Lecture Notes in Computer Science 12263

More information about this series at http://www.springer.com/series/7412

Anne L. Martel · Purang Abolmaesumi ·
Danail Stoyanov · Diana Mateus ·
Maria A. Zuluaga · S. Kevin Zhou ·
Daniel Racoceanu · Leo Joskowicz (Eds.)

Medical Image Computing and Computer Assisted Intervention – MICCAI 2020

23rd International Conference
Lima, Peru, October 4–8, 2020
Proceedings, Part III

Editors
Anne L. Martel
University of Toronto
Toronto, ON, Canada

Purang Abolmaesumi
The University of British Columbia
Vancouver, BC, Canada

Danail Stoyanov
University College London
London, UK

Diana Mateus
École Centrale de Nantes
Nantes, France

Maria A. Zuluaga
EURECOM
Biot, France

S. Kevin Zhou
Chinese Academy of Sciences
Beijing, China

Daniel Racoceanu
Sorbonne University
Paris, France

Leo Joskowicz
The Hebrew University of Jerusalem
Jerusalem, Israel

ISSN 0302-9743 ISSN 1611-3349 (electronic)
Lecture Notes in Computer Science
ISBN 978-3-030-59715-3 ISBN 978-3-030-59716-0 (eBook)
https://doi.org/10.1007/978-3-030-59716-0

LNCS Sublibrary: SL6 – Image Processing, Computer Vision, Pattern Recognition, and Graphics

This Springer imprint is published by the registered company Springer Nature Switzerland AG
The registered company address is: Gewerbestrasse 11, 6330 Cham, Switzerland

Preface

The 23rd International Conference on Medical Image Computing and Computer-Assisted Intervention (MICCAI 2020) was held this year under the most unusual circumstances, due to the COVID-19 pandemic disrupting our lives in ways that were unimaginable at the start of the new decade. MICCAI 2020 was scheduled to be held in Lima, Peru, and would have been the first MICCAI meeting in Latin America. However, with the pandemic, the conference and its program had to be redesigned to deal with realities of the "new normal", where virtual presence rather than physical interactions among attendees, was necessary to comply with global transmission control measures. The conference was held through a virtual conference management platform, consisting of the main scientific program in addition to featuring 25 workshops, 8 tutorials, and 24 challenges during October 4–8, 2020. In order to keep a part of the original spirit of MICCAI 2020, SIPAIM 2020 was held as an adjacent LatAm conference dedicated to medical information management and imaging, held during October 3–4, 2020.

The proceedings of MICCAI 2020 showcase papers contributed by the authors to the main conference, which are organized in seven volumes of *Lecture Notes in Computer Science* (LNCS) books. These papers were selected after a thorough double-blind peer-review process. We followed the example set by past MICCAI meetings, using Microsoft's Conference Managing Toolkit (CMT) for paper submission and peer reviews, with support from the Toronto Paper Matching System (TPMS) to partially automate paper assignment to area chairs and reviewers.

The conference submission deadline had to be extended by two weeks to account for the disruption COVID-19 caused on the worldwide scientific community. From 2,953 original intentions to submit, 1,876 full submissions were received, which were reduced to 1,809 submissions following an initial quality check by the program chairs. Of those, 61% were self-declared by authors as Medical Image Computing (MIC), 6% as Computer Assisted Intervention (CAI), and 32% as both MIC and CAI. Following a broad call to the community for self-nomination of volunteers and a thorough review by the program chairs, considering criteria such as balance across research areas, geographical distribution, and gender, the MICCAI 2020 Program Committee comprised 82 area chairs, with 46% from North America, 28% from Europe, 19% from Asia/Pacific/Middle East, 4% from Latin America, and 1% from Australia. We invested significant effort in recruiting more women to the Program Committee, following the conference's emphasis on equity, inclusion, and diversity. This resulted in 26% female area chairs. Each area chair was assigned about 23 manuscripts, with suggested potential reviewers using TPMS scoring and self-declared research areas, while domain conflicts were automatically considered by CMT. Following a final revision and prioritization of reviewers by area chairs in terms of their expertise related to each paper, over 1,426 invited reviewers were asked to bid for the papers for which they had been suggested. Final reviewer allocations via CMT took account of reviewer bidding, prioritization of area chairs, and TPMS scores, leading to allocating about 4 papers per reviewer. Following an initial double-blind review phase by reviewers, area chairs provided a meta-review summarizing key points of reviews and a recommendation for each paper. The program chairs then evaluated the reviews and their scores, along with the recommendation from the area chairs, to directly accept 241 papers (13%) and reject 828 papers (46%); the remainder of the papers were sent for rebuttal by the authors. During the rebuttal phase, two additional area chairs were assigned to each paper using the CMT and TPMS scores while accounting for domain conflicts. The three area chairs then independently scored each paper to accept or reject, based on the reviews, rebuttal, and manuscript, resulting in clear paper decisions using majority voting. This process resulted in the acceptance of a further 301 papers for an overall acceptance rate of 30%. A virtual Program Committee meeting was held on July 10, 2020, to confirm the final results and collect feedback of the peer-review process.

For the MICCAI 2020 proceedings, 542 accepted papers have been organized into seven volumes as follows:

- Part I, LNCS Volume 12261: Machine Learning Methodologies
- Part II, LNCS Volume 12262: Image Reconstruction and Machine Learning
- Part III, LNCS Volume 12263: Computer Aided Intervention, Ultrasound and Image Registration
- Part IV, LNCS Volume 12264: Segmentation and Shape Analysis
- Part V, LNCS Volume 12265: Biological, Optical and Microscopic Image Analysis
- Part VI, LNCS Volume 12266: Clinical Applications
- Part VII, LNCS Volume 12267: Neurological Imaging and PET

For the main conference, the traditional emphasis on poster presentations was maintained; each author uploaded a brief pre-recorded presentation and a graphical abstract onto a web platform and was allocated a personal virtual live session in which they talked directly to the attendees. It was also possible to post questions online allowing asynchronous conversations – essential to overcome the challenges of a global conference spanning many time zones. The traditional oral sessions, which typically included a small proportion of the papers, were replaced with 90 "mini" sessions where all of the authors were clustered into groups of 5 to 7 related papers; a live virtual session allowed the authors and attendees to discuss the papers in a panel format.

We would like to sincerely thank everyone who contributed to the success of MICCAI 2020 and the quality of its proceedings under the most unusual circumstances of a global pandemic. First and foremost, we thank all authors for submitting and presenting their high-quality work that made MICCAI 2020 a greatly enjoyable and successful scientific meeting. We are also especially grateful to all members of the Program Committee and reviewers for their dedicated effort and insightful feedback throughout the entire paper selection process. We would like to particularly thank the MICCAI society for support, insightful comments, and continuous engagement with organizing the conference. Special thanks go to Kitty Wong, who oversaw the entire process of paper submission, reviews, and preparation of conference proceedings. Without her, we would have not functioned effectively. Given the "new normal", none of the workshops, tutorials, and challenges would have been feasible without the true leadership of the satellite events organizing team led by Mauricio Reyes: Erik Meijering (workshops), Carlos Alberola-López (tutorials), and Lena Maier-Hein (challenges). Behind the scenes, MICCAI secretarial personnel, Janette Wallace and Johanne Langford, kept a close eye on logistics and budgets, while Mehmet Eldegez and his team at Dekon Congress and Tourism led the professional conference organization, working tightly with the virtual platform team. We also thank our sponsors for financial support and engagement with conference attendees through the virtual platform. Special thanks goes to Veronika Cheplygina for continuous engagement with various social media platforms before and throughout the conference to publicize the conference. We would also like to express our gratitude to Shelley Wallace for helping us in Marketing MICCAI 2020, especially during the last phase of the virtual conference organization.

The selection process for Young Investigator Awards was managed by a team of senior MICCAI investigators, led by Julia Schnabel. In addition, MICCAI 2020 offered free registration to the top 50 ranked papers at the conference whose primary authors were students. Priority was given to low-income regions and Latin American students. Further support was provided by the National Institutes of Health (support granted for MICCAI 2020) and the National Science Foundation (support granted to MICCAI 2019 and continued for MICCAI 2020) which sponsored another 52 awards for USA-based students to attend the conference. We would like to thank Marius Linguraru and Antonion Porras, for their leadership in regards to the NIH sponsorship for 2020, and Dinggang Shen and Tianming Liu, MICCAI 2019 general chairs, for keeping an active bridge and engagement with MICCAI 2020.

Marius Linguraru and Antonion Porras were also leading the young investigators early career development program, including a very active mentorship which we do hope, will significantly catalize young and briliant careers of future leaders of our scientific community. In link with SIPAIM (thanks to Jorge Brieva, Marius Linguraru, and Natasha Lepore for their support), we also initiated a Startup Village initiative, which, we hope, will be able to bring in promissing private initiatives in the areas of MICCAI. As a part of SIPAIM 2020, we note also the presence of a workshop for Peruvian clinicians. We would like to thank Benjaming Castañeda and Renato Gandolfi for this initiative.

MICCAI 2020 invested significant efforts to tightly engage the industry stakeholders in our field throughout its planning and organization. These efforts were led by Parvin Mousavi, and ensured that all sponsoring industry partners could connect with the conference attendees through the conference's virtual platform before and during the meeting. We would like to thank the sponsorship team and the contributions of Gustavo Carneiro, Benjamín Castañeda, Ignacio Larrabide, Marius Linguraru, Yanwu Xu, and Kevin Zhou.

We look forward to seeing you at MICCAI 2021.

October 2020

Anne L. Martel
Purang Abolmaesumi
Danail Stoyanov
Diana Mateus
Maria A. Zuluaga
S. Kevin Zhou
Daniel Racoceanu
Leo Joskowicz

Organization

General Chairs

Daniel Racoceanu — Sorbonne Université, Brain Institute, France
Leo Joskowicz — The Hebrew University of Jerusalem, Israel

Program Committee Chairs

Anne L. Martel — University of Toronto, Canada
Purang Abolmaesumi — The University of British Columbia, Canada
Danail Stoyanov — University College London, UK
Diana Mateus — Ecole Centrale de Nantes, LS2N, France
Maria A. Zuluaga — Eurecom, France
S. Kevin Zhou — Chinese Academy of Sciences, China

Keynote Speaker Chair

Rene Vidal — The John Hopkins University, USA

Satellite Events Chair

Mauricio Reyes — University of Bern, Switzerland

Workshop Team

Erik Meijering (Chair) — The University of New South Wales, Australia
Li Cheng — University of Alberta, Canada
Pamela Guevara — University of Concepción, Chile
Bennett Landman — Vanderbilt University, USA
Tammy Riklin Raviv — Ben-Gurion University of the Negev, Israel
Virginie Uhlmann — EMBL, European Bioinformatics Institute, UK

Tutorial Team

Carlos Alberola-López (Chair) — Universidad de Valladolid, Spain
Clarisa Sánchez — Radboud University Medical Center, The Netherlands
Demian Wassermann — Inria Saclay Île-de-France, France

Challenges Team

Lena Maier-Hein (Chair)	German Cancer Research Center, Germany
Annette Kopp-Schneider	German Cancer Research Center, Germany
Michal Kozubek	Masaryk University, Czech Republic
Annika Reinke	German Cancer Research Center, Germany

Sponsorship Team

Parvin Mousavi (Chair)	Queen's University, Canada
Marius Linguraru	Children's National Institute, USA
Gustavo Carneiro	The University of Adelaide, Australia
Yanwu Xu	Baidu Inc., China
Ignacio Larrabide	National Scientific and Technical Research Council, Argentina
S. Kevin Zhou	Chinese Academy of Sciences, China
Benjamín Castañeda	Pontifical Catholic University of Peru, Peru

Local and Regional Chairs

Benjamín Castañeda	Pontifical Catholic University of Peru, Peru
Natasha Lepore	University of Southern California, USA

Social Media Chair

Veronika Cheplygina	Eindhoven University of Technology, The Netherlands

Young Investigators Early Career Development Program Chairs

Marius Linguraru	Children's National Institute, USA
Antonio Porras	Children's National Institute, USA

Student Board Liaison Chair

Gabriel Jimenez	Pontifical Catholic University of Peru, Peru

Submission Platform Manager

Kitty Wong	The MICCAI Society, Canada

Conference Management

DEKON Group
Pathable Inc.

Program Committee

Ehsan Adeli	Stanford University, USA
Shadi Albarqouni	ETH Zurich, Switzerland
Pablo Arbelaez	Universidad de los Andes, Colombia
Ulas Bagci	University of Central Florida, USA
Adrien Bartoli	Université Clermont Auvergne, France
Hrvoje Bogunovic	Medical University of Vienna, Austria
Weidong Cai	The University of Sydney, Australia
Chao Chen	Stony Brook University, USA
Elvis Chen	Robarts Research Institute, Canada
Stanley Durrleman	Inria, France
Boris Escalante-Ramírez	National Autonomous University of Mexico, Mexico
Pascal Fallavollita	University of Ottawa, Canada
Enzo Ferrante	CONICET, Universidad Nacional del Litoral, Argentina
Stamatia Giannarou	Imperial College London, UK
Orcun Goksel	ETH Zurich, Switzerland
Alberto Gomez	King's College London, UK
Miguel Angel González Ballester	Universitat Pompeu Fabra, Spain
Ilker Hacihaliloglu	Rutgers University, USA
Yi Hong	University of Georgia, USA
Yipeng Hu	University College London, UK
Heng Huang	University of Pittsburgh and JD Finance America Corporation, USA
Juan Eugenio Iglesias	University College London, UK
Madhura Ingalhalikar	Symbiosis Center for Medical Image Analysis, India
Pierre Jannin	Université de Rennes, France
Samuel Kadoury	Ecole Polytechnique de Montreal, Canada
Bernhard Kainz	Imperial College London, UK
Marta Kersten-Oertel	Concordia University, Canada
Andrew King	King's College London, UK
Ignacio Larrabide	CONICET, Argentina
Gang Li	University of North Carolina at Chapel Hill, USA
Jianming Liang	Arizona State University, USA
Hongen Liao	Tsinghua University, China
Rui Liao	Siemens Healthineers, USA
Feng Lin	Nanyang Technological University, China
Mingxia Liu	University of North Carolina at Chapel Hill, USA
Jiebo Luo	University of Rochester, USA
Xiongbiao Luo	Xiamen University, China
Andreas Maier	FAU Erlangen-Nuremberg, Germany
Stephen McKenna	University of Dundee, UK
Bjoern Menze	Technische Universität München, Germany
Mehdi Moradi	IBM Research, USA

Dong Ni	Shenzhen University, China
Marc Niethammer	University of North Carolina at Chapel Hill, USA
Jack Noble	Vanderbilt University, USA
Ipek Oguz	Vanderbilt University, USA
Gemma Piella	Pompeu Fabra University, Spain
Hedyeh Rafii-Tari	Auris Health Inc., USA
Islem Rekik	Istanbul Technical University, Turkey
Nicola Rieke	NVIDIA Corporation, USA
Tammy Riklin Raviv	Ben-Gurion University of the Negev, Israel
Hassan Rivaz	Concordia University, Canada
Holger Roth	NVIDIA Corporation, USA
Sharmishtaa Seshamani	Allen Institute, USA
Li Shen	University of Pennsylvania, USA
Feng Shi	Shanghai United Imaging Intelligence Co., China
Yonggang Shi	University of Southern California, USA
Michal Sofka	Hyperfine Research, USA
Stefanie Speidel	National Center for Tumor Diseases (NCT), Germany
Marius Staring	Leiden University Medical Center, The Netherlands
Heung-Il Suk	Korea University, South Korea
Kenji Suzuki	Tokyo Institute of Technology, Japan
Tanveer Syeda-Mahmood	IBM Research, USA
Amir Tahmasebi	CodaMetrix, USA
Xiaoying Tang	Southern University of Science and Technology, China
Tolga Tasdizen	The University of Utah, USA
Pallavi Tiwari	Case Western Reserve University, USA
Sotirios Tsaftaris	The University of Edinburgh, UK
Archana Venkataraman	Johns Hopkins University, USA
Satish Viswanath	Case Western Reserve University, USA
Hongzhi Wang	IBM Almaden Research Center, USA
Linwei Wang	Rochester Institute of Technology, USA
Qian Wang	Shanghai Jiao Tong University, China
Guorong Wu	University of North Carolina at Chapel Hill, USA
Daguang Xu	NVIDIA Corporation, USA
Ziyue Xu	NVIDIA Corporation, USA
Pingkun Yan	Rensselaer Polytechnic Institute, USA
Xin Yang	Huazhong University of Science and Technology, China
Zhaozheng Yin	Stony Brook University, USA
Tuo Zhang	Northwestern Polytechnical University, China
Guoyan Zheng	Shanghai Jiao Tong University, China
Yefeng Zheng	Tencent, China
Luping Zhou	The University of Sydney, Australia

Mentorship Program (Mentors)

Ehsan Adeli	Stanford University, USA
Stephen Aylward	Kitware, USA
Hrvoje Bogunovic	Medical University of Vienna, Austria
Li Cheng	University of Alberta, Canada
Marleen de Bruijne	University of Copenhagen, Denmark
Caroline Essert	University of Strasbourg, France
Gabor Fichtinger	Queen's University, Canada
Stamatia Giannarou	Imperial College London, UK
Juan Eugenio Iglesias Gonzalez	University College London, UK
Bernhard Kainz	Imperial College London, UK
Shuo Li	Western University, Canada
Jianming Liang	Arizona State University, USA
Rui Liao	Siemens Healthineers, USA
Feng Lin	Nanyang Technological University, China
Marius George Linguraru	Children's National Hospital, George Washington University, USA
Tianming Liu	University of Georgia, USA
Xiongbiao Luo	Xiamen University, China
Dong Ni	Shenzhen University, China
Wiro Niessen	Erasmus MC - University Medical Center Rotterdam, The Netherlands
Terry Peters	Western University, Canada
Antonio R. Porras	University of Colorado, USA
Daniel Racoceanu	Sorbonne University, France
Islem Rekik	Istanbul Technical University, Turkey
Nicola Rieke	NVIDIA, USA
Julia Schnabel	King's College London, UK
Ruby Shamir	Novocure, Switzerland
Stefanie Speidel	National Center for Tumor Diseases Dresden, Germany
Martin Styner	University of North Carolina at Chapel Hill, USA
Xiaoying Tang	Southern University of Science and Technology, China
Pallavi Tiwari	Case Western Reserve University, USA
Jocelyne Troccaz	CNRS, Grenoble Alpes University, France
Pierre Jannin	INSERM, Université de Rennes, France
Archana Venkataraman	Johns Hopkins University, USA
Linwei Wang	Rochester Institute of Technology, USA
Guorong Wu	University of North Carolina at Chapel Hill, USA
Li Xiao	Chinese Academy of Science, China
Ziyue Xu	NVIDIA, USA
Bochuan Zheng	China West Normal University, China
Guoyan Zheng	Shanghai Jiao Tong University, China
S. Kevin Zhou	Chinese Academy of Sciences, China
Maria A. Zuluaga	EURECOM, France

Additional Reviewers

Alaa Eldin Abdelaal
Ahmed Abdulkadir
Clement Abi Nader
Mazdak Abulnaga
Ganesh Adluru
Iman Aganj
Priya Aggarwal
Sahar Ahmad
Seyed-Ahmad Ahmadi
Euijoon Ahn
Alireza Akhondi-asl
Mohamed Akrout
Dawood Al Chanti
Ibraheem Al-Dhamari
Navid Alemi Koohbanani
Hanan Alghamdi
Hassan Alhajj
Hazrat Ali
Sharib Ali
Omar Al-Kadi
Maximilian Allan
Felix Ambellan
Mina Amiri
Sameer Antani
Luigi Antelmi
Michela Antonelli
Jacob Antunes
Saeed Anwar
Fernando Arambula
Ignacio Arganda-Carreras
Mohammad Ali Armin
John Ashburner
Md Ashikuzzaman
Shahab Aslani
Mehdi Astaraki
Angélica Atehortúa
Gowtham Atluri
Kamran Avanaki
Angelica Aviles-Rivero
Suyash Awate
Dogu Baran Aydogan
Qinle Ba
Morteza Babaie
Hyeon-Min Bae
Woong Bae
Wenjia Bai
Ujjwal Baid
Spyridon Bakas
Yaël Balbastre
Marcin Balicki
Fabian Balsiger
Abhirup Banerjee
Sreya Banerjee
Sophia Bano
Shunxing Bao
Adrian Barbu
Cher Bass
John S. H. Baxter
Amirhossein Bayat
Sharareh Bayat
Neslihan Bayramoglu
Bahareh Behboodi
Delaram Behnami
Mikhail Belyaev
Oualid Benkarim
Aicha BenTaieb
Camilo Bermudez
Giulia Bertò
Hadrien Bertrand
Julián Betancur
Michael Beyeler
Parmeet Bhatia
Chetan Bhole
Suvrat Bhooshan
Chitresh Bhushan
Lei Bi
Cheng Bian
Gui-Bin Bian
Sangeeta Biswas
Stefano B. Blumberg
Janusz Bobulski
Sebastian Bodenstedt
Ester Bonmati
Bhushan Borotikar
Jiri Borovec
Ilaria Boscolo Galazzo

Alexandre Bousse
Nicolas Boutry
Behzad Bozorgtabar
Nadia Brancati
Christopher Bridge
Esther Bron
Rupert Brooks
Qirong Bu
Tim-Oliver Buchholz
Duc Toan Bui
Qasim Bukhari
Ninon Burgos
Nikolay Burlutskiy
Russell Butler
Michał Byra
Hongmin Cai
Yunliang Cai
Sema Candemir
Bing Cao
Qing Cao
Shilei Cao
Tian Cao
Weiguo Cao
Yankun Cao
Aaron Carass
Heike Carolus
Adrià Casamitjana
Suheyla Cetin Karayumak
Ahmad Chaddad
Krishna Chaitanya
Jayasree Chakraborty
Tapabrata Chakraborty
Sylvie Chambon
Ming-Ching Chang
Violeta Chang
Simon Chatelin
Sudhanya Chatterjee
Christos Chatzichristos
Rizwan Chaudhry
Antong Chen
Cameron Po-Hsuan Chen
Chang Chen
Chao Chen
Chen Chen
Cheng Chen
Dongdong Chen
Fang Chen
Geng Chen
Hao Chen
Jianan Chen
Jianxu Chen
Jia-Wei Chen
Jie Chen
Junxiang Chen
Li Chen
Liang Chen
Pingjun Chen
Qiang Chen
Shuai Chen
Tianhua Chen
Tingting Chen
Xi Chen
Xiaoran Chen
Xin Chen
Yuanyuan Chen
Yuhua Chen
Yukun Chen
Zhineng Chen
Zhixiang Chen
Erkang Cheng
Jun Cheng
Li Cheng
Xuelian Cheng
Yuan Cheng
Veronika Cheplygina
Hyungjoo Cho
Jaegul Choo
Aritra Chowdhury
Stergios Christodoulidis
Ai Wern Chung
Pietro Antonio Cicalese
Özgün Çiçek
Robert Cierniak
Matthew Clarkson
Dana Cobzas
Jaume Coll-Font
Alessia Colonna
Marc Combalia
Olivier Commowick
Sonia Contreras Ortiz
Pierre-Henri Conze
Timothy Cootes

Luca Corinzia
Teresa Correia
Pierrick Coupé
Jeffrey Craley
Arun C. S. Kumar
Hui Cui
Jianan Cui
Zhiming Cui
Kathleen Curran
Haixing Dai
Xiaoliang Dai
Ker Dai Fei Elmer
Adrian Dalca
Abhijit Das
Neda Davoudi
Laura Daza
Sandro De Zanet
Charles Delahunt
Herve Delingette
Beatrice Demiray
Yang Deng
Hrishikesh Deshpande
Christian Desrosiers
Neel Dey
Xinghao Ding
Zhipeng Ding
Konstantin Dmitriev
Jose Dolz
Ines Domingues
Juan Pedro Dominguez-Morales
Hao Dong
Mengjin Dong
Nanqing Dong
Qinglin Dong
Suyu Dong
Sven Dorkenwald
Qi Dou
P. K. Douglas
Simon Drouin
Karen Drukker
Niharika D'Souza
Lei Du
Shaoyi Du
Xuefeng Du
Dingna Duan
Nicolas Duchateau
James Duncan
Jared Dunnmon
Luc Duong
Nicha Dvornek
Dmitry V. Dylov
Oleh Dzyubachyk
Mehran Ebrahimi
Philip Edwards
Alexander Effland
Jan Egger
Alma Eguizabal
Gudmundur Einarsson
Ahmed Elazab
Mohammed S. M. Elbaz
Shireen Elhabian
Ahmed Eltanboly
Sandy Engelhardt
Ertunc Erdil
Marius Erdt
Floris Ernst
Mohammad Eslami
Nazila Esmaeili
Marco Esposito
Oscar Esteban
Jingfan Fan
Xin Fan
Yonghui Fan
Chaowei Fang
Xi Fang
Mohsen Farzi
Johannes Fauser
Andrey Fedorov
Hamid Fehri
Lina Felsner
Jun Feng
Ruibin Feng
Xinyang Feng
Yifan Feng
Yuan Feng
Henrique Fernandes
Ricardo Ferrari
Jean Feydy
Lucas Fidon
Lukas Fischer
Antonio Foncubierta-Rodríguez
Germain Forestier

Reza Forghani
Nils Daniel Forkert
Jean-Rassaire Fouefack
Tatiana Fountoukidou
Aina Frau-Pascual
Moti Freiman
Sarah Frisken
Huazhu Fu
Xueyang Fu
Wolfgang Fuhl
Isabel Funke
Philipp Fürnstahl
Pedro Furtado
Ryo Furukawa
Elies Fuster-Garcia
Youssef Gahi
Jin Kyu Gahm
Laurent Gajny
Rohan Gala
Harshala Gammulle
Yu Gan
Cong Gao
Dongxu Gao
Fei Gao
Feng Gao
Linlin Gao
Mingchen Gao
Siyuan Gao
Xin Gao
Xinpei Gao
Yixin Gao
Yue Gao
Zhifan Gao
Sara Garbarino
Alfonso Gastelum-Strozzi
Romane Gauriau
Srishti Gautam
Bao Ge
Rongjun Ge
Zongyuan Ge
Sairam Geethanath
Yasmeen George
Samuel Gerber
Guido Gerig
Nils Gessert
Olivier Gevaert
Muhammad Usman Ghani
Sandesh Ghimire
Sayan Ghosal
Gabriel Girard
Ben Glocker
Evgin Goceri
Michael Goetz
Arnold Gomez
Kuang Gong
Mingming Gong
Yuanhao Gong
German Gonzalez
Sharath Gopal
Karthik Gopinath
Pietro Gori
Maged Goubran
Sobhan Goudarzi
Baran Gözcü
Benedikt Graf
Mark Graham
Bertrand Granado
Alejandro Granados
Robert Grupp
Christina Gsaxner
Lin Gu
Shi Gu
Yun Gu
Ricardo Guerrero
Houssem-Eddine Gueziri
Dazhou Guo
Hengtao Guo
Jixiang Guo
Pengfei Guo
Yanrong Guo
Yi Guo
Yong Guo
Yulan Guo
Yuyu Guo
Krati Gupta
Vikash Gupta
Praveen Gurunath Bharathi
Prashnna Gyawali
Stathis Hadjidemetriou
Omid Haji Maghsoudi
Justin Haldar
Mohammad Hamghalam

Bing Han
Hu Han
Liang Han
Xiaoguang Han
Xu Han
Zhi Han
Zhongyi Han
Jonny Hancox
Christian Hansen
Xiaoke Hao
Rabia Haq
Michael Hardisty
Stefan Harrer
Adam Harrison
S. M. Kamrul Hasan
Hoda Sadat Hashemi
Nobuhiko Hata
Andreas Hauptmann
Mohammad Havaei
Huiguang He
Junjun He
Kelei He
Tiancheng He
Xuming He
Yuting He
Mattias Heinrich
Stefan Heldmann
Nicholas Heller
Alessa Hering
Monica Hernandez
Estefania Hernandez-Martin
Carlos Hernandez-Matas
Javier Herrera-Vega
Kilian Hett
Tsung-Ying Ho
Nico Hoffmann
Matthew Holden
Song Hong
Sungmin Hong
Yoonmi Hong
Corné Hoogendoorn
Antal Horváth
Belayat Hossain
Le Hou
Ai-Ling Hsu
Po-Ya Hsu
Tai-Chiu Hsung
Pengwei Hu
Shunbo Hu
Xiaoling Hu
Xiaowei Hu
Yan Hu
Zhenhong Hu
Jia-Hong Huang
Junzhou Huang
Kevin Huang
Qiaoying Huang
Weilin Huang
Xiaolei Huang
Yawen Huang
Yongxiang Huang
Yue Huang
Yufang Huang
Zhi Huang
Arnaud Huaulmé
Henkjan Huisman
Xing Huo
Yuankai Huo
Sarfaraz Hussein
Jana Hutter
Khoi Huynh
Seong Jae Hwang
Emmanuel Iarussi
Ilknur Icke
Kay Igwe
Alfredo Illanes
Abdullah-Al-Zubaer Imran
Ismail Irmakci
Samra Irshad
Benjamin Irving
Mobarakol Islam
Mohammad Shafkat Islam
Vamsi Ithapu
Koichi Ito
Hayato Itoh
Oleksandra Ivashchenko
Yuji Iwahori
Shruti Jadon
Mohammad Jafari
Mostafa Jahanifar
Andras Jakab
Amir Jamaludin

Won-Dong Jang
Vincent Jaouen
Uditha Jarayathne
Ronnachai Jaroensri
Golara Javadi
Rohit Jena
Todd Jensen
Won-Ki Jeong
Zexuan Ji
Haozhe Jia
Jue Jiang
Tingting Jiang
Weixiong Jiang
Xi Jiang
Xiang Jiang
Jianbo Jiao
Zhicheng Jiao
Amelia Jiménez-Sánchez
Dakai Jin
Taisong Jin
Yueming Jin
Ze Jin
Bin Jing
Yaqub Jonmohamadi
Anand Joshi
Shantanu Joshi
Christoph Jud
Florian Jug
Yohan Jun
Alain Jungo
Abdolrahim Kadkhodamohammadi
Ali Kafaei Zad Tehrani
Dagmar Kainmueller
Siva Teja Kakileti
John Kalafut
Konstantinos Kamnitsas
Michael C. Kampffmeyer
Qingbo Kang
Neerav Karani
Davood Karimi
Satyananda Kashyap
Alexander Katzmann
Prabhjot Kaur
Anees Kazi
Erwan Kerrien
Hoel Kervadec
Ashkan Khakzar
Fahmi Khalifa
Nadieh Khalili
Siavash Khallaghi
Farzad Khalvati
Hassan Khan
Bishesh Khanal
Pulkit Khandelwal
Maksym Kholiavchenko
Meenakshi Khosla
Naji Khosravan
Seyed Mostafa Kia
Ron Kikinis
Daeseung Kim
Geena Kim
Hak Gu Kim
Heejong Kim
Hosung Kim
Hyo-Eun Kim
Jinman Kim
Jinyoung Kim
Mansu Kim
Minjeong Kim
Seong Tae Kim
Won Hwa Kim
Young-Ho Kim
Atilla Kiraly
Yoshiro Kitamura
Takayuki Kitasaka
Sabrina Kletz
Tobias Klinder
Kranthi Kolli
Satoshi Kondo
Bin Kong
Jun Kong
Tomasz Konopczynski
Ender Konukoglu
Bongjin Koo
Kivanc Kose
Anna Kreshuk
AnithaPriya Krishnan
Pavitra Krishnaswamy
Frithjof Kruggel
Alexander Krull
Elizabeth Krupinski
Hulin Kuang

Serife Kucur
David Kügler
Arjan Kuijper
Jan Kukacka
Nilima Kulkarni
Abhay Kumar
Ashnil Kumar
Kuldeep Kumar
Neeraj Kumar
Nitin Kumar
Manuela Kunz
Holger Kunze
Tahsin Kurc
Thomas Kurmann
Yoshihiro Kuroda
Jin Tae Kwak
Yongchan Kwon
Aymen Laadhari
Dmitrii Lachinov
Alexander Ladikos
Alain Lalande
Rodney Lalonde
Tryphon Lambrou
Hengrong Lan
Catherine Laporte
Carole Lartizien
Bianca Lassen-Schmidt
Andras Lasso
Ngan Le
Leo Lebrat
Changhwan Lee
Eung-Joo Lee
Hyekyoung Lee
Jong-Hwan Lee
Jungbeom Lee
Matthew Lee
Sangmin Lee
Soochahn Lee
Stefan Leger
Étienne Léger
Baiying Lei
Andreas Leibetseder
Rogers Jeffrey Leo John
Juan Leon
Wee Kheng Leow
Annan Li
Bo Li
Chongyi Li
Haohan Li
Hongming Li
Hongwei Li
Huiqi Li
Jian Li
Jianning Li
Jiayun Li
Junhua Li
Lincan Li
Mengzhang Li
Ming Li
Qing Li
Quanzheng Li
Shulong Li
Shuyu Li
Weikai Li
Wenyuan Li
Xiang Li
Xiaomeng Li
Xiaoxiao Li
Xin Li
Xiuli Li
Yang Li (Beihang University)
Yang Li (Northeast Electric Power University)
Yi Li
Yuexiang Li
Zeju Li
Zhang Li
Zhen Li
Zhiyuan Li
Zhjin Li
Zhongyu Li
Chunfeng Lian
Gongbo Liang
Libin Liang
Shanshan Liang
Yudong Liang
Haofu Liao
Ruizhi Liao
Gilbert Lim
Baihan Lin
Hongxiang Lin
Huei-Yung Lin

Jianyu Lin
C. Lindner
Geert Litjens
Bin Liu
Chang Liu
Dongnan Liu
Feng Liu
Hangfan Liu
Jianfei Liu
Jin Liu
Jingya Liu
Jingyu Liu
Kai Liu
Kefei Liu
Lihao Liu
Luyan Liu
Mengting Liu
Na Liu
Peng Liu
Ping Liu
Quande Liu
Qun Liu
Shengfeng Liu
Shuangjun Liu
Sidong Liu
Siqi Liu
Siyuan Liu
Tianrui Liu
Xianglong Liu
Xinyang Liu
Yan Liu
Yuan Liu
Yuhang Liu
Andrea Loddo
Herve Lombaert
Marco Lorenzi
Jian Lou
Nicolas Loy Rodas
Allen Lu
Donghuan Lu
Huanxiang Lu
Jiwen Lu
Le Lu
Weijia Lu
Xiankai Lu
Yao Lu
Yongyi Lu
Yueh-Hsun Lu
Christian Lucas
Oeslle Lucena
Imanol Luengo
Ronald Lui
Gongning Luo
Jie Luo
Ma Luo
Marcel Luthi
Khoa Luu
Bin Lv
Jinglei Lv
Ilwoo Lyu
Qing Lyu
Sharath M. S.
Andy J. Ma
Chunwei Ma
Da Ma
Hua Ma
Jingting Ma
Kai Ma
Lei Ma
Wenao Ma
Yuexin Ma
Amirreza Mahbod
Sara Mahdavi
Mohammed Mahmoud
Gabriel Maicas
Klaus H. Maier-Hein
Sokratis Makrogiannis
Bilal Malik
Anand Malpani
Ilja Manakov
Matteo Mancini
Efthymios Maneas
Tommaso Mansi
Brett Marinelli
Razvan Marinescu
Pablo Márquez Neila
Carsten Marr
Yassine Marrakchi
Fabio Martinez
Antonio Martinez-Torteya
Andre Mastmeyer
Dimitrios Mavroeidis

Jamie McClelland
Verónica Medina Bañuelos
Raghav Mehta
Sachin Mehta
Liye Mei
Raphael Meier
Qier Meng
Qingjie Meng
Yu Meng
Martin Menten
Odyssée Merveille
Pablo Mesejo
Liang Mi
Shun Miao
Stijn Michielse
Mikhail Milchenko
Hyun-Seok Min
Zhe Min
Tadashi Miyamoto
Aryan Mobiny
Irina Mocanu
Sara Moccia
Omid Mohareri
Hassan Mohy-ud-Din
Muthu Rama Krishnan Mookiah
Rodrigo Moreno
Lia Morra
Agata Mosinska
Saman Motamed
Mohammad Hamed Mozaffari
Anirban Mukhopadhyay
Henning Müller
Balamurali Murugesan
Cosmas Mwikirize
Andriy Myronenko
Saad Nadeem
Ahmed Naglah
Vivek Natarajan
Vishwesh Nath
Rodrigo Nava
Fernando Navarro
Lydia Neary-Zajiczek
Peter Neher
Dominik Neumann
Gia Ngo
Hannes Nickisch
Dong Nie
Jingxin Nie
Weizhi Nie
Aditya Nigam
Xia Ning
Zhenyuan Ning
Sijie Niu
Tianye Niu
Alexey Novikov
Jorge Novo
Chinedu Nwoye
Mohammad Obeid
Masahiro Oda
Thomas O'Donnell
Benjamin Odry
Steffen Oeltze-Jafra
Ayşe Oktay
Hugo Oliveira
Marcelo Oliveira
Sara Oliveira
Arnau Oliver
Sahin Olut
Jimena Olveres
John Onofrey
Eliza Orasanu
Felipe Orihuela-Espina
José Orlando
Marcos Ortega
Sarah Ostadabbas
Yoshito Otake
Sebastian Otalora
Cheng Ouyang
Jiahong Ouyang
Cristina Oyarzun Laura
Michal Ozery-Flato
Krittin Pachtrachai
Johannes Paetzold
Jin Pan
Yongsheng Pan
Prashant Pandey
Joao Papa
Giorgos Papanastasiou
Constantin Pape
Nripesh Parajuli
Hyunjin Park
Sanghyun Park

Seyoun Park
Angshuman Paul
Christian Payer
Chengtao Peng
Jialin Peng
Liying Peng
Tingying Peng
Yifan Peng
Tobias Penzkofer
Antonio Pepe
Oscar Perdomo
Jose-Antonio Pérez-Carrasco
Fernando Pérez-García
Jorge Perez-Gonzalez
Skand Peri
Loic Peter
Jorg Peters
Jens Petersen
Caroline Petitjean
Micha Pfeiffer
Dzung Pham
Renzo Phellan
Ashish Phophalia
Mark Pickering
Kilian Pohl
Iulia Popescu
Karteek Popuri
Tiziano Portenier
Alison Pouch
Arash Pourtaherian
Prateek Prasanna
Alexander Preuhs
Raphael Prevost
Juan Prieto
Viswanath P. S.
Sergi Pujades
Kumaradevan Punithakumar
Elodie Puybareau
Haikun Qi
Huan Qi
Xin Qi
Buyue Qian
Zhen Qian
Yan Qiang
Yuchuan Qiao
Zhi Qiao
Chen Qin
Wenjian Qin
Yanguo Qin
Wu Qiu
Hui Qu
Kha Gia Quach
Prashanth R.
Pradeep Reddy Raamana
Jagath Rajapakse
Kashif Rajpoot
Jhonata Ramos
Andrik Rampun
Parnesh Raniga
Nagulan Ratnarajah
Richard Rau
Mehul Raval
Keerthi Sravan Ravi
Daniele Ravì
Harish RaviPrakash
Rohith Reddy
Markus Rempfler
Xuhua Ren
Yinhao Ren
Yudan Ren
Anne-Marie Rickmann
Brandalyn Riedel
Leticia Rittner
Robert Robinson
Jessica Rodgers
Robert Rohling
Lukasz Roszkowiak
Karsten Roth
José Rouco
Su Ruan
Daniel Rueckert
Mirabela Rusu
Erica Rutter
Jaime S. Cardoso
Mohammad Sabokrou
Monjoy Saha
Pramit Saha
Dushyant Sahoo
Pranjal Sahu
Wojciech Samek
Juan A. Sánchez-Margallo
Robin Sandkuehler

Rodrigo Santa Cruz
Gianmarco Santini
Anil Kumar Sao
Mhd Hasan Sarhan
Duygu Sarikaya
Imari Sato
Olivier Saut
Mattia Savardi
Ramasamy Savitha
Fabien Scalzo
Nico Scherf
Alexander Schlaefer
Philipp Schleer
Leopold Schmetterer
Julia Schnabel
Klaus Schoeffmann
Peter Schueffler
Andreas Schuh
Thomas Schultz
Michael Schwier
Michael Sdika
Suman Sedai
Raghavendra Selvan
Sourya Sengupta
Youngho Seo
Lama Seoud
Ana Sequeira
Saeed Seyyedi
Giorgos Sfikas
Sobhan Shafiei
Reuben Shamir
Shayan Shams
Hongming Shan
Yeqin Shao
Harshita Sharma
Gregory Sharp
Mohamed Shehata
Haocheng Shen
Mali Shen
Yiqiu Shen
Zhengyang Shen
Luyao Shi
Xiaoshuang Shi
Yemin Shi
Yonghong Shi
Saurabh Shigwan
Hoo-Chang Shin
Suprosanna Shit
Yucheng Shu
Nadya Shusharina
Alberto Signoroni
Carlos A. Silva
Wilson Silva
Praveer Singh
Ramandeep Singh
Rohit Singla
Sumedha Singla
Ayushi Sinha
Rajath Soans
Hessam Sokooti
Jaemin Son
Ming Song
Tianyu Song
Yang Song
Youyi Song
Aristeidis Sotiras
Arcot Sowmya
Rachel Sparks
Bella Specktor
William Speier
Ziga Spiclin
Dominik Spinczyk
Chetan Srinidhi
Vinkle Srivastav
Lawrence Staib
Peter Steinbach
Darko Stern
Joshua Stough
Justin Strait
Robin Strand
Martin Styner
Hai Su
Pan Su
Yun-Hsuan Su
Vaishnavi Subramanian
Gérard Subsol
Carole Sudre
Yao Sui
Avan Suinesiaputra
Jeremias Sulam
Shipra Suman
Jian Sun

Liang Sun
Tao Sun
Kyung Sung
Chiranjib Sur
Yannick Suter
Raphael Sznitman
Solale Tabarestani
Fatemeh Taheri Dezaki
Roger Tam
José Tamez-Peña
Chaowei Tan
Jiaxing Tan
Hao Tang
Sheng Tang
Thomas Tang
Xiongfeng Tang
Zhenyu Tang
Mickael Tardy
Eu Wern Teh
Antonio Tejero-de-Pablos
Paul Thienphrapa
Stephen Thompson
Felix Thomsen
Jiang Tian
Yun Tian
Aleksei Tiulpin
Hamid Tizhoosh
Matthew Toews
Oguzhan Topsakal
Jordina Torrents
Sylvie Treuillet
Jocelyne Troccaz
Emanuele Trucco
Vinh Truong Hoang
Chialing Tsai
Andru Putra Twinanda
Norimichi Ukita
Eranga Ukwatta
Mathias Unberath
Tamas Ungi
Martin Urschler
Verena Uslar
Fatmatulzehra Uslu
Régis Vaillant
Jeya Maria Jose Valanarasu
Marta Vallejo
Fons van der Sommen
Gijs van Tulder
Kimberlin van Wijnen
Yogatheesan Varatharajah
Marta Varela
Thomas Varsavsky
Francisco Vasconcelos
S. Swaroop Vedula
Sanketh Vedula
Harini Veeraraghavan
Gonzalo Vegas Sanchez-Ferrero
Anant Vemuri
Gopalkrishna Veni
Ruchika Verma
Ujjwal Verma
Pedro Vieira
Juan Pedro Vigueras Guillen
Pierre-Frederic Villard
Athanasios Vlontzos
Wolf-Dieter Vogl
Ingmar Voigt
Eugene Vorontsov
Bo Wang
Cheng Wang
Chengjia Wang
Chunliang Wang
Dadong Wang
Guotai Wang
Haifeng Wang
Hongkai Wang
Hongyu Wang
Hua Wang
Huan Wang
Jun Wang
Kuanquan Wang
Kun Wang
Lei Wang
Li Wang
Liansheng Wang
Manning Wang
Ruixuan Wang
Shanshan Wang
Shujun Wang
Shuo Wang
Tianchen Wang
Tongxin Wang

Wenzhe Wang
Xi Wang
Xiangxue Wang
Yalin Wang
Yan Wang (Sichuan University)
Yan Wang (Johns Hopkins University)
Yaping Wang
Yi Wang
Yirui Wang
Yuanjun Wang
Yun Wang
Zeyi Wang
Zhangyang Wang
Simon Warfield
Jonathan Weber
Jürgen Weese
Donglai Wei
Dongming Wei
Zhen Wei
Martin Weigert
Michael Wels
Junhao Wen
Matthias Wilms
Stefan Winzeck
Adam Wittek
Marek Wodzinski
Jelmer Wolterink
Ken C. L. Wong
Jonghye Woo
Chongruo Wu
Dijia Wu
Ji Wu
Jian Wu (Tsinghua University)
Jian Wu (Zhejiang University)
Jie Ying Wu
Junyan Wu
Minjie Wu
Pengxiang Wu
Xi Wu
Xia Wu
Xiyin Wu
Ye Wu
Yicheng Wu
Yifan Wu
Zhengwang Wu
Tobias Wuerfl
Pengcheng Xi
James Xia
Siyu Xia
Yingda Xia
Yong Xia
Lei Xiang
Deqiang Xiao
Li Xiao (Tulane University)
Li Xiao (Chinese Academy of Science)
Yuting Xiao
Hongtao Xie
Jianyang Xie
Lingxi Xie
Long Xie
Xueqian Xie
Yiting Xie
Yuan Xie
Yutong Xie
Fangxu Xing
Fuyong Xing
Tao Xiong
Chenchu Xu
Hongming Xu
Jiaofeng Xu
Kele Xu
Lisheng Xu
Min Xu
Rui Xu
Xiaowei Xu
Yanwu Xu
Yongchao Xu
Zhenghua Xu
Cheng Xue
Jie Xue
Wufeng Xue
Yuan Xue
Faridah Yahya
Chenggang Yan
Ke Yan
Weizheng Yan
Yu Yan
Yuguang Yan
Zhennan Yan
Changchun Yang
Chao-Han Huck Yang
Dong Yang

Fan Yang (IIAI)
Fan Yang (Temple University)
Feng Yang
Ge Yang
Guang Yang
Heran Yang
Hongxu Yang
Huijuan Yang
Jiancheng Yang
Jie Yang
Junlin Yang
Lin Yang
Xiao Yang
Xiaohui Yang
Xin Yang
Yan Yang
Yujiu Yang
Dongren Yao
Jianhua Yao
Jiawen Yao
Li Yao
Chuyang Ye
Huihui Ye
Menglong Ye
Xujiong Ye
Andy W. K. Yeung
Jingru Yi
Jirong Yi
Xin Yi
Yi Yin
Shihui Ying
Youngjin Yoo
Chenyu You
Sahar Yousefi
Hanchao Yu
Jinhua Yu
Kai Yu
Lequan Yu
Qi Yu
Yang Yu
Zhen Yu
Pengyu Yuan
Yixuan Yuan
Paul Yushkevich
Ghada Zamzmi
Dong Zeng
Guodong Zeng
Oliver Zettinig
Zhiwei Zhai
Kun Zhan
Baochang Zhang
Chaoyi Zhang
Daoqiang Zhang
Dongqing Zhang
Fan Zhang (Yale University)
Fan Zhang (Harvard Medical School)
Guangming Zhang
Han Zhang
Hang Zhang
Haopeng Zhang
Heye Zhang
Huahong Zhang
Jianpeng Zhang
Jinao Zhang
Jingqing Zhang
Jinwei Zhang
Jiong Zhang
Jun Zhang
Le Zhang
Lei Zhang
Lichi Zhang
Lin Zhang
Ling Zhang
Lu Zhang
Miaomiao Zhang
Ning Zhang
Pengfei Zhang
Pengyue Zhang
Qiang Zhang
Rongzhao Zhang
Ru-Yuan Zhang
Shanzhuo Zhang
Shu Zhang
Tong Zhang
Wei Zhang
Weiwei Zhang
Wenlu Zhang
Xiaoyun Zhang
Xin Zhang
Ya Zhang
Yanbo Zhang
Yanfu Zhang

Yi Zhang
Yifan Zhang
Yizhe Zhang
Yongqin Zhang
You Zhang
Youshan Zhang
Yu Zhang
Yue Zhang
Yulun Zhang
Yunyan Zhang
Yuyao Zhang
Zijing Zhang
Can Zhao
Changchen Zhao
Fenqiang Zhao
Gangming Zhao
Haifeng Zhao
He Zhao
Jun Zhao
Li Zhao
Qingyu Zhao
Rongchang Zhao
Shen Zhao
Tengda Zhao
Tianyi Zhao
Wei Zhao
Xuandong Zhao
Yitian Zhao
Yiyuan Zhao
Yu Zhao
Yuan-Xing Zhao
Yue Zhao
Zixu Zhao
Ziyuan Zhao
Xingjian Zhen
Hao Zheng
Jiannan Zheng
Kang Zheng
Yalin Zheng
Yushan Zheng
Jia-Xing Zhong
Zichun Zhong
Haoyin Zhou
Kang Zhou
Sanping Zhou
Tao Zhou
Wenjin Zhou
Xiao-Hu Zhou
Xiao-Yun Zhou
Yanning Zhou
Yi Zhou (IIAI)
Yi Zhou (University of Utah)
Yuyin Zhou
Zhen Zhou
Zongwei Zhou
Dajiang Zhu
Dongxiao Zhu
Hancan Zhu
Lei Zhu
Qikui Zhu
Weifang Zhu
Wentao Zhu
Xiaofeng Zhu
Xinliang Zhu
Yingying Zhu
Yuemin Zhu
Zhe Zhu
Zhuotun Zhu
Xiahai Zhuang
Aneeq Zia
Veronika Zimmer
David Zimmerer
Lilla Zöllei
Yukai Zou
Gerald Zwettler
Reyer Zwiggelaa

Contents – Part III

CAI Applications

Image Registration

Instrumentation and Surgical Phase Detection

Navigation and Visualization

Ultrasound Imaging

Video Image Analysis

CAI Applications

Reconstructing Sinus Anatomy from Endoscopic Video – Towards a Radiation-Free Approach for Quantitative Longitudinal Assessment

Xingtong Liu[1](✉), Maia Stiber[1], Jindan Huang[1], Masaru Ishii[2], Gregory D. Hager[1], Russell H. Taylor[1], and Mathias Unberath[1]

[1] The Johns Hopkins University, Baltimore, USA
{xingtongliu,unberath}@jhu.edu
[2] Johns Hopkins Medical Institutions, Baltimore, USA

Abstract. Reconstructing accurate 3D surface models of sinus anatomy directly from an endoscopic video is a promising avenue for cross-sectional and longitudinal analysis to better understand the relationship between sinus anatomy and surgical outcomes. We present a patient-specific, learning-based method for 3D reconstruction of sinus surface anatomy directly and only from endoscopic videos. We demonstrate the effectiveness and accuracy of our method on *in* and *ex vivo* data where we compare to sparse reconstructions from Structure from Motion, dense reconstruction from COLMAP, and ground truth anatomy from CT. Our textured reconstructions are watertight and enable measurement of clinically relevant parameters in good agreement with CT. The source code is available at https://github.com/lppllppl920/DenseReconstruction-Pytorch.

1 Introduction

The prospect of reconstructing accurate 3D surface models of sinus anatomy directly from endoscopic videos is exciting in multiple regards. Many diseases are defined by aberrations in human geometry, such as laryngotracheal stenosis, obstructive sleep apnea, and nasal obstruction in the head and neck region. In these diseases, patients suffer significantly due to the narrowing of the airway. While billions of dollars are spent to manage these patients, the outcomes are not exclusively satisfactory. An example: The two most common surgeries for nasal obstruction, septoplasty and turbinate reduction, are generally reported to *on average* significantly improve disease-specific quality of life [1], but evidence suggests that these improvements are short term in more than 40% of

---

Electronic supplementary material The online version of this chapter (https://doi.org/10.1007/978-3-030-59716-0_1) contains supplementary material, which is available to authorized users.

A. L. Martel et al. (Eds.): MICCAI 2020, LNCS 12263, pp. 3–13, 2020.
https://doi.org/10.1007/978-3-030-59716-0_1

cases [2,3]. Some hypotheses attribute the low success rate to anatomical geometry but there are no objective measures to support these claims. The ability to analyze longitudinal geometric data from a large population will potentially help to better understand the relationship between sinus anatomy and surgical outcomes. In current practice, CT is the gold standard for obtaining accurate 3D information about patient anatomy. However, due to its high cost and use of ionizing radiation, CT scanning is not suitable for longitudinal monitoring of patient anatomy. Endoscopy is routinely performed in outpatient and clinic settings to qualitatively assess treatment effect, and thus, constitutes an ideal modality to collect longitudinal data. In order to use endoscopic video data to analyze and model sinus anatomy in 3D, methods for 3D surface reconstruction that operate solely on endoscopic video are required. The resulting 3D reconstructions must agree with CT and allow for geometric measurement of clinically relevant parameters, e.g., aperture and volume.

Contributions. To address these challenges, we propose a patient-specific learning-based method for 3D sinus surface reconstruction from endoscopic videos. Our textured reconstructions are watertight and enable measurement of clinically relevant parameters in good agreement with CT. We extensively demonstrate the effectiveness and accuracy of our method on *in* and *ex vivo* data where we compare to sparse reconstructions from Structure from Motion (SfM), dense reconstruction from COLMAP [4], and ground truth anatomy from CT.

Related Work. Many methods to estimate surface reconstruction from endoscopic videos have been proposed. SfM-based methods aim for texture smoothness [5–7] and provide a sparse or dense reconstructed point cloud that is then processed by a surface reconstruction method, such as Poisson reconstruction [8]. Unfortunately, there are no guarantees that this approach will result in reasonable surfaces specifically when applied to anatomically complex structures, such as the nasal cavity in Fig. 4. Shape-from-Shading methods are often combined with fusion techniques, such as [9–11], and often require careful photometric calibration to ensure accuracy. Reconstruction with tissue deformation are handled in [12–14]. In intra-operative scenarios, SLAM-based methods [13–15] are preferable as they optimize for near real-time execution. Learning-based methods [15,16] take advantage of deep learning advancements in depth and pose estimation to improve model quality.

2 Methods

Overall Pipeline. The goal of our proposed pipeline is to automatically reconstruct a watertight textured sinus surface from an unlabeled endoscopic video. The pipeline, shown in Fig. 1, has three main components: 1) SfM based on dense point correspondences produced by a learning-based descriptor; 2) depth estimation; and 3) volumetric depth fusion with surface extraction. SfM identifies corresponding points across the video sequence and uses these correspondences

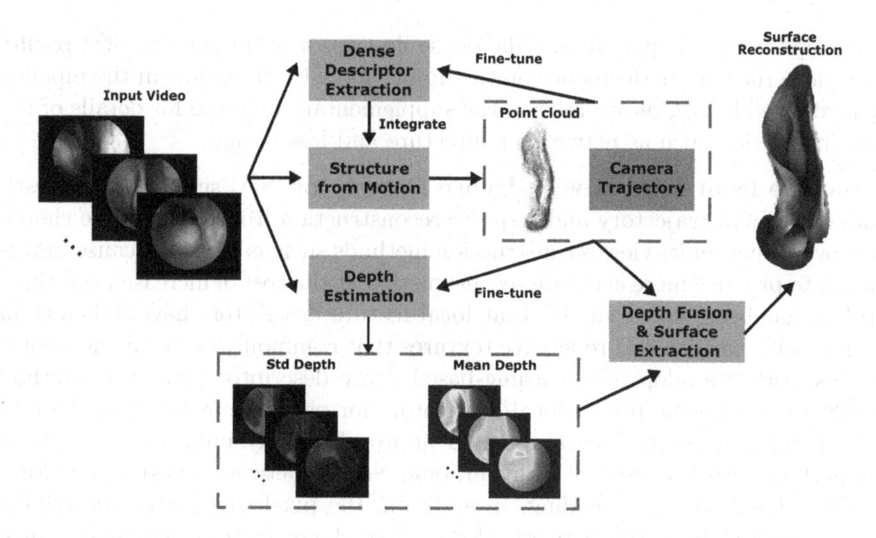

Fig. 1. Overall pipeline. Note that part of the surface reconstructions is removed in the figure to display internal structures.

to calculate both, a sparse 3D reconstruction of these points and the camera trajectory generating the video. By replacing local with learning-based descriptors during the point correspondence stage of SfM, we are able to improve the density of sparse reconstruction and the completeness of the estimated camera trajectory. We refer to this process as *Dense Descriptor Extraction*. Reliable and complete reconstructions directly from SfM are important, because these results are subsequently used for two purposes: 1) they provide self-supervisory signals for fine-tuning two learning-based modules, i.e., *Dense Descriptor Extraction* and *Depth Estimation*; 2) they are used in the *Depth Fusion & Surface Extraction* module to guide the fusion procedure. *Depth Estimation* provides dense depth measurements for all pixels in every video frame that are then aggregated over the whole video sequence using *Depth Fusion & Surface Extraction*.

Training Procedure. The two learning-modules used in our approach, namely *Dense Descriptor Extraction* and *Depth Estimation*, are both self-supervised in that they can be trained on video sequences with corresponding SfM results obtained using a conventional hand-crafted feature descriptor. This training strategy is introduced in [17,18]. Before training the complete pipeline here, we assume that both the aforementioned modules were pre-trained using such a self-supervised strategy. Then, the training order is as follows. First, the pre-trained dense descriptor extraction network is first used to establish correspondences that produce an SfM result. If the result is unsatisfactory, the dense descriptor extraction network will be fine-tuned with this SfM result for bootstrapping. This process can be repeated if necessary. We found the pre-trained dense descriptor extraction network to generalize well to unseen videos so that iterative fine-tuning was not required. Then, the depth estimation network is fine-tuned using the patient-specific dense descriptor extractor and SfM results to achieve the best performance on the input video. Each module in the pipeline is introduced below, please refer to the supplementary material for details of the implementation such as network architecture and loss design.

Structure from Motion with Dense Descriptor. SfM simultaneously estimates a camera trajectory and a sparse reconstruction from a video. We choose SfM over other multi-view reconstruction methods such as SLAM because SfM is known to produce more accurate reconstruction at the cost of increased run time. Still, it has been shown in [18] that local feature descriptors have difficulty in dealing with smooth and repetitive textures that commonly occur in endoscopy. In this work, we adopt the learning-based dense descriptor extraction method in [18] to replace the role of local descriptors for pair-wise feature matching in SfM. Intuitively, such a learning-based approach largely improves the matching performance because the Convolutional Neural Network-based architecture enables global context encoding. In addition, the pixel-wise feature descriptor map generated from the network also enables dense feature matching, which eliminates the reliance on repeatable keypoint detections. With a large number of correct pair-wise point correspondences being found, the density of the sparse reconstruction and the completeness of the camera trajectory estimate are largely improved compared with SfM with local descriptors. For each query location in the source image that is suggested by a keypoint detector, a matching location in the target image is searched for. By comparing the feature descriptor of the query location to the pixel-wise feature descriptor map of the target image, a response map is generated. In order to achieve subpixel matching accuracy, we further apply bicubic interpolation to the response map and use the position with the maximum response as the matching location. A comparison of descriptors and the impact on surface reconstruction is shown in Fig. 2.

Depth Estimation. Liu *et al.* [17] proposed a method that can train a depth estimation network in a self-supervised manner with sparse guidance from SfM results and dense inter-frame geometric consistency. In this work, we adopt a similar self-supervision scheme and the network architecture as [17] but assume that depth estimates should be probabilistic because poorly illuminated areas will likely not allow for precise depth estimates. Consequently, we model depth as a pixel-wise independent Gaussian distribution that is represented by a mean depth and its standard deviation. The related training objective is to maximize the joint probability of the training data from SfM given the predicted depth distribution. The probabilistic strategy provides some robustness to outliers from SfM, as shown in Fig. 5a. We also add an appearance consistency loss [19], which is commonly used in self-supervised depth estimation for natural scenes where photometric constancy assumptions are reasonable. This assumption, however, is invalid in endoscopy and cannot be used for additional self-supervision. Interestingly, the pixel-wise descriptor map from *Dense Descriptor Extraction* module is naturally illumination-invariant and provides a dense signal. It can thus be interpreted in analogy to appearance consistency, where appearance is now defined in terms of descriptors rather than raw intensity values. This seemed to further improve the performance, which is qualitatively shown in Fig. 5a. Based on the sparse supervision from SfM together with the dense constraints of geometric and appearance consistency, the network learns to predict accurate dense depth maps with uncertainty estimates for all frames, which are fused to form a surface reconstruction in the next step.

Depth Fusion and Surface Extraction. We apply a depth fusion method [20] based on truncated signed distance functions [21] to build a volumetric representation of the sinus surface. Depth measurements are propagated to a 3D volume using ray-casting from the corresponding camera pose and the corresponding uncertainty estimates determine the slope of the truncated signed distance function for each ray. We used SfM results to re-scale all depth estimates before the fusion to make sure all estimates are scale-consistent. To fuse all information correctly, the camera poses estimated from SfM are used to propagate the corresponding depth estimates and color information to the 3D volume. Finally, the Marching Cubes method [22] is used to extract a watertight triangle mesh surface from the 3D volume.

3 Experiments

Experiment Setup. The endoscopic videos used in the experiments were acquired from eight consenting patients and five cadavers under an IRB approved protocol. The anatomy captured in the videos is the nasal cavity. The total time duration of videos is around 40 min. Because this method is patient-specific, all data are used for training. All processing related to the proposed pipeline used 4-time spatially downsampled videos, which have a resolution of 256×320. SfM was first applied with SIFT [23] to all videos to generate sparse reconstructions and camera trajectories. Results of this initial SfM run were used to pre-train the depth estimation and dense descriptor extraction networks until convergence. Note that the pre-trained depth estimation network was not trained with appearance consistency loss. For evaluation of each individual video sequence, SfM was applied again with the pre-trained dense descriptor extraction network to generate a denser point cloud and a more complete camera trajectory. The depth estimation network, now with appearance consistency loss, were fine-tuned with the updated and sequence-specific SfM results. Note that if the pre-trained descriptor network cannot produce satisfactory SfM results on the new sequence, descriptor network fine-tuning and an extra SfM run with fine-tuned descriptor are required. All experiments were conducted on one NVIDIA TITAN X GPU. The registration algorithm used for evaluation is based on [24] to optimize over similarity transformation.

Agreement with SfM Results. Because our method is self-supervised and SfM results are used to derive supervisory signals, the discrepancy between the surface and sparse SfM reconstruction should be minimal. To evaluate the consistency between our surface reconstruction and the sparse reconstruction, we calculated the point-to-mesh distance between the two. Because scale ambiguity is intrinsic for monocular-based surface reconstruction methods, we used the

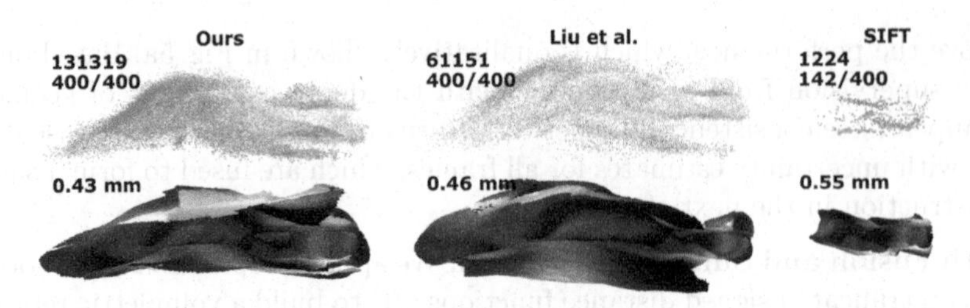

Fig. 2. Comparison of reconstruction with different descriptors. We compared the sparse and surface reconstruction using our proposed dense descriptor with those using the descriptor from Liu *et al.* [18] and SIFT [23]. The first row shows sparse reconstructions from SfM using different descriptors. The second row displays surface reconstructions estimated using the proposed method based on the sparse reconstruction. For each column, from top to bottom, the three numbers correspond to the number of points in the sparse reconstruction, number of registered views out of the total ones, and the point-to-mesh distance between the sparse and surface reconstruction. The first row shows that our sparse reconstruction is two times denser than [18]. Surface reconstruction with SIFT covers much less area and has high point-to-mesh distance, which shows the importance of having a dense enough point cloud and complete camera trajectory.

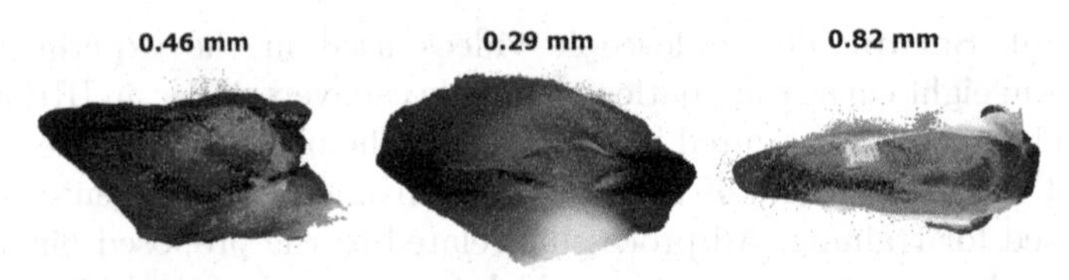

Fig. 3. Overlay of sparse and surface reconstruction. Sparse reconstruction from SfM is overlaid with surface reconstruction from the pipeline. The number in each column represents the average point-to-mesh distance.

CT surface models to recover the actual scale for all individuals where CT data are available. For those that do not have corresponding CT data, we used the average statistics of the population to recover the scale. The evaluation was conducted on 33 videos of 13 individuals. The estimated point-to-mesh distance was 0.34 ($\pm$0.14) mm. Examples of the sparse and surface reconstruction overlaid with point-to-mesh distance are shown in Fig. 3.

Consistency Against Video Variation. Surface reconstruction methods should be insensitive to variations in video capture, such as camera speed. To evaluate the sensitivity of our method, we randomly sub-sampled frames from the original video to mimic camera speed variation. The pipeline was run for each sub-sampled video and we evaluated the model consistency by aligning surface reconstructions estimated from different subsets. To simulate camera speed variation, out of every 10 consecutive video frames, only 7 frames were randomly selected. We evaluated the model consistency on 3 video sequences that cover the entire nasal cavity of three individuals, respectively. Five reconstructions that

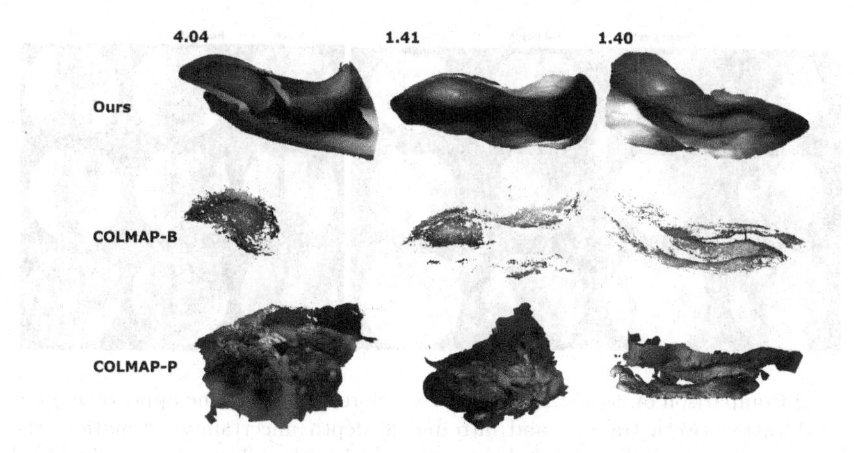

Fig. 4. Comparison of surface reconstruction from ours and COLMAP. The number in each column is the ratio of surface area between our reconstruction and COLMAP with ball pivoting [25] (COLMAP-B). Ratios are underestimated because many redundant invalid surfaces are generated in the second row. COLMAP with Poisson [8] (COLMAP-P) is shown in the last row with excessive surfaces removed already.

were computed from random subsets of each video were used for evaluation. The average residual distance after registration between different surface reconstructions was used as the metric for consistency. The scale recovery method is the same as above. The residual error was 0.21 ($\pm$0.10) mm.

Agreement with Dense Reconstruction from COLMAP. We used the ball pivoting [25] method to reconstruct surfaces in COLMAP instead of built-in Poisson [8] and Delaunay [26] methods because these two did not produce reasonable results. Three videos from 3 individuals were used in this evaluation. The qualitative comparison is shown in Fig. 4. The same scale recovery method as above was used. The average residual distance after registration between the surface reconstructions from the proposed pipeline and COLMAP is 0.24 ($\pm$0.08) mm. In terms of the runtime performance, given that a pre-trained generalizable descriptor network and depth estimation network exist, our method requires running sparse SfM with a learning-based feature descriptor, fine-tuning depth estimation network, depth fusion, and surface extraction. For the three sequences, the average runtime for the proposed method is 127 min, whereas the runtime for COLMAP is 778 min.

Agreement with CT. Model accuracy was evaluated by comparing surface reconstructions with the corresponding CT models. In this evaluation, two metrics were used: average residual error between the registered surface reconstruction and the CT model, and the average relative difference between the corresponding cross-sectional areas of the CT surface models and the surface reconstructions. The purpose of this evaluation is to determine whether our reconstruction can be used as a low-cost, radiation-free replacement for CT when calculating clinically relevant parameters. To find the corresponding cross-section of

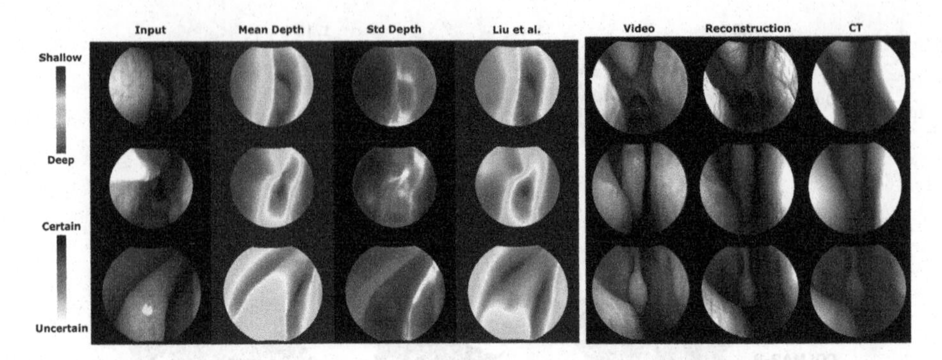

Fig. 5. (a) Comparison of depth estimation. By enforcing inter-frame appearance consistency during network training and introducing depth uncertainty estimation, the depth predictions seem to be visually better compared to those from the model trained with the settings in [17]. As can be seen in the third column, higher depth uncertainties were predicted for deeper regions and those where training data is erroneous, such as the specularity in the last row. (b) Visualization of aligned endoscopic video, dense reconstruction, and CT surface model. To produce such visualization, our dense reconstruction is first registered to the CT surface to obtain the transformation between the two coordinate systems.

two models, the surface reconstruction was first registered to the CT model. The registered camera poses from SfM were then used as the origins and orientations of the cross-sectional planes. The relative differences of all cross-sectional areas along the registered camera trajectory were averaged to obtain the final statistics. This evaluation was conducted on 7 video sequences from 4 individuals. The residual error after registration was 0.69 (±0.14) mm. As a comparison, when the sparse reconstructions from SfM are directly registered to CT models, the residual error was 0.53 (±0.24) mm. The smaller error is due to the sparsity and smaller region coverage of the sparse reconstruction compared to ours. In Fig. 5b, a visualization of the video-reconstruction-CT alignment is shown. The cross-sectional surface areas are estimated with an average relative error of 7 (±2)%. This error mainly originates from regions that were not sufficiently visualized during scoping, such as the inferior, middle, and superior meatus. These regions are included in our analysis due to the automation of cross-sectional measurements. In practice, these regions are not commonly inspected as they are hidden beneath the turbinates; if a precise measurement of these areas is desired, small modifications to video capture would allow for improved visualization. Similar to [15], such adjustments can be guided by our surface reconstruction, since the occupancy states in the fusion volume can indicate explicitly what regions were not yet captured with endoscopic video.

4 Discussion

Choice of Depth Estimation Method. In this work, a monocular depth estimation network is used to learn the complex mapping between the color appearance of a video frame and the corresponding dense depth map. The method in [17] has been shown to generalize well to unseen cases. However, the patient-specific training in this pipeline may allow for higher variance mappings since it does not need to generalize to other unseen cases. Therefore, a more complex network architecture could potentially further improve the depth estimation accuracy, leading to more accurate surface reconstruction. For example, a self-supervised recurrent neural network that predicts the dense depth map of a video frame based on the current observation and the previous frames in the video could potentially have more expressivity and be able to learn a more complex mapping, such as the method proposed by Wang *et al.* [27].

Limitations. First, this pipeline will not work if SfM fails. This could happen in some cases, such as in the presence of fast camera movement, blurry images, or tissue deformation. The latter may potentially be tackled by non-rigid SfM [28]. Second, the pipeline currently does not estimate geometric uncertainty of the surface reconstruction. A volume-based surface uncertainty estimation method may need to be developed for this purpose.

5 Conclusion

In this work, we proposed a learning-based surface reconstruction pipeline for endoscopy. Our method operates directly on raw endoscopic videos and produces watertight textured surface models that are in good agreement with anatomy extracted from CT. While this method so far has only been evaluated on videos of the nasal cavity, the proposed modules are generic, self-supervised, and should thus be applicable to other anatomies. Future work includes uncertainty estimation on the reconstructed surface models and prospective acquisition of longitudinal endoscopic video data in the clinic.

References

1. Bezerra, T.F.P., et al.: Quality of life assessment septoplasty in patients with nasal obstruction. Braz. J. Otorhinolaryngol. **78**(3), 57–62 (2012)
2. Hytönen, M., Blomgren, K., Lilja, M., Mäkitie, A.: How we do it: septoplasties under local anaesthetic are suitable for short stay surgery; the clinical outcomes. Clin. Otolaryngol.: Official J. ENT-UK; Official J. Netherlands Soc. Oto-Rhino-Laryngol. Cervico-Facial Surg. **31**(1), 64–68 (2006)
3. Hytönen, M.L., Lilja, M., Mäkitie, A.A., Sintonen, H., Roine, R.P.: Does septoplasty enhance the quality of life in patients? Eur. Arch. Otorhinolaryngol. **269**(12), 2497–2503 (2012)
4. Schönberger, J.L., Frahm, J.: Structure-from-motion revisited. In: 2016 IEEE Conference on Computer Vision and Pattern Recognition (CVPR), pp. 4104–4113, June 2016
5. Phan, T., Trinh, D., Lamarque, D., Wolf, D., Daul, C.: Dense optical flow for the reconstruction of weakly textured and structured surfaces: application to endoscopy. In: 2019 IEEE International Conference on Image Processing (ICIP), pp. 310–314, September 2019

6. Widya, A.R., Monno, Y., Okutomi, M., Suzuki, S., Gotoda, T., Miki, K.: Whole stomach 3D reconstruction and frame localization from monocular endoscope video. IEEE J. Transl. Eng. Health Med. **7**, 1–10 (2019)
7. Qiu, L., Ren, H.: Endoscope navigation and 3D reconstruction of oral cavity by visual slam with mitigated data scarcity. In: 2018 IEEE/CVF Conference on Computer Vision and Pattern Recognition Workshops (CVPRW), pp. 2278–22787, June 2018
8. Kazhdan, M., Bolitho, M., Hoppe, H.: Poisson surface reconstruction. In: Proceedings of the Fourth Eurographics Symposium on Geometry Processing, vol. 7 (2006)
9. Turan, M., Pilavci, Y.Y., Ganiyusufoglu, I., Araujo, H., Konukoglu, E., Sitti, M.: Sparse-then-dense alignment-based 3D map reconstruction method for endoscopic capsule robots. Mach. Vis. Appl. **29**(2), 345–359 (2018)
10. Tokgozoglu, H.N., Meisner, E.M., Kazhdan, M., Hager, G.D.: Color-based hybrid reconstruction for endoscopy. In: 2012 IEEE Computer Society Conference on Computer Vision and Pattern Recognition Workshops, pp. 8–15, June 2012
11. Karargyris, A., Bourbakis, N.: Three-dimensional reconstruction of the digestive wall in capsule endoscopy videos using elastic video interpolation. IEEE Trans. Med. Imaging **30**(4), 957–971 (2011)
12. Zhao, Q., Price, T., Pizer, S., Niethammer, M., Alterovitz, R., Rosenman, J.: The endoscopogram: a 3D model reconstructed from endoscopic video frames. In: Ourselin, S., Joskowicz, L., Sabuncu, M.R., Unal, G., Wells, W. (eds.) MICCAI 2016. LNCS, vol. 9900, pp. 439–447. Springer, Cham (2016). https://doi.org/10.1007/978-3-319-46720-7_51
13. Lamarca, J., Parashar, S., Bartoli, A., Montiel, J.: DefSLAM: tracking and mapping of deforming scenes from monocular sequences. arXiv preprint arXiv:1908.08918 (2019)
14. Song, J., Wang, J., Zhao, L., Huang, S., Dissanayake, G.: MIS-SLAM: real-time large-scale dense deformable slam system in minimal invasive surgery based on heterogeneous computing. IEEE Robot. Autom. Lett. **3**(4), 4068–4075 (2018)
15. Ma, R., Wang, R., Pizer, S., Rosenman, J., McGill, S.K., Frahm, J.-M.: Real-time 3D reconstruction of colonoscopic surfaces for determining missing regions. In: Shen, D., et al. (eds.) MICCAI 2019. LNCS, vol. 11768, pp. 573–582. Springer, Cham (2019). https://doi.org/10.1007/978-3-030-32254-0_64
16. Chen, R.J., Bobrow, T.L., Athey, T., Mahmood, F., Durr, N.J.: Slam endoscopy enhanced by adversarial depth prediction. arXiv preprint arXiv:1907.00283 (2019)
17. Liu, X., et al.: Dense depth estimation in monocular endoscopy with self-supervised learning methods. IEEE Trans. Med. Imaging **39**, 1438–1447 (2019)
18. Liu, X., et al.: Extremely dense point correspondences using a learned feature descriptor. In: Proceedings of the IEEE/CVF Conference on Computer Vision and Pattern Recognition, pp. 4847–4856 (2020)
19. Zhou, T., Brown, M., Snavely, N., Lowe, D.G.: Unsupervised learning of depth and ego-motion from video. In: CVPR (2017)
20. Curless, B., Levoy, M.: A volumetric method for building complex models from range images. In: Proceedings of the 23rd Annual Conference on Computer Graphics and Interactive Techniques, pp. 303–312 (1996)
21. Zach, C., Pock, T., Bischof, H.: A globally optimal algorithm for robust TV-L1 range image integration. In: ICCV. IEEE, pp. 1–8 (2007)
22. Lorensen, W.E., Cline, H.E.: Marching cubes: a high resolution 3D surface construction algorithm. ACM SIGGRAPH Comput. Graph. **21**, 163–169 (1987)

23. Lowe, D.G.: Distinctive image features from scale-invariant keypoints. Int. J. Comput. Vis. **60**(2), 91–110 (2004)
24. Billings, S., Taylor, R.: Generalized iterative most likely oriented-point (G-IMLOP) registration. IJCARS **10**(8), 1213–1226 (2015)
25. Bernardini, F., Mittleman, J., Rushmeier, H., Silva, C., Taubin, G.: The ball-pivoting algorithm for surface reconstruction. IEEE Trans. Vis. Comput. Graph. **5**(4), 349–359 (1999)
26. Cazals, F., Giesen, J.: Delaunay triangulation based surface reconstruction. In: Boissonnat, J.D., Teillaud, M. (eds.) Effective Computational Geometry for Curves and Surfaces, pp. 231–276. Springer, Heidelberg (2006). https://doi.org/10.1007/978-3-540-33259-6_6
27. Wang, R., Pizer, S.M., Frahm, J.M.: Recurrent neural network for (un-) supervised learning of monocular video visual odometry and depth. In: Proceedings of the IEEE Conference on Computer Vision and Pattern Recognition, pp. 5555–5564 (2019)
28. Khan, I.: Robust sparse and dense nonrigid structure from motion. IEEE Trans. Multimed. **20**(4), 841–850 (2018)

Inertial Measurements for Motion Compensation in Weight-Bearing Cone-Beam CT of the Knee

Jennifer Maier[1,2(✉)], Marlies Nitschke[2], Jang-Hwan Choi[3], Garry Gold[4], Rebecca Fahrig[5], Bjoern M. Eskofier[2], and Andreas Maier[1]

[1] Pattern Recognition Lab, Friedrich-Alexander-Univeristät Erlangen-Nürnberg (FAU), Erlangen, Germany
jennifer.maier@fau.de
[2] Machine Learning and Data Analytics Lab, Friedrich-Alexander-Univeristät Erlangen-Nürnberg (FAU), Erlangen, Germany
[3] College of Engineering, Ewha Womans University, Seoul, Korea
[4] Department of Radiology, Stanford University, Stanford, CA, USA
[5] Siemens Healthcare GmbH, Forchheim, Germany

Abstract. Involuntary motion during weight-bearing cone-beam computed tomography (CT) scans of the knee causes artifacts in the reconstructed volumes making them unusable for clinical diagnosis. Currently, image-based or marker-based methods are applied to correct for this motion, but often require long execution or preparation times. We propose to attach an inertial measurement unit (IMU) containing an accelerometer and a gyroscope to the leg of the subject in order to measure the motion during the scan and correct for it. To validate this approach, we present a simulation study using real motion measured with an optical 3D tracking system. With this motion, an XCAT numerical knee phantom is non-rigidly deformed during a simulated CT scan creating motion corrupted projections. A biomechanical model is animated with the same tracked motion in order to generate measurements of an IMU placed below the knee. In our proposed multi-stage algorithm, these signals are transformed to the global coordinate system of the CT scan and applied for motion compensation during reconstruction. Our proposed approach can effectively reduce motion artifacts in the reconstructed volumes. Compared to the motion corrupted case, the average structural similarity index and root mean squared error with respect to the no-motion case improved by 13–21% and 68–70%, respectively. These results are qualitatively and quantitatively on par with a state-of-the-art marker-based method we compared our approach to. The presented study shows the feasibility of this novel approach, and yields promising results towards a purely IMU-based motion compensation in C-arm CT.

Electronic supplementary material The online version of this chapter (https://doi.org/10.1007/978-3-030-59716-0_2) contains supplementary material, which is available to authorized users.

A. L. Martel et al. (Eds.): MICCAI 2020, LNCS 12263, pp. 14–23, 2020.
https://doi.org/10.1007/978-3-030-59716-0_2

Keywords: Motion compensation · Inertial measurements · CT reconstruction

1 Introduction

Osteoarthritis is a disease affecting articular cartilage in the joints, leading to a higher porosity and eventually to loss of tissue [2]. The structural change of cartilage also has an influence on its mechanical properties, i.e. its behavior when put under stress [18]. To analyze how diseased cartilage in the knee joint changes under stress compared to healthy tissue, it can be imaged under load conditions. This can be realized by scanning the knee joint in a weight-bearing standing position using a flexible C-arm cone-beam Computed Tomography (CT) system rotating on a horizontal trajectory, as depicted in Fig. 1a [15]. However, standing subjects will show more involuntary motion due to body sway when naturally standing compared to the conventional supine scanning position [22]. Since standard CT reconstruction assumes stationary objects, this motion leads to streaking artifacts, double contours, and blurring in the reconstructed volumes making them unsuitable for clinical diagnosis.

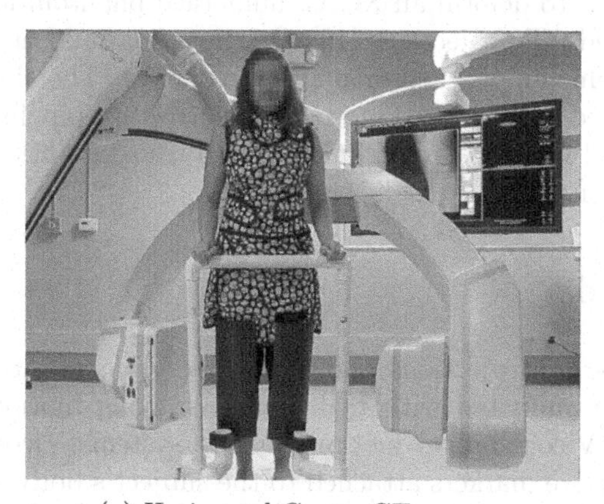

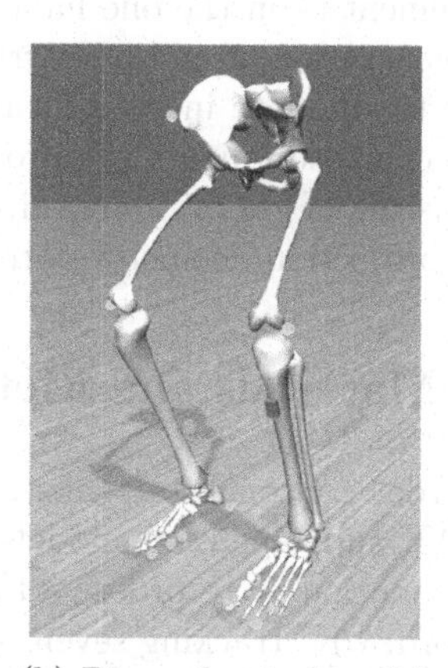

(a) Horizontal C-arm CT scan. (b) Biomechanical model.

Fig. 1. (a) Setting of a weight-bearing C-arm CT acquisition during which subjects move involuntarily. (b) Biomechanical model used for motion modeling in OpenSim and for simulation of inertial measurements. The tracked markers are shown in pink and the simulated sensor is shown in green. (Color figure online)

Multiple approaches to correct for this motion have been proposed in literature. There exist purely image-based methods, like 2D/3D registration [3] or the use of a penalized image sharpness criterion [21], that show very good motion compensation results but are computationally expensive. It is also possible to use epipolar consistency conditions for motion compensation, but for knee imaging this has so far only been applied for estimating translation and not rotation

[4]. Recently, also deep learning methods were applied in CT motion correction. Bier et al. proposed a neural network to detect anatomical landmarks in projection images, but the approach was not robust when other objects were present and was only evaluated for tracking motion [5]. Its feasibility for compensating motion was not investigated. An approach requiring external hardware is using range cameras to track motion during the scan, which so far worked well on purely simulated data [6]. The gold standard method for knee motion compensation is based on small metallic markers attached to the scanned leg, and was proposed in [8,9]. The markers tracked in the projections can be used to iteratively estimate motion. However, the placement of the markers is tedious and the metal produces artifacts in the reconstructions.

Inertial measurement units (IMUs) containing an accelerometer and a gyroscope have found use in C-arm CT for navigation [13] and calibration [14] purposes. We propose to use these small and lightweight sensors for motion compensation in C-arm CT. For this purpose, an IMU is attached to the leg of the subject to measure motion during the scan. To show the feasibility of this approach, we present a simulation study using real 3D human swaying motion recorded with an optical tracking system. These measurements are used to animate an OpenSim biomechanical model by inverse kinematics computation [11,20]. The model's movement is on the one hand used to deform an XCAT numerical phantom for the generation of motion corrupted CT projections [19]. On the other hand, it is used to simulate inertial measurements of a sensor placed on the leg. The simulated measurements are processed in a multi-stage motion correction pipeline consisting of gravity removal, local velocity computation, global transformation, and projection geometry correction.

2 Materials and Methods

In order to generate realistic X-ray projections and IMU measurements, the XCAT and OpenSim models are animated with real human swaying motion. This motion is recorded with a Vicon optical motion capture system (Vicon, Oxford, UK) tracking seven reflective markers attached to the subject's body at a sampling rate of 120 Hz. The markers are placed on the sacrum, and on the right and left anterior superior iliac spine, lateral epicondyle of the knee and malleolus lateralis. Seven healthy subjects are recorded holding a squat at 30 and 60° of knee flexion. Afterwards, a biomechanical model of the human lower body based on the model presented in [12] is scaled to each subject's anthropometry using the software OpenSim, see Fig. 1b [11,20]. With this model and the measured 3D marker positions, the inverse kinematics are computed in order to find the generalized coordinates (i.e. global position and orientation of the pelvis and the joint angles) that best represent the measured motion. Before further processing, the generalized coordinates are filtered with a moving average filter with a span of 60 in order to remove system noise from the actual movement. The 3D positions of the sacrum, and the left and right hip joint center, knee joint center and ankle joint center over time are extracted from the animated model and used for the XCAT CT projection generation (Sect. 2.1). A virtual sensor is placed on the animated model's shank in order to simulate IMU measurements (Sect. 2.2), which are used for the motion compensated reconstruction (Sect. 2.3).

2.1 Generation of Motion Corrupted Projections

The XCAT numerical phantom is a model of the human body with its legs consisting of the bones tibia, fibula, femur and patella including bone marrow and surrounding body soft tissue [19]. The shapes of these structures are defined by non-uniform rational B-Splines (NURBS) and by the positions of their control points. By shifting these control points for each time step of the simulated CT scan based on the hip, knee and ankle joint centers of the OpenSim model, the upper leg and lower leg of the XCAT model are animated individually leading to a non-rigid motion. With the deformed model for each time step, X-ray projection images of a simulated CT scan are generated. The detector of size 620 × 480 pixels with isotropic pixel resolution of 0.616 mm rotates on a virtual circular trajectory with a source detector distance of 1198 mm and a source isocenter distance of 780 mm. The angular increment between projections is 0.8° and in total 248 projections are generated, corresponding to a sampling rate of 31 Hz. Forward projections of the deformed model are created as described in [17]. Since a healthy human's knees in a natural standing position are too far apart to both fit on the detector, the rotation center of the scan is placed in the center of the left leg.

2.2 Simulation of Inertial Measurements

An IMU is a small lightweight device that measures its acceleration and angular velocity on three perpendicular axes. Additionally, the accelerometer always also senses the earth's gravitational field distributed on its three axes depending on the current orientation. Such a sensor is virtually placed on the shank 14 cm below the left knee joint aligned with the shank segment. The simulated acceleration $\mathbf{a}_i$ and angular velocity $\boldsymbol{\omega}_i$ at time point i are computed as follows [7,10]:

$$\mathbf{a}_i = \mathbf{R}_i^\top (\ddot{\mathbf{r}}_{Seg,i} + \ddot{\mathbf{R}}_i \mathbf{p}_{Sen,i} - \mathbf{g}), \tag{1}$$

$$\boldsymbol{\omega}_i = (\omega_{x,i}, \omega_{y,i}, \omega_{z,i})^\top, \tag{2}$$

$$[\boldsymbol{\omega}_i]_\times = \mathbf{R}_i^\top \dot{\mathbf{R}}_i = \begin{pmatrix} 0 & -\omega_{z,i} & \omega_{y,i} \\ \omega_{z,i} & 0 & -\omega_{x,i} \\ -\omega_{y,i} & \omega_{x,i} & 0 \end{pmatrix}. \tag{3}$$

The 3 × 3 rotation matrix $\mathbf{R}_i$ describes the orientation of the sensor at time point i in the global coordinate system, $\dot{\mathbf{R}}_i$ and $\ddot{\mathbf{R}}_i$ are its first and second order derivatives with respect to time. The position of the segment the sensor was mounted on in the global coordinate system at time point i is described by $\mathbf{r}_{Seg,i}$, with $\ddot{\mathbf{r}}_{Seg,i}$ being its second order derivative. $\mathbf{p}_{Sen,i}$ is the position of the sensor in the local coordinate system of the segment the sensor was mounted on. All required parameters are obtained by computing the forward kinematics of the biomechanical model. $\mathbf{g} = (0, -9.80665, 0)^\top$ is the global gravity vector.

2.3 Motion Compensated Reconstruction

The simulated IMU measurements are used to estimate a rigid motion describing the 3D change of orientation and position from each time step to the next in the global coordinate frame. The sensor's coordinate system $\mathbf{S}_i$ at each time step i, i.e. its orientation and position in the global frame is described by the affine matrix

$$\mathbf{S}_i = \left(\begin{array}{c|c} \hat{\mathbf{R}}_i & \hat{\mathbf{t}}_i \\ \hline \mathbf{0} & 1 \end{array}\right), \tag{4}$$

where $\hat{\mathbf{R}}_i$ is a 3×3 rotation matrix, and $\hat{\mathbf{t}}_i$ is a 3×1 translation vector. The initial pose $\mathbf{S}_0$ is assumed to be known.

The gyroscope measures the angular velocity, which is the change of orientation over time on the three axes of the sensor's local coordinate system. Therefore, this measurement can directly be used to rotate the sensor from each time step to the next. The measured accelerometer signal, however, needs to be processed to obtain the positional change over time. First, the gravity measured on the sensor's three axes is removed based on its global orientation. If $\mathbf{G}_i$ is a 3D rotation matrix containing the rotation change measured by the gyroscope at time step i, the global orientation of the sensor is described by

$$\hat{\mathbf{R}}_{i+1} = \hat{\mathbf{R}}_i \mathbf{G}_i. \tag{5}$$

The global gravity vector $\mathbf{g}$ is transformed to the sensor's local coordinate system at each time step i using

$$\mathbf{g}_i = \hat{\mathbf{R}}_i^\top \mathbf{g}. \tag{6}$$

To obtain the gravity-free acceleration $\bar{\mathbf{a}}_i$, the gravity component then is removed from $\mathbf{a}_i$ by adding the local gravity vector $\mathbf{g}_i$ at each time step i. To obtain the sensor's local velocity $\mathbf{v}_i$, i.e. its position change over time, the integral of the gravity-free acceleration is computed. The integration must be performed considering the sensor's orientation changes.

$$\mathbf{v}_{i+1} = \mathbf{G}_i^\top (\bar{\mathbf{a}}_i + \mathbf{v}_i). \tag{7}$$

In this study, we assume that the sensor's initial velocity $\mathbf{v}_0$ is known.

The local rotational change $\boldsymbol{\omega}_i$ and positional change $\mathbf{v}_i$ for each time step i are linearly resampled to the CT scan's sampling frequency and rewritten to an affine matrix

$$\boldsymbol{\Delta}_{l,i} = \left(\begin{array}{c|c} \mathbf{G}_i & \mathbf{v}_i \\ \hline \mathbf{0} & 1 \end{array}\right). \tag{8}$$

To obtain the change in the global coordinate system, $\boldsymbol{\Delta}_{l,i}$ is transformed for each time step using its pose $\mathbf{S}_i$:

$$\boldsymbol{\Delta}_{g,i} = \mathbf{S}_i \boldsymbol{\Delta}_{l,i} \mathbf{S}_i^{-1}, \tag{9}$$

$$\mathbf{S}_{i+1} = \boldsymbol{\Delta}_{g,i} \mathbf{S}_i. \tag{10}$$

For the motion compensated reconstruction, the first projection is used as reference without motion, so the affine matrix containing rotation and translation is the identity matrix $\mathbf{M}_0 = \mathbf{I}$. For each subsequent time step, the motion matrix is computed as

$$\mathbf{M}_{i+1} = \mathbf{M}_i \boldsymbol{\Delta}_{g,i}. \tag{11}$$

These motion matrices are applied to the projection matrices of the system's geometry to correct for motion.

2.4 Evaluation

We reconstruct the projections as volumes of size 512^3 with isotropic spacing of 0.5 mm using GPU accelerated filtered back-projection in the software framework CONRAD [16].

To evaluate the performance of the proposed method, the resulting reconstructions are compared to a reconstruction without motion correction, and to a ground truth reconstruction from projections where the initial pose of the subject was kept static throughout the scan. Furthermore, we compare to reconstructions from the gold standard marker-based approach. For this purpose, small highly attenuating circular markers on the skin are simulated and tracked as proposed in [9]. All volumes are scaled from 0 to 1 and registered to the reference reconstruction without motion for comparison. As metrics for image quality evaluation we compute the root mean squared error (RMSE) and the structural similarity (SSIM). The SSIM index considers differences in luminance, contrast and structure for comparison and ranges from 0 (no similarity) to 1 (identical images) [24].

3 Results

Exemplary slices of the resulting reconstructions are shown in Fig. 2. A visual comparison of the results shows a similar reduction in streaking and blurring in the reconstructions of our proposed approach and the reference marker-based approach compared to the uncorrected case. The average RMSE and SSIM values of the proposed approach and the marker-based reference approach compared to ground truth excluding background voxels are similar, see Table 1. Both methods resulted in a high average SSIM value of 0.99 (proposed) and 0.98 (reference) for 30° squats and 0.98 (proposed) and 0.97 (reference) for 60° squats. The average RMSE of 0.02 for the proposed method was slightly lower than for the marker-based approach with on average 0.03. Table 2 shows the average improvement in percent compared to the uncorrected case. While the SSIM values show an improvement of 12–21% for both proposed and reference approach, the RMSE improved on average by 68–70% for the proposed method, and on average by 57% for the marker-based approach.

4 Discussion and Conclusion

The results presented in Fig. 2 and Table 1 and 2 show that the proposed method is able to estimate and correct for involuntary subject motion during a standing acquisition. Compared to the case without motion, some artifacts are still visible and some double edges could not be restored. A reason for this is that the motion applied for projection generation is a non-rigid motion, where the upper and lower leg can move individually. The inertial sensor placed on the shank, however, is only able to estimate a rigid motion consisting of a 3D rotation and translation. Therefore it is not possible to entirely restore image quality with one sensor, even though a clear improvement compared to the uncorrected case is observable. To overcome this limitation, a second sensor could be placed on the thigh and both measurements could be combined to account for non-rigid motion.

The reference approach tracking small metallic markers in the projection images for motion compensation also estimates a 3D rigid motion, thereby allowing for a fair comparison to our proposed approach. The presented evaluation even shows slightly better results for the proposed approach in a qualitative as

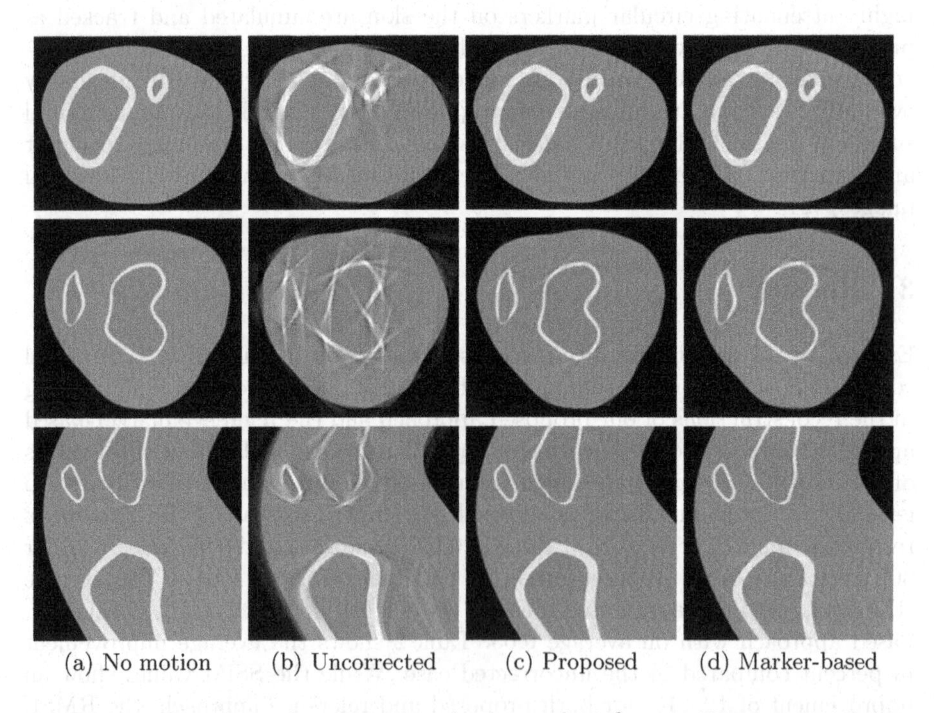

Fig. 2. Exemplary slices of a reconstructed volume of a 30° squat. First row: axial slice through shank, second row: axial slice through thigh, third row: sagittal slice. (a) scan without motion, (b) uncorrected case, (c) proposed method, (d) marker-based reference method. The motion artifacts clearly visible in the uncorrected case can be reduced by both the proposed and the reference method.

Table 1. Average structural similarity (SSIM) index and root mean squared error (RMSE) over all subjects for 30° and 60° squats. Best values are printed bold.

	SSIM		RMSE	
	30°	60°	30°	60°
Uncorrected	0.881 ± 0.053	0.821 ± 0.098	0.070 ± 0.015	0.081 ± 0.020
Proposed	**0.989 ± 0.005**	**0.984 ± 0.013**	**0.022 ± 0.004**	**0.024 ± 0.009**
Marker-based	0.980 ± 0.008	0.969 ± 0.017	0.029 ± 0.004	0.034 ± 0.008

Table 2. Average improvement in percent of structural similarity (SSIM) index and root mean squared error (RMSE) compared to the uncorrected case over all subjects for 30° and 60° squats. Best values are printed bold.

	SSIM		RMSE	
	30°	60°	30°	60°
Proposed	**12.6 ± 6.7**	**21.2 ± 14.3**	**67.6 ± 7.1**	**70.4 ± 5.5**
Marker-based	11.6 ± 6.3	19.3 ± 14.0	56.9 ± 7.0	57.1 ± 7.4

well as in a quantitative comparison. Compared to the marker-based approach, where a sufficient number of markers has to be carefully placed around the knee, for our approach, only one or two sensors per leg are necessary to track the subject's motion. This can help to facilitate and speed up the clinical process of weight-bearing imaging.

In this initial study, an optimal sensor in a well-controlled setting is assumed, however, in a real setting, some challenges will arise. One potential issue will be noise in the sensor signals. Furthermore, the gravitational signal is considerably larger than the motion to be estimated, which could pose a problem for an accurate estimation. To evaluate the influence of measurement errors on the proposed method, in a subsequent study, errors like noise or gyroscope bias will be included in the simulation.

For this study, we assumed that the initial pose and velocity of the sensor at the beginning of the C-arm scan are known, while in a real setting, these values have to be estimated. An approach for estimating the initial pose from the X-ray projection images was presented in [23]. If the sensor is placed close enough to the knee joint, it is also visible in the projection images without affecting image quality in the area of interest, and its metal components can be tracked. Using the system geometry, an average position over the scan can be obtained, which then can be refined by the first projection images. The initial velocity of the sensor could be estimated by assuming a zero mean velocity over the whole scan, since it has been shown that standing persons sway around their center of mass [1].

We presented a novel approach for motion compensation in weight-bearing C-arm CT using inertial measurements. This initial simulation is a first step towards a purely IMU-based compensation of motion in C-arm CT. We showed the feasibility of our approach which was able to improve image quality and achieve results similar to a state-of-the-art marker-based motion compensation.

Acknowledgment. Bjoern Eskofier gratefully acknowledges the support of the German Research Foundation (DFG) within the framework of the Heisenberg professorship programmme (grant number ES 434/8-1). The authors acknowledge funding support from NIH 5R01AR065248-03 and NIH Shared Instrument Grant No. S10 RR026714 supporting the zeego@StanfordLab.

References

1. Abrahamova, D., Hlavačka, F.: Age-related changes of human balance during quiet stance. Physiol. Res. **57**(6) (2008)
2. Arden, N., Nevitt, M.C.: Osteoarthritis: epidemiology. Best Pract. Res. Clin. Rheumatol. **20**(1), 3–25 (2006)
3. Berger, M., et al.: Marker-free motion correction in weight-bearing cone-beam CT of the knee joint. Med. Phys. **43**(3), 1235–1248 (2016)
4. Bier, B., et al.: Epipolar consistency conditions for motion correction in weight-bearing imaging. In: Maier-Hein, K.H., Deserno, T.M., Handels, H., Tolxdorff, T. (eds.) Bildverarbeitung für die Medizin 2017. I, pp. 209–214. Springer, Heidelberg (2017). https://doi.org/10.1007/978-3-662-54345-0_47
5. Bier, B., et al.: Detecting anatomical landmarks for motion estimation in weight-bearing imaging of knees. In: Knoll, F., Maier, A., Rueckert, D. (eds.) MLMIR 2018. LNCS, vol. 11074, pp. 83–90. Springer, Cham (2018). https://doi.org/10.1007/978-3-030-00129-2_10
6. Bier, B., et al.: Range imaging for motion compensation in C-arm cone-beam CT of knees under weight-bearing conditions. J. Imaging **4**(1), 1–16 (2018)
7. van den Bogert, A.J., Read, L., Nigg, B.M.: A method for inverse dynamic analysis using accelerometry. J. Biomech. **29**(7), 949–954 (1996)
8. Choi, J.H., et al.: Fiducial marker-based correction for involuntary motion in weight-bearing C-arm CT scanning of knees. Part I. Numerical model-based optimization. Med. Phys. **40**(9), 091905–1–091905–12 (2013)
9. Choi, J.H., et al.: Fiducial marker-based correction for involuntary motion in weight-bearing C-arm CT scanning of knees. Part II. Experiment. Med. Phys. **41**(6), 061902–1–061902–16 (2014)
10. De Sapio, V.: Advanced Analytical Dynamics: Theory and Applications. Cambridge University Press, Cambridge (2017)
11. Delp, S.L., et al.: OpenSim: open-source software to create and analyze dynamic simulations of movement. IEEE Trans. Biomed. Eng. **54**(11), 1940–1950 (2007)
12. Hamner, S.R., Seth, A., Delp, S.L.: Muscle contributions to propulsion and support during running. J. Biomech. **43**(14), 2709–2716 (2010)
13. Jost, G., Walti, J., Mariani, L., Cattin, P.: A novel approach to navigated implantation of S-2 alar iliac screws using inertial measurement units. J. Neurosurg.: Spine **24**(3), 447–453 (2016)
14. Lemammer, I., Michel, O., Ayasso, H., Zozor, S., Bernard, G.: Online mobile C-arm calibration using inertial sensors: a preliminary study in order to achieve CBCT. Int. J. Comput. Assist. Radiol. Surg. **15**, 213–224 (2019)

15. Maier, A., et al.: Analysis of vertical and horizontal circular C-arm trajectories. SPIE Med. Imaging **7961**, 796123–1–796123–8 (2011)
16. Maier, A., et al.: CONRAD - a software framework for cone-beam imaging in radiology. Med. Phys. **40**(11), 111914 (2013)
17. Maier, A., Hofmann, H., Schwemmer, C., Hornegger, J., Keil, A., Fahrig, R.: Fast simulation of X-ray projections of spline-based surfaces using an append buffer. Phys. Med. Biol. **57**(19), 6193–6210 (2012)
18. Powers, C.M., Ward, S.R., Fredericson, M., Guillet, M., Shellock, F.G.: Patellofemoral kinematics during weight-bearing and non-weight-bearing knee extension in persons with lateral subluxation of the patella: a preliminary study. J. Orthop. Sports Phys. Ther. **33**(11), 677–685 (2003)
19. Segars, W.P., Sturgeon, G., Mendonca, S., Grimes, J., Tsui, B.M.W.: 4D XCAT phantom for multimodality imaging research. Med. Phys. **37**(9), 4902–4915 (2010)
20. Seth, A., et al.: OpenSim: simulating musculoskeletal dynamics and neuromuscular control to study human and animal movement. PLoS Comput. Biol. **14**(7), 1–20 (2018)
21. Sisniega, A., Stayman, J.W., Yorkston, J., Siewerdsen, J.H., Zbijewski, W.: Motion compensation in extremity cone-beam CT using a penalized image sharpness criterion. Phys. Med. Biol. **62**(9), 3712–3734 (2017)
22. Sisniega, A., Stayman, J., Cao, Q., Yorkston, J., Siewerdsen, J., Zbijewski, W.: Image-based motion compensation for high-resolution extremities cone-beam CT. Proc. SPIE Int. Soc. Opt. Eng. **9783**, 97830K (2016)
23. Thies, M., et al.: Automatic orientation estimation of inertial sensors in C-Arm CT Projections. Curr. Dir. Biomed. Eng. **5**, 195–198 (2019)
24. Wang, Z., Bovik, A.C., Sheikh, H.R., Simoncelli, E.P.: Image quality assessment: from error visibility to structural similarity. IEEE Trans. Image Process. **13**(4), 600–612 (2004)

Feasibility Check: Can Audio Be a Simple Alternative to Force-Based Feedback for Needle Guidance?

Alfredo Illanes[1(✉)], Axel Boese[1], Michael Friebe[1], and Christian Hansen[1,2]

[1] University of Magdeburg, Universitätsplatz 2, 39106 Magdeburg, Germany
alfredo.illanes@ovgu.de
[2] Research Campus STIMULATE, Sandtorstrasse 23, 39106 Magdeburg, Germany

Abstract. Accurate needle placement is highly relevant for puncture of anatomical structures. The clinician's experience and medical imaging are essential to complete these procedures safely. However, imaging may come with inaccuracies due to image artifacts. Sensor-based solutions have been proposed for acquiring additional guidance information. These sensors typically require to be embedded in the instrument tip, leading to direct tissue contact, sterilization issues, and added device complexity, risk, and cost. Recently, an audio-based technique has been proposed for "listening" to needle tip-tissue interactions by an externally placed sensor. This technique has shown promising results for different applications. But the relation between the interaction event and the generated audio excitation is still not fully understood. This work aims to study this relationship, using a force sensor as a reference, by relating events and dynamical characteristics occurring in the audio signal with those occurring in the force signal. We want to show that dynamical information that a well-known sensor as force can provide could also be extracted from a low-cost and simple sensor such as audio. In this aim, the Pearson coefficient was used for signal-to-signal correlation between extracted audio and force indicators. Also, an event-to-event correlation between audio and force was performed by computing features from the indicators. Results show high values of correlation between audio and force indicators in the range of 0.53 to 0.72. These promising results demonstrate the usability of audio sensing for tissue-tool interaction and its potential to improve telemanipulated and robotic surgery in the future.

Keywords: Audio guidance · Force feedback · Needle interventions

1 Introduction

Percutaneous needle insertion is one of the most common minimally invasive procedures. The experience of the clinician is an important requirement for accurate

A. L. Martel et al. (Eds.): MICCAI 2020, LNCS 12263, pp. 24–33, 2020.
https://doi.org/10.1007/978-3-030-59716-0_3

placement of needles, given the reduced visual and tactile information transmitted to the clinician via the instruments. Imaging techniques such as magnetic resonance, computed tomography, or ultrasound can support clinicians in this type of procedure, but the accuracy can still not be fully assured because of artifacts present in the images [14,20].

Sensor-based solutions have been proposed for providing haptic feedback during the procedure [1,3,5,7,8,10–12,18,19,23]. However, most of these solutions require sophisticated sensors that sometimes need to be embedded in the instrument tip or shaft, leading to direct contact with human organs, sterilization issues, the use of non-standard and quality-reduced tools, added device complexity, risk and cost. This imposes serious design limitations, and therefore they have encountered difficulties in being adopted for regular clinical use.

Recently, an audio-based technique has been proposed in [9] for *listening* to the needle tip-tissue interaction dynamics using a sensor placed at the proximal end of the tool. The authors of this work has shown promising preliminary results for monitoring medical interventional devices such as needles [9], guide wires [15], and laparoscopic tools [4]. However, even if audio has proved to be a tool with potential for providing guidance information such as tissue-tissue passage, puncture and perforation events or palpation information, the generated audio dynamics are still not fully understood.

The aim of this work is to investigate the audio dynamics generated from needle-tissue interactions during needle insertion to have a better understanding of the generation of the audio excitation using the audio-based guidance technique. In this purpose, force is used as a reference since it has been widely employed in the literature to understand interactions between needles and tissue. The main idea is to relate events and dynamical characteristics extracted from the audio signal with those extracted from the force signal through indicators and event features computed by processing both signals. The audio signal is processed by extracting its homomorphic envelope. Indicators related to local event intensity, derivative, and curvature are computed from the force signal. The Pearson coefficient is used for signal-to-signal correlation between audio and force indicators. Then, event-to-event correlation between audio and force events is performed by computing features from the indicators.

Results show values of Pearson coefficient between audio and force indicators in the range of 0.53 to 0.72, being the highest one the correlation of audio with force curvature. Additionally, events of high correlated indicators exhibit a clear relationship that can be important for understanding audio behavior. Both analyses show that audio, acquired non invasively with a simple and low-cost sensor, can contain significant information that can be used as additional feedback to clinicians.

2 Method

Needle insertion and its interaction with soft tissue has been widely studied using force sensors, being possible to distinguish three phases of interaction [2,6,17].

During the first phase or pre-puncture phase, the needle tip deforms the surface in contact with the tissue producing an increase in the force. The second phase starts with the puncture event or tissue breakage, characterized by a peak in the force, followed by a sharp decrease. The third phase corresponds to the post-puncture phase, where the force can vary due to friction, collision with interior structures, or due to the puncture of a new tissue boundary. During the first phase, when audio is acquired, no audio excitation occurs since there is no tissue breakage or structure collision. The puncture during the second phase and the collisions, friction, and new punctures during the second and third phases can produce significant and complex audio excitation dynamics. However, even if an audio response is complicated, its dynamics should be related to dynamical characteristics of the force during the second and third phases. This is what we want to explore in this work.

Our aim is to extract characteristics or *feature indicators* from the force that can be related to dynamical characteristics of the audio excitation. The first indicator that we want to explore is the local intensity of the force or detrended force, which aims to emphasize the increase of force from a local deflection (contact of needle tip with the tissue) passing through its peak (puncture event) and coming back to a steady stage. We also believe that the cumulative energy stored during the boundary displacement and the fast drop in force after the puncture also influences the audio excitation, and this is why derivative and curvature indicators are also extracted.

The idea of this work is not to explain mechanical properties and fundamentals of needle insertion in soft tissue, but to demonstrate that characteristics of audio and force, even resulting from sensors of entirely different nature, can be strongly related through a sort of transfer function between both sensor modalities. Through this relationship, we also want to show the wealth of information that an audio signal can contain concerning tip-tissue interaction dynamics.

Figure 1 displays a block diagram with the main steps to relate audio and force characteristics. First, the audio signal and the force signal are processed in order to compute the different indicators extracted for enhancing the signal features that want to be compared: one audio indicator (IA), and four force indicators, related to the local intensity (IF_i), to the curvature (IF_c), to the derivative (IF_d), and one indicator that integrates curvature and intensity (IF_{ci}). Then, a signal-to-signal correlation is performed between audio and force indicators in order to assess similarity between features and also for optimizing the parameters of the processing algorithms. Finally, an event-to-event correlation is performed by computing features from the extracted indicators.

2.1 Experimental Setup and Data Acquisition

For evaluating the presented approach, the dataset generated in [9] was used, where audio signals were recorded using a stethoscope connected to a microphone attached to the proximal end of a needle via a 3D printed adapter (see Fig. 2a). This dataset consists of 80 audio recordings acquired during automatic insertion of an 18G 200mm length biopsy needle (ITP, Germany) into an ex-vivo porcine

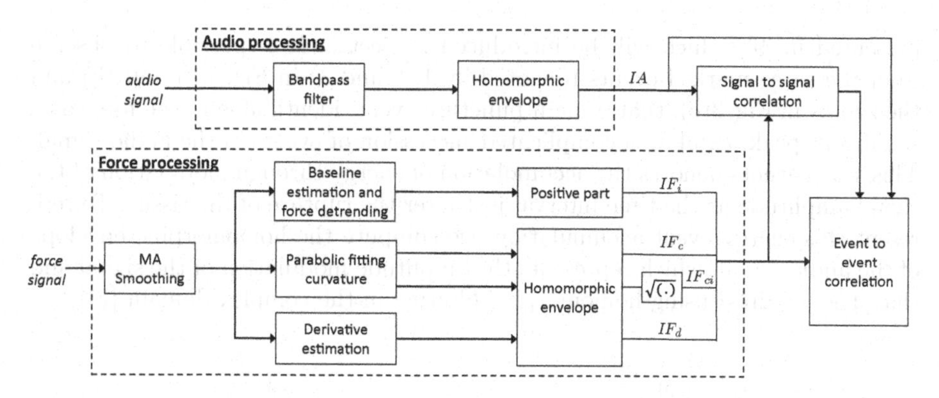

Fig. 1. General block diagram with the methodology to relate force and audio signals.

tissue phantom (see Fig. 2b). The insertion was performed automatically using a testing machine (Zwicki, Zwick GmbH & Co.KG, Ulm) at an insertion velocity of 3 mm/s that also recorded the axial needle insertion force. The audio frequency sampling was 44100 Hz, and one force sample was acquired every 0.03 mm. If we assume the velocity nearly constant, the force sampling frequency can be estimated at around 100 Hz. The acquisition of force and audio was synchronized using a trigger event visible in both the force and audio signals.

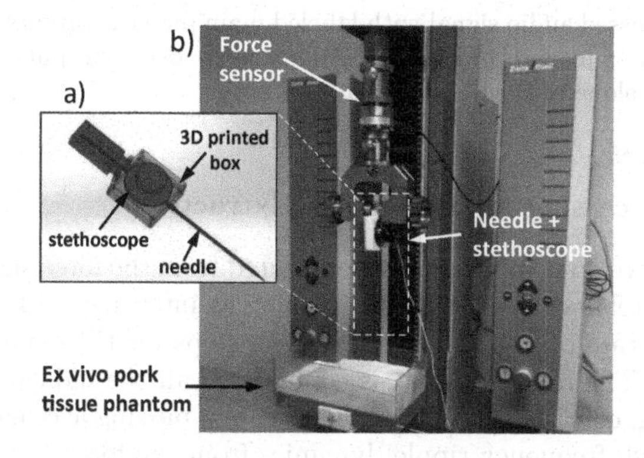

Fig. 2. Experimental setup for needle insertion in fat tissue (adapted from [9]).

2.2 Audio Indicator Extraction

The audio signal is first pre-processed using a 7th-order Butterworth bandpass filter (3–6 KHz) in order to enhance the needle tip tissue interaction information, as shown in [9].

Figure 3 displays the filtered audio signal together with the synchronized force signal during needle insertion into soft tissue, using a recording from the dataset presented in [9], which will be introduced in Sect. 3. It is possible to observe from the three marked events (denoted by 1, 2 and 3) in Fig. 3(a) and (b) and the zooms in Fig. 3(c), that a main puncture event, identifiable in the force as a significant peak, exhibits a complicated succession of events in the audio signal. This set of events denotes the accumulation of energy through modulation of the signal amplitude at the time interval just after the rupture of the tissue. To represent this energy event accumulation, we compute the homomorphic envelope of the audio signal, which represents the amplitude modulation of the signal and that it is obtained using homomorphic filtering in the complex domain [21].

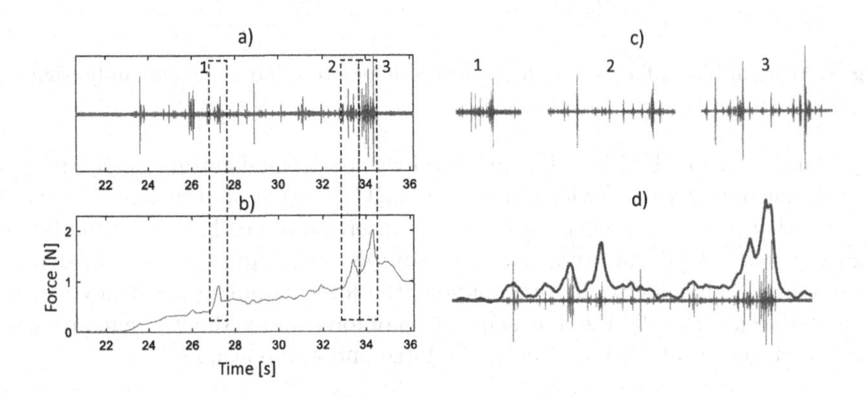

Fig. 3. (a) Bandpassed audio signal with labeled main puncture events. (b) force signal with labeled main puncture events. (c) Audio zoom over the puncture events. (d) Homomorphic envelope of the audio signal.

2.3 Force Processing for Indicators Extraction

As explained above, four indicators are extracted from the force signal by enhancing information related to characteristics such as intensity, derivative, and curvature. The diagram of Fig. 1 shows the main steps for the computation of the force indicators. The first step that is common to all the indicators is the force signal smoothing, consisting of the application of a moving average filter in order to reduce the high-frequency ripple dynamics from the force.

The force indicator IF_i intends to extract the information concerning the local intensity of the force. In this aim, a force detrending has to be performed to attenuate the very low-frequency progression present in the force signal during needle insertion. For that, the signal baseline is first estimated using a two-stage median filter [22] that is then subtracted to the smoothed force signal. Finally, the positive part of the resulting signal is extracted.

For the computation of the force indicator related to the derivative IF_d, a derivative filter in series with a smoothing filter is first applied following [13]. To only keep the most important positive characteristics of the derivative, the homomorphic envelope is extracted.

The curvature is estimated with the algorithm proposed in [16] that enhances curvature and intensity in signals. A 2nd-degree polynomial is used to fit the force signal inside a sliding window. The 2nd-degree coefficient is related to the curvature of the signal. The force indicator IF_c is finally computed by applying a homomorphic envelope to the resulting curvature.

As explained in [16], an indicator that enhance intensity and curvature jointly can be extracted by computing the product between the constant polynomial coefficient and the 2nd-degree coefficient of the polynomial. IF_{ci} is computed as the square root of the joint indicator.

Figure 4 shows the four extracted indicators, including also the original force signal and the baseline estimation (in red).

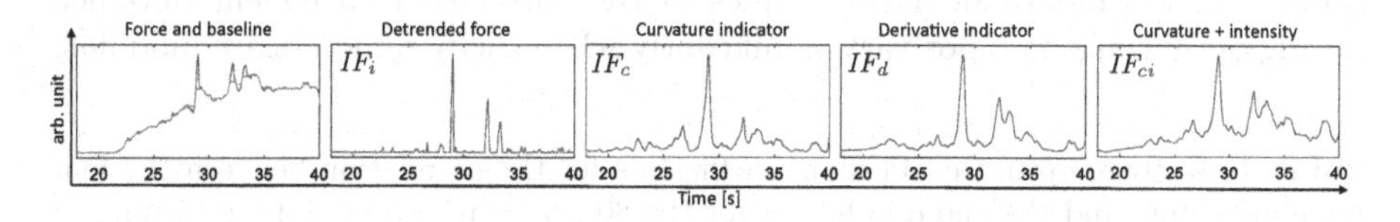

Fig. 4. Four indicators extracted from the force signal. In red color the estimated force baseline is displayed. (Color figure online)

2.4 Force and Audio Information Correlation Methodology

Two approaches are applied for putting in relation the indicators extracted from the force with the one extracted from the audio. A signal-to-signal correlation using the Pearson coefficient is first performed. This step allows assessing the similarity between both types of indicators and also to optimize the parameters involved in the computation of the indicators. In fact, the extraction of the force and audio indicators requires to set some parameters:

- The frequency cut-off of the low-pass filter to be used in the envelope extraction for the audio and force signals, denoted as lpf_a and lpf_f, respectively.
- The length of the sliding window used for computing the polynomial fitting of the force signal, denoted as h_{win}.
- The first and second stage averaging window length of the median filter applied for the force detrending, denoted by L_1 and L_2, respectively.

Each parameter is optimized by maximizing the Pearson coefficient between the force indicator and the audio indicator.

Using the optimized parameters, we perform an event-to-event correlation analysis. Significant puncture events in the force are first detected using a standard peak detector algorithm. Then, a window W is defined around the detected peak instant in the audio signal. Finally, for each detected event, we relate the maximal value of the force indicator inside W to the energy of the audio indicator, also inside W.

3 Results

Table 1 shows the optimized results of the correlation between the four force indicators and the audio indicator. The table also shows the values of the parameters for the extraction of the indicators. In bold are marked the parameters that required optimization (lpf_a, L_1, L_2, h_{win}, and lpf_f). The span of the MA smoother used for the computation of the force indicators was set to 10 samples, equivalent to 0.1 s for a sampling frequency of 100 Hz. The last column of the table shows the average optimized Pearson coefficient value ($\overline{\rho}$) over the 80 recordings of the dataset. It is possible to observe that the best correlation between audio and force is obtained with the force curvature indicator. It is important to notice in the table, in the second column from the end, that the average and standard deviation values of the correlation coefficient obtained during optimization do not vary significantly when the parameters are modified.

Table 1. Average optimized Pearson coefficients for the correlation between the four force indicators and the audio indicator for the 80 needle insertion audio recordings.

Comparison	Audio parameters	Force parameters					$mean \pm std$	$\overline{\rho}$
	$\mathbf{lpf_a}$	MAspan	$\mathbf{L_1}$	$\mathbf{L_2}$	$\mathbf{h_{win}}$	$\mathbf{lpf_f}$		
IA vs IF_i	4	10	100	100	n/a	n/a	0.489 ± 0.038	0.531
IA vs IF_c	1	10	n/a	n/a	7	1	0.611 ± 0.038	0.717
IA vs IF_d	1	10	n/a	n/a	n/a	1	0.511 ± 0.061	0.664
IA vs IF_{ci}	1	10	n/a	n/a	13	1	0.578 ± 0.050	0.672

Figure 5 displays a further analysis concerning the obtained correlation Pearson coefficients. In Fig. 5(a), the histograms of the Pearson coefficients between the four force indicators and the audio indicator are displayed. It is possible to verify that the correlation values range in general between 0.3 and 0.9, but that for curvature and the joint curvature and intensity indicators, 50% of the recording has a correlation value over 0.6, which is high considering the completely different nature between audio and force sensors. Figure 5(b), which shows the accumulative histogram of the Pearson coefficient, confirms the analysis made previously. It is possible to observe that the best correlation between audio and force is obtained with the curvature indicator IF_c, followed by the joint curvature and intensity indicator IF_{ci}. The derivative of the force IF_d also provides high values of correlation, while the local intensity indicator IF_i is the weakest indicator influencing the audio.

Figure 6 shows examples of high correlations between each force indicator and the audio indicator for different recordings belonging to the dataset. It is possible to see how the main dynamics extracted from the audio indicator can follow the dynamics obtained from the force indicator, i.e., many of the information involve in force it is somehow visible in the audio indicator signal.

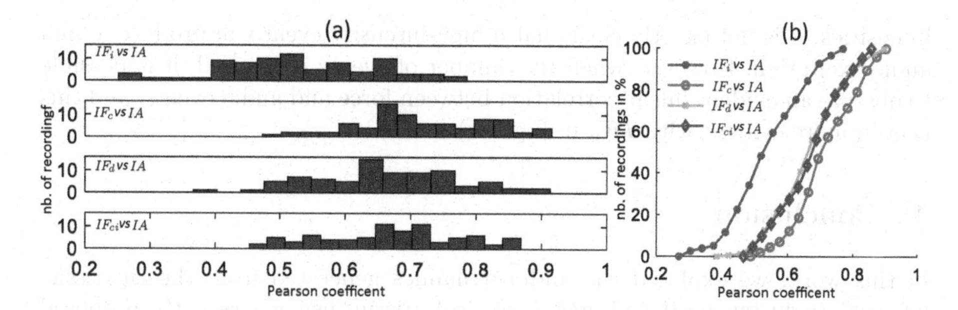

Fig. 5. Standard and cumulative histograms of the Pearson coefficients of the four force indicators with audio, for the 80 recordings.

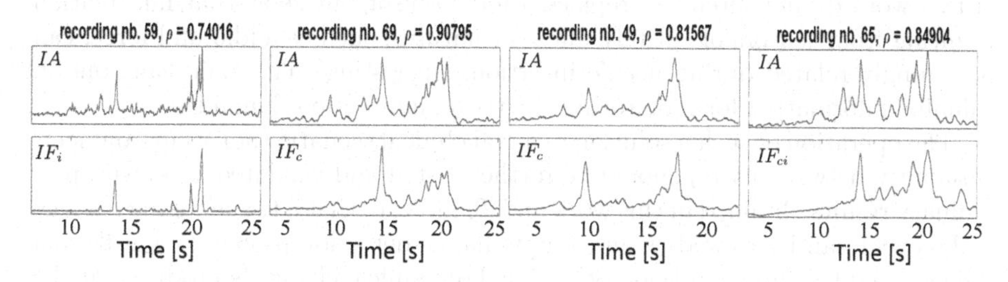

Fig. 6. Examples of force and audio indicators where high correlation were obtained.

Figure 7 shows the results of the event-to-event correlation of the two best correlations obtained in the signal-to-signal analysis: curvature IF_c indicator versus the audio indicator IA and the joint curvature and intensity indicator IF_{ci} and IA. We explore two scenarios, the first one (Fig. 7(a) and (b)) by taking into account a large number of puncture events and the second one (Fig. 7(c) and (d)) by taking into account only the most important puncture events occurring during the needle insertion process. This is done by only modifying a simple threshold in the peak detector; higher is the threshold more events will be taken into account. For both force indicators, a clear event correlation can be observed using both types of event thresholding. When a large number of events are taken into account, we can see that in the range of force events presenting low

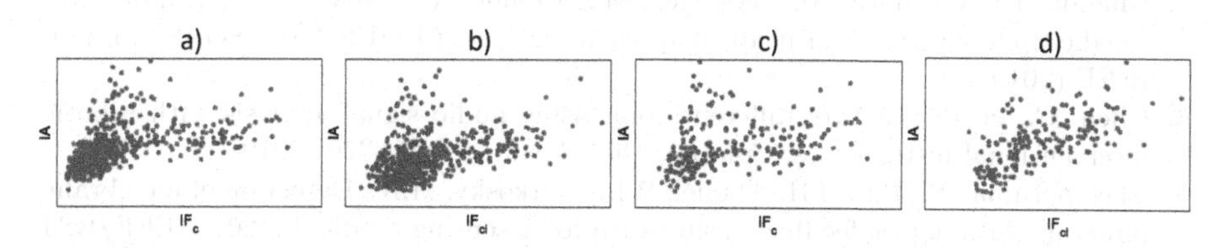

Fig. 7. Event-to-event correlation for (a) IF_c vs. IA with a low intensity-event threshold, (b) IF_{ci} vs. IA with a low intensity-event threshold, (c) IF_c vs. IA with a high intensity-event threshold, and (d) IF_{ci} vs. IA with a high intensity-event threshold.

intensities, it is not exactly clear that a high-intensity event will produce a high audio excitation. However, when the number of events is reduced, it is possible to observe an evident linear correlation between force and audio events, and this is even more evident with IF_{ci} in Fig. 7(c).

4 Conclusion

In this work, we explored the audio dynamics generated from the tip/tissue interaction during needle insertion into soft tissue using a recently proposed audio guidance technique. The main idea was to observe the effect of different characteristics of force measurement on the audio excitation. The results of this work do not intend to replace a force sensor but show that information resulting from sensors of entirely different natures, such as audio and force, can be strongly related during needle insertion, suggesting that audio can contain valuable information for monitoring tip/tissue interaction dynamics.

The operation of such a solution in a real clinical scenario (noisy environment, variability between users) should be further tested and validated. However, preliminary results obtained in this work and [9] indicate that a tissue/tissue passage audio event can be viewed as an abrupt change detection problem regardless of the user and the insertion velocity (a signal dynamical change is produced at the border of two tissues). Additionally, a puncture results in a high energy audio excitation that may simplify the processing of the signal in noisy environments.

The next steps of this work involve the exploration of non-linear dynamical relationships between force and audio using not only time-domain energy-based features but also time-variant audio signatures using frequency, scale, or modal based features.

References

1. Abayazid, M., Kemp, M., Misra, S.: 3D flexible needle steering in soft-tissue phantoms using fiber bragg grating sensors. In: 2013 IEEE International Conference on Robotics and Automation, pp. 5843–5849. IEEE (2013)
2. Abolhassani, N., Patel, R., Moallem, M.: Needle insertion into soft tissue: a survey. Med. Eng. Phys. **29**(4), 413–431 (2007)
3. Chadda, R., Wismath, S., Hessinger, M., Schäfer, N., Schlaefer, A., Kupnik, M.: Needle tip force sensor for medical applications. In: 2019 IEEE SENSORS, pp. 1–4. IEEE (2019)
4. Chen, C., et al.: Texture differentiation using audio signal analysis with robotic interventional instruments. Comput. Biol. Med. **112**, 103370 (2019)
5. Elayaperumal, S., Bae, J.H., Daniel, B.L., Cutkosky, M.R.: Detection of membrane puncture with haptic feedback using a tip-force sensing needle. In: 2014 IEEE/RSJ International Conference on Intelligent Robots and Systems, pp. 3975–3981. IEEE (2014)
6. van Gerwen, D.J., Dankelman, J., van den Dobbelsteen, J.J.: Needle-tissue interaction forces-a survey of experimental data. Med. Eng. Phys. **34**(6), 665–680 (2012)

7. Henken, K., Van Gerwen, D., Dankelman, J., Van Den Dobbelsteen, J.: Accuracy of needle position measurements using fiber bragg gratings. Minim. Invasive Ther. Allied Technol. **21**(6), 408–414 (2012)
8. Ho, S.C.M., Razavi, M., Nazeri, A., Song, G.: FBG sensor for contact level monitoring and prediction of perforation in cardiac ablation. Sensors **12**(1), 1002–1013 (2012)
9. Illanes, A., et al.: Novel clinical device tracking and tissue event characterization using proximally placed audio signal acquisition and processing. Sci. Rep. **8**(1), 1–11 (2018)
10. Iordachita, I., et al.: A sub-millimetric, 0.25 mn resolution fully integrated fiber-optic force-sensing tool for retinal microsurgery. Int. J. Comput. Assist. Radiol. Surg. **4**(4), 383–390 (2009)
11. Kalvøy, H., Frich, L., Grimnes, S., Martinsen, Ø.G., Hol, P.K., Stubhaug, A.: Impedance-based tissue discrimination for needle guidance. Physiol. Meas. **30**(2), 129 (2009)
12. Kumar, S., Shrikanth, V., Amrutur, B., Asokan, S., Bobji, M.S.: Detecting stages of needle penetration into tissues through force estimation at needle tip using fiber bragg grating sensors. J. Biomed. Opt. **21**(12), 127009 (2016)
13. Laguna, P., et al.: New algorithm for QT interval analysis in 24-hour Holter ECG: performance and applications. Med. Biol. Eng. Comput. **28**(1), 67–73 (1990)
14. Lal, H., Neyaz, Z., Nath, A., Borah, S.: CT-guided percutaneous biopsy of intrathoracic lesions. Korean J. Radiol. **13**(2), 210–226 (2012)
15. Mahmoodian, N., Schaufler, A., Pashazadeh, A., Boese, A., Friebe, M., Illanes, A.: Proximal detection of guide wire perforation using feature extraction from bispectral audio signal analysis combined with machine learning. Comput. Biol. Med. **107**, 10–17 (2019)
16. Manriquez, A.I., Zhang, Q.: An algorithm for QRS onset and offset detection in single lead electrocardiogram records. In: 2007 29th Annual International Conference of the IEEE Engineering in Medicine and Biology Society, pp. 541–544. IEEE (2007)
17. Okamura, A.M., Simone, C., O'leary, M.D.: Force modeling for needle insertion into soft tissue. IEEE Trans. Biomed. Eng. **51**(10), 1707–1716 (2004)
18. Park, Y.L., et al.: Real-time estimation of 3-D needle shape and deflection for MRI-guided interventions. IEEE/ASME Trans. Mechatron. **15**(6), 906–915 (2010)
19. Ravali, G., Manivannan, M.: Haptic feedback in needle insertion modeling and simulation. IEEE Rev. Biomed. Eng. **10**, 63–77 (2017)
20. Reusz, G., Sarkany, P., Gal, J., Csomos, A.: Needle-related ultrasound artifacts and their importance in anaesthetic practice. Br. J. Anaesth. **112**(5), 794–802 (2014)
21. Rezek, I., Roberts, S.J.: Envelope extraction via complex homomorphic filtering. Technical Report TR-98-9 Technical report (1998)
22. Sameni, R., Shamsollahi, M.B., Jutten, C.: Model-based Bayesian filtering of cardiac contaminants from biomedical recordings. Physiol. Meas. **29**(5), 595 (2008)
23. Xu, R., Yurkewich, A., Patel, R.V.: Curvature, torsion, and force sensing in continuum robots using helically wrapped FBG sensors. IEEE Robot. Autom. Lett. **1**(2), 1052–1059 (2016)

A Graph-Based Method for Optimal Active Electrode Selection in Cochlear Implants

Erin Bratu[1](✉), Robert Dwyer[2], and Jack Noble[1]

[1] Department of Electrical Engineering and Computer Science, Vanderbilt University, Nashville, TN 37235, USA
{erin.l.bratu,jack.noble}@vanderbilt.edu

[2] Department of Hearing and Speech Sciences, Vanderbilt University Medical Center, Nashville, TN 37232, USA

Abstract. The cochlear implant (CI) is a neural prosthetic that is the standard-of-care treatment for severe-to-profound hearing loss. CIs consist of an electrode array inserted into the cochlea that electrically stimulates auditory nerve fibers to induce the sensation of hearing. Competing stimuli occur when multiple electrodes stimulate the same neural pathways. This is known to negatively impact hearing outcomes. Previous research has shown that image-processing techniques can be used to analyze the CI position in CT scans to estimate the degree of competition between electrodes based on the CI user's unique anatomy and electrode placement. The resulting data permits an algorithm or expert to select a subset of electrodes to keep active to alleviate competition. Expert selection of electrodes using this data has been shown in clinical studies to lead to significantly improved hearing outcomes for CI users. Currently, we aim to translate these techniques to a system designed for worldwide clinical use, which mandates that the selection of active electrodes be automated by robust algorithms. Previously proposed techniques produce optimal plans with only 48% success rate. In this work, we propose a new graph-based approach. We design a graph with nodes that represent electrodes and edge weights that encode competition between electrode pairs. We then find an optimal path through this graph to determine the active electrode set. Our method produces results judged by an expert to be optimal in over 95% of cases. This technique could facilitate widespread clinical translation of image-guided cochlear implant programming methods.

Keywords: Cochlear implants · Graph search · Image guided cochlear implant programming

1 Introduction

In the United States, it is estimated that 2 to 3 out of every 1000 children are born with some degree of hearing loss, and 37.5 million adults experience some degree of hearing loss [1]. The cochlear implant (CI) is a neural prothesis that, over the last two decades, has become the standard treatment for severe-to-profound hearing loss [2]. As of 2012, an estimated 324,000 CIs have been implanted worldwide, and in the United

A. L. Martel et al. (Eds.): MICCAI 2020, LNCS 12263, pp. 34–43, 2020.
https://doi.org/10.1007/978-3-030-59716-0_4

States, approximately 58,000 adults and 38,000 children have received a CI [1]. The CI is primarily used in cases of sensorineural hearing loss, where damage or defects affecting hearing are in the cochlea, a.k.a., the inner ear. In a subject without hearing loss, sounds reaching the cochlea would be transduced to electrical impulses that stimulate auditory nerve fibers. The nerve fibers are tonotopically organized, meaning that activation of nerve fibers located in different regions of the cochlea create the sensation of different sound frequencies. The frequency for which a nerve fiber is activated in natural hearing is called its characteristic frequency. As such, in natural hearing nerve fibers are activated when their characteristic frequencies are present in the incoming sound.

In a patient with hearing loss, sounds no longer properly activate auditory nerve fibers. The purpose of CI is to bypass the natural transduction mechanisms and provide direct electrical stimulation of auditory nerve fibers to induce hearing sensation. A CI consists of an electrode array that is surgically implanted in the cochlea (see Fig. 1a) and an external processor. The external processor translates auditory signals to electrical impulses that are distributed to the electrodes in the array according to the patient's MAP, which is the set of processor instructions determined by an audiologist in an attempt to produce optimal hearing outcomes. Tunable parameters in a patient's MAP include the active set of electrodes, the stimulation level of each electrode, and a determination of which electrodes should be activated when a particular frequency of sound is detected by the processor. Research indicates that the locations of electrodes within the cochlea impact the quality of hearing outcomes [3–9]. Most patients have less-than-optimal electrode array placement [5], so customizing the patient's MAP is critical for optimizing hearing outcomes. Previous studies have proposed methods for segmenting cochlear anatomy and electrode arrays from pre- and post-operative computed tomography (CT) images, permitting creation of 3D models of cochlear structures [10–12].

Research has also shown that the spatial information garnered from these methods can be used to estimate channel interactions between electrodes [12, 13]. Channel interaction occurs when nerves, which naturally are activated for a finely tuned sound frequency, receive overlapping stimulation from multiple electrodes, corresponding to multiple frequency channels. This creates spectral smearing artifacts that lead to poorer hearing outcomes. Manipulating a subject's MAP to modify the active electrode set and the stimulation patterns can reduce these effects. The spatial relationship between electrodes and neural sites is a driving factor for channel interaction. Modeling an electrode as a point charge in a homogenous medium has been shown to yield similar electric field estimates to more sophisticated finite element models when using plausible tissue resistivity values within known ranges for the human cochlea [14]. Using a point charge model, Coulomb's law mandates that electric field strength, $E(\vec{x})$, at location $\vec{x}$ is inversely proportional with squared distance between $\vec{x}$ and the electrode location, $\vec{c}$.

$$E(\vec{x}) \propto \frac{1}{\|\vec{x} - \vec{c}\|^2} \tag{1}$$

Thus, as shown in Fig. 1b, when an electrode is close to neural sites (e.g., E8–E10), relatively little current is needed to activate nearby nerves, resulting in relatively little spread of excitation. However, when an electrode is distant to neural sites, neural activation requires broad stimulation patterns due to electrical current spread at greater distance. When two electrodes are close together and both distant to neural sites (e.g., E5–6), they create substantial stimulation overlap, resulting in channel interaction.

One method of visualizing the spatial relationship between electrodes and neural sites to determine when channel interaction occurs is to use distance vs. frequency (DVF) curves. These curves represent the distance from the auditory nerve spiral ganglion (SG) cells, which are the most likely target of electrical stimulation, to nearby electrodes (see Fig. 1c). The characteristic frequencies of the nerve fiber SG sites are shown on the horizontal axis, and the distance to electrodes near to those neural sites are shown with the height of an individual curve for each electrode on the vertical axis. This simplifies the process of determining which nerve pathways are likely to be stimulated by a given electrode and where two electrodes might stimulate the same region. An electrode is most likely to stimulate the nerves it is closest to, as indicated by the horizontal position of the minimum of the curve. We refer to these nerves as having SGs located in the peak activation region (PAR) for the electrode. Determining which nerves are stimulated by multiple electrodes requires making additional assumptions about the spread of excitation of each contact. In this work, we use Eq. (1) to estimate electric field strength, and we assume the activation region for an electrode includes any nerves with SG sites $\vec{x}$ that satisfy:

$$\frac{E(\vec{x})}{E(\text{PAR})} = \frac{\|\text{PAR} - \vec{c}\|^2}{\|\vec{x} - \vec{c}\|^2} \geq \tau, \tag{2}$$

which requires that the strength of the electric field in SGs must be greater than a certain fraction, τ, of the electric field in the PAR for those nerves to be considered active. This is equivalent to ensuring the ratio of squared distance from the PAR to the electrode to the square distance from another nerve SG site to the electrode is greater than τ. The DVF curves (see Fig. 1c) permit visually assessing the activation region of each contact. The activation region is defined by the width of the curve for which $\sqrt{\tau}$ times the curve height, $\sqrt{\tau}\|\vec{x} - \vec{c}\|$, is less than or equal to the minimum curve height, $\|\text{PAR} - \vec{c}\|$.

If substantial overlap of activation regions exists between neighboring electrodes, some electrodes may be selected for deactivation to reduce overlap. This is one approach for image-guided CI programming (IGCIP), i.e., a method that uses image information to assist audiologists optimize programming of CIs. The original technique for selecting the deactivated set required an expert to manually review each case and determine the optimal solution based on the information in the DVF curves. This process is not ideal for clinical translation as it can be time-consuming and requires expertise. Automated methods have been developed to eliminate the need for expert review, which either rely on an exhaustive search to optimize a cost function that relies on shape features of the DVF curves [15] or attempt to learn to replicate expert deactivation patterns using DVF curve template matching [16]. However, as presented below, the current state-of-the-art method [15] leads to optimal results in only 48% of cases, which is insufficiently reliable for widespread clinical translation. In this paper, we present an automated method for determining the active electrode set as a minimum cost path in a custom-designed graph. As our results will show, our method is significantly more robust in finding optimal deactivation plans compared to the state-of-the-art method and could facilitate automated clinical translation of IGCIP methods.

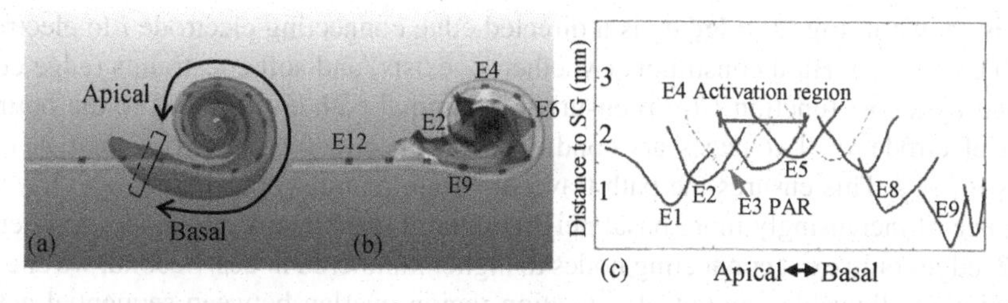

Fig. 1. (a) A 3D representation of cochlea (red) and the electrode array (gray). (b) The modiolus containing the SG cells of the auditory nerve is shown in green with estimated spread of excitation from the CI electrodes multiple colors. (c) DVF curves for the same case, showing a deactivation plan. Active electrodes are represented by solid blue lines, and deactivated electrodes by dashed gray lines. (Color figure online)

2 Methods

The dataset in this study consists of 83 cases for which we have patient-specific anatomical data that is used to generate the DVF curves and electrode deactivation plans from the current state-of-the-art technique [15] and our proposed method. All cases use an implant from one of three manufacturers: MED-EL (MD) (Innsbruck, Austria), Advanced Bionics (AB) (Valencia, California), and Cochlear (CO) (New South Wales, Australia). Of these cases, 24 used an implant from MD, 32 from AB, and 27 from CO.

2.1 Graph Definition

We propose a graph, $G = \{N, E\}$, as a set of nodes N and edges E. Each node in N represents an electrode in the array. An optimal path resulting in minimum cumulative edge cost traverses edges connecting nodes corresponding to electrodes recommended for activation. Electrodes corresponding to nodes not in the path will be recommended for deactivation. Using this approach, we (1) select the start and end nodes of the path, (2) identify valid edges using hard constraints, (3) calculate edge costs using soft constraints, and (4) use Dijkstra's algorithm [17] to find the globally optimal path.

The start node of the path is chosen to be the most apical contact (see Fig. 1a) because deactivating the most apical contact reduces stimulation of the lowest frequency nerves, creating perceived frequency upshifts that are generally bad for hearing outcomes [18]. Thus, it is desirable for this electrode to always be active. Similarly, the end node should be the electrode with PAR among the highest frequency nerves that can be effectively stimulated near the basal end of the cochlea. It is well known that electrodes outside the cochlea and those near the entrance of the cochlea are typically ineffective in stimulating auditory nerves. We use the active-shape-model based segmentation approach proposed by [19] to segment the cochlea and rely on the one-to-one point correspondence between the segmentation in the patient image and an atlas image to define a decision plane (see Fig. 1a). The plane corresponds to nerves with characteristic frequencies of 15 kHz. The first electrode apical to this plane is the end node of the path.

The edges E are defined to permit finding a minimum edge cost path from the start to the end node that represents the optimal set of active electrodes. Our proposed structure of E is shown in Fig. 2. Edge e_{ij} is a directed edge connecting electrode i to electrode j with cost $C(e_{ij})$. Hard constraints (whether e_{ij} exists) and soft constraints (edge costs defined by a cost function $C(e_{ij})$) ensure the minimal path corresponds to the optimal active electrode set. Two necessary conditions must be met for e_{ij} to exist. First, e_{ij} only exists if $i < j$. This ensures the path traverses from the most apical electrode, E1, to a sequence of increasingly more basal neighbors until reaching the end node. As seen in Fig. 3, edges only exist connecting nodes to higher numbered nodes. Second, we encode a maximum allowable amount of activation region overlap between sequential active electrodes in the path as a hard constraint. We use Eq. 2 to define the activation region for each electrode and let $\tau = 0.5$ in our experiments. We found heuristically that this value of τ selected a similar rate of active electrodes as reported in studies of the number of effectively independent electrodes as a function of electrode distance [20], and it also matched behavior of experts when selecting electrodes for deactivation. We then define overlap acceptable if the activation region for electrode j does not include the PAR for electrode i and vice versa. Thus, if the region most likely to be activated by an electrode (its PAR) is also activated by another electrode, too much overlap is occurring, in which case e_{ij} does not exist. For example, in Fig. 1c the PAR for E3 falls within the activation region for E4, therefore $\nexists$ e_{34}. An example of such a scenario is shown in our example graph in Fig. 2 with edge e_{12}. With n_1 and n_2 exhibiting too much overlap, $\nexists$ e_{12}, and thus a path from n_1 must skip n_2 and instead traverse directly to n_3 or n_4. One example allowable path in this graph is shown in red.

Soft constraints are encoded in an edge cost function,

$$C(e_{ij}) = \alpha d_i + (1 - \alpha)\beta^{(j-i-1)}, \tag{3}$$

where $d_i = \left\| \mathrm{PAR}_i - \vec{c_i} \right\|$ is the distance from electrode i to its PAR, and α and β are parameters. The second term in the cost function ensures as many electrodes are active as allowable by the hard constraints, since when $j = i + 1$, no electrodes are deactivated, but when $j > i + 1$, some electrodes are skipped in the path, indicating they will be deactivated, and, assuming $\beta > 1$, a higher cost is associated with this. Further, a larger cost is assigned when deactivating multiple electrodes in sequence, i.e., when $j \gg i + 1$, to discourage deactivations that result in large gaps in neural sites where little stimulation occurs. Larger values of β result in greater values for this penalty. The first term in Eq. (3) rewards active electrodes that tend to have lower distance to SG sites. The parameter α controls the relative contribution of the two terms.

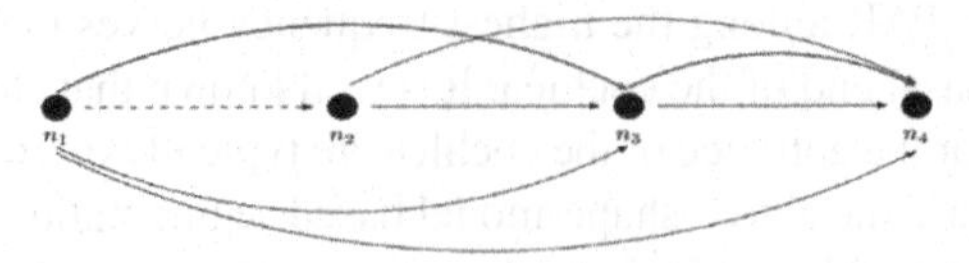

Fig. 2. Visualization of our graph design. Each node is indicated with a black circle and is labeled with a node number n_i. Dotted lines indicate invalid edges, and solid lines indicate valid edges. Red lines indicate an example path through the graph. (Color figure online)

From this graph, Dijkstra's shortest-path algorithm can determine the global cost minimizing path. The resulting path represents the set of electrodes that should remain active, while electrodes not in the path will be recommended for deactivation.

2.2 Validation Study

Ideally, we would have an expert determine the optimal deactivation plan for each of the 83 cases in our dataset and measure the rate at which the algorithm produces the optimal plan. However, for a given case, it is possible there are multiple deactivation configurations that could be considered equally optimal, and it is difficult to determine a complete set of equally optimal plans. Thus, to assess the performance of our method, we instead implemented a masked expert review study to assess optimality of the results of our algorithm compared to the current state-of-the-art algorithm and control plans for each case. In this study, an expert reviewer was presented with a graphical representation of the DVF curves for each case, showing the planned active and deactivated electrodes, similarly to Fig. 1c. The reviewer was instructed to determine whether each plan was optimal, i.e., the reviewer would not adjust anything in the presented plan. Three sets of plans for each case were presented in this study. The first set consists of the results from our proposed method using parameters $\alpha = 0.5$ and $\beta = 4$. These parameter values were determined heuristically using DVF curves from 10 cases not included in the validation set. The second set is the deactivation result from the method described in [15]. The final set includes control plans, manually created by a second expert, where the active set is close to acceptable, but suboptimal. The inclusion of control plans is used to indicate if the reviewer has a bias toward rating plans as optimal, e.g., if numerous control cases are rated as optimal, the reviewer likely has such a bias. The three sets of plans were presented one at a time in random order. The reviewer was masked to the source of each plan in order to prevent bias towards any method.

2.3 Parameter Sensitivity Analysis

We performed a parameter sweep to assess the sensitivity of the parameters in our cost function across a set of values around the heuristically determined values of $\alpha = 0.5$ and $\beta = 4$ used above in the validation study. We used our proposed method to determine the active electrode set with parameter α in the range [0.1, 0.9] with step size of 0.1 and β in the range [2, 6] with a step size of 0.5. This resulted in 81 different parameter combinations for each case. We then used the Hamming distance metric to compare the resulting plan to the plan evaluated in the validation study. Large differences would indicate greater sensitivity of the method to the parameters.

3 Results and Discussion

The results of our validation study are shown in Table 1. Our reviewer judged 79 of the plans generated using our proposed method to be optimal, rejecting only four cases. Only 40 of the plans from the previous method described in [15] were rated optimal, and none of the control plans were marked optimal. Accepting none of the control plans indicates that our expert reviewer is not biased toward accepting configurations and can accurately distinguish between optimal and close-to-optimal plans. We used McNemar mid-p tests to assess the accuracy of our plan to produce an optimal result versus that of the current state-of-the-art method in [15] as well as the control method. We found that the difference in success rates between the two methods and between the proposed and control method were highly statistically significant ($p < 10^{-9}$).

Table 1. Validation study results. Each row indicates the number of optimal and non-optimal plans generated by the given method in each column.

	Proposed	Algorithm from [14]	Control
Optimal	79	40	0
Non-optimal	4	43	83

Inspecting the four cases where the proposed deactivation plan was rejected, the reviewer noted that the plans for these cases were actually optimal, and the rejection in each case was due to erroneous reading of the DVF curves when the amount of activation region overlap between electrodes was very close to the acceptable overlap decision threshold. DVF curves for one such case are shown in Fig. 3a, along with the deactivation plan suggested by [15] in panel (b). The plan from the proposed algorithm in (a) was rejected because the reviewer mistakenly believed the PAR for E5 (green) fell within the activation region for E4 (red). Note that the hard constraints imposed by our proposed method guarantee plans that are free of this type of error, which is a significant benefit of graph-based, compared to other optimization methods. The plan from algorithm [15] in (b) is also borderline but was correctly judged to be unacceptable because the PAR for E2 (green) falls just outside the activation region for E3 (red), meaning E2 should be active.

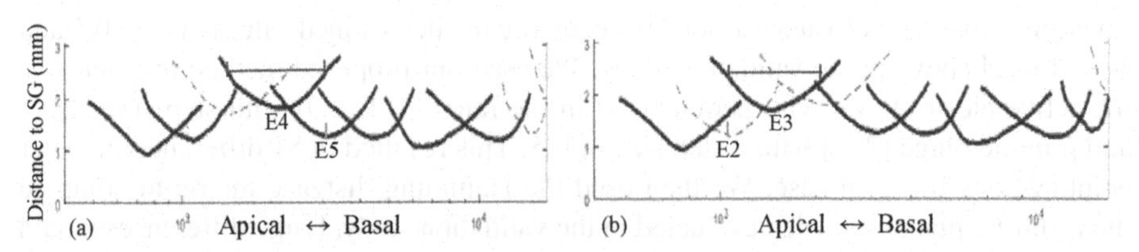

Fig. 3. (a) A rejected plan generated using our proposed method. (b) A rejected plan generated using the method from [15], showing the same case as (a).

The results of our parameter sensitivity study are shown in Fig. 4. We found that our method was relatively insensitive to low values for α and high values for β, i.e., the deactivation plan did not change from the $\alpha = 0.5$, $\beta = 4$ solution used in the validation study in this region of the parameter space. However, large numbers of plans changed when α was high or β was low. Since our validation study revealed that $\alpha = 0.5$, $\beta = 4$ produced optimal solutions, changes in many plans indicates that those configurations likely produce sub-optimal results. This finding is reasonable since, when β is low or α is high, deactivating numerous electrodes in sequence is not properly penalized in the cost function.

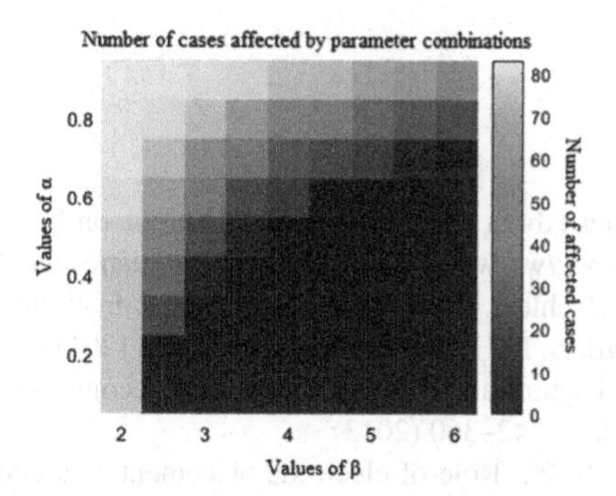

Fig. 4. Parameter sensitivity test results. Higher values indicate a greater number of plans with differences from the originally generated plans, where $\alpha = 0.5$ and $\beta = 4$.

4 Conclusion

In this study, we presented an automated graph-based approach for selecting active electrode sets in CIs. Automated selection methods reduce the time required to develop a patient-specific plan and remove the necessity for an expert reviewer to manually select the active electrodes from a set of DVF curves. Clinical translation of IGCIP techniques requires that our developed methods be robust and reliable to maximize positive hearing outcomes in patients. Our approach utilized spatial information available from previous techniques for segmenting cochlear structures and electrode arrays. We used this information to develop a graph-based solution for selecting an optimal active electrode set. To validate our results, we asked an expert reviewer to rate electrode configurations as optimal or non-optimal, where for a plan to be considered optimal, the reviewer would make no changes to that plan. 95.2% of plans created from our method were accepted as optimal, compared to only 48.2% of plans generated using the current state-of-the-art technique. Further, post-evaluation review revealed that the four rejected plans from our proposed method were actually optimal. These results suggest that our method is significantly more robust than the current state-of-the-art method and could facilitate widespread, automated clinical translation of IGCIP methods for CI programming. In the future, we plan to evaluate our method in a clinical study to confirm that the results of our method produce improved hearing outcomes for CI recipients in practice. This study would examine improvements in hearing outcomes for subjects relative to their current implant configuration over the course of several weeks by collecting data before reprogramming and again after a 3 to 6-week adjustment period to the new electrode configuration. Following successful clinical confirmation of our method, we will perform a multi-site study to assess clinically translating this method to other institutions.

Acknowledgements. This work was supported in part by grants R01DC014037, R01DC014462, and R01DC008408 from the NIDCD and by training grant T32EB021937 from the NIBIB. This content is solely the responsibility of the authors and does not necessarily represent the official views of the National Institutes of Health.

References

1. NIDCD Quick Statistics About Hearing. National Institute on Deafness and Other Communication Disorders. https://www.nidcd.nih.gov/health/statistics/quick-statistics-hearing
2. NIDCD Fact Sheet: Cochlear Implants. 2011 National Institute on Deafness and Other Communication Disorders. NIH Publication No. 11-4798 (2011)
3. Holden, L.K., et al.: Factors affecting open-set word recognition in adults with cochlear implants. Ear Hear. **34**(3), 342–360 (2013)
4. Finley, C.C., Skinner, M.W.: Role of electrode placement as a contributor to variability in cochlear implant outcomes. Otol Neurotol. **29**(7), 920–928 (2008)
5. Chakravorti, S., et al.: Further evidence of the relationship between cochlear implant electrode positioning and hearing outcomes. Otol Neurotol. **40**(5), 617–624 (2019)
6. Aschendorff, A., et al.: Quality control after cochlear implant surgery by means of rotational tomography. Otol Neurotol. **26**(1), 34–37 (2005)
7. Wanna, G.B., et al.: Impact of electrode design and surgical approach on scalar location and cochlear implant outcomes. Laryngoscope **124**(S6), S1–S7 (2014)
8. Wanna, G.B., et al.: Assessment of electrode placement and audiologic outcomes in bilateral cochlear implantation. Otol Neurotol. **32**(3), 428–432 (2011)
9. O'Connell, B.P., et al.: Electrode location and angular insertion depth are predictors of audiologic outcomes in cochlear implantation. Otol Neurotol. **37**(8), 1016–1023 (2016)
10. Zhao, Y., et al.: Automatic graph-based method for localization of cochlear implant electrode arrays in clinical CT with sub-voxel accuracy. Med. Image Anal. **52**, 1–12 (2019)
11. Noble, J.H., Dawant, B.M.: Automatic graph-based localization of cochlear implant electrodes in CT. In: Navab, N., Hornegger, J., Wells, W.M., Frangi, A.F. (eds.) MICCAI 2015. LNCS, vol. 9350, pp. 152–159. Springer, Cham (2015). https://doi.org/10.1007/978-3-319-24571-3_19
12. Noble, J.H., Labadie, R.F., Gifford, R.H., Dawant, B.M.: Image-guidance enables new methods for customizing cochlear implant stimulation strategies. IEEE Trans. Neural Syst. Rehabil. Eng. **21**(5), 820–829 (2013)
13. Noble, J.H., Gifford, R.H., Labadie, R.F., Dawant, B.M.: Statistical shape model segmentation and frequency mapping of cochlear implant stimulation targets in CT. In: Ayache, N., Delingette, H., Golland, P., Mori, K. (eds.) MICCAI 2012. LNCS, vol. 7511, pp. 421–428. Springer, Heidelberg (2012). https://doi.org/10.1007/978-3-642-33418-4_52
14. Rattay, F., Leao, R.N., Felix, H.: A model of the electrically excited human cochlear neuron. II. Influence of the three-dimensional cochlear structure on neural excitability. Hear Res. **153**(1–2), 64–79 (2001)
15. Zhao, Y., Dawant, B.M., Noble, J.H.: Automatic selection of the active electrode set for image-guided cochlear implant programming. J. Med. Imaging **3**(3), 035001 (2016)
16. Zhang, D., Zhao, Y., Noble, J.H., Dawant, B.M.: Selecting electrode configurations for image-guided cochlear implant programming using template matching. J. Med. Imaging **5**(2), 021202 (2018)
17. Dijkstra, E.W.: A note on two problems in connexion with graphs. Numer. Math. **1**, 269–271 (1959)

18. Stakhovskaya, O., Sridhar, D., Bonham, B.H., Leake, P.A.: Frequency map for the human cochlear spiral ganglion: implications for cochlear implants. J. Assoc. Res. Otolaryngol. **8**(2), 220–233 (2007)
19. Noble, J.H., Labadie, R.F., Majdani, O., Dawant, B.M.: Automatic segmentation of intra-cochlear anatomy in conventional CT. IEEE Trans. Biomed. Eng. **58**(9), 2625–2632 (2011)
20. Berg, K.A., Noble, J.H., Dwyer, R.T., Labadie, R.F., Gifford, R.H.: Speech recognition as a function of the number of channels in perimodiolar electrode recipients. J. Acoust. Soc. Am. **145**(3), 1556–1564 (2019)

Improved Resection Margins in Surgical Oncology Using Intraoperative Mass Spectrometry

Amoon Jamzad[1(✉)], Alireza Sedghi[1], Alice M. L. Santilli[1], Natasja N. Y. Janssen[1], Martin Kaufmann[2], Kevin Y. M. Ren[3], Kaitlin Vanderbeck[3], Ami Wang[3], Doug McKay[4], John F. Rudan[4], Gabor Fichtinger[1], and Parvin Mousavi[1]

[1] School of Computing, Queen's University, Kingston, ON, Canada
a.jamzad@queensu.ca
[2] Department of Medicine, Queen's University, Kingston, ON, Canada
[3] Department of Pathology, Queen's University, Kingston, ON, Canada
[4] Department of Surgery, Queen's University, Kingston, ON, Canada

Abstract. PURPOSE: Incomplete tumor resections leads to the presence of cancer cells on the resection margins demanding subsequent revision surgery and poor outcomes for patients. Intraoperative evaluations of the tissue pathology, including the surgical margins, can help decrease the burden of repeat surgeries on the patients and healthcare systems. In this study, we propose adapting multi instance learning (MIL) for prospective and intraoperative basal cell carcinoma (BCC) detection in surgical margins using mass spectrometry. METHODS: Resected specimens were collected and inspected by a pathologist and burnt with iKnife. Retrospective training data was collected with a standard cautery tip and included 63 BCC and 127 normal burns. Prospective data was collected for testing with both the standard and a fine tip cautery. This included 130 (66 BCC and 64 normal) and 99 (32 BCC and 67 normal) burns, respectively. An attention-based MIL model was adapted and applied to this dataset. RESULTS: Our models were able to predict BCC at surgical margins with AUC as high as 91%. The models were robust to changes in cautery tip but their performance decreased slightly. The models were also tested intraoperatively and achieved an accuracy of 94%. CONCLUSION: This is the first study that applies the concept of MIL for tissue characterization in perioperative and intraoperative REIMS data.

Keywords: Surgical margin detection · Multiple instance learning · Rapid evaporative ionization mass spectrometry · Intraoperative tissue characterization · Non-linear analysis · Basal Cell Carcinoma

A. Jamzad, A. Sedghi and A.M.L. Santilli—Joint first authors.

A. L. Martel et al. (Eds.): MICCAI 2020, LNCS 12263, pp. 44–53, 2020.
https://doi.org/10.1007/978-3-030-59716-0_5

1 Introduction

A main step in the clinical management of major cancers includes surgical resection of the tumor. Incomplete resection of tumors and the presence of cancer cells at the resection margins, otherwise known as "positive margins", often demands repeat surgery [10]. In some cancers, such as breast cancer, the positive surgical margin rates can be as high as 20% [4]. The subsequent revision surgeries burden the health care system with extra costs and wait times. They can also affect the cosmetic outcome of the patient, causing distress and potentially delaying life saving treatments such as chemotherapy or radiation therapy. Intraoperative evaluation of the tissue, including surgical margins, is currently a challenging task. Recent efforts have resulted in innovative perioperative and intraoperative technologies that can assess tissue in a high throughput manner, and provide surgeons with critical information on tissue pathology. The Intelligent Knife, iKnife (Waters Corp., MA), a mass spectrometry-based technology, is one such modality [2]. This technology is able to provide enriched feedback about the chemical properties of the tissue at the surgical tooltip, in real time [7,9,11,12]. The smoke created by the surgical electrocautery device is used and its molecular profiles such as lipids, fatty acids and small molecules, are analyzed through rapid evaporative ionization mass spectrometry (REIMS) [5]. iKnife can be seamlessly integrated into surgical workflows, as REIMS does not require sample preparation [13], and only a connection to the exhaust of the electrocautery device is needed for molecular profiling of the tissue.

Due to the destructive nature of electrocautery that generates the smoke for iKnife, pathology validated labels are difficult to attain. In practice, the histopathology of the surrounding tissue to a "burn" is analyzed and the burn is labelled based on an educated estimate of a pathologist. Since the data labels are not conclusively determined, they are referred to as weak labels. The problem of weakly annotated data is common to pathology where images and reports are either grossly outlined, or annotations of collected data are vague. Introduced two decades ago, multiple instance learning (MIL) [3], is a strategy for dealing with weakly labeled data. Here, a single label is assigned to a "bag" of multiple instances. The bag label is positive if it contains at least one positive instance, and negative otherwise. Using bags with different proportions of positive instances, MIL methods learn signatures of positive instances. As weak annotations result in noisy instance labels, considering a *bag* of instances, rather than each individually, helps compensate for the effect of the weak labels. It is important to identify instances that play a prominent role in predicting the overall label of a bag. This is referred to as the *attention* of an instance-[8]. Recently, attention-based MIL has been used with deep learning for whole-slide annotation of histopathology images of breast and colon cancer [6].

In this paper, for the first time, we propose to extend the concept of attention-based MIL to learn from weakly labelled REIMS data for detection of perioperative surgical margins. To create surgical smoke, the iKnife burns the tissue in contact with the tool tip. The data created in a constant stream lends itself well to the concept of *bags*. Each mass spectrum from a burn is considered as an

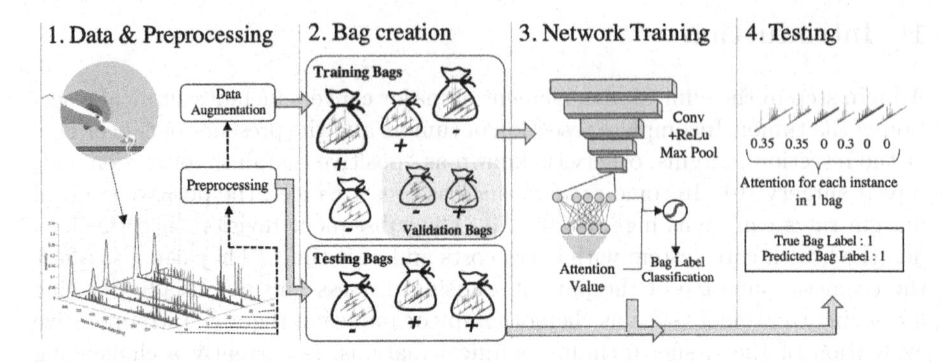

Fig. 1. Overview: Molecular profiles of the cautery smoke aspirated at the tip of the iKnife is preprocessed and augmented. The data is randomly divided into *bags* which are labelled based on instances they contain. MIL models are trained for margin classification using the training and validation bag sets. Using the trained model, the test bags are then predicted and evaluated. (Color figure online)

instance, and the stream of data from multiple burns are packed in bags. The prediction of a positive bag is, hence, an indication of the presence of positive margins. We demonstrate the accuracy of our models in prospective peroperative data, and their robustness to changes in surgical cautery tips. Finally, we investigate the feasibility of our developed approach to real time tissue typing, intraoperatively. Our methods are presented in the context of surgical margin detection for Basal Cell Carcinoma (BCC). BCC is the most commonly diagnosed cancer with a worldwide incidence rate of 2.75 million cases, and low metastasis rate [14]. Therefore, it is an ideal application for evaluation of our proposed surgical margin detection approach. Methods built on BCC data, can be translated to other cancers where surgical margins are crucial to patient outcomes.

2 Materials and Methods

Figure 1 depicts an overview of the proposed workflow. The cautery smoke aspirated at the tip of iKnife is collected for each specimen. Its spectra are selected, labelled, preprocessed, and augmented. The data is then represented as bags of instances and used for training of a deep model that is capable of predicting the bag label as well as the attention values of each instance of input spectra.

2.1 Data

Data was collected from 65 patients in 8 surgical clinic days, over a period of 10 months. Patients were recruited from the skin clinic at our institution, according to a protocol approved by the institutional HREB. BCC lesions presented on patients' head, neck or back. The suspected BCC region was first outlined by the surgeon on the skin and then resected. The resected specimen was inspected by a pathologist for perioperative point based data collection. Point burns were acquired from a cross section of the specimen containing BCC, by contact of the cautery tip with the tissue. Each burn was labelled by a derma-pathologist based on a visual inspection of the tissue at the location of the burn. A standard cautery tip was used for most of the data acquisition. To increase the specificity of the burn location, a non-insulated fine tip was used for some of the resected specimens. In Sect. 2.5, we describe how the data is divided for training and prospective testing. Experiments were performed in a controlled environment with the same operator, pathologist and surgeon at every clinic. An external lock mass of leucine enkephalin (1 ng/μl) was used for iKnife calibration (mass in negative-ion mode m/z 554.2615). The electrocautery was used on cut mode with the generator at 35 W. A sample iKnife recording consisting of five burns is shown in step 1 of Fig. 1. The chromatogram, in black, represents the total ion current (z axis), recorded over the acquisition time (x axis). Each peak in the chromatogram represents a burn. From each burn, the scan with the highest signal to noise ratio is chosen as the representative of that burn. Each scan is a mass spectral profile where the ion count, a measure of intensity, is plotted along ion mass charge ratios m/z. Five mass spectral scans are also shown in color.

2.2 Preprocessing

Using the Offline Model Builder (OMB), a Waters Corp. software platform, each scan was individually processed by normalizing, lock mass correcting and binning the intensity values. REIMS typically ionizes small molecules with mass to charge ratio of less than m/z 1200. Previous literature has reported that the majority of the total ion current to be present below m/z 900 [2]; we determined that 85% of the total ion current was between m/z 100–900 in our data. Therefore, we focused on this range for further analysis. Max binning was performed on this region with a bin size of 0.1, meaning that for a spectrum, the maximum intensity value for a m/z bin of size 0.1 is chosen to represent that range. For the range of m/z 100–900, each spectra is represented by 8000 peaks. To reduce the number of trainable parameters in the final model, we further applied max pooling with window and stride size of 10 to reduce the number of peaks (features of the spectra), to 800.

2.3 Intensity-Aware Augmentation

To avoid overfitting models to training data, it is essential to have a large number of data samples. This is not always clinically feasible. We propose a new data augmentation method for REIMS spectra that uses the inherent calibration error and background noise to create new data. First, a random shift sampled from a uniform distribution is added to the location of each peak in a spectrum. To increase the variability between multiple augmentations from the same spectrum, the shift range is also selected randomly from 0 to 3 bin widths. Next, random high-frequency Gaussian noise, multiplied by a random spline-smoothed low-frequency envelope is generated to mimic background noise. The standard deviation of the Gaussian is randomly selected proportional to the standard deviation of original spectrum. The generated noise is only added to the low-intensity peaks in the data while for the high-intensity a different Gaussian noise with half the standard deviation of the initial one is added. This intensity-aware approach ensures that the inherent molecular signatures and peak-ratios of the spectrum are preserved during augmentation.

2.4 MIL Model and Attention Mechanism

In the formulation of the MIL, a bag of instances is defined as $X = \{x_1, ..., x_n\}$, where the instances are not ordered or related to one another, and the size of the bag n may vary. Every instance in a bag has an individual binary label y, where $X_{labels} = \{y_1, ..., y_n\}$. Each bag is also assigned a single binary label $Y = max(X_{labels})$. Positive bags, $Y = 1$, have one or more instances of the target class while negative bags, $Y = 0$, have none. Considering the goal of margin classification for BCC, all BCC spectra are labeled 1 as the target class. In our architecture, similar to [6], we use a weighted average of instances where the weights are determined by a neural network and indicate each instance's attention value. The weights sum up to 1 and are invariant to the size of the bag. The proposed structure allows the network to discover similarities and differences among the instances. Ideally, instances within a positive bag that are assigned high attention values are those most likely to have label $y_i = 1$. The attention mechanism allows for easy interpretation of the bag label predictions by the models.

Every bag is fed to the network with all of its instances. The overall structure of the attention based MIL network consists of convolutional layers followed by fully connected dense layers. As visualized in step 3 of Fig. 1, every bag is passed through 5 convolutional layers (kernel size of 10) of 10, 20, 30, 40 and 50 filters. ReLu activation and max pooling is performed between every convolution. The final array is then flattened and passed through two dense layers of size 350 and 128 to get the final attention for each instance. The first dense layer and the generated attention weights are then combined in final layer and using Sigmoid activation the prediction of the bag label is outputted.

To define the sensitivity of the bag labels, we explored the minimum number of cancerous instances that would be required in a bag during training to learn the distinction between BCC and normal burns. Sweeping 2 parameters, we adjusted mean length of the bag, between 3 and 10 (standard deviation of 1), as well as the maximum involvement of cancerous instances in the positive bags between 0.1 and 0.8. This created 64 trained models and their performances are discussed in Sect. 3.1.

2.5 Experiments

To evaluate the models, we used the data from the first four clinic days as retrospective set and stratified them into 5 training/validation folds, all collected

Table 1. Table displaying the division of data into training/validation before augmentation and the separate testing folds.

	Fold	#BCC Burns	#Normal Burns	**Total Burns**	#Patients
Train	fold-1	14	24	**38**	9
	fold-2	6	30	**36**	7
	fold-3	14	20	**34**	5
	fold-4	12	25	**37**	6
	fold-5	17	28	**45**	7
Test	standard cautery	66	64	**130**	17
	fine tip cautery	32	67	**99**	11

using the standard cautery tip. The division of this data and the separation of the test sets is displayed in Table 1. The complete training set consisted of 63 BCC and 127 normal scans from 34 patients. All scans from a particular patient were kept within the same fold. Before augmentation, each fold contained approximately 1:2 ratio of cancer to normal spectra. Employing the proposed data augmentation technique, the size of each cross-validation fold increased to around 500 BCC and 500 normal scans.

For all of the experiments, the training folds were converted to a collection of 600 bags (300 negative and 300 positive bags). Bags were randomly formed in a way that an instance within a fold may be placed in more than one bag, but no two bags may have the same combination of instances. An ensemble of the 5 models was used to predict the labels of test data. Each model in the ensemble used 4 folds of the data for training and one for validation. The final label of a test bag was predicted by averaging the bag probabilities over these five models.

The data collected from clinic days 5 through 8 was used to generate two prospective test sets. The first test set contained burns collected with standard cautery tip. The second set contained burns collected with the fine cautery tip.

Intraoperative: Intraoperative data was collected from patients recruited similarly to the others in the study. During intraoperative resection, the surgeon only uses the standard cautery blade. The iKnife was connected to the cautery and smoke was collected throughout the procedure. To assess the feasibility of real time deployment and the performance of our model, an intraoperative case of neck lesion removal with continuous cut duration of 1.5 min was selected. The data was processed similar to perioperative burns and was used to the test our trained models.

3 Results and Discussion

3.1 Model Performance

The 64 trained models were tested on the prospective datasets of 500 bags generated with same parameters as their training equivalents. The AUC for each

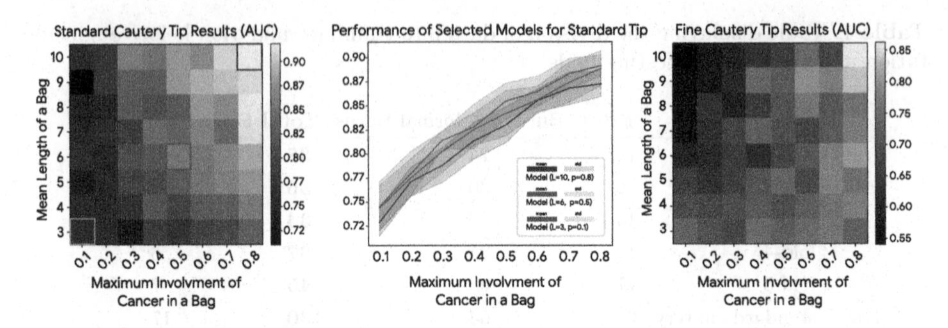

Fig. 2. *Left:* Exploring the two parameters of mean bag length and maximum involvement of cancerous instances in a positive bag, 64 models were trained on the augmented standard tip data. These models were tested on 500 standard tip bags and the mean AUC of the ensemble models are displayed. Metrics from the 3 outlined models can be seen in Table 1. *Middle:* Three outlined models from the left colormap, (L = length, p = max involvement), tested on data with mean length of 7 and standard deviation of 3 with varying max cancer involvement. *Right:* Same trained models from the left graph but tested on 500 bags of data from fine tip cautery. Mean AUC of the ensemble models are displayed.

model is visualized in the color maps of Fig. 2. The left color map displays the 64 models tested against the standard cautery tip test set. This figure demonstrates that as the bag length and cancer involvement increases, the models better learn the underlying patterns of the instances. Three models were selected (outlined) from the left color map and their performance metrics including accuracy, specificity, sensitivity, and AUC are listed in Table 2.

To mimic the uncertainty of intraoperative predictions, we tested the three selected models on bags of greater variability. All of the test bags in this case had a mean length of size 7 with a standard deviation of 3, therefore ranging from bags of size 1 to 15. The AUC, seen in the middle figure of Fig. 2, has the same upwards trend with proportion of cancerous instances as in the left colormap, even with the increased complexity of the test set. Finally, to determine the robustness of the model against potential changes in the input spectra, we examined the performance against the test dataset collected with the fine tip cautery. The fine tip was used during data collection to ideally be more precise with our burns and therefore attain better pathology guided labels. However, we do not conclusively know how the tip change may affect the signal recorded and therefore the ability of a model to perform. Trained on the same augmented standard cautery tip as both previous experiments, the right colormap in Fig. 2 visualizes the results of this experiment. The AUC trend is similar to that of the prospective standard cautery tip. However, the overall AUC is lower for most models suggesting that there may be a difference in the recorded mass spectral signal between the two tips.

As a comparison to baseline, we also implemented a standard MIL model known as multi instance support vector machine (mi-SVM). This model was

Table 2. Performance metrics of the 3 highlighted models from the left graph of Fig. 2, with comparisons to the mi-SVM baseline model [1]. Each model was tested with the prospective test set taken with the standard cautery tip. To reduce the sensitivity of the results to bag selection, each evaluation was performed 10 times with 500 randomly generated test bags.

	Accuracy	Sensitivity	Specificity	AUC
MIL (L = 3, p = 0.1)	0.72 ± 0.01	0.53 ± 0.02	0.92 ± 0.01	0.76 ± 0.02
MIL (L = 6, p = 0.5)	0.81 ± 0.01	0.71 ± 0.03	0.90 ± 0.01	0.86 ± 0.01
MIL (L = 10, p = 0.8)	0.81 ± 0.02	0.90 ± 0.01	0.72 ± 0.03	0.91 ± 0.01
mi-SVM (L = 3, p = 0.1)	0.69 ± 0.02	0.52 ± 0.04	0.85 ± 0.02	0.77 ± 0.02
mi-SVM (L = 6, p = 0.5)	0.71 ± 0.02	0.53 ± 0.06	0.89 ± 0.03	0.83 ± 0.01
mi-SVM (L = 10, p = 0.8)	0.75 ± 0.02	0.51 ± 0.03	0.99 ± 0.01	0.91 ± 0.01

presented by Andrews *et al.* in 2003 [1], based on an alternative generalization of the maximum margin idea used in SVM classification. The goal of mi-SVM, is to find both the optimal instance labelling as well as the optimal hyperplane. The performance of this baseline on the same 3 models and set of test bags is also listed in Table 2.

To demonstrate the true weakness of our labels, we performed a supervised method of principal component and linear discriminant analysis, PCA/LDA. Using the same training set, a PCA/LDA model was trained and tested on each individual scan in the standard cautery testing set. This linear approach performed at an accuracy of 75.7%. Although this method cannot be compared directly to our results as it does not utilize the bagging approach, it demonstrated the drop in performance when trying to use the instance level labels for direct classification.

3.2 Attention

Ideally a positive bag alert would root from a positive instance being given a high attention value. In a practical setting, this would alert the surgeon of a positive margin. To quantify the performance of the attention network, we evaluated the accuracy of correctly placing the highest attention value in a positive bag on a positive instance. Using the model with the highest AUC on the standard cautery tip test data (mean bag length of 10 and maximum cancerous involvement of 0.8), we were able to reach attention accuracy of 0.88 ± 0.04 on 500 test bags.

3.3 Intraoperative Trial

We demonstrated the applicability of our model on intraoperative data. Before deploying the model in the operating room, we wanted to evaluate its performance and sensitivity. The intraoperative case selected was comprised of 83 scans, including burns and no-burns spectra acquired continuously. Prospective pathology validation of the excised specimen labelled all of the margins negative, implying the absence of BCC in the scans. We created bags using a sliding window of size 10 to mimic bag creation from a continuous stream of data. For a model trained on bags with a mean size of 10 and positive bag's cancer portion of 0.8, the test on intraoperative data resulted an accuracy of 94% with a standard deviation of 6%.

4 Conclusion

In this study we adapted the concept of attention-based MIL to REIMS data analysis for perioperative margin evaluation for the first time. The framework consisted of preprocessing, intensity-aware augmentation, instance/bag representation of mass spectrmetry data, and model training. Training on retrospective BCC data, the performance of models with different bagging parameters was investigated on prospective data collected with standard and fine tip cautery blades. The feasibility of using the trained model on itraoperative data for margin assessment was also demonstrated. For future work, we plan to acquire more fine tip data to investigate the effect of transfer learning on improving the model predictive power. In practice, we are also looking at adaptive bag length selection during intraoperative data stream using the chromatogram signal. Another challenge to address is the presence of non-burn spectra, recorded during time intervals where the surgeon is not burning any tissue, along with burn signals intermixed in the interaoperative data stream. Implementation of a real-time burn screening algorithm to disregard the non-burn periods will increase the model accuracy in margin detection.

References

1. Andrews, S., Tsochantaridis, I., Hofmann, T.: Support vector machines for multiple-instance learning. In: Advances in Neural Information Processing Systems 15, pp. 577–584. MIT Press (2003)
2. Balog, J., et al.: Intraoperative tissue identification using rapid evaporative ionization mass spectrometry. Sci. Transl. Med. **5**(194), 194 (2013)
3. Dietterich, T.G., Lathrop, R.H., Lozano-Pérez, T.: Solving the multiple instance problem with axis-parallel rectangles. Artif. Intell. **89**(1), 31–71 (1997)
4. Fisher, S.L., Yasui, Y., Dabbs, K., Winget, M.D.: Re-excision and survival following breast conserving surgery in early stage breast cancer patients: a population-based study. BMC Health Serv. Res. **18**, 94 (2018)
5. Genangeli, M., Heeren, R., Porta Siegel, T.: Tissue classification by rapid evaporative ionization mass spectrometry (REIMS): comparison between a diathermic knife and CO2 laser sampling on classification performance. Anal. Bioanal. Chem. **411**, 7943–7955 (2019)
6. Ilse, M., Tomczak, J., Welling, M.: Attention-based deep multiple instance learning. In: Proceedings of the 35th International Conference on Machine Learning, vol. 80, pp. 2127–2136 (2018)

7. Kinross, J.M., et al.: iKnife: rapid evaporative ionization mass spectrometry (REIMS) enables real-time chemical analysis of the mucosal lipidome for diagnostic and prognostic use in colorectal cancer. Cancer Res. **76**(14 Suppl.), 3977 (2016)
8. Liu, G., Wu, J., Zhou, Z.H.: Key instance detection in multi-instance learning. In: Asian Conference on Machine Learning, vol. 25, pp. 253–268 (2012)
9. Marcus, D., et al.: Endometrial cancer: can the iknife diagnose endometrial cancer? Int. J. Gynecol. Cancer **29**, A100–A101 (2019)
10. Moran, M.S., et al.: Society of surgical oncology-American society for radiation oncology consensus guideline on margins for breast-conserving surgery with whole-breast irradiation in stages I and II invasive breast cancer. J. Clin. Oncol. **32**(14), 1507–1515 (2014)
11. Phelps, D.L., et al.: The surgical intelligent knife distinguishes normal, borderline and malignant gynaecological tissues using rapid evaporative ionisation mass spectrometry (REIMS). Br. J. Cancer **118**(10), 1349–1358 (2018)
12. St John, E.R., et al.: Rapid evaporative ionisation mass spectrometry of electrosurgical vapours for the identification of breast pathology: towards an intelligent knife for breast cancer surgery. Breast Cancer Res. **19**(59) (2017)
13. Strittmatter, N., Jones, E.A., Veselkov, K.A., Rebec, M., Bundy, J.G., Takats, Z.: Analysis of intact bacteria using rapid evaporative ionisation mass spectrometry. Chem. Commun. **49**, 6188–6190 (2013)
14. Verkouteren, J., Ramdas, K., Wakkee, M., Nijsten, T.: Epidemiology of basal cell carcinoma: scholarly review. Br. J. Dermatol. **177**(2), 359–372 (2017)

Self-Supervised Domain Adaptation for Patient-Specific, Real-Time Tissue Tracking

Sontje Ihler[1(✉)], Felix Kuhnke[2], Max-Heinrich Laves[1], and Tobias Ortmaier[1]

[1] Institut für Mechatronische Systeme, Leibniz Universität Hannover, Hanover, Germany
sontje.ihler@imes.uni-hannover.de

[2] Institut für Informationsverarbeitung, Leibniz Universität Hannover, Hanover, Germany

Abstract. Estimating tissue motion is crucial to provide automatic motion stabilization and guidance during surgery. However, endoscopic images often lack distinctive features and fine tissue deformation can only be captured with dense tracking methods like optical flow. To achieve high accuracy at high processing rates, we propose fine-tuning of a fast optical flow model to an unlabeled patient-specific image domain. We adopt multiple strategies to achieve unsupervised fine-tuning. First, we utilize a teacher-student approach to transfer knowledge from a slow but accurate teacher model to a fast student model. Secondly, we develop self-supervised tasks where the model is encouraged to learn from different but related examples. Comparisons with out-of-the-box models show that our method achieves significantly better results. Our experiments uncover the effects of different task combinations. We demonstrate that unsupervised fine-tuning can improve the performance of CNN-based tissue tracking and opens up a promising future direction.

Keywords: Patient-specific models · Motion estimation · Endoscopic surgery

1 Introduction

In (robot-assisted) minimally invasive surgery, instruments are inserted through small incisions and observed by video endoscopy. Remote control of instruments is a complex task and requires a trained operator. Computer assistance/guidance during surgery - in the form of automatic interpretation of endoscopic images using tracking and pose estimation - not only enables robotic surgery but can also help surgeons to operate more safely and precisely. Despite the maturity

Electronic supplementary material The online version of this chapter (https://doi.org/10.1007/978-3-030-59716-0_6) contains supplementary material, which is available to authorized users.

A. L. Martel et al. (Eds.): MICCAI 2020, LNCS 12263, pp. 54–64, 2020.
https://doi.org/10.1007/978-3-030-59716-0_6

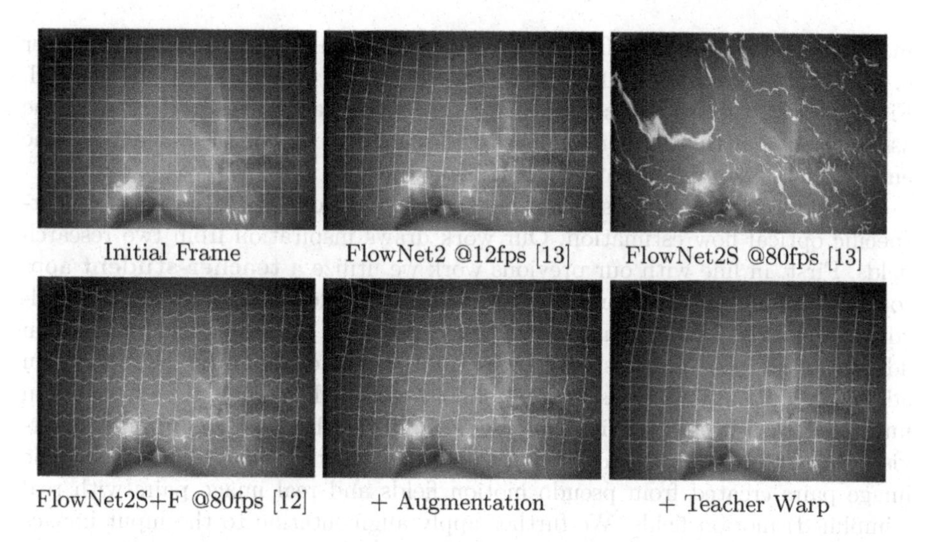

Fig. 1. Tracking liver tissue with very sparse texture and respiratory motion for 280 consecutive frames. Additional "regularization" in teacher-student fine-tuning improves smoothness of tracking grid, i.e. less drift on sparse tissue surface.

of many computer vision methods, visual motion estimation of moving tissue is still a challenging problem. Accurate motion estimation at fast feedback rates, is critical for intra-operative guidance and patient safety.

Visual motion can be estimated with sparse tracking, e.g. based on feature matching, or dense tracking algorithms like optical flow estimation. Because tissue deformation can only be fully captured with dense tracking, we focus this work on motion estimation with dense optical flow (OF). Traditionally, OF algorithms are based on conventional image processing with engineered features (an overview is provided in [19]). We will further focus on end-to-end deep learning models [5,13,15,21,22,24], as these outperform the conventional methods on public OF benchmarks [3,7,16]. Unfortunately, these models show the common speed vs. accuracy trade-off. High accuracy is achieved by high complexity, which leads to slow processing rates. On the other hand, faster models lack the capability to generalize well and provide lower accuracy. The goal is simultaneous fast and accurate flow estimation.

In a previous work [12], we propose patient-specific fine-tuning of a fast OF model based on an unsupervised teacher-student approach. A high-accuracy, but slow teacher model (FlowNet2 @12 fps [13]) is used to compute pseudo-labels for endoscopic scenes, which can then be used to fine-tune a fast, but imprecise student model (FlowNet2S @80 fps [13]) to a patient-specific domain[1]. Supervised fine-tuning with pseudo-labels improved the accuracy of the student

[1] Training samples should at best be identical to application samples. We therefore also propose to obtain training samples directly prior to the surgical intervention in the operation room. Intra-operative training time was on average 15 min.

model on the patient-specific target domain drastically at a speed up of factor 6, results are shown in Fig. 1. However, there are two drawbacks to this method. First, at best, the student can only become as good as the teacher. Second, the assumption that the teacher model provides good labels for the patient-specific endoscopic scene might be overly optimistic.

In this work, we propose a significantly extended method for fast patient-specific optical flow estimation. Our work draws inspiration from two research fields. First, in line with our previous work we utilize a **teacher-student** approach. Teacher-student approaches are found in model compression and knowledge distillation [2,10] and have attracted attention in other areas such as domain adaptation [6]. Second, inspired by **self-supervised learning** [4] we design additional optical flow tasks for the model to solve. This enables us to train on unlabeled data in a supervised manner. Joining both ideas, we propose supervised fine-tuning on pseudo-labels computed from real image pairs, synthetic image pairs created from pseudo motion fields and real image pairs with real (simplified) motion fields. We further apply augmentation to the input images during fine-tuning to pose a higher challenge for the student model. Increasing difficulty for the student is commonly used in classification to boost performance.

Our contributions are as follows: A completely unsupervised fine-tuning strategy for endoscopic sequences without labels. An improved teacher-student approach (student can outperform teacher) and novel self-supervised optical flow tasks for use during unsupervised fine-tuning.

In the following section we will explain our method. Then we describe the experimental setup and implementation details that are followed by our results. The paper concludes with a discussion and an outlook.

2 Related Work

Tissue tracking is a challenging task due to often-times sparse texture and poor lighting conditions, especially regarding stable, long-term tracking or real-time processing [23]. The issue of sparse texture was recently tackled for ocular endoscopy, by supervised fine-tuning FlowNet2S to retinal images with virtual, affine movement [9]. This was the first time the authors obtained an estimation accuracy sufficient for successfully mosaicking small scans to retinal panoramas. In our pre-experiments, a fine-tuned FlowNet2S with virtual, affine motion was not able to track tissue deformations, which are more complex the affine motion. Unsupervised training based on UnFlow [15], which does not have the restriction of a simplified motion model, did not converge on sparse textured image pairs in our experiments. In 2018, Armin et al. proposed an unsupervised method (EndoReg) to learn interframe correspondences for endoscopic images [1]. However, we have shown that both FlowNet2 and FlowNet2S outperform EndoReg, achieving higher structural similarity indices (SSI) [12].

3 Self-Supervised Teacher-Student Domain Adaptation

Our goal is fast and accurate OF estimation of a patient-specific, endoscopic target domain $T = (X, Y)$ with sample $((x_1, x_2), y) \in T$ - the tuple $(x_1, x_2) \in X$ being the input image pair sample and $y \in Y$ the corresponding motion field. Our fast student model has been trained for OF estimation on a source domain S which is disjoint from T. Unfortunately, as we don't have labels, true samples from the target domain are unknown, so we cannot directly deploy supervised domain adaptation. Our aim is therefore to obtain good approximations of target samples. First, we will revisit optical flow estimation and then introduce our proposed three training schemes: pseudo-labels, teacher-warp and zero-flow.

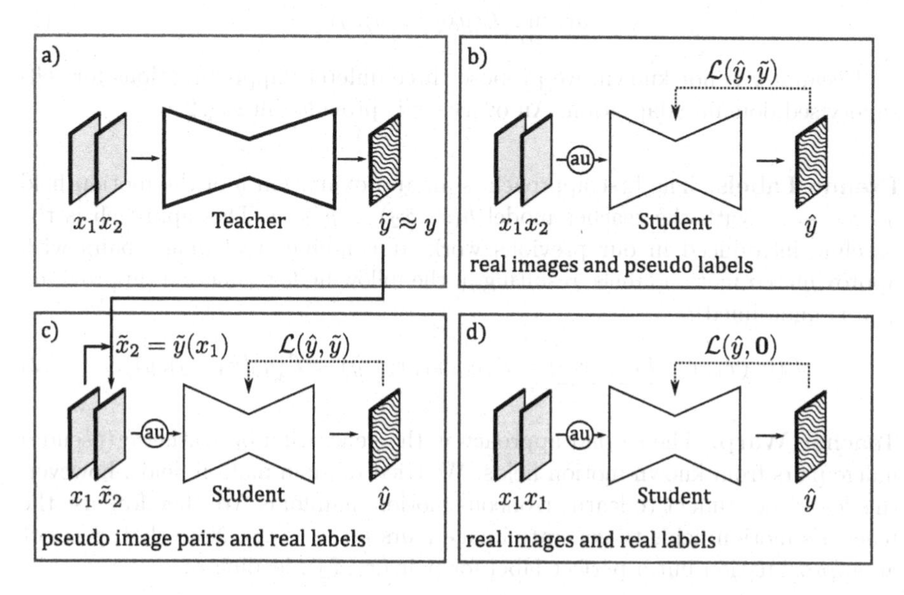

Fig. 2. Overview of method and training tasks to fine-tune a student model. a) A teacher model is used to produce pseudo labels $\tilde{y}$. b) The student can train on these pseudo labels. c) Teacher Warp: x_1 is warped using $\tilde{y}$ to create a pseudo image pair with real label. d) Zero-flow: Given the same image twice, the real flow is $\mathbf{0}$. Image augmentation (au) can be applied to increase the difficulty for the student model.

3.1 Optical Flow Estimation

Optical flow (OF) estimation is a regression problem to determine visible motion between two images. The displacement of corresponding pixels is interpreted as a vector field describing the movement of each image component [11].

The primary assumption is *brightness consistency*. If we deploy the motion field y as a mapping function $y : x_1 \rightarrow x_2$ to map the pixel values of x_1, we obtain a reconstruction $y(x_1)$ that is identical to x_2 (neglecting occlusion effects between the images). In practice, an estimated flow field $\hat{y} \approx y$ is generally not a perfect mapping. The difference in pixel values of x_2 and its reconstruction $\hat{x}_2 = \hat{y}(x_1)$ can be used to asses the quality of $\hat{y}$.

Another assumption for OF is *cycle consistency*, in other words the concatenation of a forward mapping $y : x_1 \rightarrow x_2$ and its reverse mapping $y^{-1} : x_2 \rightarrow x_1$ cancel each other out and result in zero motion: $x_1 = y^{-1}(y(x_1))$.

3.2 Teacher-Student Domain Adaptation

Our teacher-student knowledge transfer approach requires a fast student model $g_\phi : (x_1, x_2) \rightarrow \hat{y}$ parameterized with ϕ and an accurate teacher model h with fixed parameters. We optimize the student model with the following supervised objective function

$$\arg\min_\phi \mathcal{L}\big(g_\phi(x_1, x_2), y\big) \tag{1}$$

Because y is not known, we propose three different approximations for self-supervised domain adaptation. An overview is provided in Fig. 2.

Pseudo Labels. The first approach is an approximation $\tilde{y}$ of the motion field $y : x_1 \rightarrow x_2$ with the teacher model $h(x_1, x_2) = \tilde{y} \approx y$. This approach is the baseline introduced in our previous work. It combines real image pairs with approximated motion labels, resulting in the following loss function (approximation is underlined):

$$\mathcal{L}\big(g_\phi(x_1, x_2), \underline{h(x_1, x_2)}\big) = \mathcal{L}\big(g_\phi(x_1, x_2), \tilde{y}\big) \approx \mathcal{L}\big(g_\phi(x_1, x_2), y\big). \tag{2}$$

Teacher Warp. The second approach is the generation of synthetic (pseudo) image pairs from known motion fields. We tried random motion fields, however, this leads the student to learn erroneous motion manifolds. We therefore use the teacher's motion fields to generate image pairs $\tilde{y} : x_1 \rightarrow \tilde{x}_2$. Note that, $\tilde{y}$ is not an approximation but a perfect label for pair $(x_1, \tilde{x}_2)$, leading to

$$\mathcal{L}\big(g_\phi(\underline{x_1, \tilde{x}_2}), h(x_1, x_2)\big) = \mathcal{L}\big(g_\phi(x_1, \tilde{x}_2), \tilde{y}\big) = \mathcal{L}\big(g_\phi(x_1, \tilde{x}_2), y\big). \tag{3}$$

Zero-Flow Regularization. Our third and final approach is the combination of real images with real motion. This can be achieved with Zero-Flow, in other words $x_1 = x_2$ and $y = \mathbf{0}$, leading to the following loss

$$\mathcal{L}\big(g_\phi(x_1, x_1), \mathbf{0}\big). \tag{4}$$

Due to the high simplification of these samples, this loss is not suitable to be used as a sole training loss but can be used in combination with the other losses as a form of regularization (or minimum requirement of OF estimation).

Chromatic Augmentation. To increase the learning challenge for the student, we globally alter contrast γ and brightness β of $x'_v = \gamma \cdot x_v + \beta$ and add Gaussian noise n for each pixel individually (applied to value channel in HSV colour space).

4 Experiments

In our experiments we compare the proposed self-supervised domain adaptation techniques introduced in Sect. 3: teacher labels, teacher warp, zero-flow. We test the approaches individually, as well as combinations. For combinations, loss functions are summed with equal weights. All combinations were tested with and without image augmentation. A list of combinations can be seen in Table 1. We do not focus on occlusion, therefore no occlusion handling is implemented.

Table 1. Average cycle consistency error (CCE), average relative end point error EPE*. Best EPE* is achieved with a combined training scheme of teacher labels, teacher warp and zero-flow. CCE for training schemes including zero-flow is not conclusive and is therefore grayed out. († not adapted to medical image domain; +F: after fine tuning)

	Fine-Tuning Scheme				Average CCE				Average EPE*			
Model	t-labels	t-warp	zero	augm.	deform.	sparse	rot.	scale	deform.	sparse	rot.	scale
Reference FlowNet2†	-	-	-	-	0.087	0.057	0.058	0.027	0	0	0	0
Reference FlowNet2S†	-	-	-	-	0.584	0.893	0.515	0.506	0.799	0.837	0.446	0.439
Baseline FlowNet2S+F	x				0.178	**0.064**	0.138	0.100	0.303	0.059	0.193	0.102
proposed FlowNet2S+F		x			0.188	8.474	**0.115**	0.080	0.405	5.434	0.218	0.117
proposed FlowNet2S+F	x	x			**0.161**	0.067	0.118	**0.068**	**0.300**	**0.054**	**0.188**	**0.091**
proposed FlowNet2S+F	x		x		0.124	0.055	0.089	0.044	0.301	0.055	0.212	0.086
proposed FlowNet2S+F	x	x	x		0.146	0.074	0.125	0.047	**0.293**	**0.053**	0.225	**0.081**
experim. FlowNet2S+F	x			x	0.291	0.139	0.341	0.132	0.329	0.081	0.200	0.117
experim. FlowNet2S+F		x		x	0.191	0.223	1.118	0.497	0.367	0.819	2.979	0.716
experim. FlowNet2S+F	x	x		x	0.200	0.103	0.118	0.084	0.341	0.070	0.213	0.111
experim. FlowNet2S+F	x		x	x	0.152	0.067	0.114	0.052	0.344	0.077	0.248	0.115
experim. FlowNet2S+F	x	x	x	x	0.149	0.079	0.138	0.065	0.325	0.075	0.244	0.114

4.1 Setup

For our experiments we follow a similar setup to our previous work [12]. The main difference is the extension of the loss function and corresponding variations of the training samples that were added to the datasets.

Datasets. For our experiments we use four endoscopic video sequences from the Hamlyn datasets: **scale** and **rotation**: both in vivo porcine abdomen [17], **sparse** texture (and respiratory motion): in vivo porcine liver [17], strong **deformation** from tool interaction (and low texture): In vivo lung lobectomy procedure [8]. All sequences show specular highlights. The sparse-texture sequence shows very challenging lighting conditions. Each dataset represents an individual patient. To simulate intra-operative sampling from a patient-specific target domain we split the sequences into disjoint sets. Training and validation set represent the sampling phase (preparation stage), while the application phase during intervention is represented by the test set. The splits of the subsets were chosen manually, so that training and test data both cover dataset-specific motion.

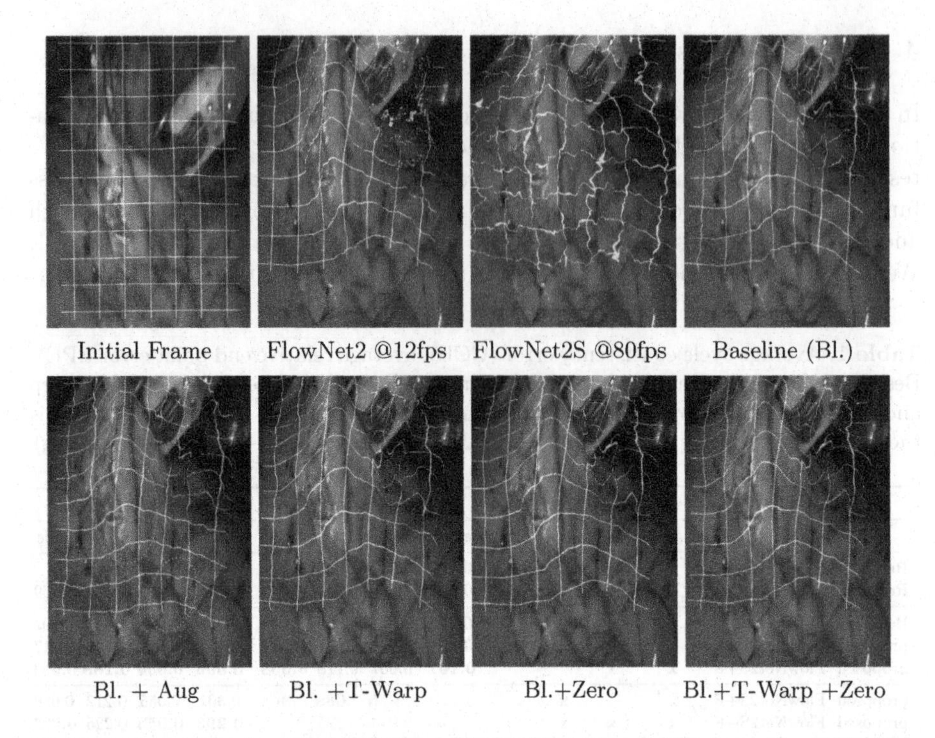

Fig. 3. Deformation dataset: Tracking results after 80 frames. The tissue is undergoing strong deformation due to tool interaction. Almost all fine-tuned models show very comparable results to the teacher FlowNet2. Unfortunately, the teacher shows drift in the bottom right and the centre of the tracked grid. Our model fine-tuned with added image augmentation produces more robust tracking results in this area. Interestingly, adding image augmentation to any other fine-tuning scheme did not improve the tracking results (also for different augmentation parameters we tried).

The number of samples for training/validation/test set are: scale - 600/240/397; rotation - 329/110/161; sparse - 399/100/307; deformation - 600/200/100. The left camera was used for all datasets.

Model. For all our experiments we used the accurate, high-complexity FlowNet2 framework [13] running at approx. 12 fps as our teacher model h and its fast FlowNet2S component running at approx. 80 fps as our low-complexity, fast student model g. FlowNet2 achieves very high accuracy on the Sintel benchmark [3], which includes elastic deformation, specular highlights and motion blur, which are common effects in endoscopy. FlowNet2S has been shown to handle sparse textures well [9]. We utilized the implementation and pretrained models from [18].

Training and Augmentation Parameters: For training we used Adam optimizer with a constant learning rate of 1e−4. Batch size was set to 8. Training was performed until convergence of validation loss, maximum of 120 epochs. Extension and augmentation of the training data set increases training times. This was not focus of this work but should be addressed in future work. We follow Sun et al.'s recommendation to fine-tune FlowNet2S with the multi-scale L1-loss [13,20]. For illumination augmentation, we sampled brightness $\beta \sim \mathcal{N}(0, 3)$ and contrast $\gamma \sim \mathcal{N}(1, 0.15)$ for each image individually. Noise n was sampled from $\mathcal{N}(0, 2)$. Value range of input images was between 0 and 255. No normalization was performed on input images.

4.2 Results

We provide the relative endpoint error ($\text{EPE}^* = ||(g(x_1, x_2) - \tilde{y})||_2$) as well as the cycle consistency error ($\text{CCE} = ||\mathbf{p} - \hat{y}^{-1}(\hat{y}(\mathbf{p}))||_2$, $\mathbf{p}$: image coordinates) for all our experiments in Table 1. We compare our fine-tuned models with the teacher and the student model, as well as our baseline (only teacher labels). All results were obtained using the test sets described in Sect. 4.1.

The best EPE* is achieved with a combined training scheme of teacher labels, teacher warp and zero-flow. The best EPE* is not necessarily indicating the best model though. A low EPE* indicates high similarity to the teacher's estimation. Almost all fine-tuned models achieve a better cycle consistency than the original student model which indicates an improved model on the target domain. A low CCE, however, is not a guarantee for a good model, best CCE is achieved by estimating zero-flow (which is not necessarily a correct estimation).

Due to the lack of annotated ground truth, we also evaluate accuracy with a tracking algorithm over consecutive frames. Small errors between two frames are not visible. However, during tracking, small errors add up for consecutive frames, resulting in drift and making small errors visible over time. We see that in most cases the advanced domain adaptation improved tracking qualities compared to the baseline, see Figs. 1, 3, 4, and 5. In some cases it even outperforms the teacher model FlowNet2. Detailed analysis is provided in the captions. For Videos see supplementary material.

Overall, adding teacher-warp to the training samples always improved the EPE*, however it should not be used on its own. As expected, training with added zero-flow almost always improved cycle consistency. Adding image augmentation seems to enforce conservative estimation. In Fig. 3 and 4 the model predicted less motion than all other training schemes. This may be beneficial in some cases e.g. very noisy data, however, the hyper-parameters for chromatic augmentation need to be chosen carefully to match the target domain.

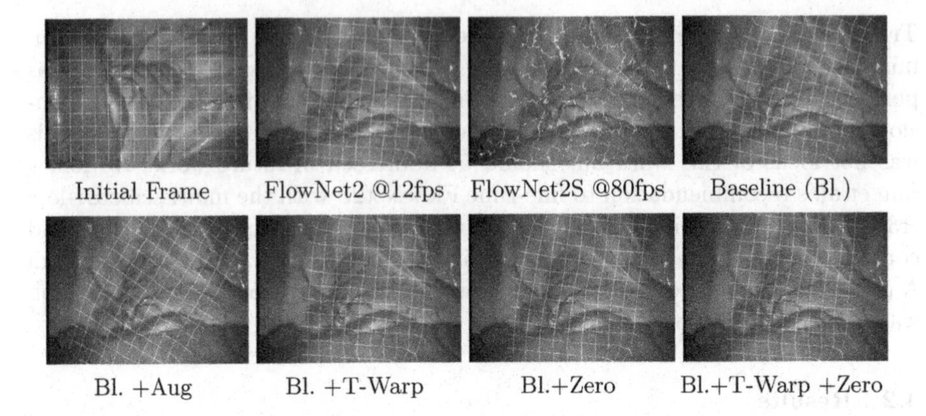

Fig. 4. Rotation dataset: Tracking results after 160 frames. The orientation of the tissue changes due to camera rotation. In contrary to Fig. 3 the tracking results come closer to the teacher model by the extended fine-tuning schemes, while the image augmentation breaks the tracking performance. The jump of the average CCE for this combination is comparably high, which might be a way to detect such erroneous models (subject to future work). Change of orientation generally seems to be a more difficult task then scale change, presumably due the lack of rotational invariance of convolutions.

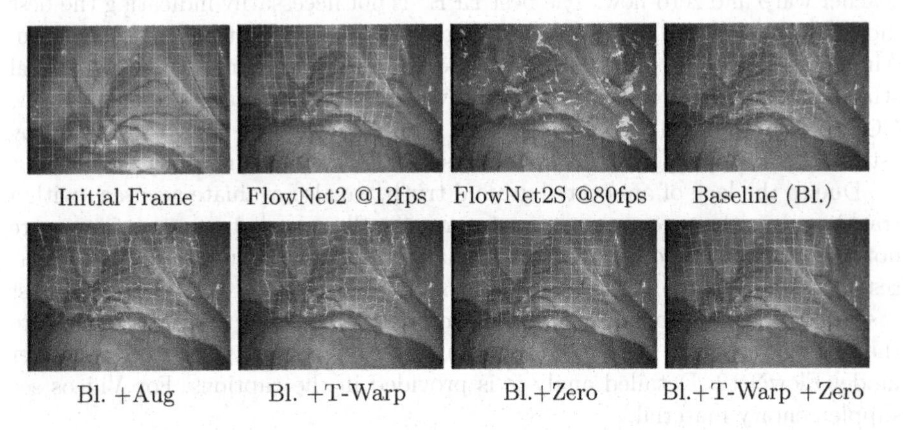

Fig. 5. Scale dataset: Tracking results after 396 frames. The scale of the tissue changes due to camera going in and out of the situs. The tissue becomes lighter with higher proximity to the camera). This sequence contains sections with temporary, static camera pose. Interestingly, drift predominantly occurs during low motion sections. The zero-flow fine-tuning scheme improves the smoothness of the tracking grid. Image augmentation reduces drift in the bottom of the mesh. However, for the upper part of the mesh, drift is lower for all other models. Please note the extended length of this sequence compared to the other sequences.

5 Conclusion

We proposed advanced, self-supervised domain adaptation methods for tissue tracking based on a teacher-student approach. This tackles the problem of lacking annotations for real tissue motion to train fast OF networks. The advanced methods improve the baseline in many cases, in some cases even outperform the complex teacher model.

Further studies are required to determine a best overall strategy. We plan to extend the training scheme with further domain knowledge. Pseudo image pairs can be created using affine transformation, which is equivalent to synthetic camera movement. Cycle consistency can and has been used for training and in estimating occlusion maps [15]. DDF-FLow learns occlusion maps using an unsupervised teacher-student approach [14]. This seems like a very promising extension for our approach. Occlusion maps are an essential next step for safe tissue tracking.

Acknowledgements. This work has received funding from the European Union as being part of the EFRE OPhonLas project.

References

1. Armin, M.A., Barnes, N., Khan, S., Liu, M., Grimpen, F., Salvado, O.: Unsupervised learning of endoscopy video frames' correspondences from global and local transformation. In: Stoyanov, D., et al. (eds.) CARE/CLIP/OR 2.0/ISIC -2018. LNCS, vol. 11041, pp. 108–117. Springer, Cham (2018). https://doi.org/10.1007/978-3-030-01201-4_13
2. Buciluă, C., Caruana, R., Niculescu-Mizil, A.: Model compression. In: ACM SIGKDD, pp. 535–541 (2006)
3. Butler, D.J., Wulff, J., Stanley, G.B., Black, M.J.: A naturalistic open source movie for optical flow evaluation. In: IEEE ECCV, pp. 611–625 (2012). https://doi.org/10.1007/978-3-642-33783-3_44
4. Doersch, C., Zisserman, A.: Multi-task self-supervised visual learning. In: IEEE ICCV, pp. 2051–2060 (2017)
5. Dosovitskiy, A., et al.: FlowNet: learning optical flow with convolutional networks. In: IEEE ICCV (2015). https://doi.org/10.1109/ICCV.2015.316
6. French, G., Mackiewicz, M., Fisher, M.: Self-ensembling for visual domain adaptation. arXiv:1706.05208 (2017)
7. Geiger, A., Lenz, P., Urtasun, R.: Are we ready for autonomous driving? The Kitti vision benchmark suite, pp. 3354–3361, May 2012. https://doi.org/10.1109/CVPR.2012.6248074
8. Giannarou, S., Visentini-Scarzanella, M., Yang, G.Z.: Probabilistic tracking of affine-invariant anisotropic regions. IEEE TPAMI **35**(1), 130–143 (2013). https://doi.org/10.1109/TPAMI.2012.81
9. Guerre, A., Lamard, M., Conze, P.H., Cochener, B., Quellec, G.: Optical flow estimation in ocular endoscopy videos using flownet on simulated endoscopy data. In: IEEE International Symposium on Biomedical Imaging (ISBI), pp. 1463–1466 (2018). https://doi.org/10.1109/ISBI.2018.8363848

10. Hinton, G., Vinyals, O., Dean, J.: Distilling the knowledge in a neural network. arXiv:1503.02531 (2015)
11. Horn, B.K.P., Schunck, B.G.: Determining optical flow. Artif. Intell. **17**, 185–203 (1981). https://doi.org/10.1016/0004-3702(81)90024-2
12. Ihler, S., Laves, M.H., Ortmaier, T.: Patient-specific domain adaptation for fast optical flow based on teacher-student knowledge transfer. arXiv:2007.04928 (2020)
13. Ilg, E., Mayer, N., Saikia, T., Keuper, M., Dosovitskiy, A., Brox, T.: FlowNet 2.0: evolution of optical flow estimation with deep networks. In: IEEE CVPR, July 2017. https://doi.org/10.1109/CVPR.2017.179
14. Liu, P., King, I., Lyu, M.R., Xu, J.: DDFlow: learning optical flow with unlabeled data distillation. In: AAAI, vol. 33, pp. 8770–8777 (2019). https://doi.org/10.1609/aaai.v33i01.33018770
15. Meister, S., Hur, J., Roth, S.: UnFlow: unsupervised learning of optical flow with a bidirectional census loss. In: AAAI, New Orleans, Louisiana, pp. 7251–7259, February 2018. arXiv:1711.07837
16. Menze, M., Geiger, A.: Object scene flow for autonomous vehicles. In: IEEE CVPR (2015). https://doi.org/10.1109/CVPR.2015.7298925
17. Mountney, P., Stoyanov, D., Yang, G.: Three-dimensional tissue deformation recovery and tracking. IEEE Signal Process. Mag. **27**(4), 14–24 (2010). https://doi.org/10.1109/MSP.2010.936728
18. Reda, F., Pottorff, R., Barker, J., Catanzaro, B.: flownet2-pytorch: pytorch implementation of flownet 2.0: evolution of optical flow estimation with deep networks (2017). https://github.com/NVIDIA/flownet2-pytorch
19. Sun, D., Roth, S., Black, M.J.: Secrets of optical flow estimation and their principles. In: IEEE CVPR, pp. 2432–2439, June 2010. https://doi.org/10.1109/CVPR.2010.5539939
20. Sun, D., Yang, X., Liu, M.Y., Kautz, J.: Models matter, so does training: an empirical study of CNNs for optical flow estimation. arXiv:1809.05571 (2018)
21. Sun, D., Yang, X., Liu, M.Y., Kautz, J.: PWC-Net: CNNs for optical flow using pyramid, warping, and cost volume. In: IEEE CVPR, pp. 8934–8943 (2018). https://doi.org/10.1109/CVPR.2018.00931
22. Wulff, J., Black, M.J.: Efficient sparse-to-dense optical flow estimation using a learned basis and layers. In: IEEE CVPR, pp. 120–130 (2015). https://doi.org/10.1109/CVPR.2015.7298607
23. Yip, M.C., Lowe, D.G., Salcudean, S.E., Rohling, R.N., Nguan, C.Y.: Tissue tracking and registration for image-guided surgery. IEEE Trans. Med. Imaging **31**(11), 2169–2182 (2012). https://doi.org/10.1109/TMI.2012.2212718
24. Yu, J.J., Harley, A.W., Derpanis, K.G.: Back to basics: unsupervised learning of optical flow via brightness constancy and motion smoothness. In: Hua, G., Jégou, H. (eds.) ECCV 2016. LNCS, vol. 9915, pp. 3–10. Springer, Cham (2016). https://doi.org/10.1007/978-3-319-49409-8_1

An Interactive Mixed Reality Platform for Bedside Surgical Procedures

Ehsan Azimi[1(✉)], Zhiyuan Niu[1], Maia Stiber[1], Nicholas Greene[1], Ruby Liu[2], Camilo Molina[3], Judy Huang[3], Chien-Ming Huang[1], and Peter Kazanzides[1]

[1] Department of Computer Science, Johns Hopkins University, Baltimore, MD, USA
eazimi1@jhu.edu
[2] Department of Biomedical Engineering, Johns Hopkins University, Baltimore, MD, USA
[3] Department of Neurosurgery, Johns Hopkins University, Baltimore, MD, USA

Abstract. In many bedside procedures, surgeons must rely on their spatiotemporal reasoning to estimate the position of an internal target by manually measuring external anatomical landmarks. One particular example that is performed frequently in neurosurgery is ventriculostomy, where the surgeon inserts a catheter into the patient's skull to divert the cerebrospinal fluid and alleviate the intracranial pressure. However, about one-third of the insertions miss the target.

We, therefore, assembled a team of engineers and neurosurgeons to develop an interactive surgical navigation system using mixed reality on a head-mounted display that overlays the target, identified in preoperative images, directly on the patient's anatomy and provides visual guidance for the surgeon to insert the catheter on the correct path to the target.

We conducted a user study to evaluate the improvement in the accuracy and precision of the insertions with mixed reality as well as the usability of our navigation system. The results indicate that using mixed reality improves the accuracy by over 35% and that the system ranks high based on the usability score.

Keywords: Surgical navigation · Neurosurgery · Augmented reality

1 Introduction

In many surgical procedures, surgeons rely on their general knowledge of anatomy and relatively crude measurements, which have inevitable uncertainties in locating internal anatomical targets. One example procedure that is frequently performed in neurosurgery is ventriculostomy (also called external ventricular

---

Electronic supplementary material The online version of this chapter (https://doi.org/10.1007/978-3-030-59716-0_7) contains supplementary material, which is available to authorized users.

A. L. Martel et al. (Eds.): MICCAI 2020, LNCS 12263, pp. 65–75, 2020.
https://doi.org/10.1007/978-3-030-59716-0_7

drainage), where the surgeon inserts a catheter into the ventricle to drain cerebrospinal fluid (CSF). In this procedure, often performed bedside and therefore without image guidance, the surgeon makes measurements relative to cranial features to determine where to drill into the skull and then attempts to insert a catheter as perpendicular to the skull as possible. Although it is one of the most commonly performed neurosurgical procedures, about one quarter to one third of catheters are misplaced or require multiple attempts [15,19], potentially resulting in brain injury and increasing healthcare costs [2].

We propose a portable navigation system, based on mixed reality (MR) on a head mounted display (HMD), specifically HoloLens (Microsoft, Redmond, WA) to provide image guidance in less structured environments, such as at the patient bedside in an intensive care unit, where it is not practical to install a separate tracking camera, computer, and display. Furthermore, the mixed reality guidance can provide an ergonomic benefit because it is visually overlaid in the surgeon's field of view, thus avoiding the need to look away from the patient to observe an external monitor.

2 Previous Work

HMDs have been used in the medical domain for treatment, education, rehabilitation, and surgery [7,9,16]. In [11], the researchers presented the use of a head-mounted display to visualize volumetric medical data for neurosurgery planning. We have previously adopted a picture-in-picture visualization for neurosurgery navigation with a custom built HMD [3,20]. With the advent of Google Glass, around 2013, many research groups started to explore using an HMD as a replacement for traditional radiology monitors [1,22]. In other research, HMDs give the surgeon an unobstructed view of the anatomy, which is rendered inside the patient's body [21]. In a recent study, Gsaxner et al. [13,18] presented a markerless registration method for head and neck surgery that matches facial features measured by the HoloLens depth camera to the CT scan. Their proposed system does not have a reference frame and instead relies on the HoloLens self-localization (SLAM), which is one source of error. With a mean TRE of about 10 mm, their reported error is too high for our application. Besides, the system's dependence on an external computer for registration makes it unsuitable for our bedside setting. Additional inaccuracies caused by network lag and head movement will add to the errors in the detection of the facial features which can be obscured by the surgeon's hands, surgical tools, and the necessary draping. Similarly, in another study, Van Doormaal et al. [10] measured HoloLens tracking accuracy, but relied on the HoloLens self-localization rather than using a reference frame. In a different study using HoloLens for neuronavigation, Frantz et al. [12] demonstrated that HoloLens self-localization is not accurate enough by comparing it to a tracked reference frame.

The most similar work to our own was reported by Li et al. [17], where they used the Microsoft HoloLens HMD to provide augmented reality guidance for catheter insertion in ventriculostomy. However, they did not track the skull and

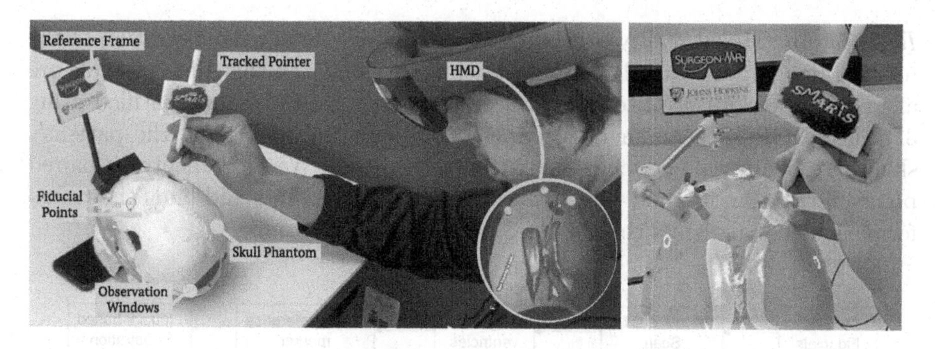

Fig. 1. Left: User wearing HMD and holding tracked pointer to perform registration. Right: Updated skull phantom (new fiducials, reference frame attachment and borescope cameras for measurement).

instead relied on preventing patient motion, which increases the risk of misplacement due to undetected motion. More recently, we proposed an ecosystem for performing surgical tasks in mixed reality [4]. This study is a representation of the immersive practice and planning module of our proposed ecosystem.

3 System Description

As a surgical navigation system, our system (Fig. 1) and workflow (Fig. 2) comprises of the following necessary components, including tracking, image segmentation, registration, guidance, and visualization:

Marker Tracking: Our system tracks the patient using a marker (reference frame) mounted on the patient's skull. We initially used a headband to attach the reference frame, as in Fig. 1-left, but based on surgeon feedback have changed to a post that is screwed into the skull, as in Fig. 1-right (in a clinical procedure, this post would be located closer to the burr hole), with an adjustable linkage to enable the marker to be positioned in the field-of-view of the HMD tracking camera. At the same time, we need to be able to identify points for surgical path planning or to select fiducials for registration. Therefore, we also track a second marker that is affixed to a pointer tool, shown in Fig. 1-left. We thus designed two image target markers that can be tracked using Vuforia Engine.

Segmentation: Our MR navigation requires a model that matches the subject of interest. To that end, we segment the CT scan of the patient. The ventricles are segmented using the connected threshold filter of SimpleITK (www.simpleitk.org), which uses user-specified coordinates, or "seeds," as well as the expected threshold, to create a binary labelmap from the medical image data. The skull, on the other hand, can be easily segmented in 3D Slicer (www.slicer.org) using its built-in threshold filter. Both the ventricle and skull segmentations are then used to create 3D models for the navigation system.

Registration: Mixed reality overlay of the 3D models requires registration of the medical imaging data to the actual patient's anatomy. To this aim, fiducials are affixed on the skull prior to the CT scan and their positions are identified using 3D Slicer. The surgeon then selects the corresponding points on the patient's skull by touching the fiducials using the tracked tool. We implemented a paired point registration method that uses these two sets of points and finds the transformation that registers the CT data to the actual anatomy.

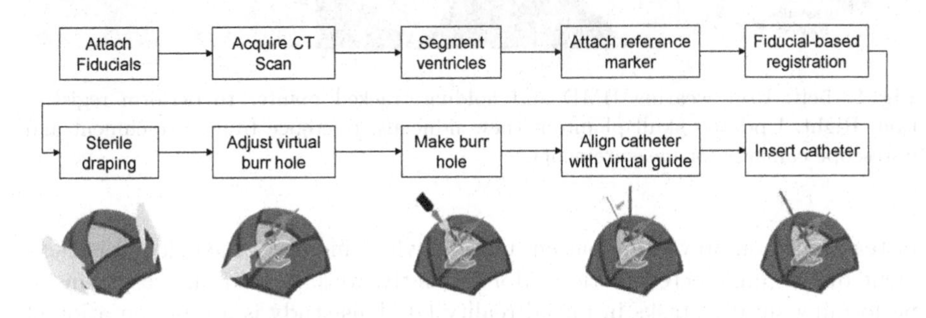

Fig. 2. Procedure workflow for ventriculostomy

User Interface and Visualization: The user interface was designed in Unity 3D (www.unity.org) and supports the workflow shown in Fig. 2, where registration fiducials are attached to the patient prior to acquisition of a CT scan.

Registration is the first procedural task that benefits from mixed reality visualization. The surgeon holds the tracked pointer tool and the HMD overlays a red sphere at the tip of the tool. This provides the surgeon with visual verification that the system is well calibrated (if not, the calibration can be repeated). The surgeon uses the pointer tool to touch each fiducial, with a voice command to trigger position capture. The system then overlays a green sphere at the captured position, which provides visual feedback that the voice command was recognized and also that the captured position is correctly aligned with the physical fiducial (if not, the point can be recollected). After collecting the positions of at least three fiducials, the surgeon issues a voice command to perform the registration. The surgeon can issue another voice command to "show skull", at which point the skull model (segmented from CT) is overlayed on the patient, enabling visual confirmation of the registration result. The surgeon can issue the command "hide skull" to turn off this overlay. In addition to this visual confirmation, the system rejects registration for which the Fiducial Registration Error (FRE) is beyond a threshold (currently 2.5 mm), and the surgeon has to register again.

The next use of mixed reality visualization occurs prior to making the burr hole. Here, the system overlays a virtual circle at the nominal position of the burr hole, based on the registered CT scan, but the surgeon can use the tracked pointer to adjust this position, if necessary. After the surgeon makes the burr hole, the system overlays a virtual line that passes through the burr hole and

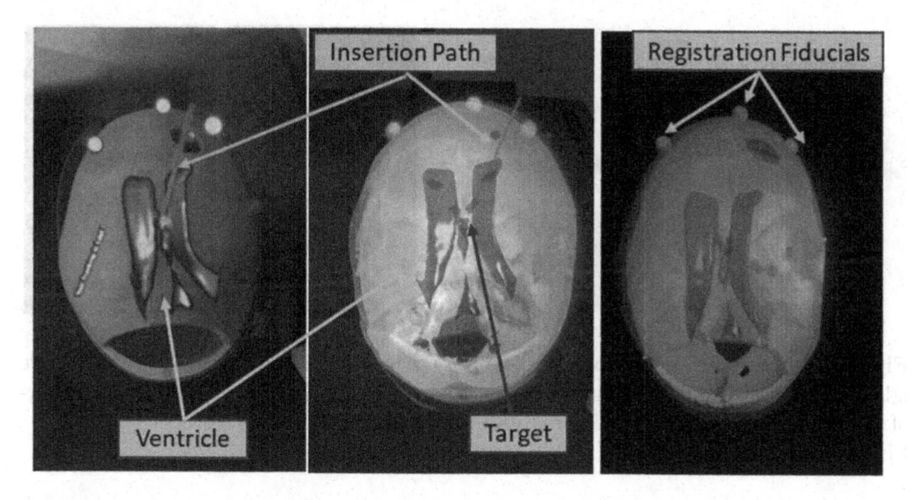

Fig. 3. Guidance and visualization in mixed reality captured from the user's view. The virtual skull is overlaid on the real skull and 3 registration points are shown. The red line is the virtual guidance path for the catheter.

to the intended target (Foramen of Monro). The surgeon then aligns the real catheter with the virtual line and advances to the target, thereby completing the procedure. Sample mixed reality visualizations are shown in Fig. 3.

In the supplementary material, we provide a video which shows registration, planning and insertion of a catheter using mixed reality.

4 Experiments

We created an experimental setup to evaluate our system in a user study.

4.1 Phantom Design

The phantom is constructed from a plastic skull with a cranial cap that is magnetically attached (Fig. 4). A clear acrylic box is inserted to hold gel that mimics the brain tissue. We determined that 1.25 teaspoons of SuperClear gelatin powder (Custom Collagen, Addison, IL) in 1 cup water provided a gel that, according to the neurosurgeons, approximated the feel of brain tissue.

We placed three spheres near the bottom of the acrylic box to use as targets. One sphere was located at the nominal position of the Foramen of Monro to represent normal anatomy. The other two spheres were offset to represent abnormal anatomy. Because our focus is to evaluate MR guidance for inserting the catheter, we created a large burr hole in the skull, thereby skipping the steps in Fig. 2 where the subject creates the burr hole. Note, however, that our burr hole is significantly larger than one created clinically so that the subject has some flexibility to adjust the catheter entry point.

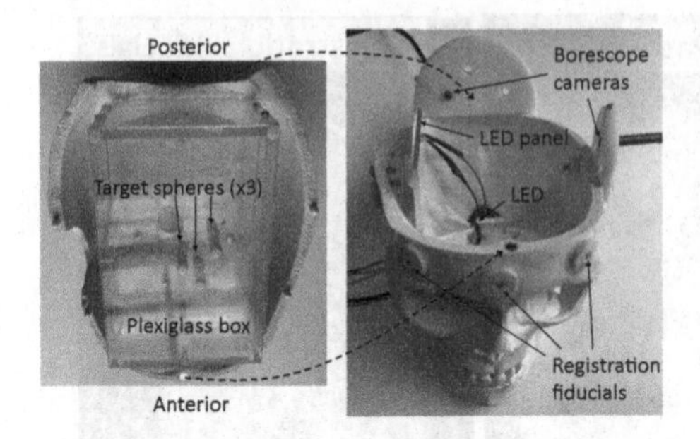

Fig. 4. Left: top of skull showing plexiglass box and targets. Right: bottom of skull showing borescope cameras and LEDs.

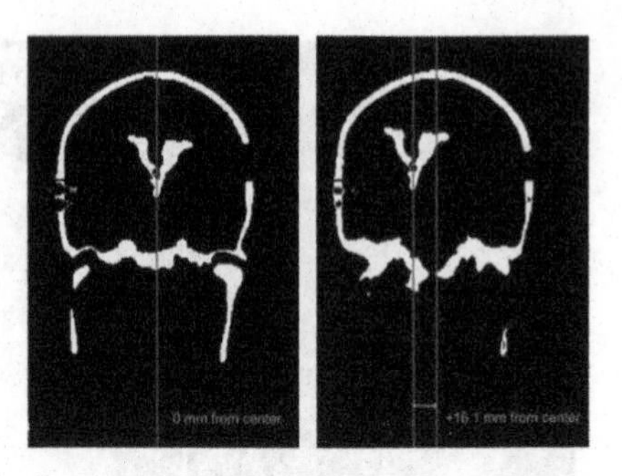

Fig. 5. Coronal views of two synthetic CT scans (left: nominal; right: abnormal).

4.2 CT Generation

After acquiring a CT scan of our phantom, we used 3D Slicer to extract the model of the skull, the positions of the fiducials, and the positions of the target spheres. Using data from another CT scan with a ventricle phantom, we created synthetic CT scans, as shown in Fig. 5. Specifically, we digitally removed the spherical targets from the CT scan and then used 3D Slicer's transform module to place the ventricle model such that its Foramen of Monro was coincident with each of the three targets. These synthetic CT scans are provided to subjects in the control group, who do not have the benefit of mixed reality visualization and would normally consult the CT scan.

4.3 Embedded Optical Measurement

We constructed a computer vision system to measure the 3D coordinates of the catheter and target, as well as to record videos (Fig. 4). One camera is fixed on the left side of the skull, and the other is on the back. Due to the different refractive indices between the gel and air, the intrinsic calibration of each camera is separately performed with a checkerboard in the gel. The extrinsic parameters, which do not change with the medium, are calibrated in air. The accuracy of the optical measurement system was obtained by identifying each of the three spherical targets inside the skull and computing the distances between them, which are 10.57 mm, 18.54 mm, and 18.43 mm. These three distances are compared to the same distances computed in the CT scans (0.68 mm axial dimension 1 mm slice spacing), which are 11.05 mm, 18.18 mm, and 18.53 mm. This results in distance errors of 0.48 mm, −0.36 mm, and 0.10 mm, which are all within one CT voxel.

Figure 6 shows images from the measurement software for a catheter insertion. During the experimental procedure, once the catheter reaches the guided position, two images are captured. The target, catheter tip and a second point on the catheter (to determine its orientation) are identified in both images. Consequently, the distances between the spherical target and the catheter tip, as well as the catheter line, are obtained to evaluate the accuracy of the catheter insertion. Our mixed reality guidance shows a virtual path to the users, so they can align their catheter along it, but does not show the insertion depth (i.e., when to stop). Therefore, we expect that mixed reality can better mitigate the distance between the target and the catheter line. Although our system does not currently track the catheter to provide feedback on the insertion depth to the user, thus requiring the user to stop insertion based on visual markings on the catheter, in the actual clinical scenario, as long as the catheter is directed close enough ($\sim$3 mm) to the target and reaches the ventricle, the surgeon would see the flow of CSF coming out of the catheter and stop further insertion.

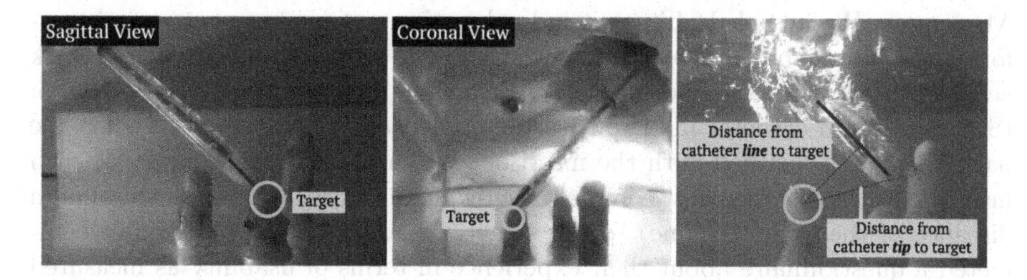

Fig. 6. Left: Sample images from two borescope cameras, showing measurements. Right: Distances from the target to the catheter tip and catheter direction

4.4 User Study

Experimental Design and Participants. We designed a within-subjects study in which each participant performed the catheter insertion task for three targets both with (MR condition) and without (baseline condition) mixed reality guidance. Following IRB approval, we recruited 10 participants. Participant ages ranged from 21 to 35 ($M = 25.44, SD = 5.11$). All had an engineering or medical background. Participants reported that they were somewhat familiar with the mixed reality devices ($M = 2.7, SD = 0.82$) on a 5-point scale, with 5 being very familiar. The majority of participants had never viewed the ventriculostomy procedure; however, we note that one participant is a neurosurgeon, who is experienced in this procedure. All participants were unpaid volunteers.

Procedure. Each participant completed two questionnaires, a pre-task survey and demographics, to provide a baseline and assess their familiarity with the procedure and with mixed reality. Then, each participant watched an instructional video on how to calibrate the HMD and perform the task. For the trials with mixed reality guidance, the user wears the HMD and calibrates the system. Next, using the tracked pointer, the participant touches each of the three fiducials and uses a voice command to acquire its 3D position. After selecting the fiducial points, the registration and alignment is performed using a voice command. The user then plans a path for catheter insertion from the entry point to the designated target. Afterwards, the user inserts the catheter by aligning it with the virtual 3D guide visualized on the HMD and then advancing to the designated depth (~6 cm). The user determines the insertion depth by reading labels on the catheter. For the trials without mixed reality, to avoid confounding factors, each participant viewed the annotated hard-copy of CT images of the phantom (coronal view as shown in Fig. 5 and saggital view) with the corresponding target, which provided the measured distance of the target with respect to its nominal position (centerline), and then inserted the catheter to reach the target. Therefore, no medical imaging knowledge was required to perform the task. The entire experiment took an average of about 45 min for each participant which included instruction video, calibration, 6 trials, and filling the questionnaires.

Measures. We included objective and subjective metrics to measure task performance and usability. Objectively, we sought to assess the participant's task accuracy. Each trial was video recorded using the optical measurement system (Sect. 4.3) when catheter insertion started and recording was stopped when the participant was satisfied with the insertion. We measured the accuracy of each insertion as the distance between the catheter tip and target and as the minimum distance of the catheter (line) to the target. In addition, the participant completed a questionnaire about their experience in terms of usability as measured by the System Usability Scale (SUS) [8] and perceived workload as measured by the NASA TLX [14].

5 Results

We used one-way repeated-measures analyses of variance (ANOVA) where the condition (either *baseline* or *mixed reality*) was set as a fixed effect, and the participant was set as a random effect. Quantitative and qualitative results are presented in Fig. 7.

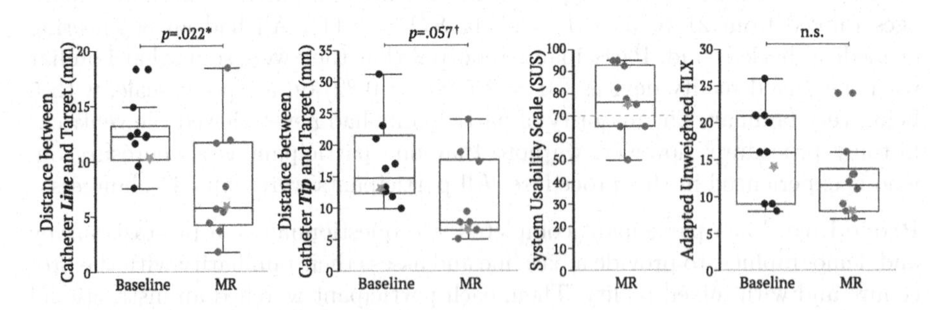

Fig. 7. Experimental results (blue star represents surgeon's data). We note that for the distance and TLX metrics lower values indicate better performance. For the SUS index, higher is better. (Color figure online)

Task Accuracy. The ANOVA test suggests that there is a significant difference in the task accuracy, measured as the distance between the catheter line and the target, $F(1, 18) = 6.24, p = .022$. The participants with the mixed reality aid were able to maintain a shorter distance to the target ($M = 7.63, SD = 5.00$) when comparing to using the baseline setup ($M = 12.21, SD = 2.93$). However, we only observed a marginal difference in measured tip distance, $F(1, 18) = 4.14, p = .057$, between the MR condition ($M = 10.96, SD = 6.61$) and the baseline condition ($M = 16.93, SD = 6.52$). Altogether, the results show more than 35% improvement in catheter tip accuracy and more than 37% improvement in catheter direction accuracy using our mixed reality navigation system.

Usability. Our data revealed that participants thought that the mixed reality method ($M = 12.1, SD = 5.04$) required less mental workload than the baseline surgical method ($M = 16.00, SD = 6.07$), although the difference was not statistically significant, $F(1, 18) = 2.44, p = .136$. Moreover, the average scored usability (SUS) of the mixed reality system ($M = 77.25, SD = 14.69$) is above the suggested usability score of 70 [8], indicating that our mixed reality system is reasonably usable for performing the ventriculostomy procedure.

Surgeon's Performance. As we have identified the surgeon's results with blue star in Fig. 7, all the surgeon's data including performance, usability, task completion time, and TLX scores are in the same interval as other participants. Specifically, the surgeon's performance without MR (baseline) in targeting accuracy for tip and line distances were 13.3 mm (M = 16.93, SD = 6.52). and 10.4 mm (M = 12:21; SD = 2:93), respectively, which with MR improved to 9.3 mm (M = 10.96, SD = 6.61) and 7.7 mm (M = 7.63; SD = 5.00). Likewise, perceived usability by the surgeon is 75 (M = 77.25; SD = 14.69). As one can see, all surgeon's quantitative and qualitative data are close to the average (within a standard deviation) and therefore, the surgeon is not an outlier.

6 Conclusion and Future Work

We proposed a new interactive portable navigation system in mixed reality for performing bedside neurosurgical procedures with a novel evaluation method. In this system, an HMD-based application is implemented which allows the surgeon to plan and perform ventriculostomy, which involves inserting a catheter through the skull into the ventricles to divert cerebrospinal fluid.

Through a user study, we show that our MR-based navigation system offered high perceived usability and improved targeting accuracy by more than 35%, suggesting clinical impact for a procedure where about one third of catheter insertions miss the target. Our future work includes integrating catheter tracking so that the system can provide additional feedback during the insertion, such as the catheter depth. We will also use a more accurate display calibration to reduce the targeting error [5,6]. We also plan to conduct a user study where all participants are intended users—neurosurgeons—with various level of expertise, to further assess the usefulness and usability of the system. This study shows the high potential of mixed reality for improving bedside surgical procedures.

Acknowledgments. The authors thank Sam Kamran, Adarsh Malapaka, and Nikhil Dave for their help in preparing the skull phantom and Kai Ding for the CT scan. Ehsan Azimi is supported by the Link Foundation fellowship.

References

1. Armstrong, D.G., Rankin, T.M., Giovinco, N.A., Mills, J.L., Matsuoka, Y.: A heads-up display for diabetic limb salvage surgery: a view through the google looking glass. J. Diab. Sci. Technol. **8**(5), 951–956 (2014)
2. Aten, Q., Killeffer, J., Seaver, C., Reier, L.: Causes, complications, and costs associated with external ventricular drainage catheter obstruction. World Neurosurg. **134**, 501–506 (2020)
3. Azimi, E., Doswell, J., Kazanzides, P.: Augmented reality goggles with an integrated tracking system for navigation in neurosurgery. In: Virtual Reality Short Papers and Posters (VRW), pp. 123–124. IEEE (2012)
4. Azimi, E., Molina, C., Chang, A., Huang, J., Huang, C.-M., Kazanzides, P.: Interactive training and operation ecosystem for surgical tasks in mixed reality. In: Stoyanov, D., et al. (eds.) CARE/CLIP/OR 2.0/ISIC -2018. LNCS, vol. 11041, pp. 20–29. Springer, Cham (2018). https://doi.org/10.1007/978-3-030-01201-4_3
5. Azimi, E., Qian, L., Kazanzides, P., Navab, N.: Robust optical see-through head-mounted display calibration: taking anisotropic nature of user interaction errors into account. In: Virtual Reality (VR). IEEE (2017)
6. Azimi, E., Qian, L., Navab, N., Kazanzides, P.: Alignment of the virtual scene to the tracking space of a mixed reality head-mounted display. arXiv preprint arXiv:1703.05834 (2017)
7. Azimi, E., et al.: Can mixed-reality improve the training of medical procedures? In: IEEE Engineering in Medicine and Biology Conference (EMBC), pp. 112–116, July 2018
8. Brooke, J.: SUS-a quick and dirty usability scale. Usabil. Eval. Ind. **189**(194), 4–7 (1996)
9. Chen, L., Day, T., Tang, W., John, N.W.: Recent developments and future challenges in medical mixed reality. In: IEEE International Symposium on Mixed and Augmented Reality (ISMAR), pp. 123–135 (2017)
10. van Doormaal, T.P., van Doormaal, J.A., Mensink, T.: Clinical accuracy of holographic navigation using point-based registration on augmented-reality glasses. Oper. Neurosurg. **17**(6), 588–593 (2019)
11. Eck, U., Stefan, P., Laga, H., Sandor, C., Fallavollita, P., Navab, N.: Exploring visuo-haptic augmented reality user interfaces for stereo-tactic neurosurgery planning. In: Zheng, G., Liao, H., Jannin, P., Cattin, P., Lee, S.-L. (eds.) MIAR 2016. LNCS, vol. 9805, pp. 208–220. Springer, Cham (2016). https://doi.org/10.1007/978-3-319-43775-0_19
12. Frantz, T., Jansen, B., Duerinck, J., Vandemeulebroucke, J.: Augmenting Microsoft's HoloLens with Vuforia tracking for neuronavigation. Healthc. Technol. Lett. **5**(5), 221–225 (2018)
13. Gsaxner, C., Pepe, A., Wallner, J., Schmalstieg, D., Egger, J.: Markerless image-to-face registration for untethered augmented reality in head and neck surgery. In: Shen, D., et al. (eds.) MICCAI 2019. LNCS, vol. 11768, pp. 236–244. Springer, Cham (2019). https://doi.org/10.1007/978-3-030-32254-0_27
14. Hart, S.G.: NASA-Task Load Index (NASA-TLX); 20 years later (2006)

15. Huyette, D.R., Turnbow, B.J., Kaufman, C., Vaslow, D.F., Whiting, B.B., Oh, M.Y.: Accuracy of the freehand pass technique for ventriculostomy catheter placement: retrospective assessment using computed tomography scans. J. Neurosurg. (JNS) **108**(1), 88–91 (2008)
16. Kersten-Oertel, M., Jannin, P., Collins, D.L.: DVV: a taxonomy for mixed reality visualization in image guided surgery. IEEE Trans. Visual Comput. Graph. **18**(2), 332–352 (2012)
17. Li, Y., et al.: A wearable mixed-reality holographic computer for guiding external ventricular drain insertion at the bedside. J. Neurosurg. (JNS) **131**, 1599–1606 (2018)
18. Pepe, A., et al.: A marker-less registration approach for mixed reality-aided maxillofacial surgery: a pilot evaluation. J. Digit. Imaging **32**(6), 1008–1018 (2019). https://doi.org/10.1007/s10278-019-00272-6
19. Raabe, C., Fichtner, J., Beck, J., Gralla, J., Raabe, A.: Revisiting the rules for freehand ventriculostomy: a virtual reality analysis. J. Neurosurg. **128**, 1–8 (2017)
20. Sadda, P., Azimi, E., Jallo, G., Doswell, J., Kazanzides, P.: Surgical navigation with a head-mounted tracking system and display. Stud. Health Technol. Inform. **184**, 363–369 (2012)
21. Sauer, F., Khamene, A., Bascle, B., Rubino, G.J.: A head-mounted display system for augmented reality image guidance: towards clinical evaluation for iMRI-guided neurosurgery. In: Niessen, W.J., Viergever, M.A. (eds.) MICCAI 2001. LNCS, vol. 2208, pp. 707–716. Springer, Heidelberg (2001). https://doi.org/10.1007/3-540-45468-3_85
22. Yoon, J.W., Chen, R.E., Han, P.K., Si, P., Freeman, W.D., Pirris, S.M.: Technical feasibility and safety of an intraoperative head-up display device during spine instrumentation. Int. J. Med. Robot. Comput. Assist. Surg. **13**(3), e1770 (2017)

Ear Cartilage Inference for Reconstructive Surgery with Convolutional Mesh Autoencoders

Eimear O' Sullivan[1](✉), Lara van de Lande[2], Antonia Osolos[2], Silvia Schievano[2], David J. Dunaway[2], Neil Bulstrode[2,3], and Stefanos Zafeiriou[1]

[1] Department of Computing, Imperial College London, London, UK
e.o-sullivan16@imperial.ac.uk
[2] UCL Great Ormond Street Institute of Child Health, London, UK
[3] Craniofacial Unit, Great Ormond Street Hospital for Children, London, UK

Abstract. Many children born with ear microtia undergo reconstructive surgery for both aesthetic and functional purposes. This surgery is a delicate procedure that requires the surgeon to carve a "scaffold" for a new ear, typically from the patient's own rib cartilage. This is an unnecessarily invasive procedure, and reconstruction relies on the skill of the surgeon to accurately construct a scaffold that best suits the patient based on limited data. Work in stem-cell technologies and bioprinting present an opportunity to change this procedure by providing the opportunity to "bioprint" a personalised cartilage scaffold in a lab. To do so, however, a 3D model of the desired cartilage shape is first required. In this paper we optimise the standard convolutional mesh autoencoder framework such that, given only the soft tissue surface of an unaffected ear, it can accurately predict the shape of the underlying cartilage. To prevent predicted cartilage meshes from intersecting with, and protruding through, the soft tissue ear mesh, we develop a novel intersection-based loss function. These combined efforts present a means of designing personalised ear cartilage scaffold for use in reconstructive ear surgery.

Keywords: Mesh autoencoders · Ear reconstruction · Cartilage

1 Introduction

Microtia is a congenital condition characterized by underdevelopment of the external ear and affects 1 in 2,000 births [9]. Microtia can be unilateral or bilateral, and occurs with varying degrees of severity, from mild abnormalities in the shape of the external ear to a complete absence of both the ear and the auricular canal [3]. Typically, microtia is treated between the ages of 9 and 12, for both functional and aesthetic purposes.

The current gold standard surgical intervention for ear reconstruction in microtia cases is a two-stage process. The first stage requires the implantation of a "scaffold" to provide the frame for a new external ear. This scaffold is

A. L. Martel et al. (Eds.): MICCAI 2020, LNCS 12263, pp. 76–85, 2020.
https://doi.org/10.1007/978-3-030-59716-0_8

constructed from the patients own rib cartilage and placed under the skin where the underdeveloped ear used to be. It remains here for a number of weeks to accommodate the stretc.h of the surrounding skin. A second procedure is then performed to release the scaffold from the skin to form a normal ear shape [14]. The drawbacks of this approach include the increased co-morbidity and potential for chest wall depression where the cartilage is harvested. Tissue engineered scaffolds, and non-absorbable auricular frameworks have also been used [8,33]. These scaffolds are made by molding biodegradable polymers into the mirror image shape of the intact contralateral ear and are then seeded with cartilage excised from the patient to encourage biocompatibility prior to subcutaneous implantation [24]. Ongoing work with stem cell tissue engineering and biofabrication indicate that it could soon be possible to bioprint 3D ear scaffolds or human patients, reducing the requirements for invasive cartilage harvesting and complex tissue engineered solutions [17,20]. While it is unclear when these advances will be available for human adoption, promising results have been seen in nude mice [20]. With this in mind, we aim to address the following question; given the ideal shape of a reconstructed ear, can we predict the underlying scaffold required that would need to be bioprinted to produce this shape?

Recent works in the field of geometric deep learning [6] have given rise to a wide variety of approaches that can be used for analysis and learning on 3D mesh structures. Mesh autoencoders [28] are of particular interest, as they allow 3D shapes to be encoded to a single latent vector representation. Highly accurate reconstructions can then be obtained by decoding these latent representations. Most, if not all, recent works on mesh autoencoders focus on compressing and then reconstructing a given shape, typically the human body, face and head [5,16,28]. Here, we aim to adapt the standard autoencoder framework to enable the encoding of one shape class and the decoding of another, thereby providing a means of inferring the structure of the cartilage lying beneath the soft tissue of the human ear. To enforce the intuition that the reconstructed cartilage mesh should lie beneath the surface of the skin, or soft tissue, of the ear, we propose a new loss function, which we refer to as an *intersection loss.*

In summary, we present a method to determine the optimal scaffold shape required for high quality ear reconstruction by adapting the standard mesh autoencoder framework such that it can be used to encode a specific structure (the ear soft tissue), and decode an alternative structure (the ear cartilage). We introduce a combined ear-and-cartilage dataset. In addition, we propose a novel intersection loss function that aims to ensure that all cartilage is within the outer surface of the soft tissue mesh. Given a lack of ground truth data for the ear scaffolds that correspond to certain ears, we present our method using actual ear cartilage and demonstrate that it is highly accurate, even with relatively small number of samples. Going forward, this makes a strong case for the use of such a method to predict the required scaffold to construct a desired ear shape.

2 Related Work

Geometric Deep Learning: Geometric deep learning is an umbrella term for the emerging techniques that facilitate the generalisation of deep neural networks and machine learning approaches to non-Euclidean domains, such as high-dimensional graphs and manifolds, and 3D mesh structures [6]. Common applications include node classification [22], molecule prediction [30], and determining mesh correspondences [4]. Multiple approaches have been presented to generalise standard 2D convolutional operators for higher-dimensional domains and irregular graph structures, [7,15,23,29]. New methods for mesh pooling, or graph coarsening, have also been developed [10,12,32].

Mesh Autoencoders: Convolutional mesh autoencoders have proven to be a highly effective means of modelling 3D data. Among the first of these was the COMA model proposed by Ranjan *et al.* [28]. By storing the barycentric coordinates of vertices removed during mesh pooling, or encoding, and using these during the unpoooling, or decoding, stage, the pooled mesh could be restored to its original topology. Truncated Chebyshev spectral convolutions were used [10]. Bouritsas *et al.* [5] later improved on this by replacing the initial spectral convolutions with aniosotropic spiral convolutions to achieve more accurate mesh reconstructions. The spiral convolution was further adapted by [16], providing a mesh convolution operator that enables fast and highly efficient training. The combination of convolutional mesh autoencoders with the spiral convolution operator has been shown to be well suited to small datasets; this was demonstrated both [5] and [16], where high reconstruction accuracies were achieved when only 80 samples were used for training.

3D Loss Functions: There exist a variety of loss functions that are typically used for 3D mesh reconstruction in geometric deep learning. The standard L1 loss function can be applied to mesh vertices, and is common when processing meshes that are in correspondence [28]. A normal consistency loss can be applied to encourage reconstructed normals to match those of the original mesh. Where exact correspondences are unavailable, the chamfer distance is often used to encourage similarity between two distinct point clouds [13]. A Laplacian objective can be used to encourage the reconstruction of smooth meshes [11,25], while the purpose of a mesh edge loss is to encourage edge length regularization. Where the reconstruction of sharp features, such as edges and corners, is desirable, a quadric loss has been shown to be beneficial [1].

While these losses have proven effective in many applications, they were not designed to enforce a particular relationship between two given surfaces. We aim to address this by proposing an intersection loss. This loss aims to ensure that points in the reconstructed sample lie on the desired side of a given plane. In the case of ear cartilage reconstruction, this helps to enforce a fundamental human understanding; that the ear cartilage should lie *within* the surface of the soft tissue of the ear.

3 Dataset

While a number of datasets for the human ear soft tissue exist [18,27,31], there are, to the best of our knowledge, none that capture the paired relationship between ear cartilage and soft tissue. To address this, we construct a database

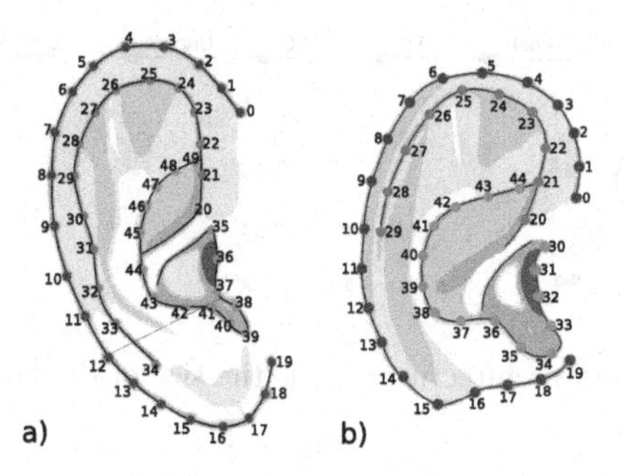

Fig. 1. a) Soft tissue (adapted from [34]) and b) Cartilage landmark sets.

of these paired samples. Data was collected from patients aged between 9 and 12 referred for a Computed Tomography (CT) scan and were reported without any abnormalities. The mean CT slice thickness and pixel size was 1 mm. The soft tissue and ear cartilage segmentations were acquired using Mimics InPrint[1]. The range used for soft tissue segmentation was −200 to +200 Hounsfield units. The ear cartilage, which has a density between that of bone and soft tissue and which can vary depending on the age of the patient, was segmented by an expert clinician. The best range for ear cartilage segmentation was found to be between −30 and −10 Hounsfield units. Although difficulties regarding the segmentation of cartilage from a CT scan limited the number of available samples, a dataset of 100 paired ear-and-cartilage instances from 68 individuals was collected.

All left meshes were mirrored so that they lay in the plane of the right ear samples. Landmarks were manually applied to both soft tissue and cartilage meshes using a set of 50 and 45 landmarks respectively, as shown in Fig. 1. The ear landmarks from [34] were used for the soft tissue, and adapted to create the cartilage landmark set. Soft tissue meshes were rigidly aligned with a soft tissue ear template mesh. The template mesh used was a non-watertight triangular mesh, consisting of 2800 vertices, and was the same as that used in [27]. The same rigid transformation was applied to the cartilage, to maintain the spatial relationship between paired cartilage and soft tissue meshes. A cartilage template mesh, containing the same number of vertices and an identical mesh topology to that of the soft tissue, was then rigidly aligned with the raw cartilage mesh. A Non-rigid Iterative Closest Point (NICP) procedure [2] was used to obtain mesh correspondences for both cartilage and soft tissue. Throughout the data processing, care was taken to ensure the spatial relationship between the soft tissue and cartilage meshes was maintained.

4 Network Architecture

Our mesh autoencoder architecture is based on that of Gong *et al.* [16]. The encoder is comprised of three convolutional layers, followed by mesh pooling.

[1] Mimics Inprint, Materialise, Leuven, Belgium.

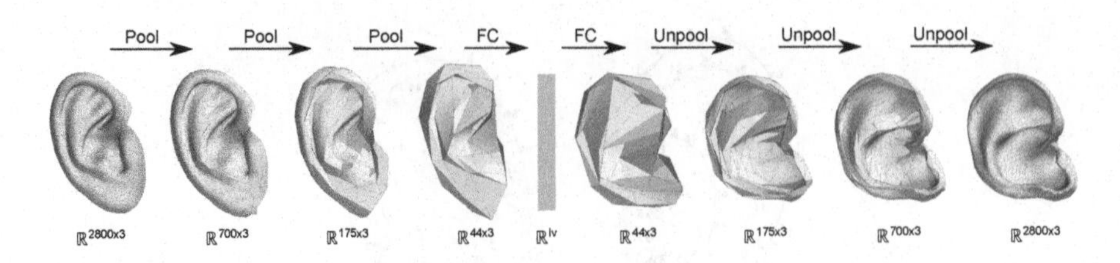

Fig. 2. Network architecture. $\mathbb{R}^{lv}$ is the size of the latent vector.

The decoder architecture is identical to that of the encoder, but in reverse. 16 latent vectors were used for the encoding. Spiral convolutionals with a length of 9 are used in all layers [16]. An overview of the architecture is shown in Fig. 2.

Mesh autoencoder frameworks typically use the barycentric coordinates of mesh vertices removed during the pooling stage to restore these vertices during the unpooling stage [28]. For computational efficiency, these values are calculated once prior to training, and then used throughout the training process. The autoencoder output is therefore a mesh with an identical topology to the input. When using an autoencoder to predict the shape of something other than the original input, however, the assumptions of identical mesh topology and barycentric coordinate locations do not necessarily hold true. Therefore, prior to training, we pool the cartilage mesh template independent of the soft tissue ear mesh and acquire the barycentric vertex locations specific to the cartilage. These are then used during the unpooling stage in place of those of the soft tissue ear mesh. This approach could be used to allow the decoder to return an arbitrary mesh toplogy, however, as the cartilage and soft tissue ear template have identical topologies, we make no such changes here.

4.1 Loss Functions

An L1 loss was applied to all reconstructed cartilage vertices. In addition to this, we also apply a loss to penalise the reconstruction if cartilage vertices are seen to protrude through the skin. This loss can be calculated using the normal direction of the vertices in cartilage and soft tissue meshes. For each vertex in the reconstructed cartilage, we find the closest vertex (nearest neighbour) in the corresponding soft tissue mesh. As the position of the predicted cartilage vertices with respect to their closest soft tissue vertex is likely to change many times during the training process, this cannot be pre-computed. For each point in the set of cartilage vertices, S_C, the closest vertex, x_s, in the set of soft tissue vertices, S_S, is instead calculated.

To determine whether the cartilage vertex intersects the surface of the soft tissue, the following formula can be applied:

$$d = n_s \cdot (x_c - x_s) \tag{1}$$

where d is the distance from the point to the soft tissue surface, x_c is a vertex in the cartilage mesh, x_s is the closest vertex in the soft tissue mesh, and n_s is the corresponding normal. The sign of d gives additional insight into the cartilage point location with respect to the soft tissue; if $d = 0$, the cartilage point lies on the plane of the soft tissue vertex. When $d < 0$, the vertex lies on the opposite side of the plane into which the normal is directed and the vertex is within the soft tissue surface. If $d > 0$, the vertex lies on the side of the plane into which the normal is directed and an intersection has occurred. In this case, the value of d can be used to determine the extent of the intersection.

The L1 loss and Intersection loss were combined as in Eq. 2, where λ_i and λ_{int} are the weighting values applied to the L1 and intersection losses respectively. Optimal performance was observed for $\lambda_i = 1$ and $\lambda_{int} = 10$.

$$l_{total} = \lambda_1 l_1 + \lambda_{int} l_{int} \tag{2}$$

As the nearest neighbour search for each vertex in the input point cloud is independent, the calculations can be trivially parallelised and efficiently computed on a GPU.

4.2 Training

The autoencoder framework is implemented in PyTorch [26]. Models were trained with a batch size of 4. The Adam optimiser, initialised with a learning rate of 0.001, and learning rate decay of 0.99, was used [21]. Mesh downsampling was achieved using the QSlim approach, while mesh upsampling in the decoder was achieved as in [28]. Early stopping was used to prevent the model from overfitting to the training data. Spiral convolutions, with a fixed length of 9, are used in each layer of the encoder-decoder framework [16]. With 80 samples in the training database, and running for up to 300 epochs, training takes 91 s on average, equating to 0.327 s per epoch on an Nvidia Titan Xp GPU.

5 Experiments

A five-fold cross-validation procedure was used. If both the right and left ear data from a given subject were used, care was taken to ensure that both sides were in the same dataset to prevent data leakage between the two datasets. Beyond this, the data was split randomly between training and testing sets. Each training and test set therefore consisted of 80 and 20 samples respectively. Mean accuracies over all data splits are reported below. Ground truth in this study was considered to be the registered cartilage meshes. Errors were calculated as the Euclidean distance between corresponding vertices in the ground truth and predicted cartilage meshes.

5.1 Ablation Study

In this section we analyse the effects of the proposed loss functions on the reconstruction quality. Three cases were tested; using just the L1 loss, using a combination of the L1 loss and chamfer distance, and using the L1 loss with the

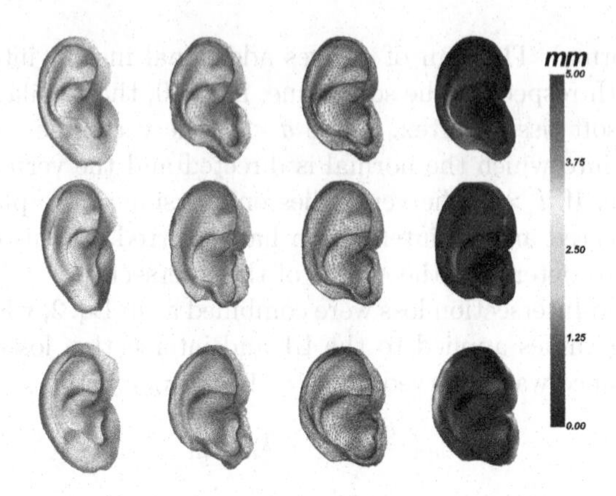

Fig. 3. Predicted cartilage for a given soft tissue sample. Rows show the soft tissue ear mesh, ground truth cartilage, and predicted cartilage meshes. The final column displays the heatmap for the Euclidean error in millimeters (mm).

proposed intersection loss. For a baseline comparison, we compare a scaled version of the soft tissue to the ground truth cartilage mesh. All soft tissue meshes were scaled by approximately 93%, equivalent to reducing the mesh by 1.5 mm to alow for skin thickness [19]. The mean vertex error, and mean number of vertex intersections per mesh, are given in Table 1. Due to the small size of the human ear, results are given in micrometers (μm). Simply scaling the soft tissue does not give a close estimate of the ground truth cartilage mesh, and results in the greatest number of intersections of the cartilage though the surface of the soft tissue. As shown in the table, the introduction of the intersection loss plays a role in reducing the mean vertex error, in addition to reducing the number of cartilage vertices that intersect the soft tissue ear mesh. Interestingly, the worst results are observed when using the L1 loss in conjunction with the chamfer distance. Examples of the predicted cartilage meshes for a given soft tissue mesh are shown Fig. 3.

Table 1. Ablation Study: The effect of different loss functions on reconstruction accuracy and number of intersections. Errors are in micrometers (μm). The mean error, standard deviation, and mean number of intersections are on a per sample basis. *Chamfer Loss* is abbreviated as *Ch.*, *Intersection Loss* is abbreviated as *Int.*

	Scaling		L1		L1 + Ch.		L1 + Int.	
	Mean	Ints.	Mean	Ints.	Mean	Ints.	Mean	Ints.
Test	324.6 ± 42.7	1683	135.1 ± 122.8	27	135.3 ± 123.0	28	$\mathbf{134.7 \pm 121.9}$	**17**
Train	–	–	578.4 ± 578.4	212	583.8 ± 413.2	215	$\mathbf{571.5 \pm 419.9}$	**202**

Table 2. Mean cartilage landmark error. A latent vector of size 16 was used. All values are reported in micrometers (μm).

	L1	L1 + Ch.	L1 + Int.
Test	210.3 ± 154.3	211.5 ± 154.7	208.9 ± 155.7
Train	712.9 ± 522.9	716.6 ± 528.4	702.6 ± 530.3

5.2 Landmark Error

Table 2 reports the landmark error for the inferenced cartilage meshes for each of the loss combinations. Minor discrepancies are noted between mean loss values for each of the implemented loss combinations, however using the intersection loss again leads to an increase in the overall accuracy. This indicates that the intersection loss discourages the predicted cartilage from protruding through the ear soft tissue, without negatively impacting the result of the reconstruction.

6 Discussion and Conclusion

Future work would look to expanding the combined ear-cartilage database to include a greater number of samples from more subjects. Neural networks tend to perform better with large quantities of data, and one of the main limitations of the system is the relatively small number of samples that can be used for training purposes; this is highlighted by the discrepancy between the reconstruction accuracies for the training and testing datasets. Though we have chosen a network architecture that has been shown to perform well on small datasets and have endeavoured to keep the number of parameters in the autoencoder small to reduce overfitting, this remains an area where further improvements could be achieved. In particular, the network has difficulties in accurately reconstructing the sharper ridges and external regions of the cartilage. This can be seen most clearly in the rows 1 and 3 of Fig. 3.

Determining the shape of the cartilage scaffold required for ear reconstruction is by no means trivial and it takes a trained and experienced surgeon to carve a cartilage scaffold from autogolous rib cartilage. The ability to predict the shape of the required cartilage scaffold, paired with the continued advancements in bioprinting technologies presents an alternative approach to creating this scaffold. With the reduced requirements for cartilage to be harvested from the ribs of the patient, morbidity and operating times would likely decrease, and the risk to chest wall deformity would be reduced.

We have presented an autoencoder framework that can be used to infer the ear cartilage shape given that of the ear soft tissue. We adapt the conventional autoencoder framework so that the encoder framework can be used to encode and decode meshes with different topologies. Additionally, we present a novel loss function, the intersection loss, and show how this can be used to reduce the number of cartilage vertices that intersect, or protrude through, the soft tissue ear surface, thereby encouraging the inference of more realistic ear cartilage samples. The loss functions can be easily optimized for GPU, facilitating fast training speeds. This approach can also allow for patient specific cartilage design; in the case of unilateral microtia, the shape of the soft tissue from the contralateral ear can be used for cartilage inference.

Acknowledgements. Stefanos Zafeiriou acknowledges support from EPSRC Fellowship DEFORM (EP/S010203/1).

References

1. Agarwal, N., Yoon, S., Gopi, M.: Learning embedding of 3D models with quadric loss (2019)
2. Amberg, B., Romdhani, S., Vetter, T.: Optimal step nonrigid ICP algorithms for surface registration, pp. 1–8. IEEE (2007)
3. Bly, R., Bhrany, A., Murakami, C., Sie, K.: Microtia reconstruction. Facial Plast. Surg. Clin. North Am. **24**(4), 577–591 (2016)
4. Boscaini, D., Masci, J., Rodolà, E., Bronstein, M.M.: Learning shape correspondence with anisotropic convolutional neural networks (2016). https://arxiv.org/abs/1605.06437
5. Bouritsas, G., Bokhnyak, S., Ploumpis, S., Zafeiriou, S., Bronstein, M.: Neural 3D morphable models: spiral convolutional networks for 3D shape representation learning and generation, pp. 7212–7221. IEEE (2019)
6. Bronstein, M.M., Bruna, J., LeCun, Y., Szlam, A., Vandergheynst, P.: Geometric deep learning: going beyond Euclidean data. IEEE Signal Process. Mag. **34**(4), 18–42 (2017)
7. Bruna, J., Zaremba, W., Szlam, A., LeCun, Y.: Spectral networks and locally connected networks on graphs. In: ICLR 2014 (2013). https://arxiv.org/abs/1312.6203. First proposal of convolution operations on graphs in the spectral domain
8. Cao, Y., Vacanti, J.P., Paige, K.T., Upton, J., Vacanti, C.A.: Transplantation of chondrocytes utilizing a polymer-cell construct to produce tissue-engineered cartilage in the shape of a human ear. Plast. Reconstr. Surg. **100**(2), 297–302 (1997)
9. Cubitt, J.J., Chang, L., Liang, D., Vandervord, J., Marucci, D.D.: Auricular reconstruction. J. Paediatr. Child Health **55**(5), 512–517 (2019)
10. Defferrard, M., Bresson, X., Vandergheynst, P. (2016). http://arxiv.org/abs/1606.09375
11. Desbrun, M., Meyer, M., Schröder, P., Barr, A.: Implicit fairing of irregular meshes using diffusion and curvature flow. In: SIGGRAPH 1999, pp. 317–324. ACM Press/Addison-Wesley Publishing Co. (1999)
12. Diehl, F., Brunner, T., Truong Le, M., Knoll, A.: Towards graph pooling by edge contraction (2019). https://graphreason.github.io/papers/17.pdf
13. Fan, H., Su, H., Guibas, L.: A point set generation network for 3D object reconstruction from a single image, pp. 2463–2471. IEEE (2017)
14. Fattah, A., Sebire, N.J., Bulstrode, N.W.: Donor site reconstitution for ear reconstruction. J. Plast. Reconstr. Aesthetic Surg. **63**(9), 1459–1465 (2009)
15. Fey, M., Lenssen, J.E., Weichert, F., Muller, H.: SplineCNN: fast geometric deep learning with continuous b-spline kernels. In: 2018 IEEE/CVF Conference on Computer Vision and Pattern Recognition, pp. 869–877 (2018)

16. Gong, S., Chen, L., Bronstein, M., Zafeiriou, S. (2019). https://arxiv.org/abs/1911.05856
17. Griffin, M.F., Ibrahim, A., Seifalian, A.M., Butler, P.E.M., Kalaskar, D.M., Ferretti, P.: Argon plasma modification promotes adipose derived stem cells osteogenic and chondrogenic differentiation on nanocomposite polyurethane scaffolds; implications for skeletal tissue engineering. Mater. Sci. Eng. C **105**, 110085 (2019)
18. Jin, C.T., et al.: Creating the Sydney York morphological and acoustic recordings of ears database. IEEE Trans. Multimed. **16**(1), 37–46 (2014)
19. Jung, B.K., et al.: Ideal scaffold design for total ear reconstruction using a three-dimensional printing technique. J. Biomed. Mater. Res. Part B Appl. Biomater. **107**(4), 1295–1303 (2019)
20. Kang, H.W., Lee, S.J., Ko, I.K., Kengla, C., Yoo, J.J., Atala, A.: A 3D bioprinting system to produce human-scale tissue constructs with structural integrity. Nat. Biotechnol. **34**(3), 312–319 (2016)
21. Kingma, D.P., Ba, J.: Adam: a method for stochastic optimization (2014). https://arxiv.org/abs/1412.6980
22. Kipf, T.N., Welling, M.: Semi-supervised classification with graph convolutional networks. In: International Conference on Learning Representations (ICLR) (2016). https://arxiv.org/abs/1609.02907
23. Levie, R., Monti, F., Bresson, X., Bronstein, M.M.: Cayleynets: graph convolutional neural networks with complex rational spectral filters. IEEE Trans. Signal Process. **67**(1), 97–109 (2019)
24. Liu, Y., et al.: In vitro engineering of human ear-shaped cartilage assisted with CAD/CAM technology. Biomaterials **31**(8), 2176–2183 (2009)
25. Nealen, A., Igarashi, T., Sorkine, O., Alexa, M.: Laplacian mesh optimization. In: GRAPHITE 2006, pp. 381–389. ACM (2006)
26. Paszke, A., et al.: Automatic differentiation in PyTorch (2017)
27. Ploumpis, S., et al.: Towards a complete 3D morphable model of the human head (2019). https://arxiv.org/abs/1911.08008
28. Ranjan, A., Bolkart, T., Sanyal, S., Black, M.J.: Generating 3D faces using convolutional mesh autoencoders (2018). http://arxiv.org/abs/1807.10267
29. Tang, S., Li, B., Yu, H.: Chebnet: efficient and stable constructions of deep neural networks with rectified power units using Chebyshev approximations (2019). https://arxiv.org/abs/1911.05467
30. Veselkov, K., et al.: Hyperfoods: machine intelligent mapping of cancer-beating molecules in foods. Sci. Rep. **9**(1), 9237 (2019)
31. Yan, P., Bowyer, K.W.: Biometric recognition using 3D ear shape. IEEE Trans. Pattern Anal. Mach. Intell. **29**(8), 1297–1308 (2007)
32. Ying, R., You, J., Morris, C., Ren, X., Hamilton, W.L., Leskovec, J.: Hierarchical graph representation learning with differentiable pooling (2018). https://arxiv.org/abs/1806.08804
33. Zhou, G., et al.: In vitro regeneration of patient-specific ear-shaped cartilage and its first clinical application for auricular reconstruction. EBioMedicine **28**(C), 287–302 (2018)
34. Zhou, Y., Zaferiou, S.: Deformable models of ears in-the-wild for alignment and recognition, pp. 626–633. IEEE (2017)

Robust Multi-modal 3D Patient Body Modeling

Fan Yang[1,2], Ren Li[1], Georgios Georgakis[1,3], Srikrishna Karanam[1(✉)], Terrence Chen[1], Haibin Ling[4], and Ziyan Wu[1]

[1] United Imaging Intelligence, Cambridge, MA, USA
{fan.yang,ren.li,georgios.georgakis,srikrishna.karanam, terrence.chen,ziyan.wu}@united-imaging.com
[2] Temple University, Philadelphia, PA, USA
[3] George Mason University, Fairfax, VA, USA
[4] Stony Brook University, Stony Brook, NY, USA
hling@cs.stonybrook.edu

Abstract. This paper considers the problem of 3D patient body modeling. Such a 3D model provides valuable information for improving patient care, streamlining clinical workflow, automated parameter optimization for medical devices *etc.* With the popularity of 3D optical sensors and the rise of deep learning, this problem has seen much recent development. However, existing art is mostly constrained by requiring specific types of sensors as well as limited data and labels, making them inflexible to be ubiquitously used across various clinical applications. To address these issues, we present a novel robust dynamic fusion technique that facilitates flexible multi-modal inference, resulting in accurate 3D body modeling even when the input sensor modality is only a subset of the training modalities. This leads to a more scalable and generic framework that does not require repeated application-specific data collection and model retraining, hence achieving an important flexibility towards developing cost-effective clinically-deployable machine learning models. We evaluate our method on several patient positioning datasets and demonstrate its efficacy compared to competing methods, even showing robustness in challenging patient-under-the-cover clinical scenarios.

Keywords: 3D patient pose and shape · Multi-modal

1 Introduction

We consider the problem of 3D patient body modeling. Given an image of a patient, the aim is to estimate the pose and shape parameters of a 3D mesh that digitally models the patient body (Fig. 1). Such a 3D representation can help augment existing capabilities in several applications. For instance, for CT

Fan Yang and Ren Li are joint first authors. This work was done during the internships of Fan Yang, Ren Li, and Georgios Georgakis with United Imaging Intelligence.

A. L. Martel et al. (Eds.): MICCAI 2020, LNCS 12263, pp. 86–95, 2020.
https://doi.org/10.1007/978-3-030-59716-0_9

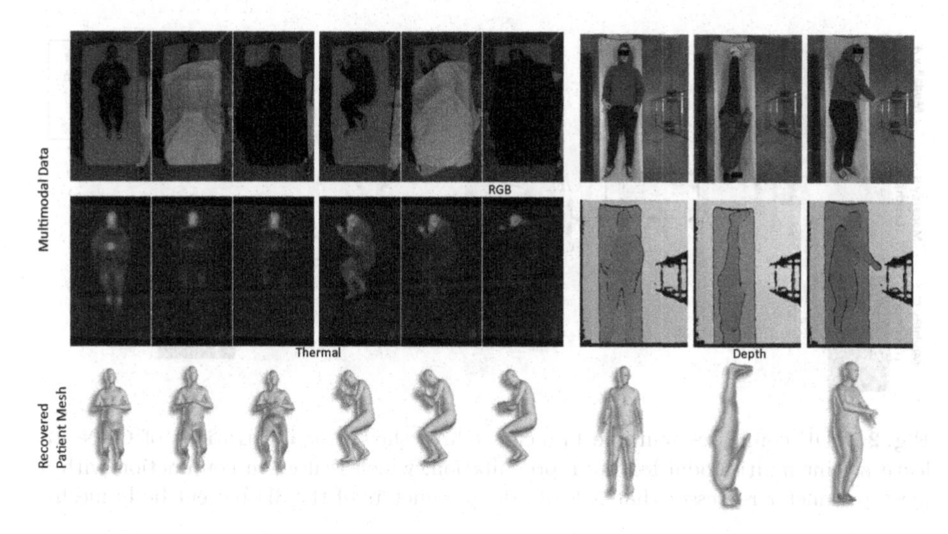

Fig. 1. We present a new approach for 3D patient body modeling that facilitates mesh inference even when the input data is a subset of all the modalities used in training.

isocentering, the 3D mesh can provide an accurate estimate of thickness for automated patient positioning [1] and radiation dose optimization [2]. In X-ray, the 3D mesh can enable automated radiographic exposure factor selection [3], ensuring optimal radiation dosage to the patient based on patient thickness estimates. Consequently, patient body modeling has seen increasing utility in healthcare [4–7].

Much recent work [8–10] has focused on estimating the 2D or 3D keypoint locations on the patient body. Such keypoints represent only a very sparse sampling of the full 3D mesh in the 3D space that defines the digital human body. The applications noted above necessitate that we go beyond just predicting keypoints and estimate the full 3D mesh representing the patient body. To address this issue, Singh *et al.* [11] presented a technique, using depth sensor data, to retrieve a full 3D patient mesh. However, this method is limited to CT-specific poses and requires depth data. If we change either the application (*e.g.*, X-ray poses and protocols) or even the sensor (*e.g.*, some applications may need RGB-only sensor), this method will need (a) fresh collection and annotation of data, and (b) retraining the model with this new data, both of which may be prohibitively expensive to do repeatedly for each application separately. These issues raise an important practical question: can we design *generic* models that can be trained *just once* and universally used across various scan protocols and application domains? Each application has its own needs and this can manifest in the form of the sensor choice (*e.g.*, RGB-only or RGB-thermal) or specific data scenario (*e.g.*, patient under the cover). To learn a model that can be trained just once and have the capability to be applied across multiple such applications requires what we call *dynamic multi-modal inference* capability. For instance, such a model trained with both RGB and thermal data can now be applied to

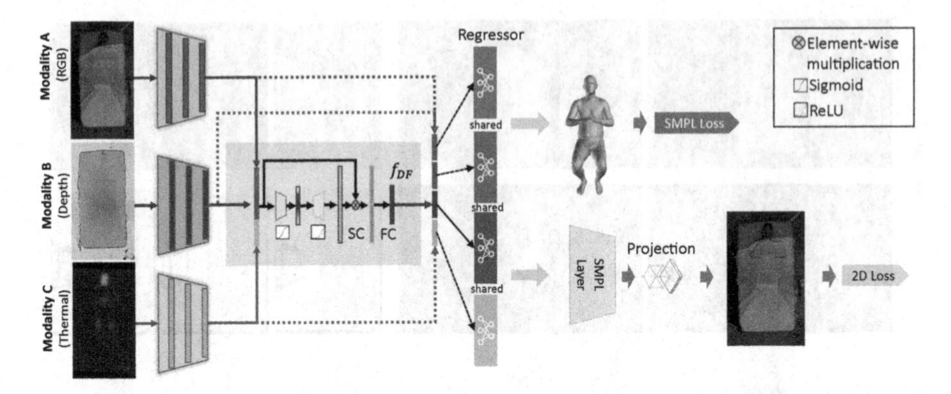

Fig. 2. RDF comprises multiple branches (three shown for illustration) of CNNs to learn a joint multi-modal feature representation, which is used in conjunction with a mesh parameter regressor that outputs the parameters of the 3D patient body mesh.

the following three scenarios without needing any application-specific retraining: RGB-only, thermal-only, or RGB-thermal. This ensures flexibility of the trained model to be used in applications that can have an RGB-only sensor, thermal-only sensor or an RGB-thermal sensor. A useful byproduct of such multi-modal inference capability is built-in redundancy to ensure system robustness. For instance, in an application with an RGB-thermal input sensor, even if one of the sensor modalities fails (*e.g.*, thermal stops working), the model above will still be able to perform 3D patient body inference with the remaining RGB-only data. These considerations, however, are not addressed by existing methods, presenting a crucial gap in clinically-deployable and scalable algorithms.

To address the aforementioned issues, we present a new *robust dynamic fusion* (RDF) algorithm for 3D patient body modeling. To achieve the multi-modal inference capability discussed above, RDF comprises a multi-modal data fusion strategy along with an associated training policy. Upon training, our RDF model can be used for 3D patient body inference under any of the possible multi-modal data modality combinations. We demonstrate these aspects under two different two-modality scenarios: RGB-depth and RGB-thermal. In both cases, we evaluate on clinically-relevant patient positioning datasets and demonstrate efficacy by means of extensive experimental comparisons with competing methods.

2 Method

The proposed *robust dynamic fusion* (RDF) framework for 3D patient body modeling comprises several key steps, as summarized in Fig. 2. Given multi-modal data input, RDF first generates features in a *joint* multi-modal feature space. While our discussion below assumes two modalities, RDF can be extended to many more modalities as well (Fig. 2 shows the scenario with three modalities). Furthermore, to make RDF robust to the absence of any particular modality during testing, we present a probabilistic scheme to perturb the input data at various multi-modal permutation levels. Our hypothesis with this training policy is that the resulting model will have been trained to predict the 3D patient model even in the absence of any particular input data modality (*e.g.*, if the thermal sensor breaks down, leading to the availability of only the RGB modality data).

Given inputs $\mathbf{I}_{m_1}$ and $\mathbf{I}_{m_2}$ from two modalities m_1 (*e.g.*, RGB) and m_2 (*e.g.*, thermal), RDF first generates feature representations for each modality $\mathbf{f}_{m_1}$ and $\mathbf{f}_{m_2}$ with two separate branches of convolutional neural networks (CNN). These individual feature vectors are then fused with our dynamic feature fusion module to give the feature representation $\mathbf{f}_{\text{DF}}$ of $\mathbf{I}_{m_1}$ and $\mathbf{I}_{m_2}$ in the joint multi-modal feature space. Given $\mathbf{f}_{\text{DF}}$, RDF generates the parameters of the 3D mesh that best describe (as measured by an objective function $L_{\text{mesh}}^{\text{DF}}$ on the mesh parameters) the patient in the input data. These parameters are then used in conjunction with an image projection operation to predict the 2D keypoints, whose error is penalized by means of an objective function L_{2D}^{DF} measuring distance to ground-truth keypoints. To strengthen the representation capability of features in each modality, RDF also computes mesh parameters directly from each of $\mathbf{f}_{m_1}$ and $\mathbf{f}_{m_2}$, each of which are penalized with objective functions ($L_{\text{mesh}}^{m_1}, L_{2D}^{m_1}$) and ($L_{\text{mesh}}^{m_2}, L_{2D}^{m_2}$) respectively. RDF is then trained with the overall loss function:

$$L = L_{\text{mesh}}^{\text{DF}} + L_{2D}^{\text{DF}} + \sum_{i=1}^{M} \left(L_{\text{mesh}}^{m_i} + L_{2D}^{m_i} \right) \tag{1}$$

where M represents the number of input modalities ($M = 2$, *e.g.*, RGB and thermal, in the context above). Note that our proposed approach is substantially different than existing state-of-the-art mesh estimation methods such as HMR [12]. While HMR also regresses mesh parameters from feature representations, it shares the same limitation as Singh *et al.* [11], *i.e.*, it can be trained only for one modality. Consequently, even if one were to use HMR in a multi-modal scenario, it would have to be in a standard two-branch fashion that assumes the availability of data from both modalities during both training and testing, leading to the same limitations and considerations discussed in Sect. 1. We next discuss each component of our RDF approach in greater detail.

Multi-modal Training. To ensure multi-modal inference flexibility discussed above, given $\mathbf{I}_{m_1}$ and $\mathbf{I}_{m_2}$, during training, we simulate several inference-time scenarios with a probabilistic data and training policy, which we achieve by adding noise to our input data streams probabilistically. Specifically, we randomly select one of the two streams m_1/m_2 with a probability p, and replace the input data array of this stream with an array of zeros. With this strategy, as training progresses, the model will have observed all the following three modality possibilities: m_1 only ($\mathbf{I}_{m_2}$ set to zero), m_2 only ($\mathbf{I}_{m_1}$ set to zero), and both m_1 and m_2, thereby "teaching" the model how to infer under any of these scenarios. Given $\mathbf{I}_{m_1}$ and $\mathbf{I}_{m_2}$ (with or without the zero changes as described above), we first extract their individual feature representations $\mathbf{f}_{m_1}$ and $\mathbf{f}_{m_2}$ with their corresponding CNN branches. We then concatenate these two feature vectors, giving $\mathbf{f}_{\text{cat}}$. Inspired by [13], we process $\mathbf{f}_{\text{cat}}$ with our feature fusion module. This fusion operation, also shown in Fig. 2, essentially generates a new feature representation, $\mathbf{f}_{DF}$, that captures interdependencies between different channels and modalities of the input feature representation. Specifically, through a series of fully connected and non-linear activation operations, we produce a vector **sc** which can be thought of as a vector of weights highlighting the importance of each channel in the input feature vector $\mathbf{f}_{\text{cat}}$. We then element-wise multiply $\mathbf{f}_{\text{cat}}$ and **sc**, which is then followed by one more fully connected unit to give $\mathbf{f}_{DF}$.

Mesh Recovery. Given $\mathbf{f}_{DF}$, RDF comprises a mesh parameter regressor module (a set of fully connected units) that estimates the parameters of the 3D patient mesh (we use Skinned Multi-Person Linear (SMPL) [14]). SMPL is a statistical model parameterized by shape $\beta \in \mathbb{R}^{10}$ and pose parameters $\theta \in \mathbb{R}^{72}$. The mesh parameter regressor module takes $\mathbf{f}_{DF}$ as input and produces the parameter estimates $\hat{\theta}$ and $\hat{\beta}$, which are penalized by an l_1 distance loss with the ground-truth parameters θ and β:

$$L_{\text{mesh}} = \left\| [\theta, \beta] - [\hat{\theta}, \hat{\beta}] \right\|_1 \tag{2}$$

Keypoints Estimation. To ensure accurate estimation of keypoints on the image, our method projects the 3D joints from the estimated mesh to image points. This is achieved using a weak-perspective projection operation [12] that consists of a translation $\rho \in \mathbb{R}^2$ and a scale $t \in \mathbb{R}$. The 2D keypoints are then computed as $\hat{\mathbf{x}}_i = s \prod(\mathbf{X}_i) + \rho$, where $\mathbf{X}_i$ is the ith 3D joint. We then supervise these predictions using an l_1 loss:

$$L_{2D} = \sum_i \left\| \mathbf{x}_i - \hat{\mathbf{x}}_i \right\|_1 \tag{3}$$

where $\mathbf{x}_i$ is the corresponding 2D ground truth.

3 Experiments

Preliminaries. As noted previously, our proposed RDF framework can in-principle be used with any number of input modalities (we only need to increase the number of input streams in Fig. 2). However, for simplicity, we demonstrate results with two separate two-modality scenarios: (m_1 = RGB, m_2 = thermal) and (m_1 = RGB, m_2 = depth). In each case, we empirically show the flexibility of RDF in inferring the 3D patient body when any subset of (m_1, m_2) modalities is available at test time. To evaluate the performance of our proposed RDF algorithm, we compare it to a competing state-of-the-art mesh recovery algorithm, HMR [12]. Note that the crux of our evaluation is in demonstrating RDF's flexibility with multi-modal inference. HMR, by design, can be used with only one data modality at a time. Consequently, the only way it can process two data modalities is by means of a two-stream architecture with data from both modalities as input. For this two-stream HMR, note that we use the concatenated features to regress the mesh parameters.

Table 1. Results on SLP, SCAN, CAD, and PKU. "T": thermal, "D": depth.

SLP	Train	Test	2D MPJPE ↓	3D MPJPE↓
HMR[12]	RGB	RGB	37.2	155
	T	T	34.2	149
	RGB-T	RGB-T	34.1	143
RDF	RGB-T	RGB	36.6	144
		T	34.7	138
		RGB-T	**32.7**	**137**

SCAN	Train	Test	2D MPJPE↓	3D MPJPE↓
HMR[12]	RGB	RGB	25.6	168
	D	D	23.7	150
	RGB-D	RGB-D	21.8	144
RDF	RGB-D	RGB	17.8	117
		D	21.6	116
		RGB-D	**16.2**	**103**

CAD	Train	Test	2D MPJPE ↓	3D MPJPE↓
HMR[12]	RGB	RGB	7.9	120
	D	D	9.2	118
	RGB-D	RGB-D	6.7	103
RDF	RGB-D	RGB	6.1	106
		D	7.2	104
		RGB-D	**5.7**	**97**

PKU	Train	Test	2D MPJPE↓	3D MPJPE↓
HMR[12]	RGB	RGB	8.8	127
	D	D	13.2	150
	RGB-D	RGB-D	8.2	118
RDF	RGB-D	RGB	7.7	123
		D	11.8	133
		RGB-D	**8.1**	**106**

Table 2. Results on SLP and CAD with three mod. "T": thermal, "D": depth.

SLP	Train	Test	2D MPJPE ↓	3D MPJPE↓
RDF	RGB-D-T	RGB	37.7	144
		T	35.5	135
		RGB-T	**34.0**	**138**

CAD	Train	Test	2D MPJPE ↓	3D MPJPE↓
RDF	RGB-D-T	RGB	6.7	108
		D	7.0	107
		RGB-D	**5.9**	**93**

Datasets, Implementation Details, and Evaluation Metrics. We use the SLP [10] dataset with images of multiple people lying on a bed for the RGB-thermal experiments. These images correspond to 15 poses collected under three different cloth coverage conditions: uncover, "light" cover (referred to as cover1), and "heavy" cover (cover2). We use PKU [15], CAD [16] and an internally-collected set of RGB-D images from a medical scan patient setup (SCAN) for the RGB-depth experiments. PKU and CAD contain a set of complex human activities recorded in daily environment, whereas the SCAN dataset has 700 images of 12 patients lying on a bed in 8 different poses. For SCAN, we create an equal 350-image/6-patient train and test split, and follow the standard protocol for other datasets. In the RDF pipeline, both modality-specific encoder networks are realized with a ResNet50 [17] architecture, which, along with the mesh parameter regressor network, is pretrained with the Humans3.6M dataset [18]. We set an initial learning rate to 0.0001, which is multiplied by 0.9 every 1,000 iterations. We use the Adam optimizer with a batch size of 64 (input image size is 224×224) and implement all code in PyTorch. All loss terms in our objective function have an equal weight of 1.0. For evaluation, we use standard metrics [18]: 2D mean per joint position error (MPJPE) in pixels and 3D MPJPE in millimeters.

Bi-modal Inference Evaluation. Table 1 shows RGB-T results on the SLP dataset and RGB-D results on the SCAN, CAD, and PKU datasets. In the "HMR" row (in both tables), "RGB" indicates training and testing on RGB-only data (similarly for thermal "T" and depth "D"). The "RGB-T"

Table 3. 3D MPJPE (mm) results of SLP evaluation under different cover scenarios.

Test modality	RGB			Thermal			RGB-T		
Cover condition	uncover	cover1	cover2	uncover	cover1	cover2	uncover	cover1	cover2
HMR [12]	139	150	154	145	149	151	141	145	143
RDF	**137**	**146**	**150**	**135**	**138**	**140**	**134**	**137**	**141**

(and similarly "RGB-D") row indicates a two-stream baseline with the two modality data streams as input. On the other hand, our proposed algorithm processes, during training, both RGB and thermal (or depth) streams of data. However, a key difference between our method and the baseline is how these algorithms are used in inference. During testing, HMR can only process data from the same modality as in training. On the other hand, RDF can infer the mesh with any subset of the input training modalities. One can note from the RGB-T results that RDF with RGB data (144 mm 3D MPJPE) is better than the baseline (155 mm 3D MPJPE) since it has access to the additional thermal modality data, thereby improving the inference results with the RGB-only modality. A similar observation can be made for the performance comparison on thermal data. RDF (137 mm) performs better than the baseline (143 mm) in the RGB-T scenario as well, substantiating the role of our feature fusion operation. Similar observations can be made from the evaluation on the SCAN/CAD/PKU datasets too.

Tri-Modal Inference Evaluation. We also evaluate our method with three modalities- RGB, depth, and thermal (RGB-D-T). Since aligned and annotated RGB-D-T data is not available, we instead use our multi-modal training policy to train with pairs of RGB-D and RGB-T data by combining the RGB-T dataset (SLP) with one RGB-D dataset (CAD). The results are shown in Table 2, where one can note our three-branch model is quite competitive when compared to the corresponding separately trained two-branch baselines (3D MPJPE 93 mm vs. two-branch RDF 97 mm on CAD RGB-D data, 138 mm vs. two-branch RDF 137 mm on SLP RGB-T data).

Under-the-Cover Evaluation. In Table 3, we evaluate the impact of patient cloth coverage on the final performance. To this end, we use "uncover", "cover1", and "cover2" labels of SLP dataset and report individual performance numbers. One can note that increasing the cloth coverage generally reduces the performance, which is not surprising. Furthermore, since there is only so much information the RGB modality can access in the covered scenarios, as opposed to the thermal modality, the performance with RGB data is also on the lower side. However, RDF generally performs better than the baseline across all these conditions, providing further evidence for the benefits of our method. Finally, some qualitative results from the output of our method are shown in Fig. 3.

Noise Robustness. We also evaluate the noise robustness of RDF and compare to HMR. In this experiment, with probability p, we replace a particular branch's

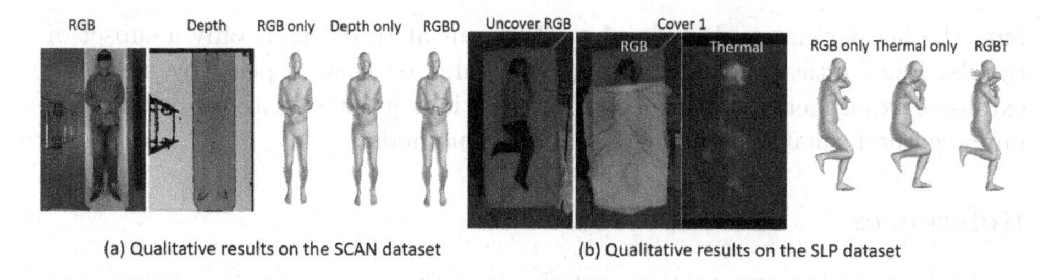

Fig. 3. Qualitative results of the proposed approach on the SCAN and SLP datasets.

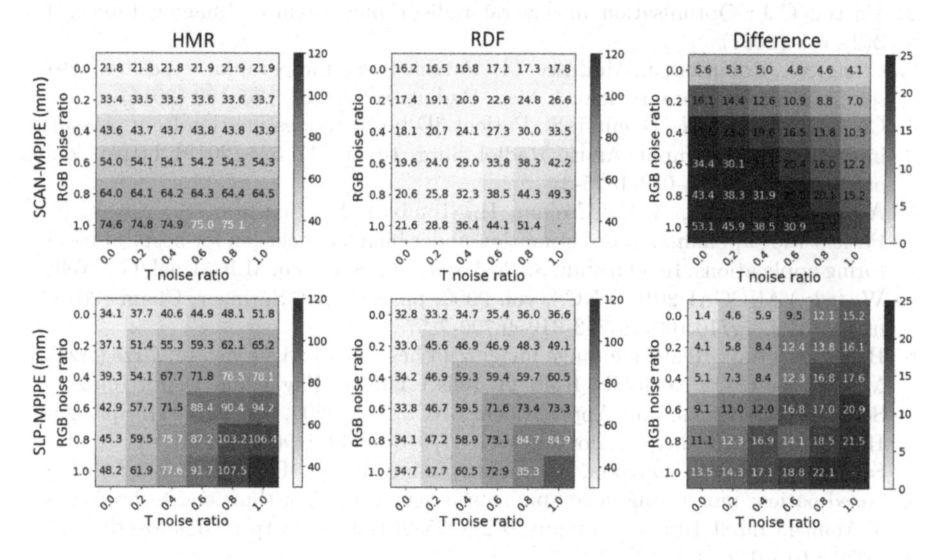

Fig. 4. HMR vs. RDF at various noise levels. "T": thermal, "D": depth.

input with an array of zeros, thus simulating the probabilistic absence of any modality's input during inference (note we ignore the case where both the inputs of both branches are zeros). In Fig. 4, we show a matrix representation of the 2D MPJPE of our method as well as baseline HMR, where one can note that with increasing noise level, both methods' performance reduces. Crucially, this performance reduction is lower for our method when compared to the baseline (see difference figure), providing evidence for improved robustness of our method.

4 Summary

We presented a new approach, called robust dynamic fusion (RDF), for 3D patient body modeling. RDF was motivated by a crucial gap of scalability and generality in existing methods, which was addressed by means of RDF's multi-modal inference capability. This was achieved by means of a novel multi-modal fusion strategy, along with an associated training policy, which enabled RDF to infer the 3D patient mesh even when the input at test time is only a subset of the data modalities used in training. We evaluated these aspects by means of extensive experiments on various patient positioning datasets and demonstrated improved performance compared to existing methods.

References

1. Li, J., et al.: Automatic patient centering for MDCT: effect on radiation dose. Am. J. Roentgenol. **188**(2), 547–552 (2007)
2. Martin, C.J.: Optimisation in general radiography. Biomed. Imaging Interv. J. **3**(2), e18 (2007)
3. Ching, W., Robinson, J., McEntee, M.: Patient-based radiographic exposure factor selection: a systematic review. J. Med. Radiat. Sci. **61**(3), 176–190 (2014)
4. Casas, L., Navab, N., Demirci, S.: Patient 3D body pose estimation from pressure imaging. Int. J. Comput. Assist. Radiol. Surg. **14**(3), 517–524 (2019). https://doi.org/10.1007/s11548-018-1895-3
5. Achilles, F., Ichim, A.-E., Coskun, H., Tombari, F., Noachtar, S., Navab, N.: Patient MoCap: human pose estimation under blanket occlusion for hospital monitoring applications. In: Ourselin, S., Joskowicz, L., Sabuncu, M.R., Unal, G., Wells, W. (eds.) MICCAI 2016. LNCS, vol. 9900, pp. 491–499. Springer, Cham (2016). https://doi.org/10.1007/978-3-319-46720-7_57
6. Bauer, S., et al.: Real-time range imaging in health care: a survey. In: Grzegorzek, M., Theobalt, C., Koch, R., Kolb, A. (eds.) Time-of-Flight and Depth Imaging. Sensors, Algorithms, and Applications. LNCS, vol. 8200, pp. 228–254. Springer, Heidelberg (2013). https://doi.org/10.1007/978-3-642-44964-2_11
7. Sathyanarayana, S., Satzoda, R.K., Sathyanarayana, S., Thambipillai, S.: Vision-based patient monitoring: a comprehensive review of algorithms and technologies. J. Ambient Intell. Human. Comput. **9**(2), 225–251 (2018). https://doi.org/10.1007/s12652-015-0328-1
8. Srivastav, V., Issenhuth, T., Kadkhodamohammadi, A., de Mathelin, M., Gangi, A., Padoy, N.: Mvor: a multi-view RGB-D operating room dataset for 2D and 3D human pose estimation. arXiv preprint arXiv:1808.08180 (2018)
9. Srivastav, V., Gangi, A., Padoy, N.: Human pose estimation on privacy-preserving low-resolution depth images. In: Shen, D., et al. (eds.) MICCAI 2019. LNCS, vol. 11768, pp. 583–591. Springer, Cham (2019). https://doi.org/10.1007/978-3-030-32254-0_65
10. Liu, S., Ostadabbas, S.: Seeing under the cover: a physics guided learning approach for in-bed pose estimation. In: Shen, D., et al. (eds.) MICCAI 2019. LNCS, vol. 11764, pp. 236–245. Springer, Cham (2019). https://doi.org/10.1007/978-3-030-32239-7_27
11. Singh, V., et al.: DARWIN: deformable patient avatar representation with deep image network. In: Descoteaux, M., Maier-Hein, L., Franz, A., Jannin, P., Collins, D.L., Duchesne, S. (eds.) MICCAI 2017. LNCS, vol. 10434, pp. 497–504. Springer, Cham (2017). https://doi.org/10.1007/978-3-319-66185-8_56
12. Kanazawa, A., Black, M.J., Jacobs, D.W. and Malik, J.: End-to-end recovery of human shape and pose. In: Computer Vision and Pattern Recognition (CVPR) (2018)
13. Hu, J., Shen, L., Sun, G.: Squeeze-and-excitation networks. In: Proceedings of the IEEE Conference on Computer Vision and Pattern Recognition, pp. 7132–7141 (2018)

14. Loper, M., Mahmood, N., Romero, J., Pons-Moll, G., Black, M.J.: SMPL: a skinned multi-person linear model. ACM Trans. Graph. (TOG) **34**(6), 1–16 (2015)
15. Liu, C., Hu, Y., Li, Y., Song, S., Liu, J.: PKU-MMD: a large scale benchmark for continuous multi-modal human action understanding. arXiv preprint arXiv:1703.07475 (2017)
16. Sung, J., Ponce, C., Selman, B., Saxena, A., Human activity detection from RGBD images. In: Workshops at the Twenty-Fifth AAAI Conference on Artificial Intelligence (2011)
17. He, K., Zhang, X., Ren, S., Sun, J.: Deep residual learning for image recognition. In: Proceedings of the IEEE Conference on Computer Vision and Pattern Recognition, pp. 770–778 (2016)
18. Ionescu, C., Papava, D., Olaru, V., Sminchisescu, C.: Human3.6m: large scale datasets and predictive methods for 3D human sensing in natural environments. IEEE Trans. Pattern Anal. Mach. Intell. **36**(7), 1325–1339 (2014)

A New Electromagnetic-Video Endoscope Tracking Method via Anatomical Constraints and Historically Observed Differential Evolution

Xiongbiao Luo(✉)

Department of Computer Science, Xiamen University, Xiamen 361005, China
xiongbiao.luo@gmail.com

Abstract. This paper develops a new hybrid electromagnetic-video endoscope 3-D tracking method that introduces anatomical structure constraints and historically observed differential evolution for surgical navigation. Current endoscope tracking approaches still get trapped in image artifacts, tissue deformation, and inaccurate sensor outputs during endoscopic navigation. To deal with these limitations, we spatially constraint inaccurate electromagnetic sensor measurements to the centerlines of anatomical tubular organs (e.g., the airway trees), which can keep the measurements physically inside the tubular organ and tackle the inaccuracy problem caused by respiratory motion and magnetic field distortion. We then propose historically observed differential evolution to precisely fuse the constrained sensor outputs and endoscopic video sequences. The new hybrid tracking framework was evaluated on clinical data, with the experimental results showing that our proposed method fully outperforms current hybrid approaches. In particular, the tracking error was significantly reduced from (5.9 mm, 9.9°) to (3.3 mm, 8.6°).

Keywords: Image-guidance intervention · Endoscopy · Surgical tracking and navigation · Differential evolution · Stochastic optimization

1 Endoscope Tracking

Endoscope three-dimensional (3-D) motion tracking is the key of image-guided minimally-invasive procedures that use endoscopes to visually examine and treat diseases in the interior of the organ cavity (tubular structures) through a natural orifice or transcutaneous port. Such a tracking aims to synchronize pre- and intra-operative image spaces in real time to provide surgeons with on-line 3-D visualization of tumor targets precisely located in the body and multimodal images. Numerous tracking methods have been discussed in the literature.

Image-based tracking aligns 2-D endoscopic video images to 3-D pre-operative volumetric data, especially performing 2-D/3-D registration to locate

A. L. Martel et al. (Eds.): MICCAI 2020, LNCS 12263, pp. 96–104, 2020.
https://doi.org/10.1007/978-3-030-59716-0_10

the endoscope [1]. Shen et al. [2] proposed to recognize typical structures to navigate the endoscope, while Byrnes et al. [3] automatically summarized endoscopic video images to indicate the endoscope's position in the body. More recently, Luo et al. [4] used multiple instance learning and a discrepancy similarity to estimate the endoscope motion, while Shen et al. [5] recovered the context-aware depth to predict the endoscopic camera pose. On the other hand, electromagnetic (EM) tracking is widely used to endoscope 3-D tracking [6–8]. Hofstad et al. [9] proposed a global to local registration method to improve the performance of EM tracking, while Attivissimo et al. [10] discussed a new EM tracking system for computer assisted surgery. Kugler et al. [11] established an experimental environment to precisely evaluate the performance of the EM tracking techniques. While image-based and EM-based endoscope tracking methods have their own advantages, they still suffer from limitations [4,12], e.g., image uncertainties, magnetic field distortion, tissue deformation, and patient movements (e.g., breathing, coughing, and beating). To deal with these limitations, a joint strategy of EM and image-based tracking provides a promising solution.

This work proposes a new hybrid strategy for smooth and accurate endoscope tracking that incorporates three new principles. First, we propose an anatomical constraint to spatially or physically limit EM sensor outputs inside anatomical tubular structures (e.g., bronchus, colon, and nasal sinus). Based on the segmented information from preoperative CT images, i.e., the centerline of the organ cavity, we constrain the current EM sensor output to be on this centerline. Such a constraint potentially keeps the EM sensor outputs inside the cavity interior under respiratory motion and tissue deformation. Next, we modify differential evolution and present a historically observed differential evolution (HODE) method to stochastically optimize EM sensor outputs and simultaneously fuse endoscopic video images. Additionally, a selective structural measure is introduced to evaluate the HODE optimization, which is robust to image artifacts or uncertainties and can accurately describe individual during iteration.

2 Approaches

The new joint EM-video endoscope tracking method consists of two main steps: (1) anatomical constraint and (2) historically observed differential evolution with parameter and fitness computation, all of which are discussed as follows.

Without loss of generality, we introduce some notations. Let $\mathbf{I}_i$ and $(\mathbf{P}_i, \mathbf{Q}_i)$ be the endoscopic video image and EM sensor tracked camera 3-D pose (including position and orientation) at time or frame i. The endoscopic camera's position is $\mathbf{P}_i$ and its orientation is represented by a unit quaternion $\mathbf{Q}_i$.

2.1 Anatomical Constraint

The idea of the anatomical constraint is to use the centerline of the hollow or tubular organ to limit or project the EM tracked position and orientation on the centerline. Hence, we first segment preoperative CT images to obtain a set of

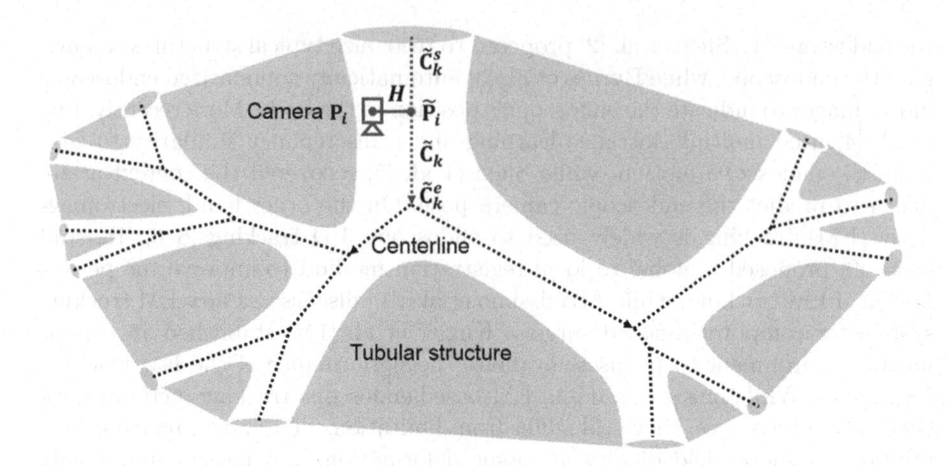

Fig. 1. Anatomical centerlines constrain the EM tracked pose inside the tubular structure to obtain the camera's projected position and updated orientation.

centerlines (curves) $\mathcal{C} = \{\mathbf{C}_k = (\mathbf{C}_k^s, \mathbf{C}_k^e)\}_{k=1}^K$ (K is the number of the extracted centerlines, $\mathbf{C}_k^s$ and $\mathbf{C}_k^e$ are the centerline's start and end points, respectively), and actually $\mathbf{C}_k$ consists of a set of 3-D points $\mathbf{C}_k = \{\mathbf{C}_k^s, \cdots, \mathbf{C}_k^e\}$.

We then assign the closest centerline $\hat{\mathbf{C}}_k$ to the position $\mathbf{P}_i$ by minimizing the distance between the 3-D position or point $\mathbf{P}_i$ and centerline set $\mathcal{C}$ (Fig. 1)

$$\hat{\mathbf{C}}_k = \arg \min_{\mathbf{C}_k \in \mathcal{C}} H(\mathbf{P}_i, \mathbf{C}_k), \tag{1}$$

where the distance $H(\mathbf{P}_i, \mathbf{C}_k)$ from the position $\mathbf{P}_i$ to $\mathbf{C}_k$ is calculate by

$$H(\mathbf{P}_i, \mathbf{C}_k) = \begin{cases} ||\mathbf{P}_i - \mathbf{C}_k^s|| & \gamma < 0 \\ ||\mathbf{P}_i - \mathbf{C}_k^e|| & \gamma > L_k \\ \sqrt{||\mathbf{P}_i - \mathbf{C}_k^s||^2 - \gamma^2} & \text{otherwise} \end{cases}, \tag{2}$$

where the centerline length $L_k = ||\mathbf{C}_k^e - \mathbf{C}_k^s||$ and γ denotes the length of the vector $(\mathbf{P}_i - \mathbf{C}_k^s)$ that is projected on the centerline $\mathbf{C}_k$, which is computed by

$$\gamma = \langle \mathbf{P}_i - \mathbf{C}_k^s, \ \mathbf{C}_k^e - \mathbf{C}_k^s \rangle / ||\mathbf{C}_k^e - \mathbf{C}_k^s||, \tag{3}$$

where $\gamma < 0$ and $\gamma > L_k$ indicate that projected point $\hat{\mathbf{P}}_i$ is located on the parent and child centerlines of the centerline $\mathbf{C}_k$, respectively; otherwise, it will be located on the centerline $\mathbf{C}_k$, and $\langle , \rangle$ denotes the dot product.

Note that we possibly obtain several closest centerlines $\{\hat{\mathbf{C}}_k\}_{k=1,2,3,\cdots}$ that have the same distance to the point $\mathbf{P}_i$ after the minimization procedure (Eq. 1). We compute the angle between the orientation $\mathbf{Q}_i$ in z−direction $\mathbf{Q}_i^z$ and centerline direction $(\hat{\mathbf{C}}_k^e - \hat{\mathbf{C}}_k^s)||\hat{\mathbf{C}}_k^e - \hat{\mathbf{C}}_k^s||^{-1}$ to determine optimal centerline $\tilde{\mathbf{C}}_k$:

$$\tilde{\mathbf{C}}_k = \arg \min_{\{\hat{\mathbf{C}}_k\}_{k=1,2,3,\cdots}} \arccos \left\langle \frac{\hat{\mathbf{C}}_k^e - \hat{\mathbf{C}}_k^s}{||\hat{\mathbf{C}}_k^e - \hat{\mathbf{C}}_k^s||}, \frac{\mathbf{Q}_i^z}{||\mathbf{Q}_i^z||} \right\rangle. \tag{4}$$

After finding the optimal centerline $\tilde{\mathbf{C}}_k$, we project the EM sensor tracked position $\mathbf{P}_i$ on the centerline $\tilde{\mathbf{C}}_k$ and obtain the projected point $\tilde{\mathbf{P}}_i$ by

$$\tilde{\mathbf{P}}_i = \tilde{\mathbf{C}}_k^s + \frac{\left\langle \mathbf{P}_i - \tilde{\mathbf{C}}_k^s, \tilde{\mathbf{C}}_k^e - \tilde{\mathbf{C}}_k^s \right\rangle}{||\tilde{\mathbf{C}}_k^e - \tilde{\mathbf{C}}_k^s||} \frac{(\tilde{\mathbf{C}}_k^e - \tilde{\mathbf{C}}_k^s)}{||\tilde{\mathbf{C}}_k^e - \tilde{\mathbf{C}}_k^s||}. \quad (5)$$

We also update the camera orientation in $z-$direction by the optimal centerline

$$\tilde{\mathbf{Q}}_i^z = (\tilde{\mathbf{C}}_k^e - \tilde{\mathbf{C}}_k^s)/||\tilde{\mathbf{C}}_k^e - \tilde{\mathbf{C}}_k^s||. \quad (6)$$

Finally, we obtain the constrained camera position $\tilde{\mathbf{P}}_i$ and orientation $\tilde{\mathbf{Q}}_i$ that are further updated by the following stochastic optimization procedure.

2.2 Historically Observed Differential Evolution

Differential evolution (DE) is a powerful tool for various stochastic optimization problems and applications [13]. Basically, DE performs three main operators: (1) mutation, (2) crossover, and (3) selection. The mutation and crossover operators play an essential role in DE, and balance convergence and computational efficiency. Improper mutation and crossover potentially result in premature convergence and local minima. The selection operator should be appropriate to precisely evaluate the quality of the solution during stochastic optimization.

DE has no any strategies to recall historical solutions that are the best solutions in the previous generation, while it also does not incorporate the current observation (the current endoscopic image and camera pose) in mutation and crossover. Moreover, accurate initialization of a population of target vectors implies that all potential solutions are powerful and fast to find the optimal solution. Unfortunately, inaccurate initialization and improper mutation or crossover possibly lead to get trapped in local minima with premature convergence. Our idea is to recall historical best solutions and simultaneously introduce the current observation into DE and presents a new version of DE discussed as follows.

Accurately Observed Initialization. Let $\mathbf{X}_{i,j}^g$ be an individual or target vector in a population of vector cloud $\mathcal{X}_i^g = \{\mathbf{X}_{i,j}^g \in \mathcal{R}^D\}_{j=1}^J$ (J is the population size) at generation or iteration g ($g = 1, 2, 3, \cdots, G$) at frame i, and $\mathbf{X}_{i,j}^g = (X_{i,j,1}^g, X_{i,j,2}^g, X_{i,j,d}^g, \cdots, X_{i,j,D}^g)$ in a D-dimensional space.

Before the target vector propagation (mutation and crossover), $\mathbf{X}_{i,j}^g$ is usually initialized in accordance with the search space constrained by the prescribed minimal and maximal bounds ($\mathbf{X}_{i,min}^g$ and $\mathbf{X}_{i,max}^g$):

$$X_{i,j,d}^g = X_{i,min,d}^g + Rand[0,1](X_{i,max,d}^g - X_{i,min,d}^g), \; d = 1, 2, \cdots, D, \quad (7)$$

where $Rand[0,1]$ is a random number that yields uniform distribution between 0 and 1. This initialization has two potential problems: first is difficult to find or determine the bounds of $\mathbf{X}_{i,j}^g$ in practice and second is that the population gets trapped in a impoverishment problem which implies that the population lose the diversity with weak exploration after several iteration.

To avoid the impoverishment problem, we introduce the previous global best $\mathbf{X}_{i-1,best}^g$ at generation g and current observation (pose), which is the constrained camera position $\tilde{\mathbf{P}}_i$ and orientation $\tilde{\mathbf{Q}}_i$ obtained above, to initialize $\mathbf{X}_{i,j}^g$

$$\mathbf{X}_{i,j}^g = \underbrace{\mathbf{X}_{i-1,best}^g}_{History} + Rand[0,1]\Delta, \Delta = \underbrace{[\tilde{\mathbf{P}}_i - \tilde{\mathbf{P}}_{i-1}, \tilde{\mathbf{Q}}_i - \tilde{\mathbf{Q}}_{i-1}]^T}_{Incremental\ observation}, \quad (8)$$

which is a seven-dimensional vector ($D = 7$) and the transpose T.

Cloud Mutation. The mutation operator aims to generate mutant or perturbed vectors in accordance with $\mathbf{X}_{i,j}^g$ difference vectors and evolutionary factors, i.e., explore more potential solutions in the search space. Five mutation strategies are most commonly used to generate the mutant vector $\mathbf{V}_{i,j}^g$ [13]. Basically, the two-difference-vector based mutation strategies provide better perturbation than others but their performance requires to be further investigated.

To exploratively propagate $\mathbf{X}_{i,best}^g$, this work proposes a new mutation strategy that integrates the cognitive and social elements defined as follows:

$$\mathbf{V}_{i,j}^g = \mathbf{X}_{i,j}^g + F_1(\underbrace{\mathbf{X}_{i,local}^g - \mathbf{X}_{i,j}^g}_{Cognitive\ element}) + F_2(\underbrace{\mathbf{X}_{i,global}^g - \mathbf{X}_{i,j}^g}_{Social\ element}), \quad (9)$$

where $\mathbf{X}_{i,local}^g$ and $\mathbf{X}_{i,global}^g$ are the local and global best individuals at generation g. While the cognitive part aims to make individuals exploring most competitive solutions in the previous iteration, the social part establishes a good solution where the population tends to reach. Two evolutionary factors F_1 and F_2 control the perturbation of two differential variations and are adaptively calculated by

$$F_1 = \mathcal{W}(\mathbf{I}_i, \mathbf{X}_{i,local}^g)/\mathcal{W}(\mathbf{I}_i, \mathbf{X}_{i,low}^g),\ F_2 = \mathcal{W}(\mathbf{I}_i, \mathbf{X}_{i,global}^g)/\mathcal{W}(\mathbf{I}_i, \mathbf{X}_{i,low}^g), \quad (10)$$

where $\mathcal{W}$ is a fitness function related to the current observed endoscopic image $\mathbf{I}_i$ and $\mathbf{X}_{i,lowest}^g$ denotes the target vector with the lowest fitness value.

Cloud Crossover. The crossover operator aims to improve the population diversity and avoid premature convergence and local minima by creating a set of trial vector $\mathbf{U}_{i,j}^g = \{U_{i,j,1}^g, U_{i,j,2}^g, U_{i,j,d}^g, \cdots, U_{i,j,D}^g\}$. Traditional crossover operators like binomial distribution provide the target and mutant vectors without any perturbation, resulting in a faster but premature convergence.

Based on our concept of using the current observation, we introduce a new crossover operator with the incremental observation Δ (Eq. 8) to obtain $\mathbf{U}_{i,j}^g$.

$$\mathbf{U}_{i,j}^g = \begin{cases} \mathbf{V}_{i,j}^g + Rand[0,1]\Delta & \text{if } (Rand[0,1] \leq C_r) \text{ or } (d = d_r) \\ \mathbf{X}_{i,j}^g + Rand[0,1]\Delta & \text{if } (Rand[0,1] \geq C_r) \text{ or } (d \neq d_r) \end{cases}, \quad (11)$$

where C_r is the crossover rate and d_r is randomly selected from $\{1, 2, d, \cdots, D\}$.

History Update and Recall. After the mutation and crossover, we need to update $\mathbf{X}_{i,j}^g$ by computing and comparing the fitness $\mathcal{W}$ of $\mathbf{U}_{i,j}^g$ and $\mathbf{X}_{i,j}^g$.

Algorithm 1: New hybrid electromagnetic-video endoscope tracking

Data: Endoscopic video sequences, EM sensor outputs, and 3-D CT images
Result: Endoscope 3-D position and orientation in the CT coordinate system

① Preprocessing: Segment CT images to obtain the centerline of the airway tree;
for $i = 1$ **to** N (frame or output number) **do**
  ② Perform anatomical constraint (Eqs. 5∼6): Obtain $\tilde{\mathbf{P}}_i$ and $\tilde{\mathbf{Q}}_i$;
  ③ Accurately observed initialization (Eq. 8): Obtain target vector set $\mathcal{X}_i^g$;
  for $g = 1$ **to** G (iteration number) **do**
    for $j = 1$ **to** J (population number) **do**
      ④ Cloud mutation (Eqs. 9∼10): Obtain mutant vector $\mathbf{V}_{i,j}^g$;
      ⑤ Cloud crossover (Eq. 11): Obtain trial vector $\mathbf{U}_{i,j}^g$;
      j=j+1;
    end
    ⑥ Calculate the fitness of each vector in two clouds $\mathcal{X}_i^g$ and $\mathcal{U}_i^g$ (Eq. 12);
    ⑦ Find the best vector $\mathbf{X}_i^g$ from $\mathcal{X}_i^g$ and $\mathcal{U}_i^g$ and store $\mathbf{X}_i^g$ in set $\mathcal{X}_i$;
    ⑧ History update and recall: Generate new target vector set $\mathcal{X}_i^{g+1}$;
    $g = g + 1$;
  end
  ⑨ Determine the estimated camera pose $[\tilde{\mathbf{P}}_i^*, \tilde{\mathbf{Q}}_i^*]^T$ at frame i (Eq. 13) ;
  $i = i + 1$;
end

For each vector in two clouds $\mathcal{U}_i^g = \{\mathbf{U}_{i,j}^g\}_{j=1}^J$ and $\mathcal{X}_i^g = \{\mathbf{X}_{i,j}^g\}_{j=1}^J$, we calculate its fitness $\mathcal{W}(\mathbf{I}_i, \mathbf{X}_{i,j}^g)$ that is defined as the similarity between the current endoscopic image $\mathbf{I}_i$ and 2-D virtual image generated by the camera pose (i.e., $\mathbf{X}_{i,j}^g$ or $\mathbf{U}_{i,j}^g$) using volume rendering techniques:

$$\mathcal{W}(\mathbf{I}_i, \mathbf{X}_{i,j}^g) = \frac{1}{M} \sum_{\{\Omega_i^m\}_{m=1}^M} \frac{1}{|\Omega_i^m|} \sum_{\Omega_i^m} \frac{(2\mu_i\mu_x + C_1)(2\sigma_{i,x} + C_2)}{(\mu_i^2 + \mu_x^2 + C_1)(\sigma_i^2 + \sigma_x^2 + C_2)}, \quad (12)$$

where $\sigma_{i,x}$ is the correlation between images $\mathbf{I}_i$ and $\mathbf{I}_x(\mathbf{X}_{i,j}^g)$ at region Ω_i^m with $|\Omega_i^m|$ pixels ($m = 1, 2, \cdots, M$); μ_i and μ_x are the mean intensity, σ_i and σ_x are the intensity variance, and C_1 and C_2 are constants. Our fitness function selects specific structural regions Ω_i^m on $\mathbf{I}_i$ to precisely compute the fitness value.

While we ascendingly sort all the vectors in $\mathcal{U}_i^g$ and $\mathcal{X}_i^g$ in accordance with their fitness values, we choose the half better-fitness vectors from $\mathcal{X}_i^g$ (recall the history) and select the half better-fitness vectors from $\mathcal{U}_i^g$ (update the history) to generate new target vector $\mathcal{X}_i^{g+1} = \{\mathbf{X}_{i,j}^{g+1}\}_{j=1}^J$ for the next iteration.

After each iteration (generation), we select the best vector $\mathbf{X}_i^g$ with the best fitness from $\mathcal{U}_i^g$ and $\mathcal{X}_i^g$ at iteration g. This implies that we obtain the best vector set $\mathcal{X}_i = \{\mathbf{X}_i^1, \cdots, \mathbf{X}_i^g, \cdots, \mathbf{X}_i^G\}$ after iteration G. Therefore, the endoscope's position and orientation $[\tilde{\mathbf{P}}_i^*, \tilde{\mathbf{Q}}_i^*]^T$ at frame i are determined by

$$\mathbf{X}_i^* = \arg \max_{\mathbf{X}_i^g \in \mathcal{X}_i} \mathcal{W}(\mathbf{I}_i, \mathbf{X}_i^g), \quad \mathbf{X}_i^* \mapsto [\tilde{\mathbf{P}}_i^*, \tilde{\mathbf{Q}}_i^*]^T \quad (13)$$

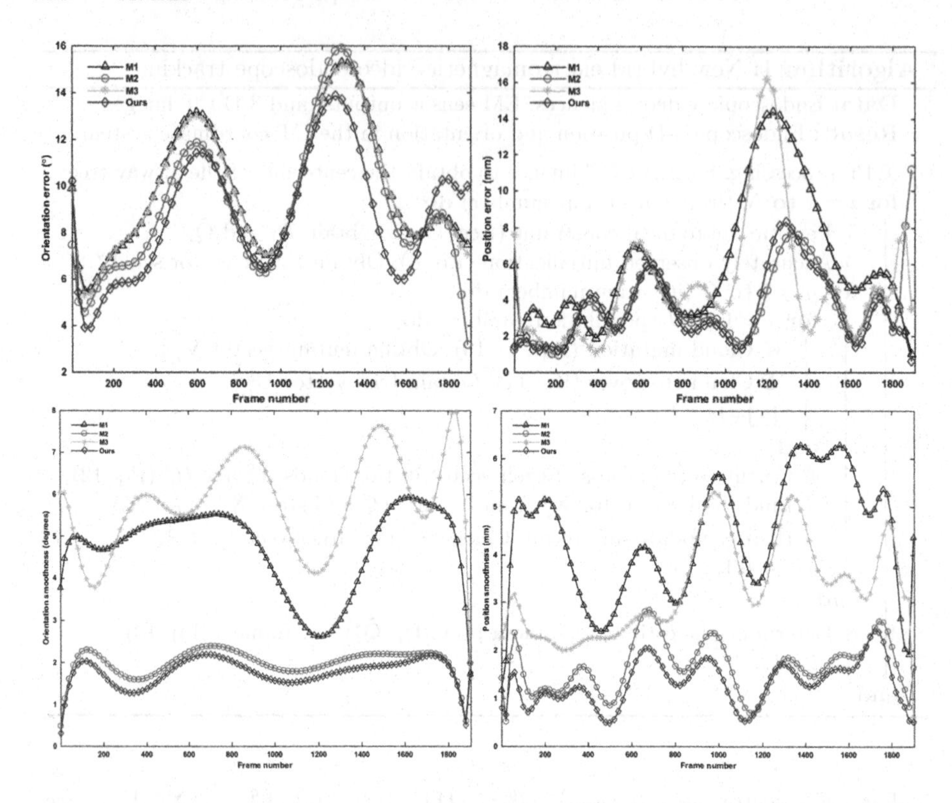

Fig. 2. Tracking error and smoothness of different methods on Case 6

Eventually, we summarized our new tracking method that combines anatomical constraint with historically observed differential evolution in Algorithm 1.

3 Results

We validate our method on six patient datasets more than 9,000 frames and compare it to several methods: (1) M1 [6], a hybrid method based on Kalman filtering, (2) M2 [7]: a stochastic optimization based method, and (3) M3 [9]: a global-to-local registration method. We manually generated ground truth and calculate the position and orientation errors, while we define the smoothness as the average Euclidean distance of the estimated positions among continuous frames as well as the orientations to evaluate if a tracking is large jitter or jump. Additionally, we investigate the fitness (similarity) of the compared methods.

Figure 2 shows the tracking error and smoothness of using the different methods. Table 1 quantitatively summarizes all the tracking results. The tracking error was reduced from (5.9 mm, 9.9°) to (3.3 mm, 8.6°). Figure 3 visually compares the tracked results of using the four methods. Based on volume rendering

Table 1. Comparison of tracking error, smoothness, fitness, and runtime of using the methods (the units of position and orientation are millimeter and degree)

Methods	M1 [6]	M2 [7]	M3 [9]	Ours
Error	(5.9 mm, 9.9°)	(3.2 mm, 9.2°)	(4.9 mm, 9.8°)	(3.3 mm, 8.6°)
Smoothness	(4.3 mm, 4.7°)	(1.6 mm, 1.9°)	(3.3 mm, 5.8°)	(1.3 mm, 1.7°)
Fitness	0.61	0.71	0.65	0.77
Runtime	1.8 s	2.0 s	2.5 s	2.6 s

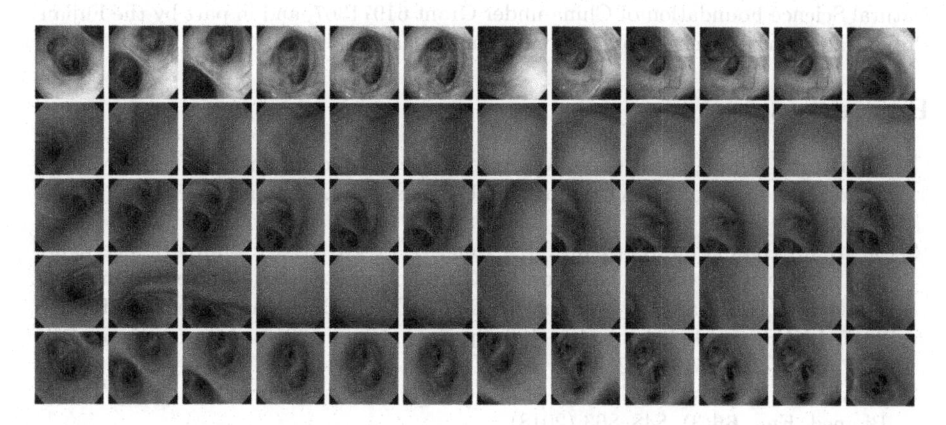

Fig. 3. Visual comparison of the results on Case 2. The first row shows real images and other rows illustrate 2-D virtual images generated by the estimated positions and orientations of using M1 [6], M2 [7], M3 [9], and ours, respectively.

methods and the estimated position and orientation, we can generate 2-D virtual images and check if they resemble real endoscopic video images: the more closely resembling, the more accurately tracking. All the experimental results demonstrate that our method significantly outperforms others.

4 Discussion and Conclusion

The objective of this work is to deal with the problems of image artifacts, patient movements, and inaccurate EM sensor outputs in endoscope tracking. Our idea to tackle these problems is to constrain the outputs and stochastically optimize the endoscope pose based on a new algorithm of historically observed differential evolution, and the experimental results demonstrate its effectiveness.

Although our proposed method is a promising and effective strategy, it still has potential limitations. First, the anatomical constraint depends on the segmentation accuracy of the bronchial tree. Next, low-quality endoscopic video images with artifacts increase the error of the fitness computation. Moreover, we further investigate the iteration and population size to improve the performance of historically observed differential evolution. Additionally, the computational time of our method was 2.6 s per frame. We will further improve the computational efficiency to meet real-time requirement of clinical applications.

In summary, this work develops a new hybrid endoscope tracking method that combines anatomical constraint and historically observed differential evolution. The experimental results demonstrate that our methods outperforms others, especially it reduced the tracking error from (5.9 mm, 9.9°) to (3.3 mm, 8.6°).

Acknowledgment. This work was supported in part by the Fundamental Research Funds for the Central Universities under Grant 20720180062, in part by the National Natural Science Foundation of China under Grant 61971367, and in part by the Fujian Provincial Natural Science Foundation under Grant 2020J01112133.

References

1. Merritt, S., Khare, R., Higgins, W.: Interactive CT-video registration for the continuous guidance of bronchoscopy. IEEE Trans. Med. Imaging **32**(8), 1376–1396 (2013)
2. Shen, M., Giannarou, S., Shah, P.L., Yang, G.-Z.: BRANCH: bifurcation recognition for airway navigation based on struCtural cHaracteristics. In: Descoteaux, M., Maier-Hein, L., Franz, A., Jannin, P., Collins, D.L., Duchesne, S. (eds.) MICCAI 2017. LNCS, vol. 10434, pp. 182–189. Springer, Cham (2017). https://doi.org/10.1007/978-3-319-66185-8_21
3. Byrnes, P., Higgins, W.: Efficient bronchoscopic video summarization. IEEE Trans. Biomed. Eng. **66**(3), 848–863 (2018)
4. Luo, X., Zeng, H.-Q., Du, Y.-P., Cheng, X.: Towards multiple instance learning and Hermann Weyl's discrepancy for robust image-guided bronchoscopic intervention. In: Shen, D., et al. (eds.) MICCAI 2019. LNCS, vol. 11768, pp. 403–411. Springer, Cham (2019). https://doi.org/10.1007/978-3-030-32254-0_45
5. Shen, M., Gu, Y., Liu, N., Yang, G.-Z.: Context-aware depth and pose estimation for bronchoscopic navigation. IEEE Robot. Autom. Lett. **4**(2), 732–739 (2019)
6. Soper, T., Haynor, D., Glenny, R., Seibel, E.: In vivo validation of a hybrid tracking system for navigation of an ultrathin bronchoscope within peripheral airways. IEEE Trans. Biomed. Eng. **57**(3), 736–745 (2010)
7. Luo, X., Wan, Y., He, X., Mori, K.: Observation-driven adaptive differential evolution and its application to accurate and smooth bronchoscope three-dimensional motion tracking. Med. Image Anal. **24**(1), 282–296 (2015)
8. Sadjadi, H., Hashtrudi-Zaad, K., Fichtinger, G.: Simultaneous localization and calibration for electromagnetic tracking systems. Int. J. Med. Robot. Comput. Assist. Surg. **12**, 189–198 (2016)
9. Hofstad, E., et al.: Intraoperative localized constrained registration in navigated bronchoscopy. Med. Phys. **44**(8), 4204–4212 (2017)
10. Attivissimo, F., Lanzolla, A., Carlone, S., Larizza, P., Brunetti, G.: A novel electromagnetic tracking system for surgery navigation. Comput. Assist. Surg, **23**(1), 42–45 (2018)
11. Kugler, D., et al.: High-precision evaluation of electromagnetic tracking. Int. J. Comput. Assist. Radiol. Surg. **14**, 1127–1135 (2019)
12. Sorriento, A., et al.: Optical and electromagnetic tracking systems for biomedical applications: a critical review on potentialities and limitations. IEEE Rev. Biomed. Eng. **13**, 212–232 (2020)
13. Das, S., Mullick, S., Suganthan, P.-N.: Recent advances in differential evolution - an updated survey. Swarm Evol. Comput. **27**, 1–30 (2016)

Malocclusion Treatment Planning via PointNet Based Spatial Transformation Network

Xiaoshuang Li[1], Lei Bi[2], Jinman Kim[2], Tingyao Li[1], Peng Li[3], Ye Tian[3], Bin Sheng[1](✉), and Dagan Feng[2]

[1] Shanghai Jiao Tong University, Dongchuan Road 800, Shanghai 200240, China
shengbin@sjtu.edu.cn
[2] The University of Sydney, Sydney, NSW 2006, Australia
[3] AceDental Software Technology (Shanghai) Co., LTD., Xiupu Road 3188, Shanghai 201315, China

Abstract. Orthodontic malocclusion treatment is a procedure to correct dental and facial morphology by moving teeth or adjusting underlying bones. It concentrates on two key aspects: the treatment planning for dentition alignment; and the plan implementation with the aid of external forces. Existing treatment planning requires significant time and effort for orthodontists and technicians. At present, no work successfully automates the process of tooth movement in orthodontics. In this study, we leverage state-of-the-art deep learning methods and propose an automated treatment planning process to take advantage of the spatial interrelationship between different teeth. Our method enables to exploit a 3-dimensional spatial transformation architecture for malocclusion treatment planning with 4 steps: (1) sub-sampling the dentition point cloud to get a critical point set; (2) extracting local features for each tooth and global features for the whole dentition; (3) obtaining transformation parameters conditioned on the features refined from the combination of both the local and global features and, (4) transforming initial dentition point cloud to the parameter-defined final state. Our approach achieves 84.5% cosine similarity accuracy (CSA) for the transformation matrix in the non-augmented dataset, and 95.3% maximum CSA for the augmented dataset. Our approach's outcome is proven to be effective in quantitative analysis and semantically reasonable in qualitative analysis.

Keywords: Deep learning · 3D point cloud · Orthodontic treatment · Spatial transformer network

1 Introduction

Malocclusion, the misalignment condition of teeth when the jaws close, is a primary oral disease; the global prevalence of malocclusion among children is

A. L. Martel et al. (Eds.): MICCAI 2020, LNCS 12263, pp. 105–114, 2020.
https://doi.org/10.1007/978-3-030-59716-0_11

currently >60% and growing each year [4,5,14,17]. Treatment based on conventional fixed appliances requires complex manual processes to straighten crooked teeth, adjust bad bites, and align the jaws correctly [9,13]. Besides, the performance of these solutions is heavily dependent on orthodontists' experience and skill. In contrast, the removable appliance requires less visit time and offers more comfort and freedom to patients [8,10]. Computer-aided modeling techniques with 3-dimensional (3D) printing has made a profound impact in the treatment planning of malocclusion. They offer solutions for teeth modeling and planning with 3D oral images [6]. Determination of a desired dental occlusion is a critical step in orthodontic treatment planning. When the patient is treated with fixed appliances, the orthodontist adjusts the appliances during every visit to approach the desired occlusion state. For the removable appliance, what orthodontist does is obtain oral scan images or computed tomography (CT) images to get 3-dimension reconstruction of the oral scan, determine the desired occlusion morphology and hand this strategy to technicians, who assist in establishing the intermediate movement states. Lack of information such as the relationship between the root and jaw bone leads to inaccuracy in the technicians' preliminary arrangement. To deal with this, orthodontists will give out several necessary revises, including step number, tooth number, and which finetune measure to take. This communication period is convoluted and time-consuming, leading to a long preparation time (up to several months).

Unfortunately, there is a limited number of works that attempt to address these challenges [2,3,12,18–21]. Murata *et al.* [12] proposed to use a classification network to make automatic morphology assessment based on 2D facial images of malocclusion patients. Unfortunately, this classification network only provides different types of facial distortion, which thereby cannot be adapted for treatment planning. Deng *et al.* [3] proposed an automatic approach for post-operational maxilla and mandible alignment by minimizing the distance between key points and landmarks. However, this alignment was applied at a global level, which thereby does not apply to the alignment of individual teeth. Wu *et al.* [19] used physically force feedback simulation to detect collisions, which is also an indispensable procedure in dentition alignment as a constrain of displacement. However, the loss function of geometry collision is discrete and non-derivable. It prevents the collision detection module from being integrated into the learnable network. Cui *et al.* [2] and Xia *et al.* [20] presented individual teeth segmentation methods based on 2-dimension CT images. Unfortunately, the performance of these methods is heavily dependent on detailed information of teeth boundaries and interrelations. To overcome these limitations, we propose a PointNet based Spatial Transformer Network (PSNT) network for malocclusion treatment planning. PointNet *et al.* [15,16] has achieved the state-of-the-art performance in the point set feature learning, where it enables to extract hierarchical features for individual point set. We introduce the following contributions to the field:

1. To the best of our knowledge, we introduce the first deep learning based method for orthodontic treatment planning, which provides simplification to the process.

2. We combine the explicit local features of each tooth and the implicit local features refined from hierarchical PointNet, to perceive the boundary and adjacency of teeth for transformation correction.
3. We propose an individual displacement augmentation approach for underfitting rectification. The introduction of this approach enables us to handle samples with large displacement.

2 Method

2.1 PSTN: PointNet Spatial Transformer Network

Spatial Transformer Network (STN) [7] was initially designed to strengthen the convolutional neural networks (CNNs) in its ability to be spatial invariant for the input images. The first part of STN, f_{loc}, called *localization network* is trained to derive a set of transformation parameters θ from the input features U, in which $\theta = f_{loc}(U)$. Then a sampling grid is created according to these parameters by a module named *grid generator*. During this period a transformation T_θ is determined by θ. After that, a *sampler* generates the sampled output features $V = T_\theta U$ to achieve the purpose of transforming the input features into an intuitively meaningful posture. The majority of the applications take STN as a differentiable plug-in for tasks such as in image classification or segmentation. To solve the clinical problems in orthodontic strategy making, we propose ***PointNet based Spatial Transformer Network*** (PSTN), a deduction of the STN concept in high-dimension data space. Figure 1 shows the framework of the proposed PSTN.

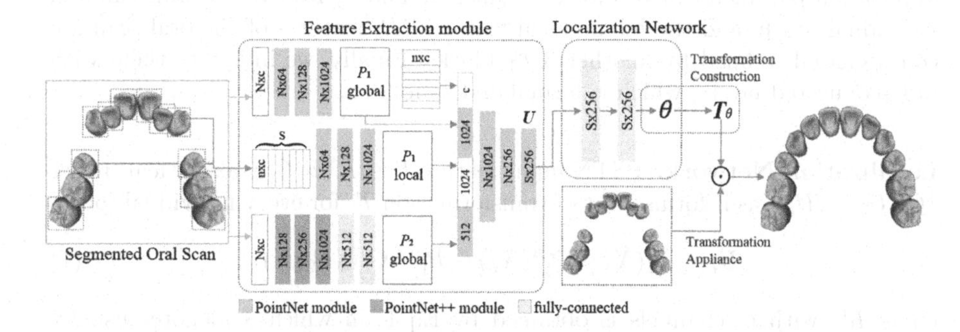

Fig. 1. An overview of the proposed PointNet based Spatial Transformer Network. The feature extraction module passes the compound feature U to the localization module, which generates parameter θ for spatial transformation construction. The mark $\odot$ refers to the transformation appliance procedure. The shape of the input point cloud is $N \times C$ with C channels, and S denotes the total number of teeth.

Instead of operating on the computational-expensive rasterized data, the feature U is extracted from the raw point cloud. Besides, the point cloud contains more details, e.g., normals and curvatures of curve patches, than the volumetric representation. This section will describe the feature extraction module as a variant of PointNet [15,16], the localization network, and construction of parameterized transformation.

Feature Extraction Modules. To obtain the input features U of PSTN from an oral scan, we adopted network structure specialized for raw point cloud, the PointNet. The vanilla PointNet [15] and its upgraded version PointNet++ [16] extract global features and hierarchical local features of unordered point set respectively. As the contour information of individual tooth plays a crucial role, we assume that the oral scan's point set was segmented into individual teeth. Thus our raw point set is partially ordered, which can be described as $X = \{X_1, X_2, ..., X_S\}$ where $X_I = \{x_1, x_2, ..., x_n\}$ and $x_i \in R^d$. X_I indicates the sampled point set of the $\boldsymbol{I}$th tooth according to the **Universal Number System** of dental notion. By default, we set $S = 32$ as the total number of teeth, including opsigenes, and $d = 3$ when x_i is presented as 3D coordinates. Each X_I is sampled unorderly from an individual tooth point set, and the X is arranged into ordered X_Is. A hybrid structure of the vanilla and the hierarchical version of PointNet is built to preserve all prior knowledge, such as segmentation information. We denote the vanilla PointNet module as $P_1(x_1, x_2, ..., x_n) = \max_{1,...,n}\{h(x_i)\}$, and the hierarchical module which implicitly learns local features as $P_2(x_1, x_2, ..., x_n)$. We achieve the feature extraction mainly with permutations and combinations of these two modules. There should be m P_1 blocks for the P_1 feature extraction, m P_2 blocks and 2 fully connection layers for P_2 feature extraction. As is shown in Fig. 1, the module duplication parameter $m = 3$ for experiments. The P_1 features of oral scan and individual tooth, and the hierarchical features (P_2 feature) of the oral scan are concatenated, refined by another 3 P_1 blocks. Finally, we integrate them with fully-connected layers, which is formulated as follows.

Localization Network. PSTN input feature $U \in R^S \times C$ is also a feature set $\{U_1, U_2, ..., U_S\}$ as it forms a set of transformation T_I for every individual tooth.

$$U_I = P_1(X_I \oplus P_1^m(X_I) \oplus P_1^m(X) \oplus P_2(X)) \tag{1}$$

$U_I \in R^C$ with C channels is obtained by Eq. 1, in which $\oplus$ denotes usually vector concatenation and γ a multi-layer perceptron. This equation can be clearly divided into 2 main steps. First we get local feature $P_1^m(X_I)$ and $P_2^m(X)$ and global feature $P_1(X)$, then another layer of P_1 is impose on the concatenation of the original point cloud X_I and these 3 features. We adopt the formulation of θ in [7] as $\theta = f_{loc}(U)$. As the applicable transformation type for orthodontic purpose is 3D rigid affine transformation including translation and rotation, the size of output θ is $S \times 16$ or $S \times 4 \times 4$ in matrix manner. In experiment the f_{loc} is layers of fully-connected network.

Parameterized Transformation and Loss Function. The source point cloud $P_s \in R^{N \times C}$ has S point set in consistent with tooth number where $N = S \times n$, and C channels including coordination and, if possible, normal vector or texture information. The parameterized transformation matrix T_θ is defined as $\theta + I_4 + bias$. Then the transformed point cloud P' is build up from the source point cloud P_s following the principle of 3D spatial transformation $P' = T_\theta(P_s) = T_\theta \cdot P_s$. Here $\cdot$ refers to matrix multiplication. Equation 2 illustrates the calculation process of a single point.

$$\begin{pmatrix} x_i^t \\ y_i^t \\ z_i^t \\ 1 \end{pmatrix} = T_\theta(P_s) = (I_4 + \begin{bmatrix} \theta_{11} & \theta_{12} & \theta_{13} & \theta_{14} \\ \theta_{21} & \theta_{22} & \theta_{23} & \theta_{24} \\ \theta_{31} & \theta_{32} & \theta_{33} & \theta_{34} \\ \theta_{41} & \theta_{42} & \theta_{43} & \theta_{44} \end{bmatrix} + bias) \begin{pmatrix} x_i^s \\ y_i^s \\ z_i^s \\ 1 \end{pmatrix} \tag{2}$$

The residual error is used as loss function to regulate the learning process. Besides the spatial residual error between P' and target P_t, the transformation matrix T_θ is also under control to be as close as possible to an orthogonal matrix. Based on the property of orthogonal matrix as $Q^T = Q^{-1}$ where Q is a general orthogonal matrix, we can calculate the residual error between $T_\theta \cdot T_\theta^T$ and identity matrix I_4. The final loss function is defined as:

$$L = \frac{1}{N} \sum |P' - P_t| + \frac{1}{16} \sum \|T_\theta \cdot T_\theta^T - I_4\|^2 \tag{3}$$

2.2 Data Augmentation

In the experiments, there were two main requirements for data augmentation. Firstly, obtaining an oral scan with strictly consistent orientation in a clinical setting is non-trivial. Factors such as different operators or scanner types can result in data acquired with minor orientation differences. Secondly, clinically collected data are cases mainly with mild crowding, and the average displacement of teeth centroid is less than 1 mm. This problem has a high free degree, and the deficit in data representativeness results in underfitting during the experiments. Two types of data augmentation were carried out to tackle these requirements: integral dentition rotation and individual tooth displacement. Here we want to emphasize the data processing for transformation matrices during data augmentation.

In this section, we use a different symbol system from the other parts of this manuscript to avoid confusion. We denote the source point cloud as X_0, target point cloud as X, and the original transformation matrix as A, so we have $X = A \cdot X_0$. Similarly, we denote the augmented source point cloud as $X_0{}'$, target point cloud as X'. The target transformation matrix A' is updated regarding the data augment.

Integral Dentition Rotation (IDR). Y-axis centered rotation R is imposed on the source and target point cloud to reinforce the rotation invariance of PSTN.

Thus we have $X_0' = RX_0$, $X' = RX$ and $A'X_0' = X'$. Given that $AX_0 = X$, we get $(AR^{-1}RX_0) = X$ and then $(RAR^{-1})RX_0 = RX$, from which we get $A' = RAR^{-1}$.

Individual Tooth Displacement (ITD). Given the constraint in orthodontic treatment that the absolute displacement of a tooth should not be larger than 12 mm, the randomly generated Displacement signal vector d in coordinate space is subjected to a normal distribution $\sqrt[3]{12}N(\mu, \sigma^2)$ with the expectation as 0 and variance as 0.1 to reduce the probability of extreme situations. d with length 3 takes the displacement position of an identity matrix, the first 3 rows of the 4th column to form the displacement matrix D.

When we augment data with individual tooth displacement, only source point cloud is impacted by D, so $X_0' = DX_0$. The target point cloud remains unchanged as $X' = X$. Given that $AX_0 = X$, we got $(AD^{-1})DX_0 = X$, from which we get $A' = AD^{-1}$.

3 Experiments and Results

3.1 Data Pre-processing and Experiment Setup

The experiments were conducted on a clinical malocclusion data set comprising 178 sample studies derived from 206 patients. Only cases without teeth extraction are considered, and each sample has pre- and post-orthodontic oral scans. For key point set selection, a commonly used *Farthest Point Sampling* (FPS) [11] method was used. The maximum sampling number for an individual tooth was set to $n = 128$, and for whole dentition, it was set to $N = 4096$. Zero-padding was used to deal with the missing teeth cases resulting from the extraction or the congenital absence of wisdom teeth. The transformation matrix from source to target point cloud was calculated with *Iterative Closest Point* (ICP) algorithm [1]. The FPS-sampled source point cloud was viewed as the input, and the label was regenerated from the input with the transformation matrix to circumvent the sampling randomness.

We randomly split the data into 150 for training and 28 samples for testing. Our method took the reorganized $4096 \times d$ input point cloud where we set d = 6 when including normals. The experiments were divided into two groups according to whether the input was augmented or not, as shown in Table 1. The vanilla PSTN (vPSTN) is a version without hierarchical modules. Two augmented experiments were carried out on PSTN structure: integral dentition rotation (IDR) and individual tooth displacement (ITD). All experiments were conducted on input with normal vectors, except for vPSTN.

3.2 Experimental Evaluation

Quantitative Evaluation. The 3D transformation of teeth can be described by the spatial transformation matrix derived by our models. Thus, the measurement of

Table 1. Cosine similarity accuracy (CSA) of generated transformation compared to the ground truth, as well as target registration error (TRE) between transformed and target point cloud.

Methods		CSA%				Mean TRE (mm^2)
		Training	Mean	Minimum	Maximum	
Non-augmented	vanilla PSTN	85.36	84.485	75.685	95.09	0.3686
	PSTN	**85.93**	**84.523**	75.665	95.17	**0.3452**
Augmented	IDR	84405	84.51	75.23	**95.31**	**0.3108**
	ITD	76.605	84.46	**75.71**	95.13	0.3894

matrix similarity can act as an accuracy index. We used two metrics to evaluate the performance of the proposed method: cosine similarity accuracy (CSA) and target registration error (TRE), as shown in Table 1. The commonly used CSA was used to measure the difference between the generated transformation matrix and its ground truth. The matrix was flattened to a vector that can be applied to cos similarity. Then the value range was mapped from [−1, 1] to [0, 1]. TRE measures the absolute distance between the transformed point cloud and its reference target, which provides an intuitive impression of the planning error.

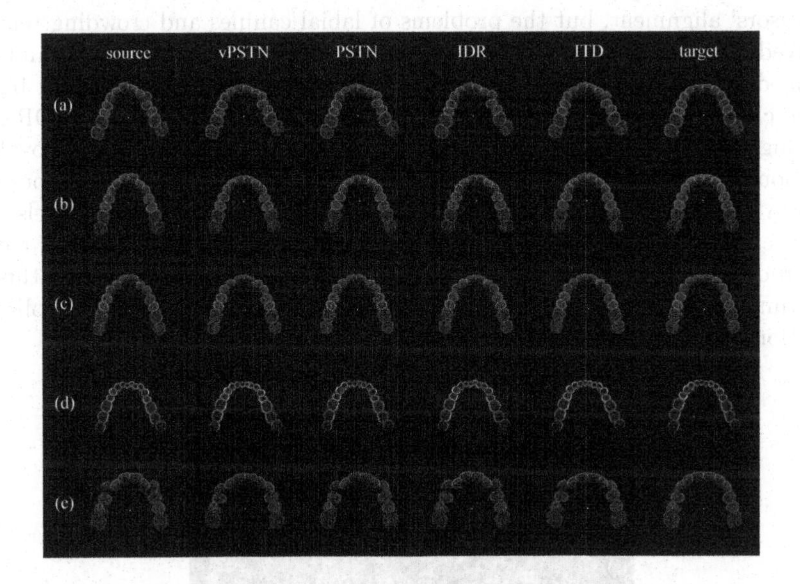

Fig. 2. Result visualization for orthodontic strategy making models of 5 cases. Among them, $(a), (b), (c)$ and (e) is maxilla while (d) is mandible.

The comparison results are presented in Table 1. It shows that the proposed PSTN achieves the best average performance among all the methods.

As expected, methods based on data augmentation had larger CSA maximum. This can be explained by the inherent data distribution. The PSTN was trained to avoid massive displacement by the orthogonal matrix constraint. Therefore, during testing, samples with milder dentition deformity had maximum CSA score and vice-versa. Integral dentition rotation augmentation strengthened the model with orientation invariance ability at the cost of inducing uncertainties for complicated cases. Individual tooth displacement explicitly increased the deformity level by disturbing every tooth's position, which enabled the model to work better with some complicated cases. However, this induces more uncertainty or even erroneous data, resulting in lower accuracy during training and more error during testing.

Qualitative Evaluation. Figure 2 shows testing results of different methods on non-augmented inputs. The target state of case (d) was not in a final state of malocclusion treatment, but was included in the data set and provided an example where our model obtained satisfactory results. The mandibular incisor in the source state of the (e) row shows an uneven arch line. In the target state, the problem remains as the left mandibular central incisor shows a lingual deviation from the arch line. Among our architecture, the ITD-augmented structure achieved the best result and is competitive to the given reference. The PSTN and its vanilla version also generated good results, but its effectiveness is less. Row (e) is another sample where ITD-augmentation outperforms other structures on incisors' alignment, but the problems of labial canines and crowding remain unsolved. Figure 3 illustrates 2 examples of the ITD data augmentation and the intermediary rearrangement during training. It produces erroneous input based on the clinical standard. In contrast, PSTN shows better results than IDR and ITD-augmentation in most cases, as shown in row (c), Fig. 2, eases the crowding situation by distal movement of maxillary lateral incisors. Our method does not perform well in severe cases with massive displacement or crowding levels and may occasionally generate overlapped dentition. In row (a) and (b), our methods produce planning that cannot reach the target point. We attribute this to the primal data set distribution. We suggest that this can be solved by collision detection based post-processing methods such as in [19].

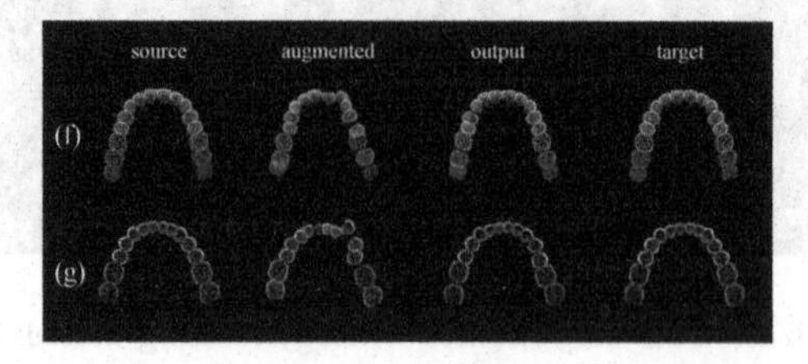

Fig. 3. Visualization of the ITD based training results. Row (f) is maxilla and row (g) is mandible.

4 Conclusions and Future Work

We proposed a PointNet based spatial transformation strategy. We demonstrate its effectiveness in malocclusion treatment planning. Our experiments show that our PointNet based feature extraction can better discriminate essential features for the 3D spatial transformer network. For future work, we will further improve the performance of the proposed method on its capability to deal with translation and rotation and seek potential clinical applications. Also, We noticed that the differences in the ability to deal with the rotation and displacement of individual teeth. The accuracy of the translation and rotation in the transformation matrix should be assessed separately, which will be handled in our next work.

Acknowledgement. Thanks are due to AceDental Software Technology for providing the clinical data and the valuable discussion. This work was supported in part by the National Natural Science Foundation of China under Grant 61872241 and Grant 61572316, in part by the Science and Technology Commission of Shanghai Municipality under Grant 18410750700, Grant 17411952600, and Grant 16DZ0501100.

References

1. Chetverikov, D., Svirko, D., Stepanov, D., Krsek, P.: The trimmed iterative closest point algorithm. In: Object Recognition Supported by User Interaction for Service Robots, vol. 3, pp. 545–548. IEEE (2002)
2. Cui, Z., Li, C., Wang, W.: Toothnet: automatic tooth instance segmentation and identification from cone beam CT images. In: Proceedings of the IEEE Conference on Computer Vision and Pattern Recognition, pp. 6368–6377 (2019)
3. Deng, H., et al.: An automatic approach to reestablish final dental occlusion for 1-piece maxillary orthognathic surgery. In: Shen, D., et al. (eds.) MICCAI 2019. LNCS, vol. 11768, pp. 345–353. Springer, Cham (2019). https://doi.org/10.1007/978-3-030-32254-0_39
4. Dimberg, L., Lennartsson, B., Arnrup, K., Bondemark, L.: Prevalence and change of malocclusions from primary to early permanent dentition: a longitudinal study. Angle Orthod. **85**(5), 728–734 (2015)
5. Dimberg, L., Lennartsson, B., Söderfeldt, B., Bondemark, L.: Malocclusions in children at 3 and 7 years of age: a longitudinal study. Eur. J. Orthod. **35**(1), 131–137 (2013)
6. Djeu, G., Shelton, C., Maganzini, A.: Outcome assessment of invisalign and traditional orthodontic treatment compared with the American board of orthodontics objective grading system. Am. J. Orthod. Dentofac. Orthop. **128**(3), 292–298 (2005)
7. Jaderberg, M., Simonyan, K., Zisserman, A., et al.: Spatial transformer networks. In: Advances in Neural Information Processing Systems, pp. 2017–2025 (2015)
8. Kravitz, N.D., Kusnoto, B., BeGole, E., Obrez, A., Agran, B.: How well does invisalign work? A prospective clinical study evaluating the efficacy of tooth movement with invisalign. Am. J. Orthod. Dentofac. Orthop. **135**(1), 27–35 (2009)
9. Lima, R.M.A., Lima, A.L.: Case report: long-term outcome of class II division 1 malocclusion treated with rapid palatal expansion and cervical traction. Angle Orthod. **70**(1), 89–94 (2000). https://doi.org/10.1043/0003-3219(2000)070⟨0089:CRLTOO⟩2.0.CO;2

10. Mavreas, D., Athanasiou, A.E.: Factors affecting the duration of orthodontic treatment: a systematic review. Eur. J. Orthod. **30**(4), 386–395 (2008). https://doi.org/10.1093/ejo/cjn018
11. Moenning, C., Dodgson, N.A.: Fast marching farthest point sampling. Technical report University of Cambridge, Computer Laboratory (2003)
12. Murata, S., Lee, C., Tanikawa, C., Date, S.: Towards a fully automated diagnostic system for orthodontic treatment in dentistry. In: 2017 IEEE 13th International Conference on e-Science (e-Science), pp. 1–8. IEEE (2017)
13. Nielsen, I.L.: Vertical malocclusions: etiology, development, diagnosis and some aspects of treatment. Angle Orthod. **61**(4), 247–260 (1991)
14. Proffit, W.R., Fields Jr., H.W., Sarver, D.M.: Contemporary Orthodontics. Elsevier Health Sciences. Elsevier, Amsterdam (2006)
15. Qi, C.R., Su, H., Mo, K., Guibas, L.J.: PointNet: deep learning on point sets for 3D classification and segmentation. In: Proceedings of the IEEE Conference on Computer Vision and Pattern Recognition, pp. 652–660 (2017)
16. Qi, C.R., Yi, L., Su, H., Guibas, L.J.: PointNet++: deep hierarchical feature learning on point sets in a metric space. In: Advances in Neural Information Processing Systems, pp. 5099–5108 (2017)
17. Thilander, B., Pena, L., Infante, C., Parada, S.S., de Mayorga, C.: Prevalence of malocclusion and orthodontic treatment need in children and adolescents in Bogota, Colombia. An epidemiological study related to different stages of dental development. Eur. J. Orthod. **23**(2), 153–168 (2001)
18. Tian, S., Dai, N., Zhang, B., Yuan, F., Yu, Q., Cheng, X.: Automatic classification and segmentation of teeth on 3D dental model using hierarchical deep learning networks. IEEE Access **7**, 84817–84828 (2019)
19. Wu, W., Chen, H., Cen, Y., Hong, Y., Khambay, B., Heng, P.A.: Haptic simulation framework for determining virtual dental occlusion. Int. J. Comput. Assist. Radiol. Surg. **12**(4), 595–606 (2016). https://doi.org/10.1007/s11548-016-1475-3
20. Xia, Z., Gan, Y., Chang, L., Xiong, J., Zhao, Q.: Individual tooth segmentation from CT images scanned with contacts of maxillary and mandible teeth. Comput. Methods Programs Biomed. **138**, 1–12 (2017)
21. Xu, X., Liu, C., Zheng, Y.: 3D tooth segmentation and labeling using deep convolutional neural networks. IEEE Trans. Visual Comput. Graph. **25**(7), 2336–2348 (2018)

Simulation of Brain Resection for Cavity Segmentation Using Self-supervised and Semi-supervised Learning

Fernando Pérez-García[1,2(✉)], Roman Rodionov[3,4], Ali Alim-Marvasti[1,3,4], Rachel Sparks[2], John S. Duncan[3,4], and Sébastien Ourselin[2]

[1] Wellcome EPSRC Centre for Interventional and Surgical Sciences (WEISS), University College London, London, UK
fernando.perezgarcia.17@ucl.ac.uk

[2] School of Biomedical Engineering and Imaging Sciences (BMEIS), King's College London, London, UK

[3] Department of Clinical and Experimental Epilepsy, UCL Queen Square Institute of Neurology, London, UK

[4] National Hospital for Neurology and Neurosurgery, Queen Square, London, UK

Abstract. Resective surgery may be curative for drug-resistant focal epilepsy, but only 40% to 70% of patients achieve seizure freedom after surgery. Retrospective quantitative analysis could elucidate patterns in resected structures and patient outcomes to improve resective surgery. However, the resection cavity must first be segmented on the postoperative MR image. Convolutional neural networks (CNNs) are the state-of-the-art image segmentation technique, but require large amounts of annotated data for training. Annotation of medical images is a time-consuming process requiring highly-trained raters, and often suffering from high inter-rater variability. Self-supervised learning can be used to generate training instances from unlabeled data. We developed an algorithm to simulate resections on preoperative MR images. We curated a new dataset, EPISURG, comprising 431 postoperative and 269 preoperative MR images from 431 patients who underwent resective surgery. In addition to EPISURG, we used three public datasets comprising 1813 preoperative MR images for training. We trained a 3D CNN on artificially resected images created on the fly during training, using images from 1) EPISURG, 2) public datasets and 3) both. To evaluate trained models, we calculate Dice score (DSC) between model segmentations and 200 manual annotations performed by three human raters. The model trained on data with manual annotations obtained a median (interquartile range) DSC of 65.3 (30.6). The DSC of our best-performing model, trained with no manual annotations, is 81.7 (14.2). For comparison, inter-rater agreement between human annotators was 84.0 (9.9). We demonstrate a training method for CNNs using simulated resection cavities that can accurately segment real resection cavities, without manual annotations.

Electronic supplementary material The online version of this chapter (https://doi.org/10.1007/978-3-030-59716-0_12) contains supplementary material, which is available to authorized users.

A. L. Martel et al. (Eds.): MICCAI 2020, LNCS 12263, pp. 115–125, 2020.
https://doi.org/10.1007/978-3-030-59716-0_12

Keywords: Neurosurgery · Segmentation · Self-supervised learning

1 Introduction

Only 40% to 70% of patients with refractory focal epilepsy are seizure-free after resective surgery [12]. Retrospective studies relating clinical features and resected brain structures (such as amygdala or hippocampus) to surgical outcome may provide useful insight to identify and guide resection of the epileptogenic zone. To identify resected structures, first, the resection cavity must be segmented on the postoperative MR image. Then, a preoperative image with a corresponding brain parcellation can be registered to the postoperative MR image to identify resected structures.

In the context of brain resection, the cavity fills with cerebrospinal fluid (CSF) after surgery [26]. This causes an inherent uncertainty in resection cavity delineation when adjacent to sulci, ventricles, arachnoid cysts or oedemas, as there is no intensity gradient separating the structures. Moreover, brain shift can occur during surgery, causing regions outside the cavity to fill with CSF (see examples in supplementary document).

Decision trees have been used for brain cavity segmentation from T_2-weighted, FLAIR, and pre- and post-contrast T_1-weighted MRI in the context of glioblastoma surgery [10,16]. Relatedly, some methods have simulated or segmented brain lesions to improve non-linear registration with missing correspondences. Brett et al. [1] propagated lesions manually segmented from pathological brain images to structurally normal brain images by registering images to a common template space. Removing the lesion from consideration when computing the similarity metric improved non-linear registration. Methods to directly compute missing correspondences during registration, which can give an estimate of the resection cavity, have been proposed [3,5,7]. Pezeshk et al. [21] trained a series of machine learning classifiers to detect lesions in chest CT scans. The dataset was augmented by propagating lesions from pathological lungs to healthy lung tissue, using Poisson blending. This data augmentation technique improved classification results for all machine learning techniques considered.

In traditional machine learning, data is represented by hand-crafted features which may not be optimal. In contrast, deep learning, which has been successfully applied to brain image segmentation [13,15], implicitly computes a problem-specific feature representation. However, deep learning techniques rely on large annotated datasets for training. Annotated medical imaging datasets are often small due to the financial and time burden annotating the data, and the need for highly-trained raters. Self-supervised learning generates training instances using unlabeled data from a source domain to learn features that can be transferred to a target domain [11]. Semi-supervised learning uses labeled as well as unlabeled data to train models [8]. These techniques can be used to leverage unlabeled medical imaging data to improve training in instances where acquiring annotations is time-consuming or costly.

We present a fully-automatic algorithm to simulate resection cavities from preoperative T_1-weighted MR images, applied to self-supervised learning for brain resection cavity segmentation. We validate this approach by comparing models trained with and without manual annotations, using 200 annotations from three human raters on 133 postoperative MR images with lobectomy or lesionectomy (133 annotations to test models performance and 67 annotations to assess inter-rater variability).

2 Methods

2.1 Resection Simulation

We generate automatically a training instance $(\boldsymbol{X}_R, \boldsymbol{Y}_R)$ representing a resected brain $\boldsymbol{X}_R$ and its corresponding cavity segmentation $\boldsymbol{Y}_R$ from a preoperative image $\boldsymbol{X}$ using the following approach.

Resection Label. A geodesic polyhedron with frequency f is generated by subdividing the edges of an icosahedron f times and projecting each vertex onto a parametric sphere with unit radius. This polyhedron models a spherical surface $S = \{V, F\}$ with vertices $V = \left\{\boldsymbol{v}_i \in \mathbb{R}^3\right\}_{i=1}^{n_V}$ and faces $F = \{\boldsymbol{f}_k\}_{k=1}^{n_F}$. Each face $\boldsymbol{f}_k = \{i_1^k, i_2^k, i_3^k\}$ is defined as a sequence of three non-repeated vertex indices.

S is perturbed with simplex noise [20], a smooth noise generated by interpolating pseudorandom gradients defined on a multidimensional simplicial grid. Simplex noise was selected as it is often used to simulate natural-looking textures or terrains. The noise at point $\boldsymbol{p} \in \mathbb{R}^3$ is computed as a weighted sum of the noise contribution of ω different octaves, with weights $\gamma^{n-1} : n \in \{1, 2, \dots, \omega\}$ controlled by the persistence parameter γ. The displacement $\delta : \mathbb{R}^3 \to [-1, 1]$ is proportional to the noise function $\phi : \mathbb{R}^3 \to [0, 1]$:

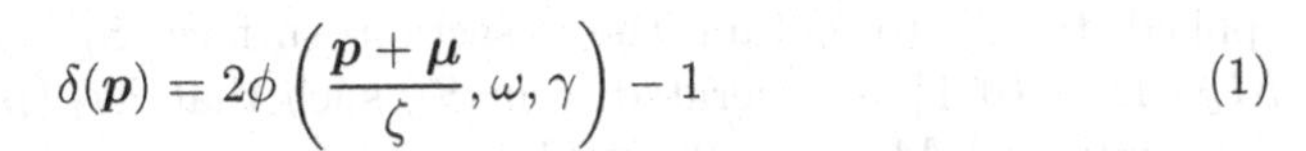

$$\delta(\boldsymbol{p}) = 2\phi\left(\frac{\boldsymbol{p} + \boldsymbol{\mu}}{\zeta}, \omega, \gamma\right) - 1 \quad (1)$$

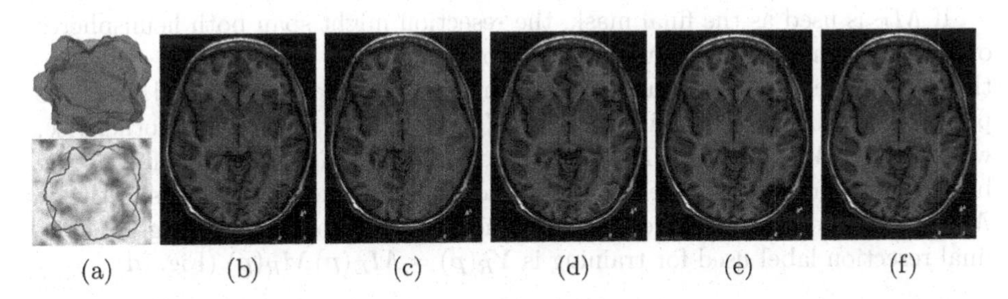

Fig. 1. Resection simulation. (a) Sphere surface mesh before (S, green) and after (S_δ, magenta) perturbation. S and S_δ (top); intersection of S and S_δ with a plane of the simplex noise volume, generated only for visualization purposes, with values between -1 (blue) and 1 (red). Radial displacement is proportional to the noise at each vertex $\boldsymbol{v}_i \in V$ (b) Transformed mesh S_E (c) Resectable hemisphere mask $\boldsymbol{M}_R$ (d) Simulated resection label $\boldsymbol{Y}_R$ (e) Simulated resected image $\boldsymbol{X}_R$ (f) Original image $\boldsymbol{X}$. (Color figure online)

where ζ is a scaling parameter to control smoothness and $\boldsymbol{\mu}$ is a shifting parameter that adds stochasticity (equivalent to a random number generator seed). Each vertex $\boldsymbol{v}_i \in V$ is displaced radially by:

$$\boldsymbol{v}_{\delta i} = \boldsymbol{v}_i + \delta(\boldsymbol{v}_i)\frac{\overrightarrow{\boldsymbol{v}_i}}{\|\overrightarrow{\boldsymbol{v}_i}\|}, \qquad \forall i \in \{1, 2, \ldots, n_V\} \tag{2}$$

to create a perturbed sphere $S_\delta = \{V_\delta, F\}$ with vertices $V_\delta = \{\boldsymbol{v}_{\delta i}\}_{i=1}^{n_V}$ (Fig. 1a).

A series of transforms is applied to S_δ to modify its volume, shape and position. Let $T_T(\boldsymbol{p})$, $T_S(\boldsymbol{s})$ and $T_R(\boldsymbol{\theta})$ be translation, scaling and rotation transforms.

Perturbing $\boldsymbol{v}_i \in V$ shifts the centroid of S_δ off the origin. S_δ is recentered at the origin by applying the translation $T_T(-\boldsymbol{c})$ to each vertex, where $\boldsymbol{c} = \frac{1}{n_V}\sum_{i=1}^{n_V} \boldsymbol{v}_{\delta i}$ is the centroid of S_δ.

Random rotations around each axis are applied to S_δ with the rotation matrix $T_R(\boldsymbol{\theta}_r) = R_x(\theta_x) \circ R_y(\theta_y) \circ R_z(\theta_z)$, where $\circ$ indicates a transform composition, $R_i(\theta_i)$ is a rotation of θ_i radians around axis i, and $\theta_i \sim \mathcal{U}(0, 2\pi)$.

A scaling transform $T_S(\boldsymbol{r})$ is applied to S_δ, where $(r_1, r_2, r_3) = \boldsymbol{r}$ are the semi-axes of an ellipsoid with volume v modeling the cavity shape. The semi-axes are computed as $r_1 = r$, $r_2 = \lambda r$ and $r_3 = r/\lambda$, where $r = (3v/4)^{1/3}$ and λ controls the semi-axes length ratios.

S_δ is translated such that it is centered at a voxel in the cortical gray matter as follows. A T_1-weighted MR image is defined as $\boldsymbol{I}_{MRI} : \Omega \to \mathbb{R}$, where $\Omega \in \mathbb{R}^3$. A full brain parcellation $\boldsymbol{G} : \Omega \to Z$ is generated for $\boldsymbol{I}_{MRI}$ using geodesical information flows [2], where Z is the set of segmented brain structures. A cortical gray matter mask $\boldsymbol{M}_{GM}^h : \Omega \to \{0, 1\}$ of hemisphere h is extracted from $\boldsymbol{G}$, where h is randomly chosen from $H = \{\text{left}, \text{right}\}$ with equal probability. A random gray matter voxel $\boldsymbol{g} \in \Omega$ is selected such that $\boldsymbol{M}_{GM_h}(\boldsymbol{g}) = 1$.

The transforms are composed as $T_E = T_T(\boldsymbol{g}) \circ T_S(\boldsymbol{r}) \circ T_R(\boldsymbol{\theta}_r) \circ T_T(-\boldsymbol{c})$ and applied to S_δ to obtain the resection surface $S_E = T_E \circ S_\delta$. A mask $\boldsymbol{M}_E : \Omega \to \{0, 1\}$ is generated from S_E such that $\boldsymbol{M}_E(\boldsymbol{p}) = 1$ for all $\boldsymbol{p}$ within the cavity and $\boldsymbol{M}_E(\boldsymbol{p}) = 0$ outside.

If $\boldsymbol{M}_E$ is used as the final mask, the resection might span both hemispheres or include non-realistic tissues such as bone or scalp (Fig. 1b). To eliminate this unrealistic scenario, a 'resectable hemisphere mask' is generated from the parcellation as $\boldsymbol{M}_R(\boldsymbol{p}) = 1$ if $\boldsymbol{G}(\boldsymbol{p}) \neq \{\boldsymbol{M}_{BG}, \boldsymbol{M}_B, \boldsymbol{M}_C, \boldsymbol{M}_{\bar{H}}\}$ and 0 otherwise, where $\boldsymbol{M}_{BG}$, $\boldsymbol{M}_B$, $\boldsymbol{M}_C$ and $\boldsymbol{M}_{\bar{H}}$ are the sets of labels in Z corresponding to the background, brainstem, cerebellum and contralateral hemisphere, respectively. $\boldsymbol{M}_R$ is smoothed using a series of binary morphological operations (Fig. 1c). The final resection label used for training is $\boldsymbol{Y}_R(\boldsymbol{p}) = \boldsymbol{M}_E(\boldsymbol{p})\boldsymbol{M}_R(\boldsymbol{p})$ (Fig. 1d).

Resected Image. To mimic partial volume effects near cavity boundaries, a Gaussian filter is applied to $\boldsymbol{M}_R(\boldsymbol{p})$ to smooth the alpha channel $\boldsymbol{A} : \Omega \to [0, 1]$, defined as $\boldsymbol{A}(\boldsymbol{p}) = \boldsymbol{M}_R(\boldsymbol{p}) * \boldsymbol{G}_{\mathcal{N}}(\boldsymbol{\sigma}), \forall \boldsymbol{p} \in \Omega$, where $*$ is the convolution operator and $\boldsymbol{G}_{\mathcal{N}}(\boldsymbol{\sigma_A})$ is a Gaussian kernel with standard deviations $\boldsymbol{\sigma_A} = (\sigma_x, \sigma_y, \sigma_z)$.

To generate a realistic CSF texture, we create a ventricle mask $\boldsymbol{M}_V : \Omega \to \{0, 1\}$ from $\boldsymbol{G}$, such that $\boldsymbol{M}_V(\boldsymbol{p}) = 1$ for all $\boldsymbol{p}$ within the ventricles and $\boldsymbol{M}_V(\boldsymbol{p}) = 0$ outside. Intensity values within ventricles are assumed to have a normal distribution [9] with a mean μ_{CSF} and standard deviation σ_{CSF} calculated from voxel intensity values in $\boldsymbol{I}_{MRI}(\boldsymbol{p}) : \forall \boldsymbol{p} \in \Omega, \boldsymbol{M}_V(\boldsymbol{p}) = 1$. A CSF-like image $\boldsymbol{I}_{CSF} : \Omega \to \mathbb{R}$ is then generated as $\boldsymbol{I}_{CSF}(\boldsymbol{p}) \sim \mathcal{N}(\mu_{CSF}, \sigma_{CSF}), \forall \boldsymbol{p} \in \Omega$, and the resected image (Fig. 1e) is the convex combination:

$$\boldsymbol{X}_R(\boldsymbol{p}) = \boldsymbol{A}(\boldsymbol{p})\boldsymbol{I}_{CSF}(\boldsymbol{p}), + [1 - \boldsymbol{A}(\boldsymbol{p})]\, \boldsymbol{I}_{MRI}(\boldsymbol{p}) \qquad \forall \boldsymbol{p} \in \Omega \tag{3}$$

Examples of simulations are presented in the supplementary document.

2.2 Dataset Description

$T1$-weighted MR images were collected from publicly available datasets Information eXtraction from Images (IXI) (566), Alzheimer's Disease Neuroimaging Initiative (ADNI)(467), and Open Access Series of Imaging Studies (OASIS) (780), for a total of 1813 images. EPISURG was obtained from patients with refractory focal epilepsy who underwent resective surgery at the National Hospital for Neurology and Neurosurgery (NHNN), London, United Kingdom. This was an analysis of anonymized data that had been previously acquired as a part of clinical care, so individual patient consent was not required. In total there were 431 patients with postoperative T_1-weighted MR images, 269 of which had a corresponding preoperative MR image (see supplementary document). All images were registered to a common template space using NiftyReg [17].

Three human raters annotated a subset of the postoperative images in EPISURG. Rater A segmented the resection cavity in 133 images. These annotations were used to test the models. This set was randomly split into 10 subsets, where the distribution of resection types (e.g. temporal, frontal, etc.) in each subset is similar. To quantify inter-rater variability, Rater B annotated subsets 1 and 2 (34 images), and Rater C annotated subsets 1 and 3 (33 images).

2.3 Network Architecture and Implementation Details

We used the PyTorch deep learning framework, training with automatic mixed precision on two 32-GB TESLA V100 GPUs. We implemented a variant of 3D U-Net [6] using two downsampling and upsampling blocks, trilinear interpolation for the synthesis path, and 1/4 of the filters for each convolutional layer. This results in a model with 100 times fewer parameters than the original 3D U-Net, reducing overfitting and computational burden. We used dilated convolutions [4], starting with a dilation factor of one, then increased or decreased in steps of one after each downsampling or upsampling block, respectively. Batch normalization and PReLU activation functions followed each convolutional layer. Finally, a dropout layer with probability 0.5 was added before the output classifier. We used an Adam optimizer [14] with an initial learning rate of 10^{-3} and weight decay of 10^{-5}. Training occurred for 60 epochs, and the learning rate was divided by 10 every 20 epochs. A batch size of 8 (4 per GPU) was used for training. 90% of the images were used for training and 10% for validation.

We wrote and used TorchIO [19] to process volumes on the fly during training. The preprocessing and random augmentation transforms used were 1) simulated resection (see Sect. 2.1), 2) MRI k-space motion artifact [23], 3) histogram standardization [18], 4) MRI bias field artifact [25], 5) normalization to zero-mean and unit variance of the foreground voxels, computed using the intensity mean as a threshold [18], 6) Gaussian noise, 7) flipping in the left-right direction, 8) scaling and rotation, and 9) B-spline elastic deformation. The resection simulation was implemented as a TorchIO [19] transform and the code is available online[1].

The following parameters were used to generate simulated resections (see Sect. 2.1): $f = 16$, $\omega = 4$, $\gamma = 0.5$, $\zeta = 3$, $\boldsymbol{\mu} \sim \mathcal{U}(0, 1000)$, $\lambda \sim \mathcal{U}(1, 2)$, and $\boldsymbol{\sigma_A} \sim \mathcal{U}(0.5, 1)$. The ellipsoid volume v is sampled from volumes of manually segmented cavities from Rater A (see Sect. 2.2).

3 Experiments and Results

We trained models with seven different dataset configurations to assess how simulated resection cavities impact model accuracy a) using datasets of similar size and scanner, b) using datasets of similar size and different scanner, c) using much larger datasets (10× increase) and d) combined with semi-supervised learning. See the supplementary document for a table summarizing the experiments and quantitative results.

All overlap measurements are expressed as 'median (interquartile range)' Dice score (DSC) with respect to the 133 annotations obtained from Rater A. Quantitative results are shown in Fig. 2.

Differences in model performance were analyzed by a one-tailed Mann-Whitney U test with a significance threshold of $\alpha = 0.05$, with Bonferroni correction for the seven experiments evaluated $\left(\frac{\alpha}{7 \times (7-1)} \approx 0.002\right)$.

3.1 Small Datasets

We trained and tested on the 133 images annotated by Rater A, using 10-fold cross-validation, obtaining a DSC of 65.3 (30.6). We refer to this dataset as *EpiPost*. For all other models, we use data without manual annotations for training and *EpiPost* for testing.

EpiPre comprised 261 preoperative MR images from patients scanned at NHNN who underwent epilepsy surgery but are not in *EpiPost*. The model trained with *EpiPre* gave a DSC of 61.6 (36.6), which was not significantly different compared to training with *EpiPost* ($p = 0.216$).

We trained a model using *PubSmall*, i.e. 261 images randomly chosen from the publicly available datasets. This model had a DSC of 69.5 (27.0).

Although there was a moderate increase in DSC, training with either *EpiPre* or *PubSmall* was not significantly superior compared to *EpiPost* after Bonferroni correction ($p = 0.009$ and $p = 0.035$, respectively).

[1] https://github.com/fepegar/resector.

3.2 Large Datasets

We trained a model using the full public dataset (*PubFull*, 1813 images), obtaining a DSC of 79.6 (17.3), which was significantly superior to *PubSmall* ($p \approx 10^{-8}$) and *EpiPost* ($p \approx 10^{-13}$). Adding *EpiPre* to *PubFull* for training did not significantly increase performance ($p = 0.173$), with a DSC of 80.5 (16.1).

For an additional training dataset, we created the *PubMed* dataset by replacing 261 images in *PubFull* with *EpiPre*. Training with *PubMed* + *EpiPre* was not significantly different compared to training with *PubFull* ($p = 0.378$), with a DSC of 79.8 (17.1).

3.3 Semi-supervised Learning

We evaluated the ability of semi-supervised learning to improve model performance by generating pseudo-labels for all unlabeled postoperative images in EPISURG (297). Pseudo-labels were generated by inferring the resection cavity label using the model trained on *PubFull* and *EpiPre*. The pseudo-labels and corresponding postoperative images were combined to create the *Pseudo* dataset.

We trained a model using *PubFull*, *EpiPre* and *Pseudo* (2371 images), obtaining a DSC of 81.7 (14.2). Adding the pseudo-labels to *PubFull* and *EpiPre* did not significantly improve performance ($p = 0.176$), indicating our semi-supervised learning approach provided no advantage. Predictions from this model are shown in Fig. 3.

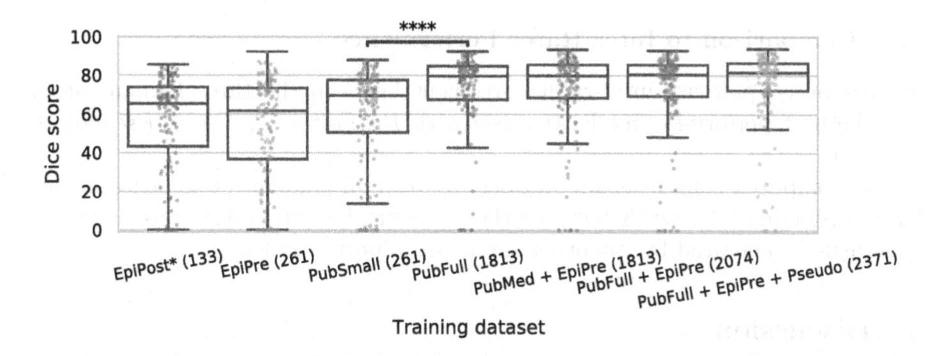

Fig. 2. DSC values between manual annotations from Rater A and segmentations for models. Values in brackets indicate number of training subjects. Note that only the first model was trained with manual annotations. *EpiPost*: postoperative images in EPISURG with manual annotations (the asterisk * indicates fully supervised training with 10-fold cross-validation); *EpiPre*: preoperative images from subjects not contained in *EpiPost*); *PubFull*: public datasets; *PubSmall*, *PubMed*: subsets of *PubFull*; *Pseudo*: pseudo-labeled postoperative images in EPISURG.

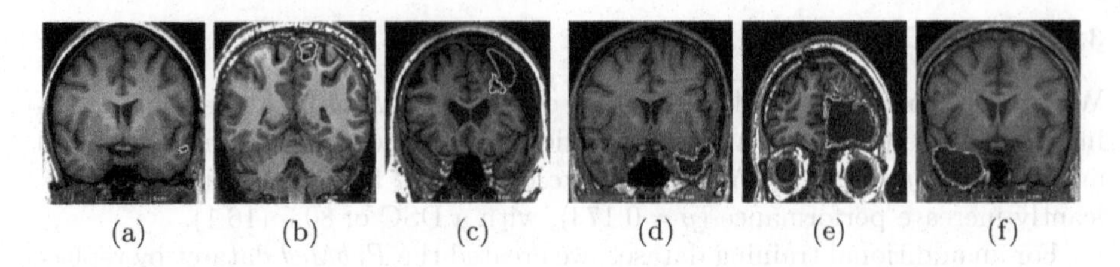

Fig. 3. Manual labels from Rater A (green) and Rater D, the model trained with *PubFull* + *EpiPre* + *Pseudo* (magenta). Errors caused by a (a) small resection, (b) blood clot in cavity and (c) brain shift; segmentations corresponding to the (d) 50th, (e) 75th and (f) 100th percentiles giving a DSC of 81.7, 86.5 and 93.8, respectively. (Color figure online)

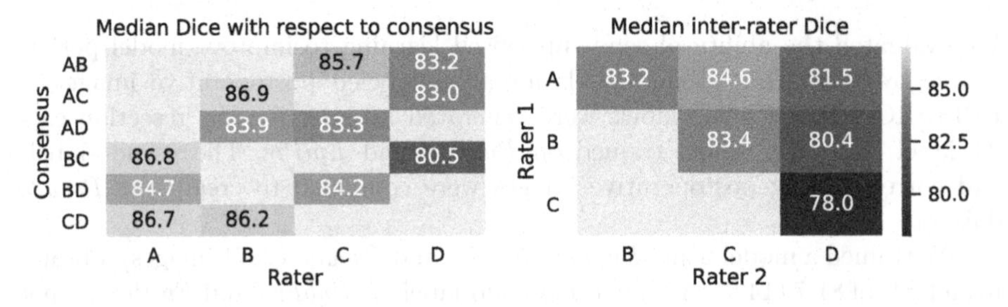

Fig. 4. Left: median DSC between segmentations by a rater and consensuses from two other raters; right: median DSC between each rater segmentations. Rater D corresponds to the model trained with *PubFull*, *EpiPre* and *Pseudo*.

3.4 Comparison to Inter-Rater Performance

We computed pairwise inter-rater agreement between the three human raters and the best performing model (trained with *PubFull* + *EpiPre* + *Pseudo*) as Rater D.

We computed consensus annotations between all pairs of raters using shape-based averaging [22]. DSCs between the segmentations from each rater and the consensuses generated by the other raters are reported in Fig. 4.

4 Discussion

We developed a method to simulate resection cavities on preoperative T_1-weighted MR images and performed extensive validation using datasets of different provenance and size. Our results demonstrate that, when the dataset is of a sufficient size, simulating resection from unlabeled data can provide more accurate segmentations compared to a smaller manually annotated dataset. We found that the most important factor for convolutional neural network (CNN) performance is using a training dataset of sufficient size (in this example, 1800+ samples). The inclusion of training samples from the same scanner or with pseudo-labels only marginally improved the model performance. However, we did not post-process the automatically-generated pseudo-labels, nor did we exclude predictions with higher uncertainty. Further improvements may be obtained by using more advanced semi-supervised learning techniques to appropriately select pseudo-labels to use for training.

Predictions errors are mostly due to 1) resection of size comparable to sulci (Fig. 3a), 2) unanticipated intensities, such as those caused by the presence of blood clots in the cavity (Fig. 3b), 3) brain shift (Fig. 3c) and 4) white matter hypointensities (Fig. 3e). Further work will involve using different internal and external cavity textures, carefully sampling the resection volume, simulating brain shift using biomechanical models, and quantifying epistemic and aleatoric segmentation uncertainty to better assess model performance [24].

The model has a lower inter-rater agreement score compared to between-human agreement values, however, this is well within the interquartile range of all the agreement values computed (Fig. 4). EPISURG will be made available, so that it may be used as a benchmark dataset for brain cavity segmentation.

Acknowledgments. The authors wish to thank Luis García-Peraza Herrera and Reuben Dorent for the fruitful discussions.

This work is supported by the UCL EPSRC Centre for Doctoral Training in Medical Imaging (EP/L016478/1). This publication represents in part independent research commissioned by the Wellcome Trust Health Innovation Challenge Fund (WT106882). The views expressed in this publication are those of the authors and not necessarily those of the Wellcome Trust.

This work uses data provided by patients and collected by the National Health Service (NHS) as part of their care and support.

References

1. Brett, M., Leff, A.P., Rorden, C., Ashburner, J.: Spatial normalization of brain images with focal lesions using cost function masking. NeuroImage **14**(2), 486–500 (2001). https://doi.org/10.1006/nimg.2001.0845. http://www.sciencedirect.com/science/article/pii/S1053811901908456
2. Cardoso, M.J., et al.: Geodesic information flows: spatially-variant graphs and their application to segmentation and fusion. IEEE Trans. Med. Imaging **34**(9), 1976–1988 (2015). https://doi.org/10.1109/TMI.2015.2418298
3. Chen, K., Derksen, A., Heldmann, S., Hallmann, M., Berkels, B.: Deformable image registration with automatic non-correspondence detection. In: Aujol, J.-F., Nikolova, M., Papadakis, N. (eds.) SSVM 2015. LNCS, vol. 9087, pp. 360–371. Springer, Cham (2015). https://doi.org/10.1007/978-3-319-18461-6_29
4. Chen, L.C., Papandreou, G., Kokkinos, I., Murphy, K., Yuille, A.L.: DeepLab: semantic image segmentation with deep convolutional nets, atrous convolution, and fully connected CRFs. arXiv:1606.00915 [cs], May 2017
5. Chitphakdithai, N., Duncan, J.S.: Non-rigid registration with missing correspondences in preoperative and postresection brain images. In: Jiang, T., Navab, N., Pluim, J.P.W., Viergever, M.A. (eds.) MICCAI 2010. LNCS, vol. 6361, pp. 367–374. Springer, Heidelberg (2010). https://doi.org/10.1007/978-3-642-15705-9_45

6. Çiçek, Ö., Abdulkadir, A., Lienkamp, S.S., Brox, T., Ronneberger, O.: 3D U-Net: learning dense volumetric segmentation from sparse annotation. arXiv:1606.06650 [cs], June 2016. http://arxiv.org/abs/1606.06650
7. Drobny, D., Carolus, H., Kabus, S., Modersitzki, J.: Handling non-corresponding regions in image registration. In: Handels, H., Deserno, T.M., Meinzer, H.-P., Tolxdorff, T. (eds.) Bildverarbeitung für die Medizin 2015. I, pp. 107–112. Springer, Heidelberg (2015). https://doi.org/10.1007/978-3-662-46224-9_20
8. van Engelen, J.E., Hoos, H.H.: A survey on semi-supervised learning. Mach. Learn. **109**(2), 373–440 (2020). https://doi.org/10.1007/s10994-019-05855-6
9. Gudbjartsson, H., Patz, S.: The rician distribution of noisy MRI data. Magn. Reson. Med. **34**(6), 910–914 (1995). https://www.ncbi.nlm.nih.gov/pmc/articles/PMC2254141/
10. Herrmann, E., et al.: Fully automated segmentation of the brain resection cavity for radiation target volume definition in glioblastoma patients. Int. J. Radiat. Oncol. Biol. Phys. **102**(3), S194 (2018). https://doi.org/10.1016/j.ijrobp.2018.07.087. https://www.redjournal.org/article/S0360-3016(18)31492-5/abstract
11. Jing, L., Tian, Y.: Self-supervised visual feature learning with deep neural networks: a survey. arXiv:1902.06162 [cs], February 2019
12. Jobst, B.C., Cascino, G.D.: Respective epilepsy surgery for drug-resistant focal epilepsy: a review. JAMA **313**(3), 285–293 (2015). https://doi.org/10.1001/jama.2014.17426
13. Kamnitsas, K., et al.: Efficient multi-scale 3D CNN with fully connected CRF for accurate brain lesion segmentation. Med. Image Anal. **36**, 61–78 (2017). https://doi.org/10.1016/j.media.2016.10.004. arXiv:1603.05959
14. Kingma, D.P., Ba, J.: Adam: a method for stochastic optimization. arXiv:1412.6980 [cs], December 2014
15. Li, W., Wang, G., Fidon, L., Ourselin, S., Cardoso, M.J., Vercauteren, T.: On the compactness, efficiency, and representation of 3D convolutional networks: brain parcellation as a pretext task. arXiv:1707.01992, **10265**, 348–360 (2017). https://doi.org/10.1007/978-3-319-59050-9_28
16. Meier, R., et al.: Automatic estimation of extent of resection and residual tumor volume of patients with glioblastoma. J. Neurosurg. **127**(4), 798–806 (2017). https://doi.org/10.3171/2016.9.JNS16146
17. Modat, M., Cash, D.M., Daga, P., Winston, G.P., Duncan, J.S., Ourselin, S.: Global image registration using a symmetric block-matching approach. J. Med. Imaging, **1**(2) (2014). https://doi.org/10.1117/1.JMI.1.2.024003. https://www.ncbi.nlm.nih.gov/pmc/articles/PMC4478989/
18. Nyúl, L.G., Udupa, J.K., Zhang, X.: New variants of a method of MRI scale standardization. IEEE Trans. Med. Imaging **19**(2), 143–150 (2000). https://doi.org/10.1109/42.836373
19. Pérez-García, F., Sparks, R., Ourselin, S.: TorchIO: a Python library for efficient loading, preprocessing, augmentation and patch-based sampling of medical images in deep learning. arXiv:2003.04696 [cs, eess, stat], March 2020
20. Perlin, K.: Improving noise. ACM Trans. Graph. (TOG) **21**(3), 681–682 (2002). https://doi.org/10.1145/566654.566636
21. Pezeshk, A., Petrick, N., Chen, W., Sahiner, B.: Seamless lesion insertion for data augmentation in CAD training. IEEE Trans. Med. Imaging **36**(4), 1005–1015 (2017). https://doi.org/10.1109/TMI.2016.2640180. https://www.ncbi.nlm.nih.gov/pmc/articles/PMC5509514/
22. Rohlfing, T., Maurer, C.R.: Shape-based averaging. IEEE Trans. Image Process. **16**(1), 153–161 (2007). https://doi.org/10.1109/TIP.2006.884936

23. Shaw, R., Sudre, C., Ourselin, S., Cardoso, M.J.: MRI k-space motion artefact augmentation: model robustness and task-specific uncertainty. In: International Conference on Medical Imaging with Deep Learning, pp. 427–436, May 2019. http://proceedings.mlr.press/v102/shaw19a.html
24. Shaw, R., Sudre, C.H., Ourselin, S., Cardoso, M.J.: A heteroscedastic uncertainty model for decoupling sources of MRI image quality. arXiv:2001.11927 [cs, eess], January 2020
25. Sudre, C.H., Cardoso, M.J., Ourselin, S.: Longitudinal segmentation of age-related white matter hyperintensities. Med. Image Anal. **38**, 50–64 (2017). https://doi.org/10.1016/j.media.2017.02.007. http://www.sciencedirect.com/science/article/pii/S1361841517300257
26. Winterstein, M., Münter, M.W., Burkholder, I., Essig, M., Kauczor, H.U., Weber, M.A.: Partially resected gliomas: diagnostic performance of fluid-attenuated inversion recovery MR imaging for detection of progression. Radiology **254**(3), 907–916 (2010). https://doi.org/10.1148/radiol09090893

Local Contractive Registration for Quantification of Tissue Shrinkage in Assessment of Microwave Ablation

Dingkun Liu, Tianyu Fu, Danni Ai(✉), Jingfan Fan, Hong Song, and Jian Yang

Beijing Engineering Research Center of Mixed Reality and Advanced Display, School of Optics and Photonics, Beijing Institute of Technology, Beijing 100081, China
danni@bit.edu.cn

Abstract. Microwave ablation is an effective minimally invasive surgery for the treatment of liver cancer. The safety margin assessment is implemented by mapping the coagulation in the postoperative image to the tumor in the preoperative image. However, an accurate assessment is a challenging task because the tissue shrinks caused by dehydration during microwave ablation. This paper proposes a fast automatic assessment method to compensate for the underestimation of the coagulation caused by the tissue shrinks and precisely quantify the tumor coverage. The proposed method is implemented on GPU including two main steps: (1) a local contractive nonrigid registration for registering the liver parenchyma around the coagulation, and (2) the fast Fourier transform-based Helmholtz-Hodge decomposition for quantifying the location of the shrinkage center and the volume of the original coagulation. The method was quantificationally evaluated on 50 groups of synthetic datasets and 9 groups of clinical MR datasets. Compared with five state-of-the-art methods, the lowest distance to the true deformation field (1.56 ± 0.74 mm) and the highest precision of safety margin (88.89%) are obtained. The mean computation time is 111 ± 13 s. Results show that the proposed method efficiently improves the accuracy of the safety margin assessment and is thus a promising assessment tool for the microwave ablation.

Keywords: Microwave ablation assessment · Nonrigid registration · Tissue shrinkage

1 Introduction

The microwave (MW) ablation is a minimally invasive treatment of hepatocellular carcinoma (HCC) by creating enough heat to destroy the tumor cells [3], making sure the tumor is completely enclosed by the coagulation with a safety margin 5 mm or larger. Generally, the safety margin is calculated by comparing the relative position between the tumor and the coagulation. Traditional global

Supported by the National Key R&D Program of China (2019YFC0119300).

A. L. Martel et al. (Eds.): MICCAI 2020, LNCS 12263, pp. 126–134, 2020.
https://doi.org/10.1007/978-3-030-59716-0_13

registration methods align the postoperative image to preoperative image based on the intensity values. However, because of the similar intensity distribution between tumor and coagulation, traditional global registration tends to wrongly match the both lesion shapes, leading to an incorrect safety margin [7,10,15,17]. To address mismatching between the tumor and coagulation, two kinds of local registration methods have been proposed recently.

In *local rigid registration* methods, transformation at the coagulation zone is considered to be rigid to prevent incorrect deformation of the coagulation to the tumor. Tanner et al. [18] performed a global nonrigid registration with local deformations constrained to an average displacement. Rieder et al. [16] presented an automatic rigid registration on region of interest (ROI) with tumor and coagulation masked out to minimize the effect of global liver deformations. Luu et al. [11] proposed an automatic nonrigid registration method with a local penalty on the deformation field. While effective, these methods neglect however the coagulation elastic deformation.

In *local incompressible registration* methods, the coagulation is considered as the incompressible tissue. Passera et al. [15] replaced the tumor ROI and the coagulation ROI with the same synthetic pattern before nonrigid registration to maintain incompressibility of the deformation. However, the image content is heavily changed and the local deformation is not estimated because of the synthetic pattern. Kim et al. [9] proposed an local incompressible nonrigid registration to constrain the incompressibility of the coagulation for preventing the inaccurate matching. Fu et al. [8] considered the coagulation as incompressible tissue and proposed an automatic workflow for estimating the nonrigid motion caused by respiratory movement and edema separately.

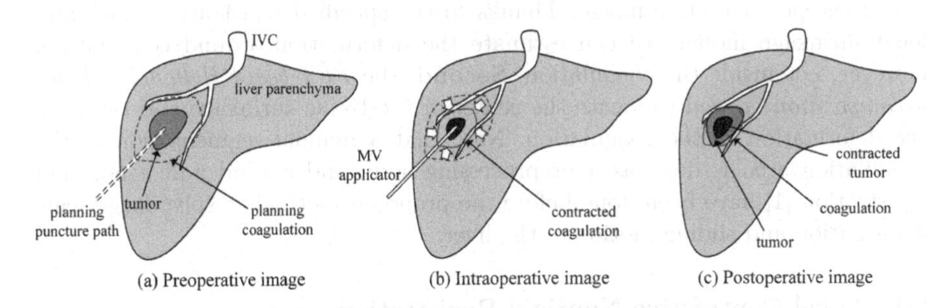

Fig. 1. The smaller safety margin caused by the tissue shrinkage. The dotted red contour and solid blue contour are the planning and contracted coagulation, respectively. The gray and black areas represent the tumor before and after shrinkage. (Color figure online)

However, it was found that MV ablation causes significant shrinkage of liver parenchyma because of water vaporization [2,5,6,10]. The above methods constrained the volume of the coagulation and neglected the shrinkage deformation of the coagulation, leading to a inaccurate safety margin. As shown in Fig. 1, the coagulation (Fig. 1(a) dotted red contour) which completely covers the tumor, shrinks into smaller size (Fig. 1(c) solid blue contour) due to the dehydration, which leading to an incorrect assessment of the inability to completely enclose the tumor (Fig. 1(a) dark gray area). In this paper, we consider the coagulation and its surrounding tissues as the contractive tissues and propose a fast workflow to automatically compensate for the tissue shrinkage and precisely quantify the safety margin. The deformation of liver parenchyma around the coagulation is estimated by a *local contractive nonrigid registration*. And then, the estimated deformation is decomposed by the *fast Fourier transform-based (FFT-based) Helmholtz-Hodge decomposition* to locate the center of tissue shrinkage and compensate the shrinkage. Based on the compensation of tissue shrinkage, the underestimation of coagulation is eliminated and the safety margin is quantified. The main contributions of this study are summarized as follows:

- The proposed local contractive registration can estimate the tissue shrinkage caused by the dehydration during MV ablation only using the preoperative and postoperative images.
- The tissue shrinkage is quantified for the assessment of MV ablation and a more comprehensive and accurate safety margin can be calculated.

2 Methods

In this paper, we estimates the deformation around coagulation and compensates for the tissue shrinkage to get an accurate safety margin assessment. **First**, a *local contractive nonrigid registration* is proposed to align the livers and model the shrinkage motion of the tissue around the coagulation with coagulation masked out of the postoperative image. Thanks to the specified registration mask and local shrinkage model, we can estimate the deformation around coagulation, however, not inside the coagulation. **Second**, the *FFT-based Helmholtz-Hodge decomposition* is used to locate the center of the tissue shrinkage and estimate the deformation in the coagulation. Note that a manual segmentation of the coagulation area is done as a preprocessing step, and a rigid and a nonrigid registration [1] have been done before the proposed method to solve large-scale deformation and sliding motion of the liver.

2.1 Local Contractive Nonrigid Registration

To obtain a smooth and diffeomorphic deformation field, we use the log-domain diffeomorphic demons [4] as the global deformation. In this deformable model, the registration is formulated as an optimization problem in which the energy function E is minimized w.r.t velocity field v:

$$E(I_F, I_M, \boldsymbol{v}) = -S(I_F, I_M \circ exp(\boldsymbol{v})) + \lambda R(exp(\boldsymbol{v})) \quad (1)$$

where I_M and I_F are the preoperative and postoperative images, respectively; $\boldsymbol{v}$ is the static velocity field and $exp(\cdot)$ is the exponential function. The parameter λ weights similarity $S(\cdot)$ against regularity $R(\cdot)$.

Consider that the images used for assessment of MV ablation are monomodal, we choose the sum of squared differences (SSD) as the similarity for efficient computation. However, this assumption sometimes does not hold because the bias field in the MR images may cause the variation of gray level in the same tissue. Therefore, we perform a histogram matching [14] between the fixed and moving images to normalize the grayscale values before registration. Moreover, we register the preoperative image to postoperative image with a mask of fixed image, to prevent compensation of both tumor and coagulation shapes instead of the surrounding anatomy. The registration mask is the invertion of coagulation segmentation. Thus, given the domain of the fixed mask Ω_F, the similarity is formulated as follow:

$$S(I_F, I_M \circ exp(\boldsymbol{v})) = \sum_{\boldsymbol{x} \in \Omega_F} (I_F(\boldsymbol{x}) - I_M \circ exp(\boldsymbol{v})(\boldsymbol{x})) \tag{2}$$

Inspired by Mansi et al. [13], we propose a local shrinkage model as the regularity by constraining local velocity field to be curl-free. Helmholtz decomposition demonstrates that any velocity $\boldsymbol{v}$ can be uniquely decomposed into a curl-free field and a divergence-free field: $\boldsymbol{v} = \boldsymbol{v}_c + \boldsymbol{v}_d$, where $\boldsymbol{v}_c$ is the curl-free field ($\nabla \times \boldsymbol{v}_c = \boldsymbol{0}$) and $\boldsymbol{v}_d$ is the divergence-free field ($\nabla \cdot \boldsymbol{v}_d = 0$). These properties lead to the following equation:

$$\nabla \cdot \boldsymbol{v}_c = \nabla \cdot \boldsymbol{v} \tag{3}$$

To compute the curl-free field for estimation of tissue deformation, the component $\boldsymbol{v}_c$ is represented by the gradient of a scalar field C. Therefore, we can get the Poisson equation by substituting $\boldsymbol{v}_c = \nabla C$ in Eqs. (3):

$$\Delta C = \nabla \cdot \nabla C = \nabla \cdot \boldsymbol{v}_c = \nabla \cdot \boldsymbol{v} \tag{4}$$

In order to remain local shrinkage for arbitrary shapes of the tissue around coagulation, we impose the 0-Dirichlet boundary condition around coagulation, and the Poisson equation is numerically solved by finite differences (FD) method in the space domain instead of frequency domain. More specifically, the 0-Dirichlet boundary condition Ω_D is represented by the coagulation segmentation dilated utilizing a morphological operation.

Given the fixed mask and 0-Dirichlet boundary condition, the local shrinkage deformation can be estimated on the ROI around the coagulation by intersection between Ω_F and Ω_D.

2.2 Quantification of the Tissue Shrinkage

Since the local shrinkage model only estimate the deformation of surrounding anatomy around coagulation, the deformation inside the coagulation is unknown. It is necessary to calculate the deformation in the coagulation area for quantification of the tissue shrinkage.

Unlike the previously defined ROI, the focus here is the entire coagulation area, so the boundary conditions are not necessary. We use an integral kernel called the *Green's function* to solve the Eqs. (4) for better computational performance. The scalar field C is calculated as follow:

$$C(\boldsymbol{x}_0) = \int_{\Omega} (\nabla \cdot \boldsymbol{v}(\boldsymbol{x})) G_{\infty}(\boldsymbol{x}_0 - \boldsymbol{x}) d\boldsymbol{x}, \boldsymbol{x}_0 \in \Omega \tag{5}$$

where G_{∞} is the *free-space Green's function* defined as $G_{\infty}(\boldsymbol{x}) = \frac{log(|\boldsymbol{x}|)}{2\pi}, \boldsymbol{x} \in \mathbb{R}^2$ and $G_{\infty}(\boldsymbol{x}) = -\frac{1}{4\pi|\boldsymbol{x}|}, \boldsymbol{x} \in \mathbb{R}^3$.

In Eqs. (5), each $C_(\boldsymbol{x})$ is computed by looping over voxels in whole image, which is computationally demanding if the domain Ω is large. In this paper, we proposed a *FFT-based Helmholtz-Hodge decomposition* to directly compute the C in parallel to improve computational efficiency. Applying the trapezoidal method, the Eqs. (5) can be discretized as follows:

$$\int_{\Omega} f(\boldsymbol{x}) d\boldsymbol{x} \approx \begin{cases} \frac{s_x s_y}{4}(\sum_{\boldsymbol{x} \in C} f_{\boldsymbol{x}} + 2\sum_{\boldsymbol{x} \in E} f_{\boldsymbol{x}} + 4\sum_{\boldsymbol{x} \in S} f_{\boldsymbol{x}}), & \boldsymbol{x} \in \mathbb{R}^2 \\ \frac{s_x s_y s_z}{8}(\sum_{\boldsymbol{x} \in C} f_{\boldsymbol{x}} + 2\sum_{\boldsymbol{x} \in E} f_{\boldsymbol{x}} + 4\sum_{\boldsymbol{x} \in S} f_{\boldsymbol{x}} + 8\sum_{\boldsymbol{x} \in I} f_{\boldsymbol{x}}), & \boldsymbol{x} \in \mathbb{R}^3 \end{cases} \tag{6}$$

where $f(\boldsymbol{x}) = \nabla \cdot \boldsymbol{v}(\boldsymbol{x})) G_{\infty}(\boldsymbol{x}_0 - \boldsymbol{x})$. $\boldsymbol{C}$, $\boldsymbol{E}$, $\boldsymbol{S}$ and $\boldsymbol{I}$ are the set of vertices at the corners, at the edges, at the surfaces, and in the interior of Ω.

Substituting the Eqs. (6) in Eqs. (5), we can get a simplified equation as follow:

$$C = k <W, \nabla \cdot \boldsymbol{v} * D> \tag{7}$$

where k is the factor and $k = \frac{s_x s_y}{8\pi}$ when $\Omega \subset \mathbb{R}^2$, $k = -\frac{s_x s_y s_z}{32\pi}$ when $\Omega \subset \mathbb{R}^3$. Given the $\Omega = [n_x \times n_y(\times n_z)]$ region of the $\boldsymbol{v}$, W is the weight matrix ($\Omega_W = [n_x \times n_y(\times n_z)]$) with values 1, 2, 4 and 8 at the corners, edges, surfaces and interior, respectively; D is the distance matrix with dimensions $\Omega_D = [(2n_x - 1)\times(2n_y - 1)]$ and values $D(i,j) = log(\sqrt{(i-n_x)^2 s_x^2 + (j-n_y)^2 s_y^2})$ when $\Omega \subset \mathbb{R}^2$, and with dimensions $\Omega_D = [(2n_x - 1)\times(2n_y - 1)\times(2n_z - 1)]$ and values $D(i,j,k) = \frac{1}{\sqrt{(i-n_x)^2 s_x^2 + (j-n_y)^2 s_y^2 + (k-n_z)^2 s_z^2}}$. $< \cdot >$ is Hadamard product.

Given the velocity field $\boldsymbol{v}$, the scalar field C can be directly computed by solving Eqs. (7). In order to further speed up the calculation, we use fast Fourier transform-based (FFT) to transform the spatial domain to the frequency domain. The center of tissue shrinkage is located at the minimum of the C, which corresponds to the position of the antenna of the MV applicator during the operation. By calculating the gradient of C, the shrinkage deformation of the coagulation can be estimated. The coagulation is mapped into the preoperative image by inverse transformation ($exp(-\boldsymbol{v}_c)$). A safety margin assessment with compensation for tissue shrinkage can be achieved.

3 Experiments and Results

In this section, we compared the proposed method with five state-of-the-art registration methods [8,9,11,16,17]. 50 groups of synthetic datasets and 9 groups of clinical MR datasets are used for quantificational evaluation. The distance to the ground-truth deformation field (DTF) and the mean surface distance (MSD) between preoperative and warped postoperative livers are used to evaluate the registration accuracy. The safety margin is used to evaluate the accuracy of compensation for tissue shrinkage. A 3-level multi-resolution is used for each registration and the max iteration number of each resolution is 300.

3.1 Evaluation on Synthetic Datasets

The accuracy and robust of the proposed method are quantified on 50 groups of synthetic datasets. We select one patient's preoperative 3D MR image ($220 \times 170 \times 73$ slices, $1.09 \times 1.09 \times 2.50\,\text{mm}^3$ voxel spacing) as the fixed image (Fig. 2(a)). Then, we create the test image by padding an artificial coagulation (Fig. 2(b)). The size of the coagulation is obtained by dilating the tumor segmentation with the ground-truth 6.74 mm safety margin. The coagulation pattern is obtained from the same patient's postoperative image. The test image is then warped by 50 ground-truth deformation fields (Fig. 2(f)) to get the warped images (Fig. 2(c)). In order to simulate the motion between preoperative and postoperative images, the ground-truth deformation field contains two part: tissue shrinkage (L_2-norm: 0.1, 0.2, ..., 5.0 mm) caused by water dehydration (Fig. 2(d)) and incompressible deformation (L_2-norm: 5.00 mm, Jacobian determinant: 1.004 ± 0.001) caused by respiration (Fig. 2(e)). We register 50 warped images to the fixed image. The estimated deformation fields are compared with the ground-truth deformation fields, and the results of safety margin are compared with ground-truth safety margin.

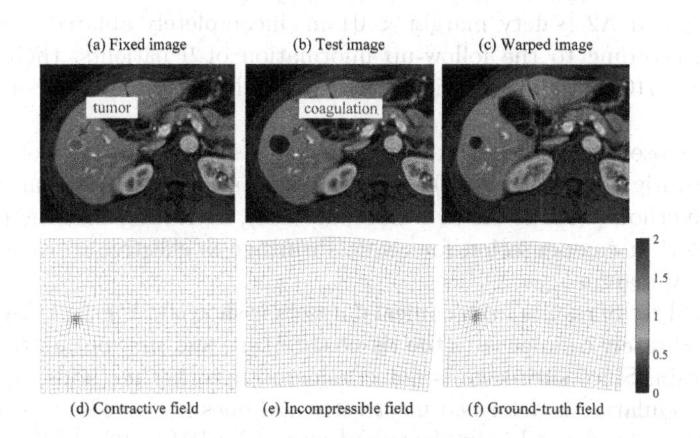

Fig. 2. Synthetic image with a coagulation pattern warped by the combination of contractive and random incompressible deformation fields.

Figure 3 provides the results of DTF and safety margin for all methods. Compared with other five methods, the proposed method exhibit the best performance for almost cases with the lowest DTF. The lowest DTF indicates that the proposed method accurately compensates the tissue shrinkage around the coagulation. As a result, the safety margin obtained by proposed method is closest to ground-truth 6.74 mm.

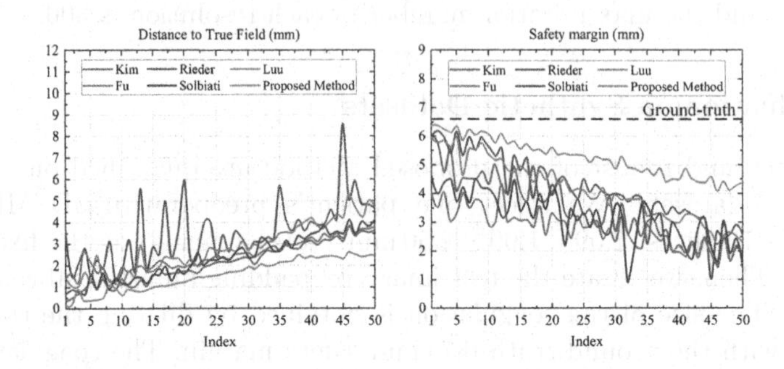

Fig. 3. The comparison results of DTF (a) and safety margin (b) by changing the amplitudes of shrinkage (0.1, 0.2,. . . , 5.0 mm) using six registration methods on synthetic datasets.

3.2 Evaluation on Clinical Datasets

The proposed method is evaluated on 9 groups of clinical datasets from 9 patients. In clinical, the safety margin is divided into three levels: A0 (safety margin $\geq$ 5 mm, completely ablated with enough safety margin, considered as cured), A1 (0 $\leq$ safety margin $<$ 5 mm, completely covered but has a risk of recurrence) and A2 (safety margin $<$ 0 mm, incompletely ablated, need reoperation). According to the follow-up information of 9 patients, there was no recurrence of HCC in all the patients, indicating that the safe boundary level is A0 or A1.

The proposed method shows a comparable registration accuracy of the livers as shown in Fig. 4(a). Figure 4(b) shows the most accurate safety margin of the proposed method (8 A0 or A1 in 9 cases, 88.89%), indicating that the proposed methods with a compensation for tissue shrinkage is effective to calculate the actual safety margin.

A typical set of results from clinical datasets is shown in Fig. 5. The proposed method accurately compensates the tissue shrinkage and thus obtain the correct safety margin. Since shrinkage is not considered by other methods, it is found that the coagulation is deformed incorrectly and does not enclose the tumor.

The proposed method is implemented on the NVIDIA 1080 GPU using the ArrayFire platform [12]. For each patient, safety margin assessment takes 111 $\pm$ 13 s with the proposed method, which is considered efficient for the assessment of MV ablation.

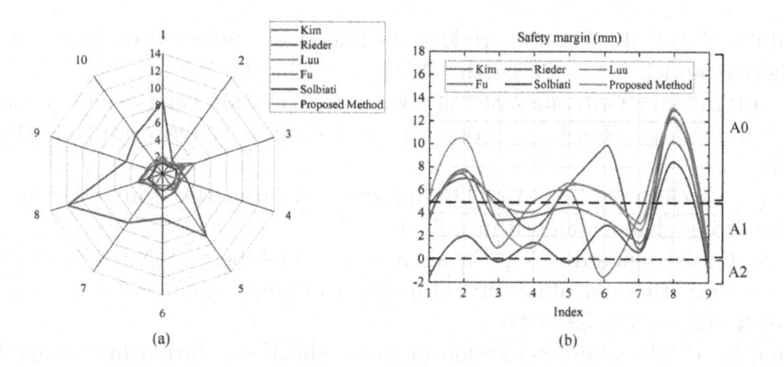

Fig. 4. The comparison results of MSD (a) and safety margin (b) using six registration methods on clinical datasets.

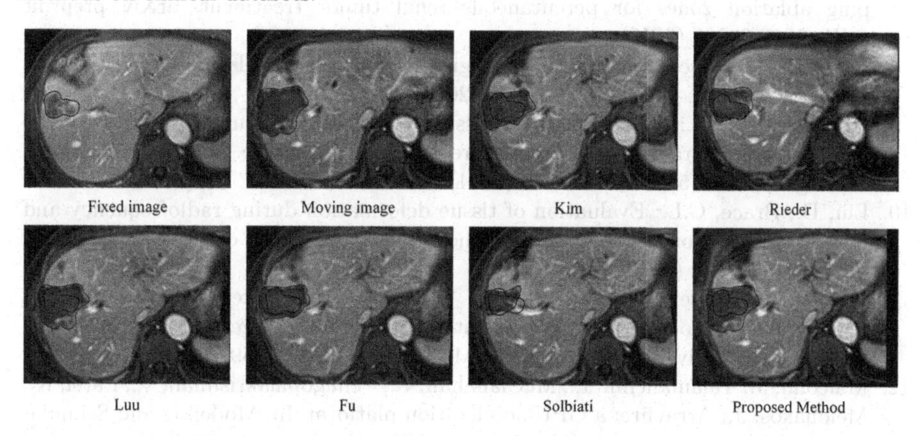

Fig. 5. Typical registration results from clinical datasets. The red and blue contours are tumor and coagulation, respectively. (Color figure online)

4 Conclusion

In this paper, tissue shrinkage caused by water dehydration is quantified in the assessment of MV ablation. A fast automatic assessment method is proposed to compensate for the shrinkage of the coagulation and quantify the safety margin. The proposed method is evaluated on 50 groups of synthetic datasets and 9 groups of clinical datasets. The experimental results indicates that the proposed method accurately quantifies and compensates the tissue shrinkage, and the most accurate safety margin is obtained. The proposed method with the quantification of tissue shrinkage has the potential to improve current assessment of MV ablation.

References

1. Ai, D., et al.: Nonrigid registration for tracking incompressible soft tissues with sliding motion. Med. Phys. **46**(11), 4923–4939 (2019)

2. Amabile, C., et al.: Tissue shrinkage in microwave ablation of liver: an ex vivo predictive model. Int. J. Hyperth. **33**(1), 101–109 (2017)
3. Dou, J.P., et al.: Outcomes of microwave ablation for hepatocellular carcinoma adjacent to large vessels: a propensity score analysis. Oncotarget **8**(17), 28758 (2017)
4. Dru, F., Vercauteren, T.: An ITK implementation of the symmetric log-domain diffeomorphic demons algorithm (2009)
5. Farina, L., Nissenbaum, Y., Cavagnaro, M., Goldberg, S.N.: Tissue shrinkage in microwave thermal ablation: comparison of three commercial devices. Int. J. Hyperth. **34**(4), 382–391 (2018)
6. Farina, L., et al.: Characterisation of tissue shrinkage during microwave thermal ablation. Int. J. Hyperth. **30**(7), 419–428 (2014)
7. Franz, A., et al.: An open-source tool for automated planning of overlapping ablation zones for percutaneous renal tumor treatment. arXiv preprint arXiv:1912.09966 (2019)
8. Fu, T., et al.: Local incompressible registration for liver ablation surgery assessment. Med. Phys. **44**(11), 5873–5888 (2017)
9. Kim, K.W., et al.: Safety margin assessment after radiofrequency ablation of the liver using registration of preprocedure and postprocedure CT images. Am. J. Roentgenol. **196**(5), W565–W572 (2011)
10. Liu, D., Brace, C.L.: Evaluation of tissue deformation during radiofrequency and microwave ablation procedures: influence of output energy delivery. Med. Phys. **46**(9), 4127–4134 (2019)
11. Luu, H.M., Niessen, W., van Walsum, T., Klink, C., Moelker, A.: An automatic registration method for pre-and post-interventional CT images for assessing treatment success in liver RFA treatment. Med. Phys. **42**(9), 5559–5567 (2015)
12. Malcolm, J., Yalamanchili, P., McClanahan, C., Venugopalakrishnan, V., Patel, K., Melonakos, J.: Arrayfire: a GPU acceleration platform. In: Modeling and Ssimulation for Defense Systems and Applications VII, vol. 8403, p. 84030A. International Society for Optics and Photonics (2012)
13. Mansi, T., Pennec, X., Sermesant, M., Delingette, H., Ayache, N.: iLogDemons: a demons-based registration algorithm for tracking incompressible elastic biological tissues. Int. J. Comput. Vis. **92**(1), 92–111 (2011)
14. Nyúl, L.G., Udupa, J.K., Zhang, X.: New variants of a method of MRI scale standardization. IEEE Trans. Med. Imaging **19**(2), 143–150 (2000)
15. Passera, K., Selvaggi, S., Scaramuzza, D., Garbagnati, F., Vergnaghi, D., Mainardi, L.: Radiofrequency ablation of liver tumors: quantitative assessment of tumor coverage through CT image processing. BMC Med. Imaging **13**(1), 3 (2013)
16. Rieder, C., et al.: Automatic alignment of pre-and post-interventional liver CT images for assessment of radiofrequency ablation. In: Medical Imaging 2012: Image-Guided Procedures, Robotic Interventions, and Modeling, vol. 8316, p. 83163E. International Society for Optics and Photonics (2012)
17. Solbiati, M., et al.: A novel software platform for volumetric assessment of ablation completeness. Int. J. Hyperth. **36**(1), 337–343 (2019)
18. Tanner, C., et al.: Volume and shape preservation of enhancing lesions when applying non-rigid registration to a time series of contrast enhancing MR breast images. In: Delp, S.L., DiGoia, A.M., Jaramaz, B. (eds.) MICCAI 2000. LNCS, vol. 1935, pp. 327–337. Springer, Heidelberg (2000). https://doi.org/10.1007/978-3-540-40899-4_33

Reinforcement Learning of Musculoskeletal Control from Functional Simulations

Emanuel Joos, Fabien Péan, and Orcun Goksel(✉)

Computer-assisted Applications in Medicine, ETH Zurich, Zurich, Switzerland
ogoksel@ethz.ch

Abstract. To diagnose, plan, and treat musculoskeletal pathologies, understanding and reproducing muscle recruitment for complex movements is essential. With muscle activations for movements often being highly redundant, nonlinear, and time dependent, machine learning can provide a solution for their modeling and control for anatomy-specific musculoskeletal simulations. Sophisticated biomechanical simulations often require specialized computational environments, being numerically complex and slow, hindering their integration with typical deep learning frameworks. In this work, a deep reinforcement learning (DRL) based inverse dynamics controller is trained to control muscle activations of a biomechanical model of the human shoulder. In a generalizable end-to-end fashion, muscle activations are learned given current and desired position-velocity pairs. A customized reward functions for trajectory control is introduced, enabling straightforward extension to additional muscles and higher degrees of freedom. Using the biomechanical model, multiple episodes are simulated on a cluster simultaneously using the evolving neural models of the DRL being trained. Results are presented for a single-axis motion control of shoulder abduction for the task of following randomly generated angular trajectories.

Keywords: Shoulder · FEM · Deep Reinforcement Learning

1 Introduction

Biomechanical tissue models have been proposed for several different anatomical structures such as the prostate, brain, liver, and muscles; for various computer-assisted applications including preoperative planning, intraoperative navigation and visualization, implant optimization, and simulated training. Musculoskeletal biomechanical simulations that use muscle activation models are used in orthopedics for functional understanding of complex joints and movements as well as for patient-specific surgical planning. Shoulder is the most complex joint in the body,

Electronic supplementary material The online version of this chapter (https://doi.org/10.1007/978-3-030-59716-0_14) contains supplementary material, which is available to authorized users.

A. L. Martel et al. (Eds.): MICCAI 2020, LNCS 12263, pp. 135–145, 2020.
https://doi.org/10.1007/978-3-030-59716-0_14

offering the greatest range-of-motion. The upper arm is actively controlled and stabilized with over 10 anatomical muscles [10] subdivided in several parts [6]. With high range-of-motion and redundancy, the shoulder is regularly exposed to forces larger than the body weight, making the area particularly prone to soft tissue damages [8,30]. Consequent surgical interventions and corresponding planing and decisions could benefit from simulated functional models.

For simulating complex biomechanical models, sophisticated computational and simulation environments are often needed, such as SOFA [11] and Artisynth [17]. Due to many tortuous effects such as time-dependent, nonlinear behaviour of muscle fibres and soft tissue and the bone and muscle contacts and collisions, the control of muscle activations required for a desired movement is not trivial. Linearized control schemes [29] easily become unstable for complex motion and anatomy, e.g. the shoulder, despite tedious controller parametrization and small simulation time-steps leading to lengthy computations [22]. Machine learning based solutions for such biomechanical models would not only enable simple, fast, stable, and thus effective controllers, but could also facilitate studying motor control paradigms such as neural adaptation and rehabilitation efficacy after orthopedic surgeries, e.g. muscle transfers.

Reinforcement learning (RL) is a machine learning technique for model control of complex behaviour, in a black-box manner from trial-and-error of input-output combinations, i.e. not requiring information on the underlying model nor its environment. With Deep Reinforcement Learning (DRL), impressive examples have been demonstrated such as for playing Atari console games [19] and the game of Go [27], for control of industrial robots [14,31], and for animating characters [13,15,16], e.g. learning how to walk. Despite DRL applications with simpler rigid and multi-body dynamics as above, soft-body structures and complex material, activation, geometry, and contact models have not been well studied. Furthermore, sophisticated simulations required for complex models are not trivial to couple with DRL strategies. In this paper, we present the DRL control of a Finite Element Method (FEM) based musculoskeletal shoulder model, while investigating two DRL approaches comparatively.

2 Materials and Methods

Musculoskeletal Shoulder Model. We herein demonstrate DRL-based control with a simplified model of the shoulder joint. We used segmentations from the BodyParts3D dataset [18], cf. Fig. 1-left. The shoulder complex consists of three bones: the humerus, the scapula and the clavicle; as well as multiple muscles, tendons and ligaments. Our model involves surface triangulations for the three bones; a manual surface fit to the ribs imitating the trunk; and B-spline based quadrilateral thin-shell meshes to model large, relatively flat muscles via FEM [21]. The bones are rigid objects and the muscles are displacement-constrained on the bones at tendon origins and insertions. Muscle fibres are modeled nonlinearly with respect to deformation based on [4]. Fibres are embedded within a linear co-rotational FE background material model, coupled at FE

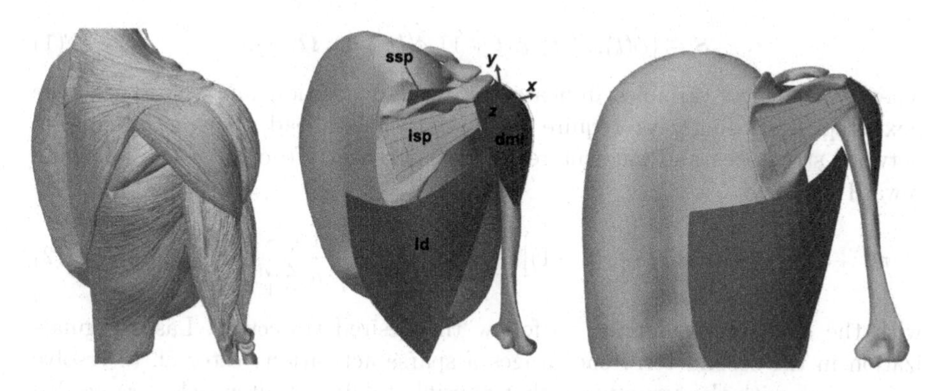

Fig. 1. Shoulder segmentation [18] (left) and our functional musculoskeletal model of the shoulder with four muscles relevant to abduction (center-right).

integration nodes. A normalized activation signal $\Omega \in [0, 1]$ sent homogeneously to all fibres of a muscle segment linearly generate internal stresses, contracting them against their background soft-tissue material, while pulling on the attached bones [21]. In this paper, we focus on the abduction motion, being a standard reference movement for diagnosis and in clinical studies [7,12,33]. Accordingly, four muscles relevant for abduction and adduction [23] supraspinatus (ssp), infraspinatus (isp), deltoid middle part (dmi), and latissimus dorsi (ld) are simulated herein, cf. Fig. 1-center.

Learning Muscle Control. Consider that at each time step t, an agent with a current *state* s_t executes an *action* a_t according to some *policy* $\pi(a_t|s_t)$, makes an observation o_t, and receives a scalar reward r_t. RL aims to find the policy π for an optimal outcome that maximizes the cumulative sum of current and discounted future rewards. This is predicted either based only on the current state, i.e. *state value* function $V_\pi(s_t) = \mathbb{E}_\pi \left[r_t + \gamma r_{t+1} + \gamma^2 r_{t+2} + ...|s_t \right]$, or based on the current state and action together, i.e. *action value* function $Q_\pi(s_t, a_t) = \mathbb{E}_\pi \left[r_t + \gamma r_{t+1} + \gamma^2 r_{t+2} + ...|s_t, a_t \right]$ also known as *Q function*. $\gamma < 1$ is the discounting factor to ensure future rewards are worth lower.

The shoulder model and its forward simulation is herein considered as black-box with its input being discrete muscles activation's and the output being the (angular) pose $\phi \in R^d$ and velocity ϕ' of the humerus, where d is the degrees of freedom to control. Using full 100% activation range of the muscles as the potential RL action set makes it difficult to predict small activation changes precisely, as well as leading potentially to large activation jumps at failed predictions. Therefore we use an action space of differential activation changes $\omega \in R^n$ where n is the number of muscles controlled. We thus additionally provide the current muscle activations $\boldsymbol{\Omega} \in \mathbb{R}^n$ as input to the agent so that it can infer the incremental effect of ω on top. To formalize a solution, typically a Markov Reward Process (MRP) [3] is defined as a quintuple with the set of possible states S, the set of possible actions A, the reward function r, the transition probability matrix P, and a discounting factor γ. Given the above, our state set is then

$$S \in \{\phi(t), \phi'(t), \hat{\phi}(t+1), \hat{\phi}'(t+1), \boldsymbol{\Omega}(t)\}, \quad (1)$$

where the hatted variables indicate the desired position and velocity at the next step. Accordingly, we require only a short look-ahead, allowing for simpler network structures and efficient real-time inference. We employ the following reward strategy

$$r(t+1) = -\left|\phi(t+1) - \hat{\phi}(t+1)\right| - \alpha \sum_{i=1}^{n} |\omega_i(t)| - \frac{1}{n} \sum_{i=1}^{n} 1_{\{|\omega_i(t)| > \omega_{\max}\}}. \quad (2)$$

with the first term enforcing to follow the desired trajectory. Lasso regularization in the second term encourages a sparse activation vector ω, to resolve redundancy with the assumption that physiologically not all muscles are needed at the same time. More sophisticated muscle recruitment strategies extensively studied in the literature can also be introduced in this reward function. The last term prevents the agent from learning a so-called "bang-bang" solution [2], where a controller alternately switches between two extreme states, e.g. $\omega_{\min}$ and $\omega_{\max}$ herein. This term then ensures a sufficient exploration of the whole interval $[\omega_{\min}, \omega_{\max}]$ during learning. P is inherent to the system being modeled, in our case the shoulder model and its forward simulation. γ is a hyperparameter defining the discounting factor, set to be 0.99 herein. To find an optimal policy, we comparatively study two following DRL strategies:

Deep Q-learning [32] is a common approach to find an optimal policy π by maximizing the action value function, i.e. solving $Q^*(s_t, a_t) = \max_\pi Q_\pi(s_t, a_t) = r_t + \gamma \sum_{s_{t+1} \in S} P_{s_t s_{t+1}} \max_{a_{t+1}} Q^*(s_{t+1}, a_{t+1})$, where s_{t+1} and a_{t+1} are respectively the next state and action, and $P_{s_t s_{t+1}}$ is the transition probability matrix for transitioning from state s_t to s_{t+1} for action alternatives. P can be populated by a so-called *replay* buffer of past experiences, e.g. earlier simulations or games. *Deep Q-Learning* (DQL) approximates the Q-value function via a neural network (NN) and outputs discrete actions, typically converging relatively fast. As the action space of DQL, we quantized [-1,1]% differential activation range in 21 steps, i.e. $A_{\text{DQL}} = \{-1, -0.9, ..., 0.9, 1\}\%$. In other words, between two simulation steps any muscle activation cannot change more than 1%.

Policy Gradient Methods work by estimating the policy gradient in order to utilize simple gradient-based optimizers such as Stochastic Gradient Descent for optimal policy decisions. To prevent large policy changes based on small sets of data, Trust Region Policy Optimization (TRPO) [24] proposes to regularize the policy update optimization by penalizing the KL divergence of policy change. This helps update the policy in an incremental manner to avoid detrimental large policy changes, also utilizing any (simulated) training data optimally. Q_π, V_π, and π_θ are estimated by different NNs, some parameters of which may be shared. Policy gradient loss can be defined as $L^{\text{PG}}(\theta) = \hat{\mathbb{E}}_t \left[\log \pi_\theta \left(a_t|s_t\right) D_t\right]$, where $D_t = Q_\pi(s_t, a_t) - V_\pi(s_t)$ is the *advantage* function [25], which represents the added value (advantage) from taking the given action a_t at state s_t.

Despite the constraint by TRPO, large policy updates may still be observed; therefore Proximal Policy Optimization (PPO) [26] proposes to clip the gradient updates around 1 within a margin defined by an hyperparameter ϵ as follows:

$$L^{\mathrm{CL}}(\theta)=\hat{\mathbb{E}}_t\Big[\min\Big(\rho_t(\theta)D_t, \mathrm{clip}(\rho_t(\theta),1-\epsilon,1+\epsilon)D_t\Big)\Big], \text{ where } \rho_t(\theta)=\frac{\pi_\theta(a_t|s_t)}{\pi_{\theta_{\mathrm{old}}}(a_t|s_t)}$$

is the policy change ratio, with θ and θ_{old} being the policy network parameter vectors, respectively, before and after the intended network update. The minimum makes sure that the change in policy ratio only effects the objective when it makes it worse. NN parameters of the policy π and the value function V are shared, allowing to also introduce a value function error L_t^{VF}. Additionally, to ensure policy exploration, an entropy bonus term is introduced [26] as follows:

$$L(\theta)=\hat{\mathbb{E}}\left[L^{\mathrm{CL}}(\theta)-c_1 L_t^{\mathrm{VF}}(\theta)+c_2 S\left[\pi_\theta\right](s_t)\right], \tag{3}$$

where c_1 and c_2 are weights, $L_t^{\mathrm{VF}}=(V_{\mathrm{new}}(s_t)-V\mathrm{old}(s_t))^2$ is the change in value function before and after the NN update, and S denotes the entropy. PPO can have continuous action spaces, so as its action set we used $A_{\mathrm{PPO}}\in[-1,1]\,\%$ corresponding to same range for our DQL implementation. In contrast to Q-learning with a replay buffer, PPO is an on-policy algorithm, i.e. it learns on-line via trial and error.

Implementation. We implemented[1] PPO [20] in Pytorch. For DQL we used its OpenAI implementation [9]. For single-muscle control DQL and PPO were both implemented as simple networks of one hidden layer with 256 neurons. For PPO with four muscles (PPO4), 3 hidden layers each with 250 neurons were used. We used ReLu activations and the Adam optimizer. For multibody biomechanical simulation, we used Artisynth [17], a framework written in Java on CPU. For training, the simulation runtime is the main computational bottleneck, with the network back-propagation taking negligible time. For speed-up, we used a CPU cluster of 100 concurrent Artisynth simulations, each running a separate simulation episode and communicating with a DRL agent over a custom TCP interface based on [1]. For simulation, an integration time-step of 100 ms was chosen for a stability and performance tradeoff. During training, at each simulation time step t (DRL *frame*), a simulation provides the respective RL agent with a state s_t as in (2) including the simulated position $\phi(t)$ and velocity $\phi'(t)$ of the humerus. The agent then calculated the respective reward $r(t)$ and, according to the current policy, executes an action, i.e. sends an update of muscles activations back to the simulation. This is repeated for a preset episode length (herein 10 s), or until the simulation "crashes" prematurely due to numerical failure, which is recorded as a high negative reward. Convergence was ascertained visually in the reward curves.

3 Experiments and Results

Herein we demonstrate experiments showing a single-axis control of the shoulder. The glenohumeral joint was thus modeled as a revolute joint allowing rotation

[1] https://github.com/CAiM-lab/PPO.

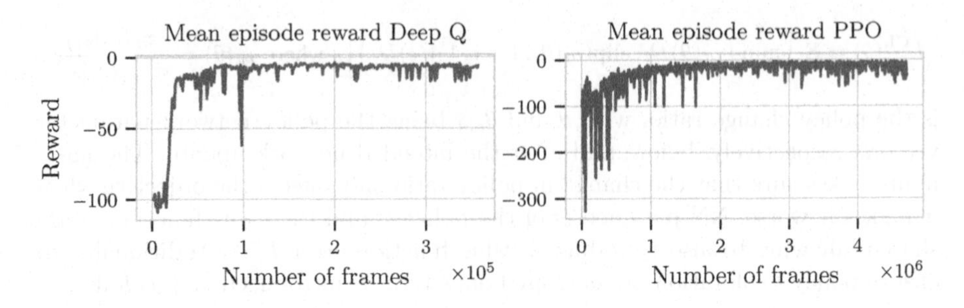

Fig. 2. Mean episode reward over last 10 episodes during training for DQL and PPO.

only around the z axis seen in Fig. 1-center. We conducted two sets of experiments: In a preliminary experiment with only one muscle (ssp), we compared the presented RL algorithms for our problem setting. A more sophisticated scenario with 4 muscles shows feasibility and scalability of the method to higher number of muscles. For training and testing, we used random trajectories. Using 5th-order polynomials [28] as $\hat{\phi}(t) = \sum_{i=0}^{5} a_i\, t^i$, we generate 5 s random sections. We set end-point velocity and acceleration constraints of zero, with end-point positions randomly sampled from [30, 90]° during training, and from [20, 100]° for testing. Using a different and wider range for the latter was to test for generalizability. By stacking such random sections while matching their end-point conditions, longer complex motions were constructed. With this, for each training episode a 10 s trajectory was generated on-the-fly, i.e. an episode being 100 frames given the integration time step of 0.1 s. Note that a trained RL agent can control an arbitrary trajectory length. For testing, a set of 100 random trajectories of each 20 s was generated once, and used to compare all presented RL agents; using root mean square error (RMSE) and mean average error (MAE) for tracking accuracy of desired trajectories.

Control of Single Muscle Activation. With this experiment we aim to comparatively study DQL and PPO. The exploration term in the reward (2) is irrelevant for DQL. In order to have a fair comparison, we thus removed this term from the reward for this experiment. Note that given a single muscle, the Lasso regularization of multiple activations in reward (2) also becomes unnecessary. Accordingly, this experiment employs a straight-forward reward function as the absolute tracking error, i.e. $r(t+1) = -\left|\phi(t+1) - \hat{\phi}(t+1)\right|$.

Episode reward is defined as $\sum_{t=0}^{T} r(t)$ were T is the episode length. Mean episode reward over last 10 episodes during training is depicted in Fig. 2 for DQL and PPO. Both algorithms are observed to show a similar learning behaviour overall, although PPO requires approximately 10 times more samples than DQL to converge, due to A_{PPO} being a continuous range. In Fig. 3 both models are shown while controlling the ssp activation in the forward simulation for a sample trajectory. It is seen that the discrete action space of DQL results in a sawtooth-like pattern in activations Ω_{ssp}, and hence a relatively oscillatory trajectory ϕ.

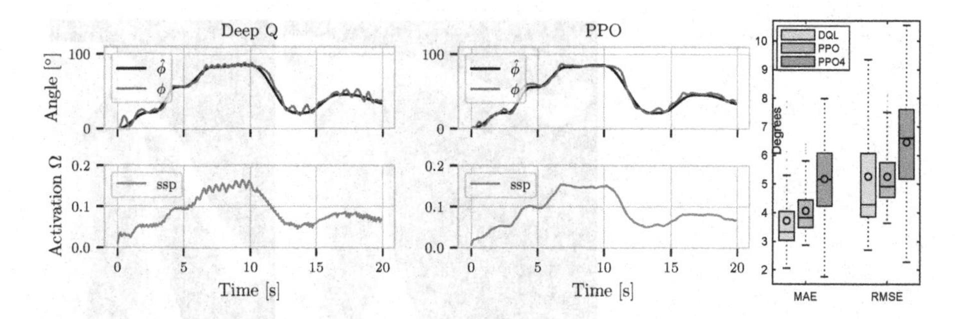

Fig. 3. A random target trajectory (black) and its tracking by RL using DQL (left) and PPO (center) with the muscle activation (bottom) and resulting angular upper arm motion (top). (right) Distribution of tracking errors for 100 random trajectories.

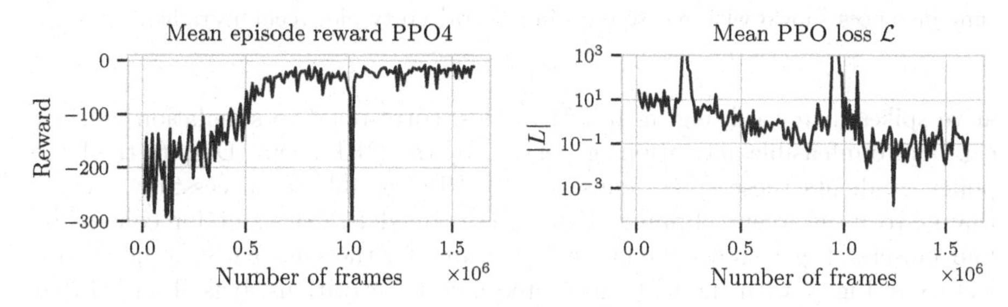

Fig. 4. Evolution of mean reward and loss of last 10 episodes during PPO4 training.

Note that for small abduction angles the moment from humerus mass is minimal, and due to this lack of a counter-acting torsional load, the control becomes difficult, i.e. any slight changes in activations Ω_{ssp} may lead to large angular changes, visible in Fig. 3 for small abduction angles.

Over 100 trajectories, DQL has an MAE of 3.70° and RMSE of 5.78°, while PPO has an MAE of 4.00° and RMSE of 5.36°. MAE and RMSE distributions of both methods over all tested trajectories can be seen in Fig. 3-right.

Muscle Control with Redundancy. In this scenario, all the four muscles relevant for abduction with redundancy are controlled at the same time. Given similar action space quantization of 21 steps, DQL for 4 muscles would require a 4^{21} dimensional discrete action space A_{DQL}, which is computationally unfeasible. Indeed, this is a major drawback of DQL preventing its extension to high dimensional input spaces. In contrast, a continuous action space for PPO is easily defined for the four muscles. Given the simulation with 4-muscles, a PPO agent (PPO4) was trained for 1.6 M frames, taking a total of 1 h including overheads for communication and Artisynth resets after crashes. Mean episode reward and loss (3) averaged over last 10 episodes are plotted in Fig. 4. Note that high gradients in policy updates due, e.g., to crashes, is a challenge Despite the 4 times higher action space dimension, a feasible learning curve is observed. Large neg-

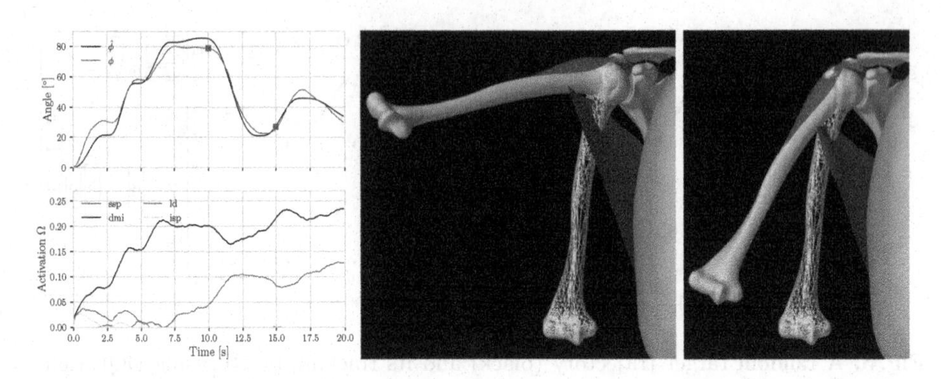

Fig. 5. Controlling four muscle activations for tracking the random abduction trajectory in Fig. 3-left, along with the activation patterns (left). Simulation frames at two time instances shown with red squares in the trajectory plot (center&right).

ative spikes in reward, e.g. near 1 M frames, correspond to simulation crashes, e.g. due to infeasible activations generated by the DRL agent. Despite the large policy gradients these spikes cause in (3), PPO is able to successfully recover thanks to its gradient clipping. Using the trained PPO4 agent for controlling four muscles, Fig. 5 shows the humerus tracking for the same earlier random trajectory in Fig. 3, with the PPO-generated muscle activations. It is observed that ssp and isp help to initiate the abduction motion – a well-known behaviour [23]. Beyond initial abduction, their activation however diminishes with the rest of the motion mainly carried out by dmi and ld, which have stronger moment arms. PPO control of 4-muscles during 100 randomly-generated trajectories results in an MAE of 5.15° and RMSE of 6.64°, with their distributions shown in Fig. 3-right. The slightly higher tracking error compared to single-muscle case is likely due to higher network capacity, training episodes, and thus time for convergence required for a higher dimensional action space.

We further tested an in-vivo trajectory from the public motion-tracking dataset of [5]. We used a *combing* motion of 17.5 s, involving lifting up the arm twice and combing while up. Using the angle between the humerus and vertical axis, the 3D tracked motion was converted to an abduction angle as our tracking target (varying between 20 and 100°). Applying our earlier-trained PPO4 agent on this shows good tracking visually, with an RMSE and MAE of 7.67° and 6.57°. These results being comparable with the earlier ones show that our method generalizes well to in-vivo trajectories even with synthetic training.

4 Conclusions

We have studied two DRL approaches demonstrating the successful application for single-axis control of a functional biomechanical model of the human shoulder. PPO was implemented in a way that allows for multiple environments to run simultaneously using network based sockets. This is indispensable for scalability to higher dimensions within reasonable computational time-frames. Inference of our NN-based DRL agents are near real-time, enabling fast control of complex functional simulations. Any constraints that make tracking suboptimal or simulation infeasible are implicitly learned with DRL, as a remedy to occasional simulation crashes occurring with conventional analytical controllers.

A main bottleneck for scalability to sophisticated models is the limitation with action spaces. In contrast to the discrete action space of DQL exploding exponentially with the curse of dimensionality, it is shown herein that the continuous action space and corresponding policy optimization of PPO enables its extension to multiple muscles with redundancy. Given the generalizable form of the learning scheme and the reward function with the proposed approach, extensions to more muscles and additional degrees-of-freedom is straightforward. This opens up the potential for full control of the shoulder and other musculoskeletal structures. This also enables neuroplasticity studies after corrective surgeries such as muscle transfers: After major orthopedic interventions, the patients may not easily adjust to postop configurations, therewith recovery expectancy and rehabilitation time-frames varying widely. Networks trained on preop settings and tested on simulated postop scenarios could provide insight into operative choices, e.g. for faster rehabilitation and improved outcomes.

References

1. Abdi, A.H., Saha, P., Srungarapu, V.P., Fels, S.: Muscle excitation estimation in biomechanical simulation using NAF reinforcement learning. In: Computational Biomechanics for Medicine, pp. 133–141 (2020)
2. Artstein, Z.: Discrete and continuous bang-bang and facial spaces or: look for the extreme points. Siam Rev. **22**(2), 172–185 (1980)
3. Bertsekas, D.P.: Dynamic Programming and Optimal Control, vol. 2. Athena Scientific, Belmont (2012)
4. Blemker, S.S., Pinsky, P.M., Delp, S.L.: A 3D model of muscle reveals the causes of nonuniform strains in the biceps brachii. J. Biomech. **38**(4), 657–665 (2005)
5. Bolsterlee, B., Veeger, H.E.J., van der Helm, F.C.T.: Modelling clavicular and scapular kinematics: from measurement to simulation. Med. Biol. Eng. Comput. **52**(3), 283–291 (2013). https://doi.org/10.1007/s11517-013-1065-2
6. Brown, J.M.M., Wickham, J.B., McAndrew, D.J., Huang, X.F.: Muscles within muscles: coordination of 19 muscle segments within three shoulder muscles during isometric motor tasks. J. Electromyogr. Kinesiol. **17**(1), 57–73 (2007)
7. Contemori, S., Panichi, R., Biscarini, A.: Effects of scapular retraction/protraction position and scapular elevation on shoulder girdle muscle activity during glenohumeral abduction. Hum. Mov. Sci. **64**, 55–66 (2019)
8. Craik, J.D., Mallina, R., Ramasamy, V., Little, N.J.: Human evolution and tears of the rotator cuff. Int. Orthop. **38**(3), 547–552 (2013). https://doi.org/10.1007/s00264-013-2204-y
9. Dhariwal, P., et al.: OpenAI baselines (2017). https://github.com/openai/baselines
10. Di Giacomo, G., Pouliart, N., Costantini, A., De Vita, A.: Atlas of Functional Shoulder Anatomy. Springer, New York (2008)

11. Faure, F., et al.: SOFA: a multi-model framework for interactive physical simulation. In: Payan, Y. (ed.) Soft Tissue Biomechanical Modeling for Computer Assisted Surgery. SMTEB, vol. 11, pp. 283–321. Springer, Heidelberg (2012). https://doi.org/10.1007/8415_2012_125
12. Gerber, C., Snedeker, J.G., Baumgartner, D., Viehöfer, A.F.: Supraspinatus tendon load during abduction is dependent on the size of the critical shoulder angle: a biomechanical analysis. J. Orthop. Res. **32**(7), 952–957 (2014)
13. Heess, N., et al.: Emergence of locomotion behaviours in rich environments. arXiv:1707.02286 (2017)
14. James, S., Johns, E.: 3D simulation for robot arm control with deep Q-learning. arXiv:1609.03759 (2016)
15. Kidziński, Ł., et al.: Learning to run challenge: synthesizing physiologically accurate motion using deep reinforcement learning. In: Escalera, S., Weimer, M. (eds.) The NIPS '17 Competition: Building Intelligent Systems. TSSCML, pp. 101–120. Springer, Cham (2018). https://doi.org/10.1007/978-3-319-94042-7_6
16. Lillicrap, T.P., et al.: Continuous control with deep reinforcement learning. arXiv:1509.02971 (2015)
17. Lloyd, J.E., Stavness, I., Fels, S.: ArtiSynth: a fast interactive biomechanical modeling toolkit combining multibody and finite element simulation. Soft Tissue Biomechanical Modeling for Computer Assisted Surgery. SMTEB, vol. 11, pp. 355–394. Springer, Heidelberg (2012). https://doi.org/10.1007/8415_2012_126
18. Mitsuhashi, N., Fujieda, K., Tamura, T., Kawamoto, S., Takagi, T., Okubo, K.: Bodyparts3D: 3D structure database for anatomical concepts. Nucleic Acids Res. **37**, D782–D785 (2008)
19. Mnih, V., Kavukcuoglu, K., Silver, D., Rusu, A.A., Veness, J., et al.: Human-level control through deep reinforcement learning. Nature **518**(7540), 529–533 (2015)
20. Paszke, A., et al.: PyTorch: an imperative style, high-performance deep learning library. In: Advances in Neural Information Processing Systems, vol. 32, pp. 8024–8035 (2019)
21. Pean, F., Goksel, O.: Surface-based modeling of muscles: functional simulation of the shoulder. Med. Eng. Phys. **82**, 1–12 (2020)
22. Péan, F., Tanner, C., Gerber, C., Fürnstahl, P., Goksel, O.: A comprehensive and volumetric musculoskeletal model for the dynamic simulation of the shoulder function. Comput. Methods Biomech. Biomed. Eng. **22**(7), 740–751 (2019)
23. Reed, D., Cathers, I., Halaki, M., Ginn, K.: Does supraspinatus initiate shoulder abduction? J. Electromyogr. Kinesiol. **23**(2), 425–429 (2013)
24. Schulman, J., Levine, S., Moritz, P., Jordan, M., Abbeel, P.: Trust region policy optimization. In: International Conference on Machine Learning (ICML), vol. PMLR 37, pp. 1889–1897 (2015)
25. Schulman, J., Moritz, P., Levine, S., Jordan, M.I., Abbeel, P.: High-dimensional continuous control using generalized advantage estimation. In: International Conference on Learning Representations (ICLR) (2016)
26. Schulman, J., Wolski, F., Dhariwal, P., Radford, A., Klimov, O.: Proximal policy optimization algorithms. arXiv:1707.06347 (2017)
27. Silver, D., Huang, A., Maddison, C.J., Guez, A., et al.: Mastering the game of go with deep neural networks and tree search. Nature **529**(7587), 484–489 (2016)
28. Spong, M.W., Hutchinson, S., Vidyasagar, M.: Robot Modeling and Control. Wiley, Hoboken (2020)
29. Stavness, I., Lloyd, J.E., Fels, S.: Automatic prediction of tongue muscle activations using a finite element model. J. Biomech. **45**(16), 2841–2848 (2012)

30. Streit, J.J., et al.: Pectoralis major tendon transfer for the treatment of scapular winging due to long thoracic nerve palsy. J. Shoulder Elbow Surg. **21**(5), 685–690 (2012)
31. Tsurumine, Y., Cui, Y., Uchibe, E., Matsubara, T.: Deep reinforcement learning with smooth policy update: application to robotic cloth manipulation. Robot. Autonom. Syst. **112**, 72–83 (2019)
32. Van Hasselt, H., Guez, A., Silver, D.: Deep reinforcement learning with double Q-learning. In: AAAI Conference on Artificial Intelligence, pp. 2094–2100 (2016)
33. Wickham, J., Pizzari, T., Stansfeld, K., Burnside, A., Watson, L.: Quantifying 'normal' shoulder muscle activity during abduction. J. Electromyogr. Kinesiol. **20**(2), 212–222 (2010)

Image Registration

MvMM-RegNet: A New Image Registration Framework Based on Multivariate Mixture Model and Neural Network Estimation

Xinzhe Luo and Xiahai Zhuang(✉)

School of Data Science, Fudan University, Shanghai, China
zxh@fudan.edu.cn

Abstract. Current deep-learning-based registration algorithms often exploit intensity-based similarity measures as the loss function, where dense correspondence between a pair of moving and fixed images is optimized through backpropagation during training. However, intensity-based metrics can be misleading when the assumption of intensity class correspondence is violated, especially in cross-modality or contrast-enhanced images. Moreover, existing learning-based registration methods are predominantly applicable to pairwise registration and are rarely extended to groupwise registration or simultaneous registration with multiple images. In this paper, we propose a new image registration framework based on multivariate mixture model (MvMM) and neural network estimation. A generative model consolidating both appearance and anatomical information is established to derive a novel loss function capable of implementing groupwise registration. We highlight the versatility of the proposed framework for various applications on multimodal cardiac images, including single-atlas-based segmentation (SAS) via pairwise registration and multi-atlas segmentation (MAS) unified by groupwise registration. We evaluated performance on two publicly available datasets, i.e. MM-WHS-2017 and MS-CMRSeg-2019. The results show that the proposed framework achieved an average Dice score of 0.871 ± 0.025 for whole-heart segmentation on MR images and 0.783 ± 0.082 for myocardium segmentation on LGE MR images (Code is available from https://zmiclab.github.io/projects.html).

1 Introduction

The purpose of image registration is to align images into a *common* coordinate space, where further medical image analysis can be conducted, including image-guided intervention, image fusion for treatment decision, and atlas-based segmentation [14]. In the last few decades, intensity-based registration

Electronic supplementary material The online version of this chapter (https://doi.org/10.1007/978-3-030-59716-0_15) contains supplementary material, which is available to authorized users.

A. L. Martel et al. (Eds.): MICCAI 2020, LNCS 12263, pp. 149–159, 2020.
https://doi.org/10.1007/978-3-030-59716-0_15

has received considerable scholarly attention. Commonly used similarity measures comprise intensity difference and correlation-based methods for intra-modality registration, and information-theoretic metrics for inter-modality registration [10,14,18,22,23].

Recently, deep learning (DL) techniques have formulated registration as a parameterized mapping function, which not only made registration in one shot possible but achieved state-of-the-art accuracies [4,8,11,24]. de Vos et al. [24] computed dense correspondence between two images by optimizing normalized cross-correlation between intensity pairs. While intensity-based similarity measures are widely used for intra-modality registration, there are circumstances when no robust metric, solely based on image appearance, can be applied. Hu et al. [11] therefore resorted to weak labels from corresponding anatomical structures and landmarks to predict voxel-level correspondence. Balakrishnan et al. [4] proposed leveraging both intensity- and segmentation-based metrics as loss functions for network optimization. More recently, Dalca et al. [8] developed a probabilistic generative model and derived a framework that could incorporate both of the intensity images and anatomical surfaces.

Meanwhile, in the literature several studies have suggested coupling registration with segmentation, in which image registration and tissue classification are performed simultaneously within the same model [2,6,20,25]. However, the search for the optimal solution of these methods usually entails computationally expensive iterations and may suffer from problems of parameter tuning and local optimum. A recent study attempted to leverage registration to perform Bayesian segmentation on brain MRI with an unsupervised deep learning framework [9]. Nevertheless, it can still be difficult to apply unsupervised intensity-based approaches to inter-modality registration or to datasets with poor imaging quality and obscure intensity class correspondence. Besides, previous DL-integrated registration methods have mainly focused on pairwise registration and are rarely extended to groupwise registration or simultaneous registration with multiple images.

In this paper, we consider the scenario in which multiple images from various modalities need to be co-registered simultaneously onto a *common* coordinate space, which is set onto a reference subject or can be implicitly assumed during groupwise registration. To this end, we propose a probabilistic image registration framework based on both multivariate mixture model (MvMM) and neural network estimation, referred to as MvMM-RegNet. The model incorporates both types of information from the appearance and anatomy associated with each image subject, and explicitly models the correlation between them. A neural network is then employed to estimate likelihood and achieve efficient optimization of registration parameters. Besides, the framework provides posterior estimation for MAS on novel test images.

The main contribution of this work is four-fold. First, we extend the conventional MvMM for image registration with multiple subjects. Second, a DL-integrated groupwise registration framework is proposed, with a novel loss function derived from the probabilistic graphical model. Third, by modelling the

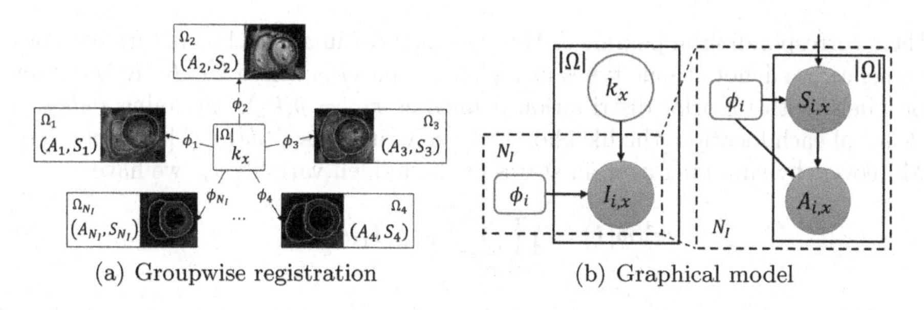

(a) Groupwise registration (b) Graphical model

Fig. 1. (a) Groupwise registration framework, (b) Graphical representation of the proposed generative model, where random variables are in circles, deterministic parameters are in boxes, observed variables are shaded and plates indicate replication.

relationship between appearance and anatomical information, our model outperforms previous ones in terms of segmentation accuracy on cardiac medical images. Finally, we investigate two applications of the proposed framework on cardiac image segmentation, i.e. SAS via pairwise registration and MAS unified by groupwise registration, and achieve state-of-the-art results on two publicly available datasets.

2 Methods

Groupwise registration aims to align every subject in a population to a *common* coordinate space Ω [5,7], referred to as the *common space* [25]. Assume we have N_I *moving* subjects $\boldsymbol{I} = \{I_i\}_{i=1}^{N_I}$, of which each is defined on spatial domain Ω_i. For each subject I_i, we can observe its appearance A_i from medical imaging as well as labels of anatomical structures S_i in various cases for image registration tasks. Thus, we can formulate $I_i = (A_i, S_i)$ as a pair of appearance and anatomical observations for each subject.

Associated with the *moving* subjects is a set of spatial transforms $\boldsymbol{\phi}$ that map points from the common space to counterparts in each subject space:

$$\boldsymbol{\phi} = \{\phi_i : y_i = \phi_i(x),\ i = 1, \ldots, N_I\}, \tag{1}$$

where $x \in \Omega$, $y_i \in \Omega_i$. The framework is demonstrated in Fig. 1(a).

2.1 Multivariate Mixture Model

The proposed method builds on a generative model of the appearance and anatomical information over a population of subjects. The likelihood function is computed as a similarity measure to drive the groupwise registration process.

For spatial coordinates in the common space, an exemplar atlas can be determined *a priori*, providing anatomical statistics of the population regardless of their corresponding appearances through medical imaging. For notational convenience, we denote tissue types using label values k_x, where $k \in K$, K is the set of labels, with its prior distribution defined as $\pi_{kx} = p(k_x)$. Assuming *independence* of each location, the likelihood can be written as $\mathcal{L}(\boldsymbol{\phi}|\boldsymbol{I}) = \prod_{x\in\Omega} p(\boldsymbol{I}_x|\boldsymbol{\phi})$. Moreover, by summing over all states of the hidden variable k_x, we have

$$\mathcal{L}(\boldsymbol{\phi}|\boldsymbol{I}) = \prod_{x\in\Omega}\sum_{k\in K} \pi_{kx}\, p(\boldsymbol{I}_x|k_x, \boldsymbol{\phi}). \tag{2}$$

Given the common-space anatomical structures, the multivariate mixture model assumes *conditional independence* of the moving subjects, namely

$$p(\boldsymbol{I}_x|k_x, \boldsymbol{\phi}) = \prod_{i=1}^{N_I} p(I_{i,x}|k_x, \phi_i) = \prod_{i=1}^{N_I} p(I_{i,y_i}|k_x), \tag{3}$$

where I_{i,y_i} denotes a patch of observations centred at $y_i = \phi_i(x)$. Given anatomical structures of each subject, one can further assume its appearance is *conditional independent* of the groupwise anatomy, i.e. $p(A_{i,y_i}|S_{i,y_i}, k_x) = p(A_{i,y_i}|S_{i,y_i})$. Hence, we can further factorize the conditional probability into

$$p(I_{i,y_i}|k_x) = p(A_{i,y_i}|S_{i,y_i})\, p(S_{i,y_i}|k_x). \tag{4}$$

Accordingly, the log-likelihood is given by

$$l(\boldsymbol{\phi}|\boldsymbol{I}) = \sum_{x\in\Omega} \log \left\{ \sum_{k\in K} \pi_{kx} \prod_{i=1}^{N_I} p(A_{i,y_i}|S_{i,y_i})\, p(S_{i,y_i}|k_x) \right\}. \tag{5}$$

In practice, we optimize the negative log-likelihood as a dissimilarity measure to obtain the desired spatial transforms $\hat{\boldsymbol{\phi}}$. The graphical representation of the proposed model is shown in Fig. 1(b).

2.2 The Conditional Parameterization

In this section, we specify in detail the conditional probability distributions (CPDs) for a joint distribution that factorizes according to the Bayesian network structure represented in Fig. 1(b).

Spatial Prior. One way to define the common-space spatial prior is to average over a cohort of subjects [2], and the resulting probabilistic atlas is used as a reference. To avoid bias from a fixed reference and consider the population as a whole, we simply adopt a flat prior over the common space, i.e. $\pi_{kx} = c_k$, $\forall x \in \Omega$ satisfying $\sum_{k\in K} c_k = 1$, where c_k is the weight to balance each tissue class.

Label Consistency. Spatial alignment of a group of subjects can be measured by their label consistency, defined as the joint distribution of the anatomical information $p(\boldsymbol{S}_{i,y_i}, k_x)$, where $\boldsymbol{S}_{i,y_i} = \{S_{i,y_i}\}_{i=1}^{N_I}$. Each CPD $p(S_{i,y_i}|k_x)$ gives the likelihood of the anatomical structure around a subject location being labelled as k_x, conditioned on the transform that maps from the common space to each subject space. We model it efficiently by a local Gaussian weighting:

$$p(S_{i,y_i}|k_x) = \sum_{z \in \mathcal{N}_{y_i}} w_z \cdot \delta(S_i(z) = k_x), \tag{6}$$

where $\delta(\cdot)$ is the Kronecker delta function, $\mathcal{N}_{y_i}$ defines a neighbourhood around y_i of radius r_s and w_z specifies the weight for each voxel within the neighbourhood. This formulation is equivalent to applying Gaussian filtering using an isotropic standard deviation σ_s to the segmentation mask [11], where we set $r_s = 3\,\sigma_s$.

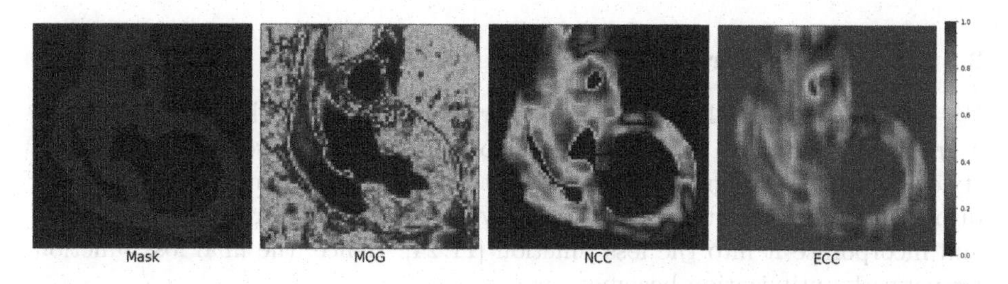

Fig. 2. Visualization of different appearance models computed from a coronal view of a whole heart MR image subject at background areas, where "Mask", "MOG", "NCC" and "ECC" denote appearance model using ROI mask, mixture of Gaussians, normalized cross correlation and entropy cross correlation, respectively. For comparison, values are normalized to intervals between 0 and 1.

Appearance Model. Finally, we seek to specify the term $p(A_{i,y_i}|S_{i,y_i})$. A common approach adopted by many tissue classification algorithms [2,9,13,16,25] is to model this CPD as a mixture of Gaussians (MOG), where intensities of the same tissue type should be clustered and voxel locations are assumed independent. Nevertheless, we hypothesize that using such an appearance model can mislead the image registration when the assumption of intensity class correspondence is violated, due to poor imaging quality, particularly in cross-modality or contrast enhanced images [19]. Let $\nabla(\cdot)$ and $\|\cdot\|$ be the voxel-wise gradient and Euclidean-norm operators, respectively. A vanilla means is to use a mask around the ROI boundaries:

$$p(A_{i,y_i}|S_{i,y_i}) = \begin{cases} 1 & \text{if } \exists\ z \in \mathcal{N}_{y_i} \text{ s.t. } \nabla\|S_i(z)\| > 0 \\ 0 & \text{otherwise,} \end{cases} \tag{7}$$

which ignores the appearance information. However, we argue that a reasonable CPD design should reflect fidelities of medical imaging and serve as a voxel-wise weighting factor for likelihood estimation. Thus, we formalize a CPD that 1) is defined with individual subjects, 2) is zero on voxels distant to the ROIs, 3) has increasing values at regions where appearance and anatomy have consistent rate of change. Therefore, we speculate that voxels with concordant gradient norms between appearance and anatomy are more contributory to determining the spatial correspondence. Based on these principles, one can estimate the CPD as a Gibbs distribution computed from an energy function or negative similarity measure between gradient-norm maps of appearance and anatomy, i.e.

$$p(A_{i,y_i}|S_{i,y_i}) = \frac{1}{Z}\exp\{-E(\|\nabla A_{i,y_i}\|, \|\nabla S_{i,y_i}\|)\}, \tag{8}$$

where Z is the normalization factor and $E(\cdot)$ can be the negative normalized cross-correlation (NCC) [3] or negative entropy correlation coefficient (ECC) [17]. Figure 2 visualises the different appearance models.

2.3 Neural Network Estimation

We formulate a neural network $g_{\boldsymbol{\theta}}(\cdot)$ parameterized by $\boldsymbol{\theta}$ that takes as input a group of N_I images to predict the deformation fields, based on a 3D UNet-style architecture designed for image registration [11]. To discourage non-smooth displacement, we resort to bending energy as a deformation regularization term and incorporate it into the loss function [11,24]. Hence, the final loss function for network optimization becomes

$$Loss(\boldsymbol{\theta}; \boldsymbol{I}) = -l(\boldsymbol{\phi}_{\boldsymbol{\theta}}|\boldsymbol{I}) + \lambda \cdot R(\boldsymbol{\phi}_{\boldsymbol{\theta}}), \tag{9}$$

where $R(\cdot)$ denotes the deformation regularization term and λ is a regularization coefficient.

2.4 Applications

In this section, we present two applications from the proposed MvMM-RegNet framework, which are validated in our experiments.

Pairwise MvMM-RegNet for SAS. Pairwise registration can be considered as a specialization of groupwise registration where the number of subjects equals two and one of the spatial coordinate transforms is the identity mapping. We will demonstrate the registration capacity of our model by performing pairwise registration on a real clinical dataset, referred to as pMvMM-RegNet.

Groupwise MvMM-RegNet for MAS. During multi-atlas segmentation (MAS), multiple expert-annotated images with segmented labels, called atlases, are co-registered to a target space, where the warped atlas labels are combined by label fusion [12]. Delightfully, our model provides a unified framework for this procedure through groupwise registration, denoted as gMvMM-RegNet. By setting the *common space* onto the target as the *reference space*, we can derive the following segmentation formula:

$$\hat{S}_T(x) = \underset{k \in K}{\operatorname{argmax}}\, p(k_x | \boldsymbol{I}_x, \boldsymbol{\phi}) = \underset{k \in K}{\operatorname{argmax}} \left\{ \pi_{kx} \prod_{i=1}^{N_I} p(I_{i,x} | k_x, \phi_i) \right\}. \tag{10}$$

In practice, the MAS result with $N_I \times t$ atlases can be generated from t times of groupwise registration over N_I subjects followed by label fusion using Eq. (10).

3 Experiments and Results

In this section, we investigate two applications of the proposed framework described in Sect. 2.4. In both of the two experiments, the neural networks were trained on a NVIDIA® RTX™ 2080 Ti GPU with the spatial transformer module adapted from open-source code in VoxelMorph [4], implemented in TensorFlow [1]. The Adam optimizer was adopted [15], with a cyclical learning rate bouncing between 1e−5 and 1e−4 to accelerate convergence and avoid shallow local optima [21].

Table 1. Average substruture Dice and Hausdorff distance (HD) of MR-to-MR and CT-to-MR inter-subject registration, with * indicating statistically significant improvement given by a Wilcoxon signed-rank test ($p < 0.001$).

Methods	MR-to-MR		CT-to-MR	
	Dice	HD (mm)	Dice	HD (mm)
WeaklyReg	0.834 ± 0.031	17.45 ± 2.482	0.842 ± 0.033	17.99 ± 2.681
Baseline-MOG	0.832 ± 0.027	19.65 ± 2.792	0.851 ± 0.028*	19.03 ± 2.564
Baseline-Mask	0.840 ± 0.028*	16.91 ± 2.374*	0.851 ± 0.032	17.51 ± 2.687*
Baseline-ECC	0.844 ± 0.026*	16.69 ± 2.355*	0.850 ± 0.032	17.70 ± 2.659
Baseline-NCC	0.847 ± 0.028*	16.83 ± 2.422	0.850 ± 0.032	17.78 ± 2.721
MVF-MvMM	Dice = 0.871 ± 0.025		HD (mm) = 17.21 ± 4.408	

3.1 pMvMM-RegNet for SAS on Whole Heart MRI

Materials and Baseline. This experiment was performed on the MM-WHS challenge dataset, which provides 120 multi-modality whole-heart images from multiple sites, including 60 cardiac CT and 60 cardiac MRI [26,27], of which 20 subjects from each of the modalities were selected as training data. Intra- (MR-to-MR) and inter-modality (CT-to-MR) but inter-subject registration tasks were

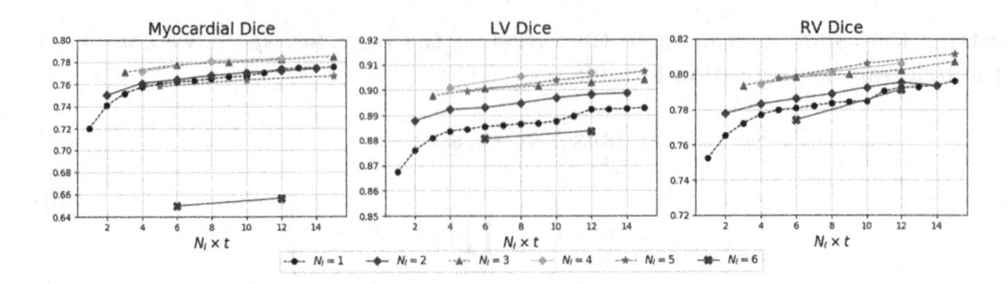

Fig. 3. Dice scores of MAS results using $N_I \times t$ atlases, where N_I denotes the number of subjects used in each groupwise registration and t counts the number of groupwise registrations performed before label fusion.

explored on this dataset, resulting in 800 propagated labels in total for 40 test MR subjects.

An optimal weighting of bending energy could lead to a low registration error, when maintaining the global smoothness of the deformations. To be balanced, we set $\lambda = 0.5$ as the default regularization strategy[1]. We analysed different variants of the appearance model described in Sect. 2.2, i.e. "MOG", "Mask", "NCC" and "ECC", and compared with a reimplementation of [11], known as "WeaklyReg", which exploited the Dice similarity metric for weakly-supervised registration. In addition, with the propagated labels obtained from pairwise registrations, we evaluated the performance of MAS by applying a simple majority vote to the results, denoted as "MVF-MvMM".

Results and Discussion. Table 1 presents the Dice statistics of both intra- and inter-modality registration tasks on the MM-WHS dataset. With increasingly plausible modelling of the relationship between appearance and anatomy, we have observed better registration accuracy especially for MR images, indicating efficacy of the proposed framework. Fusing labels by majority vote ("MVF-MvMM") can produce a better segmentation accuracy, reaching an average Dice score of 0.871 ± 0.025[2], comparable to the inter-observer variability of 0.878 ± 0.014 reported in [26].

3.2 gMvMM-RegNet for MAS on LGE CMR

Materials and Baseline. In this experiment, we explored MAS with the application of Eq. (10) on MS-CMRSeg challenge dataset [25]. The dataset consists of 45 patients scanned using three CMR sequences, i.e. the LGE, T2 and bFFSP, from which 20 patients were chosen in random for training, 5 for validation and 20 for testing. We implemented inter-subject and inter-modality groupwise registration and evaluated the MAS results on LGE CMR images.

[1] See Fig. 1 in the supplementary material for an empirical result.

[2] See Fig. 2 in the supplementary material for evaluation statistics on all cardiac substructure.

A 2D version of the network architecture described in Sect. 2.3 was devised to jointly predict the deformation fields for N_I atlases by optimizing Eq. (9). The MAS result was generated by t times of groupwise registration over N_I randomly sampled subjects followed by label fusion using Eq. (10).

Results and Discussion. The comparison between SAS and MAS highlights that more accurate and realistic segmentation is generated by groupwise registration than pairwise registration, especially for apical and basal slices[3]. Figure 3 further reports the mean Dice scores for each cardiac substructure obtained from MAS using t times of groupwise registration with N_I subjects. With a fixed total number of atlases, label fusion on 2D slices resulting from groupwise registration outperforms those from conventional pairwise registration, reaching the average myocardium Dice score of 0.783 ± 0.082. However, we also observe decline in accuracy when having a large number of subjects ($N_I \geq 5$) to be groupwise registered. This discrepancy could be attributed to the lack of network parameters compromising the predicted deformations.

4 Conclusion

In this work, we propose a probabilistic image registration framework based on multivariate mixture model and neural network estimation, coupling groupwise registration and multi-atlas segmentation in a unified fashion. We have evaluated two applications of the proposed model, i.e. SAS via pairwise registration and MAS unified by groupwise registration, on two publicly available cardiac image datasets and compared with state-of-the-art methods. The proposed appearance model along with MvMM has shown its efficacy in realizing registration on cardiac medical images characterizing inferior intensity class correspondence. Our method has also proved its superiority over conventional pairwise registration algorithms in terms of segmentation accuracy, highlighting the advantage of groupwise registration as a subroutine to MAS.

Acknowledgement. This work was supported by the National Natural Science Foundation of China (grant no. 61971142).

References

1. Abadi, M., et al.: Tensorflow: large-scale machine learning on heterogeneous distributed systems. ArXiv arXiv:1603.04467 (2015)
2. Ashburner, J., Friston, K.J.: Unified segmentation. NeuroImage **26**, 839–851 (2005)
3. Avants, B.B., Epstein, C.L., Grossman, M., Gee, J.C.: Symmetric diffeomorphic image registration with cross-correlation: evaluating automated labeling of elderly and neurodegenerative brain. Med. Image Anal. **12**(1), 26–41 (2008)

[3] See Fig. 3 in the supplementary material for visualization of the segmentation results.

4. Balakrishnan, G., Zhao, A., Sabuncu, M.R., Guttag, J.V., Dalca, A.V.: VoxelMorph: a learning framework for deformable medical image registration. IEEE Trans. Med. Imaging **38**, 1788–1800 (2019)
5. Balci, S.K., Golland, P., Shenton, M.E., Wells, W.M.: Free-form b-spline deformation model for groupwise registration. In: Medical Image Computing and Computer-assisted Intervention: MICCAI. International Conference on Medical Image Computing and Computer-Assisted Intervention 10 WS, pp. 23–30 (2007)
6. Bhatia, K.K., et al.: Groupwise combined segmentation and registration for atlas construction. In: Ayache, N., Ourselin, S., Maeder, A. (eds.) MICCAI 2007. LNCS, vol. 4791, pp. 532–540. Springer, Heidelberg (2007). https://doi.org/10.1007/978-3-540-75757-3_65
7. Bhatia, K.K., Hajnal, J., Hammers, A., Rueckert, D.: Similarity metrics for groupwise non-rigid registration. In: Ayache, N., Ourselin, S., Maeder, A. (eds.) MICCAI 2007. LNCS, vol. 4792, pp. 544–552. Springer, Heidelberg (2007). https://doi.org/10.1007/978-3-540-75759-7_66
8. Dalca, A.V., Balakrishnan, G., Guttag, J.V., Sabuncu, M.R.: Unsupervised learning of probabilistic diffeomorphic registration for images and surfaces. Med. Image Anal. **57**, 226–236 (2019)
9. Dalca, A.V., Yu, E., Golland, P., Fischl, B., Sabuncu, M.R., Eugenio Iglesias, J.: Unsupervised deep learning for bayesian brain MRI segmentation. In: Shen, D., et al. (eds.) MICCAI 2019. LNCS, vol. 11766, pp. 356–365. Springer, Cham (2019). https://doi.org/10.1007/978-3-030-32248-9_40
10. Hill, D.L.G., Batchelor, P.G., Holden, M., Hawkes, D.J.: Medical image registration. Phys. Med. Biol. **46**(3), R1–45 (2001)
11. Hu, Y., et al.: Weakly-supervised convolutional neural networks for multimodal image registration. Med. Image Anal. **49**, 1–13 (2018)
12. Iglesias, J.E., Sabuncu, M.R.: Multi-atlas segmentation of biomedical images: a survey. Med. Image Anal. **24**(1), 205–219 (2014)
13. Iglesias, J.E., Sabuncu, M.R., Leemput, K.V.: A unified framework for cross-modality multi-atlas segmentation of brain MRI. Med. Image Anal. **17**, 1181–1191 (2013)
14. Khalil, A., Ng, S.C., Liew, Y.M., Lai, K.W.: An overview on image registration techniques for cardiac diagnosis and treatment. Cardiol. Res. Pract. (2018)
15. Kingma, D.P., Ba, J.: Adam: a method for stochastic optimization. CoRR abs/1412.6980 (2014)
16. Leemput, K.V., Maes, F., Vandermeulen, D., Suetens, P.: Automated model-based tissue classification of MR images of the brain. IEEE Trans. Med. Imaging **18**, 897–908 (1999)
17. Maes, F., Collignon, A., Vandermeulen, D., Marchal, G., Suetens, P.: Multimodality image registration by maximization of mutual information. IEEE Trans. Med. Imaging **16**, 187–198 (1997)
18. Mäkelä, T., et al.: A review of cardiac image registration methods. IEEE Trans. Med. Imaging **21**, 1011–1021 (2002)
19. Pluim, J.P.W., Maintz, J.B.A., Viergever, M.A.: Mutual-information-based registration of medical images: a survey. IEEE Trans. Med. Imaging **22**, 986–1004 (2003)
20. Pohl, K.M., Fisher, J.W., Grimson, W.E.L., Kikinis, R., Wells, W.M.: A bayesian model for joint segmentation and registration. NeuroImage **31**, 228–239 (2006)
21. Smith, L.N.: Cyclical learning rates for training neural networks. 2017 IEEE Winter Conference on Applications of Computer Vision (WACV), pp. 464–472 (2015)

22. Sotiras, A., Davatzikos, C., Paragios, N.: Deformable medical image registration: a survey. IEEE Trans. Med. Imaging **32**, 1153–1190 (2013)
23. Viergever, M.A., Maintz, J.B.A., Klein, S., Murphy, K., Staring, M., Pluim, J.P.W.: A survey of medical image registration - under review. Med. Image Anal. **33**, 140–144 (2016)
24. de Vos, B.D., Berendsen, F.F., Viergever, M.A., Sokooti, H., Staring, M., Išgum, I.: A deep learning framework for unsupervised affine and deformable image registration. Med. Image Anal. **52**, 128–143 (2018)
25. Zhuang, X.: Multivariate mixture model for myocardial segmentation combining multi-source images. IEEE Trans. Pattern Anal. Mach. Intell. **41**, 2933–2946 (2019)
26. Zhuang, X., et al.: Evaluation of algorithms for multi-modality whole heart segmentation: an open-access grand challenge. Med. Image Anal. **58**, 101537 (2019)
27. Zhuang, X., Shen, J.: Multi-scale patch and multi-modality atlases for whole heart segmentation of MRI. Med. Image Anal. **31**, 77–87 (2016)

Database Annotation with Few Examples: An Atlas-Based Framework Using Diffeomorphic Registration of 3D Trees

Pierre-Louis Antonsanti[1,2(✉)], Thomas Benseghir[1], Vincent Jugnon[1], and Joan Glaunès[2]

[1] GE Healthcare, 78530 Buc, France
{pierrelouis.antonsanti,thomas.benseghir,vincent.jugnon}@ge.com
[2] MAP5, Universite de Paris, 75006 Paris, France
alexis.glaunes@parisdescartes.fr

Abstract. Automatic annotation of anatomical structures can help simplify workflow during interventions in numerous clinical applications but usually involves a large amount of annotated data. The complexity of the labeling task, together with the lack of representative data, slows down the development of robust solutions. In this paper, we propose a solution requiring very few annotated cases to label 3D pelvic arterial trees of patients with benign prostatic hyperplasia. We take advantage of Large Deformation Diffeomorphic Metric Mapping (LDDMM) to perform registration based on meaningful deformations from which we build an atlas. Branch pairing is then computed from the atlas to new cases using optimal transport to ensure one-to-one correspondence during the labeling process. To tackle topological variations in the tree, which usually degrades the performance of atlas-based techniques, we propose a simple bottom-up label assignment adapted to the pelvic anatomy. The proposed method achieves 97.6% labeling precision with only 5 cases for training, while in comparison learning-based methods only reach 82.2% on such small training sets.

1 Introduction

The automatic annotation of tree-like structures has many clinical applications, from workflow simplification in cardiac disease diagnosis [1,5,7,26] to intervention planning in arterio-venous malformations [3,10,21,25] and lesion detection in pneumology [8,11,14]. This task is particularly important in the context of interventional radiology, where minimally invasive procedures are performed by navigating small tools inside the patient's arteries under X-ray guidance. For example, benign prostatic hyperplasia symptoms can be reduced by embolizing arteries feeding the prostate to reduce its size [19].

Electronic supplementary material The online version of this chapter (https://doi.org/10.1007/978-3-030-59716-0_16) contains supplementary material, which is available to authorized users.

A. L. Martel et al. (Eds.): MICCAI 2020, LNCS 12263, pp. 160–170, 2020.
https://doi.org/10.1007/978-3-030-59716-0_16

Identifying the correct arteries to treat during the procedure - along with their neighbours - is crucial for the safety of the patient and the effectiveness of the treatment. However, this task is very challenging because of the arterial tree complexity and its topological changes induced by frequent anatomical variations. Having a large representative annotated database is also a challenge, especially in medical imaging where the sensitivity of the data and the difficulty to annotate slow down the development of learning based techniques [13]. In this context, solutions to the tree labeling problem should ideally work from a limited number of annotated samples.

Learning-Based Labeling. With the prevalence of machine learning, many articles address the problem of automatic anatomical tree labeling by extracting features from the tree to feed a learning algorithm that predicts labels probabilities. Authors of [11] and [1] first proposed Gaussian mixtures models to learn from branch features (geometrical and topological) and predict the label probabilities. In [1], labels are assigned following clinical a priori defining a set of rules. In [14], k-Nearest-Neighbours predicts label probabilities that are used in a bottom-up assignment procedure, searching through all the existing branch relationships in the training data. Since numerous features leads to high dimensional space, such techniques often show poor generalization capacity.

To reduce dimensionality, boosting algorithms and tree classifiers are used in [12,16,21] to select the most discriminant features. Assignment procedure is enriched by topological rules to improve coherence along the tree, instead of considering branches independently. Label assignment was further refined in [10] and [25] by adding a Markovian property - each branch label depending on its direct neighbours - with Markov chains parameters learned during training. Such assignment strongly depends on the database size and the task complexity in term of anatomical variability.

While previous articles had access to limited size databases (around 50 cases), authors of [26] trained a recurrent neural network preserving the topology of the coronary tree on a database composed of 436 annotated trees. The labels predictions rely on a multi-layer perceptron and a bidirectional tree-structural long short-term memory network. This interesting approach using deep learning for vessel classification is less common because it requires a lot more training data to be able to capture the anatomical variability.

Atlas-Based Labeling. Contrary to learning-based techniques, atlas-based ones can offer robust annotation even with few annotated cases [26]. An atlas is defined as a reference model that can be built from prior knowledge [3,5] or from an available annotated database [4,7,8]. Most of the atlas-based methods follow a four steps framework: the choice of the atlas, the registration onto the target, the estimation of the label probabilities and finally the assignment. In [7], authors focused on the annotation procedure, relying on previously established atlas and a manual registration. Labeling is done through a branch-and-bound algorithm extending the best partial labeling with respect to a function designed to compare observations to the atlas.

Later in [4], label probabilities are computed at the level of branches by a voting procedure, each point along a branch voting for a label. This solution can adapt to missing branches but still does not guarantee anatomical consistency of the labeling along the tree and is not robust to topological variation with respect to the atlas. To take this variability into account [3] proposes to create one atlas per known topology in the anatomy of interest, an interesting approach if the number of anatomical variants is limited. Similarly [5,8] use multi-atlas approaches, taking advantage of a distance to the atlas that quantifies topological differences. In [5], the reference case is selected as the best example in a training set following a leave-one-in cross-validation design.

While a lot of advanced methods have been proposed for medical structure registration [23], few efforts have been made to use advanced deformation models in the context of vascular tree labeling [15]. In our work, we propose an atlas-based algorithm illustrated in Fig. 1 relying on state of the art LDDMM to build the atlas and estimate realistic deformations. It is combined with an Optimal Transport based matching, to compute a relevant assignment between atlas and target branches. In the end, to handle the topological changes between the atlas and the target, a bottom-up label assignment is performed to achieve optimal results for our pelvic vasculature labeling problem.

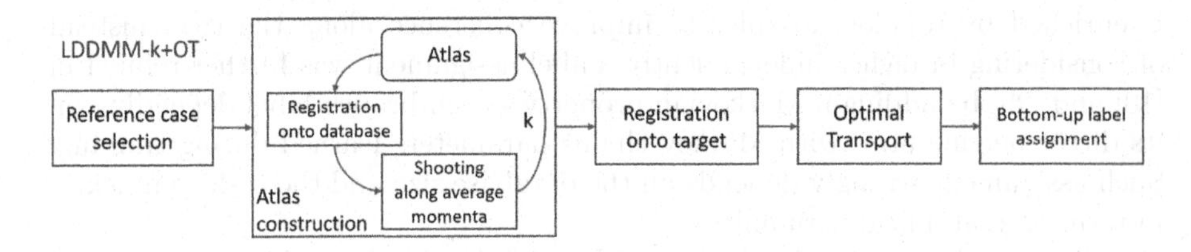

Fig. 1. The proposed atlas-based vascular tree annotation pipeline.

2 Method

Let a labeled vascular tree $T = (\{(B_\alpha, l_\alpha)\}_\alpha, M)$ be a set of branches B_α labeled $l_\alpha \in \mathbb{N}$ with connections to other branches stored in an adjacency matrix M. Two labeled branches (B_α, l_α) and (B_β, l_β) are connected if $M_{\alpha,\beta} = 1$. Each branch B_α is a polygonal curve composed of ordered points $B_\alpha[k] \in \mathbb{R}^3$ representing a vessel centerline. We also denote $\{q_i\}_i$ the unordered set of all points $B_\alpha[k]$ of the tree for all indices α and k.

The Large Deformation Diffeomorphic Metric Mapping Framework. We compute the deformation φ of a *source* shape S (in our case the atlas, a centerlines tree) onto a *target* shape T using LDDMM. This state of the art framework, detailed in [27] (Chaps. 8 to 11), allows to analyze differences between shapes via the estimation of invertible deformations of the ambient space that act upon them. In practice, the diffeomorphism φ is estimated by minimizing a cost function $\mathcal{J}(\varphi) = \mathcal{E}(\varphi) + \mathcal{A}(\varphi(S), T)$ where $\mathcal{E}$ is the deformation cost, and $\mathcal{A}$ is a data attachment term that penalizes mismatch between the deformed source $\varphi(S)$ and the target T. We chose the Normal Cycles model proposed in [22] as data attachment since it has shown good results at areas of high curvature and singular points as bifurcations or curves endpoints.

The deformation map $\varphi : \mathbb{R}^3 \to \mathbb{R}^3$ is defined as the solution at time $t = 1$ of a flow equation $\partial_t \phi(t, x) = v(t, \phi(t, x))$ with initial condition $\phi(0, x) = x$, where $v(t, \cdot)$ are time varying vector fields assumed to belong to a Hilbert space V, ensuring regularity properties. As shown in [17], in a discrete setting, denoting x_i, $i \in [1, ..., n_S]$ the discretization points of the source shape, one may derive optimality equations that must be satisfied by the trajectories $q_i(t) = \phi(t, x_i)$ when considering deformations that minimize the cost function. These equations take the following form:

$$\begin{cases} \dot{p}_i(t) = -\frac{1}{2} \nabla_{q_i(t)} \left(\sum_{j=1}^{n_S} \sum_{l=1}^{n_S} \langle p_j(t) \ , \ K_V(q_j(t), q_l(t)) p_l(t) \rangle \right) \\ \dot{q}_i(t) = \sum_{j=1}^{n_S} K_V(q_i(t), q_j(t)) p_j(t) \end{cases} \quad (1)$$

where $p_i(t)$ are auxiliary dual variables called momenta. They correspond to geodesic equations with respect to a specific Riemannian metric, written in Hamiltonian form. The deformations are then fully parametrized by the set of initial momenta $(p_i(0))_i \in \mathbb{R}^{3n_S}$.

To model deformations, we define the kernel to be a sum of Gaussian kernels $K_V(x, y) = \sum_s \exp\left(-\|x - y\|^2 \ / \ (\sigma_0/s)^2\right)$, where $s \in [1, 4, 8, 16]$ and σ_0 is half the size of our vascular trees bounding box. This multi-scale approach was introduced by [20]. Normal Cycles also require to choose specific kernels; we take a sum of constant and linear kernels for the spherical part and a Gaussian kernel for the spatial part, at two scales σ_0 and $\sigma_0/4$, using the output $(p_i(0))_i$ of the optimization process at scale σ_0 as initialization of the optimization at scale $\sigma_0/4$.

Building the Atlas. In the context of vascular tree labeling, the tree topology can be highly variable (in our database we have one topology every two cases). Trying to build atlases that take into account these variations is unrealistic. Yet we still want the automatic annotation to be robust to the choice of the reference case.

The LDDMM framework allows to derive methods for computing such atlases. More precisely, the optimal deformations generated by LDDMM are fully parameterized by the initial momenta, and their representation in the Euclidean space $\mathbb{R}^{3n_S}$ allows to perform classical linear statistics. Following [24], in order to build the atlas we select one available annotated case and compute its deformations onto a set of N targets (the selected case included). These registrations provide a collection of initial momenta $\{(p_i^k(0))_i, k \in [1, ..., N]\}$. The reference case is selected using a leave-one-in method as in [5] with respect to the labeling procedure. The atlas is then obtained by shooting via geodesic equations to deform the selected case along the average of the initial momenta $\overline{p_i(0)} = (1/N) \sum_k p_i^k(0)$. This procedure does not require the targets to be annotated, and seems suited to build an atlas of the whole database.

In addition it can be iterated by replacing the selected case with the deformed one. We refer to **LDDMM-k** for the k-th iteration of the atlas construction. Consequently LDDMM-0 refers to the case of using the reference case directly as an atlas. As illustrated in Fig. 2 the atlas converges along the iterations to an average position representative of the set of targets. Through this spatial normalization, we limit the sensitivity to the choice of the initial case as atlas. To illustrate the impact of LDDMM-k on the labeling, we chose a simple assignment procedure described in [4]: each point of the target tree votes for the label of its closest point in the labeled atlas. Then, we compute the vote per branch B_β^T in the target: $\pi(B_\beta^T, l) = \text{Card}(\{q_i^T \in B_\beta^T,\ \hat{l}(q_i^T) = l\})\ /\ \text{Card}(B_\beta^T)$. This label probability estimation does not guarantee anatomical consistency and deeply relies on the registration. Additionally, each point vote is independent from the others, allowing to characterize the quality of the registration from a labeling point of view.

Optimal Transport for a Better Assignment. During the LDDMM atlas construction and registration, each tree is seen as one shape. Consequently, there is no assumption over branch matching and topological changes. In order to provide a relevant label assignment that takes the mutual information into account we propose to use Optimal Transport. It is convenient to compute the optimal one-to-one assignment between branches of the deformed source and the target with respect to a given distance.

Based on the work of [9], each branch is re-sampled with 20 points and the distance matrix D between each branch $B_\alpha^{\varphi(S)}$ and B_β^T is given by: $D_{\alpha,\beta} = \|B_\alpha^{\varphi(S)} - B_\beta^T\|_{\mathbb{R}^{3d}}$. We tried different numbers of points d per branch ranging from 20 to 500 with no significant impact over the matching results. Considering that our problem is of limited size (17 branches per tree), a simple Kuhn-Munkres algorithm (also called Hungarian algorithm) was used to compute the assignment solution. It consists in finding minimum weight matching in bipartite graphs by minimizing the function $\sum_{\alpha,\beta} D_{\alpha,\beta}.X_{\alpha,\beta}$ with $(X_{\alpha,\beta}) \in \{0,1\}^{17 \cdot 17}$ the output boolean matrix with 1 if the branches $B_\alpha^{\varphi(S)}$ is assigned to B_β^T, 0 otherwise. To be consistent with the alternative simple label probability estimation of our LDDMM-k pipeline, we similarly define here $\pi(B_\beta^T, l_\alpha^S) := X_{\alpha,\beta}$, although these "probabilities" are always 0 or 1 in this case. This assignment process is complementary to the LDDMM-k process since it focuses on assignment between branches while LDDMM-k focuses on the atlas construction and the registration. We will call this pipeline **LDDMM-k+OT**. We will also experiment the Optimal Transport assignment without any registration (i.e. taking $\varphi = id$), which we denote **OT**.

Label Assignment Post-processing. A first label assignment procedure directly takes the highest label probability for each branch: $\hat{l}(B_\alpha^T) = \arg\max_l(\pi(B_\alpha^T, l))$, π being the output of the OT procedure or the voting probabilities estimation. This *direct assignment* is not based on a priori knowledge and directly reflects the performance of the prediction methods.

In practice, the vessels are labeled by the expert accordingly to the anatomy or area they irrigate [2]. In the application to pelvic vascular tree, when two branches of different labels share a parent, this parent is called *"Common Artery"*. This is the only clinical a priori we introduce in the method. To limit the effect of the topological variations between the atlas and the target, we propose a *bottom-up assignment* procedure: first for all B_α^T leaf of T, $\hat{l}(B_\alpha^T) = \arg\max_l(\pi(B_\alpha^T, l))$ then recursively every parent branch B_α^T is assigned a label with the rule:

$$\hat{l}(B_\alpha^T) = \begin{cases} l & \text{if } \hat{l}(B_\beta^T) = l \quad \text{for every branch } B_\beta^T \text{ child of } B_\alpha^T \\ 0 & (i.e.\text{"Common Artery"}), \text{ otherwise.} \end{cases}$$

This recursive assignment procedure, although specific to this anatomy, is quite adaptable. In fact, in most of the structured tree-shaped anatomies (coronary [1,5], airway tree [14], pelvic [2]) the branches names are also conditioned by the leaves labels. When two arteries of different labels share a common parent, this

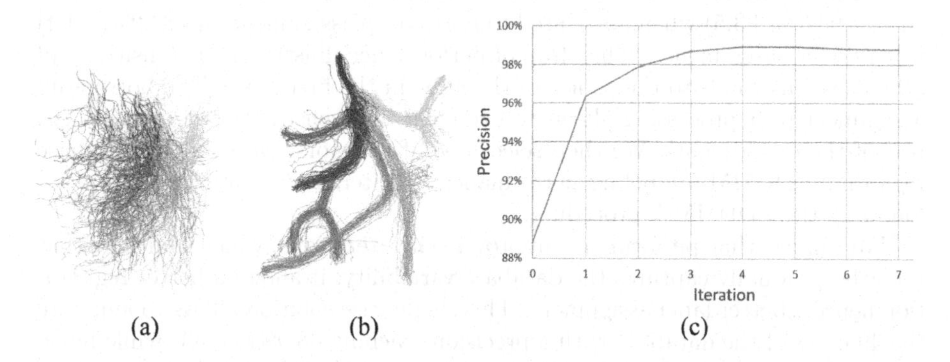

Fig. 2. Building atlases via LDDMM registration. The colors represent the ground truth labels. (a) The initial trees; (b) reference atlases at LDDMM-1 (c) Precision of one atlas over LDDMM-k iterations. (Color figure online)

parent is either unnamed (as in our application), or named by a convention provided by the experts. The latter situation corresponds to additional conditions (as in [14]) during the assignment.

3 Results

We conducted experiments on a dataset of 50 pelvic vascular trees corresponding to 43 different patients, some trees being the left and right vasculature of a single patient. The centerlines composing the vascular tree are constructed from 3D volumes (injected Cone Beam Computed Tomography). While the entire vascular tree is composed of up to 300 different branches, we manually extracted a simplified tree composed of the main arteries documented in the literature [2]. This allowed us to reduce the problem to the annotation of a 17-branches binary tree that corresponds to the typical size of trees found in the literature [5,25]. The selected arteries of interest are: the prostatic, the superior vesicle, the obturator (2), the pudendal, the inferior gluteals (2) and the superior gluteals (2). Unlabeled branches in the tree correspond to proximal common portions of the arterial tree that we label "common" arteries for a total of 7 labels. This simplified representation still captures the anatomical variability described in [2], as we found 28 different tree labels arrangements among the 50 cases with high variability of branch shapes and positions as is illustrated in Fig. 2. Registrations are computed with the library KeOps [6] allowing fast GPU computing and automatic differentiation through PyTorch [18] backends. The optimization of the functional $\mathcal{J}$ is performed using Limited-memory Broyden–Fletcher–Goldfarb–Shanno (L-BFGS) algorithm. In Fig. 2 (c) we plot the precision of the LDDMM-k at each iteration k.

A first experiment was carried out to illustrate the contributions of the atlas construction on the labeling of the database. We computed LDDMM-0 (pure registration) and LDDMM-1 (atlas construction) by registering each of the 50 available cases onto the others. It is illustrated in Fig. 2 (a, b). The average precision of one reference case used in LDDMM-0 to annotate the 49 other trees is 93.3%(±3.5) when associated to bottom-up assignment and 84.2%(±4.4) using direct assignment. This drop of performance illustrates the sensitivity of atlas-based methods to the choice of the atlas in the first place. The bottom-up assignment post processing allows to overcome this sensitivity: we use it in the rest of the experiments. We then select one of the worst cases in the database regarding LDDMM-0 labeling performance and iteratively build the new atlas following the LDDMM-k procedure.

We can see that performance improves with iterations, which indicates that the atlas gradually captures the database variability: it allows a better registration hence a better label assignment. This single-case solution allows to annotate the 49 cases of the database with a precision reaching 98.9%(±0.33) while being one of the worst with LDDMM-0. It must be pointed out that the atlas construction did not rely on any other annotated case than the one initially selected. We also computed LDDMM-1 using each tree as reference case and obtained an average score of 96.8%(±0.34), improving the average precision by 3.5%. In addition the standard deviation for LDDMM-1 confirms that iterating to build an atlas makes the LDDMM-k method robust to the initial choice of the atlas. In Fig. 2 (c) we observe that one iteration of the atlas construction is enough to greatly improve the labeling of the entire database, then the performance slowly

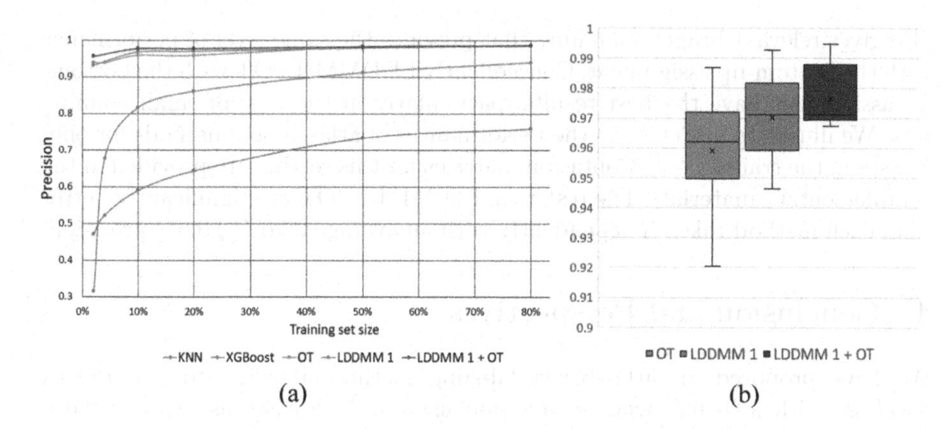

Fig. 3. (a) Comparison of the performance according to the training set size. (b) Box plot at training size 10% (5 cases).

increases until iteration 4. Therefore, and to limit the computational cost, we chose to use LDDMM-1 for the rest of the experiments.

To demonstrate that atlas based techniques described in Sect. 2 perform well compared to learning based ones in case of small size database, we implemented two classification algorithms working on branch features, KNN and XGboost, inspired from the work of [14] and [25]. To be close to the work of [14] and [16], each branch is represented by a vector composed of 28 *branch features* such as direction, length or geometrical characteristics of the centerline curves, and 13 *tree features* involving branch's relationships to the root and to its children to introduce topological information. These features are listed in the complementary material. Tuning the XGBoost parameters had no impact so we kept parameters provided in [25]. We performed 8-neigbours KNN in the space of the 10 most informative features selected by XGboost.

To compare all methods, precision has been evaluated using a cross-validation over the 50 cases with a training set and test set of varying size: from 2% (1 case) to 80% (40 cases) of the total dataset. While the notion of cross-validation is well defined for learning-based techniques, in the case of atlas-based methods, it follows the leave-one-in procedure described in Sect. 2 to select the best atlas among the training data. This case is then used to annotate the test set.

For each method we used the assignment technique giving the highest precision (direct assignment for training-based methods and bottom-up for atlas-based ones). Results are presented in Fig. 3. As expected, under 20% of training data (10 cases) precision of learning-based methods drastically drops. KNN is outperformed by other approaches, and XGBoost seems to asymptotically reach the atlas-based performances. On the other hand the atlas-based methods with bottom-up assignment perform with very little influence of the size of the training set.

Despite topological variations, the LDDMM-1 approach generates meaningful registrations showing good results when coupled with bottom-up assignment. OT also gives relevant branch matching that provides the same level of performance with the bottom-up assignment. Consequently LDDMM-1+OT with the bottom-up assignment have the best results, particularly in the case of small training sets. We illustrate in Fig. 3 (b) the performances of atlas-based methods for only 5 cases in the training set (Confusion matrices for this setting are provided in the complementary material). The results of LDDMM-1+OT are significantly better than each method taken independently with an average 97.6%(±0.97) precision.

4 Conclusion and Perspectives

We have proposed an atlas-based labeling method allowing to annotate a database with a 97.6% average precision using only 5 cases as training data. This level of precision isn't achieved by learning-based approaches even with 8 times more training data. Our method takes advantage of the LDDMM realistic deformations to build a meaningful atlas and register it onto the cases to annotate. Optimal Transport computes a global branch matching that, combined with a bottom-up label assignment, provides an anatomically consistent labeling. The bottom-up assignment procedure allows to tackle the anatomical variations that usually degrades atlas-based methods precision. This procedure may be specific to pelvic anatomy, however we believe it could be easily adapted to other ones [1,2,14]. In addition, the LDDMM framework allows to perform statistics over deformations that could be exploited to generate new realistic data or to detect anomalies. In further work we would like to extend our method to more complex cases such as missing labels or branches. In this perspective a promising lead would be to take topological changes into account as in [8] for both the deformations and the data attachment.

Acknowledgement. We would like to thank Arthur Rocha, engineer at Sao Paulo Hospital das Clinicas for his help and expertise on the annotation of pelvic vascular trees.

References

1. Akinyemi, A., Murphy, S., Poole, I., Roberts, C.: Automatic labelling of coronary arteries, pp. 1562–1566, August 2009
2. de Assis, A.M., et al.: Pelvic arterial anatomy relevant to prostatic artery embolisation and proposal for angiographic classification. Cardiovasc. Intervent. Radiol. **38**(4), 855–861 (2015). https://doi.org/10.1007/s00270-015-1114-3
3. Bogunović, H., Pozo, J.M., Cárdenes, R., Román, L.S., Frangi, A.F.: Anatomical labeling of the circle of willis using maximum a posteriori probability estimation. IEEE Trans. Med. Imaging **32**(9), 1587–1599 (2013)
4. Bülow, T., Lorenz, C., Wiemker, R., Honko, J.: Point based methods for automatic bronchial tree matching and labeling, vol. 6143 (2006)
5. Cao, Q., et al.: Automatic identification of coronary tree anatomy in coronary computed tomography angiography. Int. J. Cardiovasc. Imaging **33**(11), 1809–1819 (2017). https://doi.org/10.1007/s10554-017-1169-0

6. Charlier, B., Feydy, J., Glaunés, J.A., Collin, F.D., Durif, G.: Kernel operations on the GPU, with autodiff, without memory overflows (2020). https://www.kernel-operations.io/keops/index.html
7. Ezquerra, N., Capell, S., Klein, L., Duijves, P.: Model-guided labeling of coronary structure. IEEE Trans. Med. Imaging **17**(3), 429–441 (1998)
8. Feragen, A., Petersen, J., de Bruijne, M., et al.: Geodesic atlas-based labeling of anatomical trees: application and evaluation on airways extracted from CT. IEEE Trans. Med. Imaging **34**, 1212–1226 (2015)
9. Feydy, J., Roussillon, P., Trouvé, A., Gori, P.: Fast and scalable optimal transport for brain tractograms. In: Shen, D., et al. (eds.) MICCAI 2019. LNCS, vol. 11766, pp. 636–644. Springer, Cham (2019). https://doi.org/10.1007/978-3-030-32248-9_71
10. Ghanavati, S., Lerch, J.P., Sled, J.G.: Automatic anatomical labeling of the complete cerebral vasculature in mouse models. NeuroImage **95**, 117–128 (2014)
11. van Ginneken, B., Baggerman, W., van Rikxoort, E.M.: Robust segmentation and anatomical labeling of the airway tree from thoracic CT scans. In: Metaxas, D., Axel, L., Fichtinger, G., Székely, G. (eds.) MICCAI 2008. LNCS, vol. 5241, pp. 219–226. Springer, Heidelberg (2008). https://doi.org/10.1007/978-3-540-85988-8_27
12. Hoang, B.H., Oda, M., Mori, K., et al.: A study on automated anatomical labeling to arteries concerning with colon from 3D abdominal CT images, vol. 7962 (2011)
13. Lee, C.H., Yoon, H.J.: Medical big data: promise and challenges. Kidney Res. Clin. Pract. **36**(1), 3 (2017)
14. Lo, P., van Rikxoort, E.M., Goldin, J.G., Abtin, F., de Bruijne, M., Brown, M.R.: A bottom-up approach for labeling of human airway trees (2011)
15. Matl, S., Brosig, R., Baust, M., Navab, N., Demirci, S.: Vascular image registration techniques: a living review. Med. Image Anal. **35**, 1–17 (2017). https://doi.org/10.1016/j.media.2016.05.005
16. Matsuzaki, T., Oda, M., Kitasaka, T., Hayashi, Y., Misawa, K., Mori, K.: Automated anatomical labeling of abdominal arteries and hepatic portal system extracted from abdominal CT volumes. Med. Image Anal. **20**, 152–161 (2014)
17. Miller, M.I., Trouvé, A., Younes, L.: Geodesic shooting for computational anatomy. J. Math. Imaging Vis. **24**(2), 209–228 (2006). https://doi.org/10.1007/s10851-005-3624-0
18. Paszke, A., Gross, S., Chintala, S., et al.: Pytorch: an imperative style, high-performance deep learning library. In: Advances in Neural Information Processing Systems, vol. 32, pp. 8024–8035. Curran Associates, Inc. (2019). http://papers.neurips.cc/paper/9015-pytorch-an-imperative-style-high-performance-deep-learning-library.pdf
19. Ray, A.F., Powell, J., Hacking, N., et al.: Efficacy and safety of prostate artery embolization for benign prostatic hyperplasia: an observational study and propensity-matched comparison with transurethral resection of the prostate (the UK-rope study). BJU Int. **122**(2), 270–282 (2018)
20. Risser, L., Vialard, F., Rueckert, D., et al.: Simultaneous multi-scale registration using large deformation diffeomorphic metric mapping. IEEE Trans. Med. Imaging **30**(10), 1746–1759 (2011). https://doi.org/10.1109/TMI.2011.2146787
21. Robben, D., et al.: Simultaneous segmentation and anatomical labeling of the cerebral vasculature. Med. Image Anal. **32**, 201–215 (2016)
22. Roussillon, P., Glaunès, J.A.: Kernel metrics on normal cycles and application to curve matching. SIAM J. Imaging Sci. **9**(4), 1991–2038 (2016). http://dx.doi.org/10.1137/16M1070529

23. Sotiras, A., Davatzikos, C., Paragios, N.: Deformable medical image registration: a survey. IEEE Trans. Med. Imaging **32**(7), 1153–1190 (2013). https://doi.org/10.1109/TMI.2013.2265603
24. Vaillant, M., Miller, M., Younes, L., Trouvé, A.: Statistics on diffeomorphisms via tangent space representations. NeuroImage **23**, S161–S169 (2004). Mathematics in Brain Imaging
25. Wang, X., et al.: Automatic labeling of vascular structures with topological constraints via HMM. In: Descoteaux, M., Maier-Hein, L., Franz, A., Jannin, P., Collins, D.L., Duchesne, S. (eds.) MICCAI 2017. LNCS, vol. 10434, pp. 208–215. Springer, Cham (2017). https://doi.org/10.1007/978-3-319-66185-8_24
26. Wu, D., Wang, X., Yin, Y., et al.: Automated anatomical labeling of coronary arteries via bidirectional tree LSTMS. Comput. Assist. Radiol. Surg. **14**, 271–280 (2019)
27. Younes, L.: Shapes and Diffeomorphisms. AMS, vol. 171. Springer, Heidelberg (2019). https://doi.org/10.1007/978-3-662-58496-5. https://books.google.fr/books?id=SdTBtMGgeAUC

Pair-Wise and Group-Wise Deformation Consistency in Deep Registration Network

Dongdong Gu[1,2], Xiaohuan Cao[1], Shanshan Ma[1], Lei Chen[1], Guocai Liu[2], Dinggang Shen[1(✉)], and Zhong Xue[1(✉)]

[1] Shanghai United Imaging Intelligence Co., Ltd., Shanghai, China
Dinggang.Shen@gmail.com, zhong.xue@ieee.org
[2] Hunan University, Changsha, China

Abstract. Ideally the deformation field from one image to another should be invertible and smooth to register images bidirectionally and preserve topology of anatomical structures. In traditional registration methods, differential geometry constraints could guarantee such topological consistency but are computationally intensive and time consuming. Recent studies showed that image registration using deep neural networks is as accurate as and also much faster than traditional methods. Current popular unsupervised learning-based algorithms aim to directly estimate spatial transformations by optimizing similarity between images under registration; however, the estimated deformation fields are often in one direction and do not possess inverse-consistency if swapping the order of two input images. Notice that the consistent registration can reduce systematic bias caused by the order of input images, increase robustness, and improve reliability of subsequent data analysis. Accordingly, in this paper, we propose a new training strategy by introducing both *pair-wise* and *group-wise* deformation consistency constraints. Specifically, losses enforcing both inverse-consistency for image pairs and cycle-consistency for image groups are proposed for model training, in addition to conventional image similarity and topology constraints. Experiments on 3D brain magnetic resonance (MR) images showed that such a learning algorithm yielded consistent deformations even after switching the order of input images or reordering images within groups. Furthermore, the registration results of longitudinal elderly MR images demonstrated smaller volumetric measurement variability in labeling regions of interest (ROIs).

Keywords: Medical image registration · Deep learning · Inverse consistency · Cycle consistency

1 Introduction

Deformable image registration determines voxel-wise anatomical correspondences between a pair of images. In clinical applications, intra-subject registration can be used in follow-up studies for multi-time-point images, while inter-subject registration can be applied in quantitative population analysis of different subjects. Although accuracy is a

A. L. Martel et al. (Eds.): MICCAI 2020, LNCS 12263, pp. 171–180, 2020.
https://doi.org/10.1007/978-3-030-59716-0_17

key factor to evaluate the registration performance, consistency is also crucial for analyzing the variation of subtle anatomies. Registration consistency herein means that the deformations estimated between a pair of images should be smooth and invertible, and also their topology should be well preserved. Further, when analyzing a group of images, the deformations between multiple image pairs should also preserve group consistency. For example, deformable registration is often applied in exploring brain degenerative disease such as Alzheimer's disease (AD), and it has been found that the variation of specific brain anatomies (*e.g.*, hippocampus) is highly related to AD progression [1, 2]. Therefore, the inverse- and cycle- consistent registration of subtle anatomies is crucial for accurately evaluating such kind of brain variation, which can preserve brain topology and support more meaningful analysis.

Traditional registration approaches typically define objective functions consisting of a transformation model and a similarity measure, and apply iterative optimization. These algorithms, *e.g.*, LDDMM [3], SyN [4], and diffeomorphic Demons [5], utilize rigorous geometric theory to ensure smooth and consistent deformation, but optimization and parameter tuning are often computationally intensive and time-consuming.

Recently, deep-learning-based registration methods have been investigated to speed up the registration procedure [6–8]. Unsupervised learning is the most popular strategy since it can relieve the need of ground-truth deformations, which cannot be obtained by manual annotation. However, most unsupervised registration approaches have focused on accuracy and efficiency, but do not pay attention to registration consistency. Dalca *et al.* [9] predicted stationary velocity fields by using neural network and integrated an additional layer to obtain diffeomorphic-like registration results. Kim *et al.* [10] imposed an inverse-consistency loss by using two different registration networks. However, additional network parameters are needed, and increased registration errors could be introduced by using the warped image as new template in subsequent network. Zhang *et al.* [11] developed an inverse-consistency constraint to encourage symmetric registration by simply using a negative field, which does not strictly follow the definition of field inversing. Kuang [12] proposed a consistent training strategy to reduce the deformation folding. In summary, the aforementioned algorithms do not precisely follow the nature of inverse-consistency, and cycle-consistency has not been considered.

In this paper, we propose a new deep registration network, called Symmetric Cycle Consistency Network (SCC-Net). It takes full consideration of the consistent property of deformation fields without introducing additional network parameters. First, the inverse-consistency for image pairs under registration is introduced by performing bidirectional registration. Then, the cycle-consistency is employed for image groups by extending the idea of consistency in multiple images. Both pair-wise and group-wise consistency terms can be well incorporated into the SCC-Net.

Five public datasets of 3D brain MR images were used for evaluating SCC-Net. Besides the Dice similarity coefficient, we also introduced the consistency Dice and reproducibility distance to comprehensively evaluate registration performance. Experimental results showed that, compared with state-of-the-art registration algorithms, the proposed SCC-Net achieved comparable accuracy and significantly improved the efficiency and deformation consistency. Finally, registration for longitudinal brain images of AD patients [13] also indicated that SCC-Net can provide more consistent registration results especially for the subtle anatomies in the follow-up assessment.

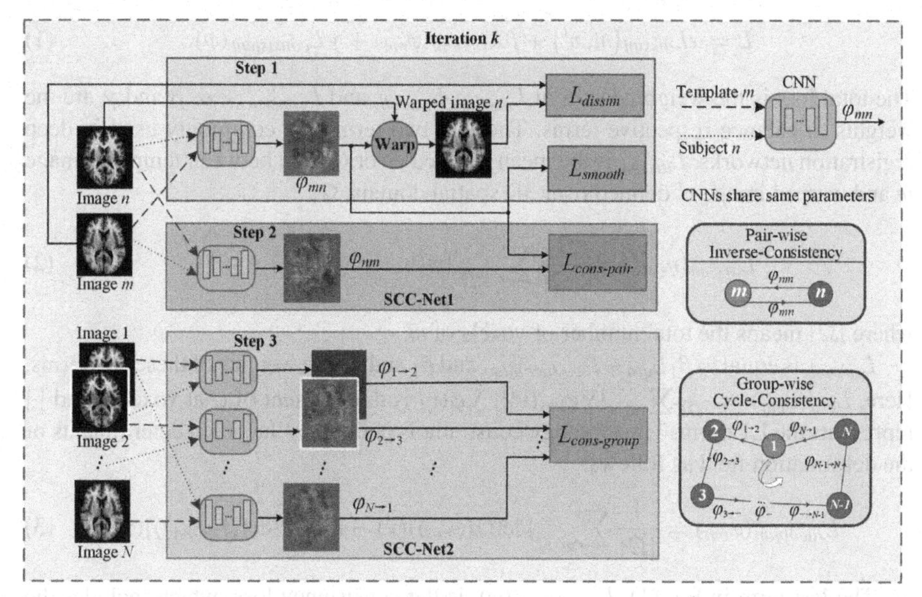

Fig. 1. An overview of the proposed method and the losses used in the kth optimization iteration. $\{m, n, 1, 2, \ldots, N\}$ represent images from the training data. The dotted and dashed lines denote subject and template images, respectively. The CNNs share the same parameters. The losses in Step 1 include image dissimilarity L_{dissim} and deformation regularization L_{smooth}. SCC-Net1 (inverse-consistency) and SCC-Net2 (cycle-consistency) have different inputs and losses.

2 Method

The goal of medical image registration is to obtain anatomical correspondences between images, and the warped subject image $S' = S(\varphi(v) + v)$ should be similar to the template image $T(v)$, where v is a voxel T, and φ denotes the deformation field. Most existing methods use image similarity and field regularization as losses for network training. To maintain consistency property, SCC-Net adopts new consistency losses in the progressive training procedure.

As shown in Fig. 1, three main steps are utilized for SCC-Net training. In Step 1, the inputs of Convolutional Neural Network (CNN) consist of an image pair (m and n), and the output is the *forward* deformation field φ_{mn}. Spatial transformation is introduced to warp image n by φ_{mn}, so the dissimilarity loss can be calculated after registration. In Step 2, we switch the input images to obtain the *backward* deformation φ_{nm}. An inverse-consistency constraint is introduced by composing the forward and backward deformations φ_{mn} and φ_{nm}. In Step 3, a group of images are randomly picked from the training set, and the training can be performed by forming an image cycle. After obtaining the deformations between any adjacent image pairs using SCC-Net, a cycle-consistency constraint is applied along the cycle. In general, the total loss can be defined as:

$$L = \alpha L_{dissim}(m, n') + \beta L_{smooth}(\varphi_{mn}) + \gamma L_{consistent}(\varphi). \quad (1)$$

The total loss is the weighted sum of L_{dissim}, L_{smooth} and $L_{consistent}$. α, β and γ are the weights to balance respective terms. The first two terms are commonly used in deep registration networks. L_{dissim} is the mean squared error (MSE) between template image m and warped image n' defined over 3D spatial domain Ω,

$$L_{dissim}(m, n') = \frac{1}{|\Omega|}\sum\nolimits_{v\in\Omega}\|m(v) - n(\varphi_{mn}(v) + v)\|_2, \quad (2)$$

where $|\Omega|$ means the total number of voxels of m.

L_{smooth} is equal to $\beta_1 L_{gra} + \beta_2 L_{Jacobian}$, and β_1 and β_2 are weights of these two terms. Here, $L_{gra}(\varphi_{mn}) = \frac{1}{|\Omega|}\sum_{v\in\Omega}|\nabla\varphi_{mn}(v)|$, $\nabla\varphi(v)$ is the gradient of φ at voxel v, and $|\cdot|$ represents the L1 norm. The topology constraint is defined by Jacobian determinants of the deformation field as follows:

$$L_{Jacobian}(\varphi_{mn}) = \frac{1}{|\Omega|}\sum\nolimits_{v\in\Omega}|\det(\mathrm{J}(\varphi_{mn}))(v) - ReLU[\det(\mathrm{J}(\varphi_{mn}))](v)|. \quad (3)$$

The last term in Eq. (1), $L_{consistent}$ (φ), is the consistency loss, which includes the inverse-consistency constraint $L_{cons\text{-}pair}$ and the cycle-consistency constraint $L_{cons\text{-}group}$. We will elaborate their definition and implementation in the following subsections.

2.1 Inverse-Consistency Constraint

Inverse-consistent registration means the bidirectional deformations estimated between an image pair should share the same pathway. That is, the composition of forward and backward deformations should be identity or close to identity. As shown in Fig. 1, for image pair m and n, we first obtain the *forward* deformation φ_{mn} that aligns subject n to template m. Then, the *backward* deformation φ_{nm} is obtained through the same network by switching m and n. If the forward and backward deformations are inverse-consistent, ideally the composition of these two deformations should be an identical transform. Therefore, the loss $L_{cons\text{-}pair}$ can be defined by:

$$L_{cons-pair}(\varphi_{mn}, \varphi_{nm}) = \frac{1}{|\Omega|}\sum\nolimits_{v\in\Omega}\|(\varphi_{mn}\circ\varphi_{nm})(v)\|_2, \quad (4)$$

where $\|\cdot\|_2$ represents L2 norm . "$\circ$" means deformation composition defined as follows:

$$(\varphi_{mn}\circ\varphi_{nm})(v) = \sum\nolimits_{w\in\mathcal{N}(u)}\varphi_{nm}(w)\prod\nolimits_{d\in\{x,y,z\}}(1 - |u_d - w_d|) + \varphi_{mn}(v). \quad (5)$$

Given a voxel $v \in \Omega$, the composition means first deforming a point v using φ_{mn} and obtaining a new point $u = v + \varphi_{mn}(v)$, and then applying deformation φ_{nm} to u. Here, we use tri-linear interpolation based on eight-grid neighborhood as shown in

Eq. (5). $\mathcal{N}(u)$ represents the eight-grid neighborhood of u. u_d and w_d represent the dth dimension of points u and w, respectively. The composition operation is performed in three-dimensional space of the deformation field, respectively.

It can be seen from Eq. (5) that the composition operation is differentiable, thus the inverse-consistency loss can be back-propagated for optimization. During training, the composition results would be gradually approaching to identity, resulting smooth and invertible deformations.

2.2 Cycle-Consistency Constraint

Inverse-consistency constraint introduced in the last section is the *pair-wise* penalty. G*roup-wise* consistency should be taken into account when considering the consistency of registration for more than two images. In general, for an image group with N images properly ordered, the adjacent two images can be aligned sequentially to get a series of deformations, *e.g.*, $\{\varphi_{1\to2}, \varphi_{2\to3}, \ldots, \varphi_{N-1\to N}\}$. The group consistency means that, when sequentially composing all the deformations, the result would be equal to the deformation estimated by directly registering the Nth and the 1st images. Therefore, we employ a *cycle-consistency constraint* as an additional loss to train the SCC-Net based on the closed deformation loop, as Step 3 shown in Fig. 1. Specifically, during training, we get pair-wise deformation outputs from SCC-Net by using losses L_{dissim} and L_{smooth} for N image pairs and the composition of deformations, $\varphi_{1\to2} \circ \varphi_{2\to3} \cdots \varphi_{N-1\to N} \circ \varphi_{N\to1}$, should be close to an identity deformation field.

To prove this idea, we design the cycle-consistency loss $L_{cons\text{-}group}$ based on a minimum image group with three images. Given any 3 images $\{m, n \text{ and } l\}$ during training, the result of registering n with m through l, should be the same as directly registering n with m. Thus, the composition of φ_{ml} and φ_{ln} should be the same as φ_{mn}. We define the cycle-consistency loss as:

$$L_{cons-group}(\varphi_{mn}, \varphi_{ml}, \varphi_{ln}) = \frac{1}{|\Omega|}\sum\nolimits_{v\in\Omega} \|(\varphi_{ml} \circ \varphi_{ln})(v) - \varphi_{mn}(v)\|_2. \quad (6)$$

In summary, we first use L_{dissim} and L_{smooth} terms in Step 1 and then introduce inverse-consistency training in Step 2. Finally, we incorporate the cycle-consistency in Step 3 to gradually enforce pair-wise and group-wise consistency in registration without introducing additional network parameters.

2.3 Network Implementation

The CNN registration network follows a similar architecture proposed in [15] and [16], which consists of encoder and decoder paths with skip-connections. In the encoder path, the concatenated template and subject images go through two convolutional layers with 3 × 3 × 3 kernels followed by *ReLU* activation. A 2 × 2 × 2 max pooling layer is then used to down-sample the extracted features. This three-layer structure are repeated to capture multi-resolution features. The decoder path applies the feature map up-sampling twice by using 2 × 2 × 2 deconvolution layers with *ReLU* activation. Two skip connection operators are applied to fuse the features from the encoder and decoder. The deformation field is obtained by a convolutional layer without activation function. The number of the final output channels is three, corresponding to three dimensions of deformation φ.

The inputs are image patches with size $100 \times 100 \times 100$, and outputs are the corresponding deformation field with size $60 \times 60 \times 60$. The output is smaller than the input due to convolution and pooling boundary effects. The network was implemented using Pytorch with Adam optimization and trained in NVIDIA Geforce RTX 2080 Ti for 200,000 iterations with batch size 1. The learning rate was initially set to 1e−5, with 0.5 decay after every 50k iterations.

3 Experiments

We evaluated the performance of SCC-Net for T1 MR image registration in five public datasets [13, 14]: ADNI, LPBA40, IBSR18, CUMC12 and MGH10. We chose 60 images from ADNI dataset and 40 images from LPBA40 and IBSR18 for training, and the rest were used for testing. To evaluate the performance of the proposed algorithm, we compared the registration model trained with and without consistency constraints: 1) deep network without consistency constraint (DL-WOC) is considered as the baseline method; 2) SCC-Net1 is the baseline model with *inverse-consistency constraint*; 3) SCC-Net2 ($N = 3$) is the SCC-Net1 model with *cycle-consistency constraint.* In SCC-Net1, 50×99 image pairs can be randomly drawn from 100 training samples, while C_{100}^{N} ($N = 3$) possible groups can be sampled for training in SCC-Net2. CUMC12 and MGH10 were only used for evaluation. The weights of L_{dissim}, L_{gra}, $L_{Jacobian}$ and $L_{consistent}$ were set as 1, 8, 10 and 10, respectively, according to our experience.

SyN [4] and Demons [5] registration algorithms were also compared with our proposed method. All the images were preprocessed by four steps: 1) skull stripping; 2) alignment to the same orientation; 3) histogram matching and intensity normalization to 0–255; 4) affine registration between each image pair under registration.

3.1 Experiments for Evaluating Registration Performance

To evaluate the registration consistency, for each pair of images A and B, both forward and backward fields φ_{AB} and φ_{BA} were calculated using our network. Then, image B (or its segmentation) can be warped using φ_{AB} to get B', and B' can be further warped by φ_{BA} and get B''. The consistency Dice is then calculated between B and B''. Similar strategy is used to evaluate the cycle-consistency computed in image pairs by applying the model trained with three-images group.

Table 1 reports the average Dice similarity coefficient (DSC) for all the methods calculated on brain ROIs [14] after registration. Bold font with * indicate significant improvement in terms of consistency Dice after registration. We also compared the accuracy Dice in Table 1, and the results marked with Δ indicate no significant difference between our algorithm and SyN [4] or Demons [5] were found (p-value > 0.05). As we know, SyN is based on diffeomorphic framework and possesses inverse consistency. By using consistent constraint in deep learning framework, we almost achieve the ideal consistent property. Boxplot of Consistency Dice in 54 brain ROIs across all the testing cases from LPBA40 dataset is shown in Fig. 2. The results showed that both the proposed

Table 1. DSC values (%) on LPBA40, IBSR18, CUMC12 and MGH10 datasets.

Method		Affine	SyN	Demons	DL-WOC	SCC-Net1	SCC-Net2
LPBA40	Consistency	–	–	94.1 ± 1.8	96.2 ± 1.1	**99.7 ± 0.2***	**99.2 ± 0.4***
	Accuracy	63.5 ± 6.2	71.0 ± 5.4$^{\Delta}$	70.6 ± 5.3	69.9 ± 5.6	69.7 ± 5.6	69.1 ± 5.7$^{\Delta}$
IBSR18	Consistency	–	–	85.1 ± 4.1	87.1 ± 3.6	**98.6 ± 0.8***	**97.0 ± 0.1***
	Accuracy	36.8 ± 9.4	49.1 ± 10.3$^{\Delta}$	49.0 ± 10.2	48.7 ± 10.4	48.3 ± 10.3	46.8 ± 10.3$^{\Delta}$
CUMC12	Consistency	–	–	78.7 ± 2.6	81.3 ± 2.6	**89.7 ± 0.8***	**88.6 ± 2.8***
	Accuracy	38.5 ± 8.7	48.9 ± 9.0	49.6 ± 9.0$^{\Delta}$	49.3 ± 9.2	48.9 ± 9.1	47.6 ± 9.1$^{\Delta}$
MGH10	Consistency	–	–	89.1 ± 3.6	89.7 ± 3.6	**98.7 ± 0.6***	**97.7 ± 0.9***
	Accuracy	40.3 ± 10.5	53.5 + 11.1	53.5 ± 11.0$^{\Delta}$	51.4 ± 11.2	51.3 ± 11.2	50.8 ± 11.2$^{\Delta}$

pair-wise and group-wise registration training strategies can improve the consistency performance.

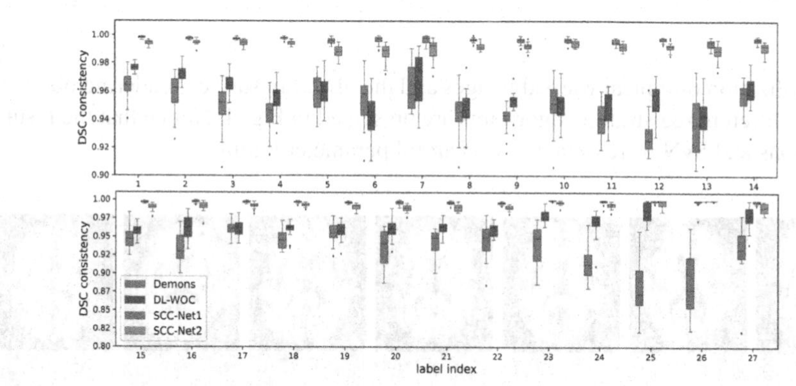

Fig. 2. Boxplot of consistency Dice of 54 brain regions in LPBA40 dataset. DSCs of all bilateral ROI pairs are averaged into one score for visualization.

Figure 3 shows the histograms of the consistency distances using different algorithms. By using the SCC-Net with pair-wise and group-wise consistent constraints, the reproducibility distance is smaller than other algorithms, thus SCC-Net1 and SCC-Net2 yielded better consistency performance. In addition, SCC-Net only takes about 12 s for one registration task, which is much faster than Demons (~1 min) and SyN (~40 min). Therefore, SCC-Net is more robust and feasible for clinical applications.

Besides consistency, to further evaluate the deformation smoothness of SCC-Net, we compared the mean number of folding across all registration results, as shown in Table 2. Less folding indicates better topology preservation and better morphology allowable performance. "J" denotes the Jacobian anti-folding term $L_{Jacobian}$. DL-WOCJ and SCC-Net-WOJ denote the DL-WOC network without J and the SCC-Net without J, respectively. The number of foldings decreases greatly in SCC-Net even without J. The results indicate that the SCC-Net yeilded more reliable deformations for our datasets.

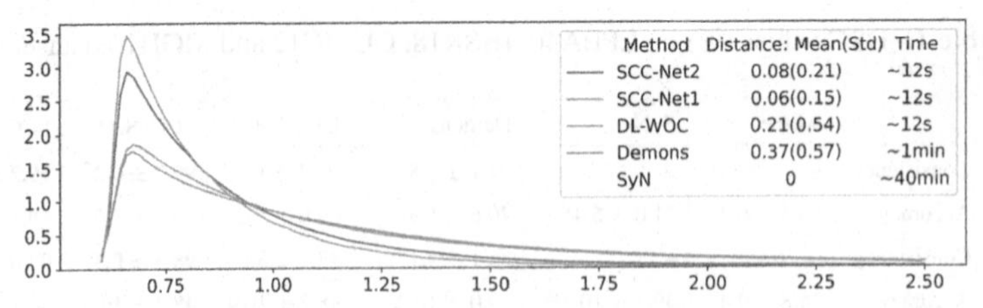

Fig. 3. Distribution of reproducibility distance in range between 0.6 mm and 2.5 mm. The values < 0.6 mm are truncated for better visualization as the majority of the error is close to 0. The legend gives the mean distance values of all testing cases with standard deviation in parentheses and the implementation time of different methods.

Table 2. Mean folding quantities and permillage across all testing sets.

Method	DL-WOCJ	DL-WOC	SCC-Net-WOJ	SCC-Net
No. of folding/‰	27416/3.80	21411/2.90	1440/0.20	1204/0.17

Visualization of typical warped images and the 3D brain surface rendering obtained by using different registration algorithms are presented in Fig. 4. Notice that the results for Demons and SyN were obtained via careful parameter tuning.

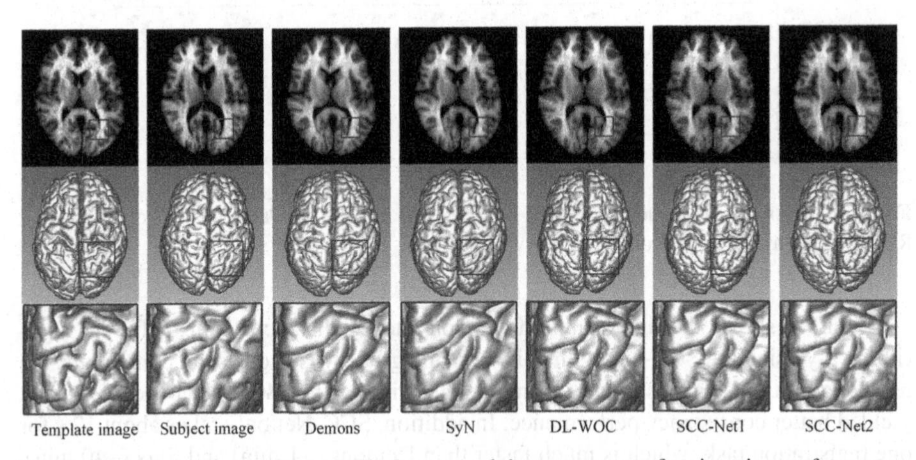

Fig. 4. Typical case from LPBA40. The boxes mark improvements of registration performance.

3.2 Experiments for Evaluating the Evolution of AD Patients

SCC-Net was also applied for longitudinal brain MR images with multiple time-points using ADNI dataset [13]. Brain atrophy occurs with AD progression, and consistent measurement of brain shrinkage is important for AD analysis. We particularly evaluated longitudinal consistency on hippocampus. Specifically, for each subject, by registering serial images and propagating the hippocampus label from the 1st time-point to the subsequent time-points, hippocampal volumes can be measured automatically.

Figure 5 shows hippocampal volumes of 10 subjects using SCC-Net and DL-WOC. Generally, the curves of SCC-Net are smoother than DL-WOC. SCC-Net can well capture longitudinal deformations and yield more temporally-consistent results, as the curves of other algorithms are often fluctuated. Notice that the curves of subject 9 are not stable in the 8th and 10th time-points. We visually inspected the results and found that the disunity in skull-strip preprocessing may cause registration error. This problem could be solved by extracting and registering image patches around the hippocampus. In summary, the results indicate that hippocampal volume variation can be captured in a more stable and reliable way by applying consistency constraint in deep-learning-based registration.

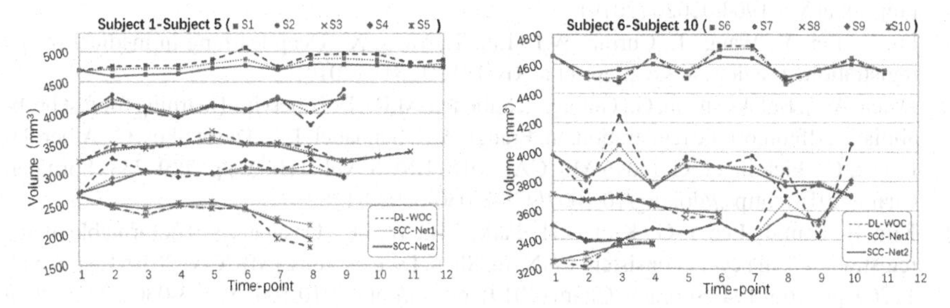

Fig. 5. Hippocampal volume measures over time for 10 subjects.

4 Conclusion

In this paper, a new deep-learning-based registration method, *i.e.*, SCC-Net, is designed to improve registration consistency. SCC-Net is trained to employ both pair-wise and group-wise consistency without introducing any extra network parameters. Robust registration results are obtained, while the computational complexity is the same as the baseline deep-learning-based registration models. Also, the experimental results indicate that the proposed method exhibits higher registration consistency and competitive registration accuracy, compared with state-of-the-art registration methods. The proposed SCC-Net is more reliable for analyzing imaging data in real clinical scenarios.

Acknowledgement. This work was partially supported by the National Key Research and Development Program of China (2018YFC0116400) and the National Natural Science Foundation of China (61671204).

References

1. Courchesne, E., et al.: Normal brain development and aging: quantitative analysis at in vivo MR imaging in healthy volunteers. Radiology **216**(3), 672–682 (2000)

2. Jack, C.R., et al.: Rates of hippocampal atrophy correlate with change in clinical status in aging and AD. Neurology **55**(4), 484–490 (2000)
3. Hart, G.L., Zach, C., Niethammer, M.: An optimal control approach for deformable registration. In: Conference on Computer Vision and Pattern Recognition (CVPR), pp. 9–16 (2009)
4. Avants, B.B., Epstein, C.L., Grossman, M., Gee, J.C.: Symmetric diffeomorphic image registration with cross-correlation: evaluating automated labeling of elderly and neurodegenerative brain. Med. Image Anal. **12**(1), 26–41 (2008)
5. Vercauteren, T., Pennec, X., Perchant, A., Ayache, N.: Diffeomorphic demons: efficient nonparametric image registration. NeuroImage **45**(1), S61–S72 (2009)
6. Cao, X., Fan, J., Dong, P., Ahmad, S., Yap, P.-T., Shen, D.: Image registration using machine and deep learning. In: Handbook of Medical Image Computing and Computer-Assisted Intervention, pp. 319–342. Academic Press (2020)
7. Haskins, G., Kruger, U., Yan, P.: Deep learning in medical image registration: a survey. arXiv preprint arXiv:1903.02026 (2019)
8. Fu, Y., Lei, Y., Wang, T., Curran, W.J., Liu, T., Yang, X.: Deep learning in medical image registration: a review. arXiv preprint arXiv:1912.12318 (2019)
9. Dalca, A.V., Balakrishnan, G., Guttag, J., Sabuncu, M.R.: Unsupervised learning for fast probabilistic diffeomorphic registration. In: Frangi, A.F., Schnabel, J.A., Davatzikos, C., Alberola-López, C., Fichtinger, G. (eds.) MICCAI 2018. LNCS, vol. 11070, pp. 729–738. Springer, Cham (2018). https://doi.org/10.1007/978-3-030-00928-1_82
10. Kim, B., Kim, J., Lee, J.-G., Kim, D.H., Park, S.H., Ye, J.C.: Unsupervised deformable image registration using cycle-consistent CNN. In: Shen, D., et al. (eds.) MICCAI 2019. LNCS, vol. 11769, pp. 166–174. Springer, Cham (2019). https://doi.org/10.1007/978-3-030-32226-7_19
11. Zhang, J.: Inverse-consistent deep networks for unsupervised deformable image registration. arXiv preprint arXiv:1809.03443 (2018)
12. Kuang, D.: Cycle-consistent training for reducing negative jacobian determinant in deep registration networks. In: Burgos, N., Gooya, A., Svoboda, D. (eds.) SASHIMI 2019. LNCS, vol. 11827, pp. 120–129. Springer, Cham (2019). https://doi.org/10.1007/978-3-030-32778-1_13
13. Mueller, S.G., Weiner, M.W., Thal, L.J., Petersen, R.C.: Ways toward an early diagnosis in Alzheimer's disease: The Alzheimer's Disease Neuroimaging Initiative (ADNI). Alzheimer's Dement. **1**(1), 55–66 (2005)
14. Klein, A., et al.: Evaluation of 14 nonlinear deformation algorithms applied to human brain MRI registration. Neuroimage **46**(3), 786–802 (2009)
15. Fan, J., Cao, X., Xue, Z., Yap, P.-T., Shen, D.: BIRNet: brain image registration using dual-supervised fully convolutional networks. Med. Image Anal. **54**, 193–206 (2019)
16. Balakrishnan, G., Zhao, A., Sabuncu, M.R., Guttag, J., Dalca, A.V.: VoxelMorph: a learning framework for deformable medical image registration. IEEE Trans. Med. Imaging **38**(8), 1788–1800 (2019)

Semantic Hierarchy Guided Registration Networks for Intra-subject Pulmonary CT Image Alignment

Liyun Chen[1,2], Xiaohuan Cao[1], Lei Chen[1], Yaozong Gao[1], Dinggang Shen[1], Qian Wang[2], and Zhong Xue[1](✉)

[1] Shanghai United Imaging Intelligence Co. Ltd., Shanghai, China
Dinggang.Shen@gmail.com, zhong.xue@ieee.org

[2] Institute for Medical Imaging Technology, School of Biomedical Engineering, Shanghai Jiao Tong University, Shanghai, China
wang.qian@sjtu.edu.cn

Abstract. CT scanning has been widely used for diagnosis, staging and follow-up studies of pulmonary nodules, where image registration plays an essential role in follow-up assessment of CT images. However, it is challenging to align subtle structures in the lung CTs often with large deformation. Unsupervised learning-based registration methods, optimized according to the image similarity metrics, become popular in recent years due to their efficiency and robustness. In this work, we consider segmented tissues, *i.e.*, airways, lobules, and pulmonary vessel structures, in a hierarchical way and propose a multi-stage registration workflow to predict deformation fields. The proposed workflow consists of two registration networks. The first network is the label alignment network, used to align the given segmentations. The second network is the vessel alignment network, used to further predict deformation fields to register vessels in lungs. By combining these two networks, we can register lung CT images *not only* in the semantic level *but also* in the texture level. In experiments, we evaluated the proposed algorithm on lung CT images for clinical follow-ups. The results indicate that our method has better performance especially in aligning critical structures such as airways and vessel branches in the lung, compared to the existing methods.

Keywords: Medical image registration · Convolution neural network · Lung CT follow-up

1 Introduction

Lung cancer is *not only* the most common cancer *but also* the leading cause of cancer deaths in the world [1]. Early computerized tomography (CT) screening can detect benign or malignant pulmonary nodules effectively in the lung, and follow-up assessment of pulmonary nodules plays an important role for early diagnosis and treatment and can significantly reduce mortality for patients [2]. During follow-up examination, radiologists need to locate the corresponding nodules and assess nodule growth from at

A. L. Martel et al. (Eds.): MICCAI 2020, LNCS 12263, pp. 181–189, 2020.
https://doi.org/10.1007/978-3-030-59716-0_18

least two time-point CT images. Because different positions and respiratory phases at each time-point of the CT series will cause a large and nonlinear deformation, an effective registration algorithm is desirable to deliver reliable anatomical correspondences across time-points.

Registration is the process to transform one or more images onto the coordinate of another image. For decades, optimization-based methods that solve the transformation parameters by maximizing the image similarity have successively received many applications. However, they are limited in high computing cost and low efficiency. Recently, learning-based image registration methods have attracted great attention due to their high efficiency and robustness. In the literature, many deep -learning based registration algorithms have been introduced for solving chest image registration problems. Sokooti *et al.* [3] proposed RegNet trained in a supervised way to register 3D follow-up chest CT images. Eppenhof *et al.* [4] applied a 3D convolutional neural network (CNN) to predict the deformation field between inhale-exhale images using the randomly generated ground-truth for training, which requires a large number of deformation fields to be prepared or generated and is hard to acquire realistic ones. VoxelMorph [5] is a popular unsupervised training framework by using spatial transformer networks (STN) [6] to recover image deformations based on similarity metric losses. Christodoulidis *et al.* [7] proposed a CNN architecture to simultaneously estimate linear and deformable registration for the pair of inhale-exhale lung MR images. The methods in [5] and [7] take intensity information of the whole images into account, not focusing on specific regions. This may cause the model to learn larger surfaces and organs and ignore small structures in the lungs. Hansen *et al.* [8] used dynamic graphs in CNNs to register two feature point sets extracted from lung CT images; however, the performance is restricted by the size and quality of point sets.

In this work, we tackle intra-subject lung CT registration and follow the unsupervised learning framework in VoxelMorph, emphasizing its performance and high efficiency. Our idea is to *not only* use semantic information to guide the registration *but also* perform network training hierarchically in a multi-stage workflow: (1) Semantic information is introduced both as inputs and as supervision. Due to the outperformance of deep learning on medical image segmentation, we generate the lung lobules and airway masks via a segmentation network and then directly take the segmentation results of organs, *i.e.*, lobules and airways, as additional information into our subsequent registration network to distinguish region shapes from simple intensity similarities. (2) A multi-stage learning registration workflow is proposed to register the segmentation surfaces and vessels in the lungs. In the first stage, we predict a rough deformation field that aligns every given segmentation region. Then, a second registration network is employed to register structures such as vessels of the corresponding lung regions. (3) We design two specific objective functions for training of each registration network. The advantage of the proposed algorithm is that it can precisely align lung structures after registering the entire chest images.

To evaluate our algorithm, we register follow-up CT scans from clinical datasets and measure its performance in aligning lung structures such as lobules, airways and vessels. Lung nodule correspondences after registration are also computed to illustrate the application of our proposed algorithm in follow-up studies. The results show that our proposed algorithm has better performance on aligning tiny vessel structures than other methods.

2 Method

A multi-stage registration framework is proposed as shown in Fig. 1. At the beginning, two images, namely fixed and moving images, go through a segmentation network to generate lung lobule and airway masks. Then, two registration networks are designed to align the lung anatomy in a hierarchical manner, *i.e.*, using 1) the *label alignment network* to align the *global structures* such as lung lobules and airways and 2) the *vessel alignment network* to further align more *local anatomies* such as vessels. Specifically, the input of *label alignment network* is the original image of the moving image, along with its lung lobe and airway masks. Then, the registration result of the moving image and the fixed image are fed into the *vessel alignment network* to further refine the vessel alignment. We jointly train these two networks in a two-step training fashion.

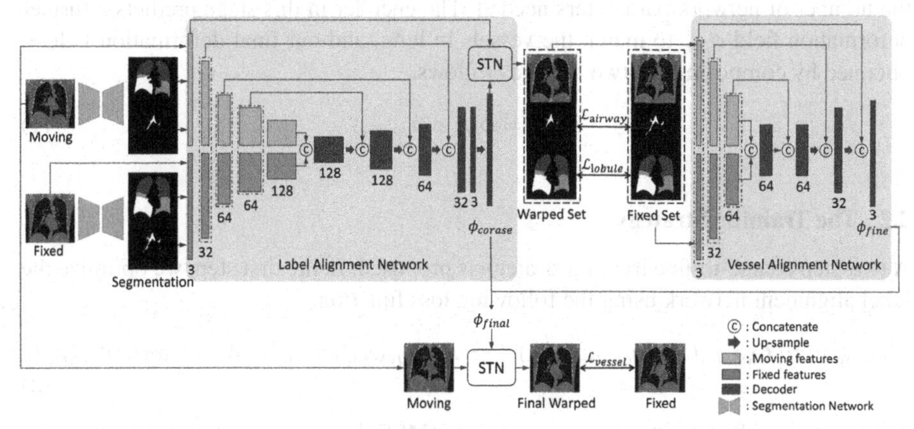

Fig. 1. The pipeline of the proposed method includes three sub-networks: segmentation, label alignment, and vessel alignment networks. For the sake of simplicity, we illustrate 2D diagrams here, while the actual implementations are in 3D.

2.1 Network Structures

Segmentation Network. Fissures and bronchus are two important structures to locate nodules based on their relative locations, but these structures are hard to register due to their low contrasts, small volumes, and complex structures. As an enhancement and a guidance for the subsequent registration networks, we train a 3D U-Net [9] to segment lung lobules S and airways A from each input image in a supervised manner. The network does not participate in the subsequent training.

Label Alignment Network. We use a Siamese encoder to compute two convolutional feature pyramids for a pair of images. Each encoder takes three channels of image pairs (lung CT images I_M and I_F, lung lobule segmentations S_M and S_F, and airway segmentations A_M and A_F) as inputs. The decoder fuses the extracted features and predicts a rough deformation field ϕ_{coarse} to align the labels of two image sets. We use skip connections to combine low-level features between the encoder and the decoder. Before concatenation, we merge features of fixed image set and moving image set via a $1 \times 1 \times 1$ convolution operation for feature fusion and dimensionality reduction. Finally, we warp the moving image and its segmentation masks according to ϕ_{coarse} and feed them together with the fixed image and its segmentation masks into the next vessel alignment network.

Vessel Alignment Network. This network has a similar architecture to the label alignment network, but it is shallower. As vessels are elongated tubular shapes, we only use the down-sampling convolutional block twice at the encoder side, hoping to keep texture features. We share the encoder's weights between two networks at the same part so that our encoder will learn to extract both semantic and texture features. It can also reduce the number of network parameters needed. The encoder in this stage predicts a further deformation field ϕ_{fine} to match the vessels in lung, and our final deformation field is obtained by composing the two fields as follows,

$$\phi_{final} = \phi_{fine} \circ \phi_{coarse}. \tag{1}$$

2.2 The Training Strategy

A two-step coarse-to-fine training strategy is proposed. In the first step, we optimize the label alignment network using the following loss function:

$$\mathcal{L}_{coarse} = \alpha_1 \mathcal{L}_{lobules}(S_M(\phi_{coarse}), S_F) + \alpha_2 \mathcal{L}_{airway}(A_M(\phi_{coarse}), A_F) + \alpha_3 \mathcal{R}(\phi_{coarse}). \tag{2}$$

The first term calculates the mean square error (MSE) between lung lobules. The second term is an airway loss, which is defined by the symmetric average closest distance between two airway segments. The last term is a regulation term to keep the predicted deformation field smooth. The airway loss is further detailed as,

$$\mathcal{L}_{airway} = \frac{1}{2}\left(\frac{1}{N_M}\sum\nolimits_{y \in A_M} \min\nolimits_{x \in A_F} d(x, y) + \frac{1}{N_F}\sum\nolimits_{x \in A_F} \min\nolimits_{y \in A_M} d(y, x)\right), \tag{3}$$

where $d(x, y)$ is the Euclidian distance between two points, and N_M and N_F are the numbers of voxels for the moving airways and the fixed airways. Compared to MSE or Dice coefficients, this function can better describe the goodness of matching between the two tubular shapes.

In the second step, we optimize the vessel alignment network. Our goal is to register the vessel structures in lung and ignore large deformations outside the lungs, and the loss function is defined as,

$$\mathcal{L}_{fine} = \beta_1 \mathcal{L}_{vessel}\big(I_M(\phi_{final}), I_F\big) + \beta_2 \mathcal{R}(\phi_{final}). \tag{4}$$

Here, we introduce a vessel-awareness loss $\mathcal{L}_{vessel}$ to measure the similarity of tubular-shaped tissues in the lung regions. We extract the Hessian-based vesselness features V [10] in an off-line way and calculate the horizontal and vertical normalized gradient correlation coefficient (GCC) of two vesselness feature maps according to Eq. (5) and Eq. (6),

$$\cos(\theta_1 - \theta_2) = \frac{g_{y,1}g_{y,2} + g_{x,1}g_{x,2}}{g_{xy,1}g_{xy,2}}, \tag{5}$$

$$\cos(\varphi_1 - \varphi_2) = \frac{g_{z,1}g_{z,2} + g_{xy,1}g_{xy,2}}{\sqrt{g_{x,1}^2 + g_{y,1}^2 + g_{z,1}^2}\sqrt{g_{x,2}^2 + g_{y,2}^2 + g_{z,2}^2}}, \tag{6}$$

where $G_i = [g_{x,i}, g_{y,i}, g_{z,i}]$ consists of the gradient vectors of the fixed vesselness map $V_{i=1}$ and the warped vesselness map $V_{i=2}$, and g_{xy} is the norms of g_x and g_y. The normalized GCC describes the angle between two gradient vectors. As for thin tubular structure, the gradient vector is approximate to the tube direction. The range of GCC values are from -1 to 1, and it will be 1 if vessels in two images overlap exactly to each other. Besides, we add intensity constraint for training the remaining areas without enhancement, by defining the vessel loss as,

$$\mathcal{L}_{vessel} = \frac{1}{N}\left(\left(I_M\left(\phi_{final}\right) - I_F\right)^2 + (1 - \cos(\theta_1 - \theta_2)) + (1 - \cos(\varphi_1 - \varphi_2))\right), \tag{7}$$

where N is image size. In addition, α_1, α_2, α_3 in Eq. (2) and β_1, β_2 in Eq. (4) are hyper-parameters.

3 Experiments

3.1 Datasets and Experiment Setting

We evaluated our proposed method using our in-house follow-up CT images. Images of 42 subjects were used for training, and images of other 6 subjects were used for testing. The proposed method was implemented using PyTorch and trained on 2 NVIDIA Titan GPUs with batch size of 2. Two preprocessing steps were used before feeding the images into the network. First, affine registration based on the ANTs [11] package was used for linear registration. Second, image intensities were clipped to [− 1000, 600] and then normalized to a range of [0, 1]. For data augmentation, we randomly cropped the images and resampled them into size 224 × 224 × 128 in the training stage. Adam optimization was adopted with an initial learning rate of 1e − 4 and a learning rate decay of 0.01.

3.2 Results

We compared our method with VoxelMorph [5] and the traditional registration toolbox elastix [12]. For VoxelMorph, two kinds of inputs (images only, and images + segmentations) were used to measure the effect of segmentation, for which we adjusted the network to have comparable size as our model for fair comparison. As for elastix, we used the images with their lung masks as inputs and chose B-Spline registration with mutual information optimization.

Results of Segmentation Network. The pre-trained segmentation network was trained on 660 pulmonary CT images and evaluated on 40 images. The results achieved an average Dice score of 96.80% for all segments of the lung.

Image-Based Analysis. Figure 2 shows the results of affine registration, elastix, VoxelMorph, and our algorithm, respectively. From the first row, it can be seen that, when tiny vessel structures are close to the rib area, our method pays more attention to vessel structures and ignores other large rib deformations, while other methods register bones and drive the vessels away.

Quantitatively, we used two major metrics to evaluate our model: (1) the mean absolute error (MAE) of intensities in the lung regions; (2) Dice coefficients for the airway and five lung lobules, including the left upper lobe (LUL), the left lower lobe (LLL), the right upper lobe (RUL), the right middle lobe (RML), and the right lower lobe (RLL). Table 1 shows the results of these methods. Our method yields an average Dice score of 96.84% for lobules and also Dice score of 80.22% for airways, which outperforms elastix as well as VoxelMorph with or without segmentation. For the mean absolute error, our method did not outperform elastix. One reason is that our method ignores large nodule deformations (which have a greater share in lung regions) and focuses more on tiny tubular structures, *i.e.*, shown in the last row of Fig. 2. In real follow-up studies, the lung nodules should keep real size/shape in each time point, so that the assessment can be more meaningful.

Nodule-Based Analysis. We further evaluated our proposed method in determining corresponding nodules from follow-up images. For each subject, two images were captured at two different time-points. Radiologists annotated the nodules at different time-points and marked corresponding pair of each nodule in different time points (with totally 59 nodule pairs manually annotated).

For each image pair, after registering them using different registration methods under comparison, we compared the nodule matching ability of different methods. If nodules in the fixed image and in the moving image are the closest to each other, we consider that they correspond each other. We show the average precision and distance between corresponding nodules, as well as the computation time of different methods in

Table 2 Our method is able to determine all nodule pairs and achieves the smallest average distance compared with all other methods. The computational time of our method is comparable to VoxelMorph but much faster than elastix. Figure 3 illustrates the distribution of distances of nodule pairs for each patient. It can be seen that our method shows a competitive ability on identifying corresponding nodules, with shorter distance and smaller variance.

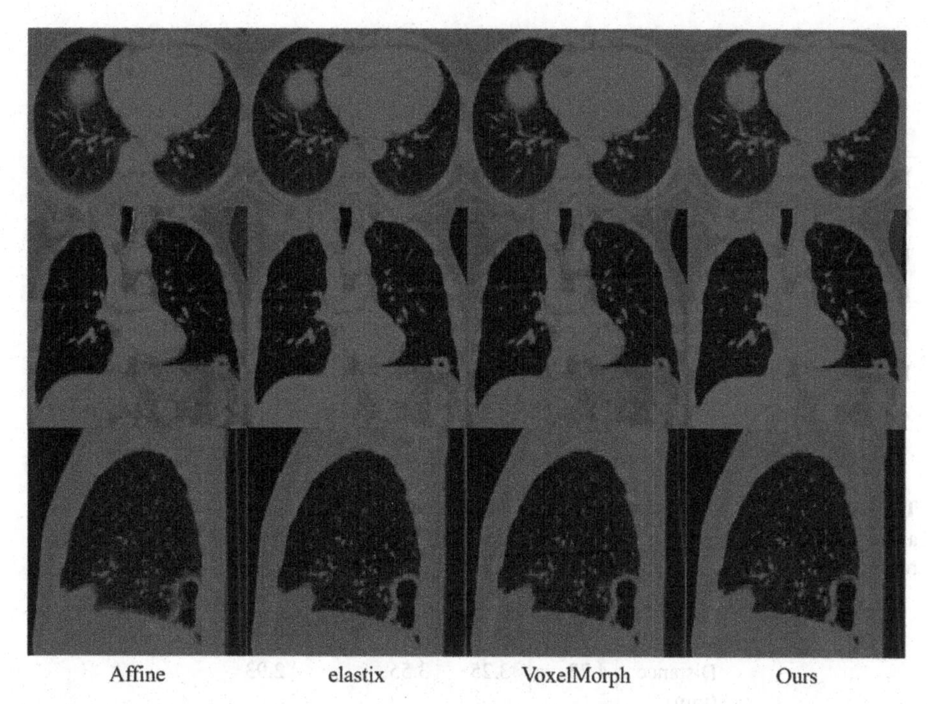

Fig. 2. Typical results of different registration methods. From left to right: Affine, elastix, VoxelMorph, and our method. Each image shows the overlay of the fixed image (orange) and the warped image (blue); when they overlap well, the image looks like greyscale image. (Color figure online)

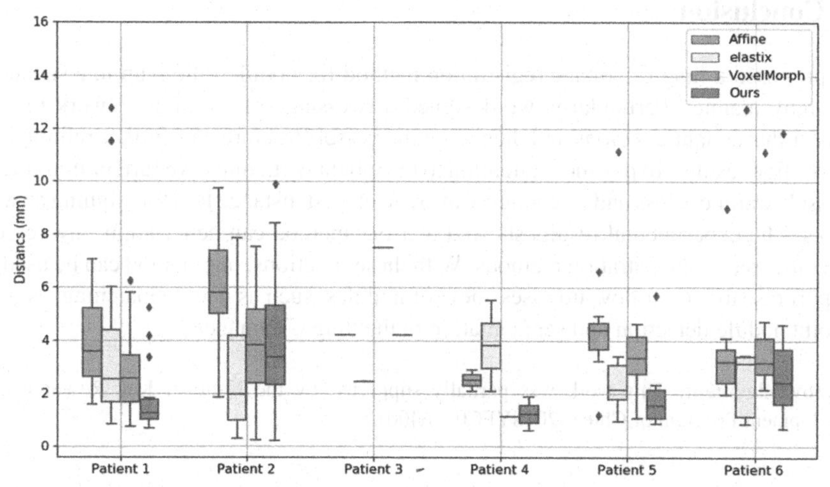

Fig. 3. Boxplots of distances for corresponding nodules by Affine, elastix, VoxelMorph and our method. The numbers of nodule pairs for the patients are 16, 25, 1, 2, 8 and 7, respectively.

Table 1. Mean average error and dice scores of different methods for follow-up CT images.

	Affine	Elastix	VoxelMorph	VoxelMorph (with segmentation)	Ours
MAE (HU)	137.65	**81.05**	87.68	91.79	85.68
LUL (%)	94.17 ± 1.89	96.48 ± 0.48	96.36 ± 1.18	96.97 ± 0.56	**97.30 ± 0.57**
LLL (%)	91.29 ± 2.38	95.54 ± 0.71	95.26 ± 1.34	96.31 ± 0.66	**96.93 ± 0.62**
RUL (%)	94.48 ± 1.32	95.77 ± 0.97	95.82 ± 0.99	96.93 ± 0.55	**97.22 ± 0.56**
RML (%)	90.77 ± 3.58	93.26 ± 2.99	93.25 ± 2.80	95.04 ± 1.76	**95.56 ± 1.74**
RLL (%)	91.28 ± 3.01	95.38 ± 0.97	95.24 ± 0.96	96.56 ± 0.76	**97.20 ± 0.62**
Airway (%)	63.49 ± 9.73	78.81 ± 2.58	77.98 ± 3.11	77.33 ± 4.03	**80.22 ± 3.26**

Table 2. Results of corresponding nodule matching for 59 nodule pairs. The first row shows the average distance of all the annotated corresponding nodule pairs after registration. The second row shows the average computation time for the subjects.

	Affine	Elastix	VoxelMorph	Ours
Distance (mm)	4.73	3.25	3.55	2.93
Time (s)	–	94.6	2.4	2.6

4 Conclusion

We proposed a lung CT image registration method for nodule follow-up in a semantic hierarchy manner. Particularly, we designed a two-stage registration network to first register the semantic regions and then align the vessel structures for further refining the registration results. To pay more attention to the tubular structures, we further introduced a vessel awareness loss and a symmetric average closest distance loss for aligning airway masks. The experimental results showed that our method can better align tiny vessels than other related registration methods. With these functions, our method can be used to support research on follow-up assessment of nodules, such as nodule matching, as well as joint nodule detection and segmentation in the lung CT images.

Acknowledgement. This work was partially supported by the National Key Research and Development Program of China (2018YFC0116400).

References

1. Bray, F., Ferlay, J., Soerjomataram, I., Siegel, R.L., Torre, L.A., Jemal, A.: Global cancer statistics 2018: GLOBOCAN estimates of incidence and mortality worldwide for 36 cancers in 185 countries. CA Cancer J. Clin. **68**(6), 394–424 (2018)

2. The National Lung Screening Trial Research Team: Reduced lung-cancer mortality with low-dose computed tomographic screening. New England J. Med. **365**(5), 395–409 (2011)
3. Sokooti, H., de Vos, B., Berendsen, F., Lelieveldt, B.P.F., Išgum, I., Staring, M.: Nonrigid image registration using multi-scale 3D convolutional neural networks. In: Descoteaux, M., Maier-Hein, L., Franz, A., Jannin, P., Collins, D.L., Duchesne, S. (eds.) MICCAI 2017. LNCS, vol. 10433, pp. 232–239. Springer, Cham (2017). https://doi.org/10.1007/978-3-319-66182-7_27
4. Eppenhof, K.A.J., Pluim, J.P.W.: Pulmonary CT registration through supervised learning with convolutional neural networks. IEEE Trans. Med. Imaging **38**(5), 1097–1105 (2019)
5. Balakrishnan, G., Zhao, A., Sabuncu, M.R., Guttag, J., Dalca, A.V.: An unsupervised learning model for deformable medical image registration. In: 2018 IEEE/CVF Conference on Computer Vision and Pattern Recognition, pp. 9252–9260 (2018)
6. Jaderberg, M., Simonyan, K., Zisserman, A., Kavukcuoglu, K.: Spatial transformer networks. In: Cortes, C., Lawrence, N.D., Lee, D.D., Sugiyama, M., Garnett, R. (eds.) Advances in Neural Information Processing Systems. pp. 2017–2025. Curran Associates, Inc. (2015)
7. Stergios, C., et al.: linear and deformable image registration with 3D convolutional neural networks. In: Stoyanov, D., et al. (eds.) RAMBO/BIA/TIA -2018. LNCS, vol. 11040, pp. 13–22. Springer, Cham (2018). https://doi.org/10.1007/978-3-030-00946-5_2
8. Hansen, L., Dittmer, D., Heinrich, M.P.: Learning deformable point set registration with regularized dynamic graph CNNs for large lung motion in COPD patients. In: Zhang, D., Zhou, L., Jie, B., Liu, M. (eds.) GLMI 2019. LNCS, vol. 11849, pp. 53–61. Springer, Cham (2019). https://doi.org/10.1007/978-3-030-35817-4_7
9. Çiçek, Ö., Abdulkadir, A., Lienkamp, S.S., Brox, T., Ronneberger, O.: 3D U-Net: learning dense volumetric segmentation from sparse annotation. In: Ourselin, S., Joskowicz, L., Sabuncu, M.R., Unal, G., Wells, W. (eds.) MICCAI 2016. LNCS, vol. 9901, pp. 424–432. Springer, Cham (2016). https://doi.org/10.1007/978-3-319-46723-8_49
10. Sato, Y., et al.: Three-dimensional multi-scale line filter for segmentation and visualization of curvilinear structures in medical images. Med. Image Anal. **2**(2), 143–168 (1998)
11. Avants, B.B., Tustison, N.J., Song, G., Cook, P.A., Klein, A., Gee, J.C.: A reproducible evaluation of ANTs similarity metric performance in brain image registration. NeuroImage **54**(3), 2033–2044 (2011)
12. Klein, S., Staring, M., Murphy, K., Viergever, M.A., Pluim, J.: elastix: a toolbox for intensity-based medical image registration. IEEE Trans. Med. Imaging **29**(1), 196–205 (2010)

Highly Accurate and Memory Efficient Unsupervised Learning-Based Discrete CT Registration Using 2.5D Displacement Search

Mattias P. Heinrich(✉) and Lasse Hansen

Institute of Medical Informatics, Universität zu Lübeck, Lübeck, Germany
heinrich@imi.uni-luebeck.de

Abstract. Learning-based registration, in particular unsupervised approaches that use a deep network to predict a displacement field that minimise a conventional similarity metric, has gained huge interest within the last two years. It has, however, not yet reached the high accuracy of specialised conventional algorithms for estimating large 3D deformations. Employing a dense set of discrete displacements (in a so-called correlation layer) has shown great success in learning 2D optical flow estimation, cf. FlowNet and PWC-Net, but comes at excessive memory requirements when extended to 3D medical registration. We propose a highly accurate unsupervised learning framework for 3D abdominal CT registration that uses a discrete displacement layer and a contrast-invariant metric (MIND descriptors) that is evaluated in a probabilistic fashion. We realise a substantial reduction in memory and computational demand by iteratively subdividing the 3D search space into orthogonal planes. In our experimental validation on inter-subject deformable 3D registration, we demonstrate substantial improvements in accuracy (at least ≈10% points Dice) compared to widely used conventional methods (ANTs SyN, NiftyReg, IRTK) and state-of-the-art U-Net based learning methods (VoxelMorph). We reduce the search space 5-fold, speed-up the run-time twice and are on-par in terms of accuracy with a fully 3D discrete network.

Keywords: Deformable registration · Deep learning · Discrete optimisation

1 Motivation and Related Work

Medical image registration aims to align two or more 3D volumes of different patients, time-points or modalities. In many practical cases the transformation is highly deformable, which poses a complex task of regressing a continuous displacement field. Conventional registration methods aim to capture larger deformations with multi-resolution strategies iterative warping, discrete displacement

A. L. Martel et al. (Eds.): MICCAI 2020, LNCS 12263, pp. 190–200, 2020.
https://doi.org/10.1007/978-3-030-59716-0_19

search or a combination of these. Many clinical tasks that could benefit from automatic and accurate image registration are time-sensitive making many conventional algorithms inapt for practical use. Deep learning based image registration (DLIR) has the promise to reduce computation times from minutes to sub-seconds as already realised in the field of image segmentation (exemplified by the wide adaptation of the U-Net [21]). In particular, unsupervised methods [2] that minimise a classical cost function are of interest, because they can reduce inference run-times without relying on extensive manual annotations of e.g. corresponding landmarks. However, up to date no boost in accuracy has been achieved in DLIR and its application remains restricted to less complex tasks. For intra-patient lung 4D-CT registration conventional approaches yield highly accurate motion estimation [22] with target registration errors of less than 1 mm (on the DIRLAB dataset), while reaching sub-second computation time when parallelised on GPU hardware [4], whereas state-of-the-art DLIR methods still exhibit large residual errors of $\approx$ 3.5 mm [1,2,9]. This is most likely due to the limited or at least ineffective capture range of multi-scale feed forward architectures (U-Nets) that were designed for segmentation. Multi-resolution strategies [15], multi-stage networks [28] or progressive hierarchical training [8] can alleviate the limitations to some degree and reduce errors for 4DCT to $\approx$ 2.5 mm.

Discrete Displacements: Exploring multiple potential displacements that are quantised as labels in MRF-based discrete optimisation has seen great success in image registration with large deformations, both for 2D images and medical volumes [10,12,18,24]. This idea is also reflected in the top-performing 2D DLIR methods for optical flow, among others FlowNet [7] and PWC-Net [26]. Instead of directly regressing a continuous displacement value, discrete methods quantise the range of expected displacements and estimate probabilities (or costs) for a (spatially regularised) label assignment. Therefore, discrete methods achieve remarkable results even when they are employed without iterative or multi-resolution strategies. Learning probabilistic and uncertainty-aware registration models has been studied in [6,17] yet in a continuous variational setting very different to discrete probabilistic methods.

In [11] a discrete probabilistic dense displacement network was presented that was trained with label supervision and substantially exceeded the accuracy of continuous (U-Net based) DLIR methods for 3D CT registration. It used a large number of control points and a densely quantised 3D displacement space. This approach comes at the cost of substantial memory use, since each feature tensor within the network requires >1200 MByte of GPU memory per channel - in our experiments it required $\approx$22 GByte GPU RAM (which is only available in few high-end cards). The authors, hence, restricted the network to few trainable weights, single channel features within their mean-field inference and made extensive use of checkpointing. We expect that a fully sampled 3D displacement space is not necessary to reach state-of-the-art results and hence a reduction in labels and subsequently memory will lead to wider adoption of discrete DLIR.

Subdivision of Label Spaces: Exploding numbers of parameters have been an issue in many conventional MRF-based registration methods and strategies to reduce the computational burden include: simplified graph models (e.g. minimum-spanning trees [27]), fewer nodes in the transformation model [10] and reduction from the dense 2D displacement space to a sparse setting where only displacements along the orthogonal 1D axes are sampled [24]. This subdivision of displacement labels, which enables a decoupled MRF-regularisation, was also used in the popular SIFT flow approach [18] and the 3D medical MRF registration of [10]. However, as shown in [12] this very sparse sampling of a 3D space leads to a significant reduction of registration accuracy. Applying CNNs to 1D signals is also expected to limit their ability to learn meaningful patterns within the displacement space. Here, we propose an intermediate strategy that decomposes the 3D space into three orthogonal 2D planes and thereby more accurately approximates the full space. A related strategy was considered for linear 3D registration in a discrete setting [32], where 2D sub-spaces of a 12 parameter affine transform where optimised using an MRF.

2.5D Approaches in Deep Learning: Several recent segmentation and classification networks for medical volumes consider multiple 2D views of the 3D input data. In [25] a 2.5D input view is created by extracting orthogonal planes of the 3D patches and representing them as RGB colour channels for a 2D CNN, which enabled transfer learning from ImageNet pretrained models. Multiview fusion of multiple 2D CNN classifiers was proven effective for pulmonary nodule classification of 3D CT in [23]. Large 2.5D planes were used to extract context features for self-supervised learning in [3]. Many fully-convolutional segmentation models decouple the prediction for axial, coronal and sagittal planes and fuse the resulting scores or directly employ 3D separable convolutions [5]. In learning-based image registration, [16] decouples the motion prediction in three orthogonal spatial planes and fuses the continuous regression values and achieves comparable accuracy to 3D U-Nets for weakly-supervised DLIR. This is fundamentally different to our idea to decompose the 3D displacement space in a discrete registration setting.

Contributions: 1) We are the first to propose an unsupervised discrete deep learning framework for 3D medical image registration that leverages probabilistic predictions to improve the guidance of the metric loss. **2)** Furthermore, we propose a highly efficient 2.5D approximation of the quantised 3D displacement space that substantially reduces the memory burden for training and computational complexity for inference of discrete DLIR. **3)** We demonstrate that the high accuracy of a full 3D search space can be matched using two iterations of sampling three orthogonal 2D displacement maps in combination with on-the-fly instance optimisation. **4)** The method is evaluated on a challenging public inter-subject abdominal CT dataset and the source code is already released anonymously.

2 Methods

3D pdd-net: We build our efficient 2.5D discrete registration framework upon the 3D pdd-net (probabilistic dense displacement network) proposed in [11],

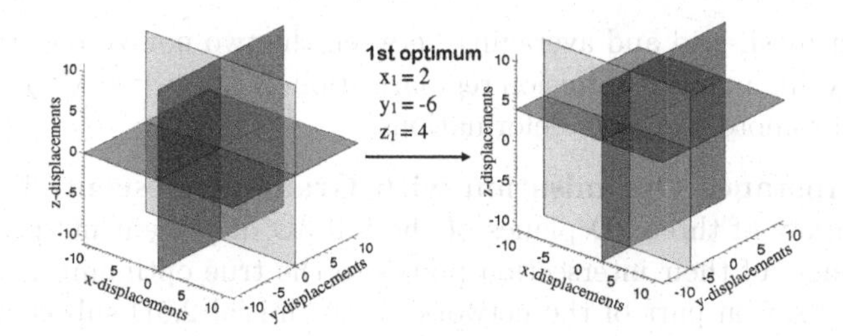

Fig. 1. Concept of 2.5D decomposition of discrete 3D displacement space. To reduce the computational complexity, displacements are only sampled along three orthogonal planes (reducing e.g. the number of quantised positions by 5× when using $|\mathcal{L}_{3D}| = 15^3 = 3375$). To compensate for the approximation error, a second pass of the same regularisation network (without image warping or additional feature computation) is performed that roots the planes at the previous optima.

which is briefly summarised below. A pair of fixed and moving scans I_F and I_M, for which we seek a spatial alignment φ, are fed through a feature extractor that uses deformable convolutions [13] and outputs a 24-channel feature map (4D tensor) with a stride of 3 voxels. Next, a B-spline transformation model with a set of $|K| \in \mathbb{R}^3$ control points on a coarser grid and a quantised 3D displacement space with linear spacing $\mathcal{L} = q \cdot \{-1, -\frac{6}{7}, -\frac{5}{7}, \ldots, +\frac{5}{7}, +\frac{6}{7}, +1\}^3$ are defined. Here q is a scalar that defines the capture range and the cardinalities are $|K| = 29^3 = 24389$ and $|\mathcal{L}| = 15^3 = 3375$. A correlation layer (cf. [7]) without any trainable weights is used to compute the matching cost of a fixed scan feature vector with all discretely displaced moving feature vectors that are within the search space spanned by $\mathcal{L}$. The second part of the network comprises several 3D max- and average pooling operations (with stride = 1) that act in alternation on either the three spatial or the three displacement dimensions and model two iterations of MRF-regularisation (approx. mean-field inference as found in [31]).

Decoupled 2.5D Subspaces: As mentioned above these dense 3D displacement computation enable highly accurate registration, but come with large memory and computational costs. We thus approximate the dense 3D space with three 2.5D subspaces and define $\mathcal{L}_{2D} = \{\mathcal{L}_{xy}, \mathcal{L}_{xz}, \mathcal{L}_{yz}\}$ where each subspace comprises a planar grid of 2D displacements: $\mathcal{L}_{xy} = q \cdot \{-1, -\frac{6}{7}, -\frac{5}{7}, \ldots, +\frac{5}{7}, +\frac{6}{7}, +1\}^2$ and a constant value z_1 for the third dimension. This step greatly reduces the memory requirements for learning a better feature extraction in the deformable convolution part, since $|\mathcal{L}_{2D}|$ is now only $3 \cdot 15 \cdot 15 = 625$, five-fold smaller than a full 3D space. The feature dissimilarities computations within the correlation layer are reduced from 4 GFlops to 790 MFlops. The spatial smoothing remains to operate on 3D (yet with a much smaller number of channels 625 instead of 3375), while the operations that regularise the displacement dimensions are now in 2D. In order to estimate a 3D field, the output of the network is converted into three 2D pseudo-probability maps (using the softmax) for each control point (see Fig. 2). The 3D vectors φ are then found by multiplying the probabilities with the displacement mesh-grid and averaging between the two non-zero elements of all three intersecting maps. A diffusion regularisation, $\lambda\cdot(|\nabla\varphi_1|^2+|\nabla\varphi_2|^2+|\nabla\varphi_3|^3)$, is added to promote plausible deformations.

Two-Step Instance Optimisation with Gradient Descent: The approximation accuracy of three 2D planes of the full 3D displacement space depends on the closeness of their intersection points to the true optimum. A single pass of the regularisation part of the network on the initial 2.5D subset might yield inaccuracies in both training and inference. Therefore, a two-step approach is proposed that to alleviate limited capabilities of a feed-forward network to find the optimal compromise between metric loss and diffusion regularisation. Thus an iterative on-the-fly instance optimisation on the intermediate 2.5D displacement probabilities is performed similar to [2] and [11]. For this purpose a continuous 3D B-spline transformation model is considered and optimised for an improved $\varphi^* := (x+\Delta x^*, y+\Delta y^*, z+\Delta z^*)$ per instance (test registration pair) using Adam. We start with the feed-forward predicted discrete cost tensor $\mathcal{C} \in H\times W\times D\times 15\times 15\times 3$, where the last dimensions describes the three subplanes (2.5D) of a full 3D displacement search region. We minimise the following loss function: $L_{\text{instance}} = L_{xy} + L_{xz} + L_{yz} + L_{\text{diff.-reg}}$. For each sub-dimension we define $L_{xy} = \mathcal{C}(x, y, z, \Delta x^*, \Delta y^*, 0)$, where differentiable bilinear sampling is used for $\Delta x^*, \Delta y^*$. We, thus iteratively update the deformation and optimise a related cost function (the sum over the three 2D displacement metric values and a diffusion regularisation) in a continuous manner.

A similar principle to combine the complementary strengths of discrete and continuous optimisation was used in [22] and [7]. Once the instance optimisation is completed, the second pass of the discrete network with more accurate initial placement of 2.5D subplanes is performed (see Fig. 1 right for visualisation), followed again by a continuous refinement.

Non-local Metric Loss: The approach in [11] was restricted to weakly-supervised learning with segmentation labels. This may limit the applicability to learn from large unlabelled datasets and introduce an bias towards the chosen labels, cf. [2]. We thus introduce a novel unsupervised non-local metric loss for discrete DLIR. Due to its high performance in other CT registration tasks e.g. [30] modality independent neighbourhood descriptors (MIND) are extracted with self-similar context as proposed in [14] yielding 12-channel tensors. Instead to directly employing a warping loss as done in [2] and most related DLIR methods, we make full use of the 2.5D probabilistic prediction and compute the warped MIND vectors of the moving scan implicitly by a weighted average of the underlying features within the search region (averaged again for the three orthogonal probability maps): $\text{MIND}_{\text{warped}} = \frac{1}{3}\text{MIND}_{\text{xy}} + \frac{1}{3}\text{MIND}_{\text{xz}} + \frac{1}{3}\text{MIND}_{\text{yz}}$.

Here the probabilistic displacements are defined as $\mathcal{P}_{xy}(x, y, z, \Delta x', \Delta y') = \frac{\exp(-\alpha\mathcal{C}(x,y,z,\Delta x',\Delta y'))}{\sum_{\Delta x',\Delta y'}\exp(-\alpha\mathcal{C}(x,y,z,\Delta x',\Delta y'))}$ and the discretely warped MIND features (single channel for brevity) as: $\text{MIND}_{\text{xy}} = \sum_{\Delta x',\Delta y'} \mathcal{P}_{xy}(x, y, z, \Delta x', \Delta y') \cdot \text{MIND}(x + \Delta x', y + \Delta y', z)$, where $\Delta x', \Delta y'$ are local coordinates.

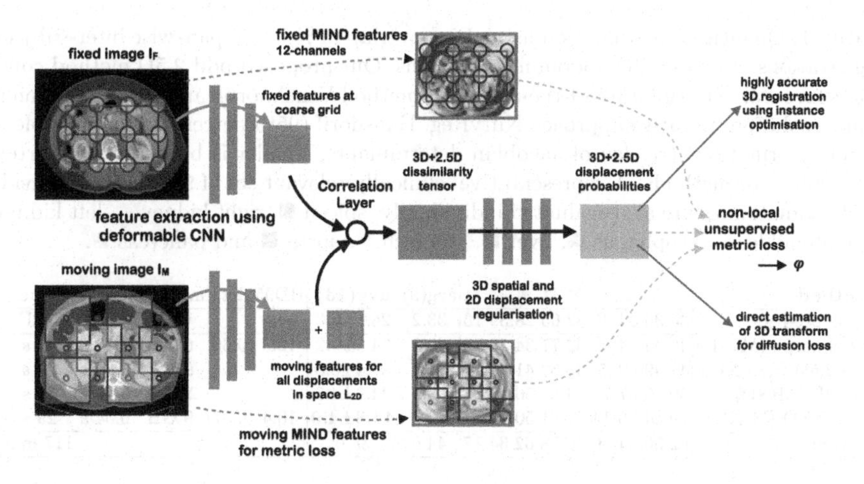

Fig. 2. Novel concept of 2.5D unsupervised dense displacement network. Deformable convolution layers firstly extract features for both fixed and moving image. Secondly, the correlation layer evaluates 2.5D displacement space for each 3D grid point yielding three 5D dissimilarity maps. These maps are spatially smoothed by filters acting on dimensions 1–3 and regularity within dimensions 4–5 (2D displacement planes) is obtained using approx. min-convolutions. The learning is supervised without any annotations using our proposed non-local MIND metric loss. The 2.5D probabilistic prediction is obtained using a softmax (over each 2D displacement plane) and either converted to continuous 3D displacements for a diffusion regularisation or further refined using instance optimisation to warp scans (or label images) with high accuracy.

We also compare the benefits of this discrete probabilistic loss with a traditional warping loss (denoted as "w/o NL" in Table 1). The source code of our complete implementation is publicly available at [1].

3 Experiments and Results

Many state-of-the art DLIR registration methods have been evaluated on private datasets with unknown complexity. The public "beyond the cranial vault" abdominal CT dataset described in [30] us used here, which was used in a MICCAI 2015 challenge and is available for download. We pre-process the scans using the following steps: First, resampling to isotropic resolution of 2 mm and automatic cropping to a similar field-of-view. Second, affine pre-registration was performed between all pairs of scans using the discrete registration tool linearBCV[2] an unbiased mean transform estimated to bring all scans into a canonical space (the run-time per registration is ≈ 2 s.)[3]. Despite these reasonable efforts they are still very complex and challenging deformation to be compensated as evident from the low initial average Dice overlap of 28.1% (see Table 1).

[1] https://github.com/multimodallearning/pdd2.5/.
[2] https://github.com/mattiaspaul/deedsBCV.
[3] https://learn2reg.grand-challenge.org/Dataset/ (Task 3).

Table 1. Quantitative evaluation using Dice overlap in % for 90 pair-wise inter-subject registrations of unseen 3D abdominal CT scans. Our proposed pdd 2.5D method compares favourable to the state-of-the-art DLIR method Voxelmorph and the conventional multiresolution, iterative approach NiftyReg. Hausdorff (95th percentile) and complexity of deformations (stddev of Jacobian determinants, smaller is better) are reported for a subset of methods. A representative, clinically relevant set of 9 out of the considered 13 anatomies are also evaluated individually: spleen ■, right kidney ■, left kidney ■, gallbladder ■, esophagus ■, liver ■, stomach ■, aorta ■ and pancreas ■.

Method	■	■	■	■	■	■	■	■	■	avg(9)	**avg(13)**	HD95	detJ	memory	infer.	time
affine pre-reg.	36	36	34	6	32	65	28	39	16	33.2	28.1±8.3	14.6		train	GPU	CPU
Voxelmorph (MIND)	49	44	44	8	32	77	34	48	19	40.0	34.0±9.2	12.8	0.66	6 GB	0.12 s	60 s
pdd 2.5D (w/o NL)	45	49	52	5	38	69	41	44	21	41.1	34.4±5.8			9 GB	0.54 s	29 s
pdd 3D (MIND)	60	65	67	12	43	81	50	56	35	52.6	44.7±4.6			22 GB	0.73 s	73 s
pdd 2.5D (MIND)	60	64	65	14	41	81	50	57	34	52.2	**44.8±4.9**	10.4	0.57	9 GB	0.54 s	29 s
NiftyReg	62	50	54	3	36	78	62	34	17	44.5	35.0					117 min

We evaluate the registration accuracy using manual segmentations for ten scans (90 registrations) that were not used during unsupervised training and compute the Dice score for all 13 labels (see Fig. 3 and Table 1 for details). We compare the state-of-the-art unsupervised Voxelmorph network [2] and use the same MIND implementation as similarity metric that yields a Dice of 34.0% (results using the default MSE loss are ≈4% points worse and the NCC loss failed to converge). A small ablation study of variants of our proposed method is performed using either fully 3D displacement spaces (pdd 3D, 44.7%), the novel 2.5D subdivision (**pdd 2.5D**, 44.8%) and a conventional warping loss (pdd 2.5D (w/o NL), 34.4%). In addition, we include results of NiftyReg [20] from [30] with an average Dice of 35.0% that performed substantially better than IRTK and ANTs SyN (28% and 27%) and only slightly worse than deeds (49%). Note, that a larger subset of test pairs was used for these methods and their initial affine alignment was likely worse. Our networks were each trained with AdamW [19] (weight decay =0.01, initial learning rate =0.005 and exponential decay with $\gamma = 0.99$) for 250 epochs (1000 iterations with a mini-batch size of 4), using affine augmentation and a weighting $\lambda = 0.025$ for the diffusion regularisation loss (a higher λ was employed during a warm-up phase for 100 iterations to stabilise the training) within ≈25 min. The instance optimisation uses Adam with learning rate 0.02, $\lambda = 5$ and 30 iterations. It was also employed during training (without gradient tracking) to enable the sampling from 2D displacement planes that are not rooted at the origin (central voxel) and thus increasing the coverage for feature learning. We repeated the training three times and report the accuracies from the worst run (Voxelmorph was trained once for 50000 iterations requiring about 8 h).

Clear improvements of 10% points accuracy gains are achieved with the discrete setting compared to the state-of-the-art in unsupervised DLIR. The gains over Voxelmorph are most visible for medium-sized organs (kidneys ■, ■) and highly deformable anatomies (stomach ■ and pancreas ■). The sorted Dice

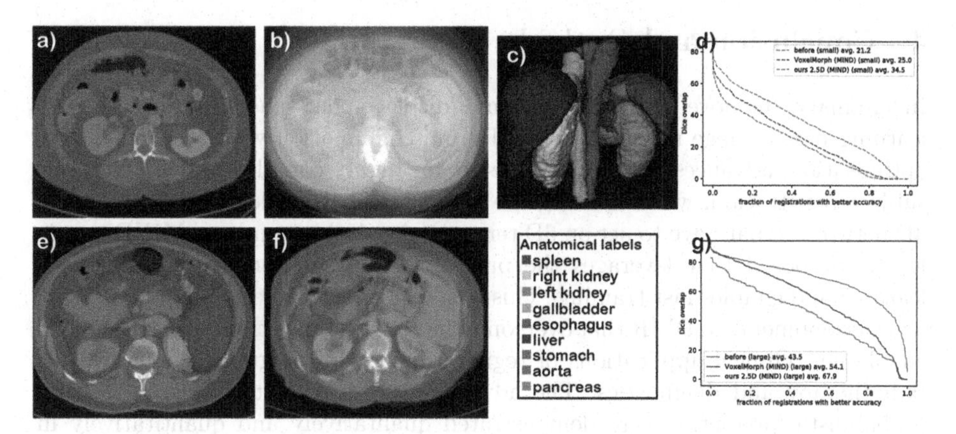

Fig. 3. Visual and numerical results for registration of unseen test scans with proposed **pdd 2.5D** method. a) random moving scan with ground truth annotation b) intensity mean of all test scans (after affine pre-registration) c) automatically propagated segmentation from training scans (stapled [29]) d) resulting Dice scores for four larger organs (spleen, kidneys, liver) e) propagated segmentation from other test scan f) intensity mean after our pair-wise registration g) resulting Dice scores for 8 smaller organs (gallbladder, esophagus, stomach, aorta, inferior and portal veins, glands) (Color figure online)

scores in Fig. 3 g) show that our 2.5D network compensates larger deformations (small initial Dice values) especially well. Our results without using the non-local MIND loss are similar to Voxelmorph, highlighting the fact that a meaningful probabilistic prediction is achieved. The proposed 2.5D approximation matches the quality of the full 3D search space (5× bigger) and reduces the memory use for the correlation layer and regularisation part during training from 10.2 to 1.7 GByte (the total memory usage is higher due to the feature extractor and non-local loss). The CPU runtime at inference is reduced by 2.5× to only 29 s (note, this includes the two-step instance optimisation, without the second step a CPU runtime of 17 s is obtained with a moderate decrease of 2% points in Dice accuracy). The complexity of the transformations estimated using the standard deviation of Jacobian determinants was 0.567 on average, 2.5% of voxels incurred a negative Jacobian (folding), indicating a reasonable smoothness for abdominal inter-subject registration with large topological differences. This can be further improved by increasing the parameters for instance optimisation to $\lambda = 15$ and 50 iterations, yielding $\text{std}(Jac) = 0.38$, 0.4% negative Jacobians and a similar Dice accuracy of 44.0%. The visual results in Fig. 3 demonstrate highly accurate alignment of unseen test images, yielding sharp intensity mean images (f) and convincing automatic 3D segmentations (c, stapled [29]).

4 Conclusion and Outlook

In summary, the novel 2.5D subdivision of displacement spaces for discrete deep learning based image registration (DLIR), in combination with a fast instance-optimisation, advances the state-of-the-art for highly complex abdominal inter-subject registration, while limiting the computational burden in comparison to 3D networks (that have to act on 6D tensors). A suitable non-local MIND metric loss is proposed that leverages the probabilistic predictions for unsupervised learning and enables fast training. This idea can further unleash its potential for new developments in DLIR that overcome the current limitations in accuracy and enable new clinical applications of registration for image-guided interventions, radiotherapy and diagnostics. The advantages of a discrete search space and probabilistic predictions are demonstrated qualitatively and quantitatively in terms of highly accurate automatic propagation of segmentations between unseen images (with improvements of 10% points Dice over state-of-the-art), lower run times and by robustly capturing of large deformations. Future work could yield further gains of the memory-efficient 2.5D displacement space by employing more powerful regularisation networks and by incorporating contextual loss terms.

References

1. Anonymous: Tackling the problem of large deformations in deep learning based medical image registration using displacement embeddings. Medical Imaging with Deep Learning, pp. 1–5 (2020, under reviewed). https://openreview.net/pdf?id=kPBUZluVq
2. Balakrishnan, G., Zhao, A., Sabuncu, M.R., Guttag, J., Dalca, A.V.: Voxelmorph: a learning framework for deformable medical image registration. IEEE Trans. Med. Imag. (2019)
3. Blendowski, M., Nickisch, H., Heinrich, M.P.: How to learn from unlabeled volume data: self-supervised 3D context feature learning. In: Shen, D., et al. (eds.) MICCAI 2019. LNCS, vol. 11769, pp. 649–657. Springer, Cham (2019). https://doi.org/10.1007/978-3-030-32226-7_72
4. Budelmann, D., König, L., Papenberg, N., Lellmann, J.: Fully-deformable 3D image registration in two seconds. In: Handels, H., Deserno, T., Maier, A., Maier-Hein, K., Palm, C., Tolxdorff, T. (eds.) BVM, pp. 302–307. Springer, Heidelberg (2019). https://doi.org/10.1007/978-3-658-25326-4_67
5. Chen, W., Liu, B., Peng, S., Sun, J., Qiao, X.: S3D-UNet: separable 3D U-Net for brain tumor segmentation. In: Crimi, A., Bakas, S., Kuijf, H., Keyvan, F., Reyes, M., van Walsum, T. (eds.) BrainLes 2018. LNCS, vol. 11384, pp. 358–368. Springer, Cham (2019). https://doi.org/10.1007/978-3-030-11726-9_32
6. Dalca, A.V., Balakrishnan, G., Guttag, J., Sabuncu, M.R.: Unsupervised learning for fast probabilistic diffeomorphic registration. In: Frangi, A.F., Schnabel, J.A., Davatzikos, C., Alberola-López, C., Fichtinger, G. (eds.) MICCAI 2018. LNCS, vol. 11070, pp. 729–738. Springer, Cham (2018). https://doi.org/10.1007/978-3-030-00928-1_82
7. Dosovitskiy, A., et al.: Flownet: learning optical flow with convolutional networks. In: Proceedings of ICCV, pp. 2758–2766 (2015)

8. Eppenhof, K.A., Lafarge, M.W., Veta, M., Pluim, J.P.: Progressively trained convolutional neural networks for deformable image registration. IEEE Trans. Med. Imag. (2019)
9. Eppenhof, K.A., Pluim, J.P.: Pulmonary CT registration through supervised learning with convolutional neural networks. IEEE Trans. Med. Imag. **38**(5), 1097–1105 (2018)
10. Glocker, B., Komodakis, N., Tziritas, G., Navab, N., Paragios, N.: Dense image registration through MRFs and efficient linear programming. Med. Image Anal. **12**(6), 731–741 (2008)
11. Heinrich, M.P.: Closing the gap between deep and conventional image registration using probabilistic dense displacement networks. In: Shen, D., et al. (eds.) MICCAI 2019. LNCS, vol. 11769, pp. 50–58. Springer, Cham (2019). https://doi.org/10.1007/978-3-030-32226-7_6
12. Heinrich, M.P., Jenkinson, M., Brady, M., Schnabel, J.A.: MRF-based deformable registration and ventilation estimation of lung CT. IEEE Trans. Med. Imag. **32**(7), 1239–1248 (2013)
13. Heinrich, M.P., Oktay, O., Bouteldja, N.: OBELISK-Net: fewer layers to solve 3D multi-organ segmentation with sparse deformable convolutions. Med. Image Anal. **54**, 1–9 (2019)
14. Heinrich, M.P., Jenkinson, M., Papież, B.W., Brady, S.M., Schnabel, J.A.: Towards realtime multimodal fusion for image-guided interventions using self-similarities. In: Mori, K., Sakuma, I., Sato, Y., Barillot, C., Navab, N. (eds.) MICCAI 2013. LNCS, vol. 8149, pp. 187–194. Springer, Heidelberg (2013). https://doi.org/10.1007/978-3-642-40811-3_24
15. Hering, A., van Ginneken, B., Heldmann, S.: mlVIRNET: multilevel variational image registration network. In: Shen, D., et al. (eds.) MICCAI 2019. LNCS, vol. 11769, pp. 257–265. Springer, Cham (2019). https://doi.org/10.1007/978-3-030-32226-7_29
16. Hering, A., Kuckertz, S., Heldmann, S., Heinrich, M.P.: Memory-efficient 2.5 D convolutional transformer networks for multi-modal deformable registration with weak label supervision applied to whole-heart CT and MRI scans. Int. J. Comput. Assist. Radiol. Surg. **14**(11), 1901–1912 (2019)
17. Krebs, J., Delingette, H., Mailhé, B., Ayache, N., Mansi, T.: Learning a probabilistic model for diffeomorphic registration. IEEE Trans. Med. Imag. **38**(9), 2165–2176 (2019)
18. Liu, C., Yuen, J., Torralba, A.: SIFT flow: dense correspondence across scenes and its applications. IEEE Trans. Patt. Anal. Mach. Intell. **33**(5), 978–994 (2010)
19. Loshchilov, I., Hutter, F.: Decoupled weight decay regularization. arXiv preprint arXiv:1711.05101 (2017)
20. Modat, M., et al.: Fast free-form deformation using graphics processing units. Comput. Methods Programs Biomed. **98**(3), 278–284 (2010)
21. Ronneberger, O., Fischer, P., Brox, T.: U-Net: convolutional networks for biomedical image segmentation. In: Navab, N., Hornegger, J., Wells, W.M., Frangi, A.F. (eds.) MICCAI 2015. LNCS, vol. 9351, pp. 234–241. Springer, Cham (2015). https://doi.org/10.1007/978-3-319-24574-4_28
22. Rühaak, J., et al.: Estimation of large motion in lung CT by integrating regularized keypoint correspondences into dense deformable registration. IEEE Trans. Med. Imag. **36**(8), 1746–1757 (2017)
23. Setio, A.A.A., et al.: Pulmonary nodule detection in CT images: false positive reduction using multi-view convolutional networks. IEEE Trans. Med. Imag. **35**(5), 1160–1169 (2016)

24. Shekhovtsov, A., Kovtun, I., Hlaváč, V.: Efficient MRF deformation model for non-rigid image matching. Comput. Vis. Image Und. **112**(1), 91–99 (2008)
25. Shin, H.C., et al.: Deep convolutional neural networks for computer-aided detection: CNN architectures, dataset characteristics and transfer learning. IEEE Trans. Med. Imag. **35**(5), 1285–1298 (2016)
26. Sun, D., Yang, X., Liu, M.Y., Kautz, J.: PWC-Net: CNNs for optical flow using pyramid, warping, and cost volume. In: Proceedings of CVPR, pp. 8934–8943 (2018)
27. Veksler, O.: Stereo correspondence by dynamic programming on a tree. In: Proceedings of CVPR, vol. 2, pp. 384–390. IEEE (2005)
28. de Vos, B.D., Berendsen, F.F., Viergever, M.A., Sokooti, H., Staring, M., Išgum, I.: A deep learning framework for unsupervised affine and deformable image registration. Med. Image Anal. **52**, 128–143 (2019)
29. Warfield, S.K., Zou, K.H., Wells, W.M.: Simultaneous truth and performance level estimation (STAPLE): an algorithm for the validation of image segmentation. IEEE Trans. Med. Imag. **23**(7), 903–921 (2004)
30. Xu, Z., et al.: Evaluation of 6 registration methods for the human abdomen on clinically acquired CT. IEEE Trans. Biomed. Eng. **63**(8), 1563–1572 (2016)
31. Zheng, S., et al.: Conditional random fields as recurrent neural networks. In: Proceedings of ICCV, pp. 1529–1537 (2015)
32. Zikic, D., et al.: Linear intensity-based image registration by Markov random fields and discrete optimization. Med. Image Anal. **14**(4), 550–562 (2010)

Unsupervised Learning Model for Registration of Multi-phase Ultra-Widefield Fluorescein Angiography

Gyoeng Min Lee[1], Kwang Deok Seo[1], Hye Ju Song[1], Dong Geun Park[2], Ga Hyung Ryu[2], Min Sagong[2], and Sang Hyun Park[1(✉)]

[1] Department of Robotics Engineering, DGIST, Daegu, South Korea
{rud557,shpark13135}@dgist.ac.kr
[2] Department of Ophthalmology, Yeungnam University College of Medicine, Daegu, South Korea

Abstract. Registration methods based on unsupervised deep learning have achieved good performances, but are often ineffective on the registration of inhomogeneous images containing large displacements. In this paper, we propose an unsupervised learning-based registration method that effectively aligns multi-phase Ultra-Widefield (UWF) fluorescein angiography (FA) retinal images acquired over the time after a contrast agent is applied to the eye. The proposed method consists of an encoder-decoder style network for predicting displacements and spatial transformers to create moved images using the predicted displacements. Unlike existing methods, we transform the moving image as well as its vesselness map through the spatial transformers, and then compute the loss by comparing them with the target image and the corresponding maps. To effectively predict large displacements, displacement maps are estimated at multiple levels of a decoder and the losses computed from the maps are used in optimization. For evaluation, experiments were performed on 64 pairs of early- and late-phase UWF retinal images. Experimental results show that the proposed method outperforms the existing methods.

Keywords: Registration · Unsupervised learning · Deep learning · Vesselness map

1 Introduction

Ultra-Widefield (UWF) retinal imaging plays an essential role in the diagnosis and treatment of eye diseases with the peripheral retinal changes. This imaging device has multimodal capabilities including fundus photographs, fluorescein angiography (FA), indocyanine green angiography, and autofluorescence images.

Electronic supplementary material The online version of this chapter (https://doi.org/10.1007/978-3-030-59716-0_20) contains supplementary material, which is available to authorized users.

A. L. Martel et al. (Eds.): MICCAI 2020, LNCS 12263, pp. 201–210, 2020.
https://doi.org/10.1007/978-3-030-59716-0_20

Among them, the FA shows pathological changes in blood vessels such as leaks, new blood vessels, and ischemia, which can be distinguished through phase-by-phase comparison. Thus, for accurate identification and quantitative evaluation of these lesions, the registration of multi-phase images is inevitable. However, the registration is non-trivial since each image has peripheral distortion in the process of projecting a 3D fundus as a 2D image and intensity distributions of the early- and late-phase images are very different and the displacements with respect to eyeball movements are often large as shown in Fig. 1.

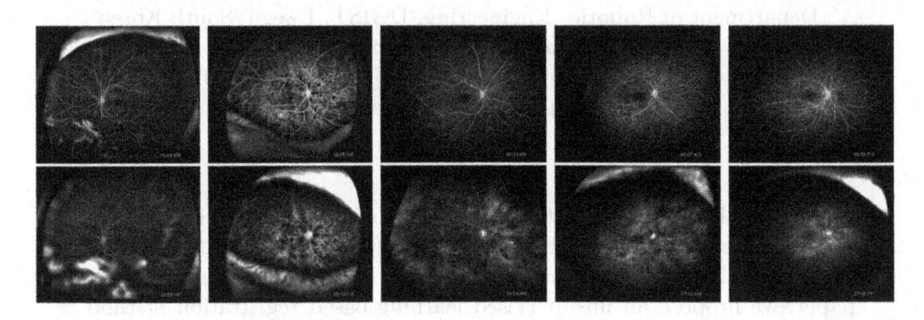

Fig. 1. Early-phase UWF FA images (Top) and late-phase UWF FA images (Bottom). It shows that the intensity distribution is different and the position of the vessel is changed by the movement between phases.

To address this problem, we propose an unsupervised learning model for aligning the multi-phase UWF FA retinal images. Recently, deep learning-based registration methods [1,15,18,21,22,27,28] that do not require ground truth displacements have been proposed, but they mostly performed intensity-based non-rigid registration to align relatively local displacements between moving and fixed images. However, these methods often fail to perform the registration of images with different intensity distribution and large displacements. We introduce a novel method that consists of an encoder-decoder style network for estimating displacements and spatial transformers to reconstruct the moved image as well as a vesselness map generated from the moving image. The moved image and vesselness map are estimated from multiple levels of a decoder and all losses from the multiple estimations are considered to address large displacements.

The main contributions of this work are as follows: (1) The proposed method provides an effective way to align inhomogeneous images, while taking advantages of the proposed unsupervised learning model by adding the loss computed with the vesselness feature map. (2) The proposed method can integrate any suitable features according to image characteristics. Though we use vesselness maps as features in this work, any advanced feature extractors that enable one to extract consistent features from inhomogeneous images can be used. (3) Large displacements are precisely estimated without pre-processing (*e.g.*, affine registration) or post-processing steps. We empirically show the benefit of using multi-level decoder predictions with improvements in performance. (4) Lastly, in the UWF FA with multiple phases, changes between images can be identified and progress can be evaluated over time. To the best of our knowledge, this is the first work to address the registration of UWF FA images.

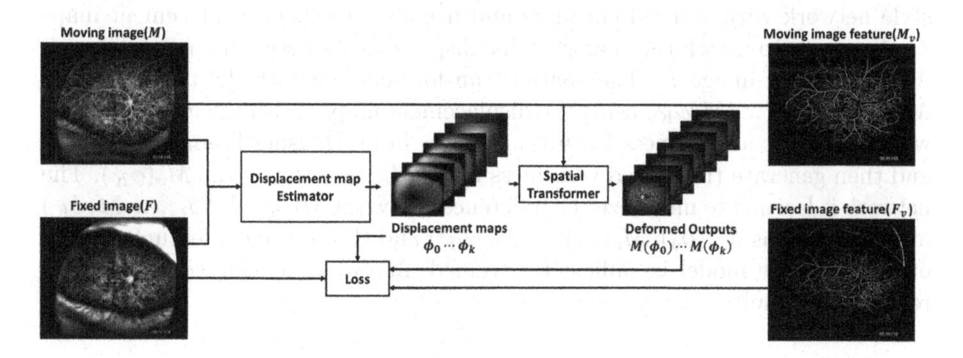

Fig. 2. Overview of our proposed method.

1.1 Related Works

Retinal Image Registration: Most of retinal image registration methods were feature-based approaches [4,7,13,20,26]. For example, vessels and bifurcation point detectors [20], SIFT [23], or edge-based Harris corner detector [6] were used to find correspondences, and then the registration was performed using iterative closest point or spline-based methods. However, these methods often fail to find robust correspondences. Recently, a deep learning-based method [14] has been proposed to find correspondences in inhomogeneous images and improve the registration performance. However, this method could not predict a dense displacement map in a single end-to-end deep learning framework.

Registration via Unsupervised Deep Learning Model: Several deep learning frameworks have been proposed for registration of medical images. Among them, supervised learning-based methods [5,12,19,24] often achieved limited results since it was difficult to produce the ground truth of displacements. Recently, registration methods [15,18,21,22] that do not require training data have been proposed. For example, Vos et al. [22] predicted a sparse displacement grid and then performed interpolation using a third-order B-spline kernel. Balakrishnan et al. first proposed a CNN network [1] that predicts dense displacement map end-to-end using spatial transformers [10]. Zhao et al. [28] proposed a cascaded framework [27] to address large displacements by gradually registering the images. However, these methods often did not work properly on the registration of images with different intensity distributions. The proposed method mainly addresses this limitation by minimizing a feature-based loss.

2 Method

Our proposed network consisting of a displacement map estimator and spatial transformers is shown in Fig. 2. In the displacement map estimator, a U-net style network with K-level encoders and decoders predicts displacement maps $\phi_0, \phi_1, ..., \phi_K$ to match the coarse to fine displacements between a moving image M and a fixed image F. The spatial transformers generate the moved images $M(\phi_0), M(\phi_1), ..., M(\phi_K)$ using the displacement maps. With the moved images, we also extract a vesselness feature maps M_v from M using Frangi filtering [8] and then generate the moved vesselness maps $M_v(\phi_0), M_v(\phi_1), ..., M_v(\phi_K)$. The network is learned to minimize the differences between $M(\phi_0), M(\phi_1), ..., M(\phi_K)$ and F as well as $M_v(\phi_0), M_v(\phi_1), ..., M_v(\phi_K)$ and the vesselness feature map F_v of F. After the model is sufficiently trained, $M(\phi_0)$ is considered as the final registration result.

2.1 Unsupervised Learning Model for Registration

Inspired by VoxelMorph [1], we follow the overall structure of that model using the spatial transformer. However, since the U-Net structure [17] is not designed to adjust large displacements, we predict the displacement maps at each level of decoder as shown in Fig. 3. M and F pass through 6 encoders and decoders (i.e., $K = 6$) with [16, 32, 32, 32, 64, 64] channels for encoding block and its reversed order for decoding block. The feature maps of each level of the decoder pass through a 3×3 convolution layer with LeakyReLU activation function to estimate the displacements with different resolutions. A convolution layer with 4×4 kernel size and 2×2 stride is used without pooling in the conv block of encoder, while a $\times 2$ upsampling is used in the conv block of decoder. The displacement maps in low resolution, i.e., ϕ_K or ϕ_{K-1}, are upsampled to the size of F by the linear interpolation.

In optimization using gradient-descent, the parameters should be differentiable with respect to the loss function. Thus, the spatial transformer proposed in [10] is used for this purpose. The spatial transformer uses bilinear interpolation to calculate values between pixels in an image. Then, the pixel values of $M(\phi)$ are sampled using ϕ. By using linear interpolation, we can transform images with discontinuities into a continuous grid. This makes it possible to perform differentiable transformations.

2.2 Loss Function

Spatial transformers allow the use of any deformation field in any of the derivative fields. Thus, we define the loss function using the feature maps obtained from the input image along with a similarity loss. Thus, similarity loss (L_{sim}) between $M(\phi)$ and F is defined as the negative cross-correlation(CC) of local regions of $M(\phi)$ and F. In particular, let $\hat{F}(p)$ and $\hat{M}(\phi(p))$ denote the mean

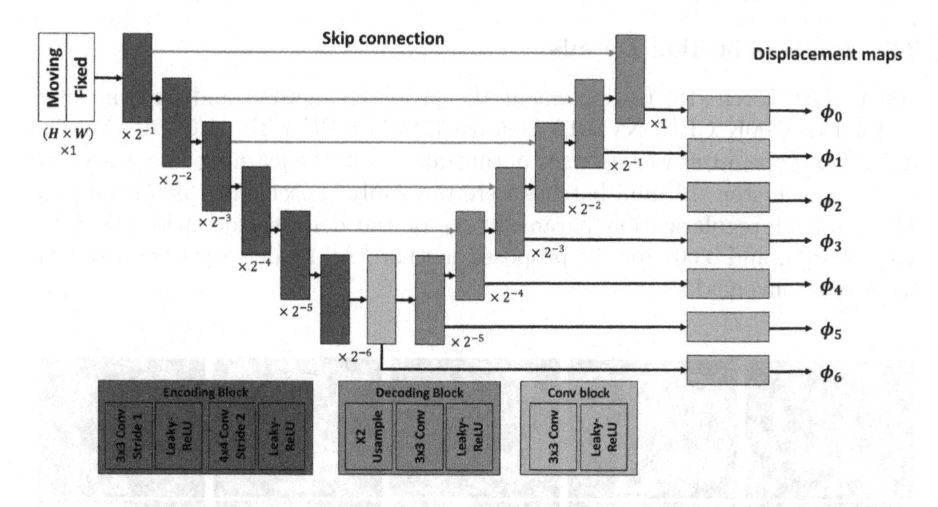

Fig. 3. Architecture of the proposed deformation field estimator network.

intensity in a $n \times n$ patch, the local cross-correlation of F and $M(\phi)$ is defined as:

$$CC(F, M(\phi)) = \sum_{p \in \Omega} \frac{(\sum_{p_i}(F(p_i) - \hat{F}(p))(M(\phi(p_i)) - \hat{M}(\phi(p))))^2}{(\sum_{p_i}(F(p_i) - \hat{F}(p)))(\sum_{p_i}(M(\phi(p_i)) - \hat{M}(\phi(p))))}, \quad (1)$$

where p_i iterates over a $n \times n$ patch around p. We set $n = 9$ in our experiments. The higher the CC value, the better the alignment. The loss function is used in the form of $L_{sim}(F, M(\phi)) = -CC(F, M(\phi))$. Furthermore, to match $M(\phi)$ with F with different characteristics, CC between $M_v(\phi)$ and F_v is also added to the loss function. For the feature maps, Frangi filter is applied to M and F and each vesselness map is extracted and scaled to [0, 1] range. Displacement maps obtained by network are applied to ϕ to generate $M_v(\phi)$. CC between $M_v(\phi)$ and F_v was calculated as: $L_{vessel}(F_v, M_v(\phi)) = -CC(F_v, M_v(\phi))$.

With L_{sim} and L_{vessel}, we also use the smoothness loss L_{smooth} to regularize unnatural displacements. It is based on the gradient of the deformation field as:

$$L_{smooth}(\phi) = \sum_{p \in \Omega} \| \nabla \phi(p) \|^2, \quad (2)$$

Finally, the final loss is defined as:

$$L_{total} = \sum_{k=0}^{K} (L_{sim}(F, M(\phi_k)) + L_{vessel}(F_v, M_v(\phi_k)) + \lambda_1 L_{smooth}(\phi_k)), \quad (3)$$

where λ is the regularization parameter.

2.3 Implementation Details

We used PyTorch [16] to implement the proposed method and experimented on Intel i7-8700K CPU, NVIDIA Geforce 1080Ti GPU with 64 GB RAM. The ADAM [11] optimizer was used for optimization and the learning rate was set to $1e^{-4}$. In each step, 16 mini-batches were randomly chosen from all 64 samples. The gradient regularization parameter λ_1 of the deformation field was 1 for VoxelMorph, and 0.001 for the proposed method. All models were trained until the losses converged.

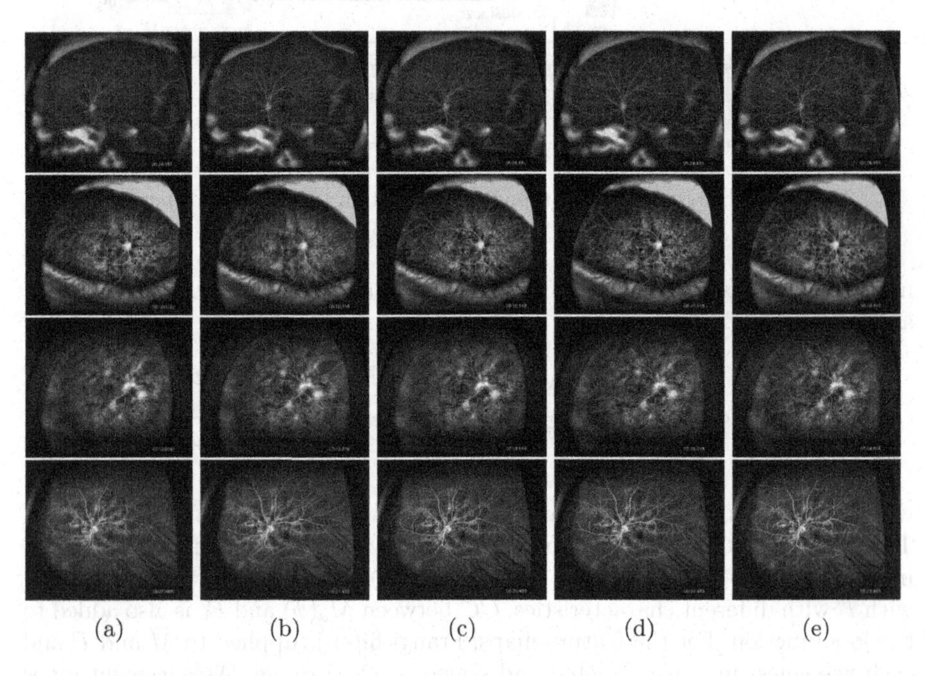

Fig. 4. Qualitative results. Each image shows overlap of a deformed early-phase (purple) and a late-phase image (green). Well overlapped blood vessels appear white. (a) Affine, (b) Affine + B-Spline, (c) VoxelMorph, (d) Ours, (e) Ours+vesselness map (Color figure online)

3 Experimental Results

Dataset. For evaluation, we used 1) 30 pairs of early- and late-phase UWF FA images with size of 3900×3076 from 30 patients with diabetic retinopathy and 2) 5∼10 multi-phase UWF FA images with the size of 4000×4000 from 34 patients with other retinal vascular diseases. All these images were acquired from a university hospital. We resized these images to a same size and performed histogram equalization to adjust the intensity range in [0, 1]. We extracted vessels and bifurcation points from the pair of images to measure the registration accuracy. For the vessel extraction, we made a binary segmentation by thresholding

Table 1. Mean recall, precision, DSC, distance scores of the proposed method and the related methods. Moving denotes the accuracy scores between moving images and fixed images.

Method	Precision	Recall	DSC	Distance
Moving	0.11	0.19	0.12 ± 0.07	14.6 ± 19.19
Affine	0.21	0.23	0.21 ± 0.16	10.65 ± 15.27
Affine + B-Spline	0.18	0.15	0.16 ± 0.07	23.72 ± 28.55
SIFT-RANSAC	0.23	0.27	0.24 ± 0.16	50.60 ± 104.7
VoxelMorph	0.40	0.44	0.41 ± 0.18	10.13 ± 20.08
+Vesselness	0.41	0.45	0.42 ± 0.18	10.25 ± 20.32
Ours	0.43	0.46	0.43 ± 0.20	8.99 ± 20.46
+Vesselness	0.45	0.48	0.45 ± 0.19	8.25 ± 20.59

of vesselness scores obtained by Frangi filtering [8] and then manually corrected errors. The bifurcation points were also annotated manually.

Experimental Settings. To confirm the superiority of the proposed method, we compared our method with affine registration, B-spline, SIFT with RANSAC and VoxelMorph. The imregister function, a Matlab built-in function, was used for the affine registration with the mutual information similarity measurement. The transformation matrix was obtained by evolutionary optimization method. After the affine transformation, a B-spline non-rigid registration model implemented in SimpleITK [25] was used and optimized with a gradient based L-BFGS-B [3] optimization algorithm which minimizes the mutual information between a moved and fixed image. OpenCV [2] was used for implementing SIFT descriptor and RANSAC [9] with the perspective transform matrix. For VoxelMorph, we used the code provided by the authors. We predict the results after the model is learned until the losses converged to a certain level around 35,000 epochs. To verify the effect of each element of the proposed method, we also generated the results using the proposed method without feature loss, i.e., in other words, VoxelMorph with multi-level displacement estimation. For evaluation, we compared the average distance between corresponding bifurcation points and the precision, recall, and DSC scores between the vascular masks in 64 pairs of early- and late-phase.

Quantitative Results. Table 1 shows the average precision, recall, DSC between masks and the average pixel distances between correspondences of the proposed method and the related methods. The distances between correspondences of affine transform were not significantly different with the distances before transformation. In the case of affine+B-spline, there were many cases where matching failed due to a large intensity distribution difference. VoxelMorph achieved better performance both in distance and DSC scores, but the

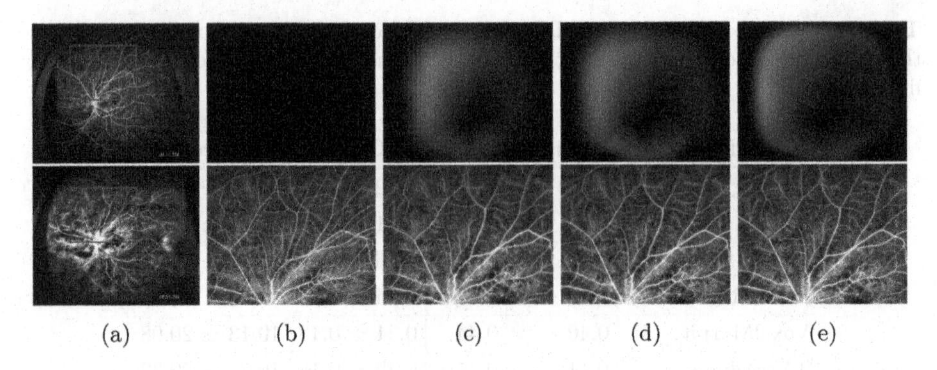

Fig. 5. Example of the results of the multi-level displacement map of the proposed method. (a) shows early-phase image (top) and late-phase image (bottom) (b) is not a deformed image and (c)–(e) are images in which the deformed image and the late image are overlapped on the red rectangular region shown in (a) with different resolutions of displacement maps (ϕ_4, ϕ_3, and ϕ_0). (Color figure online)

improvement on the distances between correspondences was relatively low. On the other hand, our proposed network achieved the best performances for most cases. Note that the reduced distance between the branching points indicates that the result of our method obtained natural displacement maps. When displacement estimation was performed at multi-levels of the decoder, performance improved by 2% for DSC and 1 for distance compared to VoxelMorph. Furthermore, the performance improved by 3% for DSC and 2 for distance when the loss computed by vesselness was used.

Qualitative Results. Figure 4 shows images of early-phase (moving) and late-phase (fixed), and their qualitative results. The result of affine registration was limited since robust features could not be often extracted and it was difficult to consider non-rigid changes. VoxelMorph aligned most of the vessels well in the pupil area by performing non-rigid registration, but some errors occurred near the periphery with large displacements. On the other hand, the proposed method achieved robust results in most areas. Figure 5 shows the example of multi-level displacement map. In addition, it was shown that elaborate matching was performed as the resolution was changed from low to high resolution.

4 Conclusions

We have proposed a novel registration method for aligning early- and late-phase UWF FA retinal images with different characteristics. The proposed method does not only take advantage of unsupervised learning-based registration methods, but also works effectively for the registration of inhomogeneous images with the addition of a feature loss. Furthermore, we introduce a way to effectively estimate large displacements by predicting the displacement map at each level of the decoder. The performance improvement of proposed method was significant on the evaluation of 64 pairs of UWF FA retinal images. We believe that the proposed technique will contribute the diagnosis and quantification of eye diseases since it is easy to find leaks, new blood vessels, and ischemia if multi-phase UWF FA images are well aligned.

Acknowledgement. This work was supported by the National Research Foundation of Korea (NRF) grant funded by the Korean Government (MSIT) (No. 2019R1C1C1008727), and the grant of the medical device technology development program funded by the Ministry of Trade, Industry and Energy (MOTIE, Korea)(20006006).

References

1. Balakrishnan, G., Zhao, A., Sabuncu, M.R., Guttag, J., Dalca, A.V.: Voxelmorph: a learning framework for deformable medical image registration. IEEE Trans. Med. Imaging **38**(8), 1788–1800 (2019)
2. Bradski, G., Kaehler, A.: Learning OpenCV: Computer Vision with the OpenCV Library. O'Reilly Media, Inc. (2008)
3. Byrd, R.H., Lu, P., Nocedal, J., Zhu, C.: A limited memory algorithm for bound constrained optimization. SIAM J. Sci. Comput. **16**(5), 1190–1208 (1995)
4. Can, A., Stewart, C.V., Roysam, B., Tanenbaum, H.L.: A feature-based, robust, hierarchical algorithm for registering pairs of images of the curved human retina. IEEE Trans. Pattern Anal. Mach. Intell. **24**(3), 347–364 (2002)
5. Cao, X., et al.: Deformable image registration based on similarity-steered CNN regression. In: Descoteaux, M., Maier-Hein, L., Franz, A., Jannin, P., Collins, D.L., Duchesne, S. (eds.) MICCAI 2017. LNCS, vol. 10433, pp. 300–308. Springer, Cham (2017). https://doi.org/10.1007/978-3-319-66182-7_35
6. Chen, J., Tian, J., Lee, N., Zheng, J., Smith, R.T., Laine, A.F.: A partial intensity invariant feature descriptor for multimodal retinal image registration. IEEE Transact. Biomed. Eng. **57**(7), 1707–1718 (2010)
7. Choe, T.E., Cohen, I.: Registration of multimodal fluorescein images sequence of the retina. In: Tenth IEEE International Conference on Computer Vision (ICCV 2005), vol. 1, pp. 106–113. IEEE (2005)
8. Frangi, A.F., Niessen, W.J., Vincken, K.L., Viergever, M.A.: Multiscale vessel enhancement filtering. In: Wells, W.M., Colchester, A., Delp, S. (eds.) MICCAI 1998. LNCS, vol. 1496, pp. 130–137. Springer, Heidelberg (1998). https://doi.org/10.1007/BFb0056195
9. Hartley, R., Zisserman, A.: Multiple View Geometry in Computer Vision. Cambridge University Press, Cambridge (2003)
10. Jaderberg, M., Simonyan, K., Zisserman, A., et al.: Spatial transformer networks. In: Advances in Neural Information Processing Systems, pp. 2017–2025 (2015)
11. Kingma, D.P., Ba, J.: Adam: A method for stochastic optimization. arXiv preprint arXiv:1412.6980 (2014)
12. Krebs, J., et al.: Robust non-rigid registration through agent-based action learning. In: Descoteaux, M., Maier-Hein, L., Franz, A., Jannin, P., Collins, D.L., Duchesne, S. (eds.) MICCAI 2017. LNCS, vol. 10433, pp. 344–352. Springer, Cham (2017). https://doi.org/10.1007/978-3-319-66182-7_40

13. Laliberté, F., Gagnon, L., Sheng, Y.: Registration and fusion of retinal images-an evaluation study. IEEE Trans. Med. Imaging **22**(5), 661–673 (2003)
14. Lee, J.A., Liu, P., Cheng, J., Fu, H.: A deep step pattern representation for multimodal retinal image registration. In: Proceedings of the IEEE International Conference on Computer Vision, pp. 5077–5086 (2019)
15. Li, H., Fan, Y.: Non-rigid image registration using fully convolutional networks with deep self-supervision. arXiv preprint arXiv:1709.00799 (2017)
16. Paszke, A., et al.: PyTorch: an imperative style, high-performance deep learning library. In: Advances in Neural Information Processing Systems 32, pp. 8024–8035. Curran Associates, Inc. (2019). http://papers.neurips.cc/paper/9015-pytorch-an-imperative-style-high-performance-deep-learning-library.pdf
17. Ronneberger, O., Fischer, P., Brox, T.: U-Net: convolutional networks for biomedical image segmentation. In: Navab, N., Hornegger, J., Wells, W.M., Frangi, A.F. (eds.) MICCAI 2015. LNCS, vol. 9351, pp. 234–241. Springer, Cham (2015). https://doi.org/10.1007/978-3-319-24574-4_28
18. Sentker, T., Madesta, F., Werner, R.: GDL-FIRE^{4D}: deep learning-based fast 4D CT image registration. In: Frangi, A.F., Schnabel, J.A., Davatzikos, C., Alberola-López, C., Fichtinger, G. (eds.) MICCAI 2018. LNCS, vol. 11070, pp. 765–773. Springer, Cham (2018). https://doi.org/10.1007/978-3-030-00928-1_86
19. Sokooti, H., de Vos, B., Berendsen, F., Lelieveldt, B.P.F., Išgum, I., Staring, M.: Nonrigid image registration using multi-scale 3D convolutional neural networks. In: Descoteaux, M., Maier-Hein, L., Franz, A., Jannin, P., Collins, D.L., Duchesne, S. (eds.) MICCAI 2017. LNCS, vol. 10433, pp. 232–239. Springer, Cham (2017). https://doi.org/10.1007/978-3-319-66182-7_27
20. Stewart, C.V., Tsai, C.L., Roysam, B.: The dual-bootstrap iterative closest point algorithm with application to retinal image registration. IEEE Trans. Med. Imaging **22**(11), 1379–1394 (2003)
21. de Vos, B.D., Berendsen, F.F., Viergever, M.A., Sokooti, H., Staring, M., Išgum, I.: A deep learning framework for unsupervised affine and deformable image registration. Med. Image Anal. **52**, 128–143 (2019)
22. de Vos, B.D., Berendsen, F.F., Viergever, M.A., Staring, M., Išgum, I.: End-to-end unsupervised deformable image registration with a convolutional neural network. In: Cardoso, M.J., et al. (eds.) DLMIA/ML-CDS -2017. LNCS, vol. 10553, pp. 204–212. Springer, Cham (2017). https://doi.org/10.1007/978-3-319-67558-9_24
23. Yang, G., Stewart, C.V., Sofka, M., Tsai, C.L.: Alignment of challenging image pairs: refinement and region growing starting from a single keypoint correspondence. IEEE Trans. Pattern Anal. Mach. Intell. **23**(11), 1973–1989 (2007)
24. Yang, X., Kwitt, R., Styner, M., Niethammer, M.: Quicksilver: fast predictive image registration-a deep learning approach. NeuroImage **158**, 378–396 (2017)
25. Yaniv, Z., Lowekamp, B.C., Johnson, H.J., Beare, R.: Simpleitk image-analysis notebooks: a collaborative environment for education and reproducible research. J. Digit. Imaging **31**(3), 290–303 (2018)
26. Zana, F., Klein, J.C.: A registration algorithm of eye fundus images using a Bayesian Hough transform. In: 7th International Conference on Image Processing and its Applications (1999)
27. Zhao, S., Dong, Y., Chang, E.I., Xu, Y., et al.: Recursive cascaded networks for unsupervised medical image registration. In: Proceedings of the IEEE International Conference on Computer Vision, pp. 10600–10610 (2019)
28. Zhao, S., Lau, T., Luo, J., Eric, I., Chang, C., Xu, Y.: Unsupervised 3D end-to-end medical image registration with volume tweening network. IEEE J. Biomed. Health Inform. (2019)

Large Deformation Diffeomorphic Image Registration with Laplacian Pyramid Networks

Tony C. W. Mok(✉) and Albert C. S. Chung

Lo Kwee-Seong Medical Image Analysis Laboratory, Department of Computer Science and Engineering, The Hong Kong University of Science and Technology, Kowloon, Hong Kong
cwmokab@connect.ust.hk

Abstract. Deep learning-based methods have recently demonstrated promising results in deformable image registration for a wide range of medical image analysis tasks. However, existing deep learning-based methods are usually limited to small deformation settings, and desirable properties of the transformation including bijective mapping and topology preservation are often being ignored by these approaches. In this paper, we propose a deep Laplacian Pyramid Image Registration Network, which can solve the image registration optimization problem in a coarse-to-fine fashion within the space of diffeomorphic maps. Extensive quantitative and qualitative evaluations on two MR brain scan datasets show that our method outperforms the existing methods by a significant margin while maintaining desirable diffeomorphic properties and promising registration speed.

Keywords: Image registration · Diffeomorphic registration · Deep Laplacian pyramid networks

1 Introduction

Deformable registration is the process of computing a non-linear transformation to align a pair of images or image volumes by maximizing certain similarity metric between the images. Deformable registration is crucial in a variety of medical image analysis, including diagnostic tasks, radiotherapy and image-guided surgery. Conventional image registration methods [2,18,23,24] often rely on the multi-resolution strategy and estimate the target transformation iteratively along with a smoothness regularization. Although conventional image registration methods excel in registration accuracy and diffeomorphic properties (i.e., invertible and topology preserving), the running time of the registration process is dependent on the degree of misalignment between the input images and can be time-consuming with high-resolution 3D image volumes. Recent unsupervised deep learning-based image registration (DLIR) methods [4,5,20,26] have demonstrated promising registration speed and quality in a variety of deformable image

A. L. Martel et al. (Eds.): MICCAI 2020, LNCS 12263, pp. 211–221, 2020.
https://doi.org/10.1007/978-3-030-59716-0_21

registration tasks. They treat the image registration problem as the pixel-wise image translation problem, which attempt to learn the pixel-wise spatial correspondence from a pair of input images by using convolutional neural networks (CNN). This significantly speeds up the registration process and shows immense potential for time-sensitive medical studies such as image-guided surgery and motion tracking. However, these approaches may not be good solutions to unsupervised large deformation image registration for two reasons. First, the gradient of the similarity metric at the finest resolution is rough in general, as many possible transformations of the moving image could yield similar measurements of similarity. Second, the optimization problem without the initialized transformation at the finest resolution is difficult due to the large degrees of freedom in the transformation parameters.

To address this challenge, a preliminary study [25] proposes to stack multiple CNNs for direct affine and deformable image registrations, which are optimized separately. Zhao et al. [30] leverage an end-to-end recursive cascaded network to refine the registration result progressively, which is identical to breaking down a large deformation into multiple small deformations. But, both methods are only optimized at the finest level using gradient descent and therefore the results can be sub-optimal as the gradient of the similarity metric can be rough at the finest resolution. Moreover, the recursive cascaded networks consume tremendous extra GPU memory, which limits the possible degree of refinement in 3D settings, resulting in minimal improvement over the brain MR registration tasks. A recent paper [9] avoids these pitfalls and utilizes multiple separated CNNs to mimic the conventional multi-resolution strategy. However, the multiple networks are trained separately and the non-linearity of feature maps in each network are collapsed into a warped image before feeding into the next level. Furthermore, these methods completely ignore the desirable diffeomorphic properties of the transformation, which can further limit their potential for clinical usage.

In this paper, we address the above challenges and present a new deep Laplacian Pyramid Image Registration Network (LapIRN) for large deformation image registration. The main contributions of this work are as follows. We

- present a novel LapIRN for large deformable image registration that utilizes the advantages of a multi-resolution strategy while maintaining the non-linearity of the feature maps throughout the coarse-to-fine optimization scheme;
- propose a new pyramid similarity metric for a pyramid network to capture both large and small misalignments between the input scans, which helps to avoid local minima during the optimization; and
- present an effective diffeomorphic setting of our method and show that our method guarantees desirable diffeomorphic properties, including the invertibility and topology preservation, of the computed transformations.

2 Methods

The Laplacian pyramid network has demonstrated its efficiency and effectiveness in a variety of computer vision tasks, including high-resolution image synthetic

[6,28], super-resolution [13] and optical flow estimation [29], in constructing high-resolution solutions, stabilizing the training and avoiding the local minima. Motivated by the successes of Laplacian pyramid networks, we propose LapIRN that naturally integrates the conventional multi-resolution strategy while maintaining the non-linearity of the feature maps throughout different pyramid levels. In the following sections, we describe the methodology of our proposed LapIRN, including the Laplacian pyramid architecture, coarse-to-fine training scheme, the loss function and, finally, we describe the diffeomorphic settings of our method.

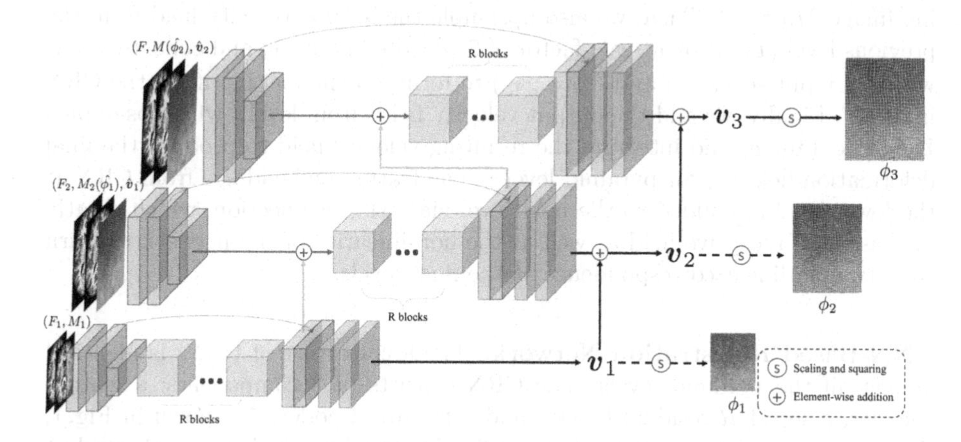

Fig. 1. Overview of the proposed 3-level deep Laplacian pyramid image registration networks in 2D settings. We utilize three identical CNN-based registration networks to mimic the registration with the multi-resolution schema. The feature maps from feature encoder, a set of R residual blocks, and feature decoder are colored with blue, green and red, respectively. The dotted paths are only included in the training phase. We highlight that all registrations are done in 3D throughout this paper. For clarity and simplicity, we depict the 2D formulation of our method in the figure. (Color figure online)

2.1 Deep Laplacian Pyramid Image Registration Networks

Given a fixed 3D scan F and a moving 3D scan M, the objective of our method is to estimate a time 1 diffeomorphic deformation field $\phi^{(1)}$ such that the warped moving scan $M(\phi^{(1)})$ is aligned with F, subject to the smoothness regularization on the predicted velocity field $\boldsymbol{v}$. Specifically, we parametrize the deformable registration problem as a function $f_\theta(F, M) = \phi^{(1)}$ with the Laplacian pyramid framework, where θ represents the learning parameters in the networks.

Network Architecture. We implement our LapIRN using a L-level Laplacian pyramid framework to mimic the conventional multi-resolution strategy. For simplicity, we set L to 3 throughout this paper. The overview of LapIRN is illustrated in Fig. 1. Specifically, we first create the input image pyramid by downsampling the input images with trilinear interpolation to obtain $F_i \in \{F_1, F_2, F_3\}$ (and $M_i \in \{M_1, M_2, M_3\}$), where F_i denotes the downsampled F with a scale factor $0.5^{(L-i)}$ and $F_3 = F$. We employ a CNN-based registration network (CRN) to solve the optimization problem for each pyramid level. For the first pyramid level, CRN captures the non-linear misalignment from the concatenated input scans with the coarsest resolution (F_1 and M_1) and outputs the 3-channel dense vector fields $\boldsymbol{v}_1$ and deformation fields ϕ_1. For pyramid level $i > 1$, we first upsample the output deformation field from the previous pyramid level (ϕ_{i-1}) by a factor of 2 to obtain $\hat{\phi}_{i-1}$ and warp M_i with $\hat{\phi}_{i-1}$ to obtain a warped moving image $M_i(\hat{\phi}_{i-1})$. Then, we also upsample the output velocity field from the previous level ($\boldsymbol{v}_{i-1}$) by a scale factor of 2 (denoted as $\hat{\boldsymbol{v}}_{i-1}$) and concatenate it with the input scans (F_i and $M_i(\hat{\phi}_{i-1})$) to form a 5-channel input for the CRN in level i. Finally, we add the output velocity fields from level i with upsampled $\hat{\boldsymbol{v}}_{i-1}$ to obtain $\boldsymbol{v}_i$ and integrate the resulting velocity field to produce the final deformation fields ϕ_i for pyramid level i. The feature embeddings from CRN at the lower level are added to the next level via a skip connection, which greatly increases the receptive field as well as the non-linearity of the network to learn complex non-linear correspondence at the finer levels.

CNN-Based Registration Network. The architecture of CRNs is identical among all the pyramid levels. The CRN consists of 3 components: a feature encoder, a set of R residual blocks, and a feature decoder. As shown in Fig. 1, the feature encoder is comprised of two 3^3 3D convolutional layers with stride 1 and one 3^3 3D convolutional layer with stride 2. In our implementation, we use 5 residual blocks for each CRN, each containing two 3^3 3D convolutional layers with pre-activation structure [8] and skip connection. Finally, a feature decoder module with one transpose convolutional layer and two consecutive 3^3 3D convolutional layers with stride 1, followed by SoftSign activation, is appended at the end to output the target velocity fields $\boldsymbol{v}$. A skip connection from the feature encoder to the feature decoder is added to prevent the vanishing of low-level features when learning the target deformation fields. In CRN, each convolution layer has 28 filters and is followed by a leaky rectified linear unit (LeakyReLU) activation [15] with a negative slope of 0.2, except for the output convolution layers.

Coarse-to-fine Training. Intuitively, our proposed LapIRN can be trained in an end-to-end manner, which is identical to learning a multi-resolution registration with deep supervision [14]. However, we found that end-to-end training for LapIRN is not an ideal training scheme as it is difficult to balance the weight of multiple losses between different resolutions. To address this issue, we propose to train LapIRN using a coarse-to-fine training scheme with a stable warm start, which is similar to [11,28]. Specifically, we first train the CRN from the coarsest level alone and then we progressively add the CRN from the next level to learn the image registration problem at a finer resolution. To avoid an unstable warm start, we freeze the learning parameters for all the pre-trained CRNs for a constant M steps whenever a new CRN is added to the training. We set M to 2000 and repeat this training scheme until the finest level is completed.

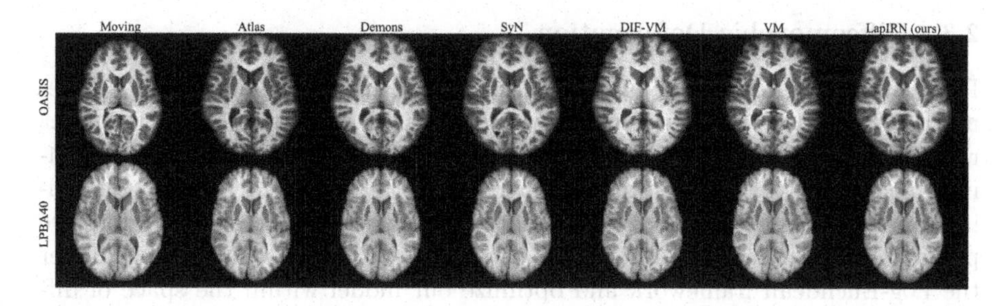

Fig. 2. Example axial MR slices from the moving, atlas and resulting warped images from Demons, SyN, DIF-VM, VM and LapIRN. The caudate and putamen are colored in red and blue respectively. Major artifacts are pointed out by yellow arrows. (Color figure online)

2.2 Similarity Pyramid

Solving the image registration problem with a intensity-based similarity metric on the finest resolution often results in local minimal solutions. By leveraging the fact that perfectly aligned image pair will yield high similarity values among all resolutions, we propose a similarity pyramid framework to address this challenge. Although the proposed similarity pyramid framework applies to a multitude of similarity measurements, we formulate it using local normalized cross-correlation (NCC) as seen in [4] for simplicity. The proposed similarity pyramid is then formulated as:

$$\mathcal{S}^K(F, M) = \sum_{i \in [1..K]} -\frac{1}{2^{(K-i)}} NCC_w(F_i, M_i), \tag{1}$$

where $\mathcal{S}^K(\cdot, \cdot)$ denotes the similarity pyramid with K levels, NCC_w represents the local normalized cross-correlation with windows size w^3, and (F_i, M_i) denotes the images in the image pyramid (i.e., F_1 is the image with the lowest resolution). A lower weight is assigned to the similarity value with lower resolution to prevent the domination of the similarity from lower level. We set w to $1 + 2i$ in our implementation. The proposed similarity pyramid captures the similarity in a multi-resolution fashion. Since the similarity metric is smoother and less sensitive to noise in the coarser resolution, integrating the similarity metric from a lower level helps to avoid local minima during the optimization problem in high-resolution.

$$\mathcal{L}_p(F, M(\phi), \boldsymbol{v}) = \mathcal{S}^p(F, M(\phi)) + \frac{\lambda}{2^{(L-p)}} ||\nabla \boldsymbol{v}||_2^2, \tag{2}$$

where $p \in [1..L]$ denotes the current pyramid level, the second terms is the smoothness regularization on the velocity field $\boldsymbol{v}$, and λ is a regularization parameter.

2.3 Diffeomorphic Deformation

Recent DLIR methods often parameterize the deformation model using displacement field u such that the dense deformation field $\phi(x) = x + u(x)$, where x represents the identity transformation. Although this parameterization is common and intuitive, the desirable properties of the predicted solution, including topology preservation and invertibility, cannot be guaranteed. Therefore, we parameterize our deformation model using the stationary velocity field under the Log-Euclidean framework and optimize our model within the space of diffeomorphic maps. Specifically, the diffeomorphic deformation field ϕ is defined as $\frac{d\phi_t}{dt} = \boldsymbol{v}(\phi^t)$, subject to $\phi^{(0)} = Id$. We follow [1,5] to integrate the (smooth) stationary velocity field $\boldsymbol{v}$ over unit time using the scaling and squaring method with time step $T = 7$ to obtain the time 1 deformation field $\phi^{(1)}$ such that $\phi^{(1)}$ is approximated to $exp(\boldsymbol{v})$, which is a member of the Lie group. Apart from that, we also report the results of LapIRN$_{disp}$, which is a variant of LapIRN parameterizing the deformation model with displacement fields instead.

3 Experiments

Data and Pre-processing. We have evaluated our method on brain atlas registration tasks using 425 T1-weighted brain MR scans from the OASIS [16] dataset and 40 brain MR scans from the LPBA40 [22] dataset. In the OASIS dataset, it includes subjects aged from 18 to 96 and 100 of the included subjects suffered from very mild to moderate Alzheimer's disease. We carry out standard preprocessing steps, including skull stripping, spatial normalization and subcortical structures segmentation, for each MR scan using FreeSurfer [7]. For OASIS, we utilize the subcortical segmentation maps of the 26 anatomical structures as the ground truth in the evaluation. In the LPBA40 dataset, the MR scans in atlas space and its subcortical segmentation map of 56 structures, which are manually delineated by experts, are used in our experiments. We resample all MR scans with isotropic voxel sizes of 1^3 mm and center cropped all the preprocessed MRI scans to $144 \times 192 \times 160$. We randomly split the OASIS dataset into 255, 20 and 150 volumes and split the LPBA40 dataset into 28, 2 and 10 volumes for training, validation and test sets, respectively. We randomly select 5 MR scans and 2 MR scans from the test sets as the atlas in OASIS and LPBA40, respectively. Finally, we register each subject to an atlas using different deformable registration methods and list the results in Table 1. In total, there are 745 and 18 combinations of test scans from OASIS and LPBA40, respectively, included in the evaluation.

Measurement. While recent DLIR methods [4,10,30] evaluate their method solely based on the Dice score between the segmentation maps in warped moving scans and the atlas, the quality of the predicted deformation fields, as well as the desirable diffeomorphic properties, are by no means to be ignored. Therefore, we evaluate our method using a sequence of measurements, including the Dice score of the subcortical segmentation maps (DSC), the percentage of voxels with non-positive Jacobian determinant ($|J_\phi|_{\leq 0}$), the standard deviation of the Jacobian determinant on the deformation fields (std($|J_\phi|$)), the volume change between the segmentation maps before and after transformation (TC) [21], and the average running time to register each pair of MR scans in seconds (Time), to provide a comprehensive evaluation on registration accuracy and the quality of solutions.

Implementation. Our proposed method LapIRN and its variants LapIRN$_{disp}$ are implemented with PyTorch [19]. We employ an Adam optimizer with a fixed learning rate $1e^{-4}$. We set λ to 4 for LapIRN, which is just enough to guarantee the smoothness of the velocity fields, and λ to 1 for LapIRN$_{disp}$. We train our networks from scratch and select the model with the highest Dice score on the validation set.

Baseline Methods. We compare our method with two conventional approaches (denoted as Demons [27] and SyN [2]) and two cutting edge DLIR methods (denoted as VM [4] and DIF-VM [5]). Demons and SyN are the top-performing registration among 14 classical non-linear deformation algorithms [12]. Both Demons and SyN utilize a 3-level multi-resolution strategy to capture large deformation. VM employs a "U" shape CNN structure to learn the dense non-linear correspondence between input scans, while DIF-VM is a probabilistic diffeomorphic variant of VM. For Demons, we use the official implementation in the ITK toolkit [17]. For SyN, we adopt the official implementation in the ANTs software package [3]. The parameters in Demons and SyN are carefully tuned to balance the tradeoff between registration accuracy and runtime. For the DLIR methods (VM and DIF-VM), we use their official implementation online with default parameters. All DLIR methods are trained from scratch.

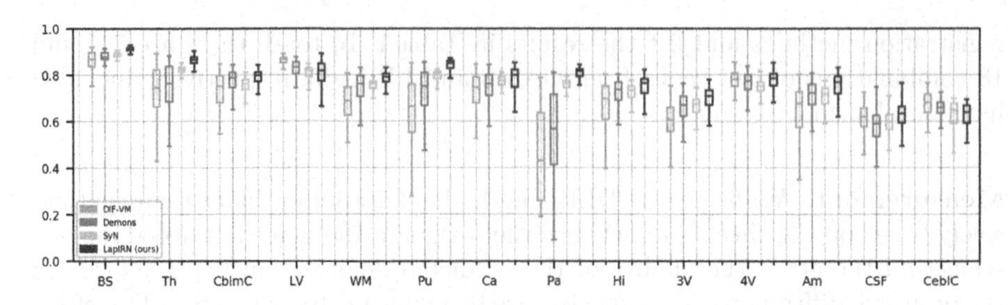

Fig. 3. Boxplots depicting the average Dice scores of each anatomical structure in OASIS for DIF-VM, SyN, Demons and our method. The left and right hemispheres of the brain are combined into one structure for visualization. The brain stem (BS), thalamus (Th), cerebellum cortex (CblmC), lateral ventricle (LV), cerebellum white matter (WM), putamen (Pu), caudate (Ca), pallidum (Pa), hippocampus (Hi), 3rd ventricle (3V), 4th ventricle (4V), amygdala (Am), CSF (CSF), and cerebral cortex (CeblC) are included.

Table 1. Quantitative evaluation of the results from OASIS and LPBA40 dataset. DSC indicates registration accuracy. $|J_\phi|_{\leq 0}$ represents the average percentage of folding voxels in the deformation fields. std($|J_\phi|$) indicates the smoothness of the deformation fields (lower is better). *TC* indicates the topology change of the anatomical structure (closer to 1 is better). *Time* indicates the average running time to register each pair of MR scans in seconds. Initial: spatial normalization.

Method	OASIS					LPBA40				
	DSC	$\lvert J_\phi\rvert_{\leq 0}$	std($\lvert J_\phi\rvert$)	TC	Time	DSC	$\lvert J_\phi\rvert_{\leq 0}$	std($\lvert J_\phi\rvert$)	TC	Time
Initial	0.567	-	-	-	-	0.586	-	-	-	-
Demons	0.715	0.000	0.259	1.102	192	0.720	0.048	0.174	1.004	190
SyN	0.723	0.000	0.357	1.109	1439	0.725	0.000	0.241	1.069	1225
DIF-VM	0.693	0.008	0.592	1.086	0.695	0.680	0.970	0.414	0.986	0.683
VM	0.727	2.626	0.611	1.054	0.517	0.705	0.884	0.319	1.025	0.519
LapIRN$_{disp}$	0.808	3.031	0.651	1.161	0.312	0.756	3.110	0.728	1.033	0.310
LapIRN	0.765	0.007	0.319	1.101	0.331	0.736	0.008	0.301	1.032	0.334

Results. Table 1 gives a comprehensive summary of the results. The variant of our method LapIRN$_{disp}$ achieves 0.808 Dice on a large scale MR brain dataset (OASIS), which outperforms both conventional methods and DLIR methods, Demons, SyN, DIF-VM, VM, by a significant margin of 13%, 18%, 17% and 11% of Dice score respectively. Nevertheless, similar to methods that work with displacement fields (i.e., VM), the solutions from LapIRN$_{disp}$ and VM cannot guarantee to be smooth and locally invertible as indicated by the high standard deviation of Jacobian determinant (0.65 and 0.61 respectively) and a large percentage of folding voxels (3% and 2.6%) in both datasets. Our proposed method LapIRN alleviates this issue and achieves the best registration performance over all the baseline methods, yielding plausible and smooth deformation fields with a standard deviation of the Jacobian determinants of 0.319 and $< 0.01\%$ folding voxels. Furthermore, the inference time of LapIRN is only 0.33 s, which is significantly faster than the conventional methods (Demons and SyN). We also highlight that our methods outperform the conventional methods even on the small-scale LPBA40 dataset, which has limited training data. Figure 2 illustrates the example of MR slices with large initial misalignment from all methods. The qualitative result shows that LapIRN is capable of large deformation, while the results from VM and DIF-VM are considered to be sub-optimal. Figure 3 depicts the average DSC for each anatomical structure in OASIS dataset. Compare to methods with diffeomorphic properties, our proposed method LapIRN achieves consistently better registration performance among 14 anatomical structures.

4 Conclusion

In this paper, we have presented a novel deep Laplacian pyramid networks for deformable image registration with the similarity pyramid, which mimics the conventional multi-resolution strategy to capture large misalignments between input scans. To guarantee the desirable diffeomorphic properties of the deformation fields, we formulate our method with diffeomorphism using the stationary vector fields under the Log-Euclidean framework. Extensive experiments have been carried out and the results showed that not only does our method achieve the state-of-the-art registration accuracy with very efficient running time (0.3 s), our methods also guarantee desirable diffeomorphic properties of the deformation fields. The formulation of our method can be easily transferred to various applications with minimum effort and has demonstrated immense potentials for time-sensitive medical studies.

References

1. Arsigny, V., Commowick, O., Pennec, X., Ayache, N.: A log-Euclidean framework for statistics on diffeomorphisms. In: Larsen, R., Nielsen, M., Sporring, J. (eds.) MICCAI 2006. LNCS, vol. 4190, pp. 924–931. Springer, Heidelberg (2006). https://doi.org/10.1007/11866565_113
2. Avants, B.B., Epstein, C.L., Grossman, M., Gee, J.C.: Symmetric diffeomorphic image registration with cross-correlation: evaluating automated labeling of elderly and neurodegenerative brain. Med. Image Anal. **12**(1), 26–41 (2008)
3. Avants, B.B., Tustison, N., Song, G.: Advanced normalization tools (ANTs). Insight j **2**(365), 1–35 (2009)
4. Balakrishnan, G., Zhao, A., Sabuncu, M.R., Guttag, J., Dalca, A.V.: An unsupervised learning model for deformable medical image registration. In: Proceedings of the IEEE Conference on Computer Vision and Pattern Recognition, pp. 9252–9260 (2018)
5. Dalca, A.V., Balakrishnan, G., Guttag, J., Sabuncu, M.R.: Unsupervised learning for fast probabilistic diffeomorphic registration. In: Frangi, A.F., Schnabel, J.A., Davatzikos, C., Alberola-López, C., Fichtinger, G. (eds.) MICCAI 2018. LNCS, vol. 11070, pp. 729–738. Springer, Cham (2018). https://doi.org/10.1007/978-3-030-00928-1_82

6. Denton, E.L., Chintala, S., Fergus, R., et al.: Deep generative image models using a Laplacian pyramid of adversarial networks. In: Advances in Neural Information Processing Systems, pp. 1486–1494 (2015)
7. Fischl, B.: Freesurfer. Neuroimage **62**(2), 774–781 (2012)
8. He, K., Zhang, X., Ren, S., Sun, J.: Identity mappings in deep residual networks. In: Leibe, B., Matas, J., Sebe, N., Welling, M. (eds.) ECCV 2016. LNCS, vol. 9908, pp. 630–645. Springer, Cham (2016). https://doi.org/10.1007/978-3-319-46493-0_38
9. Hering, A., van Ginneken, B., Heldmann, S.: mLVIRNET: multilevel variational image registration network. In: She, D., et al. (eds.) MICCAI 2019, pp. 257–265. Springer, Heidelberg (2019). https://doi.org/10.1007/978-3-030-32226-7_29
10. Hu, X., Kang, M., Huang, W., Scott, M.R., Wiest, R., Reyes, M.: Dual-stream pyramid registration network. In: Shen, D., et al. (eds.) MICCAI 2019. LNCS, vol. 11765, pp. 382–390. Springer, Cham (2019). https://doi.org/10.1007/978-3-030-32245-8_43
11. Karras, T., Aila, T., Laine, S., Lehtinen, J.: Progressive growing of GANs for improved quality, stability, and variation. arXiv preprint arXiv:1710.10196 (2017)
12. Klein, A., Andersson, J., Ardekani, B.A., et al.: Evaluation of 14 nonlinear deformation algorithms applied to human brain MRI registration. Neuroimage **46**(3), 786–802 (2009)
13. Lai, W.S., Huang, J.B., Ahuja, N., Yang, M.H.: Fast and accurate image super-resolution with deep Laplacian pyramid networks. IEEE Trans. Pattern Anal. Mach. Intell. **41**(11), 2599–2613 (2018)
14. Lee, C.Y., Xie, S., Gallagher, P., Zhang, Z., Tu, Z.: Deeply-supervised nets. In: Artificial Intelligence and Statistics, pp. 562–570 (2015)
15. Maas, A.L., Hannun, A.Y., Ng, A.Y.: Rectifier nonlinearities improve neural network acoustic models. In: International Conference on Machine Learning (ICML), vol. 30, p. 3 (2013)
16. Marcus, D.S., Wang, T.H., Parker, J., Csernansky, J.G., Morris, J.C., Buckner, R.L.: Open access series of imaging studies (OASIS): cross-sectional MRI data in young, middle aged, nondemented, and demented older adults. J. Cogn. Neurosci. **19**(9), 1498–1507 (2007)
17. McCormick, M.M., Liu, X., Ibanez, L., Jomier, J., Marion, C.: ITK: enabling reproducible research and open science. Front. Neuroinform. **8**, 13 (2014)
18. Ou, Y., Sotiras, A., Paragios, N., Davatzikos, C.: Dramms: deformable registration via attribute matching and mutual-saliency weighting. Med. Image Anal. **15**(4), 622–639 (2011)
19. Paszke, A., Gross, S., Chintala, S., et al.: Automatic differentiation in PyTorch. In: NIPS-W (2017)
20. Rohé, M.-M., Datar, M., Heimann, T., Sermesant, M., Pennec, X.: SVF-Net: learning deformable image registration using shape matching. In: Descoteaux, M., Maier-Hein, L., Franz, A., Jannin, P., Collins, D.L., Duchesne, S. (eds.) MICCAI 2017. LNCS, vol. 10433, pp. 266–274. Springer, Cham (2017). https://doi.org/10.1007/978-3-319-66182-7_31
21. Rohlfing, T., Maurer, C.R., Bluemke, D.A., Jacobs, M.A.: Volume-preserving non-rigid registration of MR breast images using free-form deformation with an incompressibility constraint. IEEE Trans. Med. Imaging **22**(6), 730–741 (2003)
22. Shattuck, D.W., et al.: Construction of a 3D probabilistic atlas of human cortical structures. Neuroimage **39**(3), 1064–1080 (2008)
23. Thirion, J.P.: Image matching as a diffusion process: an analogy with Maxwell's demons. Med. Image Anal. **2**(3), 243–260 (1998)

24. Vercauteren, T., Pennec, X., Perchant, A., Ayache, N.: Diffeomorphic demons: efficient non-parametric image registration. NeuroImage **45**(1), S61–S72 (2009)
25. de Vos, B.D., Berendsen, F.F., Viergever, M.A., Sokooti, H., Staring, M., Išgum, I.: A deep learning framework for unsupervised affine and deformable image registration. Med. Image Anal. **52**, 128–143 (2019)
26. de Vos, B.D., Berendsen, F.F., Viergever, M.A., Staring, M., Išgum, I.: End-to-end unsupervised deformable image registration with a convolutional neural network. In: Cardoso, M.J., et al. (eds.) DLMIA/ML-CDS -2017. LNCS, vol. 10553, pp. 204–212. Springer, Cham (2017). https://doi.org/10.1007/978-3-319-67558-9_24
27. Wang, H., et al.: Validation of an accelerated 'demons' algorithm for deformable image registration in radiation therapy. Phys. Med. Biol. **50**(12), 2887 (2005)
28. Wang, T.C., Liu, M.Y., Zhu, J.Y., Tao, A., Kautz, J., Catanzaro, B.: High-resolution image synthesis and semantic manipulation with conditional GANs. In: Proceedings of the IEEE Conference on Computer Vision and Pattern Recognition, pp. 8798–8807 (2018)
29. Xiang, X., Zhai, M., Zhang, R., Lv, N., El Saddik, A.: Optical flow estimation using spatial-channel combinational attention-based pyramid networks. In: IEEE International Conference on Image Processing (ICIP), pp. 1272–1276. IEEE (2019)
30. Zhao, S., Dong, Y., Chang, E.I., Xu, Y., et al.: Recursive cascaded networks for unsupervised medical image registration. In: Proceedings of the IEEE International Conference on Computer Vision, pp. 10600–10610 (2019)

Adversarial Uni- and Multi-modal Stream Networks for Multimodal Image Registration

Zhe Xu[1,2], Jie Luo[2,3], Jiangpeng Yan[1], Ritvik Pulya[2], Xiu Li[1], William Wells III[2], and Jayender Jagadeesan[2(✉)]

[1] Shenzhen International Graduate School, Tsinghua University, Shenzhen, China
[2] Brigham and Women's Hospital, Harvard Medical School, Boston, USA
jayender@bwh.harvard.edu
[3] Graduate School of Frontier Sciences, The University of Tokyo, Tokyo, Japan

Abstract. Deformable image registration between Computed Tomography (CT) images and Magnetic Resonance (MR) imaging is essential for many image-guided therapies. In this paper, we propose a novel translation-based unsupervised deformable image registration method. Distinct from other translation-based methods that attempt to convert the multimodal problem (e.g., CT-to-MR) into a unimodal problem (e.g., MR-to-MR) via image-to-image translation, our method leverages the deformation fields estimated from both: (i) the translated MR image and (ii) the original CT image in a dual-stream fashion, and automatically learns how to fuse them to achieve better registration performance. The multimodal registration network can be effectively trained by computationally efficient similarity metrics without any ground-truth deformation. Our method has been evaluated on two clinical datasets and demonstrates promising results compared to state-of-the-art traditional and learning-based methods.

Keywords: Multimodal registration · Generative adversarial network · Unsupervised learning

1 Introduction

Deformable multimodal image registration has become essential for many procedures in image-guided therapies, e.g., preoperative planning, intervention, and diagnosis. Due to substantial improvement in computational efficiency over traditional iterative registration approaches, learning-based registration approaches are becoming more prominent in time-intensive applications.

Related Work. Many learning-based registration approaches adopt fully supervised or semi-supervised strategies. Their networks are trained with ground-truth

A. L. Martel et al. (Eds.): MICCAI 2020, LNCS 12263, pp. 222–232, 2020.
https://doi.org/10.1007/978-3-030-59716-0_22

deformation fields or segmentation masks [5,12,13,16,19], and may struggle with limited or imperfect data labeling. A number of unsupervised registration approaches have been proposed to overcome this problem by training unlabeled data to minimize traditional similarity metrics, e.g., mean squared intensity differences [4,11,15,17,21,26]. However, the performances of these methods are inherently limited by the choice of similarity metrics. Given the limited selection of multimodal similarity metrics, unsupervised registration approaches may have difficulties outperforming traditional multimodal registration methods as they both essentially optimize the same cost functions. A recent trend for multimodal image registration takes advantage of the latent feature disentanglement [18] and image-to-image translation [6,20,23]. Specifically, translation-based approaches use Generative Adversarial Network (GAN) to translate images from one modality into the other modality, thus are able to convert the difficult multimodal registration into a simpler unimodal task. However, being a challenging topic by itself, image translation may inevitably produce artificial anatomical features that can further interfere with the registration process.

In this work, we propose a novel translation-based fully unsupervised multimodal image registration approach. In the context of Computed Tomography (CT) image to Magnetic Resonance (MR) image registration, previous translation-based approaches would translate a CT image into an MR-like image (tMR), and use tMR-to-MR registration to estimate the final deformation field ϕ. In our approach, the network estimates two deformation fields, namely ϕ_s of tMR-to-MR and ϕ_o of CT-to-MR, in a dual-stream fashion. The addition of the original ϕ_o enables the network to implicitly regularize ϕ_s to mitigate certain image translation problems, e.g., artificial features. The network further automatically learns how to fuse ϕ_s and ϕ_o towards achieving the best registration accuracy.

Contributions and advantages of our work can be summarized as follows:

1. Our method leverages the deformation fields estimated from the original multimodal stream and synthetic unimodal stream to overcome the shortcomings of translation-based registration;
2. We improve the fidelity of organ boundaries in the translated MR by adding two extra constraints in the image-to-image translation model Cycle-GAN.

We evaluate our method on two clinically acquired datasets. It outperforms state-of-the-art traditional, unsupervised and translation-based registration approaches.

2 Methods

In this work, we propose a general learning framework for robustly registering CT images to MR images in a fully unsupervised manner.

First, given a moving CT image and a fixed MR image, our improved Cycle-GAN module translates the CT image into an MR-like image. Then, our dual-stream subnetworks, UNet_o and UNet_s, estimate two deformation fields ϕ_o and

ϕ_s respectively, and the final deformation field is fused via a proposed fusion module. Finally, the moving CT image is warped via Spatial Transformation Network (STN) [14], while the entire registration network aims to maximize the similarity between the moved and the fixed images. The pipeline of our method is shown in Fig. 1.

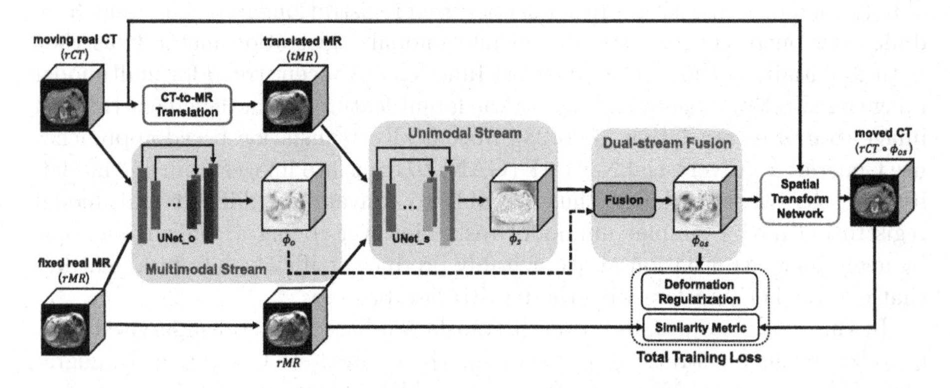

Fig. 1. Illustration of the proposed method. The entire unsupervised network is mainly guided by the image similarity between rCT $\circ$ ϕ_{os} and rMR.

2.1 Image-to-Image Translation with Unpaired Data

The CT-to-MR translation step consists of an improved Cycle-GAN with additional structural and identical constraints. As a state-of-the-art image-to-image translation model, Cycle-GAN [28] can be trained without pairwise aligned CT and MR datasets of the same patient. Thus, Cycle-GAN is widely used in medical image translation [1,9,25].

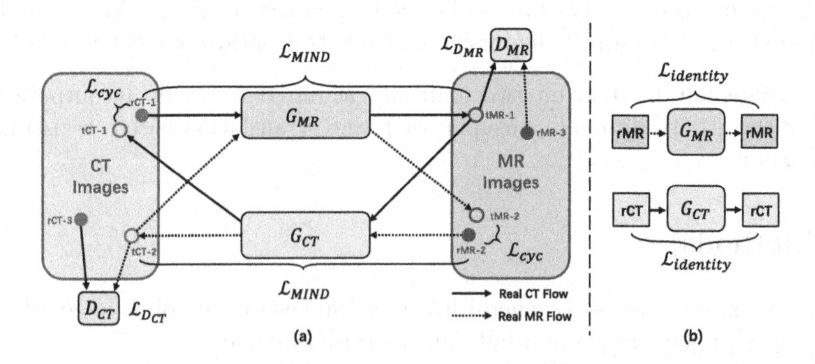

Fig. 2. Schematic illustration of Cycle-GAN with strict constraints. (a) The workflow of the forward and backward translation; (b) The workflow of identity loss.

Our Cycle-GAN model is illustrated in Fig. 2. The model consists of two generators G_{MR} and G_{CT}, which can provide CT-to-MR and MR-to-CT translation respectively. Besides, it has two discriminators D_{CT} and D_{MR}. D_{CT} is used to distinguish between translated CT(tCT) and real CT(rCT), and D_{MR} is for translated MR(tMR) and real MR(rMR). The training loss of original Cycle-GAN only adopts two types of items: adversarial loss given by two discriminators ($\mathcal{L}_{D_{CT}}$ and $\mathcal{L}_{D_{MR}}$) and cycle-consistency loss $\mathcal{L}_{cyc}$ to prevent generators from generating images that are not related to the inputs (refer to [28] for details).

However, training a Cycle-GAN on medical images is difficult since the cycle-consistency loss is not enough to enforce structural similarity between translated images and real images (as shown in the red box in Fig. 3(b)). Therefore, we introduce two additional losses, structure-consistency loss $\mathcal{L}_{MIND}$ and identity loss $\mathcal{L}_{identity}$, to constrain the training of Cycle-GAN.

MIND (Modality Independent Neighbourhood Descriptor) [8] is a feature that describes the local structure around each voxel. Thus, we minimize the difference in MIND between translated images $G_{CT}(I_{rMR})$ or $G_{MR}(I_{rCT})$ and real images I_{rMR} or I_{rCT} to enforce the structural similarity. We define $\mathcal{L}_{MIND}$ as follows:

$$\begin{aligned} L_{MIND}(G_{CT}, G_{MR}) = \frac{1}{N_{MR}|R|}\sum\nolimits_{x} ||M(G_{CT}(I_{rMR})) - M(I_{rMR})||_1 \\ + \frac{1}{N_{CT}|R|}\sum\nolimits_{x} ||M(G_{MR}(I_{rCT})) - M(I_{rCT})||_1 \end{aligned} \quad (1)$$

where M represents MIND features, N_{MR} and N_{CT} denote the number of voxels in I_{rMR} and I_{rCT}, and R is a non-local region around voxel x.

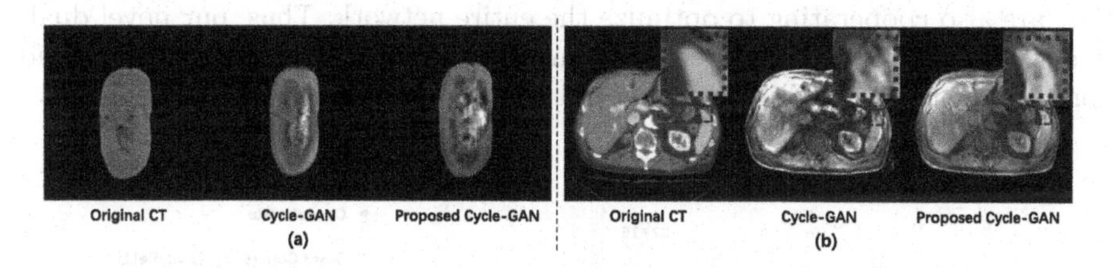

Fig. 3. CT-to-MR translation examples of original Cycle-GAN and proposed Cycle-GAN tested for (a) pig ex-vivo kidney dataset and (b) abdomen dataset.

The identity loss (as shown in Fig. 2(b)) is included to prevent images already in the expected domain from being incorrectly translated to the other domain. We define it as:

$$\mathcal{L}_{identity} = ||G_{MR}(I_{MR}) - I_{MR}||_1 + ||G_{CT}(I_{CT}) - I_{CT}||_1 \quad (2)$$

Finally, the total loss $\mathcal{L}$ of our proposed Cycle-GAN is defined as:

$$\mathcal{L} = \mathcal{L}_{D_{MR}} + \mathcal{L}_{D_{CT}} + \lambda_{cyc}\mathcal{L}_{cyc} + \lambda_{identity}\mathcal{L}_{identity} + \lambda_{MIND}\mathcal{L}_{MIND} \quad (3)$$

where λ_{cyc}, $\lambda_{identity}$ and λ_{MIND} denotes the relative importance of each term.

2.2 Dual-Stream Multimodal Image Registration Network

As shown in Fig. 3, although our improved Cycle-GAN can better translate CT images into MR-like images, the CT-to-MR translation is still challenging for translating "simple" CT images to "complex" MR images. Most image-to-image translation methods will inevitably generate unrealistic soft-tissue details, resulting in some mismatch problems. Therefore, the registration methods that simply convert multimodal to unimodal registration via image translation algorithm are not reliable.

In order to address this problem, we propose a dual-stream network to fully use the information of the moving, fixed and translated images as shown in Fig. 1. In particular, we can use effective similarity metrics to train our multimodal registration model without any ground-truth deformation.

Network Details. As shown in Fig. 1, our dual-stream network is comprised of four parts: multimodal stream subnetwork, unimodal stream subnetwork, deformation field fusion, and Spatial Transformation Network.

In ***Multimodal Stream*** subnetwork, original CT(rCT) and MR(rMR) are represented as the moving and fixed images, which allows the model to propagate original information to counteract mismatch problems in translated MR(tMR).

Through image translation, we obtain the translated MR(tMR) with similar appearance to the fixed MR(rMR). Then, in ***Unimodal Stream***, tMR and rMR are used as moving and fixed images respectively. This stream can effectively propagate more texture information, and constrain the final deformation field to suppress unrealistic voxel drifts from the multimodal stream.

During the network training, the two streams constrain each other, while they are also cooperating to optimize the entire network. Thus, our novel dual-stream design allows us to benefit from both original image information and homogeneous structural information in the translated images.

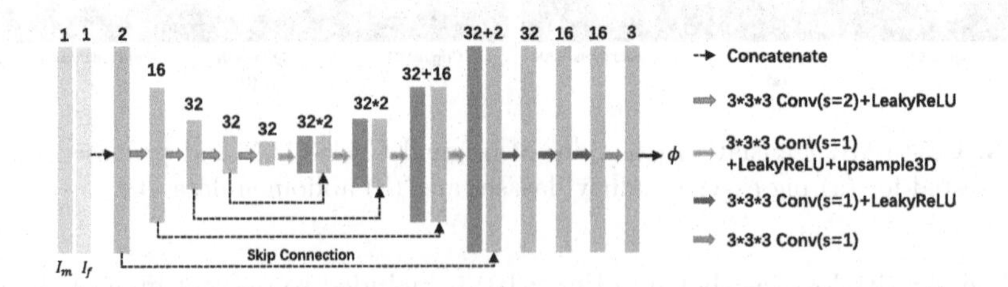

Fig. 4. Detailed architecture of UNet-based subnetwork. The encoder uses convolution with stride of 2 to reduce spatial resolution, while the decoder uses 3D upsampling layers to restore the spatial resolution.

Specifically, UNet_o and UNet_s adopt the same UNet architecture used in VoxelMorph [4] (shown in Fig. 4). The only difference is that UNet_o is with *multimodal inputs* but UNet_s is with *unimodal inputs*. Each UNet takes a single 2-channel 3D image formed by concatenating I_m and I_f as input, and outputs a volume of deformation field with 3 channels.

After Uni- and Multi-model Stream networks, we obtain two deformation fields, ϕ_o (for rCT and rMR) and ϕ_s (for tMR and rMR). We stack ϕ_o and ϕ_s, and apply a 3D convolution with size of $3 \times 3 \times 3$ to estimate the final deformation field ϕ_{os}, which is a 3D volume with the same shape of ϕ_o and ϕ_s.

To evaluate the dissimilarity between moved and fixed images, we integrate spatial transformation network (STN) [14] to warp the moving image using ϕ_{os}. The loss function consists of two components as shown in Eq. (4).

$$\mathcal{L}_{total}(I_{rMR}, I_{rCT}, \phi_{os}) = \mathcal{L}_{sim}(I_{rMR}, I_{rCT} \circ \phi_{os}) + \lambda \mathcal{L}_{smooth}(\phi_{os}) \quad (4)$$

where λ is a regularization weight. The first loss $\mathcal{L}_{sim}$ is similarity loss, which is to penalize the differences in appearance between fixed and moved images. Here we adopt SSIM [22] for experiments. Suggested by [4], deformation regularization $\mathcal{L}_{smooth}$ adopts a L2-norm of the gradients of the final deformation field ϕ_{os}.

3 Experiments and Results

Dataset and Preprocessing. We focus on the application of abdominal CT-to-MR registration.We evaluated our method on two proprietary datasets since there is no designated public repository.

1) *Pig Ex-vivo Kidney CT-MR Dataset.* This dataset contains 18 pairs of CT and MRI kidney scans from pigs. All kidneys are manually segmented by experts. After preprocessing the data, e.g., resampling and affine spatial normalization, we cropped the data to $144 \times 80 \times 256$ with 1 mm isotropic voxels and arbitrarily divided it into two groups for training (15 cases) and testing (3 cases).
2) *Abdomen (ABD) CT-MR Dataset.* This 50-patient dataset of CT-MR scans was collected from a local hospital and annotated with anatomical landmarks. All data were preprocessed into $176 \times 176 \times 128$ with the same resolution ($1\,\text{mm}^3$) and were randomly divided into two groups for training (45 cases) and testing (5 cases).

Implementation. We trained our model using the following settings: (1) The Cycle-GAN for CT-MR translation network is based on the existing implementation [27] with changes as discussed in Sect. 2.1. (2) The Uni- and Multi-modal stream registration networks were implemented using Keras with the Tensorflow backend and trained on an NVIDIA Titan X (Pascal) GPU.

3.1 Results for CT-to-MR Translation

We extracted 1792 and 5248 slices from the transverse planes of the Pig kidney and ABD dataset respectively to train the image translation network. Parameters λ_{cyc}, $\lambda_{identity}$ and λ_{MIND} were set to 10, 5, and 5 for training.

Table 1. Quantitative results for image translation.

Pig Kidney	Method	PSNR	SSIM	ABD	Method	PSNR	SSIM
	Cycle-GAN	32.07	0.9025		Cycle-GAN	22.95	0.7367
	Ours	**32.74**	**0.9532**		Ours	**23.55**	**0.7455**

Since our registration method is for 3D volumes, we apply the pre-trained CT-to-MR generator to translate moving CT images into MR-like images slice-by-slice and concatenate 2D slices into 3D volumes. The qualitative results are visualized in Fig. 3. In addition, to quantitatively evaluate the translation performance, we apply our registration method to obtain aligned CT-MR pairs and utilize SSIM [22] and PSNR [10] to judge the quality of translated MR (shown in Table 1). In our experiment, our method predicts better MR-like images on both datasets.

3.2 Registration Results

Affine registration is used as the baseline method. For traditional method, only mutual information (MI) based **SyN** [2] is compared since it is the only metric (available in ANTs [3]) for multimodal registration. In addition to SyN, we implemented the following learning-based methods: 1) **VM_MIND** and **VM_SSIM** which extends VoxelMorph with similarity metrics MIND [8] and SSIM [22]. 2) **M2U** which is a typical translation-based registration method. It generates tMR from CT and converts the multimodal problem to tMR-to-MR registration. It's noteworthy that the parameters of all methods are optimized to the best results on both datasets.

Two examples of the registration results are visualized in Fig. 5, where the red and yellow contours represent the ground truth and registered organ boundaries respectively. As shown in Fig. 5, the organ boundaries aligned by the traditional SyN method have a considerable amount of disagreement. Among all learning-based methods, our method has the most visually appealing boundary alignment for both cases. **VM_SSIM** performed significantly worse for the kidney. **VM_MIND** achieved accurate registration for the kidney, but its result for the ABD case is significantly worse. Meanwhile, **M2U** suffers from artificial features in the image translation, which leads to an inaccurate registration result.

The quantitative results are presented in Table 2. We compare different methods by the Dice score [7] and target registration error (TRE) [24]. We also provide the average run-time for each method. As shown in Table 2, our method consistently outperformed other methods and was able to register a pair of images in less than 2 s (when using GPU).

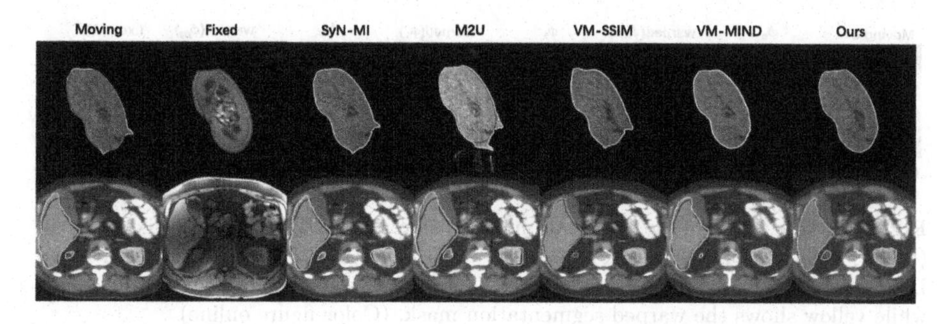

Fig. 5. Visualization results of our model compared to other methods. Upper: Pig Kidney. Bottom: Abdomen (ABD). The red contours represent the ground truth organ boundary while the yellow contours are the warped contours of segmentation masks. (Color figure online)

Table 2. Quantitative results for Pig Kidney Dataset and Abdomen (ABD) Dataset.

Metric	Organ		Affine	SyN	M2U	VM_SSIM	VM_MIND	Ours
Dice (%)	Pig	Kidney	89.53	89.87	90.21	93.75	96.48	**98.57**
	ABD	Kidney	80.03	82.36	78.96	82.21	84.58	**85.66**
		Spleen	79.58	80.38	77.76	81.79	83.11	**87.01**
		Liver	78.74	79.13	78.83	82.05	81.98	**83.34**
TRE (mm)	ABD	spleen	4.16	4.20	3.76	3.58	3.65	**2.47**
		Liver	6.55	5.61	5.91	4.72	4.87	**3.64**
Time(s) GPU/CPU	Pig		−/103	−/121	1.08/20	1.06/20	1.07/20	1.12/21
	ABD		−/108	−/137	1.23/24	1.22/22	1.21/23	1.27/24

3.3 The Effect of Each Deformation Field

In order to validate the effectiveness of the deformation field fusion, we compare ϕ_s, ϕ_o and ϕ_{os} together with warped images (shown in Fig. 6). The qualitative result shows that ϕ_s from the unimodal stream alleviates the voxel drift effect from the multimodal stream. While ϕ_o from the multimodal stream uses the original image textures to maintain the fidelity and reduce artificial features for the generated tMR image. The fused deformation field ϕ_{os} produces better alignment than both streams alone, which demonstrates the effectiveness of the joint learning step.

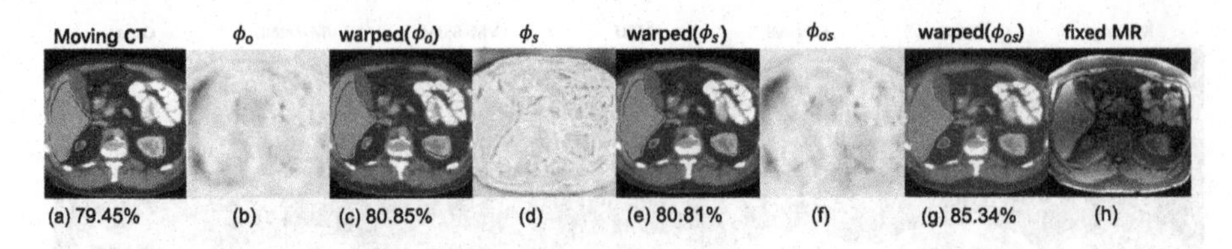

Fig. 6. Visualizations of the deformation field fusion. (a) moving image; (h) fixed image; (b/d/f) deformation fields; (c/e/g) images warped by (b/d/f), corresponding average Dice scores (%) of all organs are calculated. The contours in red represent ground truth, while yellow shows the warped segmentation mask. (Color figure online)

4 Conclusion

We proposed a fully unsupervised uni- and multi-modal stream network for CT-to-MR registration. Our method leverages both CT-translated-MR and original CT images towards achieving the best registration result. Besides, the registration network can be effectively trained by computationally efficient similarity metrics without any ground-truth deformation. We evaluated the method on two clinical datasets, and it outperformed state-of-the-art methods in terms of accuracy and efficiency.

Acknowledgement. This project was supported by the National Institutes of Health (Grant No. R01EB025964, R01DK119269, and P41EB015898) and the Overseas Cooperation Research Fund of Tsinghua Shenzhen International Graduate School (Grant No. HW2018008).

References

1. Armanious, K., Jiang, C., Abdulatif, S., Küstner, T., Gatidis, S., Yang, B.: Unsupervised medical image translation using cycle-MedGAN. In: 2019 27th European Signal Processing Conference (EUSIPCO), pp. 1–5. IEEE (2019)
2. Avants, B.B., Epstein, C.L., Grossman, M., Gee, J.C.: Symmetric diffeomorphic image registration with cross-correlation: evaluating automated labeling of elderly and neurodegenerative brain. Med. Image Anal. **12**(1), 26–41 (2008)
3. Avants, B.B., Tustison, N.J., Song, G., Cook, P.A., Klein, A., Gee, J.C.: A reproducible evaluation of ants similarity metric performance in brain image registration. NeuroImage **54**, 2033–2044 (2011)
4. Balakrishnan, G., Zhao, A., Sabuncu, M.R., Guttag, J., Dalca, A.V.: An unsupervised learning model for deformable medical image registration. In: Proceedings of the IEEE Conference on Computer Vision and Pattern Recognition, pp. 9252–9260 (2018)
5. Cao, X., et al.: Deformable image registration based on similarity-steered CNN regression. In: Descoteaux, M., Maier-Hein, L., Franz, A., Jannin, P., Collins, D.L., Duchesne, S. (eds.) MICCAI 2017. LNCS, vol. 10433, pp. 300–308. Springer, Cham (2017). https://doi.org/10.1007/978-3-319-66182-7_35

6. Cao, X., Yang, J., Wang, L., Xue, Z., Wang, Q., Shen, D.: Deep learning based inter-modality image registration supervised by intra-modality similarity. In: Shi, Y., Suk, H.-I., Liu, M. (eds.) MLMI 2018. LNCS, vol. 11046, pp. 55–63. Springer, Cham (2018). https://doi.org/10.1007/978-3-030-00919-9_7
7. Dice, L.R.: Measures of the amount of ecologic association between species. Ecology **26**(3), 297–302 (1945)
8. Heinrich, M.P., et al.: Mind: modality independent neighbourhood descriptor for multi-modal deformable registration. Med. Image Anal. **16**(7), 1423–1435 (2012)
9. Hiasa, Y., et al.: Cross-modality image synthesis from unpaired data using Cycle-GAN. In: Gooya, A., Goksel, O., Oguz, I., Burgos, N. (eds.) SASHIMI 2018. LNCS, vol. 11037, pp. 31–41. Springer, Cham (2018). https://doi.org/10.1007/978-3-030-00536-8_4
10. Hore, A., Ziou, D.: Image quality metrics: PSNR vs. SSIM. In: 2010 20th International Conference on Pattern Recognition, pp. 2366–2369. IEEE (2010)
11. Hu, X., Kang, M., Huang, W., Scott, M.R., Wiest, R., Reyes, M.: Dual-stream pyramid registration network. In: Shen, D., et al. (eds.) MICCAI 2019. LNCS, vol. 11765, pp. 382–390. Springer, Cham (2019). https://doi.org/10.1007/978-3-030-32245-8_43
12. Hu, Y., et al.: Label-driven weakly-supervised learning for multimodal deformable image registration. In: 2018 IEEE 15th International Symposium on Biomedical Imaging (ISBI 2018), pp. 1070–1074. IEEE (2018)
13. Hu, Y., et al.: Weakly-supervised convolutional neural networks for multimodal image registration. Med. Image Anal. **49**, 1–13 (2018)
14. Jaderberg, M., Simonyan, K., Zisserman, A., et al.: Spatial transformer networks. In: Advances in Neural Information Processing Systems, pp. 2017–2025 (2015)
15. Kuang, D., Schmah, T.: FAIM – a ConvNet method for unsupervised 3D medical image registration. In: Suk, H.-I., Liu, M., Yan, P., Lian, C. (eds.) MLMI 2019. LNCS, vol. 11861, pp. 646–654. Springer, Cham (2019). https://doi.org/10.1007/978-3-030-32692-0_74
16. Liu, C., Ma, L., Lu, Z., Jin, X., Xu, J.: Multimodal medical image registration via common representations learning and differentiable geometric constraints. Electron. Lett. **55**(6), 316–318 (2019)
17. Mahapatra, D., Ge, Z., Sedai, S., Chakravorty, R.: Joint registration and segmentation of Xray images using generative adversarial networks. In: Shi, Y., Suk, H.-I., Liu, M. (eds.) MLMI 2018. LNCS, vol. 11046, pp. 73–80. Springer, Cham (2018). https://doi.org/10.1007/978-3-030-00919-9_9
18. Qin, C., Shi, B., Liao, R., Mansi, T., Rueckert, D., Kamen, A.: Unsupervised deformable registration for multi-modal images via disentangled representations. In: Chung, A.C.S., Gee, J.C., Yushkevich, P.A., Bao, S. (eds.) IPMI 2019. LNCS, vol. 11492, pp. 249–261. Springer, Cham (2019). https://doi.org/10.1007/978-3-030-20351-1_19
19. Sedghi, A., et al.: Semi-supervised image registration using deep learning. In: Medical Imaging 2019: Image-Guided Procedures, Robotic Interventions, and Modeling, vol. 10951, p. 109511G. International Society for Optics and Photonics (2019)
20. Tanner, C., Ozdemir, F., Profanter, R., Vishnevsky, V., Konukoglu, E., Goksel, O.: Generative adversarial networks for mr-ct deformable image registration. arXiv preprint arXiv:1807.07349 (2018)
21. de Vos, B.D., Berendsen, F.F., Viergever, M.A., Sokooti, H., Staring, M., Išgum, I.: A deep learning framework for unsupervised affine and deformable image registration. Med. Image Anal. **52**, 128–143 (2019)

22. Wang, Z., Bovik, A.C., Sheikh, H.R., Simoncelli, E.P.: Image quality assessment: from error visibility to structural similarity. IEEE Trans. Image Process. **13**(4), 600–612 (2004)
23. Wei, D., et al.: Synthesis and inpainting-based MR-CT registration for image-guided thermal ablation of liver tumors. In: Shen, D., et al. (eds.) MICCAI 2019. LNCS, vol. 11768, pp. 512–520. Springer, Cham (2019). https://doi.org/10.1007/978-3-030-32254-0_57
24. West, J.B., et al.: Comparison and evaluation of retrospective intermodality image registration techniques. In: Medical Imaging 1996: Image Processing, vol. 2710, pp. 332–347. International Society for Optics and Photonics (1996)
25. Yang, H., et al.: Unpaired brain MR-to-CT synthesis using a structure-constrained CycleGAN. In: Stoyanov, D., et al. (eds.) DLMIA/ML-CDS -2018. LNCS, vol. 11045, pp. 174–182. Springer, Cham (2018). https://doi.org/10.1007/978-3-030-00889-5_20
26. Zhao, S., Lau, T., Luo, J., Eric, I., Chang, C., Xu, Y.: Unsupervised 3D end-to-end medical image registration with volume tweening network. IEEE J. Biomed. Health Inform. (2019)
27. Zhu, J.Y., Park, T., Isola, P., Efros, A.A.: Cyclegan (2017). https://github.com/xhujoy/CycleGAN-tensorflow
28. Zhu, J.Y., Park, T., Isola, P., Efros, A.A.: Unpaired image-to-image translation using cycle-consistent adversarial networks. In: Proceedings of the IEEE International Conference on Computer Vision, pp. 2223–2232 (2017)

Cross-Modality Multi-atlas Segmentation Using Deep Neural Networks

Wangbin Ding[1], Lei Li[2,3,4], Xiahai Zhuang[2(✉)], and Liqin Huang[1(✉)]

[1] College of Physics and Information Engineering, Fuzhou University, Fuzhou, China
hlq@fzu.edu.cn

[2] School of Data Science, Fudan University, Shanghai, China
zxh@fudan.edu.cn

[3] School of Biomedical Engineering, Shanghai Jiao Tong University, Shanghai, China

[4] School of Biomedical Engineering and Imaging Sciences, King's College London, London, UK

Abstract. Both image registration and label fusion in the multi-atlas segmentation (MAS) rely on the intensity similarity between target and atlas images. However, such similarity can be problematic when target and atlas images are acquired using different imaging protocols. High-level structure information can provide reliable similarity measurement for cross-modality images when cooperating with deep neural networks (DNNs). This work presents a new MAS framework for cross-modality images, where both image registration and label fusion are achieved by DNNs. For image registration, we propose a consistent registration network, which can jointly estimate forward and backward dense displacement fields (DDFs). Additionally, an invertible constraint is employed in the network to reduce the correspondence ambiguity of the estimated DDFs. For label fusion, we adapt a few-shot learning network to measure the similarity of atlas and target patches. Moreover, the network can be seamlessly integrated into the patch-based label fusion. The proposed framework is evaluated on the MM-WHS dataset of MICCAI 2017. Results show that the framework is effective in both cross-modality registration and segmentation.

Keywords: MAS · Cross-modality · Similarity

1 Introduction

Segmentation is an essential step for medical image processing. Many clinical applications rely on an accurate segmentation to extract specific anatomy or compute some functional indices. The multi-atlas segmentation (MAS) has proved to be an effective method for medical image segmentation [22]. Generally, it contains two steps, i.e., a pair-wise registration between target image

X. Zhuang and L. Huang are co-seniors. This work was funded by the National Natural Science Foundation of China (Grant No. 61971142), and Shanghai Municipal Science and Technology Major Project (Grant No. 2017SHZDZX01).

A. L. Martel et al. (Eds.): MICCAI 2020, LNCS 12263, pp. 233–242, 2020.
https://doi.org/10.1007/978-3-030-59716-0_23

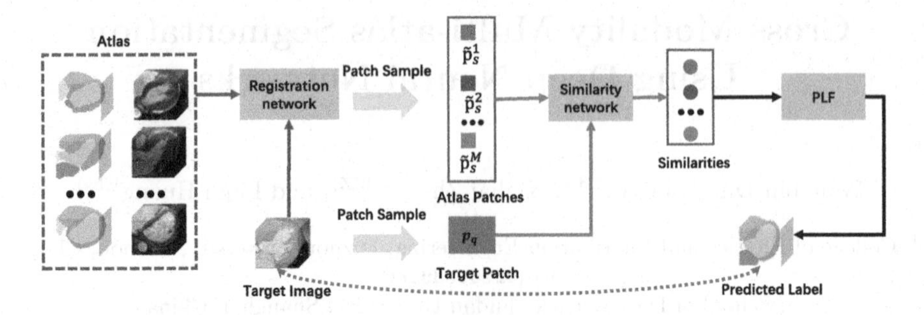

Fig. 1. The pipeline of the cross-modality MAS framework. The atlases are first warped to the target image by the registration model (see Sect. 2.1). Thus, the warped atlas label becomes a candidate segmentation of the target image simultaneously. Then, each voting patch sampled from warped atlases is weighted according to its similarity to the corresponding target patch (see Sect. 2.2). Based on the weight, one can obtain a final label for the target image using the PLF strategy.

and atlases, and a label fusion among selected reliable atlases. Conventional MAS methods normally process images from single modality, but in many scenarios they could benefit from cross-modality image processing [7]. To obtain such a method, registration and label fusion algorithms that can adapt to cross-modality data are required.

To achieve cross-modality registration, a common approach is to design a modality-invariance similarity as the registration criterion, such as mutual information (MI) [9], normalized mutual information (NMI) [15]. An alternative way is to employ structural representations of images, which are supposed to be invariant across multi-modality images [5,17]. Recently, several deep learning (DL) based multi-modality registration algorithms are developed. For example, Hu et al. proposed a weakly-supervised multi-modality registration network by exploring the dense voxel correspondence from anatomical labels [6]. Qin et al. designed an unsupervised registration network based on disentangled shape representations, and then converted the multi-modality registration into a mono-modality problem in the latent shape space [11].

For label fusion, there are several widely utilized strategies, such as majority voting (MV), plurality voting, global or local weighted voting, joint label fusion (JLF) [18], statistical modeling approach [19], and patch-based label fusion (PLF) [3]. To use cross-modality atlas, Kasiri et al. presented a similarity measurement based on un-decimated wavelet transform for cross-modality atlas fusion [8]. Furthermore, Zhuang et al. proposed a multi-scale patch strategy to extract multi-level structural information for multi-modality atlas fusion [23]. Recently, learning methods are engaged to improve the performance of label fusion. Ding et al. proposed a DL-based label fusion strategy, namely VoteNet, which can locally select reliable atlases and fuse atlas labels by plurality voting [4]. To enhance PLF strategy, Sanroma et al. and Yang et al. attempted to achieve a better deep feature similarity between target and atlas patches through deep neural networks (DNN) [13,21]. Similarly, Xie et al. incorporated a DNN to predict the weight of voting patches for the JLF strategy [20]. All these learning-based label fusion works assumed that atlas and target images come from the same modality.

This work is aimed at designing a DNN-based approach to achieve accurate registration and label fusion for cross-modality MAS. Figure 1 presents the pipeline of our proposed MAS method. The main contributions of this work are summarized as follows: (1) We present a DNN-based MAS framework for cross-modality segmentation, and validate it using the MM-WHS dataset [22]. (2) We propose a consistent registration network, where an invertible constraint is employed to encourage the uniqueness of transformation fields between cross-modality images. (3) We introduce a similarity network based on few-shot learning, which can estimate the patch-based similarity between the target and atlas images.

2 Method

2.1 Consistent Registration Network

Network Architecture: Suppose given N atlases $\{(I_a^1, L_a^1), \cdots, (I_a^N, L_a^N)\}$ and a target (I_t, L_t), for each pair of I_a^i and I_t, two registration procedures could be performed by switching the role of I_a^i and I_t. We denote the dense displacement field (DDF) from I_a^i to I_t as U^i, and vice versa as V^i. For convenience, we abbreviate I_a^i, L_a^i, U^i and V^i as I_a, L_a, U and V when no confusion is caused. Consider the label as a mapping function from common spatial space to label space: $\Omega \rightarrow \mathbb{L}$, so that

$$\tilde{L}_a(x) = L_a(x + U(x)), \tag{1}$$

$$\tilde{L}_t(x) = L_t(x + V(x)), \tag{2}$$

where $\tilde{L}_a$ and $\tilde{L}_t$ denote the warped L_a and L_t, respectively.

We develop a new registration network which can jointly estimate the forward (U) and inverse (V) DDF for a pair of input images. The advantage of joint estimation is that it can reduce the ambiguous correspondence in DDFs (see next subsection). Figure 2 shows the overall structure of the registration network. The backbone of the network is based on the U-Shape registration model [6]. Instead of using voxel-level ground-truth transformations, which is hard to obtain in cross-modality scenarios, the Dice coefficients of anatomical labels are used to train the network. Since the network is design to produce both U and V, pairwise registration errors caused by those two DDFs should been taken into account in the loss function. Thus, a symmetric Dice loss of the network is designed by

$$\mathcal{L}oss_{Dice} = \mathcal{D}ice(L_a, \tilde{L}_t) + \mathcal{D}ice(L_t, \tilde{L}_a)) + \lambda_1(\Psi(U) + \Psi(V)), \tag{3}$$

where λ_1 is the hyperparameter, $\Psi(U)$ and $\Psi(V)$ are smoothness regularizations for DDFs.

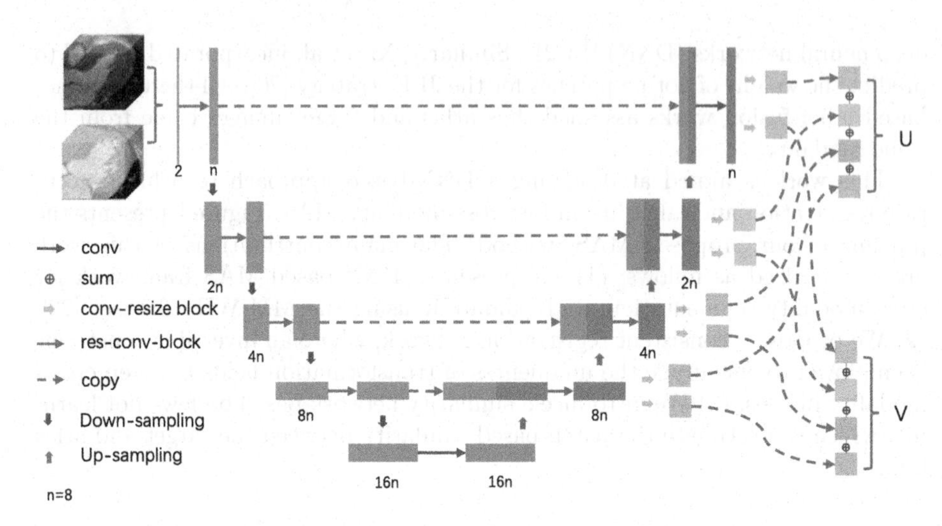

Fig. 2. The architecture of the consistent registration network.

Consistent Constraint: The $\mathcal{L}oss_{Dice}$ only provides voxel-level matching criterion for transformation field estimation. It is easily trapped into a local maximum due to the ambiguous correspondence in the voxel-level DDF. Inspired by the work of Christensen et al. [2], a consistent constraint is employed to encourage the uniqueness of the field. i.e., each voxel in L_a is mapped to only one voxel in L_t, and vice versa. To achieve this, an invertible loss $\mathcal{L}oss_{inv}$ is engaged to force the restored warped label $L_a^{'}$ (or $L_t^{'}$) to be identical to its original label L_a (or L_t),

$$\mathcal{L}oss_{inv} = \mathcal{D}ice(L_a^{'}, L_a) + \mathcal{D}ice(L_t^{'}, L_t), \tag{4}$$

where $L_a^{'}(x) = \tilde{L}_a(x + V(x)))$ and $L_t^{'}(x) = \tilde{L}_t(x + U(x)))$. Ideally, $\mathcal{L}oss_{inv}$ is equal to 0 when U and V are the inverse of each other. Therefore, it can constrain the network to produce invertible DDFs. Finally, the total trainable loss of the registration model is

$$\mathcal{L}oss_{reg} = \mathcal{L}oss_{Dice} + \lambda_2 \mathcal{L}oss_{inv}. \tag{5}$$

Here, λ_2 is the hyperparameter of the model. As only anatomical labels are needed to train the network, the consistent registration network is naturally applicable to cross-modality registration.

2.2 Similarity Network

Network Architecture: Based on the registration network, (I_a, L_a) can be deformed toward I_t and become the warped atlas $(\tilde{I}_a, \tilde{L}_a)$, where $\tilde{L}_a$ is a candidate segmentation of I_t. Given N atlases, the registration network will produce N corresponding segmentations. Then, the target label of I_t is derived by

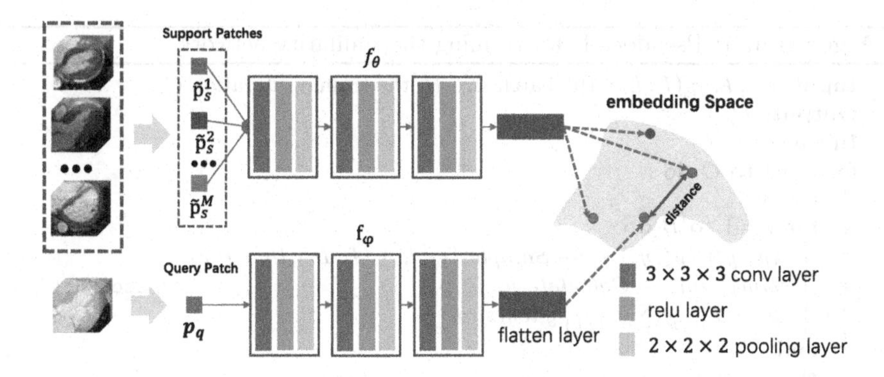

Fig. 3. The architecture of the similarity network.

combining the contribution of each warped atlas via PLF strategy. For a spatial point x, the optimal target label $\ddot{L}_t(x)$ is defined as

$$\ddot{L}_t(x) = \underset{l=\{l_1,l_2,\ldots..l_k\}}{\arg\max} \sum_{i=1}^{N} w^i(x)\delta(\tilde{L}_a^i(x), l), \tag{6}$$

where $\{l_1, l_2, \ldots..l_k\}$ is the label set, $w^i(x)$ is the contribution weight of i-th warped atlas, and $\delta(\tilde{L}_a^i(x), l)$ is the Kronecker delta function. Usually, $w^i(x)$ is measured according to the intensity similarity among local patches. Inspired by the idea of prototypical method [14], there exists an embedding that can capture more discriminative features for similarity measurement. We design a convolution network to map the original patches into a more distinguishable embedding space, and similarities (contribution weights) can be computed according to the distance between the embedded atlas and target patches.

Figure 3 shows the architecture of the similarity network. It contains two convolution ways (f_φ and f_θ), which can map the target and atlas patches into a embedding space separately. According to the prototypical method, we define the patch from target image as the query patch (p_q), and define the patches sampled from warped atlases as support patches ($p_s = \{\tilde{p}_s^1, \tilde{p}_s^2, \ldots, \tilde{p}_s^M\}$). The similarity sim_i of p_q and $\tilde{p}_s^i$ is calculated based on a softmax over the Euclidean distance of embedded atlas $f_\theta(\tilde{p}_s^i)$ and target patch $f_\varphi(p_q)$,

$$sim_i = \frac{exp(-d(f_\varphi(p_q), f_\theta(\tilde{p}_s^i)))}{\sum_{m=1}^{M} exp(-d(f_\varphi(p_q), f_\theta(\tilde{p}_s^m)))}. \tag{7}$$

Training Algorithm: We explore to train the similarity network by using the anatomical label information. Let y_i denotes the ground-truth similarity between p_q and $\tilde{p}_s^i$. The parameters of f_θ and f_φ can be optimized by minimizing the cross-entropy loss (J) of the predicted and ground-truth similarities,

Algorithm 1: Pseudocode for training the similarity network

```
Input: (Ĩ_a, L̃_a); (I_t, L_t); the batch size B; the training iteration C
Output: θ,φ
Init θ,φ
for c = 1 to C do
    J ← 0
    for b = 1 to B do
        (p̃_s^j, y_j), (p̃_s^k, y_k), p_q ← Sample((Ĩ_a, L̃_a), (I_t, L_t), thr_1, thr_2)
        sim_j, sim_k ← Calculate(p̃_s^j, p̃_s^k, p_q)          // see Eq.(7)
        J ← J + (1/B)(Σ_{i={j,k}} y_i log(sim_i))
    end
    θ^new ← θ^old − ε ∇_θ J
    φ^new ← φ^old − ε ∇_φ J
end
```

$$J = -\sum_{i=1}^{M} y_i log(sim_i). \tag{8}$$

However, y_i is hard to obtain in cross-modality scenarios. To train the network, the support patches ($\tilde{p}_s^i$) which have significant shape difference or similarity to the query patch (p_q) are used, and their corresponding y_i is decided by using the anatomical labels,

$$y_i = \begin{cases} 1 \; \mathcal{D}ice(l_q, \tilde{l}_s^i) > thr_1 \\ 0 \; \mathcal{D}ice(l_q, \tilde{l}_s^i) < thr_2 \end{cases}. \tag{9}$$

where thr_1 and thr_2 are hard thresholds, l_q and $\tilde{l}_s^i$ denote the anatomical label of p_q and $\tilde{p}_s^i$, respectively. The network is trained in a fashion of few-shot learning, each training sample is compose of a query patch (p_q) and two support patches ($\tilde{p}_s^j, \tilde{p}_s^k$) with significant shape differences ($y_j \neq y_k$). In this way, the convolution layers can learn to capture discriminative features for measuring similarity of cross-modality. Algorithm 1 provides the pseudocode. For the conciseness, the code only describe one atlas and one target setup here, while the reader can easily extend to N atlas and K targets in practice.

3 Experiment

Experiment Setup: We evaluated the framework by myocardial segmentation of the MM-WHS dataset [22]. The dataset provides 40 (20 CT and 20 MRI) images with corresponding manual segmentations of whole heart. For cross-modality setup, MR (CT) images with their labels are used as the atlases and CT (MR) images are treated as the targets. We randomly selected 24 (12 CT and 12 MR) images for training the registration network, 8 (4 CT and 4 MR) images for training the similarity network. The remaining 8 (4 CT and 4 MR) images were used as test data. Form each image, a $96 \times 96 \times 96$ sub-image around LV myocardium was cropped, and all the sub-images were normalized to zero-mean with unit-variance. In order to improve the performance, both the affine and deformable transformation were adopted for data augmentation.

For Training the Registration Network: In each training iteration, a pair of CT-MR intensity images is fed into the registration network (see Fig. 2). Then the network produce U and V, with which the MR and CT label can be warped to each other. By setting the hyperparameter λ_1 and λ_2 to 0.3 and 0.2, the total trainable loss (see Eq.(5)) of the network can be calculated. Finally, Adam optimizer is employed to train the parameters of network.

For Training the Similarity Network: For training the network, we extracted patches along the boundary of LV myocardium (which usually cover different anatomical structure). In each training iteration, the size of patch is set to $15 \times 15 \times 15$ voxels, while the thr_1 and thr_2 are set to 0.9 and 0.5, respectively (see Eq. (9)). Training sample $(\tilde{p}_s^j, \tilde{p}_s^k, p_q)$ is randomly selected and then mapped into the embedding space. Finally, the loss can be accumulated and backpropagated to optimize the parameters of f_φ and f_θ (see Algorithm 1).

Table 1. Comparison between the proposed registration network and other state-of-the-art methods.

Method	Dice (Myo)
Demons CT-MR [16]	36.1 ± 10.9%
Demons MR-CT [16]	36.9 ± 12.8%
SyNOnly CT-MR [1]	52.6 ± 13.9%
SyNOnly MR-CT [1]	55.3 ± 10.8%
CT-MR [6]	70.5 ± 4.8%
MR-CT [6]	73.4 ± 4.7%
Our CT-MR	**74.4 ± 5.2%**
Our MR-CT	**76.4 ± 4.7%**

Table 2. Comparison between the proposed MAS and other state-of-the-art methods.

Method	Dice (Myo)
U-Net [12]	86.1 ± 4.2%
Seg-CNN [10]	**87.2 ± 3.9** %
MV MR-CT	84.4 ± 3.6%
NLWV MR-CT	84.9 ± 4.0%
Our MR-CT	84.7 ± 3.9%
U-Net [12]	68.1 ± 25.3%
Seg-CNN [10]	75.2 ± 12.1%
MV CT-MR	80.8 ± 4.8%
NLWV CT-MR	81.6 ± 4.7%
Our CT-MR	**81.7 ± 4.7** %

Results: The performance of the registration network is evaluated by using the Dice score between the warped atlas label and the target gold standard label. Table 1 shows the average Dice scores over 48 (12 CT $\times$ 4 MR or 12 MR $\times$ 4 CT) LV myocardium registrations. CT-MR (MR-CT) indicates when CT (MR) images are used as atlas and MR (CT) images are used as target. Compared to the U-shape registration network [6], the proposed network achieves almost 3.5% improvement of Dice score. Additionally, our method outperforms the conventional methods (SyNOnly [1] and Demons [16]). This is reasonable as our method takes advantage of the high-level information (anatomical label) to train the registration model, which makes it more suitable for the challenging dataset of MM-WHS.

Table 2 shows the result of three different MAS methods based on our registration network. ie, MV, non-local weighted voting (NLWV) [3] and the proposed framework. Compared to other state-of-the-art methods [10,12], our framework can achieve promising performance in cross-modality myocardial segmentation. Especially in MR images, compared to the Seg-CNN [10] who won the first place of MM-WHS Challenge, our framework improves the Dice score by almost 6%. However, our MR-CT result, which is set up to use CT atlases to segment an MR target, is worse than other state-of-the-art methods. This is because the quality of atlas will affect MAS performance. Generally, MR is considered more challenging data (lower quality) compared to CT [22]. The use of low-quality MR atlases limits the segmentation accuracy of our MR-CT. Thus, the Seg-CNN [10], which is trained on purely CT data, can obtain almost 3% better Dice score than our MR-CT method. In addition, Fig. 4 demonstrates a series of intermediate results and segmentation details.

Figure 5 visualizes the performance of similarity network. The target patch is randomly selected from CT image, and the atlas patches are randomly cropped from MR images. Since Dice coefficient computes similarity of patches by using golden standard labels, it can be considered as the golden standard for cross-modality similarity estimation. Results show that the estimated similarities are well correlated to the Dice coefficient.

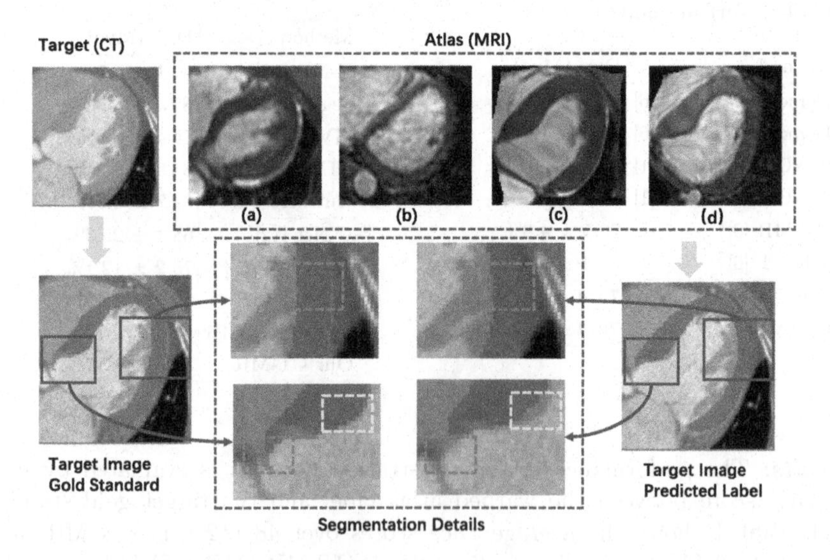

Fig. 4. Visualization of the proposed framework. Image of (a) and (b) are atlas images, (c) and (d) are corresponding warp atlas images. All the images are from axial-view. The segmentation details show different slices, where the region in the yellow box shows the error by our method, while the region in the blue boxes indicate that the proposed method performs not worse than the golden standard. (The reader is referred to the colourful web version of this article) (Color figure online)

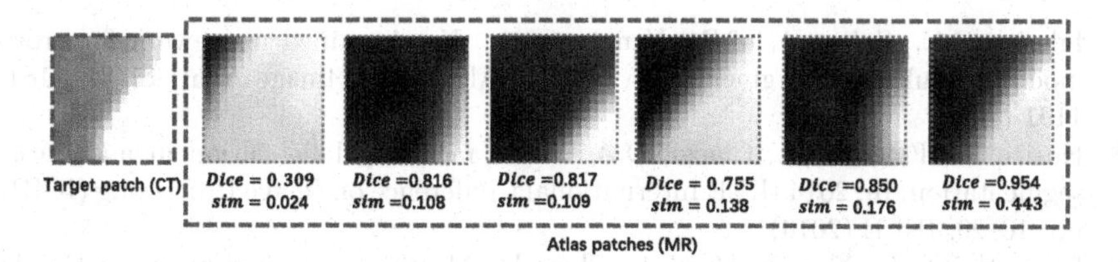

Fig. 5. Visualization of estimated similarities from the proposed network. The Dice scores (*Dice*) are calculated by the gold standard label, while the similarities (*sim*) are estimated by feeding intensity patches into the similarity network. Please note that the *sim* are normalized by the softmax function of similarity network (see Eq.(7)). Ideally, the *sim* should be positively related to the *Dice*. This figure shows both the correct (green box) and failed (red box) cases of our similarity estimation method. (Color figure online)

4 Conclusion

We have proposed a cross-modality MAS framework to segment a target image using the atlas from another modality. Also, we have described the consistent registration and similarity estimation algorithm based on DNN models. The experiment demonstrates that the proposed framework is capable of segmenting myocardium from CT or MR images. Future research aims to extend the framework to other substructure of the whole heart, and investigate the performance on different datasets.

References

1. Avants, B.B., Tustison, N., Song, G.: Advanced normalization tools (ANTs). Insight j **2**(365), 1–35 (2009)
2. Christensen, G.E., Johnson, H.J.: Consistent image registration. IEEE Trans. Med. Imaging **20**(7), 568–582 (2001)
3. Coupé, P., Manjón, J.V., Fonov, V., Pruessner, J., Robles, M., Collins, D.L.: Nonlocal patch-based label fusion for hippocampus segmentation. In: Jiang, T., Navab, N., Pluim, J.P.W., Viergever, M.A. (eds.) MICCAI 2010. LNCS, vol. 6363, pp. 129–136. Springer, Heidelberg (2010). https://doi.org/10.1007/978-3-642-15711-0_17
4. Ding, Z., Han, X., Niethammer, M.: VoteNet: a deep learning label fusion method for multi-atlas segmentation. In: Shen, D., et al. (eds.) MICCAI 2019. LNCS, vol. 11766, pp. 202–210. Springer, Cham (2019). https://doi.org/10.1007/978-3-030-32248-9_23
5. Heinrich, M.P., Jenkinson, M., Papież, B.W., Brady, S.M., Schnabel, J.A.: Towards realtime multimodal fusion for image-guided interventions using self-similarities. In: Mori, K., Sakuma, I., Sato, Y., Barillot, C., Navab, N. (eds.) MICCAI 2013. LNCS, vol. 8149, pp. 187–194. Springer, Heidelberg (2013). https://doi.org/10.1007/978-3-642-40811-3_24
6. Hu, Y., et al.: Weakly-supervised convolutional neural networks for multimodal image registration. Med. Image Anal. **49**, 1–13 (2018)

7. Iglesias, J.E., Sabuncu, M.R., Van Leemput, K.: A unified framework for cross-modality multi-atlas segmentation of brain MRI. Med. Image Anal. **17**(8), 1181–1191 (2013)
8. Kasiri, K., Fieguth, P., Clausi, D.A.: Cross modality label fusion in multi-atlas segmentation. In: 2014 IEEE International Conference on Image Processing (ICIP), pp. 16–20. IEEE (2014)
9. Luan, H., Qi, F., Xue, Z., Chen, L., Shen, D.: Multimodality image registration by maximization of quantitative-qualitative measure of mutual information. Pattern Recogn. **41**(1), 285–298 (2008)
10. Payer, C., Štern, D., Bischof, H., Urschler, M.: Multi-label whole heart segmentation using CNNs and anatomical label configurations. In: Pop, M., et al. (eds.) STACOM 2017. LNCS, vol. 10663, pp. 190–198. Springer, Cham (2018). https://doi.org/10.1007/978-3-319-75541-0_20
11. Qin, C., Shi, B., Liao, R., Mansi, T., Rueckert, D., Kamen, A.: Unsupervised deformable registration for multi-modal images via disentangled representations. In: Chung, A.C.S., Gee, J.C., Yushkevich, P.A., Bao, S. (eds.) IPMI 2019. LNCS, vol. 11492, pp. 249–261. Springer, Cham (2019). https://doi.org/10.1007/978-3-030-20351-1_19
12. Ronneberger, O., Fischer, P., Brox, T.: U-Net: convolutional networks for biomedical image segmentation. In: Navab, N., Hornegger, J., Wells, W.M., Frangi, A.F. (eds.) MICCAI 2015. LNCS, vol. 9351, pp. 234–241. Springer, Cham (2015). https://doi.org/10.1007/978-3-319-24574-4_28
13. Sanroma, G., et al.: Learning non-linear patch embeddings with neural networks for label fusion. Med. Image Anal. **44**, 143–155 (2018)
14. Snell, J., Swersky, K., Zemel, R.: Prototypical networks for few-shot learning. In: Advances in neural Information Processing Systems, pp. 4077–4087 (2017)
15. Studholme, C., Hill, D.L., Hawkes, D.J.: An overlap invariant entropy measure of 3D medical image alignment. Pattern Recogn. **32**(1), 71–86 (1999)
16. Thirion, J.: Image matching as a diffusion process: an analogy with Maxwell's demons. Med. Image Anal. **2**(3), 243–260 (1998)
17. Wachinger, C., Navab, N.: Entropy and Laplacian images: structural representations for multi-modal registration. Med. Image Anal. **16**(1), 1–17 (2012)
18. Wang, H., Suh, J.W., Das, S.R., Pluta, J.B., Craige, C., Yushkevich, P.A.: Multi-atlas segmentation with joint label fusion. IEEE Trans. Pattern Anal. Mach. Intell. **35**(3), 611–623 (2012)
19. Warfield, S.K., Zou, K.H., Wells, W.M.: Simultaneous truth and performance level estimation (STAPLE): an algorithm for the validation of image segmentation. IEEE Trans. Med. Imaging **23**(7), 903–921 (2004)
20. Xie, L., Wang, J., Dong, M., Wolk, D.A., Yushkevich, P.A.: Improving multi-atlas segmentation by convolutional neural network based patch error estimation. In: Shen, D., et al. (eds.) MICCAI 2019. LNCS, vol. 11766, pp. 347–355. Springer, Cham (2019). https://doi.org/10.1007/978-3-030-32248-9_39
21. Yang, H., Sun, J., Li, H., Wang, L., Xu, Z.: Neural multi-atlas label fusion: application to cardiac MR images. Med. Image Anal. **49**, 60–75 (2018)
22. Zhuang, X., et al.: Evaluation of algorithms for multi-modality whole heart segmentation: an open-access grand challenge. Med. Image Anal. **58**, 101537 (2019)
23. Zhuang, X., Shen, J.: Multi-scale patch and multi-modality atlases for whole heart segmentation of MRI. Med. Image Anal. **31**, 77–87 (2016)

Longitudinal Image Registration with Temporal-Order and Subject-Specificity Discrimination

Qianye Yang[1(✉)], Yunguan Fu[1,2], Francesco Giganti[3,4], Nooshin Ghavami[1,5], Qingchao Chen[6], J. Alison Noble[6], Tom Vercauteren[5], Dean Barratt[1], and Yipeng Hu[1,6]

[1] Centre for Medical Image Computing and Wellcome/EPSRC Centre for Interventional and Surgical Sciences, University College London, London, UK
qianye.yang.19@ucl.ac.uk

[2] InstaDeep, London, UK

[3] Division of Surgery and Interventional Science, University College London, London, UK

[4] Department of Radiology, University College London Hospital NHS Foundation Trust, London, UK

[5] School of Biomedical Engineering and Imaging Sciences, Kings College London, London, UK

[6] Institute of Biomedical Engineering, University of Oxford, Oxford, UK

Abstract. Morphological analysis of longitudinal MR images plays a key role in monitoring disease progression for prostate cancer patients, who are placed under an active surveillance program. In this paper, we describe a learning-based image registration algorithm to quantify changes on regions of interest between a pair of images from the same patient, acquired at two different time points. Combining intensity-based similarity and gland segmentation as weak supervision, the population-data-trained registration networks significantly lowered the target registration errors (TREs) on holdout patient data, compared with those before registration and those from an iterative registration algorithm. Furthermore, this work provides a quantitative analysis on several longitudinal-data-sampling strategies and, in turn, we propose a novel regularisation method based on maximum mean discrepancy, between differently-sampled training image pairs. Based on 216 3D MR images from 86 patients, we report a mean TRE of 5.6 mm and show statistically significant differences between the different training data sampling strategies.

Keywords: Medical image registration · Longitudinal data · Maximum mean discrepancy

1 Introduction

Multiparametric MR (mpMR) imaging has gained increasing acceptance as a noninvasive diagnostic tool for detecting and staging prostate cancer [10]. Active

A. L. Martel et al. (Eds.): MICCAI 2020, LNCS 12263, pp. 243–252, 2020.
https://doi.org/10.1007/978-3-030-59716-0_24

surveillance recruits patients with low-grade cancers that exhibit low-to-medium risk to long-term outcome [10], where mpMR imaging has been adopted to follow regions within the prostate glands and to recognise or even predict the disease progression [7]. As outlined by the panel of experts who drafted the PRECISE criteria for serial MR reporting in patients on active surveillance for prostate cancer [10], assessing radiological changes of morphological MR features is a key component when reporting longitudinal MR images. For individual regions of pathological interest, these morphological features include volume, shape, boundary, extension to neighbouring anatomical structures and the degree of conspicurity in these features. Pairwise medical image registration quantifies the morphological correspondence between two images, potentially providing an automated computational tool to measure these changes, without time-consuming and observer-biased manual reporting. This voxel-level analysis is particularly useful in developing imaging biomarkers, when the ground-truth of the disease progression are still under debate, as in this application, and cannot be reliably used to train an end-to-end progression classifier.

Registration algorithms designed for longitudinal images have been proposed for several other applications [5,8,14], such as those utilising temporal regularisation [13] when applied to a data set acquired at three or more time points. Most algorithms are still based on or derived from the basic pairwise methodologies. In this work, registration of a pair of longitudinal prostate MR images from the same patient is investigated. Classical algorithms iteratively optimise a spatial transformation between two given images without using population data. For example, a fixed set of parameters in an iterative registration algorithm may work well for one patient, but unlikely to be optimal for other patients. Substantial inter-patient variation leads to the lack of common intensity patterns or structures between different prostates. Ad hoc benign foci, varying zonal anatomy and highly patient-specific pathology are frequently observed in MR images, especially within the prostate gland. This is particularly problematic for classical iterative registration algorithms, when the regions of interest, smaller and non-metastasis tumours, are confined within the varying prostate glands, such as those from the active surveillance patient cohort considered in this study. We provide such an example in the presented results using an iterative registration algorithm.

In this paper, we propose an alternative method that uses recently-introduced deep-learning-based non-iterative registration for this application. Based on the results on holdout patients, we argue that learning population statistics in patient-specific intensity patterns [2] and weak segmentation labels [6] can overcome the difficulties due to the large inter-patient variation, for aligning intra-patient prostate MR images. In order to efficiently and effectively utilise the often limited longitudinal data for training the registration network, we compare several methods to sample the time-ordered image pairs from the same patients and those from different patients, and propose a new regularisation method to discriminate time-forward image pairs versus time-backward image pairs and/or subject-specific image pairs versus inter-subject image pairs.

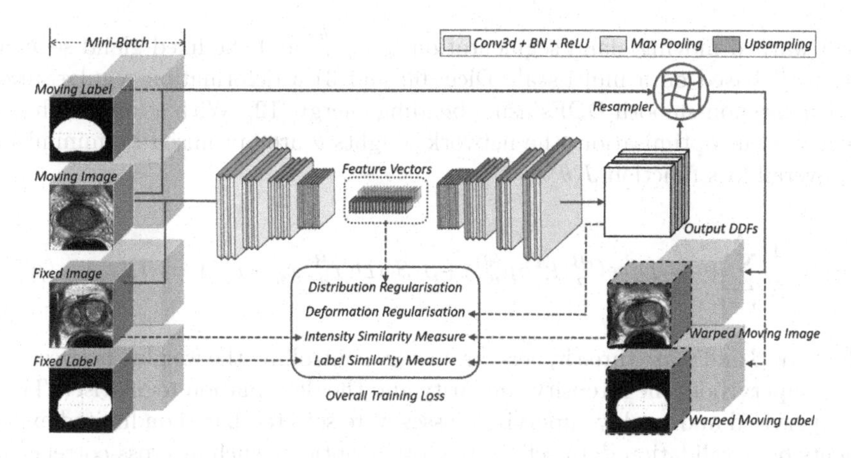

Fig. 1. Demonstration of the proposed image registration framework, with the dotted lines indicating the data flow only at the training.

We summarise the contributions in this work: 1) We developed an end-to-end deep-learning registration method tested on longitudinal images. To our knowledge, this is the first study for longitudinal MR Image registration for prostate cancer patients; 2) We present a quantitative analysis on longitudinal data sampling strategies for registration network training, with and without the new regularisation method; and 3) We report a set of rigorous results based on real clinical longitudinal data, demonstrating statistically significant differences between these sampling methods. This provides practically useful evidences for further development of the registration tools in longitudinal image analysis for active surveillance patients.

2 Methods

2.1 Learning-Based Image Registration

In this work, the pairwise registration paradigm based on deep learning is adopted for registering longitudinal MR image pairs. Denote $\{(x_n^A, x_n^B)\}, n = 1...N$, as a set of paired images to register, x_n^A and x_n^B being the moving- and fixed images, respectively. For each pair n in the set, let a pair of corresponding prostate gland anatomical segmentation, represented with binary masks, be (l_n^A, l_n^B). During training a registration network $f(\theta)$ with network parameters θ, the nth pair of images is fed into the network to predict a dense displacement field (DDF) $\mu_n^{(\theta)}$. An overview of the network training is illustrated in Fig. 1.

The training loss is comprised of three measures of the goodness-of-predicted-DDF: 1) an intensity-based similarity measure between the DDF-warped moving image $x_n^A \circ \mu_n^{(\theta)}$ and the fixed image x_n^B, the sum-of-square differences (SSD) in intensity values being used in this study; 2) an overlap measure between the

DDF-warped moving gland segmentation $l_n^A \circ \mu_n^{(\theta)}$ and the fixed gland segmentation l_n^B based on a multi-scale Dice [6]; and 3) a deformation regularisation to penalise non-smooth DDFs using bending energy [12]. With a minibatch gradient descent optimisation, the network weights θ are optimised by minimising the overall loss function $J(\theta)$:

$$J(\theta) = \frac{1}{N}\sum_{n=1}^{N}(-\alpha \cdot Dice(l_n^B, l_n^A \circ \mu_n^{(\theta)}) + \beta \cdot SSD(x_n^B, x_n^A \circ \mu_n^{(\theta)}) + \gamma \cdot \Omega_{bending}(\mu_n^{(\theta)})) \tag{1}$$

where, α, β and γ are three hyper-parameters controlling the weights between the weak supervision, the intensity similarity and the deformation regulariser. These unsupervised and weakly-supervised losses were selected based on limited experiments on a validation data set, among other options, such as cross-correlation, Jaccard and DDF gradient norms. The three weights in Eq. 1 are co-dependent with the learning rate and potentially can be reduced to two. Therefore, the fine-tuning of these hyper-parameters warrants further investigation in future studies.

2.2 Training Data Distribution

Temporal-Order and Subject-Specificity. Figure 2 illustrates example longitudinal images from individual subjects (prostate cancer patients) at multiple visits in order of time. Without loss of generality, we aim to model the morphological changes to quantify a chronological deformation field, i.e. from a baseline T2-weighted image to a follow-up, acquired at time points t_1 and t_2, $t_1 < t_2$, respectively. To train a registration network for this task, one can sample *task-specific training data*, i.e. intra-subject, time-forward image pairs. Given sufficient training data, i.e. the empirical training data distribution adequately representing the population data distribution, there is little reason to add other types of image pairs, i.e. time-backward or inter-subject pairs.

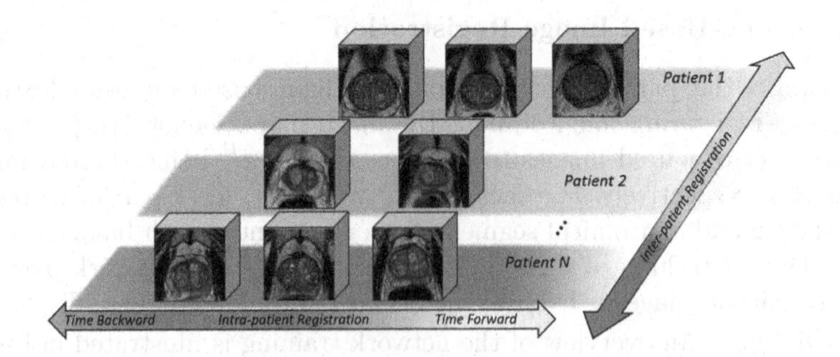

Fig. 2. Demonstration of a longitudinal image data set from active surveillance.

However, as is common in the field of medical image computing, the acquisition of training data is limited by various practical or clinical reasons. This leads to an empirical risk minimisation (ERM) on a sparse empirical training distribution [16]. ERM is known to be prone to overfitting [16]. Data augmentation strategies such as using affine/nonrigid transformation and the "mixup" [17], in geometric transformation and intensity spaces, respectively, have been applied to overcome overfitting. The former is also applied in this work.

Particularly interesting to longitudinal pairwise image registration, image pairs with reversed temporal-order, the time-backward pairs, and those with different subject-specificity, the inter-subject pairs, can be considered as augmenting data for this task to potentially improve generalisability. This is of practical importance in training longitudinal image registration networks and is the first hypothesis to test in this study.

Discriminating Prior for Regularising Network Training. Furthermore, a data augmentation strategy becomes effective if the augmented empirical data distribution better represents population distribution or "aligns" empirical feature vectors with a potentially new population distribution in feature vector space [15]. Directly mixing the intra-/inter-subject and/or time-forward/-backward image pairs does not guarantee such alignment. To utilise the prior knowledge of the known temporal order and subject specificity, we test a new regularisation approach based on empirical maximum mean discrepancy (MMD) [4]. By penalising the divergence between the feature vectors generated from different empirical data distributions. As illustrated in Fig. 1, an MMD square is computed over the two sets of high-dimensional feature vectors from the network encoder, $\{v_i\}_{i=1}^{I}$ and $\{v_j\}_{j=1}^{J}$, generated from different types of image pairs:

$$\Omega_{mmd} = \frac{1}{I^2}\sum_{i \neq i'}^{I} k_{II}(v_i, v_{i'}) - \frac{2}{IJ}\sum_{i,j=1}^{I,J} k_{IJ}(v_i, v_j) + \frac{1}{J^2}\sum_{j \neq j'}^{J} k_{JJ}(v_j, v_{j'}), \quad (2)$$

where $k(\mu, \nu) = \exp(-||\mu - \nu||^2/2\sigma)$ is a Gaussian kernel with a parameter σ [3]. I and J are the sample numbers of the feature vectors, within a minibatch of size $N = I + J$. The weighted MMD regularising term in Eq. 2 is added to the original loss in Eq. 1:

$$J^*(\theta) = J(\theta) + \lambda \cdot \Omega_{mmd}(\{v_i\}, \{v_j\}). \quad (3)$$

With the new loss in Eq. 3, we test the second hypothesis that encoding the discrimination of temporal order and subject specificity can further improve the network generalisability.

2.3 Validation

Gerneralisability on Holdout Data. The patients and their data is randomly assigned into training, validation and holdout sets. The networks are developed with training and validation sets, including hyper-parameter tuning. The generalisability is measured on the holdout set using three metrics: 1) the binary Dice similarity coefficient (DSC) between the fixed label l^B and the warped moving label $l^A \circ \mu$; 2) the MSE between the fixed image x^B and the wrapped moving image $x^A \circ \mu$; and 3) the centroid distance between the aligned prostate glands l^B and $l^A \circ \mu$. Results from paired t-tests with a significance level of $\alpha = 0.05$ are reported when comparing these metrics.

Registration Performance. Also on the holdout set, pairs of corresponding anatomical and pathological landmarks are manually identified on moving and fixed images, including patient-specific fluid-filled cysts, calcification and centroids of zonal boundaries. The target registration errors (TREs) between the corresponding pairs of landmarks, from the warped moving and those from the fixed images, are computed to quantify the registration performance. Other experiment details are provided in Sect. 3.

3 Experiments

3.1 Data and Preprocessing

216 longitudinal prostate T2-weighted MR images were acquired from 86 patients at University College London Hospitals NHS Foundation. For each patient 2–4 images were available, with intervals between consecutive visits ranging from 11 to 28 months. All the image volumes were resampled to $0.7 \times 0.7 \times 0.7$ mm^3 isotropic voxels with a normalised intensity range of $[0, 1]$. For computational consideration, all images were also cropped from the image center to $128 \times 128 \times 102$ voxels, such that the prostate glands are preserved. The prostate glands were manually segmented for the weak supervision in training and for validation. The images were split into 70, 6 and 10 patients for training, validation and holdout sets.

3.2 Network Training

A previously proposed DDF-predicting encoder-decoder architecture was used [6], which is an adapted 3D U-Net [11], with more densely connected skip layers and residual shortcuts. The network training was implemented with TensorFlow 2 [1] and made open source https://github.com/DeepRegNet/DeepReg. The Adam optimizer with an initial learning rate of 10^{-5} was used with the hyper-parameters α, β, γ and λ empirically set to 1, 1, 50 and 0.01, respectively, via qualitative assessment on the validation set. The networks were trained on Nvidia Tesla V100 GPUs with a minibatch of 4 sets of image-label data, each containing a pair of T2 MR images and a corresponding binary mask pair of prostate gland segmentation. Each network run for 272,000 iterations, approximately 64 h.

3.3 Training Image Pair Sampling

To test the first hypothesis in Sect. 2.2, three different training data sets were sampled, resulting in three networks: a network trained with only intra-subject, time-forward image pairs (IF); a second network (IF+IB) trained using all possible intra-subject pairs regardless of temporal order; and the third network (IT+IF+IB) with added inter-subject samples. All the networks were trained with the registration loss function in Eq. 1. Generalisability and TREs were computed on all the intra-subject, time-forward image pairs from the same holdout patient data.

To test the second hypothesis, two more networks were trained with the loss in Eq. 3, with respective training data sets IF+IB and IT+IF+IB. For intra-patient IF+IB pairs, two images were randomly sampled without replacement from a single patient. MMD may be sensitive to minibatch size and sample size imbalance [4], a two-stage sampling was adapted to ensure every minibatch samples 2 IF and 2 IB pairs during training the IF+IB network; and samples 2 IF/IB pairs and 2 IT pairs during training the IT+IF+IB network. For comparison purposes, the same sampling was adopted when the MMD was not used. When testing the IT+IF+IB network, with or without the MMD regulariser, results were computed on all the intra-patient pairs in the holdout set.

3.4 Comparison of Networks with an Iterative Algorithm

To test an iterative intensity-based registration, the widely-adopted nonrigid method using B-splines was tested on the same holdout images. For comparison purposes, the sum-of-square difference in intensity values was used as similarity measure with all other parameters set to default in the NiftyReg [9] package. The reported registration performance was to demonstrate its feasibility. Although the default configuration is unlikely to perform optimally, tuning this method is considered out of scope for the current study.

3.5 Results

Sampling Strategies. Networks with different training data sampling methods, described in Sect. 3.3, are summarised in Table 1. Adding time-backward and inter-subject image pairs in the training significantly improved the performance, both in network generalisability and registration performance. For example, the TREs decreased from 6.456 ± 6.822 mm to 5.801 ± 7.104 mm (p-value = 0.0004), when adding time-backward data, to 5.482 ± 5.589 mm (p-value = 0.0332), when further inter-subject data was added in training. The same conclusion was also observed in DSCs and gland CDs.

Regularisation Effect. Table 2 summarises the comparison between networks trained with MMD regularisation (Eq. 3) and without (Eq. 1). Although improved results were consistently observed in expected DSCs and TREs, statistical significance was not found on holdout set. However, we report a statistically significant improvement on the validation set, due to the introduction of the MMD regulariser, e.g. p-values = 0.016 in lowered MSEs. This was probably limited by the small holdout set and the under-optimised hyperparameters specifically for the MMD regulariser in current study.

Table 1. Registration performance with NiftyReg and networks with different sampling strategies. *See text for details including the explanation of the inferior NiftyReg result.

Methods	DSC	CD	MSE	TRE
NiftyReg*	0.270 ± 0.304	22.869 ± 11.761 mm	0.041 ± 0.019	21.147 ± 15.841 mm
w/o registration	0.701 ± 0.097	8.842 ± 4.067 mm	0.051 ± 0.090	10.736 ± 7.718 mm
IF	0.861 ± 0.042	2.910 ± 1.756 mm	0.049 ± 0.097	6.456 ± 6.822 mm
IF+IB	0.870 ± 0.033	2.257 ± 1.503 mm	0.043 ± 0.096	5.801 ± 7.104 mm
IT+IF+IB	0.885 ± 0.024	2.132 ± 0.951 mm	0.053 ± 0.014	5.482 ± 5.589 mm

Table 2. The comparison between networks trained with and without MMD regularisation. See text for more details.

Method	Test	MMD	DSC	CD	MSE	TRE
IF+IB	IF	×	0.870 ± 0.033	2.257 ± 1.503 mm	0.043 ± 0.096	5.801 ± 7.104 mm
IF+IB	IF	√	0.876 ± 0.027	2.300 ± 1.007 mm	0.042 ± 0.094	5.847 ± 6.360 mm
IT+IF+IB	IF+IB	×	0.890 ± 0.019	1.928 ± 0.797 mm	0.048 ± 0.010	5.638 ± 6.021 mm
IT+IF+IB	IF+IB	√	0.893 ± 0.023	1.527 ± 0.832 mm	0.049 ± 0.010	6.048 ± 6.721 mm

Registration Performance. As shown in Table 1, the proposed registration network, with any training data sampling strategies, produced significantly lower TREs on holdout data than the TREs before registration or that from default NiftyReg algorithms. With the overall inferior NiftyReg results summarised in the table, we report that the best-performed registration from NiftyReg achieved a comparable DSC of 0.81 and a TRE of 4.75 mm, better than the network-predicted. This provides an example of highly variable registration performance from an iterative algorithm, frequently encountered in our experiment. However, a comprehensive comparison is still needed to draw further conclusions. On average, the inference time was 0.76 s for the registration network, compared with 50.4 s for NiftyReg.

A Case Study for Longitudinal Analysis of Prostate Cancer. Figure 3 qualitatively illustrates a 60-year-old man from active surveillance with a baseline and three follow-up visits, subject to a biopsy of Gleason 3+3. The yellow arrows indicate the evolution of a marked adenomatous area within the left transitional zone. The registration network was able to track the changes of the suspicious regions between consecutive visits, in an automated, unbiased and consistent manner over 3 years. Ongoing research is focused on analysis using the registration-quantified changes.

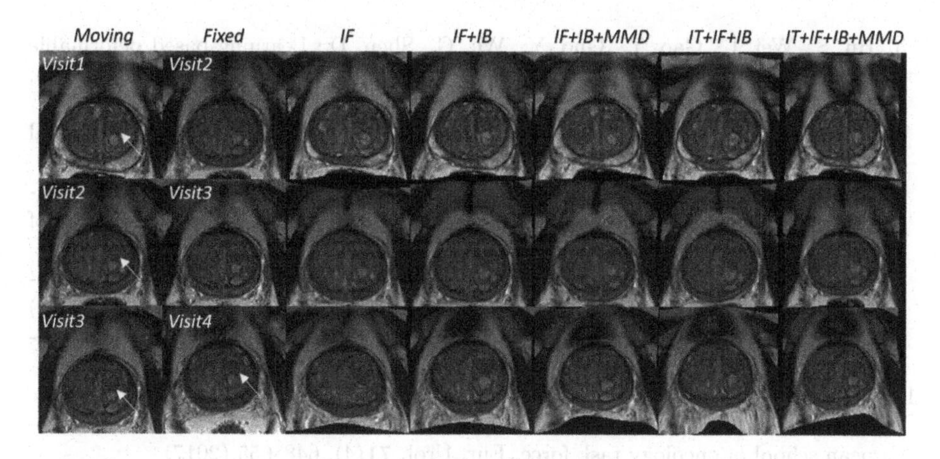

Fig. 3. Example registration results of a patient with 4 visits. The 1st and 2nd columns are the moving and fixed images. The remainder represent the network-warped images.

4 Conclusion

For the first time, we have developed a deep-learning-based image registration method and validated the network using clinical longitudinal data from prostate cancer active surveillance patients. We have also shown that adopting different training strategies significantly changes the network generalisability on holdout data.

Acknowledgment. This work is supported by the Wellcome/EPSRC Centre for Interventional and Surgical Sciences (203145Z/16/Z), Centre for Medical Engineering (203148/Z/16/Z; NS/A000049/1), the EPSRC-funded UCL Centre for Doctoral Training in Intelligent, Integrated Imaging in Healthcare (i4health) (EP/S021930/1) and the Department of Health's NIHR-fundedBiomedical Research Centre at UCLH. Francesco Giganti is funded by the UCL Graduate Research Scholarship and the Brahm Ph.D. scholarship in memory of Chris Adams.

References

1. Abadi, M., et al.: Tensorflow: Large-scale machine learning on heterogeneous distributed systems. arXiv preprint arXiv:1603.04467 (2016)
2. Balakrishnan, G., Zhao, A., Sabuncu, M.R., Guttag, J., Dalca, A.V.: Voxelmorph: a learning framework for deformable medical image registration. IEEE Trans. Med. Imaging **38**(8), 1788–1800 (2019)
3. Bousmalis, K., Trigeorgis, G., Silberman, N., Krishnan, D., Erhan, D.: Domain separation networks. In: Advances in Neural Information Processing Systems, pp. 343–351 (2016)
4. Gretton, A., Borgwardt, K.M., Rasch, M.J., Schölkopf, B., Smola, A.: A kernel two-sample test. J. Mach. Learn. Res. **13**(Mar), 723–773 (2012)

5. Hu, S., Wei, L., Gao, Y., Guo, Y., Wu, G., Shen, D.: Learning-based deformable image registration for infant MR images in the first year of life. Med. Phys. **44**(1), 158–170 (2017)
6. Hu, Y., et al.: Weakly-supervised convolutional neural networks for multimodal image registration. Med. image anal. **49**, 1–13 (2018)
7. Kim, C.K., Park, B.K., Lee, H.M., Kim, S.S., Kim, E.: MRI techniques for prediction of local tumor progression after high-intensity focused ultrasonic ablation of prostate cancer. Am. J. Roentgenol. **190**(5), 1180–1186 (2008)
8. Liao, S., et al.: A novel framework for longitudinal atlas construction with groupwise registration of subject image sequences. NeuroImage **59**(2), 1275–1289 (2012)
9. Modat, M., et al.: Fast free-form deformation using graphics processing units. Comput. Methods Programs Biomed. **98**(3), 278–284 (2010)
10. Moore, C.M., et al.: Reporting magnetic resonance imaging in men on active surveillance for prostate cancer: the precise recommendations-a report of a european school of oncology task force. Eur. Urol. **71**(4), 648–655 (2017)
11. Ronneberger, O., Fischer, P., Brox, T.: U-Net: convolutional networks for biomedical image segmentation. In: Navab, N., Hornegger, J., Wells, W.M., Frangi, A.F. (eds.) MICCAI 2015. LNCS, vol. 9351, pp. 234–241. Springer, Cham (2015). https://doi.org/10.1007/978-3-319-24574-4_28
12. Rueckert, D., Sonoda, L.I., Hayes, C., Hill, D.L., Leach, M.O., Hawkes, D.J.: Nonrigid registration using free-form deformations: application to breast MR images. IEEE Trans. Med. Imaging **18**(8), 712–721 (1999)
13. Schwartz, E., Jakab, A., Kasprian, G., Zöllei, L., Langs, G.: A locally linear method for enforcing temporal smoothness in serial image registration. In: Durrleman, S., Fletcher, T., Gerig, G., Niethammer, M., Pennec, X. (eds.) STIA 2014. LNCS, vol. 8682, pp. 13–24. Springer, Cham (2015). https://doi.org/10.1007/978-3-319-14905-9_2
14. Simpson, I.J.A., Woolrich, M.W., Groves, A.R., Schnabel, J.A.: Longitudinal brain MRI analysis with uncertain registration. In: Fichtinger, G., Martel, A., Peters, T. (eds.) MICCAI 2011. LNCS, vol. 6892, pp. 647–654. Springer, Heidelberg (2011). https://doi.org/10.1007/978-3-642-23629-7_79
15. Tzeng, E., Hoffman, J., Zhang, N., Saenko, K., Darrell, T.: Deep domain confusion: Maximizing for domain invariance. arXiv preprint arXiv:1412.3474 (2014)
16. Vapnik, V.N.: An overview of statistical learning theory. IEEE Trans. Neural Netw. **10**(5), 988–999 (1999)
17. Zhang, H., Cisse, M., Dauphin, Y.N., Lopez-Paz, D.: mixup: Beyond empirical risk minimization. arXiv preprint arXiv:1710.09412 (2017)

Flexible Bayesian Modelling for Nonlinear Image Registration

Mikael Brudfors[1](✉), Yaël Balbastre[1], Guillaume Flandin[1], Parashkev Nachev[2], and John Ashburner[1]

[1] Wellcome Centre for Human Neuroimaging, UCL, London, UK
mikael.brudfors.15@ucl.ac.uk
[2] UCL Institute of Neurology, London, UK

Abstract. We describe a diffeomorphic registration algorithm that allows groups of images to be accurately aligned to a common space, which we intend to incorporate into the SPM software. The idea is to perform inference in a probabilistic graphical model that accounts for variability in both shape and appearance. The resulting framework is general and entirely unsupervised. The model is evaluated at inter-subject registration of 3D human brain scans. Here, the main modeling assumption is that individual anatomies can be generated by deforming a latent 'average' brain. The method is agnostic to imaging modality and can be applied with no prior processing. We evaluate the algorithm using freely available, manually labelled datasets. In this validation we achieve state-of-the-art results, within reasonable runtimes, against previous state-of-the-art widely used, inter-subject registration algorithms. On the unprocessed dataset, the increase in overlap score is over 17%. These results demonstrate the benefits of using informative computational anatomy frameworks for nonlinear registration.

1 Introduction

This paper presents a flexible framework for registration of a population of images into a common space, a procedure known as spatial normalisation [1], or congealing [2]. Depending on the quality of the common space, accurate pairwise alignments can be produced by composing deformations that map two subjects to this space. The method is defined by a joint probability distribution that describes how the observed data can be generated. This *generative model* accounts for both *shape* and *appearance* variability; its conditional dependences producing a more robust procedure. Shape is encoded by a tissue *template*, that is deformed towards each image by a subject-specific composition of a rigid and a

M. Brudfors and Y. Balbastre—These authors contributed equally to this work.

Electronic supplementary material The online version of this chapter (https://doi.org/10.1007/978-3-030-59716-0_25) contains supplementary material, which is available to authorized users.

A. L. Martel et al. (Eds.): MICCAI 2020, LNCS 12263, pp. 253–263, 2020.
https://doi.org/10.1007/978-3-030-59716-0_25

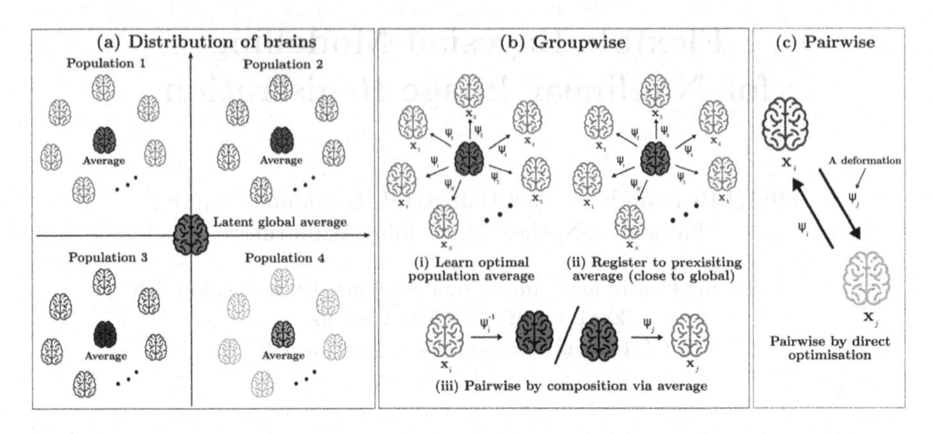

Fig. 1. **(a)** Multiple populations of brain scans have their individual averages; the assumption in this paper is that there exists a latent global average. **(b)** A groupwise method can either learn the optimal population-specific average (i), or use an already learned average (ii), the closer the learned average is to the global, the better the method should generalise to unseen test data. In both cases, all population scans are deformed towards the average. Pairwise deformations are then obtained by composing deformations via this average (iii). The proposed approach belongs here, and can be used for both (i) and (ii). **(c)** A pairwise method directly deforms one image towards another, usually by optimising some similarity metric or by applying a learned function. The common space then consists of just the two images to be registered.

diffeomorphic transform. Performing registration on the tissue level, rather than intensity, has been shown to be a more robust method of registering medical images [3]. Appearance is encoded by subject-specific Gaussian mixture models, with prior hyper-parameters shared across the population. A key assumption of the model is that there exists a latent average representation (*e.g.*, brain), this is illustrated in Fig. 1a.

Images of human organs differ in their morphology, the goal of spatial normalisation is to deform individual organs so that anatomical locations correspond between different subjects (a selective removal of the inter-individual anatomical variance). The deformations that are computed from this inter-subject registration therefore capture meaningful individual shape information. Although *not* constrained to a specific organ, our method will here be applied to spatially normalise brain magnetic resonance images (MRIs). Spatial normalisation is a critical first step in many neuroimaging analyses, *e.g.*, the comparison of tissue composition [4] or functional MRI activation [5] across individuals; shape mapping [6]; the extraction of predictive features for machine learning tasks [7]; or the identification of lesions [8]. The success of these tasks is therefore fundamentally coupled with the quality of the inter-individual alignment. Neuroimaging meta-analysis [9] is another research area that relies on spatial normalisation. Currently, statistical maps are coarsely registered into the MNI space. Better normalisation towards a more generic, multi-modal, high-resolution space could greatly improve the power and spatial specificity of such meta-analyses.

In general, registration tasks can be classified as either pairwise or groupwise (Fig. 1b–c). Pairwise methods optimise a mapping between two images, and only their two spaces exist. Groupwise methods aim to align several images into an optimal common space. Spatial normalisation aims to register a group of images into a pre-existing common space, defined by some average. Most nonlinear registration methods optimise an energy that comprises two terms: one that measures the similarity between a deformed and a fixed image and one that enforces the smoothness of the deformation. Two main families emerge, whether they penalise the displacement fields (inspired by solid physics) or their infinitesimal rate of change (inspired by fluid physics), allowing for large diffeomorphic deformations [10,11]. Concerning the optimisation scheme, a common strategy is to work with energies that allow for a probabilistic interpretation [12–16]. The optimisation can in this case be cast as an inference problem, which is the approach taken in this paper. More recently, it has been proposed to use deep neural networks to learn the normalisation function [17–20]. At training time, however, these approaches still use a two-term loss function that enforces data consistency while penalising non-smoothness of the deformations. These models have demonstrated remarkable speed-ups in runtime for volumetric image registration, with similar accuracies to the more classical methods.

Note that all of the above methods either require some sort of prior image processing or are restricted to a specific MR contrast. The method presented in this paper is instead agnostic to the imaging modality and can be applied directly to the *raw* data. This is because it models many features of the imaging process (bias field, gridding, etc.), in order not to require any processing such as skull-stripping, intensity normalisation, affine alignment or reslicing to a common grid. These properties are important for a general tool that should work 'out-of-the-box', given that imaging protocols are far from standardised – restricting a method to a particular intensity profile considerably restricts its practical use. In addition, our method allows for a user to chose the resolution of the common space. We validate our approach on a pairwise registration task, comparing it against state-of-the-art methods, on publicly available data. We achieve favourable results outperforming all other methods, within reasonable runtimes.

2 Methods

Generative Model. In this work, computing the nonlinearly aligned images is actually a by-product of doing inference on a joint probability distribution. This generative model consists of multiple random variables, modelling various properties of the observed data. It is defined by the following distribution:

$$p(\mathcal{F}, \mathcal{A}, \mathcal{S}) = p(\mathcal{F} \mid \mathcal{A}, \mathcal{S})\ p(\mathcal{A}, \mathcal{S}), \tag{1}$$

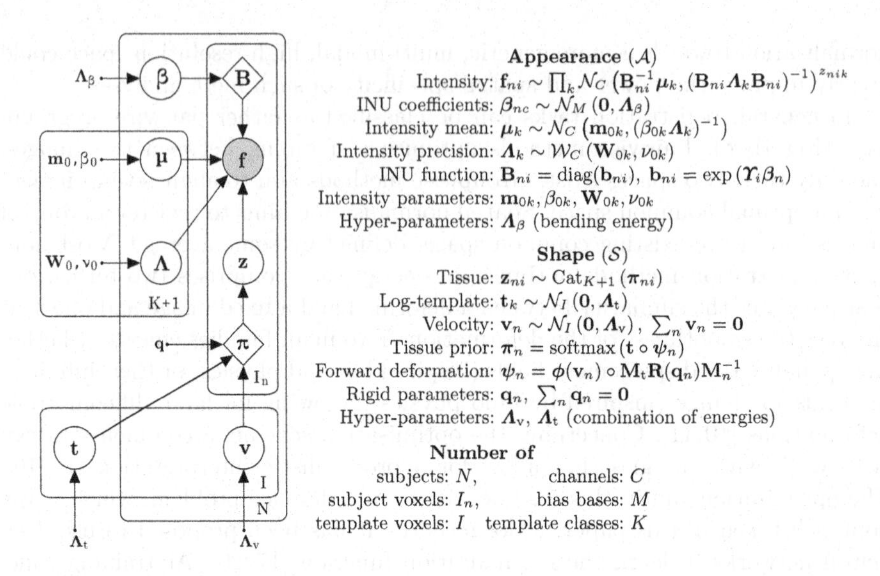

Appearance ($\mathcal{A}$)
Intensity: $\mathbf{f}_{ni} \sim \prod_k \mathcal{N}_C\left(\mathbf{B}_{ni}^{-1}\boldsymbol{\mu}_k, (\mathbf{B}_{ni}\boldsymbol{\Lambda}_k\mathbf{B}_{ni})^{-1}\right)^{z_{nik}}$
INU coefficients: $\boldsymbol{\beta}_{nc} \sim \mathcal{N}_M(\mathbf{0}, \boldsymbol{\Lambda}_\beta)$
Intensity mean: $\boldsymbol{\mu}_k \sim \mathcal{N}_C\left(\mathbf{m}_{0k}, (\beta_{0k}\boldsymbol{\Lambda}_k)^{-1}\right)$
Intensity precision: $\boldsymbol{\Lambda}_k \sim \mathcal{W}_C(\mathbf{W}_{0k}, \nu_{0k})$
INU function: $\mathbf{B}_{ni} = \mathrm{diag}(\mathbf{b}_{ni})$, $\mathbf{b}_{ni} = \exp(\boldsymbol{\Upsilon}_i\boldsymbol{\beta}_n)$
Intensity parameters: $\mathbf{m}_{0k}, \beta_{0k}, \mathbf{W}_{0k}, \nu_{0k}$
Hyper-parameters: $\boldsymbol{\Lambda}_\beta$ (bending energy)
Shape ($\mathcal{S}$)
Tissue: $\mathbf{z}_{ni} \sim \mathrm{Cat}_{K+1}(\boldsymbol{\pi}_{ni})$
Log-template: $\mathbf{t}_k \sim \mathcal{N}_I(\mathbf{0}, \boldsymbol{\Lambda}_\mathrm{t})$
Velocity: $\mathbf{v}_n \sim \mathcal{N}_I(\mathbf{0}, \boldsymbol{\Lambda}_\mathrm{v})$, $\sum_n \mathbf{v}_n = \mathbf{0}$
Tissue prior: $\boldsymbol{\pi}_n = \mathrm{softmax}(\mathbf{t} \circ \boldsymbol{\psi}_n)$
Forward deformation: $\boldsymbol{\psi}_n = \boldsymbol{\phi}(\mathbf{v}_n) \circ \mathbf{M}_\mathrm{t}\mathbf{R}(\mathbf{q}_n)\mathbf{M}_n^{-1}$
Rigid parameters: $\mathbf{q}_n$, $\sum_n \mathbf{q}_n = \mathbf{0}$
Hyper-parameters: $\boldsymbol{\Lambda}_\mathrm{v}$, $\boldsymbol{\Lambda}_\mathrm{t}$ (combination of energies)
Number of
subjects: N, channels: C
subject voxels: I_n, bias bases: M
template voxels: I template classes: K

Fig. 2. The joint probability distribution over N images. Random variables are in circles, observed are shaded, plates indicate replication, hyper-parameters have dots, diamonds indicate deterministic functions. The distributions in this figure are the Normal ($\mathcal{N}$), Wishart ($\mathcal{W}$) and Categorical (Cat). Note that $K+1$ mutually exclusive classes are modelled, but as the final class can be determined by the initial K, we do not represent it (improving runtime, memory usage and stability). The hyper-parameters ($\boldsymbol{\Lambda}_\beta$, $\boldsymbol{\Lambda}_\mathrm{v}$, $\boldsymbol{\Lambda}_\mathrm{t}$) encode a combination of absolute, membrane and bending energies. $\boldsymbol{\Lambda}_\mathrm{v}$ further penalises linear-elasticity. The sum of the shape parameters ($\mathbf{v}_n$, $\mathbf{q}_n$) are constrained to zero, to ensure that the template remains in the average position [21].

where $\mathcal{F} = \{\mathbf{F}_n\}_{n=1}^N$, $\mathbf{F}_n \in \mathbb{R}^{I_n \times C}$ are the N observed images (*e.g.*, MRI scans), each with I_n voxels and C channels (*e.g.*, MR contrasts). The two sets $\mathcal{A}$ and $\mathcal{S}$ contain the appearance and shape variables, respectively. The distribution in (1) is unwrapped in detail in Fig. 2, showing its graphical model and constituent parts. The inversion of the model in (1) is performed using a variational expectation-maximisation (VEM) algorithm. In this algorithm, each parameter (or its probability distribution, in the case of the mixture parameters) is updated whilst holding all others fixed, in an alternating manner [22]. The individual update equations are obtained from the evidence lower bound (ELBO):

$$\mathcal{L} = \sum_{\mathcal{A},\mathcal{S}} q(\mathcal{A},\mathcal{S}) \ln\left[\frac{p(\mathcal{F},\mathcal{A},\mathcal{S})}{q(\mathcal{A},\mathcal{S})}\right], \tag{2}$$

where the variational distribution is assumed to factorise as $q(\mathcal{A},\mathcal{S}) = q(\mathcal{A})q(\mathcal{S})$. The appearance updates have been published in previous work: the inference of the intensity parameters in [23]; the mode estimates of the intensity non-uniformity (INU) parameters in [12].

The contribution of this paper is to unify the shape and appearance parts as (1), providing a flexible and unsupervised image registration framework.

In particular, this framework relies on: parameterising the shape model using a combined rigid and diffeomorphic registration in the space of the template, introduction of a multi-scale optimisation method, and a novel way of computing a Hessian of the categorical data term. These will next be explained in more detail.

Spatial Transformation Model. For maximum generalisability, the model should handle image data defined on arbitrary lattices with arbitrary orientations (*i.e.*, any well formatted NIfTI file). The forward deformation ψ_n, warping the template to subject space, is the composition of a diffeomorphic transform ϕ_n, defined over the template field of view, and a rigid transform $\mathbf{R}_n$, defined in world space. The template ($\mathbf{M}_\mathrm{t}$) and subject ($\mathbf{M}_n$) orientation matrices describe the mapping from voxel to world space. Therefore, $\psi_n = \phi_n \circ \mathbf{M}_\mathrm{t} \circ \mathbf{R}_n \circ \mathbf{M}_n^{-1}$. The diffeomorphism is encoded by the initial velocity of the template 'particles' [24], and recovered by geodesic shooting [25]: $\phi_n = \mathrm{shoot}\,(\mathbf{v}_n)$. $\mathbf{R}_n$ is encoded by its projection $\mathbf{q}_n$ on the tangent space of rigid transformation matrices, and recovered by matrix exponentiation [26]. $\mathbf{R}_n$ could have included scales and shears, but keeping it rigid allows us to capture these deformations in the velocities.

Multi-scale Optimisation. Registration is a non-convex problem and is therefore highly sensitive to local minima. Multi-scale optimisation techniques can be used to circumvent this problem [2,10,20]. The proposed approach implements such a multi-scale method to help with several difficulties: local minima (especially in the rigid parameter space), slow VEM convergence, and slow runtime. The way we parameterise the spatial transformation model is what enables our multi-scale approach. If we drop all terms that do not depend on the template, velocities or rigid parameters, the ELBO in (2) reduces to:

$$\mathcal{L} \stackrel{c}{=} \sum_n \Big\{ \ln \mathrm{Cat}\,(\tilde{\mathbf{z}}_n \mid \mathrm{softmax}\,(\mathbf{t} \circ \psi_n)) + \ln p(\boldsymbol{v}_n) \Big\} + \ln p(\mathbf{t})\ , \tag{3}$$

where $\tilde{\mathbf{z}}_n$ denotes the latent class posterior probabilities (responsibilities). The two prior terms originate from the realm of PDEs, where they take the form of integrals of continuous functions. When discretised, these integrals can be interpreted as negative logs of multivariate Normal distributions (up to a constant):

$$\frac{\lambda}{2} \int_\Omega \langle f(\mathbf{x}), (\Lambda f)(\mathbf{x}) \rangle d\mathbf{x} \xrightarrow{\text{discretise}} \frac{\lambda}{2} \left(\mathbf{f}^\mathrm{T} \boldsymbol{\Lambda} \mathbf{f} \right) \,\Delta_x. \tag{4}$$

Here, $\mathbf{f}^\mathrm{T}\boldsymbol{\Lambda}\mathbf{f}$ computes the sum-of-squares of the (discrete) image gradients and Δ_x is the volume of one discrete element. Usually, Δ_x would simply be merged into the regularisation factor λ. In a multi-scale setting, it must be correctly set at each scale. In practice, the template and velocities are first defined over a very coarse grid, and the VEM scheme is applied with a suitable scaling. At convergence, they are trilinearly interpolated to a finer grid, and the scaling parameter is changed accordingly for a new iteration of VEM.

Böhning Bound. We use a Newton-Raphson algorithm to find mode estimates of the variables $\mathbf{t}$, $\mathbf{v}_n$ and $\mathbf{q}_n$, with high convergence rates. This requires the gradient and Hessian of the categorical data term. If the gradient and Hessian with respect to $\mathbf{t}_n = \mathbf{t} \circ \boldsymbol{\psi}_n$ are known, then those with respect to the variables of interest $\mathbf{t}$, $\mathbf{v}_n$ and $\mathbf{q}_n$ can be obtained by application of the chain rule (with Fisher's scoring [24]). However, the true Hessian is not well-behaved and the Newton-Raphson iterates may overshoot. Therefore, some precautions must be taken such as ensuring monotonicity using a backtracking line search [27]. Here, we make use of Böhning's approximation [28] to bound the ELBO and improve the stability of the update steps, without the need for line search. This approximation was introduced in the context of multinomial logistic regression, which relies on a similar objective function. Because this approximation allows the true objective function to be bounded, it ensures the sequence of Newton-Raphson steps to be monotically improving. However, this bound is not quite tight, leading to slower convergence rates. In this work, we therefore use a weighted average of Böhning's approximation and the true Hessian that leads to both fast and stable convergence; *e.g.*, the template Hessian becomes:

$$\frac{\partial^2 \mathcal{L}}{\partial t_{nik} \partial t_{nil}} \approx w \underbrace{\pi_{nik} \left(\delta_k^l - \pi_{nil}\right)}_{\text{True Hessian}} + (1-w) \underbrace{\frac{1}{2}\left(\delta_k^l - \frac{1}{K}\right)}_{\text{Böhning bound}}, \quad w \in [0,1]. \tag{5}$$

3 Validation

Experiments. Brain scans where regions-of-interests have been manually labelled by human experts can be used to assess the accuracy of a registration method. By warping the label images from one subject onto another, overlap scores can be computed, without the need to resample the groundtruth annotations. The labels parcelate the brain into small regions, identifying the same anatomical structures between subjects. As the labels are independent from the signal used to compute the deformations, they are well suited to be used for validation. Such a validation was done in a seminal paper [29], where 14 methods were compared at nonlinearly registering pairs of MR brain scans. Two datasets used in [29] were[1]:

- **LPBA40:** T1-weighted (T1w) MRIs of 40 subjects with cortical and subcortical labels, of which 56 were used in the validation in [29]. The two top-performing methods, from $N = 1,560$ pairwise registrations, were ART's 3dwarper [30] and ANTs' SyN [11]. The MRIs have been processed by skull-stripping, non-uniformity correction, and rigid reslicing to a common space.
- **IBSR18:** T1w MRIs of 18 subjects with cortical labels, where 96 of the labelled regions were used in the validation in [29]. The two top-performing methods, from $N = 306$ pairwise registrations, were SPM's Dartel [13] and ANTs' SyN [11]. The MRIs have non-isotropic voxels and are unprocessed; IBSR18 are therefore more challenging to register than LPBA40.

[1] nitrc.org/projects/ibsr, resource.loni.usc.edu/resources.

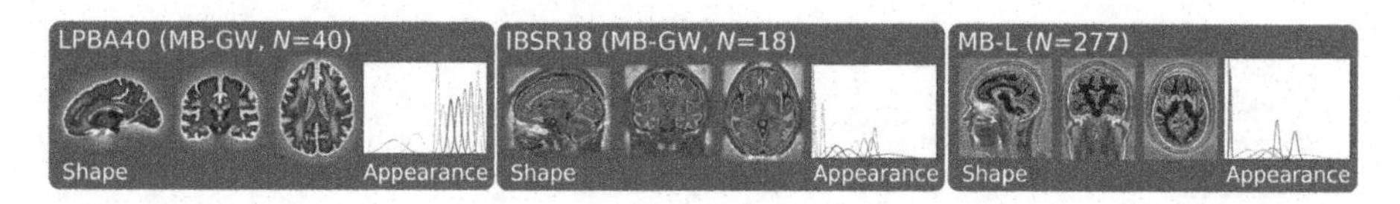

Fig. 3. Learned shape and appearance priors from fitting MB-GW to LPBA40 (left) and IBSR18 (middle); and MB-L to a training dataset (right). Colours correspond to clusters found, unsupervised, by fitting the model. Appearance densities show the expectations of the Gaussians drawn from the Gauss-Wishart priors (using 3 σ). (Color figure online)

We now compare our method, denoted MultiBrain (MB), with the top-performing methods in [29], on IBSR18 and LPBA40. The same overlap metric is used: the volume over which the deformed source labels match the target labels, divided by the total volume of the target labels (*i.e.*, the true positive rate (TPR)). Two additional registration methods are included: one state-of-the-art group-wise model, SPM's Shoot [31]; and one state-of-the-art deep learning model, the CVPR version of VoxelMorph[2] (VXM) [17]. Pairwise registrations between all subjects (in both directions) are computed using MB, SPM's Shoot and VXM. For MB, we (i) learned the optimal average from each dataset (MB-GW), and (ii) learned the optimal average from a held-out training set (MB-L); as described in Fig. 1b. These two tasks are similar to either finding the optimal template for a specific neuroimaging group study (MB-GW) or using a predefined common space for the same task (MB-L). Shoot's registration process resembles MB-GW, whilst VXM resembles MB-L. MB-L was trained on $N = 277$ held-out T1w MRIs from five different datasets: three publicly available[3]: IXI ($N_1 = 200$), MICCAI2012 ($N_2 = 35$) and MRBrainS18 ($N_3 = 7$); and two hospital curated ($N_4 = 19$, $N_5 = 16$). Training took two days on a modern workstation.

The shape and appearance models that were learned when fitting MB are shown in Fig. 3. $K = 11$ classes were used, 1 mm isotropic template voxels and the priors were initialised as uninformative. The initial template and velocity dimensions were set 8 mm cube. Energy hyper-parameters were chosen as $\lambda_\beta = 1e5$, $\lambda_v = \{2e{-}4, 0, 0.4, 0.1, 0.4\}$ (absolute, membrane, bending, linear elasticity) and $\lambda_t = \{1e{-}2, 0.5, 0\}$ (absolute, membrane, bending). The weighting was set to $w = 0.8$. The algorithm was run for a predefined number of iterations.

Results. The label overlap scores on IBSR18 are shown in Fig. 4. The figure shows, close to, unanimous better overlap for MB, compared to the other algorithms. Result plots for LPBA40 are given in the supplementary materials, as well as samples of the best and worst registrations for MB and VXM. On both IBSR18 and LPBA40, MB performs favourably. For IBSR18, the mean and median overlaps were 0.62 and 0.63 respectively for MB-GW, and both 0.59 for MB-L. Mean and median overlaps were both 0.59 for SPM's Shoot and both 0.56 for VXM. The greatest median overlap reported in [29] was about 0.55,

[2] github.com/voxelmorph/voxelmorph.
[3] brain-development.org, mrbrains18.isi.uu.nl, my.vanderbilt.edu/masi.

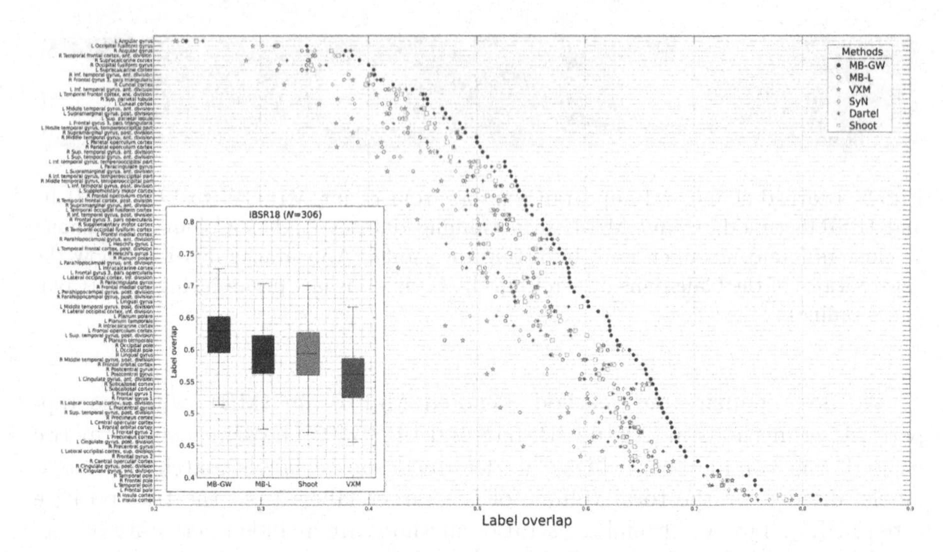

Fig. 4. Results from the validation on the IBSR18 dataset. The nonlinear registration methods include MB-GW/L, SPM's Shoot, VXM and the two top algorithms evaluated in [29]. Shown are the average label overlaps and total overlaps (the boxplot). The results in the boxplot may be compared directly with the methods of Fig. 5 in [29].

whereas the overlap from affine registration was 0.40 [32]. For LPBA40, the mean and median overlaps were both 0.76 for MB-GW and both 0.75 for MB-L. Mean and median overlaps for SPM's Shoot approach were both 0.75, and both 0.74 for VXM. The highest median overlap reported in [29] was 0.73, and that from affine registration was 0.60 [32]. Using the affine registrations as baseline, the results showed 6% to 17% greater accuracy improvements when compared to those achieved for the second most accurate nonlinear registration algorithm evaluated[4]. Computing one forward deformation took about 15 min for MB-L and 30 for MB-GW (on a modern workstation, running on the CPU).

Discussion. MB-GW does better than MB-L, this was expected as the average obtained by groupwise fitting directly on the population of interest should be more optimal than one learned from a held-out dataset, on a limited number of subjects (*e.g.*, the averages for the individual populations in Fig. 1 are more optimal than the global). Still, MB-L learned on only 277 subject does as well as, or better than, Shoot (a state-of-the-art groupwise approach). This is an exciting result that allows for groupwise accuracy spatial normalisation on small number of subjects, and to a standard common space (instead of a population-specific). With a larger and more diverse training population, accuracies are expected to improve further. One may claim that a group-wise registration scheme has unfair advantage over pairwise methods. However, as a common aim often is to spatially normalise - with the objective of making comparisons among a population of

[4] $(\text{TPR}_{\text{MB}} - \text{TPR}_{\text{Shoot}})/(\text{TPR}_{\text{MB}} - \text{TPR}_{\text{Affine}}) \times 100\%$.

scans, it would be reasonable to aim for as much accuracy as possible for this task. The purely data-driven VXM approach does better than the methods evaluated in [29]. VXM was trained on close to 4,000 diverse T1w MRIs. A larger training dataset could boost its performance. The processing that was applied to the VXM input data was done using SPM [33], whilst its training data was processed using FreeSurfer. Having used the same software could have improved its results; however, being reliant on a specific processing pipeline is inherently a weakness of any method. Furthermore, the VXM model uses a cross-correlation loss function that should be resilient to intensity variations in the T1w scans. Finally, the contrasts and fields of view in the T1w scans were slightly different from each other in the training and testing data, due to variability in field strength and scanner settings. This could have impacted the accuracy of MB-L and VXM.

4 Conclusion

This paper introduced an unsupervised learning algorithm for nonlinear image registration, which can be applied to unprocessed medical imaging data. A validation on two publicly available datasets showed state-of-the art results on registering MRI brain scans. The unsupervised, non-organ specific nature of the algorithm makes it applicable to not only brain data, but also other types of medical images. This could allow for transferring methods widely used in neuroimaging to other types of organs, *e.g.*, the liver [34]. The runtime of the algorithm is not on par with a GPU implementation of a deep learning model, but still allows for processing of a 3D brain scan in an acceptable time. The runtime should furthermore improve, drastically, by an implementation on the GPU. The proposed model could also be used for image segmentation [12] and translation [35], or modified to use labelled data, in a semi-supervised manner [23]. Finally, the multi-modal ability of the model would be an interesting avenue of further research.

Acknowledgements. MB was funded by the EPSRC-funded UCL Centre for Doctoral Training in Medical Imaging (EP/L016478/1) and the Department of Health's NIHR-funded Biomedical Research Centre at University College London Hospitals. YB was funded by the MRC and Spinal Research Charity through the ERA-NET Neuron joint call (MR/R000050/1). MB and JA were funded by the EU Human Brain Project's Grant Agreement No. 785907 (SGA2). GF and the Wellcome Centre for Human Neuroimaging is supported by core funding from the Wellcome (203147/Z/16/Z).

References

1. Friston, K.J., Ashburner, J., Frith, C.D., Poline, J.-B., Heather, J.D., Frackowiak, R.S.: Spatial registration and normalization of images. Hum. Brain Mapp. **3**(3), 165–189 (1995)
2. Zöllei, L., Learned-Miller, E., Grimson, E., Wells, W.: Efficient population registration of 3D data. In: Liu, Y., Jiang, T., Zhang, C. (eds.) CVBIA 2005. LNCS, vol. 3765, pp. 291–301. Springer, Heidelberg (2005). https://doi.org/10.1007/11569541_30

3. Heckemann, R.A., et al.: Improving intersubject image registration using tissue-class information benefits robustness and accuracy of multi-atlas based anatomical segmentation. Neuroimage **51**(1), 221–227 (2010)
4. Draganski, B., Gaser, C., Busch, V., Schuierer, G., Bogdahn, U., May, A.: Changes in grey matter induced by training. Nature **427**(6972), 311–312 (2004)
5. Fox, P.T.: Spatial normalization origins: objectives, applications, and alternatives. Hum. Brain Mapp. **3**(3), 161–164 (1995)
6. Csernansky, J.G., et al.: Hippocampal morphometry in schizophrenia by high dimensional brain mapping. PNAS **95**(19), 11406–11411 (1998)
7. Mourao-Miranda, J., et al.: Individualized prediction of illness course at the first psychotic episode: a support vector machine MRI study. Psychol. Med. **42**(5), 1037–1047 (2012)
8. Seghier, M.L., Ramlackhansingh, A., Crinion, J., Leff, A.P., Price, C.J.: Lesion identification using unified segmentation-normalisation models and fuzzy clustering. NeuroImage **41**(4), 1253–1266 (2008)
9. Yarkoni, T., Poldrack, R.A., Nichols, T.E., Van Essen, D.C., Wager, T.D.: Large-scale automated synthesis of human functional neuroimaging data. Nat. Methods **8**(8), 665 (2011)
10. Christensen, G.E., Joshi, S.C., Miller, M.I.: Volumetric transformation of brain anatomy. IEEE Trans. Med. Imaging **16**(6), 864–877 (1997)
11. Avants, B.B., Epstein, C.L., Grossman, M., Gee, J.C.: Symmetric diffeomorphic image registration with cross-correlation: evaluating automated labeling of elderly and neurodegenerative brain. Med. Image Anal. **12**(1), 26–41 (2008)
12. Ashburner, J., Friston, K.J.: Unified segmentation. NeuroImage **26**(3), 839–851 (2005)
13. Ashburner, J.: A fast diffeomorphic image registration algorithm. NeuroImage **38**(1), 95–113 (2007)
14. Andersson, J.L., Jenkinson, M., Smith, S., et al.: "Non-linear registration aka spatial normalisation FMRIB Technical report TR07JA2," FMRIB Analysis Group of the University of Oxford (2007)
15. Bhatia, K.K., et al.: Groupwise combined segmentation and registration for atlas construction. In: Ayache, N., Ourselin, S., Maeder, A. (eds.) MICCAI 2007. LNCS, vol. 4791, pp. 532–540. Springer, Heidelberg (2007). https://doi.org/10.1007/978-3-540-75757-3_65
16. Vercauteren, T., Pennec, X., Perchant, A., Ayache, N.: Diffeomorphic demons: efficient non-parametric image registration. NeuroImage **45**(1), S61–S72 (2009)
17. Balakrishnan, G., Zhao, A., Sabuncu, M.R., Guttag, J., Dalca, A.V.: VoxelMorph: a learning framework for deformable medical image registration. IEEE Trans. Med. Imaging **38**(8), 1788–1800 (2019)
18. Dalca, A., Rakic, M., Guttag, J., Sabuncu, M.: Learning conditional deformable templates with convolutional networks. In: NeurIPS, pp. 804–816 (2019)
19. Fan, J., Cao, X., Yap, P.-T., Shen, D.: BIRNet: brain image registration using dual-supervised fully convolutional networks. Med. Image Anal. **54**, 193–206 (2019)
20. Krebs, J., Delingette, H., Mailhé, B., Ayache, N., Mansi, T.: Learning a probabilistic model for diffeomorphic registration. IEEE Trans. Med. Imaging **38**(9), 2165–2176 (2019)
21. Beg, M.F., Khan, A.: Computing an average anatomical atlas using LDDMM and geodesic shooting. In: ISBI, pp. 1116–1119, IEEE (2006)
22. Bishop, C.M.: Pattern Recognition and Machine Learning. Springer, New York (2006)

23. Blaiotta, C., Freund, P., Cardoso, M.J., Ashburner, J.: Generative diffeomorphic modelling of large MRI data sets for probabilistic template construction. NeuroImage **166**, 117–134 (2018)
24. Ashburner, J., Brudfors, M., Bronik, K., Balbastre, Y.: An algorithm for learning shape and appearance models without annotations. Med. Image Anal. **55**, 197 (2019)
25. Miller, M.I., Trouvé, A., Younes, L.: Geodesic shooting for computational anatomy. J. Math. Imaging Vis. **24**(2), 209–228 (2006)
26. Woods, R.P.: Characterizing volume and surface deformations in an atlas framework: theory, applications, and implementation. NeuroImage **18**(3), 769–788 (2003)
27. Ashburner, J., Friston, K.J.: Computing average shaped tissue probability templates. NeuroImage **45**(2), 333–341 (2009)
28. Böhning, D.: Multinomial logistic regression algorithm. Ann. Inst. Stat. Math. **44**(1), 197–200 (1992)
29. Klein, A., et al.: Evaluation of 14 nonlinear deformation algorithms applied to human brain MRI registration. NeuroImage **46**(3), 786–802 (2009)
30. Ardekani, B.A., Guckemus, S., Bachman, A., Hoptman, M.J., Wojtaszek, M., Nierenberg, J.: Quantitative comparison of algorithms for inter-subject registration of 3D volumetric brain MRI scans. J. Neurosci. Methods **142**(1), 67–76 (2005)
31. Ashburner, J., Friston, K.J.: Diffeomorphic registration using geodesic shooting and Gauss-Newton optimisation. NeuroImage **55**(3), 954–967 (2011)
32. Jenkinson, M., Bannister, P., Brady, M., Smith, S.: Improved optimization for the robust and accurate linear registration and motion correction of brain images. NeuroImage **17**(2), 825–841 (2002)
33. Malone, I.B., et al.: Accurate automatic estimation of total intracranial volume: a nuisance variable with less nuisance. NeuroImage **104**, 366–372 (2015)
34. Ridgway, G., et al.: Voxel-Wise analysis of paediatric liver MRI. In: Nixon, M., Mahmoodi, S., Zwiggelaar, R. (eds.) MIUA 2018. CCIS, vol. 894, pp. 57–62. Springer, Cham (2018). https://doi.org/10.1007/978-3-319-95921-4_7
35. Brudfors, M., Ashburner, J., Nachev, P., Balbastre, Y.: Empirical bayesian mixture models for medical image translation. In: Burgos, N., Gooya, A., Svoboda, D. (eds.) SASHIMI 2019. LNCS, vol. 11827, pp. 1–12. Springer, Cham (2019). https://doi.org/10.1007/978-3-030-32778-1_1

Are Registration Uncertainty and Error Monotonically Associated?

Jie Luo[1,2](✉), Sarah Frisken[1], Duo Wang[3], Alexandra Golby[1], Masashi Sugiyama[2,4], and William Wells III[1]

[1] Brigham and Women's Hospital, Harvard Medical School, Boston, USA
jluo5@bwh.harvard.edu
[2] Graduate School of Frontier Sciences, The University of Tokyo, Tokyo, Japan
[3] Department of Automation, Tsinghua University, Beijing, China
[4] Center for Advanced Intelligence Project, RIKEN, Tokyo, Japan

Abstract. In image-guided neurosurgery, current commercial systems usually provide only rigid registration, partly because it is harder to predict, validate and understand non-rigid registration error. For instance, when surgeons see a discrepancy in aligned image features, they may not be able to distinguish between registration error and actual tissue deformation caused by tumor resection. In this case, the spatial distribution of registration error could help them make more informed decisions, e.g., ignoring the registration where the estimated error is high. However, error estimates are difficult to acquire. Probabilistic image registration (PIR) methods provide measures of registration uncertainty, which could be a surrogate for assessing the registration error. It is intuitive and believed by many clinicians that high uncertainty indicates a large error. However, the monotonic association between uncertainty and error has not been examined in image registration literature. In this pilot study, we attempt to address this fundamental problem by looking at one PIR method, the Gaussian process (GP) registration. We systematically investigate the relation between GP uncertainty and error based on clinical data and show empirically that there is a weak-to-moderate positive monotonic correlation between point-wise GP registration uncertainty and non-rigid registration error.

Keywords: Registration uncertainty · Registration error

1 Introduction

In image-guided neurosurgery (IGN), surgical procedures are often planned based on the preoperative (p-) magnetic resonance imaging (MRI). During surgery, clinicians may acquire intraoperative (i-) MRI and/or Ultrasound (US). Image registration can be used [1,2] to map the p-MRI to the intraoperative coordinate space to help surgeons locate structures or boundaries of interest during surgery (e.g., tumor margins or nearby blood vessels to be avoided) and facilitate more complete tumor resection [3–5].

A. L. Martel et al. (Eds.): MICCAI 2020, LNCS 12263, pp. 264–274, 2020.
https://doi.org/10.1007/978-3-030-59716-0_26

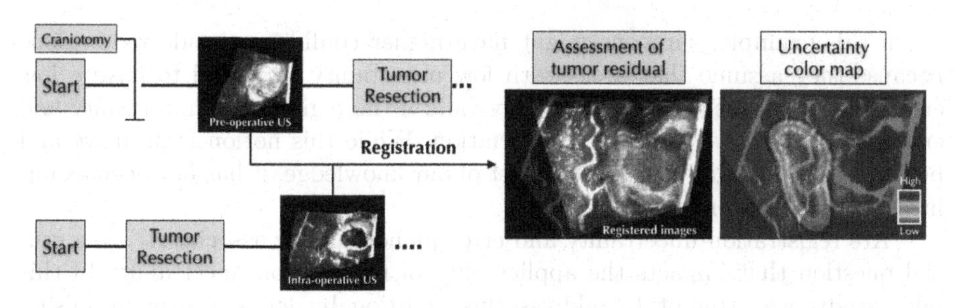

Fig. 1. An example to illustrate the usefulness of registration uncertainty in IGNs.

Even though the brain clearly undergoes non-linear deformation during surgery, rigid registration is still the standard for clinical practice [6]. Although non-rigid registration has long been a goal for IGNs, this goal is hampered because non-rigid registration error is less predictable and harder to validate than rigid registration error. In practice, if surgeons see a discrepancy between two aligned image features, they may not be able to tell if the misalignment is caused by a registration error or an actual tissue deformation caused by tumor resection. In this case, providing surgeons with a spatial distribution of the expected registration error could help them make more informed decisions, e.g., ignoring the registration where the expected error is high. However, determining this spatial distribution is difficult since:

1. Most methods that estimate registration error, such as bootstrapping [7,8], perturbed input [9–11], stereo confidence [12] and supervised learning [13–15], require multiple runs of a non-rigid registration algorithm, thus they are too time-consuming to be practical for IGNs where feedback is required within a few minutes of intraoperative image acquisition.
2. More importantly, existing methods estimate the error by detecting misaligned image features [8,12,14,15]. These methods fail in IGNs because tumor resection and retraction significantly alter the brain, particularly at the tumor margin where precision is most needed. Thus finding consistent image features near the tumor margin may be difficult.

An alternative for directly estimating the registration error is to use registration uncertainty as a surrogate. Registration uncertainty is a measure of confidence in the predicted registration and typically estimated by probabilistic image registration (PIR) [16–30]. In IGN, utilizing registration uncertainty can be helpful. As shown in Fig. 1, surgeons can inspect the residual tumor after registering the p-US image to the i-US image and decide whether to continue the resection or end the operation. An uncertainty color map overlaid on top of the registered images, where red indicated regions of low uncertainty, can be used by surgeons to dismiss clear misregistration regions where uncertainty is high and have more confidence in the registration where uncertainty is low (e.g., red regions).

In this example, surgeons might have higher confidence inside red regions because they assume that areas with low uncertainty also tend to have a low error. However, this assumption is only valid if the registration uncertainty and error have a positive monotonic association. While this notion is intuitive and believed by many clinicians, to the best of our knowledge, it has not been examined in the PIR literature.

"Are registration uncertainty and error monotonically associated?" is a crucial question that impacts the applicability of registration uncertainty. In this pilot study, we attempt to address this question by looking at a promising PIR method, Gaussian process (GP) registration [20,27,29]. We systematically investigate the GP uncertainty and error using point-wise posterior predictive checking and a patch-wise correlation test. We note that the registration uncertainty can be categorized as transformation uncertainty or label uncertainty [31]. Since the applicability of label uncertainty is still in question, this paper will focus solely on the transformation uncertainty when it refers to 'registration uncertainty'.

2 Methods

In this section, we briefly review GP registration uncertainty. Then we introduce Spearman's correlation coefficient and provide details about our point-wise and patch-wise experiments.

2.1 Review of the GP Registration Uncertainty

The stochastic GP registration approach has shown promising results in IGNs [20,27,29]. As shown in Fig. 2, a key step in the GP registration is to estimate N_* unknown displacement vectors $\mathbf{D}_*$ from N known ones $\mathbf{D}$ that were derived from automatic feature extraction and matching.

Let $\mathbf{x}$ be the grid coordinate and $\mathbf{d}(\mathbf{x}) = [d_\mathrm{x}, d_\mathrm{y}, d_\mathrm{z}]$ be the associated displacement vector. For $d(\mathbf{x})$ being one of d_x, d_y, and d_z, it is modeled as a joint Gaussian distribution $d(\mathbf{x}) \sim \mathrm{GP}(m(\mathbf{x}), \mathrm{k}(\mathbf{x}, \mathbf{x}'))$ with mean function $m(\mathbf{x}) = 0$ and covariance function $\mathrm{k}(\mathbf{x}, \mathbf{x}')$. Thus $\mathbf{D}$ and $\mathbf{D}_*$ have the following relationship:

$$\begin{bmatrix} \mathbf{D} \\ \mathbf{D}_* \end{bmatrix} \sim N\left(\mathbf{m}(\mathbf{x}), \begin{bmatrix} \mathbf{K} & \mathbf{K}_* \\ \mathbf{K}_*^T & \mathbf{K}_{**}. \end{bmatrix}\right). \quad (1)$$

In Eq. (1), $\mathbf{K} = \mathrm{k}(\mathbf{X}, \mathbf{X}) \in \mathbb{R}^{N \times N}$ and $\mathbf{K}_{**} = \mathrm{k}(\mathbf{X}_*, \mathbf{X}_*) \in \mathbb{R}^{N_* \times N_*}$ are intra-covariance matrices of $\mathbf{d}$ and $\mathbf{d}_*$ respectively. $\mathbf{K}_* = \mathrm{k}(\mathbf{X}, \mathbf{X}_*) \in \mathbb{R}^{N \times N_*}$ is the inter-covariance matrix. The interpolated displacement vector values can be estimated from the mean μ_* of the posterior distribution of $p(\mathbf{D}_* \mid \mathbf{X}_*, \mathbf{X}, \mathbf{D})$:

$$\mu_* = \mathbf{K}_*^T \mathbf{K}^{-1} \mathbf{D}. \quad (2)$$

From Eq. (1), the posterior covariance matrix can also be derived as

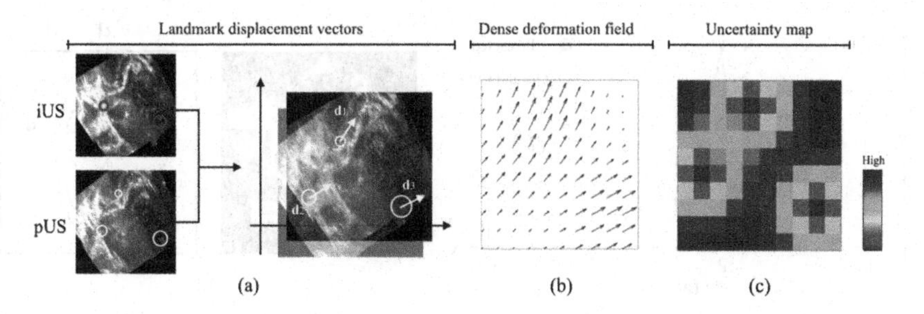

Fig. 2. (a) 3 displacement vectors; (b) A 10×10 interpolated dense deformation field; (c) A visualization of registration uncertainty. (Color figure online)

$$\mathbf{\Sigma}_* = \mathbf{K}_{**} - \mathbf{K}_*^T \mathbf{K}^{-1} \mathbf{K}_*. \tag{3}$$

Diagonal entries of $\mathbf{\Sigma}_*$ are the marginal transformation variances, and they can be used as the GP registration uncertainty. In this study, we choose the same kernel $\mathrm{k}(\mathbf{x}, \mathbf{x}') = \exp(-\frac{x^2}{a})$ for all three displacement components.

Figure 2(b) shows a 10×10 dense deformation field interpolated from three landmark displacement vectors. Each voxel is associated with an estimated displacement vector and uncertainty value. Figure 2(c) is an uncertainty color map for the displacement field.

2.2 Spearman's Rank Correlation Coefficient

Spearman's correlation coefficient, often denoted by ρ_s, is a non-parametric measure of statistical dependence between the rankings of two variables. It assesses how well their relationship can be described using a monotonic function [32].

In this study we prefer ρ_s over Pearson's correlation ρ_p for the following reasons:

1. ρ_p measures the strength of a linear relationship. To be clinically useful, registration uncertainty does not have to be linearly correlated with the error. In this sense, we prefer ρ_s which measures a less "restrictive" monotonic relationship;
2. Since ρ_s limits the influence of outliers to the value of its rank, it is less sensitive than ρ_p to strong outliers that lie in the tails of the distribution [32].

$E = \{\epsilon(1), \epsilon(2), ..., \epsilon(M)\}$ Assume there are M test points, $\mathrm{u}(i)$ and $\epsilon(i)$, which represent the uncertainty and error for point i respectively. Let U and E denote discrete random variables with values $\{\mathrm{u}(1), \mathrm{u}(2), ..., \mathrm{u}(M)\}$ and $\{\epsilon(1), \epsilon(2), ..., \epsilon(M)\}$. To measure ρ_s, we have to convert U and E to descending rank vectors rU and rE, i.e., the rank vector for $[0.2, 1.2, 0.9, 0.5, 0.1]$ would be $[2, 5, 4, 3, 1]$. Then ρ_s can be estimated as

$$\rho_\mathrm{s} = \frac{\mathrm{cov}(rU, rE)}{\sigma_{rU}\sigma_{rE}}, \tag{4}$$

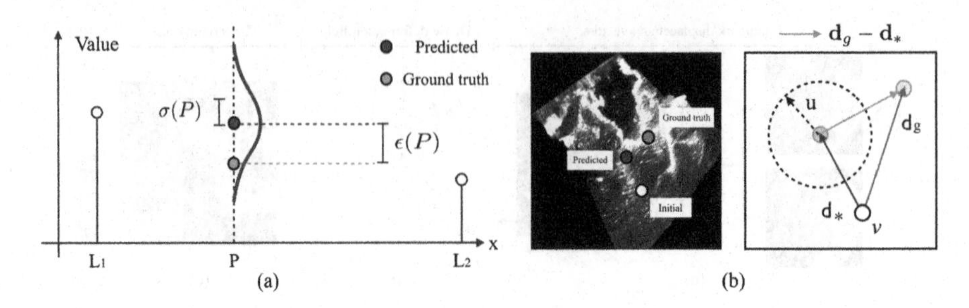

Fig. 3. (a) An illustrative example for the point-wise posterior predictive checking experiment; (b) An illustration for how to compute ϵ and u in the context of IGNs (Color figure online)

where cov is the covariance, σ's are the standard deviations. Noticing that ρ_s is by design constrained as $-1 \leq \rho_s \leq 1$, and 1 indicates a perfect positive monotonic relationship.

2.3 Point-Wise Posterior Predictive Checking

When a surgeon is removing a tumor mass near a critical structure, it is vital that s/he knows how close the predicted instrument location is from the structure and how confident the prediction is. With GP registration, we can predict the instrument location using a displacement vector. Meanwhile, we can also provide the registration uncertainty to indicate how likely the estimated instrument location to be accurate. Here, we designed a point-wise experiment to investigate whether the true location is close to the predicted location when the uncertainty is low and vice versa.

The point-wise experiment is inspired by posterior predictive checking (PPC) [33]. PPC examines the fitness of a model using the similarity between values generated by the posterior distribution and the observed ones.

In an illustrative 1D example shown in Fig. 3(a), L_1 and L_2 are two landmarks whose values are indicated by the length of vertical bars. The goal is to interpolate the value at location P. Here the blue bell-curve is the estimated posterior distribution $p(P|L_1, L_2)$ and it has a mean of p_*. Since we know the ground truth value p_g, we can compute the estimation error as $\epsilon(P) = |p_g - p_*|$. The standard deviation σ of the posterior is often used to represent the uncertainty u of the estimation.

In the context of GP registration shown in Fig. 3(b), the white circle is the initial location of voxel v on the p-US image. The green circle represents the ground truth location of deformed v on the i-US image and the blue circle is the predicted location. $\mathbf{d}_g$ and $\mathbf{d}_*$ are the ground truth and predicted displacement vectors respectively, and the registration error can be computed as $\epsilon = \|\mathbf{d}_g(i) - \mathbf{d}_*(i)\|$. As u is the registration uncertainty associated with d_*, it is visualized by a circle where the magnitude of u is the radius of the circle (a larger circle indicates a higher uncertainty).

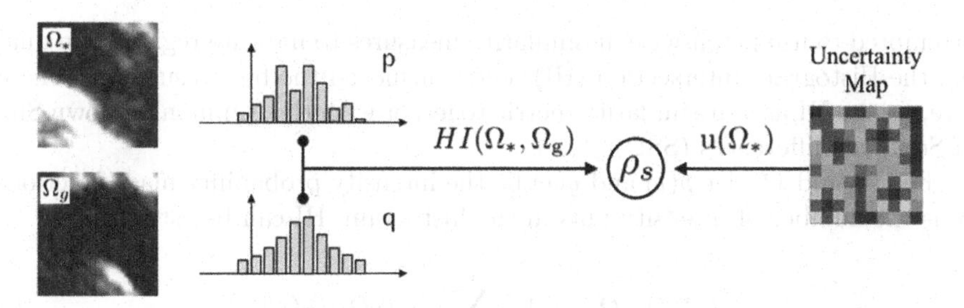

Fig. 4. An illustration for using the HI metric to compute Spearman's rank correlation coefficient for patches.

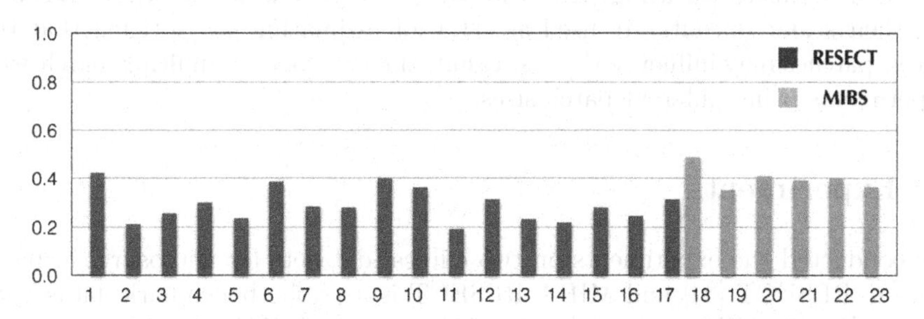

Fig. 5. The estimated ρ_s for the point-wise experiment. We can see a moderate positive monotonic relationship between u and ϵ.

In the point-wise experiment, we compute u and ϵ for every voxel-of-interest and form two discrete variables U and E. Using ρ_s, we can measure how strong their monotonic relationship is.

2.4 Patch-Wise Correlation Test

We also investigate the correlation between u and ϵ over image-patches, because surgeons may be more interested in registration errors over region of interest. We present the uncertainty to surgeons via color overlays to give them a higher level understanding of registration error.

Given a voxel v located at $\mathbf{x}_v \in \mathbb{R}^3$, we define an image patch $\Omega \subset \mathbb{R}^3$ as a sub-volume centered at $\mathbf{x}_v$, let Ω have size N. Assuming $u_\mathbf{x}$ is the voxel-wise uncertainty at location $\mathbf{x}$, we can compute the patch-wise uncertainty as the mean voxel uncertainty over Ω as $u(\Omega) = \frac{1}{N}\sum_{\mathbf{x}\in\Omega} u_\mathbf{x}$. The estimation of ϵ over a patch is not straightforward. An ideal way for measuring the patch-wise registration error would be to use the residual Euclidean distance over densely-labeled and well-distributed landmarks placed on both patches. However, to our knowledge, no existing neurosurgical dataset has such landmarks.

In this study, since all experiments are based on uni-modal registration, we use intensity-based dissimilarity metrics to measure the error between ground truth patches Ω_g's and predicted patches Ω_*'s. In a previous study that attempted to use patch-wise dissimilarity measures to indicate registration quality, the Histogram Intersection (HI) metric achieved the best result [37]. Therefore, we use HI as a dissimilarity metric together with the commonly known Sum of Squared Differences (SSD).

For $\Omega_{\rm g}$ and Ω_*, let $p(t)$ and $q(t)$ be the intensity probability mass functions. K is the number of intensity bins in the histogram. HI can be estimated as

$$HI(\Omega_*, \Omega_{\rm g}) = 1 - \sum_{i=1}^{K} \min(p(t_i), q(t_i)). \tag{5}$$

Figure 4 illustrates using HI in the patch-wise correlation test. For SSD and HI, their scalar outputs are used as $\epsilon(\Omega)$ for estimating $\rho_{\rm s}$. Noticing that the size of patches may influence the test result, thus we conduct multiple patch-wise experiments using different patch sizes.

3 Experiments

We conducted the experiments on two clinical datasets for neurosurgical registration, RESECT [34] and MIBS. RESECT is a public benchmark dataset for IGN [38], while MIBS is a proprietary dataset from BWH hospital [35,36]. Both datasets in total contain 23 sets of *p*-US and *i*-US scans that were acquired from patients with brain tumors. US data were provided as a reconstructed 3D volume. In the *p*-US to *i*-US GP registration context, we tested manually annotated landmarks in the RESECT dataset and automatically detected landmarks (around 15 landmarks per case) in MIBS, which does not have manual annotations. We note that all tested points were randomly chosen, and were excluded from the GP interpolation.

3.1 Point-Wise Experiment

In the point-wise experiment, for each landmark on the *i*-US image, GP registration estimated $\mathbf{d}_*$ and σ. Since $\mathbf{d}_{\rm g}$ is known, we can calculate ϵ and then combine all points to compute $\rho_{\rm s}$ for a pair of images.

The estimated point-wise $\rho_{\rm s}$'s are summarized in Fig. 5. For manual landmarks in RESECT, the mean value of $\rho_{\rm s}$ is 0.2899, which indicates a weak-to-moderate positive monotonic correlation. Automatically extracted landmarks achieved an average $\rho_{\rm s}$ of 0.4014, which can be categorized as a moderate-to-strong correlation. However, both scores are significantly lower than what is required for a perfect positive monotonic relationship. At this stage, it's still too early to conclude definitively whether it is safe to use GP registration uncertainty to assess the accuracy of predicted, e.g., instrument location.

We suspect that the $\rho_{\rm s}$ discrepancy between these two groups of landmarks is due to the nature of GP uncertainty and the distribution of landmarks: In GP registration, the uncertainty of a voxel depends on its distance to neighboring interpolating points, e.g., the closer to interpolating points, the lower uncertainty

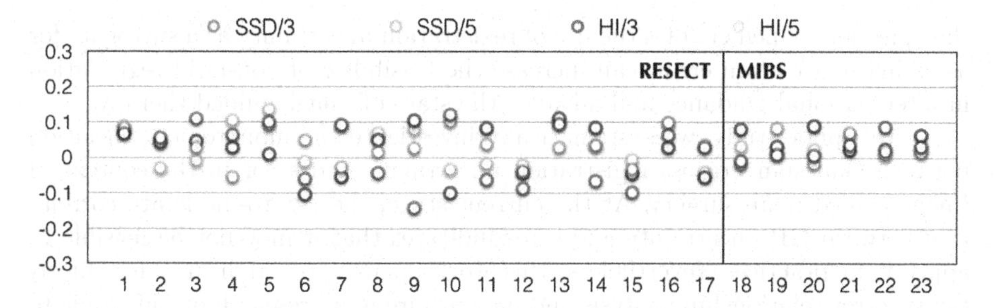

Fig. 6. The estimated $\rho_s(\Omega)$ for the patch-wise experiment. Values of $\rho_s(\Omega)$'s are consistently low for both datasets.

it has. If an annotated landmark is far away from all interpolating landmarks, it is likely to have high uncertainty. In case it happens to be located in a region with less severe deformation, that highly uncertain landmark would have a low registration error, thus lower the overall score for ρ_s.

3.2 Patch-Wise Experiment

In the patch-wise experiment, we padded $\pm k$ surrounding voxels to each landmark (same group of landmarks that is used in the point-wise experiment). For example, ± 2 padding generates a patch of the size of $5 \times 5 \times 5$. Tested values of k include 3 and 5. We calculated ϵ using SSD/HI for all patches and computed $\rho_s(\Omega)$ afterward.

The estimated patch-wise ρ_s's are shown in Fig. 6. It can be seen that values of $\rho_s(\Omega)$'s are consistently low for both datasets. We deduce the reasons for low $\rho_s(\Omega)$ values are: (1) In the presence of large deformation, e.g., tumor resection, a pair of well-matched patches may look drastically different. In this case, instead of the residual Euclidean distance over densely-labeled and well-distributed landmarks, other appearance-based dissimilarity measures become sub-optimal for estimating the registration error; (2) σ used in the calculation is transformation uncertainty, while intensities over patches is label uncertainty [31]. These two quantities may be inherently uncorrelated in GP registration. (3) Features that surgeons are interested in, e.g., tumor margins or nearby blood vessels, may be limited to a small region. It may make more sense to estimate the regional $\rho_s(\Omega)$ instead of using the whole image.

4 Conclusion

"Are registration uncertainty and error monotonically associated?" is a fundamental question that has been overlooked by researchers in the medical imaging community. There has been significant progress in the development of fast and accurate methods for performing non-rigid registration. Since all of these methods are subject to some error and rarely used in the operating room, an answer to this question, which enables the use of registration uncertainty as a surrogate for assessing registration error, can increase the feasibility of non-rigid registration in interventional guidance and advance the state of image-guided therapy.

In this pilot study, we systematically investigate the monotonic association between Gaussian process registration uncertainty and error in the context of Image-guided neurosurgery. At the current stage, the low-to-moderate correlation between GP uncertainty and error indicates that it may not be feasible to apply it in practice. Nevertheless, this work opens a research area for uncertainty/error relationship analysis and may inspire more research on this topic to verify and enhance the applicability of registration uncertainty.

Acknowledgement. MS was supported by the International Research Center for Neurointelligence (WPI-IRCN) at The University of Tokyo Institutes for Advanced Study. This work was also supported by NIH grants P41EB015898, P41EB015902 and 5R01NS049251.

References

1. Maintz, J.B.A., Viergever, M.A.: A survey of medical image registration. Med. Image Anal. **2**(1), 1–36 (1998)
2. Sotiras, A., Davatzikos, C.: Deformable medical image registration: a survey. IEEE Trans. Med. Imaging **32**(7), 1153–1190 (2013)
3. Gerard, I., Kersten-Oertel, M., Petrecca, K., Sirhan, D., Hall, J., Collins, D.L.: Brain shift in neuronavigation of brain tumors: a review. Med. Image Anal. **35**, 403–420 (2017)
4. Morin, F., et al.: Brain-shift compensation using intraoperative ultrasound and constraint-based biomechanical simulation. Med. Image Anal. **40**, 133–153 (2017)
5. Luo, M., Larson, P.S., Martin, A.J., Konrad, P.E., Miga, M.I.: An integrated multiphysics finite element modeling framework for deep brain stimulation: preliminary study on impact of brain shift on neuronal pathways. In: Shen, D., et al. (eds.) MICCAI 2019. LNCS, vol. 11768, pp. 682–690. Springer, Cham (2019). https://doi.org/10.1007/978-3-030-32254-0_76
6. Rivaz, H., Collins, D.L.: Deformable registration of preoperative MR, pre-resection ultrasound, and post-resection ultrasound images of neurosurgery. Int. J. Comput. Assist. Radiol. Surg. **10**(7), 1017–1028 (2014). https://doi.org/10.1007/s11548-014-1099-4
7. Kybic, L.: Bootstrap resampling for image registration uncertainty estimation without ground truth. IEEE Trans. Image Process. **19**(1), 64–73 (2010)
8. Shams, R., Xiao, Y., Hebert, F., Abramowitz, M., Brooks, R., Rivaz, H.: Assessment of rigid registration quality measures in ultrasound-guided radiotherapy. IEEE Trans. Med. Imaging **37**(2), 428–437 (2018)
9. Datteri, R.D., Dawant, B.M.: Automatic detection of the magnitude and spatial location of error in non-rigid registration. In: Dawant, B.M., Christensen, G.E., Fitzpatrick, J.M., Rueckert, D. (eds.) WBIR 2012. LNCS, vol. 7359, pp. 21–30. Springer, Heidelberg (2012). https://doi.org/10.1007/978-3-642-31340-0_3
10. Hub, M., Kessler, M.L., Karger, C.P.: A stochastic approach to estimate the uncertainty involved in B-spline image registration. IEEE Trans. Med. Imaging **28**(11), 1708–1716 (2009)

11. Hub, M., Karger, C.P.: Estimation of the uncertainty of elastic image registration with the demons algorithm. Phys. Med. Biol. **58**(9), 3023 (2013)
12. Saygili, G., Staring, M., Hendriks, E.A.: Confidence estimation for medical image registration based on stereo confidences. IEEE Trans. Med. Imaging **35**(2), 539–549 (2016)
13. Sokooti, H., Saygili, G., Glocker, B., Lelieveldt, B.P.F., Staring, M.: Accuracy estimation for medical image registration using regression forests. In: Ourselin, S., Joskowicz, L., Sabuncu, M.R., Unal, G., Wells, W. (eds.) MICCAI 2016. LNCS, vol. 9902, pp. 107–115. Springer, Cham (2016). https://doi.org/10.1007/978-3-319-46726-9_13
14. Sokooti, H., Saygili, G., Glocker, B., Lelieveldt, B.P.F., Staring, M.: Quantitative error prediction of medical image registration using regression forests. Med. Image Anal. **56**, 110–121 (2019)
15. Saygili G.: Local-search based prediction of medical image registration error. In: Proceedings of SPIE Medical Imaging 2018, vol. 10577 (2018). https://doi.org/10.1117/12.2293740
16. Glocker, B., Paragios, N., Komodakis, N., Tziritas, G., Navab, N.: Optical flow estimation with uncertainties through dynamic MRFs. In: Computer Vision and Pattern Recgonition (CVPR), pp. 1–8 (2008)
17. Popuri, K., Cobzas, D., Jägersand, M.: A variational formulation for discrete registration. In: Mori, K., Sakuma, I., Sato, Y., Barillot, C., Navab, N. (eds.) MICCAI 2013. LNCS, vol. 8151, pp. 187–194. Springer, Heidelberg (2013). https://doi.org/10.1007/978-3-642-40760-4_24
18. Risholm, P., Pieper, S., Samset, E., Wells, W.M.: Bayesian characterization of uncertainty in intra-subject non-rigid registration. Media **17**(5), 538–555 (2013)
19. Lotfi, T., Tang, L., Andrews, S., Hamarneh, G.: Improving probabilistic image registration via reinforcement learning and uncertainty evaluation. In: Wu, G., Zhang, D., Shen, D., Yan, P., Suzuki, K., Wang, F. (eds.) MLMI 2013. LNCS, vol. 8184, pp. 187–194. Springer, Cham (2013). https://doi.org/10.1007/978-3-319-02267-3_24
20. Wassermann, D., Toews, M., Niethammer, M., Wells, W.: Probabilistic diffeomorphic registration: representing uncertainty. In: Ourselin, S., Modat, M. (eds.) WBIR 2014. LNCS, vol. 8545, pp. 72–82. Springer, Cham (2014). https://doi.org/10.1007/978-3-319-08554-8_8
21. Yang, X., Niethammer, M.: Uncertainty quantification for LDDMM using a low-rank Hessian approximation. In: Navab, N., Hornegger, J., Wells, W.M., Frangi, A.F. (eds.) MICCAI 2015. LNCS, vol. 9350, pp. 289–296. Springer, Cham (2015). https://doi.org/10.1007/978-3-319-24571-3_35
22. Simpson, I.J.A., et al.: Probabilistic non-linear registration with spatially adaptive regularisation. Med. Image Anal. **26**(1), 203–216 (2015)
23. Heinrich, M., Simpson, I.J.A., Papiez, B., Brady, M., Schnabel, J.A.: Deformable image registration by combining uncertainty estimates from supervoxel belief propagation. Med. Image Anal. **27**, 57–71 (2016)
24. Folgoc, L.L., Delingette, H., Criminisi, A., Ayache, N.: Quantifying registration uncertainty with sparse Bayesian modelling. IEEE Trans. Med. Imaging **36**(2), 607–617 (2017)
25. Dalca, A.V., Balakrishnan, G., Guttag, J., Sabuncu, M.R.: Unsupervised learning for fast probabilistic diffeomorphic registration. In: Frangi, A.F., Schnabel, J.A., Davatzikos, C., Alberola-López, C., Fichtinger, G. (eds.) MICCAI 2018. LNCS, vol. 11070, pp. 729–738. Springer, Cham (2018). https://doi.org/10.1007/978-3-030-00928-1_82

26. Wang, J., Wells, W.M., Golland, P., Zhang, M.: Efficient Laplace approximation for Bayesian registration uncertainty quantification. In: Frangi, A.F., Schnabel, J.A., Davatzikos, C., Alberola-López, C., Fichtinger, G. (eds.) MICCAI 2018. LNCS, vol. 11070, pp. 880–888. Springer, Cham (2018). https://doi.org/10.1007/978-3-030-00928-1_99
27. Luo, J., et al.: A feature-driven active framework for ultrasound-based brain shift compensation. In: Frangi, A.F., Schnabel, J.A., Davatzikos, C., Alberola-López, C., Fichtinger, G. (eds.) MICCAI 2018. LNCS, vol. 11073, pp. 30–38. Springer, Cham (2018). https://doi.org/10.1007/978-3-030-00937-3_4
28. Sedghi, A., Kapur, T., Luo, J., Mousavi, P., Wells, W.M.: Probabilistic image registration via deep multi-class classification: characterizing uncertainty. In: Greenspan, H., et al. (eds.) CLIP/UNSURE -2019. LNCS, vol. 11840, pp. 12–22. Springer, Cham (2019). https://doi.org/10.1007/978-3-030-32689-0_2
29. Bayer, S., et al.: Investigation of feature-based nonrigid image registration using gaussian process. Bildverarbeitung für die Medizin 2020. I, pp. 156–162. Springer, Wiesbaden (2020). https://doi.org/10.1007/978-3-658-29267-6_32
30. Agn, M., Van Leemput, K.: Fast nonparametric mutual-information-based registration and uncertainty estimation. In: Greenspan, H., et al. (eds.) CLIP/UNSURE -2019. LNCS, vol. 11840, pp. 42–51. Springer, Cham (2019). https://doi.org/10.1007/978-3-030-32689-0_5
31. Luo, J., et al.: On the applicability of registration uncertainty. In: Shen, D., et al. (eds.) MICCAI 2019. LNCS, vol. 11765, pp. 410–419. Springer, Cham (2019). https://doi.org/10.1007/978-3-030-32245-8_46
32. Corder, G.W., Foreman, D.I.: Nonparametric Statistics: A Step-by-Step Approach. Wiley, Hoboken (2014)
33. Gelman, A., Carlin, J.B., Stern, H.S., Dunson, D.B., Vehtari, A., Rubin, D.: Bayesian Data Analysis, p. 653. Chapman & Hall, USA (2004)
34. Xiao, Y., Fortin, M., Unsgard, G., Rivaz, H., Reinertsen, I.: RESECT: a clinical database of pre-operative MRI and intra-operative ultrasound in low-grade glioma surgeries. Med. Phys. **44**(7), 3875–3882 (2017)
35. Machado, I., et al.: Non-rigid registration of 3D ultrasound for neurosurgery using automatic feature detection and matching. Int. J. Comput. Assist. Radiol. Surg. **13**(10), 1525–1538 (2018). https://doi.org/10.1007/s11548-018-1786-7
36. Luo, J., et al.: Using the variogram for vector outlier screening: application to feature-based image registration. Int. J. Comput. Assist. Radiol. Surg. **13**(12), 1871–1880 (2018). https://doi.org/10.1007/s11548-018-1840-5
37. Schlachter, M., et al.: Visualization of deformable image registration quality using local image dissimilarity. IEEE TMI **35**(10), 2319–2328 (2016)
38. Xiao. Y., et al.: Evaluation of MRI to ultrasound registration methods for brain shift correction: the CuRIOUS2018 challenge. IEEE TMI (2020)
39. Toews, M., Wells, W.M.: Efficient and robust model-to-image alignment using 3D scale-invariant features. Med. Image Anal. **17**, 271–282 (2013)

MR-to-US Registration Using Multiclass Segmentation of Hepatic Vasculature with a Reduced 3D U-Net

Bart R. Thomson[1(✉)], Jasper N. Smit[1], Oleksandra V. Ivashchenko[1,2], Niels F.M. Kok[1], Koert F.D. Kuhlmann[1], Theo J.M. Ruers[1,3], and Matteo Fusaglia[1]

[1] Department of Surgical Oncology, The Netherlands Cancer Institute, Amsterdam, The Netherlands
bart.thomson@icloud.com

[2] Department of Radiology, Leiden University Medical Center, Leiden, The Netherlands

[3] Department of Technical Medicine, University of Twente, Enschede, The Netherlands

Abstract. Accurate hepatic vessel segmentation and registration using ultrasound (US) can contribute to beneficial navigation during hepatic surgery. However, it is challenging due to noise and speckle in US imaging and liver deformation. Therefore, a workflow is developed using a reduced 3D U-Net for segmentation, followed by non-rigid coherent point drift (CPD) registration. By means of electromagnetically tracked US, 61 3D volumes were acquired during surgery. Dice scores of 0.77, 0.65 and 0.66 were achieved for segmentation of all vasculature, hepatic vein and portal vein respectively. This compares to inter-observer variabilities of 0.85, 0.88 and 0.74 respectively. Target registration error at a tumor lesion of interest was lower (7.1 mm) when registration is performed either on the hepatic or the portal vein, compared to using all vasculature (8.9 mm). Using clinical data, we developed a workflow consisting of multi-class segmentation combined with selective non-rigid registration that leads to sufficient accuracy for integration in computer assisted liver surgery.

Keywords: Computer assisted intervention · Liver surgery · Non-rigid registration

1 Introduction

Computer assisted intervention (i.e. CAI) aims to equip the surgeon with a "surgical cockpit", where the live position of surgical instruments, preoperative imaging and intraoperative organ position are represented within the same

B. R. Thomson and J. N. Smit—Equal contribution.

A. L. Martel et al. (Eds.): MICCAI 2020, LNCS 12263, pp. 275–284, 2020.
https://doi.org/10.1007/978-3-030-59716-0_27

coordinate system. The core process of CAI is the registration, which aims to find a geometrical mapping between the preoperative image and the intraoperative organ position.

Within computer assisted open liver surgery, the preferred method for obtaining the information about intraoperative position of the organ is via a tracked 2D ultrasound (i.e. US) probe (optically or electromagnetically tracked). Since US can visualize underlying anatomical structures (e.g. tumors, hepatic and portal vein), it is widely accepted and integrated into the surgical workflow.

By coupling a 2D US probe with a tracker, intraoperative 3D US volumes can be reconstructed. Similar to conventional tomographic images, hepatic vasculature imaged within these volumes can be automatically segmented, and used for generating a 3D model of the underlying vasculature. Such a model can be registered with its preoperative counterpart (CT or MRI-based). Accuracy of this registration is greatly dependent on the extent and accuracy of the US-based segmentation of liver vasculature.

The majority of previous work on hepatic vasculature segmentation in 2D US are based on conventional segmentation techniques. In [4,5], edge detection algorithms based on the difference of Gaussians were evaluated in phantom settings. In [11], semi-automated region growing methods were used, while in [14] dynamic texturing combined with k-nearest classification was adopted. Other methods combined extended Kalman filters with constraints on the detection of ellipsoid models [7,19] or tubular structures [1,10,18]. Despite promising results in phantom settings, these methods have proven less successful in clinical settings since they are prone to mislabeling due to their susceptibility to sub-optimal imaging conditions (e.g. artefacts, shadows, air-gaps, vessel abnormalities).

With the advances in deep learning, many convolutional neural network (CNN) based segmentation techniques that outperformed conventional algorithms, were developed for a broad range of clinical application. Similar advances are emerging in hepatic vasculature segmentation from US volumes. For example, in [16], a CNN combined with k-means clustering for hepatic vasculature extraction was proposed. The network, trained on 132 2D US images, contained a significantly smaller number of parameters compared to conventional deep learning networks and reported a segmentation accuracy, expressed as an intersection over union (IoU), of 0.696 [16]. Similar approaches were adopted in [21] and [20], where 2D [21] and 3D U-Nets [20] for hepatic vasculature segmentation were proposed. These studies reported average segmentation accuracies, expressed as Dice, of 0.5 for 2D U-Nets and 0.7 for 3D U-Nets.

In the context of registration, the segmentation results obtained in [21] were used to define a region of interest for a two-step registration procedure based on Covariance Matrix Adaptation Evolution Strategy (CMA-ES) and gradient orientation similarity. While these recent studies [16,20,21] have shown promising results, they are limited by a number of factors.

First, the aforementioned segmentation methods do not distinguish between the two major types of hepatic vasculature (i.e. hepatic and portal vein). This results in a single 3D model, where hepatic and portal veins are combined and registered to their preoperative counterpart as a single anatomical structure.

Because these vessel trees have different mechanical properties and independent mobility, joined registration may result in a larger local registration error. Additionally, depending on the tumor position with respect to the hepatic vasculature, the preservation of a hepatic or portal branch has different clinical implications and may require different degrees of accuracy. We hypothesize that a more realistic segmentation approach would distinguish between hepatic and portal vein. This will result in two separate 3D models of the hepatic vasculature which can then be registered to their preoperative counterpart independently from each other, aiming for a more accurate registration.

Second, the registration method described in [21] is based on rigid transformation between the preoperative CT and the intraoperative model of the vessels. While this methodology has been proven effective within a restricted area, it does not compensate for organ deformation throughout the entire organ. In this manuscript we will apply a non-rigid registration methodology and evaluate its accuracy in terms of two measures. Clinically, the most relevant measure is the registration accuracy that one can achieve in the tumor lesion of interest. In order to generalize the registration to the whole liver, overall registration accuracy between the vasculature is most important.

Third, previous studies are evaluated over limited clinical datasets, making it challenging to generalize to clinical use.

In this study we present a segmentation-registration pipeline, that is fully trained and validated on intraoperative imaging. By means of deep learning, we are able to fully automate the intraoperative segmentation process, which is then utilized in the automatic registration of vasculature from the pre- and intraoperative imaging.

2 Methods

The automatic non-rigid registration pipeline that is proposed in this work is schematically illustrated in Fig. 1. This pipeline enables integration of information regarding the lesions and their location with respect to the major hepatic vasculature into the surgical environment.

An initial registration is performed by recording the orientation of the US transducer and a one-point translation based on the center of the lesion. Fine registration is based on the vasculature that is present in both imaging modalities. In the preoperative imaging, vasculature is segmented semi-automatically based on the method described in [8] and refined manually. The intraoperative US vasculature is segmented automatically using a reduced filter implementation of the standard 3D U-Net architecture [3].

This architecture is used to train three different deep learning models for separate segmentation purposes; segmentation of all vasculature, solely the hepatic vein and solely the portal vein. The centerlines of both the pre- and intraoperative segmentations are then used for non-rigid registration with the coherent point drift (CPD) algorithm [17].

Following the segmentation process, both the pre- and intraoperative segmentations are resampled to isotropic spacing of 1.1 mm, increasing registration speed significantly whilst still maintaining accuracy. The registration accuracy was computed on the hepatic and portal vein by computing two measures. To measure the overlap of the vasculature, we computed the root mean squared error (RMSE) of the residual distances between the centerlines of the segmented vasculature and its preoperative registered counterpart. To measure the clinical accuracy, we computed the target registration error (TRE) as the Euclidean distance between the center of the lesion, acquired through US and manually segmented, and its preoperative registered counterpart. TRE was computed on 11 patients. For each case three TREs were found (using hepatic, portal, and all vasculature). Subsequently, the lowest TRE between the registrations using hepatic or portal vasculature was compared with the TRE found using the combined vasculature.

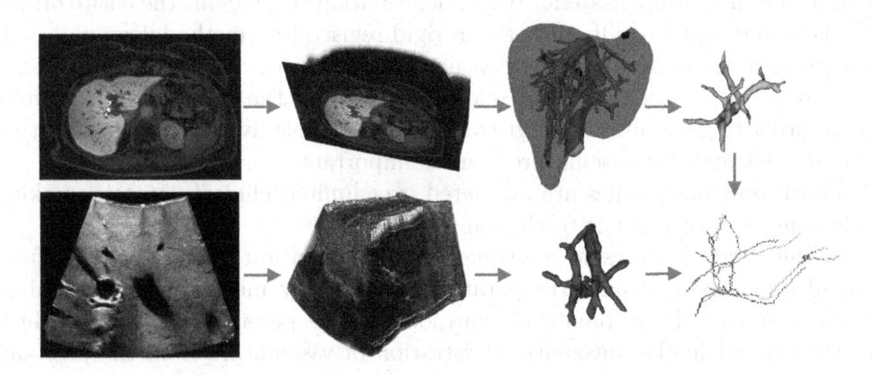

Fig. 1. Vasculature is extracted from the preoperative scan (CT or MRI) prior to surgery (top row). During surgery vasculature is extracted from a reconstructed US volume (bottom row). Centerlines from both modalities are used for registration.

2.1 Vascular Segmentation

The 3D U-Net architecture that is used is a NiftyNet [6] Tensorflow implementation similar to Çiçek et al. [3], but the amount of filters in every layer has been reduced to an eighth, to avoid bottlenecks. A learning rate of 5×10^{-3} with Adam optimizer and L1 regularization with 10^{-5} weight decay were used for training on four NVIDIA 1080 GTX GPUs with a batch size of 2. From each mean value normalized volume, 20 $144 \times 144 \times 96$ voxel patches were sampled and zero-padded with a volume of $32 \times 32 \times 32$ voxels. Data augmentation consisted of rotation between $-10°$ and $10°$, scaling between -10% and 10% and elastic deformation that is similar to [15]. The Dice loss function was used for training of the network until there was no further apparent converging of the validation loss. Segmentation performance is reported by means of the Dice score.

2.2 Gaussian Regularized Non-rigid Registration

The automatically segmented intraoperative vascular model was used for registration by means of CPD. To reduce computational cost, centerlines were extracted from the segmentations based on the method of [12]. Next, the preoperative vasculature model was mimicked as a Gaussian Mixture Model (GMM), while the intraoperative model was treated as observations from the GMM. Unlike diagnostic preoperative imaging, intraoperative US acquisition is a localized high resolution, yet noisy image of local vasculature. Therefore, point clouds of intraoperative centerlines models are fundamentally different from diagnostic imaging. CPD handles noise well and should therefore be robust to registering the complete vascular point cloud (preoperative) to a sub-set of this point cloud (intraoperative) [17].

The CPD implementation of [9] allows for tuning of two variables; α, determining the deformability of the preoperative model, to align with the intraoperative model, and β, determining the size of the Gaussian kernel that was used to find the coherent point in the intraoperative model. Both variables were optimized by means of grid search, with values in the ranges of $\alpha = 0.1, 0.2, \ldots, 0.5$ and $\beta = 50, 100, \ldots, 800$. The TRE was minimized by grid searching the amount of points that are used for registration in the range of $points = 1, 2, \ldots, 15$. The optimal combination of settings was 0.3, 550, 8 for α, β and the number of nearest points respectively.

2.3 Data

The complete dataset contained 203 stacked 2D US volumes, of which 106 volumes, acquired in 24 patients, were considered of sufficient quality. In 96 volumes, the hepatic and portal veins were delineated, of which 85 were used in training and validation of the segmentation network. The main reason of exclusion was the incorrect stacking of 2D US slices, either due to rapid turning movements of the US probe by the operator, or due to tracking or reconstruction errors. Patients scheduled for open surgery of age $\geq$18, with centrally, primary or secondary, near vasculature located liver lesions from any origin, of diameter < 8 cm were included in the dataset. Preoperative scans used for registration were no older than 2 months.

Volumes were acquired by coupling a T-shaped intraoperative US probe (T-Shaped Intraoperative I14C5T, BK Medical, Herlev Denmark) with an electromagnetic tracking system (Aurora Northern Digital, Ontario, Canada). Calibration between the tracking sensor and the US image was performed using the method described in [2]. CustusX [1] was used for acquiring the tracked images, which were then stacked in a volume using pixel nearest neighbor reconstruction. During acquisition, the US operator was instructed to acquire large volumes following a straight path from segments 4a and 8 to segments 4b and 5. Five different operators acquired the US volumes and each volume was delineated by one out of four annotators using 3D Slicer. Unclear delineations of structures have been validated by a radiologist. Five scans have been delineated by multiple annotators with the aim of setting a manual gold standard, for comparing the automatic segmentation performances. The hepatic and portal veins were segmented separately and volume sizes ranged from $293 \times 396 \times 526$ to $404 \times 572 \times 678$ pixels, depending on the zoom of the 2D slices and length of the scanning trajectory, but were down sampled to 40% prior to training, similar to [20]. Pre- and intra-operative data of 11 patients, accounting for 11 scans that contained tumor lesions, were used for evaluation of the registration pipeline.

3 Results

The reduced filter 3D U-Net obtained Dice scores of 0.77 ± 0.09, 0.65 ± 0.25 and 0.66 ± 0.13 for combined vasculature, hepatic and portal veins respectively. These values are comparable to the Dice score of the inter-observer variability (0.85 ± 0.04, 0.88 ± 0.02, 0.74 ± 0.12) for combined vasculature, hepatic and portal vasculature respectively. Figure 2 shows the segmentation result for a single case, for the different types of vasculature. The majority of mislabelling occurred on the peripheral segments of the vasculature (i.e. small vessels).

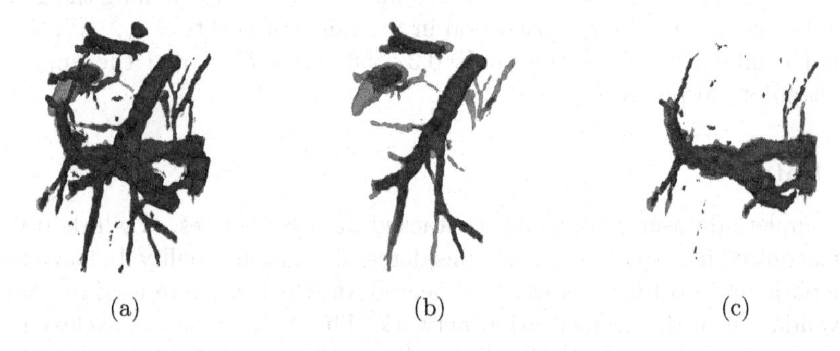

Fig. 2. Example of vascular segmentation prior to registration, with (a) all vasculature, (b) hepatic vasculature, (c) portal vasculature, with Dice scores of 0.82, 0.72 and 0.82 respectively. The ground truth delineation is indicated in green and the automatic segmentation in blue. (Color figure online)

The distribution of the RMSE of the registered vasculatures using CPD is summarized in Fig. 3a. On average, the RMSE of the combined vasculature (4.4 ± 3.9 mm) is lower than those calculated for the hepatic (7.0 ± 7.5 mm) and portal vein (4.8 ± 4.4 mm). Nevertheless, Fig. 3a shows a similar RMSE distribution for the combined, hepatic and portal vein registrations. Clinical accuracy, measured as TRE between the tumor position acquired through US and its preoperative registered counterpart, is shown in Fig. 3b. On average, selecting the lowest TRE of the hepatic or portal vasculature (7.1 ± 3.7 mm) results in a lower TRE compared with the combined vasculature (8.9 ± 5.3 mm). This can also be seen in Fig. 4a, which compares the obtained TREs for each patient. In 10 out of 11 cases, a lower TRE was found either by using only the hepatic or portal vein. In 9 out of 11 cases TRE was calculated to be below 10 mm (considered the viable clinical threshold). The total computation time of the pipeline is 62 ± 5.37 s, 68.3 ± 6.23 s and 84.5 ± 11.2 s for respectively hepatic, portal and all vasculature. Figure 4b shows a linear correlation coefficient of 0.796 and 0.331 between the

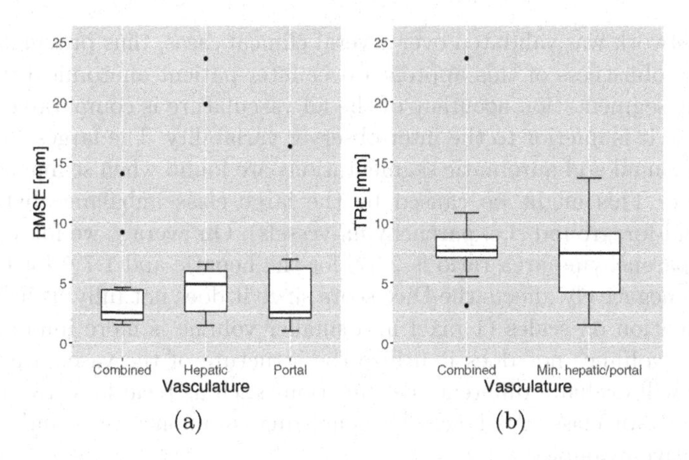

Fig. 3. (a) Registration accuracy between the vascular centerlines expressed as RMSE for the combined vasculature, hepatic and portal separately. (b) Registration accuracy of all vasculature vs the minimum between the hepatic and portal vasculature. Dots represent outliers.

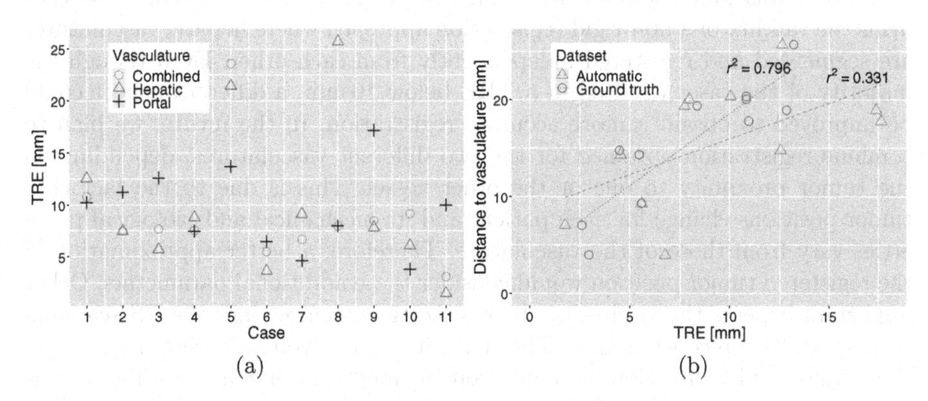

Fig. 4. (a) Lesion TRE after registration compared between the individual cases, based on whole, solely hepatic and solely portal segmentation. (b) Lesion TRE relative to distance to vasculature in ground truth segmentations.

distance of the lesion with regards to the vascular tree relative to the TRE, when registration is performed on ground truth and automatic segmentations respectively.

4 Discussion

We have presented a methodology for hepatic vasculature registration that utilizes a deep neural network to segment the hepatic and portal vein from 3D US volumes.

The network was validated over several clinical cases, thus proving the feasibility and robustness of this approach over inter-patient anatomical variations. Whilst the segmentation accuracy of the all vasculature is comparable to previous studies, it is inferior to the inter-observer variability. The largest differences between manual and automatic segmentations are found when segmenting small vasculature. This might be caused by the large class imbalance between the background-foreground (i.e. parenchyma-vessels). On average we found that the vessel-to-parenchyma area ratio is 2.3% for the hepatic and 1.7% for the portal vein. This negatively affects the Dice score since it does not fully utilize the spatial information on scales (1 pixel in a smaller volume is more important than in a larger volume) nor does it utilize the structure of the vasculature. In the future we will evaluate different cost functions such as focal loss [13]. Focal loss compensates for class imbalances by penalizing common classes and rewarding hard negative examples.

Registration accuracy, expressed as average RMS of the residual distances between the intraoperative centerlines and their registered counterparts was found to be comparable for all the three cases. Nevertheless, clinical accuracy (i.e. TRE) was found lower when using only hepatic or portal vein. This confirms the validity of a non-rigid registration approach where hepatic vasculatures are segmented and registered independently from each other. Even though the majority of the cases resulted in a TRE below 10 mm, a different aspect could be improved to obtain a more accurate registration. In the future, we plan to combine registration obtained for the two different vasculatures, depending on the tumor proximity to one or the other vessel. This is due to the fact that tumor positions change for each patient and its mechanical and biological properties vary from those of the vasculature. Therefore, a better approximation of the registered tumor position would consider these additional parameters. Other important aspects that influence the registration accuracy are the US scanning process and its reconstruction. The majority of the volumes were acquired in the cranio-caudal direction, starting from segments 4a or 8 and ending at segments 4b or 5. However, within this process, factors such as speed of acquisition, EM interference and regularity of the volume, contributed in the reconstruction accuracy of the underlying vasculature. This high variability in the parameters influencing the US volume, also resulted in large variations in the reconstructed vasculature and therefore registration accuracy.

When the TRE is determined based on ground truth segmentations, there seems to be a correlation between the lesion-to-vessel distance in the preoperative imaging and the TRE, which is not seen when using automatic segmentations. Hence, we argue that segmentation quality contributes even further to upfront prediction of which vasculature to select for the registration. We will implement this selection criteria into our pipeline, allowing us to select the most promising registration first.

Finally, the results show that both the segmentation and registration processes are highly dependent on the quality and quantity of the information contained in the US volume. Factors such as vessel-to-parenchyma ratio, scanning direction, zoom and reconstruction artefacts, strongly influence the outcome of the proposed methodology. Similar findings were also reported in [21], where only US images containing vessel-to-image ratios greater than 1% were selected for registration. In the future we will quantitatively evaluate the impact of these factors on the registration accuracy and develop deep learning methods that aim at automatically evaluating the quality of the acquired US volumes.

In conclusion, we have demonstrated that multi-class segmentation of hepatic vasculature from US volumes is feasible and, when combined with a selective non-rigid registration, accurate registration can be achieved. To our knowledge, this is the first work that utilizes deep learning based segmentation for registration purposes in hepatic ultrasound imaging where hepatic and portal vein are segmented separately. Given the promising results, validated over several patients, we are planning a prospective study in order to integrate this approach within the clinical routine of computer assisted liver interventions.

References

1. Askeland, C., et al.: CustusX: an open-source research platform for image-guided therapy. Int. J. Comput. Assist. Radiol. Surg. **11**(4), 505–519 (2016)
2. Bø, L.E., Hofstad, E.F., Lindseth, F., Hernes, T.A.: Versatile robotic probe calibration for position tracking in ultrasound imaging. Phys. Med. Biol. **60**(9), 3499 (2015)
3. Çiçek, Ö., Abdulkadir, A., Lienkamp, S.S., Brox, T., Ronneberger, O.: 3D U-Net: learning dense volumetric segmentation from sparse annotation. In: Ourselin, S., Joskowicz, L., Sabuncu, M.R., Unal, G., Wells, W. (eds.) MICCAI 2016. LNCS, vol. 9901, pp. 424–432. Springer, Cham (2016). https://doi.org/10.1007/978-3-319-46723-8_49
4. Dagon, B., Baur, C., Bettschart, V.: Real-time update of 3D deformable models for computer aided liver surgery. In: 2008 19th International Conference on Pattern Recognition, pp. 1–4. IEEE (2008)
5. Fusaglia, M., Tinguely, P., Banz, V., Weber, S., Lu, H.: A novel ultrasound-based registration for image-guided laparoscopic liver ablation. Surg. Innov. **23**(4), 397–406 (2016)
6. Gibson, E., et al.: NiftyNet: a deep-learning platform for medical imaging. Comput. Methods Programs Biomed. **158**, 113–122 (2018)
7. Guerrero, J., Salcudean, S.E., McEwen, J.A., Masri, B.A., Nicolaou, S.: Real-time vessel segmentation and tracking for ultrasound imaging applications. IEEE Trans. Med. Imaging **26**(8), 1079–1090 (2007)
8. Ivashchenko, O.V., et al.: A workflow for automated segmentation of the liver surface, hepatic vasculature and biliary tree anatomy from multiphase MR images. Magn. Reson. Imaging **68**, 53–65 (2020)
9. Khallaghi, S.: Pure numpy implementation of the coherent point drift algorithm (2017). https://github.com/siavashk/pycpd

10. Kumar, R.P., Albregtsen, F., Reimers, M., Edwin, B., Langø, T., Elle, O.J.: Three-dimensional blood vessel segmentation and centerline extraction based on two-dimensional cross-section analysis. Ann. Biomed. Eng. **43**(5), 1223–1234 (2015)
11. Lange, T., Eulenstein, S., Hünerbein, M., Lamecker, H., Schlag, P.-M.: Augmenting intraoperative 3D ultrasound with preoperative models for navigation in liver surgery. In: Barillot, C., Haynor, D.R., Hellier, P. (eds.) MICCAI 2004. LNCS, vol. 3217, pp. 534–541. Springer, Heidelberg (2004). https://doi.org/10.1007/978-3-540-30136-3_66
12. Lee, T.C., Kashyap, R.L., Chu, C.N.: Building skeleton models via 3-D medial surface axis thinning algorithms. CVGIP: Graph. Models Image Process. **56**(6), 462–478 (1994)
13. Lin, T.Y., Goyal, P., Girshick, R., He, K., Dollár, P.: Focal loss for dense object detection. In: Proceedings of the IEEE International Conference on Computer Vision, pp. 2980–2988 (2017)
14. Milko, S., Samset, E., Kadir, T.: Segmentation of the liver in ultrasound: a dynamic texture approach. Int. J. Comput. Assist. Radiol. Surg. **3**(1–2), 143 (2008)
15. Milletari, F., Navab, N., Ahmadi, S.A.: V-Net: fully convolutional neural networks for volumetric medical image segmentation. In: 2016 Fourth International Conference on 3D Vision (3DV), pp. 565–571. IEEE (2016)
16. Mishra, D., Chaudhury, S., Sarkar, M., Manohar, S., Soin, A.S.: Segmentation of vascular regions in ultrasound images: a deep learning approach. In: 2018 IEEE International Symposium on Circuits and Systems (ISCAS), pp. 1–5. IEEE (2018)
17. Myronenko, A., Song, X.: Point set registration: coherent point drift. IEEE Trans. Pattern Anal. Mach. Intell. **32**(12), 2262–2275 (2010)
18. Smistad, E., Elster, A.C., Lindseth, F.: GPU accelerated segmentation and centerline extraction of tubular structures from medical images. Int. J. Comput. Assist. Radiol. Surg. **9**(4), 561–575 (2014)
19. Song, Y., et al.: Locally rigid, vessel-based registration for laparoscopic liver surgery. Int. J. Comput. Assist. Radiol. Surg. **10**(12), 1951–1961 (2015)
20. Thomson, B.R., et al.: Hepatic vessel segmentation using a reduced filter 3D U-Net in ultrasound imaging. arXiv preprint arXiv:1907.12109 (2019)
21. Wei, W., et al.: Fast registration for liver motion compensation in ultrasound-guided navigation. In: 2019 IEEE 16th International Symposium on Biomedical Imaging (ISBI 2019), pp. 1132–1136. IEEE (2019)

Detecting Pancreatic Ductal Adenocarcinoma in Multi-phase CT Scans via Alignment Ensemble

Yingda Xia[1], Qihang Yu[1], Wei Shen[1](✉), Yuyin Zhou[1], Elliot K. Fishman[2], and Alan L. Yuille[1]

[1] Johns Hopkins University, Baltimore, USA
shenwei1231@gmail.com
[2] Johns Hopkins Medical Institutions, Baltimore, USA

Abstract. Pancreatic ductal adenocarcinoma (PDAC) is one of the most lethal cancers among the population. Screening for PDACs in dynamic contrast-enhanced CT is beneficial for early diagnosis. In this paper, we investigate the problem of automated detecting PDACs in multi-phase (arterial and venous) CT scans. Multiple phases provide more information than single phase, but they are unaligned and inhomogeneous in texture, making it difficult to combine cross-phase information seamlessly. We study multiple phase alignment strategies, *i.e.*, early alignment (image registration), late alignment (high-level feature registration), and slow alignment (multi-level feature registration), and suggest an ensemble of all these alignments as a promising way to boost the performance of PDAC detection. We provide an extensive empirical evaluation on two PDAC datasets and show that the proposed alignment ensemble significantly outperforms previous state-of-the-art approaches, illustrating the strong potential for clinical use.

Keywords: PDAC detection · Feature alignment · Pancreatic tumor segmentation

1 Introduction

Pancreatic ductal adenocarcinoma (PDAC) is the third most common cause of cancer death in the US with a dismal five-year survival of merely 9% [9]. Computed tomography (CT) is the most widely used imaging modality for the initial evaluation of suspected PDAC. However, due to the subtle early signs of PDACs in CTs, they are easily missed by even experienced radiologists.

Recently, automated PDAC detection in CT scans based on deep learning has received increasing attention [3,4,22,26], which offers great opportunities in assisting radiologists to diagnosis early-stage PDACs. But, most of these methods only unitize one phase of CT scans, and thus fail to achieve satisfying results.

Y. Xia and Q. Yu—Equally contributed to the work.

A. L. Martel et al. (Eds.): MICCAI 2020, LNCS 12263, pp. 285–295, 2020.
https://doi.org/10.1007/978-3-030-59716-0_28

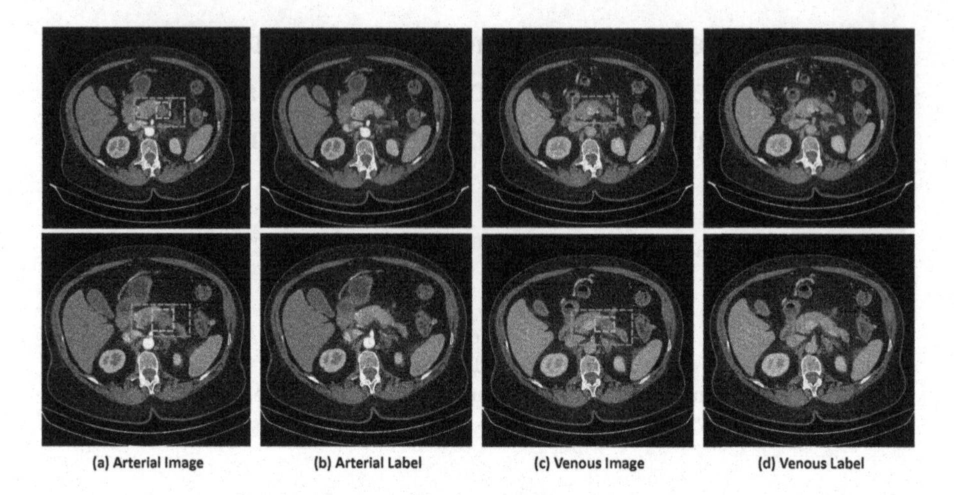

Fig. 1. Visual illustration of opportunity (top row) and challenge (bottom row) for PDAC detection in multi-phase CT scans (normal pancreas tissue - blue, pancreatic duct - green, PDAC mass - red). Top: tumor is barely visible in venous phase alone but more obvious in arterial phase. Bottom: there exist misalignment for images in these two phases given different organ size/shape and image contrast. (Color figure online)

In this paper, we aim to develop a deep learning based PDAC detection system taking multiple phases, *i.e.*, arterial and venous, of CT scans into account. This system consists of multiple encoders, each of which encodes information for one phase, and a segmentation decoder, which outputs PDAC detection results. Intuitively, multiple phases provide more information than a single phase, which certainly benefits PDAC detection. Nevertheless, how to combine this cross-phase information seamlessly is non-trivial. The challenges lie in two folds: 1) Tumor texture changes are subtle and appear differently across phases; 2) Image contents are not aligned across phases because of inevitable movements of patients during capturing multiple phases of CT scans. Consequently, a sophisticated phase alignment strategy is indispensable for detecting PDAC in multi-phase CT scans. An visual illustration is shown in Fig. 1.

We investigate several alignment strategies to combine the information across multiple phases. (1) **Early alignment** (EA): the alignment can be done in image space by performing image registration between multiple phases; (2) **Late alignment** (LA): it can be done late in feature space by performing spatial transformation between the encoded high-level features of multiple phases; (3) **Slow alignment** (SA): it can be also done step-wise in feature space by aggregating multi-level feature transformations between multiple phases. Based on an extensive empirical evaluation on two PDAC datasets [22,26], we observe that 1) All alignment strategies are beneficial for PDAC detection, 2) alignments in feature space leads to better PDAC (tumor) segmentation performance than image registration, and (3) different alignment strategies are complementary to each other, *i.e.*, an ensemble of them (**Alignment Ensemble**) significantly boosts the results, *e.g.*, approximately 4% tumor DSC score improvements over our best alignment model.

Our contributions can be summarized as follows:

- We propose late and slow alignments as two novel solutions for detecting PDACs in multi-phase CT scans and provide extensive experimental evaluation of different phase alignment strategies.
- We highlight early, late and slow alignments are complementary and a simple ensemble of them is a promising way to boost performance of PDAC detection.
- We validate our approach on two PDAC datasets [22,26] and achieve state-of-the-art performances on both of them.

2 Related Work

Automated Pancreas and Pancreatic Tumor Segmentation. With the recent advances of deep learning, automated pancreas segmentation has achieved tremendous improvements [2,10,16,17,20,21,23,25], which is an essential prerequisite for pancreatic tumor detection. Meanwhile, researchers are pacing towards automated detection of pancreatic adenocarcinoma (PDAC), the most common type of pancreatic tumor (85%) [18]. Zhu *et al.* [26] investigated using deep networks to detect PDAC in CT scans but only segmented PDAC masses in venous phase. Zhou *et al.* [22] developed the a deep learning based approach for segmenting PDACs in multi-phase CT scans, *i.e.* arterial and venous phase. They used a traditional image registration [19] approach for pre-alignment and then applied a deep network that took both phases as input. Different to their method, we also investigate how to register multiple phases in feature space.

Multi-modal Image Registration and Segmentation. Multi-modal image registration [6,7,14,19] is a fundamental task in medical image analysis. Recently, several deep learning based approaches, motivated by Spatial Transformer Networks [8], are proposed to address this task [1,13,24]. In terms of multi-modal segmentation, most of the previous works [5,11,22] perform segmentation on pre-registered multi-modal images. We also study these strategies for multi-modal segmentation, but we explore more, such as variants of end-to-end frameworks that jointly align multiple phases and segment target organs/tissues.

3 Methodology

3.1 Problem Statement

We aim at detecting PDACs from unaligned two-phase CT scans, *i.e.*, the venous phase and the arterial phase. Following previous works [22,26], venous phase is our fixed phase and arterial phase is the moving one. For each patient, we have an image $\mathbf{X}$ and its corresponding label $\mathbf{Y}$ in the venous phase, as well as an arterial phase image $\mathbf{X}'$ without label. The whole dataset is denoted as

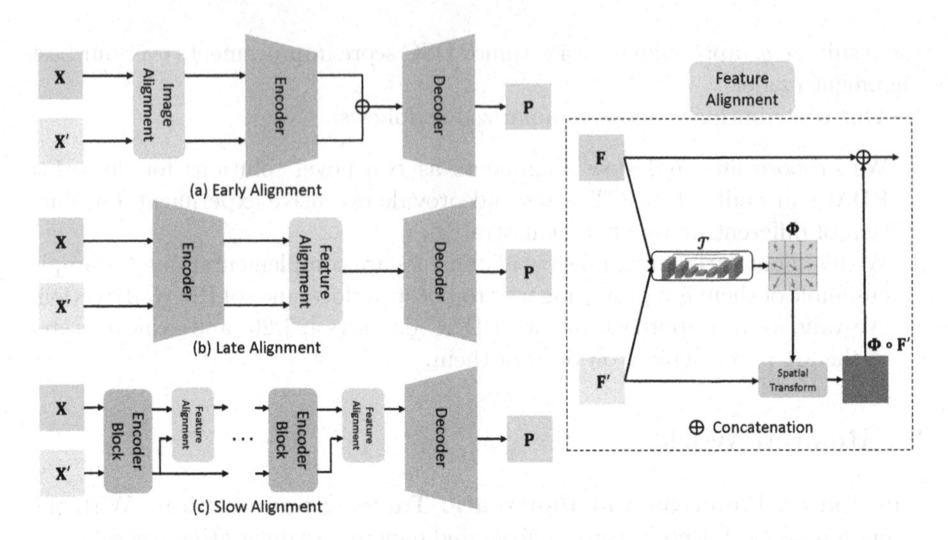

Fig. 2. An illustration of (a) early alignment (image registration) (b) late alignment and (c) slow alignment. Right: feature alignment block.

$S = \{(\mathbf{X_i}, \mathbf{X'_i}, \mathbf{Y_i}) | i = 1, 2, ..M\}$, where $\mathbf{X_i} \in \mathbb{R}^{H_i \times W_i \times D_i}$, $\mathbf{X'_i} \in \mathbb{R}^{H'_i \times W'_i \times D'_i}$ are 3D volumes representing the two-phase CT scans of the i-th patient. $\mathbf{Y_i} \in \mathcal{L}$ is a voxel-wise annotated label map, which have the same (H_i, W_i, D_i) three dimensional size as $\mathbf{X_i}$. Here, $\mathcal{L} = \{0, 1, 2, 3\}$ represents our segmentation targets, *i.e.*, background, healthy pancreas tissue, pancreatic duct (crucial for PDAC clinical diagnoses) and PDAC mass, following previous literature [22, 26]. Our goal is to find a mapping function $\mathcal{M}$ whose inputs and outputs are a pair of two-phase images $\mathbf{X}, \mathbf{X}'$ and segmentation results $\mathbf{P}$, respectively: $\mathbf{P} = \mathcal{M}(\mathbf{X}, \mathbf{X}')$. The key problem here is how to align $\mathbf{X}$ and $\mathbf{X}'$, either in image space or feature space.

3.2 Cross-Phase Alignment and Segmentation

As shown in Fig. 2, we propose and explore three types of alignment strategies, *i.e.*, early alignment, late alignment and slow alignment, for accurate segmentation.

Early (Image) Alignment. Early alignment, or image alignment strategy is adopted in [22] and some other multi-modal segmentation tasks such as BraTS challenge [11], where multiple phases (modalities) are first aligned by image registration algorithms and then fed forward into deep networks for segmentation. Here, we utilize a well-known registration algorithm, DEEDS [7], to estimate the registration field $\mathbf{\Phi}$ from an arterial image $\mathbf{X}'$ to its corresponding venous image $\mathbf{X}$. After registration, we use a network, consisting of two separtae encoders $\mathcal{F}$, $\mathcal{F}'$ and a decoder $\mathcal{G}$, to realize the mapping function $\mathcal{M}$:

$$\mathbf{P} = \mathcal{M}(\mathbf{X}, \mathbf{X}') = \mathcal{G}(\mathcal{F}(\mathbf{X}) \oplus \mathcal{F}'(\mathbf{\Phi} \circ \mathbf{X}')), \quad (1)$$

where $\oplus$ and $\circ$ denote the concatenation of two tensors and the element-wise deformation operations on a tensor, respectively.

This strategy relies on the accuracy of image registration algorithms for information alignment. If such algorithms produce errors, especially possible on subtle texture changes of PDACs, these errors will propagate and there will be no way to rescue (since alignment is only done on image level). Also, it remains a question that how much performance gain a segmentation algorithm will achieve through this separate registration procedure.

Late Alignment. An alternative way is late alignment, *i.e.*, alignment in feature space. We first encode the pair of unaligned images $(\mathbf{X}, \mathbf{X}')$ with two phase-specific encoders $(\mathcal{F}, \mathcal{F}')$, respectively. The encoded features of the two images, *i.e.*, $\mathbf{F} = \mathcal{F}(\mathbf{X})$ and $\mathbf{F}' = \mathcal{F}'(\mathbf{X}')$, are presumablely in a shared feature space. We then use a network $\mathcal{T}$ to estimate the deformable transformation field $\mathbf{\Phi}$ from arterial (moving) to venous (fixed) in the feature space by $\mathbf{\Phi} = \mathcal{T}(\mathbf{F}, \mathbf{F}')$. We apply the estimated transformation field $\mathbf{\Phi}$ to feature map $\mathbf{F}'$, then concatenate this transformed feature map $\mathbf{\Phi} \circ \mathbf{F}'$ to $\mathbf{F}$. The segmentation result $\mathbf{P}$ is obtained by feeding the concatenation to a decoder $\mathcal{G}$:

$$\mathbf{P} = \mathcal{M}(\mathbf{X}, \mathbf{X}') = \mathcal{G}(\mathbf{F} \oplus \mathbf{\Phi} \circ \mathbf{F}') = \mathcal{G}(\mathcal{F}(\mathbf{X}) \oplus \mathcal{T}(\mathbf{F}, \mathbf{F}') \circ \mathcal{F}'(\mathbf{X}')). \quad (2)$$

We name such operation as "late alignment" since the alignment is performed at the last block of feature encoders.

Slow Alignment. Late alignment performs one-off registration between two phases by only using high level features. However, it is known that the low level features of the deep network contain more image details, which motivates us to gradually align and propagate the features from multiple levels of the deep network. Following this spirit, we propose slow alignment, which leverages a stack of convolutional encoders and feature alignment blocks to iteratively align feature maps of two phases.

Let k be an integer which is not less than 1 and $(\mathbf{F}_{k-1}, \mathbf{F}'_{k-1})$ are the fused (aligned to the venous phase) feature map and the arterial feature map outputted by the $(k-1)^{th}$ convolutional encoder, respectively. First, they are encoded by a pair of convolutional encoders $(\mathcal{F}_k, \mathcal{F}'_k)$, respectively, which results in the venous feature map $\mathbf{F}_k = \mathcal{F}_k(\mathbf{F}_{k-1})$ and the arterial feature map $\mathbf{F}'_k = \mathcal{F}'_k(\mathbf{F}'_{k-1})$ at the k-th layer. Then a feature alignment block estimates a transformation field from the arterial (moving) phase to the venous (fixed) phase by

$$\mathbf{\Phi}_k = \mathcal{T}_k(\mathcal{F}_k(\mathbf{F}_{k-1}), \mathcal{F}'_k(\mathbf{F}'_{k-1})), \quad (3)$$

where $\mathcal{T}_k$ is a small U-Net. We apply the transformation field $\mathbf{\Phi}_k$ to the arterial (moving) phase, resulting in transformed arterial feature map $\mathbf{\Phi}_k \circ \mathcal{F}'_k(\mathbf{F}'_{k-1})$.

Finally, the transformed arterial feature map is concatenated with the venous feature map $\mathcal{F}_k(\mathbf{F}_{k-1})$, resulting in the fused feature map at the kth layer:

$$\mathbf{F}_k = \mathcal{F}_k(\mathbf{F}_{k-1}) \oplus \mathbf{\Phi}_k \circ \mathcal{F}'_k(\mathbf{F}'_{k-1}). \tag{4}$$

Let us rewrite the above process by a function $\mathcal{R}_k$: $\mathbf{F}_k = \mathcal{R}_k(\mathbf{F}_{k-1}, \mathbf{F}'_{k-1})$ and define $\mathbf{F}_0 = \mathbf{X}$ and $\mathbf{F}'_0 = \mathbf{X}'$, then we can iteratively derive the fused feature map at n-th convolutional encoder:

$$\mathbf{F}_n = \mathcal{R}_n\Big(\mathcal{R}_{n-1}\Big(\cdots(\mathcal{R}_1(\mathbf{F}_0, \mathbf{F}'_0), \mathbf{F}'_1), \cdots\Big), \mathbf{F}'_{n-1}\Big), \tag{5}$$

where $\mathbf{F}'_{n-1} = \mathcal{F}'_{n-1}(\mathcal{F}'_{n-2}(\cdots(\mathcal{F}'_1(\mathbf{F}'_0)))$. The final fused feature map $\mathbf{F}_n$ is fed to the decoder $\mathcal{G}$ to compute the segmentation result $\mathbf{P}$:

$$\mathbf{P} = \mathcal{M}(\mathbf{X}, \mathbf{X}') = \mathcal{G}(\mathbf{F}_n). \tag{6}$$

Alignment Ensemble. We ensemble the three proposed alignment variants by simple majority voting of the predictions. The goal of the ensemble are in two folds, where the first is to improve overall performance and the second is to see whether these three alignment methods are complementary. Usually, an ensemble of complementary approaches can lead to large improvements.

4 Experiments and Discussion

4.1 Dataset and Evaluation

We evaluate our approach on two PDAC datasets, proposed in [26] and [22] respectively. For the ease of presentation, we regard the former as PDAC dataset I and the latter as PDAC dataset II. PDAC dataset I contains 439 CT scans in total, in which 136 cases are diagnosed with PDAC and 303 cases are normal. Annotation contains voxel-wise labeled pancreas and PDAC mass. Evaluation is done by 4 fold cross-validation on these cases following [26]. PDAC dataset II contains 239 CT scans, all from PDAC patients, with pancreas, pancreatic duct (crucial for PDAC detection) and PDAC mass annotated. Evaluation are done by 3 fold cross-validation following [22].

All cases contain two phases: arterial phase and venous phase, with a spacing 0.5 mm in axial view and all annotations are verified by experienced board certified radiologists. The segmentation accuracy is evaluated using the Dice-Sørensen coefficient (DSC): DSC $(\mathcal{Y}, \mathcal{Z}) = \frac{2\times|\mathcal{Y}\cap\mathcal{Z}|}{|\mathcal{Y}|+|\mathcal{Z}|}$, which has a range of $[0, 1]$ with 1 implying a perfect prediction for each class. On dataset I, we also evaluate classification accuracy by sensitivity and specificity following a "segmentation for classification" strategy proposed in [26].

Table 1. Results on PDAC dataset I with both healthy and pathological cases. We compare our variants of alignment methods with the state-of-the-art method [26] as well as our baseline - no align (NA) version. "Misses" represents the number of cases failed in tumor detection. We also report healthy vs. pathological case classification (sensitivity and specificity) based on segmentation results. The last row is the ensemble of the three alignments.

Method	N.Pancreas	A.Pancreas	Tumor	Misses	Sens.	Spec.
U-Net [15]	86.9 ± 8.6	81.0 ± 10.8	57.3 ± 28.1	10/136	92.7	**99.0**
V-Net [12]	87.0 ± 8.4	81.6 ± 10.2	57.6 ± 27.8	11/136	91.9	99.0
MS C2F [26]	84.5 ± 11.1	78.6 ± 13.3	56.5 ± 27.2	8/136	94.1	98.5
Baseline - NA	85.8 ± 8.0	79.5 ± 11.2	58.4 ± 27.4	11/136	91.9	96.0
Ours - EA	86.7 ± 9.7	81.8 ± 10.0	60.9 ± 26.5	4/136	97.1	94.5
Ours - LA	87.5 ± 7.6	82.0 ± 10.3	62.0 ± 27.0	7/136	94.9	96.0
Ours - SA	87.0 ± 7.8	82.8 ± 9.4	60.4 ± 27.4	4/136	97.1	96.5
Ours - ensemble	**87.6 ± 7.8**	**83.3** ± 8.2	**64.4 ± 25.6**	**4/136**	**97.1**	96.0

4.2 Implementation Details

We implemented our network with PyTorch. The CT scans are first truncated within a range of HU value $[-100, 240]$ and normalized with zero mean and unit variance. In training stage, we randomly crop a patch size of 96^3 in roughly the same position from both arterial and venous phases. The optimization objective is Dice loss [12]. We use SGD optimizer with initial learning 0.005 and a cosine learning rate schedule for 40k iterations. For all our experiments, we implement the encoder and decoder architecture as U-Net [15] with 4 downsampling layers, making a total alignments of $n = 4$ in Eq. 6. The transformation fields are estimated by light-weighted U-Nets in late alignment and slow alignment, each is $\sim 8\times$ smaller than the large U-Net for segmentation, since the inputs of the small U-Nets are already the compact encoded features. The computation of EA/LA/SA is approximately 1.5/1.7/1.9 times of the computation of a single-phase U-Net. The image registration algorithm for our early alignment is DEEDS [7].

4.3 Results

Results on dataset I and II are summarized in Table 1 and Table 2 respectively, where our approach achieves the state-of-the-art performance on both datasets. Based on the results, we have three observations which leads to three findings.

Dual-Phase Alignments Are Beneficial for Detecting PDACs in Multi-phase CT Scans. On both datasets, our approaches, *i.e.* early alignment, late alignment and slow alignment, outperform single phase algorithms, i.e. U-Net [15], V-Net [12], ResDSN [25] and MS C2F [26], as well as our non-alignment dual-phase version (Baseline-NA).

Table 2. Results on PDAC dataset II with pathological cases only. We compare our variants of alignment methods with the state-of-the-art method [22]. "Misses" represents the number of cases failed in tumor detection. The last row is the ensemble of the three alignments.

Method	A.Pancreas	Tumor	Panc. duct	Misses
U-Net [15]	79.61 ± 10.47	53.08 ± 27.06	40.25 ± 27.89	11/239
ResDSN [25]	84.92 ± 7.70	56.86 ± 26.67	49.81 ± 26.23	11/239
HPN-U-Net [22]	82.45 ± 9.98	54.36 ± 26.34	43.27 ± 26.33	-/239
HPN-ResDSN [22]	85.79 ± 8.86	60.87 ± 24.95	54.18 ± 24.74	7/239
Ours - EA	83.65 ± 9.22	60.87 ± 22.15	55.38 ± 29.47	**5/239**
Ours - LA	86.82 ± 6.13	62.02 ± 24.53	64.35 ± 29.94	9/239
Ours - SA	87.13 ± 5.85	61.24 ± 24.26	64.19 ± 29.46	8/239
Ours - ensemble	$\mathbf{87.37 \pm 5.67}$	$\mathbf{64.14 \pm 21.16}$	$\mathbf{64.38 \pm 29.67}$	6/239

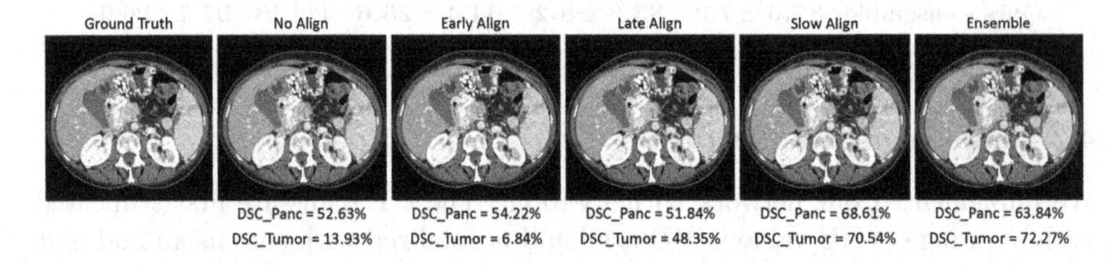

Fig. 3. An example of PDAC dataset I on venous phase. From left to right, we display ground-truth, prediction of our baseline without alignment, prediction of our early align, late align, slow align and alignment ensemble. Our feature space alignments (LA, SA) outperform no-align baseline and image registration (EA). Ensemble of the three alignment predictions also improves tumor segmentation DSC score.

Feature Space Alignments Have Larger Improvements on Segmentation Performances Than Early Alignments. Generally speaking for both datasets, our feature space alignment models (LA, SA) outperform image registration based approaches, i.e. HPN, Ours-EA, in terms of segmentation performance. Since early alignment methods apply image registration in advance, they do not guarantee a final improvement on segmentation performance. In contrast, feature space alignment methods jointly align and segment the targets in an end-to-end fashion by optimizing the final segmentation objective function, which leads to a larger improvements compared with single phase or naive dual phase methods without alignment. However, we indeed observe that early alignment leads to relatively less false negatives (misses).

An Ensemble of The Three Alignment Strategies Significantly Improve The Performances. For both dataset, Ours-Ensemble achieves the best performances, illustrating that the three alignment strategies are **complementary** to each other. An ensemble leads to significant performance gain (relatively 4% improvements on tumor segmentation DSC score compared to the best alignment model from 62.0% to 64.4%) and achieves the state-of-the-art performances on both datasets. A qualitative analysis is also shown in Fig. 3.

Last but not least, our alignment approaches also improve the sensitivity of healthy vs. pathological classification. In dataset I, we adopt the same "segmentation for classification" strategy as in [22], which classifies a case as pathological if we are able to detect any tumor mass larger than 50 voxels. Our approach can improve the overall sensitivity from 94.1% to 97.1% by reducing misses from 8 to 4, which is beneficial for the early detection of PDAC. Our approach thus has valuable potential of winning precious time for early treatments for patients.

5 Conclusion

In this paper, we study three types of alignment approaches for detecting pancreatic adenocarcinoma (PDACs) in multi-phase CT scans. Early alignment first applies registration in image space and then segment with a deep network. Late alignment and slow alignment jointly align and segment with an end-to-end deep network. The former aligns in the final encoded feature space while the latter aligns multi-stage features and propagate slowly. An ensemble of the three approaches improve the performances significantly illustrating these alignment variants are complementary to each other. We achieve the state-of-the-art performances on two PDAC datasets.

Acknowledgement. This work was supported by the Lustgarten Foundation for Pancreatic Cancer Research and also supported by NSFC No. 61672336.

References

1. Balakrishnan, G., Zhao, A., Sabuncu, M.R., Guttag, J., Dalca, A.V.: VoxelMorph: a learning framework for deformable medical image registration. IEEE Trans. Med. Imaging **38**(8), 1788–1800 (2019)
2. Cai, J., Lu, L., Zhang, Z., Xing, F., Yang, L., Yin, Q.: Pancreas segmentation in MRI using graph-based decision fusion on convolutional neural networks. In: Ourselin, S., Joskowicz, L., Sabuncu, M.R., Unal, G., Wells, W. (eds.) MICCAI 2016. LNCS, vol. 9901, pp. 442–450. Springer, Cham (2016). https://doi.org/10.1007/978-3-319-46723-8_51
3. Chu, L.C.: Utility of CT radiomics features in differentiation of pancreatic ductal adenocarcinoma from normal pancreatic tissue. Am. J. Roentgenol. **213**(2), 349–357 (2019)
4. Chu, L.C., et al.: Application of deep learning to pancreatic cancer detection: lessons learned from our initial experience. J. Am. Coll. Radiol. **16**(9), 1338–1342 (2019)
5. Dolz, J., Gopinath, K., Yuan, J., Lombaert, H., Desrosiers, C., Ayed, I.B.: Hyperdense-net: a hyper-densely connected cnn for multi-modal image segmentation. IEEE Trans. Med. Imaging **38**(5), 1116–1126 (2018)

6. Gaens, T., Maes, F., Vandermeulen, D., Suetens, P.: Non-rigid multimodal image registration using mutual information. In: Wells, W.M., Colchester, A., Delp, S. (eds.) MICCAI 1998. LNCS, vol. 1496, pp. 1099–1106. Springer, Heidelberg (1998). https://doi.org/10.1007/BFb0056299
7. Heinrich, M.P., Jenkinson, M., Brady, M., Schnabel, J.A.: MRF-based deformable registration and ventilation estimation of lung CT. IEEE Trans. Med. Imaging **32**(7), 1239–1248 (2013)
8. Jaderberg, M., Simonyan, K., Zisserman, A., et al.: Spatial transformer networks. In: Advances in Neural Information Processing Systems, pp. 2017–2025 (2015)
9. Lucas, A.L., Kastrinos, F.: Screening for pancreatic cancer. JAMA **322**(5), 407–408 (2019)
10. Man, Y., Huang, Y., Feng, J., Li, X., Wu, F.: Deep Q learning driven CT pancreas segmentation with geometry-aware U-Net. IEEE Trans. Med. Imaging **38**(8), 1971–1980 (2019)
11. Menze, B.H., et al.: The multimodal brain tumor image segmentation benchmark (BRATS). IEEE Trans. Med. Imaging **34**(10), 1993–2024 (2014)
12. Milletari, F., Navab, N., Ahmadi, S.A.: V-Net: fully convolutional neural networks for volumetric medical image segmentation. In: 2016 Fourth International Conference on 3D Vision (3DV), pp. 565–571. IEEE (2016)
13. Qin, C., Shi, B., Liao, R., Mansi, T., Rueckert, D., Kamen, A.: Unsupervised deformable registration for multi-modal images via disentangled representations. In: Chung, A.C.S., Gee, J.C., Yushkevich, P.A., Bao, S. (eds.) IPMI 2019. LNCS, vol. 11492, pp. 249–261. Springer, Cham (2019). https://doi.org/10.1007/978-3-030-20351-1_19
14. Roche, A., Malandain, G., Pennec, X., Ayache, N.: The correlation ratio as a new similarity measure for multimodal image registration. In: Wells, W.M., Colchester, A., Delp, S. (eds.) International Conference on Medical Image Computing and Computer-Assisted Intervention, vol. 1496, pp. 1115–1124. Springer, Heidelberg (1998). https://doi.org/10.1007/BFb0056301
15. Ronneberger, O., Fischer, P., Brox, T.: U-Net: convolutional networks for biomedical image segmentation. In: Navab, N., Hornegger, J., Wells, W.M., Frangi, A.F. (eds.) MICCAI 2015. LNCS, vol. 9351, pp. 234–241. Springer, Cham (2015). https://doi.org/10.1007/978-3-319-24574-4_28
16. Roth, H.R., et al.: DeepOrgan: multi-level deep convolutional networks for automated pancreas segmentation. In: Navab, N., Hornegger, J., Wells, W.M., Frangi, A.F. (eds.) MICCAI 2015. LNCS, vol. 9349, pp. 556–564. Springer, Cham (2015). https://doi.org/10.1007/978-3-319-24553-9_68
17. Roth, H.R., Lu, L., Farag, A., Sohn, A., Summers, R.M.: Spatial aggregation of holistically-nested networks for automated pancreas segmentation. In: Ourselin, S., Joskowicz, L., Sabuncu, M.R., Unal, G., Wells, W. (eds.) MICCAI 2016. LNCS, vol. 9901, pp. 451–459. Springer, Cham (2016). https://doi.org/10.1007/978-3-319-46723-8_52
18. Ryan, D.P., Hong, T.S., Bardeesy, N.: Pancreatic adenocarcinoma. N. Engl. J. Med. **371**(11), 1039–1049 (2014)
19. Vercauteren, T., Pennec, X., Perchant, A., Ayache, N.: Diffeomorphic demons: efficient non-parametric image registration. NeuroImage **45**(1), S61–S72 (2009)
20. Xia, Y., Xie, L., Liu, F., Zhu, Z., Fishman, E.K., Yuille, A.L.: Bridging the gap between 2D and 3D organ segmentation with volumetric fusion net. In: Frangi, A.F., Schnabel, J.A., Davatzikos, C., Alberola-López, C., Fichtinger, G. (eds.) MICCAI 2018. LNCS, vol. 11073, pp. 445–453. Springer, Cham (2018). https://doi.org/10.1007/978-3-030-00937-3_51

21. Yu, Q., Xie, L., Wang, Y., Zhou, Y., Fishman, E.K., Yuille, A.L.: Recurrent saliency transformation network: incorporating multi-stage visual cues for small organ segmentation. In: Proceedings of the IEEE Conference on Computer Vision and Pattern Recognition, pp. 8280–8289 (2018)
22. Zhou, Y., et al.: Hyper-pairing network for multi-phase pancreatic ductal adenocarcinoma segmentation. In: Shen, D., et al. (eds.) MICCAI 2019. LNCS, vol. 11765, pp. 155–163. Springer, Cham (2019). https://doi.org/10.1007/978-3-030-32245-8_18
23. Zhou, Y., Xie, L., Shen, W., Wang, Y., Fishman, E.K., Yuille, A.L.: A fixed-point model for pancreas segmentation in abdominal CT scans. In: Descoteaux, M., Maier-Hein, L., Franz, A., Jannin, P., Collins, D.L., Duchesne, S. (eds.) MICCAI 2017. LNCS, vol. 10433, pp. 693–701. Springer, Cham (2017). https://doi.org/10.1007/978-3-319-66182-7_79
24. Zhu, W., et al.: NeurReg: neural registration and its application to image segmentation. In: The IEEE Winter Conference on Applications of Computer Vision, pp. 3617–3626 (2020)
25. Zhu, Z., Xia, Y., Shen, W., Fishman, E., Yuille, A.: A 3D coarse-to-fine framework for volumetric medical image segmentation. In: 2018 International Conference on 3D Vision (3DV), pp. 682–690. IEEE (2018)
26. Zhu, Z., Xia, Y., Xie, L., Fishman, E.K., Yuille, A.L.: Multi-scale coarse-to-fine segmentation for screening pancreatic ductal adenocarcinoma. In: Shen, D., et al. (eds.) MICCAI 2019. LNCS, vol. 11769, pp. 3–12. Springer, Cham (2019). https://doi.org/10.1007/978-3-030-32226-7_1

Biomechanics-Informed Neural Networks for Myocardial Motion Tracking in MRI

Chen Qin[1,2(✉)], Shuo Wang[3], Chen Chen[1], Huaqi Qiu[1], Wenjia Bai[3,4], and Daniel Rueckert[1]

[1] Department of Computing, Imperial College London, London, UK
c.qin15@imperial.ac.uk
[2] Institute for Digital Communications, School of Engineering, University of Edinburgh, Edinburgh, UK
[3] Data Science Institute, Imperial College London, London, UK
[4] Department of Brain Sciences, Imperial College London, London, UK

Abstract. Image registration is an ill-posed inverse problem which often requires regularisation on the solution space. In contrast to most of the current approaches which impose explicit regularisation terms such as smoothness, in this paper we propose a novel method that can implicitly learn biomechanics-informed regularisation. Such an approach can incorporate application-specific prior knowledge into deep learning based registration. Particularly, the proposed biomechanics-informed regularisation leverages a variational autoencoder (VAE) to learn a manifold for biomechanically plausible deformations and to implicitly capture their underlying properties via reconstructing biomechanical simulations. The learnt VAE regulariser then can be coupled with any deep learning based registration network to regularise the solution space to be biomechanically plausible. The proposed method is validated in the context of myocardial motion tracking on 2D stacks of cardiac MRI data from two different datasets. The results show that it can achieve better performance against other competing methods in terms of motion tracking accuracy and has the ability to learn biomechanical properties such as incompressibility and strains. The method has also been shown to have better generalisability to unseen domains compared with commonly used L2 regularisation schemes.

1 Introduction

Medical image registration plays a crucial role in inferring spatial transformation of anatomical structures, and has been successfully used for various applications such as multi-modal image fusion, detection of longitudinal structural changes and analysis of motion patterns. Due to the intrinsic ill-posedness of the registration problem, there exist many possible solutions, i.e. spatial transformations,

Electronic supplementary material The online version of this chapter (https://doi.org/10.1007/978-3-030-59716-0_29) contains supplementary material, which is available to authorized users.

A. L. Martel et al. (Eds.): MICCAI 2020, LNCS 12263, pp. 296–306, 2020.
https://doi.org/10.1007/978-3-030-59716-0_29

to register between images. To ensure deformation to be unique and physiological plausible, regularisation techniques are often employed to convert the ill-posed problem to a well-posed one by adding a regularisation term with desired properties to the registration objective function. Specifically, image registration between a pair of images I_s, I_t can be formulated as a minimisation problem:

$$\underset{\Phi \in \mathcal{D}(\Omega)}{\operatorname{argmin}} \mathcal{L}_{sim}(I_t, I_s \circ \Phi) + \mathcal{R}(\Phi) \tag{1}$$

where $\mathcal{L}_{sim}$ stands for the image dissimilarity measure, Ω is the image domain, Φ denotes the transformation from source image I_s to target image I_t, $\mathcal{D}(\Omega)$ is the group of feasible transformations and $\mathcal{R}$ is a regularisation term. Conventionally, the regularisation is imposed explicitly on $\mathcal{R}$ or $\mathcal{D}(\Omega)$ with assumptions on the deformation field such as smoothness, diffeomorphism or incompressibility. Though these assumptions are generic, there is often a lack of incorporation of application-specific prior knowledge to inform the optimisation.

In this work, instead of imposing an explicit regularisation, we propose a biomechanics-informed neural network for image registration which can implicitly learn the behaviour of the regularisation function and incorporate prior knowledge about biomechanics into deep learning based registration. Specifically, the learning-based regularisation is formulated using a variational autoencoder (VAE) that aims to learn a manifold for biomechanical plausible deformations, and this is achieved via reconstructing biomechanically simulated deformations. The learnt regulariser is then coupled with a deep learning based image registration, which enables the parameterised registration function to be regularised by the application-specific prior knowledge, so as to produce biomechanically plausible deformations. The method is evaluated in the context of myocardial motion tracking in 2D stacks of cardiac MRI data. We show that the proposed method can achieve better performance compared to other competing approaches. It also indicates great potential in uncovering properties that are represented by biomechanics, with a more realistic range of clinical parameters such as radial and circumferential strains, as well as better generalisation ability.

Related Work. Deformable image registration has been widely studied in the field of medical image analysis. The majority of the methods explicitly assume the smoothness on the deformation, such as introducing a smoothness penalty in the form of L2 norm or bending energy to penalise the first- or second-order derivatives of the deformation [27], or parameterising the transformation via spline-based models in a relatively low dimensional representation space [25]. To take into account the invertibility of the transformation, many diffeomorphic registration methods have been proposed to parameterise the displacement field as the integral of a time-varying velocity field or stationary velocity field (SVF) [1,4], which mathematically guarantees diffeomorphism. Volumetric preservation is also a characteristic that is often desired for soft tissue tracking such as myocardium tracking to ensure that the total volume of myocardium keeps constant during image registration [12,18,26]. Such incompressible registration methods either relax $\det(J_\Phi(x)) = 1$ as a soft constraint [26] (where J_Φ is the

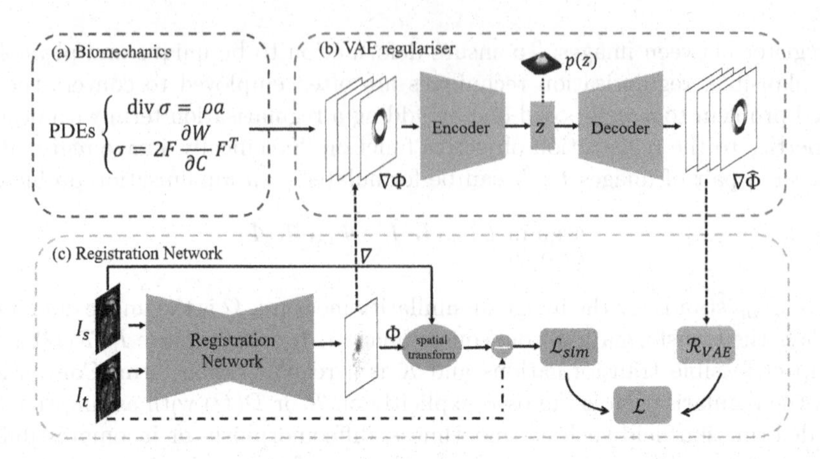

Fig. 1. Illustration of biomechanics-informed neural network for image registration: (a) Biomechanical simulations are generated according to partial differential equations (PDEs). (b) A VAE regulariser is learnt via reconstructing the first-order gradients of simulated displacement fields. (c) The registration network is regularised by biomechanics via incorporating the VAE regularisation term.

Jacobian matrix of the transformation Φ), or use specific parameterisation for the displacement field or SVF [12,18]. In addition, population statistics have also been investigated to incorporate prior knowledge for the registration process [15,24].

More recently, deep learning based image registration has been shown to significantly improve the computational efficiency. Similarly, most deep learning approaches regularise deformation fields with a smoothness penalty [10,20,21], or parameterise transformations with time-varying or stationary velocity fields [3,7,17,22]. Besides, population-based regularisation has been designed to inform registration with population-level statistics of the transformations [6]. Adversarial deformation regularisation was also proposed which penalised the divergence between the predicted deformation and the finite-element simulated data [13]. In contrast to the existing deep learning based registration work, our method proposes to implicitly learn the regularisation function via a biomechanics-informed VAE. This aims to capture deformation properties that are biomechanically plausible, with a particular focus on the application of myocardial motion tracking.

2 Method

The proposed biomechanics-regularised registration network is illustrated in Fig.1, which mainly consists of three components. First, biomechanical simulations of deformation are generated according to equations which guarantee the physical properties of the deformation (Sect. 2.1). Second, VAE is leveraged to learn the probability distribution of the simulated deformations, which implicitly captures the underlying biomechanical properties (Sect. 2.2). Finally, the learnt

VAE then acts as a regularisation function for the registration network, which regularises the solution space and helps to predict biomechanically plausible deformation without the need for any explicit penalty term (Sect. 2.3).

2.1 Biomechanical Simulations

To enable a learning-based biomechanically informed regulariser, we first propose to generate a set of deformation fields using a biomechanical simulation. From the perspective of solid mechanics, the modelling of myocardial motion can be considered as a plane-strain problem with the following governing equation [14]:

$$\text{div } \sigma = \rho a \tag{2}$$

Here σ is the stress tensor, div σ denotes the divergence of the stress tensor, ρ is the material density and a is the acceleration. This describes the equilibrium within an elastic body. To solve the deformation, the constitutive law is provided, assuming a hyper-elastic material behaviour:

$$\sigma = 2F\frac{\partial W}{\partial C}F^T \tag{3}$$

where F is the deformation gradient, C is the right Cauchy–Green deformation tensor and W stands for the strain energy density function (SEDF). We adopt a neo-Hookean constitutive model with the shear modulus of 36.75 kPa from experiments [29], which guarantees the material incompressibility. The above equations can be solved using the finite element method (FEM) with appropriate boundary conditions. For a given myocardial segmentation at the end-systolic (ES) frame, we simulate the artificial 2D deformation in a one cardiac cycle by adjusting the blood pressure (Supple. Fig. S1).

2.2 VAE Regularisation

Instead of explicitly specifying a regularisation term that is suitable for the ill-posed registration problem, here we propose to implicitly learn the prior knowledge from biomechanics as regularisation. In particular, the VAE formulation [16] is leveraged here as a learning-based regulariser to model the probability distribution of biomechanically plausible deformations. The VAE encodes an input and returns a distribution over the latent space, which enables the inverse problem to move through the latent space continuously to determine a solution.

In detail, the VAE regulariser is trained to reconstruct the first-order derivative of the biomechanically simulated deformations to remove any effects of rigid translation. Let us denote the deformation in 2D space by $\Phi = [u, v] \in \mathrm{R}^{2\times M\times N}$, where u and v denote the displacements along x- and y- directions respectively. The first-order gradients of the deformation field can be represented as $\nabla\Phi = \left[\frac{\delta u}{\delta x}, \frac{\delta u}{\delta y}, \frac{\delta v}{\delta x}, \frac{\delta v}{\delta y}\right] \in \mathrm{R}^{4\times M\times N}$, where M and N denote the spatial dimensions. The reconstruction loss of the VAE is formulated as,

$$\mathcal{R}_{\text{VAE}}(\nabla\Phi) = \|\nabla\Phi - \nabla\hat{\Phi}\|_2^2 + \beta \cdot D_{KL}(q_\theta(\mathbf{z}|\nabla\Phi)\|p(\mathbf{z})), \tag{4}$$

where $\nabla\hat{\Phi}$ denotes the reconstruction from input $\nabla\Phi$, $\mathbf{z}$ denotes the latent vector encoded by VAE, $p(\mathbf{z}) \sim \mathcal{N}(0, I)$ denotes the prior Gaussian distribution, q_θ denotes the encoder parameterised by θ and D_{KL} represents the Kullback-Leibler divergence. β is a hyperparameter that controls the trade-off between the reconstruction quality and the extent of latent space regularity.

2.3 Biomechanics-Informed Registration Network

As shown in Fig. 1(c), the registration network aims to learn a parameterised registration function to estimate dense deformation fields between a source image I_s and a target image I_t. As the ground truth dense correspondences between images are not available, the model thus learns to track the spatial features through time, relying on the temporal intensity changes as self-supervision. The image dissimilarity measure thus can be defined as $\mathcal{L}_{sim}(I_t, I_s, \Phi) = \|I_t - I_s \circ \Phi\|_2^2$, which is to minimise the pixel-wise mean squared error between the target image and the transformed source image.

To inform the registration process with prior knowledge of biomechanics, we propose to integrate the VAE regularisation into the registration network. This aims to regularise the viable solutions to fall on the manifold that is represented by the latent space of VAE, which thereby enables the network to predict biomechanically plausible deformations. This is advantageous over the explicit regularisation terms (e.g. smoothness penalty), as the data-driven regulariser implicitly captures realistic and complex properties from observational data that cannot be explicitly specified. In addition, the VAE loss $\mathcal{R}_{\text{VAE}}$ provides a quantitative metric for determining how biomechanically plausible the deformation is. Solutions close to the learnt VAE latent manifold will produce a low $\mathcal{R}_{\text{VAE}}$ score, whereas solutions far from the manifold will give a higher score. The final objective function for training the registration network is then formulated as:

$$\mathcal{L} = \mathcal{L}_{sim}(I_t, I_s \circ \Phi) + \alpha \cdot \mathcal{R}_{\text{VAE}}(\nabla\Phi \odot \mathbf{M}). \quad (5)$$

Here $\mathbf{M}$ is a binary myocardial segmentation mask which is provided to only regularise the region of interest, and $\odot$ represents the Hadamard product. A hyperparameter α is introduced here, which trades-off image similarity and the physical plausibility of the deformation.

3 Experiments and Results

Experiments were performed on 300 short-axis cardiac cine magnetic resonance image (MRI) sequences from the publicly available UK Biobank (UKBB) dataset [19]. Each scan consists of 50 frames and each frame forms a 2D stack of typically 10 image slices. Segmentation masks for the myocardium were obtained via an automated tool provided in [2]. In experiments, we randomly split the dataset into training/validation/testing with 100/50/150 subjects respectively. Image intensity was normalised between 0 and 1. Data used for biomechanical

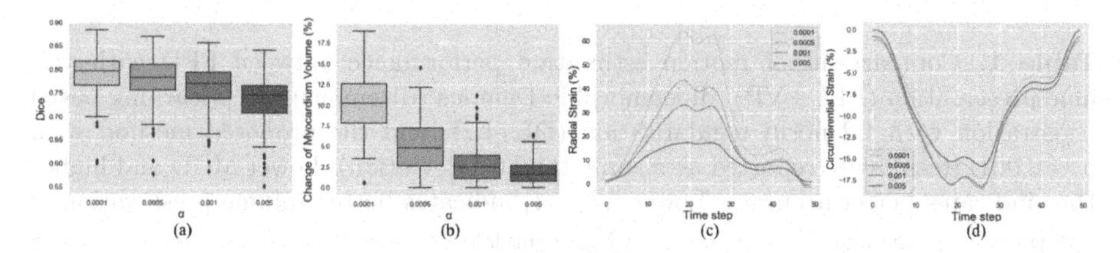

Fig. 2. Study on effects of VAE regularisation. By varying α, we show the corresponding effects on (a) Dice coefficient and (b) Change of myocardial volume between ES and ED frames; (c) Radial strain and (d) Circumferential strain across the cardiac cycle.

simulations were from a separate data set in UKBB consisting of 200 subjects. The detailed VAE architecture is described in the Supple. Fig. S2 (the latent vector dimension was set to 32, and $\beta = 0.0001$). The network architecture for registration was adopted from [20] due to its proven effectiveness in motion tracking. A learning rate of 0.0001 with Adam optimiser was employed to optimise both the VAE and the registration network.

Effect of VAE Regularisation. An ablation study to investigate the effect of α (the weight for VAE regularisation) was performed. Here α is set to values of $\{0.0001, 0.0005, 0.001, 0.005\}$, and the performance was evaluated using three quantitative measures, i.e. Dice overlap metric, change of myocardial volume and myocardial strain, as shown in Fig. 2. In detail, for results in Fig. 2(a)(b), motion was estimated between ES and end-diastolic (ED) frames of the cardiac image sequence, where ES frame is warped to ED frame with the estimated transformation. It can be observed that as α increases, the change for myocardial volume is reduced which means better volume preservation, and the observed radial and circumferential strains (Fig. 2(c)(d)) are more plausible and within the range that has been reported in the literature [11]. However, this is at the expense of the decrease of registration accuracy in terms of the Dice metric. In addition, since the motion estimation was performed in stack of 2D slices, the incompressibility of the myocardium can not be guaranteed due to the out-of-plane motion. Nevertheless, results here indicate that the proposed method has the potential and ability in implicitly capturing such biomechanical properties.

Comparison Study. The proposed registration network was then compared with other state-of-the-art motion estimation methods, including free-form deformation (FFD) with volumetric preservation (VP) [23], diffeomorphic Demons (dDemons) [28], and a deep learning based registration which uses the L2 norm to penalise displacement gradients (DL+L2) [20]. A quantitative comparison results are shown in Table 1, where the performance was evaluated in terms of the mean contour distance (MCD), Dice, and the mean absolute difference between Jacobian determinant $|J|$ and 1 over the whole myocardium, denoted as $||J|-1|$, to measure the level of volume preservation. In experiments, we separate these evaluations on different representative slices, i.e., apical, mid-ventricle, and basal slices. For fair comparison, hyperparameters in all these methods under

Table 1. Comparisons of motion estimation performance between FFD with volume preservation (FFD+VP), diffeomorphic Demons (dDemons), deep learning based registration with L2 norm regularisation (DL+L2), and the proposed method with $\alpha = 0.001$. Results are reported as mean (standard deviation). Lower MCD and higher Dice indicates better accuracy. Lower $||J| - 1|$ indicates better volume preservation.

Method	Apical			Mid-ventricle			Basal		
	MCD	Dice	$\|\|J\| - 1\|$	MCD	Dice	$\|\|J\| - 1\|$	MCD	Dice	$\|\|J\| - 1\|$
FFD+VP [23]	2.766	0.598	0.207	2.220	0.725	0.186	3.958	0.550	0.176
	(1.240)	(0.186)	(0.045)	(0.750)	(0.077)	(0.033)	(1.925)	(0.199)	(0.028)
dDemons [28]	2.178	0.651	0.133	1.603	0.784	0.148	3.568	0.560	0.132
	(1.079)	(0.186)	(0.044)	(0.629)	(0.072)	(0.036)	(1.482)	(0.176)	(0.029)
DL+L2 [20]	2.177	0.639	**0.087**	1.442	**0.806**	0.104	2.912	0.619	0.149
	(1.119)	(0.196)	(0.028)	(0.625)	(0.063)	(0.042)	(1.384)	(0.178)	0.047
Proposed	**1.734**	**0.684**	0.126	**1.417**	0.796	**0.081**	**2.280**	**0.725**	**0.124**
$\alpha = 0.001$	(0.808)	(0.160)	(0.071)	(0.402)	(0.059)	(0.020)	(1.274)	(0.149)	(0.037)

comparison are selected to ensure that they can achieve an average of $||J|-1|$ to be in a range around 0.1 or slightly higher, and α in the proposed method is thus chosen as 0.001. From Table 1, it can be observed that the proposed method can achieve better or comparable performance on all these slices in terms of both registration accuracy (MCD and Dice) and volume preservation ($||J|-1|$) compared with other methods. The performance gain is especially significant on more challenging slices such as basal slices (MCD: 2.280 (proposed) vs 2.912 (DL+L2); Dice: 0.725 vs 0.619), where prior knowledge of physically plausible deformation can have more impact on informing the alignment.

In addition, we further evaluated the performance of the proposed method on myocardial strain estimation, as shown in Table 2. The peak strain was calculated between ES and ED frames on cine MRI using the Lagrangian strain tensor formula [9], and was evaluated on apical, mid-ventricle and basal slices separately. To better understand the strains, we compared the predictions from cine MRI with reference values obtained from tagged MRI [11], which is regarded as reference modality for cardiac strain estimation. The reference values were taken from Table 2 in [11] whose results were derived also from UKBB population (but on a different subset of subjects). Compared with other methods, the proposed approach can achieve radial and circumferential strain in a more reasonable value range as defined by the reference. Particularly, basal slice strain and circumferential strain are very consistent with the reference values reported. Note that evaluation of radial strain is challenging, and there is a poor agreement in it even for commercial software packages [8]. In this case, incorporating biomechanical prior knowledge may benefit motion tracking and lead to more reasonable results. Besides, the Jacobian determinant of deformation across the entire cardiac cycle on one subject is presented in Fig. 3. Here the proposed biomechanics-informed method is compared with the one using L2 norm, and it

Table 2. Comparisons of peak strain values (%) obtained from different methods at different slices. Reference values are derived from tagged MRI based on a similar UK Biobank cohort [11]. RR: peak radial strain; CC: peak circumferential strain. Bold results indicate results that are closest to the reference values.

Method	FFD+VP [23]		dDemons [28]		DL+L2 [20]		Proposed		Reference [11]	
	RR	CC	RR	CC	RR	CC	RR	CC	RR	CC
Apical	14.9	−7.4	30.0	−14.2	**26.5**	−14.1	32.5	**−19.7**	*18.7*	*−20.8*
Mid-ventricle	12.8	−7.82	37.6	−12.7	35.9	−14.0	**33.1**	**−18.1**	*23.1*	*−19.6*
Basal	12.4	−5.9	22.8	−10.9	31.6	−13.0	**24.5**	**−16.4**	*23.8*	*−16.7*

shows better preservation of the tissue volume at different cardiac phases, with Jacobian determinant being close to 1. This implies that the proposed learning based regularisation informed by biomechanics is able to capture and represent biomechanically realistic motions. A dynamic video showing the qualitative visualisations is also presented in supplementary material.

Generalisation Study. We further performed a generalisation study of the proposed method against the L2 norm regularised network. Specifically, we deployed models trained on the UKBB dataset and directly tested them on 100 ACDC data [5]. The tested results are shown in Table 3, where MCD and Dice were compared. Though both methods achieved comparable performance on mid-ventricle slices on UKBB data (Table 1), here the proposed method generalised better on data in unseen domains, with much higher registration accuracy (Table 3). This is likely due to the benefit of the biomechanical regularisation, which enforces the generated deformations to be biomechanically plausible and less sensitive to the domain shift problem.

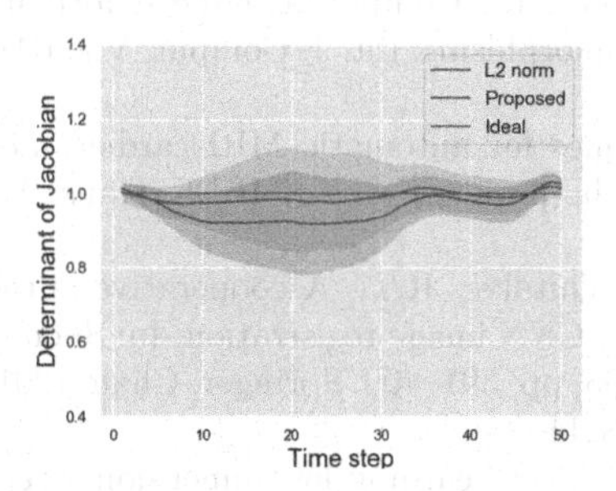

Fig. 3. Determinant of Jacobian across cardiac cycle.

Table 3. Model generalisation performance on ACDC dataset.

Method	DL+L2		Proposed	
	MCD	Dice	MCD	Dice
Apical	2.853	0.656	**2.450**	**0.707**
Mid	2.799	0.745	**2.210**	**0.783**
Basal	2.814	0.751	**2.229**	**0.789**

4 Conclusion

In this paper, we have presented a novel biomechanics-informed neural network for image registration with an application to myocardial motion tracking in cardiac cine MRI. A VAE-based regulariser is proposed to exploit the biomechanical prior knowledge and to learn a manifold for biomechanically simulated deformations. Based on that, the biomechanics-informed network can be established by incorporating the learnt VAE regularisation to inform the registration process in generating plausible deformations. Experimental results have shown that the proposed method outperforms the other competing approaches that use conventional regularisation, achieving better motion tracking accuracy with more reasonable myocardial motion and strain estimates, as well as gaining better generalisability. For future work, we will extend the method for 3D motion tracking and investigate on incorporating physiology modelling.

Acknowledgement. This work was supported by EPSRC programme grant SmartHeart (EP/P001009/1). This research has been conducted mainly using the UK Biobank Resource under Application Number 40119. The authors wish to thank all UK Biobank participants and staff.

References

1. Ashburner, J.: A fast diffeomorphic image registration algorithm. NeuroImage **38**(1), 95–113 (2007)
2. Bai, W., et al.: Automated cardiovascular magnetic resonance image analysis with fully convolutional networks. J. Cardiovasc. Magn. Reson. **20**(1), 65 (2018)
3. Balakrishnan, G., Zhao, A., Sabuncu, M.R., Guttag, J., Dalca, A.V.: VoxelMorph: a learning framework for deformable medical image registration. IEEE Trans. Med. Imaging **38**(8), 1788–1800 (2019)
4. Beg, M.F., Miller, M.I., Trouvé, A., Younes, L.: Computing large deformation metric mappings via geodesic flows of diffeomorphisms. Int. J. Comput. Vis. **61**(2), 139–157 (2005)
5. Bernard, O., et al.: Deep learning techniques for automatic MRI cardiac multi-structures segmentation and diagnosis: is the problem solved? IEEE Trans. Med. Imaging **37**(11), 2514–2525 (2018)
6. Bhalodia, R., Elhabian, S.Y., Kavan, L., Whitaker, R.T.: A cooperative autoencoder for population-based regularization of CNN image registration. In: Shen, D., et al. (eds.) MICCAI 2019. LNCS, vol. 11765, pp. 391–400. Springer, Cham (2019). https://doi.org/10.1007/978-3-030-32245-8_44
7. Bône, A., Louis, M., Colliot, O., Durrleman, S.: Learning low-dimensional representations of shape data sets with diffeomorphic autoencoders. In: Chung, A.C.S., Gee, J.C., Yushkevich, P.A., Bao, S. (eds.) IPMI 2019. LNCS, vol. 11492, pp. 195–207. Springer, Cham (2019). https://doi.org/10.1007/978-3-030-20351-1_15
8. Cao, J.J., Ngai, N., Duncanson, L., Cheng, J., Gliganic, K., Chen, Q.: A comparison of both dense and feature tracking techniques with tagging for the cardiovascular magnetic resonance assessment of myocardial strain. J. Cardiovasc. Magn. Reson. **20**(1), 26 (2018)

9. Elen, A., et al.: Three-dimensional cardiac strain estimation using spatio-temporal elastic registration of ultrasound images: a feasibility study. IEEE Trans. Med. Imaging **27**(11), 1580–1591 (2008)
10. Fan, J., Cao, X., Xue, Z., Yap, P.-T., Shen, D.: Adversarial similarity network for evaluating image alignment in deep learning based registration. In: Frangi, A.F., Schnabel, J.A., Davatzikos, C., Alberola-López, C., Fichtinger, G. (eds.) MICCAI 2018. LNCS, vol. 11070, pp. 739–746. Springer, Cham (2018). https://doi.org/10.1007/978-3-030-00928-1_83
11. Ferdian, E., et al.: Fully automated myocardial strain estimation from cardiovascular MRI-tagged images using a deep learning framework in the UK Biobank. Radiol.: Cardiothorac. Imaging **2**(1), e190032 (2020)
12. Fidon, L., Ebner, M., Garcia-Peraza-Herrera, L.C., Modat, M., Ourselin, S., Vercauteren, T.: Incompressible image registration using divergence-conforming B-splines. In: Shen, D., et al. (eds.) MICCAI 2019. LNCS, vol. 11765, pp. 438–446. Springer, Cham (2019). https://doi.org/10.1007/978-3-030-32245-8_49
13. Hu, Y., et al.: Adversarial deformation regularization for training image registration neural networks. In: Frangi, A.F., Schnabel, J.A., Davatzikos, C., Alberola-López, C., Fichtinger, G. (eds.) MICCAI 2018. LNCS, vol. 11070, pp. 774–782. Springer, Cham (2018). https://doi.org/10.1007/978-3-030-00928-1_87
14. Hunter, P.J., Smaill, B.H.: The analysis of cardiac function: a continuum approach. Prog. Biophys. Mol. Biol. **52**(2), 101–164 (1988)
15. Khallaghi, S., et al.: Statistical biomechanical surface registration: application to MR-TRUS fusion for prostate interventions. IEEE Trans. Med. Imaging **34**(12), 2535–2549 (2015)
16. Kingma, D.P., Welling, M.: Auto-encoding variational bayes. In: International Conference on Learning Representations (2014)
17. Krebs, J., Delingette, H., Mailhé, B., Ayache, N., Mansi, T.: Learning a probabilistic model for diffeomorphic registration. IEEE Trans. Med. Imaging **38**(9), 2165–2176 (2019)
18. Mansi, T., Pennec, X., Sermesant, M., Delingette, H., Ayache, N.: iLogDemons: a demons-based registration algorithm for tracking incompressible elastic biological tissues. Int. J. Comput. Vision **92**(1), 92–111 (2011)
19. Petersen, S.E., et al.: Reference ranges for cardiac structure and function using cardiovascular magnetic resonance (CMR) in Caucasians from the UK Biobank population cohort. J. Cardiovasc. Magn. Reson. **19**(1), 18 (2017)
20. Qin, C., et al.: Joint learning of motion estimation and segmentation for cardiac MR image sequences. In: Frangi, A.F., Schnabel, J.A., Davatzikos, C., Alberola-López, C., Fichtinger, G. (eds.) MICCAI 2018. LNCS, vol. 11071, pp. 472–480. Springer, Cham (2018). https://doi.org/10.1007/978-3-030-00934-2_53
21. Qin, C., et al.: Joint motion estimation and segmentation from undersampled cardiac MR image. In: Knoll, F., Maier, A., Rueckert, D. (eds.) MLMIR 2018. LNCS, vol. 11074, pp. 55–63. Springer, Cham (2018). https://doi.org/10.1007/978-3-030-00129-2_7

22. Qin, C., Shi, B., Liao, R., Mansi, T., Rueckert, D., Kamen, A.: Unsupervised deformable registration for multi-modal images via disentangled representations. In: Chung, A.C.S., Gee, J.C., Yushkevich, P.A., Bao, S. (eds.) IPMI 2019. LNCS, vol. 11492, pp. 249–261. Springer, Cham (2019). https://doi.org/10.1007/978-3-030-20351-1_19
23. Rohlfing, T., Maurer, C.R.: Intensity-based non-rigid registration using adaptive multilevel free-form deformation with an incompressibility constraint. In: Niessen, W.J., Viergever, M.A. (eds.) MICCAI 2001. LNCS, vol. 2208, pp. 111–119. Springer, Heidelberg (2001). https://doi.org/10.1007/3-540-45468-3_14
24. Rueckert, D., Frangi, A.F., Schnabel, J.A.: Automatic construction of 3-D statistical deformation models of the brain using nonrigid registration. IEEE Trans. Med. Imaging **22**(8), 1014–1025 (2003)
25. Rueckert, D., Sonoda, L.I., Hayes, C., et al.: Nonrigid registration using free-form deformations: application to breast MR images. IEEE Trans. Med. Imaging **18**(8), 712–721 (1999)
26. Shi, W., et al.: A comprehensive cardiac motion estimation framework using both untagged and 3-D tagged MR images based on nonrigid registration. IEEE Trans. Med. Imaging **31**(6), 1263–1275 (2012)
27. Sotiras, A., Davatzikos, C., Paragios, N.: Deformable medical image registration: a survey. IEEE Trans. Med. Imaging **32**(7), 1153–1190 (2013)
28. Vercauteren, T., Pennec, X., Perchant, A., Ayache, N.: Non-parametric diffeomorphic image registration with the demons algorithm. In: Ayache, N., Ourselin, S., Maeder, A. (eds.) MICCAI 2007. LNCS, vol. 4792, pp. 319–326. Springer, Heidelberg (2007). https://doi.org/10.1007/978-3-540-75759-7_39
29. Zhu, Y., Luo, X., Gao, H., McComb, C., Berry, C.: A numerical study of a heart phantom model. Int. J. Comput. Math. **91**(7), 1535–1551 (2014)

Fluid Registration Between Lung CT and Stationary Chest Tomosynthesis Images

Lin Tian[1(✉)], Connor Puett[2], Peirong Liu[1], Zhengyang Shen[1], Stephen R. Aylward[3], Yueh Z. Lee[2], and Marc Niethammer[1]

[1] Department of Computer Science, University of North Carolina at Chapel Hill, Chapel Hill, USA
lintian@cs.unc.edu

[2] Department of Radiology, University of North Carolina at Chapel Hill, Chapel Hill, USA

[3] Kitware, Inc., Clifton Park, USA

Abstract. Registration is widely used in image-guided therapy and image-guided surgery to estimate spatial correspondences between organs of interest between planning and treatment images. However, while high-quality computed tomography (CT) images are often available at planning time, limited angle acquisitions are frequently used during treatment because of radiation concerns or imaging time constraints. This requires algorithms to register CT images based on limited angle acquisitions. We, therefore, formulate a 3D/2D registration approach which infers a 3D deformation based on measured projections and digitally reconstructed radiographs of the CT. Most 3D/2D registration approaches use simple transformation models or require complex mathematical derivations to formulate the underlying optimization problem. Instead, our approach entirely relies on differentiable operations which can be combined with modern computational toolboxes supporting automatic differentiation. This then allows for rapid prototyping, integration with deep neural networks, and to support a variety of transformation models including fluid flow models. We demonstrate our approach for the registration between CT and stationary chest tomosynthesis (sDCT) images and show how it naturally leads to an iterative image reconstruction approach.

1 Introduction

Image registration is an enabling technology for image-guided interventional procedures (IGP) [7], as it can determine spatial correspondences between pre- and intra-intervention images. Conventionally, pre-interventional images are 3D CT or magnetic resonance (MR) images. Intra-interventional images are typically 2D projective X-rays or ultrasound images [11]. While fast, cheap, and easy to use, these intra-interventional images lack anatomical features and spatial information. It is therefore desirable to combine high-quality 3D pre-intervention

A. L. Martel et al. (Eds.): MICCAI 2020, LNCS 12263, pp. 307–317, 2020.
https://doi.org/10.1007/978-3-030-59716-0_30

images with sDCT [18]. In sDCT, a set of projections is acquired from spatially distributed carbon nanotube-enabled X-ray sources within a limited scanning angle. Given the stationary design, the projections can be acquired fast with high resolution and provide more spatial information than individual X-ray projections. However, this immediately raises the question of how to register a 3D CT to an sDCT volume or its 2D projections.

This may be accomplished by two approaches: (1) reconstructing a 3D image from the 2D projections and proceeding with standard 3D to 3D registration; (2) computing image similarities in the projection space (i.e., by also computing projections from the 3D volume) from which a 3D transformation is inferred. While conceptually simple, the first approach may result in dramatically different 3D image appearances as 3D reconstructions from a small set of projections with a limited scanning angle range will appear blurry and will smear anatomical structures (e.g., the ribs across the lung). Similar appearance differences have been addressed in [21] for rigid transformations via an information-theoretic similarity measure. The second approach is conceptually more attractive in a limited angle setting, as it *simulates* the limited angle acquisitions from the high-quality 3D image. Hence, sets of projection images with *similar* appearance can be compared. Such an approach is called 3D/2D registration. Different 3D/2D registration approaches have been proposed [11], but most only address rigid or affine transformations [1,4,8,9,21]. Less work considers more flexible transformations or even non-parametric models for 3D/2D registration [3,16]. A particular challenge is the efficient computation of gradients with respect to the transformation parameters. Zikic et al. [25] establish correspondences between vessels in 3D and a single 2D projection image using a diffusion registration model [13] by exploiting vessel geometry. Prümmer et al. [15,16] use curvature registration [13] and a similarity measure in projection space based on algebraic image reconstruction. The work by Flach et al. [3] relates most closely to our approach. It is based on Demons registration and digitally reconstructed radiographs (DRR) to compute image similarity in projection space. They propose a sophisticated optimization approach relying on explicit gradient computations. However, gradient computations quickly become unwieldy for complex transformation models which might explain the dearth of 3D/2D models for non-parametric registration. This complexity also hampers flexible modeling and rapid prototyping. Further, while some of the approaches above consider limited numbers of projections, they do not address limited angle acquisitions.

Our starting point for 3D/2D registration is a DRR model for the stationary design of sDCT, which we use to compute image similarity in projection space. In contrast to previous approaches, all our components (the registration model, the similarity measure, and the DRR computation) support automatic-differentiation [5,14]. This allows focusing our formulation on the forward model and supports the flexible use of our DRR component. Specifically, within one framework we can support a variety of parametric and non-parametric registration models and can also formulate an iterative image reconstruction approach.

Contributions: 1) We propose a differentiable ray-cast DRR generator consistent with sDCT; 2) We propose a 3D/2D registration framework based on these DRR projections which support automatic differentiation. This allows us to easily integrate various registration models (including non-parametric ones) without the need for complex manual gradient calculations; 3) We show how our approach allows for a straightforward implementation of a limited angle image reconstruction algorithm; 4) We demonstrate the performance of our approach for CT to simulated sDCT registration and explore its performance when varying the number of projections and the range of scanning angles.

2 Ray-Cast DRR Generator

Given a high-quality volumetric image $I_0(x)$, $x \in R^3$ (e.g., based on CT) and one or a set of projection images $I_1(y)$, $y \in R^2$, we generate DRRs by shooting virtual rays through the 3D volume to a virtual receiver plane P_r of size $W \times H$ (Fig. 1). Assume we are in the coordinate system of the receiver plane whose center is placed and will stay at the origin $(0, 0, 0)$ for all the emitters according to the stationary design of sDCT. The position of points in P_r can then be represented as $p_r = \{(w, h, 0)^T | \frac{-W}{2} \leq w \leq \frac{W}{2}, \frac{-H}{2} \leq h \leq \frac{H}{2}\}$. Given an emitter at position $e \in R^3$ and a point x on the ray from e to p_r, we represent the ray and DRR of this geometry respectively as

$$x = e + \lambda \hat{\boldsymbol{r}}, \quad \hat{\boldsymbol{r}} = \boldsymbol{r}/\|\boldsymbol{r}\|_2, \quad \boldsymbol{r} = p_r - e, \quad I_{proj} = \int I_0(x)\, dx. \tag{1}$$

We discretize the integral along the z axis of the receiver plane (Fig. 1) and obtain planes $P = z$ parallel to the receiver plane. Given the point $x_z = (0, 0, z)^T$ on the plane with the plane normal $\boldsymbol{n} = (0, 0, 1)^T$, any point x on the plane $P = z$ can be written as

$$(x - x_z) \cdot \boldsymbol{n} = 0. \tag{2}$$

Combining Eqs. 1 and 2, we obtain the intersection between the ray and the plane $P = z$:

$$x = e + \frac{(x_z - e) \cdot \boldsymbol{n}}{\boldsymbol{r} \cdot \boldsymbol{n}} \hat{\boldsymbol{r}}, \tag{3}$$

The DRR (I_{proj}) can then be written as

$$I_{proj}(w, h, e) = \sum_{z=0}^{Z} I_0(e + \frac{(x_z - e) \cdot \boldsymbol{n}}{\boldsymbol{r} \cdot \boldsymbol{n}} \hat{\boldsymbol{r}})\, dz, \ \boldsymbol{r} = (w, h, 0)^T - e, \ dz = ||\frac{\boldsymbol{r}}{\boldsymbol{r} \cdot \boldsymbol{n}}||_2. \tag{4}$$

It is possible to write Eq. 4 in tensor form where all rays (for all receiver plane coordinates (w, h)) are evaluated at once. This allows efficient computation on a GPU and only involves tensor products, interpolations, and sums. Hence, we can also efficiently implement derivatives of a DRR with respect to I_0 using automatic differentiation. This allows easy use of DRRs in general loss functions.

3 Registration Model

Our goal is to develop a generic 3D/2D registration approach that can be combined with any transformation model. In general, image registration is formulated as an energy minimization problem of the form [13]

$$\theta^* = \arg\min_{\theta} \operatorname{Reg}(\theta) + \lambda \operatorname{Sim}(I_0 \circ \Phi_\theta^{-1}, I_1),\ \lambda > 0, \tag{5}$$

where the transformation Φ_θ^{-1} is parameterized by θ, λ is a weighting factor, $\operatorname{Reg}(\theta)$ is a regularizer on the transformation parameters, $\operatorname{Sim}(A, B)$ is a similarity measure (e.g., sum of squared differences, normalized cross correlation, or mutual information) between two images A, B of the same dimension, and I_0, I_1 are the source and target images respectively. Since we are interested in 3D/2D registration, we instead have a 3D source image I_0 and a set of projection target images $\{I_1^i\}$. The similarity measure compares the DRRs of the 3D source image to their corresponding projection images. This changes Eq. 5 to

$$\theta^* = \arg\min_{\theta} \operatorname{Reg}(\theta) + \frac{\lambda}{n}\sum_{i=1}^{n} \operatorname{Sim}(P^i[I_0 \circ \Phi_\theta^{-1}], I_1^i),\ \lambda > 0, \tag{6}$$

where we now average over the similarity measures with respect to all the n projection images, and $P^i[\cdot]$ denotes the projection consistent with the geometry of the i-th projection image I_1^i (i.e., its DRR). Note that neither the regularizer nor the transformation model changed compared to the registration formulation of Eq. 5. The main difference is that the 3D transformed source image $I_0 \circ \Phi_\theta^{-1}$ undergoes different projections and the similarity is measured in the projection space. Clearly, optimizing over such a formulation might be difficult as one needs to determine the potentially rather indirect dependency between the transformation parameter θ, and the image differences measured after projection. While such dependencies have previously been determined analytically [3,16], they make using advanced registration models and rapid prototyping of registration approaches difficult. Instead, we base our entire registration approach on automatic differentiation. In such a framework, changing the transformation model becomes easy. We explore an affine transformation model ($\Phi_\theta^{-1}(x) = Ax + b$, where $\theta = \{A, b\}$, $A \in \mathbb{R}^{3\times 3}$, $b \in \mathbb{R}^{3\times 1}$), which has previously been used in 3D/2D registration. To capture large, localized deformations we also use a fluid-registration model after affine pre-registration. We use the large deformation diffeomorphic metric mapping (LDDMM) model [2,12], where a spatial transformation is parameterized by a spatio-temporal velocity field. The LDDMM formulation of 3D/2D registration in its shooting form [19,20,22] becomes

$$\theta^*(x) = \arg\min_{\theta(x)} \langle \theta, K \star \theta \rangle + \frac{\lambda}{n}\sum_{i=1}^{n} \operatorname{Sim}(P^i[I_0 \circ \Phi^{-1}(1)], I_1^i),\ \lambda > 0, \tag{7}$$

$$\Phi_t^{-1} + D\Phi^{-1} v = 0,\ \Phi^{-1}(0, x) = \Phi_0^{-1}(x), \tag{8}$$

$$m_t + div(v)m + Dv^T(m) + Dm(v) = 0,\ v = K \star m,\ m(x, 0) = \theta(x), \tag{9}$$

where θ is the initial momentum vector field which parameterizes the transformation over time via an advection equation (Eq. 8) on the transformation map Φ^{-1} controlled by the spatially-varying velocity field, v, whose evolution is governed by the Euler-Poincaré equation for diffeomorphisms (EPDiff; Eq. 9); $\Phi^{-1}(1)$ is the transformation at time $t = 1$ which warps the source image and $\Phi_0^{-1}(x)$ is its initial condition (set to the result of our affine pre-registration); K is a regularizer (we choose a multi-Gaussian [17]). This registration model can capture large deformations (as occur in lung registration), is parameterized entirely by its initial momentum field, and can guarantee diffeomorphic transformations. While technically possible, it is easy to appreciate that taking derivatives of the LDDMM 3D/2D registration model is not straightforward. However, our implementation based on automatic differentiation only requires implementation of the forward model (Eqs. 7–9). Gradient computation is then automatic.

Algorithm 1 describes our 3D/2D registration algorithm:

Algorithm 1: 3D/2D registration

Result: $\theta^*, \Phi^{-1}_{affine}$
Initialize $v = \mathbf{0}$ and $\Phi^{-1}(0) = \Phi^{-1}_{affine}$;
while *loss not converged* **do**
  Compute ϕ^{-1} and warp I_0 using ϕ^{-1};
  Compute projection images $\{P^i[I_0 \circ \Phi^{-1}]\}$ using tensor form of Eq. 4;
  Compute $\mathrm{Sim}_{NGF}(P^i[I_0 \circ \Phi^{-1}], I_1^i)$ and $\mathrm{Reg}(\theta)$;
  Back-propagate to compute $\nabla_\theta loss$ and update θ
end

---

4 Limited Angle Image Reconstruction

We have shown how our DRR model integrates within our 3D/2D registration formulation. Now we apply it for an iterative reconstruction algorithm (Algorithm 2).

Given a set of n 2D projection images $\{I^i\}$, our goal is to determine the 3D image I which is most consistent with these projections. Specifically, we solve the following minimization problem:

$$I^* = \arg\min_I \frac{1}{n} \sum_{i=1}^{n} \mathrm{Sim}(P^i[I], I^i) + \lambda_1 \mathrm{ReLU}(-I) + \frac{\lambda_2}{|\Omega|} \int_\Omega ||\nabla I||_2 \, dx, \ \lambda_1, \lambda_2 \geq 0, \tag{10}$$

where $P^i[\cdot]$ is the DRR generated for the i-th emitter (Eq. 4). Sim is the L_1 loss between $P^i[I]$ and I^i. We add two regularization terms: 1) $\mathrm{ReLU}(I)$, which encourages positive reconstructed values, and 2) $\int ||\nabla I||_2 \, dx$, which measures the total variation of I, encouraging piece-wise constant reconstructions.

Algorithm 2: Iterative reconstruction

Result: I^*
Initialize $I = 0$;
while *loss not converged* **do**
  Compute $\{P^i[I]\}$ using tensor form of Eq. 4;
  Compute $\mathrm{Sim}(P^i[I], I^i)$, $ReLU(-I)$ and $\int ||\nabla I||_2 \, dx$;
  Back-propagate to compute $\nabla_I loss$ to update I
end

5 Experiments and Results

Dataset. We use two lung datasets for evaluation: 1) a paired CT-sDCT dataset including raw sDCT projections, 2) a 3D CT dataset with landmarks.

CT-sDCT: We acquired 3 pairs of CT/sDCT images including 31 sDCT projections for each sDCT image with uniformly spaced emitters covering a scanning range of 12°. We use this dataset to test our reconstruction approach.

DIR-Lab 4DCT: This dataset contains 10 pairs of inhale/exhale lung CT images with small deformations and 300 corresponding landmarks per pair. We use this dataset to synthesize corresponding 3D/2D image sets. Specifically, we synthesize projection images from the inspiration phase CT (via our DRR model) and treat them as the target images. As we have the true 3D images with landmarks we can then perform quantitative validation.

5.1 sDCT Reconstruction

We compare our reconstruction result (using Algorithm 2 and the sDCT projections) to the given sDCT reconstructions (based on [24]) of the CT-sDCT dataset.

Figure 2 shows the resulting reconstructions. The mean squared error (MSE) between our reconstruction algorithm and the reconstruction algorithm of [24] is small, demonstrating that our method recovers image structure equally well. This result also qualitatively demonstrates the correctness of our DRR projection algorithm.

5.2 3D/2D Registrations

Given a CT image and a set of projection images we follow two 3D/2D registration approaches: 1) our approach (Algorithm 1) and 2) 3D/3D registration by first reconstructing a 3D volume (Algorithm 2) from the set of projection images. The second approach is our baseline demonstrating that the vastly different appearances of the reconstructed 3D volumes (for a low number of projections) and the CT images result in poor registration performance, which can be overcome using our approach. To quantitatively compare both approaches, we use the DIR-Lab lung CT dataset to create synthetic CT/projection set pairs, where we keep the inspiration phase CT image and create the synthetic DRRs from the expiration phase CT (which constitute our simulated raw sDCT projections). While sDCT imaging is fast, the imaging quality is lower than for CT. For comparison, we provide 3D-CT/3D-CT registration results from the literature.

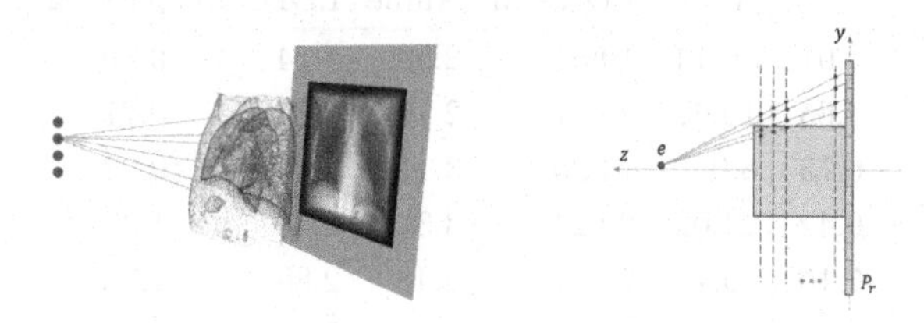

Fig. 1. Illustration of sDCT and our DRR generation. Left: Emitters are spatially distributed in sDCT [18]. Right: Side view of DRR geometry.

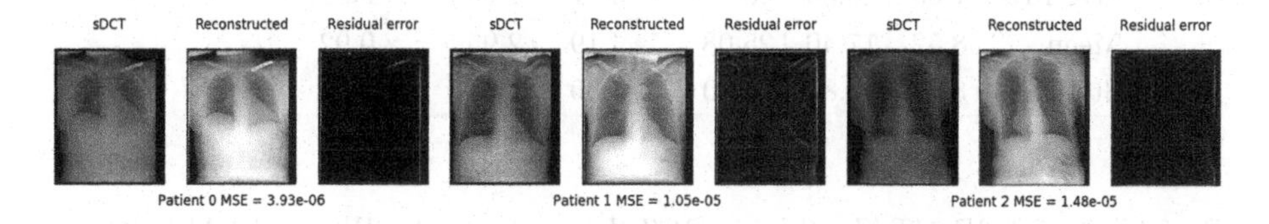

Fig. 2. Comparisons between our reconstructed volumes (Algorithm 2) and sDCT volumes (we only show one slice per case). Columns 1, 3, 6: provided sDCT images; columns 2, 5, 8: reconstructed images from our method; columns 3, 6, 9: residual error computed by subtracting our reconstructed image from the sDCT image.

Registration of 3D-CT/Synthetic 2D sDCT Projections. We resample both expiration/inspiration lung CT images to $1 \times 1 \times 1$ mm. To make our results comparable to state-of-the-art (SOTA) 3D-CT/3D-CT lung registration approaches, we segment the lung and remove all image information outside the lung mask. Then, we create DRRs from the expiration phase lung CT to simulate a set of sDCT projection images with the same emitter geometry as in our CT-sDCT dataset [18]. However, we use only 4 emitters placed evenly within 11 degrees (compared to 31 emitters in CT-sDCT dataset) to assess registration behavior for very few projections. We use $\lambda = 0.5$, the normalized gradient field (NGF) similarity measure [6], stochastic gradient descent optimization for affine registration with $\lambda = 0.01$; and NGF, lBFGS [10] for LDDMM registration. Table 1 shows the registration results. As expected, our affine and LDDMM projection results improve over the initial alignment and LDDMM improves over affine registration, indicating more complex transformations can be recovered with our 3D/2D registration approach. Due to the limited number of projections our results are not as accurate as direct 3D-CT/3D-CT registration (e.g., SOTA pTVreg [23]), which, however, would not be practical in an interventional setting.

Table 1. Mean distance of expert landmarks before and after registration with our proposed 3D/2D methods and comparison to a SOTA 3D-CT/3D-CT method.

Dataset	Initial	3D-2D registration task				3D-3D reg. task
		Reconstruction		Projection		
		Affine	LDDMM	Affine	LDDMM	pTVreg [23]
4DCT1	4.01	10.14	19.51	2.99	1.74	0.80
4DCT2	4.65	19.67	29.71	2.10	1.74	0.77
4DCT3	6.73	6.07	14.90	3.11	2.24	0.92
4DCT4	9.42	24.02	29.24	3.37	2.54	1.30
4DCT5	7.10	19.79	26.21	4.48	2.88	1.13
4DCT6	11.10	21.81	30.51	6.13	5.18	0.78
4DCT7	11.59	16.83	26.63	4.64	4.07	0.79
4DCT8	15.16	14.78	18.85	7.85	2.73	1.00
4DCT9	7.82	24.44	29.04	3.70	3.76	0.91
4DCT10	7.63	16.50	26.28	3.57	2.63	0.82
Mean	8.52	17.40	25.08	4.19	2.95	0.92
Std.	3.37	5.89	5.40	1.69	1.08	0.17

Registration of 3D-CT/Synthetic sDCT Reconstruction. We apply Algorithm 2 to reconstruct the sDCT from the synthetic projection images of the 3D-CT expiration phase image. We set $\lambda_1 = 100$, $\lambda_2 = 1$ in Eq. 10, then register the 3D-CT at inhale to the reconstructed sDCT using 3D/3D LDDMM registration.

Table 1 shows that 3D/2D registration outperforms 3D/3D registration via the reconstructed sDCT. This is because of the vast appearance difference of the sDCT image (compared to 3D-CT), because of limited-angle reconstruction, and because of the low number of projections. In fact, when using the reconstructed sDCT, registration results are worse (for affine and LDDMM) than not registering at all. Hence, using the 3D/2D approach is critical in this setting.

5.3 Influence of Angle Range and Number of Projections

We explore how our 3D/2D registration results change with the number of projections (fixing angle range of 11°) and with the angle range (fixing the number of projections at 4). Results are based on DIR-Lab case 4DCT5 (Table 1).

Figure 3 shows that both affine and LDDMM registrations improve as the angle range increases. Note that improvements are most pronounced in the Z direction, which is perpendicular to the receiver plane and thus has the least localization information in such a limited angle imaging setup. Figure 3 also shows that registration accuracy is relatively insensitive to the number of projections when restricting to a fixed angle range of 11°. This indicates that angle range is more important than the number of projections. Further, when the radiation dose is a concern, larger scanning angles with fewer emitters are preferable.

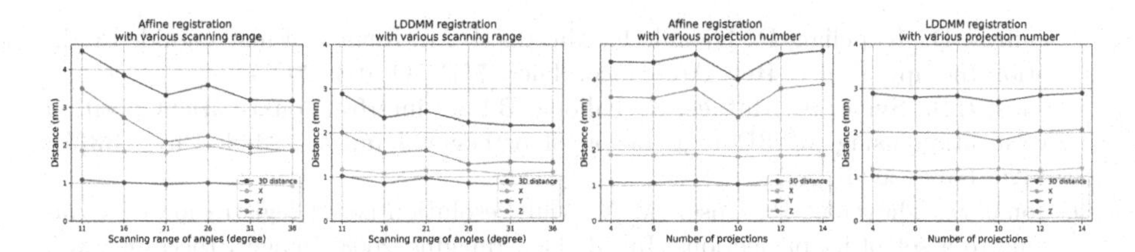

Fig. 3. Influence of angle range and number of projections. First 2 columns: Affine/LDDMM registrations accuracy improves with increasing angle range for 4 projections. Last 2 columns: Affine/LDDMM registration accuracies are relatively insensitive to the number of projections for an 11° angle range.

6 Conclusion

We proposed an sDCT-based DRR model and a 3D/2D framework for CT to sDCT registration. Our approach uses automatic differentiation and therefore easily extends to other tasks (e.g., 3D image reconstruction). We demonstrated that DRR-based 3D/2D LDDMM registration outperforms 3D/3D-reconstruction registration in a limited angle setting with small numbers of projections. Future work will extend our approach to an end-to-end 3D/2D deep network registration model. We will also explore alternative image similarity measures and quantitatively validate the approach on real CT/sDCT pairs.

Acknowledgements. Research reported in this work was supported by the National Institutes of Health (NIH) under award number NIH 1-R01-EB028283-01. The content is solely the responsibility of the authors and does not necessarily represent the official views of the NIH.

References

1. Aouadi, S., Sarry, L.: Accurate and precise 2D–3D registration based on x-ray intensity. Comput. Vis. Image Underst. **110**(1), 134–151 (2008)
2. Beg, M.F., Miller, M.I., Trouvé, A., Younes, L.: Computing large deformation metric mappings via geodesic flows of diffeomorphisms. Int. J. Comput. Vis. **61**(2), 139–157 (2005)
3. Flach, B., Brehm, M., Sawall, S., Kachelrieß, M.: Deformable 3D–2D registration for CT and its application to low dose tomographic fluoroscopy. Phys. Med. Biol. **59**(24), 7865 (2014)
4. Fu, D., Kuduvalli, G.: A fast, accurate, and automatic 2D–3D image registration for image-guided cranial radiosurgery. Med. Phys. **35**(5), 2180–2194 (2008)
5. Griewank, A., Walther, A.: Evaluating derivatives: principles and techniques of algorithmic differentiation, vol. 105. SIAM (2008)
6. Haber, E., Modersitzki, J.: Intensity gradient based registration and fusion of multi-modal images. In: Larsen, R., Nielsen, M., Sporring, J. (eds.) MICCAI 2006. LNCS, vol. 4191, pp. 726–733. Springer, Heidelberg (2006). https://doi.org/10.1007/11866763_89

7. Jaffray, D., Kupelian, P., Djemil, T., Macklis, R.M.: Review of image-guided radiation therapy. Expert Rev. Anticancer Ther. **7**(1), 89–103 (2007)
8. Jans, H.S., Syme, A., Rathee, S., Fallone, B.: 3D interfractional patient position verification using 2D–3D registration of orthogonal images. Med. Phys. **33**(5), 1420–1439 (2006)
9. Jonic, S., Thévenaz, P., Unser, M.A.: Multiresolution-based registration of a volume to a set of its projections. In: Medical Imaging 2003: Image Processing, vol. 5032, pp. 1049–1052. International Society for Optics and Photonics (2003)
10. Liu, D.C., Nocedal, J.: On the limited memory BFGS method for large scale optimization. Math. Program. **45**(1–3), 503–528 (1989)
11. Markelj, P., Tomaževič, D., Likar, B., Pernuš, F.: A review of 3D/2D registration methods for image-guided interventions. Med. Image Anal. **16**(3), 642–661 (2012)
12. Miller, M.I., Trouvé, A., Younes, L.: On the metrics and Euler-Lagrange equations of computational anatomy. Ann. Rev. Biomed. Eng. **4**(1), 375–405 (2002)
13. Modersitzki, J.: Numerical Methods for Image Registration. Oxford University Press on Demand (2004)
14. Paszke, A., et al.: PyTorch: an imperative style, high-performance deep learning library. In: Wallach, H., Larochelle, H., Beygelzimer, A., d'Alché-Buc, F., Fox, E., Garnett, R. (eds.) Advances in Neural Information Processing Systems, vol. 32, pp. 8024–8035. Curran Associates, Inc. (2019). http://papers.neurips.cc/paper/9015-pytorch-an-imperative-style-high-performance-deep-learning-library.pdf
15. Prümmer, M., Han, J., Hornegger, J.: 2D–3D non-rigid registration using iterative reconstruction. In: Workshop Vision Modeling and Visualization in Erlangen, vol. 1, pp. 187–194 (2005). http://www.vmv2005.uni-erlangen.de
16. Prümmer, M., Hornegger, J., Pfister, M., Dörfler, A.: Multi-modal 2D–3D non-rigid registration. In: Medical Imaging 2006: Image Processing, vol. 6144, p. 61440X. International Society for Optics and Photonics (2006)
17. Risser, L., Vialard, F.-X., Wolz, R., Holm, D.D., Rueckert, D.: Simultaneous fine and coarse diffeomorphic registration: application to atrophy measurement in Alzheimer's disease. In: Jiang, T., Navab, N., Pluim, J.P.W., Viergever, M.A. (eds.) MICCAI 2010. LNCS, vol. 6362, pp. 610–617. Springer, Heidelberg (2010). https://doi.org/10.1007/978-3-642-15745-5_75
18. Shan, J., et al.: Stationary chest tomosynthesis using a carbon nanotube x-ray source array: a feasibility study. Phys. Med. Biol. **60**(1), 81 (2014)
19. Shen, Z., Vialard, F.X., Niethammer, M.: Region-specific diffeomorphic metric mapping. In: Advances in Neural Information Processing Systems, pp. 1096–1106 (2019)
20. Singh, N., Hinkle, J., Joshi, S., Fletcher, P.T.: A vector momenta formulation of diffeomorphisms for improved geodesic regression and atlas construction. In: 2013 IEEE 10th International Symposium on Biomedical Imaging, pp. 1219–1222. IEEE (2013)
21. Tomazevic, D., Likar, B., Pernus, F.: 3-D/2-D registration by integrating 2-D information in 3-D. IEEE Trans. Med. Imaging **25**(1), 17–27 (2005)
22. Vialard, F.X., Risser, L., Rueckert, D., Cotter, C.J.: Diffeomorphic 3D image registration via geodesic shooting using an efficient adjoint calculation. Int. J. Comput. Vis. **97**(2), 229–241 (2012)
23. Vishnevskiy, V., Gass, T., Szekely, G., Tanner, C., Goksel, O.: Isotropic total variation regularization of displacements in parametric image registration. IEEE Trans. Med. Imaging **36**(2), 385–395 (2016)

24. Wu, G., Inscoe, C., Calliste, J., Lee, Y.Z., Zhou, O., Lu, J.: Adapted fan-beam volume reconstruction for stationary digital breast tomosynthesis. In: Medical Imaging 2015: Physics of Medical Imaging, vol. 9412, p. 94123J. International Society for Optics and Photonics (2015)
25. Zikic, D., Groher, M., Khamene, A., Navab, N.: Deformable registration of 3D vessel structures to a single projection image. In: Medical Imaging 2008: Image Processing, vol. 6914, p. 691412. International Society for Optics and Photonics (2008)

Anatomical Data Augmentation via Fluid-Based Image Registration

Zhengyang Shen(✉), Zhenlin Xu, Sahin Olut, and Marc Niethammer

Department of Computer Science, UNC Chapel Hill, Chapel Hill, USA
zyshen@cs.unc.edu

Abstract. We introduce a fluid-based image augmentation method for medical image analysis. In contrast to existing methods, our framework generates anatomically meaningful images via interpolation from the geodesic subspace underlying given samples. Our approach consists of three steps: 1) given a source image and a set of target images, we construct a geodesic subspace using the Large Deformation Diffeomorphic Metric Mapping (LDDMM) model; 2) we sample transformations from the resulting geodesic subspace; 3) we obtain deformed images and segmentations via interpolation. Experiments on brain (LPBA) and knee (OAI) data illustrate the performance of our approach on two tasks: 1) data augmentation during training and testing for image segmentation; 2) one-shot learning for single atlas image segmentation. We demonstrate that our approach generates anatomically meaningful data and improves performance on these tasks over competing approaches. Code is available at https://github.com/uncbiag/easyreg.

1 Introduction

Training data-hungry deep neural networks is challenging for medical image analysis where manual annotations are more difficult and expensive to obtain than for natural images. Thus it is critical to study how to use scarce annotated data efficiently, e.g., via data-efficient models [11,30], training strategies [20] and semi-supervised learning strategies utilizing widely available unlabeled data through self-training [3,16], regularization [4], and multi-task learning [7,31,36].

An alternative approach is data augmentation. Typical methods for medical image augmentation include random cropping [12], geometric transformations [15,18,24] (e.g., rotations, translations, and free-form deformations), and photometric (i.e., color) transformations [14,21]. Data-driven data augmentation has also been proposed, to learn generative models for synthesizing images with new appearance [9,28], to estimate class/template-dependent distributions of deformations [10,19,34] or both [6,35]. Compared with these methods, our approach focuses on a geometric view and constructs a continuous geodesic subspace as an estimate of the space of anatomical variability.

Electronic supplementary material The online version of this chapter (https://doi.org/10.1007/978-3-030-59716-0_31) contains supplementary material, which is available to authorized users.

A. L. Martel et al. (Eds.): MICCAI 2020, LNCS 12263, pp. 318–328, 2020.
https://doi.org/10.1007/978-3-030-59716-0_31

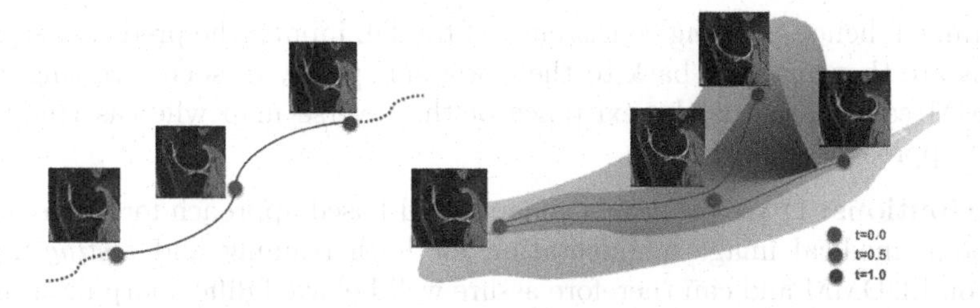

Fig. 1. Illustration of our fluid-based data augmentation using a 1D (left) and 2D (right) geodesic subspace. We assume a registration from a source to a target image in unit time. In 1D, we can sample along the geodesic path ($t \in [0,1]$) between the source ($t = 0$) and the target images ($t = 1$). We can also extrapolate $t \notin [0,1]$. In the 2D case, a source and two target images define a two-dimensional geodesic subspace.

Compared with the high dimensionality of medical images, anatomical variability is often assumed to lie in a much lower dimensional space [1]. Though how to directly specify this space is not obvious, we can rely on reasonable assumptions informed by the data itself. We assume there is a diffeomorphic transformation between two images, that image pairs can be connected via a geodesic path, and that appearance variation is implicitly captured by the appearance differences of a given image population. For longitudinal image data, we can approximate images at intermediate time points by interpolation or predict via extrapolation. As long as no major appearance changes exist, diffeomorphic transformations can provide realistic intermediate images[1]. Based on these considerations, we propose a data augmentation method based on fluid registration which produces *anatomically plausible* deformations and retains appearance differences of a given image population. Specifically, we choose the Large Deformation Diffeomorphic Metric Mapping (LDDMM) model as our fluid registration approach. LDDMM comes equipped with a metric and results in a geodesic path between a source and a target image which is parameterized by the LDDMM initial momentum vector field. Given two initial momenta in the tangent space of the same source image, we can define a geodesic plane, illustrated in Fig. 1; similarly, we can construct higher dimensional subspaces based on convex combinations of sets of momenta [22]. Our method includes the following steps: 1) we compute a set of initial momenta for a source image and a set of target images; 2) we generate an initial momentum via a convex combination of initial momenta; 3) we sample a transformation on the geodesic path determined by the momentum; and 4) we warp the image and its segmentation according to this transformation.

Data augmentation is often designed for the training phase. However, we show the proposed approach can be extended to the testing phase, e.g., a testing image is registered to a set of training images (with segmentations) and the deep learning (DL) segmentation model is evaluated in this warped space (where it

[1] In some cases, for example for lung images, sliding effects need to be considered, violating the diffeomorphic assumption.

was trained, hence ensuring consistency of the DL input); the predicted segmentations are then mapped back to their original spaces. In such a setting, using LDDMM can guarantee the existence of the inverse map whereas traditional elastic approaches cannot.

Contributions: 1) We propose a general fluid-based approach for anatomically consistent medical image augmentation for both training and *testing.* 2) We build on LDDMM and can therefore assure well-behaved diffeomorphic transformations when interpolating *and* extrapolating samples with large deformations. 3) Our method easily integrates into different tasks, such as segmentation and one-shot learning for which we show general performance improvements.

2 LDDMM Method

LDDMM [5] is a fluid-based image registration model, estimating a spatio-temporal velocity field $v(t,x)$ from which the spatial transformation φ can be computed via integration of $\partial_t\varphi(t,x) = v(t,\varphi(t,x)),\ \varphi(0,x) = x$. For appropriately regularized velocity fields [8], diffeomorphic transformations can be guaranteed. The optimization problem underlying LDDMM can be written as

$$v^* = \underset{v}{\operatorname{argmin}}\ \frac{1}{2}\int_0^1 \|v(t)\|_L^2\, \mathrm{d}dt + \mathrm{Sim}(I(1), I_1) \quad \text{s.t.} \quad \partial_t I + \langle \nabla I, v\rangle = 0,\ I(0) = I_0\ , \tag{1}$$

where ∇ denotes the gradient, $\langle\cdot,\cdot\rangle$ the inner product, and $\mathrm{Sim}(A,B)$ is a similarity measure between images. We note that $I(1,x) = I_0 \circ \varphi^{-1}(1,x)$, where φ^{-1} denotes the inverse of φ in the target image space. The evolution of this map follows $\partial_t\varphi^{-1} + D\varphi^{-1}v = 0$, where D is the Jacobian. Typically, one seeks a velocity field which deforms the source to the target image in unit time. To assure smooth transformations, LDDMM penalizes non-smooth velocity fields via the norm $\|v\|_L^2 = \langle Lv, Lv\rangle$, where L is a differential operator.

At optimality the following equations hold [33] and the entire evolution can be parameterized via the initial vector-valued momentum, $m = L^\dagger Lv$:

$$m(0)^* = \underset{m(0)}{\operatorname{argmin}}\ \frac{1}{2}\langle m(0), v(0)\rangle + \mathrm{Sim}(I_0 \circ \varphi^{-1}(1), I_1), \tag{2}$$

$$\text{s.t.} \quad \varphi_t^{-1} + D\varphi^{-1}v = 0, \quad \varphi(0,x) = x, \tag{3}$$

$$\partial_t m + \mathrm{div}(v)m + Dv^T(m) + Dm(v) = 0,\ m(0) = m_0, v = K \star m, \tag{4}$$

where we assume $(L^\dagger L)^{-1}m$ is specified via convolution $K \star m$. Equation 4 is the Euler-Poincaré equation for diffeomorphisms (EPDiff) [33], defining the evolution of the spatio-temporal velocity field based on the initial momentum m_0.

The geodesic which connects the image pair $(I_0, I_0 \circ \varphi^{-1}(1))$ and approximates the path between (I_0, I_1) is specified by m_0. We can sample along the geodesic path, assuring diffeomorphic transformations. As LDDMM assures diffeomorphic transformations, we can also obtain the inverse transformation map,

φ (defined in source image space, whereas φ^{-1} is defined in target image space) by solving

$$\varphi(1, x) = x + \int_0^1 v(t, \varphi(t, x))\, dt,\ \varphi(0, x) = x. \tag{5}$$

Computing the inverse for an arbitrary displacement field on the other hand requires the numerical minimization of $\|\varphi^{-1} \circ \varphi - id\|^2$. Existence of the inverse map cannot be guaranteed for such an arbitrary displacement field.

3 Geodesic Subspaces

We define a geodesic subspace constructed from a source image and a set of target images. Given a dataset of size N, $I_c \in \mathbb{R}^D$ denotes an individual image $c \in \{1 \dots N\}$, where D is the number of voxels. For each source image I_c, we further denote a target set of K images as $\mathbf{I_K^c}$. We define $M_K^c := \{m_0^{cj} | \mathcal{M}(I_c, I_j), \mathcal{M} : \mathbb{R}^D \times \mathbb{R}^D \to \mathbb{R}^{D \times d}, I_j \in \mathbf{I_K^c}\}$ as a set of K different initial momenta, where $\mathcal{M}$ maps from an image pair to the corresponding initial momentum via Eqs. 2–4; d is the spatial dimension. We define convex combinations of M_K^c as

$$C(M_K^c) := \left\{ \tilde{m}_0^c \middle| \tilde{m}_0^c = \sum_{j=1}^{K} \lambda_j m_0^{cj}, m_0^{cj} \in M_K^c, \lambda_j \geq 0\ \forall j, \sum_{j=1}^{K} \lambda_j = 1 \right\}. \tag{6}$$

Restricting ourselves to convex combinations, instead of using the entire space defined by arbitrary linear combinations of the momenta M_k^c allows us to retain more control over the resulting momenta magnitudes. For our augmentation strategy we simply sample an initial momentum $\tilde{m}_0^c$ from $C(M_K^c)$, which, according to the EPDiff Eq. 4, determines a geodesic path starting from I_c. If we set $K = 2$, for example, the sampled momentum parameterizes a path from a source image toward two target images, where the λ_i weigh how much the two different images drive the overall deformation. As LDDMM registers a source to a target image in unit time, we obtain interpolations by additionally sampling t from $[0, 1]$, resulting in the intermediate deformation $\varphi_{\tilde{m}_0^c}^{-1}(t)$ from the geodesic path starting at I_c and determined by $\tilde{m}_0^c$. We can also extrapolate by sampling t from $\mathbb{R} \setminus [0, 1]$. We then synthesize images via interpolation: $I_c \circ \varphi_{\tilde{m}_0^c}^{-1}(t)$.

4 Segmentation

In this section, we first introduce an augmentation strategy for general image segmentation (Sect. 4.1) and then a variant for one-shot segmentation (Sect. 4.2).

4.1 Data Augmentation for General Image Segmentation

We use a two-phase data augmentation approach consisting of (1) pre-augmentation of the training data and (2) post-augmentation of the testing data.

During the training phase, for each training image, I_c, we generate a set of new images by sampling from its geodesic subspace, $C(M_K^c)$. This results in a set of deformed images which are anatomically meaningful and retain the appearance of I_c. We apply the same sampled spatial transformations to the segmentation of the training image, resulting in a new set of warped images and segmentations. We train a segmentation network based on this augmented dataset.

During the testing phase, for each testing image, we also create a set of new images using the same strategy described above. Specifically, we pair a testing image with a set of *training* images to create the geodesic subspace for sampling. This will result in samples that come from a similar subspace that has been used for augmentation during training. A final segmentation is then obtained by warping the predicted segmentations back to the original space of the image to be segmented and applying a label-fusion strategy. Consequently, we expect that the segmentation network performance will be improved as it (1) is allowed to see multiple views of the same image and (2) the set of views is consistent with the set of views that the segmentation network was trained with.

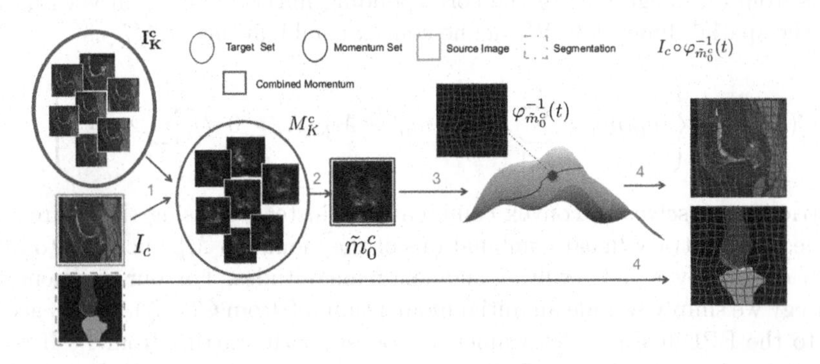

Fig. 2. Illustration of the training phase data augmentation. Given a source image I_c and a set of target images $\mathbf{I}_\mathbf{K}^\mathbf{c}$, a set of momenta M_K^c is first computed. Then a momentum $\tilde{m}_0^c$ is sampled from the convex combination of these momenta $C(M_K^c)$ defining a geodesic path starting from the source image. Lastly, a transformation $\varphi^{-1}_{\tilde{m}_0^c}(t)$ is sampled on the geodesic and used to warp the source image and its segmentation.

Figure 2 illustrates the training phase data augmentation. We first compute M_K^c by picking an image I_c, $c \in \{1 \ldots N\}$ from a training dataset of size N and a target set $\mathbf{I}_\mathbf{K}^\mathbf{c}$ of cardinality K, also sampled from the training set. We then sample $\tilde{m}_0^c \in C(M_K^c)$ defining a geodesic path from which we sample a deformation $\varphi^{-1}_{\tilde{m}_0^c}(t)$ at time point t. We apply the same strategy multiple times and obtain a new deformation set for each I_c, $c \in \{1 \ldots N\}$. The new image set $\{I_c \circ \varphi^{-1}_{\tilde{m}_0^c}(t)\}$ consisting of the chosen set of random transformations of I_c and the corresponding segmentations can then be obtained by interpolation.

Figure 3 illustrates the testing phase data augmentation. For a test image I_c and its target set $\mathbf{I}_\mathbf{K}^\mathbf{c}$ sampled from the *training set*, we obtain a set of transformations $\{\varphi^{-1}_{\tilde{m}_0^c}(t)\}$. By virtue of the LDDMM model these transformations

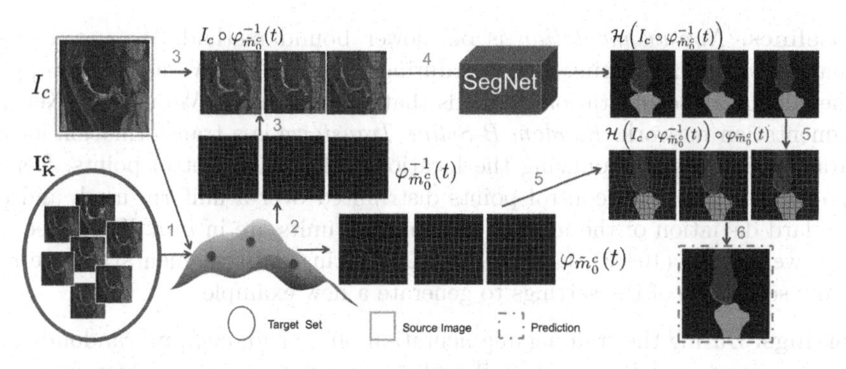

Fig. 3. Illustration of the testing phase data augmentation. Given a source image I_c and a set of target images $\mathbf{I^c_K}$, a geodesic subspace is determined first. A set of transformations $\varphi^{-1}_{\tilde{m}_0^c}(t)$ is then sampled from this space and, at the same time, the corresponding inverse transformations $\varphi_{\tilde{m}_0^c}(t)$ are obtained. A segmentation network $\mathcal{H}$ is applied to each warped image and the resulting segmentations $\mathcal{H}(I_c \circ \varphi^{-1}_{\tilde{m}_0^c}(t))$ are warped back to the source image space. A label fusion strategy is applied to obtain the final segmentation.

are invertible. For each $\varphi^{-1}_{\tilde{m}_0^c}(t)$ we can therefore efficiently obtain the corresponding inverse map $\varphi_{\tilde{m}_0^c}(t)$. We denote our trained segmentation network by $\mathcal{H} : \mathbb{R}^D \rightarrow \mathbb{R}^{D\times L}$ which takes an image as its input and predicts its segmentation labels. Here, L is the number of segmentation labels. Each prediction $\mathcal{H}\left(I_c \circ \varphi^{-1}_{\tilde{m}_0^c}(t)\right)$ is warped back to the space of I_c via $\mathcal{H}\left(I_c \circ \varphi^{-1}_{\tilde{m}_0^c}(t)\right) \circ \varphi_{\tilde{m}_0^c}(t)$. The final segmentation is obtained by merging all warped predictions via a label fusion strategy.

Dataset. The LONI Probabilistic Brain Atlas [25] (LPBA40) dataset contains volumes of 40 healthy patients with 56 manually annotated anatomical structures. We affinely register all images to a mean atlas [13], resample to isotropic spacing of 1 mm, crop them to $196 \times 164 \times 196$ and intensity normalize them to $[0, 1]$ via histogram equalization. We randomly take 25 patients for training, 10 patients for testing, and 5 patients for validation.

The Osteoarthritis Initiative [29] (OAI) provides manually labeled knee images with segmentations of femur and tibia as well as femoral and tibial cartilage [2]. We first affinely register all images to a mean atlas [13], resample them to isotropic spacing of 1 mm, and crop them to $160 \times 200 \times 200$. We randomly take 60 patients for training, 25 patients for validation, and 52 patients for testing.

To evaluate the effect of data augmentation on training datasets with different sizes, we further sample 5, 10, 15, 20, 25 patients as the training set on LPBA40 and 10, 20, 30, 40, 60 patients as the training set for OAI.

Metric. We use the average Dice score over segmentation classes for all tasks in Sect. 4.1 and Sect. 4.2.

Baselines. *Non-augmentation* is our lower bound method. We use a class-balanced random cropping schedule during training [32]. We use this cropping schedule for all segmentation methods that we implement. We use a U-Net [23] segmentation network. *Random B-Spline Transform* is a transformation locally parameterized by randomizing the location of B-spline control points. Denote $(\cdot,\cdot)$ as the number of control points distributed over a uniform mesh and the standard deviation of the normal distribution, units are in mm. The three settings we use are $(10^3, 3)$, $(10^3, 4)$, $(20^3, 2)$. During data augmentation, we randomly select one of the settings to generate a new example.

Settings. During the training augmentation phase (*pre-aug*), we randomly pick a source image and K targets, uniformly sample λ_i in Eq. 6 and then uniformly sample t. For LPBA40, we set $K = 2$ and $t \in [-1, 2]$; for the OAI data, we set $K = 1$ and $t \in [-1, 2]$. For all training sets with different sizes, for both the B-Spline and the fluid-based augmentation methods and for both datasets, we augment the training data by 1,500 cases. During the testing augmentation phase (*post-aug*), for both datasets, we set $K = 2$ and $t \in [-1, 2]$ and draw 20 synthesized samples for each test image. The models trained via the augmented training set are used to predict the segmentations. To obtain the final segmentation, we sum the softmax outputs of all the segmentations warped to the original space and assign the label with the largest sum. We test using the models achieving the best performance on the validation set. We use the optimization approach in [17] and the network of [26,27] to compute the mappings $\mathcal{M}$ on LPBA40 and OAI, respectively.

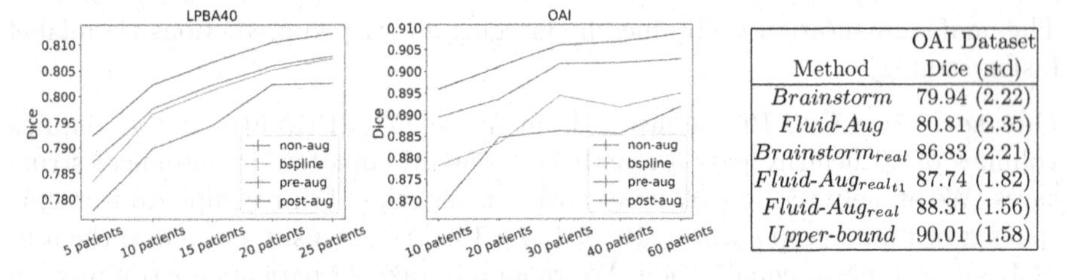

Method	OAI Dataset Dice (std)
Brainstorm	79.94 (2.22)
Fluid-Aug	80.81 (2.35)
$Brainstorm_{real}$	86.83 (2.21)
$Fluid\text{-}Aug_{real_{t1}}$	87.74 (1.82)
$Fluid\text{-}Aug_{real}$	88.31 (1.56)
Upper-bound	90.01 (1.58)

Fig. 4. Segmentation performance for segmentation tasks. The left two plots show Dice scores for the different methods with different training set sizes on the LPBA40 and OAI datasets for general segmentation. Performance increases with training set size. Fluid-based augmentation (pre-aug and post-aug) shows the best performance. The right table compares the performance for one-shot segmentation in Sect. 4.2. Fluid-based augmentation methods perform better than their *Brainstorm* counterparts.

Results. Figure 4 shows the segmentation performance on the LPBA40 and the OAI datasets. For training phase augmentation, fluid-based augmentation improves accuracy over non-augmentation and B-Spline augmentation by a large margin on the OAI dataset and results in comparable performance on the

LPBA40 dataset. This difference might be due to the larger anatomical differences in the LPBA40 dataset compared to the OAI dataset; such large differences might not be well captured by inter- and extrapolation along a few geodesics. Hence, the OAI dataset may benefit more from the anatomically plausible geodesic space. When test phase augmentation is used in addition to training augmentation, performance is further improved. This shows that the ensemble strategy used by post-aug, where the segmentation network makes a consensus decision based on different views of the image to be segmented, is effective. In practice, we observe that high-quality inverse transformations (that map the segmentations back to the test image space) are important to achieve good performance. These inverse transformations can efficiently be computed via Eq. 5 for our fluid-based approach.

4.2 Data Augmentation for One-Shot Segmentation

We explore one-shot learning. Specifically, we consider single atlas medical image segmentation, where only the atlas image has a segmentation, while all other images are unlabeled. We first review *Brainstorm* [35], a competing data augmentation framework for one-shot segmentation. We then discuss our modifications.

In *Brainstorm*, the appearance of a sampled unlabeled image is first transfered to atlas-space and subsequently spatially transformed by registering to another sampled unlabeled image. Specifically, a registration network $\mathcal{H}^r$ is trained to predict the displacement field between the atlas A and the unlabeled images. For a given image I_c, the predicted transformation to A is $\varphi_c(x) = \mathcal{H}^r(I_c, A) + x$. A set of approximated inverse transformations $\{\varphi_c^{-1}, c \in 1 \ldots N\}$ from the atlas to the image set can also be predicted by the network $\mathcal{H}^r$. These inverse transformations capture the anatomical diversity of the unlabeled set and are used to deform the images. Further, an appearance network $\mathcal{H}^a$ is trained to capture the appearance of the unlabeled set. The network is designed to output the residue r between the warped image $I_c \circ \varphi$ and the atlas, $r_c = \mathcal{H}^a(A, I_c \circ \varphi_c)$. Finally, we obtain a new set of segmented images by applying the transformations to the atlas with the new appearance: $\{(A + r_i) \circ \varphi_j^{-1}, i, j \in 1 \ldots N\}$.

We modify the *Brainstorm* idea as follows: 1) Instead of sampling the transformation φ_c^{-1}, we sample $\varphi_{\tilde{m}_0^c}^{-1}(t)$ based on our fluid-based approach; 2) We remove the appearance network and instead simply use $\{I_c \circ \varphi_c, c \in 1 \ldots N\}$ to model appearance. I.e., we retain the appearance of individual images, but deform them by going through atlas space. This results by construction in a realistic appearance distribution. Our synthesized images are $\{(I_i \circ \varphi_i) \circ \varphi_{\tilde{m}_0^j}^{-1}(t), \tilde{m}_0^j \in C(M_K^j), t \in \mathbb{R}, i, j \in 1 \ldots N\}$. We refer to this approach as *Fluid-Aug*$_{real}$.

Dataset. We use the OAI dataset with 100 manually annotated images and a segmented mean atlas [13]. We only use the atlas segmentation for our one-shot segmentation experiments.

Baseline. *Upper-bound* is a model trained from 100 images and their manual segmentations. We use the same U-net as for the general segmentation task in

Sect. 4.1. *Brainstorm* is our baseline. We train a registration network and an appearance network separately, using the same network structures as in [35]. We sample a new training set of size 1,500 via random compositions of the appearance and the deformation. We also compare with a variant replacing the appearance network, where the synthesized set can be written as $\{(I_i \circ \varphi_i) \circ \varphi_j^{-1}, i, j \in 1 \dots N\}$. We refer to this approach as $Brainstorm_{real}$.

Settings. We set $K = 2$ and $t \in [-1, 2]$ and draw a new training set with 1,500 pairs the same way as in Sect. 4.1. We also compare with a variant where we set $t = 1$ (instead of randomly sampling it), which we denote $Fluid\text{-}Aug_{real_{t=1}}$. Further, we compare with a variant using the appearance network, where the synthesized set is $\{(A+r_i) \circ \varphi_{\tilde{m}_0^j}^{-1}(t), m_0^j \in C(M_K^j), t \in \mathbb{R}, i, j \in 1 \dots N\}$. We refer to this approach as *Fluid-Aug*.

Figure 4 shows better performance for fluid-based augmentation than for *Brainstorm* when using either real or learnt appearance. Furthermore, directly using the appearance of the unlabeled images shows better performance than using the appearance network. Lastly, randomizing over the location on the geodesic ($Fluid\text{–}Aug_{real}$) shows small improvements over fixing $t = 1$ ($Fluid\text{–}Aug_{real_{t_1}}$).

5 Conclusion

We introduced a fluid-based method for medical image data augmentation. Our approach makes use of a geodesic subspace capturing anatomical variability. We explored its use for general segmentation and one-shot segmentation, achieving improvements over competing methods. Future work will focus on efficiency improvements. Specifically, computing the geodesic subspaces is costly if they are not approximated by a registration network. We will therefore explore introducing multiple atlases to reduce the number of possible registration pairs.

Acknowledgements. Research reported in this publication was supported by the National Institutes of Health (NIH) and the National Science Foundation (NSF) under award numbers NSF EECS1711776 and NIH 1R01AR072013. The content is solely the responsibility of the authors and does not necessarily represent the official views of the NIH or the NSF.

References

1. Aljabar, P., Wolz, R., Rueckert, D.: Manifold learning for medical image registration, segmentation, and classification. In: Machine Learning in Computer-Aided Diagnosis: Medical Imaging Intelligence and Analysis, pp. 351–372. IGI Global (2012)
2. Ambellan, F., Tack, A., Ehlke, M., Zachow, S.: Automated segmentation of knee bone and cartilage combining statistical shape knowledge and convolutional neural networks: data from the osteoarthritis initiative. Med. Image Anal. **52**, 109–118 (2019)

3. Bai, W., et al.: Semi-supervised learning for network-based cardiac MR image segmentation. In: Descoteaux, M., Maier-Hein, L., Franz, A., Jannin, P., Collins, D.L., Duchesne, S. (eds.) MICCAI 2017. LNCS, vol. 10434, pp. 253–260. Springer, Cham (2017). https://doi.org/10.1007/978-3-319-66185-8_29
4. Baur, C., Albarqouni, S., Navab, N.: Semi-supervised deep learning for fully convolutional networks. In: Descoteaux, M., Maier-Hein, L., Franz, A., Jannin, P., Collins, D.L., Duchesne, S. (eds.) MICCAI 2017. LNCS, vol. 10435, pp. 311–319. Springer, Cham (2017). https://doi.org/10.1007/978-3-319-66179-7_36
5. Beg, M.F., Miller, M.I., Trouvé, A., Younes, L.: Computing large deformation metric mappings via geodesic flows of diffeomorphisms. IJCV **61**(2), 139–157 (2005)
6. Chaitanya, K., Karani, N., Baumgartner, C.F., Becker, A., Donati, O., Konukoglu, E.: Semi-supervised and task-driven data augmentation. In: Chung, A.C.S., Gee, J.C., Yushkevich, P.A., Bao, S. (eds.) IPMI 2019. LNCS, vol. 11492, pp. 29–41. Springer, Cham (2019). https://doi.org/10.1007/978-3-030-20351-1_3
7. Chen, S., Bortsova, G., García-Uceda Juárez, A., van Tulder, G., de Bruijne, M.: Multi-task attention-based semi-supervised learning for medical image segmentation. In: Shen, D., et al. (eds.) MICCAI 2019. LNCS, vol. 11766, pp. 457–465. Springer, Cham (2019). https://doi.org/10.1007/978-3-030-32248-9_51
8. Dupuis, P., Grenander, U., Miller, M.I.: Variational problems on flows of diffeomorphisms for image matching. Q. Appl. Math. **LVI**, 587–600 (1998)
9. Frid-Adar, M., Diamant, I., Klang, E., Amitai, M., Goldberger, J., Greenspan, H.: GAN-based synthetic medical image augmentation for increased CNN performance in liver lesion classification. Neurocomputing **321**, 321–331 (2018)
10. Hauberg, S., Freifeld, O., Larsen, A.B.L., Fisher, J., Hansen, L.: Dreaming more data: class-dependent distributions over diffeomorphisms for learned data augmentation. In: Artificial Intelligence and Statistics, pp. 342–350 (2016)
11. Heinrich, M.P., Oktay, O., Bouteldja, N.: Obelisk-one kernel to solve nearly everything: unified 3D binary convolutions for image analysis (2018)
12. Hussain, Z., Gimenez, F., Yi, D., Rubin, D.: Differential data augmentation techniques for medical imaging classification tasks. In: AMIA Annual Symposium Proceedings, vol. 2017, p. 979. American Medical Informatics Association (2017)
13. Joshi, S., Davis, B., Jomier, M., Gerig, G.: Unbiased diffeomorphic atlas construction for computational anatomy. NeuroImage **23**, S151–S160 (2004)
14. Learned-Miller, E.G.: Data driven image models through continuous joint alignment. IEEE Trans. Pattern Anal. Mach. Intell. **28**(2), 236–250 (2005)
15. Milletari, F., Navab, N., Ahmadi, S.A.: V-net: Fully convolutional neural networks for volumetric medical image segmentation. In: 2016 Fourth International Conference on 3D Vision (3DV), pp. 565–571. IEEE (2016)
16. Nie, D., Gao, Y., Wang, L., Shen, D.: ASDNet: attention based semi-supervised deep networks for medical image segmentation. In: Frangi, A.F., Schnabel, J.A., Davatzikos, C., Alberola-López, C., Fichtinger, G. (eds.) MICCAI 2018. LNCS, vol. 11073, pp. 370–378. Springer, Cham (2018). https://doi.org/10.1007/978-3-030-00937-3_43
17. Niethammer, M., Kwitt, R., Vialard, F.X.: Metric learning for image registration. In: CVPR (2019)
18. Oliveira, A., Pereira, S., Silva, C.A.: Augmenting data when training a CNN for retinal vessel segmentation: how to warp? In: 2017 IEEE 5th Portuguese Meeting on Bioengineering (ENBENG), pp. 1–4. IEEE (2017)
19. Park, S., Thorpe, M.: Representing and learning high dimensional data with the optimal transport map from a probabilistic viewpoint. In: Proceedings of the IEEE Conference on Computer Vision and Pattern Recognition, pp. 7864–7872 (2018)

20. Paschali, M., et al.: Data augmentation with manifold exploring geometric transformations for increased performance and robustness. arXiv preprint arXiv:1901.04420 (2019)
21. Pereira, S., Pinto, A., Alves, V., Silva, C.A.: Brain tumor segmentation using convolutional neural networks in MRI images. IEEE Trans. Med. Imaging **35**(5), 1240–1251 (2016)
22. Qiu, A., Younes, L., Miller, M.I.: Principal component based diffeomorphic surface mapping. IEEE Trans. Med. Imaging **31**(2), 302–311 (2011)
23. Ronneberger, O., Fischer, P., Brox, T.: U-Net: convolutional networks for biomedical image segmentation. In: Navab, N., Hornegger, J., Wells, W.M., Frangi, A.F. (eds.) MICCAI 2015. LNCS, vol. 9351, pp. 234–241. Springer, Cham (2015). https://doi.org/10.1007/978-3-319-24574-4_28
24. Roth, H.R., et al.: Anatomy-specific classification of medical images using deep convolutional nets. In: 2015 IEEE 12th International Symposium on Biomedical Imaging (ISBI), pp. 101–104. IEEE (2015)
25. Shattuck, D.W., et al.: Construction of a 3D probabilistic atlas of human cortical structures. Neuroimage **39**(3), 1064–1080 (2008)
26. Shen, Z., Han, X., Xu, Z., Niethammer, M.: Networks for joint affine and nonparametric image registration. In: CVPR (2019)
27. Shen, Z., Vialard, F.X., Niethammer, M.: Region-specific diffeomorphic metric mapping. In: Advances in Neural Information Processing Systems, pp. 1096–1106 (2019)
28. Shin, H.C., et al.: Medical image synthesis for data augmentation and anonymization using generative adversarial networks. In: Gooya, A., Goksel, O., Oguz, I., Burgos, N. (eds.) SASHIMI 2018. LNCS, vol. 11037, pp. 1–11. Springer, Cham (2018). https://doi.org/10.1007/978-3-030-00536-8_1
29. The Osteoarthritis Initiative: Osteoarthritis initiative (OAI) dataset. https://nda.nih.gov/oai/
30. Vakalopoulou, M., et al.: AtlasNet: multi-atlas non-linear deep networks for medical image segmentation. In: Frangi, A.F., Schnabel, J.A., Davatzikos, C., Alberola-López, C., Fichtinger, G. (eds.) MICCAI 2018. LNCS, vol. 11073, pp. 658–666. Springer, Cham (2018). https://doi.org/10.1007/978-3-030-00937-3_75
31. Xu, Z., Niethammer, M.: DeepAtlas: joint semi-supervised learning of image registration and segmentation. arXiv preprint arXiv:1904.08465 (2019)
32. Xu, Z., Shen, Z., Niethammer, M.: Contextual additive networks to efficiently boost 3D image segmentations. In: Stoyanov, D., et al. (eds.) DLMIA/ML-CDS -2018. LNCS, vol. 11045, pp. 92–100. Springer, Cham (2018). https://doi.org/10.1007/978-3-030-00889-5_11
33. Younes, L., Arrate, F., Miller, M.I.: Evolutions equations in computational anatomy. NeuroImage **45**(1), S40–S50 (2009)
34. Zhang, M., Singh, N., Fletcher, P.T.: Bayesian estimation of regularization and atlas building in diffeomorphic image registration. In: Gee, J.C., Joshi, S., Pohl, K.M., Wells, W.M., Zöllei, L. (eds.) IPMI 2013. LNCS, vol. 7917, pp. 37–48. Springer, Heidelberg (2013). https://doi.org/10.1007/978-3-642-38868-2_4
35. Zhao, A., Balakrishnan, G., Durand, F., Guttag, J.V., Dalca, A.V.: Data augmentation using learned transformations for one-shot medical image segmentation. In: Proceedings of the IEEE Conference on Computer Vision and Pattern Recognition, pp. 8543–8553 (2019)
36. Zhou, Y., et al.: Collaborative learning of semi-supervised segmentation and classification for medical images. In: Proceedings of the IEEE Conference on Computer Vision and Pattern Recognition, pp. 2079–2088 (2019)

Generalizing Spatial Transformers to Projective Geometry with Applications to 2D/3D Registration

Cong Gao(✉), Xingtong Liu, Wenhao Gu, Benjamin Killeen, Mehran Armand, Russell Taylor, and Mathias Unberath

Johns Hopkins University, Baltimore, MD 21218, USA
cgao11@jhu.edu

Abstract. Differentiable rendering is a technique to connect 3D scenes with corresponding 2D images. Since it is differentiable, processes during image formation can be learned. Previous approaches to differentiable rendering focus on mesh-based representations of 3D scenes, which is inappropriate for medical applications where volumetric, voxelized models are used to represent anatomy. We propose a novel Projective Spatial Transformer module that generalizes spatial transformers to projective geometry, thus enabling differentiable volume rendering. We demonstrate the usefulness of this architecture on the example of 2D/3D registration between radiographs and CT scans. Specifically, we show that our transformer enables end-to-end learning of an image processing and projection model that approximates an image similarity function that is convex with respect to the pose parameters, and can thus be optimized effectively using conventional gradient descent. To the best of our knowledge, we are the first to describe the spatial transformers in the context of projective transmission imaging, including rendering and pose estimation. We hope that our developments will benefit related 3D research applications. The source code is available at https://github.com/gaocong13/Projective-Spatial-Transformers.

1 Introduction

Differentiable renderers that connect 3D scenes with 2D images thereof have recently received considerable attention [7,14,15] as they allow for simulating, and more importantly *inverting*, the physical process of image formation. Such approaches are designed for integration with gradient-based machine learning techniques including deep learning to, e.g., enable single-view 3D scene reconstruction. Previous approaches to differentiable rendering have largely focused on mesh-based representation of 3D scenes. This is because compared to say, volumetric representations, mesh parameterizations provide a good compromise between spatial resolution and data volume. Unfortunately, for most medical

Supported by NIH R01EB023939, NIH R21EB020113, NIH R21EB028505 and Johns Hopkins University Applied Physics Laboratory internal funds.

A. L. Martel et al. (Eds.): MICCAI 2020, LNCS 12263, pp. 329–339, 2020.
https://doi.org/10.1007/978-3-030-59716-0_32

applications the 3D scene of interest, namely the anatomy, is acquired in volumetric representation where every voxel represents some specific physical property. Deriving mesh-based representations of anatomy from volumetric data is possible in some cases [2], but is not yet feasible nor desirable in general, since surface representations cannot account for tissue variations within one closed surface. However, solutions to the differentiable rendering problem are particularly desirable for X-ray-based imaging modalities, where 3D content is reconstructed from – or aligned to multiple 2D transmission images. This latter process is commonly referred to as 2D/3D registration and we will use it as a test-bed within this manuscript to demonstrate the value of our method.

Mathematically, the mapping from volumetric 3D scene V to projective transmission image I_m can be modeled as $I_m = A(\theta)V$, where $A(\theta)$ is the system matrix that depends on pose parameter $\theta \in \mathrm{SE}(3)$. In intensity-based 2D/3D registration, we seek to retrieve the pose parameter θ such that the moving image I_m simulated from V is as similar as possible to the target fixed image I_f:

$$\min_{\theta} L_S(I_f, I_m) = \min_{\theta} L_S(I_f, A(\theta)V), \tag{1}$$

where L_S is the similarity metric loss function. Gradient decent-based optimization methods require the gradient $\frac{\partial L_S}{\partial \theta} = \frac{\partial L_S}{\partial A(\theta)} \cdot \frac{\partial A(\theta)}{\partial \theta}$ at every iteration. Although the mapping was constructed to be differentiable, analytic gradient computation is still impossible due to excessively large memory footprint of A for all practical problem sizes[1]. Traditional stochastic optimization strategies are numeric-based methods, such as CMA-ES [4]. Since the similarity functions are manually crafted, such as mutual information (MI) [16] or normalized cross correlation, these methods require an initialization which is close to the global optimum, and thus suffer from a small capture range. Recent deep learning-based methods put efforts on learning a similarity metric or regressing the pose transformation from the image observations (I_f, I_m) to extend the capture range [6,8,19]. Several researchers proposed reinforcement learning paradigms to iteratively estimate a transformation [11,13,18]. However, these learning-based methods only trained on 2D images with no gradient connection to 3D space. Spatial transformer network (STN) [9] has been applied on 3D registration problems to estimate deformation field [1,12]. Yan et al. proposed perspective transformer nets which applied STN for 3D volume reconstruction [26]. In this work, we propose an analytically differentiable volume renderer that follows the terminology of spatial transformer networks and extends their capabilities to spatial transformations in projective transmission imaging. Our specific contributions are:

- We introduce a **Pro**jective **S**patial **T**ransformer (ProST) module that generalizes spatial transformers [9] to projective geometry. This enables volumetric rendering of transmission images that is differentiable both with respect to the input volume V as well as the pose parameters θ.

[1] It is worth mentioning that this problem can be circumvented via ray casting-based implementations if one is interested in $\partial L_S / \partial V$ but not in $\partial L_S / \partial \theta$ [25].

- We demonstrate how ProST can be used to solve the non-convexity problem of conventional intensity-based 2D/3D registration. Specifically, we train an end-to-end deep learning model to approximate a convex loss function derived from geodesic distances between poses θ and enforce desirable pose updates $\frac{\partial L_S}{\partial \theta}$ via double backward functions on the computational graph.

2 Methodology

2.1 Projective Spatial Transformer (ProST)

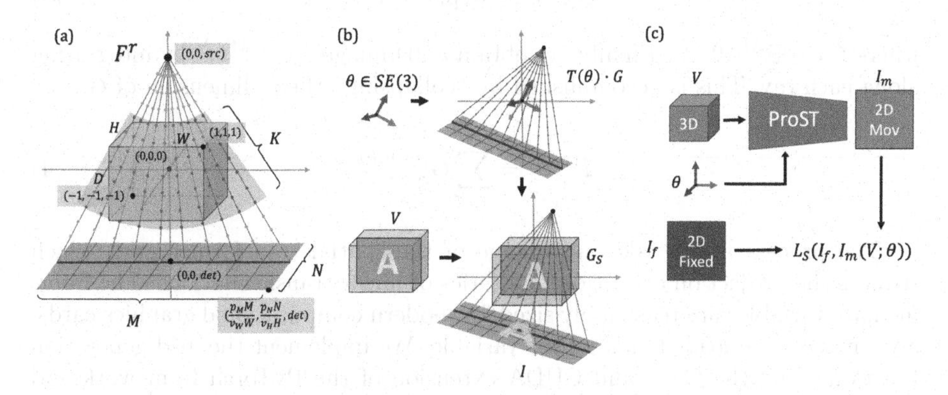

Fig. 1. (a) Canonical projection geometry and a slice of cone-beam grid points are presented with key annotations. The green fan covers the control points which are used for further reshape. (b) Illustration of grid sampling transformer and projection. (c) Scheme of applying ProST to 2D/3D registration. (Color figure online)

Canonical Projection Geometry. Given a volume $V \in \mathbb{R}^{D \times W \times H}$ with voxel size $v_D \times v_W \times v_H$, we define a reference frame F^r with the origin at the center of V. We use normalized coordinates for depth (Dv_D), width (Wv_W) and height (Hv_H), so that the points of V are contained within the unit cube $(d, w, h) \in [-1, 1]^3$. Given a camera intrinsic matrix $\mathcal{K} \in \mathbb{R}^{3\times3}$, we denote the associated source point as $(0, 0, src)$ in F^r. The spatial grid G of control points, shown in Fig. 1(a), lies on $M \times N$ rays originating from this source. Because the control points in regions where no CT voxels exist will not contribute to the line integral, we cut the grid point cloud to a cone-shape structure that covers the exact volume space. Thus, each ray has K control points uniformly spaced within the volume V, so that the matrix $G \in \mathbb{R}^{4\times(M \cdot N \cdot K)}$ of control points is well-defined, where each column is a control point in homogeneous coordinates. These rays describe a cone-beam geometry which intersects with the detection plane, centered on $(0, 0, det)$ and perpendicular to the z axis with pixel size $p_M \times p_N$, as determined by $\mathcal{K}$. The upper-right corner of the detection plane is at $(\frac{p_M M}{v_W W}, \frac{p_N N}{v_H H}, det)$.

Grid Sampling Transformer. Our Projective Spatial Transformer (ProST) extends the canonical projection geometry by learning a transformation of the control points G. Given $\theta \in \mathrm{SE}(3)$, we obtain a transformed set of control points via the affine transformation matrix $T(\theta)$:

$$G_T = T(\theta) \cdot G, \tag{2}$$

as well as source point $T(\theta) \cdot (0, 0, src, 1)$ and center of detection plane $T(\theta) \cdot (0, 0, det, 1)$. Since these control points lie within the volume V but in between voxels, we interpolate the values G_S of V at the control points:

$$G_S = \mathtt{interp}(V, G_T), \tag{3}$$

where $G_S \in \mathbb{R}^{M \times N \times K}$. Finally, we obtain a 2D image $I \in \mathbb{R}^{M \times N}$ by integrating along each ray. This is accomplished by "collapsing" the k dimension of G_S:

$$I^{(m,n)} = \sum_{k=1}^{K} G_S^{(m,n,k)} \tag{4}$$

The process above takes advantage of the spatial transformer grid, which reduces the projection operation to a series of linear transformations. The intermediate variables are reasonably sized for modern computational graphics cards, and thus can be loaded as a tensor variable. We implement the grid generation function using the C++ and CUDA extension of the PyTorch framework and embed the projection operation as a PyTorch layer with tensor variables. With the help of PyTorch autograd function, this projection layer enables analytical gradient flow from the projection domain back to the spatial domain. Figure 1(c) shows how this scheme is applied to 2D/3D registration. Without any learning parameters, we can perform registration with PyTorch's powerful built-in optimizers on large-scale volume representations. Furthermore, integrating deep convolutional layers, we show that ProST makes end-to-end 2D/3D registration feasible.

2.2 Approximating Convex Image Similarity Metrics

Following [3], we formulate an intensity-based 2D/3D registration problem with a pre-operative CT volume V, Digitally Reconstructed Radiograph (DRR) projection operator P, pose parameter θ, a fixed target image I_f, and a similarity metric loss L_S:

$$\min_{\theta \in SE(3)} L_S\Big(I_f, P(V; \theta)\Big). \tag{5}$$

Using our projection layer P, we propose an end-to-end deep neural network architecture which will learn a convex similarity metric, aiming to extend the capture range of the initialization for 2D/3D registration. Geodesic loss, L_G, which is the square of geodesic distance in $SE(3)$, has been studied for registration problems due to its convexity [17,23]. We take the implementation of [20] to

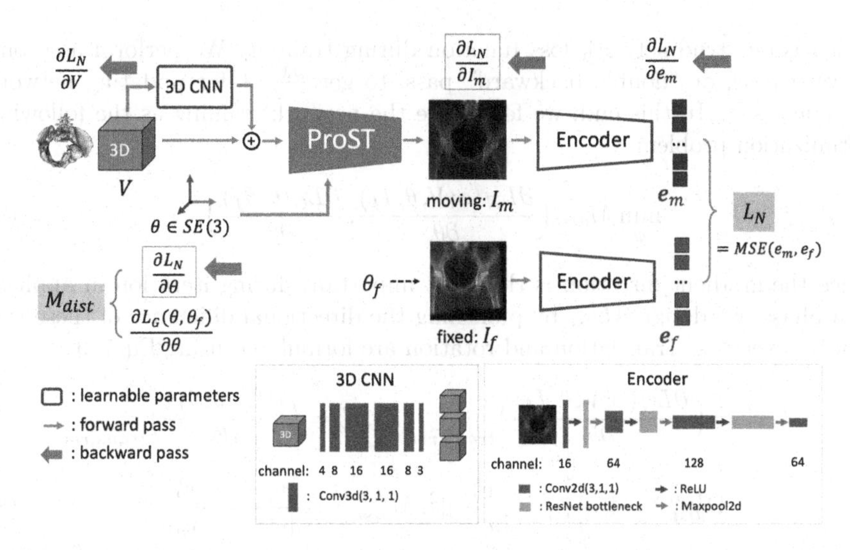

Fig. 2. DeepNet architecture. Forward pass follows the blue arrows. Backward pass follows pink arrows, where gradient input and output of ProST in Eq. 10 are highlighted with pink border. (Color figure online)

calculate the geodesic gradient $\frac{\partial L_G(\theta,\theta_f)}{\theta}$, given a sampling pose θ and a target pose θ_f. We then use this geodesic gradient to train our network, making our training objective exactly the same as our target task – learning a convex shape similarity metric.

Figure 2 shows our architecture. The input includes a 3D volume: V, a pose parameter: $\theta \in SE(3)$ and a fixed target image: I_f. All blocks which contain learnable parameters are highlighted with a red outline. The 3D CNN is a skip connection from the input volume to multi-channel expansion just to learn the residual. Projections are performed by projection layer with respect to θ, which does not have learnable parameters. The projected moving image I_m and the fixed image I_f go through two encoders, which are the same in structure but the weights are not shared, and output embedded features e_m and e_f. Our network similarity metric L_N is the mean squared error of e_m and e_f. We will then explain the design from training phase and application phase separately.

Training Phase. The goal of training is to make the gradient of our network error function w.r.t. pose parameter, $\frac{\partial L_N}{\partial \theta}$, close to the geodesic gradient $\frac{\partial L_G}{\partial \theta}$. The blue arrows in Fig. 2 show the forward pass in a single iteration. The output can be written as $L_N(\phi; V, \theta, I_f)$, where ϕ are the network parameters. We then apply back-propagation, illustrated with pink arrows in Fig. 2. This yields $\frac{\partial L_N}{\partial \theta}$ and $\frac{\partial L_N}{\partial \phi}$. Assuming L_N is the training loss, ϕ would normally be updated according to $lr \cdot \frac{\partial L_N}{\partial \phi}$, where lr is the learning rate. However, we *do not* update the network parameters during the backward pass. Instead we obtain the gradient and calculate a distance measure of these two gradient vectors, $M_{dist}(\frac{\partial L_N}{\partial \theta}, \frac{\partial L_G}{\partial \theta})$, which is our true network loss function during training. We perform a second forward pass, or "double backward" pass, to get $\frac{\partial M_{dist}}{\partial \phi}$ for updating network parameters ϕ. To this end, we formulate the network training as the following optimization problem

$$\min_{\phi} M_{dist}\Big(\frac{\partial L_N(\phi; V, \theta, I_f)}{\partial \theta}, \frac{\partial L_G(\theta, \theta_f)}{\partial \theta}\Big). \tag{6}$$

Since the gradient direction is the most important during iteration in application phase, we design M_{dist} by punishing the directional difference of these two gradient vectors. Translation and rotation are formulated using Eq. 7–9

$$v_1^t, v_1^r = \Big(\frac{\partial L_N(\phi; V, \theta, I_f)}{\partial \theta}\Big)_{trans,rot}; v_2^t, v_2^r = \Big(\frac{\partial L_G(\theta, \theta_f)}{\partial \theta}\Big)_{trans,rot} \tag{7}$$

$$M_{dist}^{trans} = ||\frac{v_1^t}{||v_1^t||} - \frac{v_2^t}{||v_2^t||}||^2, M_{dist}^{rot} = ||\frac{v_1^r}{||v_1^r||} - \frac{v_2^r}{||v_2^r||}||^2 \tag{8}$$

$$M_{dist} = M_{dist}^{trans} + M_{dist}^{rot}, \tag{9}$$

where the rotation vector is transformed into Rodrigues angle axis.

Application Phase. During registration, we fix the network parameters ϕ and start with an initial pose θ. We can perform gradient-based optimization over θ based on the following back-propagation gradient flow

$$\frac{\partial L_N}{\partial \theta} = \frac{\partial L_N}{\partial I_m} \cdot \frac{\partial I_m}{\partial G_S} \cdot \frac{\partial G_S}{\partial G_T} \cdot \frac{\partial G_T}{\partial T(\theta)} \cdot \frac{\partial T(\theta)}{\partial \theta}. \tag{10}$$

The network similarity is more effective when the initial pose is far away from the groundtruth, while less sensitive to local textures compared to traditional image-based methods, such as Gradient-based Normalized Corss Correlation (Grad-NCC) [21]. We implement Grad-NCC as a pytorch loss function L_{GNCC}, and combine these two methods to build an end-to-end pipeline for 2D/3D registration. We first detect the convergence of the network-based optimization process by monitoring the standard deviation (STD) of L_N. After it converges, we then switch to optimize over L_{GNCC} until final convergence.

3 Experiments

3.1 Simulation Study

We define our canonical projection geometry following the intrinsic parameter of a Siemens CIOS Fusion C-Arm, which has image dimensions of 1536 × 1536, isotropic pixel spacing of 0.194 mm/pixel, a source-to-detector distance of 1020 mm. We downsample the detector dimension to be 128 × 128. We train our algorithm using 17 full body CT scans from the NIH Cancer Imaging Archive [22] and leave 1 CT for testing. The pelvis bone is segmented using an automatic method in [10]. CTs and segmentations are cropped to the pelvis cubic region and downsampled to the size of $128 \times 128 \times 128$. The world coordinate frame origin is set at center of the processed volume, which 400 mm above the detector plane center.

At training iteration i, we randomly sample a pair of pose parameters, (θ^i, θ^i_f), rotation from $N(0, 20)$ in degree, translation from $N(0, 30)$ in mm, in all three axes. We then randomly select a CT and its segmentation, V_{CT} and V_{Seg}. The target fixed image is generated online from V_{CT} and θ^i_f using our ProST. V_{Seg} and θ are used as input to our network forward pass. The network is trained using SGD optimizer with a cyclic learning rate between 1e−6 and 1e−4 every 100 steps [24] and a momentum of 0.9. Batch size is chosen as 2 and we trained 100k iterations until convergence.

We performed the 2D/3D registration application by randomly choosing a pose pair from the same training distribution, (θ^R, θ^R_f). Target fixed image is generated from the testing CT and θ^R_f. We then use SGD optimizer to optimize over θ^R with a learning rate of 0.01, momentum of 0.9 for iteration. We calculate the STD of the last 10 iterations of L_N as std_{L_N}, and set a stop criterion of $std_{L_N} < 3 \times 10^{-3}$, then we switch to Gradient-NCC similarity using SGD optimizer with cyclic learning rate between 1e−3 and 3e−3, and set the stop criterion, $std_{L_{NCC}} < 1 \times 10^{-5}$. We conduct in total of 150 simulation studies for testing our algorithm.

3.2 Real X-Ray Study

We collected 10 real X-ray images from a cadaver specimen. Groundtruth pose is obtained by injecting metallic BBs 1 mm diameter into the surface of the bone and manually annotated from the X-ray images and CT scan. The pose is recovered by solving a PnP problem [5]. For each X-ray image, we randomly choose a pose parameter, rotation from $N(0, 15)$ in degree, translation from $N(0, 22.5)$ in mm, in all three axes. 10 registrations are performed for each image using the same pipeline, resulting in a total of 100 registrations.

4 Results

We compared the performance of four methods, which are Grad-NCC with SGD optimizer (GradNCC SGD), Grad-NCC with CMA-ES optimizer (GradNCC CMAES), Net only, and Net+GradNCC. The registration accuracy was used as the evaluation metric, where the rotation and translation errors are expressed in degree and millimeter, respectively. The coordinate frame F^r are used to define the origin and orientation of the pose. In Fig. 3, both qualitative and quantitative results on the testing data are shown. Numeric results are shown in Table 1. The Net+GradNCC works the best among comparisons in both studies.

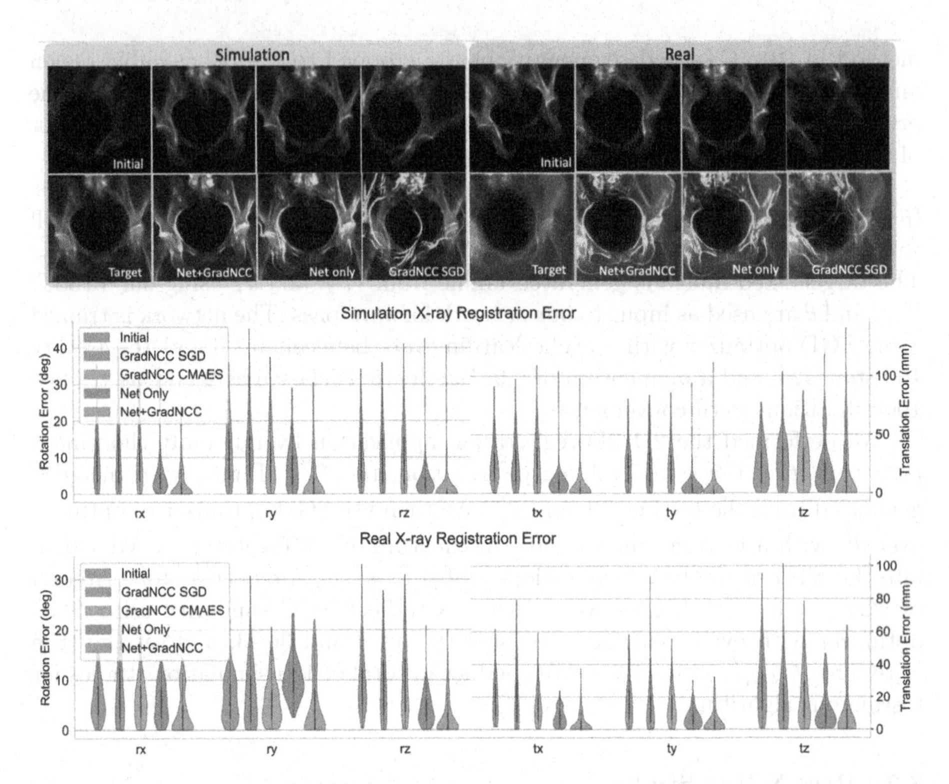

Fig. 3. The top row shows qualitative examples of Net+GradNCC, Net only, GradNCC only convergence overlap, for simulation and real X-ray respectively. The middle is the registration error distribution of simulation. The bottom is the distribution for real X-ray experiments. x, y and z-axis correspond to LR, IS and AP views.

5 Discussion

We have seen from the results that our method largely increases the capture range of 2D/3D registration. Our method follows the same iterative optimization design as the intensity-based registration methods, where the only difference is that we take advantage of the great expressivity of deep network to learn a set of more complicated filters than the conventional hand-crafted ones. This potentially makes generalization easier because the mapping that our method needs to learn is simple. In the experiment, we observed that the translation along the depth direction is less accurate than other directions in both simulation and real studies, as shown in Fig. 3, which we attribute to the current design of the network architecture and will work on that as a future direction.

Table 1. Quantitative results of 2D/3D registration

		Simulation study		Real X-ray study	
		Translation	Rotation	Translation	Rotation
Initialization		41.57 ± 18.01	21.16 ± 9.27	30.50 ± 13.90	14.22 ± 5.56
GradNCC SGD	Mean	41.83 ± 23.08	21.97 ± 11.26	29.52 ± 20.51	15.76 ± 8.37
	Median	38.30	22.12	26.28	16.35
GradNCC CMAES	Mean	40.68 ± 22.04	20.16 ± 9.32	25.64 ± 12.09	14.31 ± 6.74
	Median	37.80	20.63	23.87	13.80
Net	Mean	13.10 ± 18.53	10.21 ± 7.55	12.14 ± 6.44	13.00 ± 4.42
	Median	9.85	9.47	11.06	12.61
Net+GradNCC	Mean	7.83 ± 19.8	4.94 ± 8.78	7.02 ± 9.22	6.94 ± 7.47
	Median	0.25	0.27	2.89	3.76

6 Conclusion

We propose a novel Projective Spatial Transformer module (ProST) that generalizes spatial transformers to projective geometry, which enables differentiable volume rendering. We apply this to an example application of 2D/3D registration between radiographs and CT scans with an end-to-end learning architecture that approximates convex loss function. We believe this is the first time that spatial transformers have been introduced for projective geometry and our developments will benefit related 3D research applications.

References

1. Ferrante, E., Oktay, O., Glocker, B., Milone, D.H.: On the adaptability of unsupervised CNN-based deformable image registration to unseen image domains. In: Shi, Y., Suk, H.-I., Liu, M. (eds.) MLMI 2018. LNCS, vol. 11046, pp. 294–302. Springer, Cham (2018). https://doi.org/10.1007/978-3-030-00919-9_34
2. Gibson, E., et al.: Automatic multi-organ segmentation on abdominal CT with dense V-networks. IEEE Trans. Med. Imaging **37**(8), 1822–1834 (2018)
3. Grupp, R., et al.: Automatic annotation of hip anatomy in fluoroscopy for robust and efficient 2D/3D registration. arXiv preprint arXiv:1911.07042 (2019)
4. Hansen, N., Müller, S.D., Koumoutsakos, P.: Reducing the time complexity of the derandomized evolution strategy with covariance matrix adaptation (CMA-ES). Evol. Comput. **11**(1), 1–18 (2003)
5. Hartley, R., Zisserman, A.: Multiple View Geometry in Computer Vision. Cambridge University Press, Cambridge (2003)
6. Haskins, G., Kruger, U., Yan, P.: Deep learning in medical image registration: a survey. Mach. Vis. Appl. 1–18 (2020). https://doi.org/10.1007/s00138-020-01060-x
7. Henderson, P., Ferrari, V.: Learning single-image 3D reconstruction by generative modelling of shape, pose and shading. Int. J. Comput. Vis. **128**, 835–854 (2020). https://doi.org/10.1007/s11263-019-01219-8

8. Hou, B., et al.: Predicting slice-to-volume transformation in presence of arbitrary subject motion. In: Descoteaux, M., Maier-Hein, L., Franz, A., Jannin, P., Collins, D.L., Duchesne, S. (eds.) MICCAI 2017. LNCS, vol. 10434, pp. 296–304. Springer, Cham (2017). https://doi.org/10.1007/978-3-319-66185-8_34
9. Jaderberg, M., Simonyan, K., Zisserman, A., et al.: Spatial transformer networks. In: Advances in Neural Information Processing Systems, pp. 2017–2025 (2015)
10. Krčah, M., Székely, G., Blanc, R.: Fully automatic and fast segmentation of the femur bone from 3D-CT images with no shape prior. In: 2011 IEEE International Symposium on Biomedical Imaging: from Nano to Macro, pp. 2087–2090. IEEE (2011)
11. Krebs, J., et al.: Robust non-rigid registration through agent-based action learning. In: Descoteaux, M., Maier-Hein, L., Franz, A., Jannin, P., Collins, D.L., Duchesne, S. (eds.) MICCAI 2017. LNCS, vol. 10433, pp. 344–352. Springer, Cham (2017). https://doi.org/10.1007/978-3-319-66182-7_40
12. Kuang, D., Schmah, T.: FAIM – a ConvNet method for unsupervised 3D medical image registration. In: Suk, H.-I., Liu, M., Yan, P., Lian, C. (eds.) MLMI 2019. LNCS, vol. 11861, pp. 646–654. Springer, Cham (2019). https://doi.org/10.1007/978-3-030-32692-0_74
13. Liao, R., et al.: An artificial agent for robust image registration. In: Thirty-First AAAI Conference on Artificial Intelligence (2017)
14. Liu, S., Li, T., Chen, W., Li, H.: Soft rasterizer: a differentiable renderer for image-based 3D reasoning. In: Proceedings of the IEEE International Conference on Computer Vision, pp. 7708–7717 (2019)
15. Loper, M.M., Black, M.J.: OpenDR: an approximate differentiable renderer. In: Fleet, D., Pajdla, T., Schiele, B., Tuytelaars, T. (eds.) ECCV 2014. LNCS, vol. 8695, pp. 154–169. Springer, Cham (2014). https://doi.org/10.1007/978-3-319-10584-0_11
16. Maes, F., Collignon, A., Vandermeulen, D., Marchal, G., Suetens, P.: Multimodality image registration by maximization of mutual information. IEEE Trans. Med. Imaging **16**(2), 187–198 (1997)
17. Mahendran, S., Ali, H., Vidal, R.: 3D pose regression using convolutional neural networks. In: Proceedings of the IEEE International Conference on Computer Vision Workshops, pp. 2174–2182 (2017)
18. Miao, S., et al.: Dilated FCN for multi-agent 2D/3D medical image registration. In: Thirty-Second AAAI Conference on Artificial Intelligence (2018)
19. Miao, S., Wang, Z.J., Liao, R.: A CNN regression approach for real-time 2D/3D registration. IEEE Trans. Med. Imaging **35**(5), 1352–1363 (2016)
20. Miolane, N., Mathe, J., Donnat, C., Jorda, M., Pennec, X.: Geomstats: a python package for riemannian geometry in machine learning. arXiv preprint arXiv:1805.08308 (2018)
21. Penney, G.P., Weese, J., Little, J.A., Desmedt, P., Hill, D.L., et al.: A comparison of similarity measures for use in 2-D-3-D medical image registration. IEEE Trans. Med. Imaging **17**(4), 586–595 (1998)
22. Roth, H.R., et al.: A new 2.5D representation for lymph node detection using random sets of deep convolutional neural network observations. In: Golland, P., Hata, N., Barillot, C., Hornegger, J., Howe, R. (eds.) MICCAI 2014. LNCS, vol. 8673, pp. 520–527. Springer, Cham (2014). https://doi.org/10.1007/978-3-319-10404-1_65
23. Salehi, S.S.M., Khan, S., Erdogmus, D., Gholipour, A.: Real-time deep registration with geodesic loss. arXiv preprint arXiv:1803.05982 (2018)

24. Smith, L.N.: Cyclical learning rates for training neural networks. In: 2017 IEEE Winter Conference on Applications of Computer Vision (WACV), pp. 464–472. IEEE (2017)
25. Würfl, T., et al.: Deep learning computed tomography: learning projection-domain weights from image domain in limited angle problems. IEEE Trans. Med. Imaging **37**(6), 1454–1463 (2018)
26. Yan, X., Yang, J., Yumer, E., Guo, Y., Lee, H.: Perspective transformer nets: learning single-view 3D object reconstruction without 3D supervision. In: Advances in Neural Information Processing Systems, pp. 1696–1704 (2016)

Instrumentation and Surgical Phase Detection

TeCNO: Surgical Phase Recognition with Multi-stage Temporal Convolutional Networks

Tobias Czempiel[1(✉)], Magdalini Paschali[1], Matthias Keicher[1], Walter Simson[1], Hubertus Feussner[2], Seong Tae Kim[1], and Nassir Navab[1,3]

[1] Computer Aided Medical Procedures, Technische Universität München, Munich, Germany
tobias.czempiel@tum.de

[2] MITI, Klinikum Rechts der Isar, Technische Universität München, Munich, Germany

[3] Computer Aided Medical Procedures, Johns Hopkins University, Baltimore, USA

Abstract. Automatic surgical phase recognition is a challenging and crucial task with the potential to improve patient safety and become an integral part of intra-operative decision-support systems. In this paper, we propose, for the first time in workflow analysis, a Multi-Stage Temporal Convolutional Network (MS-TCN) that performs hierarchical prediction refinement for surgical phase recognition. Causal, dilated convolutions allow for a large receptive field and online inference with smooth predictions even during ambiguous transitions. Our method is thoroughly evaluated on two datasets of laparoscopic cholecystectomy videos with and without the use of additional surgical tool information. Outperforming various state-of-the-art LSTM approaches, we verify the suitability of the proposed causal MS-TCN for surgical phase recognition.

Keywords: Surgical workflow · Surgical phase recognition · Temporal Convolutional Networks · Endoscopic videos · Cholecystectomy

1 Introduction

Surgical workflow analysis is an integral task to increase patient safety, reduce surgical errors and optimize the communication in the operating room (OR) [1]. Specifically, surgical phase recognition can provide vital input to physicians in the form of early warnings in cases of deviations and anomalies [2] as well as context-aware decision support [3]. Another use case is automatic extraction of a surgery's protocol, which is crucial for archiving, educational and post-operative patient-monitoring purposes [4].

Computer-assisted intervention (CAI) systems based on machine learning techniques have been developed for surgical workflow analysis [5], deploying not only OR signals but also intra-operative videos, which can be captured during a

A. L. Martel et al. (Eds.): MICCAI 2020, LNCS 12263, pp. 343–352, 2020.
https://doi.org/10.1007/978-3-030-59716-0_33

laparoscopic procedure, since cameras are an integral part of the workflow. However, the task of surgical phase recognition from intra-operative videos remains challenging even for advanced CAI systems [6,7] due to the variability of patient anatomy and surgeon style [8] along with the limited availability and quality of video material [9]. Furthermore, strong similarities among phases and transition ambiguity lead to decreased performance and limited generalizability of the existing methods. Finally, most approaches dealing with temporal information, such as Recurrent Neural Networks (RNNs) [10] leverage sliding window detectors, which have difficulties capturing long-term temporal patterns.

Towards this end, we propose a pipeline utilizing dilated Temporal Convolutional Networks (TCN) [11] for accurate and fast surgical phase recognition. Their large temporal receptive field captures the full temporal resolution with a reduced number of parameters, allowing for faster training and inference time and leveraging of long, untrimmed surgical videos.

Initial approaches for surgical phase recognition [5] exploited binary surgical signals. Hidden Markov Models (HMMs) captured the temporal information with the use of Dynamic Time Warping (DTW). However, such methods relied on whole video sequences and could not be applied in an online surgery scenario. EndoNet [12] jointly performed surgical tool and phase recognition from videos, utilizing a shared feature extractor and a hierarchical HMM to obtain temporally-smoothed phase predictions. With the rise of RNNs, EndoNet was evolved to EndoLSTM, which was trained in a two-step process including a Convolutional Neural Network (CNN) as a feature extractor and an LSTM [13] for feature refinement. Endo2N2 [14] leveraged self-supervised pre-training of the feature extractor CNN by predicting the Remaining Surgery Duration (RSD). Afterwards a CNN-LSTM model was trained end-to-end to perform surgical phase recognition. Similarly, SV-RCNet [15] trained an end-to-end ResNet [16] and LSTM model for surgical phase recognition with a prior knowledge inference scheme.

MTRCNet-CL [17] approached surgical phase classification as a multi-task problem. Extracted frame features were used to predict tool information while also serving as input to an LSTM model [13] for the surgical phase prediction. A correlation loss was employed to enhance the synergy between the two tasks. The common factor of the methods mentioned above is the use of LSTMs, which retain memory of a limited sequence, that cannot span minutes or hours, which is the average duration of a surgery. Thus, they process the temporal information in a slow, sequential way prohibiting inference parallelization, which would be beneficial for their integration in an online OR scenario.

Temporal convolutions [11] were introduced to hierarchically process videos for action segmentation. An encoder-decoder architecture was able to capture both high- and low-level features in contrast to RNNs. Later, TCNs adapted dilated convolutions [18] for action localization and achieved improvement in performance due to a larger receptive field for higher temporal resolution. Multi-Stage TCNs (MS-TCNs) [19] were introduced for action segmentation and consisted of stacked predictor stages. Each stage included an individual

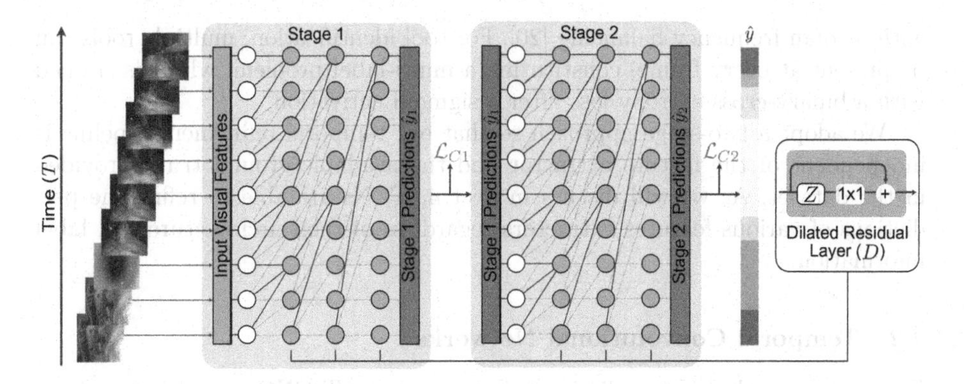

Fig. 1. Overview of the proposed TeCNO multi-stage hierarchical refinement model. The extracted frame features are forwarded to Stage 1 of our TCN, which consists of 1D dilated convolutional and dilated residual layers D. Cross-entropy loss is calculated after each stage and aggregated for the joint training of the model.

multi-layer TCN, which incrementally refined the initial prediction of the previous stages.

In this paper our contribution is two-fold: (1) We propose, for the first time in surgical workflow analysis, the introduction of causal, dilated MS-TCNs for accurate, fast and refined online surgical phase recognition. We call our method TeCNO, derived from **Te**mporal **C**onvolutional **N**etworks for the **O**perating room. (2) We extensively evaluate TeCNO on the challenging task of surgical phase recognition on two laparoscopic video datasets, verifying the effectiveness of the proposed approach.

2 Methodology

TeCNO constitutes a surgical workflow recognition pipeline consisting of the following steps: 1) We employ a ResNet50 as a visual feature extractor. 2) We refine the extracted features with a 2-stage causal TCN model that forms a high-level reasoning of the current frame by analyzing the preceding ones. The refinement 2-stage TCN model is depicted in Fig. 1.

2.1 Feature Extraction Backbone

A ResNet50 [16] is trained frame-wise without temporal context as a feature extractor from the video frames either on a single task for phase recognition or as a multi-task network when a dataset provides additional label information, for instance tool presence per frame. In the multi-task scenario for concurrent phase recognition and tool identification, our model concludes with two separate linear layers, whose losses are combined to train the model jointly. Since phase recognition is an imbalanced multi-class problem we utilize softmax activations and weighted cross entropy loss for this task. The class weights are calculated with median frequency balancing [20]. For tool identification, multiple tools can be present at every frame, constituting a multi-label problem, which is trained with a binary-cross entropy loss after a sigmoid activation.

We adopt a two-stage approach so that our temporal refinement pipeline is independent of the feature extractor and the available ground truth provided in the dataset. As we will discuss in Sect. 4, TCNs are able to refine the predictions of various features extractors regardless of their architecture and label information.

2.2 Temporal Convolutional Networks

For the temporal phase prediction task, we propose TeCNO, a multi-stage temporal convolutional network that is visualized in Fig. 1. Given an input video consisting of $x_{1:t}$, $t \in [1, T]$ frames, where T is the total number of frames, the goal is to predict $y_{1:t}$ where y_t is the class label for the current time step t. Our temporal model follows the design of MS-TCN and contains neither pooling layers, that would decrease the temporal resolution, nor fully connected layers, which would increase the number of parameters and require a fixed input dimension. Instead, our model is constructed solely with temporal convolutional layers.

The first layer of Stage 1 is a 1×1 convolutional layer that matches the input feature dimension to the chosen feature length forwarded to the next layer within the TCN. Afterwards, dilated residual (D) layers perform dilated convolutions as described in Eq. 1 and Eq. 2. The major component of each D layer is the dilated convolutional layer (Z).

$$Z_l = ReLU(W_{1,l} * D_{l-1} + b_{1,l}) \tag{1}$$

$$D_l = D_{l-1} + W_{2,l} * Z_l + b_{2,l} \tag{2}$$

D_l is the output of D (Eq. 2), while Z_l is the result of the dilated convolution of kernel $W_{1,l}$ with the output of the previous layer D_{l-1} activated by a *ReLU* (Eq. 1). $W_{2,l}$ is the kernel for the 1×1 convolutional layer, $*$ denotes a convolutional operator and $b_{1,l}, b_{2,l}$ are bias vectors.

Instead of the acausal convolutions in MS-TCN [19] with predictions $\hat{y}_t(x_{t-n}, ..., x_{t+n})$ which depend on both n past and n future frames, we use causal convolutions within our D layer. Our causal convolutions can be easily described as 1D convolutions with kernel size 3 with a dilation factor. The term causal refers to the fact that the output of each convolution is shifted and the prediction $\hat{y}$ for time step t does not rely on any n future frames but only relies on the current and previous frames i.e. $\hat{y}_t(x_{t-n}, ..., x_t)$. This allows for intra-operative online deployment of TeCNO, unlike biLSTMs that require knowledge of future time steps [21–23].

Increasing the dilation factor of the causal convolutions by 2 within the D layer for each consecutive layer we effectively increase the temporal receptive field RF of the network without a pooling operation (Eq. 3). We visualize the progression of the receptive field of the causal convolutions in Fig. 1. A single

D layer with a dilation factor of 1 and a kernel size of 3 can process three time steps at a time. Stacking 3 consecutive D layers within a stage, as seen in Fig. 1, increases the temporal resolution of the kernels to 8 time steps. The size of the temporal receptive field depends on the number of D layers $l \in [1, N]$ and is given by:

$$RF(l) = (2)^{l+1} - 1 \tag{3}$$

This results in a exponential increase of the receptive field, which significantly reduces the computational cost in comparison to models that achieve higher receptive field by increasing the kernel size or the amount of total layers [18].

Multi-stage TCN. The main idea of the multi-stage approach is to refine the output of the first stage S_1 by adding M additional stages to the network $S_{1...M}$ [24]. The extracted visual feature vectors for each frame of a surgical video $x_{1:T}$ are the input of S_1, as explained above. The output of S_1 is directly fed into the second stage S_2. As seen in Fig. 1, the outputs of S_1 and S_2 have independent loss functions and the reported predictions are calculated after S_2, where the final refinement is achieved.

After each stage $S_{1...M}$ we use a weighted cross-entropy loss to train our model, as described in Eq. 4. Here, y_t is the ground truth phase label and $\hat{y}_{mt}$ is the output prediction of each stage $m \in [1, M]$. The class weights w_c are calculated using median frequency balancing [20] to mitigate the imbalance between phases. Our TeCNO model is trained utilizing exclusively phase recognition labels without requiring any additional tool information.

$$\mathcal{L}_C = \frac{1}{M} \sum_{m}^{M} \mathcal{L}_{Cm} = -\frac{1}{M} \frac{1}{T} \sum_{m}^{M} \sum_{t}^{T} w_c y_{mt} \cdot log(\hat{y}_{mt}) \tag{4}$$

3 Experimental Setup

Datasets. We evaluated our method on two challenging surgical workflow intra-operative video datasets of laparoscopic cholecystectomy procedures for the resection of the gallbladder. The publicly available Cholec80 [25] includes 80 videos with resolutions 1920×1080 or 854×480 pixels recorded at 25 frames-per-second (fps). Each frame is manually assigned to one of seven classes corresponding to each surgical phase. Additionally, seven different tool annotation labels sampled at 1 fps are provided. The dataset was subsampled to 5 fps, amounting to ~92000 frames. We followed the split of [12,17] separating the dataset to 40 videos for training, 8 for validation, and 32 for testing.

Cholec51 is an in-house dataset of 51 laparoscopic cholecystectomy videos with resolution 1920×1080 pixels and sampling rate of 1 fps. Cholec51 includes seven surgical phases that slightly differ from Cholec80 and have been annotated by expert physicians. There is no additional tool information provided. 25 videos were utilized for training, 8 for validation and 18 for test. Our experiments for both datasets were repeated 5 times with random initialization to ensure reproducibility of the results.

Table 1. Ablative testing results for different feature extraction CNNs and increasing number of stages for Cholec80. Average metrics over multiple runs are reported (%) along with their respective standard deviation.

	AlexNet			ResNet50		
	Acc	Prec	Rec	Acc	Prec	Rec
No TCN	74.40 ± 4.30	63.06 ± 0.32	70.75 ± 0.05	82.22 ± 0.60	70.65 ± 0.08	75.88 ± 1.35
Stage I	84.04 ± 0.98	79.82 ± 0.31	79.03 ± 0.99	88.35 ± 0.30	**82.44 ± 0.46**	84.71 ± 0.71
Stage II	85.31 ± 1.02	81.54 ± 0.49	79.92 ± 1.16	**88.56 ± 0.27**	81.64 ± 0.41	**85.24 ± 1.06**
Stage III	84.41 ± 0.85	77.68 ± 0.90	79.64 ± 1.6	86.49 ± 1.66	78.87 ± 1.52	83.69 ± 1.03

Model Training. TeCNO was trained for the task of surgical phase recognition using the Adam optimizer with an initial learning rate of 5e−4 for 25 epochs. We report the test results extracted by the model that performed best on the validation set. The batch size is identical to the length of each video. Our method was implemented in PyTorch and our models were trained on an NVIDIA Titan V 12 GB GPU using Polyaxon[1]. The source code for TeCNO is publicly available[2].

Evaluation Metrics. To comprehensively measure the results of the phase prediction we deploy three different evaluation metrics suitable for surgical phase recognition [5], namely Accuracy (Acc), Precision (Prec) and Recall (Rec). Accuracy quantitatively evaluates the amount of correctly classified phases in the whole video, while Precision, or positive predictive value, and Recall, or true positive rate, evaluate the results for each individual phase [22].

Ablative Testing. To identify a suitable feature extractor for our MS-TCN model we performed experiments with two different CNN architectures, namely AlexNet [26] and ResNet50 [16]. Additionally we performed experiments with different number of TCN stages to identify which architecture is best able to capture the long temporal associations in our surgical videos.

Baseline Comparison. TeCNO was extensively evaluated against surgical phase recognition networks, namely, PhaseLSTM [12], EndoLSTM [12] and MTRC-Net [17], which employ LSTMs to encompass the temporal information in their models. We selected LSTMs over HMMs, since their superiority has been extensively showcased in the literature [14]. Moreover, MTRCNet is trained in an end-to-end fashion, while the remaining LSTM approaches and TeCNO focus on temporally refining already extracted features. Since Cholec51 does not include tool labels, EndoLSTM and MTRCNet are not applicable due to their multi-task requirement. All feature extractors for Cholec80 were trained

[1] https://polyaxon.com/.
[2] https://github.com/tobiascz/TeCNO/.

Table 2. Baseline comparison for Cholec80 and Cholec51 datasets. EndoLSTM and MTRCNet require tool labels, therefore cannot be applied for Cholec51. The average metrics over multiple runs are reported (%) along with their respective standard deviation.

	Cholec80			Cholec51		
	Acc	Prec	Rec	Acc	Prec	Rec
PhaseLSTM [12]	79.68 ± 0.07	72.85 ± 0.10	73.45 ± 0.12	81.94 ± 0.20	68.84 ± 0.11	68.05 ± 0.79
EndoLSTM [22]	80.85 ± 0.17	76.81 ± 2.62	72.07 ± 0.64	–	–	–
MTRCNet [17]	82.76 ± 0.01	76.08 ± 0.01	78.02 ± 0.13	–	–	–
ResNetLSTM [15]	86.58 ± 1.01	80.53 ± 1.59	79.94 ± 1.79	86.15 ± 0.60	70.45 ± 2.85	67.42 ± 1.43
TeCNO	$\mathbf{88.56 \pm 0.27}$	$\mathbf{81.64 \pm 0.41}$	$\mathbf{85.24 \pm 1.06}$	$\mathbf{87.34 \pm 0.66}$	$\mathbf{75.87 \pm 0.58}$	$\mathbf{77.17 \pm 0.73}$

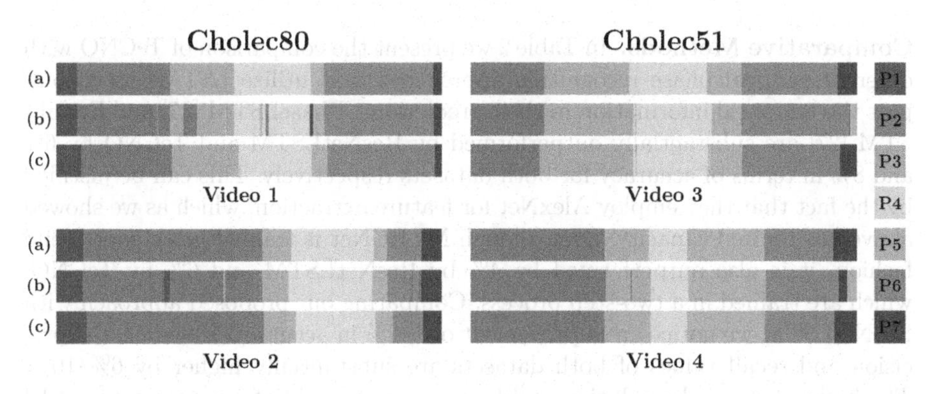

Fig. 2. Qualitative results regarding quality of phase recognition for Cholec80 and Cholec51. (a) Ground truth (b) ResNetLSTM predictions (c) TeCNO predictions. P1 to P7 indicate the phase label.

for a combination of phase and tool identification, except for the feature extractor of PhaseLSTM [25], which requires only phase labels. The CNNs we used to extract the features for Cholec51 were only trained on phase recognition since no tool annotations were available.

4 Results

Effect of Feature Extractor Architecture. As can be seen in Table 1, ResNet50 outperforms AlexNet across the board with improvements ranging from 2% to 8% in accuracy. Regarding precision and recall, the margin increases even further. For all stages ResNet50 achieves improvement over AlexNet of up to 7% in precision and 6% in recall. This increase can be attributed to the improved training dynamics and architecture of ResNet50 [16]. Thus, the feature extractor selected for TeCNO is ResNet50.

Effect of TCN and Number of Stages. Table 1 also highlights the substantial improvement in the performance achieved by the TCN refinement stages. Both AlexNet and ResNet50 obtain higher accuracy by 10% and 6% respectively with the addition of just 1 TCN Stage. Those results signify not only the need for temporal refinement for surgical phase recognition but also the ability of TCNs to improve the performance of any CNN employed as feature extractor, regardless of its previous capacity. We can also observe that the second stage of refinement improves the prediction of both architectures across our metrics. However, Stage 2 outperforms Stage 3 by 1% in accuracy for AlexNet and 2% for ResNet50. This could indicate that 3 stages of refinement lead to overfitting on the training set for our limited amount of data.

Comparative Methods. In Table 2 we present the comparison of TeCNO with different surgical phase recognition approaches that utilize LSTMs to encompass the temporal information in their predictions. PhaseLSTM [27] and EndoLSTM [27] are substantially outperformed by ResNetLSTM and TeCNO by 6% and 8% in terms of accuracy for both datasets respectively. This can be justified by the fact that they employ AlexNet for feature extraction, which as we showed above has limited capacity. Even though MTRCNet is trained in an end-to-end fashion, it is also outperformed by 4% by ResNetLSTM and 6% by TeCNO, which are trained in a two-step process. Comparing our proposed approach with ResNetLSTM we notice an improvement of 1–2% in accuracy. However, the precision and recall values of both datasets are substantially higher by 6%–10%. The higher temporal resolution and large receptive field of our proposed model allow for increased performance even for under-represented phases.

Phase Recognition Consistency. In Fig. 2 we visualize the predictions for four laparoscopic videos, two for each dataset. The results clearly highlight the ability of TeCNO to obtain consistent and smooth predictions not only within one phase, but also for the often ambiguous phase transitions. Compared against ResNetLSTM, TeCNO can perform accurate phase recognition, even for the phases with shorter duration, such as P5 and P7. Finally, TeCNO showcases robustness, since Video 3 and 4 are both missing P1. However, the performance of our model does not deteriorate.

5 Conclusion

In this paper we proposed TeCNO, a multi-stage Temporal Convolutional Neural Network, which was successfully deployed on the task of surgical phase recognition. Its full temporal resolution and large receptive field allowed for increased performance against a variety of LSTM-based approaches across two datasets. Online and fast inference on whole video-sequences was additionally achieved due to causal, dilated convolutions. TeCNO increased the prediction consistency, not only within phases, but also in the ambiguous inter-phase transitions. Future work includes evaluation of our method on a larger number of videos from a variety of laparoscopic procedures.

Acknowledgements. Our research is partly funded by the DFG research unit 1321 PLAFOKON and ARTEKMED in collaboration with the Minimal-invasive Interdisciplinary Intervention Group (MITI). We would also like to thank NVIDIA for the GPU donation.

References

1. Maier-Hein, L., et al.: Surgical data science for next-generation interventions. Nat. Biomed. Eng. **1**(9), 691–696 (2017)
2. Huaulmé, A., Jannin, P., Reche, F., Faucheron, J.L., Moreau-Gaudry, A., Voros, S.: Offline identification of surgical deviations in laparoscopic rectopexy. Artif. Intell. Med. **104**(May), 2020 (2019)
3. Padoy, N.: Machine and deep learning for workflow recognition during surgery. Minim. Invasive Ther. Allied Technol. **28**(2), 82–90 (2019)
4. Zisimopoulos, O., et al.: DeepPhase: surgical phase recognition in CATARACTS videos. In: Frangi, A.F., Schnabel, J.A., Davatzikos, C., Alberola-López, C., Fichtinger, G. (eds.) MICCAI 2018. LNCS, vol. 11073, pp. 265–272. Springer, Cham (2018). https://doi.org/10.1007/978-3-030-00937-3_31
5. Padoy, N., Blum, T., Ahmadi, S.A., Feussner, H., Berger, M.O., Navab, N.: Statistical modeling and recognition of surgical workflow. Med. Image Anal. **16**(3), 632–641 (2012)
6. Lecuyer, G., Ragot, M., Martin, N., Launay, L., Jannin, P.: Assisted phase and step annotation for surgical videos. Int. J. Comput. Assist. Radiol. Surg. **15**(4), 673–680 (2020). https://doi.org/10.1007/s11548-019-02108-8
7. Bodenstedt, S., et al.: Prediction of laparoscopic procedure duration using unlabeled, multimodal sensor data. Int. J. Comput. Assist. Radiol. Surg. **14**(6), 1089–1095 (2019). https://doi.org/10.1007/s11548-019-01966-6
8. Funke, I., Mees, S.T., Weitz, J., Speidel, S.: Video-based surgical skill assessment using 3D convolutional neural networks. Int. J. Comput. Assist. Radiol. Surg. **14**(7), 1217–1225 (2019). https://doi.org/10.1007/s11548-019-01995-1
9. Klank, U., Padoy, N., Feussner, H., Navab, N.: Automatic feature generation in endoscopic images. Int. J. Comput. Assist. Radiol. Surg. **3**(3), 331–339 (2008). https://doi.org/10.1007/s11548-008-0223-8
10. Al Hajj, H., et al.: CATARACTS: challenge on automatic tool annotation for cataRACT surgery. Med. Image Anal. **52**, 24–41 (2019)
11. Lea, C., Vidal, R., Reiter, A., Hager, G.D.: Temporal convolutional networks: a unified approach to action segmentation. In: Hua, G., Jégou, H. (eds.) ECCV 2016. LNCS, vol. 9915, pp. 47–54. Springer, Cham (2016). https://doi.org/10.1007/978-3-319-49409-8_7
12. Twinanda, A.P., Shehata, S., Mutter, D., Marescaux, J., De Mathelin, M., Padoy, N.: EndoNet: a deep architecture for recognition tasks on laparoscopic videos. IEEE Trans. Med. Imaging **36**(1), 86–97 (2017)
13. Hochreiter, S., Schmidhuber, J.: Long short-term memory. Neural Comput. **9**(8), 1735–1780 (1997)
14. Yengera, G., Mutter, D., Marescaux, J., Padoy, N.: Less is more: surgical phase recognition with less annotations through self-supervised pre-training of CNN-LSTM networks (2018)

15. Jin, Y., et al.: SV-RCNet: workflow recognition from surgical videos using recurrent convolutional network. IEEE Trans. Med. Imaging **37**(5), 1114–1126 (2018)
16. He, K., Zhang, X., Ren, S., Sun, J.: Deep residual learning for image recognition. In: 2016 IEEE Conference on Computer Vision and Pattern Recognition (CVPR), Las Vegas, NV, pp. 770–778 (2016). https://doi.org/10.1109/CVPR.2016.90
17. Jin, Y., et al.: Multi-task recurrent convolutional network with correlation loss for surgical video analysis. Med. Image Anal. **59**, 101572 (2020). https://github.com/YuemingJin/MTRCNet-CL
18. van den Oord, A.: WaveNet: a generative model for raw audio. arXiv:1609.03499 (2016)
19. Farha, Y.A., Gall, J.: MS-TCN: multi-stage temporal convolutional network for action segmentation. In: 2019 IEEE/CVF Conference on Computer Vision and Pattern Recognition (CVPR), Long Beach, CA, USA, pp. 3570–3579 (2019). https://doi.org/10.1109/CVPR.2019.00369
20. Eigen, D., Fergus, R.: Predicting depth, surface normals and semantic labels with a common multi-scale convolutional architecture. In: 2015 IEEE International Conference on Computer Vision (ICCV), pp. 2650–2658 (2015)
21. Yu, T., Mutter, D., Marescaux, J., Padoy, N.: Learning from a tiny dataset of manual annotations: a teacher/student approach for surgical phase recognition (2018)
22. Twinanda, A.P., Padoy, N., Troccaz, M.J., Hager, G.: Vision-based approaches for surgical activity recognition using laparoscopic and RBGD videos. Thesis, no. Umr 7357 (2017)
23. Graves, A., Fernández, S., Schmidhuber, J.: Bidirectional LSTM networks for improved phoneme classification and recognition. In: Duch, W., Kacprzyk, J., Oja, E., Zadrożny, S. (eds.) ICANN 2005. LNCS, vol. 3697, pp. 799–804. Springer, Heidelberg (2005). https://doi.org/10.1007/11550907_126
24. Newell, A., Yang, K., Deng, J.: Stacked hourglass networks for human pose estimation. In: Leibe, B., Matas, J., Sebe, N., Welling, M. (eds.) ECCV 2016. LNCS, vol. 9912, pp. 483–499. Springer, Cham (2016). https://doi.org/10.1007/978-3-319-46484-8_29
25. Twinanda, A.P., Mutter, D., Marescaux, J., de Mathelin, M., Padoy, N.: Single- and multi-task architectures for surgical workflow challenge at M2CAI 2016, pp. 1–7 (2016)
26. Krizhevsky, A., Sutskever, I., Hinton, G.: ImageNet classification with deep convolutional neural networks. In: Neural Information Processing Systems, vol. 25 (2012)
27. Twinanda, A.P., Mutter, D., Marescaux, J., Mathelin, M.D., Padoy, N.: Single- and multi-task architectures for surgical workflow challenge at M2CAI 2016. ArXiv, abs/1610.08844 (2016)

Surgical Video Motion Magnification with Suppression of Instrument Artefacts

Mirek Janatka[1(✉)], Hani J. Marcus[1,2], Neil L. Dorward[2], and Danail Stoyanov[1]

[1] Wellcome/EPSRC Centre for Interventional and Surgical Sciences, University College London, London, UK
mirek.janatka@ucl.ac.uk
[2] Department of Neurosurgery, The National Hospital for Neurology and Neurosurgery, Queen Square, London, UK

Abstract. Video motion magnification can make blood vessels in surgical video more apparent by exaggerating their pulsatile motion and could prevent inadvertent damage and bleeding due to their increased prominence. It could also indicate the success of restricting blood supply to an organ when using a vessel clamp. However, the direct application to surgical video could result in aberration artefacts caused by its sensitivity to residual motion from the surgical instruments and would impede its practical usage in the operating theatre. By storing the previously obtained jerk filter response of each spatial component of each image frame - both prior to surgical instrument introduction and adhering to a Eulerian frame of reference - it is possible to prevent such aberrations from occurring. The comparison of the current readings to the prior readings of a single cardiac cycle at the corresponding cycle point, are used to determine if motion magnification should be active for each spatial component of the surgical video at that given point in time. In this paper, we demonstrate this technique and incorporate a scaling variable to loosen the effect which accounts for variabilities and misalignments in the temporal domain. We present promising results on endoscopic transnasal transsphenoidal pituitary surgery with a quantitative comparison to recent methods using Structural Similarity (SSIM), as well as qualitative analysis by comparing spatio-temporal cross sections of the videos and individual frames.

Keywords: Motion magnification · Surgical visualisation · Augmented reality · Computer assisted interventions · Image guided surgery

Electronic supplementary material The online version of this chapter (https://doi.org/10.1007/978-3-030-59716-0_34) contains supplementary material, which is available to authorized users.

A. L. Martel et al. (Eds.): MICCAI 2020, LNCS 12263, pp. 353–363, 2020.
https://doi.org/10.1007/978-3-030-59716-0_34

1 Introduction

In endoscopic surgery visualising blood vessels is a common challenge as they are often beneath the tissue surface or indistinctive from the surface texture. Major complications can result from instruments causing inadvertent damage and hence bleeding [1] due to vessel imperceptibility that leads surgical error. In severe cases, such bleeding can place the patient at risk of death if it cannot be controlled or potentially lead to post-operative problems, that would require additional surgery to address [2]. The major challenge and cause for such problems in endoscopic procedures is that subsurface vessels cannot be visualised directly or detected through touch and palpation [3]. Various schemes for detecting and avoiding subsurface vasculature have been explored including Augmented Reality (AR) that fuses preoperative patient imaging with the surgical video, as well as novel optical imaging [4] and intraoperative ultrasound [5]. Similar schemes have also been used to determine if blood flow continues to pulse into a region, after being obstructed by a clamping mechanism [6]. However, all have some drawbacks in terms of surface registration accuracy, workflow additions, ergonomic problems and signal sensitivity [7]. Hence practically, the problem persists and can be a significant hurdle to the successful completion of many procedures [8].

Video motion magnification (VMM) [9–11] has previously been proposed as a mechanism to aid vessel localisation in endoscopic video, without the requirement of additional *in situ* hardware, surface registration or contrast agents [12,14,15]. VMM uses the existing variations in the endoscopic video stream that are minute and out of the range of the surgeon's perception and creates a synthetic video where such motions are perceivable. The generated video characteristics can be temporally selected, so that reoccurring motion within a certain temporal frequency range can be processed exclusively, which allows for motions of known occurrence to be selected [9,10]. In the case of vessel localisation, it is possible to isolate and amplify motion due to the heart rate (readily available in the OR), that governs the periodicity of vessel distension from the pressure wave that is generated from the heart. VMM assumes spatial consistency, effectively treating every point on the image as a time series, and requires a static view for initialisation. This creates limitations and while VMM has been demonstrated in endoscopic third ventriculostomy and robotic prostatectomy videos, it has not been effectively clinically translated or adopted [12]. VMM has also been suggested for other clinical applications, such as with video otoscopy [13].

One challenge for VMM in surgery is that other motions within the scene can cause aberrational distortions and can be disorientating to surgical view. An alternative method of using VMM was proposed in robotic partial nephrectomy, where respiratory motion is present and needs be accounted for. Rather than use VMM directly a colour map representation was generated from the raw VMM video of where pulsation was used to located vessels as an aid to assist in registration of preoperative patient data [14]. Attempts to deal with large motion presence in motion magnified videos have been suggested by segmentation which is not practical in an operation [19,20]. Yet, recent developments in temporal

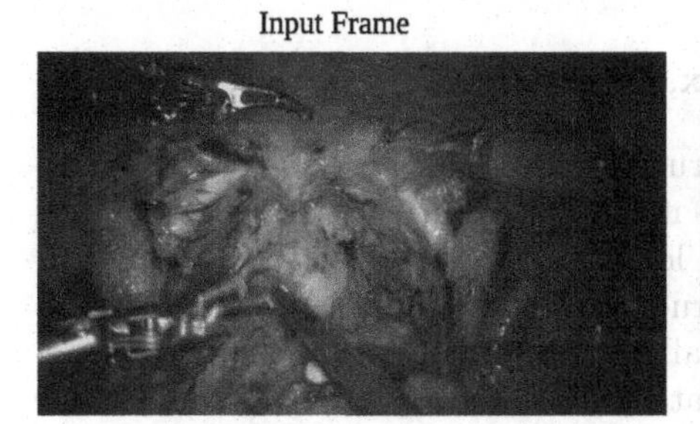

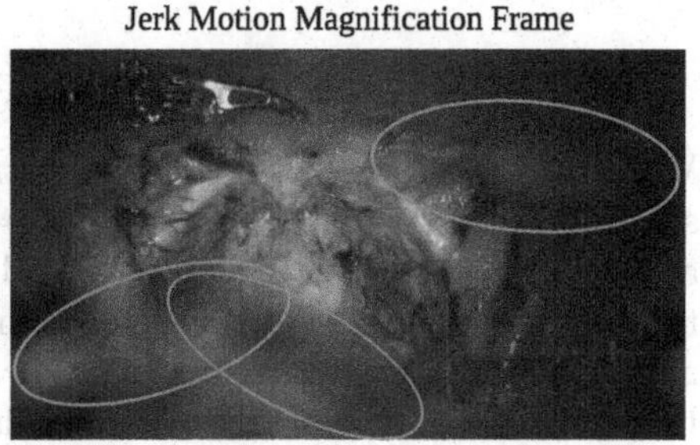

Fig. 1. Demonstrating aberrations from surgical instrument motions using video motion magnification via frame comparison. Left) Original video frame. Right) Aberrations generated jerk-based motion magnification, outlined by the green elliptical annotation [15] (Color figure online)

filtering have allowed for different components of motion, such as acceleration and jerk, to be selected based upon a principal oscillation frequency. This allows for VMM to leverage a higher band pass of frequencies than the band under investigation to exaggerate motion within video [15–18]. In our previous work, we used the third order of motion (jerk) was utilised to exclude motions from respiration and transmitted motion from larger arteries, whilst motion magnifying blood vessels in surgical video, with a filter designed around the pulse rate. This approach reduced blur distortion caused by the large motions within the scene, whilst still permitting motion magnification of vessel distension, as the pulsatile waveform contains jerk characteristic [15]. However, this jerk filter is not able to prevent the generation of motion blur aberrations from the presence of instruments moving within the scene, making the synthesised video unsuitable for surgical guidance (as shown in Fig. 1). For VMM to be a viable option for surgical intervention it must be usable with instrument motion presence or its function would be limited to just observational usage.

In this paper, we propose a technique that would allow for motion magnification to be left unaffected by the introduction of tool motion to the surgical camera's field of view. To operate, it simply requires the known heart rate and a brief sample of the view that is uninterrupted by instrument motion. It maintains the spatial consistency assumption of a fixed view point. We demonstrate this method on four cases of endoscopic video of transnasal transsphenoidal surgery, providing both qualitative and quantitative comparison to a prior method.

2 Methods

2.1 Motion Magnification

VMM operates by spatially decomposing video frames into local frequency components using Complex Steerable Pyramid (CSP) [21,22] which uses different Gabor-like wavelets (ψ) to decompose images at varying scales and orientations representations notated by S.

$$\tilde{I}(\mathbf{x}, t) = \sum S(\mathbf{x}, t) * \psi + \epsilon(\mathbf{x}, t) \tag{1}$$

Where $\tilde{I}(\mathbf{x}, t)$ represents a CSP reconstructed video frame at time t, with intensity values at $\mathbf{x} = (x, y)$ pixel value, $*$ notes a convolution operation. As well as the band-pass of S a high-pass and low-pass residual $(\epsilon(\mathbf{x}, t))$ that is unalterable is also required for the reconstruction. As the information held in $S(\cdot)$ are complex conjugates representing local frequency information, local phase can be attained. As the local motion is related to the local phase $(\phi(\mathbf{x}, t))$ via the Fourier shift theorem, motion analysis and modulation can be performed by filtering and manipulating the local phase of the video over time. As shown in our previous work [15], higher order of motion magnification utilises a temporally tuned third order Gaussian derivative to detect jerk motion $D(\mathbf{x}, t)$ within a certain pass-band of temporal frequencies. It exploits the linear relationship of convolution to gather third order of motion from local phase using a third order Gaussian derivative:

$$D(\mathbf{x}, t) := \frac{\partial^3 G_\sigma(t)}{\partial t^3} * \phi(\mathbf{x}, t) = \frac{\partial^3 \phi(\mathbf{x}, t)}{\partial t^3} * G_\sigma(t) \tag{2}$$

Where σ is the standard derivation of the Gaussian (G) derivative and is defined $\sigma = \frac{fr}{4\omega}$ [23], where ω is the temporal frequency of interest, in this case the heart rate from the electrocardiogram. fr denotes the sampling rate of the video. The Gaussian derivative convolution is applied to the time series for all scales and orientations of each pixel $(\mathbf{X})$ from the CSP representation. By detecting local jerk motion in a video, it can be exaggerated by an amplification factor α to generate $\hat{S}(\cdot)$.

$$\hat{S}(\mathbf{x}, t) = A(\mathbf{x}, t) e^{i(\phi(\mathbf{x}, t) + \alpha D(\mathbf{x}, t))} \tag{3}$$

Where A is the amplitude and ϕ is the phase of that particular local frequency at $\mathbf{x}$ at time t with respect to the S band's orientation and scale. The summations of which from the various scales and orientations can reconstruct the motion magnified frame $\hat{I}(\mathbf{x}, t)$.

2.2 Tool Motion Artefact Suppression Filter

Assuming the endoscope is statically positioned and there are no surgical instruments within the scene, the response of the jerk filter applied to the video feed can be anticipated as:

$$D(\mathbf{x}, t) = D(\mathbf{x}, t \bmod \frac{1}{\omega_c}) \tag{4}$$

Where mod is the modulo function and $\frac{1}{\omega_c}$ is the time period of the cardiac cycle, reported from the electrocardiogram. ω_c is also the ω value used to determine $G(t)$ from Eq. 2. However, in reality, due to sampling quantization and subtle variations in the heart rate, perfectly aligned repetition rarely occurs. Therefore these values are used as a guide for creating the filter with an offset range, which can then be loosened by a scaling factor. Taking these stored tool free scene readings to be $TL(\mathbf{x})$:

$$TL(\mathbf{x}) = D(\mathbf{x}, 0), D(\mathbf{x}, 1), \ldots D(\mathbf{x}, \frac{1}{\omega_c}) \tag{5}$$

To generate the offset range for the filter, the variability of $TL(\mathbf{x})$ can be found as $R(\mathbf{x})$:

$$R(\mathbf{x}) = \frac{max(TL(\mathbf{x})) - min(TL(\mathbf{x}))}{2} \tag{6}$$

This allows the comparator filter to be:

$$D(\mathbf{x}, t \bmod \frac{1}{\omega_c}) \pm \beta R(\mathbf{x}) \tag{7}$$

Where β is a scaling factor that can widen the filter further. To operate within the VMM, the comparator has to act as a switch so that the motion magnification effect can be deactivated. Therefore we define the state of $\chi(\mathbf{x}, t)$ as:

$$\begin{aligned} &1 = [D(\mathbf{x}, t) < D(\mathbf{x}, t \bmod \tfrac{1}{\omega_c}) + \beta R(\mathbf{x})] \wedge [D(\mathbf{x}, t) > D(\mathbf{x}, t \bmod \tfrac{1}{\omega_c}) - \beta R(\mathbf{x})] \\ &0 = else \end{aligned} \tag{8}$$

Where $\wedge$ is a logic *and* function. This consideration can be shown in the CSP magnified band S as $\hat{S}(\cdot)^*$, with the threshold value β being pre-assigned with χ_β.

$$\hat{S}(\mathbf{x}, t)^* = A(\mathbf{x}, t) e^{i(\phi(\mathbf{x}, t) + \alpha \chi_\beta(\mathbf{x}, t) D(\mathbf{x}, t))} \tag{9}$$

Which can be used to reconstruct $\hat{I}(\mathbf{x}, t)^*$, a motion magnified video without blur distortion from tool motion.

2.3 Synthetic Example

To better visualise how the tool motion artefact suppression filter (TMASF) works, Fig. 2 shows a demonstration on a synthetic arterial displacement profile [24], shown as a phase reading that could be taken from a single pixel from an arbitrary $S(\cdot)$, where vessel motion exists. After just under two cardiac cycles, a tool passes across the point, denoted by "tool motion" that alters the phase reading. The reading then returns to that of the vessel motion as before. The response of this displacement profile from the jerk filter [15] is shown in the middle plot of Fig. 2 in green. After a single cardiac cycle, the TMASF can be constructed, the bounds of which are shown in red. So long as the jerk filter response stays within these bounds, $\chi(\cdot)$ is equal to 1 and the amplification is performed.

However, if the jerk filter response moves outside these bounds, $\chi(\cdot)$ is equal to 0 for that particular pixel around that moment and the amplification effect is nullified. This can be seen around the time the tool motion is present. The bottom plot of Fig. 2 shows the resultant amplified signal, using both the TMASF

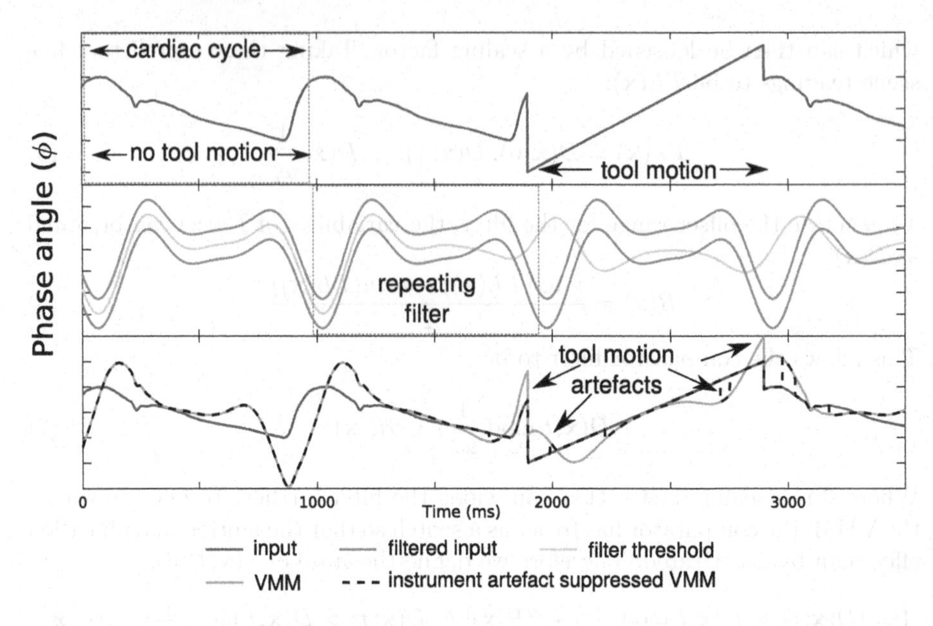

Fig. 2. Synthetic one dimensional explanation of the instrument motion suppression filter. The top plot shows the observed cardiac motion with a tool passing over it. The middle plot shows the filtered component of the motion signal and the generation of the suppression filter. The bottom plot shows the difference the suppression has on the motion magnification output signal. (Color figure online)

filter (dashed black curve) and without (magenta curve). As shown, the amplification generates tool motion artefacts on prior method where tool motion is present. The TMASF reduces such artefacts, but is not immune to them. The extent to which they are created can be reduced by decreasing the β value, which would essentially bring the red lines in Fig. 2 closer together. Yet, this could also be detrimental to the amplification effect, due to the imperfections in regularity and sampling.

3 Results

3.1 Experimental Setup

To verify if the TMASF works in surgical videos, we performed a proof of concept study using retrospective data (IDEAL Stage 0) [25], applying the filter to a series of patient cases (n = 4) that underwent endoscopic transnasal transsphenoidal surgery. The study was registered with the local institutional review board, patients provided their written consent, and videos were fully anonymised. Each video was processed with the TMASF at three β threshold scalings ($\beta = 1$, 3 and 5) and without, using the previous jerk motion magnification without the TMASF [15], using 8 orientations and quarter octave CSP, from hereafter referred to as VMM.

All samples have a brief few seconds before surgical instruments are visible in the scene for the filter to initialise. This would be a reasonable prerequisite requirement and perhaps automatically possible with a surgical robot as it is aware when the system is stationary. For all videos the magnification factor was set to $\alpha = 10$. The resolution of the capture videos were 1280×720, however were cropped to a 720×720 pixel square to account for the visible area of the video. The videos were scaled down by a half for quicker offline processing. Acquisition of the endoscope was at 25 fps. For quantitative analysis and comparison Structure Similarity Image Matrix index (SSIM) was used to compare the magnified frames to the corresponding input frames. The closer to 1 the similar the frame is to the input, meaning the less noise or motion magnification effect has been generated. To verify that the motion magnification effect is still operational, comparative spatio-temporal cross sections from places of interest are compared across all videos. Additionally, all videos are supplied in the supplementary material.

3.2 Results and Analysis

As a comparative example of how the TMASF functions, the frame-wise SSIM result for each scaling factor of the TMASF and VMM can be seen in Fig. 3, that depicts the result from case four, and is a common representation for all cases. The "tool motion" region shows how the various β values of the TMASF perform. The VMM curve shows the impact tool movement has on the generated video, with SSIM values dropping to as low as 0.55 and are volatile for this duration, with rarely reaching the lowest reading from the "no tool motion" region. Similarly to the prior "no tool motion" section, TMASF $\beta = 1$ shows that the motion magnification is impinged by the suppression filter at this β level. The other two β values show a drop in SSIM readings, compared to the "no tool motion" region, however are not as severe as the VMM curve, with $\beta = 5$ reaching an offset of -0.05 compared to $\beta = 3$ at most. These results suggest that the TMASF works, on β levels 3 and higher. To report a fuller quantitative performance of the TMASF, box plots from the SSIM reading of

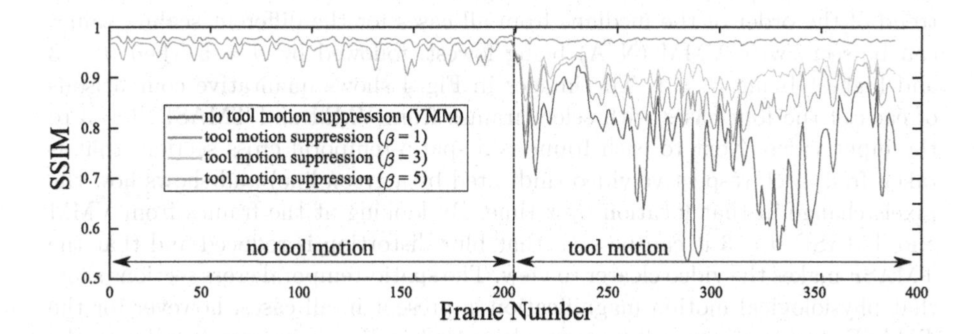

Fig. 3. Frame-wise comparison of SSIM for all four motion magnified videos

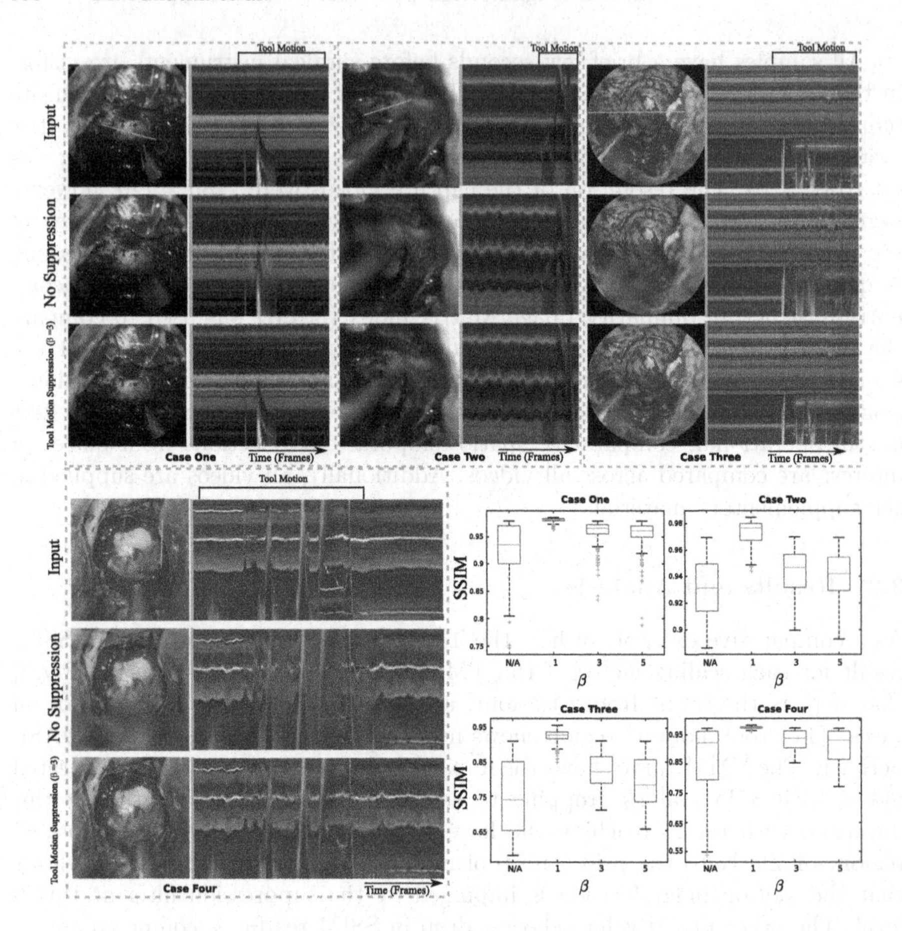

Fig. 4. Qualitative (input, VMM and TMASF β = 3 - green line on image frame indicates sample site for spatio-temporal cross section) and quantitative (using SSIM: VMM and TMASF $\beta = 1$, 3 and 5) comparisons of all four cases. (Color figure online)

the entire videos are shown on the right bottom corner of Fig. 4. A running trend of the order of the medians from all cases for the different scaling values can be seen, with VMM (N/A) being lowest, followed by β = 5, then β = 3 and highest being β = 1. The images in Fig. 4 shows qualitative comparisons of each of the four cases with select frames from VMM and TMASF β = 3 to the input video. Next to each frame is a spatio-temporal cross section, that is taken from each respective video (indicated by a green line) and shows how the pixels change in that location over time. By looking at the frames from VMM and TMASF β = 3 it can be seen that blur distortion is reduced and that the TMASF makes the video clearer to view. The spatio-temporal cross sections show that physiological motion magnification is present in all cases, however for the TMASF structure can still be seen where tool motion is present (similar to the image in the input cross section), whilst is lost in the VMM cross section. This suggests that TMASF successfully reduces aberrations from VMM videos whilst retaining the desired motion magnification effect of exaggerating the motion of the physiology. In general, the results suggest that there is a trade-off between impinging the motion magnification effect and permitting aberration occurring from instrumentation motion, depending on the β value used with TMASF.

4 Discussion

In this paper, we have proposed a filter constructed from local phase information collected prior to the insertion of surgical tools into the surgical field of view for motion magnification in endoscopic procedures. This approach can prevent aberration from being created due to instrument motion and hence allow more clinically usable surgical motion magnification augmentation, such as critical structure avoidance and assistance in vessel clamping.

We have shown that the proposed filter can reduce amplification of motion due to tools on example videos from endoscopic neurosurgery where the camera and surgical site are confined and do not move too much. This is an important consideration because our filter would need re-initialising if the endoscope is moved or after large changes to the surgical scene. Yet this could be performed quickly, as the initialisation period is the length of a heart beat but more work is needed to detect and automate any re-initialisation strategy.

The application of the TMASF is causal and is possible to combine with a real-time system as it is only reliant on past information. Further work is needed to consider user studies investigating how augmented visualisation of motion magnification can be presented to the clinical team to access the risk of cognitive overload and inattention blindness [26]. Both the motion magnification factor α and TMASF β value are variables that can be tuned to create the optimal synthesized video, however, it is not essential that the raw video is used alone. TMASF could assist in existing visualisation approaches to only show physiological motion [14]. Additionally, the ability to segregate physiological and non-physiological motion from a surgical scene could assist in surgical instrument tracking and anatomical segmentation tasks, as well as in other modalities of medical imaging that utilise motion estimation.

Acknowledgements. The work was supported by the Wellcome/EPSRC Centre for Interventional and Surgical Sciences (WEISS) [203145Z/16/Z]; Engineering and Physical Sciences Research Council (EPSRC) [EP/P027938/1, EP/R004080/1, EP/P012841/1]; The Royal Academy of Engineering [CiET1819/2/36]; Horizon 2020 FET (GA 863146) and the National Brain Appeal Innovations Fund.

References

1. Feldman, S.A., Marks, V.: The problem of haemorrhage during anaesthesia and surgery. Anaesthesia **16**(4), 410–431 (1961)

2. Marietta, M., Facchini, L., Pedrazzi, P., Busani, S., Torelli, G.: Pathophysiology of bleeding in surgery. In: Transplantation Proceedings, vol. 38, pp. 812–814. Elsevier (2006)
3. Stoyanov, D.: Surgical vision. Ann. Biomed. Eng. **40**(2), 332–345 (2012). https://doi.org/10.1007/s10439-011-0441-z
4. Clancy, N.T., Jones, G., Maier-Hein, L., Elson, D.S., Stoyanov, D.: Surgical spectral imaging. Med. Image Anal. **63**, 101699 (2020)
5. Cleary, K., Peters, T.M.: Image-guided interventions: technology review and clinical applications. Ann. Rev. Biomed. Eng. **12**, 119–142 (2010)
6. Norat, P., et al.: Application of indocyanine green videoangiography in aneurysm surgery: evidence, techniques, practical tips. Fount. Surg. **6**, 34 (2019)
7. Bernays, R.L.: Intraoperative imaging: current trends, technology and future directions. In: Transsphenoidal Surgery, pp. 56–69. Saunders, Philadelphia (2010)
8. Shander, A.: Financial and clinical outcomes associated with surgical bleeding complications. Surgery **142**(4), 20–25 (2007)
9. Wadhwa, N., Rubinstein, M., Durand, F., Freeman, W.T.: Phase-based video motion processing. ACM Trans. Graph. **32**(4), 1–10 (2013)
10. Wu, H.-Y., Rubinstein, M., Shih, E., Guttag, J., Durand, F., Freeman, W.: Eulerian video magnification for revealing subtle changes in the world. ACM Trans. Graph. **31**(4), 1–8 (2012)
11. Ngo, A.C.L., Phan, R.C.-W.: Seeing the invisible: survey of video motion magnification and small motion analysis. ACM Comput. Surv. (CSUR) **52**(6), 1–20 (2019)
12. McLeod, J.A., Baxter, J.S.H., de Ribaupierre, S., Peters, T.M.: Motion magnification for endoscopic surgery. In: Medical Imaging 2014: Image-Guided Procedures, Robotic Interventions, and Modeling, vol. 9036, p. 90360C. International Society for Optics and Photonics (2014)
13. Janatka, M., Ramdoo, K.S., Tatle, T., Pachtrachai, K., Elson, D.S., Stoyanov, D.: Examining in vivo tympanic membrane mobility using smart phone video-otoscopy and phase-based Eulerian video magnification. In: Medical Imaging 2017: Computer-Aided Diagnosis, vol. 10134, p. 101341Y. International Society for Optics and Photonics (2017)
14. Amir-Khalili, A., et al.: Automatic segmentation of occluded vasculature via pulsatile motion analysis in endoscopic robot-assisted partial nephrectomy video. Med. Image Anal. **25**(1), 103–110 (2015)
15. Janatka, M., Sridhar, A., Kelly, J., Stoyanov, D.: Higher order of motion magnification for vessel localisation in surgical video. In: Frangi, A.F., Schnabel, J.A., Davatzikos, C., Alberola-López, C., Fichtinger, G. (eds.) MICCAI 2018. LNCS, vol. 11073, pp. 307–314. Springer, Cham (2018). https://doi.org/10.1007/978-3-030-00937-3_36
16. Zhang, Y., Pintea, S.L., Van Gemert, J.C.: Video acceleration magnification. In: Proceedings of the IEEE Conference on Computer Vision and Pattern Recognition, pp. 529–537 (2017)
17. Takeda, S., Okami, K., Mikami, D., Isogai, M., Kimata, H.: Jerk-aware video acceleration magnification. In: Proceedings of the IEEE Conference on Computer Vision and Pattern Recognition, pp. 1769–1777 (2018)
18. Takeda, S., Akagi, Y., Okami, K., Isogai, M., Kimata, H.: Video magnification in the wild using fractional anisotropy in temporal distribution. In: Proceedings of the IEEE Conference on Computer Vision and Pattern Recognition, pp. 1614–1622 (2019)

19. Elgharib, M., Hefeeda, M., Durand, F., Freeman, W.T.: Video magnification in presence of large motions. In: Proceedings of the IEEE Conference on Computer Vision and Pattern Recognition, pp. 4119–4127 (2015)
20. Kooij, J.F.P., van Gemert, J.C.: Depth-aware motion magnification. In: Leibe, B., Matas, J., Sebe, N., Welling, M. (eds.) ECCV 2016. LNCS, vol. 9912, pp. 467–482. Springer, Cham (2016). https://doi.org/10.1007/978-3-319-46484-8_28
21. Portilla, J., Simoncelli, E.P.: A parametric texture model based on joint statistics of complex wavelet coefficients. Int. J. Comput. Vis. **40**(1), 49–70 (2000). https://doi.org/10.1023/A:1026553619983
22. Freeman, W.T., Adelson, E.H.: Steerable filters. In: Optical Society of America Conference on Image Understanding and Machine Vision, pp. 114–117 (1989)
23. Lindeberg, T.: Scale-space for discrete signals. IEEE Trans. Pattern Anal. Mach. Intell. **12**(3), 234–254 (1990)
24. Willemet, M., Chowienczyk, P., Alastruey, J.: A database of virtual healthy subjects to assess the accuracy of foot-to-foot pulse wave velocities for estimation of aortic stiffness. Am. J. Physiol.-Heart Circ. Physiol. **309**(4), 663–675 (2015)
25. Sedrakyan, A., Campbell, B., Merino, J.-G., Kuntz, R., Hirst, A., McCulloch, P.: IDEAL-D: a rational framework for evaluating and regulating the use of medical devices. BMJ **353**, i2372 (2016)
26. Marcus, H.J., et al.: Comparative effectiveness and safety of image guidance systems in neurosurgery: a preclinical randomized study. J. Neurosurg. **123**(2), 307–313 (2015)

Recognition of Instrument-Tissue Interactions in Endoscopic Videos via Action Triplets

Chinedu Innocent Nwoye[1(✉)], Cristians Gonzalez[2], Tong Yu[1], Pietro Mascagni[1,3], Didier Mutter[2], Jacques Marescaux[2], and Nicolas Padoy[1]

[1] ICube, University of Strasbourg, CNRS, IHU, Strasbourg, France
{nwoye,npadoy}@unistra.fr

[2] University Hospital of Strasbourg, IRCAD, IHU, Strasbourg, France

[3] Fondazione Policlinico Universitario Agostino Gemelli IRCCS, Rome, Italy

Abstract. Recognition of surgical activity is an essential component to develop context-aware decision support for the operating room. In this work, we tackle the recognition of fine-grained activities, modeled as action triplets $\langle instrument, verb, target\rangle$ representing the tool activity. To this end, we introduce a new laparoscopic dataset, *CholecT40*, consisting of 40 videos from the public dataset Cholec80 in which all frames have been annotated using 128 triplet classes. Furthermore, we present an approach to recognize these triplets directly from the video data. It relies on a module called *class activation guide*, which uses the instrument activation maps to guide the verb and target recognition. To model the recognition of multiple triplets in the same frame, we also propose a trainable *3D interaction space*, which captures the associations between the triplet components. Finally, we demonstrate the significance of these contributions via several ablation studies and comparisons to baselines on CholecT40.

Keywords: Surgical activity recognition · Action triplet · Tool-tissue interaction · Deep learning · Endoscopic video · CholecT40

1 Introduction

The recognition of the surgical workflow has been identified as a key research area in surgical data science [14], as this recognition enables the development of intra- and post-operative context-aware decision support tools fostering both surgical safety and efficiency. Pioneering work in surgical workflow recognition has mostly focused on phase recognition from endoscopic video [1,4,7,12,22,25] and from ceiling mounted cameras [2,21], gesture recognition from robotic data (kinematic [5,6], video [11,24], system events [15]) and event recognition, such as the presence of smoke or bleeding [13].

Electronic supplementary material The online version of this chapter (https://doi.org/10.1007/978-3-030-59716-0_35) contains supplementary material, which is available to authorized users.

A. L. Martel et al. (Eds.): MICCAI 2020, LNCS 12263, pp. 364–374, 2020.
https://doi.org/10.1007/978-3-030-59716-0_35

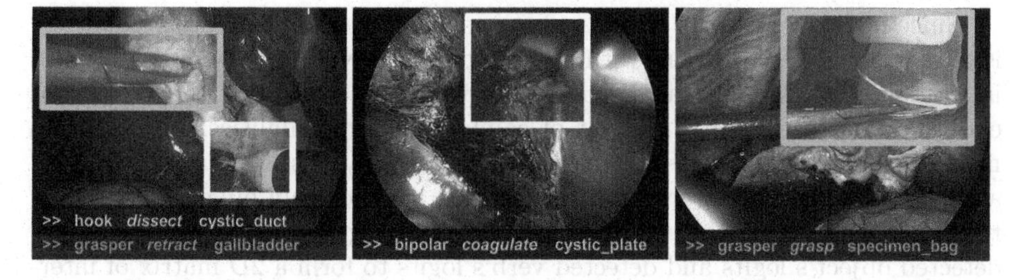

Fig. 1. Examples of action triplets from the CholecT40 dataset. The three images show four different triplets. The localization is not part of the dataset, but a representation of the weakly-supervised output of our recognition model.

In this paper, we focus on recognizing fine-grained activities representing the instrument-tissue interactions in endoscopic videos. These interactions are modeled as triplets $\langle instrument, verb, target\rangle$. Triplets represent the used instrument, the performed action, and the anatomy acted upon, as proposed in existing surgical ontologies [10,17]. The target anatomy, while more challenging to annotate, adds substantial semantics to the recognized action/instrument. Triplet information has already been used to recognize phases [10], however, to the best of our knowledge, this is the first work aiming at recognizing triplets directly from the video data. The fine-grained nature of the triplets also makes this recognition task very challenging. For comparison, the action recognition task introduced within the Endovis challenge at MICCAI 2019 targeted the recognition of 4 verbs only (*grasp, hold, cut, clip*).

To perform this work, we present a new dataset, called *CholecT40*, containing 135K action triplets annotated on 40 cholecystectomy videos from the public Cholec80 dataset [22]. The triplets belong to 128 action triplet classes, composed of 6 instruments, 8 verbs, and 19 target classes. Examples of such action triplets are $\langle grasper, retract, gallbladder\rangle$, $\langle scissor, cut, cystic\_duct\rangle$, $\langle hook, coagulate, liver\rangle$ (see also Fig. 1).

To design our recognition model, we build a multitask learning (MTL) network with three branches for the instrument, verb and target recognition. We also observe that triplets are instrument-centric: an action is only performed if an instrument is present. Indeed, clinically an action can only occur if a hand is manipulating the instrument. We therefore introduce a new module, called *class activation guide (CAG)*, which uses the weak localization information from the instrument activation maps to guide the recognition of the verbs and targets. The idea is similar to [8], which uses the human's ROI produced by FasterRCNN to inform the model on the likely location of the target. Other related works from the computer vision community [19,20,23] rely heavily on the overlap of the *subject-object* bounding boxes to learn the interactions. However, in addition to the fact that our work target triplets, our approach differs in that it does not rely on any spatial annotations in the dataset, which are expensive to generate.

Since instrument, verb, and target are multi-label classes, another challenge is to model their associations within the triplets. As will be shown in the experiments, naively assigning an ID to each triplet and classifying the IDs is not effective, due to the large amount of combinatorial possibilities. In [19,20,23] mentioned above, *human* is considered to be the only possible subject of interaction. Hence, in those works data association requires only bipartite matching to match verbs to objects. This is solvable by using the outer product of the detected object's logits and detected verb's logits to form a 2D matrix of interaction at test time [20]. Data association's complexity increases however with a triplet. Solving a triplet relationship is a tripartite graph matching problem, which is an NP-hard optimisation problem. In this work, inspired by [20], we therefore propose a *3D interaction space* to recognize the triplets. Unlike [20], where the data association is not learned, our interaction space learns the triplet relationships.

In summary, the contributions of this work are as follows:

1. We propose the first approach to recognize surgical actions as triplets of *⟨instrument, verb, target⟩* directly from surgical videos.
2. We present a large endoscopic action triplet dataset, CholecT40, for this task.
3. We develop a novel deep learning model that uses weak localization information from tool prediction to guide verb and target detection.
4. We introduce a trainable 3D interaction space to learn the relationships within the triplets.

2 Cholecystectomy Action Triplet Dataset

To encourage progress towards the recognition of instrument-tissue interactions in laparoscopic surgery, we generated a dataset consisting of 40 videos from Cholec80 [22] annotated with action triplet information. We call this dataset *CholecT40*. The cholecystectomy recordings were first annotated by a surgeon using the software *Surgery Workflow Toolbox-Annotate* from the B-com institute. For each identified *action*, the surgeon sets times for the start and end frames, then labels the *instrument*, the *verb* and the *target*. Any change in the triplet configuration marks the end of the current action and the beginning of a different one. This first step was followed by a mediation on the annotations and a class grouping carried out by another clinician. The resulting action triplets span 128 classes encompassing 6 instruments, 8 verbs, and 19 target classes. For our experiments, we downsample the videos to 1 fps yielding a total of 83.2K frames annotated with 135K action-triplet instances. Table 1 shows the frequency of occurrence of the instruments, verbs and targets in the dataset. When a tool is idle, the verb and the target are both set to *null*. Additional statistics on the co-occurence distribution of the triplets are presented in the supplementary material. The video dataset is randomly split into training (25 videos, 50.6K frames, 82.4K triplets), validation (5 videos, 10.2K frames, 15.9K triplets) and testing (10 videos, 22.5K frames, 37.1K triplets) sets.

Table 1. Dataset statistics showing the frequency of occurrence of the instruments, verbs and targets. Target ids *0...18* correspond to *null, abdominal wall/cavity, gallbladder, cystic plate, cystic artery, cystic duct, cystic pedicle, liver, adhesion, clip, fluid, specimen bag, omentum, peritoneum, gut, hepatic pedicle, tissue sampling, falciform ligament, suture.*

Instrument		Verb		Target					
Name	Count	Name	Count	ID	Count	ID	Count	ID	Count
grasper	76196	null	5807	0	5807	8	236	16	88
bipolar	5616	place/pack	273	1	1169	9	137	17	114
hook	44413	grasp/retract	74720	2	75331	10	1950	18	9
scissors	1856	clip	2578	3	5173	11	5793		
clipper	2851	dissect	42851	4	4378	12	8815		
irrigator	4522	cut	1544	5	10023	13	641		
		coagulate	4306	6	552	14	745		
		clean	3375	7	14433	15	60		

3 Methodology

To recognize the instrument-tissue interactions in the CholecT40 dataset, we build a new deep learning model, called *Tripnet*, by following a multitask learning (MTL) strategy. The principal novelty of this model is its use of the instrument's class activation guide and 3D interaction space to learn the relationships between the components of the action triplets.

Multitask Learning: Recently, multitask deep learning models have shown that correlated tasks can share deep learning layers and features to improve performance [9,16]. Following this observation, we build a MTL network with three branches for the instrument (I), verb (V), and target (T) recognition tasks. The instrument branch is a two layers convolutional network trained for instrument classification. It uses global max pooling (GMP) to learn the class activation maps (CAM) of the instruments for their weak localization, as suggested in [18]. Similarly, the verb and the target branches learn the verb and target classifications using each two convolutional layers and one fully-connected (FC)-layer. All the three branches share the same ResNet-18 backbone for feature extraction.

Class Activation Guide: The pose of the instruments is indicative of their interactions with the tissues. However, there is no bounding box annotation in the dataset that could be used to learn how to crop the action's locations, as done in [8,19,20,23]. We therefore hypothesize that the instrument's CAM from the instrument branch, learnable in a weakly supervised manner, has sufficient information to direct the verb and target detection branches towards the likely region of interest of the actions. For convenience, we regroup the three branches of the MTL into two subnets: the *instrument* subnet and the *verb-target* subnet, as illustrated in Fig. 2a. The verb-target subnet is then transformed to a *class*

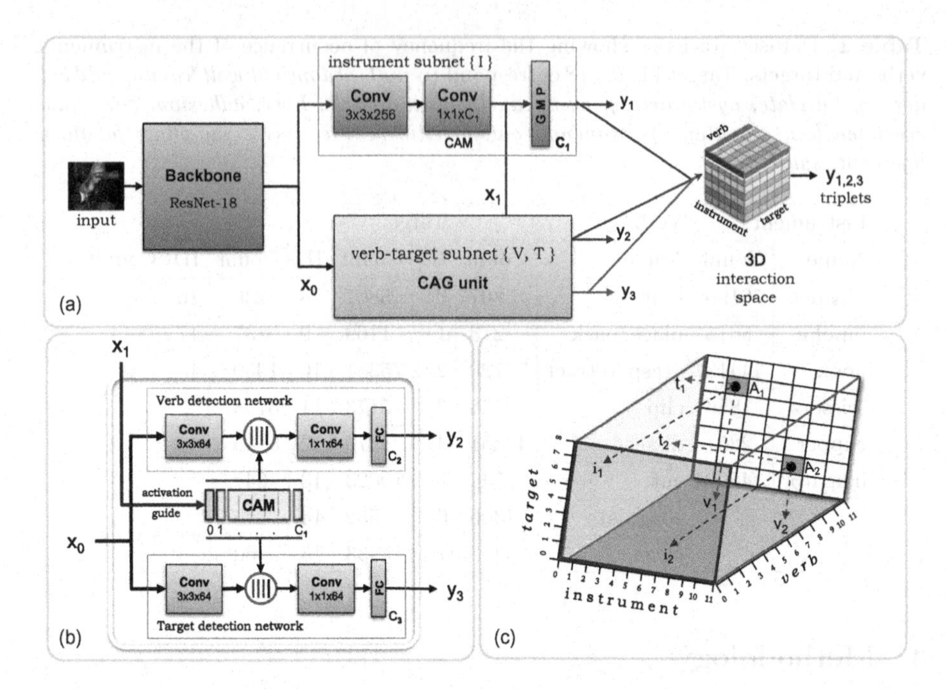

Fig. 2. Proposed model: (a) tripnet for action triplet recognition, (b) class activation guide (CAG) unit for spatially guided detection, (c) 3D interaction space for triplet association.

activation guide (CAG) unit, as shown in Fig. 2b. It receives the instrument's CAM as additional input. This CAM input is then concatenated with the verb and target features, concurrently, to guide and condition the model search space of the verb and target on the instrument appearance cue.

3D Interaction Space: Recognizing the correct action triplets involves associating the right (I, V, T) components using the raw output vectors, also called logits, of the instrument (I), verb (V) and target (T) branches. In the existing work [20], where the data association problem involves only the *object-verb* pair, the outer product of their logits is used to form a 2D matrix of component interaction at test time. In a similar manner, we propose a *3D interaction space* for associating the triplets, as shown in Fig. 2c. Unlike in [20], where the data association is not learned by the trained model, we model a trainable interaction space. Given the m-logits, n-logits and p-logits for the I, V, T respectively, we learn the triplets y using a 3D projection function Ψ as follows:

$$y = \Psi(\alpha I, \beta V, \gamma T), \tag{1}$$

where α, β, γ, are the learnable weights for projecting I, V and T to the 3D space and Ψ is an outer product operation. This gives an $m \times n \times p$ grid of logits with the three axes representing the three components of the triplets. For all $i \in I, v \in V, t \in T$ the 3D point $y_{i,v,t}$ represents a possible triplet. A 3D point with a probability above a threshold is considered a valid triplet. In practice, there are more 3D points in the space than valid triplets in the CholecT40 dataset. Therefore, we mask out the invalid points, obtained using the training set, at both train and test times.

Proposed Model: The proposed network is called *Tripnet* and shown in Fig. 2(a): it is an integration of the CAG unit and of the 3D interaction space within the MTL model. The whole model is trained end-to-end using a warm-up parameter which allows the instrument subnet to learn some semantics for a few epochs before guiding the verb-target subnet with instrument cues.

4 Experiments

Implementation Details: We perform our experiments on CholecT40. During training, we employ three types of data augmentation (rotation, horizontal flipping and patch masking) with no image preprocessing. The model is trained on images resized to $256 \times 448 \times 3$. All the individual tasks are trained for multi-label classification using the weighted sigmoid cross-entropy with logits as loss function, regularized by an L_2 norm with $1e^{-5}$ weight decay. The class weights are calculated as in [18]. The Resnet-18 backbone is pretrained on Imagenet. All the experimented models are trained using learning rates with exponential decay and initialized with the values $1e^{-3}, 1e^{-4}, 1e^{-5}$ for the subnets, backbone, and 3D interaction space, respectively. The learning rates and other hyperparameters are tuned from the validation set using grid search. Our network is implemented using TensorFlow and trained on GeForce GTX 1080 Ti GPUs.

Tasks and Metrics: To evaluate the capacity of a model to recognize correctly a triplet and its components, we use two types of metrics:

1. Instrument detection performance: This measures the average precision (AP) of detecting the correct *instruments*, as the area under the precision-recall curve per instrument (AP_I).
2. Triplet recognition performance: This measures the AP of recognizing the instrument-tissue interactions by looking at different sets of triplet components. We use three metrics: the *instrument-verb* (AP_{IV}), *instrument-target* (AP_{IT}), and *instrument-verb-target* (AP_{IVT}) metrics. All the listed components need to be correct during the AP computation. AP_{IVT} evaluate the recognition of the complete triplets.

Baselines: We build two baseline models. The naive CNN baseline is a ResNet-18 backbone with two additional 3×3 convolutional layers and a fully connected (FC) classification layer with N units, where N corresponds to the number of triplet classes ($N = 128$). The naive model learns the action-triplets using their IDs without any consideration of the components that constitute the triplets. We therefore also include an MTL baseline built with the I, V and T branches described in Sect. 3. The outputs of the three branches are concatenated and fed to an FC-layer to learn the triplets. For fair comparison, the two baselines share the same backbone as Tripnet.

Quantitative Results: Table 2 presents the AP results for the instrument detection across all triplets. The results show that the naive model does not understand the triplet components. This comes from the fact that it is designed to learn the triplets using their IDs: two different triplets sharing the same instrument or verb still have different IDs. On the other hand, the MTL and Tripnet networks, which both model the triplet components, show competing performance on instrument detection. Moreover, Tripnet outperforms the MTL baseline by 15.1% mean AP. This can be attributed to its use of CAG unit and 3D interaction space to learn better semantic information about the instrument behaviors.

Table 2. Instrument detection performance of (AP_I) across all triplets. The IDs 0...5 correspond to *grasper, bipolar, hook, scissors, clipper and irrigator*, respectively.

Model	Instrument						Mean
	0	1	2	3	4	5	
Naive CNN	75.3	04.3	64.6	02.1	05.5	06.0	27.5
MTL	96.1	**91.9**	**97.2**	55.7	30.3	76.8	74.6
Tripnet	**96.3**	91.6	**97.2**	**79.9**	**90.5**	**77.9**	**89.7**

Table 3. Action triplet recognition performance for instrument-verb (AP_{IV}), instrument-target (AP_{IT}) and instrument-verb-target (AP_{IVT}) components.

Model	AP_{IV}	AP_{IT}	AP_{IVT}	Mean
Naive CNN	7.54	6.89	5.88	6.77
MTL	14.02	7.15	6.43	9.20
Tripnet	**35.45**	**19.94**	**18.95**	**24.78**

The triplet recognition performance is presented in Table 3. The naive CNN model has again the worst performance for the AP_{IV}, AP_{IT} and AP_{IVT} metrics, as expected from the previous results. The MTL baseline model, on the other hand, performs only slightly above the naive model despite its high instrument detection performance in Table 2. This is because the MTL baseline model, after learning the components of the triplets, dilutes this semantic information by concatenating and feeding the output to an FC-layer. However, Tripnet improves over the MTL baseline by leveraging the instrument cue from the CAG unit. It also learns better triplet association by increasing the AP_{IVT} by 12.5% on average. Tripnet outperformed all the baselines in instrument-tissue interaction recognition by a minimum of 15.6%. In general, it can be observed that it is easier to learn the instrument-verb components than the instrument-target components. This is likely due to the fact that (a) a verb has a more direct association to the instrument creating the action (b) the dataset contains many more target classes than verb classes (c) many anatomical structures in the abdomen are usually discriminated with difficulty by non-medical experts.

While the action recognition performance appears to be low, it follows the same pattern as other models in the computer vision literature on action datasets of even lesser complexity. For instance, on the HICO-DET dataset [3,8] achieves 10.8%, [19] achieves 14.2% and [23] achieves 15.1% action recognition AP, also known as AP_{role}. In fact, the current state-of-the-art performance on HICO-DET dataset is 21.2% as reported on the leaderboard server. Similarly, the winner of the MICCAI 2019 subchallenge on action recognition, involving only four verb classes, scores 23.3% F1-score. This shows the challenging nature of fine-grained action recognition.

Table 4. Ablation study for the CAG unit and 3D interaction space in Tripnet model.

FC	3D (untrained)	3D (trained)	CAG	AP_I	AP_{IV}	AP_{IT}	AP_{IVT}
✓				74.6	14.02	7.15	6.43
	✓			89.3	14.28	6.99	6.03
	✓		✓	**89.7**	16.72	7.62	6.32
		✓		89.5	20.63	12.08	12.06
		✓	✓	**89.7**	**35.45**	**19.94**	**18.95**

Ablation Studies: Table 4 presents an ablation study of the novel components of the Tripnet model. The CAG unit improves the AP_{IV} and AP_{IT} by approximately 2.0% and 1.0%, respectively, justifying the need for using instrument cues in the verb and target recognition. We also observe that learning the instrument-tissue interactions is better with a trainable 3D projection than with either the untrained 3D space or with an FC-layer. This results in a large 6.0% improvement of the AP_{IVT}. We record the best performance in all four metrics by combining the CAG unit and the trained 3D interaction space. The two units complement each other and improve the results across all metrics.

Qualitative Results: To better appreciate the performance of the proposed model in understanding instrument-tissue interactions, we overlay the predictions on several surgical images in Fig. 3. The qualitative results show that Tripnet does not only improve the performance of the baseline models, but also localizes accurately the regions of interest of the actions. It is observed that the majority of incorrect predictions are due to one incorrect triplet component. Instruments are usually correctly predicted and localized. As can be seen in the complete statistics provided in the supplementary material, it is however not straightforward to predict the verb/target directly from the instrument due to the multiple possible associations. More qualitative results are included in the supplementary material.

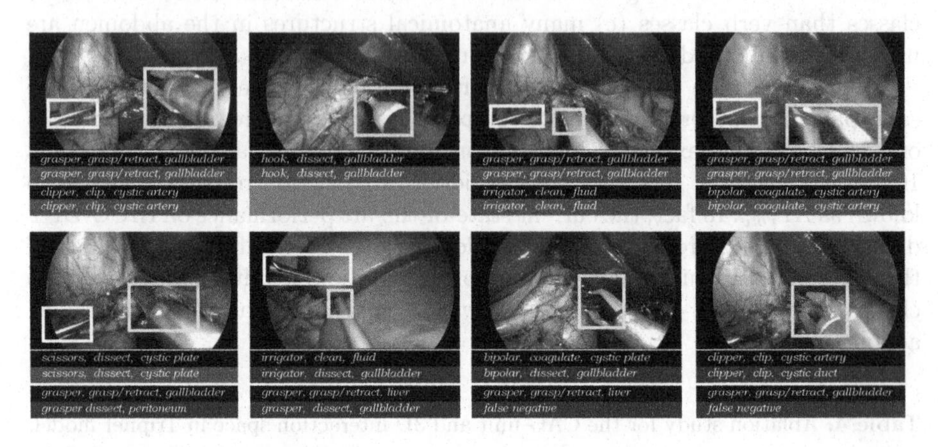

Fig. 3. Qualitative results: triplet prediction and weak localization of the regions of action (*best seen in color*). Predicted and ground-truth triplets are displayed below each image: black = ground-truth, green = correct prediction, red = incorrect prediction. A missed triplet is marked as false negative and a false detection is marked as false positive. The color of the text corresponds to the color of the associated bounding box. (Color figure online)

5 Conclusion

In this work, we tackle the task of recognizing action triplets directly from surgical videos. Our overarching goal is to detect the instruments and learn their interactions with the tissues during laparoscopic procedures. To this aim, we present a new dataset, which consists of 135k action triplets over 40 videos. For recognition, we propose a novel model that relies on instrument class activation maps to learn the verbs and targets. We also introduce a trainable 3D interaction space for learning the ⟨instrument, verb, target⟩ associations within the triplets. Experiments show that our model outperforms the baselines by a substantial margin in all the metrics, hereby demonstrating the effectiveness of the proposed approach.

Acknowledgements. This work was supported by French state funds managed within the Investissements d'Avenir program by BPI France (project CONDOR) and by the ANR (references ANR-11-LABX-0004 and ANR-16-CE33-0009). The authors would also like to thank the IHU and IRCAD research teams for their help with the data annotation during the CONDOR project.

References

1. Blum, T., Feußner, H., Navab, N.: Modeling and segmentation of surgical workflow from laparoscopic video. In: Jiang, T., Navab, N., Pluim, J.P.W., Viergever, M.A. (eds.) MICCAI 2010. LNCS, vol. 6363, pp. 400–407. Springer, Heidelberg (2010). https://doi.org/10.1007/978-3-642-15711-0_50
2. Chakraborty, I., Elgammal, A., Burd, R.S.: Video based activity recognition in trauma resuscitation. In: 2013 10th IEEE International Conference and Workshops on Automatic Face and Gesture Recognition (FG), pp. 1–8 (2013)
3. Chao, Y.W., Liu, Y., Liu, X., Zeng, H., Deng, J.: Learning to detect human-object interactions. In: 2018 IEEE Winter Conference on Applications of Computer Vision (WACV), pp. 381–389 (2018)
4. Dergachyova, O., Bouget, D., Huaulmé, A., Morandi, X., Jannin, P.: Automatic data-driven real-time segmentation and recognition of surgical workflow. Int. J. Comput. Assist. Radiol. Surg. **11**(6), 1081–1089 (2016). https://doi.org/10.1007/s11548-016-1371-x
5. DiPietro, R., et al.: Segmenting and classifying activities in robot-assisted surgery with recurrent neural networks. Int. J. Comput. Assist. Radiol. Surg. **14**(11), 2005–2020 (2019). https://doi.org/10.1007/s11548-019-01953-x
6. DiPietro, R., et al.: Recognizing surgical activities with recurrent neural networks. In: Ourselin, S., Joskowicz, L., Sabuncu, M.R., Unal, G., Wells, W. (eds.) MICCAI 2016. LNCS, vol. 9900, pp. 551–558. Springer, Cham (2016). https://doi.org/10.1007/978-3-319-46720-7_64
7. Funke, I., Jenke, A., Mees, S.T., Weitz, J., Speidel, S., Bodenstedt, S.: Temporal coherence-based self-supervised learning for laparoscopic workflow analysis. In: Stoyanov, D., et al. (eds.) CARE/CLIP/OR 2.0/ISIC -2018. LNCS, vol. 11041, pp. 85–93. Springer, Cham (2018). https://doi.org/10.1007/978-3-030-01201-4_11
8. Gkioxari, G., Girshick, R., Dollár, P., He, K.: Detecting and recognizing human-object interactions. In: Proceedings of the IEEE Conference on Computer Vision and Pattern Recognition, pp. 8359–8367 (2018)
9. Jin, Y., et al.: Multi-task recurrent convolutional network with correlation loss for surgical video analysis. Med. Image Anal. **59**, 101572 (2020)
10. Katić, D., et al.: LapOntoSPM: an ontology for laparoscopic surgeries and its application to surgical phase recognition. Int. J. Comput. Assist. Radiol. Surg. **10**(9), 1427–1434 (2015). https://doi.org/10.1007/s11548-015-1222-1
11. Kitaguchi, D., et al.: Real-time automatic surgical phase recognition in laparoscopic sigmoidectomy using the convolutional neural network-based deep learning approach. Surg. Endosc. 1–8 (2019). https://doi.org/10.1007/s00464-019-07281-0
12. Lo, B.P.L., Darzi, A., Yang, G.-Z.: Episode classification for the analysis of tissue/instrument interaction with multiple visual cues. In: Ellis, R.E., Peters, T.M. (eds.) MICCAI 2003. LNCS, vol. 2878, pp. 230–237. Springer, Heidelberg (2003). https://doi.org/10.1007/978-3-540-39899-8_29
13. Loukas, C., Georgiou, E.: Smoke detection in endoscopic surgery videos: a first step towards retrieval of semantic events. Int. J. Med. Robot. Comput. Assist. Surg. **11**(1), 80–94 (2015)
14. Maier-Hein, L., et al.: Surgical data science: enabling next-generation surgery. Nat. Biomed. Eng. **1**, 691–696 (2017)
15. Malpani, A., Lea, C., Chen, C.C.G., Hager, G.D.: System events: readily accessible features for surgical phase detection. Int. J. Comput. Assist. Radiol. Surg. **11**(6), 1201–1209 (2016). https://doi.org/10.1007/s11548-016-1409-0

16. Mondal, S.S., Sathish, R., Sheet, D.: Multitask learning of temporal connectionism in convolutional networks using a joint distribution loss function to simultaneously identify tools and phase in surgical videos. arXiv preprint arXiv:1905.08315 (2019)
17. Neumuth, T., Strauß, G., Meixensberger, J., Lemke, H.U., Burgert, O.: Acquisition of process descriptions from surgical interventions. In: Bressan, S., Küng, J., Wagner, R. (eds.) DEXA 2006. LNCS, vol. 4080, pp. 602–611. Springer, Heidelberg (2006). https://doi.org/10.1007/11827405_59
18. Nwoye, C.I., Mutter, D., Marescaux, J., Padoy, N.: Weakly supervised convolutional LSTM approach for tool tracking in laparoscopic videos. Int. J. Comput. Assist. Radiol. Surg. **14**(6), 1059–1067 (2019). https://doi.org/10.1007/s11548-019-01958-6
19. Qi, S., Wang, W., Jia, B., Shen, J., Zhu, S.C.: Learning human-object interactions by graph parsing neural networks. In: Proceedings of the European Conference on Computer Vision (ECCV), pp. 401–417 (2018)
20. Shen, L., Yeung, S., Hoffman, J., Mori, G., Fei-Fei, L.: Scaling human-object interaction recognition through zero-shot learning. In: 2018 IEEE Winter Conference on Applications of Computer Vision (WACV), pp. 1568–1576 (2018)
21. Twinanda, A.P., Alkan, E.O., Gangi, A., de Mathelin, M., Padoy, N.: Data-driven spatio-temporal RGBD feature encoding for action recognition in operating rooms. Int. J. Comput. Assist. Radiol. Surg. **10**(6), 737–747 (2015). https://doi.org/10.1007/s11548-015-1186-1
22. Twinanda, A.P., Shehata, S., Mutter, D., Marescaux, J., De Mathelin, M., Padoy, N.: EndoNet: a deep architecture for recognition tasks on laparoscopic videos. IEEE Trans. Med. Imaging **36**(1), 86–97 (2017)
23. Xu, B., Wong, Y., Li, J., Zhao, Q., Kankanhalli, M.S.: Learning to detect human-object interactions with knowledge. In: Proceedings of the IEEE Conference on Computer Vision and Pattern Recognition (2019)
24. Zia, A., Hung, A., Essa, I., Jarc, A.: Surgical activity recognition in robot-assisted radical prostatectomy using deep learning. In: Frangi, A.F., Schnabel, J.A., Davatzikos, C., Alberola-López, C., Fichtinger, G. (eds.) MICCAI 2018. LNCS, vol. 11073, pp. 273–280. Springer, Cham (2018). https://doi.org/10.1007/978-3-030-00937-3_32
25. Zisimopoulos, O., et al.: DeepPhase: surgical phase recognition in CATARACTS videos. In: Frangi, A.F., Schnabel, J.A., Davatzikos, C., Alberola-López, C., Fichtinger, G. (eds.) MICCAI 2018. LNCS, vol. 11073, pp. 265–272. Springer, Cham (2018). https://doi.org/10.1007/978-3-030-00937-3_31

AutoSNAP: Automatically Learning Neural Architectures for Instrument Pose Estimation

David Kügler[1,2(✉)], Marc Uecker[1], Arjan Kuijper[1,3], and Anirban Mukhopadhyay[1]

[1] Department of Computer Science, TU Darmstadt, Darmstadt, Germany
[2] German Center for Neuro-Degenerative Diseases (DZNE), Bonn, Germany
david.kuegler@dzne.de
[3] Fraunhofer IGD, Darmstadt, Germany

Abstract. Despite recent successes, the advances in Deep Learning have not yet been fully translated to Computer Assisted Intervention (CAI) problems such as pose estimation of surgical instruments. Currently, neural architectures for classification and segmentation tasks are adopted ignoring significant discrepancies between CAI and these tasks. We propose an automatic framework (AutoSNAP) for instrument pose estimation problems, which discovers and learns architectures for neural networks. We introduce 1) an efficient testing environment for pose estimation, 2) a powerful architecture representation based on novel Symbolic Neural Architecture Patterns (SNAPs), and 3) an optimization of the architecture using an efficient search scheme. Using AutoSNAP, we discover an improved architecture (SNAPNet) which outperforms both the hand-engineered i3PosNet and the state-of-the-art architecture search method DARTS.

Keywords: Neural architecture search · Instrument pose estimation · AutoML

1 Introduction

Deep Neural Networks (DNNs) have revolutionized Computer-Assisted Interventions (CAI) with applications ranging from instrument tracking to quality control [10,15]. However, the design of these neural architectures is a time-consuming and complex optimization task requiring extensive hyper-parameter

D. Kügler and M. Uecker—Equal contribution.

Electronic supplementary material The online version of this chapter (https://doi.org/10.1007/978-3-030-59716-0_36) contains supplementary material, which is available to authorized users.

A. L. Martel et al. (Eds.): MICCAI 2020, LNCS 12263, pp. 375–384, 2020.
https://doi.org/10.1007/978-3-030-59716-0_36

testing. Consequently, CAI researchers often adopt established neural architectures designed for other vision tasks such as large-scale image classification [5,11,13,14]. But CAI problems requiring regression instead of classification or segmentation on scarcely annotated and small datasets differ from these tasks on a fundamental level. This CAI-centric challenge is featured in instrument pose estimation for minimally-invasive temporal bone surgery [12]: Training on synthetic data is necessary, because hard-to-acquire real-world datasets with high-quality annotation are reserved for evaluation only. The state-of-the-art method (i3PosNet [7]) relies on an architecture optimized for classification. It disregards the specialization potential as described by the no-free-lunch-theorem and as realized by DNNs for registration [2] demanding a method to automatically improve the architecture.

Optimizing neural architectures for a specific problem is challenging on its own due to the following requirements: 1) an *Efficient Environment* to test candidate performance, 2) a *Succinct Representation* to describe the architecture, and 3) an *Efficient Search Algorithm* to improve candidates quickly with limited hardware. Automatic Neural Architecture Search (NAS) strategies were initially introduced in computer vision classification. Previous work [4] can be classified into two groups: discrete and continuous. The discrete strategy (e.g. NASNet) [9,19] iteratively proposes, tests and improves blocks. These blocks consist of multiple "NAS units" and are themselves combined to form full architectures. Despite being widely used in various publications, these units are not particularly efficient in terms of both optimization and functional redundancy. However, the iterative improvement scales well for distributed computing with massive computational effort (≥200 GPU days). The continuous strategy (e.g. DARTS [8]) stacks all layer options together and calculates a weighted sum, motivating the name continuous. All architectures are trained at the same time and weights are shared. This approach is more computationally efficient (4 GPU days), but very VRAM-demanding because of "stacks of layers". AutoSNAP combines the flexibility of NASNet with the speed of DARTS by introducing an intuitive yet succinct representation (instead of NAS units) and improving the efficient search and optimization strategy. The medical imaging community has recently confirmed the potential of NAS methods to segmentation [3,16–18] with adaptations for scalable [6] and resource-constrained [1] environments. We are not aware of any application of NAS to CAI.

We introduce problem-dependent learning and optimization of neural architectures to instrument pose estimation[1]. AutoSNAP implements problem-specific and limited-resources NAS for CAI with three major contributions: 1) the integration of a CAI-framework as an *efficient testing environment* for performance analysis (Fig. 1a), 2) an extensible, *succinct representation* termed Symbolic Neural Architecture Pattern (SNAP, Fig. 1b) to describe architecture blocks, and 3) an *efficient search algorithm* guided in "Optimization Space" (auto-encoder latent space) to explore and discover "new architectures" (Fig. 1c). By integrating these factors, AutoSNAP links architecture and performance

[1] Our code is available at https://github.com/MECLabTUDA/AutoSNAP.

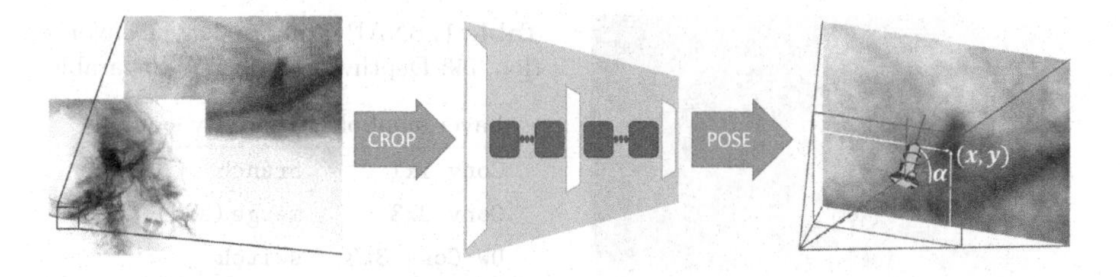

(a) *Testing Environment* from CAI: we search for a performant architecture (SNAPNet, green architecture) to estimate the pose of a surgical instrument from X-ray images.

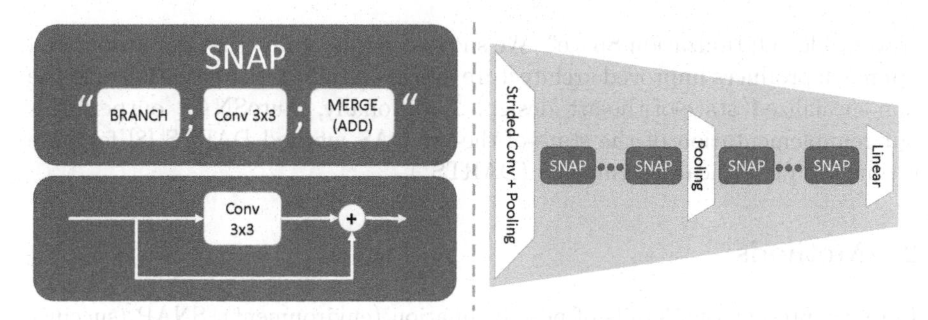

(b) *Succinct Representation*: left side: a SNAP (top) defines a corresponding neural block (bottom); right: we build the architecture (SNAPNet) by repeating this block.

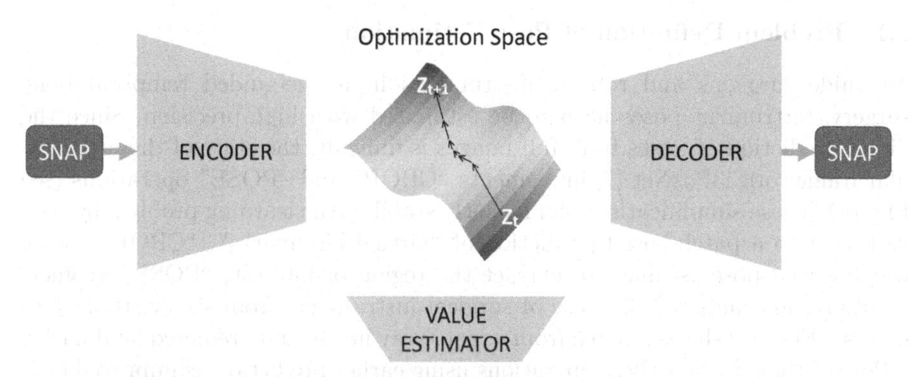

(c) *Efficient Search*: the transformation of SNAPs into a unified latent-space (auto-encoder) accelerates the search by gradient ascent on the value estimator surface.

Fig. 1. Overview of AutoSNAP components (Color figure online)

allowing for end-to-end optimization and search. We jointly train AutoSNAP's auto-encoder (Fig. 1c) using a multi-component loss. In addition to reconstruction, this loss also uses on-the-fly performance metrics from the testing environment to predict the performance of a SNAP-based architecture. In consequence, we enable the substitution of the optimization on SNAPs by the optimization in a

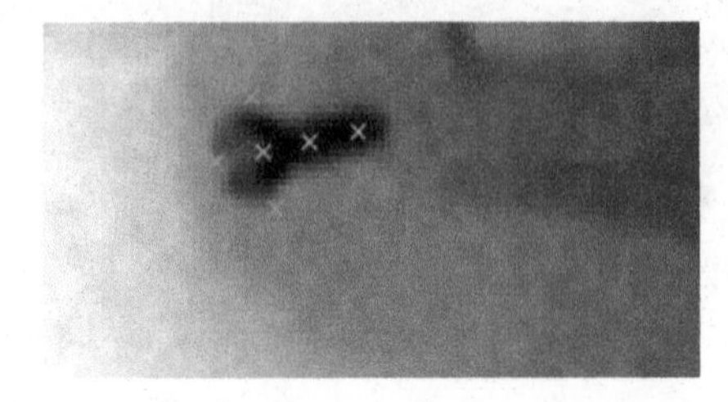

Fig. 2. X-ray image patch of a screw with "virtual landmarks".

Table 1. SNAP symbols: Conv: Convolution, DW: Depthwise, DWS: DW-Separable

Layer symbols	Topology symbols
Conv 1x1	branch
Conv 3x3	merge (add)
DW-Conv 3x3	switch
DWS-Conv 3x3	
Max-Pool 3x3 (stride 1)	

traversable "Optimization Space". We show experimentally, that our automated approach produces improved architecture designs significantly outperforming the non-specialized state-of-the-art design. Additionally, AutoSNAP outperforms our reimplementation of the state-of-the-art NAS method DARTS [8] for pose estimation of surgical instruments (DARTS*).

2 Methods

Here, we present the details of pose estimation (environment), SNAP (succinct representation), the auto-encoder and the optimization scheme (efficient search).

2.1 Problem Definition of Pose Estimation

To guide surgeons and robotic instruments in image-guided temporal bone surgery, instrument poses need to be estimated with high-precision. Since the direct prediction of poses from full images is difficult, the state-of-the-art modular framework i3PosNet [7] implements "CROP" and "POSE" operations (see Fig. 1a). These simplifications significantly stabilize the learning problem by converting it to a patch-based prediction of "virtual landmarks". "CROP" uses a rough initial pose estimate to extract the region of interest, "POSE" geometrically reconstructs the 3D pose of surgical instruments from six "virtual landmarks". Figure 2 shows a patch from a real X-ray image and predicted landmarks. i3PosNet then iterates these operations using earlier prediction as improved estimates.

Framing this prediction task as our *environment*, we search for a neural architecture (green network in Fig. 1) that minimizes the Mean-Squared Error of the point regression task (regMSE). i3PosNet, on the other hand, only adapts a non-specialized VGG-based architecture for this task. Our implementation parallelizes training and evaluation on a validation dataset on multiple machines.

2.2 Symbolic Neural Architecture Patterns (SNAPs)

With many layer types and the design of connections in DNNs, it is currently impractical to optimize the topology and layer choice of the full architecture. As a result, our full architecture (SNAPNet) repeats a block representing multiple operations as illustrated in Fig. 1b. The topology and layer choice of this block are defined by a SNAP sequence.

To automatically generate trainable models, we introduce a language to define blocks using 8 SNAP symbols (see Table 1). Each symbol corresponds to a modification of a stack of tensors which is used to build the model. Five symbols specify (trainable) layers (Table 1). Results replace the top tensor using the previous value as input. Convolutions are always preceded with BatchNormalization and ReLU activation. The three topology symbols realize modification of the stack size and order for example enabling skip connections (see Fig. 1b, left). `branch` duplicates the top element on the stack, `switch` swaps the top two elements and `merge(add)` pops the top two inputs, applies `concat` + `Conv 1x1` and pushes the result. The stack is initialized by the output of the last two blocks and SNAPs end with an implicit `merge(add)` across all tensors on the stack (ignored in Fig. 1b for simplicity, but Fig. 4 includes these).

2.3 AutoSNAP's Auto-Encoder

We introduce an auto-encoder architecture to transform the SNAP sequence into a 16-dimensional vector (latent space), since continuous vectors have favorable properties for optimization. In addition to the Encoder and Decoder, a Value Estimator predicts the architecture performance (Fig. 1c).

The **Encoder** (E) and the **Decoder** (D) use a Recurrent Neural Networks with mirrored architectures of two bi-directional LSTMs and two fully connected layers[2]. Since the last Encoder-layer uses tanh-activation, the latent space is bound to the interval of $[-1, 1]$ in each dimension. As a conditional language model, the decoder generates a sequence of symbol probabilities from a latent vector. Finally, the **Value Estimator** (V) is a linear regression layer with no activation function. Since we are interested in both high accuracy and resolution for well-performing architectures (i.e. at very low regMSE values), we estimate $-\log_{10}(\text{regMSE})$ of the candidate architecture on the validation dataset. This value metric improves the resolution and gradients of the value estimator.

We train the auto-encoder on three sets of input and target data: 1) Sequence Reconstruction, 2) Latent-space consistency and 3) Value Regression. 1) A Cross-Entropy (CE) loss enforces successful *reconstruction* on randomly generated SNAP sequences ($\hat{X}$). 2) The *consistency of the latent space* is further supported by mapping uniform random latent vectors ($\hat{z}$) to a sequence of symbol probabilities, back to the latent space under a Mean Squared Error (MSE) loss. 3) Value estimator and encoder are jointly trained to estimate the value criterion (Y) of known SNAPs (X) via MSE loss. All three loss functions are minimized simultaneously via Gradient Descent:

$$\mathcal{L} = \mathcal{L}_{\text{CE}}(D(E(\hat{X})), \hat{X}) + \mathcal{L}_{\text{MSE}}(V(E(X)), Y) + \mathcal{L}_{\text{MSE}}(E(D(\hat{z})), \hat{z}) \quad (1)$$

[2] We provide additional diagrams of the architectures in the Supplementary Materials.

Three failure modes motivate this design: 1) A disentanglement between value estimate and architecture description occurs, rendering optimization within the latent space futile. 2) The encoder only projects to a limited region of the latent space, thus some latent vectors have an unconstrained value estimate and no corresponding architecture (see a). 3) The decoder overfits to known architectures, which leaves a strong dependence on initial samples. The Latent-space consistency enforces a bijective mapping between latent and symbol spaces (cycle consistency loss) addressing issues 1 and 2. We mitigate problem 3 by only training the auto-encoder on randomly generated sequences.

2.4 Exploration and Optimization

Our approach introduces the continuous optimization of an architecture block by gradient ascent inside the latent space of the auto-encoder (Fig. 1c). The gradient for this optimization is provided by the value estimator which predicts the performance of the architecture pattern vector of a corresponding block (i.e. SNAP). By the design of the auto-encoder, the latent space is learned to represent a performance-informed vector representation. The thereby improved convexity of the latent space w.r.t. performance is the central intuition of the search.

The architecture search consists of three iterative steps: 1) Retrain the auto-encoder on all previously evaluated SNAPs (initially this is a random sample of evaluated architectures). 2) Transform a batch of best known SNAPs into the latent space, optimize them via gradient ascent on value estimator gradients until the decoded latent vector no longer maps to a SNAP with known performance. After a limit of 50 gradient steps (i.e. in vicinity of the SNAP all architectures are evaluated), a new latent vector is sampled randomly (uniform distribution) from the latent space and gradient ascent resumed. 3) Evaluate SNAPs found in step 2) using the test environment and add them to the set of known architectures. Finally, repeat the iteration at step 1).

3 Experiments

We evaluate our method on synthetic and real radiographs of surgical screws in the context of temporal bone surgery [12]. In this section, we summarize data generation, the parameters of the AutoSNAP optimization and settings specific to pose estimation as well as the evaluation metrics.

3.1 Training and Evaluation Datasets

We use the publicly available i3PosNet Dataset [7] assuming its naming conventions. Dataset A features synthetic and Dataset C real radiographs. We train networks exclusively on Dataset A, while evaluating on synthetic and real images.

Dataset A: Synthetic Images: The dataset consists of 10,000 digital images (Subjects 1 and 2) for training and 1,000 unseen images (Subject 3) for evaluation with geometrically calculated annotations. Training images are statically

Table 2. Pose evaluation (lower is better) for datasets A and C (synthetic and real X-ray images). Evaluation for one and three iterations of the i3PosNet-scheme. Mean absolute error ± standard deviation of the absolute error.

Model		Dataset A: synthetic images		Dataset C: real images	
		Position [mm]	Angle [deg.]	Position [mm]	Angle [deg.]
3 iterations	i3PosNet [7]	0.024 ± 0.016	0.92 ± 1.22	1.072 ± 1.481	9.37 ± 16.54
	DARTS* [8]	0.046 ± 0.105	1.84 ± 6.00	1.138 ± 1.199	9.76 ± 18.60
	SNAPNet-A (ours)	0.017 ± 0.012	0.52 ± 0.88	0.670 ± 1.047	7.55 ± 14.22
	SNAPNet-B (ours)	**0.016** ± 0.011	**0.49** ± 0.84	**0.461** ± 0.669	**5.02** ± 9.28
1 iteration	i3PosNet [7]	0.050 ± 0.139	1.14 ± 1.50	0.746 ± 0.818	6.59 ± 10.36
	DARTS* [8]	0.062 ± 0.146	1.81 ± 4.20	0.810 ± 0.770	7.68 ± 12.70
	SNAPNet-A (ours)	0.026 ± 0.029	0.72 ± 1.19	0.517 ± 0.678	5.32 ± 8.85
	SNAPNet-B (ours)	**0.025** ± 0.028	**0.65** ± 1.06	**0.419** ± 0.486	**4.36** ± 6.88

augmented 20-fold by random shifts and rotations to ensure similarity between training runs. For architecture search, we split the training dataset by 70/10/20 (training/online validation/testing and model selection) to identify the performance of candidate models without over-fitting to the evaluation dataset.

Dataset C: Real Images: Real X-ray images of medical screws on a phantom head are captured with a Ziehm c-arm machine (totaling 540 images). Poses are manually annotated with a custom tool.

3.2 Details of Optimization

We randomly choose 100 SNAP architectures for the initial training of the autoencoder and the value-estimator. During the search phase, 100 additional models are tested by training the models for 20 epochs and evaluation on the validation set. We stop the search after 1500 tested models. Our small-scale test environment uses 4 blocks with a pooling layer in the center. Convolutions have 24/48 features before/after the pooling layer. In total, the search takes 100 GPU hours (efficiently parallelized on two NVIDIA GeForce GTX 1080 Ti for 50 h) and requires no human interaction.

3.3 DARTS Architecture Search

We compare our results with DARTS*, our reimplementation of DARTS [8] where the * indicates our application to CAI. DARTS is an efficient, state-of-the-art NAS approach for classification from computer vision (CIFAR-10). We ported the DARTS framework to tensorflow implementing all operations as documented by the authors. This process yields a large "continuous model" with weights for the contribution of individual layers. For the evaluation and comparison with SNAPNet, we discretize and retrain the continuous model of DARTS* in analogy to the DARTS transfer from CIFAR-10 to ImageNet. Similar to DARTS on ImageNet, our DARTS* search took approximately 4 days on one

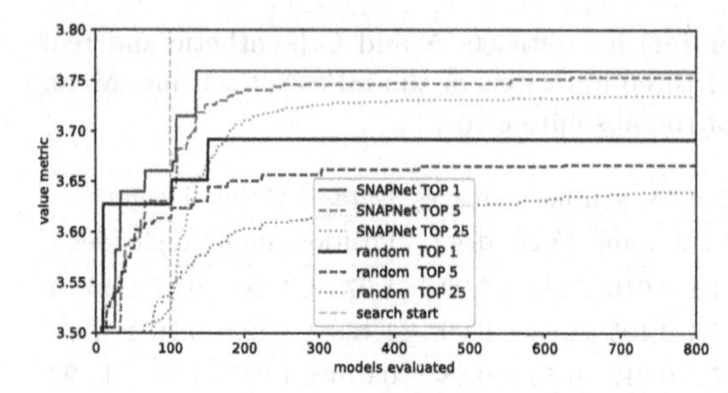

Fig. 3. Comparison of search efficiency for AutoSNAP (red) and Random Search (black) for the value metric. No significant events occur after 2 GPU days (800 models). (Color figure online)

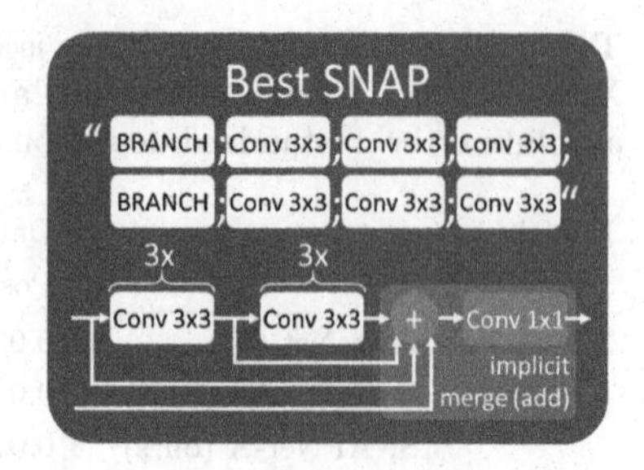

Fig. 4. Best SNAP of Auto-SNAP. Implicit merge of skip-connections, outputs of the previous 2 blocks are used as input.

GPU. Inherently, DARTS cannot efficiently be parallelized across multiple GPUs or machines because all updates are applied to the same continuous model.

3.4 Full-Scale Retraining

We retrain the full final architecture on the common training data of synthetic images from Dataset A [7]. Since efficiency is not a constraining factor for full training, we increase the number of blocks to a total of eight, four before and after the central pooling layer (see Fig. 1b). While SNAPNet-A uses 24/48 feature channels (same as the test environment), we increase the number of features to 56/112 (before/after the pooling layer) for SNAPNet-B. In consequence, the number of weights approximately quadruple from SNAPNet-A and the discrete DARTS* model to SNAPNet-B and again to i3PosNet. Like i3PosNet, we train models for 80 epochs, however using RMSProp instead of Adam. Following the spirit of automatic machine learning, we obtained hyperparameters for these models using bayesian optimization.

3.5 Evaluation Metrics

To maximize comparability and reproducibility, we follow the evaluation protocol introduced by i3PosNet [7]. Similar to i3PosNet, we report mean and standard deviation of the absolute error for position and forward angle. These are calculated w.r.t. the projection direction ignoring depth. The forward angle is the angle between the image's x-axis and the screw axis projected into the image ((Fig. 1a). The architecture optimization performance and effectiveness is reported by the value metric ($-\log_{10}$(regMSE), see Sect. 3.2).

4 Results

We compare our final architecture (SNAPNet) with two state-of-the-art methods: 1) the manually designed i3PosNet [7], and 2) an automatically identified architecture using the DARTS* [8] NAS approach.

Both AutoSNAP-based architectures outperform both reference methods by a considerable margin approximately doubling the pose estimation performance (Table 2). DARTS-based results do not even reach i3PosNet levels and show the potential of AutoSNAP for CAI applications. For synthetic images, SNAPNet consistently outperforms position and angle estimates of all other methods including a substantial increase in performance when using i3PosNet's iterative scheme. For difficult real X-ray images, on the other hand, SNAPNet can significantly reduce the instability of the iterative scheme resulting in a significant reduction of 90% and 95% confidence intervals. In general, *performance gains are slightly more pronounced for real images than for synthetic images.*

The AutoSNAP search strategy is extremely effective, discovering this best performing architecture after less than 10 GPU hours. We illustrate the top-1 SNAP and neural block used for SNAPNet-A/B in Fig. 4. Figure 3 compares the convergence of AutoSNAP to random search (a common NAS baseline), which samples architectures randomly from the search space. While additional well-performing architectures are discovered after this block confirming flexibility of the approach, even the 25th-best SNAPNet architecture outperforms the best architecture produced by random search on the validation set.

5 Conclusion

We propose AutoSNAP, a novel approach targeting efficient search for high-quality neural architectures for instrument pose estimation. While the application of neural architecture search to CAI is already a novelty, our contribution also introduces SNAPs (to represent block architectures) and an auto-encoder-powered optimization scheme (efficient search algorithm). This optimization operates on a continuous representation of the architecture in latent space. We show more than 33 % error reduction compared to two state-of-the-art methods: the hand-engineered i3PosNet and DARTS, a neural architecture search method.

The application of NAS to CAI is generally limited by a scaling of the search cost with more operations (Table 1) and limited to block optimization, e.g. no macro-architecture optimization. With respect to learning, AutoSNAP requires stable task evaluations either by good reproducibility or by experiment repetition. Especially, the identification and exploration of unexplored architectures remains a challenge. While originally designed with instrument pose estimation in CAI in mind, in the future, we will expand AutoSNAP to other CAI and medical imaging problems.

Methods like AutoSNAP enable efficient development and improvement of neural architectures. It promises to help researchers in finding well-performing architectures with little effort. In this manner, researchers can focus on the integration of deep neural networks into CAI problems.

References

1. Bae, W., Lee, S., Lee, Y., Park, B., Chung, M., Jung, K.-H.: Resource optimized neural architecture search for 3D medical image segmentation. In: Shen, D., et al. (eds.) MICCAI 2019. LNCS, vol. 11765, pp. 228–236. Springer, Cham (2019). https://doi.org/10.1007/978-3-030-32245-8_26

2. Balakrishnan, G., Zhao, A., Sabuncu, M.R., Guttag, J., Dalca, A.V.: VoxelMorph: a learning framework for deformable medical image registration. IEEE Trans. Med. Imaging **38**(8), 1788–1800 (2019)
3. Dong, N., Xu, M., Liang, X., Jiang, Y., Dai, W., Xing, E.: Neural architecture search for adversarial medical image segmentation. In: Shen, D., et al. (eds.) MICCAI 2019. LNCS, vol. 11769, pp. 828–836. Springer, Cham (2019). https://doi.org/10.1007/978-3-030-32226-7_92
4. Elsken, T., Metzen, J.H., Hutter, F.: Neural architecture search: a survey **20**, 1–21 (2019). http://jmlr.org/papers/v20/18-598.html
5. Hajj, H.A., et al.: CATARACTS: challenge on automatic tool annotation for cataract surgery. Med. IA **52**, 24–41 (2019)
6. Kim, S., et al.: Scalable neural architecture search for 3D medical image segmentation. In: Shen, D., et al. (eds.) MICCAI 2019. LNCS, vol. 11766, pp. 220–228. Springer, Cham (2019). https://doi.org/10.1007/978-3-030-32248-9_25
7. Kügler, D., et al.: i3posnet: instrument pose estimation from x-ray in temporal bone surgery. Int. J. Comput. Assist. Radiol. Surg. **15**(7), 1137–1145 (2020). https://doi.org/10.1007/s11548-020-02157-4
8. Liu, H., Simonyan, K., Yang, Y.: DARTS: differentiable architecture search. In: ICLR 2019 (2019). https://arxiv.org/pdf/1806.09055
9. Luo, R., Tian, F., Qin, T., Chen, E., Liu, T.Y.: Neural architecture optimization. In: Bengio, S., et al. (eds.) Advances in NeurIPS, vol. 31. Curran Associates, Inc. (2018)
10. Maier-Hein, L., et al.: Surgical data science for next-generation interventions. Nat. BioMed. Eng. **1**(9), 691–696 (2017)
11. Miao, S., Wang, Z.J., Liao, R.: A CNN regression approach for real-time 2D/3D registration. IEEE Trans. Med. Imaging **35**(5), 1352–1363 (2016)
12. Schipper, J., et al.: Navigation as a quality management tool in cochlear implant surgery. J. Laryngol. Otol. **118**(10), 764–770 (2004)
13. Twinanda, A.P., Shehata, S., Mutter, D., Marescaux, J., de Mathelin, M., Padoy, N.: EndoNet: a deep architecture for recognition tasks on laparoscopic videos. IEEE Trans. Med. Imaging **36**(1), 86–97 (2017)
14. Unberath, M., et al.: Enabling machine learning in x-ray-based procedures via realistic simulation of image formation. Int. J. Comput. Assist. Radiol. Surg. **14**(9), 1517–1528 (2019). https://doi.org/10.1007/s11548-019-02011-2
15. Vercauteren, T., Unberath, M., Padoy, N., Navab, N.: CAI4CAI: the rise of contextual artificial intelligence in computer assisted interventions. Proc. IEEE **108**(1), 198–214 (2020). https://doi.org/10.1109/JPROC.2019.2946993
16. Weng, Y., Zhou, T., Li, Y., Qiu, X.: NAS-Unet: neural architecture search for medical image segmentation. IEEE Access **7**, 44247–44257 (2019)
17. Yu, Q., et al.: C2FNAS: coarse-to-fine neural architecture search for 3D medical image segmentation (2019). https://arxiv.org/pdf/1912.09628
18. Zhu, Z., Liu, C., Yang, D., Yuille, A., Xu, D.: V-NAS: neural architecture search for volumetric medical image segmentation. In: 2019 International Conference on 3D Vision, pp. 240–248. IEEE Computer Society, Conference Publishing Services, Los Alamitos (2019)
19. Zoph, B., Vasudevan, V., Shlens, J., Le, V.Q.: Learning transferable architectures for scalable image recognition. In: Brown, M.S., et al. (eds.) CVPR Proceedings (2018)

Automatic Operating Room Surgical Activity Recognition for Robot-Assisted Surgery

Aidean Sharghi(✉), Helene Haugerud, Daniel Oh, and Omid Mohareri

Intuitive Surgical Inc., Sunnyvale, CA, USA
aidean.sharghikarganroodi@intusurg.com

Abstract. Automatic recognition of surgical activities in the operating room (OR) is a key technology for creating next generation intelligent surgical devices and workflow monitoring/support systems. Such systems can potentially enhance efficiency in the OR, resulting in lower costs and improved care delivery to the patients. In this paper, we investigate automatic surgical activity recognition in robot-assisted operations. We collect the first large-scale dataset including 400 full-length multi-perspective videos from a variety of robotic surgery cases captured using Time-of-Flight cameras. We densely annotate the videos with 10 most recognized and clinically relevant classes of activities. Furthermore, we investigate state-of-the-art computer vision action recognition techniques and adapt them for the OR environment and the dataset. First, we fine-tune the Inflated 3D ConvNet (I3D) for clip-level activity recognition on our dataset and use it to extract features from the videos. These features are then fed to a stack of 3 Temporal Gaussian Mixture layers which extracts context from neighboring clips, and eventually go through a Long Short Term Memory network to learn the order of activities in full-length videos. We extensively assess the model and reach a peak performance of ~88% mean Average Precision.

Keywords: Activity recognition · Surgical workflow analysis · Robotic surgery

1 Introduction

Robot-assisted surgery (RAS) has been shown to improve perioperative outcomes such as reduced blood loss, faster recovery times, and shorter hospital stays for certain procedures. However, cost, staff training requirements and increased OR workflow complexities can all be considered barriers to adoption [1]. Although human error in care delivery is inevitable in such complex environments, there is opportunity for data-driven workflow analysis systems to identify "error-prone" situations, anticipate failures and improve team efficiency in the OR [2]. To this end, there have been a few observational studies to analyze non-operative activities, robot setup workflows and team efficiency aiming

A. L. Martel et al. (Eds.): MICCAI 2020, LNCS 12263, pp. 385–395, 2020.
https://doi.org/10.1007/978-3-030-59716-0_37

to enhance total system performance in robotic surgery [1,3,4]. However, such approaches are time and cost intensive and are not scalable.

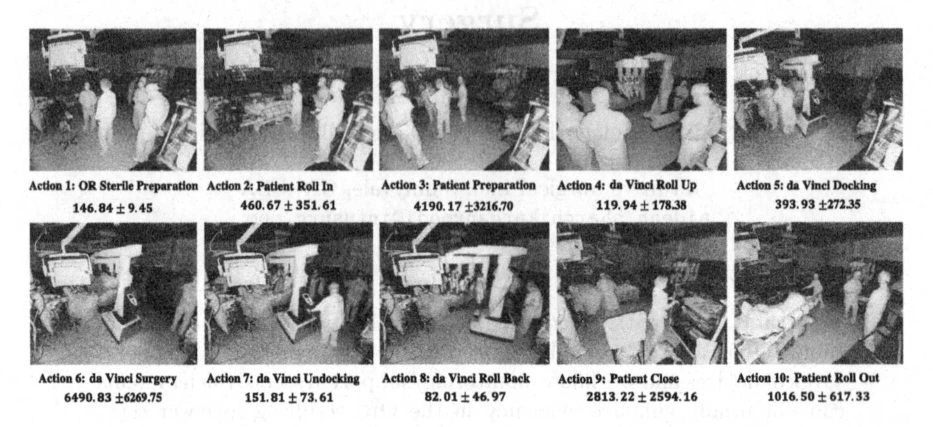

Fig. 1. We annotate each video in the dataset with its constituent actions (activities). Generally, 10 activities take place in every video. We report mean and standard deviation of number of frames per activity under the samples in this figure.

In this paper, we aim to develop a framework to automatically recognize the activity that is taking place at any moment in a full-length video captured in an OR. Almost all state-of-the-art approaches to activity recognition are based on data-driven approaches (in machine learning and computer vision), i.e., models are trained on large-scale datasets. Hence, in order to take advantage of them, we collect a *large-scale robot-assisted operation* dataset, including 400 full-length videos captured from 103 surgical cases performed by the da Vinci Xi surgical system. This is achieved by placing two vision carts, each with two Time-of-Flight (ToF) cameras inside operating rooms to create a multi-view video dataset. The reason behind using ToF cameras instead of high resolution RGB cameras is twofold: 1) to capture 3D information about the scene, 2) to preserve the privacy of patients and staff involved. Our dataset includes cases from several types of operations (see Fig. 2). After collecting the data, a trained user annotated each video with 10 clinically relevant activities, illustrated by Fig. 1.

Next, we develop an effective method to perform surgical activity recognition, i.e., automatically identifying activities that occur at any moment in the video. This problem is different from existing action recognition studied in the computer vision community. Firstly, our activities of interest have high inter/intra-class length variations. Secondly, surgical activities follow an order; i.e., first a sterile table is set up with sterile instruments (sterile preparation), then the patient is rolled in and prepared, etc. If the order is not accounted for, existing state-of-the-art methods fail to reach reliable performance due to inherent resemblance between certain activities such as robot docking and undocking. Thirdly, the videos in our dataset are full-length surgery cases and some last over 2 h. Effective learning on such long videos is difficult and precautions must be taken.

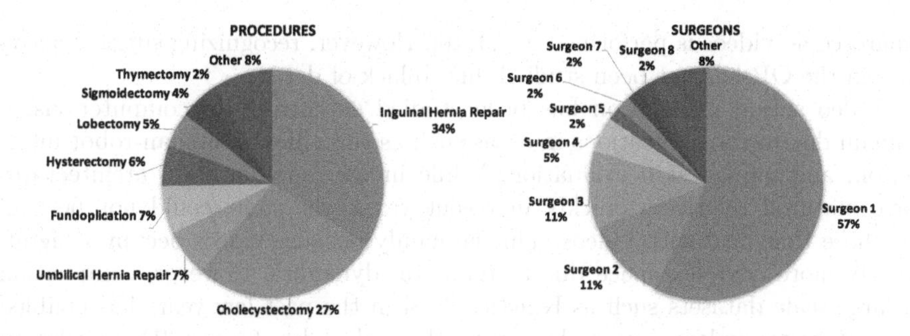

Fig. 2. A breakdown of the videos in our dataset based on type of procedure and surgeons.

Given a video, we first partition it uniformly into short clips (16 consecutive frames), and then feed them to the model in sequential order. Once the model has seen all the clips, it predicts the class of activities for every clip in the video. The right panel of Fig. 3 shows our proposed framework. Our model consists of two sub-networks. The first sub-network, shown on the left in Fig. 3, is the Inflated 3D ConvNet (I3D) [9] that is commonly used to extract discriminative features from the videos. We first fine-tune I3D (pretrained on ImageNet) on our dataset, and then use it to extract spatio-temporal features for the videos. Next, we employ 3 Temporal Gaussian Mixture (TGM) [18] layers, each with several per-class temporal filters. These layers help with extraction of features from neighboring clips, enabling the model to eventually classify each clip with contextual knowledge. Finally, we process the clips in temporal order using a Long Short Term Memory (LSTM) [17] to learn the order of activities performed. We comprehensively assess the generalization capability of our approach by splitting our videos into train and test according to different criteria, and are able to reach peak performance as high as ~88% mean Average Precision (mAP).

When dealing with robot-assisted surgeries, event and kinematics data generated by the surgical robot can be used to identify certain activities. However, this cannot replace the need for additional sensing as other activities do not involve interactions with the robot (e.g., patient in/out etc.). A smart OR integration depends on reliable detection of such activities. While in this work we study robot-assisted surgeries, detecting activities in the OR is still important in open and laparoscopic cases where such system data is not available. By not relying on such meta-data, we develop a more general solution to the problem.

2 Related Work

Automatic activity recognition has gained a lot of attention recently. Depth sensors are used in hospitals to monitor hand hygiene compliance using CNNs [12] and detect patient mobility [13,14]. In [15], a super-resolution model is used to enhance healthcare assist in smart hospitals, and surgical phase recognition in laparoscopic videos is performed in [11,16]. However, recognizing surgical activities in the OR has not been studied due to lack of data.

Video action recognition has been studied extensively in computer vision domain due to its applications in areas such as surveillance, human-robot interaction, and human skill evaluation. While image representation architectures have matured relatively quickly in recent years, the same could not be said for those that deal with videos. This is mainly because videos occupy a significantly more complex space due to temporal dynamics. However, compilation of large-scale datasets such as Kinetics [5–8] in the past few years has enabled researchers to explore deep architectures that take advantage of 3D convolution kernels leading to very deep, naturally spatio-temporal classifiers [9,19].

Action recognition methods fall under two categories depending on their input video. In action recognition on *trimmed* videos, the goal is to classify a video that is trimmed such that it contains a single action [20,24,25]. On the other hand, in an *untrimmed* video there are two key differences: 1) one or several activities may be observed, 2) some parts of the video may contain no activity. Therefore, in the latter, the objective is to find the boundaries of each action as well as classify them in their correct category [21–23].

3 Dataset

The dataset used in this work has been captured from a single medical facility with two robotic ORs and da Vinci Xi systems. Institutional Review Board approvals have been obtained to enable our data collection. Two imaging carts were placed in each room, for a total of four carts, each equipped with two ToF cameras. The baseline between the cameras on each cart is 70 cm and their orientation is fixed. This results in slightly different view in videos captured by the cameras from the same cart. However, different carts in the same room are set in strategic positions, such that if a cart's view is blocked due to clutter in the scene, the other cart can successfully capture the activities. With this setup, we captured 400 videos from 103 robotic surgical cases. Our dataset includes a variety of procedures performed by 8 surgeons/teams with different skill levels. A detailed breakdown of surgery types and surgeons is shown in Fig. 2.

Following data collection, raw data from the ToF cameras was processed to extract 2D intensity frames which are then used to create videos. These videos were manually annotated into separate cases and their constituent surgical activities using a video annotation tool [10]. We do not perform annotation quality analysis and rely on expertise of one person closely familiar with the data. Although there are multiple ways to analyze the surgical process in a robotic OR, we focus on 10 activities (Fig. 1) that are most relevant to robotic system utilization and non-operative workflows that are less investigated in prior studies.

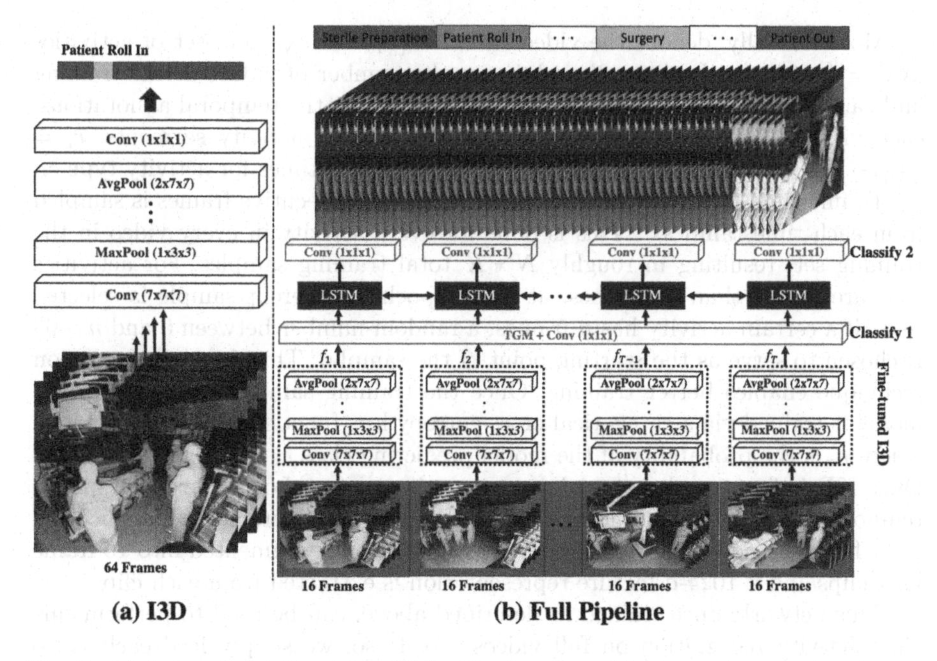

Fig. 3. (a) Inflated 3D ConvNet (I3D). We fine-tune this network such that it can classify short clips (64/16-frames during training/testing respectively) sampled from our dataset. (b) In full pipeline, we break each video uniformly into 16-frame long clips and feed each to the fine-tuned I3D to extract features $(f_1, \cdots, f_T)$. These features are then passed through TGM and convolution layers, a bidirectional LSTM, and a secondary 1D convolution layer that eventually predicts the action in each clips. Due to extreme length of the videos, using two classification layers facilitates convergence.

4 Methodology

In this section, we describe our model to perform activity recognition by breaking the problem into two subtasks; 1) extracting discriminative features for activity classes, and 2) using the features to recognize activities in full-length videos.

4.1 Inflated 3D ConvNet

Our first step to tackling full-length surgery video activity recognition is to extract discriminative features from the data. The intuition behind this task is to learn optimal representation for each class of activity such that the activities are easily distinguishable from one another. While various approaches exist, state-of-the-art models are built upon deep convolutional networks. Among them, the Inflated 3D ConvNet [9] (I3D) has become a reliable backbone to many video understanding methods, shown on the left panel of Fig. 3. I3D is a very deep network consisting of several $3D$ convolution layers and Inception V1 modules, allowing it to extract discriminative spatio-temporal features from short clips.

More formally, denote the videos as $\mathcal{V} = \{v_1, \cdots, v_N\}$ and set of activities as $\mathcal{C} = \{c_1, \cdots, c_K\}$ where N and K are the number of videos in the training and cardinality of set of activities respectively. Using the temporal annotations, each video can be partitioned into its corresponding activity segments, $v_i = \{s_{c_1}, \cdots, s_{c_K}\}$ where s_{c_i} represents the start and end times for activity type c_i.

To fine-tune I3D, a short clip consisting of 64 consecutive frames is sampled from each timestamp s_{c_i}. This is done for every activity in every video in the training set, resulting in roughly $N * K$ total training samples. For activities that are longer than 64 frames, at every epoch, a different sample is selected (e.g., if a certain activity has n frames, a random number between 0 and $n - 63$ is chosen to serve as the starting point of the sample). This data augmentation technique enables better training. Once the training samples are decided, the model is trained via a categorical cross-entropy loss function $\mathcal{L} = \sum_{i=1}^{K} c_i \log p_i$, where p_i is the probability of the model predicting class c_i for the given sample. Once I3D is fine-tuned on our data, we use it as a backbone to extract spatio-temporal features from full-length videos. Thus, we first detach the classification layer from the network. Next, every video is uniformly segmented into 16-frame long clips and a 1024-d feature representation is extracted from each clip.

This network, once trained as described above, can be used to perform surgical activity recognition on full videos. To do so, we simply feed each video one clip (16 frames) at a time to obtain a classification decision per clip. We use this approach as a baseline and report its performance. As we expect (and later confirm through experiments), this method fails to perform reliably simply because classification of each clip is done locally; the model has no memory of previously seen clips that belong to the same video. Afterall, context must help in classifying a new clip. Another major drawback of this model is its inability to account for the order of the activities. We tackle these issues in the next section.

4.2 Full-Length Surgical Activity Recognition

In order to perform in-context prediction and model the inherent order of activities, we use a recent technique [18] and add a stack of three Temporal Gaussian Mixture **(TGM)** layers before an LSTM module to model local and global temporal information in the video. Each TGM layer learns several Gaussian (temporal) filters for each class of activity and uses them to provide local context before making a classification decision. The features extracted through this layer are concatenated with I3D features and are then fed to LSTM. LSTM processes the sequence in order and stores hidden states to serve as memory of its observation. These hidden representations are used to process the input at every time step and calculate the output, therefore making LSTM a powerful tool for the purpose of full-length surgical activity recognition.

After extracting features from the training videos, each video is then represented as $v_i = \{f_1, \cdots, f_T\}$ where f_t is a feature extracted from t^{th} clip and T is the number of clips in v_i. At this stage of training, each video serves as a single training sample; the entire feature set of a video is fed to TGM layers before being processed by LSTM in sequential order to obtain $\mathcal{O}_i = \{o_1, \cdots, o_T\}$ where o_t is

Table 1. I3D fine-tuning results.

	Random			Procedure			Surgeon		
	Prec.	Rec.	F1	Prec.	Rec.	F1	Prec.	Rec.	F1
Sterile preparation	97.0 ± 1.4	96.3 ± 1.7	96.5 ± 1.3	86.44	78.46	82.26	81.08	88.24	84.51
Patient roll in	91.3 ± 3.4	91.0 ± 5.4	90.8 ± 3.4	77.94	80.3	79.1	85.0	75.0	79.69
Patient preparation	75.8 ± 3.0	78.3 ± 3.2	$770. \pm 2.6$	55.56	78.57	65.09	61.97	62.86	62.41
Robot roll up	92.8 ± 2.4	92.8 ± 2.1	92.8 ± 1.3	88.14	83.87	85.95	85.94	82.09	83.97
Robot docking	75.5 ± 6.0	80.3 ± 2.6	77.5 ± 2.6	76.19	50.0	60.38	60.29	60.29	60.29
Surgery	80.5 ± 1.7	86.0 ± 5.4	83.0 ± 1.4	69.51	81.43	75.0	73.33	74.32	73.83
Robot undocking	84.0 ± 2.4	72.5 ± 6.2	77.8 ± 4.2	75.0	72.86	73.91	68.18	62.50	65.22
Robot roll back	96.3 ± 2.2	90.8 ± 3.3	93.3 ± 2.4	94.64	80.30	86.89	98.08	76.12	85.71
Patient close	77.8 ± 1.0	79.3 ± 2.2	78.5 ± 0.6	69.44	69.44	69.44	67.16	62.5	64.75
Patient roll out	88.5 ± 3.4	90.3 ± 4.9	89.3 ± 1.5	73.53	73.53	73.53	63.64	90.0	74.56

the output after observing first t features of the video, $o_t = LSTM(\{f_1, \cdots, f_t\})$. This equation illustrates how LSTM represents each clip in a long video as a function of previous clips, hence providing context in representing each clip. This context helps to make classification decisions more robust. Furthermore, LSTM takes the order of activities into account as it processes the input sequentially. The output of the LSTM at every time step is fed to a convolution layer that serves as the classifier, classifying all the clips in the video.

To compare our model with closest work in the literature, Yengera et al. [11] uses a **frame-level** CNN combined with an LSTM to perform surgical phase recognition on laparoscopic videos. Since the loss has to be backpropagated through time, using frame-level features (on such long videos) leads to extremely high space complexity, thus, resulting in convergence issues. Unlike them, we use clip level features, significantly reducing the complexity. Moreover, we further speed up convergence while preserving the generalization capability of the model by applying the classification loss before and after LSTM. This is achieved by feeding the TGM features to a 1D convolution layer with output dimension equal to number of activity classes (hence this layer can serve as a classifier on its own), and use its prediction as input to our LSTM module combined with a secondary classification layer. The same categorical cross-entropy is applied to the outputs of both classification layers and the network is trained end-to-end.

5 Experimental Setup and Result Analysis

Train/Test Data Split. In order to comprehensively study the generalization power of the model, we perform experiments using different data split schemes.

I. *Random split.* In this experiment, we randomly select 80% of the videos for training and the rest for testing. We do not take precautions to guarantee that all 4 videos belonging to the same case are either in train or test sets.

Table 2. Full-length surgical activity recognition results (mAP).

	I3D	I3D-LSTM	I3D-TGM-LSTM
Random	62.16 ± 1.60	**88.81 ± 1.30**	88.75 ± 2.07
Procedure	47.98	75.39	**78.04**
Surgeon	50.31	78.29	**78.77**

This experiment is designed to find a meaningful upper bound to model's performance. We repeat this several times and report the mean and standard deviation of the performance to ensure an unbiased evaluation.
II. *Operating room split.* Since the entirety of our data is captured in two ORs within the same medical center, we can assess the model's generalization capability in cross-room train-test split. We train on videos from OR1 (57.5% of all the videos) and test the model on videos captured in OR2. We expect to see a drop in performance as the model is trained with less data.
III. *Procedure type split.* With this split, we assess if the model can generalize to novel procedures. In other words, we evaluate the model's performance by training on certain procedure types (Cholecystectomy, etc.) and test it on unseen categories. We used 12 out of 28 types of surgeries covered in our dataset for testing and the remaining for training. This results in a 80%–20% train-test split.
IV. *Surgeon split.* Surgeons lead the procedure and significantly affect the surgical workflow. In this experiment, we train our model on videos from select surgeons and assess the model's performance on videos from the remaining surgeons. The surgeons are split such that we obtain a 80%–20% train-test split.

Evaluation Metrics. I3D network is designed to classify short video clips and is evaluated using precision, recall, and F1 score. The full pipeline is evaluated similar to untrimmed action recognition algorithms using mean Average Precision (mAP).
Fine-Tuned I3D Results. Per class precision, recall, and F1-scores of our experiments are shown in Table 1. Following a random data split scheme, training I3D results in a robust clip-level action recognition network. We performed 4 rounds of experiments, every time with a new random data split. The model is able to maintain high average scores while keeping the standard deviation low.

When separating videos based on which OR they were captured in, the performance drops significantly (precision: 64.9, recall: 61.9, F1: 61.8% on average). The reason behind this observation is twofold: 1) the number of training videos is significantly less, 2) certain types of operations (e.g., lung Lobectomy, Thymectomy) had no instance in OR1 videos and were only seen in OR2. Thus, our clip recognition model is less robust compared to random split experiments.

Next, we test the model's generalization power when dealing with novel procedure types. This split is more difficult compared to random as the testing surgery types were never seen during the training. As shown in Table 1, F1-scores drop in every class of activity, yet retain an average F1 score of 75.15%.

Finally, when splitting the videos based on the surgeons, we observe the same trend as procedure type split. This confirms that I3D is able to maintain a decent performance when tested on unseen surgeons.

Full-Length Surgical Activity Recognition Results. As described in Sect. 4.1, we can use the baseline I3D to perform activity recognition on full-length videos. To do so, we add a 1d convolution on top of the pretrained I3D to train (only the convolution layer) on full-length videos. The performance of this baseline, in mean Average Precision (mAP), is reported under the first column of Table 2. Not surprisingly, the trend is similar to our clip-level recognition experiments; the performance is higher when using more training videos. The base I3D yields mAP of 36.91% when we split the videos by the OR they were captured in. We do not report this number in the table simply because the train-test split ratio is significantly different than the other 3 data split schemes.

When comparing I3D to our full pipeline (I3D-LSTM), we see a significant boost in performance under every split scheme. This is expected since the classification of clips is done within the context of other clips in the video, and the order of activities is learned. Peak performance of **88.81%** mAP on average was obtained. In worst case scenario (OR data split), I3D-LSTM reaches 58.40 mAP, which is significantly higher than the base I3D.

Finally, we see that TGM layer is increasing the model's generalization capability; it performs on-par with I3D-LSTM on random split, but outperforms it on all 3 other splits. I3D-TGM-LSTM reaches mAP of 63.47% on OR split, outperforming I3D-LSTM by a margin of 5%. As mentioned in previous sections, it is important to classify each clip in the video with consideration of its neighboring clips. This is the underlying reason why I3D-TGM-LSTM reaches state-of-the-art performance. The temporal filters pool class-specific features from neighboring clips. These features provide complimentary information to LSTM unit, resulting in overall increased performance.

6 Conclusion and Future Work

In this paper, we introduce the first large-scale robot-assisted operation dataset with 400 full-length videos captured from real surgery cases. We temporally annotate every video with 10 clinically relevant activities. Furthermore, we use this dataset to design a model to recognize activities throughout the video. To this end, we fine-tune the Inflated 3D ConvNet model on our dataset to learn discriminative feature representations for different types of activities. Next, using Temporal Gaussian Mixture layers and a Long Short Term Memory unit, we enable the model to keep track of activities, thus learning their order.

There are several directions for expanding our efforts in this paper. From the data perspective, we are in the process of acquiring videos from different medical centers. This allows us to: 1) assess the model's generalization ability more comprehensively, 2) explore deeper architectures. From the application stand point, our proposed framework can be used in future to provide automated workflow and efficiency metrics to surgical teams and medical institutions.

References

1. Catchpole, K., et al.: Safety, efficiency and learning curves in robotic surgery: a human factors analysis. Surg. Endosc. **30**(9), 3749–3761 (2015). https://doi.org/10.1007/s00464-015-4671-2
2. Vercauteren, T., Unberath, M., Padoy, N., Navab, N.: CAI4CAI: the rise of contextual artificial intelligence in computer-assisted interventions. Proc. IEEE **108**(1), 198–214 (2019)
3. Allers, J.C., et al.: Evaluation and impact of workflow interruptions during robot-assisted surgery. Urology **92**, 33–37 (2016)
4. Zeybek, B., Öge, T., Kılıç, C.H., Borahay, M.A., Kılıç, G.S.: A financial analysis of operating room charges for robot-assisted gynaecologic surgery: efficiency strategies in the operating room for reducing the costs. J. Turk. Ger. Gynecol. Assoc. **15**(1), 25 (2014)
5. Kay, W., et al.: The kinetics human action video dataset. arXiv preprint arXiv:1705.06950 (2017)
6. Carreira, J., Noland, E., Banki-Horvath, A., Hillier, C., Zisserman, A.: A short note about kinetics-600. arXiv preprint arXiv:1808.01340 (2018)
7. Carreira, J., Noland, E., Hillier, C., Zisserman, A.: A short note on the kinetics-700 human action dataset. arXiv preprint arXiv:1907.06987 (2019)
8. Karpathy, A., Toderici, G., Shetty, S., Leung, T., Sukthankar, R., Fei-Fei, L.: Large-scale video classification with convolutional neural networks. In: Proceedings of the IEEE Conference on Computer Vision and Pattern Recognition, pp. 1725–1732 (2014)
9. Carreira, J., Zisserman, A.: Quo vadis, action recognition? A new model and the kinetics dataset. In: Proceedings of the IEEE Conference on Computer Vision and Pattern Recognition, pp. 6299–6308 (2017)
10. Dutta, A., Zisserman, A.: The VIA annotation software for images, audio and video. In: Proceedings of the 27th ACM International Conference on Multimedia, pp. 2276–2279, October 2019
11. Yengera, G., Mutter, D., Marescaux, J., Padoy, N.: Less is more: surgical phase recognition with less annotations through self-supervised pre-training of CNN-LSTM networks. arXiv preprint arXiv:1805.08569 (2018)
12. Yeung, S., et al.: Vision-based hand hygiene monitoring in hospitals. In: AMIA (2016)
13. Ma, A.J., et al.: Measuring patient mobility in the ICU using a novel noninvasive sensor. Crit. Care Med. **45**(4), 630 (2017)
14. Yeung, S., et al.: A computer vision system for deep learning-based detection of patient mobilization activities in the ICU. NPJ Digit. Med. **2**(1), 1–5 (2019)
15. Chou, E., et al.: Privacy-preserving action recognition for smart hospitals using low-resolution depth images. arXiv preprint arXiv:1811.09950 (2018)

16. Zia, A., Hung, A., Essa, I., Jarc, A.: Surgical activity recognition in robot-assisted radical prostatectomy using deep learning. In: Frangi, A.F., Schnabel, J.A., Davatzikos, C., Alberola-López, C., Fichtinger, G. (eds.) MICCAI 2018. LNCS, vol. 11073, pp. 273–280. Springer, Cham (2018). https://doi.org/10.1007/978-3-030-00937-3_32
17. Hochreiter, S., Schmidhuber, J.: Long short-term memory. Neural Comput. **9**(8), 1735–1780 (1997)
18. Piergiovanni, A.J., Ryoo, M.S.: Temporal Gaussian mixture layer for videos. arXiv preprint arXiv:1803.06316 (2018)
19. Tran, D., Bourdev, L., Fergus, R., Torresani, L., Paluri, M.: Learning spatiotemporal features with 3D convolutional networks. In: Proceedings of the IEEE International Conference on Computer Vision, pp. 4489–4497 (2015)
20. Sudhakaran, S., Escalera, S., Lanz, O.: Gate-shift networks for video action recognition. arXiv preprint arXiv:1912.00381 (2019)
21. Liu, Y., Ma, L., Zhang, Y., Liu, W., Chang, S.F.: Multi-granularity generator for temporal action proposal. In: Proceedings of the IEEE Conference on Computer Vision and Pattern Recognition, pp. 3604–3613 (2019)
22. Xu, H., Das, A., Saenko, K.: Two-stream region convolutional 3D network for temporal activity detection. IEEE Trans. Pattern Anal. Mach. Intell. **41**(10), 2319–2332 (2019)
23. Yeung, S., Russakovsky, O., Jin, N., Andriluka, M., Mori, G., Fei-Fei, L.: Every moment counts: dense detailed labeling of actions in complex videos. Int. J. Comput. Vis. **126**(2–4), 375–389 (2018)
24. Ryoo, M.S., Piergiovanni, A.J., Tan, M., Angelova, A.: AssembleNet: searching for multi-stream neural connectivity in video architectures. arXiv preprint arXiv:1905.13209 (2019)
25. Tran, D., Wang, H., Torresani, L., Feiszli, M.: Video classification with channel-separated convolutional networks. In: Proceedings of the IEEE International Conference on Computer Vision, pp. 5552–5561 (2019)

Navigation and Visualization

Can a Hand-Held Navigation Device Reduce Cognitive Load? A User-Centered Approach Evaluated by 18 Surgeons

Caroline Brendle[1], Laura Schütz[2], Javier Esteban[1(✉)], Sandro M. Krieg[3], Ulrich Eck[1], and Nassir Navab[1]

[1] Computer Aided Medical Procedures, Technische Universität München, Boltzmannstrasse, 85748 Garching bei München, Germany
caroline.brendle@tum.de, javier.esteban@tum.de
[2] Chair for Industrial Design, Technische Universität München, Arcisstrasse 21, 80333 München, Germany
laura.schuetz@tum.de
[3] Department of Neurosurgery, Klinikum rechts der Isar, Technische Universität München, Ismaninger Strasse 22, 81675 München, Germany

Abstract. During spinal fusion surgery, the orientation of the pedicle screw in the right angle plays a crucial role for the outcome of the operation. Local separation of navigation information and the surgical situs, in combination with intricate visualizations, can limit the benefits of surgical navigation systems. The present study addresses these problems by proposing a hand-held navigation device (HND) for pedicle screw placement. The in-situ visualization of graphically reduced interfaces, and the simple integration of the device into the surgical work flow, allow the surgeon to position the tool while keeping sight of the anatomical target. 18 surgeons participated in a study comparing the HND to the state-of-the-art visualization on an external screen. Our approach revealed significant improvements in mental demand and overall cognitive load, measured using NASA-TLX ($p < 0.05$). Moreover, surgical time (One-Way ANOVA $p < 0.001$) and system usability (Kruskal-Wallis test $p < 0.05$) were significantly improved.

Keywords: Hand-held navigation device · Surgical navigation · Spinal fusion · Pedicle screw placement · Augmented reality · Cognitive load · Surgical visualization

1 Introduction

Disc degeneration, spondylolisthesis or scoliosis are exemplary indications leading to spine instability, which, in turn, can cause bone deformation or nerve

C. Brendle, L. Schütz and J. Esteban—Contributed equally to this work.
This work was partially supported by the German Federal Ministry of Research and Education (FKZ: 13GW0236B).

A. L. Martel et al. (Eds.): MICCAI 2020, LNCS 12263, pp. 399–408, 2020.
https://doi.org/10.1007/978-3-030-59716-0_38

damage [16]. In severe cases, an artificial reconstruction of the spine's stability, is performed by introducing a transpedicular screw-rod system into the spine. This treatment fixes the operated segment(s) and restores the spine's stability [5]. Screw placement in the correct angular trajectory, indicated by the pedicle and the vertebral body, is one of the most intricate tasks due to complex anatomic variants [4]. As misplacements of pedicle screws can be a source of bone breaches and neurological complications [5], precise placement and alignment of the pedicle screw instruments presents a crucial step during spinal fusion surgery.

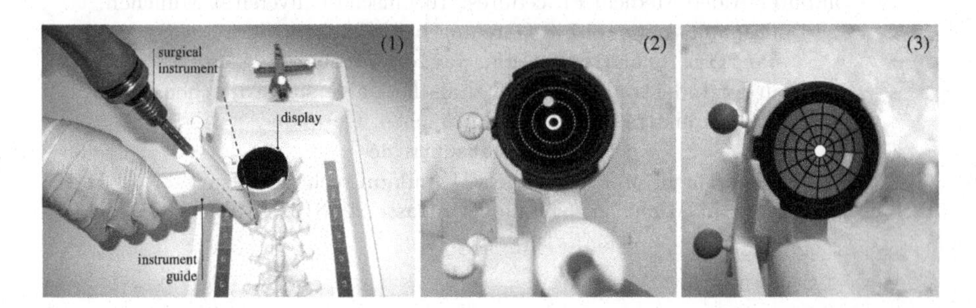

Fig. 1. (1) Hand-held navigation device, green line: instrument trajectory, red line: target trajectory, (2) Close view of Circle Display, (3) Close view of Grid Display (Color figure online)

Navigation systems assist the surgeon during placement and alignment, helping to maximize the precision of pedicle screw insertion [7]. Improved outcome and reduced radiation exposure are reasons to use these systems [15]. However, the complexity of use, and the disruption of the surgical work flow are predominant factors for the conservative percentage of navigated spine surgeries [10]. When using navigation, the operator must split attention between the navigation information, displayed on an external screen, and the situs [11], adding to the complexity of use.

We present a user-centered navigation system for spinal fusion surgery. The solution consists of a hand-held navigation device (HND) which comprises a display for navigation information and a shaft for surgical tool guidance (Figure 1). This way, the HND enables the display of meaningful information in the surgeon's field of view. The primary focus of this study is the evaluation of the HND for the task of angle alignment within the pedicle screw placement procedure. Two different visualizations for angle orientation of the surgical tool were tested with 18 physicians against the visualizations used in clinical routine. The proposed visualizations were also tested against the same visualizations implemented on an external screen. Our results show that this integration of the visualization unit into the HND, combined with a user-friendly visualization, reduces cognitive load when compared to the traditional approach. Furthermore, statistical analysis of the data showed that the proposed system presents a more time-efficient alternative to the state-of-the-art solutions.

1.1 Related Work

Moving the surgeons' attention back from an external screen to the patient is a well-researched topic in the scientific community. Carmigniani et al. [2] utilized augmented (AR) and mixed reality (MR) as well as mobile devices to overlay additional information on top of the patient, enabling the examination of the situs and organs. Likewise, several works have discussed solutions using not only head-mounted displays [1,14], but also half-silvered mirrors [18] to superimpose pre-operative images onto the patient for anatomical navigation. Léger et al. [17] compared in situ AR, desktop AR and traditional navigation with regard to attention shifts and time to completion for craniotomy planning. Their research showed that the use of AR systems resulted in less time and less attention shifts. Liebmann et al. [19] demonstrated the benefits of real-time visualizations using AR without additional intra-operative imaging for pedicle screw placement achieving the same accuracy for screw placement as state-of-the-art navigation systems. Their study examined 3D augmented views using Microsoft HoloLenses, whereas we propose visualizations on a 2D display. Navab et al. [21] developed a surgical AR technology enabling video-augmented x-ray images by extending a mobile C-arm with a video camera. As in our study, instrument axis alignment was evaluated. However, AR images are displayed using a mirror attached to the C-arm, while we propose localization of the visualization unit on the surgical instrument.

Similar approaches, utilizing mobile devices to enable in situ navigation, can be found. Kassil et al. [13] demonstrated that using a tool-mounted display can achieve better positional and angular accuracy for a drilling task. The system augments the image of a camera installed on the tool, whereas our system is a user-centered, graphic navigation interface. Experiments looked at drilling precision and completion time but did not include any evaluation of cognitive load and usability. Weber et al. [22] integrated a Navigated Image Viewer into the surgical process, and proved that this dynamic visualization helps to understand the spatial context. Gael et al. [6] investigated the potential of adding a smartphone as an interaction device. Mullaji et al. [20] presented a hand-held, iPod-based navigation approach for total knee arthroplasty. Both the smartphone and the iPod approach are displaying the traditional visualization in small size on a mobile screen. In contrast to both studies our work presents visually reduced information independent of the external monitor. An exemplary work introducing an approach for measuring cognitive load, user preference, and general usability is presented by Herrlich et al. [11]. The use of an external monitor is compared to an instrument-mounted display for the task of needle guidance. Their display reduced cognitive load while achieving the same performance in terms of time and accuracy. As opposed to their work, clinical experts were used to evaluate cognitive load and system usability during our study. Outside of the medical context, Echtler et al. [3] presented the design and implementation of a welding gun, which displays three-dimensional stud locations on the car frame relative to the current gun position. Their specific visualizations such as concentric rings and compass enabled a correct positioning of the gun tip on the surface of the frame, whereas our work is concerned with angle orientation.

2 Methods

To avoid physical de-coupling of the navigation and the surgical situs, we propose the use of a hand-held navigation device (HND). This HND is a custom-made device, developed as a result of user-centered design [12]. It consists of a handle, a shaft for tool guidance, a tracking array and a visualization unit holder. The HND is designed to hold a surgical instrument while attaching a visualization to it. This idea introduces a paradigm shift, contrasting the solutions used in clinical routine. With this approach the attention of the user is not drawn from the patient to an external screen, but kept on the surgical area.

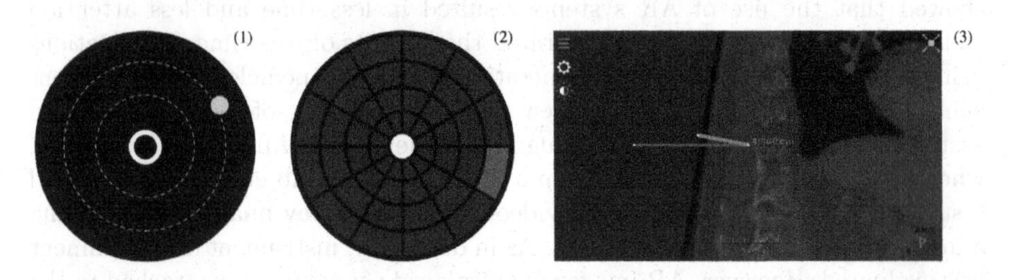

Fig. 2. (1) Circle Display, (2) Grid Display, (3) Transversal view of Traditional External

2.1 Hardware and Software Setup

The system comprises a workstation, a Polaris Vicra (NDI, Ontario, Canada), the HND, and a Ticwatch E (Mobvoi. Beijing, China). The Ticwatch is attached to the visualization unit holder to show the instrument-integrated visualization. A Polaris Vicra infrared tracking system tracks the arrays attached to both the phantoms and the instrument. The calibration between the marker and the tool is known by construction, and checked using a pivot calibration. The software uses a client-server architecture between the Ticwatch E and the workstation. The server is run as a plugin on ImFusionSuite[1], which processes the tracking information and sends it to the Ticwatch to create the visualization.

2.2 Visualizations

We propose two visually reduced interfaces (Fig. 2), displaying abstract 2D guidance for angle alignment, aiming to reduce the complexity of navigation during pedicle screw placement. Both minimalist visualizations map the real-time 3D Cartesian orientation of the HND to a polar coordinate system centered at the pre-planned insertion trajectory. This way, the relative angle distance between the planned and current trajectories is intuitively shown on the 2D screen.

[1] ImFusion GmbH, Munich, Germany (https://www.imfusion.de).

Circle Display - This interface consists of a circular element moving dynamically across the underlying background according to the HND's position calculated as described in the previous section. By orienting the HND in such a way that the circular element enters the central ring of the interface, the user achieves the target trajectory. The correct alignment is communicated to the user by additional visual feedback: change of color of the circular element from yellow to blue and the superimposition of a highlight around the inner ring.

Grid Display - The design is a grid pattern dividing the interface into 12 pie sections and four concentric circles of different radii. According to the relative orientation of the HND towards the planned trajectory, the respective grid field is highlighted in red. By moving the HND into the indicated direction of the red-marked field, the system guides the user towards the right angle. When the user reaches the target orientation, the central circle lights up in green.

Traditional External - The state-of-art visualization uses 3 slices with normal directions $tool_{X,Y,Z}$, and application point $tool_{tip}$. The target trajectory is projected on each of the planes as a red line. The HND's pose is shown as a green line. The right orientation is achieved by aligning each of the different axes individually until both lines intersect.

3 User Study

Our study used two controlled experiments to evaluate the potential benefits of the proposed approach. Eighteen experienced surgeons participated in the first experiment. It investigated the performance and usability of the HND compared to the state-of-the-art navigation system on a realistic spine phantom. A second experiment was employed to isolate and evaluate the two main factors of the solution, the in-situ visualization offered by the HND, and the developed interface itself. For this, the user was presented with an orientation task using the same pair of visualizations, both on the HND and the external screen. The distribution of the participants in both experiments followed a block randomization.

- In comparison with the state-of-the-art navigation system:

  H1. Participants using the HND for angle alignment experience reduced cognitive load and improved usability.
  H2. Participants using the HND for angle alignment achieve the planned trajectory faster and with a shorter euclidean path (Fig. 3).

- In comparison to the same visualization presented on an external screen:

  H3. Participants using the HND for angle alignment experience reduced cognitive load and improved usability.
  H4. Participants using the HND for angle alignment achieve the planned trajectory faster and with a shorter euclidean path.

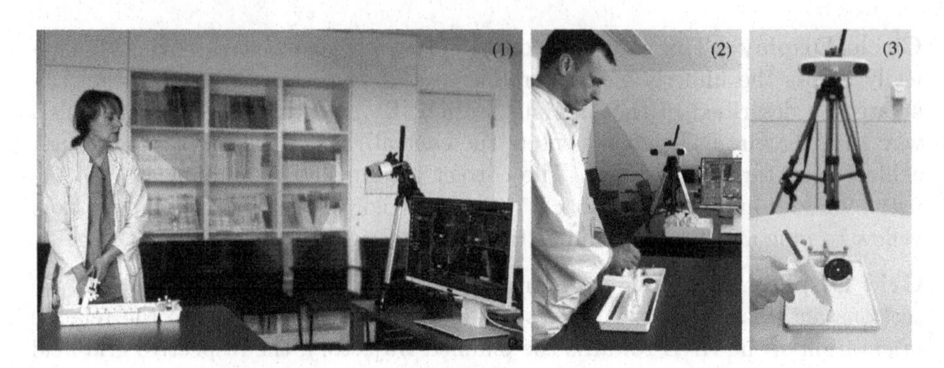

Fig. 3. (1) Setup experiment 1, neurosurgeon using the external screen, (2) Senior neurosurgeon using the visualization on the HND, (3) Setup experiment 2

3.1 Experiment 1

Participants. 18 volunteers [10 m/8 f] participated in the study. All were practicing surgeons with experience in using navigation systems and traditional techniques for visualization of 3D data (i.e., CT, MR). The mean age was 30 ± 3.7 *std*. None of them had previous experience using the proposed navigation system. 15 participants reported to have executed pedicle screw placement before.

Experiment Setup. A model of the lower lumbar spine, [levels Th11 to L5], was 3D-printed and calibrated to the tracking system. To ensure anatomical coherence, realistic pedicle screw trajectories were selected by a senior neurosurgeon on the CT of the phantom. The trajectories corresponded to the real insertion path for each right and left pedicle screw on levels Th11 to L5 of the spine, summing up to 14 different trajectories evenly distributed on both sides of the spine. The entry point of each trajectory was physically marked on the phantom. First, the user was instructed to place the HND onto one of the predefined entry points on the spine phantom. Then, the physician was asked to align it in the right anatomical, pre-planned trajectory using one of the visualization techniques. This alignment was repeated four times on two randomized trajectories for each side of the spine. The same task was repeated for all three visualizations (Circle and Grid on Display, and Traditional External).

3.2 Experiment 2

Participants. The set of participants consisted of 15 females and 27 males [age 26.8 ± 3.3]. None of them had clinical experience or previous experience using the HND system. However, 31 had experience using AR/VR.

Experiment Setup. We performed a second experiment to isolate the two main factors of our solution: the visualization techniques, and the in-situ visualization. Four different setups were evaluated as a combination of the two interface designs (Circle and Grid Display), and the two displays (integrated and external). An abstract scenario, comprising a conical 3D model with a single entry point, was built. In contrast to experiment 1, where the trajectories are anatomically right, here, random realistic trajectories were created with a fixed insertion point on top of the pyramidal model. After positioning the tip of the instrument on top of the 3D model, the user was asked to align the HND using the visualization. This alignment was repeated five times per visualization, and with the four mentioned setups.

3.3 Experimental Variables

For both experiments, time-stamped poses of the HND handler were recorded. Additionally, users had to fill out a NASA-TLX and System Usability Scale (SUS) questionnaire for measuring the cognitive load and overall usability. Cognitive load refers to the overall task load calculated using the NASA-TLX questionnaire, for six defined variables (Mental Demand, Physical Demand, Temporal Demand, Performance, Effort, Frustration) [9]. Usability was calculated using the System Usability Scale (SUS) questionnaire [8]. Further questions gathered subjective assessments of the interfaces' visual appeal, ease of use and interaction design.

4 Results

4.1 Experiment 1

Cognitive Load and Usability. The recorded data of the NASA-TLX and SUS showed a Chi-squared distribution for all the different variables ($p < 0.001$). A Kruskal Wallis test was run pairwise between visualizations, for each of the NASA-TLX variables and the SUS score. The analysis showed that the usage of both Circle Display and Grid Display resulted in a significantly lower mental demand ($p < 0.05$) and cognitive load ($p < 0.05$) when compared with the Traditional External . The usability (SUS score) for Circle Display ($79.4 \pm 14.1 std$) was significantly better ($p < 0.05$) compared to Traditional External (68.7 ± 14.7 *std*). The pairwise comparison of the two proposed designs Circle Display and Grid Display showed no significant difference ($p > 0.05$) (Fig. 4).

Experiment 1	Cognitive load	Mental demand	Usability
Circle	23.7 ± 15.4	23.6 ± 20.1	79.4 ± 14.1
Grid	25.0 ± 14.4	25.8 ± 19.0	76.7 ± 18.1
Traditional	35.3 ± 12.7	42.8 ± 18.3	68.7 ± 14.7

Experiment 2	Cognitive load	Mental demand	Usability
HND	32.2 ± 17.3	33.7 ± 25.3	80.8 ± 13.1
External Screen	33.3 ± 17.1	36.9 ± 26.9	80.2 ± 18.1

Fig. 4. Means and stds for cognitive load and mental demand (NASA-TLX, [0, 100] lower is better), and usability (SUS [0,100], higher is better) for experiment 1 and 2)

Statistical Analysis. The tool's total euclidean path was analyzed. For this, 3D time-stamped poses of the tool's shaft were saved during the experiments.

The euclidean path is calculated as the total sum of the euclidean distances between each consecutive pair of poses. This value expresses the accumulated travel distance of the tool's shaft until the alignment is successful. Intuitively, this measure provides insights into how directly the user moved the tool towards the final pose. (Fig. 5 [Chi-squared distributed ($p < 0.001$)]). Time values showed a normal distribution ($p < 0.05$). Outliers with values outside of $mean \pm 2 * std$ were excluded. A one-way ANOVA test was used to measure significance for time and Kruskal-Wallis for distance. Our approach reached significantly better results in comparison to the traditional approach regarding time ($p < 0.001$). Circle Display performed better against the traditional method, both on distance ($p = 0.0107$) and time ($p < 0.001$).

4.2 Experiment 2

Cognitive Load and Usability. The results of the NASA-TLX and SUS questionnaire showed a Chi-squared distribution of the different variables ($p < 0.001$). A Kruskal Wallis test was employed to compare the group of external visualizations (Circle External and Grid External) with the group of instrument-integrated visualizations (Circle Display and Grid Display) for all variables of NASA-TLX and the SUS score. The analysis showed no significant difference between the two groups ($p > 0.05$), neither for cognitive load, nor for usability, and consequently, pairwise comparisons were not conducted.

Statistical Analysis. To compare the different visualizations, the euclidean distances and total time of each individual task were grouped according to the visualization used (Circle Display, Grid Display, Circle External, Grid External) as shown on Fig. 5. Outliers with values outside of $mean \pm 2 * std$ were excluded from the sample. Normality was tested within the 4 groups (both for time and distance) using D'Agostino's K-squared test ($p < 0.001$). The sample has a 2.004 ratio between the larger and smaller variances. One-Way ANOVA was used to compare the results in terms of euclidean distance and time. Overall,

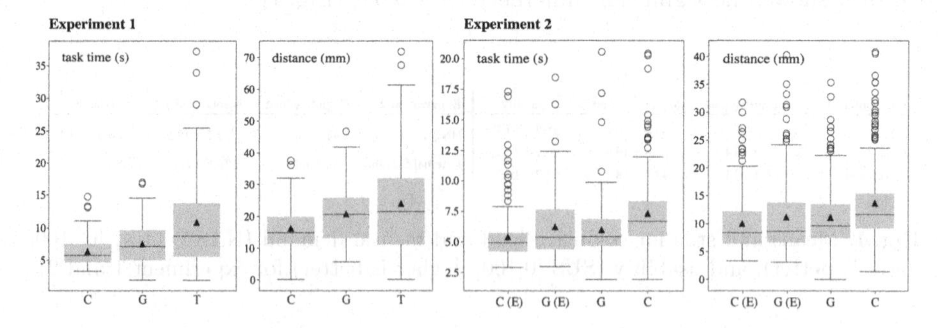

Fig. 5. Time and euclidean distance for Circle Display (C), Grid Display (G), Traditional External (T), Circle External (C(E)), Grid External (G(E))

Circle External performed best, both in terms of time and distance, followed by Circle Display and Grid Display. According to the data, the visualizations on the external screen performed significantly better than the visualizations on the device ($p < 0.001$).

5 Discussion and Conclusion

A hand-held navigation device for spinal fusion surgery was introduced. Preliminary tests revealed significant differences in favor of our approach regarding cognitive load, mental demand and usability (H1). This improvement of user ergonomics leads to a significant increase in performance, measured as total euclidean path and time (H2). Experiment 2 showed no significant results for H3. H4 has proven to be false, as the results on the external display were significantly better.

Possible factors affecting this result are the latency added to the drill visualizations, the lower resolution, and the relative small size of the screen. However, the HND still offers the advantage of in-situ navigation information, keeping the surgeon's attention on the patient. To answer the introductory question: our user-centered approach to spinal navigation enabled a significant reduction of cognitive load for the surgeon. Thus, the solution shows potential benefits for clinical application if properly integrated within the medical workflow.

Acknowledgement. Our research was initiated during MedInnovate, a graduate fellowship program for medical innovation and entrepreneurship in MedTech. We would also like to thank the Department of Neurosurgery at Klinikum Rechts der Isar for their continuous support.

References

1. Birkfellner, W., et al.: A head-mounted operating binocular for augmented reality visualization in medicine - design and initial evaluation. IEEE Trans. Med. Imaging **21**, 991–7 (2002). https://doi.org/10.1109/TMI.2002.803099
2. Carmigniani, J., Furht, B., Anisetti, M., Ceravolo, P., Damiani, E., Ivkovic, M.: Augmented reality technologies, systems and applications. Multimed. Tools Appl. **51**, 341–377 (2010). https://doi.org/10.1007/s11042-010-0660-6
3. Echtler, F., Sturm, F., Kindermann, K., Klinker, G.: 17 the intelligent welding gun: augmented reality for experimental vehicle construction, January 2003
4. Engler, J.A., Smith, M.L.: Use of intraoperative fluoroscopy for the safe placement of C2 laminar screws: technical note. Eur. Spine J. **24**(12), 2771–2775 (2015). https://doi.org/10.1007/s00586-015-4165-x
5. Fritsch, E., Duchow, J., Seil, R., Grunwald, I., Reith, W.: Genauigkeit der fluoroskopischen navigation von pedikelschrauben: Ct-basierte evaluierung der schraubenlage. Der Orthopäde **31**, 385–391 (04 2002)
6. Le Bellego, G., Bucki, M., Bricault, I., Troccaz, J.: Using a Smart phone for information rendering in computer-aided surgery. In: Jacko, J.A. (ed.) HCI 2011. LNCS, vol. 6764, pp. 202–209. Springer, Heidelberg (2011). https://doi.org/10.1007/978-3-642-21619-0_26

7. Gebhard, F., Weidner, A., Liener, U.C., Stöckle, U., Arand, M.: Navigation at the spine. Injury **35**, 35–45 (2004)
8. Grier, R., Bangor, A., Kortum, P., Peres, S.: The system usability scale. Proc. Hum. Factors Ergon. Soc. Ann. Meet. **57**, 187–191 (2013). https://doi.org/10.1177/1541931213571042
9. Hart, S., Staveland, L.: Development of NASA-TLX (task load index): results of empirical and theoretical research, pp. 139–183, April 1988. https://doi.org/10.1016/S0166-4115(08)62386-9
10. Härtl, R., Lam, K.S., Wang, J., Korge, A., Kandziora, F., Audigé, L.: Worldwide survey on the use of navigation in spine surgery. World Neurosurg. **79**, 162–172 (2013). https://doi.org/10.1016/j.wneu.2012.03.011
11. Herrlich, M., et al.: Instrument-mounted displays for reducing cognitive load during surgical navigation. Int. J. Comput. Assist. Radiol. Surg. **12**(9), 1599–1605 (2017). https://doi.org/10.1007/s11548-017-1540-6
12. Kashfi, P.: Applying a user centered design methodology in a clinical context. In: Studies in Health Technology and Informatics, vol. 160, pp. 927–31, January 2010. https://doi.org/10.3233/978-1-60750-588-4-927
13. Kassil, K., Stewart, A.: Evaluation of a tool-mounted guidance display for computer-assisted surgery, pp. 1275–1278, April 2009. https://doi.org/10.1145/1518701.1518892
14. Keller, K., State, A., Fuchs, H.: Head mounted displays for medical use. J. Disp. Technol. **4**, 468–472 (2008). https://doi.org/10.1109/JDT.2008.2001577
15. Kraus, M.D., Krischak, G., Keppler, P., Gebhard, F.T., Schuetz, U.H.: Can computer-assisted surgery reduce the effective dose for spinal fusion and sacroiliac screw insertion? Clin. Orthop. Relat. Res.® **468**, 2419–2429 (2010). https://doi.org/10.1007/s11999-010-1393-6
16. Krismer, M.: Fusion of the lumbar spine. J. Bone Joint Surg. -Br. **84**, 783–794 (2002)
17. Léger, E., Drouin, S., Collins, D.L., Popa, T., Kersten-Oertel, M.: Quantifying attention shifts in augmented reality image-guided neurosurgery. Healthcare Technol. Lett. **4**(5), 188–192 (2017). https://doi.org/10.1049/htl.2017.0062
18. Liao, H., Inomata, T., Sakuma, I., Dohi, T.: 3-D augmented reality for MRI-guided surgery using integral videography autostereoscopic image overlay. IEEE Trans. Bio-med. Eng. **57**, 1476–86 (2010). https://doi.org/10.1109/TBME.2010.2040278
19. Liebmann, F., et al.: Pedicle screw navigation using surface digitization on the Microsoft HoloLens. Int. J. Comput. Assist. Radiol. Surg. **14**(7), 1157–1165 (2019). https://doi.org/10.1007/s11548-019-01973-7
20. Mullaji, A., Shetty, G.: Efficacy of a novel iPod- based navigation system compared to traditional navigation system in total knee arthroplasty. Comput. Assist. Surg. (Abingdon Engl.) **22**, 1–13 (2016). https://doi.org/10.1080/24699322.2016.1276630
21. Navab, N., Blum, T., Wang, L., Okur, A., Wendler, T.: First deployments of augmented reality in operating rooms. Computer **45**, 48–55 (2012). https://doi.org/10.1109/MC.2012.75
22. Weber, S., Klein, M., Hein, A., Krueger, T., Lueth, T., Bier, J.: The navigated image viewer - evaluation in maxillofacial surgery. vol. 2878, pp. 762–769, November 2003

Symmetric Dilated Convolution for Surgical Gesture Recognition

Jinglu Zhang[1], Yinyu Nie[1], Yao Lyu[1], Hailin Li[2], Jian Chang[1(✉)], Xiaosong Yang[1], and Jian Jun Zhang[1]

[1] National Centre for Computer Animation, Bournemouth University, Poole, UK
jchang@bournemouth.ac.uk
[2] Communication University of China, Beijing, China

Abstract. Automatic surgical gesture recognition is a prerequisite of intra-operative computer assistance and objective surgical skill assessment. Prior works either require additional sensors to collect kinematics data or have limitations on capturing temporal information from long and untrimmed surgical videos. To tackle these challenges, we propose a novel temporal convolutional architecture to automatically detect and segment surgical gestures with corresponding boundaries only using RGB videos. We devise our method with a symmetric dilation structure bridged by a self-attention module to encode and decode the long-term temporal patterns and establish the frame-to-frame relationship accordingly. We validate the effectiveness of our approach on a fundamental robotic suturing task from the JIGSAWS dataset. The experiment results demonstrate the ability of our method on capturing long-term frame dependencies, which largely outperform the state-of-the-art methods on the frame-wise accuracy up to ∼6 points and the F1@50 score ∼6 points.

Keywords: Surgical gesture recognition · Temporal convolution network · Symmetric dilation · Self-attention

1 Introduction

Surgical gesture recognition is the process of jointly segmenting and classifying fine-grained surgical actions from surgical videos. It is crucial for surgical video understanding and building the context awareness system towards the next generation surgery [10]. However, raw surgical videos are normally untrimmed and the operation environment is particularly complicated. Consequently, detecting surgical gestures from these surgical videos with high intra-class variance and low inter-class variance is inherently quite challenging.

Recent studies apply Hidden Markov Model (HMM) [14] and its variants [8] to identify the latent state of surgical actions. The latent states transferring

Electronic supplementary material The online version of this chapter (https://doi.org/10.1007/978-3-030-59716-0_39) contains supplementary material, which is available to authorized users.

A. L. Martel et al. (Eds.): MICCAI 2020, LNCS 12263, pp. 409–418, 2020.
https://doi.org/10.1007/978-3-030-59716-0_39

among successive actions are subsequently modelled by the transition probability. Although state features in HMMs are interpretable, they only focus on few local frames hence making the model incapable of capturing the global pattern. In addition, some machine learning methods (i.e. Support Vector Machine (SVM) [15]) assemble multiple heterogeneous features (color, motion, intensity gradients, etc.) to localize and classify surgical actions. Nonetheless, these features are hand-crafted. Therefore some crucial latent features could be neglected during feature extraction procedure.

More recently, large number of approaches depend on Recurrent Neural Networks (RNN) [2,13], particularly, the Long Short Term Memory (LSTM) network, because of their notable ability of modeling sequence data in variable length. The gate mechanism of LSTM preserves temporal dependencies and drops irrelevant information during the training stage. However, LSTM-based methods only have limited ability of capturing long-term video context, due to the intrinsic vanishing gradient problem [12].

From another perspective, Lea et al. [7] introduce Temporal Convolutional Networks (TCNs) to segment and detect actions by hierarchically convolving, pooling, and upsampling input spatial features using 1-D convolutions and deconvolutions. The promising experiment results manifest that TCNs are capable of dealing with long-term temporal sequences though, the model handles information among local neighbors, thus showing incapabilities in catching global dependencies. Following this work, Farha and Gall [3] suggest a multi-stage TCN, in which each stage is composed of several dilation layers, for action segmentation. Their work demonstrates the competence of dilated convolution [11] in hierarchically collecting multi-scale temporal information without losing dimensions of data. Moreover, in order to sequentially capture the video dynamics, Funke et al. [4] randomly sample video snippets (16 consecutive frames per snippet) and utilize a 3D Convolutional Neural Network (CNN) to extract the spatial-temporal features. But still, they only consider local continuous information. Because of the huge computational cost and GPU memory expenditure of 3D-CNN, they can only train the network at the clip level rather than inputted with the whole video [18].

In this paper, we propose a symmetric dilated convolution structure embedded with self-attention kernel to jointly detect and segment fine-grained surgical gestures. Figure 1 is an overview of our framework. Taking the extracted spatial CNN features from [7] as input, the encoder captures the long temporal information with a series of 1-D dilated convolutions to enlarge the temporal receptive field, followed by an attention block to establish the one-to-one relationship across all latent representations. Symmetrically, we devise our decoder with another set of dilation layers to map the latent representations back to each frame and predict the frame-wise gesture label. Unlike 3D-CNN learning features from partial sampled clips, our network takes the whole video into consideration. Owing to the symmetric dilated convolution structure with the enclosed self-attention kernel, not only can we learn the long-range temporal information, but also we can process neighbor and global relationship simultaneously.

With the above facts, we claim our contribution as two-fold. First, we propose a symmetric dilation architecture embedded with a self-attention module. It takes into account the long-term temporal patterns and builds frame-to-frame adjacent as well as global dependencies from the surgical video sequence. Second, with the novel network architecture, our approach consistently exceeds the state-of-the-art method both on frame-level and on segmental-level metrics, improving the frame-wise accuracy ∼**6 points**, and the F1@50 score ∼**6 points**, which largely alleviates the over-segmentation error.

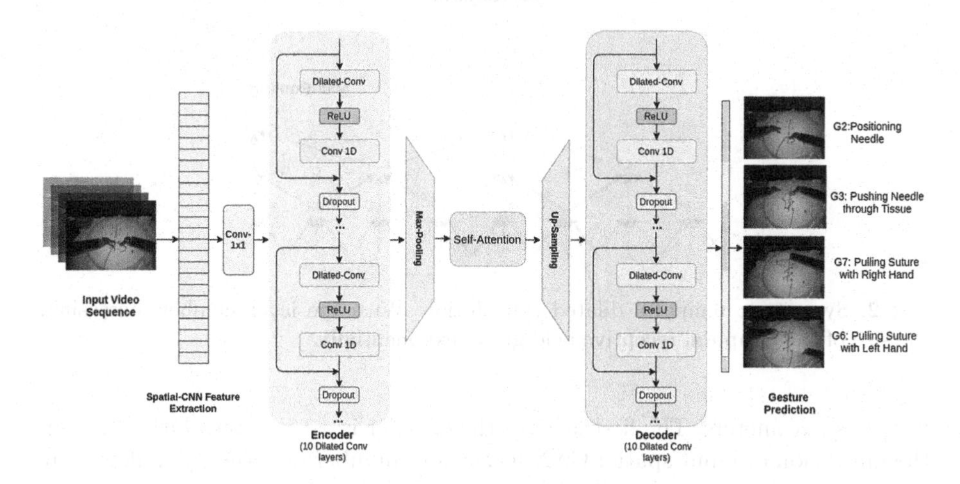

Fig. 1. Overview of our architecture. Symmetric dilation network takes frame-level spatial-CNN features as input. The architecture can be divided into five steps: 1) 1-D convolution; 2) 10 dilated convolution layers with max-pooling; 3) self-attention; 4) upsampling with 10 dilated convolution layers; 5) frame-wise prediction.

2 Methodology

The architecture of our symmetric dilation network for surgical gesture recognition is detailed in this section (see Fig. 2), which consists of two substructures: 1) the symmetric dilated Encoder-Decoder structure to capture long-term frame contents with memory-efficient connections (dilated layers) to aggregate multi-scale temporal information (see Sect. 2.1); 2) the self-attention kernel in the middle to deploy the deep frame-to-frame relations to better discriminate the similarities among different frames (see Sect. 2.2).

2.1 Symmetric Temporal Dilated Convolution

Temporal dilated convolution is a type of convolution applied on the input sequence with a defined sliding gap, which increases the temporal receptive field

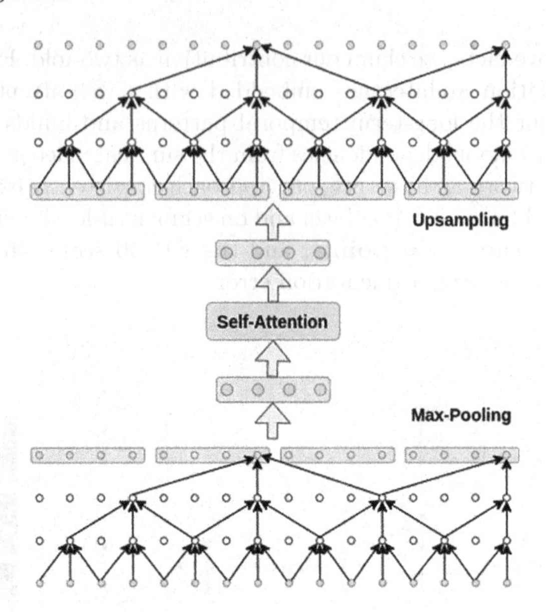

Fig. 2. Symmetric temporal dilated convolution. With the layer number increasing, the size of the temporal receptive field grows exponentially.

with less parameters. The first layer of the encoder is a 1×1 convolution to map the dimension of input spatial-CNN features to number of kernels f, followed by l layers of temporal dilated convolutions, where the dilation rates $\{s_l\}$ are set to $s_l = 2^l, l = 0, 1, ..., 9$. Because our target is the off-line recognition, we follow the details in [3] by using acausal mode with kernel size at 3. Furthermore, we apply the non-linear activation function ReLU to each dilation output followed by a residual connection between the layer input and the convolution signal. The temporal dilated procedure can be formulated as follows:

$$\hat{E}_l = \text{ReLU}(W_1 * E_{l-1} + b_1) \tag{1}$$

$$E_l = E_{l-1} + W_2 * \hat{E}_l + b_2 \tag{2}$$

where E_l is the output of l-th encoder layer, $*$ is the temporal convolutional operation, $W_1 \in \mathbf{R}^{f \times f \times 3}$, $W_2 \in \mathbf{R}^{f \times f \times 1}$ represent the weights of a dilated convolution and the weights of a 1×1 convolution with f convolutional kernels, respectively. $b_1, b_2 \in \mathbf{R}^f$ are denoted as their corresponding biases. In every dilation layer l, the receptive field R grows exponentially to capture the long range temporal pattern, expressed as: $R(l) = 2^{l+1} - 1$. By doing this, the temporal information on different scale is hierarchically aggregated while keeps the ordering of sequence. We also employ a 4×1 max-pooling layer behind the encoder dilation block to efficiently reduce the oversegmentation error (see our ablative study results in Table 2).

Our symmetric decoder has a similar structure with the encoder block, except that the max-pooling operations are replaced with a 1×4 upsampling. To get the final prediction, we use a 1×1 convolution followed by a softmax activation after the last decoder dilated convolution layer:

$$Y_t = \text{Softmax}(W * D_{L,t} + b) \quad (3)$$

where Y_t is the prediction at time t, $D_{L,t}$ is the output from the last decode dilated layer at time t, $W \in \mathbf{R}^{f \times c}$ and $b \in \mathbf{R}^c$, where $c \in [1, C]$ is the surgical gestures classes. Eventually, we use the categorical cross-entropy loss for the classification loss calculation.

2.2 Joint Frame-to-Frame Relation Learning with Self-attention

The TCNs have shown consistent robustness in handling long temporal sequences with using ***relational features*** among frames. However, current methods [1,7] only consider relations in local neighbors, which could undermine their performance in capturing relational features within a longer period. To obtain the global relationship among frames, it is essential to build frame-to-frame relational features with a non-local manner in addition to our encoder-decoder dilated convolutions. With this insight, we introduce the non-local self-attention module to extract discriminate spatial-temporal features for better prediction.

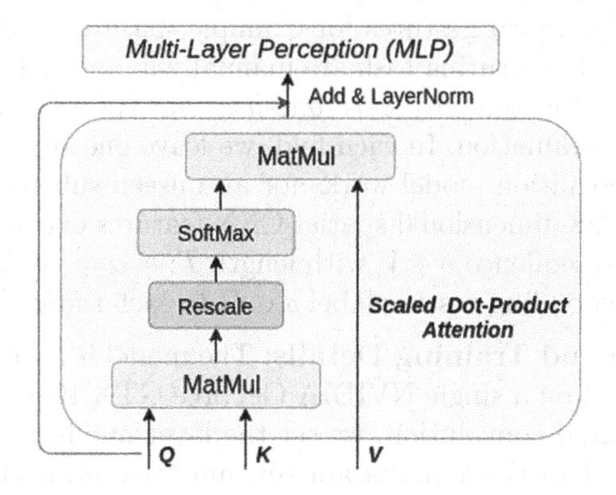

Fig. 3. Self-attention block

Self-attention, or intra-attention refers to an attention mechanism, which attends every position of the input sequence itself and build one-to-one global dependencies. This idea has been widely used in Natural Language Processing (NLP) [16], Object Detection and Segmentation [6,17], etc. The key component of self-attention is called **Scaled Dot-Product Attention**, which is calculated as:

$$\text{Attention}(Q, K, V) = \text{Softmax}(\frac{QK^T}{\sqrt{d_k}})V \quad (4)$$

where Q is a packed *Query* matrix, K and V stand for *Key-Value* pairs, and $\sqrt{d_k}$ is the feature dimension of queries and keys. The structure of the self-attention is shown in Fig. 3. In our work, the input *Queries, Keys, and Values* to the self-attention module are the same, that is the output hidden temporal states from the encoder downsampling. The first step is to take the dot product between the query and the key to calculate the similarity. This similarity determines the relevance between all other frames from the input sequence to a certain frame. Then, the dot product is rescaled by $\sqrt{d_k}$ to prevent the exploding gradient and followed by a softmax function to normalize the result. The intention of applying the softmax function here is to give relevant frames more focus and drop irrelevant ones. Eventually, the attention matrix is multiplied by the value and summed up. There is a residual connection followed by a layer normalization to feed the result to next two fully connected 1-D concolutional layers (see Fig. 3). In this manner, frame-to-frame global dependencies are constructed.

3 Evaluation

3.1 Experiment Settings

Dataset Description: We evaluate our approach on an elementary suturing task from JHU-ISI Gesture and Skill Assessment Working Set (JIGSAWS) [5], a robotic assisted bench-top model collected using *da Vinci* surgical system. There are 39 videos performed by eight surgeons with three skill levels. Ten different fine-grained surgical gestures for example, *pushing needle through tissue and oriental needle* for suturing task are manually annotated by an experienced surgeon. We follow the standard *leave-one-user-out* (LOUO), a 8-fold cross validation scheme for evaluation. In each fold, we leave one surgeon out for testing to verify if the recognition model works for an unseen subject. For the network input, we use the 128 dimensional spatial-CNN features extracted by [7] with 10 FPS. Given a video sequence $v \in V$ with length T: $v_{1:T} = (v_1, ..., v_T)$, our goal is to assign the corresponding gesture label $g \in \mathbf{G}$ to each frame: $g_{1:T} = (g_1, ..., g_T)$.

Implementation and Training Details: The model is implemented based on Pytorch and trained on a single NVIDIA GeForce GTX 1080 graphics card. For the symmetric dilated convolution, we set the layer number to 10 (see the supplementary material for the hyperparameter tuning experiment) and the channel number to 128 with the kernel size 3 followed by a dropout after each layer. In regard to the attention module, the feature dimension of queries and keys is set to 16. The network is trained for 30 epochs with the learning rate at 0.01. In addition, we apply Adam Optimizer such that $\beta_1 = 0.9$, $\beta_2 = 0.98$, and $\epsilon = 10^{-9}$.

Evaluation Metrics: We adopt three evaluation metrics in our experiments: **frame-wise accuracy**, **edit score**, and **segmented F1 score**. Frame-wise accuracy is to measure the performance in frame level. However, long gesture segments tend to have more impact than short gesture segments, and the frame-wise accuracy is not sensitive to the oversegmentation error. Therefore, we use the edit score and F1 score to assess the model at segmental level. Edit score is defined as the normalized Levenshtein distance between the prediction and the groundtruth. While F1 score is the harmonic mean of precision and recall with the threshold 10%, 25%, and 50% as defined in [7].

3.2 Comparison with the State-of-The-Arts

Table 1 compares our symmetric dilation network with other state-of-the-art methods. It can be seen that our model achieves the best performance in all three metrics. Among other approaches, the baseline model **Bi-LSTM** reaches the relative lower performance than other methods indicating that the traditional RNN-based method is incapable of handing long video sequence. **Deep Reinforcement Learning (RL)** method trains an intelligent agent with reward mechanism and achieves the high edit 87.96 and F1 score 92.0, but the low frame-wise accuracy at 81.43%, which shows its inadequacy in capturing the global similarities throughout the frames. The latest **3D-CNN** method obtains the fair frame-wise accuracy at 84.3%, but it only obtains 80.0 for the edit score. This reflects that the model based on clip-level is still inefficient in catching long temporal relationship such that it suffers from the oversegmentation error.

While our model reaches the best frame-wise accuracy at 90.1% as well as the highest edit and F1 score at 89.9 and 92.5, respectively. It demonstrates that our model is able to capture the long-range temporal information along with the frame-to-frame global dependencies.

Table 1. Comparsion with the most recent and related works for surgical gesture recognition. Acc., Edit, and F1@10, 25, 50 stand for the frame-wise accuracy, segmented edit distance, and F1 score, respectively

JIGSAWS (Suturing)	Acc.	Edit	F1@10	F1@25	F1@50
Bi-LSTM [13]	77.4	66.8	77.8	–	–
ED-TCN [7]	80.8	84.7	89.2	–	–
TricorNet [1]	82.9	86.8	–	–	–
RL [9]	81.43	87.96	92.0	90.5	82.2
3D-CNN [4]	84.3	80.0	87.0	–	–
Symmetric dilation (w. pooling) + attn	**90.1**	**89.9**	**92.5**	**92.0**	**88.2**

4 Discussion

To further investigate the functionality of each submodule in our method, we conduct ablative studies with five configurations as follows. As our network consists of a symmetric dilation structure with a self-attention kernel in the middle. We decouple it into a head dilation module, a tail dilation module, and the self-attention kernel to explore their joint effects.

(1) Self-attention module only (baseline)
(2) Baseline + head dilated convolution
(3) Baseline + tail dilated convolution

(4) Baseline + symmetric dilated convolution
(5) Baseline + symmetric dilated convolution + pooling

We apply these settings to segment and classify the surgical gestures and measure their **frame-wise accuracy**, **edit score**, and **segmented F1 score** separately. The experiment results are shown in Table 2 (see supplementary material for the visualization of ablative experiments).

(1) only: Self-attention module can achieve promising frame-wise accuracy at 87.8%, but with very low edit distance (44.0) and F1 scores. It can be concluded that attention module is robust for classification tasks while missing the long temporal information.

(1) v.s. (2) and (3): We put the temporal dilated convolution structure before and after the self-attention module and get the similar results. The results have huge improvement in edit score and F1 score with different threshold, increase around 30% in each metric. It states that temporal convolution is capable of catching long temporal patterns.

(4): The obvious improvement on the segmental level evaluation shows that the symmetric encoder-decoder dilation structure helps capture the high-level temporal features.

(5): Max-pooling and upsampling further improve the edit distance and F1 score at segmental level such that smooth the prediction and allievate the over-segmentation problem.

Above controlled experiments verify the indispensability of each component for our proposed architecture. From frame-level view, self-attention mechanism is feasible to build non-local dependencies for accurate classification. And from the segmental-level perspective, symmetric dilation with pooling is a viable solution for recognising gestures from long and complicated surgical video data.

Table 2. Ablative experiment results show the effectiveness of each submodel. Acc., Edit, and F1@{10, 25, 50}, stand for the frame-wise accuracy, segmented edit distance, and F1 score, respectively

JIGSAWS (Suturing)	Acc.	Edit	F1@10	F1@25	F1@50
Self-attn only	87.8	44.0	54.8	53.5	49.0
Head dilation + attn	90.8	76.9	82.5	81.8	79.3
Tail dilation + attn	90.5	77.9	83.4	83.4	79.7
Symmetric dilation + attn	90.7	83.7	87.7	86.9	83.6
Symmetric dilation (w. pooling) + attn	90.1	**89.9**	**92.5**	**92.0**	**88.2**

5 Conclusion

In this work, we propose a symmetric dilated convolution network with self-attention module embedded to jointly segment and classify fine-grained surgical gestures from the surgical video sequence. Evaluation of JIGSAW dataset indicates that our model can catch the long-term temporal patterns with the large temporal receptive field, which benefits from the symmetric dilation structure. In addition, a self-attention block is applied to build the frame-to-frame relationship to capture the global dependencies, while the temporal max-pooling and upsampling layer further diminish the oversegmentation error. Our approach outperforms the accuracy of state-of-the-art methods both at the frame level and segmental level. Currently, our network is designed with an off-line manner in acausal mode, and we will explore the possibility of improving and applying it for real-time surgical gesture recognition in the future work.

Acknowledgement. The authors thank Bournemouth University PhD scholarship and Hengdaoruyi Company as well as the Rabin Ezra Scholarship Trust for partly supported this research.

References

1. Ding, L., Xu, C.: Tricornet: a hybrid temporal convolutional and recurrent network for video action segmentation. arXiv preprint arXiv:1705.07818 (2017)
2. DiPietro, R., et al.: Recognizing surgical activities with recurrent neural networks. In: Ourselin, S., Joskowicz, L., Sabuncu, M.R., Unal, G., Wells, W. (eds.) MICCAI 2016. LNCS, vol. 9900, pp. 551–558. Springer, Cham (2016). https://doi.org/10.1007/978-3-319-46720-7_64
3. Farha, Y.A., Gall, J.: Ms-tcn: multi-stage temporal convolutional network for action segmentation. In: Proceedings of the IEEE Conference on Computer Vision and Pattern Recognition, pp. 3575–3584 (2019)
4. Funke, I., Bodenstedt, S., Oehme, F., von Bechtolsheim, F., Weitz, J., Speidel, S.: Using 3D convolutional neural networks to learn spatiotemporal features for automatic surgical gesture recognition in video. In: Shen, D., et al. (eds.) MICCAI 2019. LNCS, vol. 11768, pp. 467–475. Springer, Cham (2019). https://doi.org/10.1007/978-3-030-32254-0_52
5. Gao, Y., et al.: Jhu-isi gesture and skill assessment working set (jigsaws): a surgical activity dataset for human motion modeling. In: MICCAI Workshop: M2CAI, vol. 3, p. 3 (2014)
6. Hu, H., Gu, J., Zhang, Z., Dai, J., Wei, Y.: Relation networks for object detection. In: Proceedings of the IEEE Conference on Computer Vision and Pattern Recognition, pp. 3588–3597 (2018)
7. Lea, C., Flynn, M.D., Vidal, R., Reiter, A., Hager, G.D.: Temporal convolutional networks for action segmentation and detection. In: proceedings of the IEEE Conference on Computer Vision and Pattern Recognition, pp. 156–165 (2017)
8. Lea, C., Hager, G.D., Vidal, R.: An improved model for segmentation and recognition of fine-grained activities with application to surgical training tasks. In: 2015 IEEE Winter Conference on Applications of Computer Vision, pp. 1123–1129. IEEE (2015)
9. Liu, D., Jiang, T.: Deep reinforcement learning for surgical gesture segmentation and classification. In: Frangi, A.F., Schnabel, J.A., Davatzikos, C., Alberola-López, C., Fichtinger, G. (eds.) MICCAI 2018. LNCS, vol. 11073, pp. 247–255. Springer, Cham (2018). https://doi.org/10.1007/978-3-030-00937-3_29

10. Maier-Hein, L.: Surgical data science for next-generation interventions. Nat. Biomed. Eng. **1**(9), 691–696 (2017)
11. Oord, A.V.d., et al.: Wavenet: a generative model for raw audio. arXiv preprint arXiv:1609.03499 (2016)
12. Pascanu, R., Mikolov, T., Bengio, Y.: On the difficulty of training recurrent neural networks. In: International Conference on Machine Learning, pp. 1310–1318 (2013)
13. Singh, B., Marks, T.K., Jones, M., Tuzel, O., Shao, M.: A multi-stream bi-directional recurrent neural network for fine-grained action detection. In: Proceedings of the IEEE Conference on Computer Vision and Pattern Recognition, pp. 1961–1970 (2016)
14. Tao, L., Zappella, L., Hager, G.D., Vidal, R.: Surgical gesture segmentation and recognition. In: Mori, K., Sakuma, I., Sato, Y., Barillot, C., Navab, N. (eds.) MICCAI 2013. LNCS, vol. 8151, pp. 339–346. Springer, Heidelberg (2013). https://doi.org/10.1007/978-3-642-40760-4_43
15. Twinanda, A.P., Shehata, S., Mutter, D., Marescaux, J., De Mathelin, M., Padoy, N.: Endonet: a deep architecture for recognition tasks on laparoscopic videos. IEEE Trans. Med. Imaging **36**(1), 86–97 (2016)
16. Vaswani, A., et al.: Attention is all you need. In: Advances in Neural Information Processing Systems, pp. 5998–6008 (2017)
17. Wang, X., Girshick, R., Gupta, A., He, K.: Non-local neural networks. In: Proceedings of the IEEE Conference on Computer Vision and Pattern Recognition, pp. 7794–7803 (2018)
18. Zhang, S., Guo, S., Huang, W., Scott, M.R., Wang, L.: V4d: 4D convolutional neural networks for video-level representation learning. arXiv preprint arXiv:2002.07442 (2020)

Deep Selection: A Fully Supervised Camera Selection Network for Surgery Recordings

Ryo Hachiuma[1(✉)], Tomohiro Shimizu[1], Hideo Saito[1], Hiroki Kajita[2], and Yoshifumi Takatsume[2]

[1] Keio University, Yokohama, Kanagawa, Japan
{ryo-hachiuma,tomy1201,hs}@keio.jp
[2] Keio University School of Medicine, Shinjuku-ku, Tokyo, Japan
{jmrbx767,tsume}@keio.jp

Abstract. Recording surgery in operating rooms is an essential task for education and evaluation of medical treatment. However, recording the desired targets, such as the surgery field, surgical tools, or doctor's hands, is difficult because the targets are heavily occluded during surgery. We use a recording system in which multiple cameras are embedded in the surgical lamp, and we assume that at least one camera is recording the target without occlusion at any given time. As the embedded cameras obtain multiple video sequences, we address the task of selecting the camera with the best view of the surgery. Unlike the conventional method, which selects the camera based on the area size of the surgery field, we propose a deep neural network that predicts the camera selection probability from multiple video sequences by learning the supervision of the expert annotation. We created a dataset in which six different types of plastic surgery are recorded, and we provided the annotation of camera switching. Our experiments show that our approach successfully switched between cameras and outperformed three baseline methods.

Keywords: Surgery recording · Camera selection · Deep neural network

1 Introduction

Recording plastic surgeries in operating rooms with cameras has been indispensable for a variety of purposes, such as education, sharing surgery technologies and techniques, performing case studies of diseases, and evaluating medical treatment [11,13]. The targets that best depict the surgery, such as the surgical field, doctor's hands, or surgical tools, should be recorded for these purposes. The recording target differs for each surgery type and the purpose of recording.

Electronic supplementary material The online version of this chapter (https://doi.org/10.1007/978-3-030-59716-0_40) contains supplementary material, which is available to authorized users.

A. L. Martel et al. (Eds.): MICCAI 2020, LNCS 12263, pp. 419–428, 2020.
https://doi.org/10.1007/978-3-030-59716-0_40

However, it is difficult to continuously record these targets without any occlusion. If the camera is attached to the operating room environment, the targets may be occluded by the doctors, nurses, or surgical machines. If the camera is attached to the doctor's head to record video from the first-person view, the camera's field of view does not always capture the target, and the video is often affected by motion blur because of the doctor's head movements. Therefore, a first-person viewpoint camera is not suitable for recording surgery. Moreover, the doctors are disturbed by the camera which is attached to their head during the surgery which requires careful and sensitive movement.

Shimizu *et al.* [14] proposed a novel surgical lamp system with multiple embedded cameras to record surgeries. A generic surgical lamp has multiple light bulbs that illuminate the surgical field from multiple directions to reduce the shadows caused by the operating doctors. Hence, Shimizu *et al.* expected that at least one of the multiple light bulbs would almost always illuminate the surgical field. In the same way, we can expect that at least one of the multiple cameras embedded in the surgical lamp system will always record the target without occlusion.

As the cameras obtain multiple videos of a single surgery, Shimizu *et al.* also proposed a method to automatically select the image with the best view of the surgical field at each moment to generate one video. By assuming that the quality of the view is defined by the visibility (area size) of the surgical field, they used an image segmentation [8] trained for each surgery using manually annotated images. Then, they applied Dijkstra's algorithm to generate a smooth video sequence in which the selected camera does not change frequently over a period of time.

Although their method only focuses on the visibility of the surgical field, the visibility of the field is not always treated as the cue of the best view for recording surgeries. The recording target differs for each surgery type and the purpose of the recording. In some cases, the pose or the motion of the doctor's hands and surgical tools may be the most important, so the best view should not always be determined by the larger surgical field. A method based on image segmentation of predefined target objects is therefore not suitable to select the best view from the multiple videos of the surgical scene.

In this paper, we aim to establish a method to select the best view based on annotated labels by humans, rather than simply using the area size of the surgery field. We address the task in a fully supervised manner, in which the camera selection label is directly given to the prediction model during training, and the predictor learns to map the video sequences to the corresponding selected camera labels. No prior work has been proposed that addresses the task of selecting the best camera view of surgery recordings in a fully supervised manner.

The naive approach to solve this task is to formulate it as a multi-class classification problem, in which the model outputs the selected camera index. However, if the number of cameras in the recording system changes from the training time to the test time, the model cannot be applied at the test time as the number of categories (camera index) will differ. We, therefore, predict the probability of selecting each camera against each frame in the video sequences instead of predicting the selected camera index. When humans create a single video that depicts the surgery the best from multiple video sequences, they consider not only the temporal information of each video sequence but also the intra-camera context. We, therefore, present a deep neural network that aggregates information from multiple video sequences not only sequentially but also spatially (intra-camera direction) to predict the probability of selecting each camera from each frame.

As there is no dataset available to the public containing surgery recordings via multiple cameras, we record our dataset using the system proposed by Shimizu *et al.* [14]. The actual plastic surgeries are recorded at our university's school of medicine. We record six different types of surgery with five cameras attached to the surgical lamp. We validate our proposed model with this dataset, and we quantitatively evaluate our approach with three baseline methods to verify the effectiveness of our approach. The experiments show that our approach can create a video with the best views from multiple video sequences.

In summary, our contributions are as follows: (1) To the best of our knowledge, this is the first study to address the task of selecting the cameras with the best views from multiple video sequences for the purpose of recording surgery. (2) We propose an end-to-end deep learning network for the selection of cameras with the best views, which aggregates the visual context features sequentially and spatially. (3) We create a dataset of a variety of plastic surgeries recorded with multiple cameras, we provide camera-switching labels via a human expert, and we conduct extensive experimentation from qualitative and quantitative perspectives to show the effectiveness of our method. Please also refer to our accompanying video.

2 Related Work

Multiple cameras are used in many situations, such as office environments, sports stadiums, and downtown areas. Although large areas can be recorded without occlusion using multiple cameras, only the necessary information must be selected from the huge number of video sequences recorded by multiple cameras. Self-controlled camera technology, such as automatic viewpoint switching, multiple video summarization, is therefore regarded as an important issue [1].

Liu *et al.* [10] presented a rule-based camera switching method inspired by the heuristic knowledge of a professional video editor. Their method successfully switches between the viewpoints of three cameras shooting the speaker and the audience. Doubek *et al.* [4] recorded moving objects with multiple cameras embedded in an office environment. Cameras were selected based on their score, and a resistance coefficient was introduced so that the cameras were switched only when the score changed significantly. In these methods, camera switching is conducted based on the heuristic knowledge of the professional.

Recently, Chen *et al.* [2] presented a camera selection method in which the rich deep features are extracted from the neural network and online regression to broadcast soccer games. As they stated in the discussion, their method

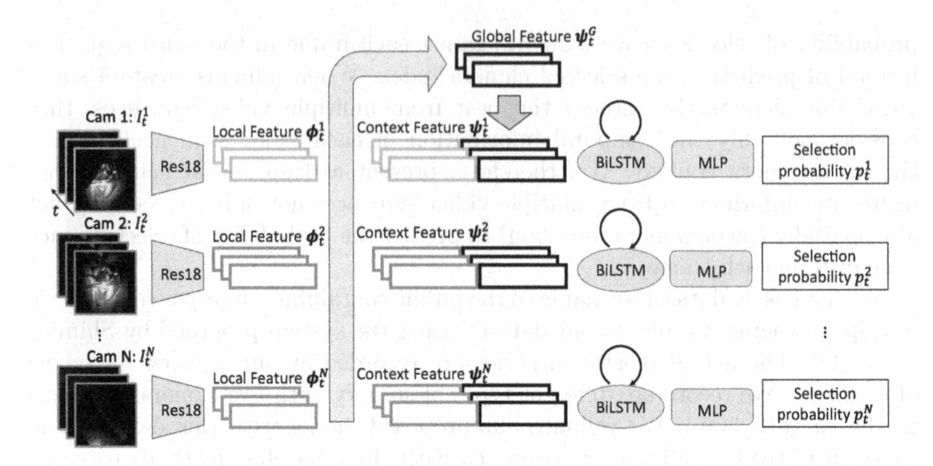

Fig. 1. The network architecture of the proposed method

is sensitive to experimental parameters, such as the range of camera duration and the weight of the regularization term. Moreover, their method requires pre-processing and post-processing of the image and camera selection indices. On the other hand, our end-to-end deep neural network directly learns to map the multiple video sequences to the camera selection indices.

The work most relevant to ours is that presented by Shimizu *et al.* [14]. They presented a surgery recording system with multiple cameras embedded in the surgery lamp, with the assumption that at least one of the cameras will always record the surgery field. They also proposed a camera selection method that can select one frame among multiple frames obtained by the cameras. They calculated the area size of the surgery field of each frame, and they applied Dijkstra's algorithm to obtain smooth camera selection indices based on the segmentation score. The aim of the current study differs from theirs. As it is not always best to record the entire surgery field, it is difficult to determine the camera selection criterion beforehand. We, therefore, provide the ground truth of camera selection directly to the predictor, and the predictor learns to map the multiple video sequences to the camera selection process in a fully supervised manner.

3 Approach

3.1 Problem Formulation

Our task is to estimate a sequence of camera labels $\mathbf{y} = \langle y_1, y_2, \ldots, y_T \rangle$, where $y_t \in [1, N]$, and T denotes the number of frames in the sequence, from image sequences $\boldsymbol{I} = \langle \boldsymbol{I}_1, \boldsymbol{I}_2, \ldots, \boldsymbol{I}_T \rangle$, where $\boldsymbol{I}_t = \langle I_t^1, I_t^2, \ldots, I_t^N \rangle$, from N cameras. The naive approach to solve this task is to treat it as an N-class classification problem similar to the image classification task: that is to say, to classify a set of images $\boldsymbol{I}_t$ into N classes for each point in time. On the other hand, our network predicts the camera selection probability $\boldsymbol{p}_t^n$ for each camera and solve the task as a binary classification problem. The predicted camera labels y_t can be obtained by calculating the maximum selection probabilities $\boldsymbol{p}_t^n$.

3.2 Network Architecture

Our network architecture is represented in Fig. 1. The network is composed of four components; visual feature extraction, spatial feature aggregation, sequential feature aggregation, and selection probability module. First, we extract the visual feature $\phi_t^n \in \mathbb{R}^{128}$ from each image I_t^n. We employ ResNet-18 [6] as the visual feature extractor.

Next, we aggregate the features from multiple cameras at each time step to consider intra-camera (spatial) context, $\psi_t^G = \mathbb{A}(\phi_t^1, \dots \phi_t^N)$, where $\mathbb{A}$ is a spatial feature aggregation module that combines the independent camera local feature ϕ_t^n into an aggregate feature ψ_t^G. The aggregated feature should be invariant to the input permutation as the number of cameras or the index of cameras can be changed at the test time. To handle unordered feature aggregation, we use the idea of neural network on 3D point cloud processing. As the order of points should not be mattered for the extraction of the point cloud's global feature, the point cloud's global feature should be invariant to the order of points. Inspired by PointNet [12], we employ max pooling as $\mathbb{A}$ to extract the global feature ψ_t^G. The global feature is concatenated to each local feature ϕ_t^n to obtain the context feature ψ_t^n.

Then, we aggregate the context features over time, $\hat{\psi}_1^n, \dots \hat{\psi}_t^n = \mathbb{B}(\psi_1^n, \dots \psi_t^n)$, where $\mathbb{B}$ is a sequential feature aggregation module that computes the sequential feature $\hat{\psi}_t^n$. In our experiments, we employ a BiLSTM recurrent neural network [5] with one hidden layer for $\mathbb{B}$ which aggregate not only the past sequence but also future sequence against the target feature.

The output feature $\hat{\psi}_1^n, \dots, \hat{\psi}_t^n$ is then fed to a multilayer perceptron (MLP) with two hidden layers and leaky rectified linear units (LeakyReLU) activation function [15] to predict the selection probability $p_1^n, \dots, p_T^n$. The sigmoid activation function is applied to the output layer. Final predicted camera labels can be obtained by calculating the maximum selection probabilities at each time step: $y_t = \arg\max_{1 \le n \le N} p_t^n$ where p_t^n denotes a camera's selection probability at the time step t of n-th camera.

3.3 Loss Function

Because we formulate the task as binary classification problem, the optimal weight of the network can be obtained using the binary cross entropy between the prediction p_t^n and the ground-truth $\hat{p}_t^n$. However, a class imbalance exists between two categories: selected and not selected. For example, the ratio of selected labels and not-selected labels is 1 : 3 in the case of four cameras. This class imbalance during training overwhelms the cross entropy loss. Easily classified negatives (not selected) comprise the majority of the loss and dominate the gradient. We therefore employ the focal loss [9], inspired by the two-stage object detection method. The focal loss L we employ is as follows:

$$L(\boldsymbol{\xi}) = -\frac{1}{NT}\sum_{n=1}^{N}\sum_{t=1}^{T}(1 - q_{t,n})^{\gamma}\log q_{t,n}, \tag{1}$$

where $q_{t,n}$ is p_t^n when the n-th camera at the time step t is selected as the best camera view ($\hat{p}_t^n = 1$), and $q_{t,n}$ is $1 - p_t^n$ when the n-th camera at the time step t is not selected as the best camera view ($\hat{p}_t^n = 0$). $\boldsymbol{\xi}$ is the parameters of the proposed model. γ is *focusing* parameter $\gamma \geq 0$.

4 Experiments

4.1 Dataset

As there is no dataset available that contains surgery recordings with multiple cameras, we use the system proposed by Shimizu *et al.* [14] to create our dataset. The surgeries are recorded at Keio University School of Medicine. Video recording of the patients is approved by the Keio University School of Medicine Ethics Committee, and written informed consent is obtained from all patients or parents. We record six different types of surgery with five cameras attached to the surgical lamp. Each surgery is 30 min long, and the video is recorded at 30 frames per second (FPS). We subsampled every five frames for each sequence. The ground-truth annotations are created by a single expert.

4.2 Network Training

We employ Adam optimizer [7] with a learning rate of 1.0×10^{-4}. When training the model, we randomly sample a data fragment of $T = 40$ frames. The model converged after 50 epochs, which takes about six hours on a GeForce Quadro GV100. We apply dropout with the 0.5 probability during training to reduce the overfitting at the first layer of MLP. We set $\gamma = 2.0$ of focal loss [9]. The weights of ResNet-18 is initialized with the pretrained ImageNet [3]. We train the model with batch size 2.

4.3 Comparison with the Other Method and Baselines

No prior work has been proposed that addresses the task of switching between cameras using deep learning for the creation of surgical videos. We therefore set three baseline methods to validate our approach. We compare the following baselines:

- **Ours w/o spa., seq.**: Our camera selection method without spatial feature aggregation and sequential feature aggregation. This method estimates the camera probability using ResNet-18 and MLP.

- **Ours w/o spa.**: Our camera selection network without the spatial feature aggregation module. In this model, the context of other cameras are not considered to predict each camera's selection probability.
- **Ours w/o seq.**: Our camera selection network without the sequential feature aggregation module. In this model, the sequential context is not considered to predict each camera's selection probability.

Table 1. Quantitative results for camera selection for sequence-out and surgery-out settings. The dice score (F value) is employed as the evaluation metric so a higher value is better. S1 to S6 denote the indices of surgeries. The accuracy of Shimizu *et al.* [14] is shown as the reference.

Method	Sequence-Out						
	S1	S2	S3	S4	S5	S6	Average
Shimizu *et al.* [14]	**0.58**	**0.59**	0.57	**0.45**	0.34	**0.56**	0.52
Ours w/o spa., seq	0.45	0.39	0.98	0.28	0.36	0.36	0.47
Ours w/o spa	0.50	0.42	0.95	0.42	0.39	0.39	0.52
Ours w/o seq	0.45	0.43	0.98	0.33	0.32	0.38	0.48
Ours	0.54	0.44	**0.98**	0.36	**0.46**	0.44	**0.54**
Method	Surgery-Out						
	S1	S2	S3	S4	S5	S6	Average
Shimizu *et al.* [14]	0.46	0.44	0.20	0.30	0.33	**0.52**	0.38
Ours w/o spa., seq	0.41	0.45	0.82	0.50	0.37	0.35	0.48
Ours w/o spa	**0.46**	0.45	0.89	0.69	0.38	0.38	0.54
Ours w/o seq	0.40	**0.48**	0.88	0.56	0.40	0.38	0.52
Ours	0.44	0.47	**0.90**	**0.70**	**0.42**	0.43	**0.56**

Moreover, we also compare the results with Shimizu *et al.* [14] as the reference. The focus of this paper is to predict the camera label which is adapted for each expert. We do not aim to generate the model generalized for different experts, but the model that represents each expert's subjective selection. In contrast, Shimizu *et al.* [14] aims to select the camera based on the generic criterion obtained by segmenting the surgical field. This means that we aim the different goals from Shimizu *et al.* [14]. However, comparing the accuracy of the method [14] evaluates how much the expert see the surgical field during the annotation. We follow the experimental setup for training the segmentation model and switching parameter as same as the original paper [14].

5 Results

5.1 Sequence-Out Evaluation

In this setting, we sequentially split six surgery videos into training and test data at the ratio of 80–20. In this setup, we adopted the sequence-out cross-validation protocol to evaluate our method. That is to say, the surgery type is known at the test-time but the sequence is unknown at the test-time. Even though the surgery type is known at the test-time, this setting is challenging as single surgery contains multiple sub-tasks so the scene drastically changes between the training and test sequence. The quantitative results are summarized in Table 1. As shown in Table 1, the prediction accuracy improves with the use of each module in the proposed network. It can be seen that Shimizu *et al.* [14] outperformed the other methods for four sequences. However, the rich segmentation annotations are needed at each surgery for Shimizu *et al.* [14].

The qualitative results of camera switching are shown in Fig. 2 (left). As Fig. 2 shows, our model successfully selects the frame with the best view from among five cameras, and the predicted camera indices are similar to the camera indices of the ground truth. For example, in the first frame (leftmost column), the surgical field is the most visible at the frame from Camera 4.

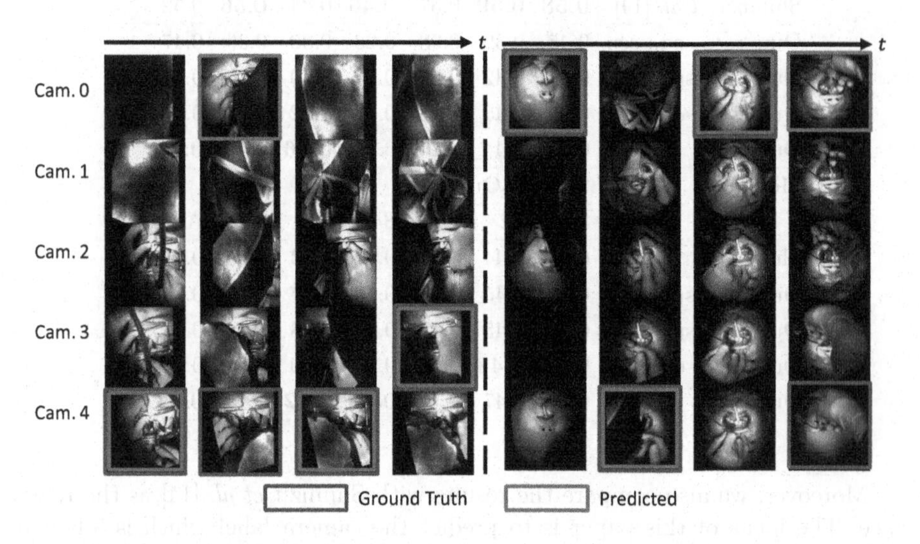

Fig. 2. Qualitative results for (left) Sequence-Out setting and (right) Surgery-Out setting. The predicted camera frame is highlighted in green, and the ground-truth camera frame is highlighted in red. The column direction indicates the time step, and the row direction indicates the camera indices (Cam.0, ..., Cam.4) (Color figure online)

5.2 Surgery-Out Evaluation

To further test the robustness of the proposed method, we perform the experiments in the surgery-out setting, in which we train our model on four surgery videos and test it on the two videos. In this setup, we adopted the surgery-out cross-validation protocol to evaluate our method. This is a more challenging setting since the surgery type is completely unknown to the model at the test-time, and a variety of surgery types exist in the dataset. The quantitative results are summarized in Table 1. As shown in Table 1, our method outperforms other baseline methods. As the unseen surgical field appeared at the test time for Shimizu *et al.* [14], the performance significantly drops for surgery-out settings.

The qualitative results of camera switching is visualized in Fig. 2 (right). Even though the type of the surgery is not included in the training dataset, our model successfully selects the best-viewed frame as like the ground-truth's. For the last frame, even though the predicted camera index is not matched to the ground-truth index, the surgery field, surgical tools, and the hand are captured within the frame.

6 Conclusion

We tackled, for the first time, the task of selecting the camera with the best view from multiple video sequences of a surgery. Our model learns to map each video sequence for selection probability while aggregating the features along the intra-camera and temporal directions in a fully supervised manner. Our experiments revealed that our method successfully selects the same camera indices as the ground truth. As the video is divided into the sub-sequences (40 frames) to input to the network, the model cannot consider the whole sequence of the long surgery video for the prediction. Therefore, we will investigate the autoregressive model which relies on the prediction of the previous output.

Acknowledgement. This research was funded by JST-Mirai Program Grant Number JPMJMI19B2, ROIS NII Open Collaborative Research 2020-20S0404, SCOPE of the Ministry of Internal Affairs and Communications and Saitama Prefecture Leading-edge Industry Design Project, Japan. We would like to thank the reviewers for their valuable comments.

References

1. Chen, J., Carr, P.: Autonomous camera systems: a survey. In: Workshops at the Twenty-Eighth AAAI Conference on Artificial Intelligence (2014)
2. Chen, J., Meng, L., Little, J.J.: Camera selection for broadcasting soccer games. In: Winter Conference on Applications of Computer Vision (WACV), pp. 427–435. IEEE (2018)
3. Deng, J., Dong, W., Socher, R., Li, L.J., Li, K., Fei-Fei, L.: ImageNet: a large-scale hierarchical image database. In: Conference on Computer Vision and Pattern Recognition (CVPR). IEEE (2009)
4. Doubek, P., Geys, I., Svoboda, T., Van Gool, L.: Cinematographic rules applied to a camera network. In: The Fifth Workshop on Omnidirectional Vision, Camera Networks and Non-Classical Cameras, pp. 17–29. Czech Technical University, Prague, Czech Republic (2004)
5. Graves, A., Fernández, S., Schmidhuber, J.: Bidirectional LSTM networks for improved phoneme classification and recognition. In: Duch, W., Kacprzyk, J., Oja, E., Zadrożny, S. (eds.) ICANN 2005. LNCS, vol. 3697, pp. 799–804. Springer, Heidelberg (2005). https://doi.org/10.1007/11550907_126

6. He, K., Zhang, X., Ren, S., Sun, J.: Deep residual learning for image recognition. In: Conference on Computer Vision and Pattern Recognition (CVPR), pp. 770–778. IEEE (Jun 2016)
7. Kingma, D.P., Ba, J.: Adam: a method for stochastic optimization. In: International Conference on Learning Representations (ICLR) (2015)
8. Li, C., Kitani, K.M.: Pixel-level hand detection in ego-centric videos. In: Conference on Computer Vision and Pattern Recognition (CVPR), pp. 3570–3577. IEEE (July 2013)
9. Lin, T., Goyal, P., Girshick, R., He, K., Dollár, P.: Focal loss for dense object detection. In: International Conference on Computer Vision (ICCV), pp. 2999–3007. IEEE, October 2017
10. Liu, Q., Rui, Y., Gupta, A., Cadiz, J.J.: Automating camera management for lecture room environments. In: Proceedings of the SIGCHI Conference on Human Factors in Computing Systems, pp. 442–449. ACM (2001)
11. Matsumoto, S., et al.: Digital video recording in trauma surgery using commercially available equipment. Scand. J. Trauma Resuscitation Emerg. Med. **21**, 27–27 (2013)
12. Qi, C.R., Su, H., Mo, K., Guibas, L.J.: Pointnet: deep learning on point sets for 3D classification and segmentation. In: Conference on Computer Vision and Pattern Recognition (CVPR), pp. 652–660. IEEE, July 2017
13. Sadri, A., Hunt, D., Rhobaye, S., Juma, A.: Video recording of surgery to improve training in plastic surgery. J. Plast. Reconstr. Aesthetic Surg. **66**(4), 122–123 (2013)
14. Shimizu, T., Oishi, K., Hachiuma, R., Kajita, H., Yoshihumi, T., Hideo, S.: Surgery recording without occlusions by multi-view surgical videos. In: International Conference on Computer Vision Theory and Applications, February 2020
15. Xu, B., Wang, N., Chen, T., Li, M.: Empirical evaluation of rectified activations in convolutional network. CoRR abs/1505.00853 (2015). http://arxiv.org/abs/1505.00853

Interacting with Medical Volume Data in Projective Augmented Reality

Florian Heinrich[1,2], Kai Bornemann[1,2], Kai Lawonn[3], and Christian Hansen[1,2(✉)]

[1] University of Magdeburg, Universitätsplatz 2, 39106 Magdeburg, Germany
hansen@isg.cs.uni-magdeburg.de
[2] Research Campus STIMULATE, Sandtorstrasse 23, 39106 Magdeburg, Germany
[3] University of Jena, Fürstengraben 1, 07743 Jena, Germany

Abstract. Medical volume data is usually explored on monoscopic monitors. Displaying this data in three-dimensional space facilitates the development of mental maps and the identification of anatomical structures and their spatial relations. Using augmented reality (AR) may further enhance these effects by spatially aligning the volume data with the patient. However, conventional interaction methods, e.g. mouse and keyboard, may not be applicable in this environment. Appropriate interaction techniques are needed to naturally and intuitively manipulate the image data. To this end, a user study comparing four gestural interaction techniques with respect to both clipping and windowing tasks was conducted. Image data was directly displayed on a phantom using stereoscopic projective AR and direct volume visualization. Participants were able to complete both tasks with all interaction techniques with respectively similar clipping accuracy and windowing efficiency. However, results suggest advantages of gestures based on motion-sensitive devices in terms of reduced task completion time and less subjective workload. This work presents an important first step towards a surgical AR visualization system enabling intuitive exploration of volume data. Yet, more research is required to assess the interaction techniques' applicability for intraoperative use.

Keywords: Interaction techniques · Medical volume data · Projective augmented reality

1 Introduction

Exploring medical image data is essential for many modern surgical procedures in terms of diagnosis, planning and image guided surgery. This volume data is usually visualized as two-dimensional (2D) slices on conventional PC monitors,

Electronic supplementary material The online version of this chapter (https://doi.org/10.1007/978-3-030-59716-0_41) contains supplementary material, which is available to authorized users.

A. L. Martel et al. (Eds.): MICCAI 2020, LNCS 12263, pp. 429–439, 2020.
https://doi.org/10.1007/978-3-030-59716-0_41

thus requiring surgeons to build mental maps of the actual patient anatomy. Three-dimensional (3D) data representations can facilitate this process [12,19]. Moreover, such visualization enables a better understanding of spatial relations and an easier identification of anatomical or pathological structures [1].

These effects can potentially be enhanced by the concept of augmented reality (AR). By showing relevant anatomy directly on the patient, surgeons no longer need to split their focus between spatially separated monitors and the operation site. Additionally, mental effort for the understanding of anatomical spatial relations can be further reduced, because of less distance between these information sources [16,21]. In the past, different technical solutions were developed to fuse both views on the patient and on the image data, e.g. augmented camera views displayed on monitors [22], optical see-through head mounted displays [17] or projector-camera-systems projecting volumetric image data directly onto the patient [21]. In comparison, projective AR does not require additional monitors in already crowded operating rooms or the user to wear head sets, which are often uncomfortable to wear for longer periods of time and could potentially compromise sterility.

To effectively explore and manipulate 3D image data, appropriate interaction methods are needed. Typical interaction tasks often require the manipulation of more degrees of freedom (DoF), than provided by conventional methods, e.g. mouse and keyboard [3]. Moreover, the need for sterility and limited space in the operating room further restrict needed interaction paradigms [15]. Therefore, more efficient techniques have been developed, that allow users to control more than two DoF simultaneously [11] and can be executed touchlessly [14]. Such intuitive methods are often based on natural human interaction in the form of gestures [6].

Related work identified advantages of using hand gestures [12,20,24] and evaluated different eligible input devices [9]. Foot gestures were identified as viable alternatives to manipulate image data even hands-freely [7,10,23]. In contrast, also interaction techniques using hand-held devices were explored [4]. Similar approaches were followed for immersive virtual reality applications using motion-sensitive controllers [2,18]. Most of these techniques were developed for the interaction with image data displayed on monitors. However, methods developed for desktop environments may not be intuitive in AR systems because of the more complex dimensionality. Additionally, in AR, position, orientation and scale of image data are fixed in space while these parameters can be modified in desktop applications. Therefore, previous research findings may not be applicable for AR environments. Hence, this work evaluates four gestural interaction techniques with respect to their applicability for exploring medical volume data in projective AR. This type of AR was chosen because of its advantages in terms of sterility and space requirements. Moreover, interactive AR system projecting image data onto the operation site have not been investigated before.

2 Materials and Methods

A user study was conducted to evaluate different gestural interaction methods for the exploration of projected medical volume data. For this study, 26 medical students (16 female, median age: 24, median year of university: 4) with basic knowledge about medical image data were recruited. In the following, details and rationales of the study are presented. A supplementary video demonstrates implemented interaction concepts and tasks of the study.

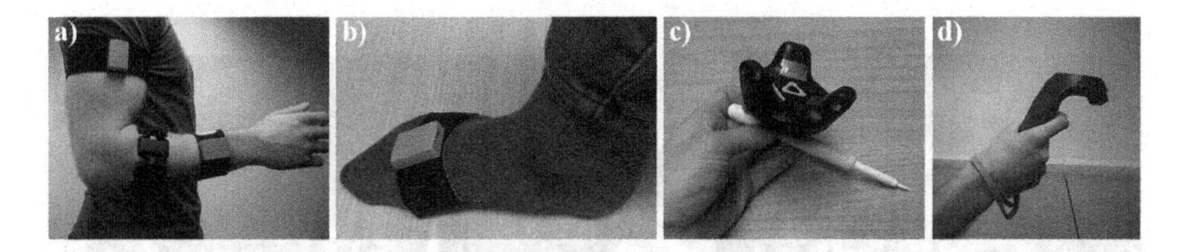

Fig. 1. Apparatus of implemented interaction techniques. a) Hand gestures. b) Foot gestures. c) Surgical instrument. d) Controller.

2.1 Interaction Techniques

Related work suggests advantages of gestural interaction techniques over conventional input modalities. To ensure, that all methods can be executed during interventions, we restricted this work's techniques to single-handed use, leaving one hand still available. Implementations should provide methods to manipulate linear parameters, as well as means to activate or deactivate these modes.

First, an interaction concept based on *hand gestures* was implemented using the Myo gesture control armband (Thalmic Labs, Canada) for gesture recognition. A positively evaluated fist gesture [8] and an easy to distinguish finger double tap gesture were used for mode activation. Two inertial measurement units (IMU; Xsens, Netherlands) were attached to the users wrist and upper arm (see Fig. 1a). Using direct kinematics, the relative hand position of the user could be determined. Changes of that position were mapped to linear parameters, resulting in an overall grabbing and moving interaction concept.

We also implemented an interaction technique based on *foot gestures.* A toe tap, i.e. lifting and lowering the forefoot [23], was included as activation gesture. Heel rotations, i.e. rotating the forefoot around the heel, were implemented to change linear parameters. Degrees of rotation were mapped to the speed at which parameters were changed. Only one DoF could be manipulated at the same time. Therefore, toe taps were used to switch between parameters. An IMU attached to the forefoot measured relative foot movements, thus recognizing gestures (see Fig. 1b).

Surgical navigation systems are commonly used in image-guided surgery [17]. Device-based gestures could by adapted by using related tracked *surgical instruments.* We attached an HTC Vive tracker (HTC Corp., Taiwan) to a 3D-printed surgical pointer (see Fig. 1c). Movements of that instrument were mapped to changes of linear parameters. Mode activations were performed using toe taps because direct means of interaction, e.g. buttons, are not available on surgical instruments.

Finally, a *controller-based interaction* method using an HTC Vive controller was implemented for our prototype (see Fig. 1d). Manipulation modes were activated by pressing and holding the trigger button at the index finger or by touching and holding the trackpad at the thumb's position. Movement of the controller in space was than mapped to changes of linear parameters.

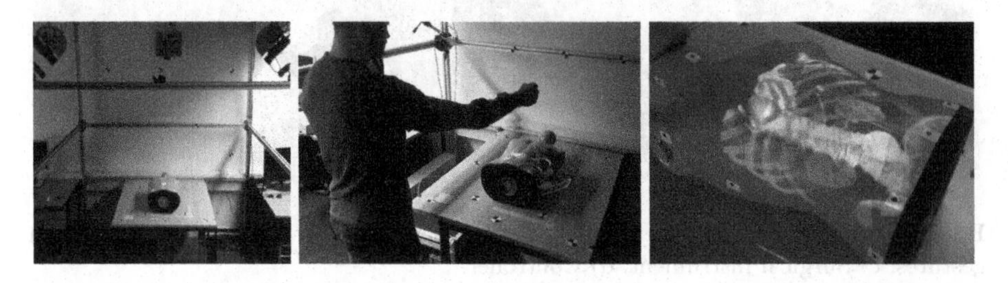

Fig. 2. The experimental setup. Projectors mounted above a table superimpose a torso phantom with a medical volume data set.

2.2 Projection of Medical Volume Data

For this work, the Panoramix DICOM example data set provided by the software 3D Slicer (Kitware, USA) was scaled to match the dimensions of a human torso phantom, on which two stereoscopic projectors with an active shutter 3D system (Barco F50 WQXGA, Barco GmbH, Germany) displayed virtual contents. Both projectors were calibrated using a photogrammetric measurement system. Calibration results included a surface model of the projection surface, as well as extrinsic and intrinsic projector parameters. The user's head position was tracked with an HTC Vive tracker attached to a helmet worn by the user. That way, both binocular and kinematic depth cues were supported. Images based on the user's spatial position were rendered for each eye using the game engine Unity (Unity Technologies, USA). Then, these images were mapped onto the projection surface scan. Rendering that textured surface from the view of individual projectors resulted in the projection of undistorted, perspectivly correct images (see Fig. 2). GPU accelerated volume ray casting was performed to determine, which parts of the data were currently visible. Thereby, current windowing width and windowing level parameters mapped the intensity values of the used DICOM data set to the range 0 to 1. Afterwards these values were used to apply color and transparency via a transfer function. Furthermore, the visualized volume could be reduced using a clipping box.

2.3 Tasks

This work's exploration tasks were limited to the manipulation of windowing parameters and the position and size of a clipping box, because in spatial projective AR systems, tasks manipulating transformation or camera parameters no longer apply. To this end, a search task was implemented that had to be completed with either windowing or clipping techniques. For that, a foreign body, i.e. a screw, was inserted into the volume data. Position, size, orientation and intensity value of the object were varied between trials to avoid bias. Only one target was present for each trial. For the windowing task, window level and window width needed to be manipulated, so that the hidden search object became visible (see Fig. 3a–b). An object was considered visible, when the windowing parameters mapped its intensity value to a specific range of the used transfer function. After the target was found, windowing parameters were changed and a new target was inserted, until a total of four objects were detected. The clipping tasks required participants to change the position and size of a clipping box so that it encapsulated the search object as closely as possible without clipping the object itself (see Fig. 3c–d). Scaling and translation of the clipping box could be performed individually for each main axis. Both tasks are demonstrated in the supplementary video. Interaction techniques' mode activation methods were used to enable changing task-specific parameters. For the windowing task, only one method was needed to start changing both windowing parameters by vertical and horizontal movements, analogous to mouse input for 2D image slices. For the clipping task, individual activation gestures were used to enable either a change in clipping box size or position. Movements were then mapped to changes in respective main axis directions. These changes were always limited to the direction of greatest change, to avoid unwilling manipulation of more than one axis. As for foot-based interaction methods, the toe tap activation gesture needed to be performed consecutively to rotate between individual parameters.

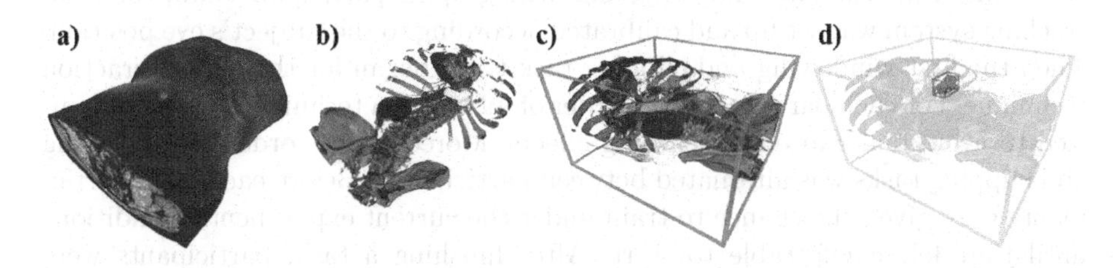

Fig. 3. Exploration task procedures. a)–b) Windowing task. c)–d) Clipping task.

2.4 Variables

Task completion time, total number of mode activations, i.e. how often the mode activation gestures were performed, and subjective workload were measured for both windowing and clipping tasks. Time measurement started after participants signalized their readiness and stopped when the individual task was considered finished. Subjective workload was estimated using the standardized raw

TLX questionnaire. For clipping tasks, accuracy was measured by calculating the Jaccard index between the user-set and a perfect clipping box. This resulted in accuracy values between 0 and 100 with higher percentages representing higher congruence. Participant performance during windowing tasks was further evaluated by a degree of efficiency that was defined as the relation between the smallest possible amount both windowing parameters needed to be changed and their total accumulated sum of changes. Results were interpreted as percentages with higher values representing higher efficiency.

Table 1. Summary of the ANOVAs' results ($\alpha < .05$).

Variable	df	F	p	Sig.	η^2	Effect	Figure
Windowing task							
Efficiency	3	2.24	0.088		0.063	Medium effect	Fig. 4a
Task completion time	3	12.40	< 0.001	*	0.271	Large effect	Fig. 4b
Mode switches	3	6.80	<0.001	*	0.169	Large effect	Fig. 4c
Subjective workload	3	5.65	0.001	*	0.145	Large effect	Fig. 4d
Clipping task							
Accuracy	3	0.47	0.705		0.014	Small effect	Fig. 5a
Task completion time	3	9.48	<0.001	*	0.221	Large effect	Fig. 5b
Mode switches	3	6.28	<0.001	*	0.159	Large effect	Fig. 5c
Subjective workload	3	8.68	<0.001	*	0.207	Large effect	Fig. 5d

2.5 Procedure

After initial instructions and collecting demographic participant data, the head tracking system was set up and calibrated according to the subject's eye position. Then the first windowing and clipping task trials began for the first interaction technique. For each participant, the order of interaction techniques were randomized to avoid bias caused by learning effects. Moreover, the order of windowing and clipping tasks was alternated between participants. Before each task, participants were given the chance to train under the current experimental conditions until they felt comfortable to start. After finishing a task, participants were asked to fill out the raw TLX questionnaire. The next interaction technique was evaluated after both tasks were fulfilled. The experiment concluded with a brief informal questionnaire about participant feedback.

3 Results

Data for the windowing and for the clipping tasks were analyzed separately. One-way ANOVAs were conducted for each task-related variable to investigate effects of the interaction technique factor. Their results are summarized by Table 1. Afterwards, post-hoc tests were conducted to analyze individual differences between factor levels. To this end, pairwise t-tests with Bonferroni correction were performed. Figure 4 and Fig. 5 visualize identified effects, as well as statistically significant post-hoc test results.

4 Discussion and Conclusion

Participants were able to fulfill both exploration tasks using all interaction techniques, as supported by similar accuracy and efficiency results. However, differences between concepts are shown for the other variables. Foot gestures generally performed worst, as indicated by higher subjective workload and longer task completion times. This may have been because the concept only allowed for the manipulation of one DoF at the same time. As a result, the concept required more mode activations for the windowing task than other methods. Moreover, some participants reported problems in successfully performing the toe tap activation gesture and found the heel rotation gesture to be exhausting, which is partially reflected by a higher subjective workload. Hand gestures performed better than foot gestures, but not as well as controller-based interaction. These differences may have been caused by the concept only mapping relative hand movement to parameters, as opposed by global hand movement implemented by the instrument and controller concepts. As a result, movement axes and changed parameter dimensions not always coincided. Implemented mode activation gestures may

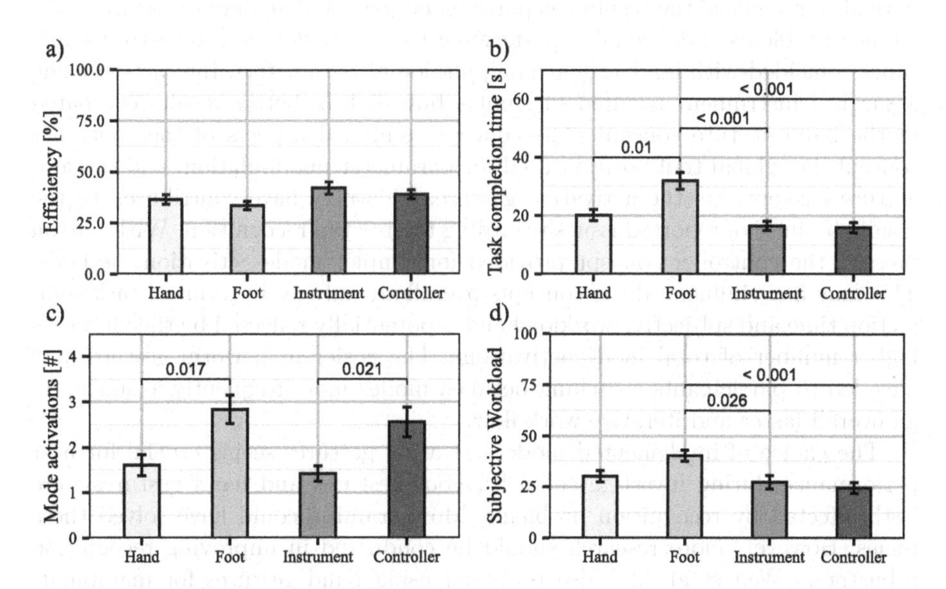

Fig. 4. Windowing task results. Effects of interaction techniques on a) efficiency, b) task completion time*, c) mode activations* and d) subjective workload*. Error bars respresent standard errors. Horizontal lines indicate statistically significant post-hoc test results.

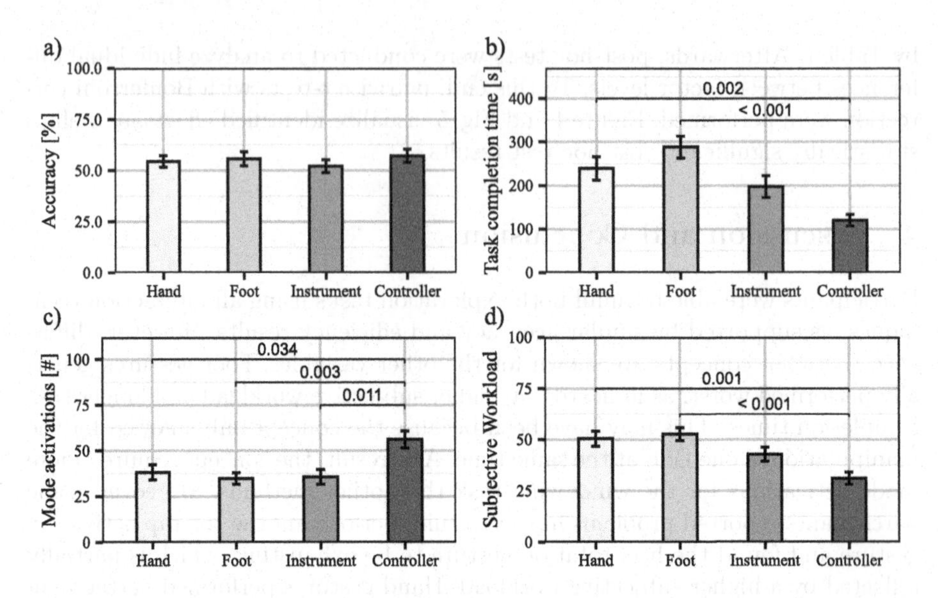

Fig. 5. Clipping task results. Effects of interaction techniques on a) accuracy, b) task completion time*, c) mode activations* and d) subjective workload*. Error bars respresent standard errors. Horizontal lines indicate statistically significant post-hoc test results.

have also influenced the results, as participants reported gesture recognition and latency problems. Additionally, performing the finger double tap gesture sometimes coincided with inadvertent manipulation of parameters. Interaction using a surgical instrument resulted in similar but slightly better results compared to the hand gesture concept. The concept combined aspects of the controller concept, i.e. global tool position used for parameter manipulation, and the foot gesture concept, i.e. the activation gesture. This may have contributed to the results because of reported issues regarding the toe tap recognition. With button presses, the controller concept provided for simpler mode activation methods. This may have influenced the concepts overall advantages in terms of task completion time and subjective workload and is potentially reflected by the method's higher number of total mode activations. The easier to perform gesture may have led to participants switching between modes more frequently, resulting in an overall faster and iterative work flow.

The choice of implemented mode activation gestures seems crucial for user performance during investigated tasks. Foot gestures and hand gestures were both affected by recognition problems. More training could have solved these issues. However, more research should be conducted in improving recognition robustness. Wen et al. [24] also proposed using hand gestures for manipulating projected radiological 3d images. They used a different set of gestures and tracking hardware, that may have performed differently in this work's evaluation. Future research could, therefore, investigate effects of different gestures and recognition systems. Since the foot gesture concept performed worst in the user study, linear parameters should be rather modified using hand gestures or hand-controlled devices because of more controllable DoF and less exhaustion. In these cases, foot gestures seem to more suited than hand gestures for mode activations because of possible inadvertent modification of linear parameters controlled by the same hand. Yet, future work should evaluate different alternatives to the evaluated toe tap. Voice commands were not included in this experiment because of low expected speech recognition rates in noisy operating rooms. However, they still may be viable mode activation alternatives [13]. Thus, voice control may complement identified limitations of the current setup and should, therefore, be investigated in a future iteration of this work.

A meaningful continuation of this work would also be a comparison of investigated methods and conventionally used approaches to interact with medical image data, i.e. mouse input and visualizations on a monitor. Compared to conventional mouse interaction, using hand gestures was shown to be advantageous for the rotation of 3D models in terms of speed and precision [12]. Hettig et al. [9] compared different hand gestures detected by the Leap Motion controller (Ultraleap, UK & USA) and the Myo armband with the conventional methods of joystick input and verbal task delegation for rotation and navigation tasks. They found that both evaluated touchless interaction methods were viable alternatives for the conventional approaches. Similar comparisons could be conducted regarding this work's methods.

The present study was conducted under laboratory conditions only. The integration of used hardware systems in clinical environments is dependent on spatial and sterility conditions. The proposed controller-based method entails some sterility issues, similar to those of mice and keyboards in the operating room. Initial findings suggest, that these problems may be solved by wrapping the controller in a sterile plastic bag. However, potential resulting performance issues would need to be examined in future research. The instrument-based method is dependent on external tracking systems. The currently used, cumbersome HTC Vive tracker would need to be replaced by different techniques. Surgical navigation systems usually involve infrared-based optical tracking cameras, which could be used for this method, as well [17]. Additionally, the implemented head tracking method using a head-worn helmet would need to be replaced for clinical environments, as well. Gierwialo et al. [5] describe an alternative solution requiring only a small marker attached to a headband, that could be more easily integrated into surgical workflow. Also, markerless methods using depth cameras, like the Microsoft Kinect sensor (Microsoft Corporation, USA) could be implemented [12,24]. Finally, dynamic patient registration methods need to be integrated to the system to correctly superimpose projected images on moving patients. Tissue deformation and respiratory movement models would also be meaningful additions. However, this topic was not focused on in this work, because first research objectives only included interaction and visualization aspects. Therefore, more research is needed to evaluate the applicability of developed techniques for the intraoperative use.

In conclusion, this work represents an important step towards the development of interaction techniques for surgical AR and has the potential to foster the clinical integration and acceptance of advanced AR visualization techniques in the operating room.

Acknowledgments. This work was funded by the German Research Foundation (HA 7819/1-2 & LA 3855/1-2).

References

1. D'Agostino, J., Diana, M., Vix, M., Soler, L., Marescaux, J.: Three-dimensional virtual neck exploration before parathyroidectomy. New Engl. J. Med. **367**(11), 1072–1073 (2012)
2. Egger, J., et al.: HTC Vive MeVisLab integration via OpenVR for medical applications. PloS One **12**(3), e0173972 (2017)
3. Gallo, L.: A study on the degrees of freedom in touchless interaction. In: SIGGRAPH Asia 2013 Technical Briefs, p. 28. ACM (2013)
4. Gallo, L., De Pietro, G., Marra, I.: 3D interaction with volumetric medical data: experiencing the wiimote. In: Proceedings of the 1st International Conference on Ambient Media and Systems. Ambi-Sys '08, ICST (Institute for Computer Sciences, Social-Informatics and Telecommunications Engineering), Brussels, BEL (2008)
5. Gierwiało, R., Witkowski, M., Kosieradzki, M., Lisik, W., Groszkowski, Ł., Sitnik, R.: Medical augmented-reality visualizer for surgical training and education in medicine. Appl. Sci. **9**(13), 2732 (2019)
6. Goth, G.: Brave NUI world. Commun. ACM **54**(12), 14–16 (2011)
7. Hatscher, B., Luz, M., Hansen, C.: Foot interaction concepts to support radiological interventions. i-com **17**(1), 3–13 (2018)
8. Hettig, J., Mewes, A., Riabikin, O., Skalej, M., Preim, B., Hansen, C.: Exploration of 3D medical image data for interventional radiology using myoelectric gesture control. In: Proceedings of the Eurographics Workshop on Visual Computing for Biology and Medicine, pp. 177–185. Eurographics Association (2015)
9. Hettig, J., Saalfeld, P., Luz, M., Becker, M., Skalej, M., Hansen, C.: Comparison of gesture and conventional interaction techniques for interventional neuroradiology. Int. J. Comput. Assist. Radiol. Surg. **12**(9), 1643–1653 (2017)
10. Jalaliniya, S., Smith, J., Sousa, M., Büthe, L., Pederson, T.: Touch-less interaction with medical images using hand & foot gestures. In: Proceedings of the 2013 ACM Conference on Pervasive and Ubiquitous Computing Adjunct Publication, pp. 1265–1274 (2013)
11. Jankowski, J., Hachet, M.: A survey of interaction techniques for interactive 3D environments. In: Eurographics (2013)
12. Kirmizibayrak, C., Radeva, N., Wakid, M., Philbeck, J., Sibert, J., Hahn, J.: Evaluation of gesture based interfaces for medical volume visualization tasks. In: Proceedings of the 10th International Conference on Virtual Reality Continuum and its Applications in Industry, pp. 69–74. ACM (2011)
13. Mentis, H.M., et al.: Voice or gesture in the operating room. In: Proceedings of the 33rd Annual ACM Conference Extended Abstracts on Human Factors in Computing Systems, pp. 773–780. ACM (2015)

14. Mewes, A., Hensen, B., Wacker, F., Hansen, C.: Touchless interaction with software in interventional radiology and surgery: a systematic literature review. Int. J. Comput. Assist. Radiol. Surg. **12**(2), 291–305 (2017)
15. Mewes, A., Saalfeld, P., Riabikin, O., Skalej, M., Hansen, C.: A gesture-controlled projection display for CT-guided interventions. Int. J. Comput. Assist. Radiol. Surg. **11**(1), 157–164 (2016)
16. Nicolau, S., Soler, L., Mutter, D., Marescaux, J.: Augmented reality in laparoscopic surgical oncology. Surg. Oncol. **20**(3), 189–201 (2011)
17. de Oliveira, M.E., Debarba, H.G., Lädermann, A., Chagué, S., Charbonnier, C.: A hand-eye calibration method for augmented reality applied to computer-assisted orthopedic surgery. Int. J. Med. Robot. Comput. Assist. Surg. **15**, e1969 (2019). https://doi.org/10.1002/rcs.1969
18. Pfeiffer, M., et al.: IMHOTEP: virtual reality framework for surgical applications. Int. J. Comput. Assist. Radiol. Surg. **13**(5), 741–748 (2018)
19. Silén, C., Wirell, S., Kvist, J., Nylander, E., Smedby, Ö.: Advanced 3D visualization in student-centred medical education. Med. Teach. **30**(5), e115–e124 (2008)
20. Silva, É.S., Rodrigues, M.A.F.: Design and evaluation of a gesture-controlled system for interactive manipulation of medical images and 3D models. SBC J. Interact. Syst. **5**(3), 53–65 (2014)
21. Sugimoto, M., et al.: Image overlay navigation by markerless surface registration in gastrointestinal, hepatobiliary and pancreatic surgery. J. Hep.-Biliary-Pancreat. Sci. **17**(5), 629–636 (2010)
22. Thomas, R.G., William John, N., Delieu, J.M.: Augmented reality for anatomical education. J. Vis. Commun. Med. **33**(1), 6–15 (2010)
23. Velloso, E., Schmidt, D., Alexander, J., Gellersen, H., Bulling, A.: The feet in human-computer interaction: a survey of foot-based interaction. ACM Comput. Surv. (CSUR) **48**(2), 21 (2015)
24. Wen, R., Nguyen, B.P., Chng, C.B., Chui, C.K.: In situ spatial AR surgical planning using projector-kinect system. In: Proceedings of the Fourth Symposium on Information and Communication Technology, pp. 164–171 (2013)

VR Simulation of Novel Hands-Free Interaction Concepts for Surgical Robotic Visualization Systems

Fang You[1,3(✉)], Rutvik Khakhar[2], Thomas Picht[2], and David Dobbelstein[1]

[1] Carl Zeiss AG, Carl-Zeiss-Strasse 22, 73447 Oberkochen, Germany
fang.you@zeiss.com, david.dobbelstein@zeiss.com

[2] Charité Universitätsmedizin Berlin, Charitéplatz 1, 10117 Berlin, Germany
rutvik.khakhar@charite.de, thomas.picht@charite.de

[3] Institute for Anthropomatics and Robotics - Intelligent Process Automation and Robotics Lab (IAR-IPR), Karlsruhe Institute of Technology (KIT), Karlsruhe, Germany
fang.you@partner.kit.edu

Abstract. In microsurgery, visualization systems such as the traditional surgical microscope are essential, as surgeons rely on the highly magnified stereoscopic view for performing their operative tasks. For well-aligned visual perspectives onto the operating field during surgery, precise adjustments of the positioning of the system are frequently required. This, however, implies that the surgeon has to reach for the device and each time remove their hand(s) from the operating field, i.e. a *disruptive event* to the operative task at hand. To address this, we propose two novel *hands-free* interaction concepts based on *head*-, and *gaze-tracking*, that should allow surgeons to efficiently control the 6D positioning of a robotic visualization system with little interruptions to the main operative task. The new concepts were purely simulated in a virtual reality (VR) environment using a HTC Vive for a *robotic* visualization system. The two interaction concepts were evaluated within the virtual reality simulation in a quantitative user study with 11 neurosurgeons at the *Charité* hospital and compared to conventional interaction with a surgical microscope. After a brief introduction to the interaction concepts in the virtual reality simulation, neurosurgeons were 29% faster in reaching a set of virtual targets (position and orientation) in simulation as compared to reaching equivalent physical targets on a 3D-printed reference object.

Keywords: Microsurgery · Virtual reality · Human-robot-interaction · Hands-free interaction · Intraoperative visualization and guidance

1 Introduction

Visualization systems, such as surgical microscopes, are necessary in microsurgery during almost the entire surgical operation as they provide the required magnification and illumination of the operating field [14] (Fig. 1). Surgeons are intensively trained to precisely coordinate their hand (and instrument) movements while looking at the magnified

Supported by BMBF (German Federal Ministry of Education and Research).

A. L. Martel et al. (Eds.): MICCAI 2020, LNCS 12263, pp. 440–450, 2020.
https://doi.org/10.1007/978-3-030-59716-0_42

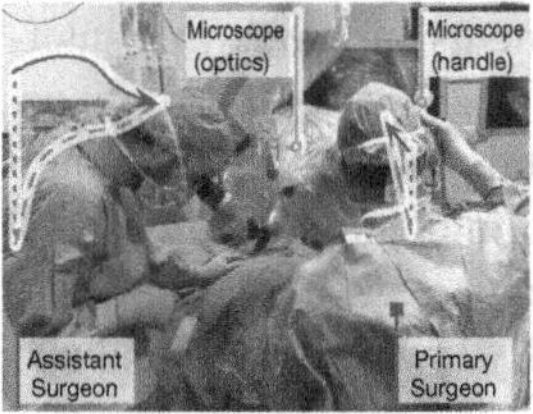

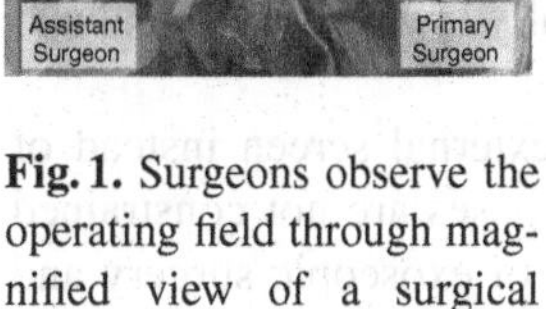

Fig. 1. Surgeons observe the operating field through magnified view of a surgical microscope with oculars.

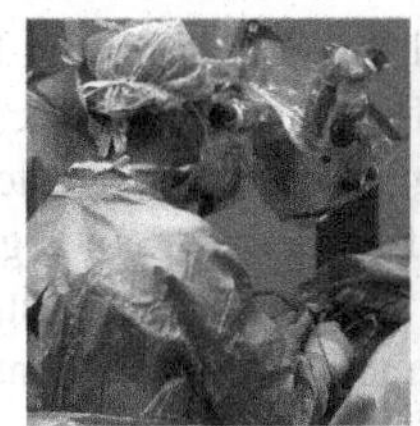
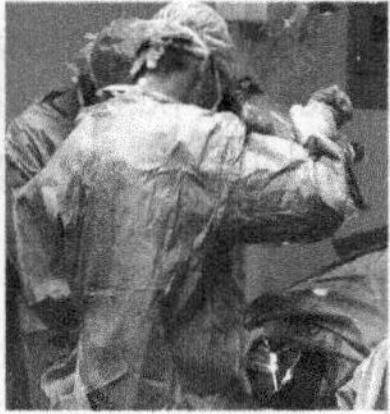

Fig. 2. The conventional adjustment to surgical microscope's FoV (position, orientation, magnification) requires the surgeon to reach for the handle and to physically interact with the device, which each time interrupts the operative task.

stereoscopic visualization. Due to the high magnification, however, only a limited view of the operating field can be provided at once, so that the perspective of the visualization system needs to be frequently adjusted (Fig. 2). This, however, each time requires the surgeon to briefly interrupt their surgical main task to reach for the system's handles, increasing the total operation time by up to 10% [9]. As these interruptions take away valuable operating time, they are often avoided by the operating surgeon despite risks introduced by working under poor visibility, e.g. at the edge of the field of view (FoV), or in unfocused areas [9].

We envision that in the future, operating surgeons can adjust the configuration of a **robotic visualization system** (**RVS**) with high precision in a *hands-free* manner [3], so that the surgeon can quickly switch between operative task and interaction with the device as they do not need to remove their hands from the operating field. Furthermore, by the introduction of quickly attainable *micro-interactions* [4], that barely interrupt the operative task, we anticipate that adjustments of the device will be performed more frequently to optimize viewing configurations over the course of varying surgical subtasks.

To achieve this, we studied the use case of surgical microscopes as a reference for visualization systems in microsurgery and simulated a RVS in virtual reality (VR). Through an iterative design process, we developed two novel interaction concepts for surgeons to control the FoV of the simulated RVS without using hands. The two novel concepts are based on a multi-modal combination of *head*-tracking, *gaze*-tracking and voice commands to interactively adjust the RVS, such as its mechanical positioning with six degrees of freedom (6DoF: 3D translation + 3D rotation), as well as its optical magnification (zoom).

Our contributions are: (1) two novel *hands-free* interaction concepts to control a robotic surgical visualization system, and (2) a user study with neurosurgeons investigating the efficiency of our proposed interaction concepts in comparison to the state-of-the-art.

2 Related Work

Eivazi et el. [9] analyzed surgical workflows and found that interruptions caused by the necessary interaction with a surgical microscope prolong the operating time by up to 10%. Afkari et el. [3] discussed the potentials and benefits of *hands-free* interaction with surgical microscopes for efficiency, ergonomics and avoiding interruptions. The paper discussed 4 hands-free surgical microscope interaction methods: via mouth switch (limited maneuverability), via foot pedal (poor ergonomic and poor mobility), via voice command (limited usability for complex adjustments, e.g. 3D motion) and gaze-tracking. It concluded that gaze-tracking could potentially lower the physical and cognitive work load.

Exoscopic visualization, i.e. providing images on an external screen instead of using oculars, can improve ergonomics, as surgeons' body poses are not constrained by the optical system. Rothe et al. [19] assessed the impact of exoscopic surgery and observed its feasibility and improvements of ergonomics. However, the hand-eye coordination for exoscopic visualization reportedly demands a higher cognitive workload [15,22].

Head-mounted displays (HMDs) are mostly envisioned as wearable interfaces that enable for a pervasive access to information [20,21] and for immersive visualizations within the environment [6]. While this is beneficial in the mostly *mobile* scenarios of wearable computing, using HMDs can also offer benefits in the *stationary* medical setting of microsurgery. Rendering the magnified view of a surgical microscope into a HMD can enable surgeons to decouple their body from the physical device and to maintain an ergonomic body posture throughout surgery. Furthermore, with sensory for the 6D pose of the head and an incorporated gaze-tracker, the user's head motion and gaze direction can be utilized for manipulating a RVS.

Head-Tracking has shown to be feasible as an interaction input for controlling robotic motion [16] and graphical interfaces [18]. Head-*motion* as a means of user input have already been explored to control a 2D cursor via head-*pointing* [5], as utilized with the Microsoft HoloLens [2], or via head-*tilting* [8], e.g. for Google Glass to wake up the embedded near-eye display [1]. Zinchenko et el. [23] has designed a VR-based control system for a robotic endoscope holder for 3D movements. The 6DoF-controlling of a RVS, however, demands for more complex head-based interaction concepts.

Gaze-tracking interaction has been widely considered for stationary and mobile interfaces [7] as it enables a quick and hands-free pointing modality. The strong link of gaze and the user's attention made it especially suited for attentive user interfaces [7, 13]. In the context of surgery, the tracked gaze point can indicate surgeon's current point of interest (POI) within the operating field. In the literature, eye-tracking systems have been integrated into traditional oculars of stereoscopic microscopes [10] to analyze the surgeon's gaze behavior [11]. It has also been used for moving a robot-held endoscope [12,13].

Our proposed interaction techniques mainly focus on controlling precise and frequent micro-adjustments to the visualization systems in micro-neurosurgery via hands-free interaction with 6DoF. We started with a workflow analysis of five neurosurgical procedures at the Charité hospital, where we could confirm Eivazi et al.'s [9] observations of frequent disruptive events during surgery by the necessity of the surgeon interrupting the surgical main task to manually reach for the microscope's handles. These interruptions occurred on average every 67 s.

3 Novel Hands-Free Interaction Concepts

In this work, we propose two novel hands-free interaction concepts that make use of a head-mounted display for adjusting the perspective of a RVS by utilizing the surgeon's 6D head pose for translation, orientation and optical magnification, and by utilizing the surgeon's gaze point for 2D lateral translation. The interaction techniques are triggered by voice commands respectively. We build upon the conceptual benefits of modern HMDs for virtual reality (VR) to visualize the RVS's imaging as a *virtual screen* floating above the operating field. This enables the surgeon to look at the virtual screen, i.e. the magnified perspective provided by the RVS, during surgery. Our proposed novel interaction concepts enable the surgeon to adjust the perspective of the RVS by hands-free interaction with the virtual screen. The novel concepts were purely developed via a VR-simulation, which allowed for iterative prototyping and refinement of the concepts with surgeons as test users.

3.1 Head-Positioning

When looking at the virtual screen, the surgeon implicitly perceives a visual perspective onto the magnified operating field. With the new concept of *head-positioning*, we allow the surgeon to naturally change their visual perspective in 6D regarding the virtual screen, by moving their head; this results in a positional adjustment of the visualization system.

A naive 1:1 mapping of head-motion to camera motion yields an non-immersive user experience as the RVS perspective and its movements would not align with the surgeon's perspective: While the surgeon visually perceives to be very close to the operating scene (due to the high magnification), the camera of the RVS is not. Depending on the zoom level, the RVS's motion is disproportionally perceived by the surgeon within the image.

To assure visual immersion during head-positioning, the calculation of the RVS adjustments considers the magnification and surgeon's perspective (6D HMD pose) with regard to the virtual screen. The intended perspective adjustment is estimated regarding the surgeon's visual perception of the 3D scene within the virtual screen. In

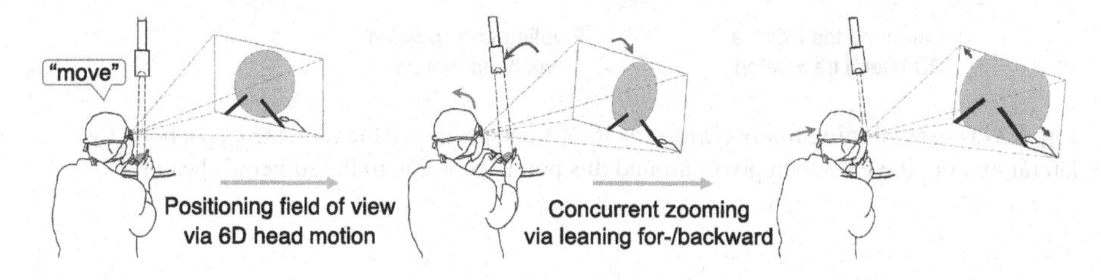

Fig. 3. Via **head-positioning**, the RVS's perspective is adjusted according to the surgeon's view point onto the virtual screen. E.g. when the surgeon is leaning towards the left, the RVS will in real-time adjust the RVS positioning to provide the visual perspective that the surgeon would see from the respective new view point. By this, the surgeon can immersively look around the operating field.

the iterative design process, this showed to greatly benefit the surgeon's predictability of movements.

The virtual screen is programmed to face towards the surgeon, so that it can be controlled as a responsive object. As an effect, the surgeon perceives the virtual screen as a *window* into the magnified operating field that dynamically follows the surgeon's motion and provides a natural change in perspective (Fig. 3). In addition, the surgeon can simultaneously zoom into or out of the scene by moving their head forward, resp. backward. Due to providing a natural change in visual perspective, we expect surgeon's to have a very fast learning curve for head-positioning and a low cognitive demand during interaction.

3.2 Gaze-Positioning

By using the gaze-tracking capabilities of an HMD integrated eye-tracker, we can infer on the surgeon's current POI within the operating field when they look at the virtual screen. We utilize this for positioning the RVS by centering the provided perspective around the surgeon's gaze point. In the iterative design process, surgeons emphasized the importance of observing the POI from varying angles. To allow to rotate the image around this POI as well, the *gaze-positioning* consists of two phases (Fig. 4): 1) upon the voice command 'center', the current gaze point is visually centered by a 2D lateral movement of the RVS in its image plane, and 2) once centered, the RVS will perform a pivot-movement, i.e. a rotation, around the reached POI, based on the surgeon tilting their head into the respective direction. As with head-positioning, the surgeon can simultaneously zoom into or out of the scene by moving their head forward, resp. backward. While the centering phase is triggered once via voice command, the pivoting-adjustment via head-rotation is available until the user stops the interaction via another voice command, e.g. 'stop'. The gaze point is not visualized to not distract the surgeon.

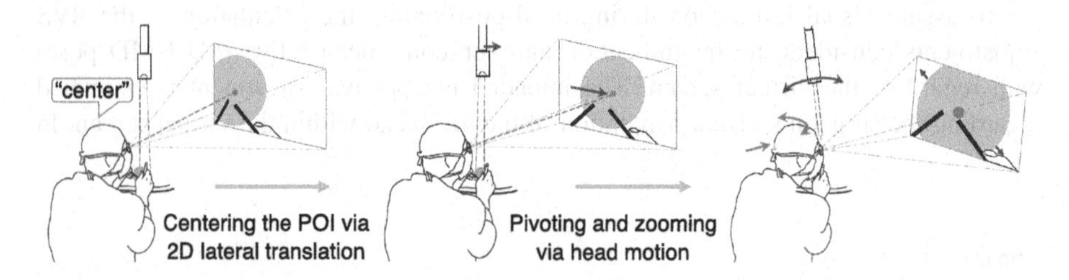

Fig. 4. Gaze-positioning firstly centers the RVS's perspective to the estimated gaze point by a 2D lateral motion, the RVS then pivots around this point according to the surgeon's head rotation.

3.3 VR - Simulation of the Novel Hands-Free Interaction Concepts

We implemented the interaction concepts as a simulated setup in VR with Unity3D. This enabled us to quickly prototype and investigate the proposed concepts in multiple sessions with surgeons at the neurosurgical department at Charité. Furthermore, it enabled us to conduct a user-centered design process to iteratively optimize the system's parameters and behavior by working closely together with surgeons as test users. We implemented the system as a VR-simulation, including the dynamic behavior of the RVS and the *virtual screen*.

In the rendered VR scene, a simulated stereo RVS is placed, and the 6D pose of its optical center is known and controlled by the proposed interaction concepts. According to the optical configuration, the 6D pose and the scale of RVS's field of view (FoV) is known and controllable through interaction. To visualize the stereo images of the RVS over the operating field, a stereo virtual screen is displayed in space, i.e. floating in front of the surgeon. The behavior of the display is dynamic: when the surgeon is operating with the stationary RVS, the screen stays stationary above the operational field; albeit, when an interaction is activated, the screen interactively moves according to the surgeon with a fixed offset. Its pose and scale are controlled during interaction.

A VR-HMD (HTC Vive Pro) was used for immersive visualization of the scene. The VR-HMD provides 6D pose data of the user's head in the virtual scene, and with the integrated eye-tracker (Pupil Labs [17]), the gaze point on the virtual screen is estimated.

Head-Positioning Interaction: The interaction of head-positioning maps and applies the real-time head motion (6D head pose) for adjusting the RVS. The calculation interprets the relative head motion as the intention of relatively adjusting the FoV. To achieve an immersive interaction, the pose of the HMD, the virtual screen (with scale) and the magnified FoV (with scale) are taken into account. The virtual screen follows the HMD with a fixed offset as soon as an interaction is activated. During *head-positioning*, the virtual screen is moved by the surgeon's head motion. As the virtual screen corresponds to the FoV of RVS, we apply screen motion to the FoV. In addition, the magnification (zooming) is controlled by the distance of the HMD to the virtual screen. Therefore, head-positioning utilizes the user's head motions to control the virtual screen's movement; By applying screen's movement to the RVS's FoV, the to-be-acquired target configuration of the RVS's camera is determined.

Gaze-Positioning Interaction: The integrated eye-tracker estimates gaze direction of the surgeon regarding the HMD. As the 6D pose of the HMD with regard to the virtual screen is known, the gaze position on the screen can be estimated. The relation between the screen and the estimated gaze point corresponds to the relation between the FoV and the observed point of interest (POI) in the operating field. The relation between the screen and the gaze indicates the intention to adjust the FoV (2D lateral translation). Upon activating the gaze-positioning, the RVS moves to put the estimated POI to the center of the FoV, equivalently moving the previous gaze point to the center of the virtual screen. The pivot movement after centering the POI is controlled by head rotation. The position of the FoV is fixed to the previously arrived POI in the mean time. The magnification of the RVS is handled the same as with head-positioning, i.e. by the distance of the user's head to the virtual screen.

4 User Studies for Evaluating Interaction Concepts

To investigate the efficiency of the newly introduced interaction concepts, we conduced a quantitative user study, where neurosurgeons would perform a target selection task over the course of two sessions: 1) in the VR-simulated setup using *head-* and *gaze-positioning* and 2) with a physical setup and *hand-positioning* (i.e. interaction via the handles) of a traditional surgical microscope as a baseline comparison.

A set of **target selection tasks** was designed to simulate operative tasks that would require adjustments to the visualization system. As shown in Fig. 5, the tasks require surgeons to touch 25 individual targets with a surgical instrument, where each target (2 mm diameter) is hidden in a tube (8 mm diameter, 10 mm height). The geometric arrangement of the tubes conceal the target so that the surgeon would need to adjust the RVS's perspective to visually expose each individual target, as the operative task requires visibility. By this, the target selection would entail a primary operative task by hand (i.e. by surgical instrument), and a secondary interaction task of adjusting the RVS.

The tubes were tilted for a maximal x- and y- rotation of 15° and 25° respectively, so that an orientation adjustment up to 58.2° was required between tasks. The sequence of task positions was randomized for each participant. The tube containing the active pointing target at its center is highlighted. When a task was completed, the next pointing target would be highlighted immediately. The tasks are identical for the VR- and for the physical-session of interaction.

VR-Session with Hands-Free Interaction: In the VR-session (Fig. 5 left) the targets are rendered in front of the surgeon. The VR-controller held in hand is rendered as a virtual surgical instrument to touch the targets. The neurosurgeons conducted the operative target selection task with *head-* and *gaze-positioning* following a within-subject design with alternating order. Each task session of 25 targets was preceded by a training session, where each participant would get familiar with the new interaction concept for a maximum of 5 min. After both task sessions were completed, we conducted an open discussion of the surgeon's user experience of the interaction. The task selection times as well as the surgeon's motion were logged within Unity.

Physical-Session with Hand-Positioning via Handles: The physical-session was conducted 2 months after the VR-session with the same participants. They conducted the identical target selection tasks on a 3D-printed target probe with a surgical instrument (Fig. 5 right), while looking through a traditional surgical microscope that they would then need to adjust with their hands to visually expose individual targets. An Arduino micro-controller is incorporated for the touch sensory, target highlighting and data logging.

4.1 Results and Discussion

11 neurosurgeons (4 attendings, 7 residents) at Charité hospital conducted the user study for both sessions. Initially, 14 neurosurgeons were recruited, but we had to exclude 3 participants as their worn glasses were incompatible to the VR-HMD. All participants stated to have no previous experience with hands-free interaction and HMDs.

Target Selection Time: Regarding time efficiency (see Fig. 6), the surgeons were more efficient with the 2 proposed hands-free interaction concepts than with the hand-based interaction of a physical device. With *head-positioning* (6.79 ± 2.35 s) surgeons

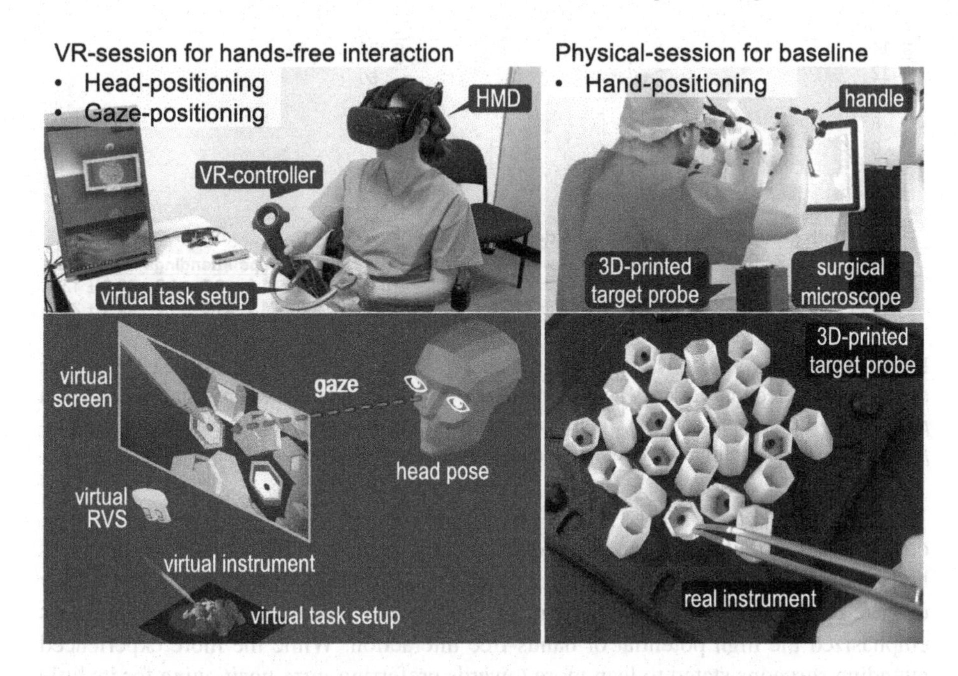

Fig. 5. Surgeons conduct the target selection tasks as: a VR-sessions with the new hands-free interaction concepts (left), and a physical-session with hand-positioning via handles on a traditional surgical microscope (right). In each session they select 25 individual targets that would require adjustment of camera perspective. The task setups share the identical geometry and dimensions. Each randomly activated target is highlighted in the VR-session via color, and in the physical-session by an LED. (Color figure online)

were 29% faster than with conventional hand-positioning via handles (9.57 ± 2.90 s), whereas with *gaze-positioning* (7.84 ± 2.76 s) they were 18% faster. The difference was higher when only taking the more experienced attending surgeons into account: they were 36% faster with *head-positioning*, and 19% with *gaze-positioning*.

Head Motion: The efficiency in movement is measured by integrating the path length of head translation (see Fig. 7). The identical tasks are achieved with 55% less head motion by *gaze-* (16.0 ± 11.9 cm) than by *head-positioning* (35.2 ± 24.2 cm). Overall, attending surgeons required about 50% of body-motion as resident surgeons for both resp. techniques (head: M = 21.3 cm; M = 42.1 cm resp.; gaze: M = 9.3 cm; M = 19.3 cm resp.).

Discussion: According to the interviews with the surgeons, the increase in efficiency of the new techniques compared to traditional interaction was largely due to the decrease of interruption time and the capability for simultaneous execution of operative and interaction tasks, avoiding the task switching time. The motion mapping of the virtual screen to the FoV showed to be effective, as the participants acknowledged the immersion of head-positioning and were able to control the system after only a brief learning phase. The centering phase of *gaze-positioning* showed to be beneficial as no continu-

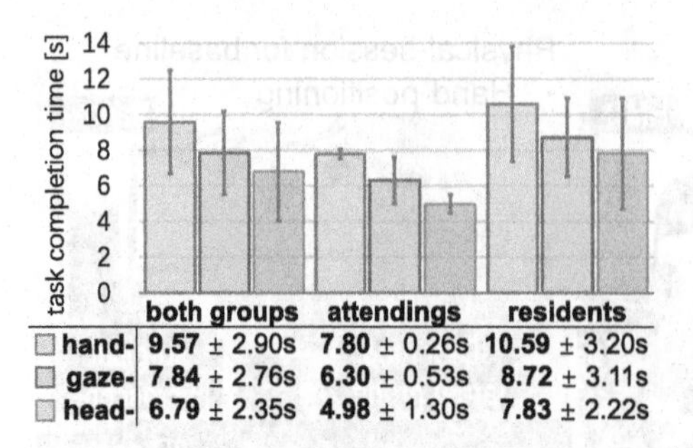

Fig. 6. Participants were 29% faster for the operative target selection tasks when using *head-positioning* in comparison to traditional *hand-positioning* for controlling the visualization system.

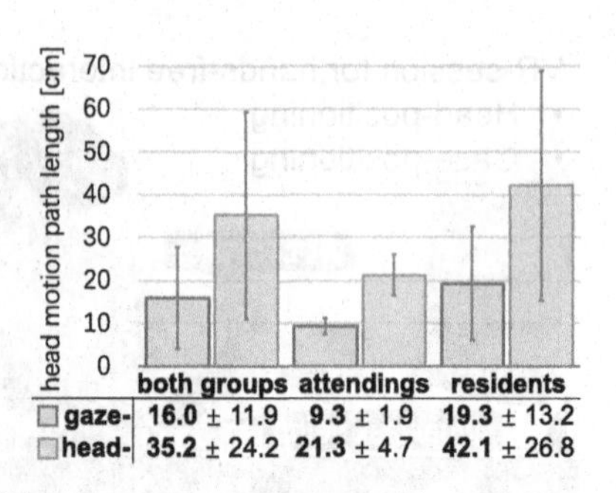

Fig. 7. The participants used 55% less head motion when using *gaze-* than *head-positioning*, measured by translation distances.

ous attention of the interaction was required. Participants also liked that it required less body movement by utilizing their gaze point. As a limitation, however, the inaccuracy of gaze-estimations can be a problem for very precise alignments. Overall, surgeons emphasized the high potential of hands-free interaction. While the more experienced attending surgeons stated to lean more towards preferring *gaze-positioning* for its little requirement in body-motion, resident surgeons leaned more towards preferring *head-positioning* for its ease to learn.

5 Conclusion and Future Work

In conclusion, two novel hands-free interaction concepts (*head-* and *gaze-positioning*) for controlling a robotic visualization system were presented and investigated in a VR-simulation. We conducted a user study with neurosurgeons at the Charité hospital and found that the novel techniques were 29%, resp. 18%, faster in comparison to positioning via handles on a traditional surgical microscope for a target selection task that required adjustment of the visualization system. The *hands-free* interaction concepts enabled surgeons to very quickly switch their tasks between operating and adjusting the visualization system, as their hands didn't have to be removed from the operational scene.

The participating surgeons acknowledged the high potential of hands-free interaction for the clinical workflow. For this, however, some safety aspects, such as avoiding potential collisions of the RVS with the surgeon or patient, still need to be addressed.

In the future, we further want to investigate whether our hands-free interaction concepts lead to more *frequent* adjustments due to little interruption time of the main operative task.

References

1. Google Glass: Glass gestures (2014). https://support.google.com/glass/answer/3064184?hl=en. Accessed 14 Feb 2019

2. Microsoft Hololens: Gaze (2016). https://docs.microsoft.com/en-us/windows/mixed-reality/gaze. Accessed 14 Feb 2019
3. Afkari, H., Eivazi, S., Bednarik, R., Mäkelä, S.: The potentials for hands-free interaction in micro-neurosurgery. In: Proceedings of the 8th Nordic Conference on Human-Computer Interaction: Fun, Fast, Foundational, pp. 401–410. ACM (2014)
4. Ashbrook, D.L.: Enabling mobile microinteractions. Ph.D. thesis, Georgia Institute of Technology (2010)
5. Bérard, F.: The perceptual window: head motion as a new input stream. In: INTERACT, pp. 238–237. Citeseer (1999)
6. Billinghurst, M., Starner, T.: Wearable devices: new ways to manage information. Computer **32**(1), 57–64 (1999)
7. Bulling, A., Gellersen, H.: Toward mobile eye-based human-computer interaction. IEEE Pervasive Comput. **4**, 8–12 (2010)
8. Crossan, A., McGill, M., Brewster, S., Murray-Smith, R.: Head tilting for interaction in mobile contexts. In: Proceedings of the 11th International Conference on Human-Computer Interaction with Mobile Devices and Services, p. 6. ACM (2009)
9. Eivazi, S., Afkari, H., Bednarik, R., Leinonen, V., Tukiainen, M., Jääskeläinen, J.E.: Analysis of disruptive events and precarious situations caused by interaction with neurosurgical microscope. Acta Neurochirurgica **157**(7), 1147–1154 (2015). https://doi.org/10.1007/s00701-015-2433-5
10. Eivazi, S., Bednarik, R., Leinonen, V., von und zu Fraunberg, M., Jääskeläinen, J.E.: Embedding an eye tracker into a surgical microscope: requirements, design, and implementation. IEEE Sensors J. **16**(7), 2070–2078 (2016)
11. Eivazi, S., Fuhl, W., Kasneci, E.: Towards intelligent surgical microscope: micro-surgeons' gaze and instrument tracking. In: Proceedings of the 22nd International Conference on Intelligent User Interfaces Companion, pp. 69–72. ACM (2017)
12. Fujii, K., Gras, G., Salerno, A., Yang, G.Z.: Gaze gesture based human robot interaction for laparoscopic surgery. Med. Image Anal. **44**, 196–214 (2018). https://doi.org/10.1016/j.media.2017.11.011
13. Gras, G., Yang, G.Z.: Intention recognition for gaze controlled robotic minimally invasive laser ablation. In: 2016 IEEE/RSJ International Conference on Intelligent Robots and Systems (IROS), pp. 2431–2437, October 2016. https://doi.org/10.1109/IROS.2016.7759379
14. Hernesniemi, J., et al.: Some collected principles of microneurosurgery: simple and fast, while preserving normal anatomy: a review. Surg. Neurol. **64**(3), 195–200 (2005)
15. Hosseini, S.M.H., et al.: Neural, physiological, and behavioral correlates of visuomotor cognitive load. Sci. Rep. **7**(1), 1–9 (2017). https://doi.org/10.1038/s41598-017-07897-z
16. Jackowski, A., Gebhard, M., Graeser, A.: A novel head gesture based interface for hands-free control of a robot. In: 2016 IEEE International Symposium on Medical Measurements and Applications (MeMeA), pp. 1–6, May 2016. https://doi.org/10.1109/MeMeA.2016.7533744
17. Kassner, M., Patera, W., Bulling, A.: Pupil: an open source platform for pervasive eye tracking and mobile gaze-based interaction. In: Proceedings of the 2014 ACM International Joint Conference on Pervasive and Ubiquitous Computing: Adjunct Publication, pp. 1151–1160. ACM (2014)
18. Plaumann, K., Ehlers, J., Geiselhart, F., Yuras, G., Huckauf, A., Rukzio, E.: Better than you think: head gestures for mid air input. In: Abascal, J., Barbosa, S., Fetter, M., Gross, T., Palanque, P., Winckler, M. (eds.) INTERACT 2015. LNCS, vol. 9298, pp. 526–533. Springer, Cham (2015). https://doi.org/10.1007/978-3-319-22698-9_36

19. Roethe, A.L., Landgraf, P., Schroeder, T., Vajkoczy, P., Picht, T.: Monitor-based exoscopic neurosurgical interventions: a task-based preparatory evaluation of 3d4k surgery. 69. Jahrestagung der Deutschen Gesellschaft fuer Neurochirurgie (DGNC) pp. Joint Meeting mit der Mexikanischen und Kolumbianischen Gesellschaft fuer Neurochirurgie- (2018). https://doi.org/10.3205/18DGNC387
20. Sauer, I.M., et al.: Mixed reality in visceral surgery: Development of a suitable workflow and evaluation of intraoperative use-cases. Ann. Surg. **266**(5), 706–712 (2017). https://doi.org/10.1097/SLA.0000000000002448
21. Starner, T.E.: Attention, memory, and wearable interfaces. IEEE Pervasive Comput. **1**(4), 88–91 (2002)
22. Werner, S., et al.: Awareness of sensorimotor adaptation to visual rotations of different size. PLOS ONE **10**(4), e0123321 (2015). https://doi.org/10.1371/journal.pone.0123321
23. Zinchenko, K., Komarov, O., Song, K.: Virtual reality control of a robotic camera holder for minimally invasive surgery. In: 2017 11th Asian Control Conference (ASCC), pp. 970–975 (2017)

Spatially-Aware Displays for Computer Assisted Interventions

Alexander Winkler[1(✉)], Ulrich Eck[1], and Nassir Navab[1,2]

[1] Computer Aided Medical Procedures, Technical University of Munich, Munich, Germany
alexander.winkler@tum.de

[2] Computer Aided Medical Procedures, Johns Hopkins University, Baltimore, USA

Abstract. We present a novel display and visual interaction paradigm, which aims at reducing the complexity of understanding the spatial transformations between the surgeon's viewpoint, the patient, the pre- or intra-operative 2D and 3D data, and surgical tools during computer assisted interventions. To the best of our knowledge, this is the first work in which the traditional interventional display, for example in surgical navigation systems, is registered both to the patient and to the surgeon's view. The closest concept is that of traditional Augmented Reality windows in which a semitransparent or video see-through display is positioned between the surgeon and the patient. In such cases, the system was providing an AR view into the patient. In the new concept introduced here, the surgeon keeps his/her own direct view to the patient without any need for additional display or direct view augmentation, but the monitor used in the operating room is now registered to the patient and surgeon's viewpoint. The display could act as fixed viewing frustum or as a mirror frustum relative to the surgeon's view. This allows the physicians to effortlessly relate their view of tools and patient to the virtual representation of the patient data. In this paper, the first realization and implementation of such a concept is presented and three clinical partners have tested the system and their first feedback is discussed in detail. They unanimously believe that this concept opens the path for facilitating interactive exploration of data and more intuitive navigation guidance in computer assisted interventions.

Keywords: Surgical visualization and mixed, augmented and virtual reality · Image-guided interventions and surgery

1 Introduction

In the very early days of X-ray fluoroscopy the radiologists used fluorescent handheld screens which they placed into the beam emanating from the X-ray

Electronic supplementary material The online version of this chapter (https://doi.org/10.1007/978-3-030-59716-0_43) contains supplementary material, which is available to authorized users.

A. L. Martel et al. (Eds.): MICCAI 2020, LNCS 12263, pp. 451–460, 2020.
https://doi.org/10.1007/978-3-030-59716-0_43

source passing through the patient. Back then the image had a direct correlation to the X-ray source, the patient, the screen and the observer, as it was observed directly where it was produced [6]. With advances in imaging and display technology, it became possible to view static or live images of the patient anatomy at any location, where a monitor could be positioned. This had the advantage of more practical patient and display positioning, and greatly reduced dose to the interventionalist, but the intuitive perception of the spatial configuration between X-ray source, patient, screen, and the viewer was lost.

CT or MRI scans, which are inherently 3D are mostly visualized as a set of three orthogonal slices along the anatomical axes of the patient or along the axis of an instrument. Only in recent years 3D images are also rendered on displays in ORs [15], commonly in addition to slice visualizations. These are often volume renderings from viewpoints, which can be controlled by the user. Surgeons often prefer the 2D images rather than 3D renderings as they have been well trained in interpreting them and thus this defines the current clinical standard [14].

Navigation in surgery has been introduced since the early 1990s and takes advantage of different visualization techniques. Many systems show slices of pre- or intraoperative data with additional annotations. These annotations relate to a surgical plan, which needs to be carried out in surgery. They are often an attempt to visualize deep seated anatomical targets and safe paths for the surgeon to reach them based on pre-operative images. In such systems, when rendering images, the position of the surgeon and location of displays are not taken into consideration. The position of the display therefore does not affect the visualization nor does the position of the user observing it.

In most medical Augmented Reality (AR) systems the focus is on augmenting the direct view of the surgeon or the view of an intra-operative imaging device [9,10]. The most related work to ours is often referred to as AR-windows. Blackwell et al. [2] presented a tracked semi-transparent display for medical in-situ augmentation, which incorporated a head tracker and stereo glasses. Their semi-transparent display consisted of a half-silvered glass pane, which reflected the image from a computer display. Such systems could create visualizations similar to Optical-See-Through Head-Mounted-Displays (OST-HMD) on the surgery site but could potentially be more comfortable. Schnaider et al. [12] addressed the same problem of creating an AR-window on patient anatomy using a semi-transparent active matrix LCD between the patient and the surgeon. Goebbels et al. [5] approached the problem again with a half-silvered glass pane, which reflected the projection of two DLP projectors generating high contrast images. Weber et al. [14] presented a system comprising of a tracked mobile opaque screen in which the position of the screen affected the visualization. Their system was again placed between the surgeon and the patient showing a slice view of the anatomy parallel to the screen. However, in this system, the user's perspective is not taken into account, i.e. the image on the screen is merely two-dimensional, independent from the viewpoint of the surgeon.

Bichlmeier et al. [1] presented an AR visualization inspired by the dentists' approach for examining the patient's mouth without changing their viewpoints.

They identified that in some AR applications rotating the object or moving around it is impossible. They suggested to generate additional virtual mirroring views to offer secondary perspectives on virtual objects within an AR view. They used a spatially tracked joystick to move a virtual mirror which reflected the virtual data like a real mirror within the AR view of an HMD.

In contrast to these systems, we propose a novel concept in which the direct view of the surgeon onto the patient and his/her working space remains completely unchanged. This includes the placement of the display behind the patient as seen by the surgeon. We make the position and orientation of the external display an integral part of the visualization pipeline. We therefore, dynamically track its pose relative to other objects of interest, such as the patient, instruments, and the surgeon's head. This serves as input to the image-guided surgery visualization user interface. Our believe is that any application in which surgical navigation is used with traditional displays can be a good use case for our technique. We introduce two novel concepts, that aim at task simplification and enhancing perception [3]. The first one uses a "Fixed View Frustum" relating the pose of the display to the patient and tools. The second one is build upon the mirror metaphor and extends the first technique to also integrate the pose of the surgeon's head as an additional parameter within the visualization pipeline, in which case the display will be associated to a "Dynamic Mirror View Frustum". This novel concept is fully implemented and its first qualitative evaluation by three expert surgeons is presented and discussed in detail in this paper.

2 Methodology

In this section, we present the technical requirements for the proposed concept, followed by a description and technical details of the visualization techniques.

2.1 Technical Requirements

We deliberately designed our concept for easy integration into existing surgical navigation solutions. While in our current implementation we use a commercial optical IR tracking system, the concept works with any sufficiently accurate 6D tracking technology. Registration is a deep-rooted principle in image-based or model-based surgical navigation systems [8]. Thus, we assume that existing systems already contain appropriate methods for registering pre-operative image data and surgical plans to the patient before surgery.

2.2 Visualization Techniques

The proposed concept only requires the following additions: First, we need to track the surgical display for example by attaching a tracking marker and calibrating the spatial relation between the physical display and the tracking marker as explained by Tuceryan et al. [13]. Second, we need to track the surgeon's viewpoint for all visualization techniques within the class of "Dynamic Mirror View

Frustum", which will be presented in the next section. Estimation of the surgeon's viewpoint can be achieved either by using a head tracking target or by mounting a camera to the surgical display in combination with existing video-based head-pose estimation algorithms. We recommend to track the pose of the display and the surgeon's viewpoint with the same surgical navigation system as the patient, not only as it is convenient in the OR, but also because then the tracking characteristics will be the same. Consequently there is no additional source of error for the visualization.

Once the required tracking information is available in a common coordinate system, we can compute the spatial relationship between the patient, the surgical display, the tools and if needed the surgeon's viewpoint.

Fixed View Frustum *(FVF)*. The first visualization method comprises a Fixed View Frustum *(FVF)* associated to the display. We render the objects according to the pinhole camera model with an on-axis perspective projection orthogonal to the middle of the display from a given distance in front of the display, e.g. one meter in our current implementation, and with the display itself as the base of the frustum. Static and dynamic objects in front of the display, e.g. patient's imaging data and tracked surgical tool representations, are projected by the frustum. The viewpoint of the rendering is on the same side of the patient as from which the surgeon also directly sees the patient. If the user moves or rotates the display, the position of the virtual camera is automatically updated. Therefore, in contrast to the standard visualization displays within surgical navigation systems, the mental mapping of the real object to its image on the screen is simplified. Examples of such FVF visualizations from two viewpoints are shown in Fig. 1.

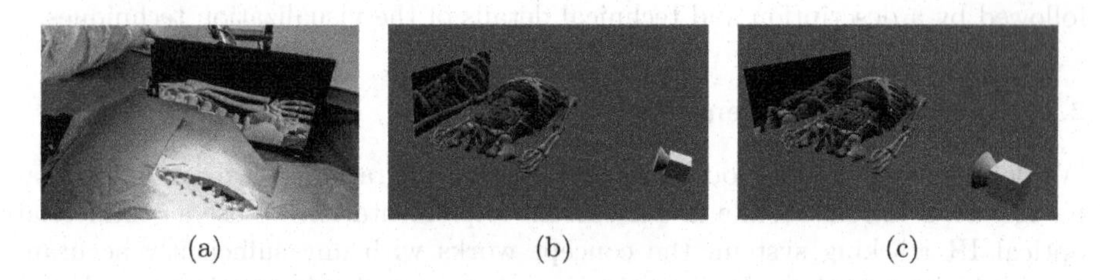

Fig. 1. Real world example (1a) and simulations (1b, 1c) from two viewpoints applying the Fixed View Frustum *(FVF)* visualization. The virtual camera is in a fixed position in front of the display.

Screen Parallel Slice Visualization *(SPSV)*. While the FVF visualization is intuitive to use and can display 3D data, surgeons often rely on slice visualization since it is the current clinical standard [14]. One can incorporate the slice view into our spatially-aware visualization. This is achieved by re-slicing the 3D volume at planes parallel to the screen. While the orientation of the slice is set automatically based on the display pose, we use a tracked tool to select the position of the slice within the volume. We render the instrument and slice using the same FVF method as before. Exemplary images can be seen in Fig. 2.

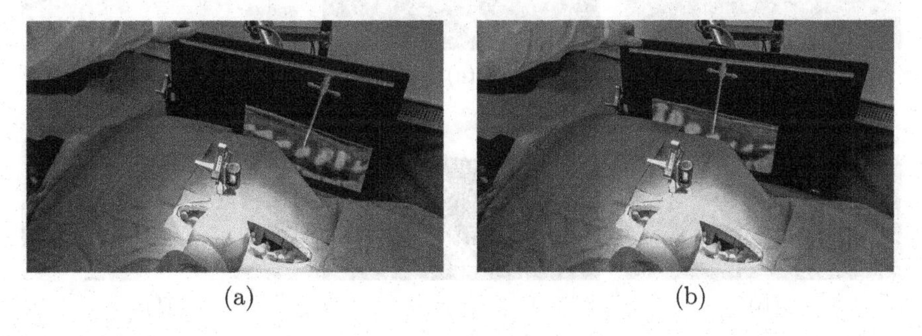

(a) (b)

Fig. 2. Slices parallel to the screen. The user is given the interaction method to rotate the screen which in turn also rotates the CT slice.

Dynamic Mirror View Frustum ***(DMVF)***. We then propose to also track the head of the surgeon relative to the display and to the patient and define the display viewing frustum as if the display was a mirror as can be seen in Fig. 3. We call this the Dynamic Mirror View Frustum *(DMVF)*.

As we use a regular screen, we do not render stereoscopically. This also simplifies the setup, as only the surgeon's head needs to be tracked not each eye.

The image is rendered with a custom projection matrix as detailed in Sect. 2.3. The user can now move left and right or back and forth and see the patient and instruments moving correspondingly on the mirror display.

When the surgeon uses this visualization concept, the screen shows the objects in front of the display just like a mirror would do, taking the poses of the screen, objects, and the viewpoint into account. Since this paradigm works similar to a real mirror, it can contribute to a more intuitive visualization of the anatomy by using motion parallax and observing structures from additional desired viewpoints. The user gains the freedom to interact with the data just by looking, without the need for an interaction device like a mouse or joystick. Voice interfaces as an alternative exhibit other difficulties, such as challenges in a noisy operating room, or the fact that voice seems not suitable to the control of continuous parameters, such as for example the rotation of an object [3]. Care must be taken, that the rendering of the mirror image is perspectively correct and even more that the user's sensory-motor system understands how and why the virtual representation moves in relationship to the user. As in this concept the user's pose also affects the visualization, inconsistencies will lead to errors in spatial understanding [4].

As the users do not constantly need to redefine their view on the anatomy, the user is enabled to toggle the interactivity on or off for example with a foot pedal. DMVF facilitates exploring and defining the best slice or projection for a given navigation task. It can then remain fixed during the surgical action

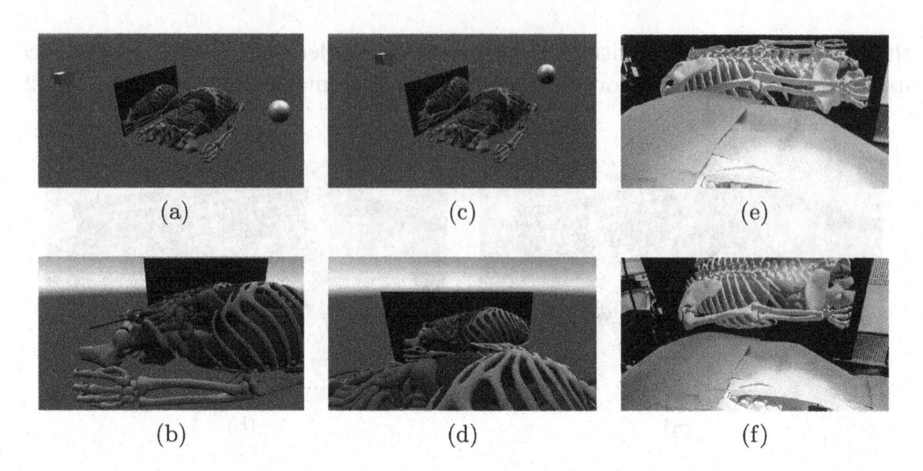

Fig. 3. Simulated mirror views (3a–3d) and a real example (3e, 3f) with the torso phantom. In the third person view (3a, 3c) it is apparent how the position of the mirror camera (left) changes in dependence to the head position (ball on the right), and the projection matrix is adjusted so that the frustum is fitted so that its corners always coincide with the corners of the screen. This results in a warped image, which appears perspectively correct as a mirror reflection from the head position (3b, 3d).

until further data exploration is desired. Furthermore it is important that the user interaction to change the perspective is safe, for both the patient and the surgeon. For example a certain distance between the display and operation site needs to be observed for sterility. Or at extreme angles the image on the screen can become too distorted so that it is hard to comprehend. To avoid this the user interaction can be restricted so that it can only be applied in a safe margin.

Viewpoint Facing Slice Visualization *(VFSV)*. If the screen and viewpoints are tracked, analogous to SPSV, a slice visualization can use both inputs to re-slice the volume. We again choose the position of the slice with a tracked tool. The orientation, however, is chosen to point towards the mirror viewpoint. This guarantees that in this Viewpoint Facing Slice Visualization *(VFSV)*, when viewed through the mirror, the slice always faces the surgeon. An example from two different viewpoints can be seen in Fig. 4.

2.3 Virtual Camera Setup

Estimation of the Mirrored Viewpoint. The mirrored world position of the viewpoint $p_{mirrored}$ can be calculated as follows: Let us define the normal to the mirror plane as $(0, 0, 1)$. In its local coordinate system the mirroring corresponds to multiplication with the matrix

$$M_{flip} = \begin{bmatrix} 1 & 0 & 0 & 0 \\ 0 & 1 & 0 & 0 \\ 0 & 0 & -1 & 0 \\ 0 & 0 & 0 & 1 \end{bmatrix}$$

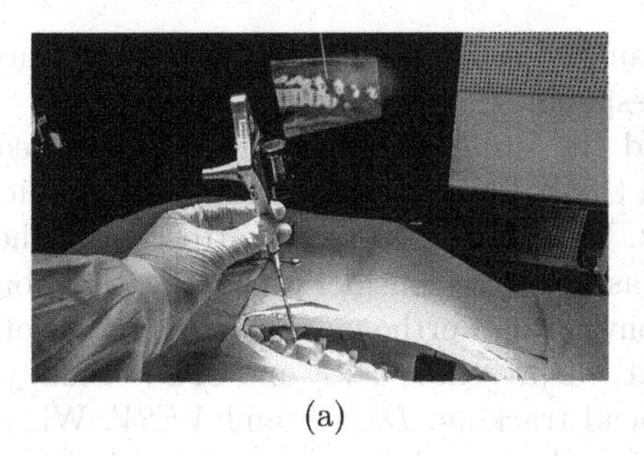

(a)

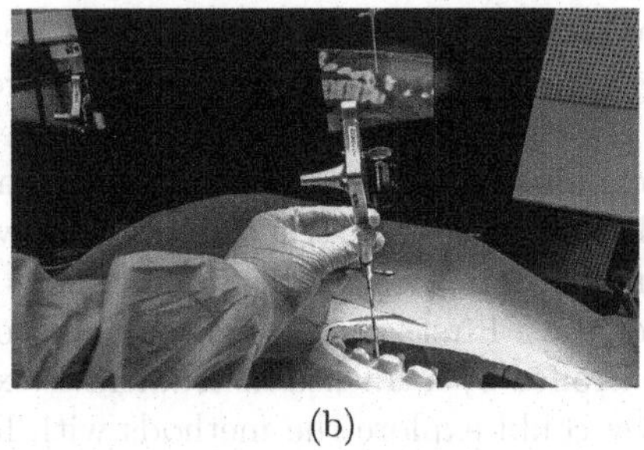

(b)

Fig. 4. The CT slice is always oriented so that it is facing the user in the Viewpoint Facing Slice Visualization *(VFSV)*

This is equivalent to scaling by −1 in z-direction. To provide the mirror view for DMVF the rendering only depends on the position of the head with respect to the mirror plane and not its orientation. We first transform the world position of the viewpoint p into local coordinates of the mirror then perform the mirroring, and finally transform it back into world coordinates, which is formulated as[1]:

$$p_{mirrored} = {}^{Mirror}T_{World} \cdot M_{flip} \cdot {}^{World}T_{Mirror} \cdot p.$$

Mirror Frustum. In order for the image on the display to render mirror views of the objects in front of it in relation to the surgeon's pose, we employ an off-axis projection matrix with the mirrored viewpoint at the apex of the frustum and the base matching the surface of the screen. Not only does this take the position of the viewpoint into account, it also warps the image, so that it appears like a projection onto the tilted plane. This frustum needs to be recomputed at every frame as the pose of the screen or the viewpoint of the user changes. The effect of the changed viewpoint on the frustum can be seen in Fig. 3a and 3c together with the warped images. This implementation is based on Robert Kooima's article "Generalized Perspective Projection" [7] and relies on a commonly available function in graphic APIs to set a projection matrix with the frustum defined by the near plane coordinates in view space, which is then rotated to be non-perpendicular and moved to be located at the mirrored viewpoint.

3 Case Study and Discussion

As a first proof of the concept, we performed a case study with three trauma surgeons (chief surgeon, attending physician, and resident), where we set up a 3D printed patient phantom, a monitor on an adjustable desk mount fixed to the OR table, and a Stryker Flashpoint 6000 tracking system to track the patient phantom, the instrument, the monitor, and the viewpoint. The 3D printed patient

[1] We define ${}^{A}T_{B}$ to represent a transformation from coordinate system A into B.

phantom is based on the Visible Korean Project data set [11], which contains segmented anatomy as well as the corresponding CT volume.

During these experiments, we asked the surgeons to explore the proposed techniques in a predefined order with a think-aloud protocol and no time limit, followed by a semi-structured interview. We asked them to pay attention to the differences in the visualization methods as well as to consider possible application scenarios. First, we let them use the conventional orthogonal slice visualization controlled by a tracked instrument. Next, we presented *FVF* and *SPSV*. Finally, they could explore the methods with head tracking: *DMVF* and *VFSV*. When using *FVF* and *DMVF*, we asked them to also switch between two modes: segmented 3D structures and direct volume rendering.

In the interviews, all surgeons stated that they would typically work with orthogonal slice views during navigated surgery, even though 3D visualization might be available. Furthermore, they stated that selecting appropriate slice views is an essential part of surgery preparation, since they typically cannot change them conveniently during the procedure. They clearly stated that this situation is not satisfactory, but the constraints of the sterile work-space would force them to do so. Imaging systems today usually offer the possibility to rotate or flip the images on the display. All participants agreed that they always set the image in the same orientation as the patient lying in front of them, so that what they do and what they see is congruent. They praised the idea to automate this and to present the images to them that match the orientation of their direct view onto the patient.

The overall feedback of the experts on the proposed concept was extremely positive. They specifically liked the fact that for *DMVF* and *VFSV*, they could interact with patient data in a sterile manner, which they considered to be important when an intervention turns out to be more complex than initially expected. While all participants typically rely on the orthogonal slice view during navigated surgery, one surgeon stated that he would utilize isosurface rendering of CT data in order to better understand spatial relations, for example when reducing complex fractures. All participants agreed that the *FVF* and *SPSV* methods could be easily integrated into the established surgical workflow.

When asked about possible applications of the proposed concept, the answers included use-cases in complex craniomaxiofacial surgery, and interventions in orthopaedics and trauma surgery, especially interventions where high precision is required, such as endoprosthetics.

Participants unanimously agree that the new visualization technique helps with orientation when it comes to getting an overview of the specific patient anatomy and how the instruments are currently positioned, which is always the first step in any procedure.

4 Conclusions

In this paper we presented a novel display and visual exploration paradigm, which aims at reducing the complexity of understanding spatial transformations between the surgeon's viewpoint, the patient, the pre/intra-operative 2D and 3D data, and surgical tools during computer assisted interventions with minimal change in the current setups. Any available surgical tracking system can be used to track the display, tool and surgeon's head supporting the integration into existing computer assisted intervention systems. The main objective is to allow the physicians to effortlessly relate their view of tools and the patient to the virtual data on surgical monitors. The users gain the possibility of interacting with the patient data just by intuitively moving their viewing position and observing it from a different perspective in relation to the patient position without the need for an interaction device like a mouse or joystick. The three surgeons experimenting with the first implementation of this concept, find it innovative, intuitive, easy to use and smooth to integrate into current image guided surgery solutions. The main objective here was to expose the community to this novel concept and allow researchers to integrate it into their clinical solutions, as the concept can only be fully evaluated once integrated within clinical setups.

Acknowledgements. This work was partially supported by Stryker Leibinger GmbH & Co. KG.

References

1. Bichlmeier, C., Heining, S.M., Feuerstein, M., Navab, N.: The virtual mirror: a new interaction paradigm for augmented reality environments. IEEE Trans. Med. Imaging **28**(9), 1498–1510 (2009). https://doi.org/10.1109/TMI.2009.2018622
2. Blackwell, M., Nikou, C., DiGioia, A.M., Kanade, T.: An image overlay system for medical data visualization. Med. Image Anal. **4**(1), 67–72 (2000). https://doi.org/10.1016/S1361-8415(00)00007-4
3. Drouin, S., Collins, D., Kersten-Oertel, M.: Interaction in augmented reality image-guided surgery. In: Mixed and Augmented Reality in Medicine, pp. 99–114. CRC Press (2018). https://doi.org/10.1201/9781315157702-7
4. Eagleson, R., de Ribaupierre, S.: Visual perception and human-computer interaction in surgical augmented and virtual reality environments. In: Mixed and Augmented Reality in Medicine, pp. 83–98. CRC Press (2018). https://doi.org/10.1201/9781315157702-6
5. Goebbels, G., et al.: Development of an augmented reality system for intra-operative navigation in maxillo-facial surgery. In: Proceedings BMBF Statustagung, Stuttgart, Germany, pp. 237–246, January 2002
6. Houston, E.J.: The x-rays. In: Young, H.W. (ed.) Popular Electricity, vol. 2, pp. 4–9. Popular Electricity Publishing Company (1909)
7. Kooima, R.: Generalized perspective projection (2009). http://csc.lsu.edu/kooima/articles/genperspective
8. Mezger, U., Jendrewski, C., Bartels, M.: Navigation in surgery. Langenbeck's Arch. Surg. **398**(4), 501–514 (2013). https://doi.org/10.1007/s00423-013-1059-4
9. Navab, N., Blum, T., Wang, L., Okur, A., Wendler, T.: First deployments of augmented reality in operating rooms. Computer **45**(7), 48–55 (2012). https://doi.org/10.1109/MC.2012.75

10. Navab, N., Traub, J., Sielhorst, T., Feuerstein, M., Bichlmeier, C.: Action-and workflow-driven augmented reality for computer-aided medical procedures. IEEE Comput. Graph. Appl. **27**(5), 10–14 (2007). https://doi.org/10.1109/MCG.2007.117
11. Park, J.S., Chung, M.S., Hwang, S.B., Lee, Y.S., Har, D.H., Park, H.S.: Visible korean human: improved serially sectioned images of the entire body. IEEE Trans. Med. Imaging **24**(3), 352–360 (2005). https://doi.org/10.1109/tmi.2004.842454
12. Schnaider, M., Schwald, B., Seibert, H., Weller, T.: Medarpa - a medical augmented reality system for minimal-invasive interventions. Stud. Health Technol. Inform. **94**, 312–314 (2003). https://doi.org/10.3233/978-1-60750-938-7-312
13. Tuceryan, M., et al.: Calibration requirements and procedures for a monitor-based augmented reality system. IEEE Trans. Vis. Comput. Graph. **1**(3), 255–273 (1995). https://doi.org/10.1109/2945.466720
14. Weber, S., Klein, M., Hein, A., Krueger, T., Lueth, T.C., Bier, J.: The navigated image viewer – evaluation in maxillofacial surgery. In: Ellis, Randy E., Peters, Terry M. (eds.) MICCAI 2003. LNCS, vol. 2878, pp. 762–769. Springer, Heidelberg (2003). https://doi.org/10.1007/978-3-540-39899-8_93
15. Wilkinson, E.P., Shahidi, R., Wang, B., Martin, D.P., Adler, J.R., Steinberg, G.K.: Remote-rendered 3D CT angiography (3DCTA) as an intraoperative aid in cerebrovascular neurosurgery. Comput. Aided Surg. **4**(5), 256–263 (1999). https://doi.org/10.3109/10929089909148178

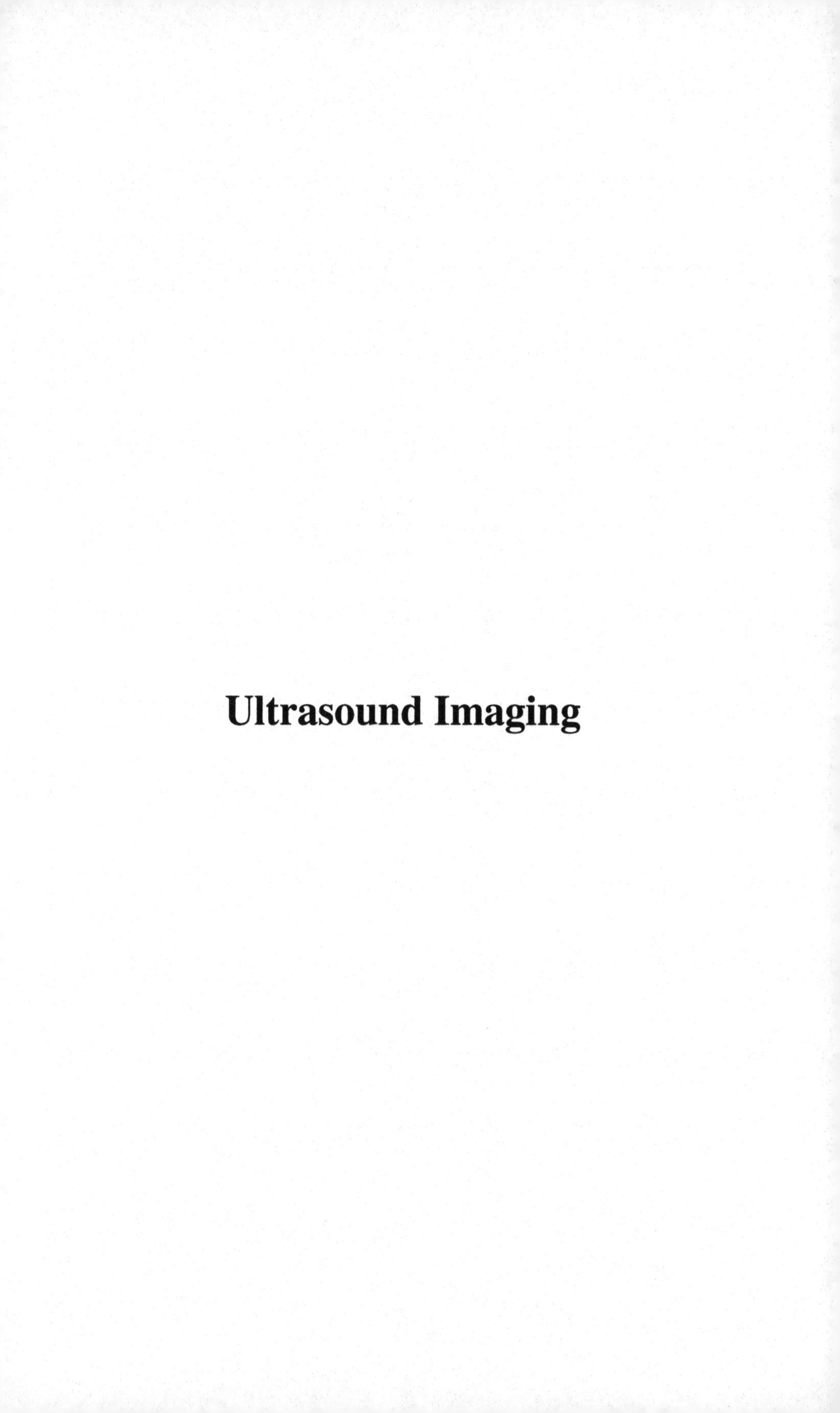

Ultrasound Imaging

Sensorless Freehand 3D Ultrasound Reconstruction via Deep Contextual Learning

Hengtao Guo[1], Sheng Xu[2], Bradford Wood[2], and Pingkun Yan[1(✉)]

[1] Department of Biomedical Engineering and Center for Biotechnology and Interdisciplinary Studies, Rensselaer Polytechnic Institute, Troy, NY 12180, USA
yanp2@rpi.edu

[2] National Institutes of Health, Center for Interventional Oncology, Radiology and Imaging Sciences, Bethesda, MD 20892, USA

Abstract. Transrectal ultrasound (US) is the most commonly used imaging modality to guide prostate biopsy and its 3D volume provides even richer context information. Current methods for 3D volume reconstruction from freehand US scans require external tracking devices to provide spatial position for every frame. In this paper, we propose a deep contextual learning network (DCL-Net), which can efficiently exploit the image feature relationship between US frames and reconstruct 3D US volumes without any tracking device. The proposed DCL-Net utilizes 3D convolutions over a US video segment for feature extraction. An embedded self-attention module makes the network focus on the speckle-rich areas for better spatial movement prediction. We also propose a novel case-wise correlation loss to stabilize the training process for improved accuracy. Highly promising results have been obtained by using the developed method. The experiments with ablation studies demonstrate superior performance of the proposed method by comparing against other state-of-the-art methods. Source code of this work is publicly available at https://github.com/DIAL-RPI/FreehandUSRecon.

Keywords: Ultrasound volume reconstruction · Deep learning · Image guided intervention

1 Introduction

Ultrasound (US) imaging has been widely used in interventional applications to monitor and trace target tissue. US possesses many advantages, such as low cost, portable setup and the capability of navigating through patient in real-time for anatomical and functional information. Transrectal ultrasound imaging (TRUS) has been commonly used for guiding prostate cancer diagnosis and can significantly reduce the false negative rate when fused with magnetic resonance imaging (MRI) [15]. However, 2D US images are difficult to be registered with 3D MRI volume, due to the differences in not only image dimension but also

A. L. Martel et al. (Eds.): MICCAI 2020, LNCS 12263, pp. 463–472, 2020.
https://doi.org/10.1007/978-3-030-59716-0_44

image appearance. In practice, a reconstructed 3D US image volume is usually required to assist such interventional tasks.

A reconstructed 3D US imaging volume visualizes a 3D region of interest (ROI) by using a set of 2D ultrasound frames [11], which can be captured by a variety of scanning techniques such as mechanical scan [5] and freehand tracked scan [17]. Among these categories, the tracked freehand scanning is the most favorable method in a number of clinical scenarios. For instance, during a prostate biopsy, the freehand scanning allows clinicians to freely move the US probe around the ROI and produces US images with much more flexibility. The tracking device, either an optical or electro-magnetic (EM) tracking system, helps to build a spatial transformation chain between the imaging planes in the world coordinate for 3D reconstruction.

US volume reconstruction from sensorless freehand scans takes a step further by removing the tracking devices attached to the US probe. The prior research on this was mainly supported by the speckle decorrelation [4,16], which maps the relative difference of position and orientation between neighboring US images to the correlation of their speckle patterns, *i.e.* higher the speckle correlation, lower the elevational distance between neighboring frames [3]. By removing the tracking devices, such sensorless reconstruction allows the clinicians to move the probe with less constraint without the concerns of blocking tracking signals. In addition, it also reduces the hardware cost. Although the speckle correlation carries information of the relative transformation between neighboring frames, relying on the decorrelation alone renders unreliable performance [1,10].

In the past decade, deep learning (DL) methods based on convolutional neural networks (CNN) have been identified as an important tool for automatic feature extraction. In the field of US volume reconstruction, a pioneer work carried out by Prevost *et al.* [14] explored the feasibility of using CNN to directly estimate the inter-frame motion between two 2D US scans. A 2D convolutional network takes two consecutive ultrasound frames and a generated optical flow field between them as a stacked input to estimate the relative rotations and translations between these two frames. However, a typical US scanning video contains rich contextual information beyond two neighboring frames. A sequence of 2D US frames can provide a more general representation of the motion trajectory of US probe. Using only two neighboring frames may lose temporal information and thus result in less accurate reconstruction. In addition, optical flow field, which is good at describing in-plane motion, may not help out-of-plane motion analysis. Besides, the prior works on decorrelation suggests that paying more attention to the speckle-rich regions can boost the reconstruction performance, which hasn't been explored.

In this paper, based on the above observations, we propose a novel Deep Contextual Learning Network (DCL-Net) for sensorless freehand 3D ultrasound reconstruction. The underlying framework takes multiple consecutive US frames as input, instead of only two neighboring frames, to estimate the trajectory of US probe by efficiently exploiting the rich contextual information. Furthermore, to make the network focus on the speckle-rich image areas to utilize the decorrelation information between frames, the attention mechanism is embedded into the network architecture. Last but not the least, we introduce a new case-wise correlation loss to enhance the discriminative feature learning to prevent overfitting the scanning style.

2 Data Materials

All TRUS scanning videos studied in this work are collected by an EM-tracking device from real clinical cases. The dataset contains 640 TRUS videos all from different subjects acquired by a Philips iU22 scanner in varied lengths. Every frame corresponds to an EM tracked vector that contains the position and orientation information of that frame. We convert this vector to a 3D homogeneous transformation matrix $M = [R\ T; 0\ 1]$, where R is a 3×3 rotation matrix and T is a 3D translation vector.

The primary task of 3D ultrasound reconstruction is to obtain the relative spatial position of two or more consecutive US frames. Without loss of generality, here we use two neighboring frames as an example for illustration. Let I_i and I_{i+1} denote two consecutive US frames with corresponding transformation matrices M_i and M_{i+1}, respectively. The relative transformation matrix M_i' can be computed as $M_i' = M_{i+1}M_i^{-1}$. By decomposing M_i' into 6 degrees of freedom $\theta_i = \{t_x, t_y, t_z, \alpha_x, \alpha_y, \alpha_z\}_i$, which contains the translations in millimeters and rotations in degrees, we can use this θ_i computed from EM tracking as the ground-truth for network training.

3 Ultrasound Volume Reconstruction

Figure 1 shows the proposed DCL-Net architecture, which is designed on top of the 3D ResNext model [18]. Our model consists of 3D residual blocks and other types of basic CNN layers [7]. The skip connections help preserve the gradients to train very deep networks. The use of the multiple pathways (cardinalities) enables the extraction of important features. In our design, 3D instead of 2D convolutional kernels are used, mainly because 3D convolutions can better extract the feature mappings along the axis of channel, which is the temporal direction in our case. Such properties enable the network to focus on the slight displacement of image features between consecutive frames. The network can thus be trained to connect these speckle correlated features to estimate the relative position and orientation.

During the training process, we stack a sequence of N frames with height and width denoted by H and W, respectively, to form a 3D input volume in the shape of $N \times H \times W$. Let $\{\theta_i | i = 1, \ldots, N-1\}$ denote the relative transform parameters between the neighboring frames. Instead of directly using these parameters as ground-truth labels for network training, the mean parameters

$$\bar{\theta} = \frac{1}{N-1} \sum_{i=1}^{N-1} \theta_i, \tag{1}$$

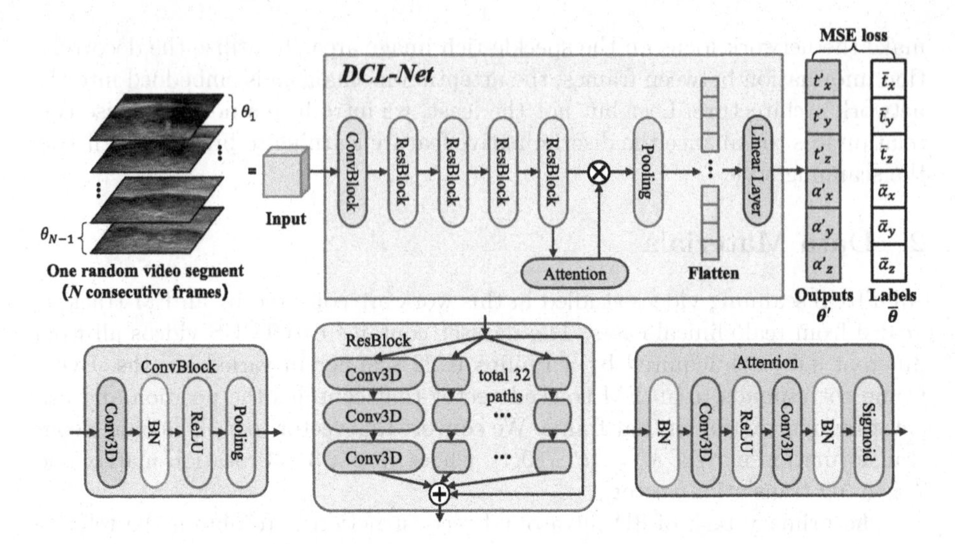

Fig. 1. An overview of the proposed DCL-Net, which takes one video segment as input volume and gives the mean motion vector as the output.

are used for the following two reasons. Most importantly, since the magnitude of motion between two frames is small, using the mean can effectively smooth the noise in probe motion. Another advantage in practice is that there is no need to modify the output layer every time when we change the number of input frames. During the test, we slide along the video sequence with a window size N. The inter-frame motion of two neighboring frames is the average motion computed in all the batches.

3.1 Attention Module

The attention mechanism in the deep learning models makes the CNN to focus on a specific region of an image, which carries salient information for the targeted task [2]. It has led to significant improvement in various computer vision tasks such as object classification [6] and segmentation [12]. In our 3D US volume reconstruction task, regions with strong speckle patterns for correlation are of high importance in estimating the transformations. Thus, we introduce a self-attention block, as shown in Fig. 1, which takes the feature maps produced by the last residual block as input and then outputs an attention map. This helps assign more weights to the highly informative regions.

3.2 Case-Wise Correlation Loss

The loss function of the proposed DCL-Net consists of two components. The first one is the mean squared error (MSE) loss, which is the most commonly used loss in deep regression problems. However, the use of MSE loss alone can

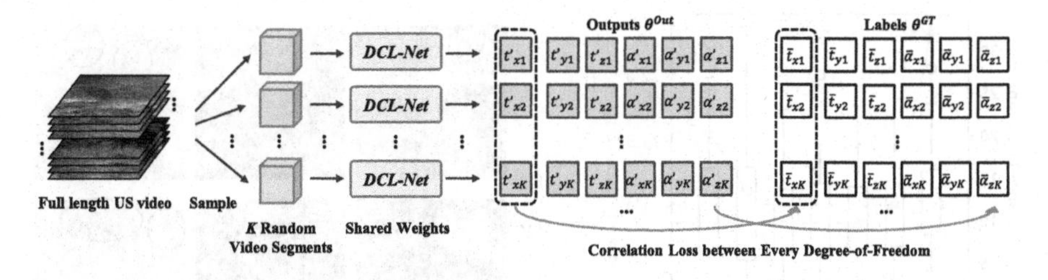

Fig. 2. Overview of the case-wise correlation loss computation.

lead to the smoothed estimation of the motion and thus the trained network tends to memorize the general style of how the clinicians move the probe, *i.e.* the mean trajectory of the ultrasound probes. This shortcoming of the MSE loss for network training has been reported before [8,19]. To deal with problem, we introduce a case-wise correlation loss based on the Pearson correlation coefficient to emphasize the specific motion pattern of a scan.

Figure 2 shows the workflow of calculating the case-wise correlation loss. K video segments with each having N frames are randomly sampled from a TRUS video. The correlation coefficients between the estimated motion and the ground truth mean are computed for every degree-of-freedom and the loss is denoted as

$$L_{corr} = 1 - \frac{1}{6} \sum_{d=1}^{6} \frac{Cov\left(\bar{\theta}_d^{GT}, \bar{\theta}_d^{Out}\right)}{\sigma\left(\bar{\theta}_d^{GT}\right) \sigma\left(\bar{\theta}_d^{Out}\right)}, \tag{2}$$

where Cov gives the covariance and σ calculates the standard deviation. The total loss is the summation of the MSE loss and the case-wise correlation loss.

4 Experiments and Results

For the experiments performed in this study, a total of 640 TRUS scanning videos (640 patients) from the Nation Institute of Health (NIH) were acquired from IRB-approved clinical trial. During an intervention, a physician used an end-firing transrectal ultrasound probe to acquire axial images by steadily sweeping through the prostate from base to apex. The positioning information given by an electromagnetic tracking device serves as the ground truth label in our training phase. The dataset is split into 500, 70 and 70 cases as training, validation and testing, respectively. Our network is trained for 300 epochs with batch size $K = 20$ using Adam optimizer [9] with initial learning rate of 5×10^{-5}, which decays by 0.9 after 5 epochs. Since the prostate US image only takes a relative small part of each frame, each frame is cropped without exceeding the imaging field and then resized to 224×224 to fit the design of ResNexts [18]. We implemented the DCL-Net using the publicly available Pytorch library [13]. The entire training phase of the DCL-Net takes about 4 h, taking 5 frames as input. During testing, it takes about 2.58 s to produce all the transformation matrix of an US video with 100 frames.

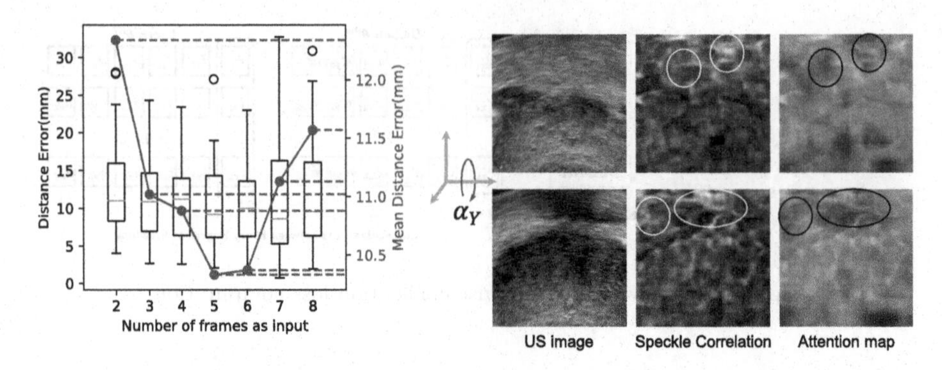

Fig. 3. (Left) effect of number of input frames. Green curve shows the mean distance error of each box. (Right) Visualization of two attention maps regarding rotation around the Y-axis. (Color figure online)

Two evaluation metrics are used for performance evaluation. The first one is the average distance between all the corresponding frame corner-points throughout a video. This distance error reveals the difference in speed and orientation variations across the entire video. The other one is the final drift [14], which is the distance between the center points of the transformed end frames of a video segment using the EM tracking data and our DCL-Net estimated motion.

4.1 Parameter Setting

We first performed experiments to determine an optimal N number of frames for each video segment. The left panel of Fig. 3 shows how the overall reconstruction performance varies as the number of consecutive frames changes. There is a decrease then increase in the error, with neighboring frame number equaling to 5 or 6 has similarly the best performance. According to our paired t-test, the calculated p-value is smaller than the confidence level of 0.05, indicating the result using 5 frames is significantly better than that using only 2 frames. The explanation is that the network takes advantage of the rich contextual information along the time-series and produces more stable trajectory estimation.

The right panel of Fig. 3 visualizes two example attention maps. The first image column shows the cropped US images. The second column is the speckle correlation map [3] between a US image and its following neighboring frame. Inside this speckle correlation map, brighter the area, longer the elevational distance to the next frame. Such pattern with dark areas at the bottom and brighter on the upper part is consistent with our TRUS scanning protocol, as there is less motion around the tip of the ultrasound probe. The third column shows the attention map regarding the rotation α_y around the Y-axis, which also indicates part of the out-of-plane rotation. The attention maps have strong activation at the bright speckle correlation regions, indicating that the attention module helps the network to focus on speckle-rich areas for better reconstruction.

Table 1. Performance of different methods on the EM-tracking dataset.

Methods	Distance error (mm)				Final drift (mm)			
	Min	Median	Max	Average	Min	Median	Max	Average
Linear motion	7.17	19.73	60.79	22.53	12.53	37.15	114.02	42.62
Decorrelation [3]	9.62	17.58	56.72	18.89	15.32	38.45	104.13	38.26
2D CNN [14]	5.66	15.8	43.35	17.42	7.05	23.13	68.87	26.81
3D CNN (NS2) [18]	2.38	10.14	31.34	12.34	1.42	19.08	68.61	21.74
Our DCL-Net	**1.97**	**9.15**	**27.03**	**10.33**	**1.09**	**17.40**	**55.50**	**17.39**

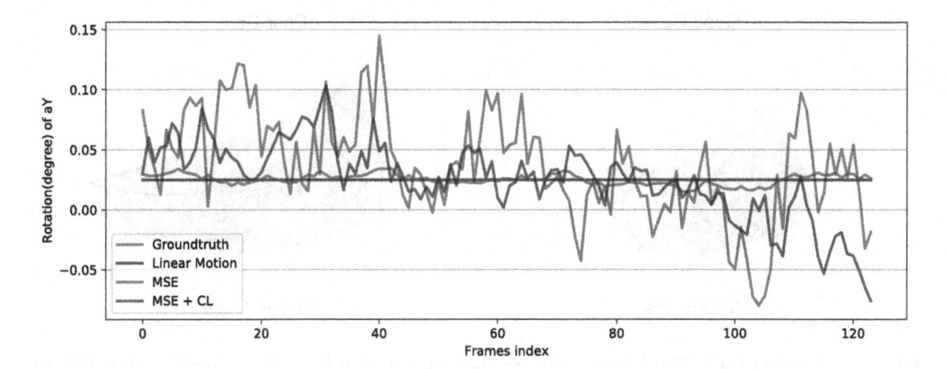

Fig. 4. Predicted rotation α_y on one video sequence with different methods. Applying correlation loss makes our prediction (blue line) more sensitive to the strongly varying speed of the ground-truth (green line). (Color figure online)

4.2 Reconstruction Performance and Discussions

Table 1 summarizes the overall comparison of the proposed DCL-Net against other existing methods. The approach of "Linear Motion" means that we first calculate the mean motion vector of the training set and then apply this fixed vector to all the testing cases. The approach of "Decorrelation" is based on the speckle decorrelation algorithm presented in [3]. "2D CNN" refers to the method presented by Prevost *et al.* [14]. "3D CNN" is the vanilla ResNext [18] architecture taking only two slices as input.

It can be seen from Table 1 that our proposed DCL-Net outperforms all the other methods. Paired t-test was performed and the performance improvement made by DCL-Net is significant in all the cases with p-value < 0.05. It is worth noting that although the average distance error of 10.33 mm achieved DCL-Net is still a large error, this is the best performance on real clinical data instead of phamtom studies. The performance of the state-of-the-art 2D-CNN method reproduced in our experiments has consistent performance compared to the accuracy reported in the paper [14]. It is a challenging task to reconstruct 3D US volume using these freehand TRUS scans and we have been making significant progress in this important area.

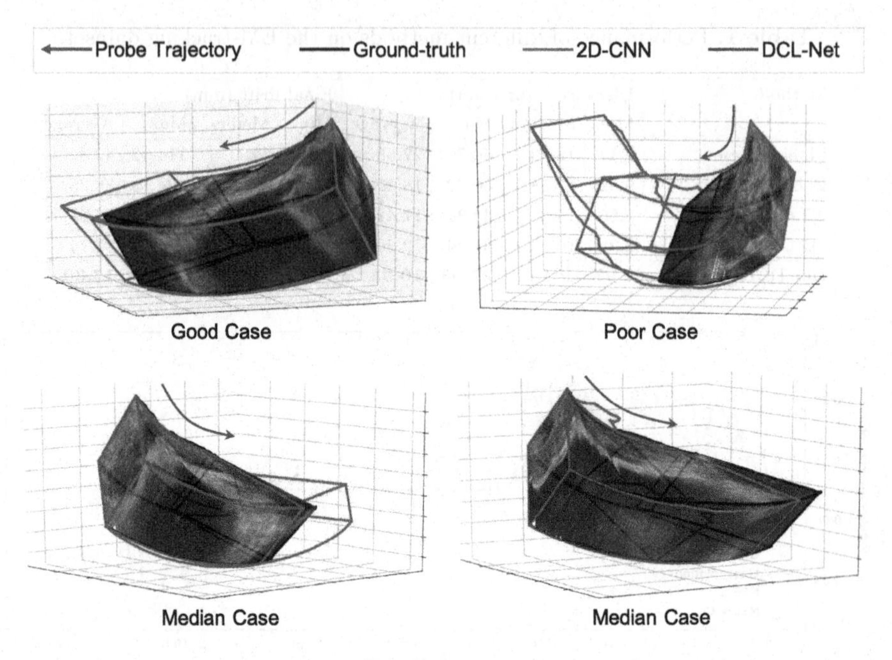

Fig. 5. Comparison of the US volume reconstruction results of four cases with different qualities.

Next, we demonstrate the effectiveness of incorporating case-wise correlation loss into the network training. Figure 4 shows the prediction of α_y along a video sequence. We can observe that the network trained with MSE loss can only produce mediocre results (red line) which are nearly constant, showing almost no sensitivity to the change in speed and orientation. Its prediction wonders around the linear motion (black line) which represents the mean value of the probe's trajectory. The correlation coefficients of all testing cases show a mean of 0.09 ± 0.03 which represents little correlation. This indicates that using MSE alone makes the network memorizing the general style of the US probe motion trajectory and fails to produce valid prediction based on image contents. By incorporating the correlation loss (CL) into the loss function (blue line), the correlation coefficients of all testing cases have a mean of 0.21 ± 0.09, representing weak correlation. Based on a paired t-test with $\alpha = 0.05$, this is found to be significantly better than the previous results. Intuitively, the network's prediction reacts more sensitively to the variation of the probe's real translation and rotation (green line).

Last but not the least, we report the volume reconstruction results using four testing cases with different reconstruction qualities as shown in Fig. 5. One good case, one bad case, and two median cases are included to offer a complete view of the performance. To reduce the clutter in the figure, we only show the comparison between our DCL-Net, the 2D-CNN [14] and the ground-truth. While producing competitive performance, the 2D-CNN method is less sensitive to the speed variations of US probe and the estimated trajectory has noisy vibration. The results sometimes even severely deviate from the ground-truth. Our proposed DCL-Net shows a much smoother trajectory estimation thanks to the contextual information provided by video segments.

5 Conclusions

This paper introduced a sensorless freehand 3D US volume reconstruction method based on deep learning. The proposed DCL-Net can well extract the information among multiple US frames to improve the US probe trajectory estimation. Experiments on a well-sized EM-tracked ultrasound dataset demonstrated that the proposed DCL-Net has benefited from the contextual learning and showed superior performance when compared to other existing methods. Further experiments on the ultrasound videos with different scanning protocols will be studied in our future work.

Acknowledgements. This work was partially supported by National Institute of Biomedical Imaging and Bioengineering (NIBIB) of the National Institutes of Health (NIH) under awards R21EB028001 and R01EB027898, and through an NIH Bench-to-Bedside award made possible by the National Cancer Institute.

References

1. Afsham, N., Rasoulian, A., Najafi, M., Abolmaesumi, P., Rohling, R.: Nonlocal means filter-based speckle tracking. IEEE Trans. Ultrason. Ferroelectr. Freq. Control **62**(8), 1501–1515 (2015)
2. Bahdanau, D., Cho, K., Bengio, Y.: Neural machine translation by jointly learning to align and translate. arXiv preprint arXiv:1409.0473 (2014)
3. Chang, R.F., et al.: 3-D US frame positioning using speckle decorrelation and image registration. Ultrasound Med. Biol. **29**(6), 801–812 (2003)
4. Chen, J.F., Fowlkes, J.B., Carson, P.L., Rubin, J.M.: Determination of scan-plane motion using speckle decorrelation: theoretical considerations and initial test. Int. J. Imaging Syst. Technol. **8**(1), 38–44 (1997)
5. Daoud, M.I., Alshalalfah, A.L., Awwad, F., Al-Najar, M.: Freehand 3D ultrasound imaging system using electromagnetic tracking. In: 2015 International Conference on Open Source Software Computing (OSSCOM), pp. 1–5. IEEE (2015)
6. Fukui, H., Hirakawa, T., Yamashita, T., Fujiyoshi, H.: Attention branch network: learning of attention mechanism for visual explanation. In: Proceedings of the IEEE Conference on Computer Vision and Pattern Recognition, pp. 10705–10714 (2019)
7. He, K., Zhang, X., Ren, S., Sun, J.: Deep residual learning for image recognition. In: Proceedings of the IEEE Conference on Computer Vision and Pattern Recognition, pp. 770–778 (2016)
8. Johnson, J., Alahi, A., Fei-Fei, L.: Perceptual losses for real-time style transfer and super-resolution. In: Leibe, B., Matas, J., Sebe, N., Welling, M. (eds.) ECCV 2016. LNCS, vol. 9906, pp. 694–711. Springer, Cham (2016). https://doi.org/10.1007/978-3-319-46475-6_43

9. Kingma, D.P., Ba, J.: Adam: a method for stochastic optimization. arXiv preprint arXiv:1412.6980 (2014)
10. Laporte, C., Arbel, T.: Learning to estimate out-of-plane motion in ultrasound imagery of real tissue. Med. Image Anal. **15**(2), 202–213 (2011)
11. Mohamed, F., Siang, C.V.: A survey on 3D ultrasound reconstruction techniques. In: Artificial Intelligence-Applications in Medicine and Biology. IntechOpen (2019)
12. Oktay, O., et al.: Attention u-net: learning where to look for the pancreas. arXiv preprint arXiv:1804.03999 (2018)
13. Paszke, A.,et al.: Automatic differentiation in PyTorch. In: NIPS 2017 Workshop Autodiff (2017)
14. Prevost, R., et al.: 3D freehand ultrasound without external tracking using deep learning. Med. Image Anal. **48**, 187–202 (2018)
15. Siddiqui, M.M., et al.: Comparison of MR/ultrasound fusion-guided biopsy with ultrasound-guided biopsy for the diagnosis of prostate cancer. JAMA **313**(4), 390–397 (2015)
16. Tuthill, T.A., Krücker, J., Fowlkes, J.B., Carson, P.L.: Automated three-dimensional us frame positioning computed from elevational speckle decorrelation. Radiology **209**(2), 575–582 (1998)
17. Wen, T., et al.: An accurate and effective FMM-based approach for freehand 3D ultrasound reconstruction. Biomed. Signal Process. Control **8**(6), 645–656 (2013)
18. Xie, S., Girshick, R., Dollár, P., Tu, Z., He, K.: Aggregated residual transformations for deep neural networks. In: Proceedings of the IEEE Conference on Computer Vision and Pattern Recognition, pp. 1492–1500 (2017)
19. Yang, Q., et al.: Low-dose CT image denoising using a generative adversarial network with wasserstein distance and perceptual loss. IEEE Trans. Med. Imaging **37**(6), 1348–1357 (2018)

Ultra2Speech - A Deep Learning Framework for Formant Frequency Estimation and Tracking from Ultrasound Tongue Images

Pramit Saha[1(✉)], Yadong Liu[2], Bryan Gick[2,3], and Sidney Fels[1]

[1] Electrical and Computer Engineering Department, University of British Columbia, Vancouver, Canada
pramit@ece.ubc.ca

[2] Department of Linguistics, University of British Columbia, Vancouver, Canada

[3] Haskins Laboratories, New Haven, CT, USA

Abstract. Thousands of individuals need surgical removal of their larynx due to critical diseases every year and therefore, require an alternative form of communication to articulate speech sounds after the loss of their voice box. This work addresses the articulatory-to-acoustic mapping problem based on ultrasound (US) tongue images for the development of a silent-speech interface (SSI) that can provide them with an assistance in their daily interactions. Our approach targets automatically extracting tongue movement information by selecting an optimal feature set from US images and mapping these features to the acoustic space. We use a novel deep learning architecture to map US tongue images from the US probe placed beneath a subject's chin to formants that we call, Ultrasound2Formant (U2F) Net. It uses hybrid spatio-temporal 3D convolutions followed by feature shuffling, for the estimation and tracking of vowel formants from US images. The formant values are then utilized to synthesize continuous time-varying vowel trajectories, via Klatt Synthesizer. Our best model achieves R-squared (R^2) measure of 99.96% for the regression task. Our network lays the foundation for an SSI as it successfully tracks the tongue contour automatically as an internal representation without any explicit annotation.

Keywords: Silent speech interface · Ultrasound tongue contour · Formant · Spatio-temporal feature · Deep neural network · Articulatory-to-acoustics

1 Introduction

Human speech is a spontaneous yet powerful mode of communication. But millions of people fail to communicate through vocalization, due to severe diseases and speech disorders. One of the key components of speech production is the

A. L. Martel et al. (Eds.): MICCAI 2020, LNCS 12263, pp. 473–482, 2020.
https://doi.org/10.1007/978-3-030-59716-0_45

vocal fold, housed within larynx, which is responsible for providing the major source excitation for speech through its vibrations. Many people need to undergo laryngectomy or surgical removal of larynx for treating laryngeal cancer, critical neck injuries and radio-necrosis of the larynx. Though laryngeal cancer accounts for only 1% of all cancers, it has around 70% 5-year survival rate [7]. Those undergoing laryngectomy are limited to speak sub-vocally by moving their vocal tract articulators including their tongue, without engaging their vocal fold. However, the loss of voice and failure to effectively communicate owing to the lack of a voice-box, can have a devastating impact in the quality of life of post-laryngectomy patients. The commercially available technologies for voice rehabilitation, including trachea-esophageal speech and electrolarynx speech, have significant limitations as discussed in [7]. Therefore, there is a need for an alternative form of personalized communication device for such patients, that do not rely on the acoustic signals to produce speech. The ability to communicate in the absence of acoustic signals can be facilitated by sensing the movement of the remaining speech articulators and converting those to speech. Such devices, also known as "silent speech interfaces" (SSI) [5], involve the extraction of speech information via electro-encephalography, surface electromyography, ultrasound imaging, electromagnetic articulography, *etc.* There are two distinct approaches of providing speech outputs from these interfaces, *viz.* 'recognition' and 'direct synthesis'. Despite the rapid progress in speech token 'recognition' led by the adoption of deep learning based classification [16], attempts in speech 'synthesis' has been less frequent due to the lack of efficient speech generation techniques.

In this paper, we introduce a novel ultrasound (US) based sound synthesis approach via a 3D convolutional neural network and formant based speech synthesis engine. The proposed Ultra2Formant Net (U2F) presents a hybrid 3D convolutional block which involves the parallel decomposition of a chunk of standard 3D convolution into individual 2D spatial and 1D temporal convolutional filters. This constrained approach takes advantage of the orthogonal nature of spatial and temporal kernels to decrease the parameter set as well as increase the strength of net spatio-temporal feature encodings. These encodings are then combined with parallel 3D convolutional output, followed by rigorous feature shuffling. The network outputs the formant frequencies, which are then fed to the Klatt speech synthesis engine for generating desired continuous speech sounds. Our codes and other details are made available at https://pramitsaha.github.io/.

2 Background and Related Works

2.1 Previous Works

The research in this field was initiated with an attempt [6] to synthesize speech based on 12 GSM vocoder parameters from tongue contour points using Multilayer-perceptrons (MLP). The input was later modified and combined with lip coordinates and mapped to 12 line spectral frequencies (LSF) using similar MLP networks [4]. The input space was further altered in [9] to additionally include *Eigentongue* features as well. Another related work [2] utilized different combinations of feature representation including eigen-tongue and correlation-based features to map to 13 MGC-LSP features using 5 layered DNN. In a follow-up work [8], the authors replaced the hand-engineered features by an autoencoder, whose bottleneck features were then fed to the DNN layers for mapping to the MGC-LSP. In order to take advantage of a shared representation, a multi-task DNN was further employed in [18] to simultaneously classify phone states and synthesize spectral parameters. The latest work in this direction has been the application of CNN-LSTMs on US images denoised via convolutional autoencoders, intended to predict 24 order MGC-LSP coefficients for synthesizing speech in Hungarian language [11]. However, the vowel transitions in the synthesized speech differ considerably from the desired speech, leading to the poor performance of the synthesized version. A promising way of achieving closer vowel trajectories is to explore the formant frequency space. The connection of articulatory space with the 2D formant space is particularly important to explore because the formant representation is a very powerful acoustic encoding that efficiently describes the essential aspects of speech using limited parameters. It also provides a lot of insight on continuous vowel trajectories, the most dynamic part of speech production, that can be utilized for better control of speech in SSI. To the best of our knowledge, this is the first investigation on deep learning based ultrasound-to-formant mapping for speech synthesis.

2.2 Formant Estimation and Tracking

Estimation and tracking of formant frequency is one of the fundamental problems in speech processing [15]. This involves determining the formant frequencies corresponding to stationary speech segment and tracking these throughout the signal. Since these formant frequencies are the direct results of resonances brought about by the tongue movement, the problem somewhat boils down to identifying and tracking the tongue contour. This is, to some extent, analogous to the applications like video object tracking, action recognition, *etc.* which utilize different spatio-temporal feature encoding schemes [1,13,14,19]. However, there are numerous other challenges encountered in ultrasound based tongue localizing and tracking. For example, we cannot use pre-trained networks popular in video processing as backbones for our application. Besides, the ultrasound images are grayscale, contain less information, possess low spatial resolution and are infested with noises and artifacts.

2.3 Challenges in tongue Tracking

Tongue is a muscular hydrostat having no conventional skeletal support, which results in its remarkably diverse and complex movements [17]. Having multiple degrees of freedom, different parts of the tongue can move simultaneously towards different directions. As such, each tiny movement or shape change of the tongue results in corresponding changes of the vocal tract resonances, which in turn, changes the formant values at that time instant. However, there is a dearth of tongue contour annotations and lack of fully-automated, generalizable

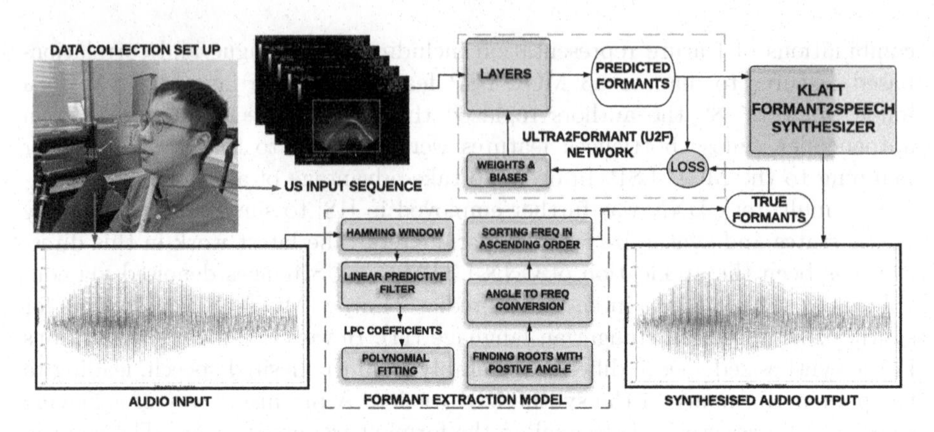

Fig. 1. Overview of proposed Ultra2Speech. The arrows indicate the data flow.

contour extraction methods that makes the U2S task much more challenging. As a result, it is crucial for a successful U2S mapping algorithm to be able to automatically track the tongue contour as a hidden representation, in order to understand the variation of formants from ultrasound.

3 Proposed Ultra2Speech (U2S) Model

We face three fundamental challenges in US-based formant estimation and tracking: (1) extracting relevant spatial information for accurate tongue contour detection; (2) encoding temporal information for understanding the dynamics of tongue movement; and (3) reaching the desired mapping between the extracted spatio-temporal features and the formant trajectories. In this section, we first introduce our **Ultra2Formant (U2F) Net**, aimed at tackling these challenges using different kernels of 3D CNNs, as illustrated in Fig. 2.

3.1 First Hidden Layer

The input video is first convolved with a set of pointwise convolutional filters with kernel of size $1 \times 1 \times 1$. Such filters are known to reduce the computational complexity before expensive $3 \times 3 \times 3$ operations. Besides, they also extract efficient low dimensional embedding and apply extra non-linear activations that help the network to model complex functions.

3.2 Hybrid Convolutional Layer

The output channels of pointwise convolutions are split into three groups, one for intra-frame spatial feature extraction, another for cross-frame temporal modeling and the other for joint spatio-temporal encoding. The spatial branch is composed of 2D CNN kernels $1 \times 3 \times 3$; the temporal branch is composed of 1D CNN time-kernels $3 \times 1 \times 1$; and the joint spatio-temporal branch is composed of 3D CNN

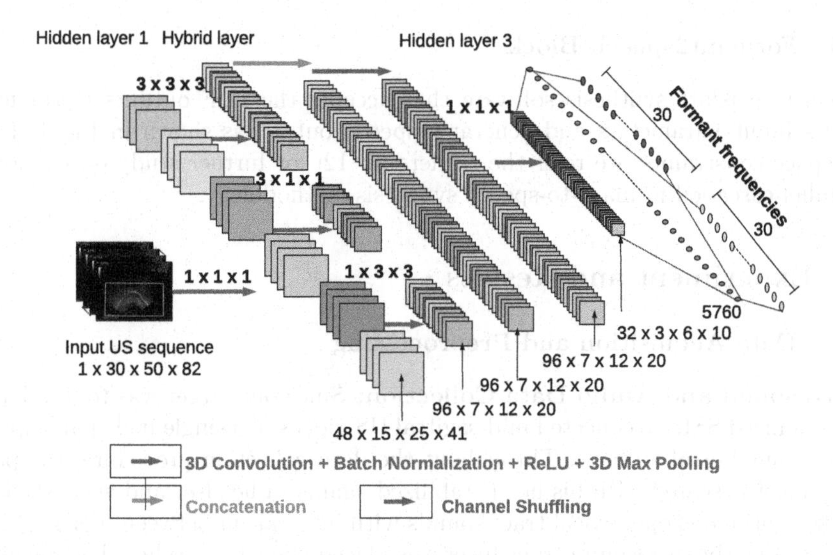

Fig. 2. Architecture of the proposed Ultra2Formant (U2F) Net

kernels $3 \times 3 \times 3$. In this way, we constrain some particular feature channels to focus more on static spatial features, while few others to focus on dynamic motion representation and remaining on encoding joint information. Factorizing part of the standard 3D convolution kernel [13] into orthogonal parallel components reduces the number of parameters thereby making it easier to train. Besides, the separation of orthogonal features also contributes towards better optimization of loss function, as reflected in the performance later. This partial decoupling of the spatial and temporal kernels of 3D CNN makes it both effective in performance and efficient in computation.

3.3 Feature Shuffling, Grouped Convolution and Fully Connected Layer

The output from three branches are concatenated together and shuffled in three groups. Consider the concatenated features with 3 groups, each having N channels. For shuffling, the output channel dimension is first reshaped into (3, N), followed by transposing and then flattening it back to its previous shape for feeding it to the next layer. This shuffling facilitates cross-group information exchange and strengthens the spatio-temporal encoding within a computational budget as shown in a different context in [20]. The feature representation is further compressed by passing it through a grouped convolutional layer of kernel size $1 \times 1 \times 1$. The output features are finally flattened and connected to two parallel sets of 30 output nodes through task-specific fully-connected layers. This joint learning paradigm aimed at estimating two sets of formant frequencies using the same network creates a shared representation beneficial for the model.

3.4 Formant2speech Block

We utilize Klatt synthesis software that accepts the U2F outputs (formants) as its input parameters and generates speech output as shown in Fig. 1. Due to space constraints, we refer the readers to [12] for further study on the joint parallel-cascaded formant-to-speech synthesis methodology.

4 Experiment and Results

4.1 Data Acquisition and Preprocessing

Ultrasound and Audio Data Collection: Since our target was to develop a personalized SSI, we collected mid-sagittal US videos of a single male participant over a number of sessions. Throughout the data collection procedure, the participant was seated with his head stabilized against a headset and was asked to make continuous open vocal tract sounds with intervals in between. For imaging the tongue, the ultrasound transducer was placed beneath the chin. The imaging was done using ALOKA SSD-5000 ultrasound system at 30 fps, with 180 9 mm radius UST-3.5 MHz convex ultrasound probe. Mono-channel audio recording was done simultaneously using Praat software, a Sennheiser MKH 416 P48 shotgun microphone and a Focusrite Scalet 2i2 preamplifier. In order to align the audio soundtrack and the video recording, participants were asked to produce sounds that involves a sudden salient acoustic change and a quick noticeable tongue movement, such as */ga/*.

Audio-Visual Alignment, Image Extraction and Pre-processing: Since the audio recording started prior to video recording, there was a time lag between audio and video recordings which ought to be calculated. For this, the frame of production of the release of */ga/* in ultrasound imaging data and the timestamp of the same event in the audio recording were identified and used for synchronization. Vowel sequences were identified and segmented from the audio recordings, and acoustic landmarks were prepared. Further, ultrasound video recording were converted to image sequences 30 fps using QuickTime 7 Pro and the frames corresponding to each vowel sequence were extracted considering the time of acoustic landmarks and the lag between audio and video recordings. The frames of spatial resolution 480×640 were cropped using a bounding box of 200×330 that contained the tongue for the entire image sequence and were further downsampled into 50×82. We also converted the images to grayscale and normalized the pixel intensities within $[0, 1]$. We chose a time window of 30 frames, resulting in a total of 13,082 videos of duration 1 s each. This time window was chosen as a trade-off between the dynamic information to be imparted to the network and the computational time required for training.

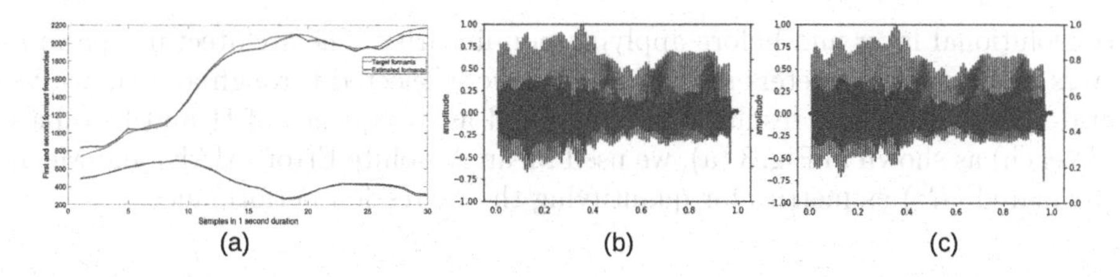

Fig. 3. (a) Time-varying formants (red indicates target and blue indicates predicted trajectories), (b) original speech signal, (c) synthesized speech signal (Color figure online)

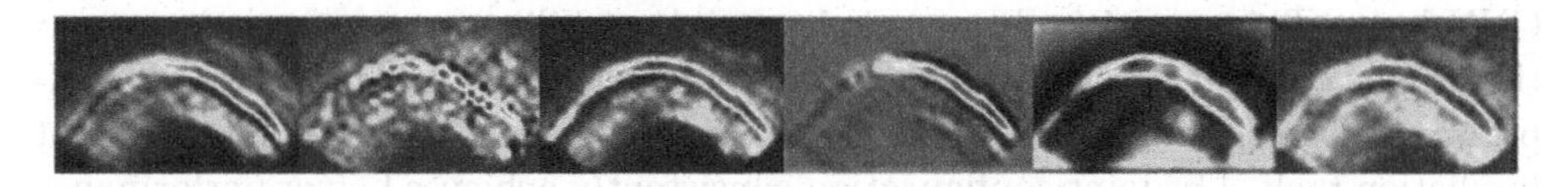

Fig. 4. Saliency maps from U2F showing internal tongue contour localization

Formant Extraction from Recorded Speech: We applied a Hamming window on frame-blocked acoustic signal of 1470 samples each. Following the traditional approach, next, we employed a 1D filter with transfer function of $1/(1 + 0.63z^-1)$ and computed the linear predictive coefficients for the filtered signal using autocorrelation method. Furthermore, we computed the roots of the predictor polynomial in order to locate the peaks in spectra of LPC filters. Only positive frequencies up to half of the sampling frequency were considered for calculations and were sorted in ascending order. In this study, we explored the first two formant frequencies - the most dominating parameters for speech trajectories.

4.2 Implementation Details and Performance Analysis

Network Architecture: The final model consists of 4 hidden layers, 3 being convolutional, with respectively 48, 96 and 32 filters and the last one being a fully-connected layer of 5760 nodes, connected parallely to the output nodes as illustrated in Fig. 2. Each 3D convolutional layer has a stride = 1 and is followed by 3D Batch Normalization[10], ReLU activation[3] and 3D max-pooling with a stride = 2. We perform a respective padding of (0, 1, 1), (1, 0, 0) and (1, 1, 1) for the spatial, temporal and spatio-temporal convolution of the hybrid layer.

Training and Evaluation: Our U2S model was implemented in PyTorch. We randomly shuffled and partitioned the data (13,082 videos) into train (80%), development (10%) and test sets (10%). The network was trained with a batch size of 10 on NVIDIA GeForce GTX 1080 Ti GPU. Mean absolute error (MAE) loss function was optimized using Adam with a learning rate of .001 for a total of 100 epochs. All the parameters were randomly initialized. In order to mitigate the problem of overfitting, we used Batch Normalization after every convolutional layer and before applying non-linearity. The architectural parameters and hyperparameters shown in Fig. 2 were selected through an exhaustive grid-search. Since the results were computed as a sequence of f1 and f2 values (30 each) as shown in Fig. 3 (a), we used Mean Absolute Error (MAE) and Mean R-squared (R^2) as metrics for quantifying the regression performance.

Results: We showcase our results on a randomly chosen sample in Fig. 3, which demonstrates that there is almost no visible distinction between the target and predicted acoustic signal. We also show the visual explanation behind the estimations made by U2F in Fig. 4 in the form of saliency maps corresponding to last CNN layer. This surprisingly reveals its striking ability to accurately represent the tongue contour internally. Table 1 presents the quantitative results and comparisons, corresponding to a joint f1–f2 prediction task Vs individual formant prediction task. The joint configuration consistently achieves better performance taking advantage of a shared representation, despite having fewer parameters. Our baseline method is the Conv-BiLSTM that has been presented as the state-of-the-art approach in [11]. The network in the fourth row of the Table 1 has exactly same architecture as U2F except that the hybrid CNN-block is replaced by regular CNN-block. The results show that the proposed U2F model outperforms the CNN-RNN baselines as well as the standard 3D CNN models even with lesser number of parameters.

Table 1. Performance comparison with baseline methods

Method	$f1$		$f2$		$f1 - f2$	
	MAE	$Mean\ R^2$	MAE	$MeanR^2$	MAE	$Mean\ R^2$
CLSTM (2-layers) + 2-FCN	.0419	86.36	.0423	85.34	.0444	86.45
CBiLSTM (2-layers) + 2-FCN	.0352	89.40	.0380	89.12	.0293	90.01
3D CNN (2-layers) + 1-FCN	.0233	96.79	.0242	96.22	.0069	98.87
3D CNN (4-layers) + 1-FCN	.0204	98.40	.0174	98.13	.0118	98.78
Our U2F (with Hybrid Conv)	**.0097**	**99.80**	**.0092**	**99.76**	**.0052**	**99.96**

Ablation Study. We conduct several ablation experiments on our dataset to analyze the contribution of different modules of U2F Net. Here, we particularly report (in Table 2) three primary variants of our model obtained by dropping the spatial layer, the temporal layer, and the channel shuffling. We can see that the network performance decreases by 1.88% and 1.44% in absence of the individual spatial and temporal encoding respectively. This shows that the hybrid block is a significant part of U2F which captures contrasting features to jointly learn the localization and tracking of tongue contour better. Similarly, the removal of shuffling block leads to an approximate decrease of the Mean R^2 by .84%. This is because channel shuffling mixes the independent as well as shared encodings and thereby enriches the input feature space for the last grouped CNN layer. All the ablation studies provide evidence in favour of our original model design.

Table 2. Ablation experiments - *Removal of spatial, temporal and shuffling blocks*

Method	$f1$		$f2$		$f1-f2$	
	MAE	$Mean\ R^2$	MAE	$MeanR^2$	MAE	$Mean\ R^2$
U2F w/o spatial kernels	.0178	97.78	.0267	97.69	.0113	98.08
U2F w/o temporal kernels	.0188	98.01	.0180	98.30	.0161	98.52
U2F w/o shuffling block	.0103	98.84	.0097	99.09	.0087	99.12

5 Discussion and Conclusion

The main contributions of our paper are four-fold:

1. We developed a novel spatio-temporal feature extraction strategy for mapping ultrasound tongue movement to formant trajectories. This involves replacing a chunk of the 3D convolutional layer by individual 2D spatial and 1D temporal convolutions for better feature encoding. A shuffling block is introduced to enable cross-feature information flow between spatial, temporal and spatio-temporal representations.
2. For the first time, we established a successful end-to-end mapping between the ultrasound tongue images and formant frequencies, that bridges the gap in SSI and opens a new dimension for articulatory speech research.
3. We provide evidences that our network has the ability to model an internal representation of tongue by optimizing a non-image based loss function. This demonstrates that the network has the potential to replace the manual selection of points for semi-automatic tongue contour extraction. This also shows the promise of using acoustic labels for tongue contour detection, thereby, replacing the need for tedious manual annotation for tongue tracing.
4. Our approach shows a striking improvement in performance over the baseline methods. We present an ablation study to explain the contribution of individual components towards better performance. Our network has the potential to encode robust spatio-temporal information in other related tasks.

Acknowledgements. This work was funded by the Natural Sciences and Engineering Research Council (NSERC) of Canada and Canadian Institutes for Health Research (CIHR).

References

1. Carreira, J., Zisserman, A.: Quo vadis, action recognition? A new model and the kinetics dataset. In: Proceedings of the IEEE Conference on Computer Vision and Pattern Recognition, pp. 6299–6308 (2017)
2. Csapó, T.G., Grósz, T., Gosztolya, G., Tóth, L., Markó, A.: DNN-based ultrasound-to-speech conversion for a silent speech interface. Proc. Interspeech **2017**, 3672–3676 (2017)

3. Dahl, G.E., Sainath, T.N., Hinton, G.E.: Improving deep neural networks for LVCSR using rectified linear units and dropout. In: 2013 IEEE International Conference on Acoustics, Speech and Signal Processing, pp. 8609–8613. IEEE (2013)
4. Denby, B., Oussar, Y., Dreyfus, G., Stone, M.: Prospects for a silent speech interface using ultrasound imaging. In: Proceedings of the 2006 IEEE International Conference on Acoustics Speech and Signal Processing, vol. 1, p. I. IEEE (2006)
5. Denby, B., Schultz, T., Honda, K., Hueber, T., Gilbert, J.M., Brumberg, J.S.: Silent speech interfaces. Speech Commun. **52**(4), 270–287 (2010)
6. Denby, B., Stone, M.: Speech synthesis from real time ultrasound images of the tongue. In: 2004 IEEE International Conference on Acoustics, Speech, and Signal Processing, vol. 1, pp. I–685. IEEE (2004)
7. Gilbert, M., et al.: Restoring speech following total removal of the larynx by a learned transformation from sensor data to acoustics. J. Acoust. Soc. Am. **141**(3), EL307–EL313 (2017)
8. Gosztolya, G., Pintér, Á., Tóth, L., Grósz, T., Markó, A., Csapó, T.G.: Autoencoder-based articulatory-to-acoustic mapping for ultrasound silent speech interfaces. In: 2019 International Joint Conference on Neural Networks (IJCNN), pp. 1–8. IEEE (2019)
9. Hueber, T., et al.: Eigentongue feature extraction for an ultrasound-based silent speech interface. In: 2007 IEEE International Conference on Acoustics, Speech and Signal Processing-ICASSP 2007, vol. 1, pp. I–1245. IEEE (2007)
10. Ioffe, S., Szegedy, C.: Batch normalization: accelerating deep network training by reducing internal covariate shift. arXiv preprint arXiv:1502.03167 (2015)
11. Juanpere, E.M., Csapó, T.G.: Ultrasound-based silent speech interface using convolutional and recurrent neural networks. Acta Acust. United Acust. **105**(4), 587–590 (2019)
12. Klatt, D.H.: Software for a cascade/parallel formant synthesizer. J. Acoust. Soc. Am. **67**(3), 971–995 (1980)
13. Luo, C., Yuille, A.L.: Grouped spatial-temporal aggregation for efficient action recognition. In: Proceedings of the IEEE International Conference on Computer Vision, pp. 5512–5521 (2019)
14. Mandal, M., Kumar, L.K., Saran, M.S., et al.: MotionRec: a unified deep framework for moving object recognition. In: The IEEE Winter Conference on Applications of Computer Vision, pp. 2734–2743 (2020)
15. O'Shaughnessy, D.: Formant estimation and tracking. In: Benesty, J., Sondhi, M.M., Huang, Y.A. (eds.) Springer Handbook of Speech Processing. SH, pp. 213–228. Springer, Heidelberg (2008). https://doi.org/10.1007/978-3-540-49127-9_11
16. Saha, P., Srungarapu, P., Fels, S.: Towards automatic speech identification from vocal tract shape dynamics in real-time MRI. Proc. Interspeech **2018**, 1249–1253 (2018)
17. Stavness, I., Lloyd, J.E., Fels, S.: Automatic prediction of tongue muscle activations using a finite element model. J. Biomech. **45**(16), 2841–2848 (2012)
18. Tóth, L., Gosztolya, G., Grósz, T., Markó, A., Csapó, T.G.: Multi-task learning of speech recognition and speech synthesis parameters for ultrasound-based silent speech interfaces. In: Interspeech, pp. 3172–3176 (2018)
19. Tran, D., Wang, H., Torresani, L., Ray, J., LeCun, Y., Paluri, M.: A closer look at spatiotemporal convolutions for action recognition. In: Proceedings of the IEEE Conference on Computer Vision and Pattern Recognition, pp. 6450–6459 (2018)
20. Zhang, X., Zhou, X., Lin, M., Sun, J.: Shufflenet: An extremely efficient convolutional neural network for mobile devices. In: Proceedings of the IEEE Conference on Computer Vision and Pattern Recognition, pp. 6848–6856 (2018)

Ultrasound Video Summarization Using Deep Reinforcement Learning

Tianrui Liu, Qingjie Meng, Athanasios Vlontzos, Jeremy Tan, Daniel Rueckert, and Bernhard Kainz(✉)

Department of Computing, Imperial College London, London, UK
b.kainz@imperial.ac.uk

Abstract. Video is an essential imaging modality for diagnostics, e.g. in ultrasound imaging, for endoscopy, or movement assessment. However, video hasn't received a lot of attention in the medical image analysis community. In the clinical practice, it is challenging to utilise raw diagnostic video data efficiently as video data takes a long time to process, annotate or audit. In this paper we introduce a novel, fully automatic video summarization method that is tailored to the needs of medical video data. Our approach is framed as reinforcement learning problem and produces agents focusing on the preservation of important diagnostic information. We evaluate our method on videos from fetal ultrasound screening, where commonly only a small amount of the recorded data is used diagnostically. We show that our method is superior to alternative video summarization methods and that it preserves essential information required by clinical diagnostic standards.

Keywords: Video summarization · Reinforcement learning · Ultrasound diagnostic

1 Introduction

Ultrasound is a popular modality in medical imaging because of its low cost, real-time capabilities, wide availability and safety. It's primary output is a video stream. It is challenging to utilize video data retrospectively since it often contains too much redundant information, is too large for easy documentation or audit and complicates remote assessment. Hence, finding a way to summarise the data without losing important information is vital.

In this paper, we present an ultrasound imaging summarization method using deep reinforcement learning. We show effectiveness for the example of fetal ultrasound screening. Given video captures from full examinations of 30 to 60 min per video, our goal is to select a small subset of frames to create a summary video that is much shorter but contains sufficient, dynamic and essential information to facilitate retrospective analysis. Our deep summarization network adopts an encoder-decoder convolutional neural network structure which first extracts visual features from frame sequence and then feeds these features into a

A. L. Martel et al. (Eds.): MICCAI 2020, LNCS 12263, pp. 483–492, 2020.
https://doi.org/10.1007/978-3-030-59716-0_46

bi-directional long short-term memory network (Bi-LSTM) for sequential modeling. The reinforcement learning (RL) network interprets the summarization task as a decision making process and takes actions on whether a frame should be selected for the summary set or not. The RL network maximizes expected rewards computed on the quality of the selected frames in terms of their representativeness, diversity, as well as the likelihood of being a standard diagnostic view plane [1]. The video summarization method can be trained in either a supervised or unsupervised way. Hence, in case the training process of our summarization network does not have clinical annotations, the proposed method can be trained in a fully unsupervised way which can still achieve encouraging performance.

Contribution: The contribution of this paper is three-fold: (1) We discuss a deep RL-based framework for ultrasound video summarization. To the best of our knowledge, this is the first method to use RL for this task. (2) We propose a novel diagnostic view plane reward for the RL network which encourages agents to select essential clinical information. (3) We take fetal screening as an example from the clinical practice and show experimental evidence for the effectiveness of our approach.

Related Work: There has been much work done in computer vision on general video summarization techniques. Early works adopt low-level or mid-level visual features to locate important segments of a video with a particular strategy such as clustering [7,9,10] and sparse dictionary learning [4,17]. In [18], long short-term memory (LSTM) has been used to model the frame-level features for video summarization. In [14], Rochan *et al.* [14] demonstrated that it is possible to model the video summarization task as an element-wise segmentation problem using fully convolutional sequence networks (FCSN). Recently, Zhou and Qiao [19] propose an RL-based deep network for general video summarization. They formulate the video summarization task as a sequential decision making process and generate video summaries by predicting the probabilities of a given frame being a key-frame. Video summarization fits the ideas of RL well. RL has become increasingly popular in medical imaging research due to its effectiveness for various tasks. For example, Alansary *et al.* [2,15] have shown that RL can be successfully used for landmark detection in medical image analysis.

In many existing video summarization methods [11,14,19], the ground-truth depends on subjective human perception of frame importance. This is more variable than defining factual image classification tasks. Different annotators may provide significantly different labels for the same video sequence. It usually requires more than ten human annotators to mitigate inter-observer variance. In medical image analysis we have the advantage that importance is often defined according to diagnostically decision criteria. In [5], a summarization approach for hysteroscopy data was proposed. Based on the motion estimation of the camera capturing the hysteroscopy videos, they make use of physicians' attention on video segments for data summarization. In [12], a video summarization-based tele-endoscopy service is introduced. They compute image moments, curvature, and multi-scale contrast to obtain the saliency map of each frame for key frames

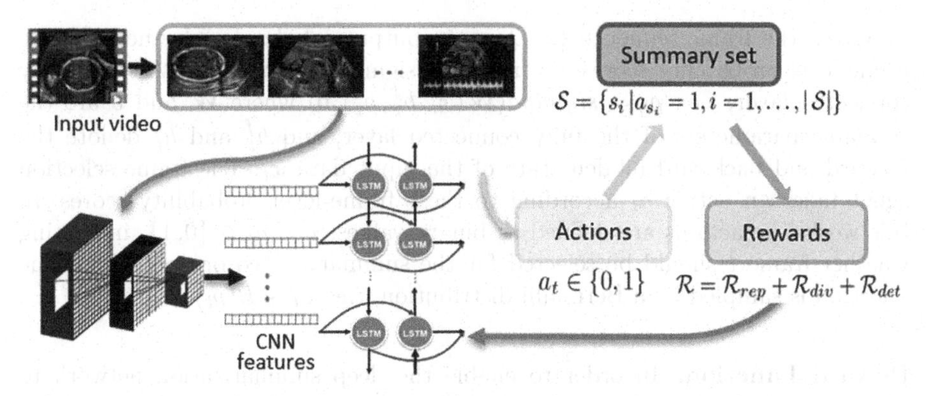

Fig. 1. Overview of the proposed ultrasound video summarization network using deep reinforcement learning. Given an input video, a feature extraction network first computes CNN feature representations for frame sequences and feeds them into a bi-directional LSTM neural network. Then, the RL network takes actions a_t on whether a frame should be selected for the summary set $\mathcal{S}$ or not. The RL network maximizes the expected rewards $\mathcal{R}$ computed on the quality of the selected frames in terms of their representativeness $\mathcal{R}_{rep}$, diversity $\mathcal{R}_{div}$, and likelihood of being a diagnostically valuable view plane $\mathcal{R}_{det}$.

selection. In this paper we explore how to effectively exploit the prior knowledge from diagnostically decision criteria for ultrasound videos.

2 Method

An overview over our RL-based ultrasound summarization network is illustrated in Fig. 1. Given an input video, a deep summarization network is used to extract deep feature representations from the input video sequence and sequentially models the frame features. It adopts an encoder-decoder convolutional network structure. Our encoder network is a diagnostic view plane detection network [3] pre-trained with ultrasound standard plane detection annotations. The decoder network takes the extracted feature maps of each input frame as input and feeds them into a Bi-LSTM to analyze features of both, past and future frames.

Following the feature extraction, the RL network interprets the diagnostic video summarization task as a decision making process, in which a decision is to include a current frame in the summary or not. The RL network accepts latent scores from the Bi-LSTM as input and takes actions a_t on whether a frame should be selected into the summary set $\mathcal{S}$ or not by maximizing the expected rewards $\mathcal{R}$. The rewards are computed on the quality of the selected frames in terms of their representativeness $\mathcal{R}_{rep}$, diversity $\mathcal{R}_{div}$, as well as the likelihood of being a standard diagnostic plane $\mathcal{R}_{det}$.

During training, the parameters of the decoder network will be learned using back-propagation, while the parameters of the encoder network are frozen.

Given the frame sequence $\{\boldsymbol{x}_t\}_{t=1}^{T}$, the outputs of the decoder network are frame-level probability scores, given by the sigmoid activation of the final fully connected layer, *i.e.*, $p_t = sigmoid(\boldsymbol{W}(\boldsymbol{x}_t, h_t^f, h_t^b), \boldsymbol{b})$ where $\boldsymbol{W}$ and $\boldsymbol{b}$ are the trainable parameters of the fully connected layer, and h_t^f and h_t^b denote the forward and backward hidden state of the input data $\boldsymbol{x}_t$. The frame selection agent takes an action a_t according to these frame-level probability scores. In this work, the actions are defined as binary values, *i.e.*, $a_t \in \{0, 1\}$, indicating whether frame t should be selected for the summary video or not. The frame selection is sampled by a Bernoulli distribution, *i.e.*, $a_t \sim B(p_t)$.

Reward Function. In order to enable the deep summarization network to select a good set of key frames for the video summary, the deep summarization network maximizes three reward terms during training:

$$\mathcal{R} = \mathcal{R}_{rep} + \mathcal{R}_{div} + \mathcal{R}_{det}, \tag{1}$$

where $\mathcal{R}_{det}$ evaluates the likelihood of a frame being a standard diagnostic plane, $\mathcal{R}_{rep}$ defines the representativeness reward and the diversity reward $\mathcal{R}_{div}$ evaluates the quality of the selected summary $\mathcal{S} = \{s_i \,|a_{s_i} = 1, i = 1, \ldots, |\,\mathcal{S}|\}$ in terms of their representativeness and diversity, respectively.

The representativeness reward $\mathcal{R}_{rep}$ is defined as

$$\mathcal{R}_{\text{rep}} = \exp(-\frac{1}{T}\sum_{t=1}^{T} \min_{t' \in \mathcal{S}} \|\boldsymbol{x}_t - \boldsymbol{x}_{t'}\|_2). \tag{2}$$

It measures how well the generated summary can represent the original video by minimizing the mean squared errors (MSE) between video frames and their nearest medoids. Maximizing $\mathcal{R}_{rep}$ can therefore help to preserve the temporal information across the entire diagnostic video.

The diversity reward $\mathcal{R}_{\text{div}}$ measures the dissimilarity between the selected frames of the summary video:

$$\mathcal{R}_{\text{div}} = \frac{1}{|\mathcal{S}|(|\mathcal{S}|-1)} \sum_{t \in \mathcal{S}} \sum_{i \in \mathcal{S}, i \neq t} d\left(\boldsymbol{x}_t, \boldsymbol{x}_i\right), \tag{3}$$

where $d(\cdot, \cdot)$ measures the cosine dissimilarity of two vectors.

We further propose a novel standard plane detection reward term $\mathcal{R}_{\text{det}}$ to encourage the agent to select essential diagnostic information.

$$\mathcal{R}_{det} = \frac{1}{|\mathcal{S}|} \sum_{i \in \mathcal{S}} s_i \left(\delta(y_i - 1) - \delta(y_i)\right), \tag{4}$$

where $\delta(\cdot)$ is a delta function, i.e. $\delta(y_i) = 1$ if $y_i = 0$ and otherwise $\delta(y_i) = 0$. $y_i = 1$ indicates that the i-th frame in $\mathcal{S}$ is classified as standard diagnostic view plane. s_i is the standard plane detection score of the i-th frame in the summary set $\mathcal{S}$ resulting from the encoder network topped up with a softmax layer.

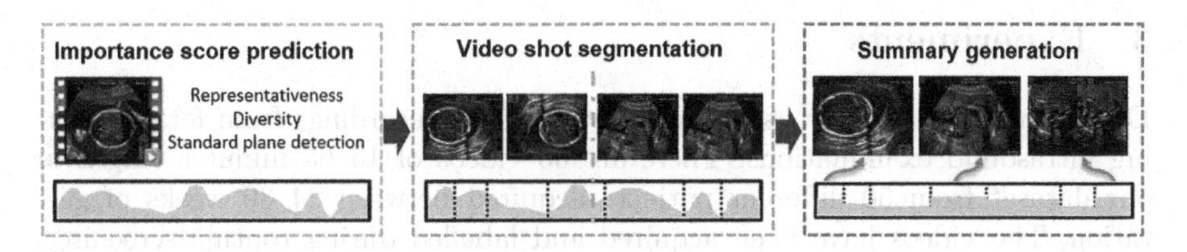

Fig. 2. Sequential summary of ultrasound video summary generation modules.

Optimization. The learning objective is to train a video summarization agent for an optimal policy π which indicates actions to take to maximize the overall reward. The expected reward $\mathcal{J}(\theta)$ is defined as $\mathcal{J}(\theta) = \mathbb{E}_{p_\theta(a|\pi)}(\mathcal{R})$ where $p_\theta(a|\pi)$ denotes the probability distribution over the actions of sequences.

The proposed ultrasound video summarization method can be trained in either a supervised or an unsupervised way. For supervised training, we utilize a loss term that promotes to minimize the MSE between the predicted frame-wise importance scores and ground-truth scores, *i.e.*,

$$\mathcal{L}_{pred} = \frac{1}{T}\sum_{t=1}^{T} \|p_t - p_t^*\|^2, \tag{5}$$

where p_t and p_t^* are the predicted frame-wise importance scores and the ground-truth user annotation scores, respectively.

A regularization term [19] is applied to penalize the selection of a large number of frames in the summary set.

$$\mathcal{L}_{reg} = \|\frac{1}{T}\sum_{t=1}^{T} p_t - \epsilon\|^2, \tag{6}$$

where ϵ is a scalar controlling the proportion of the selected frames and λ controls the relative importance of the two loss terms.

In our summarization network, the loss terms and reward terms are jointly optimized in an end-to-end manner. Thus, the total cost for the video summarization network is formulated as $\mathcal{L}_{sup} = \mathcal{L}_{pred} + \beta\mathcal{L}_{reg} - \gamma\mathcal{R}$ where β and γ are parameters to control the relevant importance of the costs.

In case that ground truth key frames through clinical annotations are limited or unavailable, our method can be used as fully unsupervised model by jointly optimizing the regularization and reward terms, *i.e.*, $\mathcal{L}_{unsup} = \beta\mathcal{L}_{reg} - \gamma\mathcal{R}$.

Video Summary Generalization. Once we get the frame-level importance scores via the deep RL network, we can generate the video summary as given in Fig. 1. First, the input video is segmented into *shots* using Kernel temporal segmentation (KTS) [13]. Then, we generate the video summaries by selecting the shots with the highest scores while keeping the duration of summaries below a threshold (*e.g.*, 15% duration of the original video). The importance score of an shot equals to the average score of the frames in that shot.

3 Experiments

Data. In this paper, we use screen capture video recordings from fetal screening ultrasound examinations. There are 50 videos of 13–65 minutes length in our dataset from 50 different patients acquired between 24–30 weeks of gestation. The videos have been acquired and labelled during routine screenings according to the guidelines in the UK National Health Service (NHS) FASP handbook [1]. The feature extraction network is trained on annotations indicating the type of standard ultrasound diagnostic plane. From all available FASP planes we have selected Brain (Cb.), Brain (Tv.), Profile, Lips, Abdominal, Kidneys, Femur, Spine (Cor.), Spine (Sag.), 4CH, 3VV, RVOT, LVOT as the most frequent exemplars.

For ultrasound video summarization, we take the freeze-frame images which are saved by the sonongraphers during the scan as the ground-truth key frames. We follow the steps in [6,18] to convert key frame annotations into frame-level scores. The videos are temporally segmented into disjoint intervals using KTS [13]. If an interval (*i.e.*, a shot) contains at least one key frame, we take this shot as a key shot and mark all the frames of it with score 1 and otherwise 0.

Evaluation Metrics. Following the protocols proposed for general video summarization methods [6,8,16], we compute the precision (P) and recall (R) according to the temporal overlap between a user annotated summary $\mathcal{A}$ and a predicted summary $\mathcal{B}$, *i.e.*,

$$P = \frac{\mathcal{A} \cup \mathcal{B}}{|A|}, \qquad R = \frac{\mathcal{A} \cup \mathcal{B}}{|B|}, \tag{7}$$

where $|\mathcal{A}|$ denotes the duration of a summary $\mathcal{A}$ and $|\mathcal{A} \cup \mathcal{B}|$ is the temporal overlap between them two. The harmonic mean F1-score against is $F = 2P \times R/(P+R)$.

Implementation. The proposed video summarization approach is implementations in PyTorch. For the RL algorithm, the number of episodes of the episodic reinforcement learning algorithm is fixed to 5. Stochastic Gradient Descent with momentum ($\rho = 0.9$) and a weight decay of $1e^{-5}$ is used to train the models. The initial learning rate is set to $1e^{-4}$ and subsequently reduced by a factor of 0.5 for every 50 epochs. The maximum training epoch is set as 300. We set $\lambda = 0.01$ and $\epsilon = 0.5$ for Eq. 6. β and γ are set as 0.01 and 1. The experiments are performed on a single TITAN RTX GPU.

3.1 Experimental Results

We conduct experiments and compare with conventional state-of-the-art video summarization methods from traditional computer vision literature, including FCSN [14] and DR-DSN [19]. We report the results in Table 1. For fair comparison, we use the same feature extraction network which has been fine-tuned

Table 1. Comparison to state-of-the-art video summarization methods. Best results in bold.

Methods	F1-scores
FCSN_{sup} [14]	59.17
DR-DSN_{sup} [19]	60.34
Proposed_{sup}(our)	**63.29**
DR-DSN_{unsup} [19]	40.92
Proposed_{unsup}(our)	**56.73**

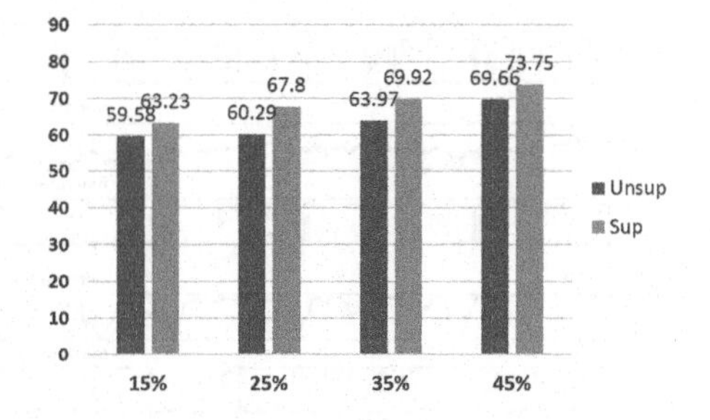

Fig. 3. Performance of supervised vs. unsupervised summarization for different summary lengths. (Colour figure online)

Table 2. Comparison of results using different combinations of reward terms.

Rewards			Learning paradigm					
$\mathcal{R}_{rep}$	$\mathcal{R}_{div}$	$\mathcal{R}_{det}$	Supervised			Unsupervised		
			F	P	R	F	P	R
×	×	×	58.41	57.43	59.52	-	-	-
✓	×	×	59.36	58.27	60.59	37.65	37.25	38.07
×	✓	×	57.34	56.41	58.40	32.57	32.03	33.15
✓	✓	×	59.20	58.22	60.31	38.82	38.42	39.24
×	×	✓	59.95	58.94	61.10	44.28	43.48	45.18
✓	×	✓	61.98	60.88	63.22	54.56	53.53	55.72
×	✓	✓	62.33	61.21	63.60	49.23	48.35	50.22
✓	✓	✓	63.23	62.08	64.54	59.58	58.56	60.74

with our ultrasound standard plane annotations for the FCSN and DR-DSN approaches.

We performed experiments on five different splits of training and testing subsets of percentages 80% and 20%, *i.e.*, 40 videos for training and 10 videos for testing each time. The averaged F1 scores for both unsupervised and supervised learning paradigm are compared in Table 1. As we can see, the proposed ultrasound video summarization approach novel can effectively exploit the prior knowledge from standard diagnostic plane, leads to significant improvements for the performance especially for unsupervised model.

Ablation Study About the Effectiveness of Rewards. We conduct ablation studies to investigate the effectiveness of the reward terms. For all the experiments in this section, we keep the same split of training and testing video.

Table 2 reports experiments for the proposed approach with different combinations of reward terms. As this table shows, As we can see, the proposed novel diagnostic view reward, *i.e.*, $\mathcal{R}_{det}$ for ultrasound video, leads to significant

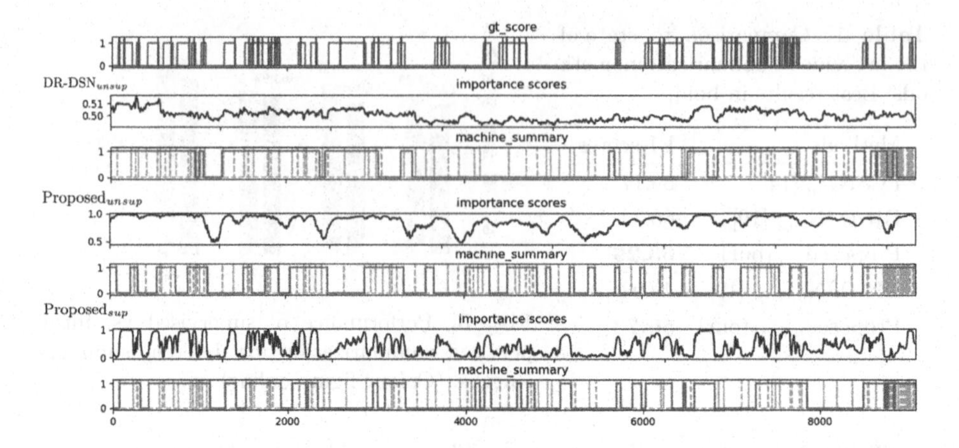

Fig. 4. Comparison of video summary result on exemplar ultrasound video.

improvements for the summarization performance especially for unsupervised model. By using the standard plane detection reward $\mathcal{R}_{det}$ alone, the summarization performance can be as good as 59.95 regarding F1-score for the supervised model and 44.28 for the unsupervised model. Compared to the results of unsupervised learning using $\mathcal{R}_{div}$ and $\mathcal{R}_{rep}$ rewards on their own, the F1-score improves 11.45 and 6.25, respectively. For supervised learning, the improvements are 2.61 and 0.6. Using the combination of all the three reward terms leads to the highest scores, which are 63.23 and 59.58 for the supervised and the unsupervised model, respectively.

Results with Different Summary Lengths. The above experiments are conducted with a constant summary length constraint of 15%, *i.e.*, the length of the summary videos are restricted to be shorter than 15% of the input video length. We also perform experiments on summaries generated with four different summary length constraints: 15%, 25%, 35% and 45% and show the results in Fig. 3. When the video summarization network is allowed to select more key shots, the F1 scores increase for both supervised models (green bars) and unsupervised models (blue bars).

Qualitative Results. Figure 4 shows a visualization of an example video summary generated by the proposed method and a baseline method [19]. The ground-truth (gt) summary is shown at the top, where the gt key frames and gt scores are shown in green and red, respectively. We observe that the summary result using our approach have higher percentage of overlap with the gt. This implies that our method is able to preserve essential information for generating optimal and meaningful summaries.

4 Conclusion

We have proposed an RL-based deep learning model for effective diagnostic video summarization. The proposed framework has the potential to save storage costs as well as to increase efficiency when browsing patient video data during retrospective analysis or audit without loosing essential information. Both the supervised and unsupervised training model of our method can achieve good performance. Hence, our ultrasound video summarization method can be used for a variety of applications also when clinical annotations are unavailable. Experiments and ablation studies show that the proposed novel diagnostic view reward leads to significant improvements for the summarization performance and is able to summarize ultrasound videos without discarding important information. Future work will focus on experiments for other kind of diagnostic videos such as endoscopy videos or physiotherapeutic movement assessment videos.

Acknowledgements. We thank the volunteers and sonographers from routine fetal screening at St. Thomas' Hospital London. This work was supported by the Wellcome Trust IEH Award [102431] and EPSRC EP/S013687/1. The research was funded/supported by the National Institute for Health Research (NIHR) Biomedical Research Center based at Guy's and St Thomas' NHS Foundation Trust, King's College London and the NIHR Clinical Research Facility (CRF) at Guy's and St Thomas'. Data access only in line with the informed consent of the participants, subject to approval by the project ethics board and under a formal Data Sharing Agreement. The views expressed are those of the author(s) and not necessarily those of the NHS, the NIHR or the Department of Health.

References

1. Fetal anomaly screening programme: handbook for ultrasound practitioners, (2015)
2. Alansary, A., et al.: Evaluating reinforcement learning agents for anatomical landmark detection. Med. Image Anal. **53**, 156–164 (2019)
3. Baumgartner, C.F., et al.: Sononet: real-time detection and localisation of fetal standard scan planes in freehand ultrasound. IEEE Trans. Med. Imaging **36**(11), 2204–2215 (2017)
4. Cong, Y., Yuan, J., Luo, J.: Towards scalable summarization of consumer videos via sparse dictionary selection. IEEE Trans. Multimedia **14**(1), 66–75 (2011)
5. Gavião, W., Scharcanski, J., Frahm, J.M., Pollefeys, M.: Hysteroscopy video summarization and browsing by estimating the physician's attention on video segments. Med. Image Anal. **16**(1), 160–176 (2012)
6. Gong, B., Chao, W.L., Grauman, K., Sha, F.: Diverse sequential subset selection for supervised video summarization. In: Ghahramani, Z., Welling, M., Cortes, C., Lawrence, N.D., Weinberger, K.Q. (eds.) Advances in Neural Information Processing Systems, Curran Associates, Inc. **27**, pp. 2069–2077 (2014), http://papers.nips.cc/paper/5413-diverse-sequential-subset-selection-for-supervised-video-summarization.pdf
7. Gygli, M., Grabner, H., Riemenschneider, H., Van Gool, L.: Creating summaries from user videos. In: ECCV, (2014)

8. Gygli, M., Grabner, H., Van Gool, L.: Video summarization by learning submodular mixtures of objectives. In: Proceedings of the IEEE Conference on Computer Vision and Pattern Recognition, pp. 3090–3098 (2015)
9. Kuanar, S.K., Panda, R., Chowdhury, A.S.: Video key frame extraction through dynamic delaunay clustering with a structural constraint. J. Visual Commun. Image Represent. **24**(7), 1212–1227 (2013)
10. Liu, T., Chan, S.: Automatic shot boundary detection algorithm using structure-aware histogram metric. In: 2014 19th International Conference on Digital Signal Processing, pp. 541–546 (2014)
11. Liu, T., Kender, J.R.: Optimization algorithms for the selection of key frame sequences of variable length. In: Heyden, A., Sparr, G., Nielsen, M., Johansen, P. (eds.) ECCV 2002. LNCS, vol. 2353, pp. 403–417. Springer, Heidelberg (2002). https://doi.org/10.1007/3-540-47979-1_27
12. Mehmood, I., Sajjad, M., Baik, S.W.: Video summarization based tele-endoscopy: a service to efficiently manage visual data generated during wireless capsule endoscopy procedure. J. Med. Syst. **38**(9), 109 (2014)
13. Potapov, D., Douze, M., Harchaoui, Z., Schmid, C.: Category-specific video summarization. In: Fleet, D., Pajdla, T., Schiele, B., Tuytelaars, T. (eds.) ECCV 2014. LNCS, vol. 8694, pp. 540–555. Springer, Cham (2014). https://doi.org/10.1007/978-3-319-10599-4_35
14. Rochan, M., Ye, L., Wang, Y.: Video summarization using fully convolutional sequence networks. In: Proceedings of the European Conference on Computer Vision (ECCV), pp. 347–363 (2018)
15. Vlontzos, A., Alansary, A., Kamnitsas, K., Rueckert, D., Kainz, B.: Multiple landmark detection using multi-agent reinforcement learning. In: Shen, D., et al. (eds.) MICCAI 2019. LNCS, vol. 11767, pp. 262–270. Springer, Cham (2019). https://doi.org/10.1007/978-3-030-32251-9_29
16. Yale Song, Vallmitjana, J., Stent, A., Jaimes, A.: Tvsum: summarizing web videos using titles. In: 2015 IEEE Conference on Computer Vision and Pattern Recognition (CVPR), pp. 5179–5187 (2015) https://doi.org/10.1109/CVPR.2015.7299154
17. Yang, M., Dai, D., Shen, L., Van Gool, L.: Latent dictionary learning for sparse representation based classification. In: The IEEE Conference on Computer Vision and Pattern Recognition (CVPR), (2014)
18. Zhang, K., Chao, W.L., Sha, F., Grauman, K.: Video summarization with long short-term memory. In: Leibe, B., Matas, J., Sebe, N., Welling, M. (eds.) ECCV 2016. LNCS, vol. 9911, pp. 766–782. Springer, Cham (2016). https://doi.org/10.1007/978-3-319-46478-7_47
19. Zhou, K., Qiao, Y., Xiang, T.: Deep reinforcement learning for unsupervised video summarization with diversity-representativeness reward. In: Thirty-Second AAAI Conference on Artificial Intelligence, (2018)

Predicting Obstructive Hydronephrosis Based on Ultrasound Alone

Lauren Erdman[1,2,3,4(✉)], Marta Skreta[1,2,4], Mandy Rickard[5], Carson McLean[1,2,3], Aziz Mezlini[1,2], Daniel T. Keefe[5], Anne-Sophie Blais[5], Michael Brudno[1,2,3,4], Armando Lorenzo[5], and Anna Goldenberg[1,2,3,6]

[1] Department of Computer Science, University of Toronto, Toronto, Canada
[2] Program in Genetics and Genome Biology, Hospital for Sick Children, Toronto, Canada
[3] Vector Institute, Toronto, Canada
[4] Center for Computational Medicine, SickKids Hospital, Toronto, Canada
lauren.erdman@sickkids.ca
[5] Department of Surgery, Division of Urology, Hospital for Sick Children, Toronto, Canada
[6] Child and Brain Development, Canadian Institute for Advanced Research (CIFAR), Toronto, Canada

Abstract. Prenatal hydronephrosis (HN) makes up nearly 30% of pediatric Urology Department visits, yet remains challenging to prognosticate without repeated ultrasounds and invasive clinical tests. We build a deep learning model, which uses still images from kidney ultrasound as input and predicts whether HN is due to an obstruction that will receive surgical intervention. We compare our custom convolutional neural network performance against other existing state-of-the-art models. Our best model predicts obstruction with an AUC of 0.93 and an AUPRC of 0.75 in a prospective test set of 89 patients (286 repeated kidney ultrasounds). We show that while maintaining a 5% false negative rate, our classifier identifies 58% of those who will have surgery due to obstruction yet received a functional renogram, indicating that this model could feasibly reduce the amount of testing done in more than half of non-surgical cases. This work demonstrates the ability of deep learning to predict obstructive HN with clinically relevant accuracy based on kidney ultrasound alone, without requiring other clinical variables as input. This algorithm has the potential to change clinical practice by stratifying HN patient risk, reducing repeated follow ups and invasive testing for less severe cases, and bringing more consistency to clinical management.

Keywords: Deep learning · Ultrasound · Pediatric urology · Hydronephrosis

1 Introduction

Prenatal hydronephrosis (HN) refers to the dilatation of the urinary tract (kidneys and/or ureters) in a fetus and occurs in up to 5% of pregnancies [1, 2]. Postnatally, HN may be

L. Erdman and M. Skreta—Co-first authors, A. Lorenzo and A. Goldenberg—Co-last authors.

Electronic supplementary material The online version of this chapter (https://doi.org/10.1007/978-3-030-59716-0_47) contains supplementary material, which is available to authorized users.

A. L. Martel et al. (Eds.): MICCAI 2020, LNCS 12263, pp. 493–503, 2020.
https://doi.org/10.1007/978-3-030-59716-0_47

self-limiting or secondary to an underlying pathology such as obstruction or retrograde flow of urine from the bladder to the kidney (vesicoureteral reflux, VUR). Differentiating between transient and pathologic HN from initial or early clinic visits is difficult as most cases will spontaneously resolve with no surgical intervention but spontaneous resolution may take up to 36 months or longer to occur [3, 4]. As a result, all infants with HN receive ongoing monitoring for obstructive HN requiring surgical intervention ("obstructive HN" here forward) regardless of dilation severity. The majority of HN patients are followed with serial ultrasounds requiring frequent clinic visits (i.e. every 3 months), typically for several years. In addition to routine monitoring with ultrasound, many infants will be subjected to invasive testing involving ionizing radiation and radioisotope exposure, requiring painful procedures such as intravenous insertion and urethral catheterization [5, 6]. An example of a standard diagnostic test for HN is a diuretic renogram, used to identify obstruction and potential loss of renal function [5]. However, there are drawbacks of functional renal testing, including the requirements of IV insertion, urethral catheterization, IV administration of medication and radioisotope and ionizing radiation exposure [7]. Therefore, a reduction in functional renograms would result in cost savings to the patient and health system as well as a reduction in harmful exposures to very young patients. Reducing this and other exposures from invasive testing motivates our desire to improve selectivity of the children who are investigated with invasive testing and ultimately reduce the amount of invasive testing at our hospital.

We hypothesize that deep learning can be used to elucidate whether HN is obstructive or will resolve on its own. Because this determination is currently made after repeated ultrasounds and invasive tests, this technology would have the potential to reduce this burden and improve the consistency regarding who receives surgery for HN obstruction. Deep learning has the ability to make accurate predictions in physiology and disease based on images alone [8]. This methodology has been particularly successful in identifying patterns in medical image datasets using various imaging modalities, including ultrasound [9–11]. Dhindsa et al. used deep learning to automatically assign Society for Fetal Urology (SFU) grades to hydronephrotic renal ultrasound without manual feature engineering. Their algorithm was able to distinguish high grade (grade 3 or 4) from low grade (grade 1 or 2) with 80% accuracy [12]. However, their work involved the automatic assignment of an SFU HN grade, which is only partially correlated with resolution [13]. Therefore, the aim of our work is to generate a systematic and automatic prediction for obstructive HN based on renal ultrasounds alone, expediting required interventions when necessary and reducing the number of invasive tests and duration of clinical follow up in infants whose HN will resolve.

2 Methods

Study Setting, Inclusion and Exclusion Criteria. This study takes place in a pediatric urology clinic in a quaternary, university affiliated pediatric teaching institution in a single-payer healthcare system. The clinic is staffed by 5 full time urology physicians, as well as pediatric nurse practitioners, medical trainees and nurses, resulting in slight variabilities of care for infants with HN between clinicians. We include infants less than

24 months of age at baseline with HN. In surgical patients with bilateral dilatation, we only include the renal unit that required surgical repair. Patient characteristics are described in Table S1. Our label of interest is obstructive HN, determined by if the kidney ultimately received surgery to resolve kidney dilation.

Dataset and Image Processing. We use serial ultrasound imaging from 294 patients. For each patient, transverse and sagittal images are selected for the right and left kidneys and labelled by view and side. Ultrasounds of left and right kidneys and ultrasounds from multiple visits from the same patient are treated as independent observations. Images are cropped to remove identifiable text annotations and ultrasound beam borders. Cropping is done by selecting a square in the centre of the image smaller than the width of the original image by a factor of 2. The contrast of the images is normalized using contrast limited adaptive histogram equalization [15]. The images are resized to 256×256 pixels. Individuals in our data set are divided into a training/validation (70%, n patients = 205, n kidneys = 1359) and held-out test set (30%, n patients = 89, n kidneys = 286) such that individuals in the test set are imaged more recently than individuals in the training set in order to evaluate our model's performance in hypothetical future patients [15]. Ultrasound images for a given patient are not split across the training and test set. Our training set shows 11.5% rate of obstruction and our test set shows 10.8%.

Model Architectures. We investigate the performance of various convolutional model architectures including models with a single kidney ultrasound view (sagittal- and transverse-specific CNN), a dual ultrasound view (Siamese sagittal and transverse CNN), architectures of off-the-shelf models that have achieved state-of-the-art (SOTA) computer vision task performance, and an ensemble of views and architectures. All our CNN models are provided a cropped ultrasound image (or two images) as input and output a single probability of obstructive HN. Our custom single-view model consists of seven convolutional layers, followed by two fully-connected layers, and a softmax binary output layer. Each layer is activated using the ReLU function. We do batch normalization on all convolution layers and max-pooling in the first, second, fifth, and seventh layers. We use dropout with a rate of 0.5 in the first fully-connected layer to reduce overfitting. Dropout is turned off during validation and testing. Our Siamese model, taking in two images with shared weights is identical to the single-model, except that two images pass through the convolutional layers instead of 1 (Fig. 1, Table S2). View-specific image features are concatenated and passed to the fully connected layers. This concatenation leads to the first fully connected layer to be double the width of the single view model.

We compare our model to three CNNs (VGG-16 with batch normalization [17], ResNet-18 [18], and DenseNet-121 [19]) that achieved SOTA performance in object recognition. These networks give us the opportunity to test whether more connections (VGG), residual connections (ResNet), or a network that has achieved SOTA performance on a medical imaging dataset would outperform our own architecture. We use the networks as implemented in PyTorch v1.0.1, but we modify the output of the network to have the same dimensionality as F1 outlined in Fig. 1 and implement layers F2-F4 to achieve the dual view architecture. We did a grid search across batch-size and learning rate to find optimal hyperparameters for each network. Finally, we generated ensemble predictions from our models by combining the training predictions in a logistic regression with the true value as the outcome. Weights combining our model outputs

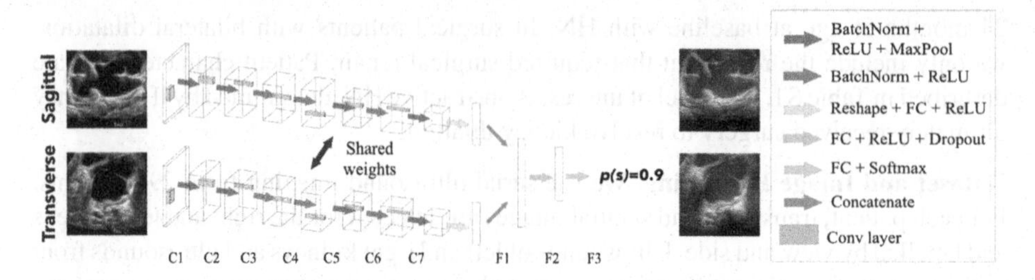

Fig. 1. Siamese model predicting obstructive HN from sagittal and transverse kidney view.

are therefore learned in a training set and applied to the validation and test set in order to create an ensemble model by view:

$$y_{view\ ens} = exp\left\{\alpha + \beta_{sag}\hat{y}_{sag} + \beta_{trans}\hat{y}_{trans}\right\} \tag{1}$$

and ensemble of our view-ensembled models:

$$y_{\text{ mod }\ ens} = exp\left\{\alpha + \sum \beta_{arch}\hat{y}_{arch}\right\} \tag{2}$$

where $y_{view\ ens}$ is the probability of obstruction as calculated by an ensemble of the predictions from each individual view: $\hat{y}_{sag}$ and $\hat{y}_{trans}$ and $y_{\text{ mod }\ ens}$ is the probability of obstruction as calculated by an ensemble of the model-specific ensemble estimates from each $\hat{y}_{arch}$, representing each of the CNN model architectures. All models described were fit with and without pre-training of all the model weights using the ImageNet benchmark data set [20]. All models were trained using PyTorch v1.0.1 on a NVIDIA® Tesla® V100 TensorCore GPU [21].

Loss Function. We train all models by minimizing the cross-entropy loss between the model output and binary obstruction labels. For each input sample, the loss has the following form:

$$L_{CE}(x_s, x_t, y) = -y\ log(\hat{y}) - (1-y) log(1-\hat{y}) \tag{3}$$

where x_s and x_t are the sagittal and transverse views, respectively, of a given kidney, y is the ground truth label, y_c is the class label, and $\hat{y} = p(y_c 1|x_s, x_t)$. For a given minibatch, m, the total loss was the average loss of the training samples.

Training and Optimization. We used stochastic gradient descent with a learning rate of 0.001, a minibatch size of 128, a weight decay of 0.005, and a momentum of 0.9. Initialization weights and biases were drawn from a zero-mean uniform distribution. We trained each model for 50 epochs and the minimum average 5-fold cross-validation loss to identify our stopping epoch. We then trained a final model on all the training data with the chosen hyperparameters and evaluated it on our held-out test set.

Model Calibration. Modern neural networks are generally poorly calibrated, leading to difficulties in interpreting reported output [22]. Calibration allows the reported model output to be framed as a confidence of the outcome, that is an output of 0.8 would represent 80% likelihood of the given kidney receiving a surgical intervention. To calibrate our output, we use Platt scaling which builds a logistic map from our model output to a calibrated output based on our validation data [23, 24].

Evaluation Metrics. We evaluate our model performances using Area Under the Receiver Operating Characteristics (AUROC) and Area Under Precision-Recall Curve (AUPRC). AUROC, which plots the true positive rate (TPR) against the false positive rate (FPR), is the probability of ranking a randomly-chosen positive sample higher than a randomly-chosen negative one. AUPRC considers both precision (PPV) and recall and does not take into account true negative information, which is useful for our imbalanced dataset since predicting the positive class correctly is more important. We do not use accuracy as a metric due to the overwhelming number of true negatives.

Empirical Confidence Intervals. To evaluate the significance of our model's performance in prospective samples, we calculate empirical confidence intervals for our AUROC and AUPRC values. This was done by resampling our predicted values and true values 999 times with replacement to generate an AUROC and AUPRC from a synthetic dataset from the same population.

Gradient Weighted Class Activation Maps. To further validate our classifier's performance, we generate Gradient-weighted Class Activation Maps (Grad-CAMs) to assess the areas of the images our classifier is highlighting as important and determine whether or not the prediction seems reasonable [25]. Grad-CAMs are able to adequately support tasks that require explanation as they are dependent on model parameters and training data [26]. We generate a Grad-CAM by passing our input ultrasound image(s) through our best performing model (the custom siamese CNN) and extracting the gradient of the predicted class with respect to feature maps from the final convolutional layer. We activate them with a ReLU function, which results in a heatmap outlining which image features positively influence the prediction for the given class. Finally, we upsample the heatmap to overlay it on the original image.

3 Results

A diverse Set of Deep Learning Algorithms All Classify Obstructive HN Based on Kidney Ultrasound Images Alone. We predict obstructive HN using only ultrasound images in a variety of models, highlighting the presence of a strong signal that is largely invariant to model selection and pre-training. With ImageNet pre-training, our custom model shows an AUROC > 0.9 and an AUPRC > 0.75 on all models containing the sagittal kidney view (Table 1). All transverse view models show an AUROC < 0.9. DenseNet-121 models produced best AUPRC values within each kidney view set, whereas our custom model produced the best test AUROC value. Model performance with randomly initialized weights was slightly lower but not significantly different (Table 2).

Class Activation Maps Help Qualify Confidence in Individual Predictions. We demonstrate that our model can localize areas in ultrasound images most indicative of an obstruction vs resolution prediction such as the parenchyma and pyramids surrounding the dilatation of the kidney, and does not merely classify more dilated kidneys

Table 1. Model performance for models pre-trained with ImageNet. Top validation and test AUROC and AUPRC indicated for each kidney ultrasound view in bold.

Model	# trainable parameters	Training AUROC (95% CI) AUPRC (95% CI)	Validation AUROC (95% CI) AUPRC (95% CI)	Test AUROC (95% CI) AUPRC (95% CI)
Sagittal view only				
VGG-16	156.3 M	0.981 (0.979,0.984) 0.725 (0.712,0.738)	0.850 (0.815,0.881) 0.764 (0.734,0.792)	0.916 (0.883,0.949) 0.767 (0.709,0.832)
ResNet-18	11.4 M	0.983 (0.980,0.986) 0.724 (0.709,0.739)	**0.870 (0.837,0.899)** 0.758 (0.731,0.789)	0.901 (0.859,0.939) 0.766 (0.699,0.836)
DenseNet-121	8.3 M	0.933 (0.924,0.942) 0.738 (0.723,0.751)	0.827 (0.792,0.859) 0.771 (0.744,0.802)	0.813 (0.722,0.894) **0.782 (0.718,0.848)**
Custom	12.5 M	1.000 (1.000, 1.000) 0.719 (0.705, 0.733)	0.675 (0.637, 0.714) **0.801 (0.772, 0.829)**	**0.932 (0.901, 0.959)** 0.758 (0.695, 0.827)
Transverse view only				
VGG-16	156.3 M	0.993 (0..991,0.994) 0.721 (0.708,0.735)	0.832 (0.796,0.866) 0.766 (0.736,0.796)	**0.868 (0.799,0.927)** 0.768 (0.708,0.844)
ResNet-18	11.4 M	0.986 (0.983,0.988) 0.723 (0.708,0.736)	**0.844 (0.808,0.876)** 0.764 (0.735,0.795)	0.798 (0.701,0.886) 0.781 (0.720,0.849)
DenseNet-121	8.3 M	0.965 (0.959,0.969) 0.728 (0.714,0.741)	0.816 (0.777,0.852) **0.772 (0.745,0.800)**	0.771 (0.667,0.868) 0.788 (0.728,0.859)
Custom	12.5 M	1.000 (0.999, 1.000) 0.719 (0.705, 0.733)	**0.844 (0.810, 0.874)** 0.765 (0.736, 0.796)	0.756 (0.668, 0.845) **0.789 (0.730, 0.852)**
Siamese Sagittal and Transverse view				
VGG-16	156.3 M	0.972 (0.968,0.976) 0.724 (0.710,0.739)	0.869 (0.842,0.896) 0.756 (0.726,0.785)	0.908 (0.869,0.945) 0.763 (0.702,0.828)
ResNet-18	11.4 M	0.980 (0.970,0.983) 0.722 (0.709,0.737)	**0.872 (0.843,0.899)** 0.756 (0.726,0.783)	0.890 (0.844,0.928) 0.767 (0.707,0.833)
DenseNet-121	8.4 M	0.989 (0.986,0.990) 0.724 (0.712,0.737)	0.858 (0.832,0.879) **0.761 (0.739,0.782)**	0.840 (0.759,0.905) **0.781 (0.720,0.850)**
Custom	12.7 M	0.999 (0.999, 0.999) 0.719 (0.706, 0.733)	0.863 (0.834, 0.891) 0.759 (0.730, 0.788)	**0.932 (0.894, 0.959)** 0.751 (0.695, 0.819)
Ensemble Sagittal and Transverse view				
VGG-16	312.6 M	0.992 (0.990,0.993) 0.723 (0.711,0.736)	0.834 (0.806,0.860) 0.766 (0.745,0.786)	0.914 (0.873,0.947) 0.761 (0.699,0.825)
ResNet-18	11.8 M	0.989 (0.986,0.990) 0.724 (0.712,0.737)	**0.858 (0.832,0.879)** 0.761 (0.739,0.782)	0.904 (0.862,0.942) 0.764 (0.705,0.828)
DenseNet-121	16.6 M	0.963 (0.958,0.968) 0.729 (0.718,0.743)	0.822 (0.795,0.847) 0.772 (0.749,0.796)	0.843 (0.769,0.903) **0.775 (0.714,0.838)**

(continued)

Table 1. *(continued)*

Model	# trainable parameters	Training AUROC (95% CI) AUPRC (95% CI)	Validation AUROC (95% CI) AUPRC (95% CI)	Test AUROC (95% CI) AUPRC (95% CI)
Custom	25 M	1.000 (1.000,1.000) 0.719 (0.704,0.734)	0.743 (0.704,0.783) **0.781 (0.754,0.811)**	**0.930 (0.897,0.959)** 0.757 (0.689,0.824)
Ensemble of all Sagittal and Transverse models				
Full ensemble	365.9 M	0.965 (0.956,0.973) 0.851 (0.841,0.860)	0.742 (0.630,0.840) 0.779 (0.724,0.848)	0.908 (0.870,0.941) 0.761 (0.700,0.826)

as having an obstruction (Fig. 2). Since our model automates curation and feature extraction from ultrasound images, text input by the ultrasound technician (e.g. "SAG LT") and edges surrounding the ultrasound beam may still remain visible in images used to classify obstructive HN. These areas of the image are noise and we find that our model successfully ignores them. In test set images that result in an incorrect prediction, we find many special clinical cases, such as a kidney with 3% function which our algorithm predicted would receive surgery due to an unresolving obstruction but the clinical team decided not to surgically intervene. Because we expect our algorithm to assist in clinical decision making, not replace it, these cases demonstrate real world scenarios in which a typical HN case would have received surgery and would rightly be predicted as such but the clinical team decided against it after obtaining auxiliary information.

Invasive Prognostic Tests in Young Children can Be Reduced with This Technology. Our data shows 142 (48%) patients with grade 4 HN during at least one of their first 3 ultrasounds. Of these patients, 126 (89%) were investigated with a functional renogram while only 66 (46%) ultimately underwent a surgical procedure to address obstruction. In our test set we are able to classify 58% (15/26) of the non-surgical patients who received a functional renogram as low-risk based on early ultrasounds while maintaining a 5% false negative rate (95% sensitivity) in our surgical cases. Therefore, even while setting a stringent threshold for foregoing a functional renogram, more than half the patients who were unnecessarily investigated with this invasive test may have been spared this procedure with additional information derived from their existing, non-invasive ultrasound.

Table 2. Model performance with randomly initialized model weights. Top validation and test AUROC and AUPRC indicated for each kidney ultrasound view in bold.

Model	# trainable parameters	Training AUROC (CI) AUPRC (CI)	Validation AUROC (CI) AUPRC (CI)	Test AUROC (CI) AUPRC (CI)
Sagittal view only				
VGG-16	156.3 M	0.919 (0.911,0.926) 0.745 (0.731,0.759)	0.844 (0.810,0.872) 0.772 (0.745,0.805)	**0.910 (0.871,0.945)** 0.766 (0.707,0.830)
ResNet-18	11.4 M	0.845 (0.827,0.861) 0.763 (0.748,0.779)	0.789 (0.746,0.832) **0.779 (0.752,0.809)**	0.901 (0.859,0.939) 0.765 (0.705,0.830)
DenseNet-121	8.3 M	0.949 (0.943,0.955) 0.735 (0.722,0.749)	0.819 (0.779,0.854) 0.771 (0.746,0.801)	0.896 (0.853,0.936) **0.770 (0.708,0.834)**
Custom	12.5 M	0.910 (0.899, 0.921) 0.742 (0.727, 0.757)	**0.852 (0.821, 0.881)** 0.761 (0.732, 0.791)	**0.910 (0.864, 0.946)** 0.766 (0.702, 0.837)
Transverse view only				
VGG-16	156.3 M	0.905 (0.893,0.915) 0.746 (0.733,0.761)	0.822 (0.785,0.856) 0.771 (0.744,0.801)	0.867 (0.795,0.926) 0.771 (0.709,0.837)
ResNet-18	11.4 M	0.781 (0.759,0.800) 0.781 (0.766,0.795)	0.713 (0.669,0.755) **0.806 (0.774,0.837)**	0.817 (0.721,0.897) **0.775 (0.718,0.840)**
DenseNet-121	8.3 M	0.954 (0.947,0.961) 0.730 (0.717,0.745)	0.794 (0.756,0.829) 0.778 (0.749,0.807)	0.883 (0.818,0.931) 0.761 (0.701,0.832)
Custom	12.5 M	0.955 (0.947, 0.962) 0.729 (0.715, 0.743)	**0.841 (0.807, 0.871)** 0.765 (0.739, 0.797)	**0.902 (0.848, 0.951)** 0.756 (0.697, 0.818)
Siamese Sagittal and Transverse view				
VGG-16	156.3 M	0.932 (0.924, 0.941) 0.737 (0.723, 0.752)	0.858 (0.829,0.887) 0.766 (0.736,0.799)	0.915 (0.878, 0.949) 0.758 (0.695, 0.823)
ResNet-18	11.4 M	0.814 (0.794,0.832) 0.768 (0.754,0.782)	0.779 (0.738,0.820) **0.780 (0.751,0.808)**	0.893 (0.849, 0.935) **0.768 (0.700, 0.835)**
DenseNet-121	8.4 M	0.962 (0.956, 0.967) 0.727 (0.713, 0.743)	0.833 (0.797, 0.865) 0.766 (0.736, 0.799)	0.909 (0.870, 0.942) 0.766 (0.701, 0.828)
Custom	12.7 M	0.943 (0.935, 0.949) 0.733 (0.719, 0.747)	**0.859 (0.826, 0.890)** 0.762 (0.734, 0.792)	**0.923 (0.887, 0.955)** 0.753 (0.692, 0.821)
Ensemble Sagittal and Transverse view				
VGG-16	312.6 M	0.926 (0.919,0.933) 0.742 (0.729,0.756)	0.836 (0.808,0.858) 0.771 (0.748,0.791)	0.916 (0.875,0.951) 0.755 (0.693,0.819)
ResNet-18	11.8 M	0.832 (0.815,0.848) 0.767 (0.754,0.781)	0.749 (0.714,0.779) **0.795 (0.773,0.816)**	0.907 (0.865,0.943) **0.760 (0.700,0.825)**
DenseNet-121	16.6 M	0.960 (0.955,0.965) 0.731 (0.719,0.744)	0.809 (0.778,0.835) 0.775 (0.753, 0.796)	0.917 (0.881,0.947) 0.757 (0.696,0.820)

(*continued*)

Table 2. (*continued*)

Model	# trainable parameters	Training AUROC (CI) AUPRC (CI)	Validation AUROC (CI) AUPRC (CI)	Test AUROC (CI) AUPRC (CI)
Custom	25 M	0.957 (0.949,0.964) 0.728 (0.713,0.742)	**0.859 (0.826,0.888)** 0.759 (0.729,0.792)	**0.926 (0.890,0.958)** 0.753 (0.692,0.817)
Ensemble of all Sagittal and Transverse models				
Full ensemble	365.9 M	0.972 (0.965,0.978) 0.758 (0.745,0.772)	0.801 (0.720,0.874) 0.781 (0.735,0.830)	0.7905 (0.666,0.894) 0.781 (0.719,0.843)

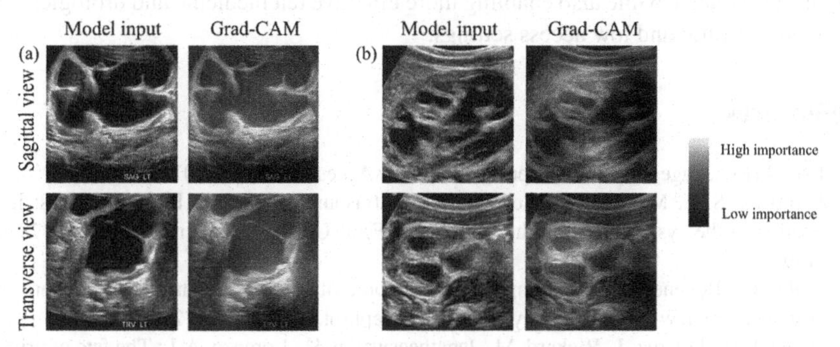

Fig. 2. Grad-CAMs visualize areas in ultrasound images deemed important by the classifier for kidneys that did and did not receive surgical intervention for obstructive HN. (a) Shows a kidney with grade 4 HN, correctly classified as obstructed. (b) Shows a healthy, non-HN kidney correctly classified as not obstructed.

4 Discussion

We have created an algorithm to automatically predict obstructive HN in infants based on ultrasound views of the kidney alone. The ability to predict HN obstruction with ultrasound images alone reduces its barrier to clinical use. For example, clinicians in more isolated communities, where ultrasounds are common but radiologists with pediatric expertise are not, can use this technology to triage high vs low risk patients, sending only the worse cases to distant pediatric centers. Automated assessment of ultrasounds for HN also has the potential to standardize clinical management. By testing a model based on HN management strategies of a single institution in other comparable institutions, there is potential to elucidate systematic differences in care across urological departments.

We are encouraged by our results but our findings come with several limitations. First, we note that our model performance was impacted by the small number of positive cases (28% of patients, 14% of kidneys) in our cohort. For this reason, we included AUPRC as one of our evaluation metrics. Second, we use surgical intervention for obstructive HN as a surrogate variable for obstructive HN however clinical practices, policies, and surgeons change over time. In an attempt to elucidate our model's performance on unseen data in this setting, we train our model on data from the earliest years (first 70% of patients) and test on patients from later years (last 30% of patients) [16]. Furthermore, we treat repeated ultrasounds from each patient and each kidney as independent samples. While this is interesting from the perspective that we can make predictions from a single set of images, we lose prior information from the patient that could be used to make a more informed decision. Future work will use repeated ultrasounds to accumulate information across patient visits and potentially improve our models' performance.

5 Conclusion

We find that prediction of obstructive HN based on ultrasound images alone is feasible. This model has the potential to reduce the reliance on invasive and repeated imaging for patient surveillance while also enabling more effective telemedicine and urological care for infants in rural and low access settings.

References

1. LNCS Homepage, http://www.springer.com/lncsAccessed 21 Nov 2016
2. Fernbach, S.K., Maizels, M., Conway, J.J.: Ultrasound grading of hydronephrosis: introduction to the system used by the Society for Fetal Urology. Pediatr. Radiol. **23**, 478–480 (1993)
3. Sidhu, G., Beyene, J., Rosenblum, N.D.: Outcome of isolated antenatal hydronephrosis: a systematic review and meta-analysis. Pediatr. Nephrol. **21**, 218–224 (2006)
4. Braga, L.H., D'Cruz, J., Rickard, M., Jegatheeswaran, K., Lorenzo, A.J.: The fate of primary nonrefluxing megaureter: a prospective outcome analysis of the rate of urinary tract infections, surgical indications and time to resolution. J. Urol. **195**, 1300–1305 (2016)
5. Bayne, C.E., Majd, M., Rushton, H.G.: Diuresis renography in the evaluation and management of pediatric hydronephrosis: what have we learned? J. Pediatr. Urol. **15**, 128–137 (2019)
6. Capone, V., et al.: Voiding cystourethrography and 99MTC-MAG3 renal scintigraphy in pediatric vesicoureteral reflux: what is the role of indirect cystography? J. Pediatr. Urol. (2019). https://doi.org/10.1016/j.jpurol.2019.06.004
7. Jacobson, D.L., et al.: The correlation between serial uultrasound and diuretic renography in children with severe unilateral hydronephrosis. J. Urol. **200**, 440–447 (2018)
8. Esteva, A., et al.: A guide to deep learning in healthcare. Nat. Med. **25**, 24–29 (2019)
9. Hosny, A., Parmar, C., Quackenbush, J., Schwartz, L.H., Aerts, H.J.W.L.: Artificial intelligence in radiology. Nat. Rev. Cancer **18**, 500–510 (2018)
10. Esteva, A., et al.: Dermatologist-level classification of skin cancer with deep neural networks. Nat. **542**, 115–118 (2017)
11. LeCun, Y., Bengio, Y., Hinton, G.: Deep learning. Nat. **521**, 436–444 (2015)
12. Dhindsa, K., Smail, L.C., McGrath, M., Braga, L.H.P., Sonnadara, R.R.: Grading prenatal hydronephrosis from ultrasound imaging using deep convolutional neural networks. In: 15th Conference on Computer and Robot Vision, (2018) https://doi.org/10.1109/crv.2018.00021
13. Braga, L.H., McGrath, M., Farrokhyar, F., Jegatheeswaran, K., Lorenzo, A.J.: Society for fetal urology classification vs urinary tract dilation grading system for prognostication in prenatal hydronephrosis: a time to resolution analysis. J. Urol. **199**, 1615–1621 (2018)
14. Wong, N.C., Koyle, M.A., Braga, L.H.: Continuous antibiotic prophylaxis in the setting of prenatal hydronephrosis and vesicoureteral reflux. Can. Urol. Assoc. J. **11**, S20–S24 (2017)
15. Zuiderveld, K.: Contrast limited adaptive histogram equalization. Graphics Gems, pp. 474–485 (1994) https://doi.org/10.1016/b978-0-12-336156-1.50061-6

16. Jung, K., Shah, N.H.: Implications of non-stationarity on predictive modeling using EHRs. J. Biomed. Inform. **58**, 168–174 (2015)
17. Simonyan, K., Zisserman, A.: Very deep convolutional networks for large-scale image recognition. arXiv [cs.CV], (2014) Available: http://arxiv.org/abs/1409.1556
18. He, K., Zhang, X., Ren, S., Sun, J.: Deep residual learning for image recognition. arXiv [cs.CV], (2015) Available: http://arxiv.org/abs/1512.03385
19. Huang, G., Liu, Z., van der Maaten, L., Weinberger, K.Q.: Densely connected convolutional networks. arXiv [cs.CV], (2016) Available: http://arxiv.org/abs/1608.06993
20. Russakovsky, Olga., et al.: ImageNet large scale visual recognition challenge. Int. J. Comput. Vision **115**(3), 211–252 (2015). https://doi.org/10.1007/s11263-015-0816-y
21. Paszke, A., et al.: PyTorch: an imperative style, high-performance deep learning library. In: Wallach H, Larochelle H, Beygelzimer A, d\textquotesingle Alché-Buc F, Fox E, Garnett R, editors. Advances in Neural Information Processing Systems 32. Curran Associates, Inc, pp. 8024–8035 (2019)
22. Guo, C., Pleiss, G., Sun, Y., Weinberger, K.Q.: On calibration of modern neural networks. arXiv [cs.LG], (2017) Available: http://arxiv.org/abs/1706.04599
23. Niculescu-Mizil, A., Caruana, R.: Predicting good probabilities with supervised learning. In: Proceedings of the 22nd international conference on Machine learning, pp. 625–632. ACM (2005)
24. Platt, J.C.: Probabilistic outputs for support vector machines and comparisons to regularized likelihood methods, p. 10 (2000) Available: http://dx.doi.org/
25. Selvaraju, R.R., Cogswell, M., Das, A., Vedantam, R., Parikh, D., Batra, D.: Grad-cam: visual explanations from deep networks via gradient-based localization. In: Proceedings of the IEEE International Conference on Computer Vision, pp. 618–626 (2017)
26. Adebayo, J., Gilmer, J., Muelly, M., Goodfellow, I., Hardt, M., Kim, B.: Sanity checks for saliency maps. arXiv [cs.CV], (2018) Available: http://arxiv.org/abs/1810.03292

Semi-supervised Training of Optical Flow Convolutional Neural Networks in Ultrasound Elastography

Ali K. Z. Tehrani(✉), Morteza Mirzaei, and Hassan Rivaz

Department of Electrical and Computer Engineering, Concordia University, Montreal, Canada
a_kafaei@encs.concordia.ca, hrivaz@ece.concordia.ca

Abstract. Convolutional Neural Networks (CNN) have been found to have great potential in optical flow problems thanks to an abundance of data available for training a deep network. The displacement estimation step in UltraSound Elastography (USE) can be viewed as an optical flow problem. Despite the high performance of CNNs in optical flow, they have been rarely used for USE due to unique challenges that both input and output of USE networks impose. Ultrasound data has much higher high-frequency content compared to natural images. The outputs are also drastically different, where displacement values in USE are often smooth without sharp motions or discontinuities. The general trend is currently to use pre-trained networks and fine-tune them on a small simulation ultrasound database. However, realistic ultrasound simulation is computationally expensive. Also, the simulation techniques do not model complex motions, nonlinear and frequency-dependent acoustics, and many sources of artifact in ultrasound imaging. Herein, we propose an unsupervised fine-tuning technique which enables us to employ a large unlabeled dataset for fine-tuning of a CNN optical flow network. We show that the proposed unsupervised fine-tuning method substantially improves the performance of the network and reduces the artifacts generated by networks trained on computer vision databases.

Keywords: Ultrasound Elastography · Convolutional Neural Networks (CNN) · Ultrasound-guid intervention · Unsupervised training

1 Introduction

Ultrasound is one of the most widely used modality in medical imaging, and is the preferred modality in image-guided interventions [1–3]. UltraSound Elastography (USE) is an imaging technique which provides relative stiffness properties of the tissue, and as such, provides additional guidance during interventions. Freehand palpation is one of the most popular methods in USE due to simplicity and

Supported by the Natural Sciences and Engineering Research Council of Canada (NSERC) RGPIN-2020-04612.

A. L. Martel et al. (Eds.): MICCAI 2020, LNCS 12263, pp. 504–513, 2020.
https://doi.org/10.1007/978-3-030-59716-0_48

availability. The basic idea of free-hand palpation is that the operator compresses the tissue by ultrasound probe, and the images before and after the compression are compared to obtain the displacement map [4]. Due to the fact that most compression is in the axial direction, axial displacement contains more available information than the lateral one. The axial displacement map is used to obtain the strain map, which is generally inversely proportional to the elastic modulus.

Convolutional Neural Networks (CNN) have been proven useful in optical flow estimation. Many network architectures such as FlowNet [5], FlowNet2 [6], PWC-Net [7] and LiteFlowNet [8] have been proposed. The displacement estimation step of the USE can be performed using optical flow CNNs [9–14]. However, computer vision images and ultrasound data are generally different in characteristics and the objectives. Computer vision images may contain small objects with a very different optical flow from the background (for example: a hand moves and the background is fixed). Whereas in USE, the movement is generally smooth and continuous. Another difference lies in the objective of the two tasks. The objective is to find sharp and accurate optical flows in computer vision, whereas in USE, the main goal is to obtain a differentiable displacement field. These differences led to the fact that the strain map generated by optical flow CNNs trained on computer vision images have lower bias but with higher variance compared to traditional elastography algorithms [11]. The lower bias of CNNs results in high contrast images but the high variance is amplified in the spatial differentiation step. Fine-tuning is a viable options to improve the network performance and reduce this variance [9–11,13].

Many researchers have tried to adopt optical flow CNNs for USE using supervised fine-tuning. The general trend among researchers is to use pre-trained networks and fine-tune them with generated simulation datasets with known ground truth [10,11,13,15]. They used pre-trained well-known optical flow CNNs such as FlowNet2, PWC-Net and LiteFlowNet, and fine-tuned them using supervised techniques. The structure of the networks are also modified to address the differences of computer vision and USE in the inputs [11].

Unsupervised fine-tuning is a more appropriate option for several reasons. First, simulation techniques entail several finite element and interpolation steps, which render the accuracy of sub-pixel ground truth displacement field inaccurate. Second, the simulation database often cannot model non-linear deformation and acoustic behaviors. Last but not least, the fine-tuning may cause forgetting effects [16], if the imaging parameters of ultrasound device is not close to the simulation data to the point that the fine-tuning with simulation deteriorates results on real data [11]. By using unsupervised techniques, the network can be fine-tuned to any target domain, i.e. different ultrasound machines and different organs.

In this paper, we propose a novel unsupervised technique to fine-tune pre-trained optical flow CNNs using real ultrasound images. Our method can be considered as a form of semi-supervised learning since the pre-trained network is trained by labeled data, whereas the fine-tuning is done using data with unknown ground truth. We use LiteFlowNet [8] since it is light and has shown good performance in optical flow. However, the proposed framework can be applied to any optical flow CNN. The network estimates 2D displacements for strain values ranging from 0.5 to 5 %, the performance of the algorithm in transient elastography [17] where displacements are very small is an area of future work. Our contribution can be summarized as follows:

1. Our results show that training on computer vision images is not enough since the statistics of ultrasound RF data and physics of the displacement field are different in these two domains.
2. We propose an unsupervised fine-tuning method in elastography.
3. We use real ultrasound images for fine-tuning, thanks to our unsupervised technique which does not need ground truth displacements.
4. We propose an automatic frame and region selection algorithm which enables the user to employ real ultrasound images without any supervision and expertise.
5. We propose a novel loss function, considering statistics of RF data and physics of the displacement field.

2 Materials and Methods

2.1 Unsupervised Training of Optical Flow Networks

A critical component of unsupervised techniques is the loss function, which can be expressed as [18–21]:

$$Loss = loss_d + loss_s + loss_c \tag{1}$$

where $loss_d$ is the data loss, $loss_s$ is the smoothness loss which can also be described as smoothness regularization, and $loss_c$ is the consistency loss, which shows how different the forward and backward flows are. Data loss can be described as the difference between the first image and the warped second image. Smoothness loss controls smoothness of the displacement and usually first- and second-order derivatives are utilized for this loss [19–21]. In [18], a combination of $L1$ norm and structural similarity ($SSIM$) is employed for data loss, an edge-aware smoothness regularizer is utilized as the smoothness loss, and L_1 norm is used for consistency loss. In [19], the robust generalized Charbonnier penalty [22] of census transform [23] is employed as the data loss. The Charbonnier penalty of forward and backward displacement as consistency loss have also been exploited in [19]. Finally, this paper also used an occlusion mask to remove the occluded pixels from the loss terms to avoid back propagation of occluded regions.

2.2 Proposed Method

Inspired by other unsupervised optical flow networks, we propose a fine-tuning strategy well-adapted to USE. Let I_1 and I_2 be the first and the second images,

w_f and w_b be forward ($I_1 \rightarrow I_2$) and backward flows ($I_2 \rightarrow I_1$). We define the data loss as:

$$loss_d = \left\langle \Phi(I_1 - \widetilde{I_2}) \right\rangle_{O_f} \tag{2}$$

where $\langle . \rangle_{O_f}$ is the mean of non-outlier pixels. $\widetilde{I_2}$ is the warped I_2 toward I_1 using w_f and Φ denotes Charbonnier penalty:

$$\Phi(x) = (x^2 + \varepsilon)^\gamma \tag{3}$$

We set γ to 0.2 similar to the fine-tuning loss in [6,7,11] and ε denotes a small number. There is no occlusion in USE. We borrow ideas from occlusion detection to find the outlier regions of displacement estimates and exclude them in all loss terms. In order to find outlier displacement estimates, we compare forward and backward displacement using the following equation:

$$O_f = \left| w^f + w^b \right| < \alpha \tag{4}$$

we set α empirically to 1 as we observed that in case of outlier the difference between forward and negative of the backward displacement is much larger than 1. The O_f image can be considered as a mask that selects reliable regions for the loss function. In unsupervised training, frame selection is critical since many pairs of frames are not suitable for displacement estimation due to a very large decorrelation between their ultrasound data, and grossly incorrect displacement estimates can back propagate wrong values to the network. In order to do frame selection during the training, image pairs wherein the O_f mask is 0 in more than 50% of pixels are excluded. The outlier mask can be considered as a hard threshold consistency loss.

Regarding the smoothness loss, the derivative of axial displacement is often the main concern. This derivative operation amplifies variance of the displacement estimates, reducing the contrast to noise ratio (CNR). In optimization-based elastography methods, smoothness constrains are imposed on the axial displacement [24,25]. Here, we enforce smoothness on both displacement and its derivative. The latter is insensitive to affine deformations and performs better in the boundaries [19]. Let axial displacement be w_a^f, and a and l denote axial and lateral direction, respectively. The first order smoothing loss can be given as:

$$loss_s^1 = \lambda_1 \left\langle \Phi \left\{ \frac{\partial}{\partial a} w_a^f - < \frac{\partial}{\partial a} w_a^f > \right\} \right\rangle_{O_f} + \lambda_2 \left\langle \Phi \left\{ \frac{\partial}{\partial l} w_a^f - < \frac{\partial}{\partial l} w_a^f > \right\} \right\rangle_{O_f} \tag{5}$$

where λ_1 and λ_2 are weights associated to the axial and lateral derivative, respectively and $\langle . \rangle$ denotes mean. The average of derivatives are subtracted from the regularization term to reduce the regularization bias [11,25]. We also consider to penalize changes in the second order axial derivative of axial displacement:

$$loss_s^2 = \lambda_3 \left\langle \Phi \left\{ \frac{\partial^2}{\partial a^2} w_a^f \right\} \right\rangle_{O_f} \tag{6}$$

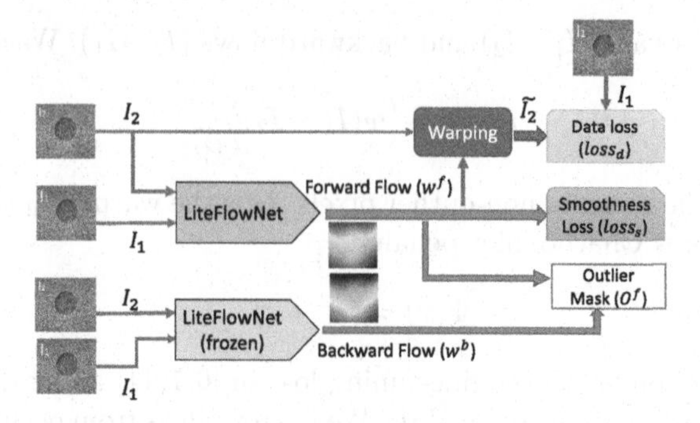

Fig. 1. Proposed network structure for unsupervised training.

The network structure for unsupervised training is shown in Fig. 1. The final loss function for training can be written as:

$$Loss = loss_d + loss_s^1 + loss_s^2 \tag{7}$$

It should be noted that LiteFlowNet is a multi-scale network with intermediate outputs. For training, intermediate loss and labels are required, but as suggested by [11], we only consider the last output since only small changes are required in fine-tuning.

2.3 Training and Practical Considerations

Unlike computer vision images, a large input size (for example, 1920 × 768 in this work) is required in USE to maintain the high frequency information of the radio frequency (RF) data. This is a limiting factor for current commercial GPUs, which generally have less than 12 GB of RAM. In addition, we estimate both forward and backward displacements in our unsupervised training framework, which further intensifies the memory limitation.

To mitigate the aforementioned problem, we employ gradient checkpointing [26], where all values of forward pass are not saved into memory. When back propagation requires the values of the forward pass, they are re computed. According to [26], it decreases memory usage up to 10 times with a computational overhead of only 20–30 %. Moreover, the network's weight are kept fixed in backward flow computation (the gray block in Fig. 1) to further reduce the memory usage.

Envelope of RF data along with RF data and imaginary parts of analytic signal were used as three separate channels. As suggested by [11], envelope along with RF data is used to compensate the loss of information in RF data by the down sampling steps in the network.

We used the pre-trained weights of [8], which was obtained by training on 72,000 pairs of simulated computer vision images with known optical flows. Fine-tuning was performed on 2200 pairs of real ultrasound RF data with unknown

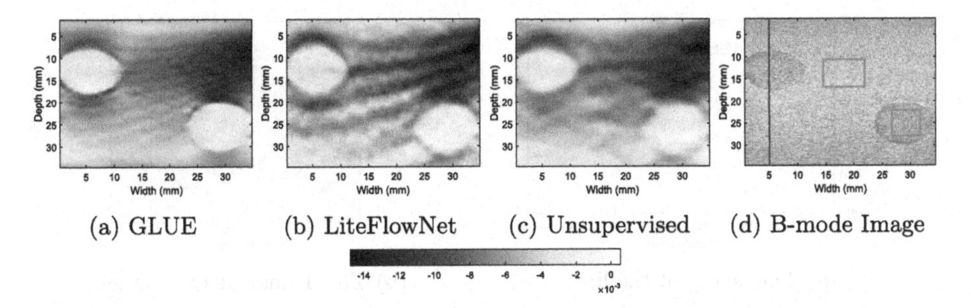

Fig. 2. Strain images of the experimental phantom with two hard inclusion. The windows used for CNR and SR computation are highlighted in the B-mode image (d). The strain of the line highlighted in red is shown in Fig. 3 (Colour figure online).

displacement maps. Strain images of the experimental seen by the network during fine-tuning. The network was trained for 20 epochs on NVIDIA TITAN V using Adam optimizer. The learning rate was set to 4e-7 and the batch size was 1 due to memory limitations. Regarding the weights of each part of the loss function, there is a trade-off between bias and variance error. Higher weights of the smoothing loss result in smoother but more biased results and lower weights lead to lower bias but higher variance. We empirically set λ_1, λ_2 and λ_3 to 0.5, 0.005, 0.2, as we observed that these weights had a good balance between bias and variance error.

3 Results

We validated our proposed unsupervised fine-tuning using an experimental phantom and *in vivo* data. We compare LiteFlowNet, our unsupervised fine-tuned LiteFlowNet and GLobal Ultrasound Elastography (GLUE) [24], which is a well known non-deep learning elastography method. Codes associated with all of these methods are available online.

3.1 Quantitative Metrics

Contrast to Noise Ratio (CNR) and Strain Ratio (SR) are two popular metrics used to assess the elastography algorithms in experimental phantoms and *in vivo* data where the ground truth is unknown. These metrics are defined as [4]:

$$SR = \frac{\overline{s}_t}{\overline{s}_b}, \qquad CNR = \sqrt{\frac{2(\overline{s}_b - \overline{s}_t)^2}{{\sigma_b}^2 + {\sigma_t}^2}}, \tag{8}$$

where $\overline{s}_b$ and $\overline{s}_t$ are average values of strain in the background and target regions of the tissue, and σ_b and σ_t denote variance values of strain in the background and target regions, respectively. CNR is a proper metric to measure a combination of bias and variance error. SR sheds light on the estimator bias in real experiments wherein the ground truth strain values are unknown [11].

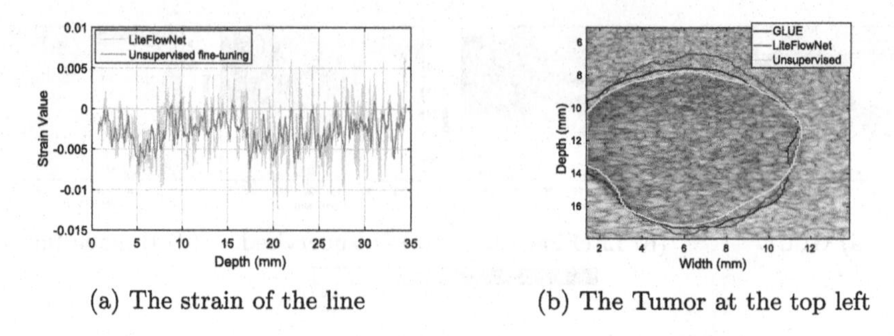

(a) The strain of the line (b) The Tumor at the top left

Fig. 3. The strain of the line specified in Fig. 2 using small smoothing window (a). The top left tumor with the edges obtained by the three USE methods (b).

3.2 Experimental Phantom

We collected ultrasound images at Concordia University's PERFORM Centre using an Alpinion E-Cube R12 research ultrasound machine (Bothell, WA, USA) with a L3-12H linear array at the center frequency of 10 MHz and sampling frequency of 40 MHz. A tissue mimicking breast phantom made by Zerdine (Model 059, CIRS: Tissue Simulation & Phantom Technology, Norfolk, VA) is used for data collection. The phantom contains several hard inclusions with elasticity values at least twice the elasticity of the tissue. The experimental phantom for test is from the same phantom but different part of the phantom is imaged. The composition of the phantom in test data is also different, where regions with only one inclusion are used for training and regions with two inclusions are used for testing. The test results are depicted in Fig. 2.

The unsupervised fine-tuning improves the strain quality of LiteFlowNet producing a smoother strain with less artifacts. The quantitative results are shown in Fig. 5. GLUE has the highest CNR which shows the high-quality strain but the SR is also the highest which indicates that it has the highest bias due to the strong regularization used in the algorithm. The unsupervised fine-tuning substantially improves the CNR of the network with very similar SR. In order to show the improvements better, the line specified in Fig. 2 (d) with small differentiation window is shown in Fig. 3 (a). The strain plots indicate that fine-tuning substantially reduces the variance error presented in the strain. We calculate the edges of the strain images using Canny edge detection and superimpose on the B-mode image to compare the size of different structures in Fig. 3 (b). Here, the top left inclusion of Fig. 2 (d) is shown. It can be seen that LiteFlowNet substantially overestimates the size of the inclusion since the red curve is well outside of the inclusion. Our unsupervised fine-tuning technique corrects this overestimation.

3.3 *In Vivo* Data

In vivo data was collected at Johns Hopkins Hospital using a research Antares Siemens system by a VF 10-5 linear array with a sampling frequency of 40 MHz and the center frequency of 6.67 MHz. Data was obtained from patients in

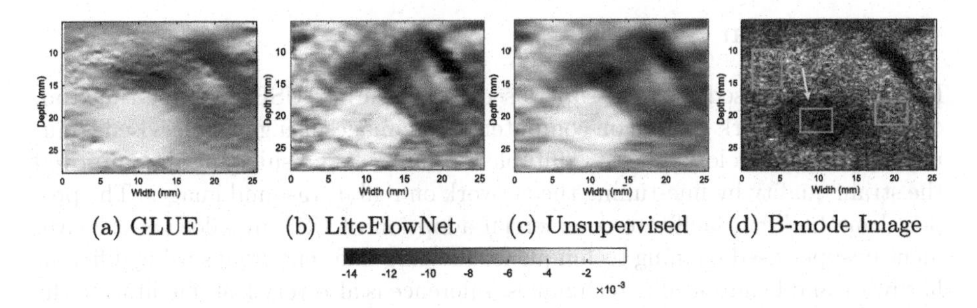

Fig. 4. Strain images of the *in vivo* data. The two background and the target windows used for CNR and SR computation are highlighted by green in the B-mode image (d). The hard tumor and soft vein are marked by yellow and red arrows, respectively (Colour figure online).

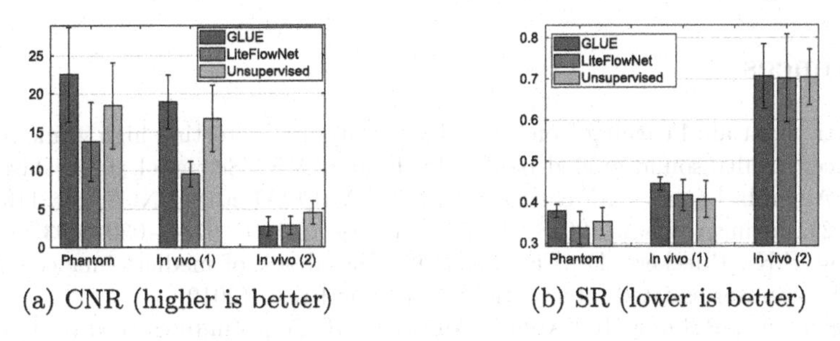

Fig. 5. CNR (a) and SR (b) of experimental phantom and *in vivo* data. *In vivo* (1) and *in vivo* (2) correspond to the CNR with background of 1 and 2 in Fig. 4 (d), respectively.

open-surgical RF thermal ablation for liver cancer [27]. The study was approved by the institutional review board with consent of all patients. The strains obtained by the compared methods are shown in Fig. 4. GLUE (a) produce high quality strain but it is over smoothed which is evident for the vein (it is marked by the orange arrow). LiteFlowNet (b) produces a strain with many artifacts and heterogeneities inside the tumor (the tumor is marked with the yellow arrow), but it preserves the vein which is a small structure. Unsupervised fine-tuning (c) not only reduces the artifacts and heterogeneity inside the tumor, but also maintains the vein similar to LiteFlowNet.

The quantitative results are given in Fig. 5. The first background window (1 in Fig. 4 (d)) is very different than the tumor region, GLUE and the fine-tuned network produce similar CNRs between the tumor and this window. The second background (2 in Fig. 4 (d)) has very high amount of artifacts and the strain value is very close to the strain of tumor. Fine-tuned network has the best CNR for this challenging background which can be confirmed by visual assessment of Fig. 4. Regarding SR, LiteFlowNet has the lowest SR but the differences with the unsupervised fine-tuned network are negligible. Among the compared methods, GLUE has the worst SR results which indicates the high bias error presented in this method.

4 Conclusion

Herein, we proposed an semi-supervised technique for USE. We fine-tuned an optical flow network trained on computer vision images using unsupervised training. We designed a loss function suitable for our task and substantially improved the strain quality by fine-tuning the network on real ultrasound images. The proposed method can facilitate commercial adoption of USE by allowing a convenient unsupervised training technique for imaging different organs using different hardware and beamforming techniques. Inference is also very fast, facilitating the use of USE in image-guided interventions.

Acknowledgement. We thank NVIDIA for the donation of the GPU. The *in vivo* data was collected at Johns Hopkins Hospital. We thank E. Boctor, M. Choti and G. Hager for giving us access to this data.

References

1. Azizi, S., et al.: Learning from noisy label statistics: detecting high grade prostate cancer in ultrasound guided biopsy. In: Frangi, A.F., Schnabel, J.A., Davatzikos, C., Alberola-López, C., Fichtinger, G. (eds.) MICCAI 2018. LNCS, vol. 11073, pp. 21–29. Springer, Cham (2018). https://doi.org/10.1007/978-3-030-00937-3_3
2. Zhou, S.K., Rueckert, D., Fichtinger, G.: Handbook of medical imagecomputing and computer assisted intervention. Academic Press, (2019)
3. Zhuang, B., Rohling, R., Abolmaesumi, P.: Region-of-interest-based closed-loop beamforming for spinal ultrasound imaging. IEEE Trans. Ultrason. Ferroelectr. Freq. Control **66**(8), 1266–1280 (2019)
4. Ophir, J., et al.: Elastography: ultrasonic estimation and imaging of the elastic properties of tissues. Proc. Inst. Mech. Eng. Part H: J. Eng. Med. **213**(3), 203–233 (1999)
5. Dosovitskiy, A., et al.: Flownet: learning optical flow with convolutional networks. In: Proceedings of the IEEE international conference on computer vision, pp. 2758–2766 (2015)
6. Ilg, E., Mayer, N., Saikia, T., Keuper, M., Dosovitskiy, A., Brox, T.: Flownet 2.0: evolution of optical flow estimation with deep networks. In: Proceedings of the IEEE conference on computer vision and pattern recognition, pp. 2462–2470 (2017)
7. Sun, D., Yang, X., Liu, M.-Y., Kautz, J.: Pwc-net: cnns for optical flow using pyramid, warping, and cost volume. In: Proceedings of the IEEE Conference on Computer Vision and Pattern Recognition, pp. 8934–8943 (2018)
8. Hui, T.W., Tang, X., Change Loy, C.: Liteflownet: a lightweight convolutional neural network for optical flow estimation. In: Proceedings of the IEEE conference on computer vision and pattern recognition, pp. 8981–8989 (2018)
9. Kibria, M.G., Rivaz, H.: GLUENet: ultrasound elastography using convolutional neural network. In: Stoyanov, D., et al. (eds.) POCUS/BIVPCS/CuRIOUS/CPM -2018. LNCS, vol. 11042, pp. 21–28. Springer, Cham (2018). https://doi.org/10.1007/978-3-030-01045-4_3
10. Peng, B., Xian, Y., Jiang, J.: A convolution neural network-based speckletracking method for ultrasound elastography. In: 2018 IEEEInternational Ultrasonics Symposium (IUS), pp. 206–212. IEEE (2018)

11. Tehrani, A.K., Rivaz, H.: Displacement estimation in ultrasound elastography using pyramidal convolutional neural network. In: IEEE Transactions on Ultrasonics, Ferroelectrics, and Frequency Control, (2020)
12. Wu, S., Gao, Z., Liu, Z., Luo, J., Zhang, H., Li, S.: Direct reconstruction of ultrasound elastography using an end-to-end deep neural network. In: Frangi, A.F., Schnabel, J.A., Davatzikos, C., Alberola-López, C., Fichtinger, G. (eds.) MICCAI 2018. LNCS, vol. 11070, pp. 374–382. Springer, Cham (2018). https://doi.org/10.1007/978-3-030-00928-1_43
13. Peng, B., Xian, Y., Zhang, Q., Jiang, J.: Neural network-based motion tracking for breast ultrasound strain elastography: an initial assessment of performance and feasibility. Ultrason. Imaging. **42**(2), 74–91 (2020)
14. Gao, Z., et al.: Learning the implicit strain reconstruction in ultrasound elastography using privileged information. Med. Image Anal. **58**, 11–18 (2019)
15. Evain, E., Faraz, K., Grenier, T., Garcia, D., De Craene, M., Bernard, O.: A pilot study on convolutional neural networks for motion estimation from ultrasound images. Ferroelectrics, and Frequency Control, IEEE Transactions on Ultrasonics, (2020)
16. Li, Z., Hoiem, D.: Learning without forgetting. IEEE Trans. Pattern Anal. Mach. Intell. **40**(12), 2935–2947 (2017)
17. Sandrin, L., et al.: Transient elastography: a new noninvasive method for assessment of hepatic fibrosis. Ultrasound Med. Biol. **29**(12), 1705–1713 (2003)
18. Godard, C., Mac Aodha, O., Brostow, G.J.: Unsupervised monocular depth estimation with left-right consistency. In: Proceedings of the IEEE Conference on Computer Vision and Pattern Recognition, pp. 270–279 (2017)
19. Meister, S., Hur, J., Roth, S.: Unflow: unsupervised learning of optical flow with a bidirectional census loss. In: Thirty-Second AAAI Conference on Artificial Intelligence, (2018)
20. Ren, Z., Yan, J., Yang, X., Yuille, A., Zha, H.: Unsupervised learning of optical flow with patch consistency and occlusion estimation. Pattern Recogn. **103**, 107191 (2020)
21. Wang, Y., Yang, Y., Yang, Z., Zhao, L., Wang, P., Xu, W.: Occlusion aware unsupervised learning of optical flow. In: Proceedings of the IEEE Conference on Computer Vision and Pattern Recognition, pp. 4884–4893 (2018)
22. Sun, D., Roth, S., Black, M.J.: A quantitative analysis of current practices in optical flow estimation and the principles behind them. Int. J. Comput. Vision **106**(2), 115–137 (2014)
23. Zabih, R., Woodfill, J.: Non-parametric local transforms for computing visual correspondence. In: Eklundh, J.-O. (ed.) ECCV 1994. LNCS, vol. 801, pp. 151–158. Springer, Heidelberg (1994). https://doi.org/10.1007/BFb0028345
24. Hashemi, H.S., Rivaz, H.: Global time-delay estimation in ultrasound elastography. IEEE Trans. Ultrason. Ferroelectr. Freq. Control **64**(10), 1625–1636 (2017)
25. Mirzaei, M., Asif, A., Rivaz, H.: Combining total variation regularization with window-based time delay estimation in ultrasound elastography. IEEE Trans. Med. Imaging **38**(12), 2744–2754 (2019)
26. Chen, T., Xu, B., Zhang, C., Guestrin, C.: Training deep nets with sublinear memory cost. arXiv preprint arXiv:1604.06174 (2016)
27. Rivaz, H., Boctor, E.M., Choti, M.A., Hager, G.D.: Real-time regularized ultrasound elastography. IEEE Trans. Med. Imaging **30**(4), 928–945 (2011)

Three-Dimensional Thyroid Assessment from Untracked 2D Ultrasound Clips

Wolfgang Wein[1](✉), Mattia Lupetti[1], Oliver Zettinig[1], Simon Jagoda[1], Mehrdad Salehi[1], Viktoria Markova[1], Dornoosh Zonoobi[2], and Raphael Prevost[1]

[1] ImFusion GmbH, Munich, Germany
wein@imfusion.com
[2] MEDO.ai, Alberta, Canada

Abstract. The diagnostic quantification of thyroid gland, mostly based on its volume, is commonly done by ultrasound. Typically, three orthogonal length measurements on 2D images are used to estimate the thyroid volume from an ellipsoid approximation, which may vary substantially from its true shape. In this work, we propose a more accurate direct volume determination using 3D reconstructions from two freehand clips in transverse and sagittal directions. A deep learning based trajectory estimation on individual clips is followed by an image-based 3D model optimization of the overlapping transverse and sagittal image data. The image data and automatic thyroid segmentation are then reconstructed and compared in 3D space. The algorithm is tested on 200 pairs of sweeps, and shows that it can provide fully automated, but also more accurate and consistent volume estimations than the standard ellipsoid method, with a median volume error of 11%.

1 Introduction

Ultrasound imaging has been for a long time the gold standard for thyroid assessment, thus replacing clinical inspection and palpation [1]. The current workflow is to perform 2D measurements on ultrasound planes and combine them into a volume using the so-called ellipsoid formula, a very coarse approximation which leads to uncertainties [11] and sub-optimal reproducibility. In this context, creating a 3D reconstruction of the thyroid would bring several benefits: (i) it would allow a more precise volume estimation of either thyroid or suspicious masses than 2D imaging [6], and (ii) it could ease matching of the same anatomy, enabling easier and more reliable regular screening of the population at risk. One way to achieve this could be 3D ultrasound solutions such as a dedicated 3D transducer, or an external tracking of a 2D probe [13]. Such approaches have not found widespread adoption though, mostly due to the high cost or cumbersome setup.

It has recently been shown that it is possible to reconstruct the 3D trajectory of a freehand ultrasound clip using deep learning [8]. Combined with a thyroid

A. L. Martel et al. (Eds.): MICCAI 2020, LNCS 12263, pp. 514–523, 2020.
https://doi.org/10.1007/978-3-030-59716-0_49

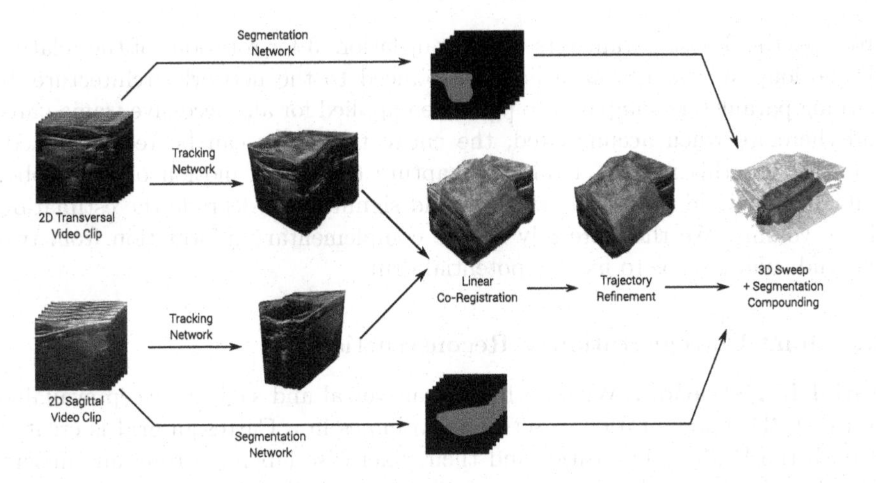

Fig. 1. Overview of the proposed method for 3D thyroid reconstruction from two perpendicular sweeps.

segmentation method (see for instance [2,5] for deep learning-based approaches, or [3] for a more generic and thorough review), this could enable a better assessment of its volume. However, this approach might still suffer from drift degrading the accuracy of such a measurement. In this paper, we therefore propose to build upon such methods by exploiting the redundancy from two perpendicular freehand acquisitions: out-of-plane distances in one sweep can be precisely recovered from the other sweep since they appear in-plane. In order to do so, our approach will aim at registering those two acquisitions while jointly refining their own trajectories. This is to our knowledge the first time a consistent 3D reconstruction of the thyroid on untracked freehand 2D ultrasound clips is presented, accurate enough to utilize volumetric lesion measurements.

2 Methods

Our overall approach takes two ultrasound clips as input, slow-swept in transversal (TRX) and sagittal (SAG) directions over one side of the thyroid gland (right or left). The output are two registered volumetric representations which are visually presented to the clinician, along with a 3D thyroid gland segmentation. An overview of the computational pipeline is shown in Fig. 1, each step of which is described below.

2.1 Deep Learning Tracking Estimation

Similarly to [9], we first train a convolutional neural network (CNN) to estimate the trajectory of the probe during the sweep based on the video clip. The network uses as input a 4-channel image (two successive ultrasound frames, and the 2D optical flow in-plane motion between them encoded as two channels) and produces the 6 pose parameters (3 for translation, 3 for rotation) of the relative 3D motion. No changes have been introduced to the network architecture or learning parameters compared to [9]. When applied for all successive frame pairs and therefore when accumulated, the entire trajectory can be reconstructed. Even though this method is able to capture the global motion of the probe, it might still yield some drift which would significantly degrade the estimation of the volume. We therefore rely on the complementary information from two perpendicular sweeps to fix this potential drift.

2.2 Joint Co-registration & Reconstruction

Initial Registration. With both the transversal and sagittal sweep available in a first 3D representation, a set of 3D volumes in a Cartesian grid is created of both the B-Mode intensities and their pixel-wise labeling, using an efficient GPU-based compounding algorithm similar to [4]. An initial rigid transformation between them is derived by assuming standardized scan directions, as well as by matching the segmented thyroid volumes with a rigid registration. This rigid transformation is further optimized by maximizing a cross-correlation metric over the image intensities. This aligns the bulk of the anatomical structures roughly, despite possibly incomplete thyroid visibility (and hence segmentation) in both sweeps. Most of the drift by the initial sensorless reconstruction method cannot really be fixed at this point, since the out-of-plane lengths of the sweeps stay fixed.

Trajectory Refinement via Co-registration. The reconstruction of both sweeps is then simultaneously refined by optimizing a number of trajectory parameters, together with the rigid transformation between the sweep centers, with respect to an image similarity on the B-Mode intensities. Since the 3D geometry within an ultrasound volume changes during this optimization, an on-the-fly multi-planar reconstruction (MPR) method is required to compare individual frames from one sweep with a compounded image from the other one. A related technique was presented in [12] for image-based optimization of the probe-to-sensor calibration for externally tracked 3D freehand ultrasound. We generalize it to optimize any parameters affecting the 3D topology of un-tracked data. A cascade of non-linear optimization using a Nelder-Mead search method are executed with increasing degrees of freedom (DOF) parameters as follows:

- **10 DOF:** Relative rigid pose plus 2 DOF per sweep, corresponding to an additional out-of-plane stretch, and rotation around the probe surface between the first and last frames of the sweep, as drift correction from the sensorless reconstruction method.
- **18 DOF:** As above, with the 2 DOF per sweep replaced by a 6-DOF rigid pose between the first and last frame.
- **30 DOF:** As above, with an additional 6 DOF pose control point in the center of the sweeps, realized through Hermite cubic spline interpolation on the individual rigid pose parameters.

– **54 DOF:** As above, with 3 control points per sweep instead of one, placed in an equidistant manner over the number of image frames.

2.3 Automated Thyroid Segmentation

Thyroid gland segmentation has been studied for many years in the literature. Following the recommendation in the review [3], we opted to segment each 2D ultrasound frame using a very standard U-Net [10] neural network. Using the same trajectory found in the previous steps, we then compound those label maps in both transversal and sagittal sweeps into 3D binary masks. After minor post-processing (morphological closing), the final segmentation is defined as the union of the two 3D segmentations, coming respectively from the transversal and corresponding sagittal sweep.

The whole pipeline is implemented in the C++ ImFusion SDK, with OpenGL shaders for similarity measure computation and image compounding, and CUDA for the deep learning models. The overall computation time on a standard desktop computer is in average 3 min, the deep learning based segmentation and tracking estimation take approximately 10 s per sweep, while the successive stages of the joint sweep reconstruction take from 20 s to 1 min. The computation time can be further reduced by running the tracking estimation and the segmentation in the background during the acquisition.

3 Experiments and Results

3.1 Data Acquisition

Our method is evaluated using a dataset of 180 ultrasound sweeps from 9 volunteers acquired from a Cicada research ultrasound machine (Cephasonics, Inc., Santa Clara, CA, USA) with a linear probe. For each volunteer, we acquired 5 transversal and 5 sagittal sweeps for each lobe of the thyroid. The aquisitions were performed by three different operators (the transversal top to bottom and the sagittal starting from the trachea) with variations in the aquisition speed, extend of the captured anatomy, tilt angles. Following guidelines in [8], the sweeps have been tracked using an optical tracking system and recorded without speckle reduction or scanline conversion.

In order to train and evaluate our method, the thyroid was manually segmented by several operators on a subset of the sweeps. The volunteers were then split into a training (5) and testing (4) subset; the first set was used to train all networks as well as fine-tune the registration parameters and contains 100 sweeps, the latter was left out for the evaluation of the method and contains 80 sweeps.

3.2 Experiments

Our evaluation design is driven by the 3D thyroid segmentation, in particular its consistency between TRX and SAG sweeps, as well as the volume estimation since this is the relevant clinical measure. We ran our whole pipeline with different configurations to test our various hypotheses. All results are summarized in Table 1 and discussed below.

Table 1. Results of the experiments on co-registered sweeps without tracking information. The Dice coefficient TRX/SAG represents the overlap between the 3D segmentation of the thyroidal gland computed from the transversal and sagittal sweeps. This number is used as quality metric for all the sweep co-registration methods presented. As reference the Dice overlap computed with the ground truth tracking is reported. All numbers are means ± standard deviation if not otherwise specified.

Exp #	Sweep Co-registration	Dice TRX/SAG Mean ± std (median)	Volume Error	Norm. vol. Error	Rel. trajectory Error	Rel. length Error
1	Ground truth tracking	0.69 ± 0.13 (0.72)	N/A	N/A	N/A	N/A
2	Rigid reg	0.61 ± 0.14 (0.63)	2.09 ± 1.44 mL	0.25 ± 0.13	0.16 ± 0.06	0.22 ± 0.13
3	10-DOF refinement	0.64 ± 0.12 (0.65)	1.22 ± 1.46 mL	0.14 ± 0.16	0.16 ± 0.06	0.17 ± 0.14
4	18-DOF refinement	0.64 ± 0.12 (0.66)	1.14 ± 1.48 mL	0.14 ± 0.16	0.17 ± 0.09	0.16 ± 0.14
5	30-DOF refinement	0.65 ± 0.12 (0.66)	1.15 ± 1.47 mL	0.14 ± 0.16	0.16 ± 0.09	0.16 ± 0.14
6	54-DOF refinement	0.66 ± 0.12 (0.68)	1.15 ± 1.45 mL	0.14 ± 0.16	0.16 ± 0.09	0.16 ± 0.14

Thyroid Segmentation The ground truth thyroid segmentations were annotated by a single non-expert operator on 58 sweeps, 50 of them from the 5 volunteer training subset and the remaining 8 from the 4 volunteer testing subset (1 pair per lobe per test volunteer). The segmentation U-Net was trained and evaluated on the first 50 sweeps from training subset. The segmentation U-Net achieves a median Dice coefficient of 0.85 on our validation set of 2D frames.

After 3D compounding of the individual frames, the Dice coefficient between manual segmentation and network output drops to an average of 0.73 ± 0.08 because of inconsistencies across slices or ambiguity near the ends of the thyroid, which is in agreement with the scores recently reported in [13]. Due to the fact that each orthogonal sweep captures a slightly different view of the thyroid, also the Dice coefficient between compounded TRX and compounded SAG sweeps with manual annotation is not 1, but rather around 0.70 ± 0.05. Since those two numbers are in the same range, we then assume that, in the context of registration evaluation and because segmentation is not the focus of the paper, metrics on the network output are a good proxy for metrics on manual segmentations. This allows us to consider all the 200 possible pairs of test sweeps in the next experiments (25 pairs per lobe per test volunteer) instead of the 8 which are manually labeled (1 pair per lobe per test volunteer).

Single Sweep vs Multi-sweep. Due to the residual drift of the tracking estimation network, evaluating the thyroid volume from a single untracked sweep produces inaccurate estimates with, according to our early experiments, a 45% error. We therefore conclude that a second perpendicular sweep is required to bring the missing out-of-plane information. In the experiments #2 to #6, we use two such sweeps, which indeed yields a significant improvement of all reported metrics.

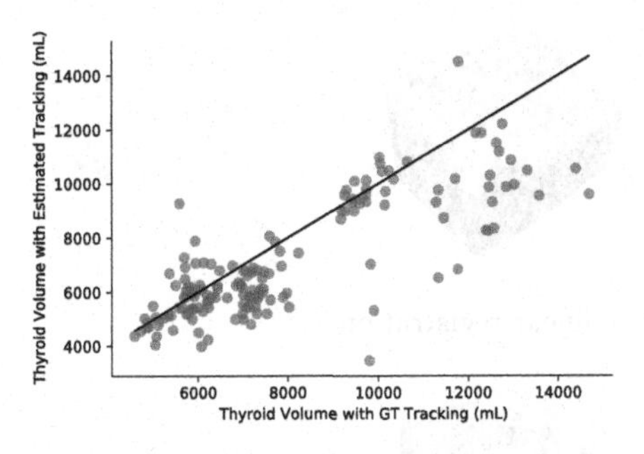

Fig. 2. Comparison of the estimated volume with ground truth tracking versus estimated trajectory on the 200 pairs of sweeps (Spearman correlation = 0.75).

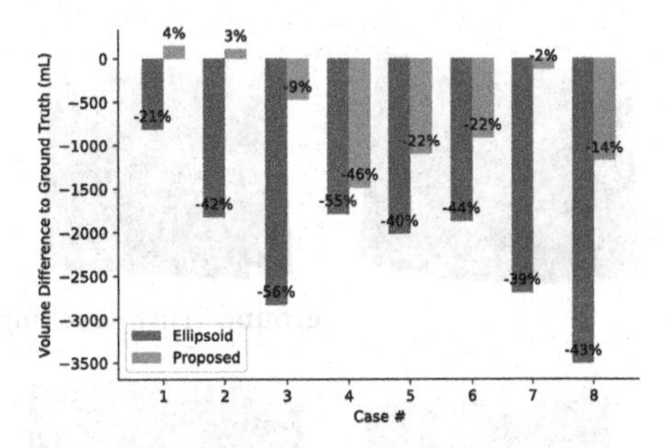

Fig. 3. Volume difference between true thyroid volume and estimated volume for both the ellipsoid method and our proposed approach on the 8 cases with manual segmentation.

Trajectory Refinement We evaluated the thyroid volume assessment at different stages of our refinement pipeline (from 10 to 54 DOF). Our experiments #3 to #6 demonstrate the benefit of further adjusting the trajectory of the two sweeps using our joint registration/trajectory refinement approach. According to a Wilcoxon statistical-test with a p-value threshold of 0.01, differences become irrelevant for motion corrections with more than 18 DOF. Furthermore, we compared the estimated sweep trajectories at each stage with their corresponding ground truth trajectories. To this purpose we defined the relative trajectory error as the cumulative in-plane translation error at each sweep frame divided by the ground truth sweep length, and similarly the relative length error as the relative error of the estimated sweep length w.r.t the ground truth length. The values of the relative trajectory and length errors are reported in the two rightmost columns of Table 1. With the same statistical-test settings as in the thyroid volume assessment, we conclude that the joint registration significantly improves the initially estimated overall length drift of the sweep, while there is no substantial change in the local trajectory estimation. The latter can be attributed to some extent to the error of the ground truth external tracking, i.e. the true local probe & tissue motion is not known with sufficient accuracy.

Figure 2 shows a comparison of the thyroid volume between the ground truth tracking and the estimated tracking. The two quantities are in strong agreement, with a Spearman correlation of 0.75, although the estimated volumes tends to be smaller than the ground truth ones. This is due to the fact that the tracking estimation network tends to underestimate the out-of-plane motion between frames on unseen sweeps, an issue that we will further investigate in future works.

To further demonstrate the reliability of our volume estimation approach, we compare it against the current clinical workflow, i.e. the ellipsoid formula, which consists in measuring the three dimensions of the thyroid in two perpendicular planes and multiplying them with a factor of 0.529 (see [11]). We report in Fig. 3 our final volume estimation as well as the ellipsoid-based estimate, compared

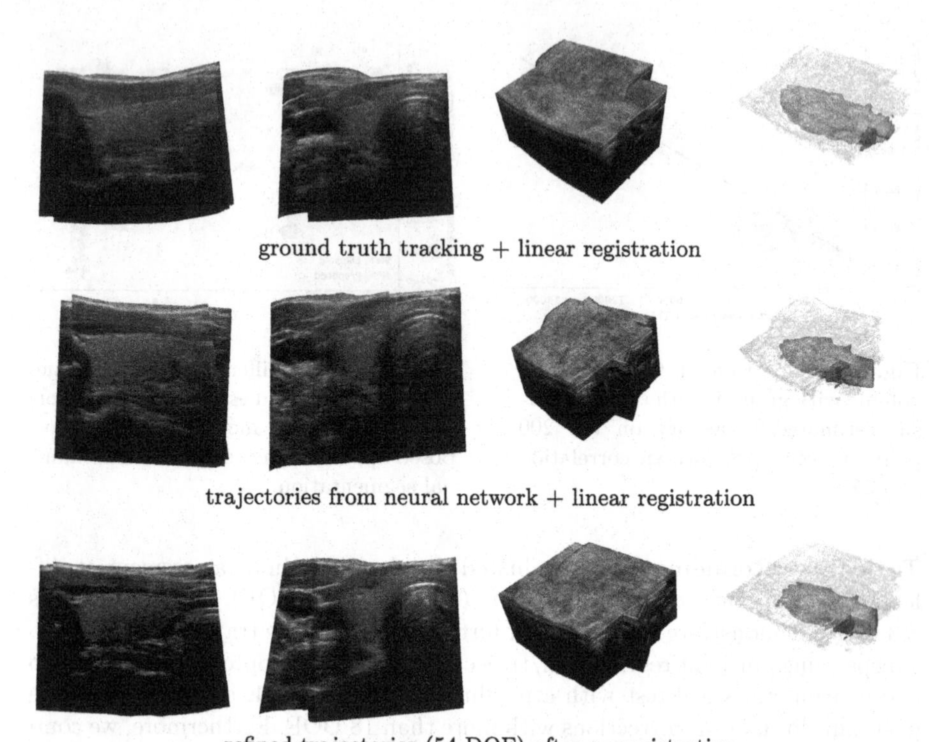

Fig. 4. 3D Reconstruction of two perpendicular sweeps of a thyroid. The four columns respectively show: (i)/(ii) blending of the two sweeps in two perpendicular planes, (iii) the registration of the two sweeps, (iv) 3D view of the resulting segmentation.

to the ground truth segmentation, on the 8 cases for which it is available. Our method yields more consistent and accurate estimates than the current clinical standard (median error of 12% vs 42%).

Finally, we show in Fig. 4 qualitative results on a representative case of our dataset, where we notice a stronger agreement between the two sweeps after our trajectory refinement. Here, the visual appearance is arguably even better than for the ground truth because internal tissue motion can be partially compensated, which is not possible with external tracking alone. This also illustrates that there is an inherent limit in ground truth accuracy for our experimental setup.

3.3 Application on Clinical Data

In an ongoing prospective study, patients undergoing screening for suspicious thyroid masses are scanned with a Philips iU22 ultrasound machine. Written consent was obtained from all patients prior to the examination. For each patient the following scans are performed:

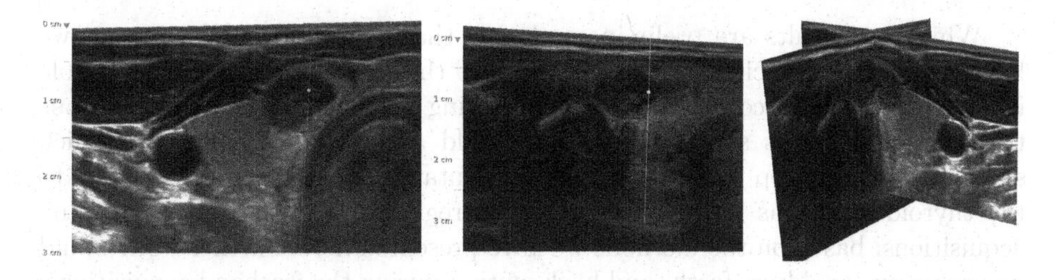

Fig. 5. Reconstructed frames from transversal and sagittal clips (left, middle), together with their 3D rendering (right) of a patient scan.

1. Conventional thyroid ultrasound by a routine clinical protocol which includes transversal and sagittal greyscale still images and cine sweep images through both thyroid lobes with a L12-5 probe.
2. Multiple 3D ultrasound volume data acquisitions with a VL13-5 transducer through the thyroid lobe, including transversal and sagittal orientations of the probe.

Three independent radiologists are measuring the thyroid nodules in three planes with the ellipsoid formula (two faculty radiologists and a fellow radiologist, all of whom had performed over 1000 thyroid ultrasound scans prior to this study). The first cases show the vast variability in the volumetric measurements of the same nodules by different readers, thus highlighting the inaccuracy of the manual method. Figure 5 shows the result of our reconstruction pipeline on one patient; instead of guessing the relationship between axial and sagittal scan planes, we provide a linked representation and a 3D rendering, using the 3D pose of the ultrasound clip frames from both sweeps as derived by our method. This alone may improve the ellipsoid method by choosing better frames to draw consistent length measurements from; in addition, an automatic segmentation method can be directly translated to 3D volume measurements. On this patient, the shown lesion was measured by three readers to be 0.85 mL, 1.06 mL, 0.76 mL, respectively. The 3D segmentation results based on our reconstruction were 0.99 mL and 0.91 mL for transversal and sagittal, respectively.

4 Conclusion

We have presented a pipeline to create accurate three-dimensional representations of the human thyroid from overlapping 2D ultrasound clips acquired in transversal and sagittal directions. Deep learning-based reconstruction and segmentation is performed individually on each sweep, then their information is combined and redundancy in the overlapping data exploited to refine the 3D reconstruction. Since neither specialized hardware or setup is required, this can have beneficial implications for many clinical applications; we therefore propose this concept as a general means to create a 3D representation from arbitrary DICOM clips, even when using inexpensive point-of-care ultrasound probes.

While our results are preliminary due to the size of our validation set, we believe they are sufficient proof to show that the presented approach is viable in general. Pending completion of our ongoing patient study with suspicious thyroid masses, it is straightforward to build a clinical software tool, which shall also allow Deep Learning-based segmentation of nodules in addition to the thyroid gland, as well as automated co-registration of repeated screening acquisitions, based on the methods we have presented here. Further work would be necessary to address forth- and back motion during the freehand acquisitions, using for instance an auto-correlation approach (as used for gating in [7]) which can remove duplicate content. On challenging clinical data, landmarks placed interactively on anatomical structures of interest may be used to constrain the joint co-registration & reconstruction, while at the same time increasing local accuracy. Last but not least, we are also working on a non-linear deformation model that is matching the skin surface of the two sweeps so that the varying pressure exerted onto the patient's neck during the scanning can be properly compensated.

References

1. Brown, M.C., Spencer, R.: Thyroid gland volume estimated by use of ultrasound in addition to scintigraphy. Acta Radiol. Oncol. Radiat. Phys. Biol. **17**(4), 337–341 (1978)
2. Chang, C.Y., Lei, Y.F., Tseng, C.H., Shih, S.R.: Thyroid segmentation and volume estimation in ultrasound images. IEEE Trans. Biomed. Eng. **57**(6), 1348–1357 (2010)
3. Chen, J., You, H., Li, K.: A review of thyroid gland segmentation and thyroid nodule segmentation methods for medical ultrasound images. Comput. Methods Program Biomed. **185**, 105329 (2020)
4. Karamalis, A., Wein, W., Kutter, O., Navab, N.: Fast hybrid freehand ultrasound volume reconstruction. In: Miga, M., Wong, I., Kenneth, H. (eds.) Proc. of the SPIE. **7261**, 726114–726118 (2009)
5. Kumar, V., et al.: Automated segmentation of thyroid nodule, gland, and cystic components from ultrasound images using deep learning. IEEE Access **8**, 63482–63496 (2020)
6. Lyshchik, A., Drozd, V., Reiners, C.: Accuracy of three-dimensional ultrasoundfor thyroid volume measurement in children and adolescents. Thyroid **14**(2), 113–120 (2004). https://doi.org/10.1089/105072504322880346. pMID:15068625
7. O'Malley, S.M., Granada, J.F., Carlier, S., Naghavi, M., Kakadiaris, I.A.: Image-based gating of intravascular ultrasound pullback sequences. IEEE Trans. Inf. Technol. Biomed. **12**(3), 299–306 (2008)
8. Prevost, R., et al.: 3D freehand ultrasound without external tracking using deep learning. Med. Image Anal. **48**, 187–202 (2018)
9. Prevost, R., Salehi, M., Sprung, J., Ladikos, A., Bauer, R., Wein, W.: Deep learning for sensorless 3D freehand ultrasound imaging. In: Descoteaux, M., Maier-Hein, L., Franz, A., Jannin, P., Collins, D.L., Duchesne, S. (eds.) MICCAI 2017. LNCS, vol. 10434, pp. 628–636. Springer, Cham (2017). https://doi.org/10.1007/978-3-319-66185-8_71

10. Ronneberger, O., Fischer, P., Brox, T.: U-Net: convolutional networks for biomedical image segmentation. In: Navab, N., Hornegger, J., Wells, W.M., Frangi, A.F. (eds.) MICCAI 2015. LNCS, vol. 9351, pp. 234–241. Springer, Cham (2015). https://doi.org/10.1007/978-3-319-24574-4_28
11. Shabana, W., Peeters, E., De Maeseneer, M.: Measuring thyroid gland volume: should we change the correction factor? Am. J. Roentgenol. **186**(1), 234–236 (2006)
12. Wein, W., Khamene, A.: Image-based method for in-vivo freehand ultrasound calibration. In: SPIE Medical Imaging 2008, San Diego, (2008)
13. Wunderling, T., Golla, B., Poudel, P., Arens, C., Friebe, M., Hansen, C.: Comparison of thyroid segmentation techniques for 3D ultrasound. In: Styner, M.A., Angelini, E.D. (eds.) Medical Imaging 2017: Image Processing. International Society for Optics and Photonics, SPIE. **10133**, 346–352 (2017) https://doi.org/10.1117/12.2254234

Complex Cancer Detector: Complex Neural Networks on Non-stationary Time Series for Guiding Systematic Prostate Biopsy

Golara Javadi[1(✉)], Minh Nguyen Nhat To[2], Samareh Samadi[1], Sharareh Bayat[1], Samira Sojoudi[1], Antonio Hurtado[3], Silvia Chang[3], Peter Black[3], Parvin Mousavi[4], and Purang Abolmaesumi[1]

[1] Department of Electrical and Computer Engineering, University of British Columbia, Vancouver, BC, Canada
golara@ece.ubc.ca
[2] Sejong University, Seoul 05006, South Korea
[3] Vancouver General Hospital, Vancouver, BC, Canada
[4] School of Computing, Queen's University, Kingston, ON, Canada

Abstract. Ultrasound is a common imaging modality used for targeting suspected cancerous tissue in prostate biopsy. Since ultrasound images have very low specificity and sensitivity for visualizing the cancer foci, a significant body of literature have aimed to develop ultrasound tissue characterization solutions to alleviate this issue. Major challenges are the substantial heterogeneity in data, and the noisy, limited number of labeled data available from pathology of biopsy samples. A recently proposed tissue characterization method uses spectral analysis of time series of ultrasound data taken during the biopsy procedure combined with deep networks. However, the real-value transformations in these networks neglect the phase information of the signal. In this paper, we study the importance of phase information and compare different ways of extracting reliable features including complex neural networks. These networks can help with analyzing the phase information to use the full capacity of the data. Our results show that the phase content can stabilize training specially with non-stationary time series. The proposed approach is generic and can be applied to several other scenarios where the phase information is important and noisy labels are present.

Keywords: Complex neural networks · Temporal enhanced ultrasound · Image guided interventions · Prostate cancer

---

This work is funded in part by the Natural Sciences and Engineering Research Council of Canada (NSERC), and the Canadian Institutes of Health Research (CIHR).
P. Black, P. Mousavi and P. Abolmaesumi—Joint senior authors.

A. L. Martel et al. (Eds.): MICCAI 2020, LNCS 12263, pp. 524–533, 2020.
https://doi.org/10.1007/978-3-030-59716-0_50

1 Introduction

Ultrasound (US) is increasingly being used for image-guided interventions, given its non-ionizing nature, wide accessibility, and low cost. This imaging technique serves many applications including confirmatory biopsies for cancer diagnosis and prognosis in the lymph nodes, breast, liver, and prostate. Annually, one million men in North America undergo US-guided prostate biopsy [8], with the global market for diagnosis and screening expected to reach $27B by 2021. An ongoing challenge facing physicians is the ability to confidently detect clinically significant prostate cancer (sPCa), i.e., distinguishing aggressive cancer that has a high likelihood of leading to metastasis *vs.* less aggressive or indolent cancer and benign tissue. Ideally, sPCa would be discovered at an early stage to maximize cure with minimal disease-related morbidity, while avoiding over-treatment of indolent cancer. Transrectal US-guided biopsy involves obtaining 8 to 12 prostate tissue samples using a systematic, yet, nontargeted approach. As such, it is blinded to individual patient intraprostatic pathology, and has a high rate of false negatives. Recent clinical practice incorporates multi-parametric magnetic resonance imaging (mp-MRI) in the diagnostic protocol to improve the overall sensitivity of cancer detection. In this approach, possible cancer lesions are outlined on mp-MRI images and the images are registered with US for targeted biopsy. However, mp-MRI has low specificity, and its fusion with US for biopsy guidance is technically challenging.

An alternative approach is to perform direct tissue characterization in US. Previous literature in this area includes spectral analysis of Radio Frequency (RF) data from single frames of US images by modeling the acoustic properties of the effective scatterers [10]. Multi-centre studies of systems that use these analyses on conventional clinical US systems have not shown significant improvements in cancer yield. Tissue elasticity imaging [23], Sheer Wave Absolute Vibro-Elastography (S-WAVE) [17] and Temporal enhanced UltraSound (TeUS) [5,22] are amongst other methods used for direct tissue characterization in US. Specifically, TeUS has demonstrated significant potential for improving the specificity of cancer detection over conventional US [4,5,13,14,16,18]. TeUS involves acquisition of a set of US frames captured over a short period of time from a stationary tissue location without mechanical excitation. This modality captures spatial variations in the scattering function, where subtle changes in nuclei configuration are the main contributors [7]. Prior work using the analysis of TeUS have assumed TeUS signals to be stationary in nature. Studies at our institution show that the US time series data acquired and used for TeUS is non-stationary in nature, which makes the previous networks sub-optimal for real-time deployment and prospective cancer detection.

In classification of non-stationary time series data, finding a representative training set of target features is a cumbersome and complicated process [15]. With this type of data, it is critical to develop methods that can unravel the full potential of the data. Complex-value neural networks have shown superior performance to their real-value counterparts for this purpose [11,19–21,24–26]. Although they have been used for real-value image classification [19], their performance is optimal when they are used with fully complex-value inputs [21]. So far, the majority of the studies on TeUS have been done with amplitude of Fast Fourier Transform (FFT) of time series. Our study shows complex-value networks are better suited at uncovering information from non-stationary time series with analysis of the complex value of FFT. A complex value network takes the correlation between the real and imaginary parts of signal into account. This includes the phase information, which can stabilize training and improve the performance of cancer detection classifier in non-stationary data [9].

In this work, we propose to use complex-value neural networks to explicitly analyze TeUS data in spectral domain (Fig. 1). We show that complex networks are better suited at uncovering information from non-stationary time series with analysis of complex value of FFT of time series data. In an *invivo* study with 188 biopsy cores from 55 patients, we achieve an area under the curve (AUC) and an accuracy (ACC) of 0.73 and 0.67, respectively. Our results show that using complex networks, we can improve tissue classification over prior art despite having highly heterogeneous and noisy labels. We believe our observations are generic and may be applied to many other applications with complex-value data or non-stationary time series.

2 Materials and Methods

2.1 Data

TeUS data were acquired from 188 biopsy cores of 55 subjects during systematic US-guided biopsy. The Institutional Research Ethics Board approved the investigation and the participants gave informed consent to be a part of this study. The systematic biopsy targets were obtained from predefined prostate zones, and the number of cores biopsies taken from one patient was based on the size of the prostate gland: 8, 10 and 12 samples were obtained for prostate volumes of less than 30 ml, 30 ml to 60 ml, and more than 60 ml, respectively. Data acquisition involved holding the ultrasound transducer still for approximately 5 s, while capturing 200 frames of RF data. The biopsy gun was then fired to collect the tissue sample.

Ground-Truth Labeling: The biopsy samples were labeled by pathologists using the Gleason scoring system (GS), which is the standard-of-care for grading prostate cancer. The aggressiveness of prostate cancer is indexed by the Gleason Grade (GG), scaled from 1 to 5, with 5 being the most aggressive. The two dominant GGs in a specimen are summed up to report the Gleason Score (GS) [12]. Non-cancerous cores include benign, atrophy, chronic inflammation, fibromuscular tissue and Prostatic Intraepithelial Neoplasia (PIN). The pathology report contains percentage of cancer in the core, often called as involvement, and the GS for the entire biopsy core. However, the spatial information on the tumor location is not specified.

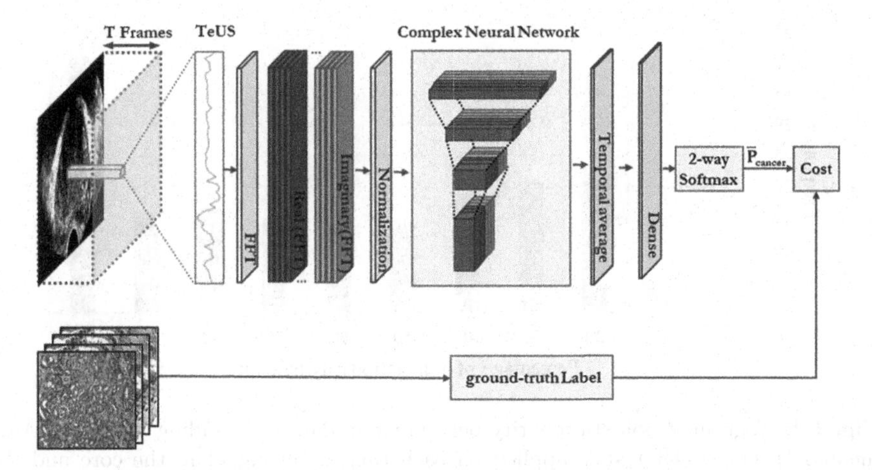

Fig. 1. Overview of the complex classification framework. The labels for the dataset come from pathology results. The network outputs a binary decision for declaring a signal as cancer or benign.

In our dataset, $n = 94$ biopsy cores are cancerous with GS 3 + 4 or higher; GS 3 + 4 (n = 37), GS 4 + 3 (n = 27), GS 4 + 4 (n = 12), and GS 4 + 5 (n = 18). We use these data for an initial evaluation of our approach to separate aggressive cancer from benign tissue through a binary labeling of the entire core as benign or cancer. The benign cores are from patients who exclusively had no cancer as confirmed by pathology, and the aggressive cancer cores are from biopsy confirmed patients with more than GS 3 + 4 in at least one core.

Preprocessing: The first step in data processing is to find the location of the biopsy sample for each core from the B-Mode data. Generally, the region of interest (ROI) is a 2×18 mm area starting 2 mm from the tip of needle. Therefore, we first detect the location of needle's tip manually. The corresponding ROI in the B-mode coordinates is then converted into the RF coordinates and the TeUS data from the mapped ROI in the RF coordinates are acquired accordingly. The data are then labeled based on the available histopathology results identifying if a core is cancerous or benign. The first 48 frequency components of RF frames acquired from this ROI with FFT creates our labeled dataset. The data are normalized to unit power. The test and train datasets have mutually exclusive patients. In multiple stages of data preprocessing, the quality of data is verified. This is to check the corresponding biopsy plane and RF frames are as close as possible without any major hand motion during the biopsy.

Stationarity Test: Non-stationarity of TeUS data implies that the statistical properties of a process generating the time series change over time. Stationarity has become a common assumption for many machine learning methods for time series analysis. There are statistical tests to check the stationarity of a signal

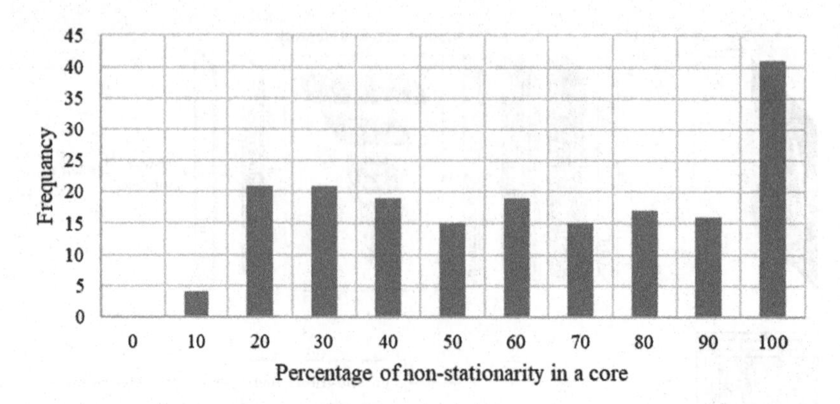

Fig. 2. Histogram of non-stationarity percentage in dataset including 188 cores. Augmented Dickey Fuller test is applied on each time series signal in the core and the percentage of the non-stationarity of each core is calculated.

such as augmented Dickey Fuller (ADF) test, and Kwiatkowski-Phillips-Schmidt-Shin (KPSS) test. ADF test examines the null hypothesis that a unit root is presented in a time series sample. To measure the non-stationarity of the dataset, ADF is applied to all the signals of an entire biopsy core and the percentage of detected non-stationary signals is calculated for each biopsy core. High portion of dataset shows non-stationarity which can be challenging for using deep learning approaches with this assumption. Figure 2 shows the histogram of percentage of non-stationarity in the dataset.

2.2 Method

Complex-Value Neural Networks: Complex equivalent of real-value convolution can be perform by convolving a complex filter matrix in Cartesian notation $\boldsymbol{W} = W_R + iW_I$ by a complex input data vector $\boldsymbol{d} = d_R + id_I$, where W_R and W_I are real matrix and d_R and d_I are real vectors. As suggested in [24],the complex convolution can be simulated using real-value entities. This operation is shown in Eq. (1) below:

$$\begin{aligned} W * d &= (W_R + iW_I) * (d_R + id_I) \\ &= (W_R * d_R - W_I * d_I) + i(W_R * d_I + W_I * d_R). \end{aligned} \tag{1}$$

Matrix notation of this complex convolution is illustrated in Fig. 3 for 1D convolution:

$$\begin{bmatrix} R(\boldsymbol{W} * \boldsymbol{d}) \\ I(\boldsymbol{W} * \boldsymbol{d}) \end{bmatrix} = \begin{bmatrix} W_R & -W_I \\ W_I & W_R \end{bmatrix} * \begin{bmatrix} d_R \\ d_I \end{bmatrix}. \tag{2}$$

The loss and activation function in a complex-value neural network must be differentiable with respect to real and imaginary parts of each component in the network. This fact restricts the choice of activation functions for a complex-value neural network. Trabelsi *et al.* [24] showed that the choice of activation

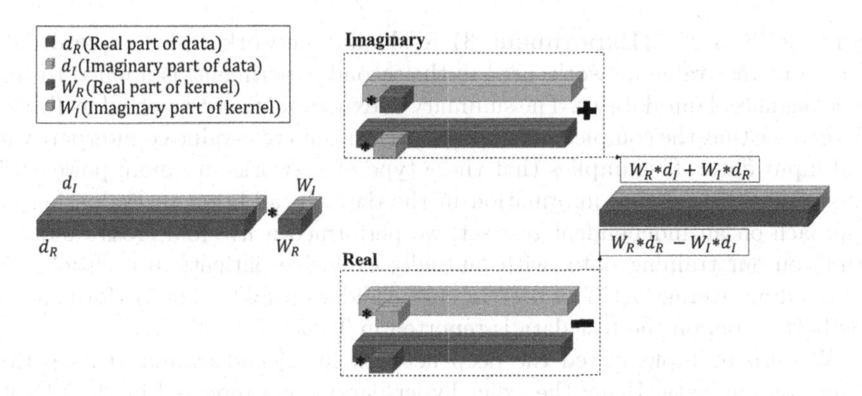

Fig. 3. Implementation details of 1D complex convolution. d_R and d_I shows the real and imaginary part of complex input. W_R and W_I show the real and imaginary kernel here.

function is also important for preserving the phase information. Many activation functions have been proposed in the literature including, ModReLU, CReLU, and zReLU [24].

Proposed Complex Classifier: The structure of the classifier network shown in Fig. 1, uses complex convolution architecture for TeUS signals. We used four layers of complex 1D convolutions with Kernel sizes of 7, 3, 3, 3, respectively. The number of filters in each layer were 4, 8, 16, and 32. Each convolution layer was followed by a complex batch normalization [24] and a MaxPooling layer. The complex part of the network connected to a real dense layer and a classifier. The network was trained using the Adam optimizer, a batch size of 256 and a learning rate of $1e-2$.

3 Experiments and Results

Data are divided into mutually exclusive training and test set with 167 cores from 41 patients for training, and 21 cores from 14 patients in test. The number of cancer and benign cores are balanced in training and test sets. We use standard evaluation metrics as prior approaches [2,6] to quantify our results. The AUC and ACC of classification are assessed for detection of benign *vs.* cancer. To evaluate the performance of complex-value classifier, we perform the following experiments: **(Experiment** 1) Complex network with complex-value FFT. We applied the complex network described in Sect. 2.2 to the first 48 components of complex-value FFT. **(Experiment** 2) Real-value networks with amplitude of FFT. A real-value network with similar structure to the complex networks of the first experiment is used as a baseline. To have a fair comparison, this network has same number of parameters as the complex network. For the input data to this real-value model, we consider the amplitude of FFT similar to previous works using TeUS [5,22]. **(Experiment 3)** Real-value network with temporal data. The same real-value network, used in the second experiment, is applied to time series signals of our dataset. The summary of experiments are reported in Table 1. We can see that the complex network outperforms its real-value counterpart with real input data. This implies that these type of networks are more powerful in uncovering the existing information in the data. In addition to evaluating our approach on an independent test set, we performed a five-fold cross-validation study on our training data, with mutually exclusive patients in each fold. We achieved an average AUC of 0.77 in cross-validation folds. The performance of the best model on the test data is reported in Table 1.

We also re-implemented the deep network in [2] and examined its performance on our data. Using the exact hyper-parameters reported in [2] (121, 406 network parameters), and performing a new hyper-parameter search to train a model similar to the size of our Complex network (21, 600 network parameters), we achieved AUC and ACC of 0.59 and 0.43, respectively. We would like to emphasize that the two studies are not directly comparable as the targeting procedure for sampling the prostate tissue in the two patient cohorts are fundamentally different. Our dataset is highly heterogeneous, since sextant biopsy involves 8–12 random cores obtained from the prostate. Conversely, the study in [2] involved likely more homogeneous tissue samples as the cores are taken from prostate regions that had suspected cancer foci pre-identified in multi-parametric MRI (mp-MRI). The population in our study is anyone suspected of prostate cancer, without any prior mp-MRI imaging and biopsied blindly with respect to pathology, hence, making the data significantly more challenging.

Table 1. Model performance for classification of cores in the test data for complex-value and real-value networks, considering the different range of involvement for the cancer cases.

Dataset	Model	No. training cores	Involvement	AUC	ACC	No. parameters
Complex-FFT	Complex	102	$\geq$ 40%	0.78	0.67	21,542
		142	$\geq$ 20%	0.78	0.64	
		167	*All*	0.76	0.81	
Amplitude-FFT	Real	102	$\geq$ 40%	0.66	0.60	21,539
		142	$\geq$ 20%	0.64	0.57	
		167	*All*	0.62	0.62	
Temporal Analysis	Real	102	$\geq$ 40%	0.60	0.52	21,539
		142	$\geq$ 20%	0.61	0.60	
		167	*All*	0.49	0.52	

We repeated the studies by considering different ranges of involvement, which is a measure of tumor in the core length. Previous works on TeUS showed using higher involvement cancers for training will result in better performance of classifier [3]. However, the number of high involvement cancer cases in our dataset is limited. The range of cancer involvement in our training dataset, as reported

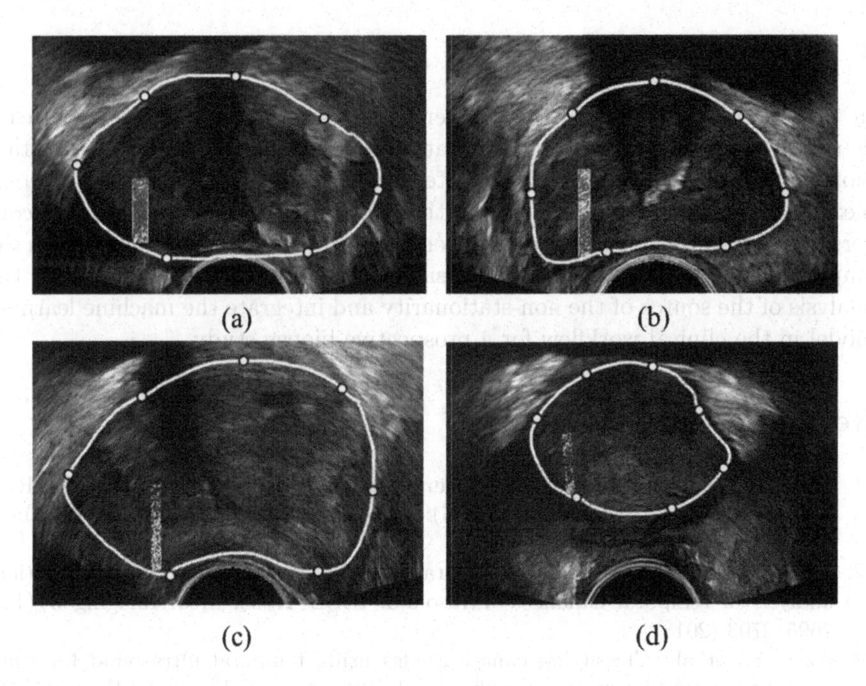

Fig. 4. Cancer likelihood maps overlaid on B-mode ultrasound images for biopsy location (ROI of 2×18 mm^2): (a) and (b) are the cancer cores with 2.5 and 1.9 mm tumor in core length. (c) and (d) are benign cores. The boundary of the prostate gland is shown in yellow. (Color figure online)

in histopathology, is between (3% – 100%) with the average of 45%. Therefore, we studied the effect of involvement starting from average involvement of our dataset for training. Table 1 shows the performance of the networks trained by considering cancer cores with different range of involvements. We can see that complex networks can perform better in all the involvement ranges. Figure 4 shows the cancer likelihood maps generated from the proposed complex network classifier overlaid on B-mode ultrasound images for a region of interest (ROI) associated with a biopsy sample taken from the prostate gland.

The primary focus of this paper is the detection of aggressive prostate cancer. According to the established clinical literature, clinically significant cancer is defined as any involvement of a Gleason grade of 4 or higher (e.g., GS 3 + 4 or 4 + 3), and not Gleason score of 3 + 3, unless it has a core length greater than 6 mm [1]. All other GS 3 + 3 cancers are considered indolent, where patients are referred to an active surveillance protocol and observed. In Table 1 we only reported the results for cores that pass the criteria for aggressive cancer. Our patient cohort does not include cores with GS 3 + 3, where tumor length is 6 mm or higher. However, we verified the performance of our trained Complex network on a subset of our data bu including the GS 3 + 3 cores, and observed an AUC of 0.76 and a balanced accuracy of 0.74.

4 Conclusion

In this paper, we addressed the problem of non-stationarity of TeUS data for detection of prostate cancer. Using data obtained during systematic prostate biopsy from 55 patients, we demonstrated that complex-value networks on complex FFT of TeUS data can help with the tissue classification and improve convergence. This type of networks can leverage the phase information as well as the amplitude of signal's spectral components. Future work will be focused on the analysis of the source of the non-stationarity and integrate the machine learning model in the clinical workflow for a prospective biopsy study.

References

1. Ahmed, H.U., et al.: Diagnostic accuracy of multi-parametric MRI and TRUS biopsy in prostate cancer (PROMIS): a paired validating confirmatory study. Lancet **389**(10071), 815–822 (2017)
2. Azizi, S., et al.: Deep recurrent neural networks for prostate cancer detection: analysis of temporal enhanced ultrasound. IEEE Trans. Med. Imaging **37**(12), 2695–2703 (2018)
3. Azizi, S., et al.: Classifying cancer grades using temporal ultrasound for transrectal prostate biopsy. In: Ourselin, S., Joskowicz, L., Sabuncu, M.R., Unal, G., Wells, William (eds.) MICCAI 2016. LNCS, vol. 9900, pp. 653–661. Springer, Cham (2016). https://doi.org/10.1007/978-3-319-46720-7_76
4. Azizi, S., et al.: Ultrasound-based detection of prostate cancer using automatic feature selection with deep belief networks. In: Navab, N., Hornegger, J., Wells, W.M., Frangi, A.F. (eds.) MICCAI 2015. LNCS, vol. 9350, pp. 70–77. Springer, Cham (2015). https://doi.org/10.1007/978-3-319-24571-3_9
5. Azizi, S.: Toward a real-time system for temporal enhanced ultrasound-guided prostate biopsy. Int. J. Comput. Assist. Radiol. Surg. **13**(8), 1201–1209 (2018). https://doi.org/10.1007/s11548-018-1749-z
6. Azizi, S.: Learning from noisy label statistics: detecting high grade prostate cancer in ultrasound guided biopsy. In: Frangi, A.F., Schnabel, J.A., Davatzikos, C., Alberola-López, C., Fichtinger, G. (eds.) MICCAI 2018. LNCS, vol. 11073, pp. 21–29. Springer, Cham (2018). https://doi.org/10.1007/978-3-030-00937-3_3
7. Bayat, S.: Investigation of physical phenomena underlying temporal-enhanced ultrasound as a new diagnostic imaging technique: theory and simulations. IEEE Trans. Ultrason. Ferroelectr. Freq. Control **65**(3), 400–410 (2017)
8. Bjurlin, M.A., Taneja, S.S.: Standards for prostate biopsy. Curr. Opin. Urol. **24**(2), 155–161 (2014)
9. Dramsch, J.S., Lüthje, M., Christensen, A.N.: Complex-valued neural networks for machine learning on non-stationary physical data. arXiv preprint (2019). arXiv:1905.12321
10. Feleppa, E., Porter, C., Ketterling, C., Dasgupta, S., Ramachandran, S., Sparks, D.: Recent advances in ultrasonic tissue-type imaging of the prostate. Acoustical Imaging, pp. 331–339. Springer, Berlin (2007)
11. Gopalakrishnan, S., Cekic, M., Madhow, U.: Robust wireless fingerprinting via complex-valued neural networks. arXiv preprint (2019). arXiv:1905.09388

12. Heidenreich, A., et al.: European association of urology: EAU guidelines on prostate cancer. part 1: screening, diagnosis, and local treatment with curative intent-update 2013. Eur. Urol. **65**(1), 124–137, January 2014
13. Imani, F.: Computer-aided prostate cancer detection using ultrasound RF time series: in vivo feasibility study. IEEE Trans. Med. Imaging **34**(11), 2248–2257 (2015)
14. Imani, F.: Augmenting MRI-transrectal ultrasound-guided prostate biopsy with temporal ultrasound data: a clinical feasibility study. Int. J. Comput. Assist. Radiol. Surg. **10**(6), 727–735 (2015)
15. Koh, B.H.D., Woo, W.L.: Multi-view temporal ensemble for classification of non-stationary signals. IEEE Access **7**, 32482–32491 (2019)
16. Moradi, M., Abolmaesumi, P., Mousavi, P.: Tissue typing using ultrasound RF time series: experiments with animal tissue samples. Med. Phys. **37**(8), 4401–4413 (2010)
17. Moradi, M.: Multiparametric 3D in vivo ultrasound vibroelastography imaging of prostate cancer: preliminary results. Med. Phys. **41**(7), 073505 (2014)
18. Nahlawi, L., et al.: Using hidden markov models to capture temporal aspects of ultrasound data in prostate cancer. In: 2015 IEEE International Conference on Bioinformatics and Biomedicine (BIBM), pp. 446–449 (2015)
19. Popa, C.A.: Complex-valued convolutional neural networks for real-valued image classification. In: 2017 International Joint Conference on Neural Networks (IJCNN), pp. 816–822. IEEE (2017)
20. Popa, C.A.: Deep hybrid real-complex-valued convolutional neural networks for image classification. In: 2018 International Joint Conference on Neural Networks (IJCNN), pp. 1–6. IEEE (2018)
21. Popa, C.A., Cernăzanu-Glăvan, C.: Fourier transform-based image classification using complex-valued convolutional neural networks. In: Huang, T., Lv, J., Sun, C., Tuzikov, A.V. (eds.) ISNN 2018. LNCS, vol. 10878, pp. 300–309. Springer, Cham (2018). https://doi.org/10.1007/978-3-319-92537-0_35
22. Sedghi, A., et al.: Deep neural maps for unsupervised visualization of high-grade cancer in prostate biopsies. Int. J. Comput. Assist. Radiol. Surg. **14**(6), 1009–1016 (2019). https://doi.org/10.1007/s11548-019-01950-0
23. Sumura, M., Shigeno, K., Hyuga, T., Yoneda, T., Shiina, H., Igawa, M.: Initial evaluation of prostate cancer with real-time elastography based on step-section pathologic analysis after radical prostatectomy: a preliminary study. Int. J. Urol. **14**(9), 811–816 (2007)
24. Trabelsi, C., et al.: Deep complex networks. arXiv preprint (2017). arXiv:1705.09792
25. Virtue, P., Stella, X.Y., Lustig, M.: Better than real: complex-valued neural nets for MRI fingerprinting. In: 2017 IEEE International Conference on Image Processing (ICIP), pp. 3953–3957. IEEE (2017)
26. Wang, S., et al.: Deepcomplexmri: exploiting deep residual network for fast parallel MR imaging with complex convolution. Magn. Reson. Imaging **68**, 136–147 (2020)

Self-Supervised Contrastive Video-Speech Representation Learning for Ultrasound

Jianbo Jiao[1(✉)], Yifan Cai[1], Mohammad Alsharid[1], Lior Drukker[2], Aris T. Papageorghiou[2], and J. Alison Noble[1]

[1] Department of Engineering Science, University of Oxford, Oxford, UK
jianbo.jiao@eng.ox.ac.uk

[2] Nuffield Department of Women's & Reproductive Health, University of Oxford, Oxford, UK

Abstract. In medical imaging, manual annotations can be expensive to acquire and sometimes infeasible to access, making conventional deep learning-based models difficult to scale. As a result, it would be beneficial if useful representations could be derived from raw data without the need for manual annotations. In this paper, we propose to address the problem of self-supervised representation learning with multi-modal ultrasound video-speech raw data. For this case, we assume that there is a high correlation between the ultrasound video and the corresponding narrative speech audio of the sonographer. In order to learn meaningful representations, the model needs to identify such correlation and at the same time understand the underlying anatomical features. We designed a framework to model the correspondence between video and audio without any kind of human annotations. Within this framework, we introduce cross-modal contrastive learning and an affinity-aware self-paced learning scheme to enhance correlation modelling. Experimental evaluations on multi-modal fetal ultrasound video and audio show that the proposed approach is able to learn strong representations and transfers well to downstream tasks of standard plane detection and eye-gaze prediction.

Keywords: Self-supervised · Representation learning · Video-audio

1 Introduction

Deep learning-based medical image analysis approaches rely heavily on annotated training data, which limits the progress in medical image analysis if a large dataset has to be manually annotated for every new task. Extracting meaningful representations directly from unlabelled data is therefore an important and interesting sub-topic in learning-based medical image analysis.

Several approaches in the literature have been explored to deal with the problem of learning from unlabelled data, which is usually termed as "self-supervised representation learning" (or unsupervised learning in some works). The common practice is to pre-train a model on unlabelled data according to a pretext

A. L. Martel et al. (Eds.): MICCAI 2020, LNCS 12263, pp. 534–543, 2020.
https://doi.org/10.1007/978-3-030-59716-0_51

task and then to fine tune the model with some specific target tasks to evaluate the learned representations. Typical pretext tasks include colourisation [16], and rotation prediction [4] for images; and tracking [13], temporal ordering [7] for videos. A recent study proposes to learn representations by contrastive predictive coding [8] and showed a powerful learning capability. Some works have also considered medical images, *e.g.*, predicting the distance between patches [12], Rubik's cube recovery [17] and anatomy-aware joint reasoning [6]. However, the above-mentioned approaches are single modality. Some recent approaches have investigated learning from natural audio and video modalities [1,9], where the pretext task is designed as video-audio alignment. In this case the audio and video are assumed to be in dense correspondence. Such multi-modal learning has not been explored for medical data before. Audio data rarely exists for medical images and even when available, it is mostly narrative diagnosis/interpretation speech, which has a sparse correlation with the visual data, making the task rather challenging.

In this paper, we propose to address the problem of self-supervised cross-modal representation learning for ultrasound video with corresponding narrative speech audio, both of which are captured on-the-fly without any manual annotations. Unlike other medical imaging modalities that are with clear anatomical structures (*e.g.*, MRI and CT), ultrasound video is much more difficult to interpret by eye for humans although experts are adept at interpreting anatomy in the acoustic patterns. As a result, learning representations automatically from unlabelled ultrasound data is rather challenging. On the other hand, in our case we have synchronised narrative speech from the sonographer accompanied with the ultrasound video. The basic assumption here is that by leveraging cross-modal correlations, a useful representation can be learned. To this end, we propose to learn the anatomical representations from ultrasound video-speech data by identifying the affinity (*i.e.*, association strength) between the two modalities. Specifically, we first randomly sample video-speech samples from our dataset, from which positive and negative pairs are generated. Unlike prior works that use straightforward simple training pairs, in this work we instead leverage hard-negative as well as hard-positive pairs for training, in order to force the model to *learn harder*. Additionally, we further introduce cross-modal contrastive learning to encourage a positive pair to have strong affinity while a negative pair will have weaker affinity in a projected shared latent space. In our diagnostic narrative speech audio, we observe that background noise (*e.g.*, button clicks, air conditioning) and unrelated conversation (*e.g.*, about the weather, travel *etc.*) exist in the raw data, which degrades the representation learning. To mitigate this problem, we propose an affinity-aware self-paced learning curriculum over the representation learning process.

To evaluate the proposed self-supervised learning framework, we consider two ultrasound-related downstream tasks: standard plane detection and eye-gaze saliency prediction. Experimental results presented show that the proposed approach significantly improves the performance of downstream tasks without referring to any manual annotations. The experiments also reveal that by

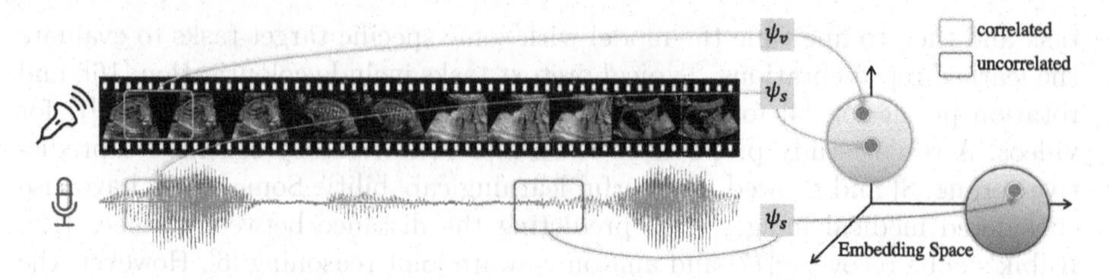

Fig. 1. Main idea of the proposed self-supervised video-speech representation learning framework. A model is trained to identify whether a sampled video-speech pair is anatomically correlated, and at the same time encourage the projected embeddings from correlated pair to lie on the same anatomical sphere (*e.g.*, the green one).(Color figure online)

leveraging speech data, useful representations can be learned when transferring to other tasks, outperforming single-modal learning methods.

The main contributions of this paper are summarised as follows:

- We propose, to our knowledge, the first self-supervised video-speech representation learning approach for ultrasound data.
- We introduce cross-modal contrastive learning and affinity-aware self-paced learning for ultrasound video-speech cross-modal representation learning.
- The proposed approach is demonstrated to be effective for two downstream tasks.

2 Method

In this section, we present the proposed self-supervised contrastive video-speech representation learning approach in detail. The main idea of the proposed approach is to build the correlation between video and narrative speech data by both explicit optimisation and implicit regularisation. In this paper, we demonstrate the idea with fetal ultrasound video synchronised with corresponding speech. A novel self-supervised learning framework is proposed accordingly. Figure 1 illustrates the main idea of the proposed framework.

2.1 Video-Speech Representation Learning

Speech audio, as an additional modality to visual data (*e.g.*, images and videos), is recorded on-the-fly without external annotations. The idea is to use the discriminative power of audio to understand visual data where ambiguity may exist. The basic assumption to learn a cross-modal (ultrasound video and corresponding speech from sonographer in our case) representation is that the two modalities share similar anatomical meanings at the same timestamp. In our case, if a model can successfully identify whether a video clip and a speech clip are correlated or not, it has learned to understand the underlying anatomical representations.

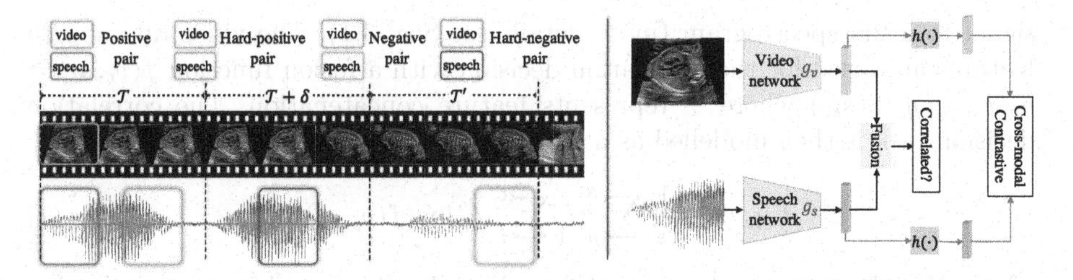

Fig. 2. Left: proposed training pair sampling scheme, where the hard-positive and hard-negative are illustrated. Right: proposed framework for self-supervised video-speech representation learning.

Based on the above assumption, we design a self-supervised learning framework to extract cross-modal representations with ultrasound video-audio data. Specifically, we randomly sample positive and negative pairs from the original video-speech data, where a positive pair indicates that the considered video and speech are correlated, while a negative pair is uncorrelated. A deep model is then trained to identify such positive/negative pairs accordingly, resembling a classification problem. Unlike natural image video and its corresponding audio where highly-dense correlations present (*e.g.*, playing the violin, cooking), the speech audio from a sonographer and the medical ultrasound video are sparsely correlated and narratively presented, making the correlation identification more challenging. To address this issue, we first propose to force the model to *learn harder* by sampling hard-negative and hard-positive pairs (as illustrated in Fig. 2 left), so as to learn a more strongly correlated representations. Suppose the ultrasound video and speech clips at time interval $\mathcal{T}$ are $\mathcal{V}_\mathcal{T}$ and $\mathcal{S}_\mathcal{T}$, the speech clip at a shifted time interval $\mathcal{T}'$ is $\mathcal{S}_{\mathcal{T}'}$, $(\mathcal{V}_\mathcal{T}, \mathcal{S}_\mathcal{T})$ is considered as a positive pair and $(\mathcal{V}_\mathcal{T}, \mathcal{S}_{\mathcal{T}'})$ a negative pair. Instead of sampling $\mathcal{T}'$ from a different scan sequence or a diverse anatomy sub-sequence, we force the learning process to be constrained by sampling $\mathcal{T}'$ from a nearby segment. This does not have to be for a different anatomy as $\mathcal{T}$, *i.e.*, $\mathcal{T}' = \mathcal{T} + \delta$ where $\delta < \mathcal{D}$ is a short random timestep and $\mathcal{D}$ is the shift range. Furthermore, we sample ultrasound video frames and speech clips to increase the generalisability: $\{v_t, s_t, s_{t'} | v \in \mathcal{V}, s \in \mathcal{S}, t \in \mathcal{T}, t' \in \mathcal{T}'\}$. In terms of the positive pair, we also make it *harder* by perturbing the alignment within $\mathcal{T}$ so that v_t and s_t do not have to be exactly at the same timestamp t. Additionally, we randomly sample a group of positive/negative pairs within each interval to represent the whole clip. Finally, the hard-positive and hard-negative pairs used for training are defined as $\{(v_{ik}, s_{jk}) | k \in [1, K]\}$ and $\{(v_{ik}, s_{t'k}) | k \in [1, K]\}$ respecctively, where $i, j \in \mathcal{T}$ and K is the number of sampling groups. An illustration of the proposed pair sampling scheme when $K = 1$ is shown in Fig. 2 (left).

With the above-mentioned sampling scheme, training pairs are fed into deep networks to extract the corresponding features. Suppose the ultrasound video and speech subnetworks as g_v and g_s, the extracted features are $g_v(v_t)$ and $g_s(\eta(s_t))$ where $\eta(\cdot)$ is a speech pre-processing function that converts the 1D signal to a 2D spectrogram (more details see Sec. 3.1). The video and speech features are fused for the correlation decision with a fusion function $f(v_t, s_t) = g_v(v_t) \oplus g_s(\eta(s_t))$ where $\oplus$ represents feature concatenation. The correlation decision task is then modelled as a classification problem with the objective:

$$\mathcal{L}_{cls} = -\frac{1}{N}\sum_{n=1}^{N}\sum_{i}^{C} c_i^n log(f(v_t, s_t)_i^n), \tag{1}$$

where N is the total number of samples while $C = 2$ for our binary classification problem, c is the label indicating if the inputs are correlated or not.

2.2 Cross-Modal Contrastive Learning

We observe that in the speech data, there exist medical-unrelated contents (*e.g.*, random talk), which deteriorate correlation learning. To address this, we introduce cross-modal contrastive learning, in addition to the afore-mentioned classification objective. The key idea here is to encourage the representations of video and speech from a (hard-)positive pair to be similar while repelling the (hard-)negative pair, at a projected embedding space. Specifically, before the feature fusion, $g_v(v_t)$ and $g_s(\eta(s_t))$ are further projected to a latent space by function $h(\cdot)$, where the cross-modal contrastive objective is applied. Suppose $y_v = h(g_v(v_t)), y_s = h(g_s(\eta(s_t)))$ are the projected embeddings, the cross-modal contrastive objective is defined as:

$$\mathcal{L}_{cont} = -log\frac{e^{sim(y_v, y_s)} - e^{sim(y_v, y_{s'})}}{\sum_{k=1}^{N}\mathbb{1}_{[k \neq v]}e^{sim(y_v, y_k)}}, \tag{2}$$

where $\mathbb{1} \in \{0, 1\}$ and $y_{s'} = h(g_s(\eta(s_{t'})))$. Function $sim(a, b) = a^\top b$ measures the similarity between two vectors. An illustration of the additional cross-modal contrastive learning is shown in Fig. 2 (right).

2.3 Affinity-Aware Self-Paced Learning

The microphone used to record the speech is placed in an open environment without any specific voice filtering, which means that all the sound is recorded including background noise (*e.g.*, air conditioner, button-click), in addition to the main narrative content. Since our main focus is the speech with meaningful descriptions, the noise inevitably affects representation learning. To this end, we further propose a self-paced learning curriculum based on the affinity between video and speech data. Specifically, we divide the multi-modal data into different affinity levels and perform specific learning schemes accordingly. For simplicity, here we choose two affinity levels, *i.e.*, low-affinity and high-affinity, where low-affinity refers to the speech audio mostly consisting of noise and the rest is called high-affinity. The affinity is automatically detected according to an energy-based voice activity detection algorithm [11].

The proposed representation learning approach is only performed on the high-affinity fragments. In terms of the low-affinity data, instead of directly discarding it, we propose to leverage the video data which provides visual clues for representation learning. As a result, we include an additional pretext task to extract representations from the whole video data. Inspired by [7], we randomly shuffle the frames in a video clip and train the model to predict the correct order. The assumption here is that the model can correct the shuffled frame order only if it understands the anatomical information within the video clip. Four frames are used to construct a clip for this task and the forward and backward sequences are considered to be the same (*e.g.*, 0–1–2–3 *v.s.* 3–2–1–0). Therefore, this pretext task is modelled as a 12-category classification problem with cross-entropy objective $\mathcal{L}_{ord}$. To avoid model cheating, the fan-shape of the ultrasound images is cropped out, keeping only the inner part.

2.4 Implementation and Training

The proposed model framework is illustrated in Fig. 2 (right). In terms of the backbone network architecture, we choose the ResNeXt-50 [14] with Squeeze-and-Excitation module [5] and dilated convolutions [15]. The video and speech subnetworks share the same architecture but are optimised separately with a joint objective function as defined in Eq. 3:

$$\mathcal{L} = \alpha\mathcal{L}_{cls} + \beta\mathcal{L}_{cont} + \gamma\mathcal{L}_{ord}, \tag{3}$$

where α, β, γ are weighting parameters and are empirically determined to be equal. The projection function $h(\cdot)$ is achieved by a multilayer perceptron (MLP) with a hidden layer and non-linear activation. The models are trained with the SGD optimizer with momentum set to 0.9 and weight decay as 5×10^{-4}. The learning rate is initialised to 10^{-3} and divided by 10 for every 20 epochs. The whole model is trained for 80 epochs. Gradient clipping is applied and the batch size is set to 32. The sampling group number $K = 2$ due to memory limitation and the interval skip range $\mathcal{D} = 5$. The whole framework is implemented using PyTorch on an NVIDIA RTX 2080Ti.

3 Experiments and Analysis

3.1 Data and Experimental Settings

The data used in this work is from a routine clinical fetal ultrasound dataset[1] with the scanned video and corresponding real-time speech as well as eye-gaze data from sonographers. In total, we have 81 scans with speech data. On average, each video scan is about 55,000 frames with frame rate of 30 fps. Each video clip $\mathcal{T}$ consists of 60 consecutive frames and we have 73,681 clips for model training. When sampling a training pair within each video clip, we extract 0.6s of the corresponding speech data and resample it to 24kHz. The speech is then converted to a 2D log-spectrogram representation of size 256 × 256, using a short-time

[1] UK Research Ethics Committee Reference 18/WS/0051.

Table 1. Evaluation results on standard plane detection (mean ± std.[%]). Best performance is marked in **bold**. Note the methods on the right side are fully-supervised using external annotations.

	Rand.Init.	Video	Ours	ImageNet Init.	SonoNet
Precision	70.4±2.3	71.9±2.0	**72.7±1.8**	74.6±1.8	82.3±1.3
Recall	64.9±1.6	71.7±3.5	**73.3±2.4**	71.2±1.9	87.3±1.1
F1-score	67.0±1.3	71.5±2.4	**72.6±1.7**	72.5±1.8	84.5±0.9

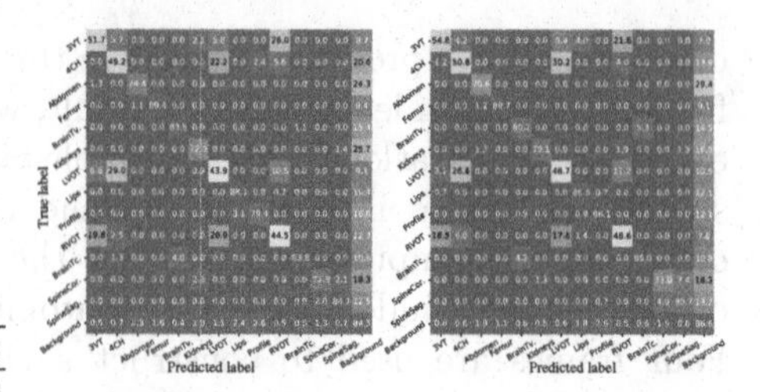

Fig. 3. Confusion matrix on standard plane detection. Left: *Video*. Right: *Ours*. (Best viewed in digital form.)

Fourier transform (STFT) with 256 frequency bands, 10ms window length and 5ms hop length. Two downstream tasks are included for the learned representation evaluation, where we use 135 scans with three-fold cross-validation (90/45 for train/test) and each scan is temporally down-sampled at a rate of 8.

3.2 Standard Plane Detection

To evaluate the learned representations, we first perform transfer learning on the standard plane detection task by fine-tuning the pre-trained weights from pretext tasks. We use 14 categories of heart three-vessel and trachea view (3VT), heart four-chamber view (4CH), abdomen, femur, brain transventricular plane (BrainTv.), kidneys, heart left ventricular outflow tract (LVOT), lips, profile, heart right ventricular outflow tract (RVOT), brain transcerebellum plane (BrainTc.), spine coronal plane (SpineCor.), spine sagittal plane (SpineSag.) and background. Standard plane labels are obtained from the dataset mentioned above. Initialisation from random weights, pre-trained weights only on video data (by the aforementioned frame order prediction task), pre-trained weights on ImageNet [10] and weights pre-trained with SonoNet [2] are included for comparison. Quantitative results are presented in Table 1. We see that the proposed self-supervised cross-modal representation approach performs better than the other alternative solutions on average. It also reveals that by leveraging the speech data with consideration of its correlation to the visual data, better performance is achieved, indicating stronger representations are learned. Note that our learned representation performs better than the ImageNet initialisation for most metrics, which also suggests that representation extracted from natural images do not transfer very well to medical data. In addition, we show the label confusion matrix in Fig. 3. It can be observed that our approach performs well on each category except the fine-grained ones like 3VT, 4CH, LVOT and RVOT (different views of fetal heart), which are challenging for sonographers too. When compared to the video-only representations (Fig. 3 left), we can see that our approach improves on almost all the categories, especially the abovementioned fine-grained ones. This is mainly due to the incorporation of visual-speech correlation where the additional speech representation reduces the ambiguity that exists in video data.

Table 2. Quantitative evaluation on eye-gaze saliency prediction. Best performance is marked in **bold**.

	KL↓	NSS↑	AUC↑	CC↑	SIM↑
Rand.Init.	3.94±0.18	1.47±0.24	0.90±0.01	0.12±0.02	0.05±0.01
Video	3.57±0.10	1.86±0.12	0.91±0.01	0.15±0.01	0.08±0.01
Ours	**3.05**±0.04	**2.68**±0.05	**0.95**±0.00	**0.22**±0.00	**0.11**±0.00
ImageNet Init.	3.95±0.28	1.72±0.25	0.89±0.02	0.14±0.02	0.08±0.01
SonoNet	3.14±0.02	2.62±0.03	0.94±0.00	0.21±0.00	0.12±0.00

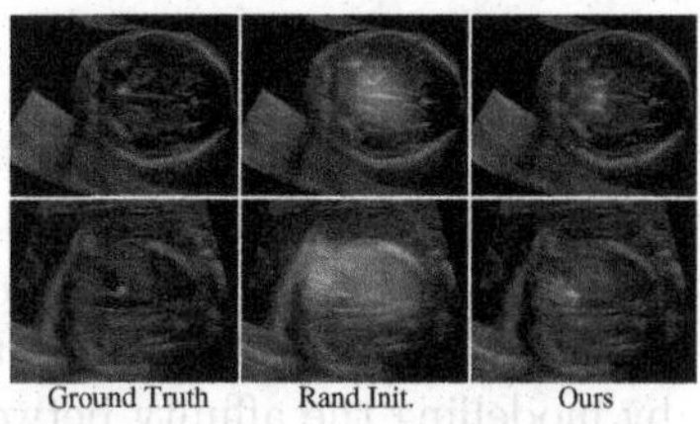

Fig. 4. Qualitative performance on eye-gaze saliency prediction.

3.3 Eye-Gaze Saliency Prediction

Since our dataset contains simultaneous eye-gaze tracking data from sonographers, in addition to the standard plane detection task as in Sec. 3.2, we further evaluate the effectiveness of the learned representations on a regression-based task, namely eye-gaze saliency prediction. Similarly, we load the pre-trained weights and fine-tune on the downstream task. A similar network architecture is used, with only the last layers modified to predict a 2D saliency map. Following [3], we use the KL divergence (KL), normalised scanpath saliency (NSS), area under curve (AUC), correlation coefficient (CC) and similarity (SIM) as the evaluation metrics. Quantitative and qualitative results are shown in Table 2 and Fig. 4 respectively, from which we can see that our approach again outperforms the alternative solutions, and even better than the approaches (ImageNet Init. and SonoNet) that were pre-trained with manual annotations.

3.4 Ablation Study

To better understand the effectiveness of the proposed learning strategies, we present an ablation study with corresponding performance shown in Table 3. We take the video-speech learning approach without cross-model contrastive learning (*CM.Contra.*) and affinity-aware self-paced learning as baseline. We can see from Table 3 that when including the cross-modal contrastive learning, the performance is improved by a large margin. Adding the affinity-aware learning scheme (*Ours*), further improves the model.

Table 3. Ablation study on each of the proposed strategies for two downstream tasks.

	Prec.[%]↑	Rec.[%]↑	F1[%]↑	KL↓	NSS↑	AUC↑	CC↑	SIM↑
Baseline	70.7±3.6	70.2±2.4	69.8±2.9	3.96±0.69	1.76±0.42	0.90±0.03	0.15±0.03	0.08±0.01
+ CM.Contra.	71.8±2.3	71.1±2.3	70.9±1.9	3.43±0.01	2.23±0.02	0.93±0.00	0.19±0.00	0.08±0.00
Ours	72.7±1.8	73.3±2.4	72.6±1.7	3.05±0.04	2.68±0.05	0.95±0.00	0.22±0.00	0.11±0.00

4 Conclusion

In this paper, we propose a self-supervised representation learning framework for ultrasound video-speech multi-modal data. To the best of our knowledge, this is the first attempt towards cross-modal representation learning without human annotations for ultrasound data. We designed a simple, yet effective, approach by modelling the affinity between these two modalities. To address the inherent sparse correlation and noise issues in the speech data, we propose a cross-modal contrastive learning scheme and an affinity-aware self-paced learning scheme. Experimental evaluation on two downstream tasks shows that the learned representations can transfer to fetal ultrasound standard plane detection and eye-gaze saliency prediction and improve the average performance accordingly. The proposed approach shows the potential to mitigate the need for laborious manual annotation work in deep learning-based applications for medical imaging via automated cross-modal self-supervision. Since the proposed approach is trained and evaluated on video data with narrative speech audio and is not specifically tailored for ultrasound, it could apply to other unseen video-audio data.

Acknowledgements. We acknowledge the EPSRC (EP/M013774/1, Project Seebibyte), ERC(ERC-ADG-2015 694581, Project PULSE), and the support of NVIDIA Corporation with the donation of the GPU.

References

1. Arandjelovic, R., Zisserman, A.: Look, listen and learn. In: ICCV (2017)
2. Baumgartner, C.F., Kamnitsas, K., Matthew, J., Fletcher, T.P., Smith, S., Koch, L.M., Kainz, B., Rueckert, D.: Sononet: real-time detection and localisation of fetal standard scan planes in freehand ultrasound. IEEE TMI **36**(11), 2204–2215 (2017)
3. Bylinskii, Z., Judd, T., Oliva, A., Torralba, A., Durand, F.: What do different evaluation metrics tell us about saliency models? IEEE TPAMI **41**(3), 740–757 (2018)
4. Gidaris, S., Singh, P., Komodakis, N.: Unsupervised representation learning by predicting image rotations. In: ICLR (2018)
5. Hu, J., Shen, L., Sun, G.: Squeeze-and-excitation networks. In: CVPR (2018)
6. Jiao, J., Droste, R., Drukker, L., Papageorghiou, A.T., Noble, J.A.: Self-supervised representation learning for ultrasound video. In: ISBI (2020)
7. Lee, H.Y., Huang, J.B., Singh, M., Yang, M.H.: Unsupervised representation learning by sorting sequences. In: ICCV (2017)
8. Oord, A.v.d., Li, Y., Vinyals, O.: Representation learning with contrastive predictive coding. arXiv preprint (2018). arXiv:1807.03748
9. Owens, A., Efros, A.A.: Audio-visual scene analysis with self-supervised multisensory features. In: Ferrari, V., Hebert, M., Sminchisescu, C., Weiss, Y. (eds.) ECCV 2018. LNCS, vol. 11210, pp. 639–658. Springer, Cham (2018). https://doi.org/10.1007/978-3-030-01231-1_39
10. Russakovsky, O., et al.: Imagenet large scale visual recognition challenge. Int. J. Comput. Vis. **115**(3), 211–252 (2015)
11. Sakhnov, K., Verteletskaya, E., Simak, B.: Approach for energy-based voicedetector with adaptive scaling factor. IAENG Int. J. Comput. Sci. **36**(4) (2009)

12. Spitzer, H., Kiwitz, K., Amunts, K., Harmeling, S., Dickscheid, T.: Improving cytoarchitectonic segmentation of human brain areas with self-supervised siamese networks. In: Frangi, A.F., Schnabel, J.A., Davatzikos, C., Alberola-López, C., Fichtinger, G. (eds.) MICCAI 2018. LNCS, vol. 11072, pp. 663–671. Springer, Cham (2018). https://doi.org/10.1007/978-3-030-00931-1_76
13. Wang, X., Gupta, A.: Unsupervised learning of visual representations using videos. In: ICCV (2015)
14. Xie, S., Girshick, R., Dollár, P., Tu, Z., He, K.: Aggregated residual transformations for deep neural networks. In: CVPR (2017)
15. Yu, F., Koltun, V.: Multi-scale context aggregation by dilated convolutions. arXiv preprint (2015). arXiv:1511.07122
16. Zhang, R., Isola, P., Efros, A.A.: Colorful image colorization. In: Leibe, B., Matas, J., Sebe, N., Welling, M. (eds.) ECCV 2016. LNCS, vol. 9907, pp. 649–666. Springer, Cham (2016). https://doi.org/10.1007/978-3-319-46487-9_40
17. Zhuang, X., Li, Y., Hu, Y., Ma, K., Yang, Y., Zheng, Y.: Self-supervised feature learning for 3D medical images by playing a rubik's cube. In: Shen, D., et al. (eds.) MICCAI 2019. LNCS, vol. 11767, pp. 420–428. Springer, Cham (2019). https://doi.org/10.1007/978-3-030-32251-9_46

Assisted Probe Positioning for Ultrasound Guided Radiotherapy Using Image Sequence Classification

Alex Grimwood[1,2,3](✉), Helen McNair[1], Yipeng Hu[2,3], Ester Bonmati[2,3], Dean Barratt[2,3], and Emma J. Harris[1]

[1] Division of Radiotherapy and Imaging, Institute of Cancer Research, Sutton, UK
alex.grimwood@ucl.ac.uk

[2] Wellcome/EPSRC Centre for Interventional and Surgical Sciences, University College London, London, UK

[3] Centre for Medical Image Computing, University College London, London, UK

Abstract. Effective transperineal ultrasound image guidance in prostate external beam radiotherapy requires consistent alignment between probe and prostate at each session during patient set-up. Probe placement and ultrasound image interpretation are manual tasks contingent upon operator skill, leading to interoperator uncertainties that degrade radiotherapy precision. We demonstrate a method for ensuring accurate probe placement through joint classification of images and probe position data. Using a multi-input multi-task algorithm, spatial coordinate data from an optically tracked ultrasound probe is combined with an image classifier using a recurrent neural network to generate two sets of predictions in real-time. The first set identifies relevant prostate anatomy visible in the field of view using the classes: outside prostate, prostate periphery, prostate centre. The second set recommends a probe angular adjustment to achieve alignment between the probe and prostate centre with the classes: move left, move right, stop. The algorithm was trained and tested on 9,743 clinical images from 61 treatment sessions across 32 patients. We evaluated classification accuracy against class labels derived from three experienced observers at 2/3 and 3/3 agreement thresholds. For images with unanimous consensus between observers, anatomical classification accuracy was 97.2% and probe adjustment accuracy was 94.9%. The algorithm identified optimal probe alignment within a mean (standard deviation) range of 3.7° (1.2°) from angle labels with full observer consensus, comparable to the 2.8° (2.6°) mean interobserver range. We propose such an algorithm could assist radiotherapy practitioners with limited experience of ultrasound image interpretation by providing effective real-time feedback during patient set-up.

Keywords: Ultrasound-guided radiotherapy · Image classification · Prostate radiotherapy

---

Electronic supplementary material The online version of this chapter (https://doi.org/10.1007/978-3-030-59716-0_52) contains supplementary material, which is available to authorized users.

A. L. Martel et al. (Eds.): MICCAI 2020, LNCS 12263, pp. 544–552, 2020.
https://doi.org/10.1007/978-3-030-59716-0_52

1 Introduction

Advanced external beam radiotherapy techniques, such as stereotactic body radiotherapy, adaptive radiotherapy and ultrahypofractionation require high precision spatiotemporal image guidance [1–3]. Ultrasound imaging is a desirable guidance technology due to the modality's excellent soft tissue contrast, high spatial resolution and ability to scan in both real-time and three dimensions. Transperineal ultrasound (TPUS) is used clinically to assist with the localization and delineation of anatomical structures for prostate radiotherapy, showing comparable precision to standard cone beam computed tomography scans (CBCT) [4]. The efficacy of ultrasound guided radiotherapy is heavily reliant upon experienced staff with a level of sonography training not typically acquired in the radiotherapy clinic [5]. With this in mind, computational technologies are being developed to assist TPUS operators in radiotherapy so that they can achieve the levels of precision demanded by state-of-the-art treatment delivery methods.

Radiotherapy patients require multiple treatment sessions. To achieve effective image guidance, anatomical structures such as the prostate must be precisely localized and registered to a patient's planning scan at each session. Automated registration methods have been investigated as a way to localize the prostate prior to treatment [6]. Such methods neglect the necessary step of ensuring probe placement is consistent with the planning scan, which can affect radiotherapy accuracy [7]. Previous studies have reported automatic identification of the optimal probe position on CT scans, however the method is computationally intensive and requires good spatial agreement between the planning CT scan and the patient many weeks later during set-up [8].

In this study, we demonstrate a method for identifying the optimal probe position by presenting the operator with actionable information in near real-time using only ultrasound images and probe tracking data obtained at patient set-up. Our method uses a supervised deep learning approach to recommend probe adjustments in response to image content and probe position.

Deep learning methods have been used extensively in other ultrasound applications [9]. Automatic segmentation of pelvic structures, such as the prostate and levator hiatus, has demonstrated performance levels exceeding manual contouring [10, 11]. Real-time classification and localization of standard scan planes in obstetric ultrasound has been demonstrated [12]. Image guidance in 3D Ultrasound target localization has also been reported for epidural needle guidance [13].

For this study, a multi-input multi-task classification approach was adopted to utilize data acquired routinely during patient set-up. The input data comprised 2D (B-scan) TPUS images and spatial coordinate information for each scan plane obtained from optical tracking. The classification labels were designed to be easily interpreted, presenting the user with simple information regarding image content and recommended probe adjustments.

2 Methods

In this section, the collection and labelling of image data is described along with the design and implementation of the classification models used in the study. The training and evaluation methods against three expert observers is also described.

2.1 Data Collection and Labelling

Clinical data was acquired using the Clarity Autoscan^{TM} system (Elekta AB, Stockholm.) as part of the Clarity Pro trial (NCT02388308) approved by the Surrey and SE Coast Region Ethics Committee and described elsewhere [14]. The system incorporated an optically tracked TPUS probe with a mechanically swept transducer array, which was used to collect 3D prostate ultrasound scans during patient set-up. According to the trial protocol, a trained radiotherapy practitioner used the Clarity probe to localize the central sagittal plane through the prostate with real-time freehand 2D imaging. Once identified, the probe was clamped in place and a static volumetric scan was acquired. Volumetric acquisition comprised recording a sequence of 2D images (B-scans) while mechanically sweeping the transducer through a 60° arc from right to left in the patient frame of reference. Each B-scan from the volume was automatically labelled with the imaging plane location and orientation in room coordinates. A total of 9,743 B-scans were acquired comprising 61 volumes across 32 patients. Images were 640×480 pixels, with pixel sizes between 0.3 mm to 0.5 mm and volumes comprising 134 to 164 B-scans depending on scan parameters.

All scan volumes were reviewed and the constituent B-scans assigned position labels by three experienced observers (2 physicists and 1 radiotherapy practitioner) as either: Centre (C) for image planes through the prostate centre; Periphery (P) for image planes through the outer prostate; Outside (O) for image planes not intersecting the prostate. The prostate centre was distinguishable from the periphery by prostate shape and anatomical features such as the urethra branching from the penile bulb. Direction labels were automatically assigned depending on the transducer motion required to position the B-scan plane at the prostate centre, as ensured manually during the acquisition. Centre images were labelled Stop (S), while Periphery and Outside images were labelled as either Left (L) or Right (R) depending on which side of the prostate centre they were located to indicate the direction of movement needed to align the transducer with the prostate. Position and direction input labels were treated as two separate class sets. Each set was encoded as an ordinal 3-vector of probabilities derived from observer consensus in each class, where: 100% consensus = 1, 66.7% (2/3) consensus = 0.667, 33.3% (1/3) consensus = 0.333.

2.2 Classifier Models

Two classifier models were developed, referred to as: MobNet and RNN (Fig. 1). MobNet, was an image classifier based upon the Keras implementation of MobileNetV2 pre-trained on the ImageNet dataset [15]. Single B-scans images were passed to the convolutional network and a custom fully connected layer was added to concatenate the last convolutional pooling layer with a 6-tuple input representing the image plane coordinates. Two dense branches each produced three class outputs. The position branch classified images according to visible prostate anatomy: Outside (O), Periphery (P) and Centre (C). The direction branch classified images according to the transducer rotation direction required to align the image plane with the prostate centre: Right (R), Stop (S) and Left (L).

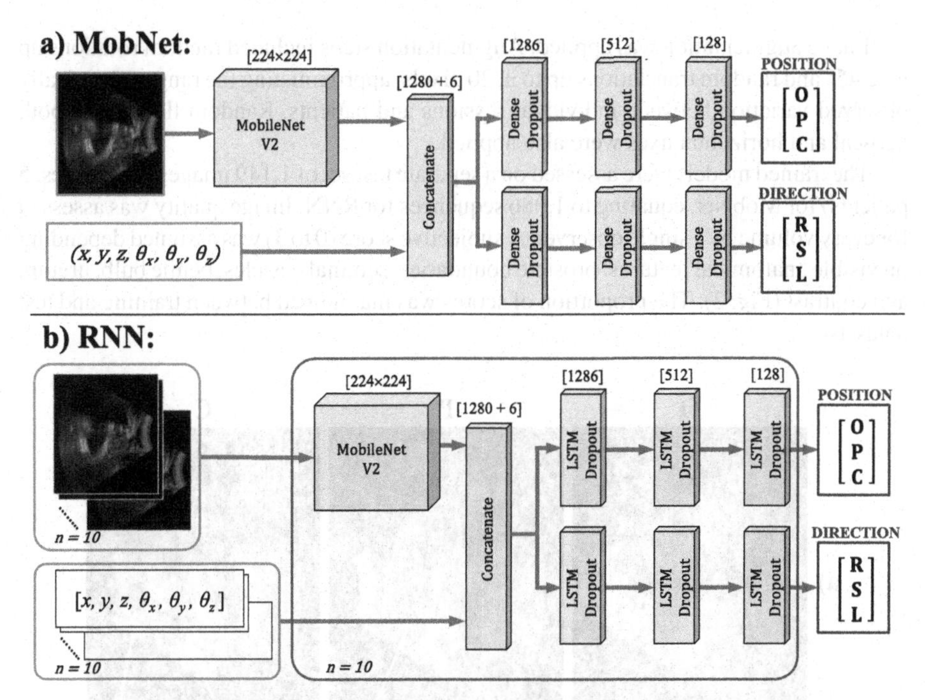

Fig. 1. MobNet (a): inputs in green are a 2D B-scan and transducer position (x, y, z) with orientation (θ_x, θ_y, θ_z). Outputs are two class sets in red: Outside, Periphery, Centre (O, P, C) for position and Right, Stop, Left (R, S, L) for direction. Input dimensions are shown in square brackets for MobileNetV2, concatenation, dense and LSTM layers. RNN (b): input sequences are generated from 10 consecutive images and transducer positions. Each is concatenated (blue box) and passed onto the LSTM layers.(Color figure online)

The second model, RNN, extended the MobNet. To improve classification of a given B-scan, P_0, the image was grouped with 9 immediately preceding B-scans from the volumetric sweep to form a 10-image sequence P_{n-10} (n =1, ..., 10). Each image was processed individually by the same MobileNetV2 backbone used for MobNet. The output from these convolutional layers was concatenated with probe orientation data and fed into the recurrent layers in sequence. The recurrent position and direction branches comprised long-short-term memory (LSTM) layers, sharing between the images to predict a single output for each 10-image sequence.

2.3 Experiments

A training set comprising 8,594 images (54 volumes, 27 patients) was used for MobNet. 8,108 image sequences were produced from the same training set for RNN. A 4-fold cross validation was used for both models over 100 epochs with an Adam optimizer (learning rate = 1×10^{-4}). Class imbalances were compensated for using sample weighting where label proportions were: [O 62%, P 23%, C 15%] and [R 43%, S 15%, L 42%]. Images were downsampled by a factor of 1.5, normalised and cropped to a fixed 224 × 224 pixel region of interest large enough to encompass the prostate where present.

Image augmentation was applied. Augmentation steps included random rotations up to ± 45° and random translations up to ± 20 pixels, approximating the range of clinically observed variation between individual sessions and patients. Random flips along both vertical and horizontal axes were also applied.

The trained models were assessed on a separate test set of 1,149 images (7 volumes, 5 patients) for MobNet, equating to 1,086 sequences for RNN. Image quality was assessed for every volume by a single observer. A subjective score (0 to 3) was assigned depending on visible anatomical features: prostate boundaries, seminal vesicles, penile bulb, urethra and contrast (Fig. 2). The proportion of scores was maintained between training and test datasets.

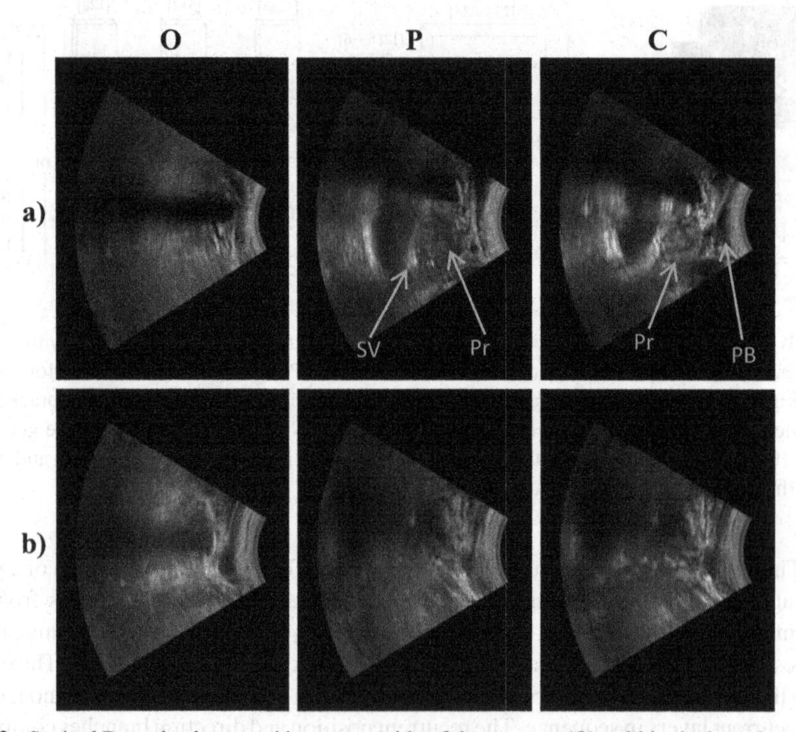

Fig. 2. Sagittal B-mode ultrasound images outside of the prostate (O), within the lateral prostate periphery (P) and through the prostate centre (C). B-scans are shown from a volume with an image quality score of 3 (a) and score of 0 (b), determined by visible anatomical features, such as: seminal vesicles (SV), penile bulb (PB) and prostate with well-defined boundary (Pr).

2.4 Analysis

MobNet and RNN model output was assessed against the observer labels for: 1) a threshold of 66.7%, where a positive label was defined as $\geq$ 2/3 consensus; and 2) a 100% consensus threshold, where only samples with full consensus were assessed.

Precision, recall, accuracy and F1-scores were calculated and confusion matrices plotted. RNN and MobNet accuracies were compared using McNemar tests, which are analogous to paired t-tests for nominal data [16]. Agreement between RNN and the observer cohort was quantified by calculating the Williams Index 95% confidence intervals [17]. Interobserver agreement was also quantified by calculating Fleiss' Kappa and Specific Agreement coefficients for each class [18, 19]. Interobserver uncertainty of the transducer angles associated with the prostate centre was quantified by calculating the mean difference between each observer and the 100% consensus labels for the 7 test volumes. RNN angular uncertainty was also calculated and compared.

3 Results and Discussion

RNN classification performance is shown in Table 1 along with the associated confusion matrices in Fig. 3. Accuracy was highest among images with 100% consensus labels, as was specificity, recall and F1-score. Maximum position and direction accuracies were 97.2% and 94.9% respectively.

Accuracy, F1-score, precision and recall were consistently higher for RNN than MobNet (Supplementary Materials). McNemar comparison tests indicated the RNN accuracy improvement over MobNet was significant for position classification ($P < 0.001$), but only marginal for direction classification ($P = 0.0049$ at 66.7%, $P = 0.522$ at 100%).

Agreement between model output and the observer cohort was quantified using Williams indices at 95% confidence intervals (CI). RNN position output significantly outperformed the observer cohort, having CI bounds between 1.03 and 1.05. RNN direction output exhibited poorer agreement, with a CI between 0.95 and 0.98; however, this is unsurprising because direction labels were assigned using information from the entire volumetric scan, rather than a 10-image sub-volume.

Table 1. RNN precision (prec), recall (rec), F1-score (F1) and accuracy for both 100% (full) and 66.7% (2/3) observer consensus label thresholds.

Class set	Class label	Full consensus threshold			2/3 Consensus threshold		
		Prec%	Rec%	F1%	Prec%	Rec%	F1%
Position	Outside	99.1	98.7	98.9	95.6	93.1	94.3
	Periphery	91.6	97.5	85.7	81.9	85.5	83.6
	Centre	100.0	85.7	92.3	79.5	83.6	79.8
Direction	Right	91.1	100.0	95.3	89.6	98.8	94.0
	Stop	93.9	74.7	83.2	77.0	81.7	79.3
	Left	98.9	93.9	96.3	98.8	89.5	93.2
Position Accuracy %		97.2			89.3		
Direction Accuracy %		94.9			92.4		

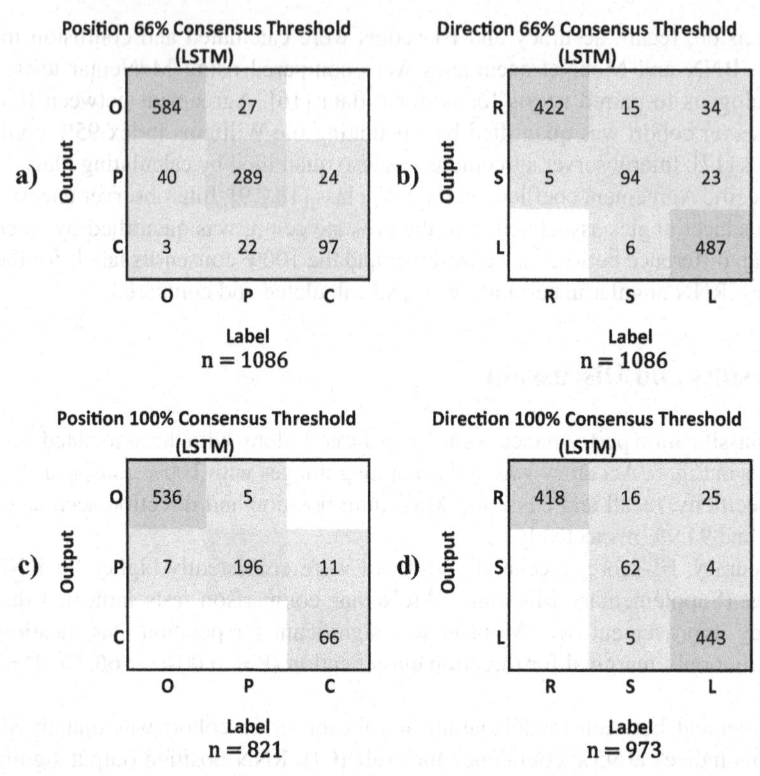

Fig. 3. RNN confusion matrices for 66.7% observer consensus thresholds (a, b) and 100% observer consensus thresholds (c, d). Total number of samples in each matrix is given by *n*.

Fleiss' Kappa scores indicated excellent interobserver agreement (> 0.8) among image labels for the position class set, with good agreement (> 0.7) for the direction class set. Specific agreement coefficients indicated consistently high interobserver agreement (> 0.8) within classes (full figures provided in Supplementary Materials). For the test dataset 100% position consensus was achieved for 884 images (821 sequences) and 100% direction consensus was achieved for 1,036 images (973 sequences). All images and sequences achieved at least 66.7% consensus in both class sets.

Mean interobserver angular uncertainty (standard deviation) was calculated to be 2.8° (2.6°). A comparable mean uncertainty of 3.7° (1.2°) was calculated for the RNN model. Finally, the RNN mean classification time was 0.24 ms per sequence on a 2.8 GHz Intel Xeon E3 CPU, establishing the possibility of incorporating a practical real-time solution on clinical systems.

4 Conclusion

A deep learning classifier incorporating recurrent layers has been shown to predict prostate anatomy and recommend probe position adjustments with high precision and accuracies comparable to expert observers. The recurrent network was significantly more accurate than a non-recurrent equivalent. This study demonstrates the possibility of enhancing ultrasound guidance precision by reducing interobserver variation and assisting ultrasound operators with finding the optimal probe position during patient set-up for ultrasound guided radiotherapy.

Acknowledgements. This work was supported by NHS funding to the NIHR Biomedical Research Centre at The Royal Marsden and The Institute of Cancer Research. The study was also supported by Cancer Research UK under Programmes C33589/A19727 and C20892/A23557, and by the Wellcome/EPSRC Centre for Interventional and Surgical Sciences (203145Z/16/Z). The study was jointly supervised by Dr Emma J. Harris, Prof. Dean Barratt and Dr. Ester Bonmati. We thank the radiographers of the Royal Marsden Hospital for their clinical support, as well as David Cooper, Martin Lachaine and David Ash at Elekta for their technical support.

References

1. Loblaw, A.: Ultrahypofractionation should be a standard of care option for intermediate-risk prostate cancer. Clin. Oncol. **32**(3), 170–174 (2020)
2. Böckelmann, F., et al.: Adaptive radiotherapy and the dosimetric impact of inter- and intrafractional motion on the planning target volume for prostate cancer patients. Strahlenther. Onkol. **196**(7), 647–656 (2020)
3. Tree, A., Ostler, P., van As, N.: New horizons and hurdles for UK radiotherapy: can prostate stereotactic body radiotherapy show the way? Clin. Oncol. (R. Coll. Radiol.) **26**(1), 1–3 (2014)
4. Li, M., et al.: Comparison of prostate positioning guided by three-dimensional transperineal ultrasound and cone beam CT. Strahlenther. Onkol. **193**(3), 221–228 (2017)
5. Hilman, S., et al.: Implementation of a daily transperineal ultrasound system as image-guided radiotherapy for prostate cancer. Clin. Oncol. (R. Coll. Radiol.) **29**(1), e49 (2017)
6. Presles, B., et al.: Semiautomatic registration of 3D transabdominal ultrasound images for patient repositioning during postprostatectomy radiotherapy. Med. Phys. **41**(12), 122903 (2014)
7. Fargier-Voiron, M., et al.: Evaluation of a new transperineal ultrasound probe for inter-fraction image-guidance for definitive and post-operative prostate cancer radiotherapy. Phys. Med. **32**(3), 499–505 (2016)
8. Camps, S.M., et al.: Automatic transperineal ultrasound probe positioning based on CT scan for image guided radiotherapy. In: Medical Imaging 2017: Image-Guided Procedures, Robotic Interventions, and Modeling (2017)
9. Liu, S., et al.: Deep learning in medical ultrasound analysis: a review. Engineering **5**(2), 261–275 (2019)
10. Yang, X., et al.: Fine-grained recurrent neural networks for automatic prostate segmentation in ultrasound images. In: Proceedings of the Thirty-First AAAI Conference on Artificial Intelligence. AAAI Press: San Francisco, California, USA, pp. 1633–1639 (2017)
11. Bonmati, E., et al.: Automatic segmentation method of pelvic floor levator hiatus in ultrasound using a self-normalizing neural network. J. Med. Imaging (Bellingham) **5**(2), 021206 (2018)
12. Baumgartner, C.F., et al.: SonoNet: real-time detection and localisation of fetal standard scan planes in freehand ultrasound. IEEE Trans. Med. Imaging **36**(11), 2204–2215 (2017)
13. Pesteie, M., et al.: Automatic localization of the needle target for ultrasound-guided epidural injections. IEEE Trans. Med. Imaging **37**(1), 81–92 (2018)
14. Grimwood, A., et al.: In vivo validation of elekta's clarity autoscan for ultrasound-based intrafraction motion estimation of the prostate during radiation therapy. Int. J. Radiat. Oncol. Biol. Phys. **102**(4), 912–921 (2018)

15. Sandler, M., et al.: MobileNetV2: inverted residuals and linear bottlenecks. In: 2018 IEEE/CVF Conference on Computer Vision and Pattern Recognition (2018)
16. Dwyer, A.: Matchmaking and McNemar in the comparison of diagnostic modalities. Radiology **178**(2), 328–330 (1991)
17. Williams, G.W.: Comparing the joint agreement of several raters with another rater. Biometrics, 619–627 (1976)
18. Fleiss, J.L., Cohen, J.: The equivalence of weighted kappa and the intraclass correlation coefficient as measures of reliability. Educ. Psychol. Measure. **33**(3), 613–619 (1973)
19. Cicchetti, D.V., Feinstein, A.R.: High agreement but low kappa: II. Resolving the paradoxes. J. Clin. Epidemiol. **43**(6), 551–558 (1990)

Searching Collaborative Agents for Multi-plane Localization in 3D Ultrasound

Yuhao Huang[1,2], Xin Yang[1,2], Rui Li[1,2], Jikuan Qian[1,2], Xiaoqiong Huang[1,2], Wenlong Shi[1,2], Haoran Dou[1,2], Chaoyu Chen[1,2], Yuanji Zhang[3], Huanjia Luo[3], Alejandro Frangi[1,4,5], Yi Xiong[3], and Dong Ni[1,2(✉)]

[1] School of Biomedical Engineering, Shenzhen University, Shenzhen, China
nidong@szu.edu.cn
[2] Medical UltraSound Image Computing (MUSIC) Lab, Shenzhen University, Shenzhen, China
[3] Department of Ultrasound, Luohu People's Hosptial, Shenzhen, China
[4] CISTIB Centre for Computational Imaging & Simulation Technologies in Biomedicine, School of Computing, University of Leeds, Leeds, UK
[5] Medical Imaging Research Center (MIRC), University Hospital Gasthuisberg, Electrical Engineering Department, KU Leuven, Leuven, Belgium

Abstract. 3D ultrasound (US) is widely used due to its rich diagnostic information, portability and low cost. Automated standard plane (SP) localization in US volume not only improves efficiency and reduces user-dependence, but also boosts 3D US interpretation. In this study, we propose a novel Multi-Agent Reinforcement Learning (MARL) framework to localize multiple uterine SPs in 3D US simultaneously. Our contribution is two-fold. First, we equip the MARL with a one-shot neural architecture search (NAS) module to obtain the optimal agent for each plane. Specifically, Gradient-based search using Differentiable Architecture Sampler (GDAS) is employed to accelerate and stabilize the training process. Second, we propose a novel collaborative strategy to strengthen agents' communication. Our strategy uses recurrent neural network (RNN) to learn the spatial relationship among SPs effectively. Extensively validated on a large dataset, our approach achieves the accuracy of 7.05°/2.21 mm, 8.62°/2.36 mm and 5.93°/0.89 mm for the mid-sagittal, transverse and coronal plane localization, respectively. The proposed MARL framework can significantly increase the plane localization accuracy and reduce the computational cost and model size.

Keywords: 3D ultrasound · NAS · Reinforcement learning

1 Introduction

Acquisition of Standard Planes (SPs) is crucial for objective and standardised ultrasound (US) diagnosis [11]. 3D US is increasingly used in clinical practice mainly because of its rich diagnostic information not contained in 2D US.

Y. Huang and X. Yang—Contribute equally to this wok.

A. L. Martel et al. (Eds.): MICCAI 2020, LNCS 12263, pp. 553–562, 2020.
https://doi.org/10.1007/978-3-030-59716-0_53

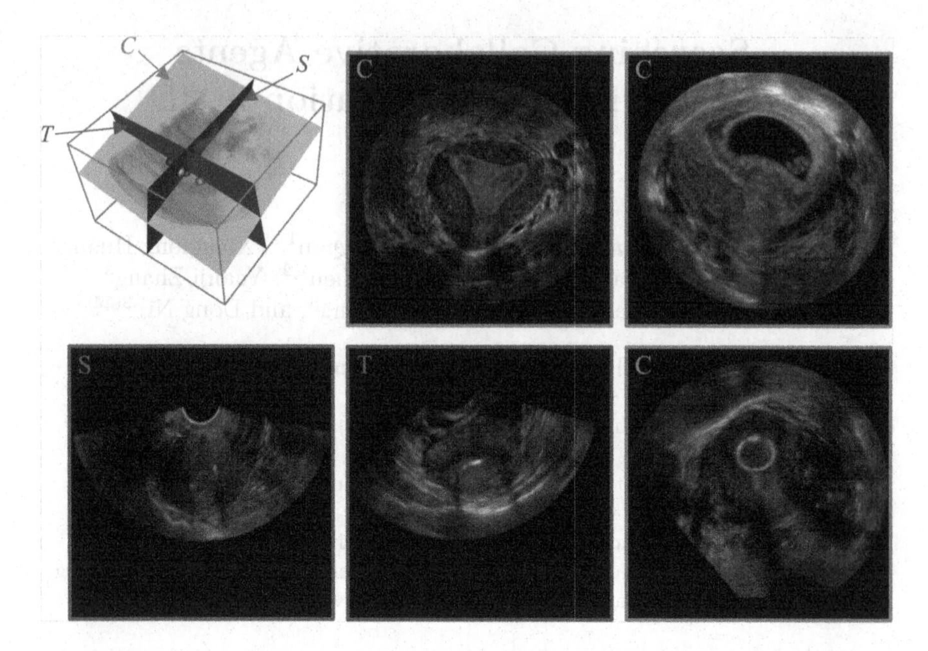

Fig. 1. Uterine SPs in 3D US (left to right). *Top row*: spatial layout of SPs, coronal SP of normal and pregnant cases. *Bottom row*: mid-sagittal (S), transverse (T) and coronal (C) SPs of one case with IUD. Red, yellow, purple and white dots are two endometrial uterine horns, endometrial uterine bottom and uterine wall bottom. (Color figure online)

Figure 1 shows coronal SPs that can only be reconstructed from 3D US. They are important for assessing congenital uterine anomalies and Intra-Uterine Device (IUD) localization [17]. 3D US can also contain multiple SPs in one shot, which improves scanning efficiency while reducing user-dependence. However, SPs are often located in 3D US manually by clinicians, which is cumbersome and time-consuming owing to a large search space and anatomical variability. Hence, automated SPs localization in 3D US is highly desirable to assist in scanning and diagnosis.

As shown in Fig. 1, automatic acquisition of uterine SPs from 3D US remains very challenging. First, the uterine SP often has extremely high intraclass variation due to the existence of IUD, pregnancy and anomaly. Second, the three SPs have very different appearance patterns, which makes designing machine learning algorithms difficult. The third challenge lies in the varied spatial relationship among planes. For 3D uterine US, in most cases, the three planes are perpendicular to each other. However, sometimes, due to uterine fibroids, congenital anomalies, etc., their spatial relationship may be different than expected.

In the literature, plane localization approaches have adopted machine learning methods to various 3D US applications. Random Forests based regression methods were first employed to localize cardiac planes [2]. Landmark alignment [9], classification [13] and regression [6,14] methods were then developed to localize fetal planes automatically. Recently, Alansary et al. [1] first proposed using a Reinforcement Learning (RL) agent for view planning in MRI volumes. Dou et al. [4] then proposed a novel RL framework with a landmark-aware alignment module for effective initialization to localize SPs in noisy US volumes. The experiments have shown that such an approach can achieve state-of-the-art results. However, these RL-based methods still have disadvantages. First, they were only designed to learn a single agent for each plane separately. The agents cannot communicate with each other to learn the inherent and invaluable spatial relationship among planes. Second, agents often use the same network structure such as the VGG model for different planes, which may lead to sub-optimal performance because SPs often have very different appearance patterns.

In this study, we propose a novel Multi-Agent RL (MARL) framework to localize multiple uterine SPs in 3D US simultaneously. We believe we are the first to employ a MARL framework for this problem. Our contribution is two-fold. First, we adopt one-shot neural architecture search (NAS) [12] to obtain the optimal agent for each plane. Specifically, Gradient-based search using Differentiable Architecture Sampler (GDAS) [3] is employed to make the training more stable and faster than Differentiable ARchiTecture Search (DARTS) [8]. Second, we propose a Recurrent Neural Network (RNN) based agent collaborative strategy to learn the spatial relationship among SPs effectively. This is a general method for establishing communication among agents.

2 Method

Figure 2 is the schematic view of our proposed method. We propose a MARL framework to localize multiple uterine SPs in 3D US simultaneously. To improve the system robustness against the noisy US environment, we first use a landmark-aware alignment model to provide a warm start for the agent [4]. Four landmarks used for the alignment are shown in Fig. 1. We further adopt the one-shot and gradient-based NAS to search the optimal agent for each plane based on the GDAS strategy [3]. Then, the RNN based agent collaborative strategy is employed to learn the spatial relationship among planes effectively.

2.1 MARL Framework for Plane Localization

To localize multiple uterine planes effectively and simultaneously, we propose a collaborative MARL framework. The framework can be defined by the *Environment*, *States*, *Actions*, *Reward Function* and *Terminal States*. In this study, we take a uterine US volume as the *Environment* and define *States* as the last nine planes predicted by three agents, with each agent obtaining three planes. A plane in Cartesian coordinate system is defined as $\cos(\zeta)x + \cos(\beta)y + \cos(\phi)z + d = 0$, where $(\cos(\zeta), \cos(\beta), \cos(\phi))$ represents the normal vector and d is the distance from the plane to the volume center. The plane parameters are updated according to its agent's *Actions* defined as $\{\pm \boldsymbol{a}_\zeta, \pm \boldsymbol{a}_\beta, \pm \boldsymbol{a}_\phi, \pm \boldsymbol{a}_d\}$. Each action taken by each agent will get its own reward signal $R \in \{-1, 0, +1\}$ calculated by the

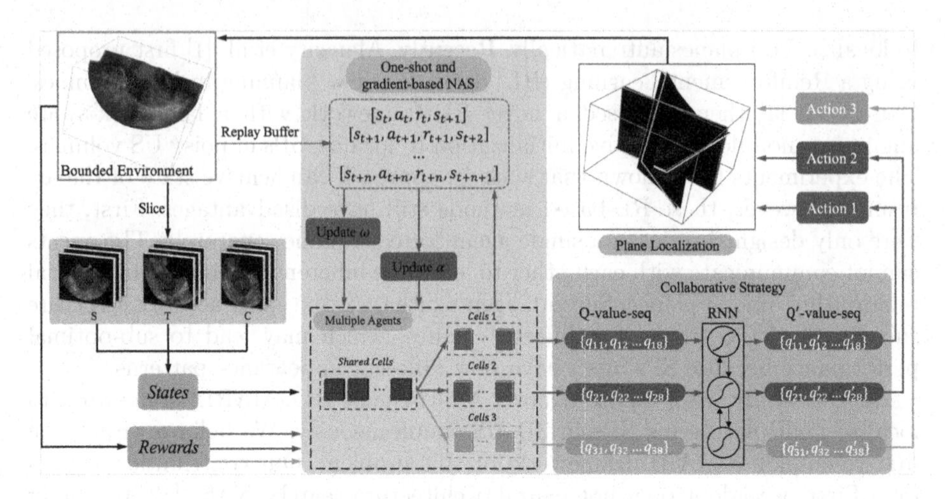

Fig. 2. Overview of the proposed framework.

Reward Function $R = sgn(D(P_{t-1} - P_g) - D(P_t - P_g))$, where D calculates the Euclidean distance from the predicted plane P_t to the ground truth plane P_g. For the *Terminal States*, we choose a fixed 50 and 30 steps during training and testing respectively based on our experiments and observation.

The agents aim to maximize both the current and future rewards and optimize their own policies for localizing corresponding SPs. To mitigate the upward bias caused by deep Q-network (DQN) [10] and stabilize the learning process, we use double DQN (DDQN) [15] and its loss function is defined as:

$$L = E[(r_t + \gamma Q(s_{t+1}, \underset{a_{t+1}}{argmax}\, Q(s_{t+1}, a_{t+1}; \omega); \tilde{\omega}) - Q(s_t, a_t; \omega))^2] \quad (1)$$

where γ is a discount factor to weight future rewards. The data sequence, including states s_t, actions a_t, rewards r_t at steps t and the next states s_{t+1}, is sampled from the experience replay buffer, which ensures the input data is independent and identically distributed during training. ω and $\tilde{\omega}$ are the weights of current and target networks. a_{t+1} is the actions in the next step predicted by the current network. Instead of outputting Q-values by fully connected layers of each agent directly and taking actions immediately, here, we propose to use an RNN based agent collaborative module to learn the spatial relationship among SPs.

2.2 GDAS Based Multi-agent Searching

In deep RL, the neural network architecture of the agent is crucial for good learning performance [7]. Previous studies [1,4] used the same VGG model for localizing different planes, which may degrade the performance because SPs often have very different appearance patterns. In this study, considering that both RL and NAS are very time-consuming, we adopt one-shot and gradient-based NAS in our MARL framework to obtain optimal network architecture for each agent.

Specifically, we use the GDAS based method [3] to accelerate the search process while keeping the accuracy. GDAS is a new gradient-based NAS method, which is stable and about 10 times faster than DARTS [8] by only optimizing the sub-graph sampled from the supernet in each iteration.

The agent search process can be defined by search space, search strategy, and performance estimation strategy. Figure 2 shows the designed network architectures, where three agents share 8 cells (5 normal cells and 3 reduce cells) and each agent has its own 4 cells (3 normal cells and 1 reduce cells). Such architecture holds the advantages that agents can not only share knowledge to collaborate with each other, but also learn their specific information as well. The cell search space is then defined as the connection mode and 10 operations including none, 3×3 convolution, 5×5 convolution, 3×3 dilated convolution, 5×5 dilated convolution, 3×3 separable convolution, 5×5 separable convolution, 3×3 max pooling, 3×3 avg pooling, and skip-connection. For the search strategy, GDAS searches by gradient descent and only updates the sub-graph sampled from the supernet in each iteration. Different batches of data are sampled from the replay buffer according to the prioritized weight and a random way to update network weights ω and architecture parameters α, respectively. We propose a new performance estimation strategy by choosing the optimal architecture parameters α^* when the sum of rewards is maximal on the validation set in all the training epochs and use it to construct the final designed agent $\mathcal{A}_{\alpha^*}$. The main reason of using this strategy is that the loss often oscillates and does not converge in RL.

2.3 RNN Based Agent Collaborative Strategy

Normally, multiple SPs in 3D US have a relatively certain spatial relationship, as shown in Fig. 1. Learning such spatial relationship is critical for automatic multi-plane localization. In [1,4], the agents were trained separately for each plane and cannot learn the invaluable spatial relationship among planes. Recently, Vlontzos et al. [16] proposed a collaborative MARL framework for multi-landmark detection by sharing weights of agent networks across the convolutional layers. However, this strategy is implicit and may not perform well on the multi-plane localization problem. In this study, we propose an RNN based collaborative strategy to learn the spatial layout among planes and improve the accuracy of localization. Specifically, the Bi-direction LSTM (BiLSTM) [5] is employed in this study, since it can combine the forward and backward information and thus strengthens the communication among agents.

Three agents can collaborate with each other through Q-values. Given the states s_t and the network's weights ω_t in step t, the agents output the Q-value sequence set $Q_t = (Q_{P_1}, Q_{P_2}, Q_{P_3})^{\mathrm{T}}$, where $Q_{P_i} = \{q_{i1}, q_{i2}, ..., q_{i8}\}$ for each plane P_i. The BiLSTM module then inputs the Q-value sequence Q_{P_i} of each plane as its hidden-state of each time-step and outputs the calibrated Q-value sequence set $Q'_t = (Q'_{P_1}, Q'_{P_2}, Q'_{P_3})$ defined by:

$$Q'_t = \mathcal{H}(Q_t, \overrightarrow{h}_{\tilde{t}-1}, \overleftarrow{h}_{\tilde{t}+1}; \theta) \quad (2)$$

where $\overrightarrow{h}_{\tilde{t}-1}$ and $\overleftarrow{h}_{\tilde{t}+1}$ are the forward and backward hidden sequences, respectively. $\mathcal{H}$ is the hidden layer function including the input gate, forget gate, output gate and cell, θ represents the parameters of BiLSTM.

3 Experimental Result

Materials and Implementation Details. We validate our proposed method on localizing three uterine SPs in 3D US. Approved by the local Institutional Review Board, 683 volumes were obtained from 476 patients by experts using a US system with an integrated 3D probe. Multiple volumes may be obtained from one patient when her uterus was abnormal or contained IUD. In our dataset, the average volume size is 261 × 175 × 277 and the unified voxel size is 0.5 × 0.5 × 0.5 mm^3. Four experienced radiologists annotated all volumes by manually localizing three SPs and four landmarks (see Fig. 1) in each volume under strict quality control. We randomly split the data into 539 and 144 volumes for training and testing at the patient level to ensure that multiple volumes of one patient belong to the same set.

In this study, we implemented our method in *Pytorch* and trained the system by Adam optimizer, using a standard PC with a NVIDIA TITAN 2080 GPU. We first obtained optimal agents by setting learning rate as $5e-5$ for the weights ω and 0.05 for the architecture parameters α in 50 epochs (about 2 days). Then we trained the MARL with learning rate = $5e-5$ and batch size = 32 in 100 epochs (about 3 days). The size of replay buffer is set as 15000 and the target network copies the parameters of the current network every 1500 iterations. We further trained the RNN module with hidden size = 64 and num layer = 2. The starting planes for training the system were randomly initialized around the ground truth within an angle range of ±20° and distance range of ±4 mm, according to the average error by landmark-aware alignment [4]. Besides, the step sizes of angle and distance in each update iteration are set as 0.5° and 0.1 mm.

Table 1. Quantitative evaluation of plane localization.

	Metrics	SARL	MARL	MARL-R	D-MARL	D-MARL-R	G-MARL	G-MARL-R
S	Ang (°)	9.68 ± 9.63	10.37 ± 10.18	8.48 ± 9.31	8.77 ± 8.83	7.94 ± 9.15	9.60 ± 8.97	7.05 ± 8.24
	Dis (mm)	2.84 ± 3.49	2.00 ± 2.38	2.41 ± 3.19	2.13 ± 3.01	2.03 ± 3.01	2.29 ± 3.03	2.21 ± 3.21
	SSIM	0.88 ± 0.06	0.89 ± 0.06	0.89 ± 0.07	0.88 ± 0.05	0.89 ± 0.06	0.87 ± 0.12	0.90 ± 0.06
T	Ang (°)	9.53 ± 8.27	9.30 ± 8.87	8.87 ± 7.37	9.03 ± 7.91	8.57 ± 7.99	8.71 ± 9.01	8.62 ± 8.07
	Dis (mm)	3.17 ± 2.58	2.99 ± 2.61	2.69 ± 2.49	2.01 ± 2.18	2.22 ± 2.31	2.37 ± 2.36	2.36 ± 2.53
	SSIM	0.75 ± 0.10	0.72 ± 0.10	0.71 ± 0.11	0.74 ± 0.12	0.75 ± 0.13	0.75 ± 0.12	0.74 ± 0.13
C	Ang (°)	8.00 ± 6.76	7.14 ± 6.67	7.17 ± 6.13	7.13 ± 1.35	6.21 ± 7.11	7.21 ± 6.60	5.93 ± 7.05
	Dis (mm)	1.46 ± 1.40	1.53 ± 1.58	1.22 ± 1.27	1.35 ± 1.42	0.96 ± 1.17	1.39 ± 1.39	0.89 ± 1.15
	SSIM	0.69 ± 0.09	0.67 ± 0.09	0.68 ± 0.09	0.68 ± 0.10	0.73 ± 0.12	0.68 ± 0.10	0.75 ± 0.12
Avg	Ang (°)	9.07 ± 8.34	8.94 ± 8.79	8.17 ± 8.13	8.31 ± 8.02	7.57 ± 8.19	8.51 ± 7.98	7.20 ± 7.89
	Dis (mm)	2.49 ± 2.74	2.17 ± 2.31	2.11 ± 2.53	1.83 ± 2.32	1.74 ± 2.37	2.02 ± 2.40	1.82 ± 2.54
	SSIM	0.77 ± 0.12	0.76 ± 0.13	0.76 ± 0.13	0.77 ± 0.12	0.79 ± 0.13	0.77 ± 0.13	0.80 ± 0.13

Quantitative and Qualitative Analysis. We evaluated the performance in terms of spatial and content similarities for localizing uterine mid-sagittal (S), transverse (T), and coronal (C) planes by the dihedral angle between two planes (Ang), difference of Euclidean distance to origin (Dis) and Structural Similarity

Table 2. Model information of compared methods.

	V-MARL	MARL	MARL-R	D-MRAL	D-MARL-R	G-MARL	G-MARL-R
FLOPs (G)	15.41	1.82	1.82	0.69	0.69	0.68	0.68
Params (M)	27.58	12.22	12.35	3.62	3.76	3.61	3.75

Index (SSIM). Ablation study was conducted by comparing the methods including Single Agent RL (SARL), MARL, MARL with RNN (MARL-R), DARTS and GDAS based MARL without and with RNN (D-MRAL, D-MARL-R, G-MARL, G-MARL-R). The landmark-aware registration provided warm starts for all the above methods and the Resnet18 served as network backbone for SARL, MARL and MARL-R instead of the VGG to reduce model parameters while keeping comparable performance.

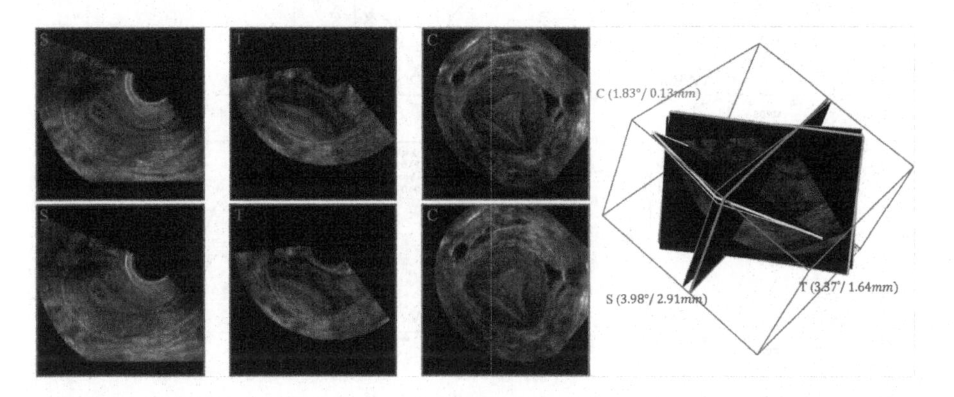

Fig. 3. One typical SPs localization result. *Left*: the ground truth (top) and prediction (bottom) of three SPs. *Right*: Spatial differences between prediction and ground truth.

Table 1 lists quantitative results for each plane and average (Avg) values for all planes by different methods. All MARL based methods perform better than the SARL because of the communication among agents. The superior performance of MARL-R, D-MARL and G-MARL compared to MARL shows the efficacy of our proposed agent searching and collaboration methods separately. Among all these methods, our proposed G-MARL-R method achieves the best results, which further illustrates the efficacy of combining these two methods. Table 2 compares the computational costs (FLOPs) and model sizes (Params) of different methods. The Restnet18-based MARL model has only 12% FLOPs and 44% parameters of the VGG-based MARL (V-MARL) model. Meanwhile, NAS-based methods save 63% FLOPs and 70% parameters of the MARL. We further compared the GDAS and the DARTS based methods in terms of training time and occupied memory. The former one G-MARL-R (0.04 days/epoch, maximum batch size = 32) is much faster and more memory efficient than the D-MARL-R (0.12 days/epoch, maximum batch size = 4).

Figure 3 shows one typical result by the G-MARL-R method. Compared from image content and spatial relationship, our method can accurately localize three SPs in one volume that are very close to the ground truth. Figure 4 illustrates the Sum of Angle and Distance (SAD) and Q-value curves of the same volume without and with the RNN. Our proposed RNN module greatly reduces the SAD and Q-value, and makes them more stable after a certain number of steps, so it can help accurately localize multiple SPs in 3D US. The optimal SAD is located around step 30, which shows that our termination method based on fixed steps is reasonable and effective.

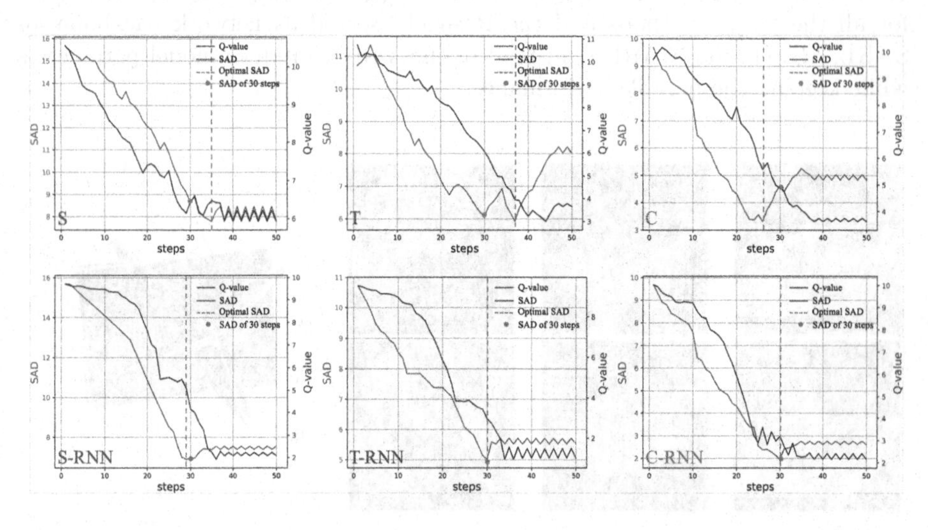

Fig. 4. SAD and Q-value curves without (top) and with (bottom) RNN module.

4 Conclusion

We propose a novel MARL framework for multiple SPs localization in 3D US. We use the GDAS-based NAS method to automatically design optimal agents with better performance and fewer parameters. Moreover, we propose an RNN based agent collaborative strategy to learn the spatial relationship among SPs, which is general and effective for establishing strong communications among agents. Experiments on our in-house large dataset validate the efficacy of our method. In the future, we will explore to search Convolutional Neural Network (CNN) and RNN modules together to improve the system performance.

Acknowledgement. This work was supported by the grant from National Key R&D Program of China (No. 2019YFC0118300), Shenzhen Peacock Plan (No. KQTD2016053112051497, KQJSCX20180328095606003) and Medical Scientific Research Foundation of Guangdong Province, China (No. B2018031).

References

1. Alansary, A., et al.: Automatic view planning with multi-scale deep reinforcement learning agents. In: Frangi, A.F., Schnabel, J.A., Davatzikos, C., Alberola-López, C., Fichtinger, G. (eds.) MICCAI 2018. LNCS, vol. 11070, pp. 277–285. Springer, Cham (2018). https://doi.org/10.1007/978-3-030-00928-1_32
2. Chykeyuk, K., Yaqub, M., Alison Noble, J.: Class-specific regression random forest for accurate extraction of standard planes from 3D echocardiography. In: Menze, B., Langs, G., Montillo, A., Kelm, M., Müller, H., Tu, Z. (eds.) MCV 2013. LNCS, vol. 8331, pp. 53–62. Springer, Cham (2014). https://doi.org/10.1007/978-3-319-05530-5_6
3. Dong, X., Yang, Y.: Searching for a robust neural architecture in four GPU hours. In: Proceedings of the IEEE Conference on Computer Vision and Pattern Recognition, pp. 1761–1770 (2019)
4. Dou, H., et al.: Agent with warm start and active termination for plane localization in 3D ultrasound. In: Shen, D., et al. (eds.) MICCAI 2019. LNCS, vol. 11768, pp. 290–298. Springer, Cham (2019). https://doi.org/10.1007/978-3-030-32254-0_33
5. Graves, A., Schmidhuber, J.: Framewise phoneme classification with bidirectional LSTM and other neural network architectures. Neural Netw. **18**(5–6), 602–610 (2005)
6. Li, Y.: Standard plane detection in 3D fetal ultrasound using an iterative transformation network. In: Frangi, A.F., Schnabel, J.A., Davatzikos, C., Alberola-López, C., Fichtinger, G. (eds.) MICCAI 2018. LNCS, vol. 11070, pp. 392–400. Springer, Cham (2018). https://doi.org/10.1007/978-3-030-00928-1_45
7. Li, Y.: Deep reinforcement learning: an overview. arXiv preprint (2017). arXiv:1701.07274
8. Liu, H., Simonyan, K., Yang, Y.: Darts: differentiable architecture search. arXiv preprint (2018). arXiv:1806.09055
9. Lorenz, C., et al.: Automated abdominal plane and circumference estimation in 3D us for fetal screening. In: Medical Imaging 2018: Image Processing, vol. 10574, p. 105740I. International Society for Optics and Photonics (2018)
10. Mnih, V., et al.: Human-level control through deep reinforcement learning. Nature **518**(7540), 529–533 (2015)
11. Ni, D., et al.: Standard plane localization in ultrasound by radial component model and selective search. Ultrasound Med. Biol. **40**(11), 2728–2742 (2014)
12. Ren, P., et al.: A comprehensive survey of neural architecture search: challenges and solutions. arXiv preprint (2020). arXiv:2006.02903
13. Ryou, H., Yaqub, M., Cavallaro, A., Roseman, F., Papageorghiou, A., Noble, J.A.: Automated 3D ultrasound biometry planes extraction for first trimester fetal assessment. In: Wang, L., Adeli, E., Wang, Q., Shi, Y., Suk, H-Il (eds.) MLMI 2016. LNCS, vol. 10019, pp. 196–204. Springer, Cham (2016). https://doi.org/10.1007/978-3-319-47157-0_24
14. Schmidt-Richberg, A., et al.: Offset regression networks for view plane estimation in 3D fetal ultrasound. In: Medical Imaging 2019: Image Processing, vol. 10949, p. 109493K. International Society for Optics and Photonics (2019)

15. Van Hasselt, H., Guez, A., Silver, D.: Deep reinforcement learning with double Q-learning. In: Thirtieth AAAI conference on artificial intelligence (2016)
16. Vlontzos, A., Alansary, A., Kamnitsas, K., Rueckert, Dl, Kainz, B.: Multiple landmark detection using multi-agent reinforcement learning. In: Shen, D., et al. (eds.) MICCAI 2019. LNCS, vol. 11767, pp. 262–270. Springer, Cham (2019). https://doi.org/10.1007/978-3-030-32251-9_29
17. Wong, L., White, N., Ramkrishna, J., Júnior, E.A., Meagher, S., Costa, F.D.S.: Three-dimensional imaging of the uterus: the value of the coronal plane. World J. Radiol. **7**(12), 484 (2015)

Contrastive Rendering for Ultrasound Image Segmentation

Haoming Li[1,2], Xin Yang[1,2], Jiamin Liang[1,2], Wenlong Shi[1,2], Chaoyu Chen[1,2], Haoran Dou[1,2], Rui Li[1,2], Rui Gao[1,2], Guangquan Zhou[3], Jinghui Fang[4], Xiaowen Liang[4], Ruobing Huang[1,2], Alejandro Frangi[1,5,6], Zhiyi Chen[4], and Dong Ni[1,2](✉)

[1] School of Biomedical Engineering, Shenzhen University, Shenzhen, China
nidong@szu.edu.cn
[2] Medical UltraSound Image Computing (MUSIC) Lab, Shenzhen University, Shenzhen, China
[3] School of Biological Sciences and Medical Engineering, Southeast University, Nanjing, China
[4] Department of Ultrasound Medicine, Third Affiliated Hospital of Guangzhou Medical University, Guangzhou, China
[5] Centre for Computational Imaging and Simulation Technologies in Biomedicine (CISTIB), School of Computing, University of Leeds, Leeds, UK
[6] Medical Imaging Research Center (MIRC), University Hospital Gasthuisberg, Electrical Engineering Department, KU Leuven, Leuven, Belgium

Abstract. Ultrasound (US) image segmentation embraced its significant improvement in deep learning era. However, the lack of sharp boundaries in US images still remains an inherent challenge for segmentation. Previous methods often resort to global context, multi-scale cues or auxiliary guidance to estimate the boundaries. It is hard for these methods to approach pixel-level learning for fine-grained boundary generating. In this paper, we propose a novel and effective framework to improve boundary estimation in US images. Our work has three highlights. First, we propose to formulate the boundary estimation as a rendering task, which can recognize ambiguous points (pixels/voxels) and calibrate the boundary prediction via enriched feature representation learning. Second, we introduce point-wise contrastive learning to enhance the similarity of points from the same class and contrastively decrease the similarity of points from different classes. Boundary ambiguities are therefore further addressed. Third, both rendering and contrastive learning tasks contribute to consistent improvement while reducing network parameters. As a proof-of-concept, we performed validation experiments on a challenging dataset of 86 ovarian US volumes. Results show that our proposed method outperforms state-of-the-art methods and has the potential to be used in clinical practice.

Keywords: Ultrasound image · Segmentation · Contrastive learning

H. Li and X. Yang—Equal contribution.

A. L. Martel et al. (Eds.): MICCAI 2020, LNCS 12263, pp. 563–572, 2020.
https://doi.org/10.1007/978-3-030-59716-0_54

1 Introduction

Ultrasound (US) is widely accepted in clinic for routine diagnosis. Automatically segmenting anatomical structures from US images is highly desired. US image segmentation witnessed its significant improvement in this deep learning era [6]. However, due to the low contrast, low resolution and speckle noise in US images, the accuracy of these methods is hampered in ambiguous boundary regions, such as the blurred boundary or shadow-occluded parts [5]. Taking the 3D ovarian US volume as an example, as shown in Fig. 1, the adverse effects of boundary ambiguity on segmentation are obvious. The boundary among follicles are blurring due to the speckle noise. Since the follicles have irregular shapes, varying volumes and complex connection status, boundary thickness of follicles are often inconsistent (Fig. 1(a–c)). In addition, no distinct boundary can be identified between the ovary and background tissues. All these factors can degrade the deep segmentation model with fuzzy and wrong predictions (Fig. 1(d)).

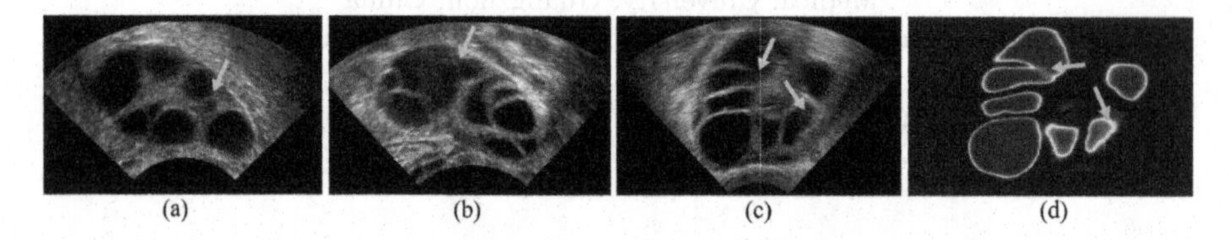

Fig. 1. Illustration for (a)(b)(c) ultrasound slices of ovary and follicle. Blurry boundary and scale-varying follicles are main challenges for segmentation (green arrows). (d) Follicle probability map of (c) generated by a U-net. Underestimation and touching boundary can be observed in the map (green arrow). (Color figure online)

In recent years, extensive attempts have been made to resolve the problems. In [8], Tu et al. proposed a classic cascaded framework, Auto-context, to revisit the context cues in probability maps for enhancement. Local semantic features were collected by RNN in [11] to refine 2D ovarian ultrasound segmentation. Making use of auxiliary guidance to refine boundary was also studied. In [1,13], edge-weighted mechanisms were proposed to guide the deep models to pay more attention to the edge of objects. Varying scales of object often make the boundary hard to be captured, especially on the small objects under limited imaging resolution, like the follicles in Fig. 1. Multi-scale architectures to fuse the context information hierarchically, like Atrous Spatial Pyramid Pooling (ASPP) [2] are explored in medical image segmentation [10]. Although effective, these methods extract global/local semantic information for boundary refinement, but cannot approach point-level learning for fine-grained boundary identification. Recently, point-wise uncertainty estimation has been investigated to enhance image segmentation by selecting informative points for loss calculation [12] and composing contour constraints [9]. These methods selected out the ambiguous points but ignored their feature enhancement to significantly modify the predictions.

In this paper, to address the aforementioned problems, we propose a Contrastive Rendering (*C-Rend*) framework to improve the boundary estimation in US images. By focusing on point-level boundary ambiguity and reducing it with fine-grained features, C-Rend is a general strategy with three highlights. *First*, inspired by recent work [4], C-Rend formulates boundary estimation as a rendering task. C-Rend can adaptively recognize ambiguous points and re-predict their boundary predictions via coarse and fine-grained feature enriched representation learning. *Second*, to further distill discriminative features for better determination of boundary location, we selectively introduce the point-wise contrastive learning [3] to maximize the divergence among different classes. Specifically, C-Rend encourages the similarity of ambiguous points from the same class and contrastively decrease the similarity of those from different classes. Boundary ambiguities are therefore further addressed. *Third*, different from previous methods which sacrifice computation overhead for performance improvement, both rendering and contrastive learning in C-Rend contribute consistent improvement while reducing the network parameters. As a proof-of-concept, we performed extensive validation experiments on a challenging dataset containing 86 3D ovarian US volumes. Results show that our proposed method outperformed strong contender and reported best results with Dice of 87.78% and 83.49% for the challenging problem of ovary and follicles, respectively.

2 Methodology

Figure 2 illustrates the flowchart of our proposed C-Rend for 3D ovarian US volume segmentation. It mainly consists of Four main components: (1) a segmentation architecture containing an asymmetric encoder-decoder structure as the backbone, (2) point selection module to select ambiguous points that need to be re-predicted, (3) a rendering head to re-predict the label of selected points based on the hybrid point-level features, (4) a contrastive learning head to further enhance the confidence on boundary. Both rendering head and contrastive learning head contribute to the update of final segmentation output.

2.1 Segmentation Architecture for Hybrid Features

As shown in Fig. 2, we build a strong segmentation backbone based on the classic 3D U-net architecture [7]. The encoder contains 4 consecutive pooling layers. Between each two pooling layers, there are two convolutional layers with dilated convolution operators. While the decoder has only 3 deconvolution layers. We discard the last deconvolution layer of the original symmetric U-net to save the computation cost and reduce model parameters. Instead, we generate two kinds of predictions from lower levels of the segmentor. The first one is the coarse probability maps (full size) for the final segmentation. The second one are the fine-grained feature maps (halved size) generated from the last two layers of our segmentor with abundant channels. These two features will jointly drive our C-Rend to recognize and render the ambiguous boundary points.

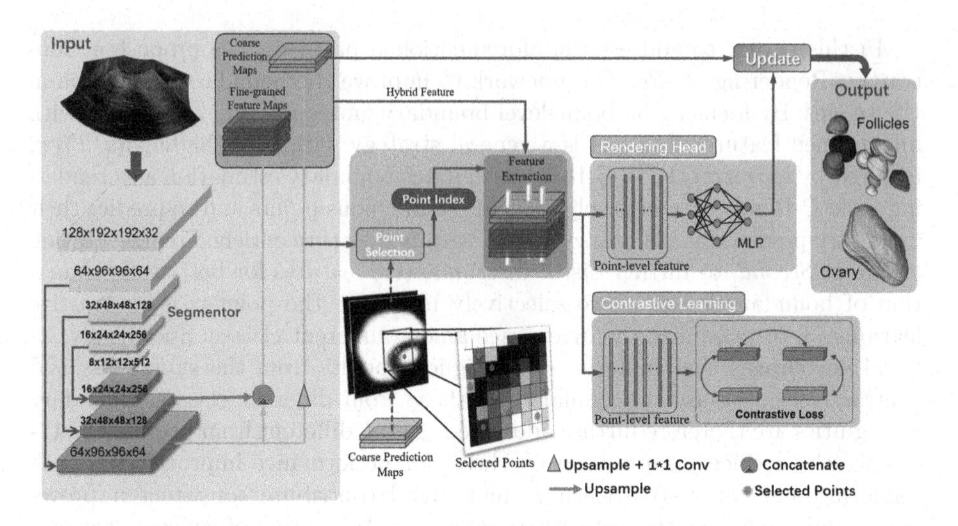

Fig. 2. The overflow of our proposed contrastive rendering (C-Rend) framework.

2.2 Rendering Head for Point-Wise Re-prediction

Point Selection Strategy. The core of rendering head is to collect point-level hybrid feature representation to re-predict ambiguous boundaries. Whereas, directly handling all the points in the entire feature map is not only computationally expensive, but also unnecessary. Most of the boundary prediction from the segmentor present high confidence. Therefore, as shown in Fig. 2, our C-Rend firstly adopts a strategy to select N points from the coarse probability map to train on [4]. Specifically, the selection strategy contains three steps. First, C-Rend randomly scatters kN seeds in the coarse prediction map. Then, based on the coarse probability maps, C-Rend randomly picks out the most βN uncertain points (e.g., those with probabilities closest to 0.5). Finally, C-Rend obtains the rest $(1-\beta)N$ points by uniformly sampling the coarse maps. During training, the selected point set is assumed to contain more points with high uncertainty than those with high confidence. In this work, we set $K=2$, N equals to the volume size, $\beta=0.7$. This configuration is desired for rendering head to learn from positive and negative samples.

Rendering Head. Figure 3(a) elaborates the details of rendering head. With the point index provided by the selection strategy, features from the coarse probability maps and fine-grained feature maps of these points are integrated into a hybrid form. In this paper, for each selected point, a 2-dimensional probability vector is extracted from the prediction map and a 96-dimensional feature vector is obtained from the fine-grained feature maps. C-Rend concatenates these two feature vectors, and sends it to the rendering head. The mission of rendering head is to re-predict the labels of selected points. Here, we use a lightweight multi-layer perceptron (MLP) with one hidden layer as the workhorse of our rendering head.

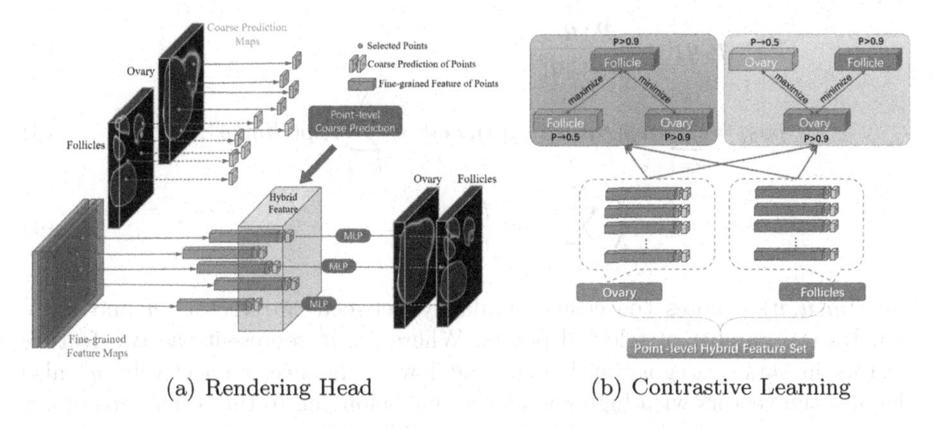

(a) Rendering Head (b) Contrastive Learning

Fig. 3. (a) Details of rendering head about hybrid point-level feature and boundary re-prediction, (b) contrastive learning on the point-level hybrid representations.

All the selected points share the same MLP (Fig. 3(a)). Rendering head outputs class label for each point, therefore its training follows the standard task-specific segmentation losses. We sum the cross-entropy loss for the rendering head over all the selected points, and define it as Rendering Loss:

$$Loss_R = -\sum_{i=1}^{N}[y_i log(\hat{y_i}) + (1 - y_i) log(1 - \hat{y_i})] \tag{1}$$

where $\hat{y_i}$ and y_i represent the re-prediction and ground truth, respectively. N denotes the number of selected points for each input volume.

2.3 Contrastive Learning

Considering the close connection among the boundary of objects belonging to different semantic class (in this paper, ovary and follicles), we adopt the point-wise contrastive loss to enhance the confidence on the selected ambiguous boundary points. Our contrastive loss roots in the concept of contrastive learning [3]. The basic idea is maximizing the divergence between feature vectors of two different classes, while minimizing those of the same class. Specifically, as shown in Fig. 3(b), based on the extracted hybrid feature vectors and probability of selected points, we enforce a doubled contrastive loss on the feature vectors from ovary and follicles. Points belonging to the follicle region with high predicted probability (>0.9) should have low similarity with those from ovary region, but should be as close as possible to the follicle points with high uncertainty (probability closest to 0.5). The case also applies to the uncertain points of ovary (Fig. 3(b) top right). We use the cosine similarity to measurement of divergence. Here we build a contrastive learning loss by maximizing the disagreement among feature vectors from different categories, meanwhile minimizing the divergence among those belonging to the same category but with different confidences. The formulation of this loss, namely $Loss_C$, is:

$$Sim(\boldsymbol{p}, \boldsymbol{q}) = \frac{\boldsymbol{p} \cdot \boldsymbol{q}}{\|\boldsymbol{p}\| \cdot \|\boldsymbol{q}\|} \tag{2}$$

$$pos_i = \sum_{k=1}^{N} exp(Sim(p^h, p^l)), neg_i = \sum_{k=1}^{N} exp(Sim(p^h, q^h)) \tag{3}$$

$$Loss_C = \gamma - \frac{1}{N} \sum_{i=1}^{N} -log(\frac{pos_i}{neg_i}) \tag{4}$$

$Sim(\boldsymbol{u}, \boldsymbol{v})$ denotes the cosine similarity between two vectors $\boldsymbol{u}$ and $\boldsymbol{v}$. N denotes the number of selected points. Where p^h, p^l represent the two feature vectors in same category with high and low confidence, respectively. q^h also denotes the vectors with high confidence but belonging to the other category. γ is a constant term and set as 2.0. The overall loss function of our network is:

$$L_{total} = Loss_{ori} + \lambda_1 Loss_R + \lambda_2 Loss_C, \tag{5}$$

where $Loss_{ori}$ denotes the main cross entropy loss on the original coarse prediction of decoder. $\lambda_1 = 0.8$ and $\lambda_2 = 0.2$ denote the weight for rendering loss and contrastive loss respectively.

3 Experimental Results

3.1 Dataset and Implementation Details

We validated our framework on the 3D ovarian US segmentation. Approved by local IRB, a total of 86 volumes were collected. The maximum original size in the dataset is $400 \times 400 \times 700$. Voxel spacing is re-sampled as $0.2 \times 0.2 \times 0.2\,\text{mm}^3$. All the ovaries and follicles were manually annotated by sonographers under the quality control of an experienced expert. All the segmentation models compared in this paper were trained on 68 volumes, the remaining 18 volumes for testing. The training dataset is further augmented with $\pm 30°$ rotation in transverse section. Limited by GPU memory, we resized the first and second dimension of the volume to 192×192, and proportionally scaled the third dimension. The segmentor finally processes the patches with size of $192 \times 192 \times 128$. 50% overlap is set during testing. In this study, all the methods were implemented in PyTorch, in which our models ran 150 epochs in a single Titan X GPU with 12 GB RAM (Nvidia Corp, Santa Clara, CA). Adam with a batch size of 1 and a fixed learning rate of 0.0001 was adopted to optimize the models. *Full code will be released soon.*

3.2 Quantitative and Quanlitative Evaluation

We adopt a set of evaluation metrics to comprehensively evaluate the segmentation results. Among them, Dice Similarity Coefficient (DSC-%), Jaccard Coefficient (JC-%), Hausdorff Distance (HD-mm) and Average Surface Distance (ASD-mm) were used to evaluate the segmentation on the ovary and follicles. In addition, False Detection (FD-%) and Missed Detection (MD-%) were utilized to assess the follicle detection. FD and MD are defined on the segmentation and IoU (Intersection over Union). A false detection (FD) of follicle means that, the segmentation hits no follicle ground truth over 30% IoU. A missed detection means that, each follicle in ground truth hits no segmentation over 30% IoU. Since the number of follicle is an important index for the ovary growth staging, the error on follicle counting was also adopted. Notably, for reproducibility, we reported two-fold cross validation results for all the compared methods in this paper, and in total 36 volumes were involved for each method.

Table 1 presents the quantitative comparison among our C-Rend and four strong re-implemented methods, including a two-stage Auto-Context [8] (*Context*), auxiliary edge assisted segmentation (*Edge*), multi-scale fusion network (*Deeplab*) [2], entropy around edge (*Entropy*) [9]. It is obvious to see that, C-Rend brings about consistent improvement over the baseline U-net (1.5% DSC on follicle, 1.2% percent DSC on ovary), while it is hard for other competitors to improve the both. C-Rend also outperforms other competitors in both follicle and ovary segmentation on almost all the metrics. Previous global contexture, edge guidance and multi-scale based designs are hence facing difficulties in handling the ambiguous boundary in fine scales. We also conduct ablation study about the modules in C-Rend. Only with the rendering head, *Rend* can already refine all the metrics over the baseline. With the constraints from contrastive learning, C-Rend gets further improvements over *Rend* with a large margin (about 0.7% in DSC for follicle and ovary). It's worth noting that, compared to the U-net baseline, our C-Rend refines the results but does not require extra computation cost (23.55M vs. 23.56M).

Table 1. Quantitative comparisons on performance and model complexity (superscript for standard deviation)

Method	Follicles				Ovary				Params
	DSC	JC	HD	ASD	DSC	JC	HD	ASD	
U-Net [7]	$82.00^{8.8}$	$70.31^{12.4}$	$12.10^{3.5}$	$4.40^{1.0}$	$86.57^{9.1}$	$76.89^{8.4}$	$6.96^{2.5}$	$1.43^{0.9}$	23.56M
Context [8]	$81.51^{9.5}$	$69.09^{12.5}$	$11.82^{2.6}$	$4.24^{1.2}$	$87.18^{6.3}$	$77.78^{9.4}$	$6.89^{2.4}$	$1.36^{0.4}$	47.12M
Edge [1]	$82.70^{9.5}$	$71.18^{10.6}$	$11.79^{2.9}$	$4.31^{1.1}$	$86.64^{8.7}$	$77.08^{7.6}$	$7.08^{3.1}$	$1.35^{0.7}$	23.59M
Deeplab [2]	$81.88^{12.3}$	$70.34^{14.1}$	$11.94^{3.3}$	$4.36^{2.1}$	$86.42^{10.2}$	$76.63^{10.4}$	$7.04^{2.1}$	$1.33^{0.3}$	25.58M
Entropy [9]	$81.64^{7.2}$	$69.13^{10.6}$	$11.96^{2.78}$	$4.24^{1.17}$	$86.76^{6.0}$	$77.22^{9.0}$	$7.05^{2.2}$	$1.35^{0.43}$	23.56M
Rend	$82.73^{7.6}$	$70.95^{8.6}$	$11.85^{2.2}$	$4.22^{1.43}$	$87.07^{8.7}$	$77.44^{9.1}$	$6.50^{2.2}$	$1.57^{0.5}$	**23.55M**
C-Rend	$\mathbf{83.49}^{7.9}$	$\mathbf{72.02}^{8.8}$	$\mathbf{11.70}^{3.0}$	$\mathbf{3.99}^{1.1}$	$\mathbf{87.78}^{10.4}$	$\mathbf{78.64}^{9.8}$	$\mathbf{6.45}^{1.9}$	$\mathbf{1.23}^{0.8}$	**23.55M**

In Table 2, we analyze the performance of different methods in detecting follicles with radius <5 mm and counting all follicles. We can observe that, embedding the rendering task and contrastive learning, C-Rend significantly reduces the FD (30%), MD (15%) and counting error (16%) when compared to the baseline. C-Rend also surpasses other competitors in all criteria. Addressing the boundary ambiguity in C-Rend from point level may profoundly contribute to the localization of small follicles.

Table 2. Evaluation about follicle detection and counting

Method	FD (<5 mm)	MD (<5 mm)	Counting error
Unet [7]	2.47 ± 9.7	16.07 ± 11.2	2.52 ± 3.8
Context [8]	3.68 ± 7.1	17.42 ± 12.1	2.27 ± 3.4
Edge [1]	5.19 ± 5.2	17.69 ± 12.3	2.47 ± 2.9
Deeplab [2]	4.88 ± 9.6	18.60 ± 10.6	2.75 ± 2.7
Entropy [9]	5.58 ± 8.7	19.60 ± 14.9	2.16 ± 3.0
Rend (ours)	4.14 ± 7.4	13.78 ± 7.1	2.51 ± 2.1
C-Rend (ours)	$\mathbf{1.71} \pm 5.4$	$\mathbf{13.62} \pm 8.6$	$\mathbf{2.10} \pm 1.9$

Figure 4 visualize the segmentation differences among the baseline U-net, *Edge*, Rend and C-Rend. In surface rendering, C-Rend gets more clear boundary prediction than other methods for follicles with different scales. In HD surface distance, Rend and C-Rend present much lower mean/max/min distance errors for both ovary and follicles than the baseline and *Edge* methods.

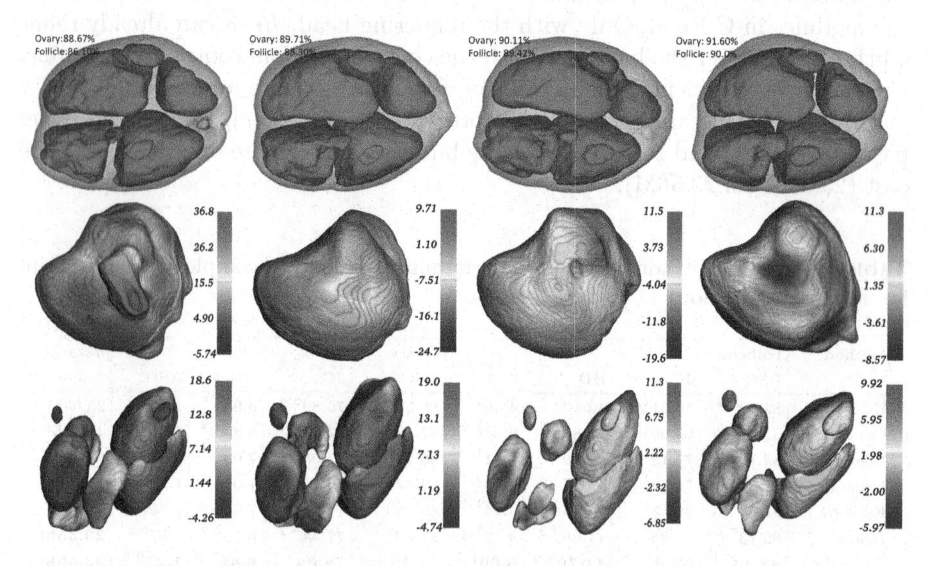

Fig. 4. Visualization results. From left to right: U-net, *Edge*, Rend and C-Rend. From Top to bottom: surface rendering of ovary (cyan) and follicles (blue), HD from ovary segmentation to ground truth, Hausdorff distance from follicle segmentation to ground truth. The colorbar is annotated with mean in the center, min and max on the ends. (Color figure online)

4 Conclusion

We proposed a general and lightweight framework for ultrasound images segmentation, which holds potentials for different deep architectures and applications. Exploiting point-level coarse prediction and fine-grained features, coupled with contrastive learning, to calibrate the ambiguous prediction on boundary are the main highlights of this work. The proposed C-Rend is extensively validated on the tough task of 3D ovarian US segmentation. In terms of efficacy and computational cost, both quantitative and qualitative evaluations support that, our C-Rend has clear advantages over other strong competitors in combating ambiguous boundary.

Acknowledgments. This work was supported by the grant from National Key R&D Program of China (No. 2019YFC0118300); Shenzhen Peacock Plan (No. KQTD2016053112051497, KQJSCX20180328095606003); Medical Scientific Research Foundation of Guangdong Province, China (No. B2018031); National Natural Science Foundation of China (No. NSFC61771130).

References

1. Chen, H., Qi, X., Yu, L., Heng, P.A.: DCAN: deep contour-aware networks for accurate gland segmentation. In: Proceedings of the IEEE Conference on Computer Vision and Pattern Recognition, pp. 2487–2496 (2016)
2. Chen, L.C., Papandreou, G., Kokkinos, I., Murphy, K., Yuille, A.L.: DeepLab: semantic image segmentation with deep convolutional nets, atrous convolution, and fully connected CRFs. IEEE Trans. Pattern Anal. Mach. Intell. **40**(4), 834–848 (2017)
3. Chen, T., Kornblith, S., Norouzi, M., Hinton, G.: A simple framework for contrastive learning of visual representations. arXiv preprint arXiv:2002.05709 (2020)
4. Kirillov, A., Wu, Y., He, K., Girshick, R.: PointRend: image segmentation as rendering. arXiv preprint arXiv:1912.08193 (2019)
5. Li, H., et al.: CR-UNet: a composite network for ovary and follicle segmentation in ultrasound images. IEEE J. Biomed. Health Inform. **24**, 974–983 (2019)
6. Liu, S., et al.: Deep learning in medical ultrasound analysis: a review. Engineering **5**, 261–275 (2019)
7. Ronneberger, O., Fischer, P., Brox, T.: U-Net: convolutional networks for biomedical image segmentation. In: Navab, N., Hornegger, J., Wells, W.M., Frangi, A.F. (eds.) MICCAI 2015. LNCS, vol. 9351, pp. 234–241. Springer, Cham (2015). https://doi.org/10.1007/978-3-319-24574-4_28
8. Tu, Z., Bai, X.: Auto-context and its application to high-level vision tasks and 3D brain image segmentation. IEEE Trans. Pattern Anal. Mach. Intell. **32**(10), 1744–1757 (2009)
9. Wang, S., Yu, L., Li, K., Yang, X., Fu, C.-W., Heng, P.-A.: Boundary and entropy-driven adversarial learning for fundus image segmentation. In: Shen, D., et al. (eds.) MICCAI 2019. LNCS, vol. 11764, pp. 102–110. Springer, Cham (2019). https://doi.org/10.1007/978-3-030-32239-7_12
10. Wang, Y., et al.: Deep attentive features for prostate segmentation in 3D transrectal ultrasound. IEEE Trans. Med. Imaging **38**(12), 2768–2778 (2019)

11. Yang, X., et al.: Towards automated semantic segmentation in prenatal volumetric ultrasound. IEEE Trans. Med. Imaging **38**(1), 180–193 (2018)
12. Yu, L., Wang, S., Li, X., Fu, C.-W., Heng, P.-A.: Uncertainty-aware self-ensembling model for semi-supervised 3D left atrium segmentation. In: Shen, D., et al. (eds.) MICCAI 2019. LNCS, vol. 11765, pp. 605–613. Springer, Cham (2019). https://doi.org/10.1007/978-3-030-32245-8_67
13. Zhu, Q., Du, B., Yan, P.: Boundary-weighted domain adaptive neural network for prostate MR image segmentation. IEEE Trans. Med. Imaging **39**, 753–763 (2019)

An Unsupervised Approach to Ultrasound Elastography with End-to-end Strain Regularisation

Rémi Delaunay[1,2](✉), Yipeng Hu[1], and Tom Vercauteren[1,2]

[1] Wellcome/EPSRC Centre for Interventional and Surgical Sciences, University College London, Gower Street, London WC1E 6BT, UK
remi.delaunay.17@ucl.ac.uk

[2] School of Biomedical Engineering and Imaging Sciences, King's College London, Strand, London WC2R 2LS, UK

Abstract. Quasi-static ultrasound elastography (USE) is an imaging modality that consists of determining a measure of deformation (i.e. strain) of soft tissue in response to an applied mechanical force. The strain is generally determined by estimating the displacement between successive ultrasound frames acquired before and after applying manual compression. The computational efficiency and accuracy of the displacement prediction, also known as time-delay estimation, are key challenges for real-time USE applications. In this paper, we present a novel deep-learning method for efficient time-delay estimation between ultrasound radio-frequency (RF) data. The proposed method consists of a convolutional neural network (CNN) that predicts a displacement field between a pair of pre- and post-compression ultrasound RF frames. The network is trained in an unsupervised way, by optimizing a similarity metric between the reference and compressed image. We also introduce a new regularization term that preserves displacement continuity by directly optimizing the strain smoothness. We validated the performance of our method by using both ultrasound simulation and *in vivo* data on healthy volunteers. We also compared the performance of our method with a state-of-the-art method called OVERWIND [17]. Average contrast-to-noise ratio (CNR) and signal-to-noise ratio (SNR) of our method in 30 simulation and 3 *in vivo* image pairs are 7.70 and 6.95, 7 and 0.31, respectively. Our results suggest that our approach can effectively predict accurate strain images. The unsupervised aspect of our approach represents a great potential for the use of deep learning application for the analysis of clinical ultrasound data.

Keywords: Ultrasound elastography · Time delay estimation · Convolutional neural network

Electronic supplementary material The online version of this chapter (https://doi.org/10.1007/978-3-030-59716-0_55) contains supplementary material, which is available to authorized users.

A. L. Martel et al. (Eds.): MICCAI 2020, LNCS 12263, pp. 573–582, 2020.
https://doi.org/10.1007/978-3-030-59716-0_55

1 Introduction

Ultrasound elastography (USE) is an imaging technique that enables the characterization of tissue mechanical properties [21]. Since its introduction in 1991 [19], strain imaging has shown usefulness in diagnostic applications where pathological alterations induce modification of tissue stiffness, such as lesion detection in liver disease [12] and tumour characterisation in the thyroid [16], breast [7] and prostate cancer [3]. This work focuses on quasi-static, free-hand strain elastography, where a time-varying axial compression is applied with an ultrasound transducer to the targeted tissue [23]. The tissue mechanical behavior is then determined by mapping the relative deformation (i.e. strain) induced by manual compression (i.e. stress).

Various methods have been proposed over the years to measure the strain. The main approach consists of determining the spatial displacement between a pair of radio-frequency (RF) ultrasound image data, acquired before and after applying an axial compression. The displacement estimation, also known as time-delay estimation, has been historically performed by maximizing a correlation function between local frame windows, either in the time or phase domain [1,2,18,20,22]. More recently, different approaches have added a regularization parameter to account for displacement discontinuity and improve displacement estimates [6,15]. Although these methods demonstrated the ability to make an accurate displacement estimation, current techniques are sensitive to noise and global decorrelation, i.e. the change of speckle appearance due to out-of-plane motion. Furthermore, real-time imaging is an important feature of elastography, and a trade-off is often made between the precision of standard approaches and their computational cost.

The recent progress of learning-based methods in computer vision for optical flow estimation have inspired new approaches for strain elastography [9]. Those methods demonstrated the ability of neural networks to exploit the ultrasound high-frequency content and to robustly produce accurate displacement estimates [14,25]. Previous networks have been trained using ultrasound simulation associated with ground truth displacement and strain labels. Accurate ground-truth images can be difficult to obtain for quasi-static elastography because the magnitude of applied stress is unknown. However, real-world ultrasound data often exhibits complex speckle patterns and echogenic features that can be quite challenging to replicate in ultrasound simulation.

In this paper, we proposed an unsupervised method for time-delay estimation that allows a neural network to be trained directly on clinical data and predict tissue displacement. Unlike previous methods, our CNN training procedure is performed without ground truth labels. Instead, the network weights are optimized by minimizing a dissimilarity function between the pre-compression image and warped compressed image. We also introduce a new regularization term that preserves displacement continuity by directly optimizing the smoothness of the strain prediction. We validated our method by applying it on both real ultrasound data and simulations. The performance of our method was evaluated by comparing our displacement field and strain prediction with the ground truth labels and a gold standard USE technique [17]. To the best of our knowledge, this is the first unsupervised deep-learning method applied to strain imaging.

2 Methods

2.1 Problem Statement

Our method follows an approach similar to a standard image registration framework, which aims to find a spatial transformation that maps a moving image into the space of a reference image. In the case of a non-rigid image registration solution, this transformation can be represented as a dense displacement field (DDF). This transformation is generally optimized in an iterative manner, by maximizing an objective function that measures the similarity between the warped moving image and the reference image, and a regularization parameter to ensure displacement continuity. From a learning-based approach perspective, the image mapping is predicted by a neural network instead of being directly optimized. An overview of the method is presented in Fig. 1.

2.2 Displacement Estimation

Our network takes as input a pair of pre- and post-compression 2D RF frames, here named *Pre* and *Post*, and predicts a DDF u. The network parameters are learned by minimizing a dissimilarity metric L_{sim} between pairs of 2D RF ultrasound data. Our training loss function also includes a regularization term, L_{reg}, acting on the predicted displacement field u and associated with a weighting hyper-parameter α, to ensure balance between the likelihood and smoothness of the predicted transformation. The optimization problem can be written as:

$$\hat{\theta} = \arg\min_{\theta} \left[L_{sim}(Pre, Post; u) \ + \ \alpha \ . \ L_{reg}(u)\right] \tag{1}$$

where θ represents the network parameters that are optimized through stochastic gradient descent.

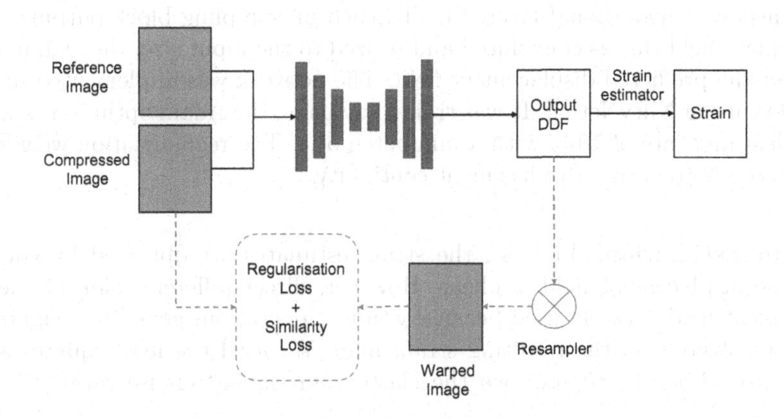

Fig. 1. Overview of the method.

The dissimilarity metric corresponds to a negative local normalized cross-correlation (NCC) which average the NCC score between sliding windows sampled from the pre-compression image and the transformed post-compression image, resampled with the predicted displacement field. The NCC between two local image windows, W_{pre} and W_{post}, with i, j pixel components can be written as:

$$NCC = \frac{1}{N} \sum_{i,j} \frac{\left[W_{pre}(i,j) - \mu_{W_{pre}}\right] \times \left[W_{post}(i,j) - \mu_{W_{post}}\right]}{\sigma_{W_{pre}} \times \sigma_{W_{post}}} \tag{2}$$

where N is the number of pixels (i, j) and μ and σ correspond to the mean and standard deviation of the images, respectively.

The regularisation term consists of the L1-norm of the second spatial derivatives of the predicted displacement field u. Given that the strain modulus corresponds to the displacement gradient, minimizing its second derivative allows to enforce the strain map smoothness. The regularisation term can be written as:

$$L_{reg} = \sum_{i,j} (\mid \partial_x^2 u_{i,j} \mid + \mid \partial_y^2 u_{i,j} \mid) \tag{3}$$

where ∂_x^2 and ∂_y^2 are the second partial derivatives in axial and lateral directions, respectively.

2.3 Implementation

Network. The architecture of our network was presented in a method for medical image registration [8], and consists in an encoder-decoder CNN with skip connections. The encoder part is composed of four down-sampling blocks, which capture the hierarchical features necessary to establish correspondence between the image pair. Each down-sampling block consists of two convolutional layers with a residual network shortcut, batch normalization, and leaky rectified linear unit (Leaky ReLU). Symmetrically, the decoder part is composed of four up-sampling blocks that consists of an additive up-sampling layer summed over a transpose convolutional layer. Finally, each up-sampling block outputs a displacement field that is convoluted and resized to the input size, then summed to output the predicted displacement field. The network was implemented in TensorFlow using NiftyNet [4]. It was trained by using the Adam optimiser, starting at a learning rate of 10-3, with a minibatch of 4. The regularisation weight was set to $\alpha = 2$ to ensure displacement continuity.

Strain Estimation. In USE, the strain estimates are obtained by computing the displacement field gradient. However, direct differentiation of the displacement field is rarely used because gradient operations generate a significant amount of noise in the resulting strain map. We used the least-squares strain estimator (LSQSE) to improve the elastogram signal-to-noise ratio [13]. The strain estimation was also implemented in TensorFlow to facilitate efficient parallel computing. In inference, the strain map prediction rate reach a total of 13 images per second on a 12 GB NVIDIA GTX-1080ti GPU.

3 Experiments

The performance of our method was evaluated on ultrasound simulations and *in vivo* data. We compared our results with a state-of-the-art strain elastography method called "tOtal Variation Regularization and WINDow-based time delay estimation" (OVERWIND) [17]. The OVERWIND results were obtained by using the publicly available MATLAB implementation, and default parameters were chosen for the simulation experiment. The regularisation coefficients were manually tuned for the *in vivo* data to globally maximize the NCC and Contrast-to-Noise Ratio (CNR) scores of the three cases. The results on simulation were also compared with strain estimates obtained by training our network with a supervised loss function. The supervised loss function was used before for time-delay estimation [25], and corresponded to the mean absolute difference (MAE) between the network prediction and the ground truth labels. The quality of the strain estimates were assessed with the CNR and Signal-to-Noise Ratio (SNR), which are metrics commonly used in USE [17,24].

Simulation Dataset. The Field-II software [10,11] was used to generate the ultrasound images. Each simulation consisted in a 3D rectangle of size 38×40 mm, containing a cylindrical-shaped inclusion with a randomly assigned diameter (from 8 to 12 mm) and position. The speckle pattern typically observed in ultrasound imaging was obtained by randomly assigning a total of 400,000 scatterers across each digital phantom. The axial compression was assigned randomly to each phantom and represented between 0.5% and 4% of the phantom total length. The Young's modulus of the inclusion was set to different values (i.e. 8, 15, 45 and 75 kPa) while the background was fixed to 25 kPa. Tissue displacements were estimated by finite element method using the Partial Differential Equation Toolbox from MATLAB, and were used to interpolate the scatterers position. A total of 192 RF lines were simulated for each image, with probe central and sampling frequency of 7 and 40 MHz, respectively. The background scatterers were associated with a random intensity value from a Gaussian distribution, to mimic homogeneous tissue. To increase the network's robustness to noise and image intensities, the inclusion scatterers intensities were either assigned to 0 or similar to the background. Moreover, white Gaussian noise with random signal power values, from 5 to 20 dBw, was added on each image to increase robustness to noise. Finally, 160 ultrasound image pairs were simulated, where 100 were used for training and 30 for validation and testing, respectively.

***In Vivo* Dataset.** We created our own *in vivo* dataset by collecting images of the arms in three human volunteers. The RF data was acquired from a Cicada 128PX system equipped with 10 MHz linear probe from Cephasonics (Cephasonics Inc., USA). The images were reconstructed using the delay-and-sum beamformer from SUPRA [5]. Experimental protocol consisted in acquiring sequences of images while slowly applying an axial compression on the volunteer's arm with the handheld ultrasound probe. The *in vivo* dataset included 1300 image pairs for training and 300 pairs for validation and testing. The three image pairs presented in the results section were taken from the testing partition and exhibit one or several blood vessels located in the arm.

3.1 Results

Figure 2 shows axial strain images estimated from three simulated image pairs and obtained with the finite element method, OVERWIND and both supervised and unsupervised models. The averaged CNR and SNR values of the entire testing dataset (i.e., 30 image pairs) are displayed in Table 1.

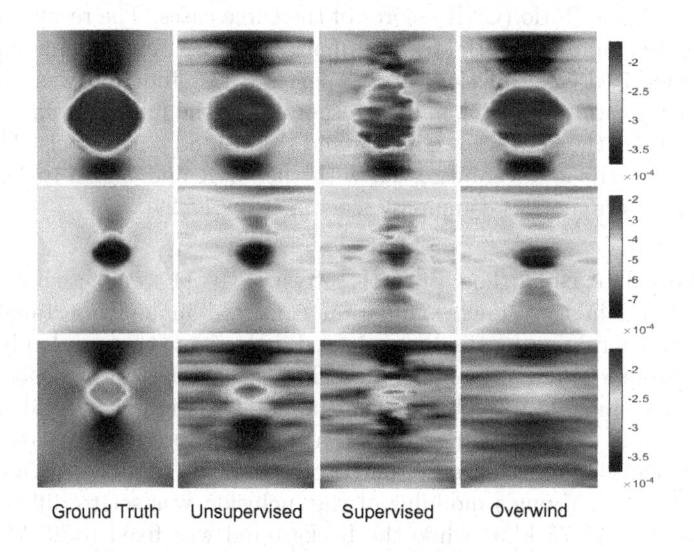

Fig. 2. Comparison of three axial strain fields computed from ultrasound simulations by finite element method (i.e., ground truth), OVERWIND, the unsupervised and supervised network. OVERWIND's regularisation parameters: $\alpha 1 = \beta 1 = 20$ and $\alpha 2 = \beta 2 = 8$.

Table 1. Mean and standard deviation of SNR and CNR for the strain images of the simulation testing dataset obtained with finite element method, OVERWIND, the unsupervised and supervised network.

	Ground truth (mean std)	Unsupervised (mean std)	Supervised (mean std)	OVERWIND (mean std)
SNR	7.51 (2.61)	6.95 (2.54)	7.79 (2.01)	9.31 (3.51)
CNR	9.15 (2.73)	7.70 (3.8)	4.22 (2.08)	6.33 (3.6)

Figure 3 shows axial strains estimated from three different *in vivo* image pairs. CNR and SNR values of each case can be found in Table 2. The local NCC values between the post-compression and resampled pre-compression images are also displayed, to indicate the quality of the predicted displacement fields. The mean of the three CNR values are 7 and 8.43 for our unsupervised model and OVERWIND, respectively. The average SNR values are 0.31 and 0.7 for our method and OVERWIND, respectively. Displacement field estimates from both the simulation and *in vivo* experiments are available in the supplementary material.

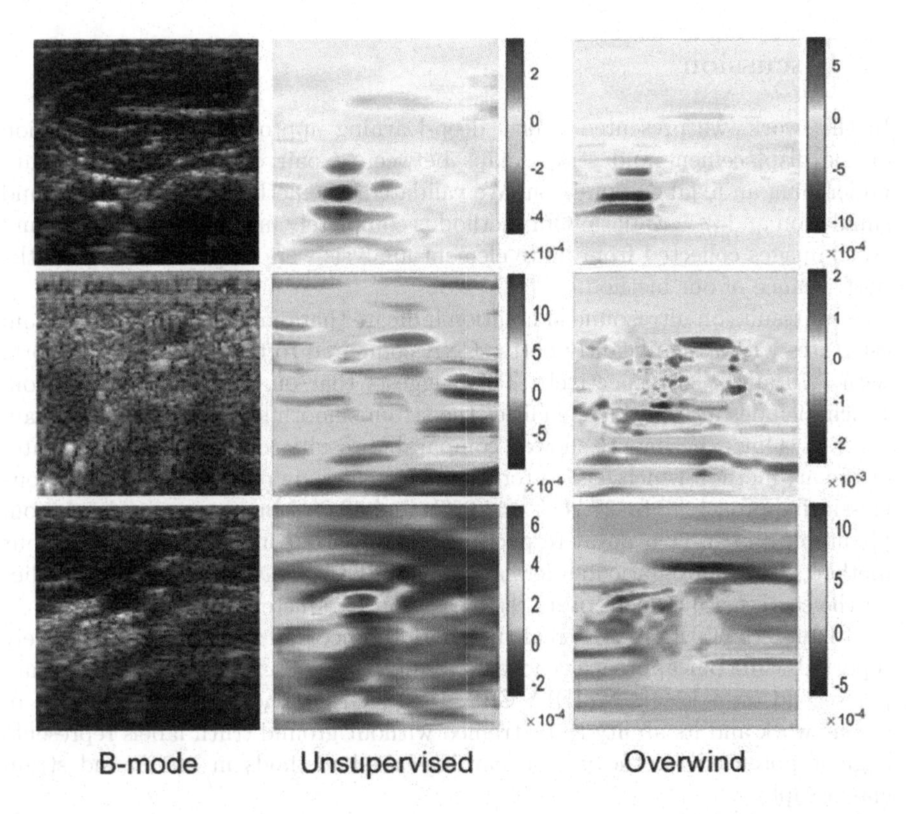

Fig. 3. Comparison of axial strains estimated by our method and OVERWIND on three image pairs from the *in vivo* testing dataset. OVERWIND's regularisation parameters: $\alpha1 = \beta1 = 0.2$ and $\alpha2 = \beta2 = 0.05$.

Table 2. SNR and CNR of strain estimates from our unsupervised method and OVERWIND, and local NCC scores for the three image pair results from the *in vivo* dataset.

	Unsupervised			OVERWIND		
	Case 1	Case 2	Case 3	Case 1	Case 2	Case 3
SNR	0.14	0.07	0.1	0.12	0.35	0.23
CNR	6.22	5.2	4.5	4.9	3.15	2.23
LNCC	0.87	0.81	0.94	0.90	0.81	0.93

4 Discussion

In this work, we presented a new deep-learning approach for the estimation of the displacement and strain maps between a pair of ultrasound RF data undergoing an axial compression. We validated our method on both ultrasound simulation and *in vivo* data. Our method is completely unsupervised and ground truth images collected from finite element analysis were only used to assess the performance of our method.

Our results on ultrasound simulation indicate that our method predicts strain estimates with a significantly better CNR compared to the supervised network, with 7.70 and 4.20 respectively. This suggests that our training loss function, which includes a term that penalizes the strain smoothness, improves the strain contrast. Our experiments showed comparable results to OVERWIND, a state-of-the-art method which has, in terms of CNR, already outperformed a previous classical approach [6] by 27.26%, 144.05%, and 49.90% on average in simulation, phantom, and in-vivo data, respectively, as reported in [17]. In addition, our method is fully automatic in inference while OVERWIND strain estimation relies on the correct adjustment of its regularisation parameters.

Finally, the OVERWIND real-time performance had not been quantitatively reported while our approach reached a strain prediction rate of about 13 frames per second on a 12 GB NVIDIA GTX-1080ti GPU. The real-time inference of our network and its ability to be trained without ground truth labels represents a great potential for the use of learning-based methods in ultrasound strain elastography.

Acknowledgement. This work was supported by the EPSRC [NS/A000049/1], [NS/A000050/1], [EP/L016478/1] and the Wellcome Trust [203148/Z/16/Z], [203145/Z/16/Z]. Tom Vercauteren is supported by a Medtronic/RAEng Research Chair [RCSRF1819/7/34].

References

1. Ara, S.R., et al.: Phase-based direct average strain estimation for elastography. IEEE Trans. Ultrason. Ferroelectr. Freq. Control **60**(11), 2266–2283 (2013). https://doi.org/10.1109/TUFFC.2013.6644732

2. Chen, H., Shi, H., Varghese, T.: Improvement of elastographic displacement estimation using a two-step cross-correlation method. Ultrasound Med. Biol. **33**(1), 48–56 (2007). https://doi.org/10.1016/j.ultrasmedbio.2006.07.022
3. Correas, J.M., Tissier, A.M., Khairoune, A., Khoury, G., Eiss, D., Hélénon, O.: Ultrasound elastography of the prostate: state of the art. Diagn. Interv. Imaging **94**(5), 551–560 (2013). https://doi.org/10.1016/j.diii.2013.01.017
4. Gibson, E., et al.: NiftyNet: a deep-learning platform for medical imaging. Comput. Methods Programs Biomed. **158**, 113–122 (2018). https://doi.org/10.1016/j.cmpb.2018.01.025
5. Göbl, R., Navab, N., Hennersperger, C.: SUPRA: open-source software-defined ultrasound processing for real-time applications. Int. J. Comput. Assist. Radiol. Surg. **13**(6), 759–767 (2018). https://doi.org/10.1007/s11548-018-1750-6
6. Hashemi, H.S., Rivaz, H.: Global time-delay estimation in ultrasound elastography. IEEE Trans. Ultrason. Ferroelectr. Freq. Control **64**(10), 1625–1636 (2017). https://doi.org/10.1109/TUFFC.2017.2717933
7. Hiltawsky, K.M., Krüger, M., Starke, C., Heuser, L., Ermert, H., Jensen, A.: Freehand ultrasound elastography of breast lesions: clinical results. Ultrasound Med. Biol. **27**(11), 1461–1469 (2001). https://doi.org/10.1016/S0301-5629(01)00434-3
8. Hu, Y., et al.: Weakly-supervised convolutional neural networks for multimodal image registration. Med. Image Anal. **49**, 1–13 (2018). https://doi.org/10.1016/j.media.2018.07.002
9. Ilg, E., Mayer, N., Saikia, T., Keuper, M., Dosovitskiy, A., Brox, T.: FlowNet 2.0: evolution of optical flow estimation with deep networks. In: Proceedings of the IEEE Conference on Computer Vision and Pattern Recognition, pp. 2462–2470 (2017). https://doi.org/10.1109/CVPR.2017.179
10. Jensen, J.A.: FIELD: a program for simulating ultrasound systems. Med. Biol. Eng. Comput. **34**, 351–352 (1996)
11. Jensen, J.A., Svendsen, N.B.: Calculation of pressure fields from arbitrarily shaped, apodized, and excited ultrasound transducers. IEEE Trans. Ultrason. Ferroelectr. Freq. Control **39**(2), 262–267 (1992). https://doi.org/10.1109/58.139123
12. Jeong, W.K., Lim, H.K., Lee, H.K., Jo, J.M., Kim, Y.: Principles and clinical application of ultrasound elastography for diffuse liver disease. Ultrasonography **33**(3), 149–160 (2014). https://doi.org/10.14366/usg.14003
13. Kallel, F., Ophir, J.: A least-squares strain estimator for elastography. Ultrason. Imaging **19**(3), 195–208 (1997). https://doi.org/10.1177/016173469701900303
14. Kibria, M.G., Rivaz, H.: GLUENet: ultrasound elastography using convolutional neural network. In: Stoyanov, D., et al. (eds.) POCUS/BIVPCS/CuRIOUS/CPM-2018. LNCS, vol. 11042, pp. 21–28. Springer, Cham (2018). https://doi.org/10.1007/978-3-030-01045-4_3
15. Kuzmin, A., Zakrzewski, A.M., Anthony, B.W., Lempitsky, V.: Multi-frame elastography using a handheld force-controlled ultrasound probe. IEEE Trans. Ultrason. Ferroelectr. Freq. Control **62**(8), 1486–1500 (2015). https://doi.org/10.1109/TUFFC.2015.007133
16. Kwak, J.Y., Kim, E.K.: Ultrasound elastography for thyroid nodules: recent advances. Ultrasonography **33**(2), 75–82 (2014). https://doi.org/10.14366/usg.13025
17. Mirzaei, M., Asif, A., Rivaz, H.: Combining total variation regularization with window-based time delay estimation in ultrasound elastography. IEEE Trans. Med. Imaging **38**(12), 2744–2754 (2019). https://doi.org/10.1109/TMI.2019.2913194

18. Ophir, J., Cespedes, I., Garra, B., Ponnekanti, H., Huang, Y., Maklad, N.: Elastography: ultrasonic imaging of tissue strain and elastic modulus in vivo. Eur. J. Ultrasound **3**(1), 49–70 (1996). https://doi.org/10.1016/0929-8266(95)00134-4
19. Ophir, J., Céspedes, I., Ponnekanti, H., Yazdi, Y., Li, X.: Elastography: a quantitative method for imaging the elasticity of biological tissues. Ultrason. Imaging **13**(2), 111–134 (1991). https://doi.org/10.1177/016173469101300201
20. Shi, H., Varghese, T.: Two-dimensional multi-level strain estimation for discontinuous tissue. Phys. Med. Biol. **52**(2), 389–401 (2007). https://doi.org/10.1088/0031-9155/52/2/006
21. Sigrist, R.M., Liau, J., Kaffas, A.E., Chammas, M.C., Willmann, J.K.: Ultrasound elastography: review of techniques and clinical applications. Theranostics **7**(5), 1303–1329 (2017). https://doi.org/10.7150/thno.18650
22. Varghese, T., Konofagou, E.E., Ophir, J., Alam, S.K., Bilgen, M.: Direct strain estimation in elastography using spectral cross-correlation. Ultrasound Med. Biol. **26**, 1525–1537 (2000). https://doi.org/10.1016/S0301-5629(00)00316-1
23. Varghese, T.: Quasi-static ultrasound elastography. Ultrasound Clin. **4**(3), 323–338 (2009). https://doi.org/10.1016/j.cult.2009.10.009
24. Wang, J., Huang, Q., Zhang, X.: Ultrasound elastography based on the normalized cross-correlation and the PSO algorithm. In: 2017 4th International Conference on Systems and Informatics (ICSAI), pp. 1131–1135. IEEE (2017). https://doi.org/10.1109/ICSAI.2017.8248455
25. Wu, S., Gao, Z., Liu, Z., Luo, J., Zhang, H., Li, S.: Direct reconstruction of ultrasound elastography using an end-to-end deep neural network. In: Frangi, A.F., Schnabel, J.A., Davatzikos, C., Alberola-López, C., Fichtinger, G. (eds.) MICCAI 2018. LNCS, vol. 11070, pp. 374–382. Springer, Cham (2018). https://doi.org/10.1007/978-3-030-00928-1_43

Automatic Probe Movement Guidance for Freehand Obstetric Ultrasound

Richard Droste[1(✉)], Lior Drukker[2], Aris T. Papageorghiou[2], and J. Alison Noble[1]

[1] Institute of Biomedical Engineering, University of Oxford, Oxford, UK
richard.droste@eng.ox.ac.uk

[2] Nuffield Department of Women's and Reproductive Health, University of Oxford, Oxford, UK

Abstract. We present the first system that provides real-time probe movement guidance for acquiring standard planes in routine freehand obstetric ultrasound scanning. Such a system can contribute to the worldwide deployment of obstetric ultrasound scanning by lowering the required level of operator expertise. The system employs an artificial neural network that receives the ultrasound video signal and the motion signal of an inertial measurement unit (IMU) that is attached to the probe, and predicts a guidance signal. The network termed US-GuideNet predicts either the movement towards the standard plane position (goal prediction), or the next movement that an expert sonographer would perform (action prediction). While existing models for other ultrasound applications are trained with simulations or phantoms, we train our model with real-world ultrasound video and probe motion data from 464 routine clinical scans by 17 accredited sonographers. Evaluations for 3 standard plane types show that the model provides a useful guidance signal with an accuracy of 88.8% for goal prediction and 90.9% for action prediction.

Keywords: Fetal ultrasound · Probe guidance · Ultrasound navigation

1 Introduction

Ultrasound scanning is an indispensable diagnostic tool in obstetrics due to its safety, real-time results and low cost. At the same time, many women in developing countries do not receive a single ultrasound examination throughout their pregnancy due to a lack of skilled operators [20]. The main tasks of ultrasound scanning are the acquisition, examination/verification and interpretation of pre-defined standard anatomical planes that enable the detection of fetal abnormalities. Systems that provide assistance for or automate these tasks have the potential of enabling worldwide access to ultrasound scanning by reducing the level of necessary expertise. Standard plane examination/verification and interpretation are largely standardized [18], can be performed remotely [3], and

A. L. Martel et al. (Eds.): MICCAI 2020, LNCS 12263, pp. 583–592, 2020.
https://doi.org/10.1007/978-3-030-59716-0_56

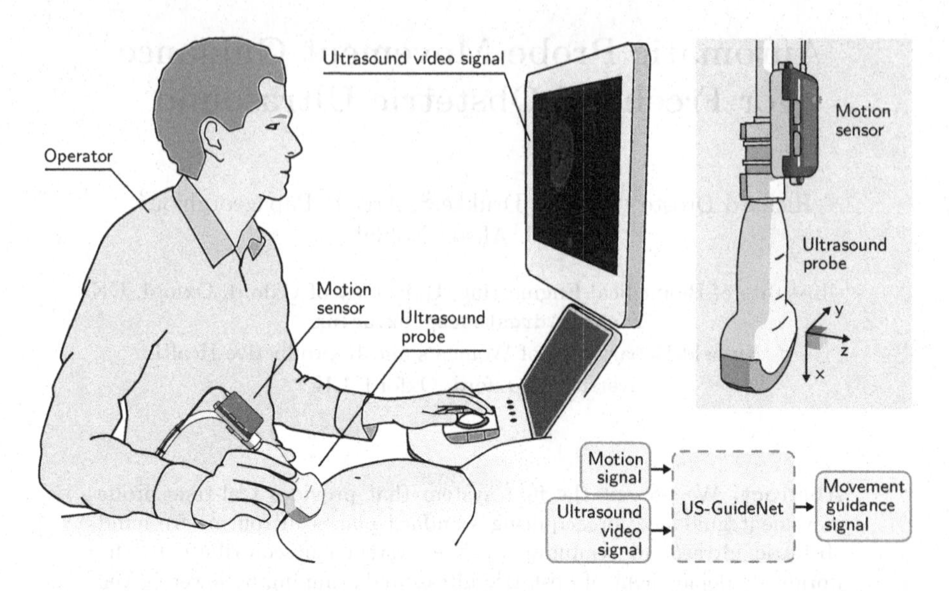

Fig. 1. System overview. *Left*: An operator performs ultrasound scanning with a routine clinical setup while the motion of the probe is recorded with an IMU. *Bottom right*: The *US-GuideNet* receives the IMU motion signal and the ultrasound video signal as inputs and outputs a real-time probe movement guidance signal. *Top right*: Attachment of the IMU to the ultrasound probe and IMU coordinate system.

can be facilitated through automated image analysis [24]. Freehand standard plane acquisition, on the other hand, is harder to facilitate/automate since it is not standardized and requires interaction with the mother. It demands years of training and is the rate-limiting step even for experienced sonographers [1].

To address this issue, we present the first system that provides real-time probe movement guidance for fetal standard plane acquisition in routine freehand obstetric ultrasound scanning. An overview of the system is presented in Fig. 1. An artificial neural network termed *US-GuideNet* receives the ultrasound video signal alongside the signal of a motion sensor that is attached to the ultrasound probe, and outputs probe movement guidance that directs the operator towards the desired standard plane. No specialized equipment is required: The motion sensor is a common inertial measurement unit (IMU) that is attached to the probe of a standard clinical ultrasound machine. Further, the *US-GuideNet* neural network is designed to be extremely lightweight and can run real time inference on a CPU. Behavioral cloning (BC), a type of imitation learning, has emerged as a powerful technique to train neural networks to perform complex real-world tasks such as autonomous driving [14]. Here, we collect 5079 demonstrations of standard plane acquisitions from 464 2nd- and 3rd-trimester scans acquired by 17 accredited sonographers and implement two settings of BC: 1) For *goal prediction*, the network predicts the movement that leads directly to the estimated position of the standard plane. 2) For *action prediction*, the network predicts the next movement that the expert would perform.

2 Related Work

Various approaches have been proposed to address the difficulty of ultrasound standard plane acquisition.

Robotic Ultrasound. Human-controlled robotic systems have been developed that allow experienced sonographers to perform obstetric ultrasound exams remotely [22]. Automated robotic systems have been proposed for highly structured tasks such as finding planes of motionless objects [8,11] or the human liver [13]. However, despite on-going efforts [23], no robotic system has been proposed that can automate the complex task of obstetric ultrasound scanning.

Simplified Acquisition Protocols. Instead of assisting operators to acquire typical freehand 2D standard planes, previous work has proposed to automatically extract standard planes from data that are acquired with a simplified protocol, such as 3D ultrasound volumes [7,17] or linear sweeps over the maternal abdomen [10]. Moreover, IMUs have been used to acquire 3D ultrasound volumes with 2D probes [5,16]. However, these methods are applicable only for a subset of standard planes (fetal abdomen and head) and the standard plane quality is not up to par with typical freehand scanning.

Phantoms and Simulated Environments. Recent studies have proposed learning based systems that are trained to acquire ultrasound planes in simplified environments. One study proposes an algorithm that learns to find a view of the adult heart in a grid of pre-acquired ultrasound images [12]. Moreover, learning-based systems have been proposed in which a robotic actuator finds predefined views of simple tissue phantoms [6] or a fetal US phantom [21]. However, a fetus in the mother's womb is a dynamic and highly variant object that can not be well-represented with static simulations or a phantom. Furthermore, these algorithms are purely image-based and therefore rely on the exact execution of the predicted actions, which is only possible within a simulation or with a robotic system. Here, we train a guidance algorithm with video and probe motion data from a large number of real-world expert demonstrations from routine scanning. Moreover, our algorithm receives the real-time probe motion signal and can therefore react to the movements of a human operator.

3 Method

Figure 1 presents an overview of the proposed system. An operator performs routine obstetric ultrasound scanning with a standard clinical machine. An inertial measurement unit (IMU) motion sensor is attached to the ultrasound probe and an on-board attitude and heading reference system (AHRS) estimates the sensor's orientation in the earth coordinate system. The motion sensor signal and

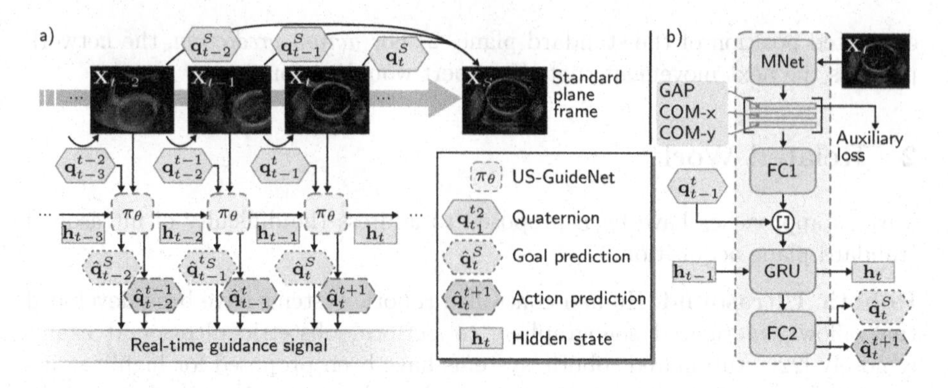

Fig. 2. a) Proposed behavioral cloning framework. b) *US-GuideNet* architecture.

the machine video signal are input into a neural network, *US-GuideNet*, that outputs a 3D rotation of the probe that guides the operator towards the standard plane. The network training method is described in Sect. 3.1, the network architecture is detailed in Sect. 3.2 and implementation details are provided in Sect. 3.3.

3.1 Learning from Expert Demonstrations

We pose the problem of training a neural network to predict a probe guidance signal as a behavioral cloning problem. That is, we record standard plane acquisition demonstrations from several experts for a large number of patients and train the network to replicate the demonstrated behavior. In general, a standard plane acquisition consists of live B-mode scanning followed by *freezing* the ultrasound video and optionally selecting a previous frame with the desired appearance from a *cine-buffer*. We define the finally selected frame as the standard plane.

Problem Formulation. Figure 2 a) presents the formulation of the learning problem. Let $\{\mathbf{X}_t \in \mathbb{R}^{H \times W} | t \in \mathcal{T}\}$ be ultrasound video frames of a standard plane acquisition, temporally downsampled to 6Hz, with resolution $H \times W$ and frame indices $\mathcal{T} = \{i\}_{i=0}^{F}$, where F is the *freeze* frame index. Moreover, let $\mathbf{X}_S$ be the standard plane with index $S \in \mathcal{T} \setminus \{0\}$. Finally, let $\mathbf{q}_t = [q_w, q_x, q_y, q_z]^\top$ be the probe orientation quaternion of frame $\mathbf{X}_t$ and $\mathbf{q}_t^{t_2} := \mathbf{q}_t^* \mathbf{q}_{t_2}$ the probe rotation quaternion from frame $\mathbf{X}_t$ to frame $\mathbf{X}_{t_2}$, where $\mathbf{q}^*$ is the conjugate. We represent orientations with quaternions since they can be smoothly interpolated without discontinuities or singularities, and are numerically stable and computationally efficient [15]. Euler angles, in contrast, another popular representation of rotations, suffer from discontinuities such as the *gimbal lock*. We do not consider probe translation in this work since the IMU is not suitable to estimate it accurately.

Behavioral Cloning. We train a policy network $\pi_\theta : s_t \mapsto u_t$ termed *US-GuideNet* with parameters θ that maps the state s_t at time step $t \in \mathcal{T}$ to an action u_t. We define the state as the tuple $s_t := (\mathbf{X}_t, \mathbf{q}_{t-1}^t, \mathbf{h}_{t-1})$, where $\mathbf{h}_{t-1}$ is the hidden state of a recurrent neural network within π_θ. We explore the two different settings for the action u_t: *goal prediction* and *action prediction.* For *goal prediction*, the policy $\pi_\theta^g : s^t \mapsto \hat{\mathbf{q}}_t^S$ estimates the rotation from the current orientation to the orientation of the standard plane. If the estimated standard plane orientation is accurate, this policy is optimal, *i.e.*, it guides the operator directly to the standard plane. However, it is not guaranteed that enough information has been seen at time t for an accurate estimation of the standard plane orientation. Therefore, we explore a second setting, *action prediction*, where the policy $\pi_\theta^a : s_t \mapsto \hat{\mathbf{q}}_t^{t+1}$ estimates the next rotation that the operator would perform. This policy aims to closely mimic the expert sonographer behavior.

Loss Function. During training, a demonstration is constructed from a subset of indices $\mathcal{T}_D \subset \mathcal{T}$ with start and end indices t_0 and T. Let $\hat{\mathbf{Q}} := [\hat{\mathbf{q}}_t^{t_2}]_{t=t_0}^T$ and $\mathbf{Q} := [\mathbf{q}_t^{t_2}]_{t=t_0}^T$ be the predicted and ground truth rotation quaternion sequences, with $t_2 \in \{S, t+1\}$ for *goal prediction* and *action prediction* respectively. We add an auxiliary output after the MNet in order to facilitate and regularize its training: Since we want the MNet to recognize the appearance of standard planes, we input the average pooled MNet features into a softmax layer that predicts the class probabilities of the SonoNet standard plane classifier [2] for each frame. Let $\hat{\mathbf{P}} = [\hat{\mathbf{p}}_t]_{t=t_0}^T$ be the auxiliary softmax output and $\mathbf{Y} = [\mathbf{y}_t]_{t=t_0}^T$ the SonoNet class probabilities, with $\hat{\mathbf{p}}_t, \mathbf{y}_t \in \mathbb{R}_{\geq 0}^{14}$. The total training loss $\mathcal{L}$ is

$$\mathcal{L} = \sum_{t=t_0+W}^{T} \Big\{ \underbrace{-\frac{1}{\|\hat{\mathbf{q}}_t^{t_2}\|} \hat{\mathbf{q}}_t^{t_2} \cdot \mathbf{q}_t^{t_2}}_{\text{Similarity loss}} + \underbrace{\alpha \left(1 - \|\hat{\mathbf{q}}_t^{t_2}\|^2\right)^2}_{\text{Norm loss}} \Big\} \underbrace{- \beta \sum_{t=t_0}^{T} \mathbf{y}_t^\top \operatorname{diag}(\mathbf{w}) \log(\mathbf{p}_t)}_{\text{Auxiliary loss}}$$

where $\cdot$ denotes the dot-product, $\alpha, \beta \in \mathbb{R}_{>0}$ are scalar weighting factors, $\mathbf{w} \in \mathbb{R}_{\geq 0}^{14}$ is a weight vector that balances the SonoNet class probabilities, and W is a warm up time for the rotation prediction.

3.2 US-GuideNet Architecture

The *US-GuideNet* policy network receives the ultrasound video and probe motion signals and outputs predicted expert probe rotations as described in Sect. 3.1. We design the architecture for small time and space computational complexity (runtime and model size) such that it can run real-time inference on the CPU of an inexpensive computer. The network architecture is illustrated in Fig. 2 b). At each time step t, the ultrasound video frame X_t is fed into a MobileNet V2 (MNet) convolutional neural network [19], which consists of a cascade of lightweight depthwise-separable and pointwise convolutions. We use MNet with a width-multiplier of 0.5, *i.e.*, 50% reduced number of channels. Next, the dimensionality of the MNet output is reduced with a custom *ConcatPool* operation that preserves both semantic and spatial information by concatenating global average pooled (GAP) features with the x and y coordinates of the centers of mass (COM-x/y) of the feature maps. After reducing the features to

128 channels with a fully-connected layer *FC1*, they are concatenated with the current probe rotation quaternion $\mathbf{q}_{t-1}^t$ and input into a gated recurrent unit [4] (GRU) with 132 input channels and 128 hidden channels. Finally *FC2*, a fully-connected layer with one 128-channel hidden layer, outputs the 4-dimensional probe rotation quaternion.

3.3 Experimental Setup

Data Acquisition. The data were acquired as part of the PULSE (Perception Ultrasound by Learning Sonographic Experience) project, a prospective study of routine fetal ultrasound scans performed in all trimesters by accredited sonographers and fetal medicine doctors at the maternity ultrasound unit, Oxford University Hospitals NHS Foundation Trust, Oxfordshire, United Kingdom. The exams were performed on a GE Voluson E8 scanner (General Electric, USA) while the video signal of the machine monitor was recorded lossless at 30 Hz. The motion of each of two curved linear array transducer (2D) probes was recorded with a NGIMU IMU/AHRS (x-io Technologies Ltd., UK). Each motion sensor was attached to the cable outlet of the probe with a custom 3D-printed mount as shown in Fig. 1. The probe orientation quaternions were sampled at 400Hz. This study was approved by the UK Research Ethics Committee (Reference 18/WS/0051) and written informed consent was given by all participating pregnant women and operators. In this paper we use ultrasound video and corresponding gaze data of 464 s and third trimester scans acquired by 17 accredited sonographers between May 2018 and February 2020.

Data Processing. We extracted the standard plane acquisitions from the ultrasound scans with a purpose-built program based on optical character recognition. For each of the 5079 resulting acquisitions, the program outputs the corresponding live B-Mode scanning segment, the *freeze* frame and the *cine-buffer*-corrected standard plane. In addition, the program labels acquisitions of the biometry standard planes: the femur standard plane (FSP), the abdominal standard plane (ASP) and the trans-ventricular plane (TVP) [18], which we use for evaluation. The acquisition duration was limited to 10s before the standard plane. We automatically corrected any lag between the video and motion signals by correlating frame differences with probe motion, and manually verified the synchronization. The video frames were cropped such that the ultrasound machine's graphical user interface was removed, and normalized to zero-mean and unit-variance. The scans are divided into 80% for training and 20% for testing.

Training. For each training epoch, a demonstration of 32 frames is randomly selected from each standard plane acquisition, which corresponds to a duration 5.3s at 6Hz. It is ensured that $\min_t\{\mathbf{q}_t \cdot \mathbf{q}_S\} \geq 0.7$ for each demonstration. The frames are augmented by randomly changing of the brightness, contrast and gamma by ±10% and randomly symmetrically cropping up to 20% of the frame border. The frames are then down-sampled to the network input resolution of 224×288. The MNet is pre-trained via the auxiliary loss with a large number of ultrasound frames. The entire *US-GuideNet* neural network is then trained from the demonstrations for 20 epochs with the AdamW optimizer [9] with weight decay of 10^{-2} and initial learning rate of 0.001, which is decayed by a factor of 0.1 every 8 epochs. The batch size is set to 8 and the warm up time for the rotation loss to 1s. After training with all demonstrations, the model is fine-tuned for each evaluation plane (FSP, ASP and TVP) separately for 16 epochs.

Evaluation and Baseline. We evaluate the trained model on the full-length standard plane acquisitions (clipped to 10s before the standard plane). For each time step, we classify the predicted probe rotation as correct if and only if $\angle(\mathbf{q}_t\hat{\mathbf{q}}_t^{t_2}, \mathbf{q}_{t_2}) < \angle(\mathbf{q}_t, \mathbf{q}_{t_2}), t_2 \in \{S, t+1\}$, *i.e.*, if applying the predicted rotation reduces the angle to the target orientation. As before, $t_2 = S$ for *goal prediction* and $t_2 = t + 1$ for *action prediction*. As a baseline rotation prediction we use $\mathbf{q}_{t-1}^{t}$, *i.e.*, continuing in the current direction of rotation at each time step.

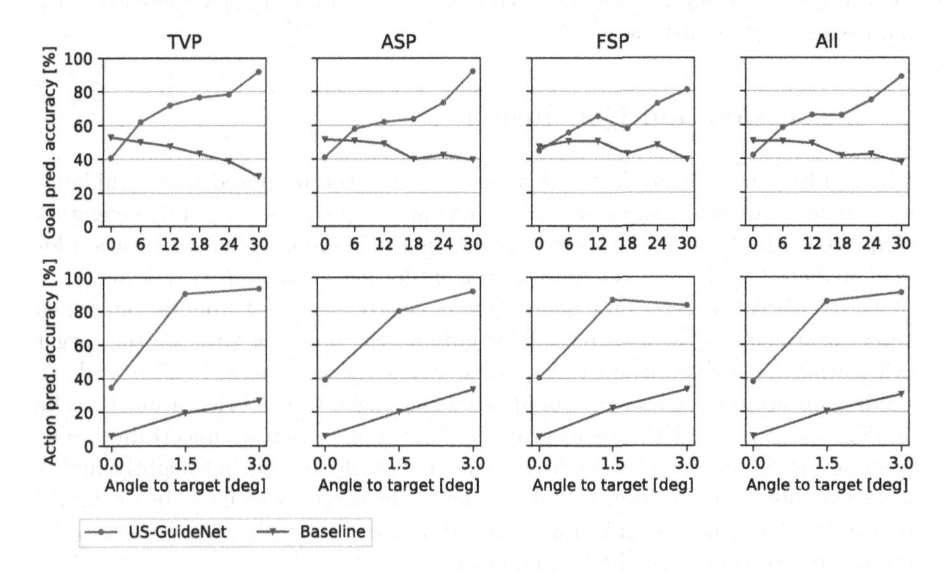

Fig. 3. Experimental results for the evaluated standard planes: TVP (head), ASP (abdomen) and FSP (femur). In addition, the overall accuracies are provided.

4 Results

The experimental results are shown in Fig. 3. The average accuracy of the guidance signal and baseline is evaluated for different ranges of the angular distance to the target (standard plane orientation for *goal prediction* or next probe position for *action prediction*). This enables the separation of the performance for coarse (large angular distance) and fine (low angular distance) adjustments. The x-axis of the individual plots provides the lower limits of the ranges, which extend to the next-higher x-axis value. Across the *action prediction* and *goal prediction* settings and all evaluated standard plane types, a common pattern can be observed that the accuracy of the guidance signal tends to increase with increasing angular distance to the standard plane.

Goal Prediction. The *goal prediction* accuracies are given in the upper row of Fig. 3. The guidance signal performs better than the baseline for any angle range >6°. The accuracy of the guidance signal increases with increasing angular distances to the standard plane, ranging from 42.2% for angles range0° to 6° to 88.8% for angles >30°, with 81.0% for the FSP and 92.1% for the ASP. The average baseline accuracy slightly declines towards higher angular distances.

Action Prediction. The guidance signal accuracy is higher than the baseline accuracy for all target distance ranges. The average guidance signal accuracy increases from 38.0% for angles 0° to 1.5° to 90.9% for distances >3°. At angles >3°, the largest accuracy is observed for the TVP with 93.3% and the lowest for the FSP with 83.3%. The average baseline accuracy slightly increases with increasing angular distances.

5 Discussion and Conclusion

The results presented in Sect. 4 demonstrate that the proposed probe guidance system for obstetric ultrasound scanning indeed provides a useful navigation signal towards the respective target, which is the standard plane orientation for *goal prediction* and the next expert movement for *action prediction*. The accuracy of *US-GuideNet* increases for larger differences to the target orientation, which shows that the algorithm is robust for guiding the operator towards the target orientation from distant starting points. For small distances, it is difficult to predict an accurate guidance signal since the exact target orientation may be subject to inter- and intra-sonographer variations or sensor uncertainty. The accuracy is similar for *goal prediction* and *action prediction* but slightly higher for action prediction at intermediate angles to the target, which can be explained by the fact that the action is always based on the previously seen frames, while the target position might be yet unknown.

The guidance signal accuracies are generally the highest for the abdominal and head standard planes (ASP and TVP). The accuracy for the femur standard plane (FSP) is slightly lower, which can be explained by the fact the femur is part of an extremity and therefore subject to more fetal movement, which can make its final position unpredictable. Moreover, the FSP is defined via two anatomical landmarks—the distal and proximal ends of the femur—while the ASP and TVP are determined by the appearance of more anatomical structures [18]. This might make it more difficult for the model to predict the FSP position that was chosen by the operator, since it is subject to more degrees of freedom.

A limitation of our study is that we test our algorithm with pre-acquired data. However, in contrast with previous work [6,12,21] which uses simulations or phantoms, our proposed system is trained and evaluated on data from real-world routine ultrasound scanning. Moreover, instead of relying on the exact execution of the probe guidance as in previous work [6,12,21], our system reacts to the actual operator probe movements that are sensed with an IMU. This suggests that the system will perform well in future tests on volunteer subjects. In general, the accuracy of the predictions of *US-GuideNet* is evident from the large improvements over the baseline of simply continuing the current direction of rotation. While probe translation is not predicted due to IMU limitations, only the through-plane sweeping translation would usually change the view of the fetus while the sideways sliding and downwards/upwards translations would shift the fetal structure within the ultrasound image. In combination with the rotation guidance, this leaves one degree of freedom to be determined by the operator.

In conclusion, this paper presents the first probe movement guidance system for the acquisition of standard planes in freehand obstetric ultrasound scanning. Moreover, it is the first guidance system for any application of ultrasound standard plane acquisition that is trained with video and probe motion data from routine clinical scanning. Our experiments have shown that the proposed *US-GuideNet* network and behavioral cloning framework result in an accurate guidance system. These results will serve as a foundation for subsequent validation studies with novice operators. The proposed algorithm is lightweight which facilitates the deployment for existing ultrasound machines.

Acknowledgements. We acknowledge the ERC (ERC-ADG-2015 694581, project PULSE) and the NIHR Oxford Biomedical Research Centre.

References

1. Bahner, D.P., et al.: Language of transducer manipulation. J. Ultrasound Med. **35**(1), 183–188 (2016)
2. Baumgartner, C.F., et al.: SonoNet: real-time detection and localisation of fetal standard scan planes in freehand ultrasound. IEEE Trans. Med. Imag. **36**(11), 2204–2215 (2017)
3. Britton, N., Miller, M.A., Safadi, S., Siegel, A., Levine, A.R., McCurdy, M.T.: Tele-ultrasound in resource-limited settings: a systematic review. Front. Public Health **7**, 244 (2019)
4. Cho, K., van Merrienboer, B., Bahdanau, D., Bengio, Y.: On the properties of neural machine translation: encoder-decoder approaches. In: Eighth Workshop on Syntax, Semantics and Structure in Statistical Translation (SSST-8), pp. 103–111 (2014)
5. Housden, R., Treece, G.M., Gee, A.H., Prager, R.W.: Calibration of an orientation sensor for freehand 3D ultrasound and its use in a hybrid acquisition system. BioMed. Eng. OnLine **7**(1), 5 (2008)
6. Jarosik, P., Lewandowski, M.: Automatic ultrasound guidance based on deep reinforcement learning. In: IEEE International Ultrasonics Symposium (IUS), pp. 475–478 (2019)
7. Li, Y., et al.: Standard plane detection in 3D fetal ultrasound using an iterative transformation network. In: Frangi, A.F., Schnabel, J.A., Davatzikos, C., Alberola-López, C., Fichtinger, G. (eds.) MICCAI 2018. LNCS, vol. 11070, pp. 392–400. Springer, Cham (2018). https://doi.org/10.1007/978-3-030-00928-1_45

8. Liang, K., Rogers, A.J., Light, E.D., von Allmen, D., Smith, S.W.: Three-dimensional ultrasound guidance of autonomous robotic breast biopsy: feasibility study. Ultrasound Med. Biol. **36**(1), 173–177 (2010)
9. Loshchilov, I., Hutter, F.: Decoupled weight decay regularization. In: International Conference on Learning Representations (ICLR) (2019)
10. Maraci, M.A., Bridge, C.P., Napolitano, R., Papageorghiou, A., Noble, J.A.: A framework for analysis of linear ultrasound videos to detect fetal presentation and heartbeat. Med. Image Anal. **37**, 22–36 (2017)
11. Mebarki, R., Krupa, A., Chaumette, F.: 2-D ultrasound probe complete guidance by visual servoing using image moments. IEEE Trans. Robot. **26**(2), 296–306 (2010)
12. Milletari, F., Birodkar, V., Sofka, M.: Straight to the point: reinforcement learning for user guidance in ultrasound. In: Wang, Q., et al. (eds.) PIPPI/SUSI -2019. LNCS, vol. 11798, pp. 3–10. Springer, Cham (2019). https://doi.org/10.1007/978-3-030-32875-7_1
13. Mustafa, A.S.B., et al.: Development of robotic system for autonomous liver screening using ultrasound scanning device. In: IEEE International Conference on Robotics and Biomimetics (ROBIO), pp. 804–809 (2013)
14. Pan, Y., et al.: Agile autonomous driving using end-to-end deep imitation learning. In: Robotics: Science and Systems (RSS) (2019)
15. Pavllo, D., Feichtenhofer, C., Auli, M., Grangier, D.: Modeling human motion with quaternion-based neural networks. Int. J. Comput. Vis. **128**, 855–872 (2020)
16. Prevost, R., et al.: 3D freehand ultrasound without external tracking using deep learning. Med. Image Anal. **48**, 187–202 (2018)
17. Rahmatullah, B., Papageorghiou, A., Noble, J.A.: Automated selection of standardized planes from ultrasound volume. In: Suzuki, K., Wang, F., Shen, D., Yan, P. (eds.) MLMI 2011. LNCS, vol. 7009, pp. 35–42. Springer, Heidelberg (2011). https://doi.org/10.1007/978-3-642-24319-6_5
18. Salomon, L.J., et al.: Practice guidelines for performance of the routine mid-trimester fetal ultrasound scan. Ultrasound Obstet. Gynecol. **37**(1), 116–126 (2011)
19. Sandler, M., Howard, A., Zhu, M., Zhmoginov, A., Chen, L.C.: MobileNetV2: inverted residuals and linear bottlenecks. In: IEEE Conference on Computer Vision and Pattern Recognition (CVPR), pp. 4510–4520 (2018)
20. Shah, S., Bellows, B.A., Adedipe, A.A., Totten, J.E., Backlund, B.H., Sajed, D.: Perceived barriers in the use of ultrasound in developing countries. Crit. Ultrasound J. **7**(1), 1–5 (2015). https://doi.org/10.1186/s13089-015-0028-2
21. Toporek, G., Wang, H., Balicki, M., Xie, H.: Autonomous image-based ultrasound probe positioning via deep learning. In: Hamlyn Symposium on Medical Robotics (2018)
22. Vilchis, A., Troccaz, J., Cinquin, P., Masuda, K., Pellissier, F.: A new robot architecture for tele-echography. IEEE Trans. Robot. Autom. **19**(5), 922–926 (2003)
23. Wang, S., et al.: Robotic-assisted ultrasound for fetal imaging: evolution from single-arm to dual-arm system. In: Althoefer, K., Konstantinova, J., Zhang, K. (eds.) TAROS 2019. LNCS (LNAI), vol. 11650, pp. 27–38. Springer, Cham (2019). https://doi.org/10.1007/978-3-030-25332-5_3
24. Yaqub, M., Kelly, B., Papageorghiou, A.T., Noble, J.A.: A deep learning solution for automatic fetal neurosonographic diagnostic plane verification using clinical standard constraints. Ultrasound Med. Biol. **43**(12), 2925–2933 (2017)

Video Image Analysis

ISINet: An Instance-Based Approach for Surgical Instrument Segmentation

Cristina González([✉]), Laura Bravo-Sánchez, and Pablo Arbeláez

Center for Research and Formation in Artificial Intelligence,
Universidad de los Andes, Bogotá, Colombia
{ci.gonzalez10,lm.bravo10,pa.arbelaez}@uniandes.edu.co

Abstract. We study the task of semantic segmentation of surgical instruments in robotic-assisted surgery scenes. We propose the Instance-based Surgical Instrument Segmentation Network (ISINet), a method that addresses this task from an instance-based segmentation perspective. Our method includes a temporal consistency module that takes into account the previously overlooked and inherent temporal information of the problem. We validate our approach on the existing benchmark for the task, the Endoscopic Vision 2017 Robotic Instrument Segmentation Dataset [2], and on the 2018 version of the dataset [1], whose annotations we extended for the fine-grained version of instrument segmentation. Our results show that ISINet significantly outperforms state-of-the-art methods, with our baseline version **duplicating** the Intersection over Union (IoU) of previous methods and our complete model **triplicating** the IoU.

Keywords: Robotic-assisted surgery · Instrument type segmentation · Image-guided surgery · Computer assisted intervention · Medical image computing

1 Introduction

In this paper, we focus on the task of semantic segmentation of surgical instruments in robotic-assisted surgery scenes. In other words, we aim at identifying the instruments in a surgical scene and at correctly labeling each instrument pixel with its class. The segmentation of surgical instruments or their type is frequently used as an intermediate task for the development of computer-assisted surgery systems [15] such as instrument tracking [6], pose estimation [31], and surgical phase estimation [28], which in turn have applications ranging from operating

C. González and L. Bravo-Sánchez—Both authors contributed equally to this work.

Electronic supplementary material The online version of this chapter (https://doi.org/10.1007/978-3-030-59716-0_57) contains supplementary material, which is available to authorized users.

A. L. Martel et al. (Eds.): MICCAI 2020, LNCS 12263, pp. 595–605, 2020.
https://doi.org/10.1007/978-3-030-59716-0_57

room optimization to personalization of procedures, and particularly, in preoperative planning [15,19,32]. Hence, developing reliable methods for the semantic segmentation of surgical instruments can advance multiple fields of research.

The task of instrument segmentation in surgical scenes was first introduced in the Endoscopic Vision 2015 Instrument Segmentation and Tracking Dataset [3]. However, the objective was not to distinguish among instrument types, but to extract the instruments from the background and label their parts. The dataset's annotations were obtained using a semi-automatic method, leading to a misalignment between the groundtruth and the images [2]. Another limitation of this pioneering effort was the absence of substantial background changes, which further simplified the task.

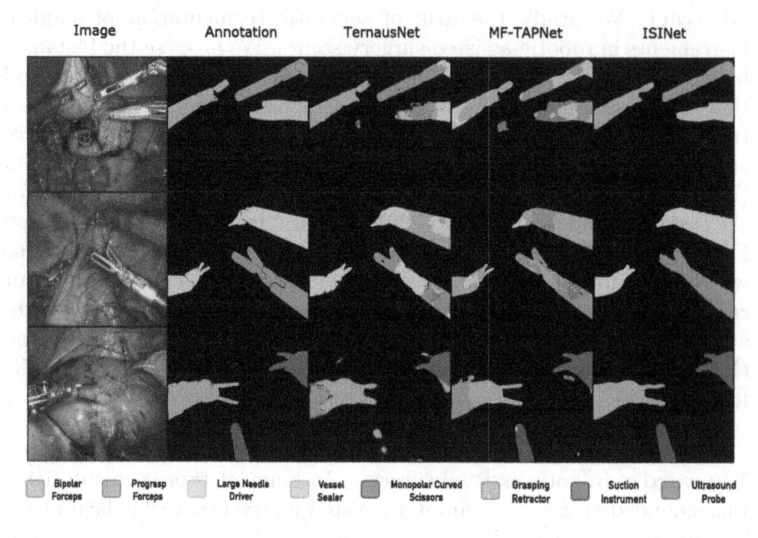

Fig. 1. Each row depicts an example result for the task of instrument type segmentation on the EndoVis 2017 and 2018 datasets. The columns from left to right: image, annotation, segmentation of TernausNet [27], segmentation of MFTAPNet [14] and the segmentation of our method ISINet. The instrument colors represent instrument types. Best viewed in color.

The Endoscopic Vision 2017 Robotic Instrument Segmentation (EndoVis 2017) Dataset [2] was developed to overcome the drawbacks of the 2015 benchmark. This dataset contains 10 robotic-assisted surgery image sequences, each composed of 225 frames. Eight sequences make up the training data and two sequences the testing data. The image sequences show up to 5 instruments per frame, pertaining to 7 instrument types. In this dataset, the task was modified to include annotations for instrument types as well as instrument parts. To date, this dataset remains the only existing experimental framework to study this fine-grained version of the instrument segmentation problem. Despite the effort put into building this dataset, it still does not reflect the general problem, mainly due to the limited amount of data, unrealistic surgery content (the videos are recorded from skills sessions), and the sparse sampling of the original videos, which limits temporal consistency.

The next installment of the problem, the Endoscopic Vision 2018 Robotic Scene Segmentation Dataset, increased the complexity of surgical image segmentation by including anatomical objects such as organs and non-robotic surgical instruments like gauze and suturing thread. In contrast to the 2017 dataset, these images were taken from surgical procedures and thus boast a large variability in backgrounds, instrument movements, angles, and scales. Despite the additional annotations, the instrument class was simplified to a general *instrument* category that encompasses all instrument types. For this reason, the 2018 dataset cannot be used for the 2017 fine-grained version of the instrument segmentation task.

State-of-the-art methods for the segmentation of surgical instruments follow a pixel-wise semantic segmentation paradigm in which the class of each pixel in an image is predicted independently. Most methods [2,3,9,14,27] modify the neural network U-Net [25], which in turn is based on Fully Convolutional Networks (FCN) [21]. Some of these methods attempt to take into account details that could differentiate the instrument as a whole, by using boundaries [8], depth perception [23], post-processing strategies [7], saliency maps [13] or pose estimation [18]. Nevertheless, these techniques have a label consistency problem in which a single instrument can be assigned multiple instrument types, that is, there is a lack of spatial consistency in class labels within objects. [17] addresses this challenge by employing an instance-based segmentation approach; however, their work on gynecological instruments was developed on a private dataset.

The second limitation of state-of-the-art models for this task is the difficulty in ensuring label consistency for an instrument through time, that is, usually the instrument classes are predicted frame by frame without considering the segmentation labels from previous frames. Recently, MF-TAPNet [14] was the first method to include a temporal prior to enhance segmentation. This prior is used as an attention mechanism and is calculated using the optical flow of previous frames. Other methods that use temporal cues have been mostly developed for surgical instrument datasets that focus on instrument tracking instead of instrument segmentation [4,16,30,31]. More recently, those methods developed in [26] employ temporal information for improving the segmentations or for data augmentation purposes. Instead of using temporal information to improve the segmentations, our method employs the redundancy in predictions across frames to correct mislabeled instruments, that is, to ensure temporal consistency.

In this paper, we address the label consistency problem by introducing an instance-based segmentation method for this task, the **I**nstance-based **S**urgical **I**nstrument Segmentation **Net**work (ISINet). Figure 1 shows examples of the segmentation of robotic-assisted surgery scenes predicted by the state-of-the-art TernausNet [27] compared to the result of ISINet. In contrast to pixel-wise segmentation methods, our approach first identifies instrument candidates, and then assigns a unique category to the complete instrument. Our model builds on the influential instance segmentation system Mask R-CNN [10], which we adapted to the instrument classes present in the datasets, and to which we added a temporal consistency module that takes advantage of the sequential nature of the data. Our temporal consistency strategy (i) identifies instrument instances over different frames in a sequence, and (ii) takes into account the class predictions of consecutive individual frames to generate a temporally consistent class prediction for a given instance.

As mentioned above, a limiting factor in solving instrument segmentation is the relative scarcity of annotated data, particularly for the fine-grained version of the task. In order to quantitatively assess the influence of this factor in algorithmic performance, we collect additional instrument type annotations for the 2018 dataset, extending thus the 2017 training data. Following the 2017 dataset annotation protocol, we manually annotate the instruments with their types and temporally consistent instance labels, with the assistance of a specialist. Thus, with our additional annotations, we augment the data available for this task with 15 new image sequences, each composed of 149 frames, and provide annotations for new instrument types. Our annotations make the EndoVis 2018 dataset the second experimental framework available for studying the instrument segmentation task.

We demonstrate the efficacy of our approach by evaluating ISINet's performance in both datasets, with and without using the additional data during training. In all settings, the results show that by using an instance-based approach, we **duplicate** and even **triplicate** the performance of the state-of-the-art pixel-wise segmentation methods.

Our main contributions can be summarized as follows:

1. We present ISINet, an instance-based method for instrument segmentation.
2. We propose a novel temporal consistency module that takes advantage of the sequential nature of data to increase the accuracy of the class predictions of instrument instances.
3. We provide new data for studying the instrument segmentation task, and empirically demonstrate the need for this additional data.

To ensure reproducibility of our results and to promote further research on this task, we provide the pre-trained models, source code for ISINet and additional annotations created for the EndoVis 2018 dataset[1].

2 ISINet

Unlike pixel-wise segmentation methods, which predict a class for each pixel in the image, instance-based approaches produce a class label for entire object instances. Our method, Instance-based Surgical Instrument segmentation Network (ISINet), builds on the highly successful model for instance segmentation in natural images, Mask R-CNN. We adapt this architecture to the fine-grained instrument segmentation problem by modifying the prediction layer to the number of classes found in the EndoVis 2017 and 2018 datasets, and develop a

[1] https://github.com/BCV-Uniandes/ISINet.

module to promote temporal consistency in the per-instance class predictions across consecutive frames. Our temporal consistency module works in two steps: first, in the *Matching Step*, for each image sequence we identify and follow the instrument instances along the sequence and then, in the *Assignment Step* we consider all the predictions for each instance and assign an overall instrument type prediction for the instance.

Initially, for the images in a sequence I from frame $t = 1$ to the final frame T, we obtain via a candidate extraction model (M), particularly Mask R-CNN, a set of n scores (S), object candidates (O) and class predictions (C) for every frame t. Where n correspond to all the predictions with a confidence score above 0.75.

$$(\{\{S_{i,t}\}_{i=1}^{n}\}_{t=1}^{T}, \{\{O_{i,t}\}_{i=1}^{n}\}_{t=1}^{T}, \{\{C_{i,t}\}_{i=1}^{n}\}_{t=1}^{T}) = M(\{I_t\}_{t=1}^{T})$$

We calculate the backward optical flow (OF), that is from frame t to $t-1$, for all the consecutive frames in a sequence. For this purpose we use FlowNet2 [12] (F) pre-trained on the MPI Sintel dataset [5] and use the PyTorch implementation [24].

$$OF_{t \to t-1} = F([I_t, I_{t-1}])$$

Matching Step. For the candidate matching step given a frame t, we retrieve the candidates $\{\{O_{i,t}\}_{i=1}^{n}\}_{t=t-f}^{t}$, scores $\{\{S_{i,t}\}_{i=1}^{n}\}_{t=t-f}^{t}$ and class $\{\{C_{i,t}\}_{i=1}^{n}\}_{t=t-f}^{t}$ predictions from the f previous frames. We use the optical flow to iteratively warp each candidate from the previous frames to the current frame t. For example, to warp frame $t-2$ into frame t, we apply the following equation from frame $t-2$ to $t-1$, and from $t-1$ to t.

$$\hat{O}_{i,t-1} = \underset{t-1 \to t}{Warp}(O_{i,t-1}, OF_{t \to t-1})$$

Once we obtain the warped object candidates $\hat{O}$, we follow an instrument instance through time by matching every warped object candidate from the f frames and the current frame t amongst themselves by finding reciprocal pairings in terms of the Intersection over Union (IoU) metric between each possible candidate pair. Additionally, we only consider reciprocal pairings that have an IoU larger than a threshold U. The end result of the matching step is a set O of m instances along frames $t-f$ to t, an instance O_k need not be present in all the frames.

$$\{\{O_{k,t}\}_{k=1}^{m}\}_{t=t-f}^{t} = Matching(\{\{\hat{O}_{i,t}\}_{i=1}^{n}\}_{t=t-f}^{t})$$

Assignment Step. The objective of this step is to update the class prediction for each instance in the current frame, by considering the predictions of the previous f frames. For this purpose, we use a function A that considers both the classes and scores for each instance k.

$$C_{k,t} = A([C_{k,t-f}, \cdots, C_{k,t}], [S_{k,t-f}, \cdots, S_{k,t}])$$

We repeat the Matching and Assignment steps for every frame t in a sequence and for all the sequences in a set. For our final method we set f to 6, U to 0 and 0.5 for the 2017 and 2018 datasets respectively, and define A as the mode of the input classes weighted by their scores, we validate these parameters experimentally in the supplementary material.

Implementation Details. For training Mask R-CNN we use the official implementation [22]. We train until convergence with the 1x schedule of the implementation, a learning rate of 0.0025, weight decay of $1e^{-4}$, and 4 images per batch on an NVIDIA TITAN-X Pascal GPU. Additionally, for all experiments we use a ResNet-50 [11] backbone pre-trained on the MS-COCO dataset [20].

3 Experiments

3.1 Additional Annotations for EndoVis 2018

We provide additional instrument-type data for the task of instrument-type segmentation in surgical scenes. For this purpose, we manually extend the annotations of the EndoVis 2018 with the assistance of a specialist. Originally, this dataset's instruments are annotated as a general *instrument* class and are labeled with their parts (*shaft*, *wrist* and *jaws*). To make this dataset available for the study of fine-grained instrument segmentation, that is, to distinguish among instrument types, we further annotate each instrument in the dataset with its type. Based on the classes presented in the 2017 version of this dataset and the DaVinci systems catalog [29], we identify 9 instrument types: Bipolar Forceps (both Maryland and Fenestrated), Prograsp Forceps, Large Needle Driver, Monopolar Curved Scissors, Ultrasound Probe, Suction Instrument, Clip Applier, and Stapler. However, we refrain from evaluating the Stapler class due to the limited amount of examples.

For the dataset's 15 image sequences, each composed of 149 frames, we manually extract each instrument from the other objects in the scene and assign it one of the 10 aforementioned types. We label each instrument by taking into account its frame and its complete image sequence to ensure a correct label in blurry or partially occluded instances. We maintain the instrument part annotations as additional information useful for grouping-based segmentation methods. Furthermore, ensure that our instance label annotations are consistent throughout the frames of a sequence to make the dataset suitable for training instance-based segmentation methods. Our annotations are compatible with the original scene segmentation task and with the MS-COCO standard dataset format.

3.2 Experimental Setup

For our experimental framework, we use the EndoVis 2018 and 2017 datasets. In both datasets the images correspond to robot-assisted surgery videos taken with the DaVinci robotic system, and the annotations are semantic and instance segmentations of robotic instrument tools.

Table 1. Comparison of ISINet against the state-of-the-art methods for this task. The results are shown in terms of the IoU per class, mean IoU across classes, challenge IoU and IoU. D stands for use of additional data and T for temporal consistency module. Best values in bold.

(a) Comparison against the state-of-the-art on the EndoVis 2017 dataset.

Method	D	T	challenge IoU	IoU	Instrument classes							mean class IoU
					Bipolar Forceps	Prograsp Forceps	Large Needle Driver	Vessel Sealer	Grasping Retractor	Monopolar Curved Scissors	Ultrasound Probe	
TernausNet[27]			35.27	12.67	13.45	12.39	20.51	5.97	1.08	1.00	**16.76**	10.17
MF-TAPNet[14]			37.35	13.49	16.39	14.11	19.01	8.11	0.31	4.09	13.40	10.77
ISINet (Ours)			53.55	49.57	36.93	37.80	47.06	24.96	**2.01**	19.99	13.90	26.92
		✓	**55.62**	**52.20**	**38.70**	**38.50**	**50.09**	**27.43**	**2.01**	**28.72**	12.56	**28.96**
ISINet (Ours)	✓		66.27	62.70	59.53	45.73	58.65	24.38	2.87	35.85	**28.33**	36.48
	✓	✓	**67.74**	**65.18**	**62.86**	**46.46**	**64.12**	**27.77**	**3.06**	**37.12**	25.18	**38.08**

(b) Comparison against the state-of-the-art on the EndoVis 2018 dataset.

Method	D	T	challenge IoU	IoU	Instrument classes							mean class IoU
					Bipolar Forceps	Prograsp Forceps	Large Needle Driver	Monopolar Curved Scissors	Ultrasound Probe	Suction Instrument	Clip Applier	
TernausNet[27]			46.22	39.87	44.20	4.67	0.00	50.44	0.00	0.00	0.00	14.19
MF-TAPNet[14]			67.87	39.14	69.23	6.10	11.68	70.24	0.57	14.00	**0.91**	24.68
ISINet (Ours)			72.99	**71.01**	73.55	**48.98**	30.38	**88.17**	**2.23**	**37.84**	0.00	40.16
		✓	**73.03**	70.97	**73.83**	48.61	**30.98**	88.16	2.16	37.68	0.00	**40.21**
ISINet (Ours)	✓		77.19	75.25	76.55	48.79	50.24	**91.50**	0.00	44.95	0.00	44.58
	✓	✓	**77.47**	**75.59**	**76.60**	**51.18**	**52.31**	91.08	0.00	**45.87**	0.00	**45.29**

For the experimental validation process, we divide the original training images of the 2018 dataset into two sets, the validation set with sequences 2, 5, 9, and 15, while the remaining sequences are part of the training set. As the 2017 dataset is smaller we use 4-fold cross-validation with the standard folds described in [27]. For the quantitative evaluation we use three metrics based on the Intersection over Union (IoU) metric, each more stringent than the preceding metric. For a prediction P and groundtruth G in a frame i, we compute the challenge IoU, the metric from the 2017 challenge, which only considers the classes that are present in a frame. A variation of the IoU (Eq. 1) averaged over all the classes C and frames N, and the mean class IoU (Eq. 2) which corresponds to the IoU per class averaged across the classes.

$$IoU = \frac{1}{N} \sum_{i=1}^{N} \left(\frac{1}{C} \sum_{c=1}^{C} \frac{P_{i,c} \cap G_{i,c}}{P_{ic} \cup G_{i,c}} \right) \tag{1}$$

$$mean\ cIoU = \frac{1}{C} \sum_{c=1}^{C} \left(\frac{1}{N} \sum_{i=1}^{N} \frac{P_{i,c} \cap G_{i,c}}{P_{i,c} \cup G_{i,c}} \right) \tag{2}$$

3.3 Experimental Validation

We compare the performance of our approach ISINet with the state-of-the-art methods TernausNet [27] and MF-TAPNet [14]. For TernausNet we use the pre-trained models provided for the 2017 dataset, and for the 2018 dataset we retrain using the official implementation with the default 2017 parameters and a learning rate of $1e^{-4}$. For MF-TAPNet, we retrain the method for both datasets with the official implementation and the default parameter setting. We present the results of this experiment for the EndoVis 2017 dataset on Table 1a and on Table 1b for the EndoVis 2018 dataset.

The results show that our baseline method, that is, without the temporal consistency module nor the additional data, outperforms TernausNet and MF-TAPNet in all the overall IoU metrics in both datasets, **duplicating** or **triplicating** the performance of the state-of-the-art depending on the metric. For some instrument classes, the improvement is more than 30.0 IoU on both datasets. We observe that the improvement correlates with the number of examples of a class, as can be seen by the performance of the Grasping Retractor and Clip Applier instruments. Figure 1 depicts the advantages of our method over the state-of-the-art in the 2017 dataset, by segmenting previously unidentified classes and recovering complete instruments. Please refer to the supplementary material for additional qualitative results, including error modes.

We design an experiment to assess the effect of training on additional data and compare the results against only training with one dataset. For the 2018 dataset the additional data is the complete 2017 dataset. However, as the EndoVis 2017 dataset uses a 4-fold validation scheme, we train on three folds and the 2018 data, and evaluate on the remaining fold. The final result is the average of the results on all folds. Considering that not all classes are present in both datasets, we only predict segmentations for each dataset's existing classes. However, we train with all the available examples. Tables 1a and 1b demonstrate that, for both datasets, training on additional data results in better performance compared to training on a single dataset. The performance of 6 out of 7 classes on the EndoVis 2017 dataset improves, and 4 out of 7 classes of the EndoVis 2018 dataset follow this trend, with some of them increasing by up to **20.0 IoU** percentage points. These results confirm the need for additional data in order to solve this task.

In order to validate our temporal consistency module, we evaluate ISINet with and without the module (T) and with and without using additional data (D). We perform these experiments on both datasets. Tables 1a and 1b show that our temporal consistency module improves the overall metrics with and without additional data on both datasets. Our module corrects outlier predictions in all the classes except the Ultrasound Probe, with some of them increasing nearly 8% points on the 2017 dataset. Despite the overall gain on the 2018 dataset, we hypothesize that the increment is less compared to the 2017 dataset due to the higher variability of the 2018 dataset.

4 Conclusions

In this paper, we address the task of instrument segmentation in surgical scenes by proposing an instance-based segmentation approach. Additionally, we propose a temporal consistency module that considers an instance's predictions across the frames in a sequence. Our method ISINet outperforms the state-of-the-art semantic segmentation methods in the benchmark dataset EndoVis 2017 and the EndoVis 2018 dataset. We extend the former dataset for this task by manually annotating the instrument types. Additionally, our results indicate that using more data during the training process improved model generalization for both datasets. Finally, we observe that our temporal consistency module enhances performance by better preserving the identity of objects across time. We will provide the code and pre-trained models for both datasets, and the instrument type annotations for the EndoVis 2018 dataset.

Acknowledgments. The authors thank Dr. Germán Rosero for his support in the verification of the instrument type annotations.

References

1. Allan, M., et al.: 2018 robotic scene segmentation challenge. arXiv preprint arXiv:2001.11190 (2020)
2. Allan, M., et al.: 2017 robotic instrument segmentation challenge. arXiv preprint arXiv:1902.06426 (2019)
3. Bodenstedt, S., et al.: Comparative evaluation of instrument segmentation and tracking methods in minimally invasive surgery. arXiv preprint arXiv:1805.02475 (2018)
4. Bouget, D., Benenson, R., Omran, M., Riffaud, L., Schiele, B., Jannin, P.: Detecting surgical tools by modelling local appearance and global shape. IEEE Trans. Med. Imaging **34**(12), 2603–2617 (2015)
5. Butler, D.J., Wulff, J., Stanley, G.B., Black, M.J.: A naturalistic open source movie for optical flow evaluation. In: Fitzgibbon, A., Lazebnik, S., Perona, P., Sato, Y., Schmid, C. (eds.) ECCV 2012. LNCS, vol. 7577, pp. 611–625. Springer, Heidelberg (2012). https://doi.org/10.1007/978-3-642-33783-3_44
6. Du, X., et al.: Articulated multi-instrument 2-D pose estimation using fully convolutional networks. IEEE Trans. Med. Imaging **37**(5), 1276–1287 (2018). https://doi.org/10.1109/tmi.2017.2787672
7. Lee, E.J., Plishker, W., Liu, X., Kane, T., Bhattacharyya, S.S., Shekhar, R.: Segmentation of surgical instruments in laparoscopic videos: training dataset generation and deep-learning-based framework, vol. 10951 (2019). https://doi.org/10.1117/12.2512994
8. García-Peraza-Herrera, L.C., et al.: ToolNet: holistically-nested real-time segmentation of robotic surgical tools. In: 2017 IEEE/RSJ International Conference on Intelligent Robots and Systems (IROS), pp. 5717–5722. IEEE (2017)
9. García-Peraza-Herrera, L.C., et al.: Real-time segmentation of non-rigid surgical tools based on deep learning and tracking. In: Peters, T., et al. (eds.) CARE 2016. LNCS, vol. 10170, pp. 84–95. Springer, Cham (2017). https://doi.org/10.1007/978-3-319-54057-3_8
10. He, K., Gkioxari, G., Dollar, P., Girshick, R.: Mask R-CNN. In: The IEEE International Conference on Computer Vision (ICCV), October 2017
11. He, K., Zhang, X., Ren, S., Sun, J.: Deep residual learning for image recognition. In: Proceedings of the IEEE Conference on Computer Vision and Pattern Recognition, pp. 770–778 (2016)

12. Ilg, E., Mayer, N., Saikia, T., Keuper, M., Dosovitskiy, A., Brox, T.: FlowNet 2.0: evolution of optical flow estimation with deep networks. In: IEEE Conference on Computer Vision and Pattern Recognition (CVPR), July 2017. http://lmb.informatik.uni-freiburg.de//Publications/2017/IMKDB17
13. Islam, M., Li, Y., Ren, H.: Learning where to look while tracking instruments in robot-assisted surgery. In: Shen, D., et al. (eds.) MICCAI 2019. LNCS, vol. 11768, pp. 412–420. Springer, Cham (2019). https://doi.org/10.1007/978-3-030-32254-0_46
14. Jin, Y., Cheng, K., Dou, Q., Heng, P.-A.: Incorporating temporal prior from motion flow for instrument segmentation in minimally invasive surgery video. In: Shen, D., et al. (eds.) MICCAI 2019. LNCS, vol. 11768, pp. 440–448. Springer, Cham (2019). https://doi.org/10.1007/978-3-030-32254-0_49
15. Joskowicz, L.: Computer-aided surgery meets predictive, preventive, and personalized medicine. EPMA J. **8**(1), 1–4 (2017)
16. Jung, I., Son, J., Baek, M., Han, B.: Real-time MDNet. In: Ferrari, V., Hebert, M., Sminchisescu, C., Weiss, Y. (eds.) ECCV 2018. LNCS, vol. 11208, pp. 89–104. Springer, Cham (2018). https://doi.org/10.1007/978-3-030-01225-0_6
17. Kletz, S., Schoeffmann, K., Benois-Pineau, J., Husslein, H.: Identifying surgical instruments in laparoscopy using deep learning instance segmentation. In: 2019 International Conference on Content-Based Multimedia Indexing (CBMI), pp. 1–6 (2019)
18. Kurmann, T., et al.: Simultaneous recognition and pose estimation of instruments in minimally invasive surgery. In: Descoteaux, M., Maier-Hein, L., Franz, A., Jannin, P., Collins, D.L., Duchesne, S. (eds.) MICCAI 2017. LNCS, vol. 10434, pp. 505–513. Springer, Cham (2017). https://doi.org/10.1007/978-3-319-66185-8_57
19. Lalys, F., Jannin, P.: Surgical process modelling: a review. Int. J. Comput. Assist. Radiol. Surg. **9**(3), 495–511 (2013)
20. Lin, T.-Y., et al.: Microsoft COCO: common objects in context. In: Fleet, D., Pajdla, T., Schiele, B., Tuytelaars, T. (eds.) ECCV 2014. LNCS, vol. 8693, pp. 740–755. Springer, Cham (2014). https://doi.org/10.1007/978-3-319-10602-1_48
21. Long, J., Shelhamer, E., Darrell, T.: Fully convolutional networks for semantic segmentation. In: Proceedings of the IEEE Conference on Computer Vision and Pattern Recognition, pp. 3431–3440 (2015)
22. Massa, F., Girshick, R.: MaskRCNN-benchmark: fast, modular reference implementation of Instance Segmentation and Object Detection algorithms in PyTorch (2018). https://github.com/facebookresearch/maskrcnn-benchmark
23. Mohammed, A., Yildirim, S., Farup, I., Pedersen, M., Hovde, Ø.: Streoscennet: surgical stereo robotic scene segmentation, vol. 10951 (2019). https://doi.org/10.1117/12.2512518
24. Reda, F., Pottorff, R., Barker, J., Catanzaro, B.: flownet2-pytorch: pytorch implementation of flownet 2.0: evolution of optical flow estimation with deep networks (2017). https://github.com/NVIDIA/flownet2-pytorch
25. Ronneberger, O., Fischer, P., Brox, T.: U-Net: convolutional networks for biomedical image segmentation. In: Navab, N., Hornegger, J., Wells, W.M., Frangi, A.F. (eds.) MICCAI 2015. LNCS, vol. 9351, pp. 234–241. Springer, Cham (2015). https://doi.org/10.1007/978-3-319-24574-4_28
26. Ross, T., et al.: Robust medical instrument segmentation challenge 2019. arXiv preprint arXiv:2003.10299 (2020)
27. Shvets, A.A., Rakhlin, A., Kalinin, A.A., Iglovikov, V.I.: Automatic instrument segmentation in robot-assisted surgery using deep learning (2018)

28. Spediel, S., et al.: Surgical workflow and skill analysis (2019). https://endovissub-workflowandskill.grand-challenge.org
29. Intuitive Surgical: Da vinci surgical system (2019). https://www.intuitive.com/en-us/products-and-services/da-vinci
30. Sznitman, R., Ali, K., Richa, R., Taylor, R.H., Hager, G.D., Fua, P.: Data-driven visual tracking in retinal microsurgery. In: Ayache, N., Delingette, H., Golland, P., Mori, K. (eds.) MICCAI 2012. LNCS, vol. 7511, pp. 568–575. Springer, Heidelberg (2012). https://doi.org/10.1007/978-3-642-33418-4_70
31. Sznitman, R., Becker, C., Fua, P.: Fast part-based classification for instrument detection in minimally invasive surgery. In: Golland, P., Hata, N., Barillot, C., Hornegger, J., Howe, R. (eds.) MICCAI 2014. LNCS, vol. 8674, pp. 692–699. Springer, Cham (2014). https://doi.org/10.1007/978-3-319-10470-6_86
32. Twinanda, A.P., Shehata, S., Mutter, D., Marescaux, J., De Mathelin, M., Padoy, N.: EndoNet: a deep architecture for recognition tasks on laparoscopic videos. IEEE Trans. Med. Imaging **36**(1), 86–97 (2017)

Reliable Liver Fibrosis Assessment from Ultrasound Using Global Hetero-Image Fusion and View-Specific Parameterization

Bowen Li[1(✉)], Ke Yan[1], Dar-In Tai[2], Yuankai Huo[3], Le Lu[1], Jing Xiao[4], and Adam P. Harrison[1]

[1] PAII Inc., Bethesda, MD 20817, USA
libowen755@paii-labs.com
[2] Chang Gung Memorial Hospital, Linkou, Taiwan, ROC
[3] Vanderbilt University, Nashville, TN 37235, USA
[4] PingAn Technology, Shenzhen, China

Abstract. Ultrasound (US) is a critical modality for diagnosing liver fibrosis. Unfortunately, assessment is very subjective, motivating automated approaches. We introduce a principled deep convolutional neural network (CNN) workflow that incorporates several innovations. First, to avoid overfitting on non-relevant image features, we force the network to focus on a clinical region of interest (ROI), encompassing the liver parenchyma and upper border. Second, we introduce global hetero-image fusion (GHIF), which allows the CNN to fuse features from any arbitrary number of images in a study, increasing its versatility and flexibility. Finally, we use "style"-based view-specific parameterization (VSP) to tailor the CNN processing for different viewpoints of the liver, while keeping the majority of parameters the same across views. Experiments on a dataset of 610 patient studies (6979 images) demonstrate that our pipeline can contribute roughly 7% and 22% improvements in partial area under the curve and recall at 90% precision, respectively, over conventional classifiers, validating our approach to this crucial problem.

Keywords: View fusion · Ultrasound · Liver fibrosis · Computer-aided diagnosis

1 Introduction

Liver fibrosis is a major health threat with high prevalence [19]. Without timely diagnosis and treatment, liver fibrosis can develop into liver cirrhosis [19] and

---

Electronic supplementary material The online version of this chapter (https://doi.org/10.1007/978-3-030-59716-0_58) contains supplementary material, which is available to authorized users.

A. L. Martel et al. (Eds.): MICCAI 2020, LNCS 12263, pp. 606–615, 2020.
https://doi.org/10.1007/978-3-030-59716-0_58

even hepatocellular carcinoma [21]. While histopathology remains the gold standard, non-invasive approaches minimize patient discomfort and danger. Elastography is a useful non-invasive modality, but it is not always available or affordable and it can be confounded by inflammation, presence of steatosis, and the patient's etiology [2,12,23]. Assessment using conventional ultrasound (US) may be potentially more versatile; however, it is a subjective measurement that can suffer from insufficient sensitivities, specificities, and high inter- and intra-rater variability [13,15]. Thus, there is great impetus for an automated and less subjective assessment of liver fibrosis. This is the goal of our work.

Although a relatively understudied topic, prior work has advanced automated US fibrosis assessment [4,14,16–18]. In terms of deep convolutional neural networks (CNNs), Meng *et al.* [16] proposed a straightforward liver fibrosis parenchyma VGG-16-based [22] classifier and tested it on a small dataset of 279 images. Importantly, they only performed image-wise predictions and do not report a method for study-wise classification. On the other hand, Liu *et al.* [14] correctly identified the value in fusing features from all US images in a study when making a prediction. However, their algorithm requires exactly 10 images. But, real patient studies may contain any arbitrary number of US scans. Their feature concatenation approach would also drastically increase computational and memory costs as more images are incorporated. Moreover, they rely on 13 manually labeled indicators as ancillary supervision, which are typically not available without considerable labor costs. Finally, their system treats all US images identically, even though a study consists of different *viewpoints* of the liver, each of which may have its own set of clinical markers correlating with fibrosis. Ideally, a liver fibrosis assessment system could learn directly from supervisory signals already present in hospital archives, *i.e.*, image-level fibrosis scores produced during daily clinical routines. In addition, a versatile system should also be able to effectively use all US images/views in a patient study with no ballooning of computational costs, regardless of their number.

We fill these gaps by proposing a robust and versatile pipeline for conventional ultrasound liver fibrosis assessment. Like others, we use a deep CNN, but with key innovations. First, informed by clinical practice [1], we ensure the network focuses only on a clinically-important region of interest (ROI), *i.e.*, the liver parenchyma and the upper liver border. This prevents the CNN from erroneously overfitting on spurious or background features. Second, inspired by hetero-modal image segmentation (HeMIS) [6], we adapt and expand on this approach and propose global hetero-image fusion (GHIF) as a way to learn from, and perform inference on, any arbitrary number of US scans within a study. While GHIF share similarities with deep feature-based multi-instance learning [10], there are two important distinctions: (1) GHIF includes variance as part of the fusion, as per HeMIS [6]; (2) GHIF is trained using arbitrary image combinations from a patient study, which is possible because, unlike multi-instance learning, each image (or instance) is strongly supervised by the same label. We are the first to propose and develop this mechanism to fuse global CNN feature vectors. Finally, we implement a view-specific parameterization (VSP) that tailors the CNN processing based on 6 common liver US views. While the majority of processing is shared, each view possesses its own

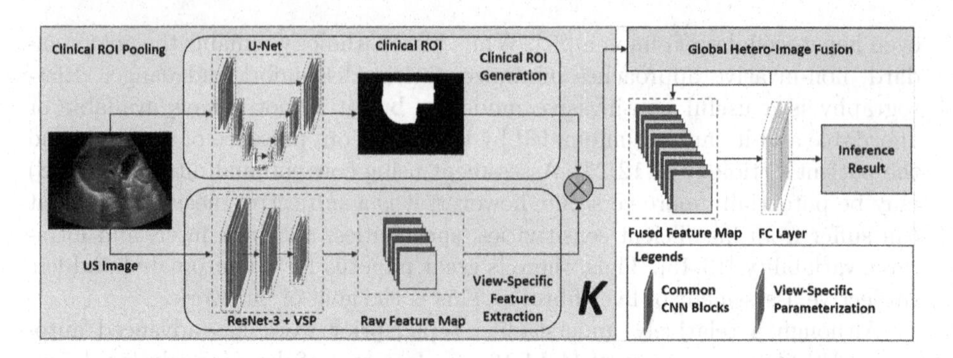

Fig. 1. Algorithmic workflow depicting the clinical ROI pooling, GHIF, and VSP. We use plate notation to depict the repeated workflow across the K images in a US study.

set of so-called "style"-based normalization parameters [9] to customize the analysis. While others have used similar ideas segmenting different anatomical structures [8], we are the first to apply this concept for clinical decision support and the first to use it in concert with a hetero-image fusion mechanism. The result is a highly robust and practical liver fibrosis assessment solution.

To validate our approach, we use a cross-validated dataset of 610 US patient studies, comprising 6976 images. We measure the ability to identify patients with moderate to severe liver fibrosis. Compared to strong classification baselines, our enhancements are able to improve recall at 90% precision by 22%, with commensurate boosts in partial areas under the curve (AUCs). Importantly, ablation studies demonstrate that each component contributes to these performance improvements, demonstrating that our liver fibrosis assessment pipeline, and its constituent clinical ROI, GHIF, and VSP parts, represents a significant advancement for this important task.

2 Methods

We assume we are given a dataset, $\mathcal{D} = \{\mathcal{X}_i, y_i\}_{i=1}^{N}$, comprised of US patient studies and ground-truth labels indicating liver fibrosis status, dropping the i when convenient. Each study $\mathcal{X}_i$, in turn, is comprised of an arbitrary number of K_i 2D conventional US scans of the patient's liver, $\mathcal{X}_i = \{\mathbf{X}^1 \ldots \mathbf{X}^{K_i}\}$. Figure 1 depicts the workflow of our automated liver assessment tool, which combines clinical ROI pooling, GHIF, and VSP.

2.1 Clinical ROI Pooling

We use a deep CNN as backbone for our pipeline. Popular deep CNNs, *e.g.*, ResNet [7], can be formulated with the following convention:

$$\hat{y}^k = f\left(g\left(\mathbf{A}^k\right); \mathbf{w}\right), \tag{1}$$

$$\mathbf{A}^k = h\left(\mathbf{X}^k; \theta\right), \tag{2}$$

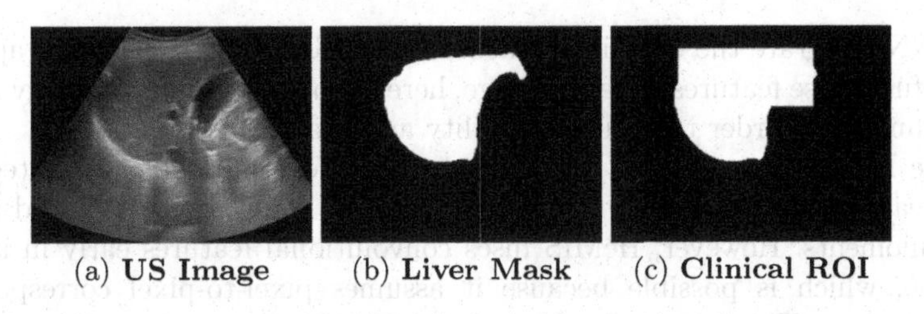

(a) **US Image** (b) **Liver Mask** (c) **Clinical ROI**

Fig. 2. (a) Depicts an US image, whose liver mask is rendered in (b). As shown in (c), the clinical ROI the mask is extended upward to cover the top liver border.

where $h(.;\theta)$ is a fully-convolutional network (FCN) feature extractor parameterized by θ, $\mathbf{A}^k$ is the FCN output, $g(.)$ is some global pooling function, *e.g.*, average pooling, and $f(.;\mathbf{w})$ is a fully-connected layer (and sigmoid function) parameterized by $\mathbf{w}$. When multiple US scans are present, a standard approach is to aggregate individual image-wise predictions, *e.g.*, taking the median:

$$\hat{y} = \text{median}(\{\hat{y}^1 \ldots \hat{y}^K\}). \tag{3}$$

This conventional approach may have drawbacks, as it is possible for the CNN to overfit to spurious background variations. However, based on clinical practice [1], we know *a priori* that certain features are crucial for assessing liver fibrosis, *e.g.*, the parenchyma texture and surface nodularity. As Fig. 2 demonstrates, to make the CNN focus on these features we use a masking technique. We first generate a liver mask for each US scan. This is done by training a simple segmentation network on a small subset of the images. Then, for each scan, we create a rectangle that just covers the top half of and 10 pixels above the liver mask, to ensure the liver border is covered. The resulting binary mask is denoted $\mathbf{M}$. Because we only need to ensure we capture enough of the liver parenchyma and upper border to extract meaningful features, $\mathbf{M}$ need not be perfect.

With a clinical ROI obtained, we formulate the pooling function in (1) as a masked version of global average pooling:

$$g(\mathbf{A};\mathbf{M}) = \text{GAP}(\mathbf{M} \odot \mathbf{A}), \tag{4}$$

where $\odot$ and GAP(.) denote the element-wise product and global average pooling, respectively. Interestingly, we found that including the zeroed-out regions within the global average pooling benefits performance [3,25]. We posit their inclusion helps implicitly capture liver size characteristics, which is another important clinical US marker for liver fibrosis [1].

2.2 Global Hetero-Image Fusion

A challenge with US patient studies is that they may consist of a variable number of images, each of a potentially different view. Ideally, all available US images would contribute to the final prediction. In (3) this is accomplished via a late fusion of independent and image-specific predictions. But, this does not allow the CNN integrate the combined features across US images. A better approach would fuse these features. The challenge, here, is to allow for an arbitrary number of US images in order to ensure flexibility and practicality.

The HeMIS approach [6] to segmentation offers a promising strategy that fuses features from arbitrary numbers of images using their first- and second-order moments. However, HeMIS fuses convolutional features early in its FCN pipeline, which is possible because it assumes pixel-to-pixel correspondence across images. This is completely violated for US images. Instead, only *global* US features can be sensibly fused together, which we accomplish through global hetero-image fusion (GHIF). More formally, we use $\mathcal{A} = \{\mathbf{A}^k\}_{k=1}^K$ and $\mathcal{M} = \{\mathbf{M}^k\}_{k=1}^K$ to denote the set of FCN features and clinical ROIs, respectively, for each image. Then GHIF modifies (1) to accept any arbitrary set of FCN features to produce a *study-wise* prediction:

$$\hat{y} = f\left(g\left(\mathcal{A};\mathcal{M}\right);\mathbf{w}\right), \tag{5}$$

$$g(\mathcal{A};\mathcal{M}) = \text{concat}\left(\text{mean}(\mathcal{G}), \text{var}(\mathcal{G}), \max(\mathcal{G})\right), \tag{6}$$

$$\mathcal{G} = \{\text{GAP}(\mathbf{M}^k \odot \mathbf{A}^k)\}_{k=1}^K, \tag{7}$$

Besides the first- and second-order moments, GHIF (6) also incorporates the max operator as a powerful hetero-fusion function [26]. All three operators can accept any arbitrary numbers of samples to produce one fused feature vector. To the best of our knowledge, we are the first to apply hetero-fusion for global feature vectors. The difference, compared to late fusion, is that features, rather than predictions are fused. Rather than always inputting all US scans when training, an important strategy is choosing random combinations of the K scans for every epoch. This provides a form of data augmentation and allows the CNN to learn from image signals that may be suppressed otherwise. An important implementation note is that training with random combinations of images can make GHIF's batch statistics unstable. For this reason, a normalization not relying on batch statistics, such as instance-normalization [24], should be used.

2.3 View-Specific Parameterization

While GHIF can effectively integrate arbitrary numbers of US images within a study, it uses the same FCN feature extractor across all images, treating them all identically. Yet, there are certain US features, such as vascular markers, that are specific to particular views. As a result, some manner of view-specific analysis could help push performance further. In fact, based on guidance from our clinical partner, US views of the liver can be roughly divided into 6 categories, which focus on different regions of the liver. These are shown in Fig. 3.

A naive solution would be to use a dedicated deep CNN for each view category. However, this would drastically reduce the training set for each dedicated CNN and would sextuple the number of parameters, computation, and memory consumption. Intuitively, there should be a great deal of analysis that is common across US views. The challenge is to retain this shared analysis, while also providing some tailored processing for each category.

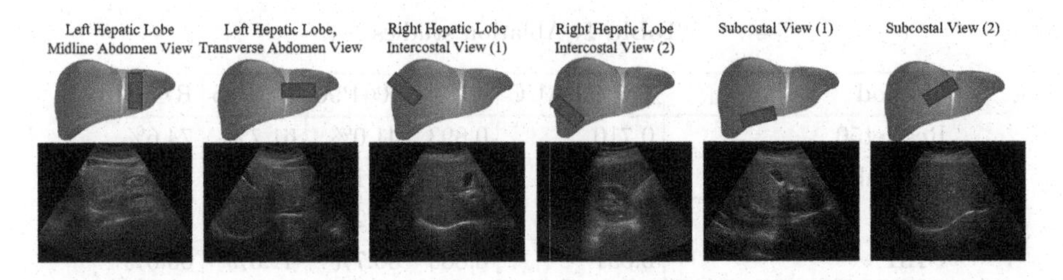

Fig. 3. Liver views in our dataset. Blue square: position of the US probe. Liver cartoons adapted from the DataBase Center for Life Science (https://commons.wikimedia.org/wiki/File:201405_liver.png), licensed under the Creative Commons Attribution 4.0 International (Color figure online)

To do this, we adapt the concept of "style" parameters to implement a view-specific parameterization (VSP) appropriate for US-based fibrosis assessment. Such parameters refer to the affine normalization parameters used in batch- [11] or instance-normalization [24]. If these are switched out, while keeping all other parameters constant, one can alter the behavior of the CNN in quite dramatic ways [8,9]. For our purposes, retaining view-specific normalization parameters allows for the majority of parameters and processing to be shared across views. VSP is then realized with a minimal number of additional parameters.

More formally, if we create 6 sets of normalization parameters for an FCN, we can denote them as $\Omega = \{\omega_1 \ldots \omega_6\}$. The FCN from (2) is then modified to be parameterized also by Ω:

$$\mathbf{A}^k = h\left(\mathbf{X}^k; \theta, \omega_{v_k}\right), \tag{8}$$

where v_k indexes each image by its view and θ now excludes the normalization parameters. VSP relies on identifying the view of each US scan in order to swap in the correct normalization parameters. This can be recorded as part of the acquisition process. Or, if this is not possible, we have found classifying the US views automatically to be quite reliable.

3 Experiments

Dataset. We test our system on a dataset of 610 US patient studies collected from the Chang Gung Memorial Hospital in Taiwan, acquired from Siemens, Philips, and Toshiba devices. The dataset comprises 232 patients, among which 95 (40.95%) patients have moderate to severe fibrosis (27 with severe liver steatosis). All patients were diagnosed with hepatitis B. Patients were scanned up to 3 times, using a different scanner type each time. Each patient study is composed of up to 14 US images, corresponding to the views in Fig. 3. The total number of images is 6 979. We use 5-fold cross validation, splitting each fold at the patient level into 70%, 20%, and 10%, for training, testing, and validation, respectively. We also manually labeled liver contours from 300 randomly chosen US images.

Table 1. Ablation studies

Method	Partial AUC	AUC	R@P90	R@P85	R@P80
ResNet50	0.710	0.893	41.0%	61.7%	74.6%
Clinical ROI	0.744	0.908	46.1%	59.2%	82.1%
Global Fusion	0.591	0.845	30.2%	36.0%	49.8%
GHIF	0.691	0.885	36.7%	41.0%	65.0%
GHIF (I-Norm)	0.762	0.907	57.1%	71.1%	80.6%
GHIF + VSP (I-Norm)	**0.783**	**0.913**	**63.4%**	**78.3%**	**84.2%**

Implementation Details and Comparisons. Experiments evaluated our workflow against several strong classification baselines, where throughout we use the same ResNet50 [7] backbone (pretrained on ImageNet [5]). For methods using the clinical ROI pooling of (4), we use a truncated version of ResNet (only the first three layer blocks) for $h(.)$ in (2). This keeps enough spatial resolution prior to the masking in (4). We call this truncated backbone "ResNet-3". To create the clinical ROI, we train a simple 2D U-Net [20] on the 300 images with masks. For training, we perform standard data augmenation with random brightness, contrast, rotations, and scale adjustments. We use the stochastic gradient descent optimizer and a learning rate of 0.001 to train all networks.

For baselines that can output only image-wise predictions, we test against a conventional ResNet50 and also a ResNet-3 with clinical ROI pooling. For these two approaches, following clinical practices, we take the median value across the image-wise predictions to produce a study-wise prediction. All subsequent study-wise baselines are then built off the ResNet-3 with clinical ROI pooling. We first test the global feature fusion of (6), but only train the ResNet-3 with all available images in a US study. In this way, it follows the spirit of Liu *et al.*'s global fusion strategy [14]. To reveal the impact of our hetero-fusion training strategy that uses different random combinations of US images per epoch, we also test two GHIF variants, one using batch-normalization and one using instance-normalization. The latter helps measure the importance of using proper normalization strategies to manage the instability of GHIF's batch statistics. Finally, we test our proposed model which incorporates VSP on top of GHIF and clinical ROI pooling.

Evaluation Protocols. The problem setup is binary classification, *i.e.*, identifying patients with moderate to severe liver fibrosis, which are the patient cohorts requiring intervention. While we report full AUCs, we primarily focus on operating points within a useful range of specificity or precision. Thus, we evaluate using partial AUCs that only consider false positive rates within 0 to 30% because higher values lose their practical usefulness. Partial AUCs are normalized to be within a range of 0 to 1. We also report recalls at a range of precision points (R@P90, R@P85, R@P80) to reveal the achievable sensitivity at high precision points. We report mean values and mean graphs across all cross-validation folds.

Results. Table 1 presents our AUC, partial AUC and recall values, whereas Fig. 4 graphs the partial receiver operating characteristics (ROCs). Several

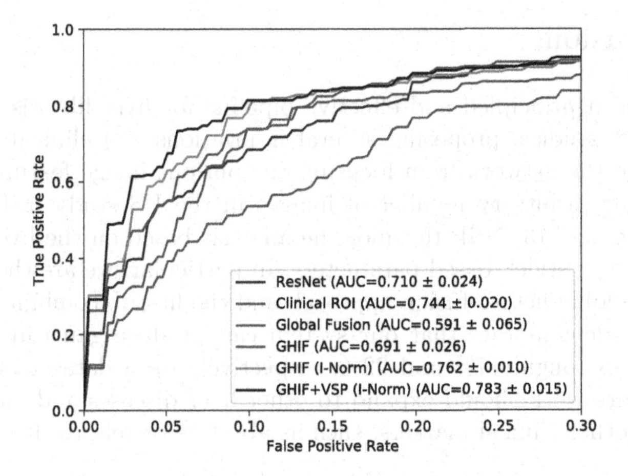

Fig. 4. Partial ROC curves are graphed, with corresponding partial AUCs found in the legend. Partial AUC scores have been normalized to range from 0–1. Both ROCs and AUCs correspond to mean measures taken across the cross-validation folds.

conclusions can be drawn. First, clinical ROI pooling produces significant boosts in performance, validating our strategy of forcing the network to focus on important regions of the image. Second, not surprisingly, global fusion without training with random combinations of images, performs very poorly, as only presenting all study images during training severely limits the data size and variability, handicapping the model. For instance, compared to variants that train on individual images, global fusion effectively reduces the training size by about a factor of 10 in our dataset. In contrast, the GHIF variants, which train with the combinatorial number of random combinations of images, not only avoids drastically reducing the training set size, but can effectively increase it. Importantly, as the table demonstrates, using an appropriate choice of instance normalization is crucial in achieving good performance with GHIF. Although not shown, switching to instance normalization did not improve performance for the image-wise or global fusion models. The boosts in performance is apparent in the partial AUC and recalls at high precision points, underscoring the need to analyze results at appropriate operating points. Finally, adding the VSP provides even further performance improvements, particularly in R@P80-R@P90 values, which see a roughly 4–7% increase over GHIF alone. This indicates that VSP can significantly enhance the recall at the very demanding precision points necessary for clinical use. In total, compared to a conventional classifier, the enhancements we articulate contribute to roughly 8% improvements in partial AUCs and 22% in R@P90 values. Table 1 of our supplementary material also presents AUCs when only choosing to input one particular view in the model during inference. We note that performance is highest when all views are inputted into the model, indicating that our pipeline is able to usefully exploit the information across views. Our supplementary also includes liver segmentation results and success and failure cases for our system.

4 Conclusion

We presented a principled and effective pipeline for liver fibrosis characterization from US studies, proposing several innovations: (1) clinical ROI pooling to discourage the network from focusing on spurious image features; (2) GHIF to manage any arbitrary number of images in the US study in both training and inference; and (3) VSP to tailor the analysis based on the liver view being presented using "style"-based parameters. In particular, we are the first to propose a deep global hetero-fusion approach and the first to combine it with VSP. Experiments demonstrate that our system can produce gains in partial AUC and R@P90 of roughly 7% and 22%, respectively on a dataset of 610 patient studies. Future work should expand to other liver diseases and more explicitly incorporate other clinical markers, such as absolute or relative liver lobe sizing.

References

1. Aubé, C., Bazeries, P., Lebigot, J., Cartier, V., Boursier, J.: Liver fibrosis, cirrhosis, and cirrhosis-related nodules: imaging diagnosis and surveillance. Diagn. Interv. Imaging **98**(6), 455–468 (2017)
2. Chen, C.J., et al.: Effects of hepatic steatosis on non-invasive liver fibrosis measurements between hepatitis b and other etiologies. Appl. Sci. **9**, 1961 (2019)
3. Chen, H., et al.: Anatomy-aware Siamese network: exploiting semantic asymmetry for accurate pelvic fracture detection in x-ray images (2020)
4. Chung-Ming, W., Chen, Y.-C., Hsieh, K.-S.: Texture features for classification of ultrasonic liver images. IEEE Trans. Med. Imaging **11**(2), 141–152 (1992)
5. Deng, J., Dong, W., Socher, R., Li, L.J., Li, K., Fei-Fei, L.: ImageNet: a large-scale hierarchical image database. In: CVPR 2009 (2009)
6. Havaei, M., Guizard, N., Chapados, N., Bengio, Y.: HeMIS: hetero-modal image segmentation. In: Ourselin, S., Joskowicz, L., Sabuncu, M.R., Unal, G., Wells, W. (eds.) MICCAI 2016. LNCS, vol. 9901, pp. 469–477. Springer, Cham (2016). https://doi.org/10.1007/978-3-319-46723-8_54
7. He, K., Zhang, X., Ren, S., Sun, J.: Deep residual learning for image recognition. In: 2016 IEEE Conference on Computer Vision and Pattern Recognition, CVPR 2016, Las Vegas, NV, USA, 27–30 June 2016, pp. 770–778 (2016)
8. Huang, C., Han, H., Yao, Q., Zhu, S., Zhou, S.K.: 3D U^2-Net: a 3D universal U-Net for multi-domain medical image segmentation. In: Shen, D., et al. (eds.) MICCAI 2019. LNCS, vol. 11765, pp. 291–299. Springer, Cham (2019). https://doi.org/10.1007/978-3-030-32245-8_33
9. Huang, X., Belongie, S.: Arbitrary style transfer in real-time with adaptive instance normalization. In: ICCV (2017)
10. Ilse, M., Tomczak, J.M., Welling, M.: Attention-based deep multiple instance learning. arXiv preprint arXiv:1802.04712 (2018)
11. Ioffe, S., Szegedy, C.: Batch normalization: accelerating deep network training by reducing internal covariate shift. In: 32nd International Conference on Machine Learning, pp. 448–456 (2015)
12. Lee, C.H., et al.: Interpretation us elastography in chronic hepatitis b with or without anti-HBV therapy. Appl. Sci. **7**, 1164 (2017)

13. Li, S., et al.: Liver fibrosis conventional and molecular imaging diagnosis update. J. Liver, **8** (2019)
14. Liu, J., Wang, W., Guan, T., Zhao, N., Han, X., Li, Z.: Ultrasound liver fibrosis diagnosis using multi-indicator guided deep neural networks. In: Suk, H.-I., Liu, M., Yan, P., Lian, C. (eds.) MLMI 2019. LNCS, vol. 11861, pp. 230–237. Springer, Cham (2019). https://doi.org/10.1007/978-3-030-32692-0_27
15. Manning, D., Afdhal, N.: Diagnosis and quantitation of fibrosis. Gastroenterology **134**(6), 1670–1681 (2008)
16. Meng, D., Zhang, L., Cao, G., Cao, W., Zhang, G., Hu, B.: Liver fibrosis classification based on transfer learning and FCNet for ultrasound images. IEEE Access **5**, 5804–5810 (2017)
17. Mojsilovic, A., Markovic, S., Popovic, M.: Characterization of visually similar diffuse diseases from b-scan liver images with the nonseparable wavelet transform. In: Proceedings of International Conference on Image Processing, vol. 3, pp. 547–550 (1997)
18. Ogawa, K., Fukushima, M., Kubota, K., Hisa, N.: Computer-aided diagnostic system for diffuse liver diseases with ultrasonography by neural networks. IEEE Trans. Nucl. Sci. **45**(6), 3069–3074 (1998)
19. Poynard, T., et al.: Prevalence of liver fibrosis and risk factors in a general population using non-invasive biomarkers (FibroTest). BMC Gastroenterol. **10**, 40 (2010)
20. Ronneberger, O., Fischer, P., Brox, T.: U-Net: convolutional networks for biomedical image segmentation. In: Navab, N., Hornegger, J., Wells, W.M., Frangi, A.F. (eds.) MICCAI 2015. LNCS, vol. 9351, pp. 234–241. Springer, Cham (2015). https://doi.org/10.1007/978-3-319-24574-4_28
21. Saverymuttu, S.H., Joseph, A.E., Maxwell, J.D.: Ultrasound scanning in the detection of hepatic fibrosis and steatosis. BMJ **292**(6512), 13–15 (1986)
22. Simonyan, K., Zisserman, A.: Very deep convolutional networks for large-scale image recognition. In: International Conference on Learning Representations (2015)
23. Tai, D.I., et al.: Differences in liver fibrosis between patients with chronic hepatitis B and C. J. Ultrasound Med. **34**(5), 813–821 (2015)
24. Ulyanov, D., Vedaldi, A., Lempitsky, V.S.: Instance normalization: the missing ingredient for fast stylization. CoRR abs/1607.08022 (2016)
25. Wang, Y., et al.: Weakly supervised universal fracture detection in pelvic X-Rays. In: Shen, D., et al. (eds.) MICCAI 2019. LNCS, vol. 11769, pp. 459–467. Springer, Cham (2019). https://doi.org/10.1007/978-3-030-32226-7_51
26. Zhou, Y., et al.: Hyper-pairing network for multi-phase pancreatic ductal adenocarcinoma segmentation. In: Shen, D., et al. (eds.) MICCAI 2019. LNCS, vol. 11765, pp. 155–163. Springer, Cham (2019). https://doi.org/10.1007/978-3-030-32245-8_18

Toward Rapid Stroke Diagnosis with Multimodal Deep Learning

Mingli Yu[1], Tongan Cai[1(✉)], Xiaolei Huang[1], Kelvin Wong[2], John Volpi[3], James Z. Wang[1(✉)], and Stephen T. C. Wong[2]

[1] The Pennsylvania State University, University Park, State College, PA, USA
{cta,jwang}@psu.edu

[2] TT and WF Chao Center for BRAIN and Houston Methodist Cancer Center, Houston Methodist Hospital, Houston, TX, USA

[3] Eddy Scurlock Comprehensive Stroke Center, Department of Neurology, Houston Methodist Hospital, Houston, TX, USA

Abstract. Stroke is a challenging disease to diagnose in an emergency room (ER) setting. While an MRI scan is very useful in detecting ischemic stroke, it is usually not available due to space constraint and high cost in the ER. Clinical tests like the Cincinnati Pre-hospital Stroke Scale (CPSS) and the Face Arm Speech Test (FAST) are helpful tools used by neurologists, but there may not be neurologists immediately available to conduct the tests. We emulate CPSS and FAST and propose a novel multimodal deep learning framework to achieve computer-aided stroke presence assessment over facial motion weaknesses and speech inability for patients with suspicion of stroke showing facial paralysis and speech disorders in an acute setting. Experiments on our video dataset collected on actual ER patients performing specific speech tests show that the proposed approach achieves diagnostic performance comparable to that of ER doctors, attaining a 93.12% sensitivity rate while maintaining 79.27% accuracy. Meanwhile, each assessment can be completed in less than four minutes. This demonstrates the high clinical value of the framework. In addition, the work, when deployed on a smartphone, will enable self-assessment by at-risk patients at the time when stroke-like symptoms emerge.

Keywords: Stroke · Emergency medicine · Computer vision · Facial video analysis · Machine learning

1 Introduction

Stroke is a common but fatal vascular disease. It is the second leading cause of death and the third leading cause of disability [14]. With acute ischemic stroke,

---

M. Yu, T. Cai, X. Huang, and J.Z. Wang are supported by Penn State University. S.T.C. Wong and K. Wong are supported by the T.T. and W.F. Chao Foundation and the John S. Dunn Research Foundation.

M. Yu and T. Cai—Made equal contributions.

A. L. Martel et al. (Eds.): MICCAI 2020, LNCS 12263, pp. 616–626, 2020.
https://doi.org/10.1007/978-3-030-59716-0_59

where parts of the brain tissue suffer from restrictions in blood supply to tissues, the shortage of oxygen needed for cellular metabolism quickly causes long-lasting damage to the areas of brain cells. The sooner a diagnosis is made, the earlier the treatment can begin and the better the outcome is expected for patients.

There is no rapid assessment approach for stroke. The gold standard test for stroke is the diffusion-weighted MRI scan that detects brain lesions, yet it is usually not accessible in the ER setting. There are two commonly adopted clinical tests for stroke in the ER, the Cincinnati Pre-hospital Stroke Scale (CPSS) [16] and the Face Arm Speech Test (FAST) [8]. Both methods assess the presence of any unilateral facial droop, arm drift, and speech disorder. The patient is requested to repeat a specific sentence (CPSS) or have a conversation with the doctor (FAST), and abnormality arises when the patient slurs, fails to organize his speech, or is unable to speak. However, the scarcity of neurologists [17] prevents such tests to be effectively conducted in all stroke emergency situations.

Recently, researchers have been striving for more accurate detection and evaluation methods for neurological disorders with the help of machine intelligence. Many researchers focused on the detection of facial paralysis. Studies have adopted 2D image analysis by facial landmarks with a criterion-based decision boundary [3,10,13]. Some used either videos [4,19], 3D information [1,15], or optical flow [20,24] as input features. Either rule-based [6,11] or learning-based [12,18] methods were applied to perform the analysis. However, existing approaches either evaluate their methods between subjects that are normal versus those with clear signs of a stroke, deal with only synthetic data, fail in capturing the spatiotemporal details of facial muscular motions, or rely on experimental settings with hard constraints on the head orientation, hindering their adoption in real clinical practice or for patient self-assessment.

We propose a new deep learning method to analyze the presence of stroke in patients with suspicion or identified with a high risk of stroke. We formulate this task as a binary classification problem, *i.e.*, stroke vs. non-stroke. With patients recorded while performing a set of vocal tests, videos are processed to extract facial-motions-only frames, and the audio is transcribed. A deep neural network is set up for the collaborative classification of the videos and transcripts.

The **main contributions** of this work are summarized in three aspects:

1. To our knowledge, this is the first work to analyze the presence of stroke among actual ER patients with suspicion of stroke using computational facial motion analysis, and is also the first attempt to adopt a natural language processing (NLP) method for the speech ability test on at-risk stroke patients. We expect this work to serve as a launchpad for more research in this area.
2. A multi-modal fusion of video and audio deep learning models is introduced with 93.12% sensitivity and 79.27% accuracy in correctly identifying patients with stroke, which is comparable to the clinical impression given by ER physicians. The framework can be deployed on a mobile platform to enable self-assessment for patients right after symptoms emerge.
3. The proposed temporal proposal of human facial videos can be adopted in general facial expression recognition tasks. The proposed multi-modal method

can potentially be extended to other clinical tests, especially for neurological disorders that result in muscular motion abnormalities, expressive disorder, and cognitive impairments.

2 Emergency Room Patient Dataset

The clinical dataset for this study was acquired in the emergency rooms (ERs) of Houston Methodist Hospital by the physicians and caregivers from the Eddy Scurlock Stroke Center at the Hospital under an IRB-approved[1] study. We took months to recruit a sufficiently large pool of patients in certain emergency situations. The subjects chosen are patients with suspicion of stroke while visiting the ER. 47 males and 37 females have been recruited in a race-nonspecific way.

Each patient is asked to perform two speech tasks: 1) to repeat the sentence "it is nice to see people from my hometown" and 2) to describe a "cookie theft" picture. The ability of speech is an important indicator to the presence of stroke; if the subject slurs, mumbles, or even fails to repeat the sentence, they have a very high chance of stroke [8,16]. The "cookie theft" task has been making great success in identifying patients with Alzheimer's-related dementia, aphasia, and some other cognitive-communication impairments [5].

The subjects are video recorded as they perform the two tasks with an iPhone X's camera. Each video has metadata information on both clinical impressions by the ER physician (indicating the doctor's initial judgement on whether the patient has a stroke or not from his/her speech and facial muscular conditions) and ground truth from the diffusion-weighted MRI (including the presence of acute ischemic stroke, transient ischemic attack (TIA), etc.). Among the 84 individuals, 57 are patients diagnosed with stroke using the MRI, 27 are patients who do not have a stroke but are diagnosed with other clinical conditions. In this work, we construct a binary classification task and only attempt to identify stroke cases from non-stroke cases, regardless of the stroke subtypes.

Our dataset is unique, as compared to existing ones [7,10], because our subjects are actual patients visiting the ERs and the videos are collected in an unconstrained, or "in-the-wild", fashion. In most existing work, the images or videos were taken under experimental settings, where good alignment and stable face pose can be assumed. In our dataset, the patients can be in bed, sitting, or standing, where the background and illumination are usually not under ideal control conditions. Apart from this, we only asked patients to focus on the picture we showed to them, without rigidly restricting their motions. The acquisition of facial data in natural settings makes our work robust and practical for real-world clinical use, and ultimately empowers our method for remote diagnosis of stroke and self-assessment in any setting.

[1] This study received IRB approval: Houston Methodist IRB protocol No. Pro00020577, Penn State IRB site No. SITE00000562.

3 Methodology

We propose a computer-aided diagnosis method to assess the presence of stroke in a patient visiting ER. This section introduces our information extraction methods, separate classification modules for video and audio, and the overall network fusion mechanism.

3.1 Information Extraction

For each raw video, we propose a spatiotemporal proposal of frames and conduct a machine speech transcription for the raw audio.

Spatiotemporal Proposal of Facial Action Video: We develop a pipeline to extract frame sequences with near-frontal facial pose and minimum non-facial motions. First, we detect and track the location of the patient's face as a square bounding box. During the same process, we detect and track the facial landmarks of the patient, and estimate the pose. Frame sequences 1) with large roll, yaw or pitch, 2) showing continuously changing pose metrics, or 3) having excessive head translation estimated with optical flow magnitude are excluded. A stabilizer with sliding window over the trajectory of between-frame affine transformations smooths out pixel-level vibrations on the sequences before classification.

Speech Transcription: We record the patient's speech and transcribe the recorded speech audio file using Google Cloud Speech-to-Text service. Each audio segment is turned into a paragraph of text in linear time, together with a confidence score for each word, ready for subsequent classification.

3.2 Deep Neural Network for Video Classification

Facial motion abnormality detection is essential to stroke diagnosis [10], but challenges remain in several approaches to this problem. First, the limited number of videos prevent us from training 3D networks (treating time as the 3rd dimension) such as C3D [21] and R3D [22], because these networks have a large number of parameters and their training can be difficult to converge with a small dataset. Second, although optical flow has been proven useful in capturing temporal changes in gesture or action videos, it is ineffective in identifying subtle facial motions due to noise [23] and can be expensive to compute.

Network Architecture: In this work, we propose the deep neural network shown in Fig. 1 for binary classification of a video to *stroke* vs. *non-stroke*. For a video consisting of N frames, we take the kth and k+1th frames and calculate the difference between their embedding features right after the first 3×3 convolutional layer. Next, a ResNet-34 [9] model outputs the class probability vector p_k for each frame based on the calculated feature differences. An overall temporal loss $\mathcal{L}$ is calculated by combining a negative log-likelihood (NLL) loss term and smoothness regularization loss terms based on the class probability vectors of three consecutive frames. From the frame-level class probabilities, a video-level

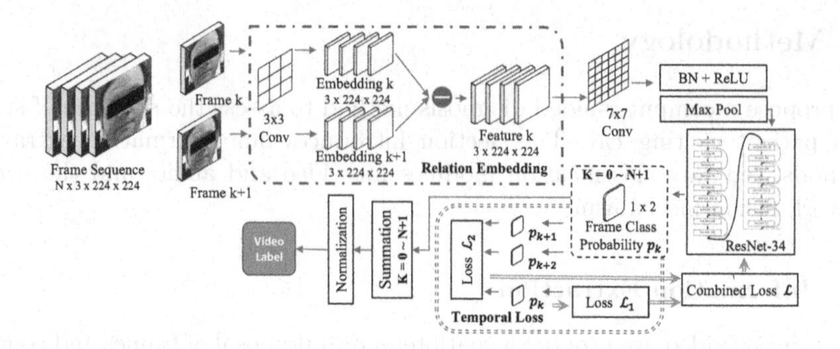

Fig. 1. Flow diagram of the video classification module.

class probability vector is obtained by averaging over all frames' probabilities, and the predicted video label will be the class with higher probability.

Relation Embedding: One novelty of our proposed video module is in the classification using feature differences between consecutive frames instead of using directly the frame features. The rationale behind this choice is in that we expect the network to learn and classify based on motion features. Features from single frames contain a lot of information about the appearance of face which are useful for face identification but not useful in characterizing facial motion abnormality.

Temporal Loss: Denote i as the frame index, y_i as the class label for frame i obtained based on the ground truth video label, and p_i as the predicted class probability. The combined loss $\mathcal{L}$ for frame i is defined with three terms: $\mathcal{L}^{(i)} = \mathcal{L}_1^{(i)} + \alpha(\mathcal{L}_{2_{(1)}}^{(i)} + \beta\mathcal{L}_{2_{(2)}}^{(i)})$, where α and β are tunable weighting coefficients. The first loss term, $\mathcal{L}_1^{(i)} = -(y_i \log p_i + (1 - y_i)\log(1 - p_i))$, is the NLL loss. Note that we sample a subset of the frames to mitigate the overfitting on shape identity. By assuming consecutive frames have continuous and similar class probabilities, we develop a $\mathcal{L}_{2_{(1)}}^{(i)}$ loss defined on the frame class probabilities for three adjacent frames where λ is a small threshold to restrict the inconsistency by penalizing those frames with large class probability differences. We also design another loss, $\mathcal{L}_{2_{(2)}}^{(i)}$, to encourage random walk around the convergence point to mitigate overfitting. Specifically,

$$\mathcal{L}_{2_{(1)}}^{(i)} = \sum_{j\in[0,1,2]} \max\{(|p_{i+j} - p_{i+1+j}| - \lambda), 0\}\ , \tag{1}$$

$$\mathcal{L}_{2_{(2)}}^{(i)} = \sum_{j\in[0,1,2]} \mathbb{1}_{p_{i+j}==p_{i+1+j}}\ . \tag{2}$$

In practice, we adopt a batch training method, and all frames in a batch are weighted equally for loss calculation.

3.3 Transcript Classification for Speech Ability Assessment

We formulate the speech ability assessment as a text classification problem. Subjects without speech disorder complete the speech task with organized sentences and maintain a good vocabulary set size, whereas patients with speech impairments either put up a few words illogically or provide mostly unrecognizable speech. Hence, we concurrently formulate a binary classification on the speech given by the subjects to determine if stroke exists.

Preprocessing: For each speech case $\mathcal{T} := \{t_i, ..., t_N\}$ extracted from the obtained transcripts where t_i is a single word and N is the number of words in the case, we first define the encoding of the words $\mathcal{E}$ over the training set by their order of appearance, $\mathcal{E}(t_i) := d_i$, $d_i \in \mathbb{I}$; $\mathcal{E}(\mathcal{T}) := \mathcal{D}$ and $\mathcal{D} = \{d_i, ..., d_N\} \in \mathbb{I}^n$. We denote the vocabulary size obtained as v. Due to the length difference between cases, we pad the sequences to the max length m of the dataset, so that $\mathcal{D}' = \{d_i, ..., d_N, p_1, ..., p_{m-n}\} \in \mathbb{I}^m$ where p_i denotes a constant padding value. We further embed the padded feature vectors to an embedding dimension E so that the final feature vector has $\mathcal{X} := \{x_i, .., x_m\}$ and $x_i \in \mathbb{R}^{E \times v}$.

Text Classification with Long Short-Term Memory (LSTM): We construct a basic bidirectional LSTM model to classify the texts. For the input $\mathcal{X} = \{x_i, .., x_m\}$, the LSTM model generates a series of hidden states $\mathcal{H} := \{h_1, .., h_m\}$ where $h_i \in \mathbb{R}^t$. We take the output from the last hidden state h_t, apply a fully-connected (FC) layer before output (class probabilities/logits) $\hat{y_i} \in \mathbb{R}^2$. For our task, we leave out the last FC layer for model fusion.

3.4 Two-Stream Network Fusion

Overall structure of the model: Figure 2 shows the data pipeline of the proposed fusion model. Videos that are collected following the protocol are

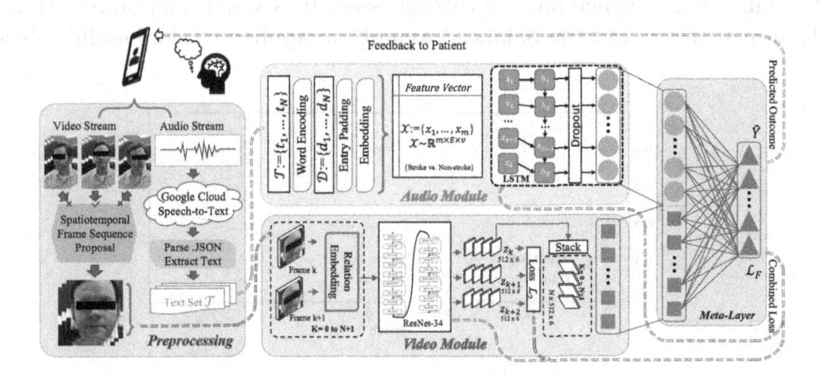

Fig. 2. Flow diagram of the two-stream network fusion process. z_k is the feature output from the ResNet-34 before the final FC layer.

uploaded to a database, while audios are extracted and forwarded to Google Cloud Speech-to-Text which generates the transcript. Meanwhile, videos are sent to the spatiotemporal proposal module to perform face detection, tracking, cropping, and stabilization. The preprocessed data are further handled per-case-wise and loaded into the audio and video modules. Finally, a "meta-layer" combines the outputs of the two modules and gives the final prediction on each case.

Fusion Scheme: We take a simple fusion scheme of the two models. For both the video and text/audio modules, we remove the last fully-connected layer before the output and concatenate the feature vectors. We construct a fully-connected "meta-layer" for the output of class probabilities. For all the N frames in a video, the frame-level class probabilities from the model are concatenated into $\hat{Y} = \{\hat{p_1}, ..., \hat{p_N}\}$. The fusion loss $\mathcal{L}_F$ is defined in a similar way as the temporal loss; instead of using only video-predicted probabilities, the fusion loss combines both video- and text-predicted probabilities. Note again that the fusion model operates at the frame level, and a case-level prediction is obtained by summing and normalizing class probabilities over all frames.

4 Experiments and Results

Implementation and Training: The whole framework[2] is running on Python 3.7 with Pytorch 1.4, OpenCV 3.4, CUDA 9.0, and Dlib 19. The model starts with a pretrained model on ImageNet. The entire pipeline runs on a computer with a quad-core CPU and one GTX 1070 GPU. To accommodate for the existing class imbalance inside the dataset and ER setting, a higher class weight (1.25) is assigned to the non-stroke class. For evaluation, we report the accuracy, specificity, sensitivity and area under the ROC curve (AUC) from 5-fold cross validation results. The loss curves for one of the folds are presented in Fig. 3. The learning rate is tuned to 0.0001 and we early stop at epoch 8 due to the quick convergence and overfitting issue. It is worth mentioning that the early stop strategy and the balanced weight are applied to the baselines below.

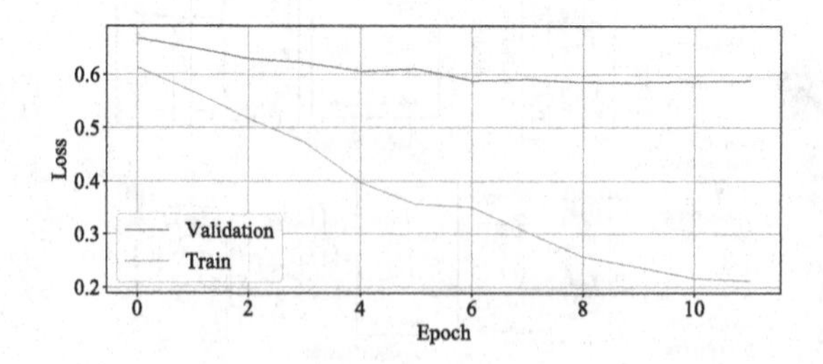

Fig. 3. Loss curves for one of the folds.

[2] Codes are available in https://github.com/0CTA0/MICCAI20_MMDL_PUBLIC.

Baselines and Comparison: To evaluate our proposed method, we construct a number of baseline models for both video and audio tasks. The ground truth for comparison is the binary diagnosis result for each video/audio. General video classification models for video tagging or action recognition are not suitable for our task since they require all frames throughout a video clip to have the same label. In our task, since stroke patients may have many normal motions, the frame-wise labels may not be equal to the video label all the time. For single frame models such as ResNet-18, we use the same preprocessed frames as input and derive a binary label for each frame. The label with more frames is then assigned as the video-level label. For multiple frame models, we simply input the same preprocessed video. We also compare with a traditional method based on identifying facial landmarks and analyzing facial asymmetry ("Landmark + Asymmetry"), which detects the mid-line of a patient's face and checks for bilateral pixel-wise differences on between-frame optical-flow vectors. The binary decision is given by statistical values including the number of peaks and average asymmetry index. We further tested our video module separately with and without using feature differences between consecutive frames. We compare the result of our audio module to that of sound wave classification with pattern recognition on spectrogram [2].

Table 1. Experimental results. Best values are in **bold**

Method	Metrics			
	Accuracy	Specificity	Sensitivity	AUC
Landmark+Asymmetry	62.35%	66.67%	60.34%	0.6350
Video Module w/o difference	70.28%	11.11%	**98.24%**	0.5467
Video Module w/ difference	76.67%	62.21%	96.42%	0.7281
ResNet-18	58.20%	42.75%	70.22%	0.5648
VGG16	52.30%	27.98%	71.22%	0.4960
Audio Module Only	70.24%	40.74%	84.21%	0.6248
Soundwave/Spectrogram	68.67%	59.26%	77.58%	0.6279
Proposed Method	**79.27%**	66.07%	93.12%	**0.7831**
Clinical Impression	72.94%	**77.78%**	70.68%	0.7423

As shown in Table 1, the proposed method outperforms all the strong baselines by achieving a 93.12% sensitivity and a 79.27% accuracy. The improvements of our proposed method from the baselines on accuracy are ranging from 10% to 30%. It is noticeable that proven image classification baselines (ResNet, VGG) are not ideal for our "in-the-wild" data. Comparing to the clinical impressions given by ER doctors, the proposed method achieves even higher accuracy and greatly improves the sensitivity, indicating that more stroke cases are correctly identified by our proposed approach. ER doctors tend to rely more on the speech abilities of the patients and may overlook subtle facial motion weaknesses.

Our objective is to identify real stroke and fake stroke cases among incoming patients, who are already identified with high risk of stroke. If the patterns are subtle or challenging to observe by humans, ER doctors may have difficulty on those cases. We infer that the video module in our framework can detect those subtle facial motions that doctors can neglect and complement the diagnosis based on speech/audio. On the other hand, the ER doctors have access to emergency imaging reports and other information in the Electronic Health Records (EHR). With more information incorporated, we believe the performance of the framework can get further improved. It is also important to note that by using feature difference between consecutive frames for classification, the performance of the video module is greatly improved, validating our hypothesis about modeling based on motion.

Through the experiments, all the methods are experiencing low specificity (*i.e.*, identifying non-stroke cases), which is reasonable because our subjects are patients with suspicion of stroke rather than the general public. False negatives would be dangerous and could lead to hazardous outcome. We also took a closer look at the false negative and false positive cases. The false negatives are due to the labeling of cases using final diagnosis given based on diffusion-weighted MRI (DWI). DWI can detect very small lesions that may not cause noticeable abnormalities in facial motion or speech ability. Such cases coincide with the failures in clinical impression. The false positives typically result from background noise in audio, varying shapes of beard, or changing illumination conditions. A future direction is to improve specificity with more robust methods on both audio and video processing.

We also evaluate the efficiency of our approach. The recording runs for a minute, the extraction and upload of audio takes half a minute, the transcribing takes an extra minute, and the video processing is completed in two minutes. The prediction with the deep models can be achieved within seconds with GTX 1070. Therefore, the entire process takes no more than four minutes per case. If the approach is deployed onto a smartphone, we can rely on Google Speech-to-Text's real-time streaming method and perform the spatiotemporal frame proposal on the phone. Cloud computing can be leveraged to perform the prediction in no more than a minute, after the frames are uploaded. In such a case, the total time for one assessment should not exceed three minutes. This is ideal for performing stroke assessment in an emergency setting and the patients can make self-assessments even before the ambulance arrives.

5 Conclusions

We proposed a multi-modal deep learning framework for on-site clinical detection of stroke in an ER setting. Our framework is able to identify stroke based on abnormality in the patient's speech ability and facial muscular movements. We construct a deep neural network for classifying patient facial video, and fuse the network with a text classification model for speech ability assessment. Experimental studies demonstrate that the performance of the proposed approach is comparable to clinical impressions given by ER doctors, with a 93.12% sensitivity and a 79.27% accuracy. The approach is also efficient, taking less than four minutes for assessing one patient case. We expect that our proposed approach will be clinically relevant and can be deployed effectively on smartphones for fast and accurate assessment of stroke by ER doctors, at-risk patients, or caregivers.

References

1. Claes, P., Walters, M., Vandermeulen, D., Clement, J.G.: Spatially-dense 3D facial asymmetry assessment in both typical and disordered growth. J. Anat. **219**(4), 444–455 (2011)
2. Dennis, J., Tran, H.D., Li, H.: Spectrogram image feature for sound event classification in mismatched conditions. IEEE Signal Process. Lett. **18**(2), 130–133 (2010)
3. Dong, J., Lin, Y., Liu, L., Ma, L., Wang, S.: An approach to evaluation of degree of facial paralysis based on image processing and pattern recognition. J. Inf. Comput. Sci. **5**(2), 639–646 (2008)
4. Frey, M., et al.: Three-dimensional video analysis of the paralyzed face reanimated by cross-face nerve grafting and free gracilis muscle transplantation: Quantification of the functional outcome. Plast. Reconstr. Surg. **122**(6), 1709–1722 (2008)
5. Giles, E., Patterson, K., Hodges, J.R.: Performance on the Boston cookie theft picture description task in patients with early dementia of the alzheimer's type: missing information. Aphasiology **10**(4), 395–408 (1996)
6. Guo, Z., et al.: An unobtrusive computerized assessment framework for unilateral peripheral facial paralysis. IEEE J. Biomed. Health Inform. **22**(3), 835–841 (2017)
7. Guo, Z., et al.: Deep assessment process: objective assessment process for unilateral peripheral facial paralysis via deep convolutional neural network. In: Proceedings of the IEEE International Symposium on Biomedical Imaging, pp. 135–138 (2017)
8. Harbison, J., Hossain, O., Jenkinson, D., Davis, J., Louw, S.J., Ford, G.A.: Diagnostic accuracy of stroke referrals from primary care, emergency room physicians, and ambulance staff using the face arm speech test. Stroke **34**(1), 71–76 (2003)
9. He, K., Zhang, X., Ren, S., Sun, J.: Deep residual learning for image recognition. In: Proceedings of the IEEE Conference on Computer Vision and Pattern Recognition, pp. 770–778 (2016)
10. He, S., Soraghan, J.J., O'Reilly, B.F.: Automatic motion feature extraction with application to quantitative assessment of facial paralysis. In: Proceedings of the IEEE International Conference on Acoustics, Speech and Signal Processing, vol. 1, pp. 441–444 (2007)
11. Horta, R., Aguiar, P., Monteiro, D., Silva, A., Amarante, J.M.: A facegram for spatial-temporal analysis of facial excursion: applicability in the microsurgical reanimation of long-standing paralysis and pretransplantation. J. Cranio-Maxillofacial Surg. **42**(7), 1250–1259 (2014)
12. Hsu, G.S.J., Chang, M.H.: Deep hybrid network for automatic quantitative analysis of facial paralysis. In: Proceedings of the IEEE International Conference on Advanced Video and Signal Based Surveillance, pp. 1–7 (2018)
13. Hu, Y., Chen, L., Zhou, Y., Zhang, H.: Estimating face pose by facial asymmetry and geometry. In: Proceedings of the IEEE International Conference on Automatic Face and Gesture Recognition (F&G), pp. 651–656 (2004)

14. Johnson, W., Onuma, O., Owolabi, M., Sachdev, S.: Stroke: a global response is needed. Bull. World Health Organ. **94**(9), 634 (2016)
15. Khairunnisaa, A., Basah, S.N., Yazid, H., Basri, H.H., Yaacob, S., Chin, L.C.: Facial-paralysis diagnostic system based on 3D reconstruction. In: AIP Conference Proceedings, vol. 1660, p. 070026. AIP Publishing (2015)
16. Kothari, R.U., Pancioli, A., Liu, T., Brott, T., Broderick, J.: Cincinnati prehospital stroke scale: reproducibility and validity. Ann. Emerg. Med. **33**(4), 373–378 (1999)
17. Leira, E.C., Kaskie, B., Froehler, M.T., Adams Jr., H.P.: The growing shortage of vascular neurologists in the era of health reform: Planning is brain!. Stroke **44**(3), 822–827 (2013)
18. Li, P., et al.: A two-stage method for assessing facial paralysis severity by fusing multiple classifiers. In: Sun, Z., He, R., Feng, J., Shan, S., Guo, Z. (eds.) CCBR 2019. LNCS, vol. 11818, pp. 231–239. Springer, Cham (2019). https://doi.org/10.1007/978-3-030-31456-9_26
19. Anping, S., Guoliang, X., Xuehai, D., Jiaxin, S., Gang, X., Wu, Z.: Assessment for facial nerve paralysis based on facial asymmetry. Aust. Phys. Eng. Sci. Med. **40**(4), 851–860 (2017). https://doi.org/10.1007/s13246-017-0597-4
20. Soraghan, J.J., O'Reilly, B.F., McGrenary, S., He, S.: Automatic facial analysis for objective assessment of facial paralysis. In: Proceedings of the 1st International Conference on Computer Science from Algorithms to Applications, Cairo, Egypt (2009)
21. Tran, D., Bourdev, L., Fergus, R., Torresani, L., Paluri, M.: Learning spatiotemporal features with 3D convolutional networks. In: Proceedings of the IEEE International Conference on Computer Vision, pp. 4489–4497 (2015)
22. Tran, D., Wang, H., Torresani, L., Ray, J., LeCun, Y., Paluri, M.: A closer look at spatiotemporal convolutions for action recognition. In: Proceedings of the IEEE Conference on Computer Vision and Pattern Recognition, pp. 6450–6459 (2018)
23. Wang, L., et al.: Temporal segment networks: towards good practices for deep action recognition. In: Leibe, B., Matas, J., Sebe, N., Welling, M. (eds.) ECCV 2016. LNCS, vol. 9912, pp. 20–36. Springer, Cham (2016). https://doi.org/10.1007/978-3-319-46484-8_2
24. Wang, S., Li, H., Qi, F., Zhao, Y.: Objective facial paralysis grading based on p_{face} and EigenFlow. Med. Biol. Eng. Comput. **42**(5), 598–603 (2004). https://doi.org/10.1007/BF02347540

Learning and Reasoning with the Graph Structure Representation in Robotic Surgery

Mobarakol Islam[1,2], Lalithkumar Seenivasan[1], Lim Chwee Ming[3], and Hongliang Ren[1(✉)]

[1] Department of Biomedical Engineering, National University of Singapore, Singapore, Singapore
{mobarakol,lalithkumar_s}@u.nus.edu, ren@nus.edu.sg, hlren@ieee.org
[2] Biomedical Image Analysis Group, Imperial College London, London, UK
[3] Department of Otolaryngology, Singapore General Hospital, Singapore, Singapore
lim.chwee.ming@singhealth.com.sg

Abstract. Learning to infer graph representations and performing spatial reasoning in a complex surgical environment can play a vital role in surgical scene understanding in robotic surgery. For this purpose, we develop an approach to generate the scene graph and predict surgical interactions between instruments and surgical region of interest (ROI) during robot-assisted surgery. We design an attention link function and integrate with a graph parsing network to recognize the surgical interactions. To embed each node with corresponding neighbouring node features, we further incorporate SageConv into the network. The scene graph generation and active edge classification mostly depend on the embedding or feature extraction of node and edge features from complex image representation. Here, we empirically demonstrate the feature extraction methods by employing label smoothing weighted loss. Smoothing the hard label can avoid the over-confident prediction of the model and enhances the feature representation learned by the penultimate layer. To obtain the graph scene label, we annotate the bounding box and the instrument-ROI interactions on the robotic scene segmentation challenge 2018 dataset with an experienced clinical expert in robotic surgery and employ it to evaluate our propositions.

1 Introduction

In modern surgical practices, minimally invasive surgeries (MIS) has become a norm as it greatly reduces the trauma of open surgery in patients and reduces their recovery period. The accuracy and reliability of the MIS, and dexterity of a surgeon improve with the introduction of robot-assisted MIS (RMIS)[13]. Although RMIS holds great potential in the medical domain, its applications and capabilities are still limited by the lack of haptic feedback, reduced field of

This work supported by NMRC Bedside & Bench under grant R-397-000-245-511.

A. L. Martel et al. (Eds.): MICCAI 2020, LNCS 12263, pp. 627–636, 2020.
https://doi.org/10.1007/978-3-030-59716-0_60

view, and the deficit of evaluation matrix to measure a surgeon's performance[2, 8,9,13]. In recent times, attempts have been made to address these limitations by introducing new techniques inspired by advancement in the computer vision field. To address the shortcomings and assist the surgeons better, surgical tool tracking [2,7] and instrument segmentation [8,14] have been introduced to allow the overlay of the medical image on top of the surgical scene. While considerable progress has been made in detecting the tools and semantic segmentation, few key problems still exist. Firstly, despite identifying the instruments, the system is unable to decide when to stimulate haptic sense. Secondly, the system cannot determine when to overlay the medical image and when not to during the surgery to ensure that it does not obstruct the surgeon's view of the surgical scene. Finally, the lack of standardized evaluation matrix for surgeons and the inability of the system to perform spatial understanding are the key reasons for these limitations.

Spatial understanding, termed as deep reasoning in the artificial intelligence (AI) domain, enables the system to infer the implicit relationship between instruments and tissues in the surgical scene. Incorporating reasoning capability in robotic surgery can empower the machine to understand the surgical scene and allows the robot to execute informed decisions such as stimulating a haptic sense or overlay medical images over to assist surgeons based on interactions between the tissue and tool. Furthermore, this would allow the system to identify the tasks performed by surgeons, which can then be used for evaluating their performance. However, the detection of tool-tissue interaction poses one major problem. Most of the advancements made in the AI domains, such as convolutional neural networks (CNNs), addresses issues in the euclidean domain[20]. However, detecting interactions between tissue and tools can fall under non-euclidean space where defective tissues can be related to multiple instruments, and each instrument could have multiple interaction types. For the machine learning problems in the non-euclidean domain, graph neural network (GNN) has been proposed, which allows the network to perform spatial reasoning based on features extracted from the connected nodes[15,20]. Inspired by graph parsing neural network (GPNN)[15], here, we present an enhanced GNN model that infers on the tool-tissue interaction graph structure to predicts the presence and type of interaction between the defective tissue and the surgical tools. Our key contributions in this work are as follows:

- Label smoothened features: Improve model performance by label smoothened node and edge features.
- Incorporate SageConv [4]: An inductive framework that performs node embedding based on local neighbourhood node features.
- Attention link function: Attention-based link function to predict the adjacent matrix of a graph.
- Surgical scene graph dataset: Annotate bounding box and tool-tissue interaction to predict scene graph during robotic surgery.

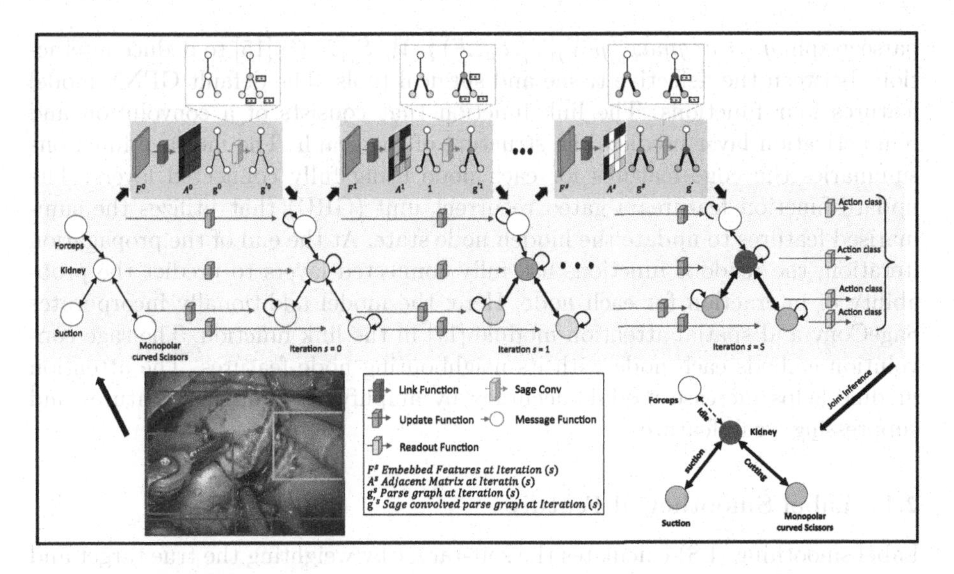

Fig. 1. Proposed architecture: Given a surgical scene, firstly, label smoothened features $\mathcal{F}$ are extracted. The network then outputs a parse graph based on the $\mathcal{F}$. The attention link function predicts the adjacent matrix of the parse graph. The thicker edge indicates possible interaction between the node. The SageConv embeds each node with its neighbouring node features. A brighter node color represents the existence of neighbouring node features within a node. The message function updates the node values based on the sum of interaction features between neighbour nodes. The hidden features of each node are updated by the update node. The readout function predicts the node labels that signify the interaction type between the connected nodes. (Color figure online)

2 Proposed Approach

To perform surgical scene understanding, the interactions between the defective tissue and surgical instruments are presented in the non-euclidean domain. Surgical tools, defective tissues and their interactions are represented as node and edge of a graph structure using the label smoothened spatial features extracted using the ResNet18 [5] (as Sect. 2.1). Inspired from GPNN [15], the complete tissue-tool interaction is modelled in the form of graph $\mathcal{G}(\mathcal{V}, \mathcal{E}, \mathcal{Y})$. The nodes $v \in \mathcal{V}$ of the graph signifies the defective tissue and instruments. The graph's edges $e \in \mathcal{E}$ denoted by $e = (x, w) \in \mathcal{V} \text{ x } \mathcal{V}$ indicates the presence of interaction. The notation $y_v \in \mathcal{Y}$ correlates to a set of labels, holding the interaction state of each node $v \in \mathcal{V}$.

Each image frame in the surgical scene is inferred as a parse graph $g = (\mathcal{V}_g, \mathcal{E}_g, \mathcal{Y}_g)$, where, g is a subgraph of $\mathcal{G}, \mathcal{E}_g \subseteq \mathcal{E}$ and $\mathcal{V}_g \subseteq \mathcal{V}$ as shown in Fig. 1. Given a surgical scene, the SSD network is first employed to detect bounding boxes. Label smoothened features $\mathcal{F}$, comprising of node features $\mathcal{F}_v$ and edge features $\mathcal{F}_e$ are then extracted based on these detected bounding boxes. Utilizing these features, an enhanced graph-based deep reasoning model then infers a parse graph, $g^* = argmax_g\ p(\mathcal{Y}_g|\mathcal{V}_g, \mathcal{E}_g, \mathcal{F})\ p(\mathcal{V}_g, \mathcal{E}_g|\mathcal{F}, \mathcal{G})$ [15] to deduce interactions between the defective tissue and surgical tools. The default GPNN model features four functions. The link function that consists of a convolution and relu activation layer predicts the structure of the graph. The message functions summaries the edge features for each node using fully connected layers. The update function features a gated recurrent unit (GRU) that utilizes the summarised features to update the hidden node state. At the end of the propagation iteration, the readout functions use fully connected layers to predict the probability of interaction for each node. Here, the model additionally incorporates SageConv and spatial attention module [17] in the link function. The sage convolution embeds each node with its neighbouring node features. The attention module helps increase module accuracy by amplifying significant features and suppressing weak features.

2.1 Label Smoothened Feature

Label smoothing (LS) calculates the soft-target by weighting the true target and use it to measure the cross-entropy loss. For example, if T_k is the true onehot target and ϵ is the smoothing factor then smoothen label for the K classes, $T_k^{LS} = T_k(1-\epsilon) + \epsilon/K$ and LS cross-entropy loss for the model prediction P_k, $CE^{LS} = \sum_{k=1}^{K} -T_k^{LS} log(P_k)$. It is observed that LS prevents the over-confident of model learning and calibrates the penultimate layer and represents the features of the same classes in a tight cluster [12]. We employ LS on ResNet18 to train with surgical object classification and extract edge and node features for graph learning. With LS, the cluster of the same classes is much tighter than the extracted features without LS. The label smoothened features are represented by $\mathcal{F} = \mathcal{F}_v, \mathcal{F}_e$, where $\mathcal{F}_v$ and $\mathcal{F}_e$ denotes the label smoothened node and edge features respectively.

2.2 Inductive Graph Learning by SageConv

SageConv[4], an inductive framework, enables node embedding by sampling and aggregating from the node's local neighbourhood. It is incorporated into the model to introduce inductive graph learning in the parse graph. Based on the adjacent matrix deduced by the link function, the SageConv further embeds each node in the parse graph by sampling and aggregating its connected neighbouring node features. Given an independent node feature $\mathcal{F}_v$, $\forall v \in \mathcal{V}$, the SageConv outputs a new embedded node features $\mathcal{F}_v \leftarrow \mathcal{F}_v^K$, $\forall v \in \mathcal{V}$ where $\mathcal{F}_v^K$ denotes the aggregated node features from the connected neighbourhood(K) nodes.

3 Experiments

3.1 Dataset

In this work, the robotic scene segmentation challenge 2018 dataset [1] is exploited to generate a graph-based tissue-tool interaction dataset. The training

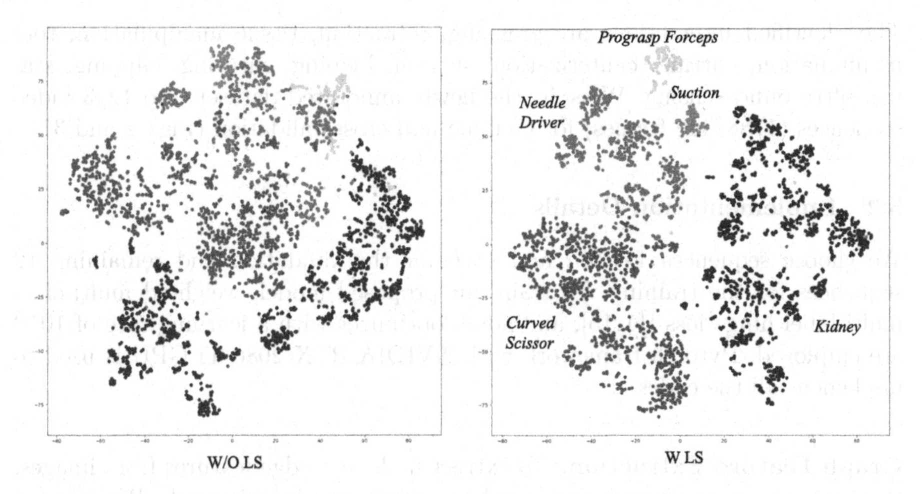

Fig. 2. Node and edge feature embedding with and without label smoothing (LS). We choose 5 semantically different classes to plot the tSNE. Based on the comparison between the two images, it is observed that the clusters containing features obtained from the same class are more compact when extracted using the LS method.

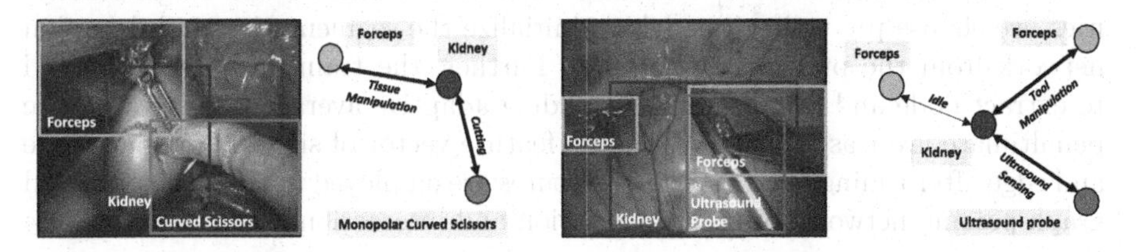

Fig. 3. Graph annotation for tool-tissue interaction in a robotic surgical scene: Each surgical scene is annotated using a graph structure. Each node is the graph denotes a defective tissue or surgical instrument. The edges between the nodes represent the type of interaction between the nodes.

subset consists of 15 robotic nephrectomy procedures captured on the da Vinci X or Xi system. There are 149 frames per video sequence, and the dimension of each frame is 1280×1024. Segmentation annotations are provided with 10 different classes, including instruments, kidneys, and other objects in the surgical scenario. The main differences with the 2017 instrument segmentation dataset [3] are annotation of kidney parenchyma, surgical objects such as suturing needles, Suturing thread clips, and additional instruments. We annotated the graphical representation of the interaction between the surgical instruments and the defective tissue in the surgical scene with the help of our clinical expertise with the da Vinci Xi robotic system. We also delineate the bounding box to identify all the surgical objects. Kidney and instruments are represented as nodes and active edges annotated as the interaction class in the graph. In total, 12 kinds of interactions were identified to generate the scene graph representation.

The identified interactions are grasping, retraction, tissue manipulation, tool manipulation, cutting, cauterization, suction, looping, suturing, clipping, staple, ultrasound sensing. We split the newly annotated dataset into 12/3 video sequences (1788/449 frames) for training and cross-validation (Figs. 2 and 3).

3.2 Implementation Details

We choose sequences $1st, 5th$ and $16th$ for the validation and remaining 12 sequences for the training. To train our proposed model, weighted multi-class multi-label hinge loss [11,15], and adam optimizer with a learning rate of 10^{-5} are employed. Pytorch framework with NVIDIA RTX 2080 Ti GPU is used to implement all the codes.

Graph Feature Extraction. To extract node and edge features from images, there are extensive experiments we have conducted in this work. We select 4 different types of deep neural networks such as classification, detection, segmentation and multitask learning models and train them with corresponding annotations for both w and w/o LS in cross-entropy loss. As instrument segmentation challenge 2017 [3] consists of similar instruments and surgical activities, we choose pre-trained models to initialize the segmentation and detection network from the previous works [6,7]. Further, the trained model is utilized to extract node and edge features by adding adaptive average pooling after the penultimate layer as [12,15]. We choose feature vector of size 200 for each node and edge after tuning. The extracted features are employed to train the proposed graph parsing network for scene generation and interaction prediction.

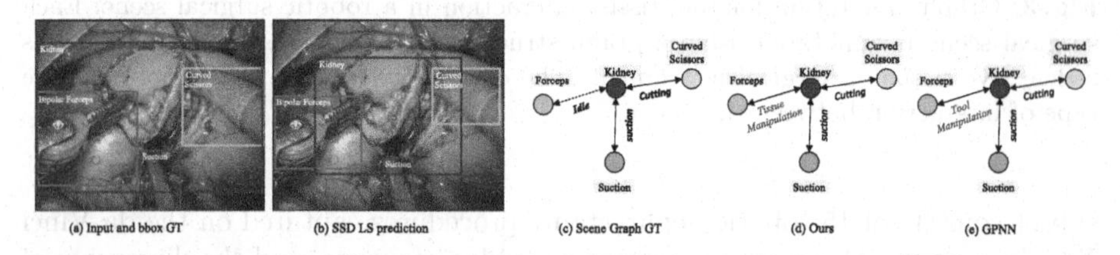

Fig. 4. Qualitative analysis: (a) Input surgical scene with true bounding box. (b) Bounding box prediction based on SSD LS model. (c) Scene graph ground truth that represents the tissue-tool interaction in the scene. (d) Predicted scene graph based on our proposed model. (e) Predicted scene graph based on GPNN model [15]. The interaction highlighted in red in (d) and (e) indicates the false prediction made by the respective models. (Color figure online)

3.3 Results

The enhanced network proposed in the model is evaluated both qualitatively and quantitatively against the state of the art networks. Fig. 4 shows the qualitative performance of the SSD LS model (b) in predicting the bounding box against the ground truth (a). The figure also shows our model's (d) scene graph prediction against the ground truth (GT) (c) and the base GPNN model (e). Although both models made a wrong interaction prediction between the kidney and the forceps, based on the spatial features observed, it is arguable that our model's prediction was more reasonable compared to the GPNN model. Qualitative comparison with GraphSage [4], GAT [18] and Hpooling [19] models were not performed as these models are unable to predict the adjacent matrix (presence of interaction) between the graph nodes.

Table 1. Comparison of our proposed graph model's performance against the state of the art model in scene graph generation. RO. denotes ReadOut module.

Detection		Graph scene prediction @ 0.5479mAP				
Models	mAP ↑	Models	mAP ↑	Hinge ↓	Recall ↑	AUC ↑
SSD [10] w LS	**0.5479**	**Ours**	**0.3732**	**1.74**	0.7989	0.8789
SSD [10] w/o LS	0.5290	GPNN [15]	0.3371	1.79	0.8001	0.8782
YoloV3 [16] w LS	0.4836	GraphSage [4] + RO	0.2274	3.10	0.6957	0.7468
YoloV3 [16] w/o LS	0.4811	GAT [18]	0.1222	16.14	0.7577	0.7866
–	–	Graph HPooling [19]	0.1883	13.93	0.7727	0.8282

Quantitative comparison results between the proposed model and the state of the art model are reported in Table 1. Firstly, the advantage of incorporating the label smoothing technique is reported under the detection. When compared with the default YoloV3[16] and SSD[10] detection models against incorporating them with the label smoothing technique, an increase in mean average precision (mAP) is reported. Secondly, the performance of our overall proposed model against graph-based models is also shown in Table 1. When compared with the default GPNN [15], GraphSage [4] (incorporated with Readout function from [15]), GAT [18] and Graph Hpooling, our enhanced model achieves a higher mAP and marginally better area under the curve. However, the GPNN [15] has a marginally better performance in terms of hinge loss and recall.

Table 2. Ablation study of proposed model while integrating different methods and modules.

Modules and Methods				mAP	Hinge	AUC
Base	LS	Attention	SageConv			
✓	✓	✓	✓	0.3732	1.74	0.8789
✓	✓	✓	×	0.3662	1.70	0.8699
✓	✓	×	×	0.3371	1.79	0.8782
✓	×	×	×	0.3145	3.31	0.8122

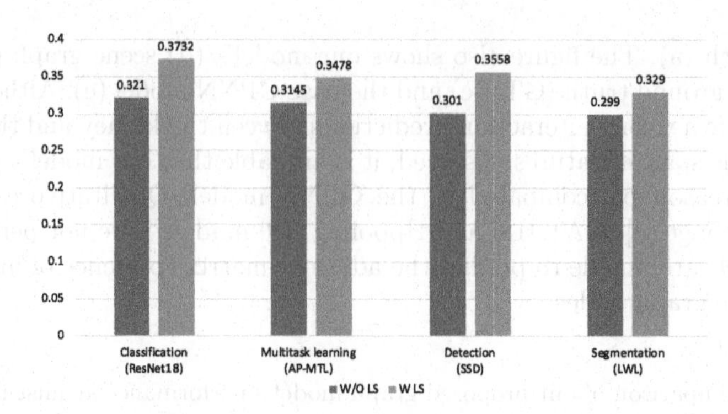

Fig. 5. Mean average precision (mAP) of the scene graph prediction with various feature embedding models such as classification, detection and segmentation. Our experiments confirm that classfication model with LS produces highest accuracy. We exploit ResNet18 [5], AP-MTL [7], SSD [10] and LWL [6] models to extract the feature from the penultimate layer.

The ablation study of the proposed model is shown in Table 2. The table relates the increase in module's performance to every module added onto the default GPNN base module. The use of label smoothened features has significantly increased the base model's performance in terms of mAP, hinge loss and AUC. Additional attention link function has further increased the accuracy of the model. Further additional of SageConv has increased both the mAP and AUC performance of the model. Figure 5 demonstrates our models performance on various type of deep neural network model such as classification (ResNet18), multitask learning (AP-MTL [7]), detection (SSD [10]) and segmentation (LWL [6]). It is interesting to see that the classification model extracts the most distinguishable features comparing to other methods. The figure also infers that extracted features with LS have better representation than without LS.

4 Discussion and Conclusion

In this work, we propose an enhanced graph-based network to perform spatial deep reasoning of the surgical scene to predict the interaction between the defective tissue and the instruments with an inference speed of 15 ms. We design and attention link function and integrated SageConv with a graph parsing network to generate a scene graph and predict surgical interactions during robot-assisted surgery. Furthermore, we demonstrate the advantage of label smoothing in graph feature extraction and the enhancement of the model performance. Spatio-temporal scene graph generation and relative depth estimation from sequential video in the surgical environment can be the future direction of robotic surgical scene understanding.

References

1. Allan, M., et al.: 2018 robotic scene segmentation challenge. arXiv preprint arXiv:2001.11190 (2020)
2. Allan, M., Ourselin, S., Thompson, S., Hawkes, D.J., Kelly, J., Stoyanov, D.: Toward detection and localization of instruments in minimally invasive surgery. IEEE Trans. Biomed. Eng. **60**(4), 1050–1058 (2012)
3. Allan, M., et al.: 2017 robotic instrument segmentation challenge. arXiv preprint arXiv:1902.06426 (2019)
4. Hamilton, W., Ying, Z., Leskovec, J.: Inductive representation learning on large graphs. In: Advances in Neural Information Processing Systems, pp. 1024–1034 (2017)
5. He, K., Zhang, X., Ren, S., Sun, J.: Deep residual learning for image recognition. In: Proceedings of the IEEE Conference on Computer Vision and Pattern Recognition, pp. 770–778 (2016)
6. Islam, M., Li, Y., Ren, H.: Learning where to look while tracking instruments in robot-assisted surgery. In: Shen, D., et al. (eds.) MICCAI 2019. LNCS, vol. 11768, pp. 412–420. Springer, Cham (2019). https://doi.org/10.1007/978-3-030-32254-0_46
7. Islam, M., VS, V., Ren, H.: AP-MTL: attention pruned multi-task learning model for real-time instrument detection and segmentation in robot-assisted surgery (2020)
8. Laina, I., et al.: Concurrent segmentation and localization for tracking of surgical instruments. In: Descoteaux, M., Maier-Hein, L., Franz, A., Jannin, P., Collins, D.L., Duchesne, S. (eds.) MICCAI 2017. LNCS, vol. 10434, pp. 664–672. Springer, Cham (2017). https://doi.org/10.1007/978-3-319-66185-8_75
9. Lee, J.Y., et al.: Ultrasound needle segmentation and trajectory prediction using excitation network. Int. J. Comput. Assist. Radiol. Surg. **15**(3), 437–443 (2020). https://doi.org/10.1007/s11548-019-02113-x
10. Liu, W., et al.: SSD: single shot multibox detector. In: Leibe, B., Matas, J., Sebe, N., Welling, M. (eds.) ECCV 2016. LNCS, vol. 9905, pp. 21–37. Springer, Cham (2016). https://doi.org/10.1007/978-3-319-46448-0_2
11. Moore, R., DeNero, J.: L1 and L2 regularization for multiclass hinge loss models. In: Symposium on Machine Learning in Speech and Language Processing (2011)
12. Müller, R., Kornblith, S., Hinton, G.E.: When does label smoothing help? In: Advances in Neural Information Processing Systems, pp. 4696–4705 (2019)
13. Okamura, A.M., Verner, L.N., Reiley, C., Mahvash, M.: Haptics for robot-assisted minimally invasive surgery. In: Kaneko, M., Nakamura, Y. (eds.) Robotics Research. Springer Tracts in Advanced Robotics, vol. 66, pp. 361–372. Springer, Heidelberg (2010). https://doi.org/10.1007/978-3-642-14743-2_30
14. Pakhomov, D., Premachandran, V., Allan, M., Azizian, M., Navab, N.: Deep residual learning for instrument segmentation in robotic surgery. In: Suk, H.-I., Liu, M., Yan, P., Lian, C. (eds.) MLMI 2019. LNCS, vol. 11861, pp. 566–573. Springer, Cham (2019). https://doi.org/10.1007/978-3-030-32692-0_65
15. Qi, S., Wang, W., Jia, B., Shen, J., Zhu, S.-C.: Learning human-object interactions by graph parsing neural networks. In: Ferrari, V., Hebert, M., Sminchisescu, C., Weiss, Y. (eds.) ECCV 2018. LNCS, vol. 11213, pp. 407–423. Springer, Cham (2018). https://doi.org/10.1007/978-3-030-01240-3_25
16. Redmon, J., Farhadi, A.: Yolov3: an incremental improvement. arXiv preprint arXiv:1804.02767 (2018)

17. Roy, A.G., Navab, N., Wachinger, C.: Concurrent Spatial and Channel 'Squeeze & Excitation' in Fully Convolutional Networks. In: Frangi, A.F., Schnabel, J.A., Davatzikos, C., Alberola-López, C., Fichtinger, G. (eds.) MICCAI 2018. LNCS, vol. 11070, pp. 421–429. Springer, Cham (2018). https://doi.org/10.1007/978-3-030-00928-1_48
18. Veličković, P., Cucurull, G., Casanova, A., Romero, A., Lio, P., Bengio, Y.: Graph attention networks. arXiv preprint arXiv:1710.10903 (2017)
19. Zhang, Z., et al.: Hierarchical graph pooling with structure learning. arXiv preprint arXiv:1911.05954 (2019)
20. Zhou, J., et al.: Graph neural networks: a review of methods and applications. arXiv preprint arXiv:1812.08434 (2018)

Vision-Based Estimation of MDS-UPDRS Gait Scores for Assessing Parkinson's Disease Motor Severity

Mandy Lu[1], Kathleen Poston[2], Adolf Pfefferbaum[2,3], Edith V. Sullivan[2], Li Fei-Fei[1], Kilian M. Pohl[2,3], Juan Carlos Niebles[1], and Ehsan Adeli[1,2](✉)

[1] Computer Science Department, Stanford University, Stanford, CA, USA
eadeli@stanford.edu
[2] School of Medicine, Stanford University, Stanford, CA, USA
[3] Center of Health Sciences, SRI International, Menlo Park, CA, USA

Abstract. Parkinson's disease (PD) is a progressive neurological disorder primarily affecting motor function resulting in tremor at rest, rigidity, bradykinesia, and postural instability. The physical severity of PD impairments can be quantified through the Movement Disorder Society Unified Parkinson's Disease Rating Scale (MDS-UPDRS), a widely used clinical rating scale. Accurate and quantitative assessment of disease progression is critical to developing a treatment that slows or stops further advancement of the disease. Prior work has mainly focused on dopamine transport neuroimaging for diagnosis or costly and intrusive wearables evaluating motor impairments. For the first time, we propose a computer vision-based model that observes non-intrusive video recordings of individuals, extracts their 3D body skeletons, tracks them through time, and classifies the movements according to the MDS-UPDRS gait scores. Experimental results show that our proposed method performs significantly better than chance and competing methods with an F_1-score of 0.83 and a balanced accuracy of 81%. This is the first benchmark for classifying PD patients based on MDS-UPDRS gait severity and could be an objective biomarker for disease severity. Our work demonstrates how computer-assisted technologies can be used to non-intrusively monitor patients and their motor impairments. The code is available at https://github.com/mlu355/PD-Motor-Severity-Estimation.

Keywords: Movement Disorder Society Unified Parkinson's Disease Rating Scale · Gait analysis · Computer vision

1 Introduction

Parkinson's disease (PD) is a progressive neurological disorder that primarily affects motor function. Early, accurate diagnosis and objective measures of disease severity are crucial for development of personalized treatment plans aimed to slow or stop continual advancement of the disease [27]. Prior works aiming

A. L. Martel et al. (Eds.): MICCAI 2020, LNCS 12263, pp. 637–647, 2020.
https://doi.org/10.1007/978-3-030-59716-0_61

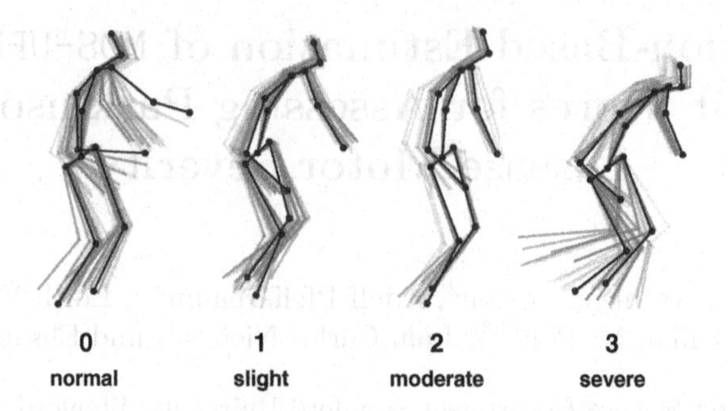

Fig. 1. Progressive PD impairments demonstrated by 3D gait (poses fade over time; left/right distinguished by color) with MDS-UPDRS gait score shown below each skeleton. Participants are taken from our clinical dataset. Classes 0 to 2 progressively decrease in mobility with reduced arm swing and range of pedal motion (*i.e.*, reduced stride amplitude and footlift) while class 3 becomes imbalanced. (Color figure online)

to objectively assess PD severity or progression are either based on neuroimages [1,4] or largely rely on quantifying motor impairments via wearable sensors that are expensive, unwieldy, and sometimes intrusive [12,13]. With the rapid development of deep learning, video-based technologies now offer non-intrusive and scalable ways of quantifying human movements [7,16], yet to be applied to clinical applications such as PD.

PD commonly causes slowing of movement, called bradykinesia, and stiffness, called rigidity, that is visible during the gait and general posture of patients. The Movement Disorder Society-Unified Parkinson's Disease Rating Scale (MDS-UPDRS) [10] is the most commonly used method in clinical and research to assess the severity of these motor symptoms. Specifically, the MDS-UPDRS gait test requires a subject to walk approximately 10 meters away from and toward an examiner. Trained specialists assess the subject's posture with respect to movement and balance (*e.g.*, 'stride amplitude/speed', 'height of foot lift', 'heel strike during walking', 'turning', and 'arm swing') by observation. MDS-UPDRS item 3.10 is scored on a 5-level scale that assesses the severity of PD gait impairment, ranging from a score of 0 indicating no motor impairments to a score of 4 for patients unable to move independently (see Fig. 1).

We propose a method based on videos to assess PD severity related to gait and posture impairments. Although there exist a few video-based methods which assess gait for PD diagnosis [8,11,30], we define a new task and a principled benchmark by estimating the standard MDS-UPDRS scores. There are several challenges to this new setting: (1) there are no baselines to build upon; (2) since it is harder to recruit patients with severe impairments, the number of participants in our dataset is imbalanced across MDS-UPDRS classes; (3) clinical datasets are typically limited in the number of participants, presenting difficulty for training deep learning models; (4) estimating MDS-UPDRS scores defines a multi-class classification problem on a scale of scores from 0 to 4, while prior work only focused on diagnosing PD *vs.* normal. To address these challenges, our 3D pose

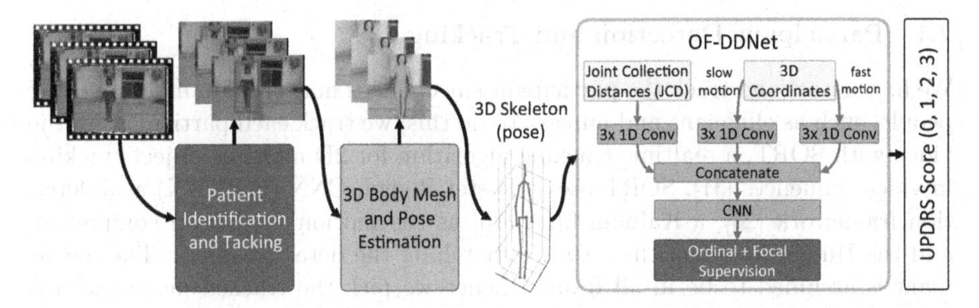

Fig. 2. The proposed framework: we first track the participant throughout the video and remove other persons, *e.g.*, clinicians. Then, we extract the identified participants' 3D body mesh and subsequently the skeletons. Finally, our proposed OF-DDNet estimates the MDS-UPDRS gait score based on only the 3D pose sequence.

estimation models are trained on large public datasets. Then, we use the trained models to extract 3D poses (3D coordinates of body joints) from our clinical data. Therefore, estimation of the MDS-UPDRS scores is only performed on low-dimensional pose data which are agnostic to the clinical environment and video background. To deal with data imbalance, we propose a model with a focal loss [20], which is coupled with an ordinal loss component [26] to leverage the order of the MDS-UPDRS scores.

Our novel approach for automatic vision-based evaluation of PD motor impairments takes monocular videos of the MDS-UPDRS gait exam as input and automatically estimates each participants's gait score on the MDS-UPDRS standard scale. To this end, we first identify and track the participant in the video. Then, we extract the 3D skeleton (a.k.a. pose) from each video frame (visualized in Fig. 1). Finally, we train our novel temporal convolutional neural network (TCNN) on the sequence of 3D poses by training a Double-Features Double-Motion Network [31] (DD-Net) with the new hybrid ordinal-focal objective, which we will refer to as hybrid Ordinal Focal DDNet (OF-DDNet) (see Fig. 2).

The novelties of our work are three-fold: (1) we define a new benchmark for PD motor severity assessment based on video recordings of MDS-UPDRS exams; (2) for the first time, we propose a framework based on 3D pose acquired from non-intrusive monocular videos to quantify movements in 3D space; (3) we propose a method with a hybrid ordinal-focal objective that accounts for the imbalanced nature of clinical datasets and leverages the ordinality MDS-UPDRS scores.

2 Method

As shown in Fig. 2, the input consists of a monocular video of each participant walking in the scene. First, we track each participant in the video using the SORT (Simple Online and Realtime Tracking) algorithm [3] and identify the bounding boxes corresponding to the participant. These bounding boxes along with the MDS-UPDRS exam video are passed to a trained 3D pose extraction model (denoted by SPIN) [18], which provides pose input to OF-DDNet.

2.1 Participant Detection and Tracking

We first detect and track the participant since videos may contain multiple other people, such as clinicians and nurses. To do this, we track each participant in the video with SORT, a realtime tracking algorithm for 2D multiple object tracking in video sequences [31]. SORT uses a Faster Region CNN (FrRCNN) as a detection framework [25], a Kalman filter [15] as the motion prediction component, and the Hungarian algorithm [19] for matching the detected boxes. The participant is assumed to be in all frames, hence we pick the tracked person who is consistently present in all frames with the greatest number of bounding boxes as the patient.

2.2 3D Body Mesh and Pose Extraction

Next, we extract the 3D pose from each frame by feeding the corresponding image and the bounding box found in the previous step as input to SPIN (SMPL oPtimization IN the loop) [18]. SPIN is a state-of-the-art neural method for estimating 3D human pose and shape from 2D monocular images. Based on a single 2D image, the Human Mesh Recovery (HMR) regressor provided by [16] generates predictions for pose parameters θ_{reg}, shape parameters β_{reg}, camera parameters Π_{reg}, 3D joints X_{reg} of the mesh and their 2D projection $J_{reg} = \Pi_{reg}(X_{reg})$. Following the optimization routine proposed in SMPLify [5], these are initial parameters for the SMPL body model [21], a function $M(\theta, \beta)$ of pose parameters θ and shape parameters β that returns the body mesh. A linear regressor W performs regression on the mesh to find 3D joints J_{smpl}. These regressed joint values are supplied to the iterative fitting routine, which encourages the 2D projection of the SMPL joints J_{smpl} to align with the annotated 2D keypoints J_{reg} by penalizing their weighted distance. The fitted model subsequently provides supervision for the regressor, forming an iterative training loop. In our proposed method, we generate 3D pose for each video frame by performing regression on the 3D mesh output from SMPL, which has been fine-tuned in the SPIN loop. SPIN was initialized with pretrained SMPL [21] and HMR pretrained on the large Human3.6M [14] and MPI-INF-3DHP [22] datasets, providing over 150k training images with 3D joint annotations, as well as large-scale datasets with 2D annotations (*e.g.*, COCO [20] and MPII [2]).

2.3 Gait Score Estimation with OF-DDNet

Our score estimation model, OF-DDNet, builds on top of DD-Net [31] by adding a hybrid ordinal-focal objective. DD-Net [31] was chosen for its state-of-the-art performance at orders of magnitude smaller in parameter size than comparable methods. OF-DDNet takes as input 3D joints and outputs the participant's `MDS-UPDRS` gait score. Our model has a lightweight TCNN-based architecture that prevents overfitting. To address the variance of 3D Cartesian joints to both location and viewpoint, two new features are calculated: (1) Joint Collection Distances (JCD) and (2) two-scale motion features. JCD is a location-viewpoint invariant feature that represents the Euclidean distances between joints as a matrix M, where $M_{ij}^k = \|J_i^k - J_j^k\|$ for joints J_i and J_j at frame k of total K frames. Since this is a symmetric matrix, only the upper triangular matrix is preserved and flattened to a dimension of $\binom{n}{2}$ for n joints. A two-scale motion feature is introduced for global scale invariance which measures temporal difference between nearby frames. To capture varying scales of global motion, we calculate slow motion (M_k^{slow}) and fast motion (M_k^{fast})

$$\begin{aligned} M_k^{slow} &= S^{k+1} - S_k, k \in \{1, 2, 3, ..., K-1\}, \\ M_k^{fast} &= S^{k+2} - S_k, k \in \{1, 3, 5, ..., K-2\}, \end{aligned} \tag{1}$$

where $S_k = \{J_1^k, J_2^k, ...J_n^k\}$ denotes the set of joints for the kth frame. The JCD and two-scale motion features are embedded into latent vectors at each frame through a series of convolutions to learn joint correlation and reduce the effect of skeleton noise. Then, the embeddings are concatenated and run through a series of 1D convolutions and pooling layers, culminating with a softmax activation on the final layer to output a probability distribution for each class.

2.4 Hybrid Ordinal-Focal Loss

To leverage the ordinal nature of MDS-UPDRS scores and to combat the natural class imbalance in clinical datasets, we propose a hybrid ordinal ($\mathcal{O}$) focal ($\mathcal{F}$) loss with a trade-off hyperparamter λ as $\mathcal{L} = \mathcal{F} + \lambda\mathcal{O}$. Although many regression or threshold-based ordinal loss functions exist [23,26], this construction allows its use in conjunction with our focal loss.

Focal Loss is introduced to combat class imbalance [20]. It was initially proposed for binary classification, but it is naturally extensible to multi-class classification (*e.g.*, $C > 2$ classes). We apply focal loss for predicting label y with probability p:

$$\mathcal{F}(y,p) = \sum_{i=1}^{C} -\alpha(1-p_i)^{\gamma} y_i log(p_i). \tag{2}$$

The modulating factor $(1 - p_i)^{\gamma}$ is small for easy negatives where the model has high certainty and close to 1 for misclassified examples. This combats class imbalance by down-weighting learning for easy negatives, while preserving basic cross-entropy loss for misclassified examples. We set the default focusing parameter of $\gamma = 2$ and weighting factor $\alpha = 0.25$ as suggested by [20].

Ordinal Loss is used to leverage the intrinsic order in the MDS-UPDRS scores. We implement a loss function that penalizes predictions more if they are violating the order. This penalization incorporates the actual labels $\bar{y} \in \{0, 1, 2, 3\}$ to indicate order instead of the probability vectors used in cross-entropy. Given the estimated label $\hat{\bar{y}} \in \{0, 1, 2, 3\}$, we calculate the absolute distance $w = |\bar{y} - \hat{\bar{y}}|$ and incorporate this with categorical cross-entropy to generate our ordinal loss:

$$\mathcal{O}(y,p) = -\frac{1+w}{C} \sum_{i=1}^{C} y_i log(p_i). \tag{3}$$

3 Experiments

3.1 Dataset

We collected video recordings from 30 research participants who met UK Brain Bank diagnostic criteria of MDS-UPDRS exams scored by a board-certified movement disorders neurologist. All videos of PD participants were recorded during the off-medication state, defined according to previously published protocols [24]. All study procedures were approved by the Stanford Institutional Review Board and written informed consent was obtained from all participants in this study. We first extracted the sections of the video documenting the gait examination, in which participants were instructed to walk directly toward and away from the camera twice. The gait clips range from 17 s to 54 s with 30 frames per second. Our dataset includes 21 exams with score 1, 4 exams with score of 2, 4 exams with score of 3 and 1 exam with score 0. Participants who cannot walk at all or without assistance from another person are scored 4, thus we exclude this class from our analysis due to the difficulty in obtaining videos recordings of their gait exam.

To augment the normal control cohort (*i.e.*, score 0), we include samples from the publicly available CASIA Gait Database A [28], a similar dataset with videos of 20 non-PD human participants filmed from different angles. We extracted corresponding videos where participants walk directly toward and away from the camera, with length of minimum 16 and maximum 53 s. The underlying differences between the datasets should not bias our analyses because all score estimation algorithms operate on pose data with similar characteristics (same view points and duration) across all classes and we normalize and center the pose per participant by aligning temporal poses based on their hip joint.

3.2 Setup

We preprocess our dataset by 1) clipping each video into samples of 200 frames each, where the number of clips per exam depends on its length, 2) supplying two additional cropped videos per exam for sparse classes 2 and 3 and 3) joint normalization and centering at the mid-hip. To address the subjective nature of MDS-UPDRS scoring by clinicians, we incorporate a voting mechanism. Each sub-clip is labeled same as the exam itself for training to independently examine each sub-part of the exam. This voting mechanism adds robustness to the overall system and allows us to augment the dataset for proper training of the TCNN. To account for the limited dataset size, all evaluations in this study were performed using a participant-based leave-one-out cross-fold validation on all 50 samples. We note that the clips and crops for each exam are *never* separated by the train/test split. Optimal hyperparameters for the gait scoring model were obtained by performing a grid search using inner leave-one-out cross validation and the Adam optimizer ($\beta_1 = 0.9, \beta_2 = 0.999$) [17]. Best performance was achieved at 600 epochs, batch size of 64, filter size of 32 and an annealing learning rate from 1^{-3} to 1^{-6}. For evaluation, we report per-class and macro average F_1, area under ROC curve (AUC), precision (Pre), and recall (Rec).

Table 1. Per-class MDS-UPDRS gait score prediction performance of our method.

Gait score	F_1	AUC	Pre	Rec
0	0.91	0.93	0.91	0.91
1	0.81	0.91	0.73	0.91
2	0.73	0.87	0.80	0.67
3	0.86	0.90	1.00	0.75
Macro average	0.83	0.90	0.86	0.81

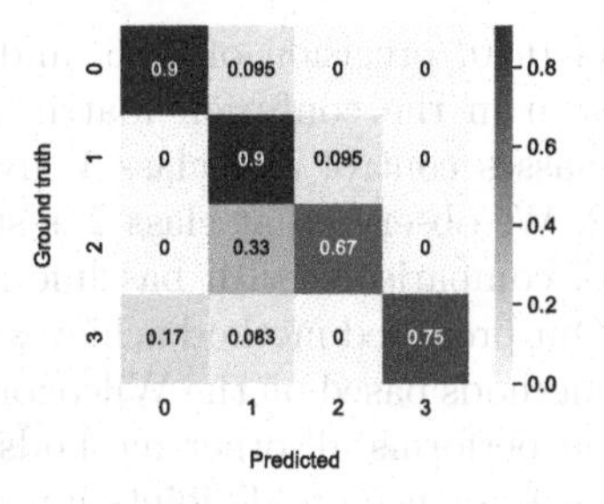

Fig. 3. Confusion matrix of OF-DDNet.

Table 2. Comparison with baseline and ablated methods. * indicates statistical difference at ($p < 0.05$) compared with our method, measured by the Wilcoxon signed rank test [29]. Best results are in bold. See text for details about compared methods.

Method	F_1	AUC	Pre	Rec
OF-DDNet (Ours)	**0.83**	**0.90**	**0.86**	**0.81**
1) Baseline CNN*	0.73	0.86	0.79	0.69
2) Baseline OF-CNN*	0.74	0.83	0.79	0.71
3) DD-Net* [31]	0.74	0.84	0.80	0.69
4) 2D joints* [6]	0.61	0.77	0.61	0.62
5) Ours w/o focal	0.79	0.83	0.83	0.76
6) Ours w/o ordinal	0.78	0.88	0.84	0.74
7) Regression*	0.67	n/a	0.70	0.65
8) DeepRank* [23]	0.74	0.80	0.79	0.71

3.3 Baseline Methods and Ablation Studies

We compare our results with several baselines: 1) we feed raw 3D joints from SPIN directly into a 1D CNN modeled after DD-Net architecture sans double features and embedding layer (see Fig. 2), 2) OF-CNN, the same as (1) but with our OF loss, and 3) the original DD-Net [31] with basic cross-entropy loss. We also conduct an ablation study on the choice of pose extraction method by 4) using 2D joints (instead of 3D) extracted with OpenPose [6] as input to OF-DDNet. To evaluate the hybrid loss function, we separately examine our method 5) without the focal loss component and 6) without the ordinal component. We further examine our ordinal component by replacing it with 7) a regression loss (MSE) for DD-Net with an extra sigmoid-activated dense output layer and finally with 8) DeepRank [23], a ranking CNN which cannot be combined with focal loss.

3.4 Results

The results of our proposed OF-DDNet are summarized in Table 1. Our method sets a new benchmark for this task with macro-average F_1-score of 0.83, AUC of 0.90, precision of 0.86, and balanced accuracy (average recall) of 81%. As seen in the confusion matrix (Fig. 3), the overall metrics for well-represented classes control and class 1 are fairly high, followed by class 3 and then class 2. We observe that class 2 is strictly misclassified as lower severity. The results of comparisons with baseline and ablated methods are summarized in Table 2. Our proposed method achieves significantly better performance than many other methods based on the Wilcoxon signed rank test [29] ($p < 0.05$), and consistently outperforms all other methods. Our results show that all methods have higher performance on 3D joints input than 2D input, as even a baseline 1D CNN has better performance than the full DD-Net model with 2D joints. This demonstrates that 3D joints provide valuable information for the prediction model, which has not been explored before. Similarly, we note that on 3D joint input, all classification methods outperformed the regression model, suggesting that classification outperforms regression at this task. Regarding the loss function, OF-DDNet significantly outperforms our baseline CNN with categorical cross-entropy. Adding ordinal (Method 5 in the Table) and focal (Method 6) losses to baseline DD-Net both improve accuracy, but their combined performance (OF-DDNet) outperforms all. DeepRank (Method 7) had high confidence on predictions and poor performance on sparse classes, suggesting an overfitting problem that encourages the use of a simple ordinal loss for our small dataset.

4 Discussion

Our method achieves compelling results on automatic vision-based assessment of PD severity and sets a benchmark for this task. We demonstrate the possibility of predicting PD motor severity using only joint data as the input to a prediction model, and the efficacy of 3D joint data in particular. Furthermore, we show the effectiveness of a hybrid ordinal-focal loss for tempering the effects of a small, imbalanced dataset and leveraging the ordinal nature of the `MDS-UPDRS`. However, it is necessary to note that there is inherent subjectivity in the `MDS-UPDRS` scale [9] despite attempts to standardize the exam through objective criterion (*e.g.*, stride amplitude/speed, heel strike, arm swing). Physicians often disagree on ambiguous cases and lean toward one score versus another based on subtle cues. Clinical context suggests our results are consistent with physician experience. As corroborated in the results of OF-DDNet, the most difficult class to categorize in clinical practice is score 2 since the `MDS-UPDRS` defines its distinction from score 1 solely by "minor" versus "substantial" gait impairment, shown in Fig. 1. Control (class 0) exhibits high arm swing and range of pedal motion while classes 1 and 2 have progressively reduced mobility and increased stiffness (*i.e.*, reduced arm swing and stride amplitude/foot lift). Class 3 exhibits high imbalance issues with stooped posture and lack of arm swing, which aids mobility, presenting a high fall risk. In practice, class 3 is easier to distinguish from the other classes because it only requires identifying that a participant requires an assisted-walking device and cannot walk independently. Likewise, our model performs well for class 3 except in challenging cases which may require human judgement, such as determining what constitutes "safe" walking.

This study presents a few notable limitations. A relatively small dataset carries risk of overfitting and uncertainty in the results. We mitigated the former through data augmentation techniques and using simple models (DD-Net) instead of deep or complex network architectures; and the latter with leave-one-out cross validation instead of the traditional train/validation/test split used in deep learning. Similarly, our classes are imbalanced with considerably fewer examples in classes 2 and 3 than in classes 0 and 1, which we attempt to address through our custom ordinal focal loss and by augmenting sparse classes through cropping. Additionally, due to a shortage of control participants in our clinical dataset, we include examples of non-PD gait from the public CASIA dataset. The data is obfuscated by converting to normalized pose, which has similar characteristics across both datasets. However, expanding the clinical dataset by recruiting more participants from underrepresented classes would strengthen the results and presents a direction for future work.

5 Conclusion

In this paper, we presented a proof-of-concept of the potential to assess PD severity from videos of gait using an automatic vision-based approach. We provide a first benchmark for estimating `MDS-UPDRS` scores with a neural model trained on 3D joint data extracted from video. This method works even with a small dataset due to data augmentation, the use of a simple model and our hybrid ordinal focal loss and has opportunity for application to similar video classification problems in the medical space. Our proposed method is simple to set up and use because it only requires a video of gait as input; thus, in remote or resource-limited regions with few experts it provides a way to form estimates of disease progression. In addition, such scalable automatic vision-based methods can help perform time-intensive and tedious collection and labelling of data for research and clinical trials. In conclusion, our work demonstrates how computer-assisted intervention (CAI) technologies can provide clinical value by reliably and unobtrusively assisting physicians by automatic monitoring of PD patients and their motor impairments.

Acknowledgment. This research was supported in part by NIH grants AA010723, AA017347, AG047366, and P30AG066515. The content is solely the responsibility of the authors and does not necessarily represent the official views of the National Institutes of Health. This study was also supported by the Stanford School of Medicine Department of Psychiatry & Behavioral Sciences 2021 Innovator Grant Program and the Stanford Institute for Human-centered Artificial Intelligence (HAI) AWS Cloud Credit.

References

1. Adeli, E., et al.: Joint feature-sample selection and robust diagnosis of Parkinson's disease from MRI data. NeuroImage **141**, 206–219 (2016)

2. Andriluka, M., Pishchulin, L., Gehler, P., Schiele, B.: 2D human pose estimation: new benchmark and state of the art analysis. In: Proceedings of the IEEE Conference on computer Vision and Pattern Recognition, pp. 3686–3693 (2014)
3. Bewley, A., Ge, Z., Ott, L., Ramos, F., Upcroft, B.: Simple online and realtime tracking. In: ICIP, pp. 3464–3468. IEEE (2016)
4. Bharti, K., et al.: Neuroimaging advances in Parkinson's disease with freezing of gait: a systematic review. NeuroImage: Clin. **24**, 102059 (2019)
5. Bogo, F., Kanazawa, A., Lassner, C., Gehler, P., Romero, J., Black, M.J.: Keep it SMPL: automatic estimation of 3D human pose and shape from a single image. In: Leibe, B., Matas, J., Sebe, N., Welling, M. (eds.) ECCV 2016. LNCS, vol. 9909, pp. 561–578. Springer, Cham (2016). https://doi.org/10.1007/978-3-319-46454-1_34
6. Cao, Z., Simon, T., Wei, S.E., Sheikh, Y.: Realtime multi-person 2D pose estimation using part affinity fields. In: CVPR (2017)
7. Chiu, H.k., Adeli, E., Wang, B., Huang, D.A., Niebles, J.C.: Action-agnostic human pose forecasting. In: 2019 IEEE Winter Conference on Applications of Computer Vision (WACV), pp. 1423–1432. IEEE (2019)
8. Cho, C.W., Chao, W.H., Lin, S.H., Chen, Y.Y.: A vision-based analysis system for gait recognition in patients with Parkinson's disease. Expert Syst. Appl. **36**(3), 7033–7039 (2009)
9. Evers, L.J., Krijthe, J.H., Meinders, M.J., Bloem, B.R., Heskes, T.M.: Measuring Parkinson's disease over time: the real-world within-subject reliability of the MDS-UPDRS. Mov. Disord. **34**(10), 1480–1487 (2019)
10. Goetz, C.G., et al.: Movement disorder society-sponsored revision of the unified Parkinson's disease rating scale (MDS-UPDRS): scale presentation and clinimetric testing results. Mov. Disord.: Off. J. Mov. Disord. Soc. **23**(15), 2129–2170 (2008)
11. Han, J., Jeon, H.S., Jeon, B.S., Park, K.S.: Gait detection from three dimensional acceleration signals of ankles for the patients with Parkinson's disease. In: Proceedings of the IEEE The International Special Topic Conference on Information Technology in Biomedicine, vol. 2628, Ioannina (2006)
12. Hobert, M.A., Nussbaum, S., Heger, T., Berg, D., Maetzler, W., Heinzel, S.: Progressive gait deficits in Parkinson's disease: a wearable-based biannual 5-year prospective study. Front. Aging Neurosci. **11**, 22 (2019)
13. Hssayeni, M.D., Jimenez-Shahed, J., Burack, M.A., Ghoraani, B.: Wearable sensors for estimation of parkinsonian tremor severity during free body movements. Sensors **19**(19), 4215 (2019)
14. Ionescu, C., Papava, D., Olaru, V., Sminchisescu, C.: Human3. 6m: large scale datasets and predictive methods for 3D human sensing in natural environments. IEEE Trans. Pattern Anal. Mach. Intell. **36**(7), 1325–1339 (2013)
15. Kalman, R.E.: A new approach to linear filtering and prediction problems (1960)
16. Kanazawa, A., Black, M.J., Jacobs, D.W., Malik, J.: End-to-end recovery of human shape and pose. In: CVPR, pp. 7122–7131 (2018)
17. Kingma, D.P., Ba, J.: Adam: a method for stochastic optimization. arXiv preprint arXiv:1412.6980 (2014)
18. Kolotouros, N., Pavlakos, G., Black, M.J., Daniilidis, K.: Learning to reconstruct 3D human pose and shape via model-fitting in the loop. In: ICCV (2019)
19. Kuhn, H.W.: The Hungarian method for the assignment problem. Naval Res. Logist. Q. **2**(1–2), 83–97 (1955)
20. Lin, T.Y., Goyal, P., Girshick, R., He, K., Dollár, P.: Focal loss for dense object detection. In: CVPR, pp. 2980–2988 (2017)
21. Loper, M., Mahmood, N., Romero, J., Pons-Moll, G., Black, M.J.: SMPL: a skinned multi-person linear model. ACM Trans. Graph. **34**(6), 1–16 (2015)

22. Mehta, D., et al.: VNect: real-time 3D human pose estimation with a single RGB camera, vol. 36 (2017). http://gvv.mpi-inf.mpg.de/projects/VNect/
23. Pang, L., Lan, Y., Guo, J., Xu, J., Xu, J., Cheng, X.: DeepRank: a new deep architecture for relevance ranking in information retrieval. In: Proceedings of the 2017 ACM on Conference on Information and Knowledge Management, pp. 257–266 (2017)
24. Poston, K.L., et al.: Compensatory neural mechanisms in cognitively unimpaired Parkinson disease. Ann. Neurol. **79**(3), 448–463 (2016)
25. Redmon, J.: DarkNet: open source neural networks in c (2013–2016). http://pjreddie.com/darknet/
26. Rennie, J.D., Srebro, N.: Loss functions for preference levels: regression with discrete ordered labels. In: IJCAI Workshop Advances in Preference Handling (2005)
27. Venuto, C.S., Potter, N.B., Ray Dorsey, E., Kieburtz, K.: A review of disease progression models of Parkinson's disease and applications in clinical trials. Mov. Disord. **31**(7), 947–956 (2016)
28. Wang, L., Tan, T., Ning, H., Hu, W.: Silhouette analysis-based gait recognition for human identification. PAMI **25**(12), 1505–1518 (2003)
29. Wilcoxon, F.: Individual comparisons by ranking methods. In: Kotz, S., Johnson, N.L. (eds.) Breakthroughs in Statistics, Springer Series in Statistics (Perspectives in Statistics), pp. 196–202. Springer, New York (1992). https://doi.org/10.1007/978-1-4612-4380-9_16
30. Xue, D., et al.: Vision-based gait analysis for senior care. arXiv preprint arXiv:1812.00169 (2018)
31. Yang, F., Wu, Y., Sakti, S., Nakamura, S.: Make skeleton-based action recognition model smaller, faster and better. In: ACM Multimedia Asia, pp. 1–6 (2019)

Searching for Efficient Architecture for Instrument Segmentation in Robotic Surgery

Daniil Pakhomov[1(✉)] and Nassir Navab[1,2]

[1] Johns Hopkins University, Baltimore, USA
dpakhom1@jhu.edu

[2] Technische Universität München, Munich, Germany

Abstract. Segmentation of surgical instruments is an important problem in robot-assisted surgery: it is a crucial step towards full instrument pose estimation and is directly used for masking of augmented reality overlays during surgical procedures. Most applications rely on accurate real-time segmentation of high-resolution surgical images. While previous research focused primarily on methods that deliver high accuracy segmentation masks, majority of them can not be used for real-time applications due to their computational cost. In this work, we design a light-weight and highly-efficient deep residual architecture which is tuned to perform real-time inference of high-resolution images. To account for reduced accuracy of the discovered light-weight deep residual network and avoid adding any additional computational burden, we perform a differentiable search over dilation rates for residual units of our network. We test our discovered architecture on the EndoVis 2017 Robotic Instruments dataset and verify that our model is the state-of-the-art in terms of speed and accuracy tradeoff with a speed of up to 125 FPS on high resolution images.

1 Introduction

Robot-assisted Minimally Invasive Surgery (RMIS) provides a surgeon with improved control, facilitating procedures in confined and difficult to access anatomical regions. However, complications due to the reduced field-of-view provided by the surgical camera limit the surgeon's ability to self-localize. Computer assisted interventions (CAI) can help a surgeon by integrating additional information. For example, overlaying pre- and intra-operative imaging with the surgical console can provide a surgeon with valuable information which can improve decision making during complex procedures [15]. Integrating this data is a complex task and involves understanding relations between the patient anatomy, operating instruments and surgical camera. Segmentation of the instruments in the camera images is a crucial component of this process and can be used to prevent rendered overlays from occluding the instruments while providing crucial input to instrument tracking frameworks [1,13].

A. L. Martel et al. (Eds.): MICCAI 2020, LNCS 12263, pp. 648–656, 2020.
https://doi.org/10.1007/978-3-030-59716-0_62

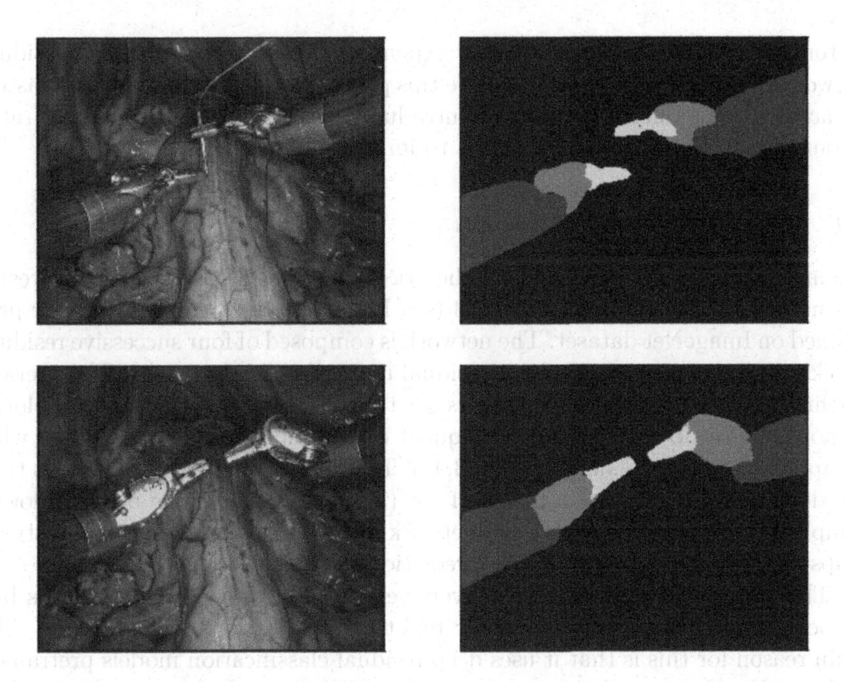

Fig. 1. Multi-class instrument segmentation results delivered by our method on sequences from validational subset of Endovis 2017 dataset.

There has been a significant development in the field of instrument segmentation [10,12,14] based on recent advancement in deep learning [3,6]. Additionally, release of new datasets for surgical tools segmentation with high-resolution challenging images further improved the field by allowing methods to be more rigorously tested [2]. While the state-of-the-art methods show impressive pixel accurate results [12,14], the inference time of these methods makes them unsuitable for real-time applications. In our work we address this problem and present a method that has a faster than real time inference time while delivering high quality segmentation masks.

2 Method

In this work, we focus on the problem of surgical instrument segmentation. Given an input image, every pixel has to be classified into one of C mutually exclusive classes. We consider two separate tasks with increasing difficulty: binary tool segmentation and multi-class tool segmentation. In the first case, each pixel has to be classified into $C = 2$ classes as belonging to surgical background or instrument. In the second task, into $C = 4$ classes, namely tool's shaft, wrist, jaws and surgical background. Our method is designed to perform these tasks efficiently while delivering high quality results. First, we discuss the previous state-of-the-art method based on dilated residual networks (Sect. 2.1) and highlight the major factor that makes it computationally expensive. Then, we present light residual networks (Sect. 2.2) that allow to solve this problem and make the method faster. To account for reduced accuracy we introduce a search for optimal dilation rates in our model (Sect. 2.3) which allows to improve its accuracy.

2.1 Dilated Residual Networks

We improve upon previous state-of-the-art approach [12] based on dilated residual network that employs ResNet-18 (see Fig. 2), a deep residual network pretrained on ImageNet dataset. The network is composed of four successive residual blocks, each one consisting of two residual units of "basic" type [6]. The average pooling layer is removed and stride is set to one in the last two residual blocks responsible for downsampling, subsequent convolutional layers are dilated with an appropriate rate as suggested in [3,12]. This allows to obtain predictions that are downsampled only by a factor of 8× (in comparison to the original downsampling of 32×) which makes the network work on a higher resolution features maps (see Fig. 2) and deliver finer predictions [3,12].

The aforementioned method delivers very accurate segmentation masks but is too computationally expensive for real-time applications (see. Table 1). The main reason for this is that it uses deep residual classification models pretrained on ImageNet that usually have great number of filters at the last layers: when the model is being transformated into dilated residual network, the downsampling operations at the last two residual blocks are removed forcing the convolutional layers to operate on the input that is spatially bigger by a factor of two and four respectively [3,12] (see Fig. 2). Even the most shallow deep residual network ResNet-18 is too computationally expensive when converted into dilated residual network. This motivates us to create a smaller deep residual network that can be pretrained on the imageNet and converted into dilated residual model that exhibits improved running time.

2.2 Light Residual Networks

In order to solve the aforementioned problem, we introduce a light residual network, a deep residual model which satisfies the requirements:

- **Low latency:** exhibits low latency inference on high-resolution images when converted into dilated residual network. This is achieved by reducing the number of filters in the last stages of the network (see Fig. 2). In the original resnet networks [6], the number of filters in the last stages is considerable: after being converted into dilated versions, these stages experience increase in the computational price by a factor of two and four [3]. Since last stages are the biggest factor responsible for the increased inference time, we decrease the number of channels in them. This significantly decreases the inference time (see Table 1). While we noticed a considerable decrease in the accuracy of the model on the ImageNet dataset, the performance on the segmentation dataset only moderately decreased. We attribute it to the fact the number of filters in the last layers is of significant importance for the imagenet classification task because it needs to differentiate between one thousand classes compared to only four in our case.

- **Low GPU memory consumption:** the memory requirements of the network allow it to be trained with optimal batch and image crop sizes on single GPU device. Dilated residual networks consume a considerable amount of memory when trained for segmentation task [3], while still requiring relatively big batch size and crop size [5]. Similar to the previous section, the biggest factor responsible for the memory consumption again involves the last stages of the network [3]. Since every activation has to be stored in memory during backpropagation [4], the layers that generate the biggest activations are contributing the most to the increased memory consumption. The last layers of the residual network work on increased spatial resolution after being converted to the dilated residual network and have more channels than other layers. By decreasing the number of channels in the last layers, we solve the problem and are able to train the model with sufficient batch size and image crop size [5].
- **ImageNet pretraining:** the network is pretrained on ImageNet dataset which was shown to be essential for good performance on small segmentation datasets like Endovis [14]. We pretrain all our models on the ImageNet dataset following the parameters suggested in [6].

We present two versions of Light ResNet-18 named Light ResNet-18-v1 and Light ResNet-18-v2 with number of channels in the last layers set to 64 and 32 respectively. Second version exhibits improved runtime speed at the expense of decreased accuracy on the segmentation task. All models were pretrained on imageNet dataset and then converted into dilated residual networks following [12]. After being converted, the network were trained on the Endovis 2017 [2] segmentation dataset for binary and multi-class instrument segmentation tasks. While being fast, the networks exhibit reduced accuracy compared to state-of-the-art methods (see. Table 1). In order to account for that and improve the accuracy without reducing the speed of the network, we search for optimal dilation rates for each layer.

2.3 Searching for Optimal Dilation Rates

In order to improve accuracy of our model further but avoid adding any new parameters or additional computational overhead [7], we search for optimal integer dilation rates for each residual unit. Since trying out all possible combinations of dilation rates and retraining a model each time is infeasible, we formulate the problem of dilation rate search as an optimization problem [17].

First, we update residual units of our model. Original residual unit [6] can be expressed in a form (see Fig. 3):

$$x_{l+1} = x_l + \mathcal{F}(x_l) \tag{1}$$

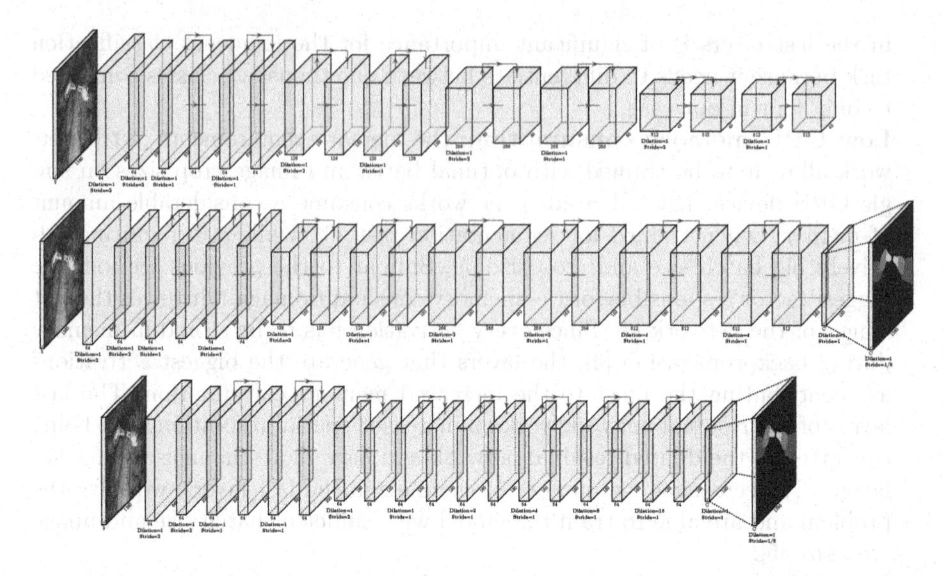

Fig. 2. Top row: simplified ResNet-18 before being converted into dilated fully convolutional network (FCN). Middle row: ResNet-18 after being converted into FCN following the method of [12]. Since the stride was set to one in two layers and the number of channels was left the same, convolutional layers are forced to work on a greater resolution, therefore, computational overhead is substantially increased, making the network unable to deliver real-time performance. Bottom row: Our light-weight ResNet-18 being converted into dilated FCN and dilation rates are set to the values found during our differentiable search. The decreased number of channels makes the network fast while the found dilation rates increase its accuracy without any additional parameters or computational overhead. Green arrows represent the residual unit of a "basic" type, black arrows represent skip connections. First two layers does not have any residual units and are simplified in the figure. Dashed lines represent the bilinear upsampling operation. The figure is better viewed in a pdf viewer in color and zoomed in for details. The figure in bigger resolution is available in the supplementary material. (Color figure online)

where x_l and x_{l+1} are input and output of the l-th unit, and $\mathcal{F}$ is a residual function.

In order to allow our network to choose residual units with different dilation rates, we introduce gated residual unit which has N different residual connections with different dilation rates:

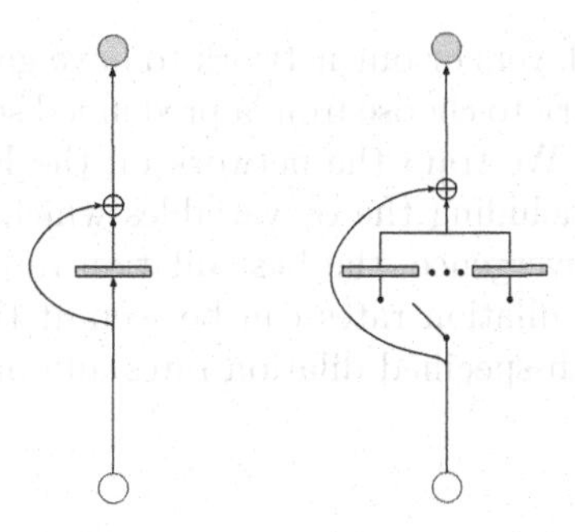

Fig. 3. Left: Simplified residual unit, a building block of deep residual networks which are usually formed as a sequence of residual units. Right: A group of residual units with different dilation rates combined with a discrete decision gate which forces the network to choose only one of residual units depending on which of them leads to a better overall performance of the network. The gate is controlled by a variable which receives gradients during training. During optimization each layer has its own set of residual units and gate variables and by the end of training only the residual units with dilation rates that perform best are left (other choices can be safely removed). We perform search for dilation rates only for the last four residual units of our network.

$$x_{l+1} = x_l + \sum_{i=0}^{N} \mathbf{Z}_i \cdot \mathcal{F}_i(x_l) \tag{2}$$

$$\sum_{i=0}^{N} \mathbf{Z}_i = 1 \tag{3}$$

$$\forall i \; \mathbf{Z}_i \in \{0, 1\} \tag{4}$$

where $\mathbf{Z}$ is a gate that decides which residual connection to choose and is represented with discrete one-hot-encoded categorical variable, N is equal to the number of dilation rates that we consider. In order to be able to search for the best dilation rates, we need to also optimize the gate variable $\mathbf{Z}$ which is not differentiable since it represents a hard decision. But it is possible to obtain estimated gradients of such hard decisions by introducing perturbations in the system in the form of noise [9,11,16,17]. To be more specific, we use recent work that introduces differentiable approximate sampling mechanism for categorical variables based on a Gumbel-Softmax distribution (also known as a concrete distribution) [9,11,16]. We use Gumbel-Softmax to relax the discrete distribution to be continuous and differentiable with reparameterization trick:

$$\bar{\mathbf{Z}}_i = softmax((\log \alpha_i + G_i)/\tau) \tag{5}$$

Where G_i is an ith Gumbel random variable, $\bar{\mathbf{Z}}$ is the softened one-hot random variable which we use in place of $\mathbf{Z}$, α_i is a parameter that controls which residual unit to select and is optimized during training, τ is the temperature of the softmax, which is steadily annealed to be close to zero during training.

We update last four layers or our network to have gated residual units. Each of them allows the network to choose from a predefined set of dilation rates which we set to $\{1, 2, 4, 8, 16\}$. We train the network on the Endovis 2017 dataset by optimizing all weights including the α_i variables which control the selection of dilation rates. Upon convergence, the best dilation rates are decoded from the α_i variables. Discovered dilation rates can be seen at the Fig. 2. Next we train the residual network with specified dilation rates and original residual units.

2.4 Training

After the optimal dilation rates are discovered we update Light ResNet-18-v1 and ResNet-18-v2 to use them. Light ResNet-18-v1 and ResNet-18-v2 were first pretrained on Imagenet. We train networks on the Endovis 2017 train dataset [2]. During training we recompute the batch normalization statistics [5,8]. We optimize normalized pixel-wise cross-entropy loss [3] using Adam optimization algorithm. Random patches are cropped from the images [3] for additional regularization. We employ crop size of 799. We use the 'poly' learning rate policy with an initial learning rate of 0.001 [5]. The batch size is set to 32.

3 Experiments and Results

We test our method on the EndoVis 2017 Robotic Instruments dataset [2]. There are 10 75-frame sequences in the test dataset that features 7 different robotic surgical instruments [2]. Samples from the dataset and qualitative results of our method are depicted in Fig. 1. We report quantitative results in terms of accuracy and inference time of our method in the Table 1 and Table 2.

As it can be seen our method is able to deliver pixel accurate segmentation while working at an extremely fast frame rate of up to 125 FPS.

Table 1. Quantitative results of our method compared to other approaches in terms of accuracy (measured using mean intersection over union metric) and latency (measured in milliseconds). Latency was measured for an image of input size 1024 × 1280 using NVIDIA GTX 1080Ti GPU. It can be seen that our light backbone allows for significantly decreased inference time, while learnt dilations help to improve decreased accuracy.

Model	Binary segmentation		Parts segmentation	
	IOU	Time	IOU	Time
TernausNet-16 [2,14]	0.888	184 ms	0.737	202 ms
Dilated ResNet-18 [12]	**0.896**	126 ms	**0.764**	126 ms
Dilated Light ResNet-18-v1	0.821	17.4 ms	0.728	17.4 ms
Light ResNet-18-v1 w/learnt dilations	0.869	17.4 ms	0.742	17.4 ms
Dilated Light ResNet-18-v2	0.805	**11.8 ms**	0.706	**11.8 ms**
Light ResNet-18-v2 w/learnt dilations	0.852	**11.8 ms**	0.729	**11.8 ms**

Table 2. Latency of our method as measured on a modern NVIDIA Tesla P100 GPU for an image of input size 1024 × 1280. We can see that fastest of our models is able to work at 125 frames per second.

Model	Binary segmentation		Parts segmentation	
	IOU	Time	IOU	Time
Light ResNet-18-v1 w/learnt dilations	0.869	11.5 ms	0.742	11.5 ms
Light ResNet-18-v2 w/learnt dilations	0.852	**7.95 ms**	0.729	**7.95 ms**

4 Discussion and Conclusion

In this work, we propose a method to perform real-time robotic tool segmentation on high resolution images. This is an important task, as it allows the segmentation results to be used for applications that require low latency, for example, preventing rendered overlays from occluding the instruments or estimating the pose of a tool [1]. We introduce a lightweight deep residual network to model the mapping from the raw images to the segmentation maps that is able to work at high frame rate. Additionally, we introduce a method to search for optimal dilation rates for our lightweight model, which improves its accuracy in binary tool and instrument part segmentation. Our results show the benefit of our method for this task and also provide a solid baseline for the future work.

References

1. Allan, M., et al.: Image based surgical instrument pose estimation with multi-class labelling and optical flow. In: Navab, N., Hornegger, J., Wells, W.M., Frangi, A.F. (eds.) MICCAI 2015. LNCS, vol. 9349, pp. 331–338. Springer, Cham (2015). https://doi.org/10.1007/978-3-319-24553-9_41
2. Allan, M., et al.: 2017 robotic instrument segmentation challenge. arXiv preprint arXiv:1902.06426 (2019)
3. Chen, L.-C., Papandreou, G., Kokkinos, I., Murphy, K., Yuille, A.L.: DeepLab: semantic image segmentation with deep convolutional nets, atrous convolution, and fully connected CRFs. arXiv preprint arXiv:1606.00915 (2016)
4. Chen, T., Xu, B., Zhang, C., Guestrin, C.: Training deep nets with sublinear memory cost. arXiv preprint arXiv:1604.06174 (2016)
5. Cheng, B., et al.: Panoptic-DeepLab. arXiv preprint arXiv:1910.04751 (2019)
6. He, K., Zhang, X., Ren, S., Sun, J.: Deep residual learning for image recognition. In: Proceedings of the IEEE Conference on Computer Vision and Pattern Recognition, pp. 770–778 (2016)
7. He, Y., Keuper, M., Schiele, B., Fritz, M.: Learning dilation factors for semantic segmentation of street scenes. In: Roth, V., Vetter, T. (eds.) GCPR 2017. LNCS, vol. 10496, pp. 41–51. Springer, Cham (2017). https://doi.org/10.1007/978-3-319-66709-6_4
8. Ioffe, S., Szegedy, C.: Batch normalization: accelerating deep network training by reducing internal covariate shift. arXiv preprint arXiv:1502.03167 (2015)

9. Jang, E., Gu, S., Poole, B.: Categorical reparameterization with gumbel-softmax. arXiv preprint arXiv:1611.01144 (2016)
10. Laina, I., et al.: Concurrent segmentation and localization for tracking of surgical instruments. In: Descoteaux, M., Maier-Hein, L., Franz, A., Jannin, P., Collins, D.L., Duchesne, S. (eds.) MICCAI 2017. LNCS, vol. 10434, pp. 664–672. Springer, Cham (2017). https://doi.org/10.1007/978-3-319-66185-8_75
11. Maddison, C.J., Mnih, A., Teh, Y.W.: The concrete distribution: a continuous relaxation of discrete random variables. arXiv preprint arXiv:1611.00712 (2016)
12. Pakhomov, D., Premachandran, V., Allan, M., Azizian, M., Navab, N.: Deep residual learning for instrument segmentation in robotic surgery. In: Suk, H.-I., Liu, M., Yan, P., Lian, C. (eds.) MLMI 2019. LNCS, vol. 11861, pp. 566–573. Springer, Cham (2019). https://doi.org/10.1007/978-3-030-32692-0_65
13. Pezzementi, Z., Voros, S., Hager, G.D.: Articulated object tracking by rendering consistent appearance parts. In: 2009 IEEE International Conference on Robotics and Automation, ICRA 2009, pp. 3940–3947. IEEE (2009)
14. Shvets, A.A., Rakhlin, A., Kalinin, A.A., Iglovikov, V.I.: Automatic instrument segmentation in robot-assisted surgery using deep learning. In: 2018 17th IEEE International Conference on Machine Learning and Applications (ICMLA), pp. 624–628. IEEE (2018)
15. Taylor, R.H., Menciassi, A., Fichtinger, G., Dario, P.: Medical robotics and computer-integrated surgery. In: Siciliano, B., Khatib, O. (eds.) Springer Handbook of Robotics, pp. 1199–1222. Springer, Heidelberg (2008). https://doi.org/10.1007/978-3-540-30301-5_53
16. Veit, A., Belongie, S.: Convolutional networks with adaptive inference graphs. arXiv preprint arXiv:1711.11503 (2017)
17. Xie, S., Zheng, H., Liu, C., Lin, L.: SNAS: stochastic neural architecture search. arXiv preprint arXiv:1812.09926 (2018)

Unsupervised Surgical Instrument Segmentation via Anchor Generation and Semantic Diffusion

Daochang Liu[1,5], Yuhui Wei[1], Tingting Jiang[1(✉)], Yizhou Wang[1,3,4], Rulin Miao[2], Fei Shan[2], and Ziyu Li[2]

[1] NELVT, Department of Computer Science, Peking University, Beijing, China
ttjiang@pku.edu.cn

[2] Peking University Cancer Hospital, Beijing, China

[3] Center on Frontiers of Computing Studies, Peking University, Beijing, China

[4] Advanced Institute of Information Technology, Peking University, Hangzhou, China

[5] Deepwise AI Lab, Beijing, China

Abstract. Surgical instrument segmentation is a key component in developing context-aware operating rooms. Existing works on this task heavily rely on the supervision of a large amount of labeled data, which involve laborious and expensive human efforts. In contrast, a more affordable unsupervised approach is developed in this paper. To train our model, we first generate anchors as pseudo labels for instruments and background tissues respectively by fusing coarse handcrafted cues. Then a semantic diffusion loss is proposed to resolve the ambiguity in the generated anchors via the feature correlation between adjacent video frames. In the experiments on the binary instrument segmentation task of the 2017 MICCAI EndoVis Robotic Instrument Segmentation Challenge dataset, the proposed method achieves 0.71 IoU and 0.81 Dice score without using a single manual annotation, which is promising to show the potential of unsupervised learning for surgical tool segmentation.

Keywords: Surgical instrument segmentation · Unsupervised learning · Semantic diffusion

1 Introduction

Instrument segmentation in minimally invasive surgery is fundamental for various advanced computer-aided intervention techniques such as automatic surgical skill assessment and intra-operative guidance systems [2]. Given its importance,

D. Liu and Y. Wei—Equal contribution.

Electronic supplementary material The online version of this chapter (https://doi.org/10.1007/978-3-030-59716-0_63) contains supplementary material, which is available to authorized users.

A. L. Martel et al. (Eds.): MICCAI 2020, LNCS 12263, pp. 657–667, 2020.
https://doi.org/10.1007/978-3-030-59716-0_63

surgical instrument segmentation has witnessed remarkable progress from early traditional methods [4,20,25] to recent approaches using deep learning [7,9–15,18,23,27]. However, such success is largely built upon supervised learning from a large amount of annotated data, which are very expensive and time-consuming to collect in the medical field, especially for the segmentation task on video data. Besides, the generalization ability of supervised methods is almost inevitably hindered by the domain gaps in real-world scenarios across different hospitals and procedure types.

In the literature, several attempts have been made to handle the lack of manual annotations [5,11,16,22,26]. Image level annotations of tool presence were utilized in [16,26] to train neural networks in a weakly-supervised manner. Jin et al. [11] propagated the ground truth across neighboring video frames using motion flow for semi-supervised learning, while Ross et al. [22] reduced the number of necessary labeled images by employing re-colorization as a pre-training task. Recently, a self-supervised approach was introduced to generate labels using the kinematic signal in the robot-assisted surgery [5]. Compared to prior works, this study steps further to essentially eliminate the demand for manual annotations or external signals by proposing an unsupervised method for the binary segmentation of surgical tools. Unsupervised learning has been successfully investigated in other surgical domains such as surgical workflow analysis [3] and surgical motion prediction [6], implying its possibility in instrument segmentation.

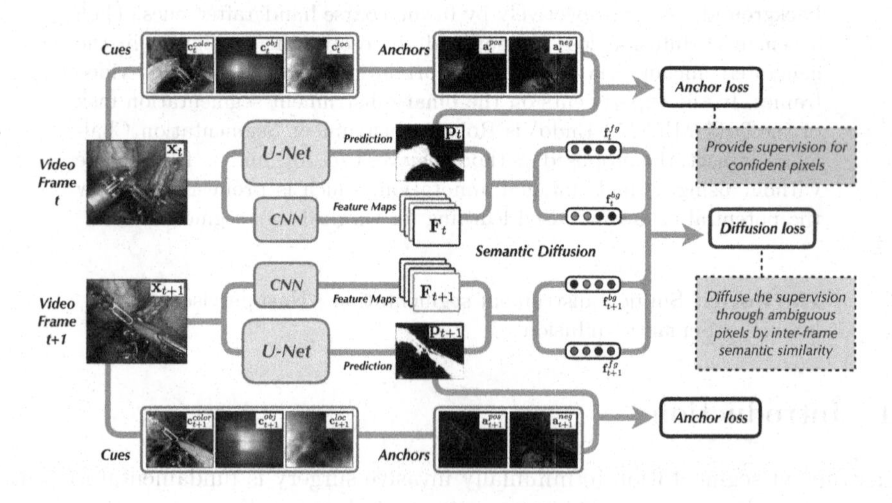

Fig. 1. Framework overview

Our method, which includes *anchor generation* and *semantic diffusion*, learns from the general prior knowledge about surgical tools. As for the anchor generation, we present a new perspective to train neural networks in the absence of labeled data, i.e., generating reliable pseudo labels from simple cues. A diverse collection of cues, including color, objectness, and location cues, are fused to be positive anchors and negative anchors, which correspond to pixels highly likely to be instruments and backgrounds respectively. Although individual cues are coarse and biased, our fusion process leverages the complementary information in these cues and thereby suppresses the noise. The segmentation model is then trained based on these anchors. However, since the anchors only cover a small portion of image pixels, a semantic diffusion loss is designed to propagate supervisory signals from anchor pixels to remaining pixels that are ambiguous to be instruments or not. The core idea of this loss is to exploit the temporal correlation in surgery videos. Specifically, adjacent video frames should share similar semantic representations in both instrument and background regions.

In the experiments on the EndoVis 2017 dataset [2], the proposed method achieves encouraging results (0.71 IoU and 0.81 Dice) on the binary segmentation task without using a single manual annotation, indicating its potential to reduce cost in clinical applications. Our method can also be easily extended to the semi-supervised setting and obtain performance comparable to the state-of-the-art. In addition, experiments on the ISIC 2016 dataset [8] demonstrate that the proposed model is inherently flexible to be applied in other domains like skin lesion segmentation. In summary, our contributions are three-fold: 1) An unsupervised approach for binary segmentation of surgical tools 2) A training strategy by generating anchor supervision from coarse cues 3) A semantic diffusion loss to explore the inter-frame semantic similarity.

2 Method

As illustrated in Fig. 1, our unsupervised framework[1] consists of two aspects, 1) generating anchors to provide initial training supervision 2) augmenting the supervision by a semantic diffusion loss. Our framework is elaborated as follows.

2.1 Anchor Generation

In conventional supervised methods, human knowledge is passed to the segmentation model through annotating large-scale databases. To be free of annotation, we encode the knowledge about surgical instruments into hand-designed cues instead and generate pseudo labels for training. The selection of such cues should adhere to two principles, i.e., *simplicity* and *diversity*. The simplicity of cues prevents virtually transferring intensive efforts from the data annotation to the design process of cues, while the diversity enriches the valuable information that we can take advantage of in different cues. Based on these principles, three cues are computed, including color, objectness and location cues. Given a video frame $\mathbf{x}_t \in \mathbb{R}^{HW}$ at time t with height H and width W, probability maps $\mathbf{c}_t^{color} \in [0,1]^{HW}$, $\mathbf{c}_t^{obj} \in [0,1]^{HW}$, $\mathbf{c}_t^{loc} \in [0,1]^{HW}$ are extracted according to the three cues respectively, which are fused as pseudo labels later.

[1] By unsupervised, we mean no manual annotation of surgical instruments is used.

Color. Color is an obvious visual characteristic to distinguish instruments from the surrounding backgrounds. Surgical instruments are mostly of grayish and plain colors, while the background tissues tend to be reddish and highly-saturated. Therefore, we multiply the inverted A channel in the LAB space, i.e., one minus the channel, and the inverted S channel in the HSV space to yield the probability map $\mathbf{c}_t^{color}$.

Objectness. Another cue can be derived from to what extent the image region is object-like. Surgical instruments are often with well-defined boundaries, while the background elements scatter around and fail to form a concrete shape. In detail, the objectness map $\mathbf{c}_t^{obj}$ is retrieved using a class-agnostic object detector [1]. Although this detector originally targets at daily scenes, we find it also give rich information in abdominal views.

Location. The third cue is based on the pixel location of instrument in the screen. Instead of a fixed location prior, an adaptive and video-specific location probability map is obtained by averaging the color maps across the whole video: $\mathbf{c}_t^{loc} = \frac{1}{T}\sum_t \mathbf{c}_t^{color}$, where T is the video length. The location map roughly highlights the image areas where instruments frequently appear in this video.

Anchor Generation. As shown in Fig. 1, the resultant cue maps are very coarse and noisy. Therefore, anchors are generated from these cues to suppress the noise. Concretely, the positive anchor $\mathbf{a}_t^{pos} \in [0,1]^{HW}$ [2] is defined as the element-wise product of all cues: $\mathbf{a}_t^{pos} = \mathbf{c}_t^{color}\mathbf{c}_t^{obj}\mathbf{c}_t^{loc}$, which captures the confident instrument regions that satisfy all the cues. Similarly, the negative anchor $\mathbf{a}_t^{neg} \in [0,1]^{HW}$ is defined as the element-wise product of all inverted cues: $\mathbf{a}_t^{neg} = (\mathbf{1}-\mathbf{c}_t^{color})(\mathbf{1}-\mathbf{c}_t^{obj})(\mathbf{1}-\mathbf{c}_t^{loc})$, which captures the confident background regions that satisfy none of the cues. As in Fig. 1, the false response is considerably minimized in the generated anchors.

Anchor Loss. The anchors are then regarded as pseudo labels to train the segmentation network, a vanilla U-Net [21] in this paper. We propose an anchor loss to encourage network activation on the positive anchor and inhibit activation on the negative anchor:

$$\mathcal{L}_{anc}(\mathbf{x}_t) = \frac{1}{HW}\sum_i -\mathbf{a}_{t,i}^{pos}\,\mathbf{p}_{t,i} - \mathbf{a}_{t,i}^{neg}(1-\mathbf{p}_{t,i}) \tag{1}$$

where $\mathbf{p}_t \in [0,1]^{HW}$ denotes the prediction map from the network and i is the pixel index. The loss is computed for each pixel and averaged over the whole image. Compared to the standard binary cross-entropy, this anchor loss only imposes supervision on the pixels that are confident to be instruments or backgrounds, keeping the network away from being disrupted by the noisy cues. However, the anchors only amount to a minority of image pixels. On the remaining ambiguous pixels outside the anchors, the network is not supervised and its behavior is undefined. Such a problem is tackled by the following semantic diffusion loss.

[2] [0, 1] means values are between 0 and 1 both inclusively.

2.2 Semantic Diffusion

Apart from the cues mentioned above, temporal coherence is another natural source of knowledge for unsupervised learning in the sequential data. We argue that the instruments in adjacent video frames usually share similar semantics, termed as *inter-frame instrument-instrument similarity*. This temporal similarity is assumed to be stronger than the semantic similarity between the instrument and the background within a single frame, i.e., the *intra-frame instrument-background similarity*. To this end, the semantic feature maps from a pre-trained convolutional neural network (CNN) are first aggregated within the instrument and background regions respectively using the prediction map:

$$\mathbf{f}_t^{fg} = \sum_i \mathbf{p}_{t,i}\mathbf{F}_{t,i} \qquad \mathbf{f}_t^{bg} = \sum_i (1 - \mathbf{p}_{t,i})\mathbf{F}_{t,i} \tag{2}$$

where $\mathbf{F}_t \in \mathbb{R}^{HW \times D}$ represents the CNN feature maps of frame $\mathbf{x}_t$, and $\mathbf{F}_{t,i} \in \mathbb{R}^D$ denotes the features at pixel i, and D is the channel number, $\mathbf{f}_t^{fg} \in \mathbb{R}^D$ and $\mathbf{f}_t^{bg} \in \mathbb{R}^D$ are the aggregated features for the instrument and the background correspondingly. Then given two adjacent frames $\mathbf{x}_t$ and $\mathbf{x}_{t+1}$, a semantic diffusion loss in a quadruplet form is proposed to constrain the *inter-frame instrument-instrument similarity* to be higher than the *intra-frame instrument-background similarities* by a margin:

$$\mathcal{L}_{dif}^{fg}(\mathbf{x}_t, \mathbf{x}_{t+1}) = \max(\phi(\mathbf{f}_t^{fg}, \mathbf{f}_t^{bg}) + \phi(\mathbf{f}_{t+1}^{fg}, \mathbf{f}_{t+1}^{bg}) - 2\phi(\mathbf{f}_t^{fg}, \mathbf{f}_{t+1}^{fg}) + m^{fg}, 0) \tag{3}$$

where $\phi(\cdot, \cdot)$ denotes the cosine similarity between two features and m^{fg} is a hyperparameter controlling the margin. Likewise, another semantic diffusion loss can be formulated to enforce the *inter-frame background-background similarity*:

$$\mathcal{L}_{dif}^{bg}(\mathbf{x}_t, \mathbf{x}_{t+1}) = \max(\phi(\mathbf{f}_t^{fg}, \mathbf{f}_t^{bg}) + \phi(\mathbf{f}_{t+1}^{fg}, \mathbf{f}_{t+1}^{bg}) - 2\phi(\mathbf{f}_t^{bg}, \mathbf{f}_{t+1}^{bg}) + m^{bg}, 0). \tag{4}$$

Lastly, the anchor loss and the semantic diffusion loss are optimized collectively:

$$\mathcal{L}_{full}(\mathbf{x}_t, \mathbf{x}_{t+1}) = \mathcal{L}_{anc}(\mathbf{x}_t) + \mathcal{L}_{anc}(\mathbf{x}_{t+1}) + \mathcal{L}_{dif}^{fg}(\mathbf{x}_t, \mathbf{x}_{t+1}) + \mathcal{L}_{dif}^{bg}(\mathbf{x}_t, \mathbf{x}_{t+1}). \tag{5}$$

Driven by the semantic diffusion loss, the initial signals on the confident anchor pixels are propagated to remaining ambiguous pixels. Our network benefits from such augmented supervision and outputs accurate and complete segmentation. Note that the semantic diffusion loss is generally not restricted to adjacent frames and can be also imposed on any image pair exhibiting inter-image similarity.

3 Experiment

Dataset. Our method is evaluated on the dataset of the 2017 MICCAI EndoVis Robotic Instrument Segmentation Challenge [2] (EndoVis 2017), which consists of 10 abdominal porcine procedures videotaped by the da Vinci Xi systems. Our work focuses on the binary instrument segmentation task, where each frame is separated into instruments and backgrounds. As our method is unsupervised, we do not use any annotations during the training process. Note that the ground truth of the test set is still held out by the challenge organizer.

Table 1. Results of the binary segmentation task from EndoVis 2017. Experimental results in the setting SS are reported.

$\mathcal{L}_{anc}$	$\mathcal{L}_{dif}^{fg}$	$\mathcal{L}_{dif}^{bg}$	IoU (%)	Dice (%)
✓			49.47	64.21
✓	✓		50.78	65.16
✓		✓	67.26	78.94
✓	✓	✓	**70.56**	**81.15**

Setup. Experiments are carried out in two different settings. 1) *Train Test (TT)*: This setting is common for supervised methods, where the learning and the inference are performed on two different sets of data. In this setting, we follow the previous convention [11] and conduct 4-fold cross-validation on the released 8 training videos of EndoVis 2017, with the same splits as prior works. Our method can attain real-time online inference speed in this setting. 2) *Single Stage (SS)*: This is a specific setting for our unsupervised method. Since the learning involves no annotation, we can directly place the learning and the inference on the same set of data, i.e., the released training set of EndoVis 2017. In application scenarios, the model needs to be re-trained when new unseen data comes, therefore this setting is more suitable for the offline batch analysis. Following previous work [11], we use intersection-over-union (IoU) and Dice coefficient to measure our performance.

Implementation Details. We extract the semantic feature maps from the *relu*5_3 layer of the VGG16 [24] pre-trained on ImageNet, which are interpolated to the same size as the prediction map. The VGG16 extractor is frozen when training our U-Net. The margin factors m^{fg} and m^{bg} are set as 0.2 and 0.8. The prediction map is thresholded to be final segmentation mask using the Otsu algorithm [17]. Our implementation uses official pre-trained CNN models and parameters in PyTorch [19]. Codes will be released to offer all details.

3.1 Results on EndoVis 2017

Results on EndoVis 2017 are reported in Table 1. Firstly we assess the network performance only using the anchor loss $\mathcal{L}_{anc}$ based on our cues, where we get the basic performance. After we combine the semantic diffusion losses, especially the background semantic diffusion loss $\mathcal{L}_{dif}^{bg}$, the performance is strikingly improved. This result proves our assumption that adjacent video frames are similar to each other in both foreground and background regions. Since the background area is relatively more similar between the video frames, it is seen from Table 1 that $\mathcal{L}_{dif}^{bg}$ brings more improvement on the performance than $\mathcal{L}_{dif}^{fg}$.

3.2 The Choice of Cues

Different combinations of cues are examined to research their effects on the network performance. Here we run the Otsu thresholding algorithm [17] not only on the network prediction map $\mathbf{p}$ but also on the corresponding positive anchor $\mathbf{a}^{pos}$ and the inverted negative anchor $\mathbf{1} - \mathbf{a}^{neg}$ to generate segmentation masks. The Otsu algorithm is adaptive to the disparate intensity level of the probabilistic maps. The resultant masks are then evaluated against the ground truth. As the results shown in Table 2, the best network performance comes from the combination of all three cues, because more kinds of cues can provide extra information from different aspects. Meanwhile, a single kind of cue may produce good results on the anchors, but may not be helpful to the final network prediction, because a single cue may contain lots of noise and it needs to be filtered out by the fusion with other useful cues. Also, different kinds of cues have varying effects on the network performance. For example, it is noticed that the $\mathbf{c}^{color}$ and $\mathbf{c}^{obj}$ cues are more important than the $\mathbf{c}^{loc}$ from the table.

Table 2. The choice of cues (setting SS)

$\mathbf{c}^{color}$	$\mathbf{c}^{obj}$	$\mathbf{c}^{loc}$	IoU (%)			Dice (%)		
			$\mathbf{a}^{pos}$	$\mathbf{1} - \mathbf{a}^{neg}$	$\mathbf{p}$	$\mathbf{a}^{pos}$	$\mathbf{1} - \mathbf{a}^{neg}$	$\mathbf{p}$
✓			55.60	55.60	45.27	69.21	69.21	60.09
	✓		16.01	16.01	14.23	26.57	26.57	23.57
		✓	16.90	16.90	21.32	28.11	28.11	33.97
	✓	✓	20.28	18.93	47.39	32.48	30.99	62.51
✓		✓	41.44	19.21	43.74	57.00	31.46	59.30
✓	✓		38.69	22.09	63.27	53.70	35.19	75.56
✓	✓	✓	38.64	18.53	**70.56**	53.73	30.56	**81.15**

3.3 Compared to Supervised Methods

At present, unsupervised instrument segmentation is still less explored, with few methods that can be directly compared with. Therefore, to provide an indirect reference, our method is adjusted to the semi-supervised and fully-supervised settings and compared with previous supervised methods in Table 3. When fully-supervised, we substitute the anchors by the ground truth on all the frames. Since our contribution is not in the network architecture and we do not use special modules such as attention beyond the U-Net, our fully-supervised performance is close to some earlier works, which can be thought of as an upper bound of our unsupervised solution. When semi-supervised, the anchors are replaced with the ground truth on 50% frames in the same periodical way as in [11]. Our method has competitive performance with the state-of-the-art in the semi-supervised setting. Lastly, in the last two rows without using any annotation, we achieve the preeminent performance. More data is exploited for learning in the setting SS than in the setting TT, which explains why the setting SS has better results.

Table 3. Comparison with supervised methods (mean ± std). Results of prior works are quoted from [11]. Not all the existing fully-supervised methods are listed due to limited space. Our network architecture is the vanilla U-Net.

Supervision	Method	Setting	IoU (%)	Dice (%)
100%	U-Net [21]	TT	75.44 ± 18.18	84.37 ± 14.58
100%	Ours ($\mathcal{L}_{full}$)	TT	81.55 ± 14.52	88.83 ± 11.50
100%	TernausNet [23]	TT	83.60 ± 15.83	90.01 ± 12.50
100%	MF-TAPNet [11]	TT	87.56 ± 16.24	93.37 ± 12.93
50%	Semi-MF-TAPNet [11]	TT	80.03 ± 16.87	**88.07** ± 13.15
50%	Ours ($\mathcal{L}_{full}$)	TT	**80.33** ± 14.69	87.94 ± 11.53
0%	Ours ($\mathcal{L}_{full}$)	TT	67.85 ± 15.94	79.42 ± 13.59
0%	Ours ($\mathcal{L}_{full}$)	SS	**70.56** ± 16.09	**81.15** ± 13.79

3.4 Qualitative Result

In this section, some visual results from our method are plotted. Firstly, as seen in Fig. 2, the three cues are very coarse, e.g., the background can still be found on the color cue maps. By the fusion of noisy cues, the anchors become purer, which are nonetheless very sparse. Then via the semantic diffusion loss, which augments the signals on the anchors, the network can find the lost instrument region in those anchor pictures, as shown in the success cases in Fig. 2a and Fig. 2b. Although in some pictures there are difficulties such as complicated scenes and lighting inconstancy, we can also get good performance in these cases. However, there are still some failure cases, such as the special probe (Fig. 2e) that is not thought of as the instrument in the ground truth. Also, the dimmed light and the dark organ (Fig. 2d) can also have negative effects on the reliability of cues. A video demo is attached in the supplementary material.

3.5 Extension to Other Domain

An exploratory experiment is conducted on the skin lesion segmentation task of ISIC 2016 benchmark [8] to inspect whether our model can be migrated to other domains. We conform to the official train-test split. Due to the dramatic color variations of lesions, the color cue is excluded. The location cue is set as a fixed 2D Gaussian center prior since ISIC is not a video dataset. In view of the background similarity shared by most images, we sample random image pairs for semantic diffusion. Our flexibility is provisionally supported by the results in Table 4. Specific cues for skin lesions can be designed in future for better results.

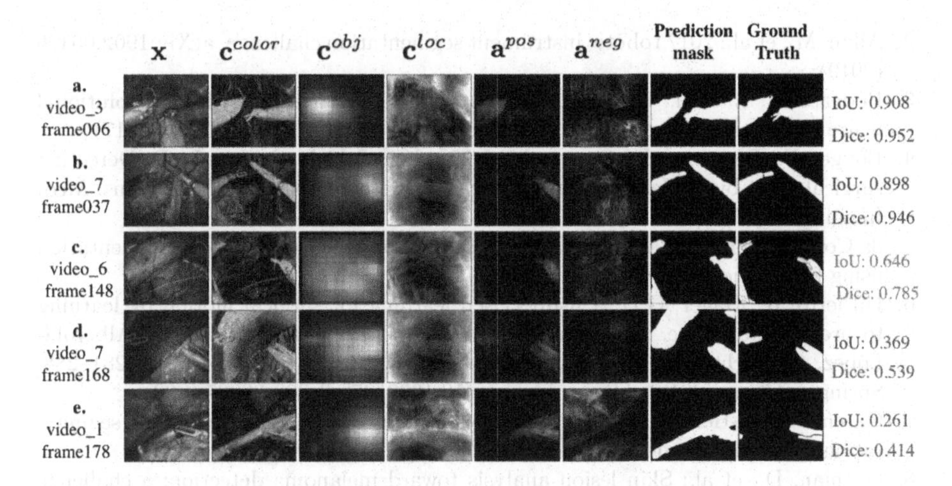

Fig. 2. Visual results for success and failure cases.

Table 4. Results on the skin lesion segmentation task of ISIC 2016

Supervision	Method	Setting	IoU (%)	Dice (%)
100%	Challenge winner [8]	TT	84.3	91.0
100%	Ours ($\mathcal{L}_{full}$)	TT	83.6	90.3
50%	Ours ($\mathcal{L}_{full}$)	TT	81.1	88.6
0%	Ours ($\mathcal{L}_{full}$)	TT	63.3	74.9
0%	Ours ($\mathcal{L}_{full}$)	SS	**64.4**	**75.7**

4 Conclusion and Future Work

This work proposes an unsupervised surgical instrument segmentation method via anchor generation and semantic diffusion, whose efficacy and flexibility are validated by empirical results. The current framework is still limited to binary segmentation. In future works, multiple class-specific anchors could be generated for multi-class segmentation, while additional grouping strategies could be incorporated as post-processing to support instance or part segmentation.

Acknowledgments. This work was partially supported by MOST-2018AAA0102004 and the Natural Science Foundation of China under contracts 61572042, 61527804, 61625201. We also acknowledge the Clinical Medicine Plus X-Young Scholars Project, and High-Performance Computing Platform of Peking University for providing computational resources. Thank Boshuo Wang for making the video demo.

References

1. Alexe, B., Deselaers, T., Ferrari, V.: Measuring the objectness of image windows. IEEE TPAMI **34**(11), 2189–2202 (2012)

2. Allan, M., et al.: 2017 robotic instrument segmentation challenge. arXiv:1902.06426 (2019)
3. Bodenstedt, S., et al.: Unsupervised temporal context learning using convolutional neural networks for laparoscopic workflow analysis. arXiv:1702.03684 (2017)
4. Bouget, D., Benenson, R., Omran, M., Riffaud, L., Schiele, B., Jannin, P.: Detecting surgical tools by modelling local appearance and global shape. IEEE Trans. Med. Imaging **34**(12), 2603–2617 (2015)
5. da Costa Rocha, C., Padoy, N., Rosa, B.: Self-supervised surgical tool segmentation using kinematic information. In: ICRA (2019)
6. DiPietro, R., Hager, G.D.: Unsupervised learning for surgical motion by learning to predict the future. In: Frangi, A.F., Schnabel, J.A., Davatzikos, C., Alberola-López, C., Fichtinger, G. (eds.) MICCAI 2018. LNCS, vol. 11073, pp. 281–288. Springer, Cham (2018). https://doi.org/10.1007/978-3-030-00937-3_33
7. García-Peraza-Herrera, L.C., et al.: ToolNet: holistically-nested real-time segmentation of robotic surgical tools. In: IROS (2017)
8. Gutman, D., et al.: Skin lesion analysis toward melanoma detection: a challenge at the international symposium on biomedical imaging (ISBI) 2016, hosted by the international skin imaging collaboration (ISIC). arXiv:1605.01397 (2016)
9. Hasan, S.K., Linte, C.A.: U-NetPlus: a modified encoder-decoder U-Net architecture for semantic and instance segmentation of surgical instruments from laparoscopic images. In: Annual International Conference of the IEEE Engineering in Medicine and Biology Society (EMBC) (2019)
10. Islam, M., Li, Y., Ren, H.: Learning where to look while tracking instruments in robot-assisted surgery. In: Shen, D., et al. (eds.) MICCAI 2019. LNCS, vol. 11768, pp. 412–420. Springer, Cham (2019). https://doi.org/10.1007/978-3-030-32254-0_46
11. Jin, Y., Cheng, K., Dou, Q., Heng, P.-A.: Incorporating temporal prior from motion flow for instrument segmentation in minimally invasive surgery video. In: Shen, D., et al. (eds.) MICCAI 2019. LNCS, vol. 11768, pp. 440–448. Springer, Cham (2019). https://doi.org/10.1007/978-3-030-32254-0_49
12. Laina, I., et al.: Concurrent segmentation and localization for tracking of surgical instruments. In: Descoteaux, M., Maier-Hein, L., Franz, A., Jannin, P., Collins, D.L., Duchesne, S. (eds.) MICCAI 2017. LNCS, vol. 10434, pp. 664–672. Springer, Cham (2017). https://doi.org/10.1007/978-3-319-66185-8_75
13. Milletari, F., Rieke, N., Baust, M., Esposito, M., Navab, N.: CFCM: segmentation via coarse to fine context memory. In: Frangi, A.F., Schnabel, J.A., Davatzikos, C., Alberola-López, C., Fichtinger, G. (eds.) MICCAI 2018. LNCS, vol. 11073, pp. 667–674. Springer, Cham (2018). https://doi.org/10.1007/978-3-030-00937-3_76
14. Ni, Z.L., et al.: BARNet: bilinear attention network with adaptive receptive field for surgical instrument segmentation. arXiv:2001.07093 (2020)
15. Ni, Z.L., Bian, G.B., Xie, X.L., Hou, Z.G., Zhou, X.H., Zhou, Y.J.: RASNet: segmentation for tracking surgical instruments in surgical videos using refined attention segmentation network. In: Annual International Conference of the IEEE Engineering in Medicine and Biology Society (EMBC) (2019)
16. Nwoye, C.I., Mutter, D., Marescaux, J., Padoy, N.: Weakly supervised convolutional LSTM approach for tool tracking in laparoscopic videos. Int. J. Comput. Assist. Radiol. Surg. **14**(6), 1059–1067 (2019). https://doi.org/10.1007/s11548-019-01958-6
17. Otsu, N.: A threshold selection method from gray-level histograms. IEEE Trans. Syst. Man Cybern. **9**(1), 62–66 (1979)

18. Pakhomov, D., Premachandran, V., Allan, M., Azizian, M., Navab, N.: Deep residual learning for instrument segmentation in robotic surgery. In: Suk, H.-I., Liu, M., Yan, P., Lian, C. (eds.) MLMI 2019. LNCS, vol. 11861, pp. 566–573. Springer, Cham (2019). https://doi.org/10.1007/978-3-030-32692-0_65
19. Paszke, A., et al.: PyTorch: an imperative style, high-performance deep learning library. In: Advances in Neural Information Processing Systems (2019)
20. Rieke, N., et al.: Real-time localization of articulated surgical instruments in retinal microsurgery. Med. Image Anal. **34**, 82–100 (2016)
21. Ronneberger, O., Fischer, P., Brox, T.: U-Net: convolutional networks for biomedical image segmentation. In: Navab, N., Hornegger, J., Wells, W.M., Frangi, A.F. (eds.) MICCAI 2015. LNCS, vol. 9351, pp. 234–241. Springer, Cham (2015). https://doi.org/10.1007/978-3-319-24574-4_28
22. Ross, T., et al.: Exploiting the potential of unlabeled endoscopic video data with self-supervised learning. Int. J. Comput. Assist. Radiol. Surg. **13**(6), 925–933 (2018). https://doi.org/10.1007/s11548-018-1772-0
23. Shvets, A.A., Rakhlin, A., Kalinin, A.A., Iglovikov, V.I.: Automatic instrument segmentation in robot-assisted surgery using deep learning. In: IEEE International Conference on Machine Learning and Applications (ICMLA) (2018)
24. Simonyan, K., Zisserman, A.: Very deep convolutional networks for large-scale image recognition. arXiv:1409.1556 (2014)
25. Speidel, S., et al.: Visual tracking of da Vinci instruments for laparoscopic surgery. In: Medical Imaging 2014: Image-Guided Procedures, Robotic Interventions, and Modeling (2014)
26. Vardazaryan, A., Mutter, D., Marescaux, J., Padoy, N.: Weakly-supervised learning for tool localization in laparoscopic videos. In: Stoyanov, D., et al. (eds.) LABELS/CVII/STENT -2018. LNCS, vol. 11043, pp. 169–179. Springer, Cham (2018). https://doi.org/10.1007/978-3-030-01364-6_19
27. Yamazaki, Y., et al.: Automated surgical instrument detection from laparoscopic gastrectomy video images using an open source convolutional neural network platform. J. Am. Coll. Surg. **230**(5), 725.e1–732.e1 (2020)

Towards Accurate and Interpretable Surgical Skill Assessment: A Video-Based Method Incorporating Recognized Surgical Gestures and Skill Levels

Tianyu Wang, Yijie Wang, and Mian Li(✉)

Shanghai Jiao Tong University, Shanghai, China
{gunnerwang27,yijiewang,mianli}@sjtu.edu.cn

Abstract. Nowadays, surgical skill assessment becomes increasingly important for surgical training, given the explosive growth of automation technologies. Existing work on skill score prediction is limited and deserves more promising outcomes. The challenges lie on complicated surgical tasks and new subjects as trial performers. Moreover, previous work mostly provides local feedback involving each individual video frame or clip that does not manifest human-interpretable semantics itself. To overcome these issues and facilitate more accurate and interpretable skill score prediction, we propose a novel video-based method incorporating recognized surgical gestures (segments) and skill levels (for both performers and gestures). Our method consists of two correlated multi-task learning frameworks. The main task of the first framework is to predict final skill scores of surgical trials and the auxiliary tasks are to recognize surgical gestures and to classify performers' skills into self-proclaimed skill levels. The second framework, which is based on gesture-level features accumulated until the end of each previously identified gesture, incrementally generates running intermediate skill scores for feedback decoding. Experiments on JIGSAWS dataset show our first framework on C3D features pushes state-of-the-art prediction performance further to 0.83, 0.86 and 0.69 of Spearman's correlation for the three surgical tasks under LOUO validation scheme. It even achieves 0.68 when generalizing across these tasks. For the second framework, additional gesture-level skill levels and captions are annotated by experts. The trend of predicted intermediate skill scores indicating problematic gestures is demonstrated as interpretable feedback. It turns out such trend resembles human's scoring process.

Keywords: Surgical skill assessment · Incorporating recognized surgical gestures and skill levels · Interpretable feedback

Electronic supplementary material The online version of this chapter (https://doi.org/10.1007/978-3-030-59716-0_64) contains supplementary material, which is available to authorized users.

A. L. Martel et al. (Eds.): MICCAI 2020, LNCS 12263, pp. 668–678, 2020.
https://doi.org/10.1007/978-3-030-59716-0_64

1 Introduction

With the society's increasingly high expectations towards modern surgery, surgical training programs have become more and more important. Better surgical skills have been shown to be strongly associated with fewer postoperative complications and lower rates of re-operation and re-admission [3,25]. Traditionally the trainee is supervised by experts who assess his/her performance and give feedback for skill improvement. However, such way of training still suffers from strong subjectivity and lack of efficiency. With the introduction of Objective Structured Assessment of Technical Skills (OSATS) [20], a grading system for medical schools, many researchers begin to focus on automatic assessment of surgical skills based on surgical trials (*i.e.*, instances of each surgical task) instead.

A lot of methods have been developed for automatic surgical skill assessment [2,5–7,9,10,19,29,32,34]. Some proposed frameworks to classify trial performers' surgical skills into self-proclaimed or score-based expertise levels (*i.e.*, discretized from exact skill scores with thresholds) [5–7,9,10,29,34]. Although considerable accuracy has been achieved by these work, measuring the quality of surgical skills with only discretized levels is far from enough. Instead, the work in [32] explored different holistic features from kinematic data to predict exact skill scores. The work in [19] recently proposed a video-based method for skill score prediction on in-vivo clinical data. However, the amount of such work is still limited and more promising progress needs to be reported. In particular, the challenges lie on complicated surgical tasks (*e.g.*, suturing) and on new subjects not previously seen in the training data. Being aware of the surgical gestures (atomic surgical activity segments) in those trials may alleviate the former issue [18,32], while knowing about performers' and even gestures' skill levels may deal with the latter one. To this end, the largely unexplored relationship among exact skill scores, surgical gestures [4,11,18,33] and skill levels is still worth studying. Moreover, exploring such relationship also helps generate more human-interpretable feedback for skill improvement in addition to frame-level or clip-level feedback only [9,19,21,32].

As kinematic data is usually unavailable in real scenarios, video-based methods are more flexible and desirable [19]. To this end, some work in Action Quality Assessment (AQA) [21–23,30] can inspire the development of surgical skill assessment. It was shown that C3D features (*i.e.*, direct visual features extracted with C3D network [28]) can be used for assessing human actions [21]. Furthermore, the work in [30] demonstrated the effectiveness of segmenting full actions into multiple stages before the assessment. State-of-the-art performance was achieved with a Multi-Task Learning (MTL) paradigm [23]. Moreover, it was indicated that there is an utility in learning a single model across multiple actions [22].

In this paper, a novel video-based method incorporating recognized surgical gestures and skill levels (for both trial performers and gestures) is proposed. Our method consists of two correlated MTL frameworks for accurate and human-interpretable surgical skill assessment, respectively. The main task of the first framework is to predict the final skill scores (*i.e.*, ranging from 6 to 30) for surgical trials and its auxiliary tasks are to recognize surgical gestures and to classify performers' skills into self-proclaimed skill levels. For the second framework, the previous skill score predictor is fine-tuned incrementally on accumulated gesture-level features to predict the running intermediate skill scores. The evaluation on gestures' skill levels is employed as the auxiliary task. By exploring the trend of predicted intermediate skill scores we can identify the problematic gestures for each surgical trial. Our proposed method is experimented on JHI-ISI Gesture and Skill Assessment Working Set (JIGSAWS) [1]. The first framework for skill score prediction on C3D features outperforms (LOUO) state-of-the-art methods on individual surgical tasks and establishes the first competitive result across the three tasks. For the second framework, we invite experts to additionally annotate the dataset. The predicted feedback is demonstrated qualitatively and evaluated quantitatively in two aspects. It turns out to resemble the scoring process of humans. In summary, our contributions are three-fold: (1) a framework for more accurate skill score prediction; (2) a framework for interpretable feedback generation; (3) additional annotations of JIGSAWS dataset for future work on more human-interpretable surgical skill assessment beyond scoring.

2 Method

In this section, two correlated frameworks are proposed to predict exact final skill scores and to generate the gesture-level feedback, respectively. The feature extractor acts as the common backbone for both frameworks.

The extracted features from RGB surgical videos should account for all frames that involve gestures. Two types of features are considered in the experiments: (1) concatenated clip-level features extracted with C3D network [28] (16-frame clips with a 8-frame overlap between two consecutive clips); (2) concatenated frame-level features extracted with ResNet-101 [12]. Pre-trained networks are utilized and fine-tuned in our MTL-VF framework. For convenience of the mathematical formulation, our method will be illustrated with C3D features. We denote features extracted from video i as $\mathbf{x}_i \in \mathbb{R}^{c_i \times 4096}$ where c_i is the number of 8-frame clips.

2.1 Multi-Task Learning with Video-Level Features (MTL-VF)

Lots of work [5–7,9,10,29,34] has focused on predicting self-proclaimed skill levels rather than exact skill scores. Although the former task itself is not good enough to supervise surgeons' training, it can guide the latter task to reach a more accurate conclusion, especially when the range of skill scores is quite large. Moreover, when human experts assess the exam performance, they usually dive into each detailed part before reaching the final score. This process also helps them further explain and analyze the exam outcomes. To this end, we believe that knowing which of and when those predefined surgical gestures occur in each surgical trial is beneficial for the assessment of that trial.

Based on the analysis above, we refer to the paradigm of MTL. It aims to facilitate the generalization of each individual task sharing a common backbone by learning them jointly. In particular, MTL was shown to improve the performance of human action assessment [23]. Therefore, we formulate the MTL-VF framework (Fig. 1a) where the main task is to predict final skill scores and the auxiliary tasks are to recognize surgical gestures and to classify performers' skills into self-proclaimed skill levels. Since surgical gestures are more diverse and vary with trials unlike segments in most actions (*e.g.*, only five fixed segments in diving [30]), such MTL-VF framework is necessary for the skill score predictor to generalize well. Besides a common feature extractor, we formulate the distinct task-specific heads below.

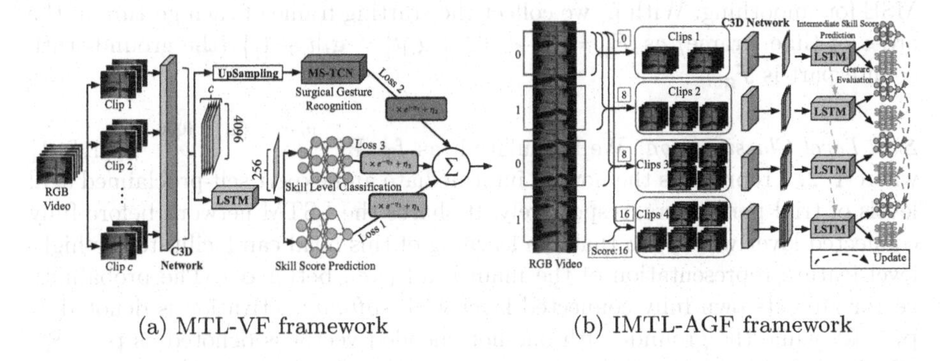

(a) MTL-VF framework (b) IMTL-AGF framework

Fig. 1. An overview of our proposed frameworks. MTL-VF consists of a common feature extractor and three task-specific heads. For IMTL-AGF, ground-truth running intermediate skill scores and final skill score are shown in grey boxes. The "update" arrow refers to back propagation on LSTM and task-specific heads.

Skill Score Prediction (Main Head). It is formulated as a mapping function from extracted features $\mathbf{x}_i$ of video i (in total N videos for training) to a final skill score $\hat{s}_i$ denoted as $f_1 : \mathbb{R}^{\max_{i\in\{1,\cdots,N\}}\{c_i\}\times 4096} \to \mathbb{R}$. A one-layer Long Short-Term Memory (LSTM) network is utilized to capture spatio-temporal relations within video-level features, rendering the intermediate output $\mathbf{o}_i \in \mathbb{R}^{256}$. Then this output goes through a fully connected layer (possibly with *sigmoid* activation) to obtain the predicted skill score $\hat{s}_i$. Denote the ground-truth skill score as s_i, then the loss function to minimize for training is defined as

$$\mathcal{L}_1 = \frac{1}{N}\sum_{i=1}^{N}\left[(\hat{s}_i - s_i)^2 + |\hat{s}_i - s_i|\right]. \tag{1}$$

It turns out placing a Mean Absolute Error (MAE) term in addition to the regular Mean Squared Error (MSE) can result in a better performance [23].

Surgical Gesture Recognition. We denote the non-empty global gesture vocabulary as $\mathcal{G}$ and number of valid frames of video i as t_i. This head f_2 identifies surgical gestures of video i (instances from the vocabulary) with input $\mathbf{x}_i^* \in \mathbb{R}^{t_i\times 4096}$

upsampled from $\mathbf{x}_i$ using nearest-neighbor interpolation and output $\hat{\mathbf{g}}_i \in \mathcal{G}^{t_i}$ (corresponding to the ground truth $\mathbf{g}_i$). The Multi-Stage Temporal Convolutional Network (MS-TCN) proposed in [8] is utilized as it works well even for long and densely annotated sequences. The loss function is

$$\mathcal{L}_2 = \frac{1}{N}\sum_{i=1}^{N}\sum_{j=1}^{M}\left(\mathcal{L}_{\text{ce}}^{i,j} + \lambda \mathcal{L}_{\text{t-mse}}^{i,j}\right), \tag{2}$$

where M is the number of stages in MS-TCN, λ is the balancing weight, $\mathcal{L}_{\text{ce}}^{i,j}$ is the Cross Entropy (CE) loss for classification, and $\mathcal{L}_{\text{t-mse}}^{i,j}$ is the truncated MSE for smoothing. With $\hat{\mathbf{g}}_i$ we collect the starting frame of each gesture in the order of time (frame) as $\mathcal{F}_{\hat{\mathbf{g}}_i} = \left\{t \in \mathbb{N}_+ \mid \hat{\mathbf{g}}_i[t] \neq \hat{\mathbf{g}}_i[t-1]\right\}$ (the ground-truth counterpart is $\mathcal{F}_{\mathbf{g}_i}$).

Skill Level Classification. We formulate it as $f_3 : \mathbb{R}^{\max_{i \in \{1,\cdots,N\}}\{c_i\} \times 4096} \rightarrow \{1, 2, 3\}$ where 1, 2, 3 represents the novice, intermediate and expert self-proclaimed skill levels of trial performers, respectively. It shares the LSTM network before fully connected layer with f_1 so that the learning of this head can facilitate the high-level feature representation of the main head (*i.e.,* better $\mathbf{o}_i$). The probability vector after its own fully connected layer with *softmax* activation is denoted as $\hat{\mathbf{p}}_i \in \mathbb{R}^3$ while the ground-truth one-hot encoded vector is denoted as $\mathbf{p}_i \in \mathbb{R}^3$. The final skill levels are therefore $\arg\max_{k=1,2,3} \hat{\mathbf{p}}_i[k]$ and $\arg\max_{k=1,2,3} \mathbf{p}_i[k]$, respectively. We use CE as the loss of this single-label multi-class classification problem,

$$\mathcal{L}_3 = -\frac{1}{N}\sum_{i=1}^{N}\sum_{k=1}^{3}\left(\mathbf{p}_i[k] \log \hat{\mathbf{p}}_i[k]\right). \tag{3}$$

Task uncertainties (with variable change for numerical stability) η_1, η_2, η_3 [14] are learned to balance the scales of $\mathcal{L}_1, \mathcal{L}_2, \mathcal{L}_3$, respectively. The final loss is thus

$$\mathcal{L} = \frac{1}{2}\left[(\mathcal{L}_1 e^{-\eta_1} + \eta_1) + (\mathcal{L}_2 e^{-\eta_2} + \eta_2) + (\mathcal{L}_3 e^{-\eta_3} + \eta_3)\right]. \tag{4}$$

2.2 Incremental Multi-Task Learning with Accumulated Gesture-Level Features (IMTL-AGF)

While MTL-VF learns to predict final skill score of each trial, it is also important to point out how intermediate details contribute to such score. Past work on skill or action quality assessment mostly derived frame-level [32] or clip-level [21] contribution values. These values can help identify problematic poses locally. However, in practice more human-interpretable feedback is also required to quickly know which parts of the trial deserve adjustment. To this end, the feedback system can benefit from a clear recognition of surgical gestures.

As the trial proceeds, human experts usually examine whether each surgical gesture is well performed. The final skill score is gradually built up with these gestures. Enlightened by the work in [21], we incrementally fine-tune the skill score predictor trained in MTL-VF to generate running intermediate skill scores for feedback decoding. For each gesture $g \in \mathcal{F}_{\mathbf{g}_i}$ in trial video i there is an annotated value $v_g \in \{0, 1\}$ indicating the skill level of the gesture, *i.e.*, pass ($v_g = 1$) or fail ($v_g = 0$). The vector involving such values of all gestures in video i is normalized with L1 norm to obtain $\mathbf{v}_i = [\bar{v}_{g_1}, \bar{v}_{g_2}, ...]$. Then the ground-truth intermediate skill scores are calculated as the cumulative sums of $s_i\mathbf{v}_i = [s_i\bar{v}_{g_1}, s_i\bar{v}_{g_2}, ...]$ until the end of each gesture (s_i is final skill score). Accordingly, we feed in accumulated gesture-level features extracted from incremental numbers of clips in video i.

Our IMTL-AGF framework is illustrated in Fig. 1b. Two tasks are formulated: intermediate skill score prediction and gesture evaluation. During the training, we initialize C3D, LSTM and skill score predictor's fully connected layer with the parameters learned by MTL-VF (C3D is frozen then). The gesture boundaries are based on the same annotations in the dataset as MTL-VF. During the testing, the output $\hat{\mathbf{g}}_i$ predicted by surgical gesture recognition head of MTL-VF is utilized to derive gesture boundaries and then generate accumulated gesture-level features. Through studying the differences between two consecutive values in $\hat{\mathbf{v}}_i$ predicted by IMTL-AGF we can tell which gestures are not performed as expected.

Intermediate Skill Score Prediction (Main Head). It fine-tunes the skill score prediction head in Sect. 2.1. Specifically, we use an unsupervised **rank loss** between two consecutive predicted intermediate skill scores in addition to supervised MSE so that the intermediate scores evolve in an non-decreasing manner.

Gesture Evaluation. We formulate it as $f_4 : \mathbb{R}^{c_i \times 4096} \rightarrow \mathbb{R}^{|\mathcal{G}|}$. Note that we evaluate gesture types from gesture vocabulary $\mathcal{G}$ here instead of gestures to relieve the difficulty and overfitting of this auxiliary task. Gesture type differs from gesture in that one gesture type can appear more than once as gestures (instances) in a video. The accumulated skill level of gesture type reaches 1 only when all of its gestures that have appeared are of skill level 1. The predicted output of this head is a vector $\hat{\mathbf{h}}_{i,g}$ containing accumulated skill level of each gesture type (after *sigmoid*) until the end of gesture g in video i. The ground truth $\mathbf{h}_{i,g}$ is thus a multi-hot encoded vector. As this head represents a multi-label classification problem, Binary Cross Entropy (BCE) is used as the loss function,

$$\mathcal{L}_4 = -\frac{1}{N}\sum_{i=1}^{N}\sum_{z=1}^{|\mathcal{G}|}\left[\mathbf{h}_{i,g}[z]\log\hat{\mathbf{h}}_{i,g}[z] + (1-\mathbf{h}_{i,g}[z])\log\left(1-\hat{\mathbf{h}}_{i,g}[z]\right)\right]. \quad (5)$$

3 Experiments

3.1 Dataset and Implementation

The JIGSAWS dataset [1] consists of kinematic and video data from 8 subjects performing three surgical tasks: Knot Tying (KT), Needle Passing (NP) and Suturing (SU). For our analysis we only use the video data. Each surgical trial is divided into atomic segments (gestures) under a common vocabulary. Moreover, the modified Global Rating Score (GRS) as final skill score and self-proclaimed skill level of the performer are assigned for each trial. We employ the standard Leave-One-Supertrial-Out (LOSO) and Leave-One-User-Out (LOUO) cross-validation schemes. To establish the cause and effect between the surgical gestures and final skill scores, skill-related annotations are necessary for each gesture. Such annotations can help provide interpretable feedback to the surgeons. To this end, several experts are invited to look at the original annotations and watch the videos. Then for each video they assign each gesture with one binary skill level (0 indicates fail and 1 indicates pass) and comment with several sentences for those gestures with skill level of 0[1].

PyTorch [24] is used to implement both frameworks[2]. The C3D network and ResNet-101 for MTL-VF were pre-trained on Sports-1M [13] and ImageNet [26], respectively. Video frames are resized to 150×112 for KT and SU, 130×112 for NP, and processed with 112×112 center crop. Both frameworks are trained for 100 epochs with the Adam optimizer [15]. The learning rates are 5×10^{-4} for MS-TCN and 1×10^{-5} for feature extractor, other task-specific heads and η's.

Table 1. Performance comparison of skill score prediction. "Across" means forming the data from surgical trials of KT, NP and SU together without distinction.

Method	Task & scheme							
	KT		NP		SU		Across	
	LOSO	LOUO	LOSO	LOUO	LOSO	LOUO	LOSO	LOUO
SMT+DCT+DFT [32]	0.70	0.73	0.38	0.23	0.64	0.10	–	–
DCT+DFT+ApEn [32]	0.63	0.60	0.46	0.25	0.75	0.37	–	–
ResNet101-LSTM	0.52	0.36	0.84	0.33	0.73	0.67	–	–
C3D-LSTM [21]	0.81	0.60	**0.84**	0.78	0.69	0.59	–	–
C3D-SVR [21]	0.71	0.33	0.75	-0.17	0.42	0.37	–	–
S3D* [30]	0.64	0.14	0.57	0.35	0.68	0.03	–	–
Our ResNet101-MTL-VF	0.63	0.72	0.73	0.48	**0.79**	0.68	0.73	0.64
Our C3D-MTL-VF	**0.89**	**0.83**	0.75	**0.86**	0.77	**0.69**	**0.75**	**0.68**

3.2 Experiment Results

MTL-VF Framework. We use the benchmark metrics Spearman's correlation ρ to measure the performance of skill score prediction (Table 1). The values of ρ are averaged on folds specified by LOSO and LOUO. We compare MTL-VF on ResNet101 and C3D features with two kinematic-based [32] and four video-based [21,30] methods. Due to lack of video-based results on JIGSAWS dataset, we implement those well-known methods. Since S3D [30] only deals

[1] Additional annotations for JIGSAWS dataset can be accessed via request.

[2] Our code is available on https://github.com/gunnerwang/MTL-VF-and-IMTL-AGF.

Table 2. Results of surgical gesture recognition under LOUO validation scheme on SU.

Method	Metrics		
	Acc	Edit	F1@{10, 25, 50}
Seg-ST-CNN [16]	74.7	66.6	–
TCN [17]	81.4	83.1	–
3D CNN [11]	**84.3**	80.0	87.0, –, –
RL (full) [18]	81.43	87.96	**92.0, 90.5, 82.2**
MS-TCN [8]	78.85	85.80	88.5, 86.6, 75.8
C3D-MTL-VF	82.05	86.63	90.6, 89.1, 80.3

Table 3. Accuracy results of self-proclaimed skill level classification under LOSO validation scheme.

Method	Task		
	KT	NP	SU
S-HMM [27]	94.4	96.2	97.4
ApEn [32]	**99.9**	100.0	100.0
ConvNet [9]	92.1	100.0	100.0
3D ConvNet [10]	95.1	100.0	100.0
C3D-MTL-VF	97.5	**100.0**	**100.0**

with fixed number of segments, we re-implement it by assigning weights in [31] to each segment. The standard deviations of these results among folds can be found in supplementary material. For the two auxiliary tasks, average values of benchmark metrics under benchmark validation schemes are reported in Table 2 and 3.

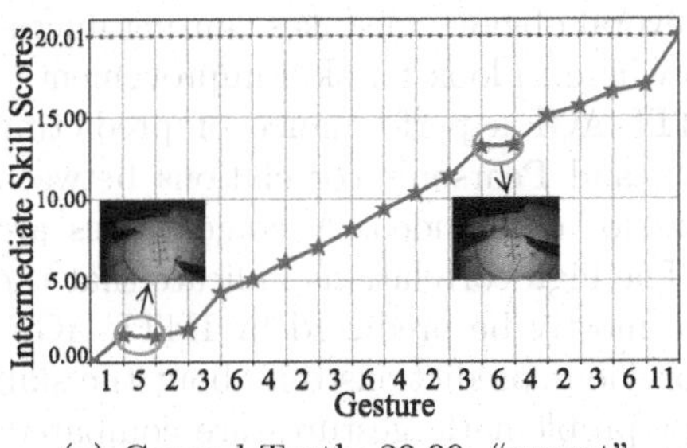

(a) Ground Truth: 20.00, "expert"

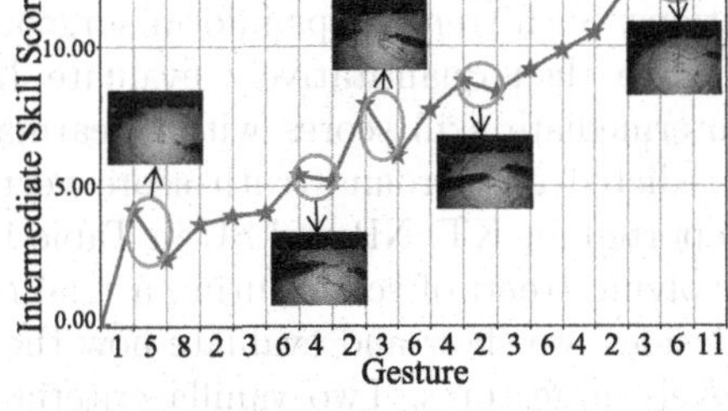

(b) Ground Truth: 12.00, "novice"

Fig. 2. Two examples of feedback in SU task. Problematic gestures are identified with green circle. Ground-truth final skill scores and skill levels are attached. (Color figure online)

Table 4. Spearman and Pearson correlations of IMTL-AGF between predicted and ground-truth intermediate scores.

		Spearman	Pearson
KT	LOSO	0.97	0.95
	LOUO	0.96	0.95
NP	LOSO	0.98	0.97
	LOUO	0.98	0.94
SU	LOSO	0.99	0.98
	LOUO	0.98	0.96

Table 5. Accuracy results of IMTL-AGF for predicting gesture-level skill levels under two different criteria.

		Criterion 1	Criterion 2
KT	LOSO	88.33	79.01
	LOUO	86.10	78.06
NP	LOSO	83.36	80.00
	LOUO	84.09	79.90
SU	LOSO	84.53	81.19
	LOUO	83.93	79.51

In general the video-based methods perform better than the kinematic-based ones, especially for NP and SU. It seems that effect-driven visual cues other than the motion trajectory are also important in skill assessment. Moreover, for existing studies there is an obvious degradation of performance under LOUO compared with LOSO, indicating a lack of generalization on new subjects. Similar degradation is observed for SU task which consists of longer and more diverse gesture sequences. Our MTL-VF alleviates the two problems by additionally learning performers' skill levels and trials' gesture sequences. It works well on both ResNet101 (compared with ResNet101-LSTM) and C3D features. The "Across" column further demonstrates that our single model can generalize well across different surgical tasks. For surgical gesture recognition and skill level classification, MTL-VF on regular C3D features achieves competitive performances compared with state-of-the-art methods employing specific spatio-temporal features.

IMTL-AGF Framework. We first qualitatively demonstrate the gesture-level feedback with two examples in SU (Fig. 2). Interestingly, the final score of trial and possibly performer's skill level are reflected on the trend of intermediate skill scores. For well-performed trials the score is built up smoothly except for some glitches, while for worse trials there are several obvious mistakes causing fluctuations. Such trend helps inform surgeons where to look for skill improvement.

We then quantitatively evaluate IMTL-AGF's performance of predicting intermediate skill scores with Spearman's and Pearson's correlations between predicted and ground-truth score vectors for each video. Average values are reported for KT, NP and SU in Table 4. The high correlations indicate that the evolving trend of score during a trial can mostly be predicted by IMTL-AGF. Besides, we show and evaluate how the previous predictions tell about the skill levels of gestures. Two vanilla criteria for problematic gestures are compared: (1) negative and (2) negative plus smallest non-negative differences between consecutive intermediate skill scores. Table 5 shows the average accuracy values regarding skill levels of gestures under these criteria. Notably, in real applications the criterion should vary with cases and the accuracy varies accordingly.

4 Conclusion and Future Work

In this work, a novel video-based method incorporating recognized surgical gestures and skill levels is proposed with two frameworks: MTL-VF and IMTL-AGF. Experiments on JIGSAWS show MTL-VF for skill score prediction outperforms state-of-the-art methods under LOUO validation scheme, while the feedback generated by IMTL-AGF resembles human's scoring process. The future work involves (1) performing ablation studies of auxiliary tasks for MTL-VF and (2) incorporating verbal captions to generate more interpretable feedback.

Acknowledgements. This work was supported in part by Science and Technology Commission of Shanghai Municipality under Grant No.: 18511105603. Special thanks go to Dr. Qiongjie Zhou's team from Obstetrics and Gynecology Hospital affiliated to Fudan University for the help on extra annotations.

References

1. Ahmidi, N., et al.: A dataset and benchmarks for segmentation and recognition of gestures in robotic surgery. IEEE Trans. Biomed. Eng. **64**(9), 2025–2041 (2017)
2. Benmansour, M., Handouzi, W., Malti, A.: A neural network architecture for automatic and objective surgical skill assessment. In: CISTEM, pp. 1–5. IEEE (2018)
3. Birkmeyer, J.D., et al.: Surgical skill and complication rates after bariatric surgery. N. Engl. J. Med. **369**, 1434–1442 (2013)
4. DiPietro, R., Hager, G.D.: Automated surgical activity recognition with one labeled sequence. In: Shen, D., et al. (eds.) MICCAI 2019. LNCS, vol. 11768, pp. 458–466. Springer, Cham (2019). https://doi.org/10.1007/978-3-030-32254-0_51
5. Ershad, M., Rege, R., Majewicz, A.: Surgical skill level assessment using automatic feature extraction methods. In: Medical Imaging: Image-Guided Procedures, Robotic Interventions, and Modeling, vol. 10576 (2018)
6. Fard, M.J., et al.: Machine learning approach for skill evaluation in robotic-assisted surgery. In: WCECS, vol. 1 (2016)
7. Fard, M.J., et al.: Automated robot-assisted surgical skill evaluation: predictive analytics approach. Int. J. Med. Robot. Comput. Assist. Surg. **14**(1), e1850 (2018)
8. Farha, Y.A., Gall, J.: MS-TCN: multi-stage temporal convolutional network for action segmentation. In: CVPR, pp. 3575–3584. IEEE (2019)
9. Ismail Fawaz, H., Forestier, G., Weber, J., Idoumghar, L., Muller, P.-A.: Evaluating surgical skills from kinematic data using convolutional neural networks. In: Frangi, A.F., Schnabel, J.A., Davatzikos, C., Alberola-López, C., Fichtinger, G. (eds.) MICCAI 2018. LNCS, vol. 11073, pp. 214–221. Springer, Cham (2018). https://doi.org/10.1007/978-3-030-00937-3_25
10. Funke, I., Mees, S.T., Weitz, J., Speidel, S.: Video-based surgical skill assessment using 3D convolutional neural networks. IJCARS **14**(7), 1217–1225 (2019)
11. Funke, I., Bodenstedt, S., Oehme, F., von Bechtolsheim, F., Weitz, J., Speidel, S.: Using 3D convolutional neural networks to learn spatiotemporal features for automatic surgical gesture recognition in video. In: Shen, D., et al. (eds.) MICCAI 2019. LNCS, vol. 11768, pp. 467–475. Springer, Cham (2019). https://doi.org/10.1007/978-3-030-32254-0_52
12. He, K., Zhang, X., Ren, S., Sun, J.: Deep residual learning for image recognition. In: CVPR, pp. 770–778. IEEE (2016)
13. Karpathy, A., et al.: Large-scale video classification with convolutional neural networks. In: CVPR, pp. 1725–1732. IEEE (2014)
14. Kendall, A., Gal, Y., Cipolla, R.: Multi-task learning using uncertainty to weigh losses for scene geometry and semantics. In: CVPR, pp. 7482–7491. IEEE (2018)
15. Kingma, D.P., Ba, J.: Adam: a method for stochastic optimization. In: ICLR (2015)
16. Lea, C., Reiter, A., Vidal, R., Hager, G.D.: Segmental spatiotemporal CNNs for fine-grained action segmentation. In: Leibe, B., Matas, J., Sebe, N., Welling, M. (eds.) ECCV 2016. LNCS, vol. 9907, pp. 36–52. Springer, Cham (2016). https://doi.org/10.1007/978-3-319-46487-9_3
17. Lea, C., Vidal, R., Reiter, A., Hager, G.D.: Temporal convolutional networks: a unified approach to action segmentation. In: Hua, G., Jégou, H. (eds.) ECCV 2016. LNCS, vol. 9915, pp. 47–54. Springer, Cham (2016). https://doi.org/10.1007/978-3-319-49409-8_7
18. Liu, D., Jiang, T.: Deep reinforcement learning for surgical gesture segmentation and classification. In: Frangi, A.F., Schnabel, J.A., Davatzikos, C., Alberola-López, C., Fichtinger, G. (eds.) MICCAI 2018. LNCS, vol. 11073, pp. 247–255. Springer, Cham (2018). https://doi.org/10.1007/978-3-030-00937-3_29

19. Liu, D., Jiang, T., Wang, Y., Miao, R., Shan, F., Li, Z.: surgical skill assessment on in-vivo clinical data via the clearness of operating field. In: Shen, D., et al. (eds.) MICCAI 2019. LNCS, vol. 11768, pp. 476–484. Springer, Cham (2019). https://doi.org/10.1007/978-3-030-32254-0_53
20. Martin, J.A., et al.: Objective structured assessment of technical skill (OSATS) for surgical residents. Br. J. Surg. **84**(2), 273–278 (1997)
21. Parmar, P., Morris, B.T.: Learning to score olympic events. In: CVPR-W, pp. 20–28. IEEE (2017)
22. Parmar, P., Morris, B.T.: Action quality assessment across multiple actions. In: WACV, pp. 1468–1476. IEEE (2019)
23. Parmar, P., Morris, B.T.: What and how well you performed? A multitask learning approach to action quality assessment. In: CVPR, pp. 304–313. IEEE (2019)
24. Paszke, A., et al.: Automatic differentiation in pytorch. In: NIPS-W (2017)
25. Regenbogen, S., et al.: Patterns of technical error among surgical malpractice claims: an analysis of strategies to prevent injury to surgical patients. Ann. Surg. **246**(5), 705–711 (2007)
26. Russakovsky, O., et al.: ImageNet large scale visual recognition challenge. Int. J. Comput. Vis. **115**, 211–252 (2015)
27. Tao, L., Elhamifar, E., Khudanpur, S., Hager, G.D., Vidal, R.: Sparse hidden markov models for surgical gesture classification and skill evaluation. In: Abolmaesumi, P., Joskowicz, L., Navab, N., Jannin, P. (eds.) IPCAI 2012. LNCS, vol. 7330, pp. 167–177. Springer, Heidelberg (2012). https://doi.org/10.1007/978-3-642-30618-1_17
28. Tran, D., Bourdev, L., Fergus, R., Torresani, L., Paluri, M.: Learning spatiotemporal features with 3D convolutional networks. In: ICCV, pp. 4489–4497. IEEE (2015)
29. Wang, Z., Majewicz Fey, A.: Deep learning with convolutional neural network for objective skill evaluation in robot-assisted surgery. Int. J. Comput. Assist. Radiol. Surg. **13**(12), 1959–1970 (2018). https://doi.org/10.1007/s11548-018-1860-1
30. Xiang, X., Tian, Y., Reiter, A., Hager, G.D., Tran, T.D.: S3D: Stacking segmental P3D for action quality assessment. In: ICIP, pp. 928–932. IEEE (2018)
31. Zhou, K., Qiao, Y., Xiang, T.: Deep reinforcement learning for unsupervised video summarization with diversity-representativeness reward. In: AAAI (2018)
32. Zia, A., Essa, I.: Automated surgical skill assessment in RMIS training. Int. J. Comput. Assist. Radiol. Surg. **13**(5), 731–739 (2018). https://doi.org/10.1007/s11548-018-1735-5
33. Zia, A., Hung, A., Essa, I., Jarc, A.: Surgical activity recognition in robot-assisted radical prostatectomy using deep learning. In: Frangi, A.F., Schnabel, J.A., Davatzikos, C., Alberola-López, C., Fichtinger, G. (eds.) MICCAI 2018. LNCS, vol. 11073, pp. 273–280. Springer, Cham (2018). https://doi.org/10.1007/978-3-030-00937-3_32
34. Zia, A., Sharma, Y., Bettadapura, V., Sarin, E.L., Essa, I.: Video and accelerometer-based motion analysis for automated surgical skills assessment. Int. J. Comput. Assist. Radiol. Surg. **13**(3), 443–455 (2018). https://doi.org/10.1007/s11548-018-1704-z

Learning Motion Flows for Semi-supervised Instrument Segmentation from Robotic Surgical Video

Zixu Zhao[1], Yueming Jin[1(✉)], Xiaojie Gao[1], Qi Dou[1,2], and Pheng-Ann Heng[1,3]

[1] Department of Computer Science and Engineering, The Chinese University of Hong Kong, Hong Kong, China
{zxzhao,ymjin}@cse.cuhk.edu.hk

[2] Shun Hing Institute of Advanced Engineering, CUHK, Hong Kong, China

[3] T Stone Robotics Institute, CUHK, Hong Kong, China

Abstract. Performing low hertz labeling for surgical videos at intervals can greatly releases the burden of surgeons. In this paper, we study the semi-supervised instrument segmentation from robotic surgical videos with sparse annotations. Unlike most previous methods using unlabeled frames individually, we propose a dual motion based method to wisely learn motion flows for segmentation enhancement by leveraging temporal dynamics. We firstly design a flow predictor to derive the motion for jointly propagating the frame-label pairs given the current labeled frame. Considering the fast instrument motion, we further introduce a flow compensator to estimate intermediate motion within continuous frames, with a novel cycle learning strategy. By exploiting generated data pairs, our framework can recover and even enhance temporal consistency of training sequences to benefit segmentation. We validate our framework with binary, part, and type tasks on 2017 MICCAI EndoVis Robotic Instrument Segmentation Challenge dataset. Results show that our method outperforms the state-of-the-art semi-supervised methods by a large margin, and even exceeds fully supervised training on two tasks (Our code is available at https://github.com/zxzhaoeric/Semi-InstruSeg/).

Keywords: Semi-supervised segmentation · Motion flow · Surgical video

1 Introduction

By providing the context perceptive assistance, semantic segmentation of surgical instrument can greatly benefit robot-assisted minimally invasive surgery towards superior surgeon performance. Automatic instrument segmentation also serves as a crucial cornerstone for more downstream capabilities such as tool pose estimation [12], tracking and control [4]. Recently, convolutional neural network

A. L. Martel et al. (Eds.): MICCAI 2020, LNCS 12263, pp. 679–689, 2020.
https://doi.org/10.1007/978-3-030-59716-0_65

(CNN) has demonstrated new state-of-the-arts on surgical instrument segmentation thanks to the effective data-driven learning [7,10,13,21]. However, these methods highly rely on abundant labeled frames to achieve the full potential. It is expensive and laborious especially for high frequency robotic surgical videos, entailing the frame-by-frame pixel-wise annotation by experienced experts.

Some studies tend to utilize extra signals to generate parsing masks, such as robot kinematic model [3,16], weak annotations of object stripe and skeleton [6], and simulated surgical scene [15]. However, additional efforts are still required for other signal access or creation. Considerable effort has been devoted to utilizing the large-scale unlabeled data to improve segmentation performance for medical image analysis [2,23,24]. For example, Bai et al. [2] propose a self-training strategy for cardiac segmentation, where the supervised loss and semi-supervised loss are alternatively updated. Yu et al. [23] raise an uncertainty-aware mean-teacher framework for 3D left atrium segmentation by learning the reliable knowledge from unlabeled data. In contrast, works focusing on the effective usage of unlabeled surgical video frames are limited. The standard mean teacher framework has recently been applied to the semi-supervised liver segmentation by computing the consistency loss of laparoscopic images [5]. Ross et al. [20] exploit a self-supervised strategy by using GAN-based re-colorization on individual unlabeled endoscopic frames for model pretraining.

Unfortunately, these semi-supervised methods propose to capture the information based on separate unlabeled video frames, failing to leverage the inherent temporal property of surgical sequences. Given 50% labeled frames with a labeling interval of 2, a recent approach [10] indicates that utilizing temporal consistency of surgical videos benefit semi-supervised segmentation. Optical flows are used to transfer predictions of unlabeled frames to adjacent position whose labels are borrowed to calculate semi-supervised loss. Yet this method heavily depends on accurate optical flow estimation and fails to provide trustworthy semi-supervision in model with some erroneous transformations.

In this paper, we propose a dual motion based semi-supervised framework for instrument segmentation by leveraging the self-contained sequential cues in surgical videos. Given sparsely annotated sequences, our core idea is to derive the motion flows for annotation and frame transformation that recover the temporal structure of raw videos to boost semi-supervised segmentation. Specifically, we firstly design a flow predictor to learn the motion between two frames with a video reconstruction task. We propose a joint propagation strategy to generate frame-label pairs with learned flows, alleviating the misalignment of pairing propagated labels with raw frames. Next, we design a flow compensator with a frame interpolation task to learn the intermediate motion flows. A novel unsupervised cycle learning strategy is proposed to optimize models by minimizing the discrepancy between forward predicted frames and backward cycle reconstructions. The derived motion flows further propagate intermediate frame-label pairs as the augmented data to enhance the sequential consistency. Rearranging the training sequence by replacing unlabeled raw frames with generated data pairs, our framework can greatly benefit segmentation performance. We extensively evaluate the method on surgical instrument binary, part, and type segmentation tasks on 2017 MICCAI EndoVis Challenge dataset. Our method consistently outperforms state-of-the-art semi-supervised segmentation methods by a large margin, as well as exceeding the fully supervised training on two tasks.

2 Method

Figure 1 illustrates our dual motion-based framework. It uses raw frames to learn dual motion flows, one for recovering original annotation distribution (top branch) and the other for compensating fast instrument motions (bottom branch). We ultimately use learned motion flows to propagate aligned frame-label pairs as a substitute for unlabeled raw frames in video sequences for segmentation training.

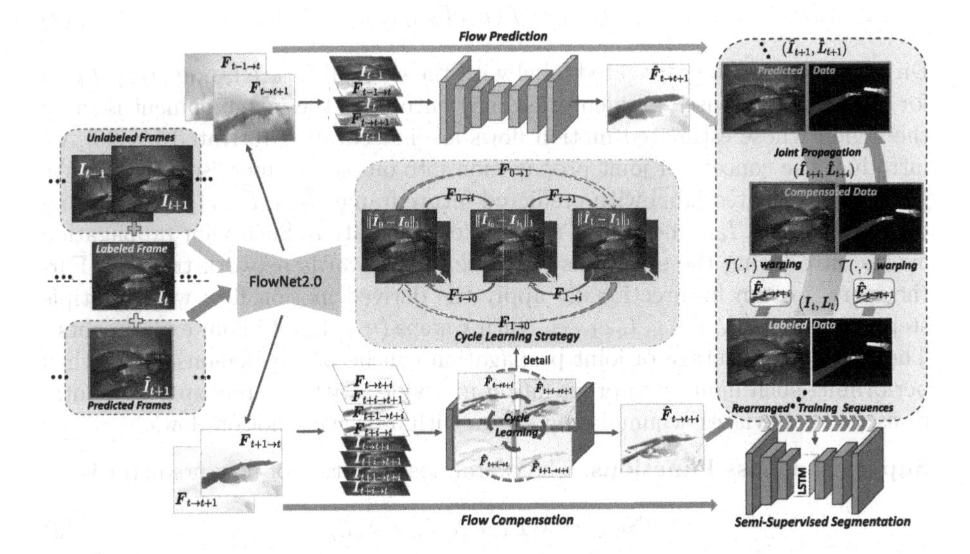

Fig. 1. The illustration of the proposed framework. We learn motion flows along *flow prediction* branch and *flow compensation* branch successively, which are used to joint propagate the aligned data pairs for semi-supervised segmentation.

2.1 Flow Prediction for Joint Propagation

With a video having T frames as $\boldsymbol{I} = \{I_0, I_1, ..., I_{T-1}\}$, we assume that $\boldsymbol{I}$ is labeled with intervals. For instance, only $\{I_0, I_5, I_{10}, ...\}$ are labeled with interval 4, accounting for 20% labeled data. This setting is reasonable to clinical practice, as it is easier for surgeons to perform low hertz labeling. Sharing the spirit with [10], we argue that the motion hidden within the continuous raw frames can be applied to corresponding instrument masks. Therefore, we first derive the motion flow from raw frames with a video reconstruction task, as shown in *Flow*

Prediction branch in Fig. 1. Given the sequence $I_{t':t+1}$, we aim to estimate the motion flow $\hat{F}_{t\to t+1}$ that can translate the current frame I_t to future frame I_{t+1}:

$$\hat{F}_{t\to t+1} = \mathcal{G}(I_{t':t+1}, F_{t'+1:t+1}), \quad \hat{I}_{t+1} = \mathcal{T}(I_t, \hat{F}_{t\to t+1}), \tag{1}$$

where $\mathcal{G}$ is a 2D CNN based flow predictor with the input $I_{t':t}$. Optical flows F_i between successive frames I_i and I_{i-1} are calculated by FlowNet2.0 [8]. $\mathcal{T}$ is a forward warping function which is differentiable and implemented with bilinear interpolation [25]. Instead of straightforwardly relying on the optical flow [10], which suffers from the undefined problem for the dis-occluded pixels, we aim to learn the motion vector (u, v) as a precise indicator for annotation propagation which can effectively account for the gap. Intuitively, the instrument mask follows the same location shift as its frame. For an unlabeled frame I_{t+1}, we can borrow the adjacent annotation L_t and use the derived flow for its label propagation:

$$\hat{L}_{t+1} = \mathcal{T}(L_t, \hat{F}_{t\to t+1}). \tag{2}$$

Directly pairing the propagated label with the original future frame $(I_{t+1}, \hat{L}_{t+1})$ for our semi-supervised segmentation may encounter the mis-alignment issue in the region whose estimated motion flows are inaccurate. Motivated by [26], we introduce the concept of joint propagation into our semi-supervised setting. We pair the propagated label with predicted future frame $(\hat{I}_{t+1}, \hat{L}_{t+1})$, while leaving the original data I_{t+1} merely for motion flow generation. Such joint propagation avoids introducing the erroneous regularization towards network training. Furthermore, we can bi-directionally apply the derived motion flow with multiple steps, obtaining $(\hat{I}_{t-k:t+k}, \hat{L}_{t-k:t+k})$ with k steps ($k = 1, 2, 4$ in our experiments). The superior advantage of joint propagation can be better demonstrated when performing such multi-step propagation in a video with severely sparse annotations, as it alleviates accumulating errors within derived motion flows.

Supervised Loss Functions. The overall loss function of flow predictor is:

$$\mathcal{L}_{Pred} = \lambda_1 \mathcal{L}_1 + \lambda_p \mathcal{L}_p + \lambda_s \mathcal{L}_s, \tag{3}$$

consisting of the primary loss, i.e., L1 loss $\mathcal{L}_1 = \|\hat{I}_{t+1} - I_{t+1}\|_1$, which can capture subtle modification rather than L2 loss [17]; perceptual loss $\mathcal{L}_p$ to retain structural details in predictions, detailed definition in [11]; smooth loss $\mathcal{L}_s = \|\nabla F_{t\to t+1}\|_1$ to encourage neighboring pixels to have similar flow values [9]. We empirically set the weights as $\lambda_1 = 0.7$, $\lambda_p = 0.2$, and $\lambda_s = 0.1$ for a more robust combination in eliminating the artifacts and occlusions than [17].

2.2 Flow Compensation via Unsupervised Cycle Learning

The fast instrument motions between successive frames always occur even in a high frequency surgical video. Smoothing large motions thus improves the sequential consistency, as well as adds data variety for semi-supervised segmentation. In this scenario, we try to compensate motion flows with a frame interpolation task. However, existing interpolation approaches are not suitable for our case. Either the optical flow based methods [9,18] rely on consistent motion frequency, or the kernel based method [14] contradicts the alignment prerequisite. Hence, we propose an unsupervised flow compensator with a novel cycle learning strategy, which forces the model to learn intermediate flows by minimizing the discrepancy between forward predicted frames and their backward cycle reconstructions.

Given two continuous frames I_0 and I_1, our ultimate goal is to learn a motion flow $\hat{F}_{0\to i}$ with a time $i \in (0,1)$ to jointly propagate the intermediate frame-label pair $(\hat{I}_i, \hat{L}_i)$. In the *Flow Compensation* branch, we first use the pretrained FlowNet2.0 to compute bi-directional optical flows $(F_{0\to 1}, F_{1\to 0})$ between two frames. We then use them to approximate the intermediate optical flows $\boldsymbol{F}$:

$$\begin{aligned} F_{i\to 0} &= -(1-i)iF_{0\to 1} + i^2 F_{1\to 0}, & F_{1\to i} &= F_{1\to 0} - F_{i\to 0}, \\ F_{i\to 1} &= (1-i)^2 F_{0\to 1} - i(1-i)F_{1\to 0}, & F_{0\to i} &= F_{0\to 1} - F_{i\to 1}. \end{aligned} \tag{4}$$

Such approximation suits well in smooth regions but poorly around boundaries, however, it can still serve as an essential initialization for subsequent cycle learning. The approximated flows $\boldsymbol{F}$ are used to generate warped frames $\hat{\boldsymbol{I}}$, including forward predicted frames $\hat{I}_{i_0}$, $\hat{I}_1$ and backward reconstructed frames $\hat{I}_{i_1}$, $\hat{I}_0$:

$$\hat{I}_{i_0} = \mathcal{T}(I_0, F_{0\to i}), \hat{I}_1 = \mathcal{T}(\hat{I}_{i_0}, F_{i\to 1}), \hat{I}_{i_1} = \mathcal{T}(\hat{I}_1, F_{1\to i}), \hat{I}_0 = \mathcal{T}(\hat{I}_{i_1}, F_{i\to 0}). \tag{5}$$

We then establish a flow compensator that based on a 5-stage U-Net to refine motion flows with cycle consistency. It takes the two frames (I_0, I_1), four initial approximations $\boldsymbol{F}$, and four warped frames $\hat{\boldsymbol{I}}$ as input, and outputs four refined flows $\hat{\boldsymbol{F}}$, where $\hat{F}_{0\to i}$ is applied on I_0 for joint frame-label pair generation.

Unsupervised Cycle Loss. The key idea is to learn the motion flow that can encourage models to satisfy cycle consistency in time domain. Intuitively, we expect that the predicted $\hat{I}_1$ and reconstructed $\hat{I}_0$ are well overlapped with the original raw data I_1 and I_0. Meanwhile, two intermediate frames warped along a cycle, i.e., $\hat{I}_{i_1}$ and $\hat{I}_{i_0}$, should show the similar representations. Keeping this in mind, we use L1 loss to primarily constrain the inconsistency of each pair:

$$\mathcal{L}_1^c = \lambda_0 \|\hat{I}_0 - I_0\|_1 + \lambda_i \|\hat{I}_{i_0} - \hat{I}_{i_1}\|_1 + \lambda_1 \|\hat{I}_1 - I_1\|_1. \tag{6}$$

To generate sharper predictions, we add the perceptual loss $\mathcal{L}_p^c$ on the three pairs (perceptual loss definition in [11]). Our overall unsupervised cycle loss is defined as $\mathcal{L}_{cycle} = \mathcal{L}_1^c + \lambda_p \mathcal{L}_p^c$, where we empirically set $\lambda_0 = 1.0$, $\lambda_i = 0.8$, $\lambda_1 = 2.0$, and $\lambda_p = 0.01$. Our cycle regularization can avoid relying on the immediate frames and learn the motion flow in a completely unsupervised way.

2.3 Semi-supervised Segmentation

For semi-supervised segmentation, we study the sparsely annotated video sequences $\boldsymbol{I} = \{I_0, I_1, ..., I_{T-1}\}$ with a label interval h. The whole dataset consists of labeled subset $\mathcal{D}_L = \{(I_t, L_t)\}_{t=hn}$ with N frames and unlabeled subset

$\mathcal{D}_U = \{I_t\}_{t \neq hn}$ with $M = hN$ frames. Using consecutive raw frames, our flow predictor learns motion flows with a video reconstruction task, which are used to transfer the adjacent annotations for the unlabeled data. With the merit of joint propagation, we pair the generated labels and frames, obtaining the re-labeled set $\mathcal{D}_R = \{\hat{I}_t, \hat{L}_t\}_{t \neq hn}$ with M frames. Subsequently, our flow compensator learns the intermediate motion flow with an unsupervised video interpolation task. We can then extend the dataset by adding $\mathcal{D}_C = \{\tilde{I}_{t_0}, \tilde{L}_{t_0}\}_{t=1}^{T-1}$ with N+M-1 compensated frames with interpolation rate as 1. Our flow predictor and compensator are designed based on U-Net, with network details in supplementary. We finally consider $\mathcal{D}_L \cup \mathcal{D}_R \cup \mathcal{D}_C$ as the training set for semi-supervised segmentation. For the network architecture, we basically adopt the same backbone as [10], i.e., U-Net11 [19] with pretrained encoders from VGG11 [22]. Excitingly, different from other semi-supervised methods, our motion flow based strategy retains and even enhances the inherent sequential consistency. Therefore, we can still exploit temporal units, such as adding convolutional long short term memory layer (ConvLSTM) at the bottleneck, to increase segmentation performance.

3 Experiments

Dataset and Evaluation Metrics. We validate our method on the public dataset of Robotic Instrument Segmentation from 2017 MICCAI EndoVis Challenge [1]. The video sequences with a high resolution of 1280 × 1024 are acquired from *da Vinci Xi* surgical system during different porcine procedures. We conduct all three sub-tasks of this dataset, i.e., binary (2 classes), part (4 classes) and type (8 classes), with gradually fine-grained segmentation for an instrument. For direct and fair comparison, we follow the same evaluation manner in [10], by using the released 8 × 225-frame videos for 4-fold cross-validation, also with the same fold splits. Two metrics are adopted to quantitatively evaluate our method, including mean intersection-over-union (IoU) and Dice coefficient (Dice).

Implementation Details. The framework is implemented in Pytorch with NVIDIA Titan Xp GPUs. The parameters of pretrained FlowNet2.0 are frozen while training the overall framework with Adam optimizer. The learning rate is set as $1e-3$ and divided by 10 every 150 epochs. We randomly crop 448 × 448 sub-images as the framework input. For training segmentation models, we follow the rules in [10]. As for the ConvLSTM based variant, the length of input sequence is 5. The initial learning rate is set as $1e-4$ for ConvLSTM layer while $1e-5$ for other network components. All the experiments are repeated 5 times to account for the stochastic nature of DNN training.

Comparison with Other Semi-supervised Methods. We implement several state-of-the-art semi-supervised segmentation methods for comparison, including ASC [14] (interpolating labels with adaptive separable convolution), MF-TAPNet [10] (propagating labels with optical flows), self-training method [2], Re-color [20] (GAN-based re-colorization for model initialization), and UA-MT [23] (uncertainty-aware mean teacher). We conduct experiments under the setting of

Table 1. Comparison of instrument segmentation results on three tasks (mean ± std).

Methods	Frames used		Binary segmentation		Part segmention		Type segmentation	
	Label	Unlabel	IoU (%)	Dice (%)	IoU (%)	Dice (%)	IoU (%)	Dice (%)
U-Net11	100%	0	82.55±12.51	89.76±9.10	64.87±14.46	76.08±13.05	36.83±26.36	45.48±28.16
U-Net11*	100%	0	83.17±12.01	90.22±8.83	64.96±14.12	76.57±12.44	40.31±24.38	49.57±25.39
TernausNet [21]	100%	0	83.60±15.83	90.01±12.50	65.50±17.22	75.97±16.21	33.78±19.16	44.95±22.89
MF-TAPNet [10]	100%	0	87.56±16.24	93.37±12.93	67.92±16.50	77.05±16.17	36.62±22.78	48.01±25.64
ASC [14]	20%	80%	78.51±13.40	87.17±9.88	59.07±14.76	70.92±13.97	30.19±17.65	41.70±20.62
ASC*	20%	80%	78.33±12.67	87.04±12.85	58.93±14.61	70.76±13.40	30.60±16.55	41.88±22.24
Self-training [2]	20%	80%	79.32±12.11	87.62±9.46	59.30±15.70	71.04±14.04	31.00±25.12	42.11±24.52
Re-color [20]	20%	80%	79.85±13.55	87.78±10.10	59.67±15.14	71.51±15.13	30.72±25.66	41.47±25.30
MF-TAPNet [10]	20%	80%	80.06±13.26	87.96±9.57	59.62±16.01	71.57±15.90	31.55±18.72	42.35±22.41
UA-MT [23]	20%	80%	80.68±12.63	88.20±9.61	60.11±14.49	72.18± 13.78	32.42±21.74	43.61±26.30
Our Dual MF	20%	80%	**83.42±12.73**	**90.34±9.25**	**61.77±14.19**	**73.22±13.25**	**37.06±25.03**	**46.55±27.10**
Our Dual MF*	20%	80%	**84.05±13.27**	**91.13±9.31**	**62.51±13.32**	**74.06±13.08**	**43.71±25.01**	**52.80±26.16**
U-Net11	30%	0	80.16±13.69	88.14±10.14	61.75±14.40	72.44±13.41	31.96±27.98	38.52±31.02
Our Single MF	30%	70%	83.70±12.47	90.46±8.95	63.02±14.80	74.49±13.76	39.38±25.54	48.49±26.92
Our Dual MF	30%	70%	**84.12±13.18**	**90.77±9.45**	**63.82±15.63**	**74.74±13.84**	**39.61±26.45**	**48.80±27.67**
Our Dual MF*	30%	70%	**84.62±13.54**	**91.63±9.13**	**64.89±13.26**	**76.33±12.61**	**45.83±21.96**	**56.11±22.33**
U-Net11	20%	0	76.75±14.69	85.75±11.36	58.50±14.65	70.70±13.95	23.53±24.84	26.74±27.17
Our Single MF	20%	80%	83.10±12.18	90.15±8.83	61.20±14.10	72.49±12.94	36.72±23.62	46.50±25.09
Our Dual MF	20%	80%	**83.42±12.73**	**90.34±9.25**	**61.77±14.19**	**73.22±13.25**	**37.06±25.03**	**46.55±27.19**
Our Dual MF*	20%	80%	**84.05±13.27**	**91.13±9.31**	**62.51±13.32**	**74.06±13.08**	**43.71±25.01**	**52.80±26.16**
U-Net11	10%	0	75.93±15.03	85.09±11.77	55.24±15.27	67.78±14.97	15.87±16.97	19.30±19.99
Our Single MF	10%	90%	82.05±14.35	89.23±10.65	57.91±14.51	69.28±14.79	30.24±21.33	40.12±24.21
Our Dual MF	10%	90%	**82.70±13.21**	**89.74±9.56**	**58.29±15.60**	**69.54±15.23**	**31.28±19.53**	**41.01±21.91**
Our Dual MF*	10%	90%	**83.10±12.45**	**90.02±8.80**	**59.36±14.38**	**70.20±13.96**	**33.64±20.19**	**43.20±22.70**

Note: * denotes that the temporal unit ConvLSTM is added at the bottleneck of the segmentation network.

20% frames being labeled with annotation interval as 4. Most above methods are difficult to gain profit from temporal units except ASC, due to the uncontinuous labeled input or network design. We use the same network backbone (U-Net11) among these methods for fair comparison. Table 1 compares our segmentation results with other semi-supervised methods. We also report fully supervised results of U-Net11 as upper bound, as well as two benchmarks TernausNet [21], and MF-TAPNet for reference. Among the semi-supervised methods, UA-MT achieves slightly better performance as it draws out more reliable information from unlabeled data. Notably, our method consistently outperforms UA-MT across three tasks by a large margin, i.e., 2.68% in IoU and 2.24% in Dice on average. After adding the temporal unit, results of ASC degrade instead on two tasks due to the inaccurate interpolated labels. As our semi-supervised method can enhance sequential consistency by expanded frame-label pairs, our results can be further improved with ConvLSTM, even surpassing the fully supervised training (U-Net11*) by 0.91% Dice on binary task and 3.23% Dice on type task.

Analysis of Our Semi-supervised Methods. For 6 × 225-frame training videos in each fold, we study the frames labeled at an interval of 2, 4, and 8, resulting in 30%, 20%, and 10% annotations. Table 1 also lists results of three ablation settings: (1) Our Single MF: U-Net11 trained by set $\{\mathcal{D}_L \cup \mathcal{D}_R\}$ with Flow Prediction; (2) Our Dual MF: U-Net11 trained by set $\{\mathcal{D}_L \cup \mathcal{D}_R \cup \mathcal{D}_C\}$ with Flow Prediction and Compensation; (3) Our Dual MF*: U-Net11 embedded with ConvLSTM and trained by set $\{\mathcal{D}_L \cup \mathcal{D}_R \cup \mathcal{D}_C\}$. It is observed that under all annotation ratios, compared with U-Net11 trained by labeled set $\mathcal{D}_L$ alone, our

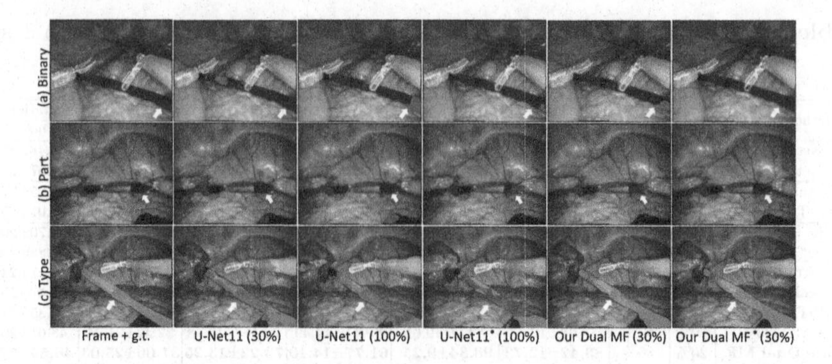

Fig. 2. Visualization of instrument (a) binary, (b) part, and (c) type segmentation. From left to right, we present frame with ground-truth, results of fully supervised training and our semi-supervised methods. * denotes incorporating ConvLSTM units.

flow based framework can progressively boost the semi-supervised performance with generated annotations. We gain the maximum benefits in the severest condition (10% labeling), where our Single MF has already largely improved the segmentation by 6.12% IoU and 4.14% Dice (binary), 2.67% IoU and 1.50% Dice (part), 14.37% IoU and 20.82% Dice (type). Leveraging compensated pairs, our Dual MF with 30% and 20% labels is even able to exceed the full annotation training by 1–3% IoU or Dice, corroborating that our method can recover and adjust the motion distribution for better network training. It can be further verified using temporal units. We only see slight improvements in fully supervised setting (the first two rows) because some motion inconsistency existed in original videos decreases the model learning capability of temporal cues. Excitingly, the increment is obvious between our Dual MF and Dual MF⋆, especially for the toughest type segmentation. For instance, IoU and Dice can be boosted by 6.65% and 6.25% in 20% labeling case. Figure 2 shows some visual results. Our Dual

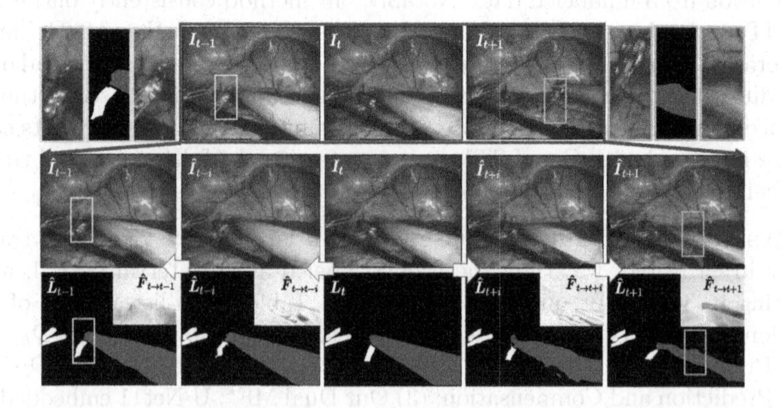

Fig. 3. Example of rearranged training sequence with propagation step $k = 1$.

MF* can largely suppress misclassified regions in Ultrasound probes for binary and part tasks, and achieve more complete and consistent type segmentation. It is even better at distinguishing hard mimics between instruments than fully supervised U-Net11*.

Analysis of Frame-Label Pairs. Our joint propagation can alleviate the misalignment issue from label propagation. In Fig. 3, labels in certain regions, like jaw (yellow) and shaft (orange) of instruments, fail to align with the original frames (first row) due to imprecision in learned flows, but correspond well with propagated frames (second row) as they experience the same transformation. The good alignment is crucial for segmentation. Besides, our learned flows can propagate instruments to a more reasonable position with smooth motion shift. The fast instrument motion is slowed down from I_t to $\hat{I}_{t+1}$ with smoother movement of Prograsp Forceps (orange), greatly benefiting ConvLSTM training.

4 Conclusions

We propose a flow prediction and compensation framework for semi-supervised instrument segmentation. Interestingly, we study the sparsely annotated surgical videos from the fresh perspective of learning the motion flow. Large performance gain over state-of-the-art semi-supervised methods demonstrates the effectiveness of our framework. Inherently our method can recover the temporal structure of raw videos and be applied to surgical videos with high motion inconsistency.

Acknowledgments. This work was supported by Key-Area Research and Development Program of Guangdong Province, China (2020B010165004), Hong Kong RGC TRS Project No.T42–409/18-R, National Natural Science Foundation of China with Project No. U1813204, and CUHK Shun Hing Institute of Advanced Engineering (project MMT-p5–20).

References

1. Allan, M., et al.: 2017 robotic instrument segmentation challenge. arXiv preprint (2019). arXiv:1902.06426
2. Bai, W., et al.: Semi-supervised learning for network-based cardiac MR image segmentation. In: Descoteaux, M., Maier-Hein, L., Franz, A., Jannin, P., Collins, D.L., Duchesne, S. (eds.) MICCAI 2017. LNCS, vol. 10434, pp. 253–260. Springer, Cham (2017). https://doi.org/10.1007/978-3-319-66185-8_29
3. da Costa Rocha, C., Padoy, N., Rosa, B.: Self-supervised surgical tool segmentation using kinematic information. In: 2019 International Conference on Robotics and Automation (ICRA), pp. 8720–8726. IEEE (2019)
4. Du, X., et al.: Patch-based adaptive weighting with segmentation and scale (pawss) for visual tracking in surgical video. Med. Image Anal. **57**, 120–135 (2019)
5. Fu, Y., et al.: More unlabelled data or label more data? A study on semi-supervised laparoscopic image segmentation. In: Wang, Q., et al. (eds.) DART/MIL3ID -2019. LNCS, vol. 11795, pp. 173–180. Springer, Cham (2019). https://doi.org/10.1007/978-3-030-33391-1_20

6. Fuentes-Hurtado, F., Kadkhodamohammadi, A., Flouty, E., Barbarisi, S., Luengo, I., Stoyanov, D.: Easylabels: weak labels for scene segmentation in laparoscopic videos. Int. J. Compu. Assist. Radiol. Surg. **14**(7), 1247–1257 (2019)
7. García-Peraza-Herrera, L.C., et al.: Toolnet: holistically-nested real-time segmentation of robotic surgical tools. In: 2017 IEEE/RSJ International Conference on Intelligent Robots and Systems (IROS), pp. 5717–5722. IEEE (2017)
8. Ilg, E., Mayer, N., Saikia, T., Keuper, M., Dosovitskiy, A., Brox, T.: Flownet 2.0: evolution of optical flow estimation with deep networks. In: Proceedings of the IEEE Conference on Computer Vision and Pattern Recognition, pp. 2462–2470 (2017)
9. Jiang, H., Sun, D., Jampani, V., Yang, M.H., Learned-Miller, E., Kautz, J.: Super slomo: high quality estimation of multiple intermediate frames for video interpolation. In: Proceedings of the IEEE Conference on Computer Vision and Pattern Recognition, pp. 9000–9008 (2018)
10. Jin, Y., Cheng, K., Dou, Q., Heng, P.-A.: Incorporating temporal prior from motion flow for instrument segmentation in minimally invasive surgery video. In: Shen, D., et al. (eds.) MICCAI 2019. LNCS, vol. 11768, pp. 440–448. Springer, Cham (2019). https://doi.org/10.1007/978-3-030-32254-0_49
11. Johnson, J., Alahi, A., Fei-Fei, L.: Perceptual losses for real-time style transfer and super-resolution. In: Leibe, B., Matas, J., Sebe, N., Welling, M. (eds.) ECCV 2016. LNCS, vol. 9906, pp. 694–711. Springer, Cham (2016). https://doi.org/10.1007/978-3-319-46475-6_43
12. Kurmann, T., et al.: Simultaneous recognition and pose estimation of instruments in minimally invasive surgery. In: Descoteaux, M., Maier-Hein, L., Franz, A., Jannin, P., Collins, D.L., Duchesne, S. (eds.) MICCAI 2017. LNCS, vol. 10434, pp. 505–513. Springer, Cham (2017). https://doi.org/10.1007/978-3-319-66185-8_57
13. Milletari, F., Rieke, N., Baust, M., Esposito, M., Navab, N.: CFCM: segmentation via coarse to fine context memory. In: Frangi, A.F., Schnabel, J.A., Davatzikos, C., Alberola-López, C., Fichtinger, G. (eds.) MICCAI 2018. LNCS, vol. 11073, pp. 667–674. Springer, Cham (2018). https://doi.org/10.1007/978-3-030-00937-3_76
14. Niklaus, S., Mai, L., Liu, F.: Video frame interpolation via adaptive separable convolution. In: Proceedings of the IEEE International Conference on Computer Vision, pp. 261–270 (2017)
15. Pfeiffer, M., et al.: Generating large labeled data sets for laparoscopic image processing tasks using unpaired image-to-image translation. In: Shen, D., et al. (eds.) MICCAI 2019. LNCS, vol. 11768, pp. 119–127. Springer, Cham (2019). https://doi.org/10.1007/978-3-030-32254-0_14
16. Qin, F., Li, Y., Su, Y.H., Xu, D., Hannaford, B.: Surgical instrument segmentation for endoscopic vision with data fusion of rediction and kinematic pose. In: 2019 International Conference on Robotics and Automation (ICRA), pp. 9821–9827. IEEE (2019)
17. Reda, F.A., et al.: Sdc-net: video prediction using spatially-displaced convolution. In: Proceedings of the European Conference on Computer Vision (ECCV), pp. 718–733 (2018)
18. Reda, F.A., et al.: Unsupervised video interpolation using cycle consistency. In: Proceedings of the IEEE International Conference on Computer Vision, pp. 892–900 (2019)
19. Ronneberger, O., Fischer, P., Brox, T.: U-net: convolutional networks for biomedical image segmentation. In: International Conference on Medical Image Computing and Computer-Assisted Intervention, pp. 234–241. Springer (2015)

20. Ross, T., Zimmerer, D., Vemuri, A., Isensee, F., Wiesenfarth, M., Bodenstedt, S., Both, F., Kessler, P., Wagner, M., Müller, B., Kenngott, H., Speidel, S., Kopp-Schneider, A., Maier-Hein, K., Maier-Hein, L.: Exploiting the potential of unlabeled endoscopic video data with self-supervised learning. Int. J. Comput. Assist. Radiol. Surg. **13**(6), 925–933 (2018). https://doi.org/10.1007/s11548-018-1772-0
21. Shvets, A.A., Rakhlin, A., Kalinin, A.A., Iglovikov, V.I.: Automatic instrument segmentation in robot-assisted surgery using deep learning. In: 2018 17th IEEE International Conference on Machine Learning and Applications (ICMLA), pp. 624–628. IEEE (2018)
22. Simonyan, K., Zisserman, A.: Very deep convolutional networks for large-scale image recognition. arXiv preprint (2014). arXiv:1409.1556
23. Yu, L., Wang, S., Li, X., Fu, C.-W., Heng, P.-A.: Uncertainty-aware self-ensembling model for semi-supervised 3D left atrium segmentation. In: Shen, D., et al. (eds.) MICCAI 2019. LNCS, vol. 11765, pp. 605–613. Springer, Cham (2019). https://doi.org/10.1007/978-3-030-32245-8_67
24. Zhang, Y., Yang, L., Chen, J., Fredericksen, M., Hughes, D.P., Chen, D.Z.: Deep adversarial networks for biomedical image segmentation utilizing unannotated images. In: Descoteaux, M., Maier-Hein, L., Franz, A., Jannin, P., Collins, D.L., Duchesne, S. (eds.) MICCAI 2017. LNCS, vol. 10435, pp. 408–416. Springer, Cham (2017). https://doi.org/10.1007/978-3-319-66179-7_47
25. Zhou, T., Tulsiani, S., Sun, W., Malik, J., Efros, A.A.: View synthesis by appearance flow. In: Leibe, B., Matas, J., Sebe, N., Welling, M. (eds.) ECCV 2016. LNCS, vol. 9908, pp. 286–301. Springer, Cham (2016). https://doi.org/10.1007/978-3-319-46493-0_18
26. Zhu, Y., et al.: Improving semantic segmentation via video propagation and label relaxation. In: Proceedings of the IEEE Conference on Computer Vision and Pattern Recognition, pp. 8856–8865 (2019)

Spectral-spatial Recurrent-Convolutional Networks for *In-Vivo* Hyperspectral Tumor Type Classification

Marcel Bengs[1(✉)], Nils Gessert[1], Wiebke Laffers[2], Dennis Eggert[3], Stephan Westermann[4], Nina A. Mueller[4], Andreas O. H. Gerstner[5], Christian Betz[3], and Alexander Schlaefer[1]

[1] Institute of Medical Technology and Intelligent Systems, Hamburg University of Technology, Hamburg, Germany
marcel.bengs@tuhh.de
[2] Departments of Otorhinolaryngology/Head and Neck Surgery, Carl von Ossietzky University Oldenburg, Oldenburg, Germany
[3] Clinic and Polyclinic for Otolaryngology, University Medical Center Hamburg-Eppendorf, Hamburg, Germany
[4] Department of Otorhinolaryngology/Head and Neck Surgery, University of Bonn, Bonn, Germany
[5] ENT-Clinic, Klinikum Braunschweig, Brunswick, Germany

Abstract. Early detection of cancerous tissue is crucial for long-term patient survival. In the head and neck region, a typical diagnostic procedure is an endoscopic intervention where a medical expert manually assesses tissue using RGB camera images. While healthy and tumor regions are generally easier to distinguish, differentiating benign and malignant tumors is very challenging. This requires an invasive biopsy, followed by histological evaluation for diagnosis. Also, during tumor resection, tumor margins need to be verified by histological analysis. To avoid unnecessary tissue resection, a non-invasive, image-based diagnostic tool would be very valuable. Recently, hyperspectral imaging paired with deep learning has been proposed for this task, demonstrating promising results on *ex-vivo* specimens. In this work, we demonstrate the feasibility of *in-vivo* tumor type classification using hyperspectral imaging and deep learning. We analyze the value of using multiple hyperspectral bands compared to conventional RGB images and we study several machine learning models' ability to make use of the additional spectral information. Based on our insights, we address spectral and spatial processing using recurrent-convolutional models for effective spectral aggregating and spatial feature learning. Our best model achieves an AUC of 76.3 %, significantly outperforming previous conventional and deep learning methods.

Keywords: Hyperspectral imaging · Head and neck cancer · Spatio-spectral deep learning

M. Bengs and N. Gessert—Both authors contributed equally.

A. L. Martel et al. (Eds.): MICCAI 2020, LNCS 12263, pp. 690–699, 2020.
https://doi.org/10.1007/978-3-030-59716-0_66

1 Introduction

Head and neck cancers are responsible for 3.6% of cancer-specific deaths while being the sixth most common type of cancer [22]. For a patient's prognosis, early detection is critical [14]. Late detection of malignancy has been reported for tumors in the head and neck area [11], suggesting that accurate diagnostic tools would be valuable. The typical diagnostic procedure for head and neck cancer starts with an endoscopic intervention where a medical expert examines tissue regions. In case malignancy is suspected for a tumor region, an invasive biopsy, followed by histological evaluation is performed. While being considered the gold standard for diagnosis, tissue removal can cause function deterioration [1]. Also, a biopsy is time-consuming and a costly procedure that needs to be performed under anesthesia, leading to a risk for the patient.

As a consequence, fast and accurate, non-invasive diagnosis or patient referral would have the potential to significantly improve the head and neck cancer diagnostic procedure both for patients and medical staff. For this purpose, several optical imaging techniques, embedded into an endoscope, have been studied over the recent years. For example, optical microscopy [18], optical coherence tomography using infrared light [2] and narrow-band imaging augmented by fluorescence [23] have been studied for detection of malignancy.

Another promising imaging modality is hyperspectral imaging (HSI) where multiple images are acquired at several different wavelengths of light. The method has also been studied in the context of head and neck cancer detection [20,21]. This includes an *in-vivo* study where conventional machine learning methods were employed for pixel-wise tissue classification using a single patient for training [16]. Very recently, deep learning methods have been employed for HSI-based head and neck cancer detection in an *ex-vivo* setting [10,12]. Also, distinguishing healthy tissue from tumor areas has been studied with different convolutional neural networks (CNNs) for spatio-spectral HSI processing [3,7]. In this work, we study the more challenging task of *in-vivo* tumor type classification using HSI and different machine learning methods.

In terms of methods, a straight-forward solution is stacking all spectral bands in the CNN's input channel dimension and assuming the spectral bands to be similar to RGB color channels. However, several studies have demonstrated that joint spatio-spectral deep learning can substantially improve classification performance. For example, Liu et al. used convolutional long short-term memory (LSTM) models instead of CNNs, adopting recurrent processing for temporal sequences to spectral bands [17]. Also, a 3D extension of the popular Inception architecture for joint spatial and spectral processing where all dimensions are treated with convolutions has been employed [12]. However, there have hardly been any studies detailing the individual importance of spectral and spatial information. Also, it is often unclear how much additional value multiple spectral bands provide over RGB images and whether machine learning methods are able to use the additional spectral information effectively.

Therefore, we compare both conventional machine learning approaches and deep learning methods using only spectral, mostly spatial or combined spatial

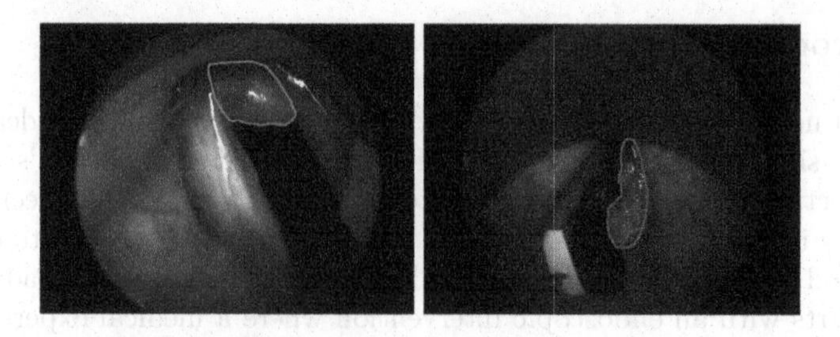

Fig. 1. Example RGB images including marked tumors. Left, a bengin tumor at the vocal folds is shown, right, a malignant tumor at the vocal folds is shown.

and spectral information. To maximize use of all available information, 3D CNNs with convolution-based processing for both spatial and spectral dimensions should be able to extract rich features. However, the approach assumes similar importance of spectral and spatial information although there is a high correlation and similarity between spectral bands. We hypothesize that learning a compact spectral representation first, followed by a focus on spatial processing is more effective. Therefore, we propose the use of recurrent-convolutional models with a convolutional gated recurrent unit (CGRU) followed by a CNN. The CGRU first aggregates the 3D hyperspectral cube into a 2D representation which is then processed by a 2D CNN. We provide an analysis of several architectural variations of the proposed model and show how changes in the number of spectral bands affect performance.

Summarized, our contributions are three-fold. First, we demonstrate the feasibility of deep learning-based *in-vivo* head and neck tumor type classification in hyperspectral images. Second, we provide an extensive analysis of different models for evaluating the spectral and spatial information content. Third, we propose a convolutional-recurrent model for efficient spectral and spatial processing.

2 Methods

2.1 Data Set

The data set consists of 98 patients, who were examined in a previous study due to mucous membrane abnormalities in the area of the upper aerodigestive tract [3]. Data acquisition was performed as described by Gerstner et al. [8] at the department of otorhinolaryngology at the University of Bonn. The study was approved by the local ethics committee (176/10 & 061/13, University of Bonn). The acquisition was performed throughout a normal procedure for diagnosis. An endoscopic device (Karl Storz GmbH & CoKG, Tuttlingen, Germany) was used for visual assessment. A Polychrome V monochromator (TillPhotonics, Gräfelfing, Germany) was used as light source for HSI and a monochromatic CCD-camera (AxioCamMRm, Carl Zeiss Microimaging GmbH, Göttingen, Germany) was employed for HSI data acquisition. The spectral bands range from

430 nm to 680 nm with a step size of 10 nm and a bandwidth of 15 nm. The spatial resolution is 1040×1388 pixels. Biopsy samples were extracted from the imaged tumor areas, followed by histopathological evaluation. After the acquisition, we use the ImageJ-implementation of the SIFT-algorithm "Linear Stack Alignment with SIFT" [19] for aligning of the HSI images. Experts performed ground-truth annotation using RGB representations derived from the HSI images and the histopathological report. Note that this method induces some label noise as tumor outlines might be slightly inaccurate. In total, there are 83 patients with benign tumor region and 15 patients with malignant tumor region. Thus, the learning problem is particularly challenging due to extreme class imbalance. Tissue regions include the larynx, oropharynx and hypopharynx. Example images and highlighted tumor regions are shown in Fig. 1. For model training and evaluation, we crop patches of size 32×32 from the marked areas, including a margin towards the marked border. Overall, we obtain 18 025 patches from all patients. For training and evaluation of our models, we define three subsets with a size of 19 patients (5 malignant/14 benign) each and apply a cross-fold scheme. Note, data from one patient does not appear in different subsets. We split each subset into a test set (3 malignant/8 benign) and validation set (2 malignant/6 benign). We perform hyperparameter optimization using grid search on the validation splits and we report performance metrics for the test splits.

2.2 Machine Learning and Deep Learning Methods

We consider several different machine learning methods for *in-vivo* head and neck tumor type classification. A first baseline is the use of a random forests (**RF**) and support vector machines (**SVM**), similar to a previous approach where a single patient was used for training [16]. We use the spectral bands as features for the RF and SVM and perform pixel-wise classification. Thus, these conventional models only use spectral information and no spatial context. In contrast, deep learning methods such as CNNs are designed for spatial processing. A first straightforward approach is to use a 2D CNN with RGB color images, which can be derived from the HSI stacks. Here, all color channels are stacked into the first layer's channel dimension which is the standard approach for natural color images. We refer to this approach as **2D CNN RGB**. This method serves as a baseline for the use of mostly spatial information and no additional spectral bands. The approach can be directly extended to the use of hyperspectral images by treating the additional spectral bands as colors channels. Thus, here, we also stack all N_S spectral bands into the CNN's first layer's input channel, resulting in an input tensor of size $x \in \mathbb{R}^{B \times H \times W \times N_S}$ where B is the batch size and H and W are spatial dimensions. This has been proposed for *ex-vivo* head and neck cancer classification [13]. We refer to this method as **2D CNN HSI**. A disadvantage of this approach is that the entire spectral information vanishes after the first layer. The CNN computes multiple weighted averages of the spectral bands which are simply treated as the CNN's feature dimension afterward.

This motivates the use of explicit processing methods of the spectral dimension. One approach is to treat the spectral dimension equivalently to a spatial

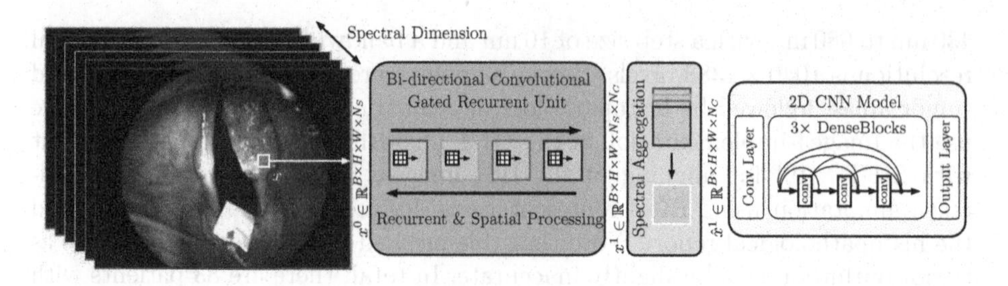

Fig. 2. The CGRU-CNN we propose for hyperspectral image data. The model's output is a binary classification into benign and malignant.

dimension and convolving over it, as employed by [12]. Thus, the CNN's input tensor changes from $x \in \mathbb{R}^{B \times H \times W \times N_S}$ to $\hat{x} \in \mathbb{R}^{B \times H \times W \times N_S \times 1}$ with a channel dimension of one. Here, the 2D CNN architecture can be directly extended to 3D by employing 3D instead of 2D convolutions. We refer to this method as **3D CNN HSI**.

An approach that has not been considered for head and neck cancer detection is the use of recurrent processing for the spectral dimension, adopted from sequential, temporal learning problems. One approach is to use a convolutional LSTM that performs joint spatial and spectral processing in its units, where each spectral image is processed sequentially. Liu et al. employed this approach in the remote sensing area using a bi-directional model [17]. We adopt a similar method using the more efficient gated recurrent units [5], also employing convolutions within its gates. We refer to this method as **CGRU HSI**.

While the CGRU HSI model performs some spatial processing with convolutional gates, high-level, abstract representations are difficult to learn due to limited network depth [17]. Therefore, we extend this model by using a 2D CNN for spatial processing after a CGRU layer. Thus, the CGRU receives $x^0 \in \mathbb{R}^{B \times H \times W \times N_S}$ as its input and outputs $x^1 \in \mathbb{R}^{B \times H \times W \times N_S \times N_C}$ where N_C is the CGRU's hidden dimension. Then, we employ a pooling or state selection function $\hat{x}^1 = f_{sel}(x^1)$ that aggregates the tensor's N_S processed spectral states into a single representation $\hat{x}^1 \in \mathbb{R}^{B \times H \times W \times N_C}$. We either select the last spectral state as the output or perform mean or max pooling over all states. Using the last state is a common approach as it captures condensed information from all states, however, state pooling might provide additional information by explicitly combining all states [4]. The resulting representation is then processed by a 2D CNN with the same structure as 2D CNN RGB or 2D CNN HSI. The architecture **CGRU-CNN** is depicted in Fig. 2.

All our CNN-based models follow the concept of densely-connected convolutional networks [15]. After an initial convolutional layer, three DenseBlocks follow. Within each DenseBlock, several convolutional layers are employed which make efficient reuse of already computed features by processing all outputs from preceding layers. In between DenseBlocks, an average pooling layer performs spatial and, in case of a 3D CNN, spectral downsampling with a stride of two.

Table 1. Results for all experiments, comparing different methods. All values are given in percent. 95 % CIs are provided in brackets. Last, Mean and Max refer to the spectral aggregation strategy.

	AUC	**Sensitivity**	**Specificity**	**F1-Score**
SVM HSI	60.5(59 – 61)	58.9(58 – 60)	61.9(61 – 63)	60.8(60 – 61)
RF HSI [16]	59.6(59 – 60)	71.7(70 – 73)	47.5(46 – 49)	62.7(62 – 64)
2D CNN RGB	61.2(60 – 62)	68.5(67 – 70)	52.2(51 – 54)	58.9(58 – 60)
2D CNN HSI [13]	62.3(61 – 64)	60.0(58 – 61)	55.1(54 – 56)	57.6(57 – 58)
CGRU HSI [17]	63.5(62 – 65)	61.0(59 – 63)	61.0(59 – 62)	61.4(60 – 62)
3D CNN HSI [12]	67.8(67 – 69)	65.7(64 – 67)	59.6(58 – 61)	62.5(62 – 64)
CNN-CGRU	69.9(69 – 71)	72.4(71 – 74)	61.9(61 – 63)	66.4(65 – 67)
CGRU-CNN (Last)	**76.3(75 – 77)**	69.4(68 – 71)	**68.9(68 – 70)**	69.5(69 – 71)
CGRU-CNN (Mean)	73.7(73 – 75)	**74.5(73 – 76)**	63.2(62 – 64)	67.8(67 – 69)
CGRU-CNN (Max)	70.9(70 – 72)	74.3(73 – 77)	66.9(65 – 68)	**70.1(69 – 71)**

After the last DenseBlock, a global average pooling layer is employed, followed by a fully-connected layer that performs binary classification. We train all models with a cross-entropy loss and Adam optimizer. We perform class balancing by reweighting the loss function with the inverse class frequency $w_i = \frac{N}{N_i}$ where N is the total number of samples and N_i the number of samples for class i. Hyperparameters such as feature map size and learning rate are optimized for the individual models based on validation performance to provide a fair comparison between all methods. Similarly, we tune RF and SVM hyperparameters including maximum tree depth, the number of trees (RF), the box constraint and kernel (SVM).

We consider the F1-Score, sensitivity, specificity and the area under the receiver operating curve (AUC) as our performance metrics. We do not consider traditional accuracy as it highly misleading for an extremely imbalanced problem such as ours. We report 95 % confidence intervals (CI) using bias corrected and accelerated bootstrapping with $n_{CI} = 10\,000$ bootstrap samples. We test for statistically significant difference in our performance metrics using a permutation test with $n_P = 10\,000$ samples and a significance level of $\alpha = 0.05$ [6].

3 Results

The results of our experiments are shown in Table 1. Using RGB and HSI in a conventional 2D CNN leads to similar performance. Notably, RF and SVM achieve performance in a similar range. When using spatio-spectral processing with a 3D CNN performance improves significantly for the AUC. Overall, variants of our proposed CGRU-CNN architecture performs best. The difference in the AUC, F1-Score and specificity is significant ($p < 0.05$) compared to the previously proposed model 3D CNN. Comparing spectral aggregation strategies,

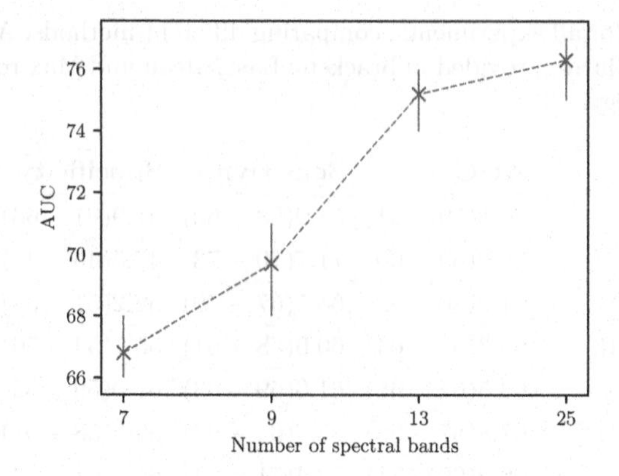

Fig. 3. Comparison of the AUC with CIs for models trained with a different number of spectral bands with our CGRU-CNN (Last) architecture. For reduction, we selected every second, third and fourth spectral band.

performance varies for different metrics but remains consistently higher than for the other methods. Performing spatial processing first with CNN-CGRU performs better than 3D CNN while being significantly outperformed by CGRU-CNN in terms of the AUC and F1-Score ($p < 0.05$). In Fig. 3 we show the effect of using a different number of spectral dimensions with our CGRU-CNN model. When using fewer spectral dimensions the AUC is slightly reduced. For half the number of spectral bands, performance remains similar to using all spectral bands.

4 Discussion

We address the problem of *in-vivo* tumor type classification with hyperspectral imaging. Previously, this task has been addressed for *ex-vivo* data with deep learning methods have shown promising results [12]. However, the *in-vivo* setting makes the task much more difficult due to high variability in terms of motion, lighting and surrounding anatomy. In terms of methods, 2D CNNs and spatio-spectral 3D CNNs have been used for tissue classification. Yet, the individual importance of spatial and spectral information as well as machine learning model's ability to use the information is still unclear, especially under challenging *in-vivo* conditions. Therefore, we employ several machine learning methods to study the information content of spatial and spectral dimensions and whether models can effectively utilize the information.

Considering our results in Table 1, we observe a similar performance of 2D CNNs and conventional methods. First, the surprisingly good performance of RFs and SVMs indicates that conventional methods are effectively using spectral information, while not taking spatial relations into account. Second, 2D CNN RGB demonstrates that using spatial information only can also lead to a similar performance level. As a result, we observe that both spectral information alone

(RF/SVM) and spatial information with few bands (2D CNN RGB) are valuable for the task at hand, suggesting that combining both is reasonable.

Using 2D CNNs with HSI in the color channels does not improve performance, indicating that other processing techniques are required. The use of CGRU only, as previously performed for remote sensing [17], does not lead to major improvements either. While simultaneously processing spatial information with convolutions and spectral information with a recurrent gating mechanism, the architecture is very shallow and therefore likely limited in terms of its capability to learn expressive spatial features. This motivates the use of 3D CNNs that treat both spatial and spectral information equally, processing both with convolutions. We observe a significant improvement in the AUC for this approach over the previously mentioned methods. However, treating spatial and spectral information completely equally is also problematic as spectral bands are highly correlated and very similar. Therefore, we propose to use a CGRU-CNN architecture that disentangles spectral and spatial processing by first learning a spatial representation from spectral bands that is then processed by a 2D CNN. This approach significantly outperforms all other methods in terms of the AUC and F1-Score. Interestingly, performing spectral processing after spatial processing with CNN-CGRU performs worse. This suggests that reducing spectral redundancy first by learning a compact representation is preferable. Comparing spectral aggregation strategies, all variations perform consistently well. Using the last state performs best for the AUC, suggesting that the CGRU's learned representation effectively captures relevant information and pooling all spectral states is not required.

Last, we investigate the value of the spectral dimension in more detail. We pose the hypothesis, that spectral bands, while adding valuable information, might be redundant due to high similarity and correlation. While learning a compact spectral representation is one approach, directly removing spectral bands should also reduce redundancy in this regard. The results in Fig. 3 show that taking only half the spectral bands still performs very well, indicating that there is indeed redundancy. When reducing the spectral dimension further, however, performance drops notably. This signifies once again that additional spectral information is helpful but too much redundancy limits performance improvement.

Overall, an AUC of 76.3 % demonstrates the feasibility of *in-vivo* tumor type classification but is likely too limited for clinical decision support yet. Still, we find valuable insights on different machine learning models' capability of utilizing spatial and spectral information. The concept of recurrent-convolutional models has been used for temporal data [9] which we adopt and extend by bi-directional units and state aggregation methods. Thus, we design a model that demonstrates the concept's efficacy for a hyperspectral learning problem which could substantially improve other medical applications of hyperspectral imaging.

5 Conclusion

We demonstrate the feasibility of *in-vivo* tumor type classification using hyperspectral imaging with spectral and spatial deep learning methods. We design and compare several different methods for processing hyperspectral images to study the value of spatial and spectral information in a challenging *in-vivo* context. Our analysis indicates that processing the spectral dimension with a convolutional gated recurrent network followed by spatial processing with a CNN is preferable. Particularly, conventional and previous deep learning methods are clearly outperformed. These results should be considered when using HSI for tissue classification in future work.

Acknowledgments. This work was partially supported by Forschungszentrum Medizintechnik Hamburg (Grant: 02fmthh2019).

References

1. Alieva, M., van Rheenen, J., Broekman, M.L.D.: Potential impact of invasive surgical procedures on primary tumor growth and metastasis. Clin. Exp. Metastasis **35**(4), 319–331 (2018). https://doi.org/10.1007/s10585-018-9896-8
2. Arens, C., Reussner, D., Woenkhaus, J., Leunig, A., Betz, C., Glanz, H.: Indirect fluorescence laryngoscopy in the diagnosis of precancerous and cancerous laryngeal lesions. Eur. Arch. Oto-rhino-laryngology. **264**(6), 621–626 (2007)
3. Bengs, M., et al.: Spatio-spectral deep learning methods forin-vivohyperspectral laryngeal cancer detection. In: SPIE Medical Imaging 2020: Computer-Aided Diagnosis. p. in print (2020)
4. Chen, Q., Ling, Z.H., Zhu, X.: Enhancing sentence embedding with generalized pooling. In: COLING (2018)
5. Cho, K., et al.: Learning phrase representations using RNN encoder-decoder for statistical machine translation. In: EMNLP (2014)
6. Efron, B., Tibshirani, R.J.: An Introduction to the Bootstrap. CRC Press, Boca Raton (1994)
7. Eggert, D., et al.: In vivo detection of laryngeal cancer by hyperspectral imaging combined with deep learning methods (conference presentation). In: Imaging, Therapeutics, and Advanced Technology in Head and Neck Surgery and Otolaryngology 2020, vol. 11213, p. 112130L. International Society for Optics and Photonics (2020)
8. Gerstner, A.O., et al.: Hyperspectral imaging of mucosal surfaces in patients. J. Biophotonics **5**(3), 255–262 (2012)
9. Gessert, N., et al.: Spatio-temporal deep learning models for tip force estimation during needle insertion. Int. J. Comput. Assist. Radiol. Surg. **14**(9), 1485–1493 (2019). https://doi.org/10.1007/s11548-019-02006-z
10. Grigoroiu, A., Yoon, J., Bohndiek, S.E.: Deep learning applied to hyperspectral endoscopy for online spectral classification. Sci. Rep. **10**(1), 1–10 (2020)
11. Habermann, W., Berghold, A., J Devaney, T.T., Friedrich, G.: Carcinoma of the larynx: predictors of diagnostic delay. Laryngoscope **111**(4), 653–656 (2001)
12. Halicek, M., et al.: Optical biopsy of head and neck cancer using hyperspectral imaging and convolutional neural networks. In: Optical Imaging, Therapeutics, and Advanced Technology in Head and Neck Surgery and Otolaryngology 2018, vol. 10469, p. 104690X. International Society for Optics and Photonics (2018)
13. Halicek, M., et al.: Deep convolutional neural networks for classifying head and neck cancer using hyperspectral imaging. J. Biomed. Opt. **22**(6), 060503 (2017)

14. Horowitz, A.M.: Perform a death-defying act: the 90-second oral cancer examination. J. Am. Den. Assoc. **132**, 36S–40S (2001)
15. Huang, G., Liu, Z., Van Der Maaten, L., Weinberger, K.Q.: Densely connected convolutional networks. In: CVPR, pp. 2261–2269 (2017)
16. Laffers, W., et al.: Early recognition of cancerous lesions in the mouth and oropharynx: automated evaluation of hyperspectral image stacks. HNO **64**(1), 27–33 (2016)
17. Liu, Q., Zhou, F., Hang, R., Yuan, X.: Bidirectional-convolutional LSTM based spectral-spatial feature learning for hyperspectral image classification. Remote Sens. **9**(12), 1330 (2017)
18. Löhler, J., Gerstner, A., Bootz, F., Walther, L.: Incidence and localization of abnormal mucosa findings in patients consulting ent outpatient clinics and data analysis of a cancer registry. Eur. Arch. Oto-Rhino-Laryngology. **271**(5), 1289–1297 (2014)
19. Lowe, D.G.: Distinctive image features from scale-invariant keypoints. Int. J. Comput. Vis. **60**(2), 91–110 (2004)
20. Regeling, B., et al.: Development of an image pre-processor for operational hyperspectral laryngeal cancer detection. J. Biophotonics **9**(3), 235–245 (2016)
21. Regeling, B., et al.: Hyperspectral imaging using flexible endoscopy for laryngeal cancer detection. Sensors **16**(8), 1288 (2016)
22. Shield, K.D., et al.: The global incidence of lip, oral cavity, and pharyngeal cancers by subsite in 2012. CA Cancer J. Clin. **67**(1), 51–64 (2017)
23. Volgger, V., et al.: Evaluation of optical coherence tomography to discriminate lesions of the upper aerodigestive tract. Head Neck **35**(11), 1558–1566 (2013)

Synthetic and Real Inputs for Tool Segmentation in Robotic Surgery

Emanuele Colleoni[1,2(✉)], Philip Edwards[1,2], and Danail Stoyanov[1,2]

[1] Wellcome/EPSRC Centre for Interventional and Surgical Sciences (WEISS), University College London, 43-45 Foley St., Fitzrovia, London W1W 7EJ, UK
[2] Department of Computer Science, University College London, Fitzrovia, UK
emanuele.colleoni.19@ucl.ac.uk

Abstract. Semantic tool segmentation in surgical videos is important for surgical scene understanding and computer-assisted interventions as well as for the development of robotic automation. The problem is challenging because different illumination conditions, bleeding, smoke and occlusions can reduce algorithm robustness. At present labelled data for training deep learning models is still lacking for semantic surgical instrument segmentation and in this paper we show that it may be possible to use robot kinematic data coupled with laparoscopic images to alleviate the labelling problem. We propose a new deep learning based model for parallel processing of both laparoscopic and simulation images for robust segmentation of surgical tools. Due to the lack of laparoscopic frames annotated with both segmentation ground truth and kinematic information a new custom dataset was generated using the da Vinci Research Kit (dVRK) and is made available.

Keywords: Instrument detection and segmentation · Surgical vision · Computer assisted interventions

1 Introduction

Robotic minimally invasive surgery is now an established surgical paradigm across different surgical specialties [18]. While the mechanical design and implementation of surgical robots can support automation and advanced features for surgical navigation and imaging, significant effort is still needed to automatically understand and infer information from the surgical site for computer assistance. Semantic segmentation of the surgical video into regions showing instruments and tissue is a fundamental building block for such understanding [13,15] and to pose estimation for robotic control [1,6] and surgical action recognition [23].

The most effective semantic segmentation approaches for surgical instruments have used deep learning and Fully Convolutional Neural Networks (FCNNs) [8]. Various architectures have been reported including novel encoder-decoders using established pre-trained feature extractors or adding attention fusion modules in the decoding part of the network [14,17]. These have

A. L. Martel et al. (Eds.): MICCAI 2020, LNCS 12263, pp. 700–710, 2020.
https://doi.org/10.1007/978-3-030-59716-0_67

dramatically improved algorithm performance compared to early methods [2]. More recently, robotic systems articles have also reported the coupling of visual information with kinematic data [22] and the possibility of using kinematic information to produce surgical tools segmentation ground truth [4]. In their work, da Costa Rocha et al. [4] employed a Grabcut-based cost function to iteratively estimate the optimal pose of the kinematic model in order to produce accurate segmentation labels through tool's model projection on the image plane. The major weakness of this method is its strong dependence to an accurate initial pose of the tool, that is not trivial in the surgical scenario. A similar approach was attempted by Qin et al. [19]: their method is based on a particle filter optimization that repeatedly updates the pose of the tool to match the silhouette projection of the surgical tool with a vision-based segmentation mask obtained using a ToolNet [8]. However, this procedure heavily rely on optical markers and on a navigation system for initial tool pose estimation. Moreover, the procedure has been proposed only for non-articulated rigid tools, that limits the field of applicability of this method. Despite progress, robust semantic segmentation for surgical scene understanding remains a challenging problem with insufficient high quality data to train deep architectures and more effort needed to exploit all the available information on instrument geometry or from robotic encoders.

In this paper, we propose a novel multi-modal FCNN architecture that exploits visual, geometric and kinematic information for robust surgical instrument detection and semantic segmentation. Our model receives two input images: one image frame recorded with a da Vinci® (Intuitive Surgical Inc, CA) system and a second image obtained loading the associated kinematic data into a virtual da Vinci Research Kit (dVRK) simulator. The global input is an image couple showing real (containing visual features) and simulated (containing geometric tools information from robot Computer-Aided Design (CAD) models) surgical tools that share the same kinematic values. We show that the simulation images obtained exploiting kinematic data can be processed in parallel with their real counterpart to improve segmentation results in presence of variable light conditions or blood. This is the first attempt that uses a deep learning framework for parallel processing of images produced using a robot simulator and a laparoscopic camera to improve surgical tool segmentation avoiding iterative shape matching. Due to a lack of a dataset annotated with both kinematic data and segmentation labels, we built a custom dataset of 14 videos for the purpose.[1]

2 Methods

2.1 Dataset Generation with dVRK

We use the dVRK system to record both video and kinematic information about the instrument motion. Because the system is robotics we can repeat movements

[1] https://www.ucl.ac.uk/interventional-surgical-sciences/davinci-segmentation-kinematic.

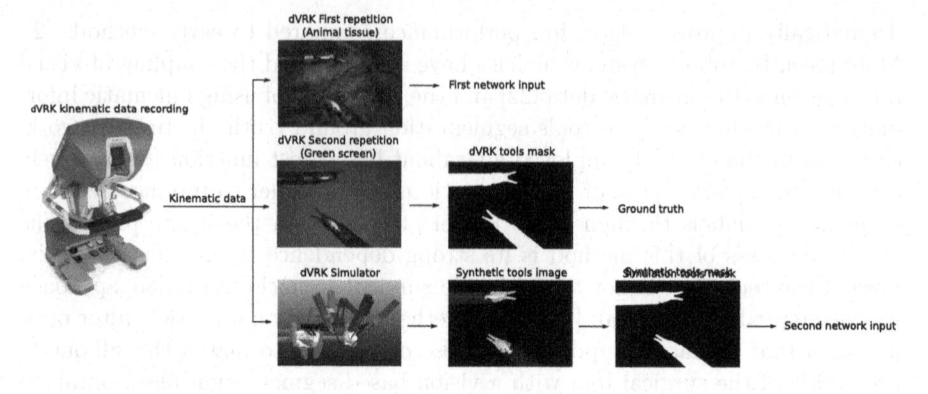

Fig. 1. The figure shows the workflow for the generation of our dataset. Once the kinematic data of a movement is recorded using the dVRK, it is first reproduced over an animal tissue background. A second repetition with the same kinematic is then performed on an OLED green screen. The ground truth for each image is the generated using background subtraction technique. The collected kinematic data are then loaded on a dVRK simulator to produce simulation images of the tools, that are successively binarized to produce the second input of the proposed FCNN.

previously executed by an operator on a da Vinci Surgical System[2] (DVSS) in clinical practice. Each instrument on the dVRK held by the Patient Side Manipulators (PSM) is defined by 7 joints (6 revolute and 1 prismatic), while the Endoscope Control Manipulator has 4 (3 revolute and 1 prismatic). In this study we only use EndoWrist® Large Needle Drivers for simplicity, although the same workflow can be extended to the whole family of articulated surgical tools if appropriate models and control information is available (currently not implemented in dVRK).

To produce each video in the dataset we followed four consecutive steps:

- First, we perform a surgical movement recording kinematic data using our dVRK;
- Then we collect image frames with animal tissue background using the recorded kinematic data stream;
- The same movement is reproduced a second time on a green screen to obtain tools ground truth segmentation masks;
- Finally, for each frame, we produced an associated image of the virtual tools obtained employing a dVRK simulator by making use of the recorded kinematic information.

Our dataset generation procedure is shown in Fig. 1.

Kinematic Data. An action is first performed on the DVSS without a background and the kinematic information of the PSMs and ECM is recorded. The

[2] https://www.intuitive.com/.

recording framework was implemented in MATLAB using the Robotic System Toolbox[3] to access robot articulations joint values from the dVRK. The result is a 7 by N matrix for each PSM and 4 by N for the ECM, where each row corresponds to one joint (starting from the ECM/PSM's base till the tip of the tool) and consecutive columns represent consecutive time steps.

Video Data. The video frames are collected by repeating the recorded action over an animal tissue background. The stored joint coordinates of all PSMs and ECM are sent to the DVSS via the dVRK using our MATLAB interface in order to have precise movement reproduction. Video images are synchronously collected every 150 ms to avoid redundancy in the data.

Ground Truth Generation. The segmentation ground truth is produced by physically replacing the animal tissue background with a green screen. We chose to use an Organic Light-Emitting Diode (OLED) screen emitting green light to avoid shadows that generally decrease segmentation performances. Once the screen is conveniently placed to entirely cover the camera Field Of View (FOV), the same recorded action employed in the previous phase is reproduced. Finally we removed the tools from the FOV and an image is collected showing only the background. The segmentation ground truth for each frame is then obtained by subtracting the background image to each frame and by applying a threshold to the *L1-norm* of the subtraction result.

The ground truth generation procedure allows us to avoid issues originated by a virtual replacement of the segmentation mask background, such as image matting and blending [3]. The reliability of our ground truth generation methodology has been tested on 7 further video couples, where a same action was reproduced twice on the green background. The resulting ground truth masks for each video couple were then compared using Intersection over Union (IoU) metric, obtaining an overall 99.8% median evaluation score with Interquartile Range (IQR) of 0.05%.

Simulation Images. We load the kinematic data collected using our dVRK into the simulation model [7] to virtually reproduce the performed movement in CoppeliaSim[4] [20]. Images were simultaneously collected at the same frame rate used for the recorded videos in order to have synchronization between simulation and dVRK information. The produced images were then thresholded to obtain a segmentation mask used to feed the proposed network.

2.2 Network Architecture

We propose a double-input FCNN for simultaneous processing of frames collected using the dVRK and segmentation masks of their simulated counterpart.

[3] https://uk.mathworks.com/products/robotics.html.
[4] https://www.coppeliarobotics.com/.

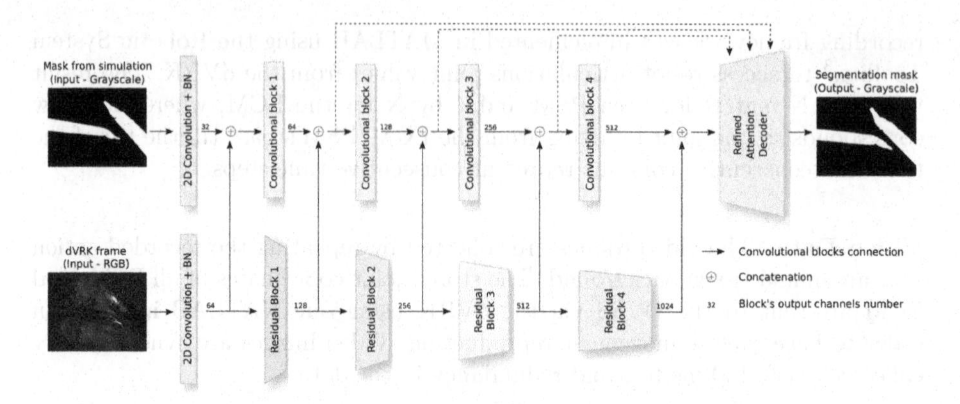

Fig. 2. Architecture of the proposed model. Two connected branches are used to extract features from dVRK frames (RGB) and simulated tools segmentation masks (Grayscale). The lower branch is made of residual blocks developed in [10] pre-trained on Imagenet dataset, while the upper branch is composed of 4 convolutional blocks, each one performing 3 convolutions + batch normalization + RELU activation. The number of output channels for all encoders' blocks is written at the end of each one of them. The decoder part is described in [17], taking encoder's output and skip connections as input.

The architecture of the proposed network is shown in Fig. 2. We a used the commonly adopted U-net structure as starting point for our model [21]. We chose to concatenate features extracted from both inputs from a very early stage in the network, that has shown to be more effective than merging them only in the decoding part [9].

The encoder branch for dVRK frames processing was implemented following work conducted in [17] and [11], where features are extracted from the image using ResNet50 [10] blocks pre-trained on Imagenet dataset [5]. Each residual block consists of multiple consecutive sub-blocks, namely 3, 4, 6 and 3 sub-blocks for residual block 1 to 4 respectively. Each sub-block is composed by 3 convolutional layers, where a Batch Normalization (BN) and a Rectified Linear Unit (RELU) activation are applied on the output of each convolution stage. The resulting output is concatenated with the input of the sub-block using a skip connection. We built a second branch parallel to ResNet50 for simulated tools' mask processing. Following ResNet50, a 7×7 convolution + BN + RELU activation operation is first performed on the image. The result is then passed through 4 consecutive convolutional blocks. At each block, the input coming from the previous convolutional block is first concatenated with the output of the relative parallel residual block. The result is then processed using 3 different convolutional layers, i.e. a first 1×1 convolution with stride 1 to double the number of channels, followed by a 3×3 convolution (stride 1×1) and a final 3×3 convolution with stride 2×2 to halve the output's height and width.

The segmentation probability map is finally obtained concatenating the outputs of both the last convolutional and residual blocks and processing them employing the decoding architecture (based on attention fusion modules and decoder blocks)developed in [17]. Each attention block takes as second input the features extracted from intermediate encoding layers of both branches using skip connection. Such methodology has shown to be particularly useful to properly recover information lost in the network's early blocks [21].

2.3 Loss Function Details

We selected the sum of per-pixel binary crossentropy [11,17] which has been previously used for instrument detection and articulation tracking [6,16], and IoU loss (to prevent the network to be sensitive to class unbalance) as loss function to train our model. The binary crossentropy is defined as:

$$L_{bce} = \frac{1}{\Omega} \sum_{n \in \Omega} [p_n log\widehat{p}_n + (1 - p_n) log(1 - \widehat{p}_n)] \quad (1)$$

where p_n and $\widehat{p}_n$ are the FCNN output and ground truth values of pixel n into mask domain Ω.

IoU loss is defined as:

$$L_{IoU} = 1 - IoU_{score} \quad (2)$$

and IoU_{score} is defined as:

$$IoU_{score} = \frac{TP}{TP + FP + FN} \quad (3)$$

where TP is the number of pixels correctly classified as tools' pixels, while FP and FN are the numbers of pixels mis-classified as tools and background respectively. Following [11], we chose a threshold value of 0.3 to binarize our output probability mask.

The resulting loss function is then defined by the sum of L_{bce} and L_{Iou}:

$$L = L_{bce} + L_{IoU} \quad (4)$$

3 Experiments and Results

Dataset. We collected 14 videos of 300 frames each (frame size = 720 × 576), for a total amount of 4200 annotated frames. In particular, 8 videos were used for the training phase, 2 for validation and the remaining 4 for testing. We employed 5 different kinds of animal tissues (chicken breast and back, lamb and pork loin, beef sirloin) for the entire dataset, changing the topology of the background and varying illumination conditions in each video to increase data variability. Lamb kidneys and blood were placed in the background and on the tools of the test set videos in order to properly test algorithms' performance on conditions not seen in the training phase. Finally, following [6], we added Fractional Brownian Motion

Table 1. Comparison with the state of the art architectures on our test set [11,17]. Results for each video are presented in terms of median Intersection over Union IoU score and Interquartile Range (IQR) over all video frames in percentages.

Median Value (%) / IQR (%) of IoU score						
	No smoke			Added smoke		
	Ternausnet	Rasnet	Proposed	Ternausnet	**Rasnet**	**Proposed**
Video 1	73.41/12.24	79.25/8.84	**81.80/7.74**	24.39/9.29	44.85/15.80	**54.15/14.16**
Video 2	73.94/2.88	78.44/3.80	**81.79/3.71**	58.87/7.15	73.98/9.41	**79.49 4.31**
Video 3	78.57/6.13	84.56/3.97	**91.04/2.79**	78.02/6.28	85.62/4.41	**91.20/3.41**
Video 4	92.07/3.82	95.09/2.15	**95.16/2.68**	49.65/10.84	78.06/17.28	**82.93/8.62**

noise[5] to simulate cauterize smoke on test set frames. Each frame has been first cropped to remove dVRK side artefacts and then resized to 256 × 320 to reduce its processing computational load. We produced a simulation segmentation mask (see Sect. 2) for each frame using the dVRK simulator in order to feed our double-input FCNN. All data will be made available for research.

Implementation and Runtime Analysis. The proposed model was implemented in Tensorflow/Keras[6] and trained on GPU NVIDIA® Tesla® V100. We chose Adam as optimizer for our network [12], with a learning rate of 0.001 and exponential decay rates $\beta 1$ and $\beta 2$ of 0.9 and 0.999 respectively. We selected the best model weights considering IoU score as evaluation metric obtained on the validation set.

Comparison Experiments. We examined the benefit introduced by adding geometric and kinematic information in the network input by comparing our results with the ones obtained using Ternausnet [11] and Rasnet [17] architectures after being trained on the proposed training set. Performances were first evaluated on the test set, repeating the prediction a second time on the same frames after adding simulated smoke. We selected IoU as the evaluation metric. As shown in Table 1, the proposed model achieved good results on all videos compared to the state of the art, with an overall median IoU score of 88.49% (IQR = 11.22%) on the test set and a score of 80.46% (IQR = 21.27%) when simulated smoke is superimposed to dVRK frames. Rasnet and Ternausnet obtained IoU scores of 82.27% (IQR = 11.55%) and 76.67% (IQR = 13.63%) respectively on raw test videos, while their median performances decreased to 75.78% (IQR = 23.48%) and 54.70% (IQR = 34.57%) with added smoke. Focusing on single videos, the lowest scores were obtained on Video 1 and Video 2 by all the considered architectures, both with or without smoke. The best performances were instead achieved by the proposed model on Video 4 (without smoke) and Video 3, with more than 95% and 91% median IoU score respectively.

[5] https://nullprogram.com/blog/2007/11/20/.
[6] https://www.tensorflow.org/guide/keras.

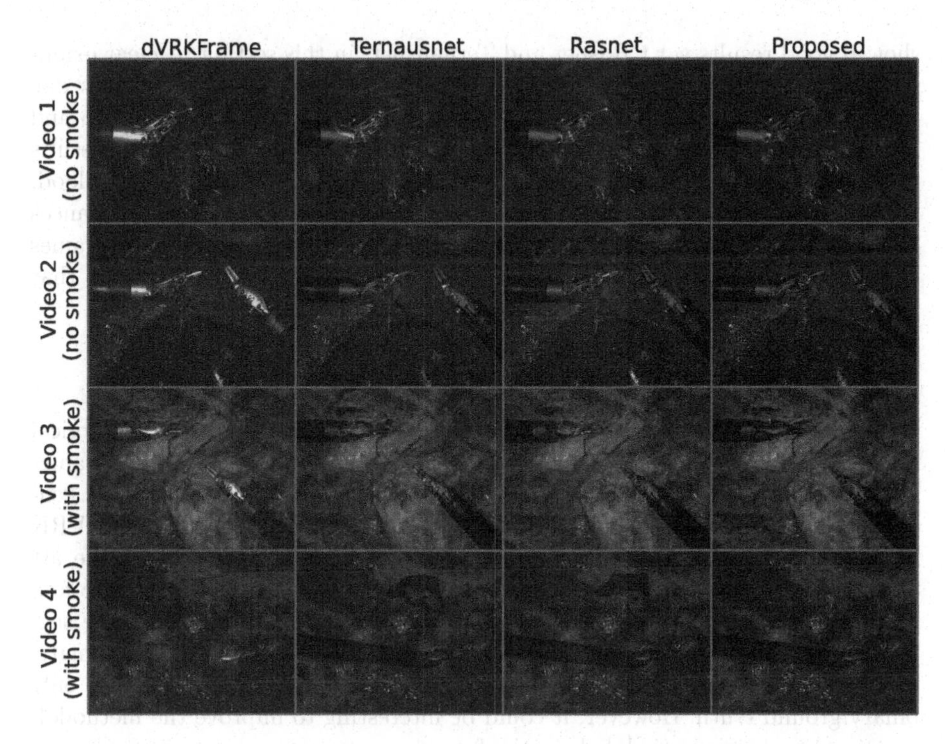

Fig. 3. Visual example of the obtained results. An image frame for each video (first column) is reported without added smoke (Video 1 and Video 2) and with smoke (Video 3 and Video 4). Segmentation results for Ternausnet, Rasnet and the proposed model are displayed in columns 2, 3 and 4 respectively.

We carried out a further analysis only on test frames that present tools occlusion to investigate models performances under this particular condition. Results were evaluated using the same evaluation metric employed during previous experiments. Even in this situation, the proposed model outperformed the other architectures, achieving a median IoU of 93.70% (IQR = 1.94%), superior to both Rasnet (median IoU = 87.45%, IQR = 2.80%) and Ternausnet (median IoU = 84.38%, IQR = 5.91%).

Robustness Discussion. Since no blood on the tools was seen during the training phase, all the architectures learned to label red pixels as background, leading to a mis-classification of the tools' portions covered by blood as seen in Fig. 3. Information extracted from segmentation masks of simulated tools helped the proposed network to better recognize the non-covered areas on the tools. Such result is highlighted when smoke is added in the image, with ΔIoU scores of 5.51% and 20.62% on Video 2 and 5.58% and 13,18% on Video 3, compared to Rasnet and Ternausnet respectively. Smoke also deteriorated segmentation performance on videos with poor initial illumination conditions but our model showed good results w.r.t Rasnet and Ternausnet, in this setting. A clear example is shown in Table 1 for Video 2, where the decrease in performances due to presence of smoke for the proposed model is only 2.3%, against 4.46% and 15.07% suffered from Rasnet and Ternausnet respectively. The worst results in presence of smoke were obtained on Video 1, where the presence of blood, darkness and smoke led all the considered models to segmentation performances below 55% (*IoU* score). Finally, the behaviour of all the architectures on frames with tool occlusions resembled the one in previous experiments, showing that such scenario do not particularly affect the results.

4 Discussion and Conclusions

In this paper, we proposed a double-input FCNN for segmentation of surgical tools from laparoscopic images. Our model takes as inputs both dVRK frames and segmentation masks produced using a dVRK simulator. Each mask is generated by projecting the simulated tools, conveniently positioned using dVRK kinematic data, onto the image plane. Our method achieved state of the art performance against image only-based models, suggesting that geometric and kinematic data can be employed by deep learning frameworks to improve segmentation. We produced a new dataset with segmentation labels and kinematic data for the purpose. Unfortunately, our procedure allows us to produce only binary ground truth. However, it could be interesting to improve the methodology to generate semantic labels in the future.

At the best of our knowledge, this is the first attempt to join visual and kinematic features for tool segmentation using a multi-modal FCNN, avoiding iterative shape-matching algorithms [4,19]. A further comparison with these methods could however be taken in consideration as future work. Several ways could be investigated as well to improve our method, e.g. using a residual-learning modelling approach to estimate the difference between simulated and estimated tool masks or trying different architectures [9] to better exploit segmentation input. Moreover, it could be interesting to study the performances of the proposed model when noise is added to the kinematic data. Finally, a scenario with tool-tissue interaction could be of great interest from the dataset generation point of view as well for further evaluation analysis in such conditions.

Acknowledgements. The work was supported by the Wellcome/EPSRC Centre for Interventional and Surgical Sciences (WEISS) [203145Z/16/Z]; Engineering and Physical Sciences Research Council (EPSRC) [EP/P027938/1, EP/R004080 /1, EP/P012841/1]; The Royal Academy of Engineering Chair in Emerging Technologies Scheme; and Horizon 2020 FET (GA 863146). We thank Intuitive Surgical Inc and the dVRK community for their support of this work.

References

1. Allan, M., Ourselin, S., Hawkes, D.J., Kelly, J.D., Stoyanov, D.: 3-D pose estimation of articulated instruments in robotic minimally invasive surgery. IEEE Trans. Med. Imaging **37**(5), 1204–1213 (2018)

2. Bouget, D., Allan, M., Stoyanov, D., Jannin, P.: Vision-based and marker-less surgical tool detection and tracking: a review of the literature. Med. Image Anal. **35**, 633–654 (2017)
3. Chuang, Y.Y., Curless, B., Salesin, D.H., Szeliski, R.: A bayesian approach to digital matting. In: Proceedings of the 2001 IEEE Computer Society Conference on Computer Vision and Pattern Recognition. CVPR 2001, vol. 2, pp. II-II. IEEE (2001)
4. da Costa Rocha, C., Padoy, N., Rosa, B.: Self-supervised surgical tool segmentation using kinematic information. In: 2019 International Conference on Robotics and Automation (ICRA), pp. 8720–8726. IEEE (2019)
5. Deng, J., Dong, W., Socher, R., Li, L.J., Li, K., Fei-Fei, L.: Imagenet: a large-scale hierarchical image database. In: 2009 IEEE Conference on Computer Vision and Pattern Recognition, pp. 248–255. IEEE (2009)
6. Du, X., Kurmann, T., Chang, P.L., Allan, M., Ourselin, S., Sznitman, R., Kelly, J.D., Stoyanov, D.: Articulated multi-instrument 2-D pose estimation using fully convolutional networks. IEEE Trans. Med. Imaging **37**(5), 1276–1287 (2018)
7. Fontanelli, G.A., Selvaggio, M., Ferro, M., Ficuciello, F., Vendittelli, M., Siciliano, B.: A v-rep simulator for the da vinci research kit robotic platform. In: 2018 7th IEEE International Conference on Biomedical Robotics and Biomechatronics (Biorob), pp. 1056–1061. IEEE (2018)
8. García-Peraza-Herrera, L.C., et al.: Toolnet: holistically-nested real-time segmentation of robotic surgical tools. In: 2017 IEEE/RSJ International Conference on Intelligent Robots and Systems (IROS), pp. 5717–5722. IEEE (2017)
9. Guo, Z., Li, X., Huang, H., Guo, N., Li, Q.: Deep learning-based image segmentation on multimodal medical imaging. IEEE Trans. Radiat. Plasma Med. Sci. **3**(2), 162–169 (2019)
10. He, K., Zhang, X., Ren, S., Sun, J.: Deep residual learning for image recognition. In: Proceedings of the IEEE conference on computer vision and pattern recognition, pp. 770–778 (2016)
11. Iglovikov, V., Shvets, A.: Ternausnet: U-net with vgg11 encoder pre-trained on imagenet for image segmentation. arXiv preprint (2018). arXiv:1801.05746
12. Kingma, D.P., Ba, J.: Adam: a method for stochastic optimization. arXiv preprint (2014). arXiv:1412.6980
13. Kurmann, T., et al.: Simultaneous recognition and pose estimation of instruments in minimally invasive surgery. In: Descoteaux, M., Maier-Hein, L., Franz, A., Jannin, P., Collins, D.L., Duchesne, S. (eds.) MICCAI 2017. LNCS, vol. 10434, pp. 505–513. Springer, Cham (2017). https://doi.org/10.1007/978-3-319-66185-8_57
14. Laina, I., Rieke, N., Rupprecht, C., Vizcaíno, J.P., Eslami, A., Tombari, F., Navab, N.: Concurrent segmentation and localization for tracking of surgical instruments. In: International Conference on Medical Image Computing and Computer-Assisted Intervention, pp. 664–672. Springer (2017)
15. Moccia, S., et al.: Uncertainty-aware organ classification for surgical data science applications in laparoscopy. IEEE Trans. Biomed. Eng. **65**(11), 2649–2659 (2018)
16. Mohammed, A., Yildirim, S., Farup, I., Pedersen, M., Hovde, Ø.: Streoscennet: surgical stereo robotic scene segmentation. In: Medical Imaging 2019: Image-Guided Procedures, Robotic Interventions, and Modeling, vol. 10951, p. 109510P. International Society for Optics and Photonics (2019)
17. Ni, Z.L., Bian, G.B., Xie, X.L., Hou, Z.G., Zhou, X.H., Zhou, Y.J.: Rasnet: segmentation for tracking surgical instruments in surgical videos using refined attention segmentation network. In: 2019 41st Annual International Conference of the IEEE Engineering in Medicine and Biology Society (EMBC), pp. 5735–5738. IEEE (2019)

18. Palep, J.H.: Robotic assisted minimally invasive surgery. J. Minimal Access Surg. **5**(1), 1 (2009)
19. Qin, F., Li, Y., Su, Y.H., Xu, D., Hannaford, B.: Surgical instrument segmentation for endoscopic vision with data fusion of rediction and kinematic pose. In: 2019 International Conference on Robotics and Automation (ICRA), pp. 9821–9827. IEEE (2019)
20. Rohmer, E., Singh, S.P., Freese, M.: V-rep: a versatile and scalable robot simulation framework. In: 2013 IEEE/RSJ International Conference on Intelligent Robots and Systems, pp. 1321–1326. IEEE (2013)
21. Ronneberger, O., Fischer, P., Brox, T.: U-Net: convolutional networks for biomedical image segmentation. In: Navab, N., Hornegger, J., Wells, W.M., Frangi, A.F. (eds.) MICCAI 2015. LNCS, vol. 9351, pp. 234–241. Springer, Cham (2015). https://doi.org/10.1007/978-3-319-24574-4_28
22. Su, Y.H., Huang, K., Hannaford, B.: Real-time vision-based surgical tool segmentation with robot kinematics prior. In: 2018 International Symposium on Medical Robotics (ISMR), pp. 1–6. IEEE (2018)
23. Tao, L., Zappella, L., Hager, G.D., Vidal, R.: Surgical gesture segmentation and recognition. In: Mori, K., Sakuma, I., Sato, Y., Barillot, C., Navab, N. (eds.) MICCAI 2013. LNCS, vol. 8151, pp. 339–346. Springer, Heidelberg (2013). https://doi.org/10.1007/978-3-642-40760-4_43

Perfusion Quantification from Endoscopic Videos: Learning to Read Tumor Signatures

Sergiy Zhuk[1(✉)], Jonathan P. Epperlein[1], Rahul Nair[1], Seshu Tirupathi[1], Pól Mac Aonghusa[1], Donal F. O'Shea[3], and Ronan Cahill[2]

[1] IBM Research, Damastown, Dublin D15 HN66, Ireland
{sergiy.zhuk,jpepperlein,rahul.nair,seshutir,aonghusa}@ie.ibm.com
[2] University College Dublin, Dublin, Ireland
ronan.cahill@ucd.ie
[3] Royal College of Surgeons, Dublin, Ireland
donalfoshea@rcsi.ie

Abstract. Intra-operative (this work was partially supported by Disruptive Technologies Innovation Fund, Ireland, project code DTIF2018 240 CA) identification of malignant versus benign or healthy tissue is a major challenge in fluorescence guided cancer surgery. We propose a perfusion quantification method for computer-aided interpretation of subtle differences in dynamic perfusion patterns which can be used to distinguish between normal tissue and benign or malignant tumors intra-operatively by using multispectral endoscopic videos. The method exploits the fact that vasculature arising from cancer angiogenesis gives tumors differing perfusion patterns from the surrounding normal tissues. Experimental evaluation of our method on a cohort of colorectal cancer surgery endoscopic videos suggests that it discriminates between healthy, cancerous and benign tissues with 95% accuracy.

Keywords: Perfusion quantification · Bio-physical modeling · Explainable features design · Tissue classification · Cancer

1 Introduction

Visible and near infrared (NIR) light sources are widely available in endoscopic cameras, providing high-resolution, multispectral videos of internal tissue and organs (e.g. Fig. 1 B), and enabling quantification of perfusion[1] by means of a

[1] Perfusion is the passage of fluid through the circulatory or lymphatic system to a capillary bed in tissue.

Electronic supplementary material The online version of this chapter (https://doi.org/10.1007/978-3-030-59716-0_68) contains supplementary material, which is available to authorized users.

A. L. Martel et al. (Eds.): MICCAI 2020, LNCS 12263, pp. 711–721, 2020.
https://doi.org/10.1007/978-3-030-59716-0_68

fluorescent dye, such as Indocyanine Green (ICG), which has become an important aid to decision making during fluorescence-guided surgery [2,8]: ICG is utilized for lymph node mapping and identification of solid tumors [15].

Intra-operative identification of malignant versus benign or healthy tissue is a major challenge in cancer surgery; surgeons typically use a combination of visual and tactile evidence to identify tissue. ICG has been observed to accumulate in cancers, [19]. In practice, however, consistent detection of sometimes subtle and complex differences in ICG perfusion has proven challenging. Intra-operative interpretation requires a surgeon to track spatial and temporal fluorescence intensity patterns simultaneously over several minutes, and it can be challenging to distinguish variations in relative fluorescence intensity due to confusing factors such as inflammation [7]. We hypothesized that observation of differences in structure of vasculature and perfusion patterns using ICG-enhanced fluorescence could be used to differentiate between benign, malignant, and healthy tissue, and that *perfusion patterns characterized by ICG inflow and outflow can serve as a marker to identify most of the benign and malignant tumors intra-operatively.* In this regard it seems natural to ask, whether *computer assisted interpretation* of perfusion could assist with perfusion quantification by approximating the judgement of a surgeon? In addressing this question we propose a method, as our key contribution, based on *bio-physical modeling* of in-vivo perfusion. Our model characterizes dynamic perfusion patterns by (i) estimating time-series of ICG fluorescence intensities, representing ICG inflow and outflow within a region of the tissue, from multispectral videos, and (ii) fitting parameters of a bio-physical perfusion model to the estimated time-series to create a compact signature of perfusion consisting of a vector of bio-physical features of ICG inflow and outflow, (iii) showing experimentally that the generated signature is discriminant for benign, malignant and normal tissue using traditional machine learning (ML) techniques.

We perform experimental validation on a corpus of 20 colorectal cancer multispectral endoscopic videos captured during surgical procedures. An experimental framework is implemented combining computer vision with a bio-physical model and basic ML. By estimating time-series of ICG intensities for a number of randomly selected Regions of Interest (ROIs) we fit parameters of our bio-physical model to time-series of ICG intensities and derive a signature for each ROI. The derived signatures are then used as features for standard supervised classification methods to attempt to differentiate normal ROIs from suspicious (benign or malignant tumors). Experiments show that our approach can match the intra-operative interpretation of an expert surgeon with 86% accuracy for ROI-based correctness, and 95% accuracy for patient-based correctness (i.e., compared with post-operative pathology findings on excised tissue).

Our choice of traditional ML technologies is deliberate, and intended to emphasize that high quality computer assisted interpretation in real-time can be provided even for limited-size datasets by combining basic ML tools with bio-physical perfusion models. Our approach represents a valuable step towards computer assisted automation of intra-operative interpretation of tissue.

Related Work. Image-based cancer detection methods have received significant attention with the advent of advanced fluorescence imaging systems and progress in image processing. We consider them in three broad categories (1) image classification using Deep Learning (DL), (2) observation by the surgeon of perfusion in the colon and, (3) quantitative analysis of bio-physical models based on perfusion patterns.

DL has been applied for cancer screening using images from mammography [11] and histology [17]. Several DL architectures like recurrent convolutional neural networks and ensembles of deep learners that include pre-trained networks have been proposed. However, DL methods typically require significant numbers of labelled training examples for each class to estimate model parameters. The cancer screening study [11] for example, required several tens of thousands of mammograms with known pathology for DL implementation.

Biophysical models of perfusion provide an alternative method. The premise is that differences in ICG inflow during a relatively short timeframe of minutes after injection, *wash-in* phase, and ICG outflow during the beginning of the venous outflow, *wash-out* phase, can serve as a marker for most benign and malignant tumors. Indeed, cancer angiogenesis has been recognized to lead to abnormal vasculature, characterized by higher internal pressure and a discontinuous basement membrane, in comparison with normal, healthy tissue [4,12]. Hence, ICG inflow over the *wash-in*, and ICG outflow over the *wash-out* could already disclose valuable discriminative signatures.

The approach taken here involves feature engineering based on bio-physical models combined with ML classifiers. This approach is better suited when training data is limited. Since we are primarily interested in perfusion dynamics, i.e. the evolution of fluorescence intensity in specific tissue over time, the task is one of video classification. This task cannot be readily translated to an image classification problem for several reasons, precluding the use of existing methods. *Although containing many individual images, each surgical video represents only a few data points*, as will become clear later. Attempting to augment the training set by, either tiling the frame with many ROIs and creating a data point for each ROI, or by extracting short time sequences of entire frames, results in duplication of data points instead of "new" ones.

Differences in vasculature are also used in [9] to build a hidden Markov model of perfusion from DCE-MRI images. A key difference between the approach in [9] and the multispectral NIR videos used here is the much higher sampling rate: 1/90 FPS for DCE-MRI [9] vs. 10 FPS in the present case.

The bio-physical model proposed here generalizes compartment models like the ones proposed in [3,6] by modeling ICG intensity time-series as a response of a generic second-order linear system with exponential input, which is a sum of one real and two potentially complex exponentials, to allow for oscillating behavior as it is observed in ICG time-series estimated from videos of human tissue perfusion (e.g., Fig. 1 and 2). The coefficients and exponents of these exponential terms form a set of features which we will call *3EXP*.

Perfusion quantification based on reading off manually a number of so called *time-to-peak* (TTP) features directly from the estimated NIR time-series is used in the surgical literature [5,18]. To exploit the natural synergy between surgical approaches and bio-physics we define a *tumor signature* as a combination of *3EXP* and *TTP* features. In our experiments this combination outperforms each set individually: 95% accuracy for patient-based correctness with 100% sensitivity and 92% specificity. To the best of our knowledge this is the first time a combination of a generalized compartment model and TTP-features has been applied to provide a discriminative signature of tumors in human tissue.

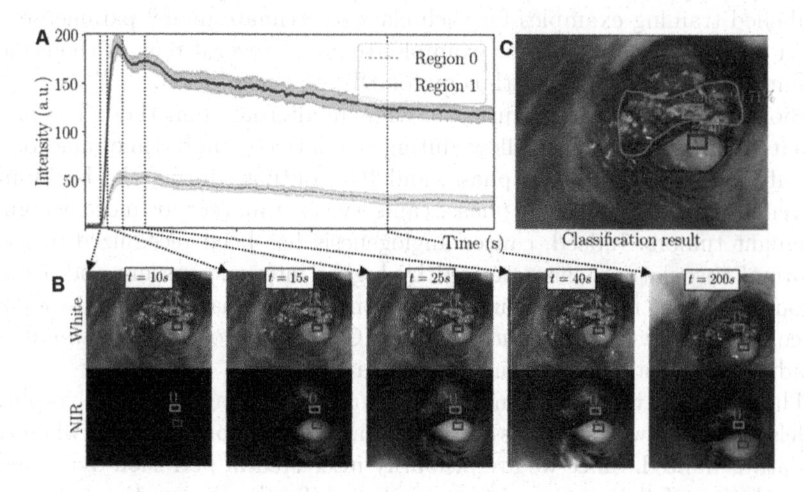

Fig. 1. Panel A: NIR-intensity time-series for two ROIs; $I_0(t)$ (ROI 0) and $I_1(t)$ (ROI 1). At time instant t the value $I_i(t)$ equals the mean of the intensity taken over ROI i, $i = 0, 1$, and the bands denote ± one standard deviation around that mean. Panel B: White – visible light video sequence, NIR – NIR light sequence. Panel C: ROI with surgical team annotation, and a classification result showing ROIs correctly classified as normal (green) and cancer (light blue). (Colour figure online)

2 Methods

Data Preprocessing: Fluorescence Intensity Estimation. The data sources here are composed of multispectral endoscopic videos (e.g. Fig. 1 B). For ICG concentrations used in the surgical procedures considered here, *fluorescence intensity at the peak wavelengths is proportional to the ICG concentration*, see e.g. [1], consequently we use the *NIR intensity* extracted from the NIR-channel video frames (see Fig. 1) as a proxy for the ICG concentration in the tissue.

The initial frame of each video has areas of suspicious and normal areas identified by a surgical team (see Fig. 1, panel C). We randomly select ROIs within those areas, e.g. ROI 0 and 1 in Fig. 1. After injection of the NIR dye, the *NIR intensity* within each of ROIs is extracted from the NIR video stream for as long as the ROI stays visible in the field of view. Data collection is straightforward only if the camera and the tissue within field of view do not move. For inter-operative surgical use, the camera is handheld and tissue contracts and expands, making acquisition very challenging. For example, ROI 0 in Fig. 1 between time $t = 1\,\mathrm{s}$ and $t = 300\,\mathrm{s}$ shows considerable drift. As this is typically the case in colorectal surgical procedures, motion stabilization techniques [16] are required to compensate for motion during collection of time-series. In what follows we assume that the data collection has already taken place.

The outcome of data preprocessing is a set of time-series estimating temporal evolution of the NIR intensity for each pixel contained in each ROI. The NIR intensity in each ROI is aggregated by taking the mean the ROI. Doing this for each time-step, we get a time-series of intensities for each ROI; let $I_{p,r}(t)$ denote the aggregated intensity of ROI r in patient p at time t. The time axis is $t \in \{0, \delta_t, 2\delta_t, \ldots, T_{p,r}\}$, where δ_t is dictated by the frame rate of the imaging equipment (here, $\delta_t = 0.1\,\mathrm{s}$) and $T_{p,r}$ denotes the time at which either ROI r was lost or data collection was terminated.

2.1 Parametric Bio-Physical Models

To create bio-physical signatures we parametrize time-series of NIR intensities $I_{p,r}(t)$ as follows:

$$y(t; \tau, D, K, \tau_i, \theta, y_{dc}) = y_{\exp}(t - \theta; \tau, D, K, \tau_i) \cdot H(t - \theta) + y_{dc} \tag{1}$$

where $y_{\exp}(t; \tau, D, K, \tau_i)$ is the response of a linear time-invariant second-order system to an exponential input (see e.g. [14]), i.e. $y_{\exp}$ solves the equation

$$\tau^2 \ddot{y}(t) + 2D\tau\dot{y}(t) + y(t) = Ke^{-t/\tau_i}, \quad y(0) = \dot{y}(0) = 0, \tag{2}$$

τ is the **time constant**, and D is the **damping**, which together govern the speed of the response and whether there are oscillations; K is known as the **gain** and responsible for the amplitude of the response, and τ_i^{-1} is the **input decay rate** which determines the rate of decay for large t. Finally, θ in (1) is a time delay, which accounts for the time it takes from the start of the video until the ICG reaches the imaged tissue, whereas y_{dc} represents the background fluorescence observed until that point (see Fig. 2, left); $H(t - \theta) = 0$ for $t \leq \theta$, and $H(t - \theta) = 1$ for $t > \theta$, hence this term ensures that $y(t; \tau, D, K, \tau_i, \theta, y_{dc}) = y_{dc}$ until fluorescence levels increase past the background level.

This parametrization includes many biological compartment models such as [3,6] as special cases: They typically model $I_{p,r}$ as a sum of two or three exponentials with **real** coefficients and exponents; as shown in (4), the response $y_{\exp}$ allows for complex values in coefficients and exponents, and hence can model oscillations observed in ICG time-series estimated from videos of human tissue perfusion (see also Fig. 2, right).

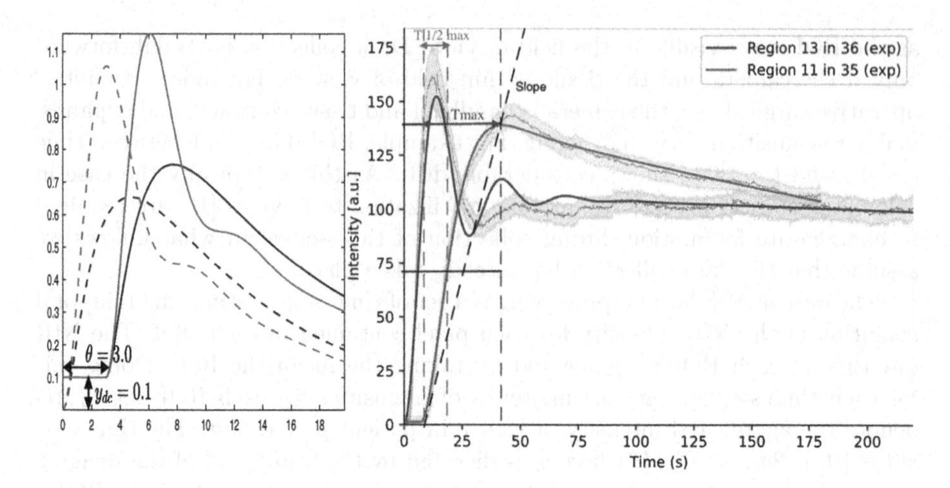

Fig. 2. Left panel: The delay θ moves the responses to the right, whereas y_{dc} moves them up. Right panel: Features $T_{\max}$, $T_{1/2\max}$, $\texttt{Slope}$, and $T_R = T_{1/2\max}/T_{\max}$ as defined in [18]. One response with oscillations (blue), corresponding to $D < 1$ in (2) and complex values for A_i and λ_i, $i = 2, 3$, in (4), and one without oscillations (orange), which corresponds to $D \geq 1$ and real A_i, λ_i. (Color figure online)

The parameters are estimated by solving the following optimization problem:

$$\begin{aligned} &\textbf{minimize } J(\tau, D, K, \tau_i, \theta, y_{dc}; I_{p,r}) = \sum_t \Big(y(t; \tau, D, K, \tau_i, \theta, y_{dc}) - I_{p,r}(t)\Big)^2 \\ &\textbf{such that } D, K, \theta, y_{dc} > 0 \quad \textbf{and} \quad 0 < \tau < \tau_i \end{aligned} \tag{3}$$

The objective in (3) is a least-squares data-misfit: minimizing J one finds parameters $\tau, D, K, \tau_i, \theta, y_{dc}$ such that (1) is as close as possible to the data $I_{p,r}$. The constraints in (3) enforce the parameters to be strictly positive. The intuition behind the constraint on τ and τ_i is that τ_i captures the slow decay during the wash-out phase, whereas τ governs the *more rapid* wash-in; a faster process has a smaller time constant.

Problem (3) can be solved by any stand-alone solver supporting box constraints: In Sect. 3, the trust-region reflective method from SciPy [10] was used. Two examples of fitted responses $y(t; \tau, D, K, \tau_i, \theta, y_{dc})$ are given in Fig. 2.

Features of Fluorescence Intensity Dynamics. For each patient and ROI, we obtain the corresponding time-series of NIR intensity, $I_{p,r}(t)$ for $t \in \{0, \dots, T_{p,r}\}$ and estimate six parameters $(\tau, D, K, \tau_i, \theta, y_{dc})$ as suggested in Sect. 2.1. Of those, only the $(\tau, D, K, \tau_i, \theta)$ are meaningful, as the offset y_{dc} depends on the background brightness and hence on the imaging equipment and the conditions under which the data is collected, but not on the tissue itself. We further derive additional features from the five basic parameters.

It is well-known (see also Suppl. material) from linear systems theory [14] that y_{exp} can also be represented as follows:

$$y_{\text{exp}}(t;\tau,D,K,\tau_i) = A_1 e^{\lambda_1 t} + A_2 e^{\lambda_2 t} + A_3 e^{\lambda_3 t}, \quad A_1 = \frac{K\tau_i^2}{\tau_i^2 + \tau^2 - 2D\tau\tau_i}$$

$$\lambda_k = -\frac{D - (-1)^k\sqrt{D^2-1}}{\tau}, \quad A_k = K\frac{D + (-1)^k\sqrt{D^2-1} - \tau/\tau_i}{2\sqrt{D^2-1}(1 - 2D\tau/\tau_i + (\tau/\tau_i)^2)}, \tag{4}$$

where $k = 2, 3$ and $\lambda_1 = -\frac{1}{\tau_i}$. Intuitively, y_{exp} can be split into slow and fast components: the slowest component is given by $A_1 e^{\lambda_1 t}$ (as per the constraint $0 < \tau < \tau_i$ enforced in (3)), it captures wash-out and the final decay of NIR-intensity and is effectively defined by τ_i (larger τ_i results in slower decay, see Fig. 6 of Suppl. material); in contrast, the wash-in phase is mainly governed by the fast components $A_2 e^{\lambda_2 t} + A_3 e^{\lambda_3 t}$, the dynamics of the second-order response (see Fig. 5 of Suppl. material).

Note that while there is a one-to-one mapping between (A_k, λ_k) and the parameters $(\tau, D, K, \tau_i, \theta, y_{dc})$ of (1), fitting (A_k, λ_k) to the data $I_{p,r}$ directly may require working numerically with complex numbers, and that the A_k have no clear physical meaning; hence it is hard to give physically meaningful constraints on λ_k and A_k. For $(\tau, D, K, \tau_i, \theta, y_{dc})$, this is straightforward, see also the discussion below (3).

The real and imaginary parts (denoted by $\Re$ and $\Im$) of A_k and λ_k, $k = 1, 2, 3$ in (4) form a set of features, which we call *3EXP*. Another popular way of quantifying perfusion in the surgical literature is based on extracting a number of so called *time-to-peak* (TTP) features $T_{\max}$, $T_{1/2\max}$, T_R and `Slope`, which are presented in Fig. 2, directly from the estimated NIR-intensity time-series $I_{p,r}$. To exploit the natural synergy between surgical approaches to perfusion quantification and bio-physics we define a *tumor signature* as a combination of *3EXP* and *TTP* features. Here we compute *TTP* features directly from the *fitted response* $y(t;\tau,D,K,\tau_i,\theta,y_{dc})$ after solving (2) and getting the parameters of (1). In summary, we obtain a bio-physical signature represented by the twelve features summarized in Table 1, all obtained from the *fitted response*.

Table 1. Signature: 3EXP+TTP features

Feature	Bio-physical meaning
$T_{\max}$	Time to reach first NIR intensity peak
$T_{\frac{1}{2}\max}$	Time to reach half the above peak value
T_R	Ratio of $T_{\frac{1}{2}\max}/T_{\max}$
`Slope`	Approximate rate of increase until NIR intensity reaches peak
$-\lambda_1$	Slowest rate of decay, i.e. the *wash-out* rate
$-\Re\lambda_2, -\Re\lambda_3$	Fastest and 2nd-fastest rates of decay; Equal if there are oscillations, see next row
$\Im\lambda_2 = -\Im\lambda_3$	If 0, then no oscillations, else defines the frequency of initial oscillations
$A_1, \Re A_2, \Re A_3, \Im A_2 = -\Im A_3$	Coefficients of the three exponentials

3 Experimental Validation: Tissue Classification

ROI Classification. The ROI classification task seeks to assign a medically relevant label to each ROI for which the signature, a vector of features described in Table 1, is available. A supervised machine learning approach is employed to perform this task as shown in Fig. 3. Two classification accuracy metrics are reported experimentally, one related to ROI-level classification and another to case-level correctness. The classifier is trained on a corpus of multispectral videos for which ROI annotations are provided by a surgical team and used as a ground truth to assess *ROI* accuracy. Additionally a single, per-video, case-level label is provided through post-operative pathology analysis. Case-level labels are used experimentally as a ground truth to assess *case* accuracy.

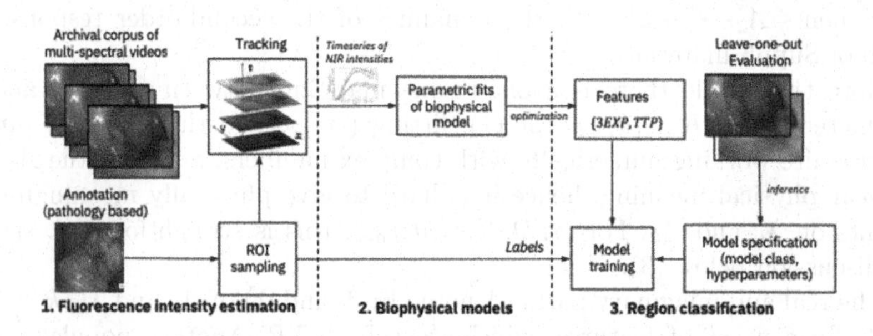

Fig. 3. Process for classifier design. Each step of the pipeline was used to assemble a labelled dataset from a corpus of multi-spectral videos. A supervised classification algorithm was trained and evaluated based on a leave-one-out test framework.

With the ROIs defined, the vector of features (see Table 1) is obtained as described in Sect. 2.1. Given the number of cases available, a Leave-One-Out (LOO) evaluation scheme was employed. There are known limitations to LOO evaluation, such as higher variance for other training data, but given our cohort size we use this approach. The model is trained with ROI and case labels for $n-1$ patients and tested on the n-th patient by selecting random ROIs from the n-th video. This process is repeated n times, hence each case video is used as the test video once.

The processing pipeline implements several practical quality criteria. First, quality checks on the input data extracted from videos are needed. Several filtering rules were implemented to ensure that the estimated parameters $\tau, D, K, \tau_i, \theta$ were meaningful. To ensure that the fits adequately represented the underlying data, a threshold of 10% was also considered on the L_1 loss.

A noisy-OR aggregation from the set of predicted ROI-level classifications is used to derive a patient-level label. Specifically, for a case with n ROIs, c of which are predicted to be cancerous, the case-level class probability is derived as $P(\text{'cancer'}) = \frac{c}{n}(1-p_{fp}) + \frac{n-c}{n}p_{fn}$ where p_{fp} and p_{fn} are the false positive and false negative rates estimated from the statistics of LOO evaluation.

Patient-level variations were observed in the data. To mitigate, a set of healthy ROIs was used as a reference to normalize the features across patients. All features were split into two classes: 'rate features' (e.g. $\lambda_1, \Re\lambda_2, \Re\lambda_3, Slope$), and 'absolute features' (e.g. $T_{\max}, T_{12\max}, T_R$). 'Rate' features were normalized as a ratio, e.g. $\lambda_1^{cancer} \mapsto \lambda_1^{cancer} / \lambda_1^{healthy}$, and 'absolutes' were subtracted from the reference value. The classifier was trained on normalized features.

Data. The proposed signature was evaluated by performing ROI classification on a dataset of 20 patients (8 with cancer) consisting of 20 multispectral endoscopic videos with annotations of suspicious and healthy ROIs (Fig. 1,C). ROI annotations were provided by a surgeon, and pathology findings per patient (normal, cancer, benign) were given from post-operative pathology analysis.
To evaluate classifier performance, several ROIs were selected at random for each patient (ranging between 11 and 40 ROIs per case) defining a total of 503 samples with labels; it was relatively balanced in terms of labels: "benign" (n = 198), "cancer" (n = 124), and "normal" (n = 181). Some ROIs were discarded as quality thresholds were not met: either the time series was too short and/or coefficient D was unrealistic (7 ROIs), or the time constant τ was too large (32 ROIs) or the L_1 loss exceeded 10% (9 ROIs). After discarding all these cases, the resulting dataset had 435 ROIs. For each of these ROIs, NIR-intensity time-series were estimated as described in Sect. 2. The length of the time-series focused on the initial *wash-in* period: time-series lasts between 40 and 100 seconds for each ROI. For each time-series the features described in Table 1 were computed by solving (3) using `curve_fit` from the `optimize` package of SciPy [10] by means of the *trust-region* reflective method (`'trf'`) and taking $\tau < 100$, $\tau_i > 150$.

Classifier. We experimented with several combinations of feature sets, filtering rules, and machine learning algorithms. The best performing feature set was based on *3EXP*-features combined with *TTP*-features (Table 1). Several standard classifiers from the Scikit-Learn package [13] were tested. The *ensemble gradient boosted tree method* was found to perform the best. This model was further refined using a grid search to fine tune the hyper-parameters. An example of a classification result is given in Fig. 1, panel C.

Results. Table 2 shows results for a set of tested classifiers. For the best performing pipeline, using the gradient boosted classifier, mean LOO accuracy score for ROI-based correctness (the fraction of ROIs correctly predicted in unseen patients) was 86.38%; case accuracy (the number of unseen patients correctly classified) was 95% (19 out of 20 cases). The best performing pipeline has a 100% cancer sensitivity and 91.67% specificity for patient-level classification. The results suggest that the signature of *3EXP- and TTP-features* is discriminant. Note that *TTP*-features offer a slightly higher performance on ROI classification accuracy compared to *3EXP*. However, *3EXP* has a 20% point improvement in case accuracy, 37.5% point improvement in sensitivity, and a 8.34% point improvement in specificity (a difference of two patients).

Table 2. Evaluation results comparing several classifiers for the same features. The last two rows show the best performing pipeline on different feature sets.

Features	Model	ROI accuracy	Case Accuracy	Sensitivity	Specificity
3EXP + TTP	Nearest Neighbors	71.49%	65.00%	50.00%	75.00%
	SVM (RBF)	76.60%	65.00%	25.00%	91.67%
	Gaussian Process	68.51%	60.00%	0.00%	**100.00%**
	Decision Tree	85.53%	90.00%	100.00%	83.33%
	Random Forest	78.30%	65.00%	50.00%	75.00%
	Naive Bayes	48.09%	50.00%	37.50%	58.33%
	QDA	54.04%	55.00%	62.50%	50.00%
	XGBoost	**86.38%**	**95.00%**	**100.00%**	91.67%
3EXP	XGBoost	76.95%	90.00%	87.50%	91.67%
TTP	XGBoost	82.55%	70.00%	50.00%	83.33%

4 Conclusions

Our experimental results suggest that the proposed method can reproduce post-operative pathology analysis and is a promising step towards automated intra-operative visual interpretation of tissue. A research priority is to scale up collection, processing and classification of videos to include more collaborators, procedures and applications. In the longer term integration with hardware platforms for robotic surgery is also a promising avenue for future research.

5 Supplementary Material

In Supplement-proofs we prove that compartment models of [3,6] are special cases of our model (2), and we show how shapes of responses, $y(t; \tau, D, K, \tau_i, \theta, y_{dc})$ given in (1), vary for different parameters $\tau, D, K, \tau_i, \theta, y_{dc}$. Screen recording Supplement-video demonstrates the classification pipeline of Fig. 3 in action.

References

1. Benson, R.C., Kues, H.A.: Fluorescence properties of indocyanine green as related to angiography. Phys. Med. Biol. **23**(1), 159–163 (1978). https://doi.org/10.1088/0031-9155/23/1/017
2. Boni, L., David, G., Dionigi, G., Rausei, S., Cassinotti, E., Fingerhut, A.: Indocyanine green-enhanced fluorescence to assess bowel perfusion during laparoscopic colorectal resection. Surg. Endosc. **30**(7), 2736–2742 (2015). https://doi.org/10.1007/s00464-015-4540-z
3. Choi, M., Choi, K., et al.: Dynamic fluorescence imaging for multiparametric measurement of tumor vasculature. J. Biomed. Optics **16**(4), 046008 (2011). https://doi.org/10.1117/1.3562956

4. De Palma, M., Biziato, D., Petrova, T.V.: Microenvironmental regulation of tumour angiogenesis. Nat. Rev. Cancer **17**(8), 457 (2017). https://doi.org/10.1038/nrc.2017.51
5. Diana, M., et al.: Enhanced-reality video fluorescence: a real-time assessment of intestinal viability. Ann. Surg. **259**(4), 700–707 (2014). https://doi.org/10.1097/SLA.0b013e31828d4ab3
6. Gurfinkel, M., et al.: Pharmacokinetics of icg and hpph-car for the detection of normal and tumor tissue using fluorescence, near-infrared reflectance imaging: a case study. Photochem. Photobiol. **72**(1), 94–102 (2000). https://doi.org/10.1562/0031-8655(2000)072
7. Holt, D., et al.: Intraoperative near-infrared imaging can distinguish cancer from normal tissue but not inflammation. PLOS ONE. **9**(7), e103342 (2014). https://doi.org/10.1371/journal.pone.0103342
8. Huh, Y.J., et al.: Efficacy of assessing intraoperative bowel perfusion with near-infrared camera in laparoscopic gastric cancer surgery. J. Laparoendosc. Adv. Surg. Tech. **29**(4), 476–483 (2019). https://doi.org/10.1089/lap.2018.0263
9. Jayender, J., et al.: Statistical learning algorithm for in situ and invasive breast carcinoma segmentation. Comput. Med. Imaging Graph **37**(4), 281–292 (2013). https://doi.org/10.1016/j.compmedimag.2013.04.003
10. Jones, E., et al.: SciPy: open source scientific tools for Python, (2001) http://www.scipy.org/
11. McKinney, S.M., et al.: International evaluation of an ai system for breast cancer screening. Nat. **577**(7788), 89–94 (2020). https://doi.org/10.1038/s41586-019-1799-6
12. Nishida, N., et al.: Angiogenesis in cancer. Vasc. Health Risk Manage. **2**(3), 213 (2006). https://doi.org/10.2147/vhrm.2006.2.3.213
13. Pedregosa, F., et al.: Scikit-learn: machine learning in Python. J. Mach. Learn. Res. **12**, 2825–2830 (2011). https://doi.org/10.5555/1953048.2078195
14. Phillips, C.L., et al.: Feedback Control Systems. Prentice Hall, 4 ed, (2000)
15. Schaafsma, B.E., et al.: The clinical use of indocyanine green as a near-infrared fluorescent contrast agent for image-guided oncologic surgery. J.Surg. Oncol. **104**(3), 323–332 (2011). https://doi.org/10.1002/jso.21943
16. Selka, F., et al.: Fluorescence-based enhanced reality for colorectal endoscopic surgery. In: Ourselin, S., Modat, M. (eds.) WBIR 2014. LNCS, vol. 8545, pp. 114–123. Springer, Cham (2014). https://doi.org/10.1007/978-3-319-08554-8_12
17. Shapcott, C.M., Rajpoot, N., Hewitt, K.: Deep learning with sampling for colon cancer histology images. Front. Bioeng. Biotech. **7**, 52 (2019). https://doi.org/10.3389/fbioe.2019.00052
18. Son, G.M., Kwon, M.S., Kim, Y., Kim, J., Kim, S.H., Lee, J.W.: Quantitative analysis of colon perfusion pattern using indocyanine green (ICG) angiography in laparoscopic colorectal surgery. Surg. Endosc. **33**(5), 1640–1649 (2018). https://doi.org/10.1007/s00464-018-6439-y
19. Veys, I., et al.: Icg-fluorescence imaging for detection of peritoneal metastases and residual tumoral scars in locally advanced ovarian cancer: a pilot study. J. Surg. Oncol. **117**(2), 228–235 (2018). https://doi.org/10.1002/jso.24807

Asynchronous in Parallel Detection and Tracking (AIPDT): Real-Time Robust Polyp Detection

Zijian Zhang[1], Hong Shang[1], Han Zheng[1], Xiaoning Wang[1], Jiajun Wang[2], Zhongqian Sun[1], Junzhou Huang[1], and Jianhua Yao[1](✉)

[1] Tencent AI Lab, Shenzhen, China
jianhuayao@tencent.com
[2] Tencent Healthcare, Shenzhen, China

Abstract. Automatic polyp detection during colonoscopy screening test is desired to reduce polyp miss rate and thus lower patients' risk of developing colorectal cancer. Previous works mainly focus on detection accuracy, however, real-time and robust polyp detection is as important to be adopted in clinical workflow. To maintain accuracy, speed and robustness for polyp detection at the same time, we propose a framework featuring two novel concepts: (1) decompose the task into detection and tracking steps to take advantage of both high resolution static images for accurate detection and the temporal information between frames for fast tracking and robustness. (2) run detector and tracker in two parallel threads asynchronously so that a heavy but accurate detector and a light tracker can efficiently work together. We also propose a robustness metric to evaluate performance in realistic clinical setting. Experiments demonstrated that our method outperformed the state-of-the-art results in terms of accuracy, robustness and speed.

Keywords: Computer-aided detection and diagnosis (CAD) · Tracking · Polyp

1 Introduction

Identifying and removing adenomatous polyps at early stage via colonoscopy screening can effectively reduce a patient's risk of developing colorectal cancer, a top leading cause of cancer-related deaths worldwide. However, it is reported that the polyp miss rate is up to 27% due to subjective colonoscope operating and physician fatigue after long duty [2]. Therefore, an automatic polyp detection system is desired to assist colonoscopists to improve polyp detection rate (PDR) for consistent and reliable screening. Besides, such a system should not slow down colonoscopy exam, which is key to get adopted in clinical workflow considering the large routine screening volume and colonoscopist's working

Z. Zhang and H. Shang—Equal contribution.

A. L. Martel et al. (Eds.): MICCAI 2020, LNCS 12263, pp. 722–731, 2020.
https://doi.org/10.1007/978-3-030-59716-0_69

habit. Therefore, a clinically viable polyp detection system should have three attributes: high accuracy (recall and precision), good robustness (stable detection on one polyp) and real-time speed. In clinical practice, undesirable alerts either as visual notice or sound alert distract the colonoscopists and slow down the workflow. Undesirable alerts occur in two scenarios: false positives on non-polyp frames (measured by precision), and unnecessary multiple alerts on one polyp due to unstable detections (measured by robustness).

Existing polyp detection solutions can be categorized into two groups, differing by whether temporal information between sequential frames is used. Most previous works belong to the first group, which detects polyps on each frame independently without using temporal information (detection-only methods). This group leverage widely-studied object detection algorithms, especially deep convolutional neural network (CNN) based algorithms, and large amount of training data, as individual images are easier to collect than video sequences [18,19]. The second group of methods utilize temporal information in colonoscopy video, introducing a tracker to refine detection results on each frame [22,23] or using 3D CNN to incorporate time as a third dimension [21]. Although detection-only methods have achieved high accuracy in both retrospective and prospective clinical studies [18], they haven't reached real-time performance and their robustness has not been studied explicitly, making them still not up to the standard for practical clinical deployment.

Aiming at superior performance on accuracy, robustness and speed at the same time, we propose a novel framework decomposing the task into detection and tracking steps, to take advantage of both high resolution static images for accurate detection and the temporal information between frames for fast tracking and robustness. This is inspired by the observation that once human detect an object, tracking the object is faster since only a small local region needs to be browsed, instead of thoroughly searching the entire field for arbitrary number of unknown objects. Most previous tracker involved methods alternate detector and tracker and run them sequentially [22,23]. We propose a novel scheme to put detector and tracker in two parallel threads and let them run at their own pace. In this way, only the tracker needs to respond to each frame in real-time, and the detector can be a large and accurate model with less speed requirement. Therefore, our method is named as Asynchronous in Parallel Detection and Tracking (AIPDT). The idea of splitting two models into two parallel threads has already been explored in computer vision, such as for visual SLAM (simultaneous localization and mapping) and tracking [7,8], however, as far as we know it has not been applied in medical applications such as detecting colorectal polyps in colonoscopy video.

2 Method

2.1 Framework

AIPDT is a general framework running a Detector and a Tracker in two parallel threads asynchronously, as shown in Fig. 1. Detector is responsible for identifying

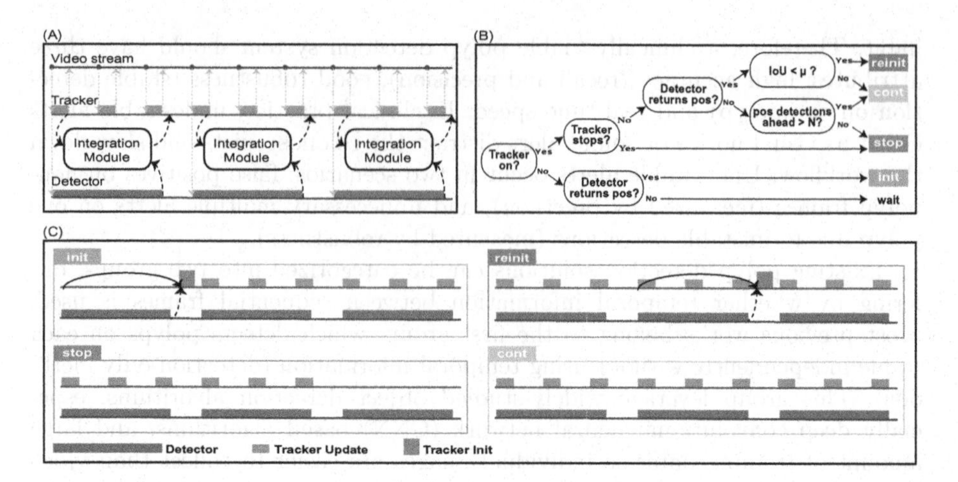

Fig. 1. Illustration of AIPDT framework. A) Overall pipeline; B) Integration module handling interaction between Detector and Tracker; C) Four states for Tracker in the integration module.

polyps with high accuracy, while Tracker ensures to return target location on each frame in real-time. Detector, as running slower than Tracker, only processes intermittent frames and provides guidance to Tracker in a lower frequency, either as initialization or for verification. Once Detector returns a result, it is compared to Tracker's result on the same frame to determine how to guide Tracker for the following frames. Note that Tracker's result is returned earlier for Detector's current frame, and thus needs to be stored for later comparison.

Integration module handles the interaction between Detector and Tracker and updates the status of Tracker as shown in Fig. 1 B and C. Tracker remains idle (wait) until Detector finds a polyp. Tracker gets initialized (init) with Detector's output bounding box and the corresponding image, which is an earlier frame considering the processing time by Detector, then Tracker updates the detection for upcoming frames. When Tracker has a large disagreement with Detector, measured by intersection over union (IoU) below a threshold μ, Tracker gets reinitialized (reinit), otherwise Tracker continues (cont). In case that Detector returns negative (no polyp) while Tracker is still tracking a polyp, a false negative possibly occurs as most of time a polyp transits smoothly. Therefore, Tracker continues (cont) until it identifies the disappearance of the target (output confidence below a threshold) and thus stops (stop), which helps to reduce jittering effect. In case Detector returns a false positive (FP), Tracker is also activated. If Detector returns at least N consecutive positive detections ahead, FP is unlikely, otherwise, Tracker is stopped if Detector returns negative (stop). Here N is a hyperparameter balancing FP and robustness. Note that reappearing polyps are considered as new ones currently, while re-detection and re-identification as in long term tracking is planned in our future work.

As a modularized framework, the building blocks of AIPDT, Detector and Tracker, can be chosen from any recent advanced algorithms, with the only requirement that Tracker is real-time (>25 frames per second (FPS)).

2.2 Detector

We chose Yolov3 as our Detector [16]. Compared to other detection algorithms, such as RCNN and its variants, Yolov3 unifies the whole detection pipeline into a single network to extract features, predict bounding boxes and classify detections concurrently on the entire image, and thus achieving a good trade-off between speed and performance.

2.3 Tracker

We compared two real-time trackers, namely, CSRDCF [15] and SiamRPN++ [12], as high performing representatives among online updating trackers and offline training trackers respectively. CSRDCF is a correlation filter based tracker with spatial/channel reliability using hand-crafted features (HoG and colour names). SiamRPN++ is built on Siamese architecture with offline training [4], thus deep models can be utilized to leverage vast labeled data which is impossible to handle with online updating. Siamese network, together with region-proposal-like bounding box regression [13,17] build the foundation of recent advance in real-time tracking algorithm [9,11]. Note CSRDCF runs on a CPU, while SiamRPN++ requires a GPU.

2.4 Implementation Details

AIPDT was implemented in Python. Yolov3 was implemented within TensorFlow [1]. For CSRDCF, we used OpenCV implementation [5]. For SiamRPN++, its original code in PyTorch was reused with an Alexnet backbone [12]. A preprocessing module was integrated before Yolov3 to filter out unqualified frames. All experiments ran on a NVIDIA Tesla M40 GPU with 24GB memory and an Intel Xeon E5-2680 v4 2.40GHz CPU.

Yolov3 was trained with momentum optimizer (momentum of 0.9), batch size of 64, weight decay of 10^{-5}, learning rate starting from 0.001 divided by 10 at steps of 120k, multiple input scales from 320 to 608, random stretching/hue/saturation/exposure for data augmentation. Its output was filtered with a threshold of 0.25 on confidence. SiamRPN++ was trained with momentum optimizer (momentum of 0.9), batch size of 128, weight decay of 0.0001, a warmup learning rate of 0.005 for first 5 epochs then logarithmically decaying from 0.01 to 0.0005, random shift/scale/blur/flip/color for data augmentation. For SiamRPN++, natural image dataset used in [12] served for both pretraining and fine tuning as additional data. AIPDT related hyperparameters were set as $N = 3$, $\mu = 0.7$.

3 Experiments and Results

3.1 Datasets

A public dataset provided at MICCAI 2018 Gastrointestinal Image ANAlysis (GIANA) challenge, referred as GIANA dataset, was used. The dataset, as part of CVC-VideoClinicDB [3], consisted of 18 video sequences and was split into training (14 videos, 9470 images, 7995/1475 of positive/negative) and testing set (4 videos, 2407 images, 2030/377 of positive/negative) following [23]. Additionally, we collected extra data from our collaborating hospital, approved by the Institutional Review Board of the hospital with exemption of ethics review ID. These private data were named as D1 and D2 with details as follow. D1 was a collection of 124726 individual images with 64932/59794 positive/negative ratio. D2 was a collection of 118 videos with 110067 positive frames. D1 augmented the training set for Detector, while D1 and D2 were used for Tracker training.

3.2 Accuracy Metrics

We used two methods to determine whether a detection is a true positive, with the first one consistent with previous works [22,23] for fair comparison in Sect. 3.3, and the second one more restrict for realistic experiments in Sect. 3.4. In the first one, a prediction is considered as a True Positive when its centroid falls inside an annotated bounding box, otherwise, as a False Positive. True Negative occurs when none detected on a normal image, and False Negative occurs when an annotated region does not contain the centroid of any detections. For the second one, centroid-based measure is replaced with the measure of IoU between the predicted and ground truth bounding box. IoU > 0.5 depicts a true positive detection, which is commonly used in object detection studies [6,14]. For both methods, accuracy metrics such as precision, recall and F_1 were evaluated.

Table 1. Accuracy comparison

Methods	Precision	Recall	F_1
Unet [23]	81.7	72.0	76.5
Unet+Opt [23]	78.1	93.7	85.2
AIPDT-Detector	98.3	70.5	82.1
AIPDT	90.6	84.5	87.5

3.3 Comparisons to State-of-the-Art

We compared our method with the state-of-the-art methods in two aspects: speed and accuracy. In terms of accuracy, [23] achieved the highest score in both polyp detection and localization task in MICCAI 2018 GIANA challenge. For a fair comparison, we used the same training and testing sets with the same accuracy metrics. As shown in Table 1, Unet / Unet+Opt represented the detector-only / detector+tracker solution in [23]. Correspondingly, AIPDT-Detector represented our detector only method. Results showed that introducing tracker improved performance for both methods, however, the degree of improvement cannot be directly compared due to a mismatch of baseline. Our method was built on a detection algorithm (Yolov3) while [23] used a segmentation algorithm (Unet). Nevertheless, both our Detector-only and Detector+Tracker solutions outperformed counterparts in [23] respectively in terms of F1.

In terms of speed, latency and FPS were measured and shown in Table 2, demonstrating that our system achieves real-time performance, much faster than previous works. Although method proposed in [19] achieved 25 fps via multi-threaded processing, its latency still exceeded real-time requirement, leading to delayed display with dragging effect.

Table 2. Speed comparison.

	AIPDT(G)[a]	AIPDT(C)[b]	[19]	[21]	[22]
Latency (ms)	13.9 ± 6.5	19.5 ± 10.8	76.8 ± 5.6	1230	154
FPS	72	51	25	0.8	6.5

[a]AIPDT with GPU-based SiamRPN++ tracker
[b]AIPDT with CPU-based CSRDCF tracker

3.4 Real-Time Evaluation

In Sect. 3.3, as well as most previous studies [19,22,23], processing time was not considered when evaluating model accuracy on colonoscopy video stream and each frame was processed independently. However, in a realistic colonoscopy exam, new frames arrive at a speed of 25 fps. If a system cannot process a frame fast enough, it has to skip some frames, causing mismatch when overlaying prediction on new frames where the target already moves. Inspired by the real-time challenge of VOT 2017 [10], we propose a new evaluation mode, named as real time mode (RT-mode), which sends images to a system at 25 fps, and takes the last reported result as output after each frame (40 ms, the elapse time for each frame), even if the system does not respond for the current frame and the output is for an earlier frame. RT-mode aligns well with the actual system setup and operator's perception, thus can better reflect the realistic performance and provide meaningful comparison among different methods.

In addition to precision, recall and F_1, we introduced a new metric, called alert times, to measure the robustness of our algorithm in RT-mode. An ideal system should generate one alert for each polyp. However, multiple alerts usually occur, especially for methods processing each frame independently without using temporal information so that one finding could be disrupted between video frames and it could be treated as a new finding next time it is detected. Such jittering effect inevitably distracts colonoscopists and slows down the workflow. However, it has not been explicitly studied yet. Therefore, alert times was proposed for quantitatively evaluating robustness by measuring number of alerts per polyp.

Detector was retrained by adding our private data D1. SiamRPN++ tracker was chosen to leverage offline training of our private data D1 and D2. Detailed comparison between CSRDCF and SiamRPN++ refers to Sect. 3.5. Comparison of AIPDT and the detector only baseline on GIANA test set was shown in Table 3. Note each run in RT-mode returns slightly different results even with identical setting, since processing time is stochastic and arbitrary frames may be skipped for each run. Therefore we repeated each evaluation 5 times and reported the mean value with standard deviation.

Compared to marginal improvement of AIPDT over baseline in terms of static accuracy (93.9 vs 94 F_1), significant improvement (88.5 vs 94.5 F_1) was observed when evaluated in RT-mode, demonstrating that AIPDT effectively improved the realistic performance by achieving high accuracy and high speed at the same time. Such gap was even enlarged (14.8% higher F_1) with a stricter metric of IoU>0.5, as it took longer time for Detector to generate outputs while the target already moved, but AIPDT enabled accurate localization with real time tracking. Alert times were also reduced by incorporating temporal information in AIPDT (2.2 vs 1.5 alert times). Two example sequences were shown in Fig. 2, where AIPDT identified the polyp on all the frames, while Detector-only suffered from missing detections especially at edge and less accurate localization due to delayed outputs.

Table 3. Real-time evaluation on GIANA test set.

Method	RT-mode	IoU > 0.5[a]	Precision	Recall	F_1	Alert #
Detector-only			97.8	90.3	93.9	NA
	✓		92.2 ± 0.1	85.1 ± 0.1	88.5 ± 0.1	2.2 ± 1.7
	✓	✓	80.6 ± 0.2	74.3 ± 0.2	77.3 ± 0.2	2.2 ± 1.7
AIPDT			95.0	93.0	94.0	NA
	✓		94.8 ± 0.1	94.3 ± 0.1	94.5 ± 0.1	1.5 ± 0.7
	✓	✓	92.3 ± 0.3	91.8 ± 0.2	92.1 ± 0.2	1.5 ± 0.7

[a] IoU > 0.5 for a correct detection, otherwise used centroid-based loose measure

3.5 Tracker Comparison

Two candidate trackers, CSRDCF and SiamRPN++, were compared in terms of intrinsic tracking performance, including both speed and the area under curve (AUC) of success plot in no-reset evaluation [20], accuracy and robustness (alert

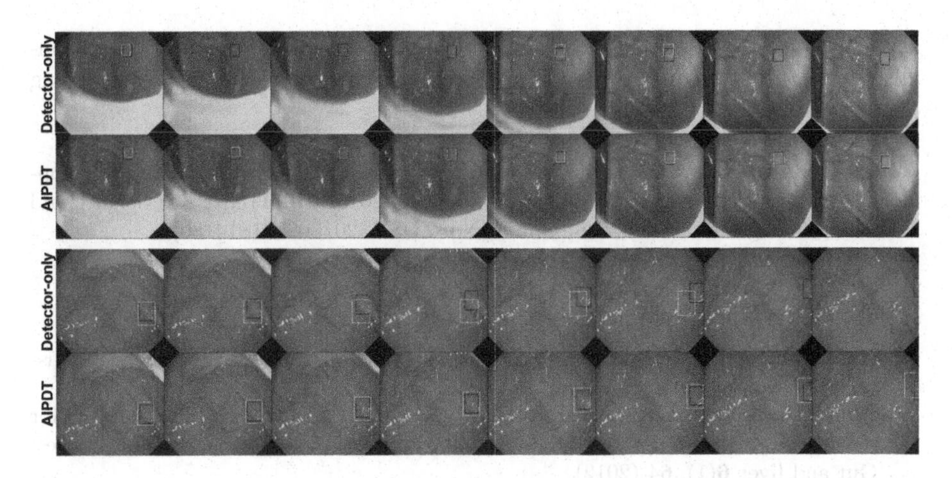

Fig. 2. Sequence examples with ground truth (black boxes) and prediction (white boxes). In the first example, Detector-only misses 5 frames and has two alerts. In the second example, Detector-only misses 2 frames. AIPDT detects every frame and has one alert for both video sequences.

times), using the GIANA test set. Results in Table 4 showed that SiamRPN++, if leveraging fine tuning with domain specific data, can outperform CSRDCF in terms of both speed and accuracy. Performance was further improved by incorporating Tracker stop mechanism, since solely relying on Detector to stop may suffer from false negative.

Table 4. Tracker Comparison. FT denotes fine-tuning with domain specific data, Stop denotes tracker stop mechanism.

Tracker	FT	Stop	FPS	AUC	Precision	Recall	F_1	Alert #
CSRDCF			50	0.323	92.1 ± 0.4	87.6 ± 0.5	89.8 ± 0.5	1.4 ± 0.6
SiamRPN++			170	0.252	67.5 ± 0.5	64.4 ± 0.4	65.9 ± 0.4	1.5 ± 0.8
SiamRPN++	✓		170	0.524	93.3 ± 0.1	88.8 ± 0.1	91.0 ± 0.1	1.5 ± 0.7
SiamRPN++	✓	✓	170	0.524	92.3 ± 0.3	91.8 ± 0.2	92.1 ± 0.2	1.5 ± 0.7

4 Conclusion

A polyp detection system was developed with AIPDT framework, which was more accurate, more robust and faster than previous works, making it more viable for clinical deployment. The improvement came from introducing tracking mechanism to utilize temporal information, and running two parallel threads asynchronously to get high accuracy and high speed simultaneously. Additionally, new evaluation metrics were proposed to address user experience during operation.

Acknowledgement. This work was founded by the Key Area Research and Development Program of Guangdong Province, China (No. 2018B010111001) and Science and Technology Program of Shenzhen, China (No. ZDSYS201802021814180).

References

1. Abadi, M., et al.: Tensorflow: large-scale machine learning on heterogeneous distributed systems. arXiv preprint arXiv:1603.04467 (2016)
2. Ahn, S.B., Han, D.S., Bae, J.H., Byun, T.J., Kim, J.P., Eun, C.S.: The miss rate for colorectal adenoma determined by quality-adjusted, back-to-back colonoscopies. Gut and liver **6**(1), 64 (2012)
3. Angermann, Q., et al.: Towards real-time polyp detection in colonoscopy videos: adapting still frame-based methodologies for video sequences analysis. In: Cardoso, M.J., et al. (eds.) CARE/CLIP -2017. LNCS, vol. 10550, pp. 29–41. Springer, Cham (2017). https://doi.org/10.1007/978-3-319-67543-5_3
4. Bertinetto, L., Valmadre, J., Henriques, J.F., Vedaldi, A., Torr, P.H.S.: Fully-convolutional siamese networks for object tracking. In: Hua, G., Jégou, H. (eds.) ECCV 2016. LNCS, vol. 9914, pp. 850–865. Springer, Cham (2016). https://doi.org/10.1007/978-3-319-48881-3_56
5. Bradski, G.: The OpenCV Library. Dr. Dobb's J. Softw. Tools. **25**, 120–125 (2000)
6. Everingham, M., Van Gool, L., Williams, C.K., Winn, J., Zisserman, A.: The pascal visual object classes (voc) challenge. Int. J. Comput. Vision **88**(2), 303–338 (2010)
7. Fan, H., Ling, H.: Parallel tracking and verifying: a framework for real-time and high accuracy visual tracking. In: Proceedings of the IEEE International Conference on Computer Vision, pp. 5486–5494 (2017)
8. Klein, G., Murray, D.: Parallel tracking and mapping for small AR workspaces. In: Proceedings of the 2007 6th IEEE and ACM International Symposium on Mixed and Augmented Reality, pp. 1–10. IEEE (2007)
9. Kristan, M., et al.: The sixth visual object tracking vot2018 challenge results. In: Proceedings of the European Conference on Computer Vision (ECCV), pp. 0–0 (2018)
10. Kristan, M., et al.: The visual object tracking vot2017 challenge results. In: Proceedings of the IEEE International Conference on Computer Vision, pp. 1949–1972 (2017)
11. Kristan, M., et al.: The seventh visual object tracking vot2019 challenge results. In: Proceedings of the IEEE International Conference on Computer Vision Workshops, pp. 0–0 (2019)
12. Li, B., Wu, W., Wang, Q., Zhang, F., Xing, J., Yan, J.: Siamrpn++: evolution of siamese visual tracking with very deep networks. In: Proceedings of the IEEE Conference on Computer Vision and Pattern Recognition, pp. 4282–4291 (2019)
13. Li, B., Yan, J., Wu, W., Zhu, Z., Hu, X.: High performance visual tracking with siamese region proposal network. In: Proceedings of the IEEE Conference on Computer Vision and Pattern Recognition, pp. 8971–8980 (2018)
14. Lin, T.Y., et al.: Microsoft COCO: common objects in context. In: Fleet, D., Pajdla, T., Schiele, B., Tuytelaars, T. (eds.) ECCV 2014. LNCS, vol. 8693, pp. 740–755. Springer, Cham (2014). https://doi.org/10.1007/978-3-319-10602-1_48

15. Lukezic, A., Vojir, T., Cehovin Zajc, L., Matas, J., Kristan, M.: Discriminative correlation filter with channel and spatial reliability. In: Proceedings of the IEEE Conference on Computer Vision and Pattern Recognition, pp. 6309–6318 (2017)
16. Redmon, J., Farhadi, A.: Yolov3: an incremental improvement. arXiv preprint arXiv:1804.02767 (2018)
17. Ren, S., He, K., Girshick, R., Sun, J.: Faster r-cnn: towards real-time object detection with region proposal networks. In: Advances in neural information processing systems, pp. 91–99 (2015)
18. Wang, P., et al.: Real-time automatic detection system increases colonoscopic polyp and adenoma detection rates: a prospective randomised controlled study. Gut. **68**(10), 1813–1819 (2019)
19. Wang, P., et al.: Development and validation of a deep-learning algorithm for the detection of polyps during colonoscopy. Nat. Biomed. Eng. **2**(10), 741 (2018)
20. Wu, Y., Lim, J., Yang, M.H.: Object tracking benchmark. IEEE Trans. Pattern Anal. Mach. Intell. **37**(9), 1834–1848 (2015)
21. Yu, L., Chen, H., Dou, Q., Qin, J., Heng, P.A.: Integrating online and offline three-dimensional deep learning for automated polyp detection in colonoscopy videos. IEEE J. Biomed. Health Inform. **21**(1), 65–75 (2016)
22. Zhang, R., Zheng, Y., Poon, C.C., Shen, D., Lau, J.Y.: Polyp detection during colonoscopy using a regression-based convolutional neural network with a tracker. Pattern Recogn. **83**, 209–219 (2018)
23. Zheng, H., Chen, H., Huang, J., Li, X., Han, X., Yao, J.: Polyp tracking in video colonoscopy using optical flow with an on-the-fly trained cnn. In: 2019 IEEE 16th International Symposium on Biomedical Imaging (ISBI 2019), pp. 79–82. IEEE (2019)

OfGAN: Realistic Rendition of Synthetic Colonoscopy Videos

Jiabo Xu[1,2](✉), Saeed Anwar[1,2], Nick Barnes[1], Florian Grimpen[3], Olivier Salvado[2], Stuart Anderson[2], and Mohammad Ali Armin[1,2]

[1] Research School of Engineering, Australian National University, Canberra, Australia
jiabo.xu@anu.edu.au
[2] Data61, CSIRO, Canberra, Australia
[3] Department of Gastroenterology and Hepatology, RBWH, Herston, Australia

Abstract. Data-driven methods usually require a large amount of labelled data for training and generalization, especially in medical imaging. Targeting the colonoscopy field, we develop the Optical Flow Generative Adversarial Network (OfGAN) to transform simulated colonoscopy videos into realistic ones while preserving annotation. The advantages of our method are three-fold: the transformed videos are visually much more realistic; the annotation, such as optical flow of the source video is preserved in the transformed video, and it is robust to noise. The model uses a cycle-consistent structure and optical flow for both spatial and temporal consistency via adversarial training. We demonstrate that the performance of our OfGAN overwhelms the baseline method in relative tasks through both qualitative and quantitative evaluation.

Keywords: Colonoscopy · Optical flow · Generative adversarial network · Domain transformation

1 Introduction

Deep learning achieves impressive performance in many machine learning tasks, such as classification [14,26], semantic segmentation [24], and object detection [13]. Those remarkable models rely on large high-quality datasets, such as ImageNet [7], Cityscape [5] and Pascal VOC [10]. However, the amount and quality of labelled data for training is often a limiting factor. In medical imaging, annotating real data is a challenging and tedious task; besides, medical data are usually subject to strict privacy rules that impose a limitation on sharing. A solution to this problem is generating synthetic data in a large quantity within

Supported by ANU and CSIRO.

Electronic supplementary material The online version of this chapter (https://doi.org/10.1007/978-3-030-59716-0_70) contains supplementary material, which is available to authorized users.

A. L. Martel et al. (Eds.): MICCAI 2020, LNCS 12263, pp. 732–741, 2020.
https://doi.org/10.1007/978-3-030-59716-0_70

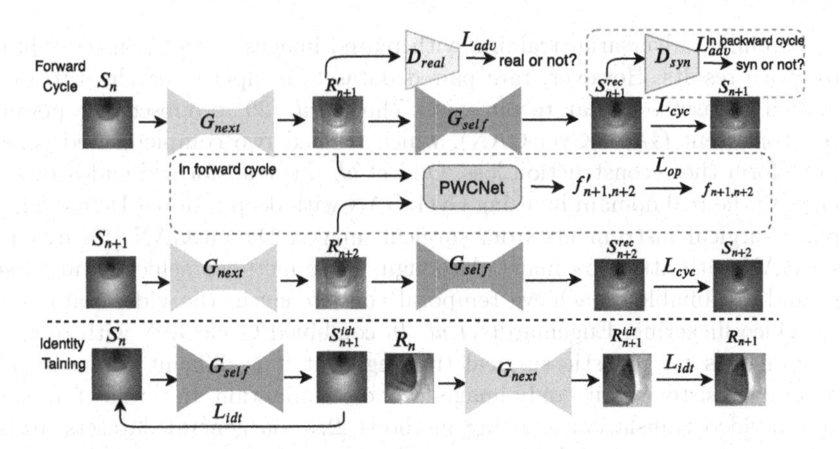

Fig. 1. Forward cycle and identity loss of OfGAN. G_{next} transforms the input to the next frame of the target domain. It is trained by the temporal-consistent loss which forces the reserved generation (G_{self}) to be same as the ground-truth synthetic frame.

controlled conditions. However, the lack of realism of synthetic data might limit the usefulness of the trained models when applied to real data. In this study, we propose a novel, ConvNet based model to increase the realism of synthetic data. We specifically work on simulated colonoscopy videos, but our approach can be expanded to other surgical assistance simulations.

Deep convolutional networks achieve remarkable performance on extracting low-dimensional features from image space [3]. Transforming one image to another requires the model to "understand" both the input and output domain in spatial domain. However, not only involving the spatial domain, our video transformation task overcomes three more challenges: (1) How to transform the input to the target domain while preserving the original annotation. (2) How to capture the temporal information between frames to form a consistent video. (3) How to synthesize near-infinite colonoscopy frames.

Generative Adversarial Networks (GANs) [12,20,23,25,28,29] fill the gap of generating and transforming high-quality images [28]. Generally, GANs consist of a generator and a discriminator. The generator is trained to generate a sample approximating the target distribution while the discriminator learns to judge the realness of the given sample. An elaborate adversarial training makes it possible to fit or transform a complex distribution.

The domain distributions play a vital role in transformation. Hence, directly transforming colonoscopy images to another domain is challenging when the distance in between is significant. Recently, Shrivastava *et al.* [25] refined synthetic small-sized grayscale images to be real-like through their S+U GAN. After that, Mahmood *et al.* [20] applied the idea of S+U GAN to remove patient-specific feature from real colonoscopy images. Both mentioned methods employ the target domain in grayscale, which dramatically reduces the training burden.

Combining adversarial training with paired images [1,15,23] usually fulfills impressive results. However, rare paired datasets compel researchers to tackle unpaired datasets, similar to our case, Zhu *et al.* [29] proposed the powerful Cycle-consistent GAN (CycleGAN), which trained two complemented generators to form the reconstruction loss. Oda *et al.* [21] transformed endoscopy CT images to the real domain by using CycleGAN with deep residual U-Net [24]. To replace surgical instruments from surgical images, DavinicGAN [18] extended CycleGAN with attention maps. Although, these methods achieve limited success and are unable to achieve temporal consistency in the video dataset. To solve video flickering, Engelhardt *et al.* [9] combined CycleGAN with temporal discriminators for realistic surgical training. But it is difficult to ensure high temporal consistency with only image-level discrimination. In terms of unpaired video-to-video translation, existing methods [2,4] on general datasets utilized similar structures as CycleGAN and novel networks for predicting future frames. However, these methods do not restrict the transformed structure to its origin; instead, they encourage novel realistic features. Our OfGAN improves CycleGAN to temporal level by forcing the generator to transform the input frame to its next real-alike frame while restricting the optical flow between two continuous output frames to be identical with their input counterparts. This setup achieves remarkable performance at pixel-level in spatial as well as temporal domain transformation.

The contributions of this exposition are:

1. **Optical Flow GAN**: Based on the standard cycle-consistent structure, we create and implement the OfGAN, which is able to transform the domain while keeping the temporal consistency of colonoscopy videos and rarely influencing the original optical flow annotation.
2. **Real-enhanced Colonoscopy Generation**: Our method can be incorporated in a colonoscopy simulator to generate near-infinite real-enhanced videos. The real generated videos possess very similar optical flow annotation with the synthetic input. Frames inside the transformed videos are consistent and smooth.
3. **Qualitative and Quantitative Evaluation**: The model is evaluated on our synthetic and a published CT colonoscopy datasets [23] both qualitatively and quantitatively. The transformation can be applied to annotation and thus create labels associated with the new realistic data.

2 Methodology

Let us consider that we are given a set of synthetic colonoscopy videos $S = \boldsymbol{s}$ and real colonoscopy videos $R = \boldsymbol{r}$, where $\boldsymbol{s} = s_1, s_2, \ldots, s_n$ and $\boldsymbol{r} = r_1, r_2, \ldots, r_m$, then s_n represents the n-th frame in the synthetic video and r_m represents the m-th frame in the real video. It should be noted that there is no real frame corresponding to any synthetic frame. Furthermore, we have ground-truth optical flow $F = \boldsymbol{f}$ for all synthetic data, where $\boldsymbol{f} = f_{1,2}, \cdots, f_{n-1,n}$ and $f_{n-1,n}$ indicates the ground-truth optical flow between frame $n-1$ and n. The goal is to learn a mapping $G : S \rightarrow R'$ where R' is a set of novel videos whose optical flow is identical to S while keeping the structure of S unchanged. To achieve this, we follow cycle-adversarial [29] training by using two generative models G_{self} and G_{next}, their corresponding discriminators D_{syn} and D_{real} as well as an optical flow estimator Op to form an optical flow cycle-consistent structure.

2.1 Temporal Consistent Loss

Different from the reconstruction loss in CycleGAN, our model tries to reconstruct the next frame of the given distribution. More specifically, forward cycle is connected by two mapping functions: $G_{next} : s_n \rightarrow r'_{n+1}$ and $G_{self} : r'_{n+1} \rightarrow s^{rec}_{n+1}$. G_{next} tries to transform a given synthetic frame s_n to be similar to a frame from the real domain, at the same time it predicts the next frame to obtain r'_{n+1}. In the reverse mapping, G_{self} transforms r'_{n+1} to s^{rec}_{n+1}. Further, our temporal consistent loss narrows the gap between s^{rec}_{n+1} and s_{n+1}. The generator G_{next} performs spatial and temporal transformation simultaneously while G_{self} only involves spatial transformation. Besides we have a backward cycle obeying the reverse mapping chain: $r_m \rightarrow G_{self} \rightarrow s'_m \rightarrow G_{next} \rightarrow r^{rec}_{m+1}$. We use ℓ_1 loss to mitigate blurring. The overall temporal consistent loss is given by:

$$\begin{aligned}\mathcal{L}_{cyc}(G_{next}, G_{self}) = & \mathrm{E}_{s\sim Pdata(S)}[||G_{self}(G_{next}(s_n)) - s_{n+1}||_1] \\ & + \mathrm{E}_{r\sim Pdata(R)}[||G_{next}(G_{self}(r_m)) - r_{m+1}||_1].\end{aligned} \quad (1)$$

2.2 Adversarial Loss

Adversarial loss [12] is utilized for both mapping functions described in the previous section. For the mapping function $G_{next} : s_n \rightarrow r'_{n+1}$ the formula of adversarial loss is:

$$\begin{aligned}\mathcal{L}_{adv}(G_{next}, D_{real}, S, R) = & \mathrm{E}_{r\sim Pdata(R)}[\log D_{real}(r)] \\ & + \mathrm{E}_{s\sim Pdata(S)}[1 - \log D_{real}(G_{next}(s))]\end{aligned} \quad (2)$$

For the reverse pair G_{self} and D_{syn}, the adversarial loss is $\mathcal{L}_{adv}(G_{self}, D_{syn}, R, S)$ where the positions of synthetic and real data are interchanged.

2.3 Perceptual Identity Loss

Nevertheless, the temporal-consistent loss itself is insufficient to force each generator to generate its targets. We use identity loss to force G_{next} to strictly generate the next frame and G_{self} to transform the current frame. Furthermore, we find measuring the distance on the perceptual level achieves better results. Finally, the formula is as follows:

$$\begin{aligned}\mathcal{L}_{idt}(G_{next}, G_{self}) = & \mathrm{E}_{r\sim Pdata(R)}[\theta(G_{next}(r_m)), \theta(r_{m+1})] \\ & + \mathrm{E}_{s\sim Pdata(S)}[(\theta(G_{self}(s_n)), \theta(s_n)]\end{aligned} \quad (3)$$

where the $\theta(\cdot)$ indicates the perceptual extractor.

2.4 Optical Flow Loss

In addition to the above operations in the unsupervised situation, the optical flow loss utilizes supervised information to preserve annotation and stabilize the training. We restrict each two continuous real-alike frames to have the same optical flow as their corresponding synthetic frames, as shown in Fig. 1. The optical flow loss is:

$$\mathcal{L}_{op}(G_{next}) = \mathrm{E}_{s\sim Pdata(S),f\sim Pdata(F)}[||Op(r'_n, r'_{n+1}) - f_{n,n+1}||_1], \tag{4}$$

where the $Op(\cdot)$ represents a non-parameteric model for optical flow estimation and $r'_n = G_{next}(s_n)$.

Therefore, the overall loss function can be presented as:

$$\begin{aligned}\mathcal{L}(G_{next}, G_{self}, D_{syn}, D_{real}) =& \mathcal{L}_{adv}(G_{next}, G_{self}) + \lambda\mathcal{L}_{cyc}(G_{next}, G_{self}) \\ &+ \beta\mathcal{L}_{idt}(G_{next}, G_{self}) + \sigma\mathcal{L}_{op}(G_{next}),\end{aligned} \tag{5}$$

where we have λ, β and σ as the importance of each term. The target is to solve the min-max problem of

$$G^*_{next}, G^*_{self} = \arg \min_{G_{next},G_{self}} \max_{D_{syn},D_{real}} \mathcal{L}(G_{next}, G_{self}, D_{syn}, D_{real}).$$

2.5 Implementation Details

To be fair with competing methods, we adopt many training parameters from CycleGAN. We use an encoder-decoder structure for the generators and PatchGAN [19] for discriminators. Both generators consist of two down-sample and two up-sample layers with six residual blocks in between. For extracting perceptual features, we use the output of the second convolution block of pre-trained VGG-13 [26] on ImageNet. Similarly, the optical flow is estimated via pre-trained PWC-Net [27]. Furthermore, to optimize the network, we employ Adam [17] optimizer with beta equal to (0.5, 0.999) and a learning rate of $2e^{-4}$. The input frames are resized to 256 × 256 while corresponding optical flow is re-scaled to the proper value. We set $\lambda = 150$, $\beta = 75$ and $\sigma = 0.1$. The framework is implemented in PyTorch [22] and trained on 4 Nvidia P100 GPUs for 100 epochs.

3 Experiments

The synthetic data we utilized is simulated by a colonoscopy simulator [6]. We extracted 8000 synthetic colonoscopy frames from five videos with ground-truth optical flow and 2741 real frames from 12 videos for training. Similarly, for testing, 2000 unknown synthetic frames are captured from two lengthy videos. The real data is captured from patients by our specialists. We perform fish-eye correction for all the real data and discard the real frames with extreme lighting conditions, wall-only images, and blurred images. Subsequently, we are left with

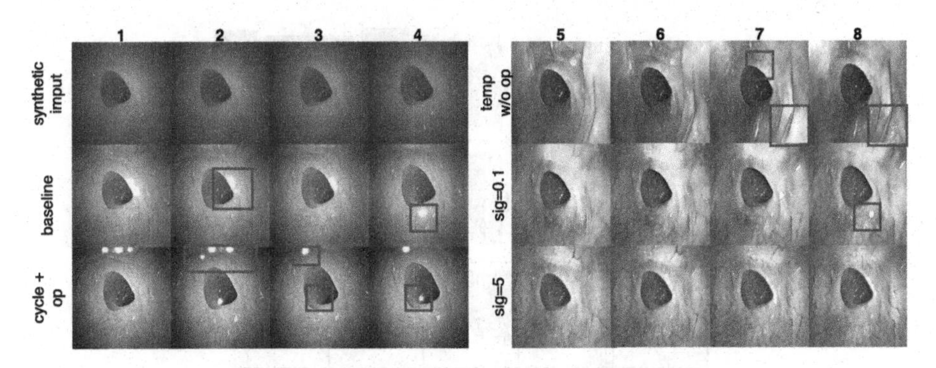

Fig. 2. Qualitative evaluation of four successive frames of each model. Each row, from top to bottom left to right, are input frames, results from the baseline, standard CycleGAN plus our optical flow loss, temporal consistent loss only, complete OfGAN with $\sigma = 0.1$ and $\sigma = 5$. Red rectangles highlight unseen features of one front frame. Differences are best viewed on zoom-in screen. (Color figure online)

1472 real images for training. Further, we also test our model on a published CT colonoscopy dataset [23] qualitatively.

We present the qualitative and the quantitative evaluation on our test results. The qualitative evaluation focuses on the single frame quality and temporal consistency in a subjective manner. For quantitative analysis, we use an auxiliary metric, Domain Temporal-Spatial score (DTS), to measure temporal and spatial quality simultaneously.

3.1 Qualitative Evaluation

The single-frame quality measures are two-fold. On the one hand, it measures if the transformed frame looks much more like real ones while, on the other hand, it evaluates if it contains less noise. For temporal consistency, we select four continuous frames and mainly concentrate on inconsistency among them. We regard the famous CycleGAN as our baseline model and furnish four models for ablation study. Results show that merely adding optical flow loss to the model does not improve rather results in worse performance on both spatial and temporal quality. The standard cycle structure does not involve in any temporal information, besides no spatial and temporal information can be learned at the same time. As a result, the black corner turns to be more obvious, and more inconsistent white spots emerge. Furthermore, only applying temporal-consistent loss (Fig. 2 *row 1, column 5–8*) intervenes in the converging of original training, which produces large scale mask-like noises. The combination of both optical flow loss and temporal-consistent loss gives much more realistic and consistent results (Fig. 2 *row 2, column 5–8*). Almost no white spots appear on any frames where the colon wall looks more bloody. A pale mask-like noise arises on the right. In terms of single frame quality (Fig. 3*b*), our method achieves better realness than the baseline method. By comparison, it is obvious that our method successfully

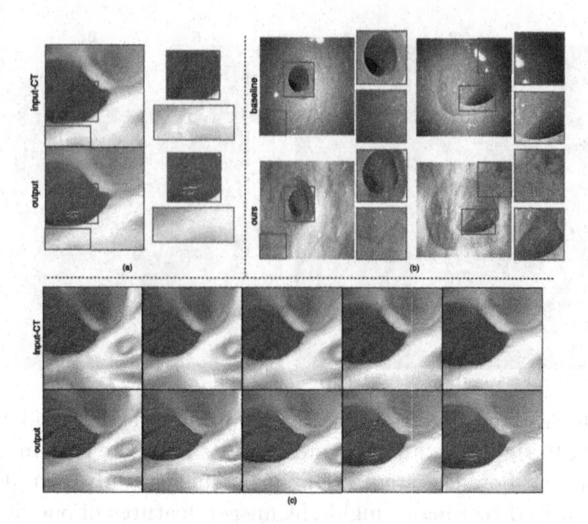

Fig. 3. (a) Qualitative evaluation of CT colonoscopy outputs, CT input (top), and our transformed results (bottom). (b) Qualitative assessment of the selected frame between two selected pairs from the baseline (top) and our method (bottom). Zoomed zones come from the detail inside the nearby red rectangles. (c) Qualitative evaluation on five continuous CT frame pairs, CT input (top), and our transformed results (bottom). (The images are best viewed on screen using zoom functionality). (Color figure online)

removes black corners and complements the detail in the deeper lumen. Besides, the white spots are rare in our results. The surface of the baseline method is so glossy that it looks far different from a human organ. On the contrary, our method has more vivid light effect.

The choice of parameter σ is a trade-off between consistency and realness. From values 0.1 to 5, results vary from the best realistic to the best consistent. Hence, we can adjust it depending on specific application scenarios.

We also test our method on CT colonoscopy videos (Fig. 3*a top row*) whose surface is coarse and texture-less compared with our synthetic data. In this case, we have no ground-truth optical flow of the input; instead, we use the estimated optical flow as the ground-truth for training. Our method successfully colors the surface to be realistic. Besides, it also removes the coarse surface and adds more realistic reflections of light inside the dark lumen. The lack of blood vessels is due to our non-blood-vessel-rich real data. Sequential frames (Fig. 3*c*) show that the innovative light reflection is consistent throughout these frames. In addition, no apparent noise nor inconsistent features appear.

3.2 Quantitative Evaluation

The quantitative evaluation should combine temporal, spatial, and domain distance measurements to overcome the trade-off problem. Hence, we utilize the DTS weighted by four normalized metrics, Average End point Error (AEPE) [8], average perceptual loss ($\mathcal{L}_{perc}$) [16], and average style loss ($\mathcal{L}_{style}$) [11] as the auxiliary metric for combining spatial and temporal dimensions. AEPE is used to measure how well two continuous outputs possess the same optical flow as the corresponding inputs, which also indicates the consistency of temporal output. We use AEPE-GT (E_{gt}) and AEPE-Pred (E_{pred}), which are AEPEs between the result and ground-truth, and estimated optical flow of the input. $\mathcal{L}_{perc}$ and $\mathcal{L}_{style}$ is for spatial quality and domain distance. The weight selection is depended on the prior of each term. The coefficients are set up empirically based on the importance of each term. To calculate the mean, we randomly select ten samples from the entire real dataset for each test data. Finally, these metrics are normalized on 36 test cases with different hyper-parameters. The smaller the DTS, the better the performance. The overall formula of DTS is:

$$DTS = \frac{3}{8}\mathcal{N}(\mathrm{E}_{gt}) + \frac{1}{8}\mathcal{N}(\mathrm{E}_{pred}) + \frac{1}{4}\mathcal{N}(\mathcal{L}_{perc}) + \frac{1}{4}\mathcal{N}(\mathcal{L}_{style}) + 0.5, \quad (6)$$

where $\mathcal{N}(\cdot)$ means normalization, adding 0.5 to make every value positive.

Table 1. Quantitative evaluation on test cases, referring to Fig. 2.

Approach	E_{gt}	E_{pred}	$\mathcal{L}_{perc}$	$\mathcal{L}_{style}$(1e−3)	DTS
Synthetic	0	0.15	3.31	11.19	1.47
Baseline	1.27	**0.24**	2.53	3.22	0.37
Cycle + op	1.53	0.81	2.43	2.82	0.40
Temp w/o op	2.85	2.35	**2.39**	**2.37**	0.77
sig=5	**1.19**	0.36	2.47	2.96	0.31
sig=0.1	1.22	0.31	2.49	2.64	**0.27**

The baseline method sacrifices the realness to achieve good consistency while only using temporal consistent loss is contrary, and both cases obtain a worse DTS (Table 1). Our method takes both advantages, even though not the best, and beats the baseline on E_{gt}, $\mathcal{L}_{perc}$, $\mathcal{L}_{style}$, and DTS. Notice that E_{gt} relies on the accuracy of the optical flow estimator, PWC-Net, as it has achieved state-of-the-art $\mathrm{E}_{gt} = 2.31$ on MPI Sintel [27]. Even though we use different dataset, we think our $\mathrm{E}_{gt} = 1.22$ (Table 1 *last row*) indicates the optical flow sufficiently identical to the ground-truth.

4 Conclusion

Our proposed OfGAN extends labeled synthetic colonoscopy video to real-alike ones. We have shown the performance of our OfGAN on our synthetic dataset and published CT datasets. The transformed dataset has outstanding temporal and spatial quality, which can be used for data augmentation, domain adaptation, and other machine learning tasks to enhance the performance. In term of the limitation, the performance of the proposed method might reduce if it fails to transform a frame correctly in a sequence. This can cause a dramatic effect on generating long videos, which needs to be dealt with in the future.

References

1. Armanious, K., et al.: Medgan: medical image translation using gans. Comput. Med. Imaging Graph. **79**, 101684 (2020)
2. Bansal, A., Ma, S., Ramanan, D., Sheikh, Y.: Recycle-gan: unsupervised video retargeting. In: Proceedings of the European conference on computer vision (ECCV), pp. 119–135 (2018)
3. Bengio, Y., Courville, A., Vincent, P.: Representation learning: a review and new perspectives. IEEE Trans. Pattern Anal. Mach. Intell. **35**(8), 1798–1828 (2013)
4. Chen, Y., Pan, Y., Yao, T., Tian, X., Mei, T.: Mocycle-gan: unpaired video-to-video translation. In: Proceedings of the 27th ACM International Conference on Multimedia, pp. 647–655 (2019)
5. Cordts, M., et al.: The cityscapes dataset for semantic urban scene understanding. In: Proceedings of the IEEE conference on computer vision and pattern recognition, pp. 3213–3223 (2016)
6. De Visser, H., et al.: Developing a next generation colonoscopy simulator. Int. J. Image Graph. **10**(02), 203–217 (2010)
7. Deng, J., Dong, W., Socher, R., Li, L.J., Li, K., Fei-Fei, L.: Imagenet: a large-scale hierarchical image database. In: 2009 IEEE conference on computer vision and pattern recognition, pp. 248–255. IEEE (2009)
8. Dosovitskiy, A., et al.: Flownet: learning optical flow with convolutional networks. In: Proceedings of the IEEE international conference on computer vision, pp. 2758–2766 (2015)
9. Engelhardt, S., De Simone, R., Full, P.M., Karck, M., Wolf, I.: Improving surgical training phantoms by hyperrealism: deep unpaired image-to-image translation from real surgeries. In: Frangi, A.F., Schnabel, J.A., Davatzikos, C., Alberola-López, C., Fichtinger, G. (eds.) MICCAI 2018. LNCS, vol. 11070, pp. 747–755. Springer, Cham (2018). https://doi.org/10.1007/978-3-030-00928-1_84
10. Everingham, M., Van Gool, L., Williams, C.K.I., Winn, J., Zisserman, A.: The pascal visual object classes (voc) challenge. Int. J. Comput. Vision **88**(2), 303–338 (2010)
11. Gatys, L.A., Ecker, A.S., Bethge, M.: Image style transfer using convolutional neural networks. In: Proceedings of the IEEE conference on computer vision and pattern recognition, pp. 2414–2423. IEEE (2016)
12. Goodfellow, I., et al.: Generative adversarial nets. In: Advances in neural information processing systems, pp. 2672–2680 (2014)
13. He, K., Gkioxari, G., Dollár, P., Girshick, R.: Mask r-cnn. In: Proceedings of the IEEE international conference on computer vision, pp. 2961–2969. IEEE (2017)
14. He, K., Zhang, X., Ren, S., Sun, J.: Deep residual learning for image recognition. In: Proceedings of the IEEE conference on computer vision and pattern recognition, pp. 770–778. IEEE (2016)
15. Isola, P., Zhu, J.Y., Zhou, T., Efros, A.A.: Image-to-image translation with conditional adversarial networks. In: Proceedings of the IEEE conference on computer vision and pattern recognition, pp. 1125–1134. IEEE (2017)
16. Johnson, J., Alahi, A., Fei-Fei, L.: Perceptual losses for real-time style transfer and super-resolution. In: Leibe, B., Matas, J., Sebe, N., Welling, M. (eds.) ECCV 2016. LNCS, vol. 9906, pp. 694–711. Springer, Cham (2016). https://doi.org/10.1007/978-3-319-46475-6_43
17. Kingma, D.P., Ba, J.: Adam: a method for stochastic optimization. arXiv preprint arXiv:1412.6980 (2014)

18. Lee, K., Jung, H.: Davincigan: unpaired surgical instrument translation for data augmentation (2018)
19. Li, C., Wand, M.: Precomputed real-time texture synthesis with markovian generative adversarial networks. In: Leibe, B., Matas, J., Sebe, N., Welling, M. (eds.) ECCV 2016. LNCS, vol. 9907, pp. 702–716. Springer, Cham (2016). https://doi.org/10.1007/978-3-319-46487-9_43
20. Mahmood, F., Chen, R., Durr, N.J.: Unsupervised reverse domain adaptation for synthetic medical images via adversarial training. IEEE Trans. Med. Imaging **37**(12), 2572–2581 (2018)
21. Oda, M., Tanaka, K., Takabatake, H., Mori, M., Natori, H., Mori, K.: Realistic endoscopic image generation method using virtual-to-real image-domain translation. Healthcare Technology Letters (2019)
22. Paszke, A., et al.: Automatic differentiation in pytorch, (2017)
23. Rau, A., et al.: Implicit domain adaptation with conditional generative adversarial networks for depth prediction in endoscopy. Int. J. Comput. Assist. Radiol. Surg. **14**(7), 1167–1176 (2019). https://doi.org/10.1007/s11548-019-01962-w
24. Ronneberger, O., Fischer, P., Brox, T.: U-Net: convolutional networks for biomedical image segmentation. In: Navab, N., Hornegger, J., Wells, W.M., Frangi, A.F. (eds.) MICCAI 2015. LNCS, vol. 9351, pp. 234–241. Springer, Cham (2015). https://doi.org/10.1007/978-3-319-24574-4_28
25. Shrivastava, A., Pfister, T., Tuzel, O., Susskind, J., Wang, W., Webb, R.: Learning from simulated and unsupervised images through adversarial training. In: Proceedings of the IEEE conference on computer vision and pattern recognition, pp. 2107–2116 (2017)
26. Simonyan, K., Zisserman, A.: Very deep convolutional networks for large-scale image recognition. arXiv preprint arXiv:1409.1556 (2014)
27. Sun, D., Yang, X., Liu, M.Y., Kautz, J.: Pwc-net: cnns for optical flow using pyramid, warping, and cost volume. In: Proceedings of the IEEE Conference on Computer Vision and Pattern Recognition, pp. 8934–8943. IEEE (2018)
28. Wang, T.C., Liu, M.Y., Zhu, J.Y., Tao, A., Kautz, J., Catanzaro, B.: High-resolution image synthesis and semantic manipulation with conditional gans. In: Proceedings of the IEEE conference on computer vision and pattern recognition, pp. 8798–8807. IEEE (2018)
29. Zhu, J.Y., Park, T., Isola, P., Efros, A.A.: Unpaired image-to-image translation using cycle-consistent adversarial networks. In: Proceedings of the IEEE international conference on computer vision, pp. 2223–2232. IEEE (2017)

Two-Stream Deep Feature Modelling for Automated Video Endoscopy Data Analysis

Harshala Gammulle(✉), Simon Denman, Sridha Sridharan, and Clinton Fookes

The Signal Processing, Artificial Intelligence and Vision Technologies (SAIVT), Queensland University of Technology, Brisbane, Australia
{pranali.gammule,s.denman,s.sridharan,c.fookes}@qut.edu.au

Abstract. Automating the analysis of imagery of the Gastrointestinal (GI) tract captured during endoscopy procedures has substantial potential benefits for patients, as it can provide diagnostic support to medical practitioners and reduce mistakes via human error. To further the development of such methods, we propose a two-stream model for endoscopic image analysis. Our model fuses two streams of deep feature inputs by mapping their inherent relations through a novel relational network model, to better model symptoms and classify the image. In contrast to handcrafted feature-based models, our proposed network is able to learn features automatically and outperforms existing state-of-the-art methods on two public datasets: KVASIR and Nerthus. Our extensive evaluations illustrate the importance of having two streams of inputs instead of a single stream and also demonstrates the merits of the proposed relational network architecture to combine those streams.

Keywords: Endoscopy image analysis · Deep networks · Relational networks.

1 Introduction

In medicine, endoscopy procedures on the Gastrointestinal (GI) tract play an important role in supporting domain experts to track down abnormalities within the GI tract of a patient. Such abnormalities may be a symptom for a life-threatening disease such as colorectal cancer. This analysis is typically carried out manually by a medical expert, and detecting critical symptoms relies solely on the experience of the practitioner, and is susceptible to human error. As such, we seek to automate the process of endoscopic video analysis, providing support to human experts during diagnosis.

Due to advancements in biomedical engineering, extensive research has been performed to support and improve the detection of anomalies via machine learning and computer vision techniques. These methods have shown great promise, and can detect abnormalities that can be easily missed by human experts

A. L. Martel et al. (Eds.): MICCAI 2020, LNCS 12263, pp. 742–751, 2020.
https://doi.org/10.1007/978-3-030-59716-0_71

[9,13,25]. Yet automated methods face multiple challenges when analysing endoscopic videos, due to overlaps between symptoms and the difficult imaging conditions.

Most previous endoscopy analysis approaches obtain a set of hand-crafted features and train models to detect abnormalities [2,16]. For example, in [16] the encoded image features are obtained through a bidirectional marginal Fisher analysis (BMFA) and classified using a support vector machine (SVM). In [18], local binary patterns (LBP) and edge histogram features are used with logistic regression. A limitation of these hand-crafted methods is that they are highly dependent on the domain knowledge of the human designer, and as such risk losing information that best describes the image. Therefore, through the advancement of deep learning approaches and due to their automatic feature learning ability, research has focused on deep learning methods. However, training these deep learning models from scratch is time consuming and requires a great amount of data. To overcome this challenge, transfer learning has been widely used; whereby a deep neural network that is trained on a different domain is adapted to the target domain through fine-tuning some or all layers. Such approaches have been widely used for anomaly detection in endoscopy videos obtained from the GI tract. The recent methods [19,20] on computer aided video endoscopy analysis predominately extract discriminative features from a pre-trained convolutional neural network (CNN), and classify them using a classifier such as a Logistic Model Tree (LMT) or SVM. In [4], a Bayesian optimisation method is used to optimise the hyper-parameters for a CNN based model for endoscopy data analysis. In [1], the authors tested multiple existing pre-trained CNN network features to better detect abnormalities.

In [14] the authors propose an architecture that consists of two feature extractors. The outputs of these are multiplied using an outer product at each location of the image and are pooled to obtain an image descriptor. This architecture models local pairwise feature interactions. The authors of [26] introduce a hierarchical bilinear pooling framework where they integrate multiple cross-layer bilinear modules to obtain information from intermediate convolution layers. In [15] several skip connections between different layers were added to detect objects in different scales and aspect ratios. In contrast, the proposed work extracts semantic features from different CNN layers and explicitly models the relationship between these through a novel relation mapping network.

In this paper, we introduce a relational reasoning approach [24] that is able to map the relationships among individual features extracted by a pre-trained deep neural network. We extract features from the mid layers of a pre-trained deep model and pass them through the relational network, which considers all possible relationships among individual features to classify an endoscopy image. Our primary evaluations are performed on the KVASIR dataset [20], containing endoscopic images and eight classes to detect. We also evaluate the proposed model on the Nerthus dataset [21] to further demonstrate the effectiveness of the proposed model. For both datasets, the proposed method outperforms the existing state-of-the-art.

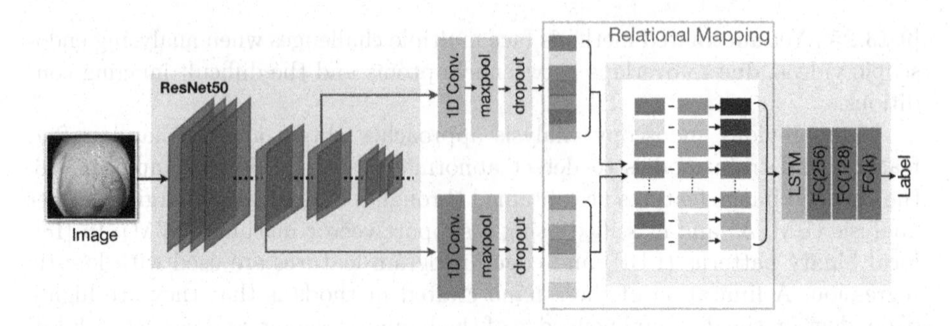

Fig. 1. Proposed Model: The semantic features of the input image are extracted through two layers of a pre-trained ResNet50 network, and the relations among the encoded feature vectors are mapped through the relational network which facilitates the final classification.

2 Method

In this paper, we propose a deep relational model that obtains deep information from two feature streams, that are combined to understand the class of the input endoscopy image. An overview of our proposed architecture is given in Fig. 1.

Training a CNN model from scratch is time consuming and requires a large dataset. Therefore, in practice it is more convenient to use a pre-trained network and adapt this to a target domain, and this has been shown to be an effective method in the computer vision [7,8] and medical domains [1,20]. To obtain the two feature streams we utilise a pre-trained ResNet50 [10] network, trained on ImageNet [23]. Training on large-scale datasets such as ImageNet [23] improves the ability of the network to capture important patterns in input images and translate them into more discriminative feature vectors, that support different computer vision tasks.

When extracting features from a pre-trained CNN model, features from earlier layers contain more local information than those from later layers; though later layers contain more semantic information [6]. Thus combining such features offers more discriminative information and facilitates our final prediction task. In this study, we combine features from an earlier layer and a later layer from the pre-trained CNN model. This allows us to capture spatial and semantic features, both of which are useful for accurate classification of endoscopy images. We avoid features from the final layers as they are over-compressed and do not contain information relating to our task, instead containing information primarily for the task the network is previously trained on (i.e. object detection). Our extracted features are further encoded through 1D convolutional and max pooling layers, and passed through a relational network to map the relationship between feature vectors, facilitating the final classification task.

2.1 Semantic Feature Extractor

The input image, X, is first passed through the Semantic Feature Extractor (SFE) module. The SFE is based on a ResNet50 pre-trained CNN, and features are extracted from two layers. We denote the respective features as,

$$\theta^1 \in \mathbb{R}^{(W^1, H^1, D^1)}, \tag{1}$$

$$\theta^2 \in \mathbb{R}^{(W^2, H^2, D^2)}, \tag{2}$$

where W^1, H^1, D^1 and W^2, H^2, D^2 denote the sizes of the respective three-dimensional vectors. We reshape these vectors to two-dimensions such that they are of shape $(L^1 = W^1 \times H^1, D^1)$ and $(L^2 = W^2 \times H^2, D^2)$.

2.2 Relational Network

The resultant two-dimensional feature vectors are passed through separate 1D convolution functions, f^{E^1} and f^{E^2}, to further encode these features from the individual streams such that,

$$\beta^1 = f^{E^1}(\theta^1), \tag{3}$$

$$\beta^2 = f^{E^2}(\theta^2). \tag{4}$$

Then through a relational network, f^{RN}, we map all possible relations among the two input feature streams. Our relational network is inspired by the model introduced in [24]. However, there exists a clear distinction between the proposed architecture and that of [24]. [24] utilises a relational network to map the relationships among the pixels in an input image. In the proposed work we illustrate that a relational network can be effectively utilised to map the correspondences among two distinct feature streams. We define the output of the relational network, γ, as,

$$\gamma = f^{RN}([\beta^1, \beta^2]), \tag{5}$$

where f^{RN} is composed of f_g and f_h which are Multi-Layer Perceptrons (MLPs), $i \in L^1$ and $j \in L^2$,

$$\gamma = f_h(\sum_i \sum_j f_g(\beta_i^1, \beta_j^2)). \tag{6}$$

The resultant vector, γ, is passed through a decoding function, f^D, which is composed of a layer of LSTM cells [11], and three fully connected layers to generate the classification of the input image,

$$y = f^D(\gamma). \tag{7}$$

3 Experiments

3.1 Datasets

We utilise two publicly available endoscopy datasets, KVASIR and Nerthus, to demonstrate the capability of our model to analyse endoscopy images and detect varying conditions within the GI tract.

The KVASIR Dataset [20] was released as part of the medical multimedia challenge presented by MediaEval [22]. It is based on images obtained from the GI tract via an endoscopy procedure. The dataset is composed of images that are annotated and verified by medical doctors, and captures 8 different classes. The classes are based on three anatomical landmarks (z-line, pylorus, cecum), three pathological findings (esophagitis, polyps, ulcerative colitis) and two other classes (dyed and lifted polyps, dyed resection margins) related to the polyp removal process. Overall, the dataset contains 8,000 endoscopic images, with 1,000 image examples per class. We utilise the standard test set released by the dataset authors, where 4,000 samples are used for model training and 4,000 for testing.

The Nerthus Dataset [21] is composed of 2,552 images from 150 colonoscopy videos. The dataset contains 4 different classes defined by the Boston Bowel Preparation Scale (BBPS) score, that ranks the cleanliness of the bowel and is an essential part of a successful colonoscopy (the endoscopy examination of the bowel). The number of examples per class lies within the range 160 to 980, and the data is annotated by medical doctors. We use the training/testing splits provided by the dataset authors.

3.2 Metrics

For the evaluations on the KVASIR dataset we utilise the metrics accuracy, precision, recall, F1-score, and matthews correlation coefficient (MCC) as suggested in [20]. The evaluations on the Nerthus dataset utilise the accuracy metric.

3.3 Implementation Details

We use a pre-trained ResNet50 [10] network and extract features from two layers: 'activation_36' and 'activation_40'. Feature shapes are $(14 \times 14 \times 1024)$ and $(14 \times 14 \times 256)$ respectively. For the encoding of each feature stream we utilise a 1D convolution layer with a kernel size of 3 and 32 filters, followed by a BatchNorm_ReLu [12] and a dropout layer, with a dropout rate of 0.25. The LSTM used has 300 hidden units and the output is further passed through three fully connected layers with the dimensionality of 256, 128 and k (number of classes) respectively. The model is trained using the RMSProp optimiser with a learning rate of 0.001 with a decay of 8×10^{-9} for 100 epochs. Implementation is completed in Keras [5] with a theano [3] backend.

3.4 Results

We use the KVASIR dataset for our primary evaluation and compare our results with recent state-of-the-art models (see Table 1). The first block of results in Table 1 are the results obtained from various methods introduced for the MediaEval Challenge [22] on the KVASIR data. In [16], a dimensionality reduction method called bidirectional marginal Fisher analysis (BMFA) which uses a Support Vector Machine (SVM) is proposed; while in [18] a method that combines 6 different features (JCD, Edge Histogram, Color Layout, AutoColor Correlogram, LBP, Haralick) and uses a logistic regressor to classify these features is presented. Aside from hand-crafted feature based methods, in [20] ResNet50 CNN features are extracted and fed to a Logistic Model Tree (LMT) classifier, and in [19], a GoogLeNet based model is employed. The authors in [2], introduce an approach where they obtained a collection of hand-crafted features (Tamura, ColorLayout, EdgeHistogram and, AutoColorCorrelogram) and deep CNN network features (VGGNet and Inception-V3 features), and train a multi-class SVM. This model records the highest performance among the previous state-of-the-art methods. However, with two streams of deep feature fusion and relation mapping, our proposed model is able to outperform [2] by 2.3% in accuracy, 5.1% in precision, 4.5% in recall, 5% in F1-score, 5.1% in MCC and 1.4% in specificity.

In [1], the authors have tested extracting features from input endoscopic images through different pre-trained networks and classifying them through a multi-class SVM. In Table 1 we show these results for ResNet50 features, MobileNet features and a combined deep feature obtained from multiple pre-trained CNN networks. In our proposed method, we also utilise features from a ResNet50 network, yet instead of naively combining features we utilise the proposed relational network to effectively attend to the feature vectors, and derive salient features for classification.

Table 1. The evaluation results on KVASIR dataset.

Method	Accuracy	Precision	Recall	F1-score	MCC	Specificity
Liu [16]	0.926	0.703	0.703	0.703	0.660	0.958
Petsch [19]	0.939	0.755	0.755	0.755	0.720	0.965
Naqvi [18]	0.942	0.767	0.774	0.767	0.736	0.966
Pogorelov [20]	0.957	0.829	0.826	0.826	0.802	0.975
Agrawal [2]	0.961	0.847	0.852	0.847	0.827	0.978
ResNet50 [1]	0.611	–	–	–	–	–
MobileNet [1]	0.817	–	–	–	–	–
Combined feat. [1]	0.838	–	–	–	–	–
Proposed	**0.984**	**0.898**	**0.897**	**0.897**	**0.878**	**0.992**

Figure 2 shows the confusion matrix for the evaluation results on the KVASIR dataset. For clarity we represent the classes as 0- 'dyed-lifted-polyps', 1- 'dyed-resection-margins', 2- 'esophagitis', 3- 'normal-cecum', 4- 'normal-pylorus', 5- 'normal-z-line', 6- 'polyps', 7- 'ulcerative-colitis'. Confusions occur primarily between the normal-z-line and esophagitis classes, and a number of classes are classified correctly for all instances.

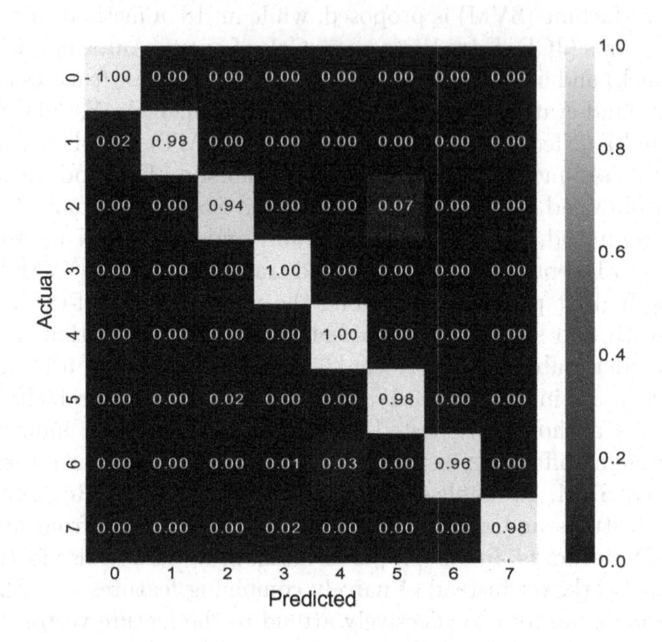

Fig. 2. Confusion matrix illustration for the KVASIR dataset.

To further illustrate the importance of our two-stream architecture and the value of the relational network for combining these feature streams, we visualise (in Fig. 3) the activations obtained from the LSTM layer of the proposed model and two ablation models, each with only one input stream. The ablation model in Fig. 3 (b) receives the feature stream θ^1 as the input, while ablation model in Fig. 3 (c) receives the feature stream θ^2 as the input. In the ablation models (b) and (c), as in [24] the relational network is used to model relationships within a single vector.

The activations are obtained for a randomly selected set of 500 images from the KVASIR test-set, and we use t-SNE [17] to plot them in two dimensions.

Considering Fig. 3 (a), we observe that samples from a particular class are tightly grouped and clear separation exists between classes. However, in the ablation models (b) and (c) we observe significant overlaps between the embeddings from different classes, indicating that the model is not capable of discriminating between those classes. These visualisations provide further evidence of the importance of utilising multiple input streams, and how they can be effectively fused

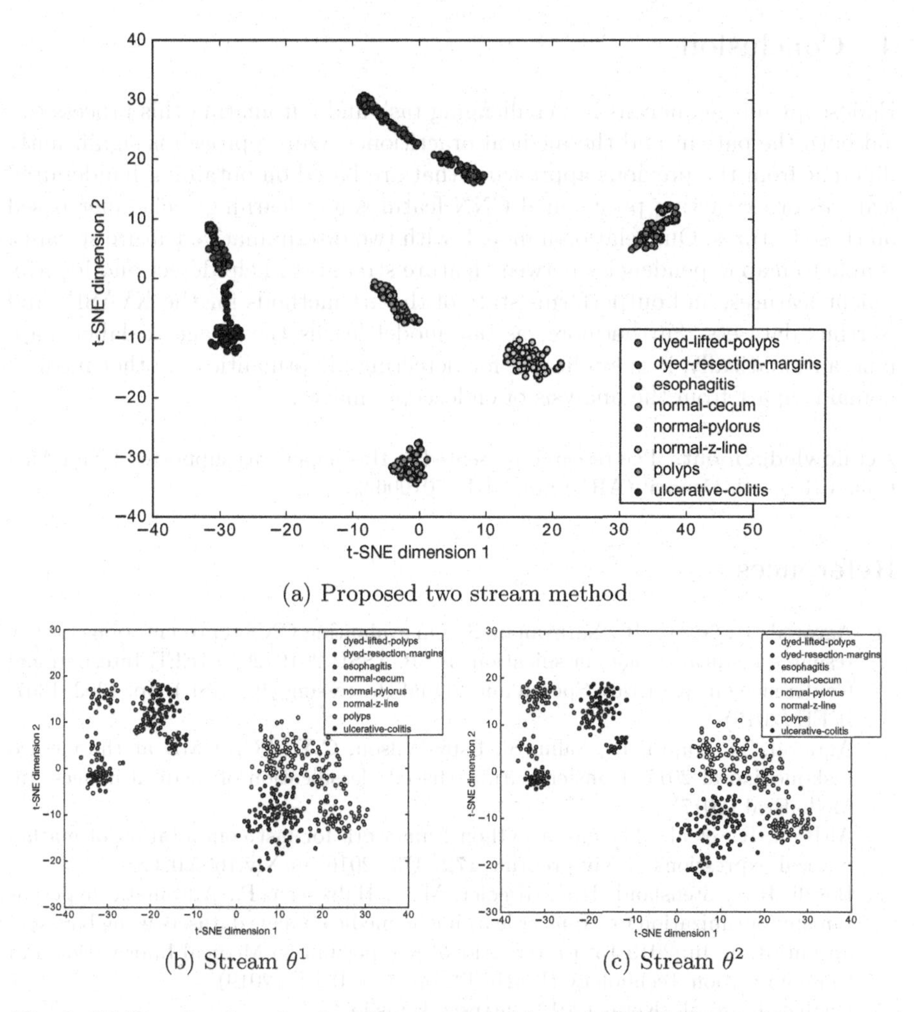

(a) Proposed two stream method

(b) Stream θ^1

(c) Stream θ^2

Fig. 3. 2D Visualisation of embeddings extracted from the LSTM layer of the proposed model (a) and two ablation models (b and c)

together with the proposed relational model to learn discriminative features to support the classification task.

To demonstrate the effectiveness of our model on different problem domains, we evaluated our model on the Nerthus dataset [21]. While the task in this dataset, measuring the cleanliness of the bowel based on the BBPS value, is less challenging compared to the abnormality classification task in the KVASIR dataset, the Nerthus dataset provides a different evaluation scenario to investigate the generalisability of the proposed approach. We obtained a 100% accuracy when predicting the BBPS value with our proposed model, while the baseline model of [21] has only achieved a 95% accuracy. This clearly illustrates the applicability of the proposed architecture for different classification tasks within the domain of automated endoscopy image analysis.

4 Conclusion

Endoscopy image analysis is a challenging task and automating this process can aid both the patient and the medical practitioner. Our approach is significantly different from the previous approaches that are based on obtaining handcrafted features or extracting pre-trained CNN features and learning a classifier based on these features. Our relational model, with two discriminative feature streams, is able to map dependencies between feature streams to help detect and identify salient features, and outperforms state-of-the-art methods for the KVASIR and Nerthus datasets. Furthermore, as our model learns the image to label mapping automatically, it is applicable for detecting abnormalities in other medical domains apart from the analysis of endoscopy images.

Acknowledgement. The research presented in this paper was supported by an Australian Research Council (ARC) grant DP170100632.

References

1. Agrawal, T., Gupta, R., Narayanan, S.: On evaluating CNN representations for low resource medical image classification. In: ICASSP 2019–2019 IEEE International Conference on Acoustics, Speech and Signal Processing (ICASSP), pp. 1363–1367. IEEE (2019)
2. Agrawal, T., Gupta, R., Sahu, S., Espy-Wilson, C.Y.: SCL-UMD at the medico task-mediaeval 2017: transfer learning based classification of medical images. In: MediaEval (2017)
3. Al-Rfou, R., et al.: Theano: a python framework for fast computation of mathematical expressions. arXiv preprint **472**, 473 (2016). arXiv:1605.02688
4. Borgli, R.J., Stensland, H.K., Riegler, M.A., Halvorsen, P.: Automatic hyperparameter optimization for transfer learning on medical image datasets using bayesian optimization. In: 2019 13th International Symposium on Medical Information and Communication Technology (ISMICT), pp. 1–6. IEEE (2019)
5. Chollet, F., et al.: Keras (2015). https://keras.io
6. Gammulle, H., Denman, S., Sridharan, S., Fookes, C.: Two stream LSTM: a deep fusion framework for human action recognition. In: 2017 IEEE Winter Conference on Applications of Computer Vision (WACV), pp. 177–186. IEEE (2017)
7. Gammulle, H., Denman, S., Sridharan, S., Fookes, C.: Forecasting future action sequences with neural memory networks. British Machine Vision Conference (BMVC) (2019)
8. Gammulle, H., Denman, S., Sridharan, S., Fookes, C.: Predicting the future: a jointly learnt model for action anticipation. In: Proceedings of the IEEE International Conference on Computer Vision, pp. 5562–5571 (2019)
9. Guo, X., Yuan, Y.: Triple ANet: adaptive abnormal-aware attention network for WCE image classification. In: Shen, D., et al. (eds.) MICCAI 2019. LNCS, vol. 11764, pp. 293–301. Springer, Cham (2019). https://doi.org/10.1007/978-3-030-32239-7_33
10. He, K., Zhang, X., Ren, S., Sun, J.: Deep residual learning for image recognition. In: Proceedings of the IEEE Conference on Computer Vision and Pattern Recognition, pp. 770–778 (2016)

11. Hochreiter, S., Schmidhuber, J.: Long short-term memory. Neural Comput. **9**(8), 1735–1780 (1997)
12. Isola, P., Zhu, J.Y., Zhou, T., Efros, A.A.: Image-to-image translation with conditional adversarial networks. In: The IEEE Conference on Computer Vision and Pattern Recognition (CVPR), July 2017
13. Kumar, N., Rajwade, A.V., Chandran, S., Awate, S.P.: Kernel generalized-gaussian mixture model for robust abnormality detection. In: Descoteaux, M., Maier-Hein, L., Franz, A., Jannin, P., Collins, D.L., Duchesne, S. (eds.) MICCAI 2017. LNCS, vol. 10435, pp. 21–29. Springer, Cham (2017). https://doi.org/10.1007/978-3-319-66179-7_3
14. Lin, T.Y., RoyChowdhury, A., Maji, S.: Bilinear CNN models for fine-grained visual recognition. In: Proceedings of the IEEE International Conference on Computer Vision, pp. 1449–1457 (2015)
15. Liu, W., et al.: SSD: single shot multibox detector. In: Leibe, B., Matas, J., Sebe, N., Welling, M. (eds.) ECCV 2016. LNCS, vol. 9905, pp. 21–37. Springer, Cham (2016). https://doi.org/10.1007/978-3-319-46448-0_2
16. Liu, Y., Gu, Z., Cheung, W.K.: HKBU at mediaeval 2017 medico: medical multimedia task (2017)
17. Maaten, L., Hinton, G., Visualizing data using t-SNE: Visualizing data using t-SNE. J. Mach. Learn. Res. **9**(Nov), 2579–2605 (2008)
18. Naqvi, S.S.A., Nadeem, S., Zaid, M., Tahir, M.A.: Ensemble of texture features for finding abnormalities in the gastro-intestinal tract. In: MediaEval (2017)
19. Petscharnig, S., Schöffmann, K., Lux, M.: An inception-like CNN architecture for GI disease and anatomical landmark classification. In: MediaEval (2017)
20. Pogorelov, K., et al.: Kvasir: a multi-class image dataset for computer aided gastrointestinal disease detection. In: Proceedings of the 8th ACM on Multimedia Systems Conference, pp. 164–169 (2017)
21. Pogorelov, K., et al.: Nerthus: a bowel preparation quality video dataset. In: Proceedings of the 8th ACM on Multimedia Systems Conference, pp. 170–174 (2017)
22. Riegler, M., et al.: Multimedia for medicine: the medico task at mediaeval **2017** (2017)
23. Russakovsky, O., et al.: Imagenet large scale visual recognition challenge. Int. J. Comput. Vis. **115**(3), 211–252 (2015). https://doi.org/10.1007/s11263-015-0816-y
24. Santoro, A., et al.: A simple neural network module for relational reasoning. In: Advances in Neural Information Processing Systems, pp. 4967–4976 (2017)
25. Wang, X., Ju, L., Zhao, X., Ge, Z.: Retinal abnormalities recognition using regional multitask learning. In: Shen, D., et al. (eds.) MICCAI 2019. LNCS, vol. 11764, pp. 30–38. Springer, Cham (2019). https://doi.org/10.1007/978-3-030-32239-7_4
26. Yu, C., Zhao, X., Zheng, Q., Zhang, P., You, X.: Hierarchical bilinear pooling for fine-grained visual recognition. In: Proceedings of the European conference on computer vision (ECCV), pp. 574–589 (2018)

Rethinking Anticipation Tasks: Uncertainty-Aware Anticipation of Sparse Surgical Instrument Usage for Context-Aware Assistance

Dominik Rivoir[1,3](✉), Sebastian Bodenstedt[1,3], Isabel Funke[1,3], Felix von Bechtolsheim[2,3], Marius Distler[2,3], Jürgen Weitz[2,3], and Stefanie Speidel[1,3]

[1] Translational Surgical Oncology, National Center for Tumor Diseases (NCT), Dresden, Germany
dominik.rivoir@nct-dresden.de

[2] Department of Visceral, Thoracic and Vascular Surgery, Faculty of Medicine and University Hospital Carl Gustav Carus, Technische Universität Dresden, Dresden, Germany

[3] Centre for Tactile Internet with Human-in-the-Loop (CeTI), TU Dresden, Dresden, Germany

Abstract. Intra-operative anticipation of instrument usage is a necessary component for context-aware assistance in surgery, e.g. for instrument preparation or semi-automation of robotic tasks. However, the sparsity of instrument occurrences in long videos poses a challenge. Current approaches are limited as they assume knowledge on the timing of future actions or require dense temporal segmentations during training and inference. We propose a novel learning task for anticipation of instrument usage in laparoscopic videos that overcomes these limitations. During training, only sparse instrument annotations are required and inference is done solely on image data. We train a probabilistic model to address the uncertainty associated with future events. Our approach outperforms several baselines and is competitive to a variant using richer annotations. We demonstrate the model's ability to quantify task-relevant uncertainties. To the best of our knowledge, we are the first to propose a method for anticipating instruments in surgery.

Keywords: Anticipation · Uncertainty · Bayesian Deep Learning · Surgical instruments · Surgical tools · Surgical workflow analysis

---

Funded by the German Research Foundation (DFG, Deutsche Forschungsgemeinschaft) as part of Germany's Excellence Strategy - EXC 2050/1 - Project ID 390696704 - Cluster of Excellence "Centre for Tactile Internet with Human-in-the-Loop" (CeTI) of Technische Universität Dresden.

Electronic supplementary material The online version of this chapter (https://doi.org/10.1007/978-3-030-59716-0_72) contains supplementary material, which is available to authorized users.

A. L. Martel et al. (Eds.): MICCAI 2020, LNCS 12263, pp. 752–762, 2020.
https://doi.org/10.1007/978-3-030-59716-0_72

1 Introduction

Anticipating the usage of surgical instruments before they appear is a highly useful task for various applications in computer-assisted surgery. It represents a large step towards understanding surgical workflow to provide context-aware assistance. For instance, it enables more efficient instrument preparation [21]. For semi-autonomous robot assistance, instrument anticipation can facilitate the identification of events that trigger the usage of certain tools and can eventually help a robotic system to decide when to intervene. Anticipating the use of certain instruments such as the irrigator can further enable early detection and anticipation of complications like bleedings.

The proposed applications require continuous anticipation estimates in long surgeries. Many instruments occur only rarely and briefly, i.e. only sparse annotations are available. Nevertheless, a useful anticipation framework should only react to these sparse occurrences and remain idle otherwise. Our approach addresses these requirements and is thus applicable to real-world scenarios. We train a neural network to regress the remaining time until the occurrence of a specific instrument within a given future horizon. Our uncertainty-quantification framework addresses the uncertainty associated with future events. This enables the identification of trigger events for instruments by measuring decreases in uncertainty associated with anticipation estimates (Fig. 1).

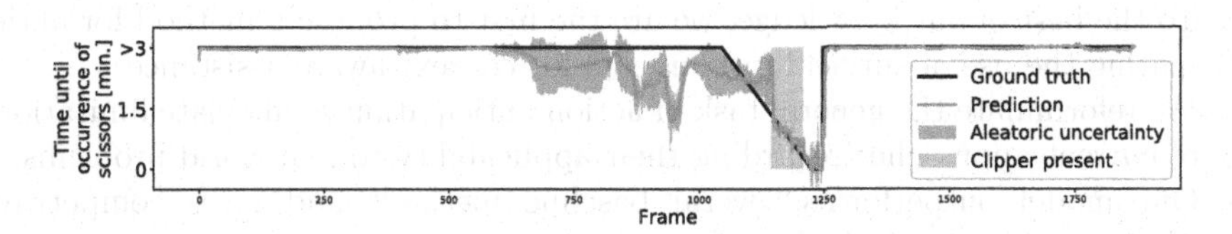

Fig. 1. Predicted remaining time until occurrence of the scissors within a horizon of three minutes on an example surgery. In cholecystectomies, the clipper is an indicator for future use of the scissors. Error and uncertainty decrease when the clipper appears.

Various works have investigated short-horizon anticipation of human actions [4,9,14,26]. However, these methods are only designed to predict actions in the immediate future based on a single frame [4,26] or short sequences of frames [9]. Most importantly, they are trained and evaluated on artificially constructed scenarios where a new action is assumed to occur within typically one second and only its correct category is unknown. In surgery however, the challenge rather lies in correctly timing predictions given the video stream of a whole surgical procedure. Our task definition considers long video sequences with sparse instrument usage and encourages models to only react to relevant cues. Jain et al. [14] address this by adding the default activity *'drive straight'* for anticipation of driver activities. For our task, we propose a similar category when no instrument is used within a defined horizon.

While methods for long-horizon anticipation exist, they require dense, rich action segmentations of the observed sequence during inference and mostly do not use visual features for anticipation [1,2,5,15,22]. This is not applicable to our task, since the information required for anticipating the usage of some instruments relies heavily on visual information. For instance, the usage of the irrigator is often triggered by bleedings and cannot be predicted solely from instrument signals. Some methods utilize visual features but nevertheless require dense action labels [19,28]. However, these labels are tedious to define and annotate and therefore not applicable to many real-world scenarios, especially surgical applications. In contrast, we propose to predict the remaining time until the occurrence of sparse events rather than dense action segmentations. During training, only sparse instrument annotations are required and inference is done solely on image data. This aids data annotation since instrument occurrence does not require complex definitions or expert knowledge [20].

The uncertainty associated with future events has been addressed by some approaches [1,26]. Similar to Farha et al. [1], we do so by learning a probabilistic prediction model. Bayesian Deep Learning through Monte-Carlo Dropout provides a framework for estimating uncertainties in deep neural networks [6]. Kendall et al. [16] identify the model and data as two relevant sources of uncertainty in machine learning. Several approaches have been proposed for estimating these quantities [8,18,23,27]. We evaluate our model's ability to quantify task-relevant uncertainties using these insights. The contributions of this work can be summarized as follows:

1. To the best of our knowledge, we are the first to propose a method for anticipating the use of surgical instruments for context-aware assistance.
2. We reformulate the general task of action anticipation to alleviate limitations of current approaches regarding their applicability to real-world problems.
3. Our model outperforms several baseline methods and gives competitive results by learning only from sparse annotations.
4. We evaluate our model's ability to quantify uncertainties relevant to the task and show that we can improve performance by filtering uncertain predictions.
5. We demonstrate the potential of our model to identify trigger events for surgical instruments through uncertainty quantification.

2 Methods

2.1 Anticipation Task.

Regression: We define surgical instrument anticipation as a regression task of predicting the remaining time until the occurrence of one of K surgical instruments within a future horizon of h minutes. Given frame x from a set of recorded surgeries and an instrument τ, the ground truth for the regression task is defined as

$$r_h(x, \tau) = \min\{t_x(\tau), h\}, \quad (1)$$

where $t_x(\tau)$ is the true remaining time in minutes until the occurrence of τ with $t_{x'}(\tau) = 0$ for frame x' where τ is present. The target value is truncated at h

minutes since we assume that instruments cannot be anticipated accurately for arbitrarily long intervals. This design choice encourages the network only to react when the usage of an instrument in the foreseeable future is likely and predict a constant otherwise. Opposed to current definitions for anticipation tasks, we do not assume an imminent action or rely on dense action segmentations.

Classification: For regularization, we add a similar classification objective to predict one of three categories $c_h(x, \tau) \in \{anticipating, present, background\}$, which correspond to an instrument appearing within the next h minutes, being present and a background category when neither is the case. In Sect. 3.2, we discuss the benefits of this regularization task for uncertainty quantification.

2.2 Bayesian Deep Learning

Due to the inherent ambiguity of future events, anticipation tasks are challenging and benefit from estimating uncertainty scores alongside model predictions. Bayesian neural networks enable uncertainty quantification by estimating likelihoods for predictions rather than point estimates [16]. Given data X with labels Y, Bayesian neural networks place a prior distribution $p(\omega)$ over parameters ω which results in the posterior $p(\omega|Y, X) = p(Y|X, \omega)p(\omega)/\int p(Y|X, \omega)p(\omega)d\omega$. Since the integration over the parameter space makes learning $p(\omega|Y, X)$ intractable, variational inference [10] is often used to approximate $p(\omega|Y, X)$ by minimizing the Kullback-Leibler divergence $KL(q_\theta(\omega)||p(\omega|Y, X))$ to a tractable distribution $q_\theta(\omega)$. Gal et al. [6] have shown that this is equivalent to training a network with dropout if $q_\theta(\omega)$ is in the family of binomial dropout distributions.

During inference, we draw $T = 10$ parameter samples $\omega_t \sim q_\theta(\omega)$ to approximate the predictive expectation $E_{p(y|x,Y,X)}(y^r)$ for regression variables y^r and the predictive posterior $p(y^c|x, Y, X)$ for classification variables y^c (Eq. 2 & 3), where $f_{\omega_t}(x)$ and $p_{\omega_t}(x)$ are the network's regression and softmax outputs parametrized by ω_t [16].

$$E_{p(y|x,Y,X)}(y^r) \approx \int f_\omega(x) q_\theta(\omega) d\omega \approx \frac{1}{T} \sum_{t=1}^{T} f_{\omega_t}(x) =: \hat{f}_\theta(x) \tag{2}$$

$$p(y^c|x, Y, X) \approx \int p_\omega(x) q_\theta(\omega) d\omega \approx \frac{1}{T} \sum_{t=1}^{T} p_{\omega_t}(x) =: \hat{p}_\theta(x) \tag{3}$$

We estimate uncertainties through the predictive variance $Var_{p(y|x,Y,X)}$. Kendall et al. [16] argue that predictive uncertainty can be divided into aleatoric (data) uncertainty, which originates from missing information in the data, and epistemic (model) uncertainty, which is caused by the model's lack of knowledge about the data. Intuitively, uncertainty regarding future events mainly corresponds to aleatoric uncertainty but likely induces model variation (epistemic) as well.

For regression, we follow Kendall et al.'s approach to estimate epistemic uncertainty (Eq. 4). Computing the predictive variance over parameter variation captures noise in the model. We omit aleatoric uncertainty for regression, since it was not effective for our task. For classification variables, we follow Kwon et al. [18] (Eq. 5). The epistemic term captures noise in the model by estimating variance over parameter samples, while the variance over the multinomial softmax distribution in the aleatoric term captures inherent sample-independent noise. In the classification case, uncertainties are averaged over all classes.

$$Var_{p(y|x,Y,X)}(y^r) \approx \underbrace{\frac{1}{T}\sum_{t=1}^{T}(f_{\omega_t}(x) - \hat{f}_\theta(x))^2}_{\sigma^2_{epistemic}} + \underbrace{\frac{1}{T}\sum_{t=1}^{T}\sigma^2_{\omega_t}(x)}_{(\sigma^2_{aleatoric})} \tag{4}$$

$$Var_{p(y|x,Y,X)}(y^c) \approx \underbrace{\frac{1}{T}\sum_{t=1}^{T}(p_{\omega_t}(x) - \hat{p}_\theta(x))^2}_{\sigma^2_{epistemic}} + \underbrace{\frac{1}{T}\sum_{t=1}^{T}p_{\omega_t}(x)(1 - p_{\omega_t}(x))}_{\sigma^2_{aleatoric}} \tag{5}$$

2.3 Model, Data and Training

The model (suppl. material) consists of a Bayesian AlexNet-style Convolutional Network [17] and a Bayesian Long Short-Term Memory network (LSTM) [12]. We sample dropout masks with a dropout rate of 20% once per video and per sample and reuse the same masks at each time step as proposed by Gal et al. [7] for recurrent architectures. The AlexNet backbone has proven effective in a similar setting [3] and empirically gave the best results. State-of-the-art architectures such as ResNet [11] performed poorly as they appeared to learn from future frames through batch statistics in batch-normalization layers [13]. Further, the AlexNet can be trained from scratch, which is beneficial for introducing dropout layers. The code is published on https://www.gitlab.com/nct_tso_public/ins_ant.

We train on the Cholec80 dataset [24] of 80 recorded cholecystectomies and anticipate $K = 5$ sparsely used instruments which are associated with specific tasks in the surgical workflow, i.e. bipolar (appears in 4.8% of frames), scissors (1.8%), clipper (3.2%), irrigator (5.3%), and specimen bag (6.2%). Grasper and hook are dropped as they are used almost constantly during procedures and hence, are not of interest for anticipation. We extract frames at 1 fps, resize to width and height of 384×216, process batches of 128 sequential frames and accumulate gradients over three batches. We use 60 videos for training and 20 for testing. We train for 100 epochs using the Adam optimizer (learning rate 10^{-4}). The loss (Eq. 6) is composed of smooth L1 [25] for the primary regression task, cross entropy (CE) for the regularizing classification task and L2-regularization, where θ are parameter estimates of the approximate distribution $q_\theta(\omega)$. We set $\lambda = 10^{-2}$ and $\gamma = 10^{-5}$.

$$L_h(x,\tau) = \sum_{\tau}(\text{Smooth}L1(\hat{f}^{\tau}_{\theta}(x), r_h(x,\tau)) + \lambda \cdot CE(\hat{p}^{\tau}_{\theta}(x), c_h(x,\tau))) + \gamma \cdot \|\theta\|^2_2 \tag{6}$$

Table 1. Comparison to baselines. Results for *OracleHist* are parenthesized since it is an offline approach. We report the mean over instrument types in minutes per metric.

	$h = 2\,\mathrm{min}$		$h = 3\,\mathrm{min}$		$h = 5\,\mathrm{min}$		$h = 7\,\mathrm{min}$	
	wMAE	pMAE	wMAE	pMAE	wMAE	pMAE	wMAE	pMAE
MeanHist	0.56	0.93	0.85	1.34	1.41	2.14	1.97	2.79
OracleHist (offline)	(0.49)	(0.83)	(0.71)	(1.18)	(1.11)	(1.73)	**(1.48)**	(2.23)
Ours non-Bayes.	0.44	0.67	**0.64**	0.93	1.13	1.64	1.58	2.21
Ours+Phase	**0.42**	**0.64**	0.67	0.94	**1.07**	**1.49**	1.61	**2.16**
Ours	0.44	0.65	**0.64**	**0.92**	1.09	1.53	1.58	**2.16**

Table 2. Instrument-wise error in minutes for a horizon of 3 min.

	Bipolar		Scissors		Clipper		Irrigator		Specimen Bag	
	wMAE	pMAE	wMAE	pMAE	wMAE	pMAE	wMAE	pMAE	wMAE	pMAE
MeanH.	0.85	1.36	0.81	1.42	0.80	1.29	0.89	1.40	0.89	1.24
OracleH.	(0.79)	(1.17)	(0.76)	(1.32)	(0.80)	(1.39)	(0.73)	(1.11)	**(0.50)**	(0.95)
O. non-B.	**0.75**	**0.92**	**0.49**	0.78	0.74	1.01	0.71	1.05	0.54	0.92
O.+Ph.	0.79	0.96	0.53	0.82	0.77	1.01	0.72	1.05	0.56	**0.88**
Ours	0.76	0.96	0.51	**0.76**	**0.71**	**0.90**	**0.70**	**1.03**	0.55	0.93

3 Evaluation

3.1 Anticipation Results

Evaluation Metrics: We evaluate frame-wise based on a weighted mean absolute error (*wMAE*). We average the MAE of 'anticipating' frames ($0 < r_h(x, \tau) < h$) and 'background' frames ($r_h(x, \tau) = h$) to compensate for the imbalance in the data. As instruments are not always predictable, a low recall does not necessarily indicate poor performance, making precision metrics popular for anticipation [9,26]. We capture the idea of precision in the *pMAE* as the MAE of predictions with $0.1h < \hat{f}_\theta^\tau(x) < 0.9h$ when the model is anticipating τ. Factors are chosen for robustness against variations during 'background' predictions.

Baselines: Since our task is not comparable to current anticipation methods, we compare to two histogram-based baselines. For instrument τ, the ith bin accumulates the occurrences of τ within the ith segments of all training video. If the bin count exceeds a learned threshold, we assume that τ occurs regularly in the ith video segment and generate anticipation values according to Eq. 1. Using 1000 bins, the thresholds are optimized to achieve best training performance w.r.t our main metric wMAE. For *MeanHist*, segments are expanded to the mean video duration. For *OracleHist*, we expand the segments to the real video duration at train and test time. This is a strong baseline as instrument usage correlates strongly with the progress of surgery, which is not known beforehand. See Fig. 4 for a visual overview of the baseline construction.

Additionally, we compare our model to a variant simultaneously trained on dense surgical phase segmentations [24] to investigate whether our model achieves competitive results using only sparse instrument labels. Surgical phases strongly correlate with instrument usage [24] and have shown to be beneficial for the related tasks of predicting the remaining surgery duration [25]. Finally, we compare to a non-Bayesian variant of our model without dropout to show that the Bayesian setting does not lead to a decline in performance.

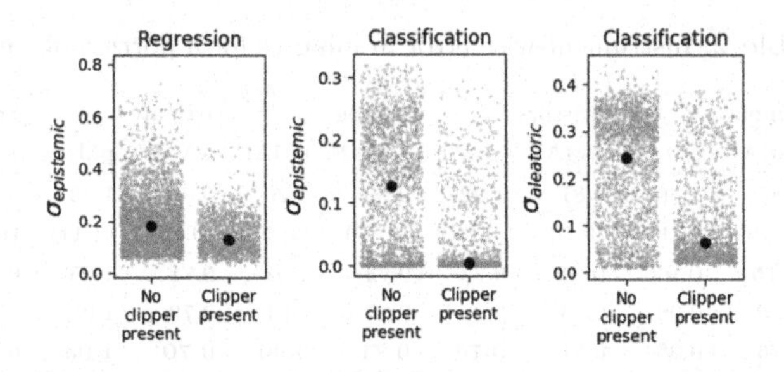

Fig. 2. Frame-wise (orange) and median (black) uncertainties for anticipating predictions of scissors depending on the clipper's presence. Since the clipper often proceeds scissors we expect lower uncertainty when the clipper is present. (Color figure online)

Discussion: We train models on horizons of 2, 3, 5 and 7 min and repeat runs for *Ours* and *Ours non-Bayes.* four times and *Ours+Phase* twice. In all settings, our methods outperform *MeanHist* by a large margin (Table 1). Compared to the offline baseline *OracleHist*, we achieve lower pMAE and comparable wMAE errors, even though knowledge about the duration of a procedure provides strong information regarding the occurrence of instruments. Further, there is no visible difference in performance with and without surgical phase regularization. This suggests that our approach performs well by learning only from sparse instrument labels. The Bayesian setting also does not lead to a drop in performance while adding the advantage of uncertainty estimation. For $h = 3$, we outperform *MeanHist* for all instruments and *OracleHist* for all except the specimen bag (Table 2). However, this instrument is easy to anticipate when the procedure duration is known since it is always used toward the end. For instrument-wise errors of other horizons, see the supplementary material.

3.2 Uncertainty Quantification Results

We analyze the model's ability to quantify uncertainties using a model for $h = 3$. For all experiments, we consider predictions which indicate that the model is anticipating an instrument. We define *anticipating predictions* as $\hat{f}_\theta^\tau(x)$ where $0.1 \cdot h < \hat{f}_\theta^\tau(x) < 0.9 \cdot h$, and $\hat{p}_\theta^\tau(x)$ with $\text{argmax}_j \hat{p}_\theta^\tau(x) = \textit{anticipating}$.

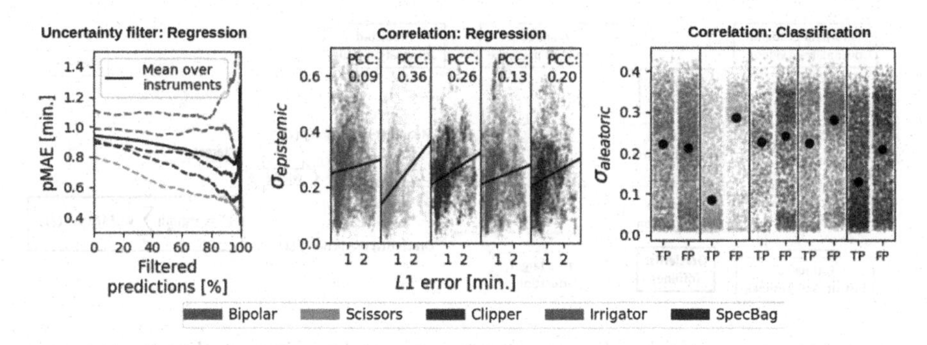

Fig. 3. *Left:* pMAE as a plot of percentiles of filtered predictions w.r.t. $\sigma_{epistemic}$. *Center:* Frame-wise error-uncertainty plot per instrument with Pearson Correlation Coefficient (PCC) and linear fit (black). *Right:* Frame-wise (colored) and median (black) uncertainty of true positives (TP) vs. false positives (FP) for class *anticipating*. (Color figure online)

Uncertainty quantification enables identification of events which trigger the usage of instruments. We evaluate the model using the known event-trigger relationship of scissors and clipper in cholecystectomies, where the cystic duct is first clipped and subsequently cut. Hence, we expect lower uncertainty for anticipating scissors when a clipper is visible. Figure 2 supports this hypothesis for epistemic uncertainty during regression. However, the difference is marginal and most likely not sufficient for identifying trigger events. Even though clipper occurrence makes usage of the scissors foreseeable, predicting the exact time of occurrence is challenging and contains noise. Uncertainties for classification are more discriminative. The classification objective eliminates the need for exact timing and enables high-certainty class predictions. Both epistemic and aleatoric estimates are promising. This is consistent with our hypothesis that uncertainty regarding future events is contained in aleatoric uncertainty but induces epistemic uncertainty as well.

We assess the quality of uncertainty estimates through correlations with erroneous predictions as high uncertainty should result in higher errors. For regression (Fig. 3, center), we observe the highest Pearson Correlation Coefficients (PCC) for scissors, clipper and specimen bag. These instruments are presumably the most predictable since they empirically yield the best results (Table 2) and are correlated with specific surgical phases ('Clipping & Cutting' and 'Gallbladder Packaging') [24]. Irrigator and Bipolar yield less reliable predictions as they are used more dynamically. For classification (Fig. 3, right), the median aleatoric uncertainty of true positive predictions for the 'anticipating' class is almost consistently lower than for false positives. Scissors and specimen bag show the largest margin. We can reduce precision errors (pMAE) by filtering uncertain predictions, shown in Fig. 3 (left) for epistemic regression uncertainty. As expected, the decrease is steeper for instruments with higher PCC. See the supplementary material for corresponding plots with other horizons.

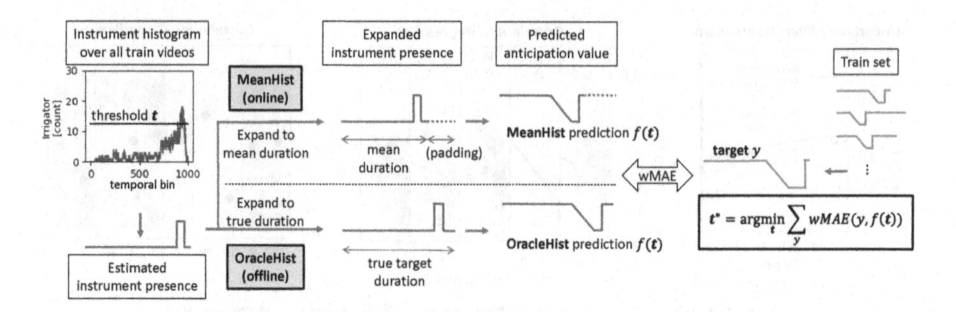

Fig. 4. Overview of the baseline construction (example: irrigator). The temporal histogram of instrument occurrences is thresholded to obtain an estimated instrument presence. The instrument presence signal is used to generate anticipation values according to Eq. 1. For each instrument, we find the optimal threshold t^* w.r.t. *wMAE* on the train set. For the online method *MeanHist*, predictions are expanded to the mean duration of procedures in the train set. For the offline method *OracleHist*, we assume the length of the target procedure to be known and expand predictions accordingly.

4 Conclusion

We propose a novel task and model for uncertainty-aware anticipation of intra-operative instrument usage. Limitations of current anticipation methods are addressed by enabling anticipation of sparse events in long videos. Our approach outperforms several baselines and matches a model variant using richer annotations, which indicates that sparse annotations suffice for this task. Since uncertainty estimation is useful for both anticipation tasks and intra-operative applications, we employ a probabilistic model. We demonstrate the model's ability to quantify task-relevant uncertainties by investigating error-uncertainty correlations and show that we can reduce errors by filtering uncertain predictions. Using a known example, we illustrate the model's potential for identifying trigger events for instruments, which could be useful for robotic applications. Future work could investigate more effective methods for uncertainty quantification where aleatoric uncertainty is especially interesting due to its link to future events.

References

1. Abu Farha, Y., Gall, J.: Uncertainty-aware anticipation of activities. In: Proceedings of the IEEE International Conference on Computer Vision Workshops (2019)
2. Abu Farha, Y., Richard, A., Gall, J.: When will you do what?-anticipating temporal occurrences of activities. In: Proceedings of the IEEE Conference on Computer Vision and Pattern Recognition, pp. 5343–5352 (2018)
3. Bodenstedt, S., et al.: Active learning using deep Bayesian networks for surgical workflow analysis. Int. J. Comput. Assist. Radiol. Surg. **14**(6), 1079–1087 (2019)
4. Damen, D., et al.: Scaling egocentric vision: the epic-kitchens dataset. In: Proceedings of the European Conference on Computer Vision (ECCV), pp. 720–736 (2018)

5. Du, N., Dai, H., Trivedi, R., Upadhyay, U., Gomez-Rodriguez, M., Song, L.: Recurrent marked temporal point processes: embedding event history to vector. In: Proceedings of the 22nd ACM SIGKDD International Conference on Knowledge Discovery and Data Mining, pp. 1555–1564 (2016)
6. Gal, Y., Ghahramani, Z.: Dropout as a Bayesian approximation: representing model uncertainty in deep learning. In: International Conference on Machine Learning, pp. 1050–1059 (2016)
7. Gal, Y., Ghahramani, Z.: A theoretically grounded application of dropout in recurrent neural networks. In: Advances in Neural Information Processing Systems, pp. 1019–1027 (2016)
8. Gal, Y., Islam, R., Ghahramani, Z.: Deep Bayesian active learning with image data. In: Proceedings of the 34th International Conference on Machine Learning, vol. 70, pp. 1183–1192. JMLR. org (2017)
9. Gao, J., Yang, Z., Nevatia, R.: Red: reinforced encoder-decoder networks for action anticipation (2017)
10. Graves, A.: Practical variational inference for neural networks. In: Advances in Neural Information Processing Systems, pp. 2348–2356 (2011)
11. He, K., Zhang, X., Ren, S., Sun, J.: Deep residual learning for image recognition. In: Proceedings of the IEEE Conference on Computer Vision and Pattern Recognition, pp. 770–778 (2016)
12. Hochreiter, S., Schmidhuber, J.: Long short-term memory. Neural Comput. **9**, 1735–80 (1997). https://doi.org/10.1162/neco.1997.9.8.1735
13. Ioffe, S., Szegedy, C.: Batch normalization: accelerating deep network training by reducing internal covariate shift. In: International Conference on Machine Learning, pp. 448–456 (2015)
14. Jain, A., Singh, A., Koppula, H.S., Soh, S., Saxena, A.: Recurrent neural networks for driver activity anticipation via sensory-fusion architecture. In: 2016 IEEE International Conference on Robotics and Automation (ICRA), pp. 3118–3125. IEEE (2016)
15. Ke, Q., Fritz, M., Schiele, B.: Time-conditioned action anticipation in one shot. In: Proceedings of the IEEE Conference on Computer Vision and Pattern Recognition, pp. 9925–9934 (2019)
16. Kendall, A., Gal, Y.: What uncertainties do we need in Bayesian deep learning for computer vision? In: Advances in Neural Information Processing Systems, pp. 5574–5584 (2017)
17. Krizhevsky, A., Sutskever, I., Hinton, G.E.: ImageNet classification with deep convolutional neural networks. In: Advances in Neural Information Processing Systems, pp. 1097–1105 (2012)
18. Kwon, Y., Won, J.H., Kim, B.J., Paik, M.C.: Uncertainty quantification using Bayesian neural networks in classification: application to biomedical image segmentation. Comput. Stat. Data Anal. **142**, 106816 (2020)
19. Mahmud, T., Hasan, M., Roy-Chowdhury, A.K.: Joint prediction of activity labels and starting times in untrimmed videos. In: Proceedings of the IEEE International Conference on Computer Vision, pp. 5773–5782 (2017)
20. Maier-Hein, L., et al.: Can masses of non-experts train highly accurate image classifiers? In: Golland, P., Hata, N., Barillot, C., Hornegger, J., Howe, R. (eds.) MICCAI 2014. LNCS, vol. 8674, pp. 438–445. Springer, Cham (2014). https://doi.org/10.1007/978-3-319-10470-6_55
21. Maier-Hein, L., et al.: Surgical data science for next-generation interventions. Nat. Biomed. Eng. **1**(9), 691–696 (2017)

22. Mehrasa, N., Jyothi, A.A., Durand, T., He, J., Sigal, L., Mori, G.: A variational auto-encoder model for stochastic point processes. In: Proceedings of the IEEE Conference on Computer Vision and Pattern Recognition, pp. 3165–3174 (2019)
23. Shridhar, K., Laumann, F., Liwicki, M.: Uncertainty estimations by softplus normalization in Bayesian convolutional neural networks with variational inference. arXiv preprint arXiv:1806.05978 (2018)
24. Twinanda, A.P., Shehata, S., Mutter, D., Marescaux, J., De Mathelin, M., Padoy, N.: EndoNet: a deep architecture for recognition tasks on laparoscopic videos. IEEE Trans. Med. Imaging **36**(1), 86–97 (2016)
25. Twinanda, A.P., Yengera, G., Mutter, D., Marescaux, J., Padoy, N.: RSDNet: learning to predict remaining surgery duration from laparoscopic videos without manual annotations. IEEE Trans. Med. Imaging **38**(4), 1069–1078 (2018)
26. Vondrick, C., Pirsiavash, H., Torralba, A.: Anticipating visual representations from unlabeled video. In: Proceedings of the IEEE Conference on Computer Vision and Pattern Recognition, pp. 98–106 (2016)
27. Wang, G., Li, W., Aertsen, M., Deprest, J., Ourselin, S., Vercauteren, T.: Aleatoric uncertainty estimation with test-time augmentation for medical image segmentation with convolutional neural networks. Neurocomputing **338**, 34–45 (2019)
28. Zhong, Y., Xu, B., Zhou, G.T., Bornn, L., Mori, G.: Time perception machine: temporal point processes for the when, where and what of activity prediction. arXiv preprint arXiv:1808.04063 (2018)

Deep Placental Vessel Segmentation for Fetoscopic Mosaicking

Sophia Bano[1(✉)], Francisco Vasconcelos[1], Luke M. Shepherd[1], Emmanuel Vander Poorten[2], Tom Vercauteren[3], Sebastien Ourselin[3], Anna L. David[4], Jan Deprest[5], and Danail Stoyanov[1]

[1] Wellcome/EPSRC Centre for Interventional and Surgical Sciences (WEISS) and Department of Computer Science, University College London, London, UK
sophia.bano@ucl.ac.uk

[2] Department of Mechanical Engineering, KU Leuven University, Leuven, Belgium

[3] School of Biomedical Engineering and Imaging Sciences, King's College London, London, UK

[4] Fetal Medicine Unit, University College London Hospital, London, UK

[5] Department of Development and Regeneration, University Hospital Leuven, Leuven, Belgium

Abstract. During fetoscopic laser photocoagulation, a treatment for twin-to-twin transfusion syndrome (TTTS), the clinician first identifies abnormal placental vascular connections and laser ablates them to regulate blood flow in both fetuses. The procedure is challenging due to the mobility of the environment, poor visibility in amniotic fluid, occasional bleeding, and limitations in the fetoscopic field-of-view and image quality. Ideally, anastomotic placental vessels would be automatically identified, segmented and registered to create expanded vessel maps to guide laser ablation, however, such methods have yet to be clinically adopted. We propose a solution utilising the U-Net architecture for performing placental vessel segmentation in fetoscopic videos. The obtained vessel probability maps provide sufficient cues for mosaicking alignment by registering consecutive vessel maps using the direct intensity-based technique. Experiments on 6 different in vivo fetoscopic videos demonstrate that the vessel intensity-based registration outperformed image intensity-based registration approaches showing better robustness in qualitative and quantitative comparison. We additionally reduce drift accumulation to negligible even for sequences with up to 400 frames and we incorporate a scheme for quantifying drift error in the absence of the ground-truth. Our paper provides a benchmark for fetoscopy placental vessel segmentation and registration by contributing the first in vivo vessel segmentation and fetoscopic videos dataset.

Keywords: Fetoscopy · Deep learning · Vessel segmentation · Vessel registration · Mosaicking · Twin-to-twin transfusion syndrome

Electronic supplementary material The online version of this chapter (https://doi.org/10.1007/978-3-030-59716-0_73) contains supplementary material, which is available to authorized users.

A. L. Martel et al. (Eds.): MICCAI 2020, LNCS 12263, pp. 763–773, 2020.
https://doi.org/10.1007/978-3-030-59716-0_73

1 Introduction

Twin-to-twin transfusion syndrome (TTTS) is a rare condition during pregnancy that affects the placenta shared by genetically identical twins [6]. It is caused by abnormal placental vascular anastomoses on the chorionic plate of the placenta between the twin fetuses that disproportionately allow transfusion of blood from one twin to the other. Fetoscopic laser photocoagulation is a minimally invasive procedure that uses a fetoscopic camera and a laser ablation tool. After insertion into the amniotic cavity, the surgeon uses the scope to identify the inter-twin anastomoses and then photocoagulates them to treat the TTTS. Limited field-of-view (FoV), poor visibility [17], unusual placenta position [13] and limited maneuverability of the fetoscope may hinder the photocoagulation resulting in increased procedural time and incomplete ablation of anastomoses leading to persistent TTTS. Automatic segmentation of placental vessels and mosaicking for FoV expansion and creation of a placental vessel map registration may provide computer-assisted interventions (CAI) support for TTTS treatment to support the identification of abnormal vessels and their ablation status.

CAI techniques in fetoscopy have concentrated efforts on visual mosaicking to create RGB maps of the placental vasculature for surgical planning and navigation [4,12,19,20,24,25]. Several approaches have been proposed for generating mosaics based on: (1) detection and matching of visual point features [12,20]; (2) fusion of visual tracking with electromagnetic pose sensing to cater for drifting error in ex vivo experiments [24]; (3) direct pixel-wise alignment of gradient orientations for a single in vivo fetoscopic video [19]; (4) deep learning-based homography estimation for fetoscopic videos captured from various sources [4,25]; and (5) detection of stable regions using R-CNN and using these regions as features for placental image registration in an underwater phantom setting [15]. While promising results have been achieved for short video sequences [4,25], long-term mapping remains a significant challenge [19] due to a variety of factors that include occlusion by the fetus, non-planar views, floating amniotic fluid particles and poor video quality and resolution. Some of these challenges can be addressed by identifying occlusion-free views in fetoscopic videos [5]. Immersion in the amniotic fluid also causes distortions due to light refraction [9,10] that are often hard to model. Moreover, ground-truth homographies are not available for in vivo fetoscopic videos making qualitative (visual) evaluation the widely used standard for judging the quality of generated mosaics. Quantitative evaluation of registration errors in fetoscopic mosaicking has been limited to ex vivo, phantom, or synthetic experiments.

Fetoscopic mosaicking aims to densely reconstruct the surface of the placenta but limited effort has been directed at identification and localisation of vessels in the fetoscopic video. Placental vasculature can be pre-operatively imaged with MRI for surgical planning [2], but there is currently no integration with fetoscopic imaging that enables intra-operative CAI navigation. Moreover, there are no publicly available annotated datasets to perform extensive supervised training. Methods based on a multi-scale vessel enhancement filter [14] have been developed for segmenting vasculature structures from ex vivo high-resolution photographs of the entire placental surface [1,11]. However, such methods fail on in vivo fetoscopy [22], where captured videos have significantly poorer visibility conditions, lower resolution, and a narrower FoV.

Identifying vessels and creating an expanded vessel map can support laser ablation but remains an open problem to date. In this paper, we propose a framework for generating placental vasculature maps that expand the FoV of the fetoscopic camera by performing vessel segmentation in fetoscopic images, registration of consecutive frames, and blending of vessel prediction probability maps from multiple views in a unified mosaic image. We use the U-Net architecture for vessel segmentation since it is robust even with limited training data [21]. Comparison with the available alternative [22] confirmed the superior performance of U-Net for placental vessel segmentation. Alignment of vessel segmentations from consecutive frames is performed via direct registration of the probability maps provided by the U-Net. Finally, multiple overlapping probability maps are blended in a single reference frame. Additionally, we propose the use of quantitative metrics to evaluate the drifting error in sequential mosaicking without relying on the ground-truth (GT). Such temporal evaluation is crucial when GT is not available in surgical video. Our contributions can be summarised as follows:

- A placental vessel segmentation deep learning network trained on 483 manually annotated in vivo fetoscopic images that significantly outperforms the available alternative [22], showing accurate results on 6 in vivo sequences from different patients, with significant changes in appearance.
- Validation of fetoscopic image registration driven exclusively from vessel segmentation maps. We show that, when vessels are visible, this approach is more reliable than performing direct image registration. Many of the visibility challenges in lighting conditions and the presence of moving occlusions are filtered out and vessels are found to have unique recognisable shapes.
- A quantitative evaluation of drift registration error for in vivo fetoscopic sequences, by analysing the similarity between overlapping warped images and predicted segmentation maps. This measures the registration consistency after sequential registration of multiple frames.
- Contribute the first placental vessel segmentation dataset (483 images from 6 subjects) and 6 in vivo video clips, useful for benchmarking results in this domain. Completely anonymised videos of TTTS fetoscopic procedure were obtained from the University College London Hospital. This dataset is made publicly available for research purpose here: https://www.ucl.ac.uk/interventional-surgical-sciences/fetoscopy-placenta-data.

2 Vessel Segmentation-Based Mosaicking

Our proposed framework consists of a segmentation block followed by the registration block as shown in Fig. 1. A U-Net architecture is used for obtaining the prediction maps for the vessels (Sect. 2.1). Vessel probability maps from two

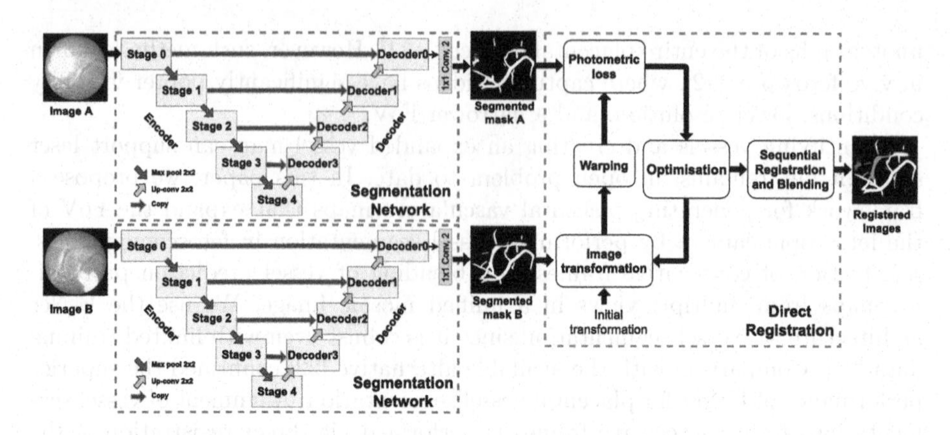

Fig. 1. An overview of the proposed framework which is composed of the segmentation block followed by the direct registration block for mosaic generation.

consecutive frames are then aligned through affine registration (Sect. 2.2). These transformations are accumulated in sequence with respect to the first frame to generate an expanded view of the placental vessels.

2.1 U-Net for Placental Vessel Segmentation

U-Net [21] provides the standard architecture for semantic segmentation and continues to be the base of many of the state-of-the-art segmentation models [3, 18]. U-Net is a fully convolutional network which is preferred for medical image segmentation since it results in accurate segmentation even when the training dataset is relatively small. Placental vessel segmentation is considered as a pixel-wise binary classification problem. Unlike [21], which used the binary cross-entropy loss (L_{bce}), we use the sum of binary cross-entropy loss and intersection over union (Jaccard) loss during training given by:

$$\begin{aligned} L(p,\hat{p}) &= L_{bce}(p,\hat{p}) + L_{iou}(p,\hat{p}), \\ &= -\frac{\sum[p(\log\hat{p}) + (1-p)\log(1-\hat{p})]}{N} + \left[1 - \frac{\sum(p\,.\,\hat{p}) + \delta}{\sum(p+\hat{p}) - \sum(p\,.\,\hat{p}) + \delta}\right], \end{aligned} \tag{1}$$

where p is the flattened ground-truth label tensor, $\hat{p}$ is the flattened predicted label tensor, N is the total number of pixels in the image and $\delta = 10^{-5}$ is arbitrary selected to avoid division by zero. We empirically found that the combined loss (Eq. 1) results in improved accuracy compared to when L_{bce} alone is used. Detailed description of the U-Net can be found in [21]. The vessel probability maps from the U-Net are then used for the vessel registration.

2.2 Vessel Map Registration and Mosaicking

Segmentations from consecutive frames are aligned via registration of the probability maps provided by the segmentation network. We perform registration on the probability maps, rather than on binary segmentation masks as this provides smoother cost transitions for iterative optimisation frameworks, facilitating convergence. We approximate the registration between consecutive frames with affine transformations. We confirmed the observations in [19] that projective registration leads to worse results. In our in vivo datasets, camera calibration is not available and lens distortion cannot be compensated, therefore the use of projective transformations can lead to excessive over-fitting to distorted patterns on the image edges. Existing literature [19] shows that direct-registration approach is robust in fetoscopy where feature-based methods fail due to lack of texture and resolution [4]. Therefore, we use a standard pyramidal Lucas-Kanade registration framework [7] that minimises the bidirectional least-squares difference, also referred to as photometric loss, between a fixed image and a warped moving image. A solution is found with the Levenberg-Marquardt iterative algorithm. Several available implementations of Levenberg-Marquardt offer the option of computing the optimisation step direction through finite differences (e. g. MATLAB's lsqnonlin implementation), however, we have observed that explicitly computing it using the Jacobian of the cost function is necessary for convergence. Given the fetoscopic images have a circular FoV, the registration is performed with a visibility mask and a robust error metric is applied to deal with complex overlap cases [8].

After multiple sequential registrations are performed, they can be blended into a single image, which not only expands the FoV of the vessel map but also filters single-frame segmentation errors whenever multiple results overlap. This is achieved by taking the average probability over all available overlapping values for each pixel, leading to the results presented in Sect. 4.

3 Experimental Setup and Evaluation Protocol

We annotated 483 sampled frames from 6 different in vivo TTTS laser ablation videos. The image appearance and quality varied in each video due to the variation in intra-operative environment among different cases, artefacts and lighting conditions, resulting in increased variability in the data (sample images are shown in Fig. 2). We first selected the non-occluded (no fetus or tool presence) frames through a separate frame-level fetoscopic event identification approach [5] since vessels are mostly visible in such frames. The videos were then down-sampled from 25 to 1 *fps* to avoid including very similar frames in the annotated samples and each frame was resized to 448×448 pixel resolution. The number of clear view samples varied in each in vivo video, hence the number of annotated images varied in each fold (Table 1). We use the pixel annotation tool[1] for annotating and creating a binary mask of vessels in each frame. A fetal medicine specialist further verified our annotations to confirm the correctness of our labels. For direct image registration, we use continuous unannotated video clips from the 6 in vivo videos.

[1] Pixel annotation tool: https://github.com/abreheret/PixelAnnotationTool.

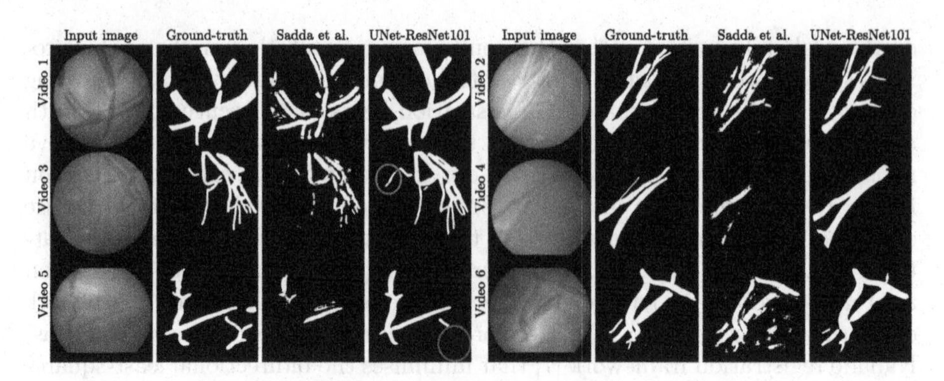

Fig. 2. Qualitative comparison of U-Net (ResNet101) with the baseline (Sadda et al. [22]) network for the placental vessel segmentation.

Table 1. Six fold cross-validation and comparison of the baseline [22] with U-Net architecture (having different backbones) for placental vessel segmentation.

No. of validation images		Fold 1	Fold 2	Fold 3	Fold 4	Fold 5	Fold 6	Overall
		121	101	39	88	37	97	483
Sadda et al. [22] (Baseline)	Dice	0.61 ± 0.17	0.57 ± 0.17	0.54 ± 0.17	0.59 ± 0.17	0.56 ± 0.16	0.50 ± 0.16	0.57 ± 0.17
	IoU	0.46 ± 0.16	0.42 ± 0.15	0.39 ± 0.16	0.43 ± 0.16	0.40 ± 0.15	0.35 ± 0.16	0.41 ± 0.16
U-Net (Vanilla) [21]	Dice	0.82 ± 0.12	0.73 ± 0.17	0.82 ± 0.09	0.70 ± 0.20	0.67 ± 0.19	0.72 ± 0.13	0.75 ± 0.15
	IoU	0.71 ± 0.13	0.60 ± 0.18	0.70 ± 0.19	0.57 ± 0.22	0.53 ± 0.19	0.58 ± 0.15	0.62 ± 0.17
U-Net (VGG16)	Dice	0.82 ± 0.12	0.69 ± 0.14	$\mathbf{0.84 \pm 0.08}$	0.73 ± 0.19	0.70 ± 0.18	0.74 ± 0.14	0.75 ± 0.14
	IoU	0.71 ± 0.14	0.55 ± 0.16	$\mathbf{0.73 \pm 0.11}$	0.61 ± 0.21	0.56 ± 0.19	0.60 ± 0.16	0.63 ± 0.16
U-Net (Resnet50)	Dice	0.84 ± 0.10	0.74 ± 0.14	0.83 ± 0.09	0.74 ± 0.19	$\mathbf{0.72 \pm 0.17}$	0.72 ± 0.16	0.77 ± 0.14
	IoU	$\mathbf{0.74 \pm 0.12}$	0.61 ± 0.16	$\mathbf{0.73 \pm 0.12}$	$\mathbf{0.62 \pm 0.21}$	$\mathbf{0.58 \pm 0.18}$	0.58 ± 0.17	0.65 ± 0.16
U-Net (Resnet101)	Dice	$\mathbf{0.85 \pm 0.07}$	$\mathbf{0.77 \pm 0.16}$	0.83 ± 0.08	$\mathbf{0.75 \pm 0.18}$	0.70 ± 0.18	$\mathbf{0.75 \pm 0.12}$	$\mathbf{0.78 \pm 0.13}$
	IoU	$\mathbf{0.74 \pm 0.10}$	$\mathbf{0.64 \pm 0.17}$	0.72 ± 0.12	$\mathbf{0.62 \pm 0.20}$	0.56 ± 0.19	$\mathbf{0.62 \pm 0.15}$	$\mathbf{0.66 \pm 0.15}$

We perform 6-fold cross-validation to compare and verify the robustness of the segmentation algorithms. Mean Intersection over Union (IoU) and Dice (F1) scores are used to evaluate the segmentation performance (reported in Table 1). We experiment with the vanilla U-Net [21] and with VGG16 [23], ResNet50 [16] and ResNet101 [16] backbones (with pre-trained Imagenet weights) to search for the best performing architecture. In each training iteration, a sub-image of size 224×224 is cropped at random after augmenting the image with rotation, horizontal or vertical flip, and illumination intensity change randomly. This helps in increasing the data and variation during training. A learning rate of $3e^{-4}$ with Adam optimiser and our combined loss (Eq. 1) is used. For each fold, training is performed for 1000 epochs with early stopping and the best performing weights on the training dataset are captured and used to validate the performance on the left-out (hold) set of frames.

In order to evaluate our segmentation driven registration approach, we compare it against standard intensity-based registration. We use the same registration pipeline (described in Sect. 2.2) for both approaches, only changing the input data. Quantification of fetoscopic mosaicking performance with in vivo data remains a difficult challenge in the absence of GT registration. We propose to indirectly characterise the drift error accumulated by consecutive registrations by aligning non-consecutive frames with overlapping FoVs and measuring their structural similarity (SSIM) and the IoU of their vessel predicted maps (results shown in Fig. 3). We use a non-overlapping sliding window of 5 frames and compute the SSIM and IoU between frame 1 and the reprojected non-consecutive frames (from 2 to 5). We highlight that this evaluation is mostly suitable for strictly sequential registration, when non-consecutive frames can provide an unbiased measure of accumulated drift.

4 Results and Discussion

We perform comparison of the U-Net (having different backbones) with the existing baseline placental vessel segmentation network [22]. The qualitative comparison is shown in Fig. 2 and 6-fold cross-validation comparison is reported in Table 1. Sadda et al. [22] implemented a 3-stage U-Net with 8 convolutions in each stage and designed a positive class weighted loss function for optimisation. Compared to the Vanilla U-Net (IoU = 0.62), the performance of [22] is lower ((IoU = 0.41). When experimenting with U-Net with different backbones, we found that the results are comparable though U-Net with ResNet101 backbone overall showed an improvement of 3% Dice (4% IoU) over the Vanilla U-Net. Therefore, we selected U-Net with ResNet101 as the segmentation architecture for obtaining vessel probability maps for completely unlabelled clips from the 6 fetoscopic videos. The robustness of the obtained vessel probability maps is further verified by the generated mosaics. From Table 1, we note that the segmentation results on fold-5 are significantly lower for all methods. This is mainly because most of the vessels were thin in this validation set and there were a few erroneous manual annotations which were rightly detected as false negative. From Fig. 2 (video 3, blue circle), we observe that U-Net (ResNet101) even managed to detect small vessels which were originally missed by the human annotator. In video 5, some vessels around the lower right side of the image were missed due to poor illumination (indicated by red circle).

The performance comparison between the vessel-based and image-based registration shown in Fig. 3 reports the SSIM and IoU scores for all image/prediction-map pair overlaps that are up to 5 frames apart. For each of the 6 unseen and unlabelled test video clips, we use the segmentation network trained on the frames from the remaining 5 videos as input for predicting the vessel maps and registration. It is worth noting that our method has usually lower SSIM values than the baseline between consecutive frames (leftmost box of plots in Fig. 3a). This is to be expected since the baseline overfits to occlusions in the scene (amniotic fluid particles, fetus occlusions, etc). However, when analysing

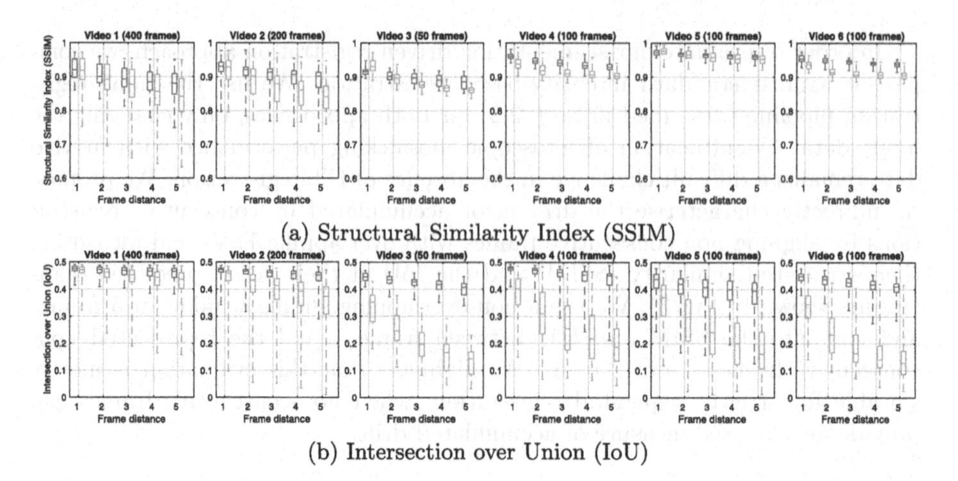

(a) Structural Similarity Index (SSIM)

(b) Intersection over Union (IoU)

Fig. 3. Registration performance while using vessel prediction maps (blue) and intensity images (red) as input over 5 frame distance. (Color figure online)

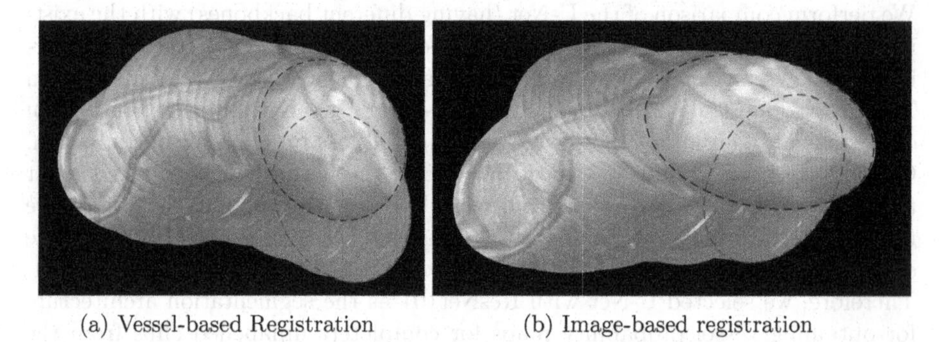

(a) Vessel-based Registration

(b) Image-based registration

Fig. 4. Image registration on video 1 (400 frames duration). First (blue) and last (red) frames are highlighted. Mosaics are generated relative to the frame closest to the centre (i. e. this frame has the same shape as the original image). (Color figure online)

the overlap 5 frames apart our method becomes more consistent. Note from Fig. 3b (Video 3 to 6) the gradually decreasing IoU over the 5 consecutive frames for the image-based registration compared to the vessel-based registration. This effect is because of the increasing drift in the image-based registration method resulting in heavy misalignment. The qualitative comparison in Fig. 4 further supports our method's better performance. Using vessel prediction maps are more robust since it overcomes challenges such as poor visibility and resolution that accounts for introducing drift in the image-based registration. Figure 5 shows the qualitative results of the vessel-based registration for the 6 leave-one-out unlabelled in vivo video clips (refer to the supplementary video for the sequential qualitative comparison). Note that vessel-based mosaicking not only

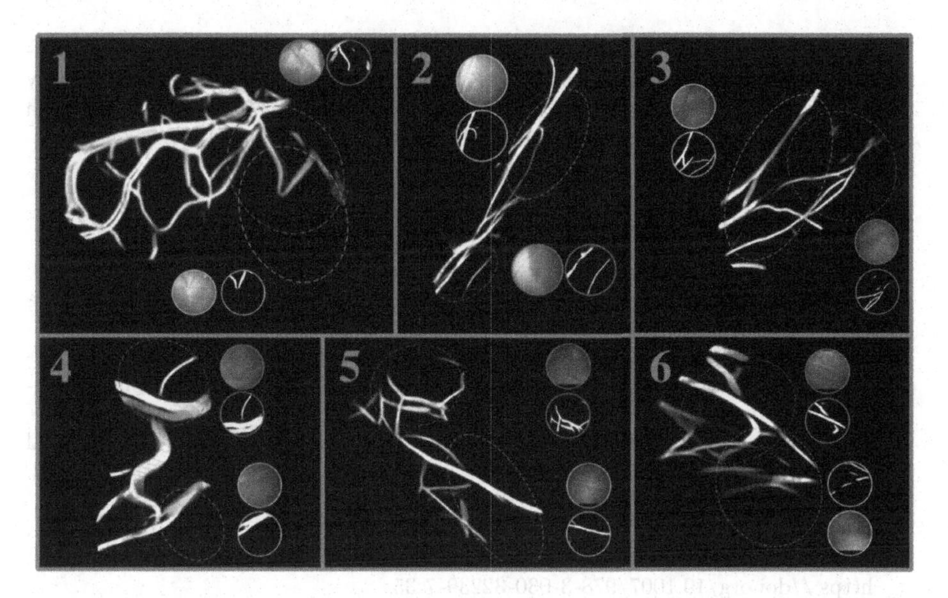

Fig. 5. Visualisation of the vessel maps generated from the segmentation predictions for the 6 in vivo clips. First (blue) and last (red) frames are highlighted. Refer to the supplementary video for the qualitative comparison. (Color figure online)

generated an increased FoV image but also helped in improving vessel segmentation (occluded or missed vessel) results by blending several registered frames.

5 Conclusion

We proposed a placental vessel segmentation driven framework for generating chorionic plate vasculature maps from in vivo fetoscopic videos. Vessel probability maps were created by training a U-Net on 483 manually annotated placental vessel images. Direct vessel-based registration was performed using the vessel probability maps which not only helped in minimising error due to drift but also corrected missing vessels that occurred due to partial occlusion in some frames, alongside providing a vascular map of the chorionic plate of the mono-chorionic placenta. The proposed framework was evaluated through both quantitative and qualitative comparison with the existing methods for validating the segmentation and registration blocks. Six different in vivo video clips were used to validate the generated mosaics. Our proposed framework along with the contributed vessel segmentation and in vivo fetoscopic videos datasets provide a benchmark for future research on this problem.

Acknowledgments. This work was supported by the Wellcome/EPSRC Centre for Interventional and Surgical Sciences (WEISS) at UCL (203145Z/16/Z), EPSRC (EP/P027938/1, EP/R004080/1, NS/A000027/1), the H2020 FET (GA 863146) and

Wellcome [WT101957]. Danail Stoyanov is supported by a Royal Academy of Engineering Chair in Emerging Technologies (CiET1819/2/36) and an EPSRC Early Career Research Fellowship (EP/P012841/1). Tom Vercauteren is supported by a Medtronic/Royal Academy of Engineering Research Chair [RCSRF1819/7/34].

References

1. Almoussa, N., et al.: Automated vasculature extraction from placenta images. In: Medical Imaging 2011: Image Processing, vol. 7962, p. 79621L. International Society for Optics and Photonics (2011)
2. Aughwane, R., Ingram, E., Johnstone, E.D., Salomon, L.J., David, A.L., Melbourne, A.: Placental MRI and its application to fetal intervention. Prenatal diagnosis (2019)
3. Badrinarayanan, V., Kendall, A., Cipolla, R.: SegNet: a deep convolutional encoder-decoder architecture for image segmentation. IEEE Trans. Pattern Anal. Mach. Intell. **39**(12), 2481–2495 (2017)
4. Bano, S., et al.: Deep sequential mosaicking of fetoscopic videos. In: Shen, D., et al. (eds.) MICCAI 2019. LNCS, vol. 11764, pp. 311–319. Springer, Cham (2019). https://doi.org/10.1007/978-3-030-32239-7_35
5. Bano, S., et al.: FetNet: a recurrent convolutional network for occlusion identification in fetoscopic videos. Int. J. Comput. Assist. Radiol. Surg. **15**(5), 791–801 (2020)
6. Baschat, A., et al.: Twin-to-twin transfusion syndrome (TTTS). J. Perinat. Med. **39**(2), 107–112 (2011)
7. Bouguet, J.Y., et al.: Pyramidal implementation of the affine Lucas Kanade feature tracker description of the algorithm. Intel Corporation **5**(1–10), 4 (2001)
8. Brunet, F., Bartoli, A., Navab, N., Malgouyres, R.: Direct image registration without region of interest. In: Vision, Modeling, and Visualization, pp. 323–330 (2010)
9. Chadebecq, F., et al.: Refractive structure-from-motion through a flat refractive interface. In: Proceedings of the IEEE International Conference on Computer Vision, pp. 5315–5323 (2017)
10. Chadebecq, F., et al.: Refractive two-view reconstruction for underwater 3D vision. Int. J. Comput. Vis. **128**, 1–17 (2019)
11. Chang, J.M., Huynh, N., Vazquez, M., Salafia, C.: Vessel enhancement with multiscale and curvilinear filter matching for placenta images. In: International Conference on Systems, Signals and Image Processing, pp. 125–128. IEEE (2013)
12. Daga, P., et al.: Real-time mosaicing of fetoscopic videos using SIFT. In: Medical Imaging 2016: Image-Guided Procedures, Robotic Interventions, and Modeling, vol. 9786, p. 97861R. International Society for Optics and Photonics (2016)
13. Deprest, J., Van Schoubroeck, D., Van Ballaer, P., Flageole, H., Van Assche, F.A., Vandenberghe, K.: Alternative technique for nd: YAG laser coagulation in twin-to-twin transfusion syndrome with anterior placenta. Ultrasound Obstet. Gynecol. Official J. Int. Soc. Ultrasound Obstet. Gynecol. **11**(5), 347–352 (1998)
14. Frangi, A.F., Niessen, W.J., Vincken, K.L., Viergever, M.A.: Multiscale vessel enhancement filtering. In: Wells, W.M., Colchester, A., Delp, S. (eds.) MICCAI 1998. LNCS, vol. 1496, pp. 130–137. Springer, Heidelberg (1998). https://doi.org/10.1007/BFb0056195
15. Gaisser, F., Peeters, S.H., Lenseigne, B.A., Jonker, P.P., Oepkes, D.: Stable image registration for in-vivo fetoscopic panorama reconstruction. J. Imaging **4**(1), 24 (2018)

16. He, K., Zhang, X., Ren, S., Sun, J.: Deep residual learning for image recognition. In: IEEE Conference on Computer Vision and Pattern Recognition, pp. 770–778 (2016)
17. Lewi, L., Deprest, J., Hecher, K.: The vascular anastomoses in monochorionic twin pregnancies and their clinical consequences. Am. J. Obstet. Gynecol. **208**(1), 19–30 (2013)
18. Litjens, G., et al.: A survey on deep learning in medical image analysis. Med. Image Anal. **42**, 60–88 (2017)
19. Peter, L., et al.: Retrieval and registration of long-range overlapping frames for scalable mosaicking of in vivo fetoscopy. Int. J. Comput. Assist. Radiol. Surg. **13**(5), 713–720 (2018). https://doi.org/10.1007/s11548-018-1728-4
20. Reeff, M., Gerhard, F., Cattin, P., Gábor, S.: Mosaicing of endoscopic placenta images. INFORMATIK 2006-Informatik für Menschen, Band 1 (2006)
21. Ronneberger, O., Fischer, P., Brox, T.: U-Net: convolutional networks for biomedical image segmentation. In: Navab, N., Hornegger, J., Wells, W.M., Frangi, A.F. (eds.) MICCAI 2015. LNCS, vol. 9351, pp. 234–241. Springer, Cham (2015). https://doi.org/10.1007/978-3-319-24574-4_28
22. Sadda, P., Imamoglu, M., Dombrowski, M., Papademetris, X., Bahtiyar, M.O., Onofrey, J.: Deep-learned placental vessel segmentation for intraoperative video enhancement in fetoscopic surgery. Int. J. Comput. Assist. Radiol. Surg. **14**(2), 227–235 (2018). https://doi.org/10.1007/s11548-018-1886-4
23. Simonyan, K., Zisserman, A.: Very deep convolutional networks for large-scale image recognition. In: International Conference on Learning Representations (2015)
24. Tella-Amo, M., et al.: Pruning strategies for efficient online globally consistent mosaicking in fetoscopy. J. Med. Imaging **6**(3), 035001 (2019)
25. Bano, S., et al.: Deep learning-based fetoscopic mosaicking for field-of-view expansion. Int. J. Comput. Assist. Radiol. Surg., 1–10 (2020)

Deep Multi-view Stereo for Dense 3D Reconstruction from Monocular Endoscopic Video

Gwangbin Bae[1(✉)], Ignas Budvytis[1], Chung-Kwong Yeung[2], and Roberto Cipolla[1]

[1] Department of Engineering, University of Cambridge, Cambridge, UK
{gb585,ib255,rc10001}@cam.ac.uk
[2] Bio-Medical Engineering (HK) Limited, Hong Kong, China
ck.yeung@nisi.hk

Abstract. 3D reconstruction from monocular endoscopic images is a challenging task. State-of-the-art multi-view stereo (MVS) algorithms based on image patch similarity often fail to obtain a dense reconstruction from weakly-textured endoscopic images. In this paper, we present a novel deep-learning-based MVS algorithm that can produce a dense and accurate 3D reconstruction from a monocular endoscopic image sequence. Our method consists of three key steps. Firstly, a number of depth candidates are sampled around the depth prediction made by a pre-trained CNN. Secondly, each candidate is projected to the other images in the sequence, and the matching score is measured using a patch embedding network that maps each image patch into a compact embedding. Finally, the candidate with the highest score is selected for each pixel. Experiments on colonoscopy videos demonstrate that our patch embedding network outperforms zero-normalized cross-correlation and a state-of-the-art stereo matching network in terms of matching accuracy and that our MVS algorithm produces several degrees of magnitude denser reconstruction than the competing methods when same accuracy filtering is applied.

Keywords: Multi-view stereo · 3D reconstruction · Endoscopy

1 Introduction

The capability of estimating depth can improve the quality and safety of the monocular endoscopy. The obtained depth information can be used to estimate the shape and size of the lesions, improving the accuracy of the visual biopsy, or to identify the safest navigation path. Numerous attempts have been made with such motivations. However, methods that require device modification (e.g. additional light source [10,13], depth sensors [2], stereo setup [3]) could not be evaluated *in vivo* due to engineering and regulatory barriers.

A. L. Martel et al. (Eds.): MICCAI 2020, LNCS 12263, pp. 774–783, 2020.
https://doi.org/10.1007/978-3-030-59716-0_74

A cheaper alternative is to perform 3D reconstruction directly from the images by using multi-view geometry. For example, structure-from-motion (SfM) [14] identifies the matches between multiple images of the scene (taken from different viewpoints) and jointly optimizes their 3D coordinates and the relative camera poses by minimizing the reprojection error. The reconstructed points can be trusted as they are geometrically verified. However, the resulting reconstruction is very sparse. For a sequence of 8 colonoscopy images (of resolution 624 × 540), SfM can only reconstruct about 200 points.

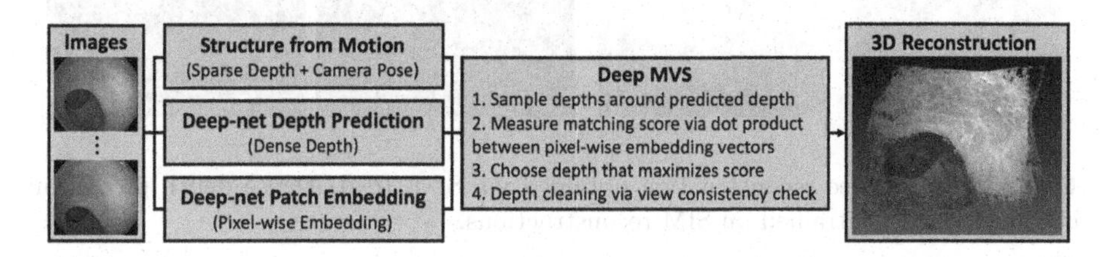

Fig. 1. Our deep multi-view stereo pipeline.

Recent works [5,8] have shown that it is possible to train deep neural networks with sparse SfM reconstruction. After training, such a network can predict dense pixel-wise depth map for a given image. Nonetheless, the predicted depth map lacks accuracy as it is generated from a single image, and is not validated from other viewpoints. For example, single-view depth prediction can be affected by the presence of motion blur, light speckles, or fluids.

A possible solution is to use multi-view stereo (MVS) algorithms. Once the camera poses are estimated (e.g. via SfM), an MVS algorithm tries to find the optimal depth each pixel should have in order for it to be projected to the visually similar pixels in the other images. However, classical MVS algorithms suffer from two weaknesses - large search space and poor matching accuracy. For example, in PatchMatch stereo [1], the depth map is initialized randomly from a uniform distribution ranging between some pre-set upper ($d_{\max}$) and lower limit ($d_{\min}$). Then, the depths with high matching score are propagated to the neighboring pixels. In BruteForce stereo [9], selected number of depths ranging from $d_{\min}$ to $d_{\max}$ are tried for each pixel, and the one with the highest score is selected. In both scenarios, finding the correct depth becomes challenging if $(d_{\max} - d_{\min})$ is large. Another problem with the conventional MVS approaches is the inaccuracy of the patch-matching. A typical matching function is the zero-normalized cross-correlation (ZNCC). Since the computational cost increases quadratically with the patch size, small patch sizes (e.g. 7 × 7) are often preferred. However, small patch can lead to ambiguous matches especially in texture-less images.

In this paper, we present a deep-learning-based MVS pipeline (see Fig. 1) that can solve the aforementioned problems. The novelty of our approach is three-fold. Firstly, we use a monocular depth estimation network to constrain the search space for the depth candidate sampling. Secondly, we introduce a

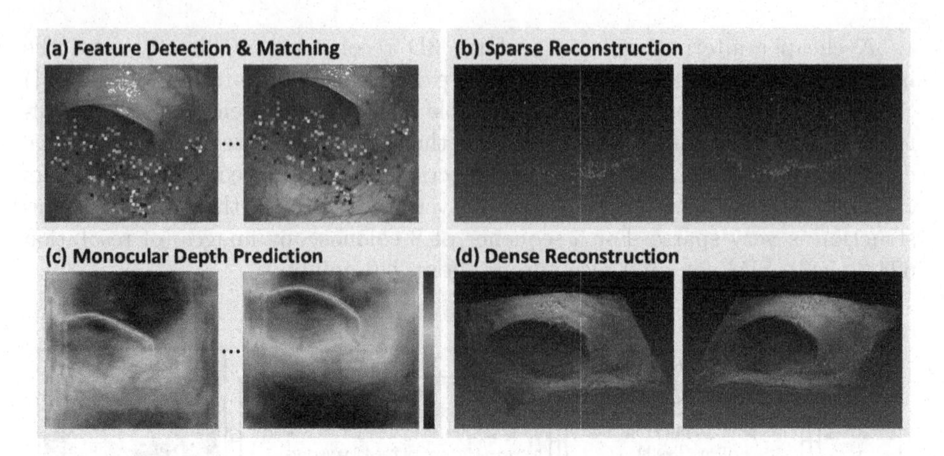

Fig. 2. (a–b) Sparse reconstruction obtained via SfM. (c–d) Dense depth prediction obtained via CNN trained on SfM reconstructions.

novel patch embedding network that significantly improves the accuracy and reduces the computation cost compared to the ZNCC and other stereo matching network. Lastly, we demonstrate that, after measuring the scores for the neighboring images in the sequence, selecting the minimum, as opposed to maximum, improves the quality of the reconstruction by enforcing the multi-view consistency. Our method is evaluated on colonoscopy videos but can be extended to any monocular endoscopy.

2 Method

Our method consists of three pre-processing steps followed by a multi-view stereo reconstruction. The three pre-processing steps are (1) sparse reconstruction via SfM, (2) monocular depth estimation, and (3) embedding vector generation via patch embedding network. Then, the MVS pipeline generates a geometrically validated 3D reconstruction. The following sections provide details of each step.

2.1 Sparse Depth and Camera Pose Estimation via SfM

Firstly, the sparse reconstruction and relative camera poses are estimated via SfM. In order to minimize the effect of the non-rigid surface deformation, the endoscopy videos are split into short sequences, consisting of 8 consecutive frames separated by 0.08 s. For each sequence, the SfM reconstruction is obtained using OpenSfM [9]. When optimizing the 3D feature coordinates and the camera poses, only the features that appear in all 8 images are considered. This ensures that the camera poses are supported by all feature coordinates and are hence accurate. See Fig. 2 for an example of a resulting sparse reconstruction.

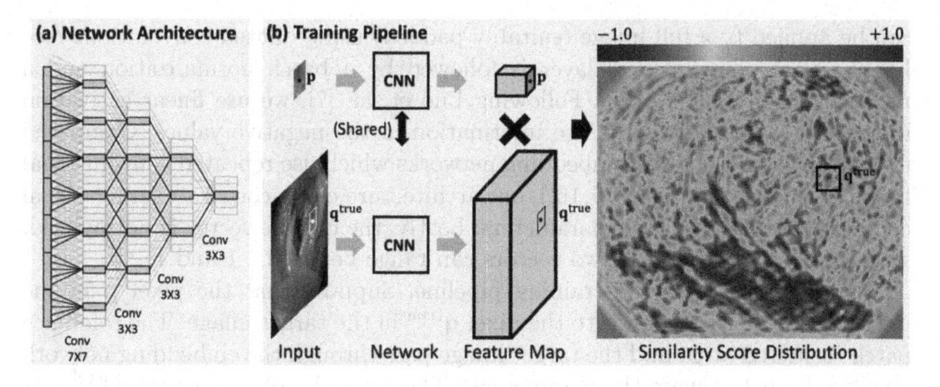

Fig. 3. (a) The architecture of our patch embedding network. (b) During training, a reference patch and the target image are passed through the network. The network is optimized so that the similarity score (i.e. dot product between patch embedding vectors) is maximized at the correct matching pixel.

2.2 Dense Depth Prediction via CNN

The sparse reconstruction obtained via SfM is then used to train a monocular depth estimation network. Due to the inherent scale ambiguity of SfM, the training loss is computed after matching the scale of the prediction to that of the ground truth. More formally, the loss is defined as:

$$L = \min_{s \in \mathcal{S}} \sum_{\mathbf{p}} \mathbb{1}\left(d_{\mathbf{p}}^{\text{true}} > 0\right) \left(d_{\mathbf{p}}^{\text{true}} - s \times d_{\mathbf{p}}^{\text{pred}}\right)^2 \quad (1)$$

where $\mathcal{S}$ is a discrete set of scaling factors (e.g. ranging from 0.5 to 2.0) and $\mathbb{1}\left(d_{\mathbf{p}}^{\text{true}} > 0\right)$ is a binary variable which is equal to 1 if a true depth is available for the pixel $\mathbf{p} = (u, v)$ and is 0 otherwise. By applying such scale-invariant loss, the network is able to learn the relative depth (i.e. the ratio between pixel-wise depths). In this paper, we use the U-Net architecture [12] with a single output channel. See Fig. 2 for an example of a predicted depth map.

2.3 Pixel-Wise Embedding via Patch Embedding Network

The goal of the patch embedding network is to map an image patch around each pixel into an embedding vector $\mathbf{f}$, so that the dot product between two vectors can be used as a measure of their patch similarity. Inspired by the SIFT feature descriptor [6], we divide the receptive field (of size 49×49 pixel) into 7×7 cells of the same size. Then, a 7×7 convolutional layer (with 64 output channels) is used to identify the low-level features in each cell. This results in a feature map of size $7 \times 7 \times 64$. Then, a set of three convolutional layers with 3×3 kernels maps the feature map into a 64-dimensional vector. The same network can be applied to a full image (suitably padded) using dilated convolutions (see Fig. 3). Each convolutional layer is followed by a batch normalization and a rectified linear unit (ReLU). Following Luo et. al. [7], we use linear activation in the last layer to preserve the information in the negative values. Compared to the conventional patch embedding networks which use repeated convolutional layers of small size (e.g. [7,15,16]), our architecture can incorporate larger visual context while having fewer parameters. Lastly, the output vector is normalized, so that the dot product of two vectors can range between −1 and 1.

Figure 3 illustrates the training pipeline. Suppose that the pixel $\mathbf{p}$ in the reference image corresponds to the pixel $\mathbf{q}^{\text{true}}$ in the target image. The reference patch (centered at $\mathbf{p}$) and the target image pass through the embedding network (the two branches share the parameters). The network outputs a vector $\mathbf{f}_{\mathbf{p}}^{\text{ref}}$ and a vector map of size $H \times W \times 64$. $\mathbf{f}_{\mathbf{p}}^{\text{ref}}$ is then multiplied (via dot product) to the embedding vector at each pixel in the target image. This generates a pixel-wise score distribution.

Ideally, we want the score to be high only near the pixel $\mathbf{q}^{\text{true}}$. To achieve this objective, we introduce a novel soft contrastive loss, which is defined as:

$$\begin{aligned} L_{\mathbf{p},\mathbf{q}^{\text{true}}} &= \sum_{\mathbf{q}} \max(w_{\mathbf{q}}, 0) \left(1 - \mathbf{f}_{\mathbf{p}}^{\text{ref}} \cdot \mathbf{f}_{\mathbf{q}}^{\text{target}}\right) \\ &\quad + \sum_{\mathbf{q}} \max(-w_{\mathbf{q}}, 0) \max\left(\mathbf{f}_{\mathbf{p}}^{\text{ref}} \cdot \mathbf{f}_{\mathbf{q}}^{\text{target}} - \alpha, 0\right) \\ \text{where} \quad w_{\mathbf{q}} &= \cos\left(\frac{||\mathbf{q} - \mathbf{q}^{\text{true}}||\pi}{5}\right) \quad \text{if } ||\mathbf{q} - \mathbf{q}^{\text{true}}|| \le 5 \text{ and } -1 \text{ otherwise.} \end{aligned} \tag{2}$$

As a result, the score is maximized if $\mathbf{q}$ is less than 2.5 pixels away from $\mathbf{q}^{\text{true}}$ and is minimized elsewhere if it is larger than the threshold, α. In this paper α is set to 0.7 empirically.

2.4 Deep Multi-view Stereo Reconstruction

Our MVS reconstruction pipeline consists of three key steps. Firstly, 50 depth candidates are sampled uniformly from $0.9 \times d_{\text{pred}}$ to $1.1 \times d_{\text{pred}}$, where d_{pred} is the predicted depth. Each candidate is then projected to all the other images in the sequence, and the score is measured for each of them. This gives $N_{\text{image}} - 1$ scores. From these, the minimum is selected and is assigned to the depth candidate. Once the score is assigned to every candidate, the one with the highest score is selected. More formally, the depth at pixel $\mathbf{p}$ in the i-th image is selected as:

$$\hat{d}_{\mathbf{p}}^{i} = \underset{d \in \mathcal{D}}{\operatorname{argmax}} \left[\min_{j} \left(\mathbf{f}_{\mathbf{p}}^{i} \cdot \mathbf{f}_{P^{j}(d)}^{j}\right)\right] \tag{3}$$

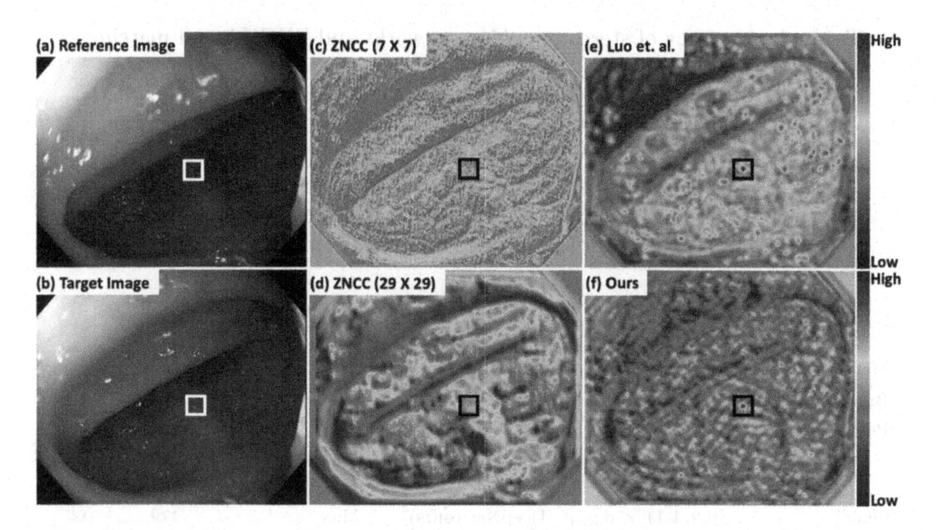

Fig. 4. Qualitative comparison between the score distribution generated by different methods shows that our method leads to significantly reduced matching ambiguity.

where $\mathcal{D}$ and $P^j(d)$ represent the set of depth candidates and the projection of the depth candidate d on the j-th image.

The resulting depth-map is then filtered via view consistency check. In this step, $\hat{d}^i_{\mathbf{p}}$ is projected to a different image in the sequence and is compared to the estimated depth value at the projected pixel. If there is less than 1% difference between the two values, the two depths are considered "consistent". If this is satisfied for all 7 images in the sequence, $\hat{d}^i_{\mathbf{p}}$ is merged into the final reconstruction. The number of survived pixels is then used as a quantitative measure of the reconstruction accuracy.

3 Experiments

3.1 Experimental Setup

Our dataset consists of 51 colonoscopy videos, each containing a full procedure (~20 min) of a different patient. 201,814 image sequences are extracted, of which 34,868 are successfully reconstructed via SfM. The sequences from 40 videos are used to train, validate and test the depth estimation network and the patch embedding network. For the training and testing of the patch embedding network, the SIFT [6] features that survived the SfM reconstruction are used to establish the ground truth matching. Lastly, the performance of our MVS pipeline is tested on the sequences from the remaining 11 videos.

The depth estimation and patch embedding networks are implemented and trained using PyTorch [11] framework. Both networks are trained for 80 epochs with a batch size of 32 using Adam optimizer [4] ($\beta_1 = 0.9$, $\beta_2 = 0.999$). The learning rate is set to 0.001 and is reduced every 20 epochs, to 0.0007, 0.0003 and 0.0001, respectively.

Table 1. Accuracy of stereo matching networks and ZNCC-based matching.

	Patch size	# Params	Median error	>3px	>5px	>10px	Runtime(s)
ZNCC	7 × 7	N/A	125.419	0.675	0.671	0.664	64.9
ZNCC	29 × 29	N/A	1.0	0.174	0.127	0.109	777.3
ZNCC	49 × 49	N/A	1.414	0.209	0.135	0.097	2029.1
Luo et. al. [7]	37 × 37	695136	**1.000**	**0.064**	0.034	0.020	**28.0**
Ours	49 × 49	120768	**1.000**	0.077	**0.028**	**0.013**	**28.0**

Table 2. Number of pixels that survive view consistency check.

Method	Search space	Matching function	Score selection	Average	Median
BruteForce [9]	$(d_{\min}, d_{\max})$	ZNCC (7 × 7)	Max	31	6
PatchMatch [9]	$(d_{\min}, d_{\max})$	ZNCC (7 × 7)	Max	17	1
Ours	$(0.9, 1.1) \times d_{\text{pred}}$	ZNCC (7 × 7)	Max	157	29
		DeepNet [7]	Max	164	39
		DeepNet (ours)	Max	479	96
		DeepNet [7]	Min	5910	3622
		DeepNet (ours)	Min	**10454**	**6452**

3.2 Accuracy of the Patch Embedding Network

The aim of this experiment is to quantitatively evaluate the matching accuracy of our patch embedding network. The accuracy is measured as follows: The embedding vector generated for the reference pixel is convolved with all the embeddings in the target image. The pixel in the target image that maximizes the score (i.e. dot product) is then selected as $\mathbf{q}^{\text{pred}}$. The error is defined as $||\mathbf{q}^{\text{pred}} - \mathbf{q}^{\text{true}}||$. We report the median error and the percentage of matches with error larger than 3, 5, and 10 pixels. We also measure the time it takes to run a stereo reconstruction with each scoring method. Table 1 shows the results.

While small patch does not contain sufficient information, large patch includes the surrounding pixels the appearance of which is highly view-dependent. This explains why the accuracy of the ZNCC matching peaks at an intermediate patch size (29 × 29). On the contrary, deep-learning-based approaches show high accuracy despite the large patch size. Compared to the state-of-the-art stereo matching network [7], our network contains fewer parameters, has larger receptive field, and achieves better accuracy except for the 3 pixel error rate. Since the patch embedding networks encode each patch into a concise representation, the matching requires less computation compared to the ZNCC, resulting in significantly reduced reconstruction runtime.

Figure 4 shows the score distribution generated by each method. For ZNCC, small patch leads to ambiguous matches, while large patch results in over-smoothed score distribution. Our network, compared to Luo et. al. [7], shows high score only near the correct pixel. This is mainly due to the soft-contrastive loss (Eq. 2) which penalizes the large scores at incorrect pixels.

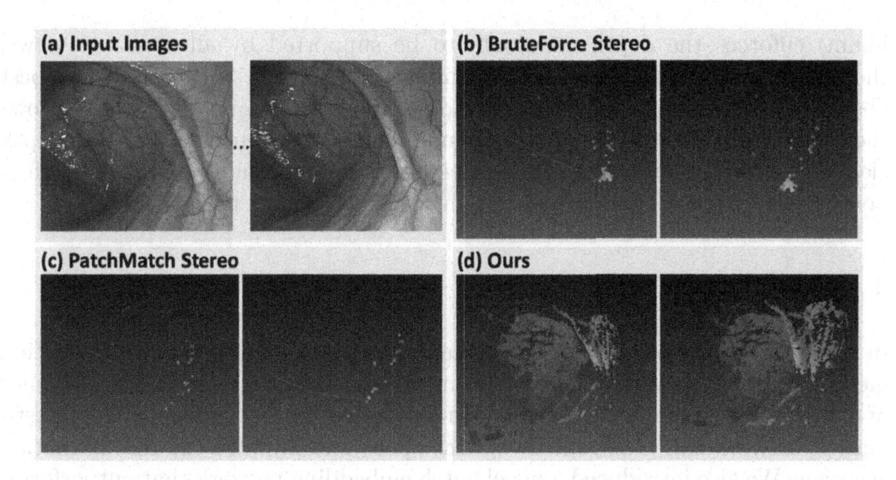

Fig. 5. 3D reconstructions obtained from different methods.

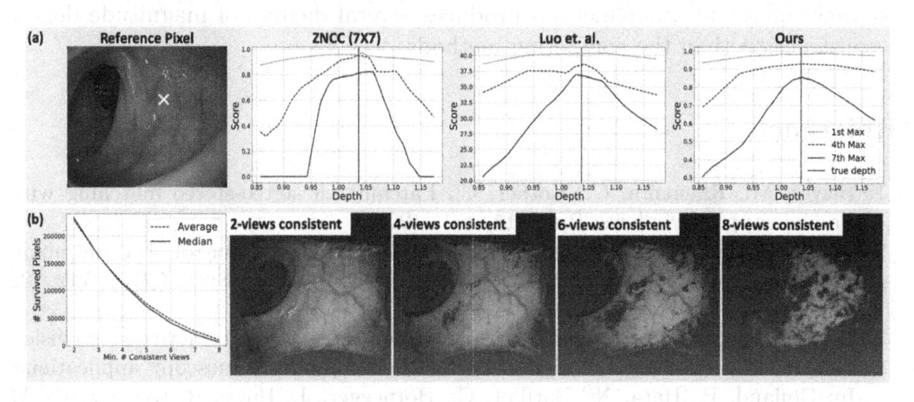

Fig. 6. (a) Visual justification for selecting the minimum score. (b) Trade-off between reconstruction density and accuracy.

3.3 Evaluation of the MVS Reconstruction

This experiment aims to compare the accuracy of our MVS pipeline to that of the competing methods. Since ground truth dense reconstruction is not available, we use the number of pixels that survive the view consistency check (see Sect. 2.4) as a quantitative measure of accuracy. Table 2 shows the obtained results. For our method, we also show the contribution of each component. Compared to the BruteForce and PatchMatch reconstruciton (implemented in OpenSfM [9]), our method produces several orders of magnitude denser reconstruction. Figure 5 provides qualitative comparison between the resulting 3D reconstructions.

Figure 6 shows the results of two additional experiments. Firstly, the score for each depth candidate is computed with different n-th best selection of the matching score. In BruteForce and PatchMatch, the maximum is selected, while we choose the minimum. The result shows that selecting the 7-th best (i.e. minimum) enforces the depth candidate to be supported by all available views, thereby suppressing the scores of the ambiguous matches. Such behavior is best observed when using our patch embedding network. Second experiment shows the relationship between the minimum number of consistent view (used in depth cleaning) and the number of survived pixels. By decreasing this parameter, it is possible to obtain denser reconstruction, while sacrificing the accuracy.

4 Conclusion

In this work, we proposed a deep-learning-based multi-view stereo reconstruction method that can produce dense and accurate 3D reconstruction from a sequence of monocular endoscopic images. We demonstrated that a pre-trained depth estimation network can constrain the search space and improve the reconstruction accuracy. We also introduced a novel patch embedding network that outperforms ZNCC and the state-of-the-art stereo matching network. For a fixed constraint on view consistency, our method produces several degrees of magnitude denser reconstruction than the competing methods.

References

1. Bleyer, M., Rhemann, C., Rother, C.: Patchmatch stereo-stereo matching with slanted support windows. Bmvc **11**, 1–11 (2011)
2. Chen, G., Pham, M., Redarce, T.: Sensor-based guidance control of a continuum robot for a semi-autonomous colonoscopy. Robot. Auton. Syst. **57**(6), 712–722 (2009)
3. Hou, Y., Dupont, E., Redarce, T., Lamarque, F.: A compact active stereovision system with dynamic reconfiguration for endoscopy or colonoscopy applications. In: Golland, P., Hata, N., Barillot, C., Hornegger, J., Howe, R. (eds.) MICCAI 2014. LNCS, vol. 8673, pp. 448–455. Springer, Cham (2014). https://doi.org/10.1007/978-3-319-10404-1_56
4. Kingma, D.P., Ba, J.: Adam: a method for stochastic optimization. arXiv preprint arXiv:1412.6980 (2014)
5. Liu, X., et al.: Self-supervised learning for dense depth estimation in monocular endoscopy. In: Stoyanov, D., et al. (eds.) CARE/CLIP/OR 2.0/ISIC -2018. LNCS, vol. 11041, pp. 128–138. Springer, Cham (2018). https://doi.org/10.1007/978-3-030-01201-4_15
6. Lowe, D.G.: Distinctive image features from scale-invariant keypoints. Int. J. Comput. Vision **60**(2), 91–110 (2004)
7. Luo, W., Schwing, A.G., Urtasun, R.: Efficient deep learning for stereo matching. In: The IEEE Conference on Computer Vision and Pattern Recognition (CVPR), June 2016
8. Ma, R., Wang, R., Pizer, S., Rosenman, J., McGill, S.K., Frahm, J.-M.: Real-time 3D reconstruction of colonoscopic surfaces for determining missing regions. In: Shen, D., et al. (eds.) MICCAI 2019. LNCS, vol. 11768, pp. 573–582. Springer, Cham (2019). https://doi.org/10.1007/978-3-030-32254-0_64
9. Mapillary: Opensfm (2017). https://github.com/mapillary/OpenSfM

10. Parot, V., et al.: Photometric stereo endoscopy. J. Biomed. Opt. **18**(7), 076017 (2013)
11. Paszke, A., et al.: Pytorch: an imperative style, high-performance deep learning library. In: Advances in Neural Information Processing Systems, pp. 8026–8037 (2019)
12. Ronneberger, O., Fischer, P., Brox, T.: U-Net: convolutional networks for biomedical image segmentation. In: Navab, N., Hornegger, J., Wells, W.M., Frangi, A.F. (eds.) MICCAI 2015. LNCS, vol. 9351, pp. 234–241. Springer, Cham (2015). https://doi.org/10.1007/978-3-319-24574-4_28
13. Schmalz, C., Forster, F., Schick, A., Angelopoulou, E.: An endoscopic 3D scanner based on structured light. Med. Image Anal. **16**(5), 1063–1072 (2012)
14. Ullman, S.: The interpretation of structure from motion. Proc. Roy. Soc. London **203**(1153), 405–426 (1979). https://doi.org/10.1098/rspb.1979.0006
15. Zagoruyko, S., Komodakis, N.: Learning to compare image patches via convolutional neural networks. In: Proceedings of the IEEE Conference on Computer Vision and Pattern Recognition, pp. 4353–4361 (2015)
16. Žbontar, J., LeCun, Y.: Stereo matching by training a convolutional neural network to compare image patches. J. Mach. Learn. Res. **17**(1), 2287–2318 (2016)

Endo-Sim2Real: Consistency Learning-Based Domain Adaptation for Instrument Segmentation

Manish Sahu[1(✉)], Ronja Strömsdörfer[1], Anirban Mukhopadhyay[2], and Stefan Zachow[1]

[1] Zuse Institute Berlin (ZIB), Berlin, Germany
Sahu@zib.de
[2] Department of Computer Science, TU Darmstadt, Darmstadt, Germany

Abstract. Surgical tool segmentation in endoscopic videos is an important component of computer assisted interventions systems. Recent success of image-based solutions using fully-supervised deep learning approaches can be attributed to the collection of big labeled datasets. However, the annotation of a big dataset of real videos can be prohibitively expensive and time consuming. Computer simulations could alleviate the manual labeling problem, however, models trained on simulated data do not generalize to real data. This work proposes a consistency-based framework for joint learning of simulated and real (unlabeled) endoscopic data to bridge this performance generalization issue. Empirical results on two data sets (15 videos of the Cholec80 and EndoVis'15 dataset) highlight the effectiveness of the proposed *Endo-Sim2Real* method for instrument segmentation. We compare the segmentation of the proposed approach with state-of-the-art solutions and show that our method improves segmentation both in terms of quality and quantity.

Keywords: Endoscopic instrument segmentation · Unsupervised domain adaptation · Self-supervised learning · Consistency learning

1 Introduction

Segmentation of surgical tools in endoscopic videos is fundamental for realizing automation in computer assisted minimally invasive surgery or surgical workflow analysis.[1] Recent research addresses the problem of surgical tool segmentation using fully-supervised deep learning.

[1] EndoVis Sub-challenges - 2015, 2017, 2018 and 2019 (URL).

Funded by the German Federal Ministry of Education and Research (BMBF) under the project COMPASS (grant no. - 16 SV 8019).

Electronic supplementary material The online version of this chapter (https://doi.org/10.1007/978-3-030-59716-0_75) contains supplementary material, which is available to authorized users.

A. L. Martel et al. (Eds.): MICCAI 2020, LNCS 12263, pp. 784–794, 2020.
https://doi.org/10.1007/978-3-030-59716-0_75

One of the barriers for employing fully-supervised Deep Neural Networks (DNNs) is the creation of a sufficiently large ground truth data set. Even though the annotation process occupies valuable time of expert surgeons, it is essential for training fully-supervised DNNs. While the research community is discussing effective ways for annotating surgical videos[2], researchers are exploring possibilities beyond purely supervised approaches [7,19,21]. A promising approach is utilizing virtual environments which provide a big synthetic dataset with clean labels for learning [19]. However, DNNs trained purely on synthetic data (source domain) perform poorly when tested on real data (target domain) due to the domain/distribution shift problem [25]. Domain adaptation [22] is commonly employed to address this performance generalization issue. In an annotation-free scenario, unsupervised domain adaptation (UDA) is applied since labels are not available for the target domain. This can be achieved, for instance, by using a multi-step approach [19], where simulated images are first translated to realistic looking ones, before a DNN is trained with these pairs of translated images and their corresponding segmentation masks. However, we argue that jointly learning from simulated and real data in an end-to-end fashion is important for adaptation of DNNs on both domains.

We propose an end-to-end, simulation to real-world domain adaptation approach for instrument segmentation, which does not require labeled (real) data for training a DNN (model). At its core, there is an interplay between consistency learning for unlabeled real images and supervised learning for labeled simulated images. The design and the loss functions of this approach are motivated by the key observation that the instrument shape is the most consistent feature between simulated and real images, while the texture and lighting may strongly vary. Unlike Generative Adversarial Network (GAN) based approaches [19] which mainly exploit style (texture, lighting) information [9], our proposed loss focuses on the consistency of tool shapes for better generalization. Moreover, our approach, being end-to-end and non-adversarial by construction, is computationally more efficient (training time of hours vs days) in comparison to other multi-step and adversarial training approaches (*I2I* and *PAT*).

We demonstrate, by quantitative and qualitative evaluation of 15 videos of the *Cholec80* dataset, that our proposed approach can either be used as an alternative or in conjunction with a GAN-based unsupervised domain adaptation method. Moreover, we show generalization ability of our approach across an additional dataset (EndoVis [1]) and unseen instrument types.

2 Related Work

Research on instrument segmentation in minimally invasive procedures is dominated by fully-supervised learning using DNN architectures with: nested design [8], recurrent layers [2], dilated convolution [18] reusing pre-trained models [24], utilizing fusion modules [10] or attention modules [15,16] to improve segmentation outcome. A comparison study on segmentation networks is, for instance, presented in [5]. Some researchers have focused on multi-task

[2] SAGES Innovation Weekend - Surgical Video Annotation Conference 2020.

Table 1. Comparison with current domain adaptation approaches. ($L \rightarrow$ *Labeled* and $UL \rightarrow$ *Unlabled*)

Approach	Source (L|UL)	Target (L|UL)	Domain adaptation (Steps)
PAT [21]	Real (L)	Real (L)	Semi-supervised (Two-step)
I2I [19]	Simulated (L)	Real (UL)	Unsupervised (Two-step)
Ours	Simulated (L)	Real (UL)	Unsupervised (End-to-end)

learning with landmark localisation [13] or attention prediction [11], while others use additional inputs, such as temporal priors from motion-flow [12] or kinematic pose information [20]. However, research in annotation efficient (semi-labeled) settings is limited. A weakly supervised learning approach is employed by [7] utilizing weak scribbles labeling. [21] proposed pre-training with auxiliary task (*PAT*) on unlabelled data via self-supervised learning, followed by fine-tuning on a subset of labeled data in a semi-supervised learning setup.

The sole research work on unlabeled target with unsupervised learning is *I2I* [19] which adopts a two-step strategy: first translating simulated images into real-looking laparoscopic images via a Cycle-GAN and then training these translated images with labels in a supervised manner.

Our work is in line with *I2I* and focuses on the distribution shift problem via *unsupervised domain adaptation*, where the DNN in not trained with labeled target data, unlike *PAT*. However, in contrast to using a two-step approach [19], we employ an end-to-end consistency-based joint learning approach. An overview of our proposed approach in contrast to *I2I* and *PAT* is provided in Table 1.

3 Method

Our proposed joint-learning framework (see Fig. 1) is based on the principles of consistency learning. It aims to bridge the domain gap between simulated

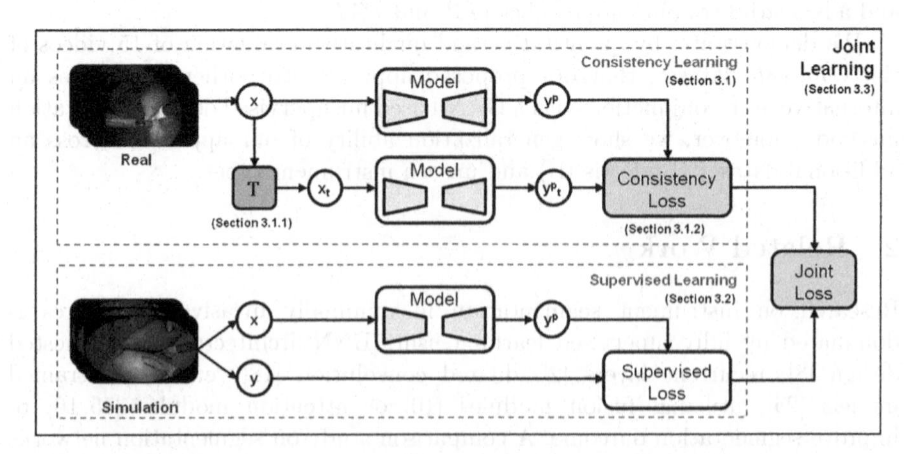

Fig. 1. The proposed joint learning approach that involves supervised learning for simulated data and consistency learning for real unlabeled data.

and real data by aligning the DNN model on both domains. For the instrument segmentation task we use TerNaus11 [24] as the DNN model as it is widely used and provides direct comparison with the previous work on domain adaptation (*I2I* [19]). The rest of the core techniques of our proposed approach are described in this section.

3.1 Consistency Learning

The core idea of consistency learning (CL) is to learn meaningful representations by forcing a model (f_θ) to produce a consistent output for an input which is *perturbed in a semantics-preserving* manner [3,4,17].

$$\min_{\theta} D\{f_\theta(x), f_\theta(x_t)\} \tag{1}$$

In a general setting, given an input x and its perturbed form x_t, the sensitivity of the model to the perturbation is minimized via a distance function $D(.,.)$. The traditional CL setting [14,23] uses mean-square-error (MSE) loss and commonly referred to as *Pi-Model.*

3.1.1. Perturbation (T)

A set of shape-preserving perturbations is necessary to learn the effective visual features in consistency learning. To facilitate learning, we implemented two data perturbation schemes: (1) *pixel-intensity* (comprising random brightness and contrast shift, posterisation, solarisation, random gamma shift, random HSV color space shift, histogram equalization and contrast limited adaptive histogram equalization) and (2) *pixel-corruption* (comprising Gaussian noise, motion blurring, image compression, dropout, random fog simulation and image embossing) and apply both of them in two ways:

- **weak**: applying one of the pixel-intensity based perturbations
- **strong**: applying weak scheme followed by one of the pixel-corruption based perturbations

All perturbations are chosen in a stochastic manner. Please refer to supplementary material for more details regarding the perturbation schemes.

3.1.2. Shape-Based Consistency Loss (S-CL)

In order to *generalize* to a real domain, the model needs to focus on generic features of an instrument, like shape, and to *adapt* to its visual appearance in an image. Therefore, we propose to utilize a shape-based measure (*Jaccard*) in conjunction with an entropy-based one (*cross-entropy*). In contrast to MSE, a shape-based loss formulation is invariant to the scale of the instrument as it depends on the overlap between prediction and ground-truth. A combination of two losses is quite common in a fully-supervised setting [24].

$$\mathcal{L}_{cl} = -\frac{1}{2}\left(\sum^{C} y_t \log y_p + log\frac{|y_t \cap y_p|}{|y_t \cup y_p| - |y_t \cap y_p|}\right) \tag{2}$$

3.2 Supervised Learning

In order to learn effectively from the simulated image and its segmentation labels, we use the same combination of *cross-entropy* and *Jaccard* (see Eq. 2) as the supervised loss function.

3.3 Joint Learning

Jointly training the model on both simulated and real domains is important for alignment on both dataset distributions and the main objective of domain adaptation. Therefore, we employ two losses: supervised loss $\mathcal{L}_{sl}$ for labeled data and consistency loss $\mathcal{L}_{cl}$ for unlabeled data which are combined together through a joint loss,

$$\mathcal{L} = \mathcal{L}_{sl} + \alpha * \mathcal{L}_{cl} \tag{3}$$

The consistency loss is employed with a time-dependent weighting factor α in order to facilitate learning. During training, the model updates the model predictions iteratively for the unlabeled data while training on both labeled and unlabeled data. Thus, our joint loss minimization can be seen as propagating labels from labeled source data to unlabeled target data.

4 Datasets and Experimental Setup

4.1 Datasets

Source. Source data is taken from [19] which is acquired by translating rendered images (from 3-D laparoscopic simulations) to Cholec80 (images from in-vivo laparoscopic videos) style images[3]. The rendered dataset contains 20 K images (2K images per patient) describing a random view of a laparoscopic scene with each tissue having a distinct texture and a presence of two tools (grasper and hook). In this paper, we mention the specific datasets (from [19]) by the following abbreviations:

- ***Sim***: rendered dataset
- ***SimRandom*** *(SR)*: I2I translated images with five random styles
- ***SimCholec*** *(SC)*: I2I translated images with five Cholec80 image styles

Real^{C}. We have built an instrument segmentation dataset comprising of 15 videos of the Cholec80 dataset [26]. The segmentation dataset is prepared by acquiring frames at five frames per second, resulting in a total of 7170 images. These images are segmented for seven tools (grasper, hook, scissors, clipper, bipolar, irrigator and specimen bag). In order to evaluate our approach, the dataset is divided into two parts: training ($\mathbf{Real}^{\mathbf{C}}_{\mathbf{UL}}$) and testing ($\mathbf{Real}^{\mathbf{C}}_{\mathbf{L}}$) dataset, containing 10 and 5 videos respectively.

[3] Please note that the rendered and real data sets are unpaired and highly unrelated.

***Real^{EV}*.** We also evaluate on EndoVis'15 dataset for rigid instruments [1] containing 160 training ($\mathbf{Real}_{\mathbf{UL}}^{\mathbf{EV}}$) and 140 testing ($\mathbf{Real}_{\mathbf{L}}^{\mathbf{EV}}$) images, acquired from 6 in-vivo video recordings. Similar to $Real^C$ dataset, seven conventional laparoscopic instrument with challenging conditions like occlusion, smoke and bleeding are present in the dataset.

4.2 Experimental Setup

Implementation. We have built a standardised implementation where the model architecture, data processing and transformation, training procedure (optimizer, regularization, learning rate schedule, number of iterations etc.) are all shared in order to provide a direct comparison with other methods. All the simulated input images and labels are first pre-processed with stochastically-varying circular masks to simulate the scene of real Cholec80 endoscopic images. We are evaluating for (binary) instrument segmentation using dice score as the metric. Also, we report all the results as an average of three training runs throughout our experiments. Our proposed framework is implemented in PyTorch and it takes four hours of training time on NVIDIA Quadro P6000 GPU.

Hyper-parameters. We use a batch size of 8 for 50 epochs and apply weight decay ($1e-6$) as standard regularization. For consistency training, we linearly increase the weight of the unlabeled loss term α (see Eq. 3) from 0 to 1 over the training. This linear increment of weight can be perceived as a warm up for the model, where it begins to understand the notion of instrument shape and linearly moves towards adapting to the real data distribution.

5 Results and Discussion

5.1 Quantitative Comparison

Joint Learning on Simulated Data. In this experiment we compared the performance of employing supervised learning on labeled simulated dataset (Sim) along with consistency learning on unlabeled real data in a traditional setting (Pi-Model) and our proposed approach for the instrument segmentation task on $Real_L^C$ and $Real_L^{EV}$ datasets. The experimental results (Table 2a) highlight the usability of our joint learning approach as an alternative to the Cycle-GAN based *I2I* framework.

Table 2. Quantitative comparison using Dice score (std).

(a) Comparison with Pi-Model

Method	$Sim \rightarrow Real_L^C$	$Sim \rightarrow Real_L^{EV}$
Pi-Model	.62 (.31)	.54 (.32)
Ours	**.75** (.29)	**.76** (.17)

(b) Comparison with I2I [19]

Method	$SR \rightarrow Real_L^C$	$SC \rightarrow Real_L^C$
I2I	.75 (.30)	.77 (.29)
Ours	**.81** (.28)	**.80** (.28)

Table 3. Qualitative analysis with respect to artifacts ($X = Real^C_L$)

	$Our_{Sim\to X}$	$Our_{SimCholec\to X}$	$I2I_{SimCholec\to X}$	$FS_{Real^C_{UL}\to X}$
Blood	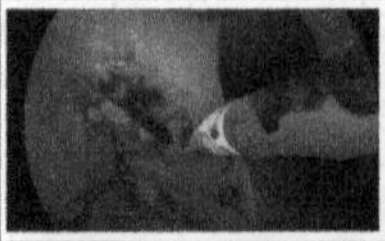	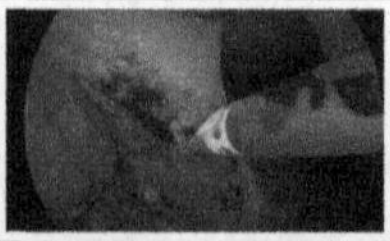	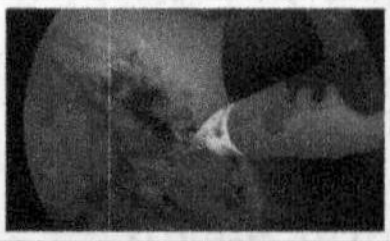	
Motion	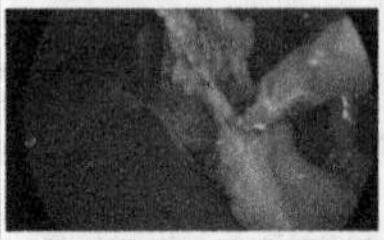	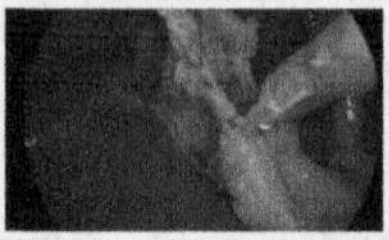	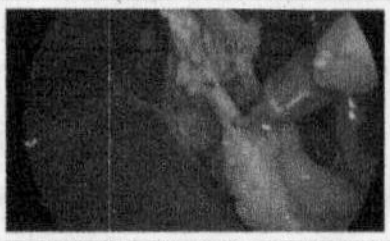	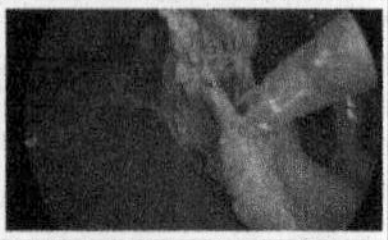
Smoke	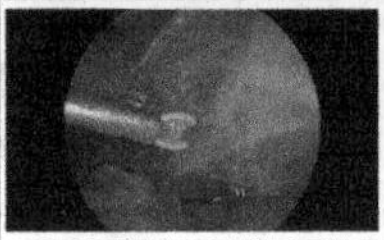	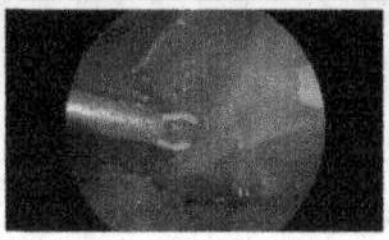	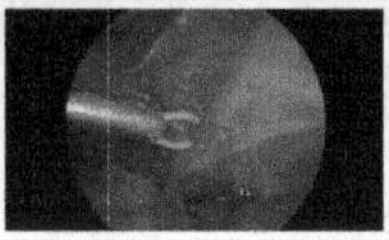	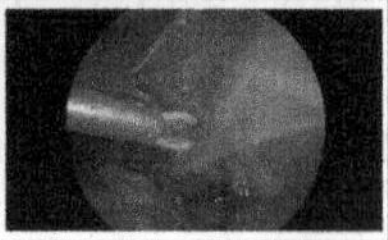

Joint Learning on Translated Data. In this experiment we employed our approach on two translated datasets of *I2I* (*SimRandom* and *SimCholec*). Our proposed approach is applied in conjunction with the *I2I* approach and the results (Table 2b) show the performance of joint learning against supervised learning on the translated images (*SimRandom* and *SimCholec*). The results indicate that our approach improves segmentation when used in conjunction with the GAN-based approach.

5.2 Qualitative Analysis

The images in Table 3 depict the segmentation quality under challenging conditions. Note that these conditions were absent in the simulated dataset. Our proposed approach performs similar to *I2I* when trained on the *simulated dataset.* However, the segmentation quality improves and looks similar to that of the fully-supervised approach when trained on the *translated dataset.* Please refer to supplementary material for more details.

Table 4. Performance comparison with respect to design choices. ($X = Real^C_L$)

(a) Ablation on simulated dataset

Loss + T	$Our_{Sim\to X}$
Supervised Baseline	.30
MSE + Weak	.56
S-CL + Weak	.69
MSE + Strong	.62
S-CL + Strong	**.75**

(b) Ablation on translated dataset

Loss + T	$SR \to X$	$SC \to X$
I2I	.75	.77
MSE + Weak	.72	.76
S-CL + Weak	.73	.77
MSE + Strong	.79	.81
S-CL + Strong	**.81**	**.80**

Table 5. Comparison of tool generalization. ($X = Real_L^C$)

Tools	$Our_{Sim \to X}$	$Our_{SimCholec \to X}$	$I2I_{SimCholec \to X}$	$FS_{Real_{UL}^C \to X}$
Grasper	.72	**.81**	.75	**.90**
Hook	.82	**.88**	.84	**.93**
Scissors	.70	**.76**	.72	**.91**
Clipper	.75	**.85**	.77	**.89**
Bipolar	.70	**.80**	.72	**.89**
Irrigator	.74	**.82**	.78	**.88**

5.3 Ablation Study on $Real^C$ Dataset

Since our consistency learning framework has multiple components, we study the effects of adding or removing a component to provide insights on reasons for the performance.

The experiments (Table 4a) show the performance on the *simulated* dataset (*Sim*) with perturbation schemes (*weak* and *strong*) and consistency loss functions (*MSE* and *S-CL*). The substantial performance gap between *Supervised Baseline* (supervised learning on source data only) and other approaches demonstrate the domain gap between simulated and real data and the effectiveness of employing our joint learning approach. Strong perturbations in general perform better than weak perturbations, and our proposed approach outperforms Pi-Model. The ablation (Table 4b) on the *translated* dataset (*SimRandom* and *SimCholec*) indicate that the consistency framework improves the overall performance of the DNN model from .75/.77 to .81/.80 on *SimRandom* and *SimCholec* datasets respectively. However, it also highlights that the choice of consistency losses (traditional and proposed loss) have low impact when the synthetic images are translated with *I2I*. Please note that the same set of perturbations are applied as data augmentations for *Supervised Baseline* (Table 4a) and *I2I* (Table 4b) in order to provide direct and fair comparison.

5.4 Generalization of Tools from Source to Target

Our experiments highlight a generalization ability of the DNN model trained on labelled simulated data and unlabeled real data for the instrument segmentation task, considering that the simulated data contains only two conventional laparoscopic tools in contrast to the six tools (excluding specimen bag) in the real data. The results (Table 5) highlight the strength of *I2I* and our approach to generalize for unseen tools in the real domain.

6 Conclusion

We introduced an efficient and end-to-end joint learning approach for the challenging problem of domain shift between synthetic and real data. Our proposed approach enforces the DNN to learn jointly from simulated and real data by employing a shape-focused consistency learning framework. It also takes only four hours of training time in contrast to multiple days for the GAN-based approach [19]. Our proposed framework has been validated for the instrument segmentation task against the baseline (supervised learning on source data only), traditional CL approach (Pi-Model) and state-of-the-art *I2I* approach. The results highlight that competitive performance (.75 vs .75/.77) can be achieved by using the proposed framework as an alternative. As a complementary method to the current GAN-based unsupervised domain adaption method, the performance further improves (to .81). Finally, the generalization capabilities of our approach across two datasets and unseen instruments is highlighted to express the feasibility of utilizing virtual environments for instrument segmentation task.

In future, we plan to study the effects of perturbations in detail. Another avenue of investigation could be the direct learning of the perturbations from the data using Automatic Machine Learning [6]. Our method being flexible and unsupervised with respect to target data, can be further used for depth estimation by exploiting depth maps from the simulated virtual environments. Since the community has not yet reached a consensus about how to efficiently label surgical videos, our end-to-end approach of joint learning provides a direction towards effective segmentation of surgical tools in the annotation scarce reality.

References

1. Endovis sub-challenge: Instrument segmentation and tracking. https://endovissub-instrument.grand-challenge.org/ (2015)
2. Attia, M., Hossny, M., Nahavandi, S., Asadi, H.: Surgical tool segmentation using a hybrid deep CNN-RNN auto encoder-decoder. In: 2017 IEEE International Conference on Systems, Man, and Cybernetics (SMC), pp. 3373–3378. IEEE (2017)
3. Bachman, P., Alsharif, O., Precup, D.: Learning with pseudo-ensembles. In: Advances in Neural Information Processing Systems, pp. 3365–3373 (2014)
4. Becker, S., Hinton, G.E.: Self-organizing neural network that discovers surfaces in random-dot stereograms. Nature **355**(6356), 161–163 (1992)
5. Bodenstedt, S., et al.: Comparative evaluation of instrument segmentation and tracking methods in minimally invasive surgery. arXiv preprint arXiv:1805.02475 (2018)
6. Cubuk, E.D., Zoph, B., Mane, D., Vasudevan, V., Le, Q.V.: Autoaugment: learning augmentation strategies from data. In: Proceedings of the IEEE Conference on Computer Vision and Pattern Recognition, pp. 113–123 (2019)
7. Fuentes-Hurtado, F., Kadkhodamohammadi, A., Flouty, E., Barbarisi, S., Luengo, I., Stoyanov, D.: Easylabels: weak labels for scene segmentation in laparoscopic videos. Int. J. Comput. Assist. Radiol. Surg. 14(7), 1–11 (2019). https://doi.org/10.1007/s11548-019-02003-2
8. García-Peraza-Herrera, L.C., et al.: Toolnet: holistically-nested real-time segmentation of robotic surgical tools. In: 2017 IEEE/RSJ International Conference on Intelligent Robots and Systems (IROS), pp. 5717–5722. IEEE (2017)
9. Huang, X., Liu, M.Y., Belongie, S., Kautz, J.: Multimodal unsupervised image-to-image translation. In: Proceedings of the European Conference on Computer Vision (ECCV), pp. 172–189 (2018)

10. Islam, M., Atputharuban, D.A., Ramesh, R., Ren, H.: Real-time instrument segmentation in robotic surgery using auxiliary supervised deep adversarial learning. IEEE Robot. Autom. Lett. **4**(2), 2188–2195 (2019)
11. Islam, M., Li, Y., Ren, H.: Learning where to look while tracking instruments in robot-assisted surgery. MICCAI 2019. LNCS, vol. 11768, pp. 412–420. Springer, Cham (2019). https://doi.org/10.1007/978-3-030-32254-0_46
12. Jin, Y., Cheng, K., Dou, Q., Heng, P.-A.: Incorporating temporal prior from motion flow for instrument segmentation in minimally invasive surgery Video. In: Shen, D., et al. (eds.) MICCAI 2019. LNCS, vol. 11768, pp. 440–448. Springer, Cham (2019). https://doi.org/10.1007/978-3-030-32254-0_49
13. Laina, I., et al.: Concurrent segmentation and localization for tracking of surgical instruments. In: Descoteaux, M., Maier-Hein, L., Franz, A., Jannin, P., Collins, D.L., Duchesne, S. (eds.) MICCAI 2017. LNCS, vol. 10434, pp. 664–672. Springer, Cham (2017). https://doi.org/10.1007/978-3-319-66185-8_75
14. Laine, S., Aila, T.: Temporal ensembling for semi-supervised learning. In: International Conference on Learning Representations (2017)
15. Ni, Z.L., Bian, G.B., Xie, X.L., Hou, Z.G., Zhou, X.H., Zhou, Y.J.: Rasnet: segmentation for tracking surgical instruments in surgical videos using refined attention segmentation network. In: 2019 41st Annual International Conference of the IEEE Engineering in Medicine and Biology Society (EMBC), pp. 5735–5738. IEEE (2019)
16. Ni, Z.-L., et al.: RAUNet: residual attention u-net for semantic segmentation of cataract surgical instruments. In: Gedeon, T., Wong, K.W., Lee, M. (eds.) ICONIP 2019. LNCS, vol. 11954, pp. 139–149. Springer, Cham (2019). https://doi.org/10.1007/978-3-030-36711-4_13
17. Oliver, A., Odena, A., Raffel, C.A., Cubuk, E.D., Goodfellow, I.: Realistic evaluation of deep semi-supervised learning algorithms. In: Advances in Neural Information Processing Systems, pp. 3235–3246 (2018)
18. Pakhomov, D., Premachandran, V., Allan, M., Azizian, M., Navab, N.: Deep residual learning for instrument segmentation in robotic surgery. In: Suk, H.-I., Liu, M., Yan, P., Lian, C. (eds.) MLMI 2019. LNCS, vol. 11861, pp. 566–573. Springer, Cham (2019). https://doi.org/10.1007/978-3-030-32692-0_65
19. Pfeiffer, M., et al.: Generating large labeled data sets for laparoscopic image processing tasks using unpaired image-to-image Translation. In: Shen, D., et al. (eds.) MICCAI 2019. LNCS, vol. 11768, pp. 119–127. Springer, Cham (2019). https://doi.org/10.1007/978-3-030-32254-0_14
20. Qin, F., Li, Y., Su, Y.H., Xu, D., Hannaford, B.: Surgical instrument segmentation for endoscopic vision with data fusion of rediction and kinematic pose. In: 2019 International Conference on Robotics and Automation (ICRA), pp. 9821–9827. IEEE (2019)
21. Ross, T., et al.: Exploiting the potential of unlabeled endoscopic video data with self-supervised learning. Int. J. Comput. Assist. Radiol. Surg. **13**(6), 925–933 (2018). https://doi.org/10.1007/s11548-018-1772-0
22. Saenko, K., Kulis, B., Fritz, M., Darrell, T.: Adapting visual category models to new domains. In: Daniilidis, K., Maragos, P., Paragios, N. (eds.) ECCV 2010. LNCS, vol. 6314, pp. 213–226. Springer, Heidelberg (2010). https://doi.org/10.1007/978-3-642-15561-1_16
23. Sajjadi, M., Javanmardi, M., Tasdizen, T.: Regularization with stochastic transformations and perturbations for deep semi-supervised learning. In: Advances in Neural Information Processing Systems, pp. 1163–1171 (2016)

24. Shvets, A.A., Rakhlin, A., Kalinin, A.A., Iglovikov, V.I.: Automatic instrument segmentation in robot-assisted surgery using deep learning. In: 2018 17th IEEE International Conference on Machine Learning and Applications (ICMLA), pp. 624–628. IEEE (2018)
25. Torralba, A., Efros, A.A.: Unbiased look at dataset bias. In: Computer Vision and Pattern Recognition, pp. 1521–1528. IEEE (2011)
26. Twinanda, A.P., Shehata, S., Mutter, D., Marescaux, J., De Mathelin, M., Padoy, N.: Endonet: a deep architecture for recognition tasks on laparoscopic videos. IEEE Trans. Med. Imaging **36**(1), 86–97 (2016)

Author Index

Printed in the United States
By Bookmasters